MATHEMATICAL PHYSICS

For BSc [Hons.] / MSc and BE / ME students of all leading Indian universities

MATHEMATICAL PHYSICS

Dr Binoy Bhattacharya PhD [Moscow], MIE

Sr Scientific Officer of Structural Engineering
Research Centre, Roorkee & Chennai
Superintending Engineer [Civil] of DVC, Kolkata

New Central Book Agency (P) Ltd

Co-Published by

Bijay Publications

E-mail : bijaypublications@gmail.com

Exclusively Distributed by

Bijay Pustak Bhandar (P) Ltd

90/6A M. G. Road, Bhawani Dutta Lane, Kolkata-700007

✆ : (033) 4605-7856, 8777898576, 8274916104

E-mail : bijaypustakbhandar@gmail.com

NCBA

REGD OFFICE
8/1 Chintamoni Das Lane, Kolkata 700 009, India
email: ncbakolkata@gmail.com

OVERSEAS
NCBA (UK) Ltd, 149 Park Avenue North, Northampton, NN3 2HY, UK
email: ncbauk@yahoo.com

CORPORATE OFFICE
501A/1 Right Side [Upper Ground Floor]
Govind Puri, Kalkaji, New Delhi 110 019
email: delhincba@gmail.com

BIJAY PUBLICATIONS
90/6A M. G. Road, Bhawani Dutta Lane, Kolkata-7
E-mail : bijaypublications@gmail.com

BIJAY PUSTAK BHANDAR (P) LTD
90/6A M. G. Road, Bhawani Dutta Lane, Kolkata-7
E-mail : bijaypustakbhandar@gmail.com

MATHEMATICAL PHYSICS • Dr Binoy Bhattacharya

First Published: 2009
Thoroughly Revised and Updated Second Edition: 2016
Reprinted: 2019, 2025

PUBLISHER • TYPESETTER • PRINTER
New Central Book Agency (P) Ltd
8/1 Chintamoni Das Lane, Kolkata 700 009

CO-PUBLISHED BY
Bijay Publications
90/6A M G Road, Bhawani Dutta Lane, Kolkata 700007

PROOF CHECKING
Dipan Roy

PROJECT TEAM
Abhijit Sen, Pradip Biswas

ISBN: 978 93 5255 021 0

Price ₹ 965.00

To
ANIRBAN
—my beloved son

Contents

Preface to the Second Edition

I am very much pleased to present this second edition of my book.

In this edition, I have attempted to rectify the typographical errors in the previous edition as far as possible.

Moreover, I have added topics on:

(i) Distribution Functions,

(ii) Infinite Series and

(iii) Functions of Two or More Variables as appendices at the end of the book. I hope, these will be very much useful to the students and readers to whom I am indebted for their constant interest in this book.

I convey my sincere thanks to Sri Amitabha Sen of M/s New Central Book Agency (P) Ltd for taking so much of care for this book.

I will remain grateful to receive comments and criticisms from any corner towards betterment of this book.

Doljatra
23 March 2016

Binoy Bhattacharyya

Preface to the First Edition

Mathematical Physics covers a wide range of topics in applied mathematics. Writing such a book is a challenge to any author. I accepted the challenge because of the present-day need of the students and the research workers in the fields of physical sciences. The main purpose of this book is to bring under one umbrella the existing mathematical methods for solving various problems in engineering and physical sciences.

I have attempted to write every chapter of this book in such a way as to cater to the needs of a wide class of readers. All the topics have been discussed in detail as far as possible and have been supported with example problems for the clear understanding of the students.

As we know, differential equations are at the root of solving problems in physics and engineering. Due to this reason, a complete chapter on 'Ordinary Differential Equations in Physical Process' has been devoted to various problems from different corners of physical sciences. The topics on partial differential equations have been discussed in two separate chapters. In one chapter, I have discussed the methods for finding the general solutions of the various types of partial differential equations. In another chapter on 'Boundary Value Problems', the Fourier method has been used to solve the partial differential equations which govern a wide class of problems in physics and engineering.

In the chapter on 'Integral Equations' I have discussed the existing methods for solving different kinds of integral equations. Numerous example problems have also been solved.

The chapter on 'Special Integrals' is a collection of 'famous' integrals which we come across in physical sciences. The elliptic integrals have been discussed in more detail; short tables for values of these integrals are also provided.

The advancement of computers and their applications in the present era demands numerical methods in solving complex problems in physics and engineering. So I have laid stress on the chapter on 'Numerical Methods' which is, however, not exhaustive. But this will help the students to develop a solid introductory base and this will be very useful for them to go ahead with advanced books on the subject, whenever needed.

I will find myself highly rewarded if this book proves useful even to a small section of the students.

I am extremely thankful to Sri Amitabha Sen of New Central Book Agency (P) Ltd for his kind consent to publish the book. I am also thankful to Sri Soumen Paul and Sri Prabhat Jas for their help and cooperation and to Sri Abhijit Sen for his sincerety and patience in performing the himalayan task of the typesetting job.

I sincerely thank Sri Dipan Roy and Sri Pradip Kumar Choudhury for their excellent proof-reading job without which the book could not have been in its present form.

Useful comments and criticisms for the improvement of this book will be gratefully accepted.

Binoy Bhattacharyya

5 September 2008
Kolkata 700 040

Chapter 1

Vector Algebra

1.1 Introduction

Some physical quantities, like mass, density, temperature, etc., are completely specified by giving their magnitude alone. These are called *scalar quantities.* But many other quantities, like displacement, force, electric intensity, etc., require direction in addition to their magnitude for the complete and comprehensive specification. These are called *vector quantities.*

A vector quantity may be graphically represented by a straight line AB, pointing in the direction from A to B such that the length of the line gives the magnitude of the vector quantity. However, in this book, bold-faced letters are used to represent vectors. Thus, **a** represents 'vector a'. The magnitude of vector quantities will be represented by ordinary letters. Thus, 'A' will represent the length of the vector **A**.

Two vectors **A** and **B** are said to be equal when their magnitudes as well as directions are identical. In that case

$$\mathbf{A} = \mathbf{B}, \quad \text{i.e.,} \quad \mathbf{A} - \mathbf{B} = \mathbf{0}. \tag{1.1}$$

The right-hand side of this vector equation is also a vector called *null vector.* Thus, a null vector has a zero magnitude.

1.2 Unit Vectors

A vector having unit scalar magnitude is defined as *unit vector.* Any vector **A** can be written as

$$\mathbf{A} = A\mathbf{k}, \tag{1.2}$$

where **k** is unit vector in the direction of the vector **A**. In the Cartesian coordinate system (*Fig 1.1*), unit vectors **i**, **j**, **k** in the directions of x, y and z axes respectively are commonly used.

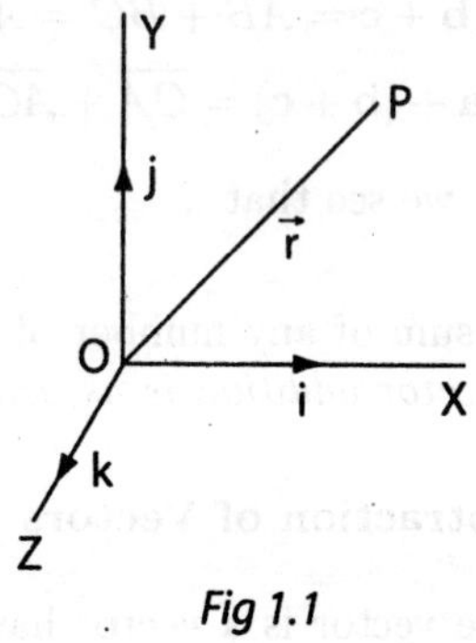

Fig 1.1

Thus, if $P(x, y, z)$ be any point in space under Cartesian coordinate system, then the vector **r** represented by OP in length and direction will be denoted by

$$\mathbf{r} = \mathbf{i}x + \mathbf{j}y + \mathbf{k}z. \tag{1.3}$$

This is known as *position vector* of P, whose length will be given by

$$r = \left(x^2 + y^2 + z^2\right)^{1/2}. \tag{1.4}$$

1.3 Addition of Vectors

Let there be any two given vectors $\mathbf{a}$ and $\mathbf{b}$ represented by the sides OA and AB of a parallelogram $OABC$ (*Fig 1.2*). Thus

$$\overrightarrow{OA} = \mathbf{a} \quad \text{and} \quad \overrightarrow{AB} = \mathbf{b}. \tag{1.5}$$

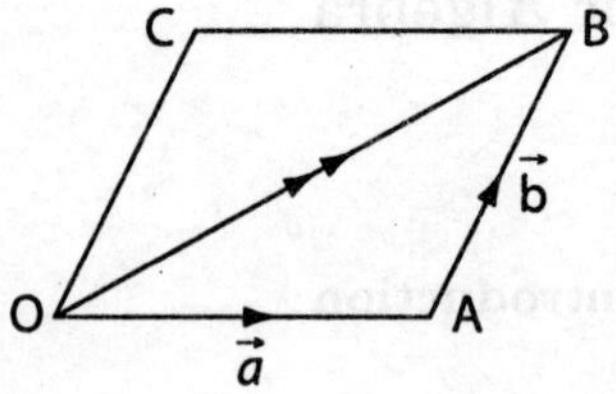

Fig 1.2

Then the vector given by $\overrightarrow{OB}$ is defined as the sum of the given vectors $\mathbf{a}$ and $\mathbf{b}$, and is written as

$$\overrightarrow{OB} = \overrightarrow{OA} + \overrightarrow{AB} = \mathbf{a} + \mathbf{b}. \tag{1.6}$$

Now, since the opposite sides of a parallelogram are equal and parallel, we can write

$$\overrightarrow{OA} = \overrightarrow{CB} = \mathbf{a} \quad \text{and} \quad \overrightarrow{AB} = \overrightarrow{OC} = \mathbf{b}. \tag{1.7}$$

$$\therefore \quad \overrightarrow{OA} + \overrightarrow{AB} = \overrightarrow{OB} = \overrightarrow{OC} + \overrightarrow{CB}, \quad \text{i.e.,} \quad \mathbf{a} + \mathbf{b} = \mathbf{b} + \mathbf{a}. \tag{1.8}$$

Hence, vector addition is commutative.

Again, let us consider vectors (*Fig 1.3*)

$$\overrightarrow{OA} = \mathbf{a}, \quad \overrightarrow{AB} = \mathbf{b} \quad \text{and} \quad \overrightarrow{BC} = \mathbf{c}.$$

Then

$$\mathbf{a} + \mathbf{b} = \overrightarrow{OA} + \overrightarrow{AB} = \overrightarrow{OB}.$$

$$\therefore \quad (\mathbf{a} + \mathbf{b}) + \mathbf{c} = \overrightarrow{OB} + \overrightarrow{BC} = \overrightarrow{OC}.$$

Similarly

$$\mathbf{b} + \mathbf{c} = \overrightarrow{AB} + \overrightarrow{BC} = \overrightarrow{AC}.$$

$$\therefore \quad \mathbf{a} + (\mathbf{b} + \mathbf{c}) = \overrightarrow{OA} + \overrightarrow{AC} = \overrightarrow{OC}.$$

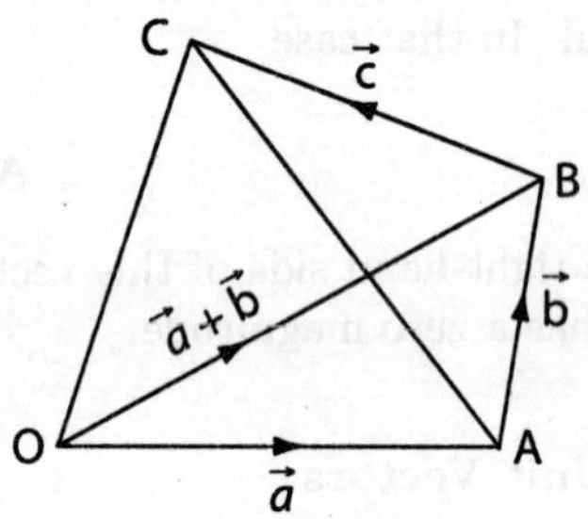

Fig 1.3

Hence, we see that

$$\mathbf{a} + (\mathbf{b} + \mathbf{c}) = (\mathbf{a} + \mathbf{b}) + \mathbf{c}. \tag{1.9}$$

Thus, sum of any number of vectors is independent of the order in which they are added. That is, *vector addition is associative.*

1.4 Subtraction of Vectors

A negative vector is a vector having same magnitude as that of a given vector, but opposite direction. Thus, if a vector $\overrightarrow{AB} = \mathbf{a}$, any vector having length equal to AB but opposite direction will be $= -\mathbf{a}$.

The operation of vector subtraction of two given vectors $\mathbf{a}$ and $\mathbf{b}$ may be regarded as the operation of addition of vectors $\mathbf{a}$ and $-\mathbf{b}$, and is written as

$$\mathbf{a} - \mathbf{b} = \mathbf{a} + (-\mathbf{b}). \tag{1.10}$$

1.5 Expression of Vector in Terms of Position Vectors of Its End Points

Let A and B are two points in Cartesian coordinates with position vectors

$$\begin{aligned} \mathbf{a} &= \mathbf{i}x_1 + \mathbf{j}y_1 + \mathbf{k}z_1 \\ \text{and} \quad \mathbf{b} &= \mathbf{i}x_2 + \mathbf{j}y_2 + \mathbf{k}z_2. \end{aligned}$$

Then, $\overrightarrow{OA} = \mathbf{a}$, $\overrightarrow{OB} = \mathbf{b}$.

Now, $\overrightarrow{OA} + \overrightarrow{AB} = \overrightarrow{OB}$.

$\therefore$ $\overrightarrow{AB} = \overrightarrow{OB} - \overrightarrow{OA} = \mathbf{b} - \mathbf{a}$, i.e., $\overrightarrow{AB} =$ position vector of $B -$ position vector of A.

$$\begin{aligned} \therefore \quad \overrightarrow{AB} &= (\mathbf{i}x_2 + \mathbf{j}y_2 + \mathbf{k}z_2) - (\mathbf{i}x_1 + \mathbf{j}y_1 + \mathbf{k}z_1) \\ &= \mathbf{i}(x_2 - x_1) + \mathbf{j}(y_2 - y_1) + \mathbf{k}(z_2 - z_1). \end{aligned}$$

$$\therefore \quad AB = \sqrt{(x_2 - x_1)^2 + (y_2 - y_1)^2 + (z_2 - z_1)^2}.$$

EXAMPLE 1.1. *The points $A(2\mathbf{i} - \mathbf{j} + \mathbf{k})$, $B(\mathbf{i} - 3\mathbf{j} - 5\mathbf{k})$ and $C(3\mathbf{i} - 4\mathbf{j} - 4\mathbf{k})$ are vertices of a triangle. Find lengths of its sides.*

Solution:

$$\begin{aligned} &\overrightarrow{AB} = (\mathbf{i} - 3\mathbf{j} - 5\mathbf{k}) - (2\mathbf{i} - \mathbf{j} + \mathbf{k}) = -\mathbf{i} - 2\mathbf{j} - 6\mathbf{k} \\ \therefore \quad &AB = \sqrt{(-1)^2 + (-2)^2 + (-6)^2} = \sqrt{41}. \\ &\overrightarrow{BC} = (3\mathbf{i} - 4\mathbf{j} - 4\mathbf{k}) - (\mathbf{i} - 3\mathbf{j} - 5\mathbf{k}) = 2\mathbf{i} - \mathbf{j} + \mathbf{k} \\ \therefore \quad &BC = \sqrt{(2)^2 + (-1)^2 + (1)^2} = \sqrt{6}. \\ &\overrightarrow{CA} = (2\mathbf{i} - \mathbf{j} + \mathbf{k}) - (3\mathbf{i} - 4\mathbf{j} - 4\mathbf{k}) = -\mathbf{i} + 3\mathbf{j} + 5\mathbf{k} \\ \therefore \quad &CA = \sqrt{(-1)^2 + (3)^2 + (5)^2} = \sqrt{35}. \end{aligned}$$

1.6 Multiplication of Vectors by Scalars

Let $\overrightarrow{OA} = \mathbf{a}$ and m is any positive real number. Then, the vector $m \cdot \overrightarrow{OA} = m\mathbf{a}$ is defined to be a vector having magnitude equal to m times that of the given vector in the same direction. Similarly, the vector $-m \cdot \overrightarrow{OA} = -m\mathbf{a}$ is a vector in the direction opposite to that of $\mathbf{a}$ and having magnitude equal to m times that of $\mathbf{a}$. Multiplication of vectors by scalars is commutative, associative and distributive. That is, for any two vectors $\mathbf{a}$ and $\mathbf{b}$,

$$\left.\begin{aligned} m\mathbf{a} &= \mathbf{a}m \\ m(n\mathbf{a}) &= (mn)\mathbf{a} = n(m\mathbf{a}) \\ (m+n)\mathbf{a} &= m\mathbf{a} + n\mathbf{a} \\ m(\mathbf{a} + \mathbf{b}) &= m\mathbf{a} + m\mathbf{b} \end{aligned}\right\}, \tag{1.11}$$

where m and n are scalar numbers.

1.7 Linear Dependence of Vectors

Let **a**, **b**, **c**, ... be vectors and $\alpha, \beta, \gamma, \ldots$ are scalars, not all of which are zero. If there exists a relation of the type

$$\alpha\mathbf{a} + \beta\mathbf{b} + \gamma\mathbf{c} + \cdots = 0, \tag{1.12}$$

then the system of vectors **a**, **b**, **c** is said to be linearly dependent.

If the system of vectors **a**, **b**, **c** are not linearly dependent, then

$$\alpha = \beta = \gamma = \cdots = 0. \tag{1.13}$$

The system of vectors in this case is said to be linearly independent.

If $\mathbf{r} = \alpha\mathbf{a} + \beta\mathbf{b} + \gamma\mathbf{c} + \ldots$, i.e., $-\mathbf{r} + \alpha\mathbf{a} + \beta\mathbf{b} + \gamma\mathbf{c} + \cdots = 0$, then system of vectors **r**, **a**, **b**, **c**, ... is linearly dependent.

Any vector **r** coplanar with any two non-collinear vectors **a** and **b** can be uniquely expressed as a linear combination of the given vectors, i.e.,

$$\mathbf{r} = \alpha\mathbf{a} + \beta\mathbf{b}, \tag{1.14}$$

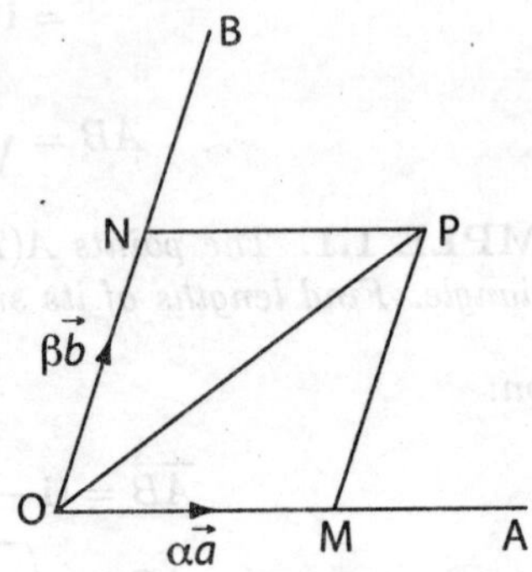

Fig 1.4

where α, β are scalars. In order to prove it, we take $\overrightarrow{OA} = \mathbf{a}$ and $\overrightarrow{OB} = \mathbf{b}$ which are non-collinear (*Fig 1.4*) and a vector $\overrightarrow{OP} = \mathbf{r}$ coplanar with **a** and **b**. Draw PM and PN parallel to OB and OA respectively.

Then, scalars α, β can be choser in such a manner that $\overrightarrow{OM} = \alpha\mathbf{a}$ and $\overrightarrow{ON} = \beta\mathbf{b}$.

$$\therefore \quad \mathbf{r} = \overrightarrow{OP} = \overrightarrow{OM} + \overrightarrow{MP} = \overrightarrow{OM} + \overrightarrow{ON} = \alpha\mathbf{a} + \beta\mathbf{b}.$$

In order to prove that this combination is unique, let us suppose that **r** can be expressed in another form, like $\mathbf{r} = \alpha'\mathbf{a} + \beta'\mathbf{b}$.

$$\therefore \quad \alpha\mathbf{a} + \beta\mathbf{b} = \alpha'\mathbf{a} + \beta'\mathbf{b}, \quad \text{or,} \quad (\alpha - \alpha')\mathbf{a} + (\beta - \beta')\mathbf{b} = 0.$$

$$\therefore \mathbf{a} = -\frac{\beta - \beta'}{\alpha - \alpha'}\mathbf{b},$$

which means that **a** is a scalar multiple of **b**, i.e., **a** and **b** are collinear. This contradicts that **a** and **b** are non-collinear. Hence, we must have $\alpha - \alpha' = 0$, i.e., $\alpha = \alpha'$. Similarly, we can prove that $\beta - \beta' = 0$, or, $\beta = \beta'$.

Hence, it is proved that the above combination is unique.

From equation (1.14) we can write, $-\mathbf{r} + \alpha\mathbf{a} + \beta\mathbf{b} = 0$, i.e., **r**, **a**, **b** vectors are linearly dependent. Hence, the necessary and sufficient condition that three vectors be linearly dependent is that they be coplanar.

EXAMPLE 1.2. *Examine whether the vectors* $\mathbf{i} - 3\mathbf{j} + 2\mathbf{k}$, $2\mathbf{i} - 4\mathbf{j} - 4\mathbf{k}$, $3\mathbf{i} + 2\mathbf{j} - \mathbf{k}$ *are linearly independent or not.*

Solution: If α, β, γ be three scalars, then for the given vectors to be linearly dependent

$$\alpha(\mathbf{i} - 3\mathbf{j} + 2\mathbf{k}) + \beta(2\mathbf{i} - 4\mathbf{j} - 4\mathbf{k}) + \gamma(3\mathbf{i} + 2\mathbf{j} - \mathbf{k}) = 0$$

$$\text{or,} \quad (\alpha + 2\beta + 3\gamma)\mathbf{i} + (-3\alpha - 4\beta + 2\gamma)\mathbf{j} + (2\alpha - 4\beta - \gamma)\mathbf{k} = 0.$$

Since **i**, **j**, **k** are non-coplanar vectors, we have

$$\begin{aligned} \alpha + 2\beta + 3\gamma &= 0 \\ -3\alpha - 4\beta + 2\gamma &= 0 \\ 2\alpha - 4\beta - \gamma &= 0 \end{aligned}$$

Solving first two equations we get

$$\frac{\alpha}{16} = \frac{\beta}{-11} = \frac{\gamma}{2} = \lambda \text{ (say)}. \qquad \therefore \quad \alpha = 16\lambda, \beta = -11\lambda, \gamma = 2\lambda.$$

Putting these values in third equation we get, $(32 + 44 - 2)\lambda = 0$, i.e., $74\lambda = 0$, i.e., $\lambda = 0$.

$$\therefore \quad \alpha = \beta = \gamma = 0.$$

Hence, the system is linearly independent.

1.8 Rectangular Resolution of Vectors

Any vector **A** can be resolved along the Cartesian coordinate axes. The components of the vector along the axes are called *rectangular components.* Let the origin O be one extremity of the vector **A**. If we draw a rectangular parallelopiped with three edges along the axes so that vector **A** is the diagonal of the solid figure (*Fig 1.5*), we can write

$$\begin{aligned} \mathbf{A} = \overrightarrow{OE} &= \overrightarrow{OP} + \overrightarrow{PB} + \overrightarrow{BE} \\ &= \overrightarrow{OP} + \overrightarrow{OC} + \overrightarrow{OG} \\ &= \mathbf{A}_x + \mathbf{A}_y + \mathbf{A}_z, \end{aligned} \qquad (1.15)$$

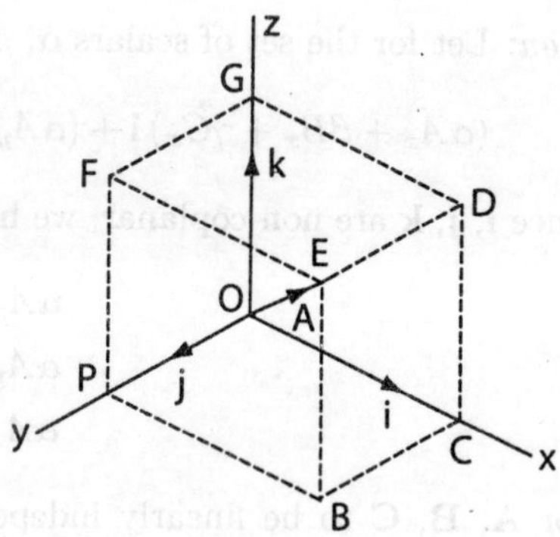

Fig 1.5

where, $\mathbf{A_x}$, $\mathbf{A_y}$, $\mathbf{A_z}$ are components of the vector **A** in the directions of x, y and z axes.

If **i**, **j**, **k** are unit vectors along the directions of x, y, z axes respectively, then

$$\mathbf{A_x} = \mathbf{i}A_x, \quad \mathbf{A_y} = \mathbf{j}A_y, \quad \mathbf{A_z} = \mathbf{k}A_z.$$

Hence

$$\mathbf{A} = \mathbf{i}A_x + \mathbf{j}A_y + \mathbf{k}A_z. \qquad (1.16)$$

If the vector **A** makes angles α, β, γ with x, y and z axes respectively, then

$$\left.\begin{aligned} A_x &= A\cos\alpha \\ A_y &= A\cos\beta \\ A_z &= A\cos\gamma \end{aligned}\right\}, \qquad (1.17)$$

where, $A = |\,\mathbf{A}\,|$.

Squaring and adding equations (1.17), we get

$$A_x^2 + A_y^2 + A_z^2 = A^2\left(\cos^2\alpha + \cos^2\beta + \cos^2\gamma\right).$$

But, $A_x^2 + A_y^2 + A_z^2 = OP^2 + PB^2 + BE^2 = A^2$.

$$\therefore \quad \cos^2\alpha + \cos^2\beta + \cos^2\gamma = 1. \tag{1.18}$$

This is known as *vector direction cosine law* and $\cos\alpha$, $\cos\beta$, $\cos\gamma$ are called the *direction cosines* of OE.

EXAMPLE 1.3. *Prove that, for vectors*

$$\begin{aligned} \mathbf{A} &= \mathbf{i}A_x + \mathbf{j}A_y + \mathbf{k}A_z, \\ \mathbf{B} &= \mathbf{i}B_x + \mathbf{j}B_y + \mathbf{k}B_z \\ \textit{and} \quad \mathbf{C} &= \mathbf{i}C_x + \mathbf{j}C_y + \mathbf{k}C_z \end{aligned}$$

to be linearly independent, the determinant

$$\begin{vmatrix} A_x & A_y & A_z \\ B_x & B_y & B_z \\ C_x & C_y & C_z \end{vmatrix} \neq 0.$$

Solution: Let for the set of scalars α, β, γ, $\alpha\mathbf{A} + \beta\mathbf{B} + \gamma\mathbf{C} = 0$, i.e.,

$$(\alpha A_x + \beta B_x + \gamma C_x)\,\mathbf{i} + (\alpha A_y + \beta B_y + \gamma C_y)\,\mathbf{j} + (\alpha A_z + \beta B_z + \gamma C_z)\,\mathbf{k} = 0.$$

Since **i**, **j**, **k** are non-coplanar, we have

$$\left.\begin{aligned} \alpha A_x + \beta B_x + \gamma C_x = 0 \\ \alpha A_y + \beta B_y + \gamma C_y = 0 \\ \alpha A_z + \beta B_z + \gamma C_z = 0 \end{aligned}\right\}. \tag{1}$$

For **A**, **B**, **C** to be linearly independent, $\alpha = \beta = \gamma = 0$. Hence, combining these with equations (1) we get the required result. The rudiments of determinants are provided in Appendix A.

1.9 Scalar Product of Two Vectors

Scalar product of two vectors **a** and **b** is defined to be equal to $ab\cos\theta$, where $a = |\mathbf{a}|$, $b = |\mathbf{b}|$ and θ =angle between the directions of a and b. Thus

$$\mathbf{a}\cdot\mathbf{b} = |\mathbf{a}||\mathbf{b}|\cdot\cos\theta = ab\cos\theta. \tag{1.19}$$

Now, from *Fig 1.6*, $b\cos\theta$ is the length of the resolved part OC of OB, i.e., $\mathbf{a}\cdot\mathbf{b} = a(b\cos\theta) = a$ (resolved part of modulus of **b** along **a**).

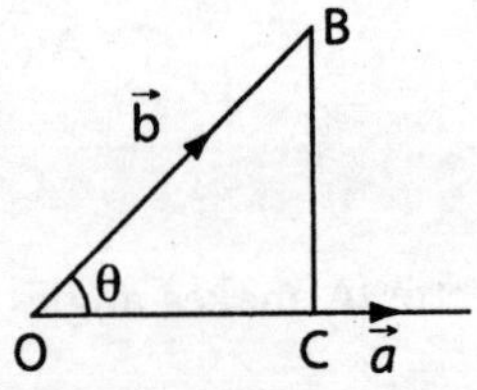

Fig 1.6

Similarly, $\mathbf{a} \cdot \mathbf{b} = b(a \cos\theta) = b$ (resolved part of modulus of **a** along the direction of **b**). Hence, scalar or dot product of two vectors is equal to the product of the length of one of them with resolved part of the length of the other in the direction of former.

From equation (1.19) it is clear that, if **a** and **b** are two non-zero vectors, then, their dot product will be zero only when the directions of vectors are perpendicular; the dot product will be equal to product of their moduli when their directions are parallel.

For the unit vectors **i**, **j**, **k** along the rectangular coordinate system

$$\mathbf{i} \cdot \mathbf{i} = \mathbf{j} \cdot \mathbf{j} = \mathbf{k} \cdot \mathbf{k} = 1 \tag{1.20}$$

$$\text{and} \quad \mathbf{i} \cdot \mathbf{j} = \mathbf{j} \cdot \mathbf{k} = \mathbf{k} \cdot \mathbf{i} = 0. \tag{1.21}$$

Hence, the scalar or dot product of two vectors

$$\mathbf{A} = \mathbf{i}A_x + \mathbf{j}A_y + \mathbf{k}A_z \tag{1.22}$$

$$\text{and} \quad \mathbf{B} = \mathbf{i}B_x + \mathbf{j}B_y + \mathbf{k}B_z \tag{1.23}$$

will be written as

$$\mathbf{A} \cdot \mathbf{B} = A_xB_x + A_yB_y + A_zB_z. \tag{1.24}$$

Scalar product of vectors obeys distributive law, i.e., for three vectors **a**, **b** and **c**, we can write

$$\mathbf{a} \cdot (\mathbf{b} + \mathbf{c}) = \mathbf{a} \cdot \mathbf{b} + \mathbf{a} \cdot \mathbf{c}. \tag{1.25}$$

For two vectors **a** and **b**, we can have the following relations:

(1) $$(\mathbf{a}+\mathbf{b}) \cdot (\mathbf{a}-\mathbf{b}) = \mathbf{a} \cdot \mathbf{a} + \mathbf{b} \cdot \mathbf{a} - \mathbf{a} \cdot \mathbf{b} - \mathbf{b} \cdot \mathbf{b}$$
$$= \mathbf{a}^2 - \mathbf{b}^2 \quad (\because\ \mathbf{a} \cdot \mathbf{b} = \mathbf{b} \cdot \mathbf{a})$$

(2) $$(\mathbf{a}+\mathbf{b})^2 = (\mathbf{a}+\mathbf{b}) \cdot (\mathbf{a}+\mathbf{b})$$
$$= \mathbf{a} \cdot \mathbf{a} + \mathbf{b} \cdot \mathbf{a} + \mathbf{a} \cdot \mathbf{b} + \mathbf{b} \cdot \mathbf{b}$$
$$= \mathbf{a}^2 + 2\mathbf{a} \cdot \mathbf{b} + \mathbf{b}^2.$$

EXAMPLE 1.4. *Prove that the triangle whose vertices are* $A(2\mathbf{i}+4\mathbf{j}-\mathbf{k})$, $B(4\mathbf{i}+5\mathbf{j}+\mathbf{k})$ *and* $C(3\mathbf{i}+6\mathbf{j}-3\mathbf{k})$ *is an isosceles right-angled triangle.*

Solution: Let O be the origin. Then

$$\overrightarrow{AB} = \overrightarrow{OB} - \overrightarrow{OA} = (4\mathbf{i}+5\mathbf{j}+\mathbf{k}) - (2\mathbf{i}+4\mathbf{j}-\mathbf{k}) = 2\mathbf{i}+\mathbf{j}+2\mathbf{k}.$$

$$\therefore \quad AB = \sqrt{(2)^2+(1)^2+(2)^2} = 3.$$

$$\overrightarrow{BC} = \overrightarrow{OC} - \overrightarrow{OB} = (3\mathbf{i}+6\mathbf{j}-3\mathbf{k}) - (4\mathbf{i}+5\mathbf{j}+\mathbf{k}) = -\mathbf{i}+\mathbf{j}-4\mathbf{k}.$$

$$\therefore \quad BC = \sqrt{(-1)^2+(1)^2+(-4)^2} = 3\sqrt{2}$$

$$\text{and} \quad \overrightarrow{CA} = \overrightarrow{OC} - \overrightarrow{OA} = (3\mathbf{i}+6\mathbf{j}-3\mathbf{k}) - (2\mathbf{i}+4\mathbf{j}-\mathbf{k}) = \mathbf{i}+2\mathbf{j}-2\mathbf{k}.$$

$$\therefore \quad CA = \sqrt{(1)^2+(2)^2+(-2)^2} = 3.$$

Hence, $AB = CA = 3, BC = 3\sqrt{2}$.

Also

$$\overrightarrow{AB} \cdot \overrightarrow{CA} = (2\mathbf{i}+\mathbf{j}+2\mathbf{k}) \cdot (\mathbf{i}+2\mathbf{j}-2\mathbf{k}) = 2+2-4 = 0.$$

$\therefore$ AB is perpendicular to CA. Hence proved.

EXAMPLE 1.5. *Find the angle between the vectors* $\mathbf{a} = 4\mathbf{i} + 3\mathbf{j} + \mathbf{k}$ *and* $\mathbf{b} = 2\mathbf{i} - \mathbf{j} + 2\mathbf{k}$. *Also, find a unit vector perpendicular to both* **a** *and* **b**.

Solution: We know, $\cos\theta = \dfrac{\mathbf{a}\cdot\mathbf{b}}{ab}$, where θ is the angle between **a** and **b**.

Here

$$a = \sqrt{(4)^2 + (3)^2 + (1)^2} = \sqrt{26},$$

$$b = \sqrt{(2)^2 + (-1)^2 + (2)^2} = 3$$

$$\text{and}\quad \mathbf{a}\cdot\mathbf{b} = (4\mathbf{i} + 3\mathbf{j} + \mathbf{k})\cdot(2\mathbf{i} - \mathbf{j} + 2\mathbf{k})$$
$$= 8 - 3 + 2 = 7.$$

$$\therefore\quad \cos\theta = \frac{7}{3\sqrt{26}},\quad \text{i.e.,}\quad \theta = \cos^{-1}\left(\frac{7}{3\sqrt{26}}\right).$$

Again, let **r** be the unit vector perpendicular to both **a** and **b** so that $\mathbf{r} = \alpha\mathbf{i} + \beta\mathbf{j} + \gamma\mathbf{k}$.

$$\therefore\quad \alpha^2 + \beta^2 + \gamma^2 = 1$$

and $\mathbf{r}\cdot\mathbf{a} = 0$ and $\mathbf{r}\cdot\mathbf{b} = 0$.

From $\mathbf{r}\cdot\mathbf{a} = 0$ we have

$$(\alpha\mathbf{i} + \beta\mathbf{j} + \gamma\mathbf{k})\cdot(4\mathbf{i} + 3\mathbf{j} + \mathbf{k}) = 0,\quad \text{i.e.,}\quad 4\alpha + 3\beta + \gamma = 0. \tag{1}$$

Similarly, $\mathbf{r}\cdot\mathbf{b} = 0$ gives

$$2\alpha - \beta + \gamma = 0. \tag{2}$$

By cross multiplication, we get

$$\frac{\alpha}{4} = \frac{\beta}{-2} = \frac{\gamma}{-10} = \frac{\sqrt{\alpha^2 + \beta^2 + \gamma^2}}{\sqrt{(4)^2 + (-2)^2 + (-10)^2}} = \frac{1}{\sqrt{120}}.$$

$$\therefore\quad \alpha = \frac{4}{\sqrt{120}} = \frac{2}{\sqrt{30}},\quad \beta = \frac{-1}{\sqrt{30}},\quad \gamma = \frac{-5}{\sqrt{30}}.$$

Hence

$$\mathbf{r} = \frac{1}{\sqrt{30}}(2\mathbf{i} - \mathbf{j} - 5\mathbf{k}).$$

EXAMPLE 1.6. *Prove by vector method the Cauchy's inequality*

$$(a_1b_1 + a_2b_2 + a_3b_3)^2 < \left(a_1^2 + a_2^2 + a_3^2\right)\left(b_1^2 + b_2^2 + b_3^2\right).$$

Solution: Let

$$\mathbf{a} = a_1\mathbf{i} + a_2\mathbf{j} + a_3\mathbf{k}$$
$$\mathbf{b} = b_1\mathbf{i} + b_2\mathbf{j} + b_3\mathbf{k}$$

be two vectors. Then

$$a^2 = a_1^2 + a_2^2 + a_3^2 \quad\text{and}\quad b^2 = b_1^2 + b_2^2 + b_3^2. \tag{1}$$

Again

$$\mathbf{a}\cdot\mathbf{b} = (a_1\mathbf{i} + a_2\mathbf{j} + a_3\mathbf{k})\cdot(b_1\mathbf{i} + b_2\mathbf{j} + b_3\mathbf{k}) = a_1b_1 + a_2b_2 + a_3b_3. \quad (2)$$

Since $\mathbf{a}\cdot\mathbf{b} = ab\cos\theta$,

$$\therefore \quad a_1b_1 + a_2b_2 + a_3b_3 = ab\cos\theta.$$

But $\cos\theta \le 1$, hence

$$a_1b_1 + a_2b_2 + a_3b_3 \le ab, \quad \text{or,} \quad (a_1b_1 + a_2b_2 + a_3b_3)^2 \le a^2b^2.$$

$$\therefore \quad (a_1b_1 + a_2b_2 + a_3b_3)^2 \le (a_1^2 + a_2^2 + a_3^2)(b_1^2 + b_2^2 + b_3^2)$$

[by using equation (1)]

1.10 Vector or Cross Product of Two Vectors

The vector or cross product of two vectors **a** and **b** is a vector **c** in a direction perpendicular to the plane formed by **a** and **b**, and has magnitude equal to $ab\sin\theta$, where θ is the angle between the directions of **a** and **b**.

Thus

$$\mathbf{c} = \mathbf{a}\times\mathbf{b} = (ab\sin\theta)\mathbf{n}, \quad (1.26)$$

where **n** is unit vector perpendicular to both **a** and **b**. The direction of vector **c** is regarded positive if the rotation from **a** to **b** appears counter-clockwise. Thus, vectors **a**, **b** and **c** form a right-handed triad.

From *Fig 1.7*, it is obvious that

$$\mathbf{a}\times\mathbf{b} = -\mathbf{b}\times\mathbf{a}, \quad (1.27)$$

i.e., cross product of two vectors is non-commutative.

From equation (1.26), we find that if $\theta = \pi/2$, i.e., if the vectors **a** and **b** are perpendicular to each other, then

$$\mathbf{c} = \mathbf{a}\times\mathbf{b} = ab\mathbf{n}. \quad (1.28)$$

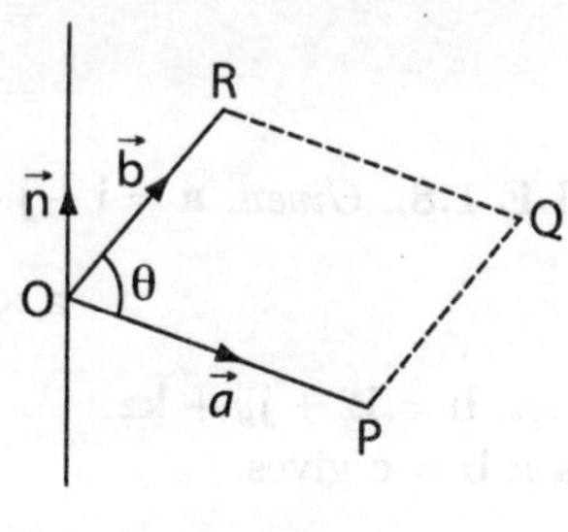

Fig 1.7

Again, if $\theta = 0$, i.e., if the vectors **a** and **b** are parallel (or antiparallel when $\theta = \pi$)

$$\mathbf{c} = \mathbf{a}\times\mathbf{b} = 0. \quad (1.29)$$

From above, it is obvious that, if **i**, **j**, **k** are unit vectors along x, y and z axes respectively of the Cartesian coordinates, then following relations exist:

$$\mathbf{i}\times\mathbf{i} = \mathbf{j}\times\mathbf{j} = \mathbf{k}\times\mathbf{k} = 0 \quad (1.30)$$

$$\mathbf{i}\times\mathbf{j} = \mathbf{k}, \quad \mathbf{j}\times\mathbf{k} = \mathbf{i}, \quad \mathbf{k}\times\mathbf{i} = \mathbf{j}. \quad (1.31)$$

Geometrically, the magnitude of the vector product of two vectors **a** and **b** (i.e., $ab\sin\theta$) is the area of the parallelogram $OPQR$ formed with sides **a** and **b**.

1.11 Vector Product in Terms of Unit Vectors

Let vectors **a** and **b** be expressed in terms of unit vectors **i**, **j**, **k** in the form

$$\left.\begin{aligned}\mathbf{a} &= a_x\mathbf{i} + a_y\mathbf{j} + a_z\mathbf{k} \\ \mathbf{b} &= b_x\mathbf{i} + b_y\mathbf{j} + b_z\mathbf{k}\end{aligned}\right\}. \tag{1.32}$$

$$\begin{aligned}\therefore \quad \mathbf{a} \times \mathbf{b} &= (a_x\mathbf{i} + a_y\mathbf{j} + a_z\mathbf{k}) \times (b_x\mathbf{i} + b_y\mathbf{j} + b_z\mathbf{k}) \\ &= (a_y b_z - a_z b_y)\,\mathbf{i} + (a_z b_x - a_x b_z)\,\mathbf{j} + (a_x b_y - a_y b_x)\,\mathbf{k}.\end{aligned} \tag{1.33}$$

Expression (1.33) may be written in the determinant form

$$\mathbf{a} \times \mathbf{b} = \begin{vmatrix} \mathbf{i} & \mathbf{j} & \mathbf{k} \\ a_x & a_y & a_z \\ b_x & b_y & b_z \end{vmatrix}. \tag{1.34}$$

EXAMPLE 1.7. *Prove that,* $(\mathbf{a} \times \mathbf{b})^2 = a^2b^2 - (\mathbf{a} \cdot \mathbf{b})^2$.

Solution: If θ be the angle between **a** and **b**,

$$\mathbf{a} \times \mathbf{b} = (ab \sin\theta)\mathbf{n}, \quad \mathbf{n} = \text{unit vector}$$

perpendicular to both **a** and **b**.

$$\begin{aligned}\therefore \quad (\mathbf{a} \times \mathbf{b})^2 &= a^2b^2 \sin^2\theta, \quad \because \mathbf{n}^2 = 1 \\ &= a^2b^2\,(1 - \cos^2\theta) \\ &= a^2b^2 - a^2b^2\cos^2\theta \\ &= a^2b^2 - (\mathbf{a} \cdot \mathbf{b})^2, \quad \because \mathbf{a} \cdot \mathbf{b} = ab\cos\theta.\end{aligned}$$

EXAMPLE 1.8. *Given,* $\mathbf{a} = \mathbf{i} + \mathbf{j} + \mathbf{k}$ *and* $\mathbf{c} = \mathbf{j} - \mathbf{k}$, *find a vector* **b** *such that*

$$\mathbf{a} \times \mathbf{b} = \mathbf{c} \quad \text{and} \quad \mathbf{a} \cdot \mathbf{b} = 3.$$

Solution: Let, $\mathbf{b} = \mathbf{i}x + \mathbf{j}y + \mathbf{k}z$.

Then, $\mathbf{a} \times \mathbf{b} = \mathbf{c}$ gives

$$\begin{vmatrix} \mathbf{i} & \mathbf{j} & \mathbf{k} \\ 1 & 1 & 1 \\ x & y & z \end{vmatrix} = \mathbf{j} - \mathbf{k}, \quad \text{or,} \quad \mathbf{i}(z - y) - \mathbf{j}(z - x) + \mathbf{k}(y - x) = \mathbf{j} - \mathbf{k}.$$

$$\therefore \quad z - y = 0, \quad x - z = 1 \quad \text{and} \quad x - y = 1,$$

$$\text{i.e.,} \quad y = z, \quad x = 1 + z = 1 + y. \tag{1}$$

Again, $\mathbf{a} \cdot \mathbf{b} = 3$ gives

$$(\mathbf{i} + \mathbf{j} + \mathbf{k}) \cdot (\mathbf{i}x + \mathbf{j}y + \mathbf{k}z) = 3, \quad \text{i.e.,} \quad x + y + z = 3. \tag{2}$$

Solving equations (1) and (2), we get

$$y = z = 2/3 \quad \text{and} \quad x = 1 + y = 5/3.$$

Hence, $b = \frac{1}{3}(5\mathbf{i} + 2\mathbf{j} + 2\mathbf{k})$.

EXAMPLE 1.9. *Find the sine of the angle between the vectors* $\mathbf{a} = \mathbf{i} + 3\mathbf{j} + 2\mathbf{k}$ *and* $\mathbf{b} = 2\mathbf{i} - 4\mathbf{j} + \mathbf{k}$.

Solution: The modulus of the given vectors are

$$a = \sqrt{1+9+4} = \sqrt{14} \quad \text{and} \quad b = \sqrt{4+16+1} = \sqrt{21}.$$

Now

$$\mathbf{a} \times \mathbf{b} = \begin{vmatrix} \mathbf{i} & \mathbf{j} & \mathbf{k} \\ 1 & 3 & 2 \\ 2 & -4 & 1 \end{vmatrix} = 11\mathbf{i} + 3\mathbf{j} - 10\mathbf{k}.$$

$$\therefore \quad (a \times b)^2 = (11)^2 + (3)^2 + (-10)^2 = 230. \qquad (1)$$

If θ be the angle between vectors **a** and **b**, then $(ab\sin\theta)^2\mathbf{n}^2 = 230$, i.e., $14 \times 21 \times \sin^2\theta = 230$ $(\because\ \mathbf{n}^2 = 1)$.

$$\therefore \quad \sin\theta = \sqrt{\frac{115}{147}}.$$

EXAMPLE 1.10. *Prove that,* $\mathbf{a} \times (\mathbf{b} + \mathbf{c}) + \mathbf{b} \times (\mathbf{c} + \mathbf{a}) + \mathbf{c} \times (\mathbf{a} + \mathbf{b}) = 0$.

Solution:

$$\begin{aligned} \text{LHS} &= \mathbf{a} \times \mathbf{b} + \mathbf{a} \times \mathbf{c} + \mathbf{b} \times \mathbf{c} + \mathbf{b} \times \mathbf{a} + \mathbf{c} \times \mathbf{a} + \mathbf{c} \times \mathbf{b} \\ &= (\mathbf{a} \times \mathbf{b} + \mathbf{b} \times \mathbf{a}) + (\mathbf{b} \times \mathbf{c} + \mathbf{c} \times \mathbf{b}) + (\mathbf{c} \times \mathbf{a} + \mathbf{a} \times \mathbf{c}) \\ &= 0 + 0 + 0 = 0 \end{aligned}$$

$$\because\ \mathbf{a} \times \mathbf{b} = -\mathbf{b} \times \mathbf{a}, \quad \mathbf{b} \times \mathbf{c} = -\mathbf{c} \times \mathbf{b}, \quad \mathbf{c} \times \mathbf{a} = -\mathbf{a} \times \mathbf{c}.$$

1.12 Surface Area as Vector

We have seen that the magnitude of the vector product of two vectors **a** and **b** represents the area of a parallelogram whose sides are **a** and **b**. Hence, area of a parallelogram can be represented by a vector **c** drawn at right angles to the plane of the area. But the vector **c** does not indicate any shape of the area. Therefore, any plane surface can be represented by a vector whose magnitude is proportional to area of the surface and whose direction is normal to the plane of the surface following the right-handed screw rule (*Fig 1.8*).

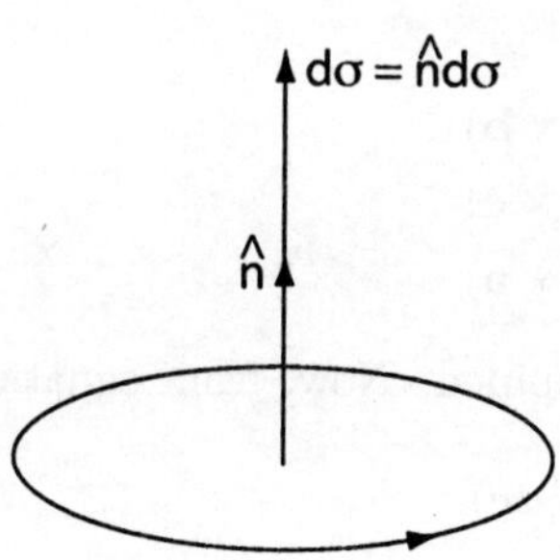

Fig 1.8

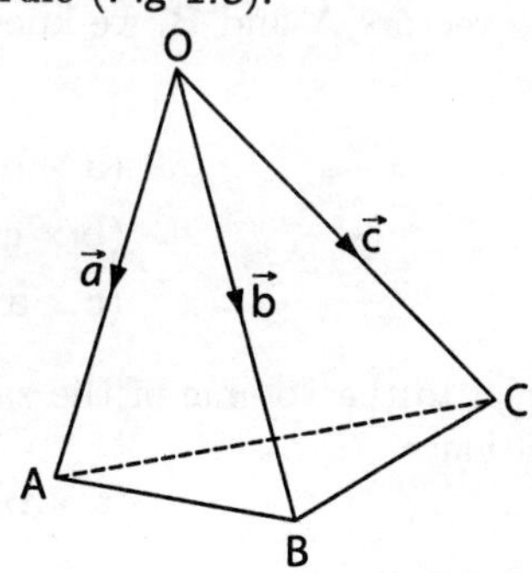

Fig 1.9

Area vector of a surface which is a part of a closed figure like tetrahedron is conventionally represented by a straight line pointing outward. The vector representing the entire surface of a closed figure is zero. We will prove it for a tetrahedron. In *Fig 1.9*, let $\overrightarrow{OA} = \mathbf{a}$, $\overrightarrow{OB} = \mathbf{b}$ and $\overrightarrow{OC} = \mathbf{c}$. Then $\overrightarrow{AC} = \mathbf{c} - \mathbf{a}$ and $\overrightarrow{AB} = \mathbf{b} - \mathbf{a}$. Area vectors of three faces passing through O are

$$\tfrac{1}{2}(\mathbf{a} \times \mathbf{b}), \quad \tfrac{1}{2}(\mathbf{b} \times \mathbf{c}) \quad \text{and} \quad \tfrac{1}{2}(\mathbf{c} \times \mathbf{a})$$

and area vector of the base ABC is

$$\tfrac{1}{2}\left[(\mathbf{c} - \mathbf{a}) \times (\mathbf{b} - \mathbf{a})\right] = \tfrac{1}{2}\left[\mathbf{c} \times \mathbf{b} - \mathbf{c} \times \mathbf{a} - \mathbf{a} \times \mathbf{b}\right] \quad (\because\ \mathbf{a} \times \mathbf{a} = 0).$$

Hence, the area vector of all the surfaces of the tetrahedron is

$$\tfrac{1}{2}\left[\mathbf{a} \times \mathbf{b} + \mathbf{b} \times \mathbf{c} + \mathbf{c} \times \mathbf{a} + \mathbf{c} \times \mathbf{b} - \mathbf{c} \times \mathbf{a} - \mathbf{a} \times \mathbf{b}\right]$$

which is equal to zero ($\because\ \mathbf{b} \times \mathbf{c} = -\mathbf{c} \times \mathbf{b}$).

1.13 Scalar Triple Product

The vector product of two vectors **b** and **c** is a vector. This resultant vector may be multiplied scalarly or vectorially by a third vector **a**. Thus, there are two types of vector triple products, namely, $\mathbf{a} \cdot (\mathbf{b} \times \mathbf{c})$ and $\mathbf{a} \times (\mathbf{b} \times \mathbf{c})$. The former is called *scalar triple product* and the latter *vector triple product*.

In the *Fig 1.10*, vectors **a**, **b**, **c** are shown. The $\mathbf{b} \times \mathbf{c}$ is a vector along the normal to the plane containing **b** and **c** whose magnitude is $bc \sin\theta$. If this vector is directed at an angle α with the plane of **a** and **c**, then

$$\mathbf{a} \cdot (\mathbf{b} \times \mathbf{c}) = \mathbf{a} \cdot \mathbf{D} = Da\cos\alpha = Dh, \quad (1.35)$$

where $\mathbf{D} = (bc\sin\theta)\mathbf{n}$, **n** being unit vector perpendicular to the plane containing **b** and **c**, and h = height of the parallelopiped with **a** and **b** as edges. Hence, right-hand side of equation (1.35) represents the volume of the parallelopiped $OBDCHGFA$ (*Fig 1.10*).

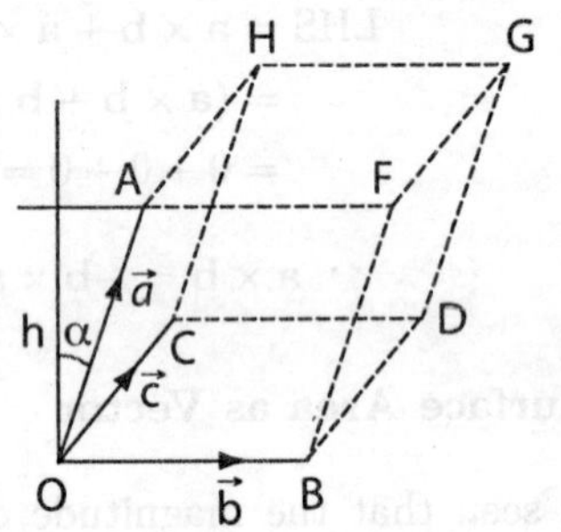

Fig 1.10

The scalar triple product is also known as *box product* and written as [**abc**].

For any two vectors **A** and **B**, we know that $\mathbf{A} \cdot \mathbf{B} = \mathbf{B} \cdot \mathbf{A}$. Hence, for three vectors **a**, **b**, **c**, we can write

$$(\mathbf{a} \times \mathbf{b}) \cdot \mathbf{c} = \mathbf{c} \cdot (\mathbf{a} \times \mathbf{b}) \quad (1.36)$$

$$(\mathbf{b} \times \mathbf{c}) \cdot \mathbf{a} = \mathbf{a} \cdot (\mathbf{b} \times \mathbf{c}) \quad (1.37)$$

$$(\mathbf{c} \times \mathbf{a}) \cdot \mathbf{b} = \mathbf{b} \cdot (\mathbf{c} \times \mathbf{a}) \quad (1.38)$$

and each is equal to the volume of the same parallelopiped. Now, from equations (1.36) and (1.37), we find that

$$(\mathbf{a} \times \mathbf{b}) \cdot \mathbf{c} = \mathbf{a} \cdot (\mathbf{b} \times \mathbf{c}).$$

Similarly, we can write

$$(\mathbf{c} \times \mathbf{a}) \cdot \mathbf{b} = \mathbf{c} \cdot (\mathbf{a} \times \mathbf{b}) \quad \text{and} \quad (\mathbf{b} \times \mathbf{c}) \cdot \mathbf{a} = \mathbf{b} \cdot (\mathbf{c} \times \mathbf{a}).$$

Thus, *in the scalar triple product, the position of dot and cross can be interchanged provided the cyclic order of the vectors* **a**, **b**, **c** *are maintained.* Therefore, we have

$$\mathbf{a}\cdot(\mathbf{b}\times\mathbf{c})=\mathbf{c}\cdot(\mathbf{a}\times\mathbf{b})=\mathbf{b}\cdot(\mathbf{c}\times\mathbf{a}) \quad \text{or,} \quad [\mathbf{abc}]=[\mathbf{cab}]=[\mathbf{bca}]. \tag{1.39}$$

For unit vectors **i**, **j**, **k** along rectangular coordinate axes, we have

$$[\mathbf{ijk}]=[\mathbf{jki}]=[\mathbf{kij}]=1. \tag{1.40}$$

If the vectors **a**, **b**, **c** are expressed in terms of unit vectors **i**, **j**, **k** in the form

$$\left.\begin{aligned}\mathbf{a}&=a_x\mathbf{i}+a_y\mathbf{j}+a_z\mathbf{k}\\ \mathbf{b}&=b_x\mathbf{i}+b_y\mathbf{j}+b_z\mathbf{k}\\ \mathbf{c}&=c_x\mathbf{i}+c_y\mathbf{j}+c_z\mathbf{k}\end{aligned}\right\}, \tag{1.41}$$

then

$$[\mathbf{abc}]=(a_x\mathbf{i}+a_y\mathbf{j}+a_z\mathbf{k})\begin{vmatrix}\mathbf{i}&\mathbf{j}&\mathbf{k}\\ b_x&b_y&b_z\\ c_x&c_y&c_z\end{vmatrix} \quad \text{i.e.,} \quad [\mathbf{abc}]=\begin{vmatrix}a_x&a_y&a_z\\ b_x&b_y&b_z\\ c_x&c_y&c_z\end{vmatrix}. \tag{1.42}$$

If the vectors **a**, **b**, **c** lie in a plane, the volume of the parallelopiped is zero. Hence, for *three vectors* **a**, **b**, **c** *to be coplanar, their scalar product vanishes.*

EXAMPLE 1.11. *Find the volume of the parallelopiped whose edges are represented by*

$$\mathbf{a}=2\mathbf{i}-3\mathbf{j}+4\mathbf{k}, \quad \mathbf{b}=\mathbf{i}+2\mathbf{j}-\mathbf{k}, \quad \mathbf{c}=3\mathbf{i}-\mathbf{j}+2\mathbf{k}.$$

Solution: Volume of the parallelopiped is

$$[\mathbf{abc}]=\begin{vmatrix}2&-3&4\\ 1&2&-1\\ 3&-1&2\end{vmatrix}=2(4-1)+3(2+3)+4(-1-6)=7 \quad (\text{taking } +).$$

EXAMPLE 1.12. *Prove that, for three vectors* **a**, **b**, **c**, $[\mathbf{a}+\mathbf{b},\mathbf{b}+\mathbf{c},\mathbf{c}+\mathbf{a}]=2[\mathbf{abc}]$.

Solution:

$$\begin{aligned}[\mathbf{a}+\mathbf{b},\mathbf{b}+\mathbf{c},\mathbf{c}+\mathbf{a}]&=(\mathbf{a}+\mathbf{b})\cdot[(\mathbf{b}+\mathbf{c})\times(\mathbf{c}+\mathbf{a})]\\ &=(\mathbf{a}+\mathbf{b})\cdot[\mathbf{b}\times\mathbf{c}+\mathbf{b}\times\mathbf{a}+\mathbf{c}\times\mathbf{c}+\mathbf{c}\times\mathbf{a}]\\ &=\mathbf{a}\cdot(\mathbf{b}\times\mathbf{c})+\mathbf{a}\cdot(\mathbf{b}\times\mathbf{a})+\mathbf{a}\cdot(\mathbf{c}\times\mathbf{a})\\ &\quad+\mathbf{b}\cdot(\mathbf{b}\times\mathbf{c})+\mathbf{b}\cdot(\mathbf{b}\times\mathbf{a})+\mathbf{b}\cdot(\mathbf{c}\times\mathbf{a}) \quad (\because\ \mathbf{c}\times\mathbf{c}=0)\\ &=[\mathbf{abc}]+[\mathbf{bca}]=2[\mathbf{abc}].\end{aligned}$$

EXAMPLE 1.13. *Find the value of* λ *for which the vectors*

$$\mathbf{a}=2\mathbf{i}-\mathbf{j}+\mathbf{k}, \quad \mathbf{b}=\mathbf{i}+2\mathbf{j}-3\mathbf{k}, \quad \mathbf{c}=3\mathbf{i}+\lambda\mathbf{j}+5\mathbf{k}$$

are coplanar.

Solution: The vectors **a**, **b**, **c** are coplanar if, $[\mathbf{abc}] = 0$

$$\text{or,}\quad \begin{vmatrix} 2 & -1 & 1 \\ 1 & 2 & -3 \\ 3 & \lambda & 5 \end{vmatrix} = 0$$

$$\text{or,}\quad 2(10+3\lambda)+(5+9)+(\lambda-6)=0$$

$$\text{or,}\quad 7\lambda+28=0,\quad \therefore\quad \lambda=-4.$$

EXAMPLE 1.14. *If* **a**, **b**, **c** *be any three non-coplanar vectors, prove that any other vector* **r** *can be expressed as*

$$\mathbf{r} = \frac{\mathbf{r}\cdot\mathbf{a}}{[\mathbf{abc}]}(\mathbf{b}\times\mathbf{c}) + \frac{\mathbf{r}\cdot\mathbf{b}}{[\mathbf{abc}]}(\mathbf{c}\times\mathbf{a}) + \frac{\mathbf{r}\cdot\mathbf{c}}{[\mathbf{abc}]}(\mathbf{a}\times\mathbf{b}).$$

Solution: Since **a**, **b**, **c** are non-coplanar vectors, we have $[\mathbf{abc}] \neq 0$.

Now, $\mathbf{b}\times\mathbf{c}$, $\mathbf{c}\times\mathbf{a}$, $\mathbf{a}\times\mathbf{b}$, each being a vector, any vector **r** can be expressed as

$$\mathbf{r} = \alpha(\mathbf{b}\times\mathbf{c}) + \beta(\mathbf{c}\times\mathbf{a}) + \gamma(\mathbf{a}\times\mathbf{b}), \tag{1}$$

where α, β, γ are constants.

Multiplying both sides of equation (1) scalarly by **a**, we have

$$\begin{aligned} \mathbf{r}\cdot\mathbf{a} &= \alpha\mathbf{a}\cdot(\mathbf{b}\times\mathbf{c}) + \beta\mathbf{a}\cdot(\mathbf{c}\times\mathbf{a}) + \gamma\mathbf{a}\cdot(\mathbf{a}\times\mathbf{b}) \\ &= \alpha[\mathbf{abc}] \quad [\because\ \mathbf{a}\cdot(\mathbf{c}\times\mathbf{a})=0, \mathbf{a}\cdot(\mathbf{a}\times\mathbf{b})=0] \end{aligned}$$

$$\therefore\quad \alpha = \frac{\mathbf{r}\cdot\mathbf{a}}{[\mathbf{abc}]}.$$

Similarly, we can find

$$\beta = \frac{\mathbf{r}\cdot\mathbf{b}}{[\mathbf{abc}]},\quad \gamma = \frac{\mathbf{r}\cdot\mathbf{c}}{[\mathbf{abc}]}.$$

Hence the result.

1.14 Vector Triple Product

The triple product of three vectors **a**, **b**, **c** is written as $\mathbf{a}\times(\mathbf{b}\times\mathbf{c})$. The vector $\mathbf{A} = \mathbf{b}\times\mathbf{c}$ is perpendicular to the plane containing **b** and **c**. Hence, the vector $\mathbf{B} = \mathbf{a}\times\mathbf{A}$ will be perpendicular to the plane containing vectors **a** and **A**, i.e., vectors **a** and $\mathbf{b}\times\mathbf{c}$. Therefore, the vector $\mathbf{B} = \mathbf{a}\times(\mathbf{b}\times\mathbf{c})$ will lie in the plane formed by vectors **b** and **c**, and at the same time, it will be perpendicular to vector **a**. Similarly, the vector $(\mathbf{a}\times\mathbf{b})\times\mathbf{c}$ will lie in the plane formed by vectors **a** and **b**, and at the same time, will be perpendicular to vector **c**. Hence

$$\mathbf{a}\times(\mathbf{b}\times\mathbf{c}) \neq (\mathbf{a}\times\mathbf{b})\times\mathbf{c}, \tag{1.43}$$

i.e., vector triple product is not associative.

Since the vector $\mathbf{B} = \mathbf{a}\times(\mathbf{b}\times\mathbf{c})$ lies in the plane formed by vectors **b** and **c**, we can write **B** as a linear combination of **b** and **c** in the form

$$\mathbf{B} = \alpha\mathbf{b} + \beta\mathbf{c}, \tag{1.44}$$

where α, β are scalar quantities to be determined. Here we have assumed that the vectors **a**, **b** are not collinear. Now, vector **B** is also perpendicular to the vector **a**. Hence, $\mathbf{a} \cdot \mathbf{B} = 0$

$$\text{i.e.,} \quad \alpha(\mathbf{a} \cdot \mathbf{b}) + \beta(\mathbf{a} \cdot \mathbf{c}) = 0, \quad \text{or,} \quad \frac{\alpha}{\mathbf{a} \cdot \mathbf{c}} = \frac{-\beta}{\mathbf{a} \cdot \mathbf{b}} = \lambda \text{ (say)}$$

$$\text{i.e.,} \quad \alpha = \lambda(\mathbf{a} \cdot \mathbf{c}), \quad \beta = -\lambda(\mathbf{a} \cdot \mathbf{b}). \tag{1.45}$$

So equation (1.44) can be written as

$$\mathbf{B} = \mathbf{a} \times (\mathbf{b} \times \mathbf{c}) = \lambda \{\mathbf{b}(\mathbf{a} \cdot \mathbf{c}) - \mathbf{c}(\mathbf{a} \cdot \mathbf{b})\}. \tag{1.46}$$

In order to determine λ, we consider a special case in which vectors **a**, **b**, **c** are taken in the form:

$$\left.\begin{aligned} \mathbf{b} &= \mathbf{i}b_x, \\ \mathbf{c} &= \mathbf{i}c_x + \mathbf{j}c_y \\ \text{and} \quad \mathbf{a} &= \mathbf{i}a_x + \mathbf{j}a_y + \mathbf{k}a_z \end{aligned}\right\}. \tag{1.47}$$

In this case

$$\mathbf{b} \times \mathbf{c} = \mathbf{i}b_x \times (\mathbf{i}c_x + \mathbf{j}c_y) = \mathbf{k}b_x c_y.$$

$$\begin{aligned} \therefore \quad \mathbf{a} \times (\mathbf{b} \times \mathbf{c}) &= (\mathbf{i}a_x + \mathbf{j}a_y + \mathbf{k}a_z) \times \mathbf{k}b_x c_y = -\mathbf{j}a_x b_x c_y + \mathbf{i}a_y b_x c_y \\ &= \mathbf{i}b_x(a_x c_x + a_y c_y) - (\mathbf{i}c_x + \mathbf{j}c_y)(a_x b_x) = \mathbf{b}(\mathbf{a} \cdot \mathbf{c}) - \mathbf{c}(\mathbf{a} \cdot \mathbf{b}). \end{aligned} \tag{1.48}$$

Equations (1.48) and (1.46) show that $\lambda = 1$. Thus, the vector triple product can be written as

$$\mathbf{a} \times (\mathbf{b} \times \mathbf{c}) = \mathbf{b}(\mathbf{a} \cdot \mathbf{c}) - \mathbf{c}(\mathbf{a} \cdot \mathbf{b}). \tag{1.49}$$

Similarly

$$\mathbf{b} \times (\mathbf{c} \times \mathbf{a}) = \mathbf{c}(\mathbf{b} \cdot \mathbf{a}) - \mathbf{a}(\mathbf{b} \cdot \mathbf{c}) \tag{1.50}$$

$$\text{and} \quad \mathbf{c} \times (\mathbf{a} \times \mathbf{b}) = \mathbf{a}(\mathbf{c} \cdot \mathbf{b}) - \mathbf{b}(\mathbf{c} \cdot \mathbf{a}). \tag{1.51}$$

Now

$$\begin{aligned} \mathbf{a} \cdot \mathbf{b} &= \mathbf{b} \cdot \mathbf{a}, \quad \therefore \quad \mathbf{c}(\mathbf{b} \cdot \mathbf{a}) = \mathbf{c}(\mathbf{a} \cdot \mathbf{b}), \\ \mathbf{a} \cdot \mathbf{c} &= \mathbf{c} \cdot \mathbf{a}, \quad \therefore \quad \mathbf{b}(\mathbf{a} \cdot \mathbf{c}) = \mathbf{b}(\mathbf{c} \cdot \mathbf{a}) \\ \text{and} \quad \mathbf{b} \cdot \mathbf{c} &= \mathbf{c} \cdot \mathbf{b}, \quad \therefore \quad \mathbf{a}(\mathbf{b} \cdot \mathbf{c}) = \mathbf{a}(\mathbf{c} \cdot \mathbf{b}) \end{aligned}$$

so by adding both sides of equations (1.49), (1.50) and (1.51), we get an important identity

$$\mathbf{a} \times (\mathbf{b} \times \mathbf{c}) + \mathbf{b} \times (\mathbf{c} \times \mathbf{a}) + \mathbf{c} \times (\mathbf{a} \times \mathbf{b}) = 0. \tag{1.52}$$

EXAMPLE 1.15. *Verify the formula for vector triple product*

$$\mathbf{a} \times (\mathbf{b} \times \mathbf{c}) = \mathbf{b}(\mathbf{a} \cdot \mathbf{c}) - \mathbf{c}(\mathbf{a} \cdot \mathbf{b})$$

by taking $\mathbf{a} = \mathbf{i} - 2\mathbf{j} + \mathbf{k}$, $\mathbf{b} = 2\mathbf{i} + \mathbf{j} + \mathbf{k}$, $\mathbf{c} = \mathbf{i} + 2\mathbf{j} - \mathbf{k}$.

Solution: Here

$$\mathbf{b} \times \mathbf{c} = \begin{vmatrix} \mathbf{i} & \mathbf{j} & \mathbf{k} \\ 2 & 1 & 1 \\ 1 & 2 & -1 \end{vmatrix} = -3\mathbf{i} + 3\mathbf{j} + 3\mathbf{k}.$$

$$\therefore \quad \mathbf{a} \times (\mathbf{b} \times \mathbf{c}) = (\mathbf{i} - 2\mathbf{j} + \mathbf{k}) \times (-3\mathbf{i} + 3\mathbf{j} + 3\mathbf{k})$$

$$= \begin{vmatrix} \mathbf{i} & \mathbf{j} & \mathbf{k} \\ 1 & -2 & 1 \\ -3 & 3 & 3 \end{vmatrix} = -9\mathbf{i} - 6\mathbf{j} - 3\mathbf{k}. \qquad (1)$$

Again

$$\mathbf{a} \cdot \mathbf{c} = (\mathbf{i} - 2\mathbf{j} + \mathbf{k}) \cdot (\mathbf{i} + 2\mathbf{j} - \mathbf{k}) = 1 - 4 - 1 = -4.$$

$$\therefore \quad \mathbf{b}(\mathbf{a} \cdot \mathbf{c}) = (2\mathbf{i} + \mathbf{j} + \mathbf{k})(-4) = -8\mathbf{i} - 4\mathbf{j} - 4\mathbf{k} \qquad (2)$$

$$\mathbf{a} \cdot \mathbf{b} = (\mathbf{i} - 2\mathbf{j} + \mathbf{k}) \cdot (2\mathbf{i} + \mathbf{j} + \mathbf{k}) = 2 - 2 + 1 = 1.$$

$$\therefore \quad \mathbf{c}(\mathbf{a} \cdot \mathbf{b}) = (\mathbf{i} + 2\mathbf{j} - \mathbf{k}) \cdot 1 = \mathbf{i} + 2\mathbf{j} - \mathbf{k}. \qquad (3)$$

From (2) and (3)

$$\mathbf{b}(\mathbf{a} \cdot \mathbf{c}) - \mathbf{c}(\mathbf{a} \cdot \mathbf{b}) = (-8\mathbf{i} - 4\mathbf{j} - 4\mathbf{k}) - (\mathbf{i} + 2\mathbf{j} - \mathbf{k}) = -9\mathbf{i} - 6\mathbf{j} - 3\mathbf{k}. \qquad (4)$$

Thus, right-hand side of (1) and (4) being equal, the formula for vector triple product is verified.

EXAMPLE 1.16. *Prove the identity* $\mathbf{a} \times [\mathbf{a} \times (\mathbf{a} \times \mathbf{b})] = (\mathbf{a} \cdot \mathbf{a})(\mathbf{b} \times \mathbf{a})$.

Solution:

$$\text{LHS} = \mathbf{a} \times [\mathbf{a}(\mathbf{a} \cdot \mathbf{b}) - \mathbf{b}(\mathbf{a} \cdot \mathbf{a})] = (\mathbf{a} \cdot \mathbf{b})\mathbf{a} \times \mathbf{a} - (\mathbf{a} \cdot \mathbf{a})(\mathbf{a} \times \mathbf{b})$$

$$= 0 + (\mathbf{a} \cdot \mathbf{a})(\mathbf{b} \times \mathbf{a}) \quad (\because \mathbf{a} \times \mathbf{a} = 0).$$

EXAMPLE 1.17. *Show that* $(\mathbf{a} \times \mathbf{b}) \times \mathbf{c} = \mathbf{a} \times (\mathbf{b} \times \mathbf{c})$ *only when* $\mathbf{a}$ *and* $\mathbf{c}$ *are collinear or* $(\mathbf{a} \times \mathbf{c}) \times \mathbf{b} = 0$.

Solution: Given, $(\mathbf{a} \times \mathbf{b}) \times \mathbf{c} = \mathbf{a} \times (\mathbf{b} \times \mathbf{c})$.

This is possible, if

$$\mathbf{b}(\mathbf{a} \cdot \mathbf{c}) + \mathbf{c}(\mathbf{a} \cdot \mathbf{b}) - \mathbf{b}(\mathbf{a} \cdot \mathbf{c}) - \mathbf{a}(\mathbf{b} \cdot \mathbf{c}) = 0$$

$$\text{or, if } \mathbf{c}(\mathbf{a} \cdot \mathbf{b}) - \mathbf{a}(\mathbf{b} \cdot \mathbf{c}) = 0$$

$$\text{or, if } (\mathbf{a} \times \mathbf{c}) \times \mathbf{b} = 0.$$

This shows that either $\mathbf{b} = 0$ or $\mathbf{a} \times \mathbf{c} = 0$.
But $\mathbf{b} \neq 0$, hence $\mathbf{a} \times \mathbf{c} = 0$.
Hence, $\mathbf{a}$ and $\mathbf{c}$ are collinear.

EXAMPLE 1.18. *Show that vectors* $\mathbf{a} \times (\mathbf{b} \times \mathbf{c})$, $\mathbf{b} \times (\mathbf{c} \times \mathbf{a})$, $\mathbf{c} \times (\mathbf{a} \times \mathbf{b})$ *are coplanar.*

Solution: Let $\mathbf{A} = \mathbf{a} \times (\mathbf{b} \times \mathbf{c}), \mathbf{B} = \mathbf{b} \times (\mathbf{c} \times \mathbf{a})$ and $\mathbf{C} = \mathbf{c} \times (\mathbf{a} \times \mathbf{b})$.

The vectors $\mathbf{A}$, $\mathbf{B}$, $\mathbf{C}$ will be coplanar if

$$[\mathbf{ABC}] = 0, \quad \text{i.e.,} \quad \mathbf{A} \cdot (\mathbf{B} \times \mathbf{C}) = 0$$

$$\text{or,} \quad [\mathbf{b}(\mathbf{a} \cdot \mathbf{c}) - \mathbf{c}(\mathbf{a} \cdot \mathbf{b})] \cdot [\{\mathbf{c}(\mathbf{b} \cdot \mathbf{a}) - \mathbf{a}(\mathbf{b} \cdot \mathbf{c})\} \times \{\mathbf{a}(\mathbf{c} \cdot \mathbf{b}) - \mathbf{b}(\mathbf{c} \cdot \mathbf{a})\}] = 0$$

$$\text{or,} \quad [\mathbf{b}(\mathbf{a} \cdot \mathbf{c}) - \mathbf{c}(\mathbf{a} \cdot \mathbf{b})] \cdot [(\mathbf{b} \cdot \mathbf{a})(\mathbf{c} \cdot \mathbf{b})\mathbf{c} \times \mathbf{a} - (\mathbf{b} \cdot \mathbf{a})(\mathbf{c} \cdot \mathbf{a})\mathbf{c} \times \mathbf{b}$$

$$- (\mathbf{b} \cdot \mathbf{c})(\mathbf{c} \cdot \mathbf{b})\mathbf{a} \times \mathbf{a} - (\mathbf{b} \cdot \mathbf{c})(\mathbf{c} \cdot \mathbf{a})\mathbf{a} \times \mathbf{b}] = 0.$$

Now, $\mathbf{b} \cdot (\mathbf{c} \times \mathbf{b}) = 0$, $\mathbf{c} \cdot (\mathbf{c} \times \mathbf{b}) = 0$, etc., and $\mathbf{a} \times \mathbf{a} = 0$.
$\therefore$ $(\mathbf{a} \cdot \mathbf{c})(\mathbf{b} \cdot \mathbf{a})(\mathbf{c} \cdot \mathbf{b})\mathbf{b} \cdot (\mathbf{c} \times \mathbf{a}) - (\mathbf{a} \cdot \mathbf{b})(\mathbf{b} \cdot \mathbf{c})(\mathbf{c} \cdot \mathbf{a})\mathbf{c} \cdot (\mathbf{a} \times \mathbf{b}) = 0$,
i.e., $(\mathbf{a} \cdot \mathbf{b})(\mathbf{b} \cdot \mathbf{c})(\mathbf{c} \cdot \mathbf{a}) \{[\mathbf{bca}] - [\mathbf{cab}]\} = 0$, which is true since $[\mathbf{bca}] = [\mathbf{cab}]$.
Hence proved.

1.15 Scalar Product of Four Vectors

If **a**, **b**, **c**, **d** are four vectors, their scalar product is written as $(\mathbf{a}\times\mathbf{b})\cdot(\mathbf{c}\times\mathbf{d})$.

Now, let us suppose that $\mathbf{c}\times\mathbf{d}=\mathbf{p}$. Then

$$\begin{aligned}(\mathbf{a}\times\mathbf{b})\cdot(\mathbf{c}\times\mathbf{d}) &= (\mathbf{a}\times\mathbf{b})\cdot\mathbf{p}=\mathbf{a}\cdot(\mathbf{b}\times\mathbf{p})=\mathbf{a}\cdot[\mathbf{b}\times\mathbf{c}\times\mathbf{d}]\\ &= \mathbf{a}\cdot[\mathbf{c}(\mathbf{b}\cdot\mathbf{d})-\mathbf{d}(\mathbf{b}\cdot\mathbf{c})]=(\mathbf{b}\cdot\mathbf{d})(\mathbf{a}\cdot\mathbf{c})-(\mathbf{b}\cdot\mathbf{c})(\mathbf{a}\cdot\mathbf{d}).\end{aligned}$$

$$\therefore\quad (\mathbf{a}\times\mathbf{b})\cdot(\mathbf{c}\times\mathbf{d})=\begin{vmatrix}\mathbf{a}\cdot\mathbf{c} & \mathbf{b}\cdot\mathbf{c}\\ \mathbf{a}\cdot\mathbf{d} & \mathbf{b}\cdot\mathbf{d}\end{vmatrix}. \tag{1.53}$$

1.16 Vector Product of Four Vectors

If **a**, **b**, **c**, **d** are four vectors, their vector product is written as $(\mathbf{a}\times\mathbf{b})\times(\mathbf{c}\times\mathbf{d})$.

Now, let $\mathbf{c}\times\mathbf{d}=\mathbf{p}$. Then

$$\begin{aligned}(\mathbf{a}\times\mathbf{b})\times(\mathbf{c}\times\mathbf{d}) &= (\mathbf{a}\times\mathbf{b})\times\mathbf{p}=\mathbf{b}(\mathbf{a}\cdot\mathbf{p})-\mathbf{a}(\mathbf{b}\cdot\mathbf{p})\\ &= [\mathbf{b}\{\mathbf{a}\cdot(\mathbf{c}\times\mathbf{d})\}-\mathbf{a}\{\mathbf{b}\cdot(\mathbf{c}\times\mathbf{d})\}]\\ &= \mathbf{b}[\mathbf{acd}]-\mathbf{a}[\mathbf{bcd}].\end{aligned} \tag{1.54}$$

Again, let us suppose that $\mathbf{a}\times\mathbf{b}=\mathbf{p}$.

$$\begin{aligned}\therefore\quad (\mathbf{a}\times\mathbf{b})\times(\mathbf{c}\times\mathbf{d}) &= \mathbf{p}\times(\mathbf{c}\times\mathbf{d})=\mathbf{c}(\mathbf{p}\cdot\mathbf{d})-\mathbf{d}(\mathbf{p}\cdot\mathbf{c})\\ &= \mathbf{c}\{(\mathbf{a}\times\mathbf{b})\cdot\mathbf{d}\}-\mathbf{d}\{(\mathbf{a}\times\mathbf{b})\cdot\mathbf{c}\}\\ &= \mathbf{c}[\mathbf{abd}]-\mathbf{d}[\mathbf{abc}].\end{aligned} \tag{1.55}$$

Thus, we get two expressions for vector product of four vectors given by equations (1.54) and (1.55).

EXAMPLE 1.19. *Show that, any four vectors* **a**, **b**, **c**, **d** *satisfy*

$$[\mathbf{abc}]\mathbf{d}=[\mathbf{bcd}]\mathbf{a}+[\mathbf{cad}]\mathbf{b}+[\mathbf{abd}]\mathbf{c}.$$

Solution: Equating the two expressions (1.54) and (1.55) for vector product of four vectors **a**, **b**, **c**, **d**, we have

$$\begin{aligned}\mathbf{b}[\mathbf{acd}]-\mathbf{a}[\mathbf{bcd}] &= \mathbf{c}[\mathbf{abd}]-\mathbf{d}[\mathbf{abc}]\\ \text{i.e.,}\quad \mathbf{d}[\mathbf{abc}] &= \mathbf{a}[\mathbf{bcd}]-\mathbf{b}[\mathbf{acd}]+\mathbf{c}[\mathbf{abd}].\end{aligned} \tag{1}$$

Since, for every change of cyclic order in a scalar triple product of vectors, there is a change of sign

$$\therefore\quad [\mathbf{acd}]=[\mathbf{dac}]=-[\mathbf{cad}].$$

$\therefore$ Equation (1) becomes

$$\mathbf{d}[\mathbf{abc}]=\mathbf{a}[\mathbf{bcd}]+\mathbf{b}[\mathbf{cad}]+\mathbf{c}[\mathbf{abd}].$$

Hence proved.

EXAMPLE 1.20. *Show that,* $(\mathbf{b}\times\mathbf{c})\times(\mathbf{c}\times\mathbf{a})=[\mathbf{abc}]\mathbf{c}$, *and deduce that*

$$[\mathbf{b}\times\mathbf{c},\mathbf{c}\times\mathbf{a},\mathbf{a}\times\mathbf{b}]=[\mathbf{abc}]^2.$$

Solution: Let, $\mathbf{c} \times \mathbf{a} = \mathbf{p}$. Then

$$\begin{aligned}(\mathbf{b} \times \mathbf{c}) \times (\mathbf{c} \times \mathbf{a}) &= (\mathbf{b} \times \mathbf{c}) \times \mathbf{p} = \mathbf{c}(\mathbf{b} \cdot \mathbf{p}) - \mathbf{b}(\mathbf{c} \cdot \mathbf{p}) \\ &= \mathbf{c}\{\mathbf{b} \cdot (\mathbf{c} \times \mathbf{a})\} - \mathbf{b}\{\mathbf{c} \cdot (\mathbf{c} \times \mathbf{a})\} = \mathbf{c}[\mathbf{cab}] - \mathbf{b}[\mathbf{cac}] \\ &= \mathbf{c}[\mathbf{abc}], \quad \because \ [\mathbf{cac}] = 0. \qquad (1)\end{aligned}$$

Now, multiplying both sides of (1) scalarly by $\mathbf{a} \times \mathbf{b}$, we get

$$\begin{aligned}(\mathbf{a} \times \mathbf{b}) \cdot [(\mathbf{b} \times \mathbf{c}) \times (\mathbf{c} \times \mathbf{a})] &= (\mathbf{a} \times \mathbf{b}) \cdot [\mathbf{abc}]\mathbf{c} = [\mathbf{abc}]\mathbf{a} \times \mathbf{b} \cdot \mathbf{c} \\ &= [\mathbf{abc}][\mathbf{abc}] = [\mathbf{abc}]^2.\end{aligned}$$

Hence proved.

EXAMPLE 1.21. *Prove that*

$$(\mathbf{b} \times \mathbf{c}) \cdot (\mathbf{a} \times \mathbf{d}) + (\mathbf{c} \times \mathbf{a}) \cdot (\mathbf{b} \times \mathbf{d}) + (\mathbf{a} \times \mathbf{b}) \cdot (\mathbf{c} \times \mathbf{d}) = 0.$$

Solution: Following the equation (1.53), we can write

$$\begin{aligned}\text{LHS} &= \begin{vmatrix} \mathbf{b} \cdot \mathbf{a} & \mathbf{b} \cdot \mathbf{d} \\ \mathbf{c} \cdot \mathbf{a} & \mathbf{c} \cdot \mathbf{d} \end{vmatrix} + \begin{vmatrix} \mathbf{c} \cdot \mathbf{b} & \mathbf{c} \cdot \mathbf{d} \\ \mathbf{a} \cdot \mathbf{b} & \mathbf{a} \cdot \mathbf{d} \end{vmatrix} + \begin{vmatrix} \mathbf{a} \cdot \mathbf{c} & \mathbf{a} \cdot \mathbf{d} \\ \mathbf{b} \cdot \mathbf{c} & \mathbf{b} \cdot \mathbf{d} \end{vmatrix} \\ &= (\mathbf{b} \cdot \mathbf{a})(\mathbf{c} \cdot \mathbf{d}) - (\mathbf{b} \cdot \mathbf{d})(\mathbf{c} \cdot \mathbf{a}) + (\mathbf{c} \cdot \mathbf{b})(\mathbf{a} \cdot \mathbf{d}) - (\mathbf{c} \cdot \mathbf{d})(\mathbf{a} \cdot \mathbf{b}) \\ &\qquad + (\mathbf{a} \cdot \mathbf{c})(\mathbf{b} \cdot \mathbf{d}) - (\mathbf{a} \cdot \mathbf{d})(\mathbf{b} \cdot \mathbf{c}) \\ &= 0.\end{aligned}$$

Hence proved.

1.17 Reciprocal System of Vectors

Let there be a set of three non-coplanar vectors **a**, **b**, **c**. The set of other three vectors **A**, **B**, **C** defined by the equations

$$\mathbf{A} = \frac{\mathbf{b} \times \mathbf{c}}{[\mathbf{abc}]}, \quad \mathbf{B} = \frac{\mathbf{c} \times \mathbf{a}}{[\mathbf{abc}]}, \quad \mathbf{C} = \frac{\mathbf{a} \times \mathbf{b}}{[\mathbf{abc}]} \qquad (1.56)$$

are called *reciprocal system to the vectors* **a**, **b**, **c**.

The concept of reciprocal vectors has got much importance in Solid State Physics in connection with reciprocal lattice.

1.18 Properties of Reciprocal Systems

Property 1. If **a**, **b**, **c** and **A**, **B**, **C** be reciprocal systems of vectors, then

$$\mathbf{a} \cdot \mathbf{A} = \mathbf{b} \cdot \mathbf{B} = \mathbf{c} \cdot \mathbf{C} = 1. \qquad (1.57)$$

Proof. We have

$$\mathbf{A} = \frac{\mathbf{b} \times \mathbf{c}}{[\mathbf{abc}]}. \qquad \therefore \quad \mathbf{a} \cdot \mathbf{A} = \mathbf{a} \cdot \frac{\mathbf{b} \times \mathbf{c}}{[\mathbf{abc}]} = \frac{[\mathbf{abc}]}{[\mathbf{abc}]} = 1.$$

Similarly

$$\mathbf{b} \cdot \mathbf{B} = \mathbf{c} \cdot \mathbf{C} = 1.$$

Hence

$$\mathbf{a}\cdot\mathbf{A}+\mathbf{b}\cdot\mathbf{B}+\mathbf{c}\cdot\mathbf{C}=3 \quad \text{and} \quad \mathbf{A}=\frac{1}{\mathbf{a}},\quad \mathbf{B}=\frac{1}{\mathbf{b}},\quad \mathbf{C}=\frac{1}{\mathbf{c}}.$$

□

Property 2. The product of any vector of one system with a vector of reciprocal system which does not correspond to it is zero

$$\text{i.e.,}\quad \mathbf{a}\cdot\mathbf{B}=0,\quad \mathbf{b}\cdot\mathbf{C}=0,\ldots,\ \text{etc.},\ldots \tag{1.58}$$

Proof. We have

$$\mathbf{a}\cdot\mathbf{B}=\mathbf{a}\cdot\frac{\mathbf{c}\times\mathbf{a}}{[\mathbf{abc}]}=\frac{[\mathbf{aca}]}{[\mathbf{abc}]}=0,$$

since $[\mathbf{aca}]$ being the scalar triple product of three vectors two of which are equal is zero. □

Property 3. The scalar triple product of any three non-coplanar vectors **a, b, c** is reciprocal to the corresponding scalar triple product of reciprocal vectors **A, B, C**.

Proof. We have

$$[\mathbf{ABC}]=\mathbf{A}\cdot(\mathbf{B}\times\mathbf{C})=\frac{(\mathbf{b}\times\mathbf{c})\cdot[(\mathbf{c}\times\mathbf{a})\times(\mathbf{a}\times\mathbf{b})]}{[\mathbf{abc}]^3}\quad\text{[using equation (1.56)]}.$$

Now

$$\begin{aligned}(\mathbf{c}\times\mathbf{a})\times(\mathbf{a}\times\mathbf{b}) &= (\mathbf{c}\times\mathbf{a})\times\mathbf{p},\quad \text{where } \mathbf{p}=\mathbf{a}\times\mathbf{b}\\ &= \mathbf{a}(\mathbf{p}\cdot\mathbf{c})-\mathbf{c}(\mathbf{p}\cdot\mathbf{a})\\ &= (\mathbf{a}\times\mathbf{b})\cdot\mathbf{ca}-(\mathbf{a}\times\mathbf{b})\cdot\mathbf{ac}\\ &= [\mathbf{abc}]\mathbf{a},\quad \because\ (\mathbf{a}\times\mathbf{b})\cdot\mathbf{a}=0.\end{aligned}$$

$$\therefore\quad [\mathbf{ABC}]=\frac{(\mathbf{b}\times\mathbf{c})\cdot[\mathbf{abc}]\mathbf{a}}{[\mathbf{abc}]^3}=\frac{(\mathbf{b}\times\mathbf{c})\cdot\mathbf{a}}{[\mathbf{abc}]^2}=\frac{[\mathbf{abc}]}{[\mathbf{abc}]^2}.$$

$$\therefore\quad [\mathbf{ABC}]=\frac{1}{[\mathbf{abc}]}. \tag{1.59}$$

□

EXAMPLE 1.22. *If*

$$\mathbf{A}=\frac{\mathbf{b}\times\mathbf{c}}{[\mathbf{abc}]},\quad \mathbf{B}=\frac{\mathbf{c}\times\mathbf{a}}{[\mathbf{abc}]},\quad \mathbf{C}=\frac{\mathbf{a}\times\mathbf{b}}{[\mathbf{abc}]},$$

then prove that

$$\mathbf{a}=\frac{\mathbf{B}\times\mathbf{C}}{[\mathbf{ABC}]},\quad \mathbf{b}=\frac{\mathbf{C}\times\mathbf{A}}{[\mathbf{ABC}]},\quad \mathbf{c}=\frac{\mathbf{A}\times\mathbf{B}}{[\mathbf{ABC}]}.$$

Solution: We have

$$\mathbf{B}\times\mathbf{C}=\frac{(\mathbf{c}\times\mathbf{a})\times(\mathbf{a}\times\mathbf{b})}{[\mathbf{abc}]^2}=\frac{(\mathbf{c}\times\mathbf{a})\cdot\mathbf{ba}-(\mathbf{c}\times\mathbf{a})\cdot\mathbf{ab}}{[\mathbf{abc}]^2}\quad(\because\ (\mathbf{c}\times\mathbf{a})\cdot\mathbf{a}=0)$$

$$=\frac{[\mathbf{cab}]\cdot\mathbf{a}}{[\mathbf{abc}]^2}=\frac{1}{[\mathbf{abc}]}\cdot\mathbf{a}.$$

Again

$$[\mathbf{ABC}]=\frac{1}{[\mathbf{abc}]}\quad\text{[by equation (1.59)]}$$

$$\therefore\quad\frac{\mathbf{B}\times\mathbf{C}}{[\mathbf{ABC}]}=\frac{1}{[\mathbf{abc}]}\cdot\mathbf{a}[\mathbf{abc}]=\mathbf{a}.$$

Similarly, we can prove other results.

1.19 Vectors in Mechanics

1. Centre of Mass

Consider particles of masses m_1, m_2, m_3, ... placed at points whose position vectors with respect to the origin O are **a**, **b**, **c**, ... respectively. Then the position vector of their centre of mass G will be

$$\overrightarrow{OG}=\mathbf{r}=\frac{m_1\mathbf{a}+m_2\mathbf{b}+m_3\mathbf{c}+\ldots}{m_1+m_2+m_3+\ldots}.\tag{1.60}$$

If the position vectors **a**, **b**, **c**, ... are given in terms of unit vectors **i**, **j**, **k** in the form

$$\mathbf{a}=x_1\mathbf{i}+y_1\mathbf{j}+z_1\mathbf{k}$$
$$\mathbf{b}=x_2\mathbf{i}+y_2\mathbf{j}+z_2\mathbf{k}$$
$$\mathbf{c}=x_3\mathbf{i}+y_3\mathbf{j}+z_3\mathbf{k},\quad\text{etc.,}$$

and the position vector of centre of mass is written in the form $\mathbf{r}=x\mathbf{i}+y\mathbf{j}+z\mathbf{k}$, then from equation (1.60) we get

$$x\mathbf{i}+y\mathbf{j}+z\mathbf{k}=\frac{\sum(m_1x_1)\mathbf{i}+\sum(m_1y_1)\mathbf{j}+\sum(m_1z_1)\mathbf{k}}{\sum m_1}.$$

Equating the coefficients of **i**, **j**, **k** on both sides, we get the coordinates of centre of mass

$$x=\frac{\sum m_1x_1}{\sum m_1},\quad y=\frac{\sum m_1y_1}{\sum m_1},\quad z=\frac{\sum m_1z_1}{\sum m_1}.\tag{1.61}$$

EXAMPLE 1.23. *Particles of masses* 4, 6, 10 *grams are placed at points* $A(2\mathbf{i}+3\mathbf{j}-\mathbf{k})$, $B(-\mathbf{i}+4\mathbf{j}+5\mathbf{k})$ *and* $C(6\mathbf{i}-\mathbf{j}+2\mathbf{k})$ *respectively. Find the position vector of the centre of masses.*

Solution: Let **r** be the position vector of the centre of masses. Then

$$\mathbf{r}=\frac{4(2\mathbf{i}+3\mathbf{j}-\mathbf{k})+6(-\mathbf{i}+4\mathbf{j}+5\mathbf{k})+10(6\mathbf{i}-\mathbf{j}+2\mathbf{k})}{4+6+10}$$

$$\text{or,}\quad\mathbf{r}=\frac{1}{20}(62\mathbf{i}-26\mathbf{j}+46\mathbf{k}).$$

2. Work Done by a Force

Let a particle be displaced from its initial position at P to final position at Q under the action of a constant force **F**. Let the position vectors of P, Q relative to an origin O be respectively **a** and **b** (*Fig 1.11*).

Then, $\overrightarrow{OP} = \mathbf{a}$, $\overrightarrow{OQ} = \mathbf{b}$.

$$\therefore \quad \overrightarrow{PQ} = \overrightarrow{OQ} - \overrightarrow{OP} = \mathbf{b} - \mathbf{a}.$$

The work done by the force **F**

$$= F\cos\theta \left|\overrightarrow{PQ}\right| = \mathbf{F}\cdot\overrightarrow{PQ} = \mathbf{F}\cdot(\mathbf{b}-\mathbf{a}). \quad (1.62)$$

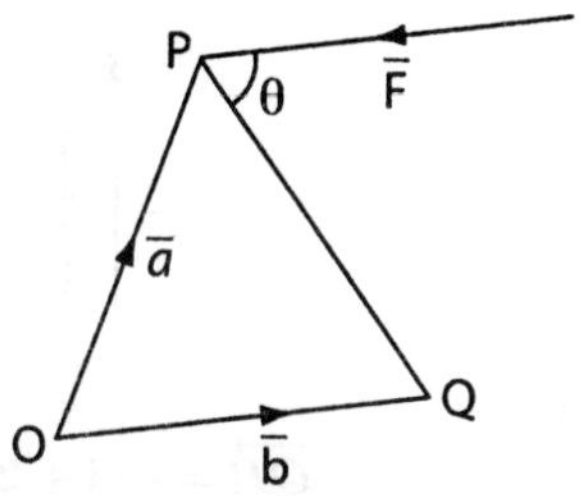

Fig 1.11

EXAMPLE 1.24. *A particle under the action of a constant force* $\mathbf{F} = 3\mathbf{i}+4\mathbf{j}+\mathbf{k}$ *is displaced from the point* $(\mathbf{i} - 3\mathbf{j} + 4\mathbf{k})$ *to* $(5\mathbf{i} + \mathbf{j} - 2\mathbf{k})$. *Calculate the work done by the force.*

Solution:

$$\text{Displacement} = (5\mathbf{i} + \mathbf{j} - 2\mathbf{k}) - (\mathbf{i} - 3\mathbf{j} + 4\mathbf{k}) = 4\mathbf{i} + 4\mathbf{j} - 6\mathbf{k}.$$

$$\therefore \quad \text{Work done} = (3\mathbf{i} + 4\mathbf{j} + \mathbf{k}) \cdot (4\mathbf{i} + 4\mathbf{j} - 6\mathbf{k}) = 12 + 16 - 6 = 22 \text{ units.}$$

3. Moment of a Force

Let a force **F** be acting at a point P having position vector **a**. Let Q be another point with position vector **b** about which the moment of the force **F** is required.

Then, $\overrightarrow{OP} = \mathbf{a}$, $\overrightarrow{OQ} = \mathbf{b}$.

$$\therefore \quad \overrightarrow{PQ} = \overrightarrow{OQ} - \overrightarrow{OP} = \mathbf{b} - \mathbf{a}.$$

QN is drawn perpendicular to the line of action of the force **F** (*Fig 1.12*).

Then moment of the force **F** about the point Q is $\mathbf{F}\cdot\overrightarrow{QN}$.

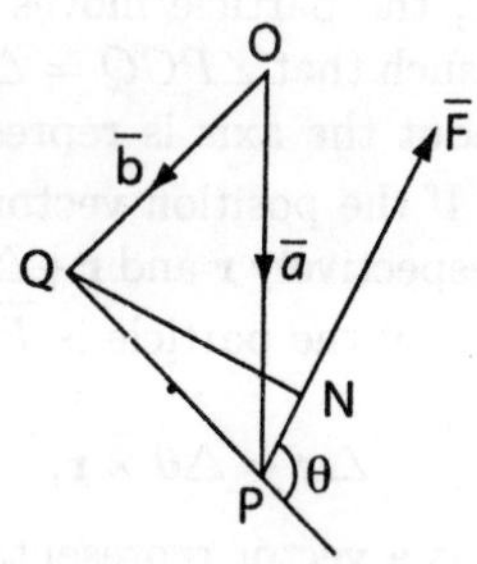

Fig 1.12

Now, $QN = PQ\sin\theta$, where θ is the angle between $\overrightarrow{PQ}$ and **F**.

$\therefore$ The magnitude of the moment is

$$\mathbf{F}\cdot PQ\sin\theta = |\,\mathbf{F}\times(\mathbf{b}-\mathbf{a})\,|. \quad (1.63)$$

In order to find moment of a force about a line, it is necessary to calculate the moment of the force about any point on that line. The resolved part of this moment along the given line will be the required moment of the force about the given line.

EXAMPLE 1.25. *A force* $(5\mathbf{i}-3\mathbf{j}+4\mathbf{k})$ *is acting at a point* $P(\mathbf{i}+2\mathbf{j}+3\mathbf{k})$. *Find the moment of the force about the point* $Q(2\mathbf{i} - 3\mathbf{j} + 5\mathbf{k})$. *Also, find the moment of the force about the line joining* Q *and* $R(-\mathbf{i} + 2\mathbf{j} + \mathbf{k})$.

Solution: Here, $\overrightarrow{OP} = \mathbf{i} + 2\mathbf{j} + 3\mathbf{k}$, $\overrightarrow{OQ} = 2\mathbf{i} - 3\mathbf{j} + 5\mathbf{k}$.

$$\therefore \quad \overrightarrow{PQ} = \overrightarrow{OQ} - \overrightarrow{OP} = \mathbf{i} - 5\mathbf{j} + 2\mathbf{k}.$$

$\therefore$ Moment of the force about Q

$$= \mathbf{F} \times \overrightarrow{PQ} = (5\mathbf{i} - 3\mathbf{j} + 4\mathbf{k}) \times (\mathbf{i} - 5\mathbf{j} + 2\mathbf{k})$$

$$= \begin{vmatrix} \mathbf{i} & \mathbf{j} & \mathbf{k} \\ 5 & -3 & 4 \\ 1 & -5 & 2 \end{vmatrix} = 14\mathbf{i} - 6\mathbf{j} - 22\mathbf{k}.$$

Again

$$\overrightarrow{QR} = (-\mathbf{i} + 2\mathbf{j} + \mathbf{k}) - (2\mathbf{i} - 3\mathbf{j} + 5\mathbf{k}) = -3\mathbf{i} + 5\mathbf{j} - 4\mathbf{k}.$$

$\therefore$ Moment of the force about the direction $\overrightarrow{QR}$

= Resolved part of moment of the force about Q along QR

$$= (14\mathbf{i} - 6\mathbf{j} - 22\mathbf{k}) \cdot (-3\mathbf{i} + 5\mathbf{j} - 4\mathbf{k})$$
$$= -42 - 30 + 88 = 16 \text{ units.}$$

4. Angular Velocity, Angular Momentum

Let a particle of mass m be rotating with a constant angular velocity $\boldsymbol{\omega}$ about a fixed axis OA (*Fig 1.13*), where O is the origin.

Suppose, the particle moves from P to Q in time Δt such that $\angle PCQ = \Delta\theta$. This small rotation about the axis is represented by the vector $\overrightarrow{CA}$. If the position vectors of points P and Q be respectively $\mathbf{r}$ and $\mathbf{r}+\Delta\mathbf{r}$, then linear displacement of the particle is $\overrightarrow{PQ} = \Delta\mathbf{r}$.

Then

$$\Delta\mathbf{r} = \Delta\boldsymbol{\theta} \times \mathbf{r}, \quad (1.64)$$

where, $\Delta\boldsymbol{\theta}$ is a vector represented by $\overrightarrow{CA}$.

Angular velocity is the rate of change of angular displacement with respect to time.

Hence

$$\boldsymbol{\omega} = \lim_{\Delta t \longrightarrow 0} \frac{\Delta\boldsymbol{\theta}}{\Delta t} = \frac{d\boldsymbol{\theta}}{dt}. \quad (1.65)$$

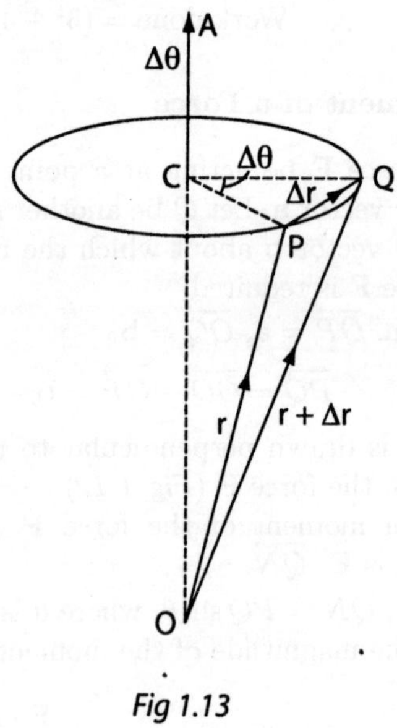

Fig 1.13

Thus, angular velocity $\boldsymbol{\omega}$ is a vector along the direction $\overrightarrow{CA}$.

Dividing both sides of equation (1.64) by Δt and proceeding to the limit as $\Delta t \longrightarrow 0$, we get

$$\lim_{\Delta t \longrightarrow 0} \frac{\Delta\mathbf{r}}{\Delta t} = \lim_{\Delta t \longrightarrow 0} \frac{\Delta\boldsymbol{\theta}}{\Delta t} \times \mathbf{r}, \quad \text{i.e.,} \quad \mathbf{v} = \boldsymbol{\omega} \times \mathbf{r}. \quad (1.66)$$

Angular momentum $\mathbf{L}$ is also regarded as vectors and represented in a similar manner.

In this case, linear momentum of the particle $\mathbf{p} = m\mathbf{v}$; the angular momentum $\mathbf{L}$ is given by

$$\mathbf{L} = \mathbf{r} \times \mathbf{p}. \quad (1.67)$$

1.20 Miscellaneous Problems

EXAMPLE 1.26. *Prove that medians of a triangle are concurrent and find the point of concurrency.*

Solution: Let **a**, **b**, **c** be the position vectors of the vertices A, B, C of the triangle ABC and D, E, F are the middle points of sides (*Fig 1.14*).

$\therefore$ Position vectors of D, E, F are $\frac{1}{2}(\mathbf{b}+\mathbf{c})$, $\frac{1}{2}(\mathbf{c}+\mathbf{a})$, $\frac{1}{2}(\mathbf{a}+\mathbf{b})$ respectively.

Now, the point G which divides AD in the ratio 2 : 1 is

$$\frac{1}{3}\left[2\frac{\mathbf{b}+\mathbf{c}}{2}+1\cdot\mathbf{a}\right]=\frac{1}{3}(\mathbf{a}+\mathbf{b}+\mathbf{c}). \qquad (1)$$

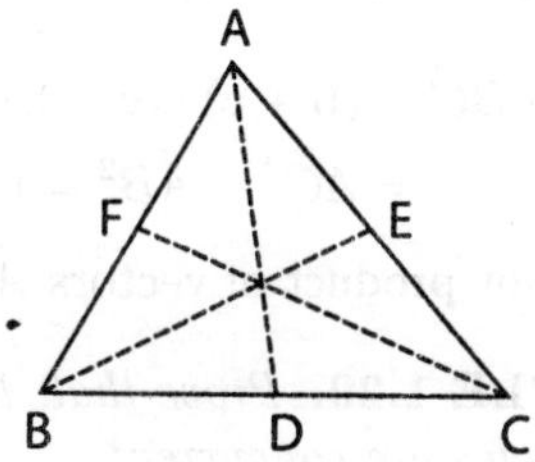

Fig 1.14

In the same manner, position vector of G can be found considering BE and CF, and in each case it will be same as (1).

Hence, three medians are concurrent and meet at the point $\frac{1}{3}(\mathbf{a}+\mathbf{b}+\mathbf{c})$ which is also centroid of the triangle.

EXAMPLE 1.27. *Find the vector equation of a straight line passing through a given point* **a** *and parallel to a given vector* **b**.

Solution: Let A be the given point having position vector **a** relative to origin O and P be any point on the line through A which is parallel to **b** (*Fig 1.15*) having position vector **r**.

Now, $\overrightarrow{AP}$ being parallel to **b**, can be written as $\overrightarrow{AP}=\lambda\mathbf{b}$, where λ is a constant, positive or negative.

Since, $\overrightarrow{OP}=\overrightarrow{OA}+\overrightarrow{AP}$

$$\therefore \quad \mathbf{r}=\mathbf{a}+\lambda\mathbf{b}, \qquad (1)$$

which is the vector equation of the required line.

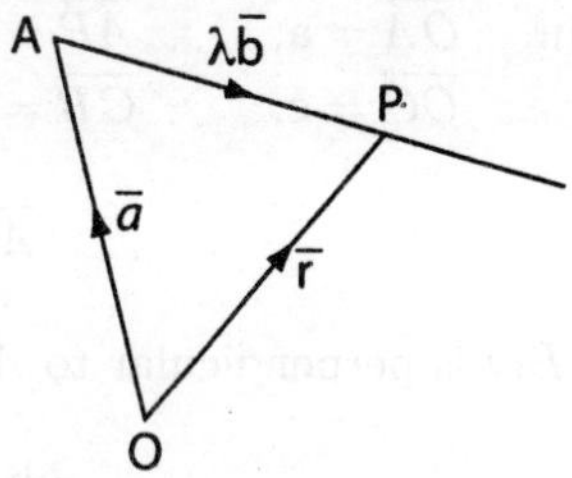

Fig 1.15

If the line passes through origin, then $\mathbf{a}=0$, and the equation of the required line becomes

$$\mathbf{r}=\lambda\mathbf{b}. \qquad (2)$$

If the line passes through two given points **a** and **b**, it becomes parallel to the vector $(\mathbf{b}-\mathbf{a})$, and hence its equation will be

$$\mathbf{r}=\mathbf{a}+\lambda(\mathbf{b}-\mathbf{a}). \qquad (3)$$

EXAMPLE 1.28. *Prove that the diagonals of a rhombus are at right angles.*

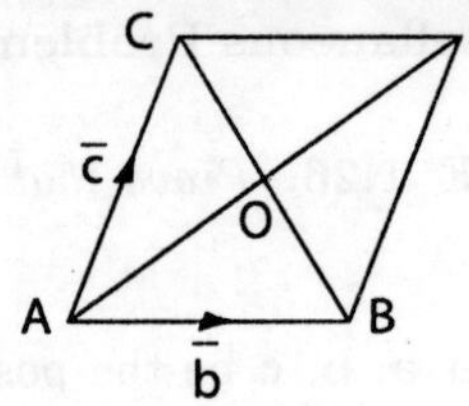

Fig 1.16

Solution: In the *Fig 1.16*, $ABCD$ is a rhombus whose diagonals are AD and BC.

$$\therefore \quad AB = AC.$$

Now, $\overrightarrow{AD} = \mathbf{b} + \mathbf{c}$ and $\overrightarrow{BC} = \mathbf{c} - \mathbf{b}$.

$$\therefore \quad \overrightarrow{AD} \cdot \overrightarrow{BC} = (\mathbf{b} + \mathbf{c}) \cdot (\mathbf{c} - \mathbf{b}) = \mathbf{c}^2 - \mathbf{b}^2$$

$$= AC^2 - AB^2 = 0.$$

Since dot product of vectors $\overrightarrow{AD}$ and $\overrightarrow{BC}$ is zero, they are perpendicular.

EXAMPLE 1.29. *Prove that, in any triangle, the perpendiculars from the vertices upon the opposite sides are concurrent.*

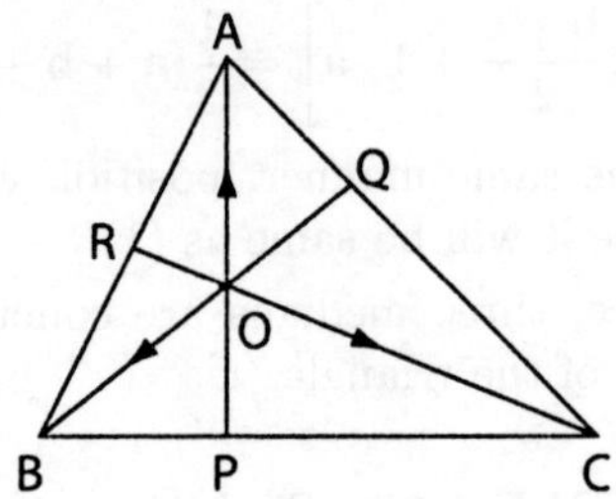

Fig 1.17

Solution: Let the perpendiculars from vertices B, C of the triangle upon opposite sides meet at the point O (*Fig 1.17*). With this point as origin, let the position vectors of vertices A, B, C be $\mathbf{a}$, $\mathbf{b}$, $\mathbf{c}$ respectively. The perpendiculars meet the opposite sides at points Q, R. Draw the line AO and extend it to meet BC at P.

$$\begin{array}{llll} \text{Now,} & \overrightarrow{OB} = \mathbf{b}, & \therefore\ \overrightarrow{BQ} = -\mu\mathbf{b}, & \text{where } \mu = \text{constant.} \\ \text{Again,} & \overrightarrow{OA} = \mathbf{a}, & \therefore\ \overrightarrow{AP} = -\lambda\mathbf{a}, & \text{where } \lambda = \text{constant} \\ \text{and} & \overrightarrow{OC} = \mathbf{c}, & \therefore\ \overrightarrow{CR} = -\nu\mathbf{c}, & \text{where } \nu = \text{constant.} \end{array}$$

$$\therefore \quad \overrightarrow{AC} = \mathbf{c} - \mathbf{a} \quad \text{and} \quad \overrightarrow{AB} = \mathbf{b} - \mathbf{a}.$$

Since BQ is perpendicular to AC,

$$\therefore \quad -\mu\mathbf{b} \cdot (\mathbf{c} - \mathbf{a}) = 0, \quad \text{i.e.,} \quad \mathbf{a} \cdot \mathbf{b} = \mathbf{b} \cdot \mathbf{c}. \tag{1}$$

Again, CR is perpendicular to AB,

$$\therefore \quad -\nu\mathbf{c} \cdot (\mathbf{b} - \mathbf{a}) = 0, \quad \text{i.e.,} \quad \mathbf{b} \cdot \mathbf{c} = \mathbf{c} \cdot \mathbf{a}. \tag{2}$$

$\therefore$ $\mathbf{a} \cdot \mathbf{b} = \mathbf{b} \cdot \mathbf{c} = \mathbf{c} \cdot \mathbf{a}$, i.e., $\lambda\mathbf{a} \cdot (\mathbf{c} - \mathbf{b}) = 0$.
Hence, AP is perpendicular to BC.

EXAMPLE 1.30. *If D, E, F are middle points of the sides of a triangle ABC, show that*

$$\Delta DEF = \tfrac{1}{4}\Delta ABC.$$

Solution: Let $\mathbf{a}$, $\mathbf{b}$, $\mathbf{c}$ be position vectors of the vertices A, B, C of the triangle (*Fig 1.14*).
Then, $\overrightarrow{BC} = \mathbf{c} - \mathbf{b}$, $\overrightarrow{BA} = \mathbf{a} - \mathbf{b}$.

$$\therefore \quad \Delta ABC = \tfrac{1}{2}\overrightarrow{BC} \times \overrightarrow{BA} = \tfrac{1}{2}(\mathbf{c} - \mathbf{b})(\mathbf{a} - \mathbf{b}). \tag{1}$$

Now, position vectors of D, E, F are: $D = \frac{1}{2}(\mathbf{b}+\mathbf{c})$, $E = \frac{1}{2}(\mathbf{c}+\mathbf{a})$, $F = \frac{1}{2}(\mathbf{a}+\mathbf{b})$.

$$\therefore \quad \overrightarrow{EF} = \tfrac{1}{2}\left[(\mathbf{a}+\mathbf{b}) - (\mathbf{c}+\mathbf{a})\right] = \tfrac{1}{2}(\mathbf{b}-\mathbf{c})$$

$$\text{and} \quad \overrightarrow{ED} = \tfrac{1}{2}\left[(\mathbf{b}+\mathbf{c}) - (\mathbf{c}+\mathbf{a})\right] = \tfrac{1}{2}(\mathbf{b}-\mathbf{a}).$$

$$\therefore \quad \triangle DEF = \tfrac{1}{2}\overrightarrow{EF} \times \overrightarrow{ED} = \tfrac{1}{2}\left[\tfrac{1}{2}(\mathbf{b}-\mathbf{c}) \times \tfrac{1}{2}(\mathbf{b}-\mathbf{a})\right] = \tfrac{1}{4}\cdot\tfrac{1}{2}(\mathbf{c}-\mathbf{b}) \times (\mathbf{a}-\mathbf{b})$$
$$= \tfrac{1}{4}\triangle ABC \quad \text{[by equation (1)]}$$

EXAMPLE 1.31. *Prove vectorially that, in any triangle ABC,*

$$a = b\cos C + c\cos B \quad \textit{and} \quad a^2 = b^2 + c^2 - 2bc\cos A.$$

Solution: Let, $\overrightarrow{BC} = \mathbf{a}$, $\overrightarrow{CA} = \mathbf{b}$ and $\overrightarrow{AB} = \mathbf{c}$ (*Fig 1.18*).

Then, $\overrightarrow{BC} + \overrightarrow{CA} + \overrightarrow{AB} = 0$.

$$\therefore \quad \overrightarrow{BC} = -\left(\overrightarrow{CA} + \overrightarrow{AB}\right). \qquad (1)$$

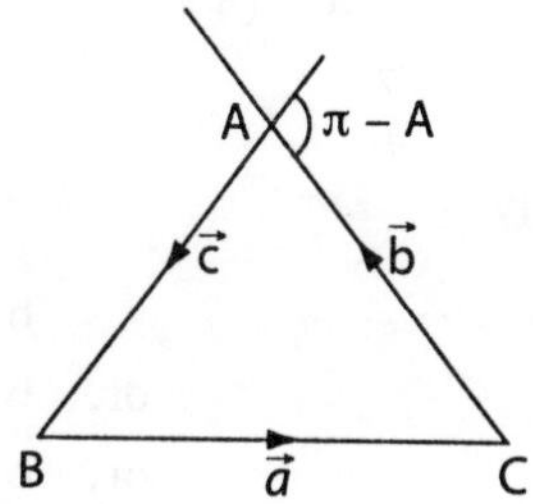

Fig 1.18

Squaring both sides

$$\overrightarrow{BC}^2 = \overrightarrow{CA}^2 + \overrightarrow{AB}^2 + 2\overrightarrow{CA}\cdot\overrightarrow{AB}$$
$$\text{or,} \quad a^2 = b^2 + c^2 + 2bc\cos(\pi - A)$$
$$= b^2 + c^2 - 2bc\cos A.$$

Again

$$\overrightarrow{BC}\cdot\overrightarrow{BC} = -\left(\overrightarrow{CA} + \overrightarrow{AB}\right)\cdot\overrightarrow{BC} \quad \text{[from (1)]}$$
$$\text{or,} \quad \overrightarrow{BC}^2 = -\overrightarrow{CA}\cdot\overrightarrow{BC} - \overrightarrow{AB}\cdot\overrightarrow{BC}$$
$$\text{or,} \quad a^2 = -ba\cos(\pi - C) - ca\cos(\pi - B)$$
$$\text{or,} \quad a = b\cos C + c\cos B.$$

EXAMPLE 1.32. *Prove by vector method,* $\cos(\alpha - \beta) = \cos\alpha\cos\beta + \sin\alpha\sin\beta$.

Solution: Let, OP and OQ be any two lines in the x-y plane making angles α and β with x-axis respectively. Let $\mathbf{i}, \mathbf{j}$ be unit vectors along x and y axes (*Fig 1.19*).

If $\overrightarrow{OA}$ and $\overrightarrow{OB}$ are unit vectors along OP and OQ, then

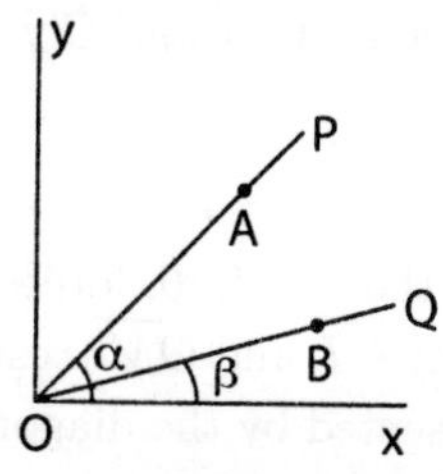

Fig 1.19

$$\overrightarrow{OA}\cdot\overrightarrow{OB} = 1\cdot 1\cdot\cos(\alpha - \beta) = \cos(\alpha - \beta). \quad (1)$$

But, $\overrightarrow{OA}$ and $\overrightarrow{OB}$ can be expressed as

$$\overrightarrow{OA} = \mathbf{i}\cos\alpha + \mathbf{j}\sin\alpha, \quad \overrightarrow{OB} = \mathbf{i}\cos\beta + \mathbf{j}\sin\beta.$$
$$\therefore \quad \overrightarrow{OA}\cdot\overrightarrow{OB} = (\mathbf{i}\cos\alpha + \mathbf{j}\sin\alpha)\cdot(\mathbf{i}\cos\beta + \mathbf{j}\sin\beta)$$
$$= \cos\alpha\cos\beta + \sin\alpha\sin\beta. \qquad (2)$$

From (1) and (2), we get

$$\cos(\alpha - \beta) = \cos\alpha\cos\beta + \sin\alpha\sin\beta.$$

EXAMPLE 1.33. *Prove by vector method, that in any triangle*

$$\frac{\sin A}{a} = \frac{\sin B}{b} = \frac{\sin C}{c},$$

the notations having usual meanings.

Solution: Let the sides of the triangle be represented by vectors **a**, **b**, **c**.
Then, $\mathbf{a} + \mathbf{b} + \mathbf{c} = 0$.

$$\therefore \quad \mathbf{a} \times (\mathbf{a} + \mathbf{b} + \mathbf{c}) = 0, \quad \text{or,} \quad \mathbf{a} \times \mathbf{b} + \mathbf{a} \times \mathbf{c} = 0, \quad \because \mathbf{a} \times \mathbf{a} = 0$$

$$\text{i.e.,} \quad \mathbf{a} \times \mathbf{b} = \mathbf{c} \times \mathbf{a}. \tag{1}$$

Similarly

$$\mathbf{b} \times (\mathbf{a} + \mathbf{b} + \mathbf{c}) = 0$$

$$\text{or,} \quad \mathbf{b} \times \mathbf{a} + \mathbf{b} \times \mathbf{c} = 0, \quad \because \mathbf{b} \times \mathbf{b} = 0$$

$$\text{or,} \quad \mathbf{b} \times \mathbf{c} = \mathbf{a} \times \mathbf{b}. \tag{2}$$

From (1) and (2), we get

$$\mathbf{b} \times \mathbf{c} = \mathbf{c} \times \mathbf{a} = \mathbf{a} \times \mathbf{b}$$

$$\text{or,} \quad bc\sin(\pi - A) = ca\sin(\pi - B) = ab\sin(\pi - C)$$

(taking magnitudes only).

Dividing throughout by *abc*,

$$\frac{\sin A}{a} = \frac{\sin B}{b} = \frac{\sin C}{c}.$$

EXAMPLE 1.34. *Forces P, Q act at O and have a resultant R. If any transversal cuts their lines of action at A, B and C respectively, prove that*

$$\frac{P}{OA} + \frac{Q}{OB} = \frac{R}{OC}.$$

Solution: In the *Fig 1.20*, forces P and Q are represented by $\overrightarrow{OL}$ and $\overrightarrow{OM}$; resultant force R will be represented by the diagonal $\overrightarrow{ON}$. Then

$$\vec{P} + \vec{Q} = \vec{R}. \tag{1}$$

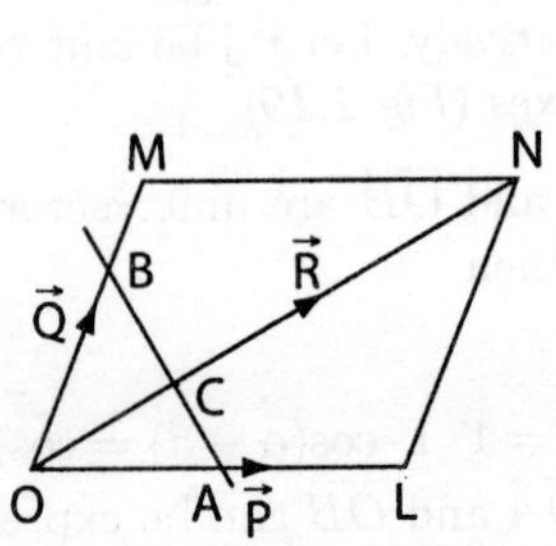

Fig 1.20

Now, let $\overrightarrow{OA} = \mathbf{a}$, $\overrightarrow{OB} = \mathbf{b}$.

$$\therefore \quad \overrightarrow{AB} = \overrightarrow{OB} - \overrightarrow{OA} = \mathbf{b} - \mathbf{a},$$

so that we can write

$$\overrightarrow{AC} = \alpha(\mathbf{b} - \mathbf{a}), \quad \alpha = \text{constant}.$$

$$\therefore \quad \overrightarrow{OC} = \overrightarrow{OA} + \overrightarrow{AC} = \mathbf{a} + \alpha(\mathbf{b} - \mathbf{a}) = (1 - \alpha)\mathbf{a} + \alpha\mathbf{b}. \tag{2}$$

Again, vectors $\vec{P}$, $\vec{Q}$ and $\vec{R}$ can be expressed as

$$\vec{P} = m \cdot \overrightarrow{OA}, \quad \vec{Q} = n \cdot \overrightarrow{OB} \quad \text{and} \quad \vec{R} = p \cdot \overrightarrow{OC}, \tag{3}$$

where m, n and p are constants.

From (3), we have

$$\frac{P}{OA} = m, \quad \frac{Q}{OB} = n, \quad \frac{R}{OC} = p.$$

Thus, we are to prove, $m + n = p$.

Now, from (1), we get

$$m \cdot \overrightarrow{OA} + n \cdot \overrightarrow{OB} = p \cdot \overrightarrow{OC}, \quad \text{or,} \quad \frac{m}{p}\mathbf{a} + \frac{n}{p}\mathbf{b} = (1 - \alpha)\mathbf{a} + \alpha\mathbf{b} \quad \text{[from (2)]}$$

Comparing coefficients of **a** and **b**, we get

$$\frac{m}{p} = 1 - \alpha \quad \text{and} \quad \frac{n}{p} = \alpha. \quad \therefore \quad \frac{m + n}{p} = 1, \quad \text{or,} \quad m + n = p.$$

Hence proved.

EXAMPLE 1.35. *A particle in equilibrium is subjected to the forces:*

$$\mathbf{F_1} = -10\mathbf{k}, \quad \mathbf{F_2} = u\left(\frac{12}{13}\mathbf{i} - \frac{12}{13}\mathbf{j} + \frac{3}{13}\mathbf{k}\right),$$

$$\mathbf{F_3} = v\left(-\frac{12}{13}\mathbf{i} - \frac{12}{13}\mathbf{j} + \frac{3}{13}\mathbf{k}\right), \quad \mathbf{F_4} = w(\cos\theta\mathbf{i} + \sin\theta\mathbf{j}).$$

Find the values of u, v, w.

Solution: Since the forces are in equilibrium, $\mathbf{F_1} + \mathbf{F_2} + \mathbf{F_3} + \mathbf{F_4} = 0$.

After addition, we will get an equation of the type, $L\mathbf{i} + M\mathbf{j} + N\mathbf{k} = 0$.

Since **i**, **j**, **k** are non-coplanar

$$\therefore \quad L = M = N = 0.$$

Here

$$L = \frac{12}{13}u - \frac{12}{13}v + w\cos\theta = 0,$$

$$M = -\frac{12}{13}u - \frac{12}{13}v + w\sin\theta = 0$$

$$\text{and} \quad N = \frac{3}{13}u + \frac{3}{13}v - 10 = 0.$$

Solving for u, v, w, we get

$$u = \frac{65}{3}(1 - \cot\theta), \quad v = \frac{65}{3}(1 + \cot\theta) \quad \text{and} \quad w = 40\,\text{cosec}\,\theta.$$

EXERCISE

1. The sides of a parallelogram are $2\mathbf{i}+4\mathbf{j}-5\mathbf{k}$ and $\mathbf{i}+2\mathbf{j}+3\mathbf{k}$. Find unit vectors parallel to the diagonals.

2. Prove that vectors $2\mathbf{i}+3\mathbf{j}-6\mathbf{k}$, $6\mathbf{i}-2\mathbf{j}+3\mathbf{k}$ and $3\mathbf{i}+6\mathbf{j}-2\mathbf{k}$ form the sides of an equilateral triangle.

3. Three forces of magnitudes a, $2a$, $3a$ meet at a point and their directions are along the diagonals of three adjacent faces of a cube. Find the magnitude and direction of the resultant force.

4. Prove that the figure formed by joining the mid-points of the sides of a quadrilateral taken in order is a parallelogram.

5. If two concurrent forces be represented by $n \cdot \overrightarrow{OP}$ and $m \cdot \overrightarrow{OQ}$ respectively, prove that their resultant is given by $(m+n) \cdot \overrightarrow{OR}$, where R divides PQ, such that $n \cdot PR = m \cdot RQ$.

6. Prove that the magnitude of the resultant of forces $\mathbf{P}$ and $\mathbf{Q}$ is $\left(P^2+Q^2+2PQ\cos\theta\right)^{1/2}$, where θ is the angle between the directions of $\mathbf{P}$ and $\mathbf{Q}$.

7. Find cosine of the angle between the vectors $\mathbf{a} = 4\mathbf{i}+3\mathbf{j}+\mathbf{k}$ and $\mathbf{b} = 2\mathbf{i}-\mathbf{j}+2\mathbf{k}$.

   Find also the unit vector perpendicular to both $\mathbf{a}$ and $\mathbf{b}$.

8. If $\mathbf{p}$, $\mathbf{q}$ are unit vectors and θ is the angle between them, show that $\sin\theta/2 = \frac{1}{2}\,|\mathbf{p}-\mathbf{q}|$.

9. Prove vectorially, that in any triangle—

   (a) $a\cos B - b\cos A = \left(a^2-b^2\right)/c$;

   (b) $2(bc\cos A + ca\cos B + ab\cos C) = a^2+b^2+c^2$.

10. If $\mathbf{a}$, $\mathbf{b}$, $\mathbf{c}$ are coplanar vectors, prove that

$$\begin{vmatrix} \mathbf{a} & \mathbf{b} & \mathbf{c} \\ \mathbf{a}\cdot\mathbf{a} & \mathbf{a}\cdot\mathbf{b} & \mathbf{a}\cdot\mathbf{c} \\ \mathbf{b}\cdot\mathbf{a} & \mathbf{b}\cdot\mathbf{b} & \mathbf{b}\cdot\mathbf{c} \end{vmatrix} = 0.$$

11. Prove that, in any triangle, $\sin(\alpha+\beta) = \sin\alpha\cos\beta + \cos\alpha\sin\beta$.

12. If $\mathbf{a} = 3\mathbf{i}-\mathbf{j}+2\mathbf{k}$, $\mathbf{b} = 2\mathbf{i}+\mathbf{j}-\mathbf{k}$, $\mathbf{c} = \mathbf{i}-2\mathbf{j}+2\mathbf{k}$, find $(\mathbf{a}\times\mathbf{b})\times\mathbf{c}$ and $\mathbf{a}\times(\mathbf{b}\times\mathbf{c})$, and hence show that $(\mathbf{a}\times\mathbf{b})\times\mathbf{c} \neq \mathbf{a}\times(\mathbf{b}\times\mathbf{c})$.

    Also, show that $\mathbf{a}\times\mathbf{b}$ is perpendicular to both $\mathbf{a}$ and $\mathbf{b}$.

13. If $\mathbf{a}\times\mathbf{b} = \mathbf{c}\times\mathbf{d}$ and $\mathbf{a}\times\mathbf{c} = \mathbf{b}\times\mathbf{d}$, show that $\mathbf{a}-\mathbf{d}$ is parallel to $\mathbf{b}-\mathbf{c}$.

14. If vectors $\mathbf{A} = \left(1,a,a^2\right)$, $\mathbf{B} = \left(1,b,b^2\right)$ and $\mathbf{C} = \left(1,c,c^2\right)$ are non-coplanar and if

$$\begin{vmatrix} a & a^2 & 1+a^2 \\ b & b^2 & 1+b^2 \\ c & c^2 & 1+c^2 \end{vmatrix} = 0,$$

    show that $abc+1=0$.

15. If $\mathbf{l}$, $\mathbf{m}$, $\mathbf{n}$ be three non-coplanar vectors, prove that

$$[\mathbf{lmn}](\mathbf{a} \times \mathbf{b}) = \begin{vmatrix} \mathbf{l}\cdot\mathbf{a} & \mathbf{l}\cdot\mathbf{b} & \mathbf{l} \\ \mathbf{m}\cdot\mathbf{a} & \mathbf{m}\cdot\mathbf{b} & \mathbf{m} \\ \mathbf{n}\cdot\mathbf{a} & \mathbf{n}\cdot\mathbf{b} & \mathbf{n} \end{vmatrix}.$$

16. If $(\mathbf{a} \times \mathbf{b}) \times \mathbf{c} = \mathbf{a} \times (\mathbf{b} \times \mathbf{c})$, then prove that, either $\mathbf{b}$ is perpendicular to both $\mathbf{a}$ and $\mathbf{c}$ or $\mathbf{a}$ and $\mathbf{c}$ are collinear.

17. Prove that, $\mathbf{d}\left[\mathbf{a} \times \{\mathbf{b} \times (\mathbf{c} \times \mathbf{d})\}\right] = (\mathbf{b}\cdot\mathbf{d})[\mathbf{acd}]$.

18. Find the value of $\mathbf{P} = \mathbf{i} \times (\mathbf{a} \times \mathbf{i}) + \mathbf{j} \times (\mathbf{a} \times \mathbf{j}) + \mathbf{k} \times (\mathbf{a} \times \mathbf{k})$.

19. Prove that, $\mathbf{a} \times [\mathbf{b} \times (\mathbf{c} \times \mathbf{d})] = \mathbf{a}(\mathbf{b}\cdot\mathbf{d}) \times \mathbf{c} - \mathbf{a}(\mathbf{b}\cdot\mathbf{c}) \times \mathbf{d} = [\mathbf{acd}]\mathbf{b} - \mathbf{c}(\mathbf{a}\cdot\mathbf{b}) \times \mathbf{d}$.

20. Prove that

$$\begin{aligned}[\mathbf{a} \times \mathbf{b}, \mathbf{c} \times \mathbf{d}, \mathbf{e} \times \mathbf{f}] &= [\mathbf{abd}][\mathbf{cef}] - [\mathbf{abc}][\mathbf{def}] \\ &= [\mathbf{abc}][\mathbf{fcd}] - [\mathbf{abf}][\mathbf{ecd}] \\ &= [\mathbf{cda}][\mathbf{bef}] - [\mathbf{cdb}][\mathbf{aef}].\end{aligned}$$

21. Prove that

$$2(\mathbf{c} \times \mathbf{d}) \times (\mathbf{a} \times \mathbf{b}) = \begin{vmatrix} \mathbf{a} & a_1 & a_2 & a_3 \\ \mathbf{b} & b_1 & b_2 & b_3 \\ -\mathbf{c} & c_1 & c_2 & c_3 \\ -\mathbf{d} & d_1 & d_2 & d_3 \end{vmatrix},$$

where, $\mathbf{a} = a_1\mathbf{i} + a_2\mathbf{j} + a_3\mathbf{k}, \mathbf{b} = b_1\mathbf{i} + b_2\mathbf{j} + b_3\mathbf{k}$, etc.

22. If $\mathbf{a}$, $\mathbf{b}$, $\mathbf{c}$ are three vectors, such that $\mathbf{a} + \mathbf{b} + \mathbf{c} = 0$, show that $\mathbf{a} \times \mathbf{b} = \mathbf{b} \times \mathbf{c} = \mathbf{c} \times \mathbf{a}$. Is the converse true?

23. Determine the values of λ so that for all real x, the vectors $\lambda x\mathbf{i} - 6\mathbf{j} + 3\mathbf{k}$ and $x\mathbf{i} + 2\mathbf{j} + 2\lambda x\mathbf{k}$ make an obtuse angle with each other.

24. In a triangle OAB, E is the mid-point of OB and D is a point on AB such that $AD : DB = 2 : 1$. If OD and OE intersect at P, show that $OP : PD = 3 : 2$.

25. If the vectors $a\mathbf{i} + \mathbf{j} + \mathbf{k}$, $\mathbf{i} + b\mathbf{j} + \mathbf{k}$ and $\mathbf{i} + \mathbf{j} + c\mathbf{k}$ are coplanar, then prove that

$$\frac{1}{1-a} + \frac{1}{1-b} + \frac{1}{1-c} = 1.$$

26. If A, B, C, D are any four points in space, prove that

$$\left|\overrightarrow{AB} \times \overrightarrow{CD} + \overrightarrow{BC} \times \overrightarrow{AD} + \overrightarrow{CA} \times \overrightarrow{BD}\right| = 4\triangle ABC.$$

27. Find all values for λ, such that $(x, y, z) \neq (0, 0, 0)$ and

$$(\mathbf{i} + \mathbf{j} + 3\mathbf{k})x + (3\mathbf{i} - 3\mathbf{j} + \mathbf{k})y + (-4\mathbf{i} + 5\mathbf{j})z = \lambda(x\mathbf{i} + y\mathbf{j} + z\mathbf{k}).$$

28. Let $\mathbf{a} = a_1\mathbf{i} + a_2\mathbf{j} + a_3\mathbf{k}$, $\mathbf{b} = b_1\mathbf{i} + b_2\mathbf{j} + b_3\mathbf{k}$ and $\mathbf{c} = c_1\mathbf{i} + c_2\mathbf{j} + c_3\mathbf{k}$ be three non-zero vectors such that **c** is a unit vector perpendicular to both **a** and **b**. If the angle between **a** and **b** is $\pi/6$, then show that

$$\begin{vmatrix} a_1 & a_2 & a_3 \\ b_1 & b_2 & b_3 \\ c_1 & c_2 & c_3 \end{vmatrix}^2 = \tfrac{1}{4}\sum a_1^2 \cdot \sum b_1^2.$$

ANSWERS

1. $\frac{1}{\sqrt{69}}(-\mathbf{i} - 2\mathbf{j} + 8\mathbf{k}), \frac{1}{7}(3\mathbf{i} + 6\mathbf{j} - 2\mathbf{k})$.

3. $R = 5a$, d.c.'s are $\frac{1}{\sqrt{2}}, \frac{4}{5\sqrt{2}}, \frac{3}{5\sqrt{2}}$.

7. $\frac{7}{3\sqrt{26}}; \frac{1}{\sqrt{185}}(7\mathbf{i} - 6\mathbf{j} - 10\mathbf{k})$.

18. $2a$.

27. $0, -1, -1$.

Chapter 2

Vector Calculus

I. DIFFERENTIATION OF VECTORS

2.1 Vector Functions of a Single Scalar Variable

The position vector of any point in space can be expressed in terms of three mutually perpendicular unit vectors $\mathbf{i}, \mathbf{j}, \mathbf{k}$ as

$$\mathbf{r} = x\mathbf{i} + y\mathbf{j} + z\mathbf{k}. \tag{2.1}$$

If the scalars x, y, z can be expressed as functions of a single scalar parameter 't' in the form

$$x = f_1(t), \quad y = f_2(t), \quad z = f_3(t) \tag{2.2}$$

then, from (2.1) we have

$$\mathbf{r} = f_1(t)\mathbf{i} + f_2(t)\mathbf{j} + f_3(t)\mathbf{k} \tag{2.3}$$

Thus, for any value of the scalar parameter t, the vector $\mathbf{r}$ is completely known in respect of its magnitude as well as direction. We say that $\mathbf{r}$ is a vector function of a scalar variable t and write it as

$$\mathbf{r} = \mathbf{f}(t) \tag{2.4}$$

For example, the equation of a parabola

$$y^2 = 4ax, \quad z = 0 \tag{2.5}$$

can be written in parametric form as

$$x = at^2, \quad y = 2at, \quad z = 0 \tag{2.6}$$

Thus, for a parabola, the vector function is

$$\mathbf{r} = \mathbf{f}(t) = at^2\mathbf{i} + 2at\mathbf{j} \tag{2.7}$$

The vector function $\mathbf{r} = \mathbf{f}(t)$ is called a *single-valued vector function* if for every value of the parameter t there corresponds one and only one value of $\mathbf{r}$.

The limit of any vector function $\mathbf{r} = \mathbf{f}(t)$ as $t \longrightarrow \alpha$ will be a vector of the form

$$\lim_{t \longrightarrow \alpha} \mathbf{r} = a\mathbf{i} + b\mathbf{j} + c\mathbf{k}, \tag{2.8}$$

where scalars a, b, c are limiting values of component functions obtained in the usual manner for scalar functions. A vector function $\mathbf{r} = \mathbf{f}(t)$ is said to be continuous at $t = \alpha$, if and only if all its component functions are continuous at $t = \alpha$.

In this chapter, we will discuss vector functions of single variable only.

2.2 Differentiation of Vectors

We consider a vector function **r** of a scalar variable t which undergoes a continuous change in magnitude and direction. Let, corresponding to an infinitesimally small change in t from t to $t+\delta t$, the function **r** becomes $\mathbf{r}+\delta\mathbf{r}$. Then, change in the vector function is (*Fig 2.1*)

$$\delta\mathbf{r} = \mathbf{f}(t+\delta t) - \mathbf{f}(t)$$

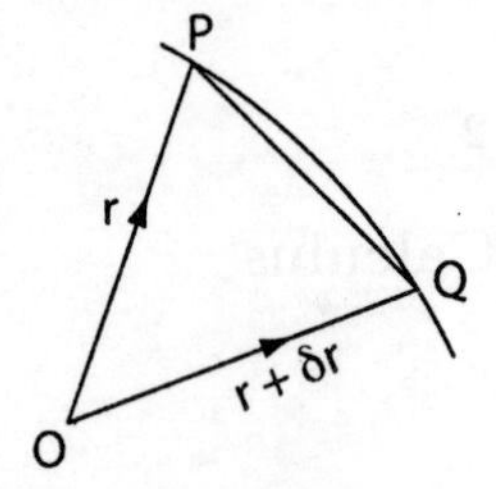

Fig 2.1

and average rate of change in the vector function is

$$\frac{\delta\mathbf{r}}{\delta t} = \frac{\mathbf{f}(t+\delta t) - \mathbf{f}(t)}{\delta t}.$$

Now, as $\delta t \longrightarrow 0$, this ratio attains a limiting value which is the rate of change of **r**,

$$\text{i.e.,} \quad \frac{d\mathbf{r}}{dt} = \lim_{\delta t \longrightarrow 0} \frac{\delta\mathbf{r}}{\delta t} \tag{2.9}$$

This is the derivative or differential coefficient of the vector function **r** with respect to scalar variable t.

But, $\mathbf{r} = f_1(t)\mathbf{i} + f_2(t)\mathbf{j} + f_3(t)\mathbf{k}$.

$$\therefore \quad \frac{d\mathbf{r}}{dt} = \frac{df_1}{dt}\mathbf{i} + \frac{df_2}{dt}\mathbf{j} + \frac{df_3}{dt}\mathbf{k} \tag{2.10}$$

provided that the derivatives of the functions $f_1(t)$, $f_2(t)$, $f_3(t)$ exist.

2.3 Rules for Differentiation

(a) *Differentiation of Sum (or Difference) of Two Vectors*: Let **r** be the sum of two vectors **a** and **b** which are vector functions of scalar variable t, so that

$$\mathbf{r} = \mathbf{a} \pm \mathbf{b}. \tag{2.11}$$

Then, for an infinitesimal change of t from t to $t+\delta t$,

$$\mathbf{r}+\delta\mathbf{r} = (\mathbf{a}+\delta\mathbf{a}) \pm (\mathbf{b}+\delta\mathbf{b}).$$

$$\therefore \quad \delta\mathbf{r} = \delta\mathbf{a} \pm \delta\mathbf{b}.$$

Hence

$$\frac{d\mathbf{r}}{dt} = \lim_{\delta t \longrightarrow 0}\left(\frac{\delta\mathbf{r}}{\delta t}\right) = \lim_{\delta t \longrightarrow 0}\left(\frac{\delta\mathbf{a}}{\delta t} \pm \frac{\delta\mathbf{b}}{\delta t}\right) = \frac{d\mathbf{a}}{dt} \pm \frac{d\mathbf{b}}{dt}. \tag{2.12}$$

(b) *Differentiation of Scalar Product of Two Vectors*: Let, $\mathbf{r} = \mathbf{a}\cdot\mathbf{b}$. Then, for change of t to $t+\delta t$,

$$\mathbf{r}+\delta\mathbf{r} = (\mathbf{a}+\delta\mathbf{a})\cdot(\mathbf{b}+\delta\mathbf{b}) = \mathbf{a}\cdot\mathbf{b} + \mathbf{a}\cdot\delta\mathbf{b} + \delta\mathbf{a}\cdot\mathbf{b} + \delta\mathbf{a}\cdot\delta\mathbf{b}$$

Neglecting the term $\delta\mathbf{a}\cdot\delta\mathbf{b}$ due to smallness we have $\delta\mathbf{r} = \mathbf{a}\cdot\delta\mathbf{b} + \delta\mathbf{a}\cdot\mathbf{b}$.

In the limit $\delta t \longrightarrow 0$, we get

$$\frac{d\mathbf{r}}{dt} = \frac{d\mathbf{a}}{dt}\cdot\mathbf{b} + \mathbf{a}\cdot\frac{d\mathbf{b}}{dt}. \tag{2.13}$$

Since, as we already know that $\mathbf{a} \cdot \mathbf{b} = \mathbf{b} \cdot \mathbf{a}$, the order of the factors in equation (2.13) can be changed at pleasure.

If $\mathbf{a} = \mathbf{b}$, then $\mathbf{r} = \mathbf{a} \cdot \mathbf{a} = \mathbf{a}^2$.

$$\therefore \quad \frac{d\mathbf{r}}{dt} = \frac{d}{dt}(\mathbf{a} \cdot \mathbf{a}) = \mathbf{a} \cdot \frac{d\mathbf{a}}{dt} + \frac{d\mathbf{a}}{dt} \cdot \mathbf{a} = 2\mathbf{a} \cdot \frac{d\mathbf{a}}{dt}. \tag{2.14}$$

(c) *Differentiation of Vector Product of Two Vectors*: Let, $\mathbf{r} = \mathbf{a} \times \mathbf{b}$.
Then

$$\mathbf{r} + \delta\mathbf{r} = (\mathbf{a} + \delta\mathbf{a}) \times (\mathbf{b} + \delta\mathbf{b}) = \mathbf{a} \times \mathbf{b} + \delta\mathbf{a} \times \mathbf{b} + \mathbf{a} \times \delta\mathbf{b} + \delta\mathbf{a} \times \delta\mathbf{b}.$$

$$\therefore \quad \delta\mathbf{r} = \delta\mathbf{a} \times \mathbf{b} + \mathbf{a} \times \delta\mathbf{b} \quad \text{(neglecting small terms)}.$$

$$\therefore \quad \frac{d\mathbf{r}}{dt} = \frac{d\mathbf{a}}{dt} \times \mathbf{b} + \mathbf{a} \times \frac{d\mathbf{b}}{dt}. \tag{2.15}$$

(d) *Differentiation of Scalar Triple Product*: Let, $\mathbf{r} = \mathbf{a} \cdot (\mathbf{b} \times \mathbf{c})$.
Then

$$\begin{aligned} \mathbf{r} + \delta\mathbf{r} &= (\mathbf{a} + \delta\mathbf{a}) \cdot [(\mathbf{b} + \delta\mathbf{b}) \times (\mathbf{c} + \delta\mathbf{c})] \\ &= (\mathbf{a} + \delta\mathbf{a}) \cdot [\mathbf{b} \times \mathbf{c} + \mathbf{b} \times \delta\mathbf{c} + \delta\mathbf{b} \times \mathbf{c}] \quad \text{(neglecting small terms)} \\ &= \mathbf{a} \cdot (\mathbf{b} \times \mathbf{c}) + \mathbf{a} \cdot (\delta\mathbf{b} \times \mathbf{c}) + \mathbf{a} \cdot (\mathbf{b} \times \delta\mathbf{c}) + \delta\mathbf{a} \cdot (\mathbf{b} \times \mathbf{c}). \\ \therefore \quad \delta\mathbf{r} &= \delta\mathbf{a} \cdot (\mathbf{b} \times \mathbf{c}) + \mathbf{a} \cdot (\delta\mathbf{b} \times \mathbf{c}) + \mathbf{a} \cdot (\mathbf{b} \times \delta\mathbf{c}). \end{aligned}$$

Dividing both sides by δt and proceeding to the limit $\delta t \longrightarrow 0$, we get

$$\frac{d\mathbf{r}}{dt} = \frac{d\mathbf{a}}{dt} \cdot (\mathbf{b} \times \mathbf{c}) + \mathbf{a} \cdot \left(\frac{d\mathbf{b}}{dt} \times \mathbf{c}\right) + \mathbf{a} \cdot \left(\mathbf{b} \times \frac{d\mathbf{c}}{dt}\right). \tag{2.16}$$

(e) *Differentiation of a Vector Triple Product*: Let $\mathbf{r} = \mathbf{a} \times (\mathbf{b} \times \mathbf{c})$.
Then

$$\begin{aligned} \mathbf{r} + \delta\mathbf{r} &= (\mathbf{a} + \delta\mathbf{a}) \times [(\mathbf{b} + \delta\mathbf{b}) \times (\mathbf{c} + \delta\mathbf{c})] \\ &= (\mathbf{a} + \delta\mathbf{a}) \times [(\mathbf{b} \times \mathbf{c} + \mathbf{b} \times \delta\mathbf{c} + \delta\mathbf{b} \times \mathbf{c}] \quad \text{(neglecting small terms)} \\ &= \mathbf{a} \times (\mathbf{b} \times \mathbf{c}) + \delta\mathbf{a} \times (\mathbf{b} \times \mathbf{c}) + \mathbf{a} \times (\mathbf{b} \times \delta\mathbf{c}) + \mathbf{a} \times (\delta\mathbf{b} \times \mathbf{c}) \\ \text{or,} \quad \delta\mathbf{r} &= \delta\mathbf{a} \times (\mathbf{b} \times \mathbf{c}) + \mathbf{a} \times (\mathbf{b} \times \delta\mathbf{c}) + \mathbf{a} \times (\delta\mathbf{b} \times \mathbf{c}). \end{aligned}$$

$$\therefore \quad \frac{d\mathbf{r}}{dt} = \frac{d\mathbf{a}}{dt} \times (\mathbf{b} \times \mathbf{c}) + \mathbf{a} \times \left(\mathbf{b} \times \frac{d\mathbf{c}}{dt}\right) + \mathbf{a} \times \left(\frac{d\mathbf{b}}{dt} \times \mathbf{c}\right). \tag{2.17}$$

2.4 Constant Vectors

A vector is said to be constant when both its magnitude and direction remain unchanged. So, for a constant vector $\mathbf{f}(t)$

$$\mathbf{f}(t + \delta t) = \mathbf{f}(t)$$

so that

$$\frac{d\mathbf{f}}{dt} = \lim_{\delta t \longrightarrow 0} \frac{\mathbf{f}(t + \delta t) - \mathbf{f}(t)}{\delta t} = 0. \tag{2.18}$$

Thus, derivative of a constant vector is zero.

A vector may also have only constant magnitude or only constant direction.

Let us suppose that the vector $\mathbf{f}(t)$ is of constant magnitude. Then

$$|\mathbf{f}(t)| = \text{constant}, \quad \text{i.e.,} \quad \frac{d}{dt}|\mathbf{f}(t)| = 0.$$

Now, $\mathbf{f}(t)\cdot\mathbf{f}(t) = |\mathbf{f}(t)|^2$.

$$\therefore \quad \frac{d}{dt}(\mathbf{f}\cdot\mathbf{f}) = \frac{d}{dt}|\mathbf{f}(t)|^2, \quad \text{or,} \quad \mathbf{f}\cdot\frac{d\mathbf{f}}{dt} + \frac{d\mathbf{f}}{dt}\cdot\mathbf{f} = 0$$

$$\text{or,} \quad 2\mathbf{f}\cdot\frac{d\mathbf{f}}{dt} = 0, \quad \text{i.e.,} \quad \mathbf{f}\cdot\frac{d\mathbf{f}}{dt} = 0. \tag{2.19}$$

But we know that dot product of two perpendicular vectors is zero. Hence, from equation (2.19) we can say that: *If* $\mathbf{f}(t)$ *is a vector function of constant magnitude (but changing direction), then vectors* $\mathbf{f}$ *and* $\frac{d\mathbf{f}}{dt}$ *are perpendicular to each other.*

Again, let us suppose that the vector $\mathbf{f}(t)$ is of constant direction but with changing magnitude. In that case we can assume

$$\mathbf{f}(t) = \phi(t)\cdot\boldsymbol{\lambda}(t), \tag{2.20}$$

where $\phi(t)$ denotes the constant magnitude of the vector $\mathbf{f}(t)$ and $\boldsymbol{\lambda}(t)$ is a unit vector function with constant direction and also of constant magnitude (i.e., unity). Thus $\boldsymbol{\lambda}(t)$ is a constant vector.

$\therefore$ From equation (2.18)

$$\frac{d\boldsymbol{\lambda}}{dt} = 0. \tag{2.21}$$

Now, differentiating equation (2.20) with respect to t,

$$\frac{d\mathbf{f}}{dt} = \phi\cdot\frac{d\boldsymbol{\lambda}}{dt} + \frac{d\phi}{dt}\boldsymbol{\lambda}, \tag{2.22}$$

$$\text{i.e.,} \quad \mathbf{f}\times\frac{d\mathbf{f}}{dt} = \mathbf{f}\times\left[\phi\frac{d\boldsymbol{\lambda}}{dt} + \frac{d\phi}{dt}\boldsymbol{\lambda}\right] = \phi\boldsymbol{\lambda}\times\left[\phi\frac{d\boldsymbol{\lambda}}{dt} + \frac{d\phi}{dt}\boldsymbol{\lambda}\right] \quad \text{[using (2.20)]}$$

$$= \phi^2\boldsymbol{\lambda}\times\frac{d\boldsymbol{\lambda}}{dt} + \phi\frac{d\phi}{dt}\boldsymbol{\lambda}\times\boldsymbol{\lambda}.$$

$$\therefore \quad \mathbf{f}\times\frac{d\mathbf{f}}{dt} = \phi^2\boldsymbol{\lambda}\times\frac{d\boldsymbol{\lambda}}{dt} \quad [\because \boldsymbol{\lambda}\times\boldsymbol{\lambda} = 0].$$

$$\therefore \quad \mathbf{f}\times\frac{d\mathbf{f}}{dt} = 0 \quad \left[\because \text{by (2.21)}, \frac{d\boldsymbol{\lambda}}{dt} = 0\right].$$

Thus, *if* $\mathbf{f}(t)$ *is a vector function of constant direction, the cross product of* $\mathbf{f}(t)$ *and* $\frac{d\mathbf{f}}{dt}$ *is equal to zero.*

2.5 Partial Differentiation of Vectors

When a vector $\mathbf{f}$ is a function of more than one scalar independent variable, say x, y, z, then the partial derivative $\frac{\partial \mathbf{f}}{\partial x}$ denotes the rate of change of $\mathbf{f}$ with respect to x when y and z remain constant.

Thus

$$\frac{\partial \mathbf{f}}{\partial x} = \lim_{\Delta x \longrightarrow 0} \frac{\mathbf{f}(x+\Delta x, y, z) - \mathbf{f}(x, y, z)}{\Delta x}.$$

Similarly

$$\frac{\partial \mathbf{f}}{\partial y} = \lim_{\Delta y \longrightarrow 0} \frac{\mathbf{f}(x, y+\Delta y, z) - \mathbf{f}(x, y, z)}{\Delta y}$$

and

$$\frac{\partial \mathbf{f}}{\partial z} = \lim_{\Delta z \longrightarrow 0} \frac{\mathbf{f}(x, y, z+\Delta z) - \mathbf{f}(x, y, z)}{\Delta z}. \tag{2.23}$$

Second and higher order partial derivatives are obtained in the usual manner. Again, if **f** and ϕ are two vector functions of scalar variables x, y, z, then

$$\left.\begin{aligned} \frac{\partial}{\partial x}(\mathbf{f} \cdot \phi) &= \mathbf{f} \cdot \frac{\partial \phi}{\partial x} + \frac{\partial \mathbf{f}}{\partial x} \cdot \phi \\ \frac{\partial}{\partial y}(\mathbf{f} \cdot \phi) &= \mathbf{f} \cdot \frac{\partial \phi}{\partial y} + \frac{\partial \mathbf{f}}{\partial y} \cdot \phi \\ \text{and} \quad \frac{\partial}{\partial z}(\mathbf{f} \cdot \phi) &= \mathbf{f} \cdot \frac{\partial \phi}{\partial z} + \frac{\partial \mathbf{f}}{\partial z} \cdot \phi \end{aligned}\right\}. \tag{2.24}$$

The total change in **f** due to simultaneous change in the variables is called the *total differential* of **f** and is given by

$$d\mathbf{f} = \frac{\partial \mathbf{f}}{\partial x} dx + \frac{\partial \mathbf{f}}{\partial y} dy + \frac{\partial \mathbf{f}}{\partial z} dz. \tag{2.25}$$

2.6 Time Derivative—Velocity and Acceleration

Let us consider a particle P moving along a curve in space. At any instant of time, its position vector is given by $\overrightarrow{OP} = \mathbf{r}$. This position vector is a continuous function of time, since it will change due to the motion of the particle (*Fig 2.2*).

Differentiating **r** with respect to time t, we get velocity of the particle as

$$\mathbf{v} = \dot{\mathbf{r}} = \frac{d\mathbf{r}}{dt}. \tag{2.26}$$

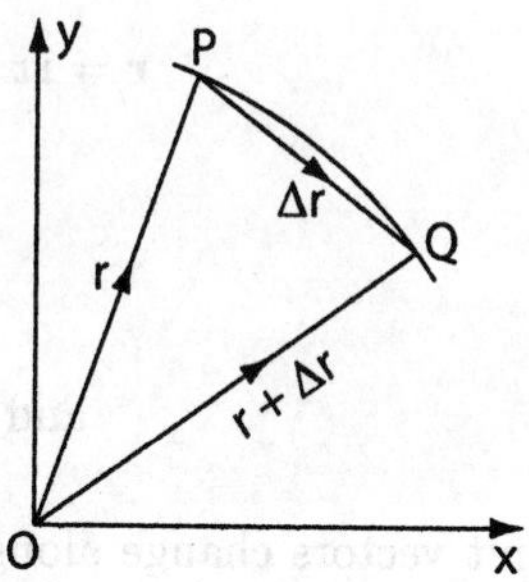

Fig 2.2

Differentiating **v** with respect to time t, we get acceleration of the particle as

$$\mathbf{a} = \dot{\mathbf{v}} = \ddot{\mathbf{r}} = \frac{d^2\mathbf{r}}{dt^2}. \tag{2.27}$$

Now, let us suppose that the position vector of the particle is given in rectangular coordinate system as

$$\mathbf{r} = x\mathbf{i} + y\mathbf{j} + z\mathbf{k}. \tag{2.28}$$

Then the velocity and acceleration of the particle are

$$\left.\begin{aligned} \mathbf{v} &= \dot{x}\mathbf{i} + \dot{y}\mathbf{j} + \dot{z}\mathbf{k} \\ \text{and} \quad \mathbf{a} &= \ddot{x}\mathbf{i} + \ddot{y}\mathbf{j} + \ddot{z}\mathbf{k} \end{aligned}\right\}. \tag{2.29}$$

Let us now consider the motion of the particle in a plane under a central force acting on the particle and directed along the radius vector towards a fixed point. The position of the particle is expressed in terms of radius vector $\mathbf{r}$ and the angle θ which it makes with x-axis. These polar coordinates are related with the Cartesian coordinates in the form (*Fig 2.3*)

$$x = r\cos\theta, \quad y = r\sin\theta. \tag{2.30}$$

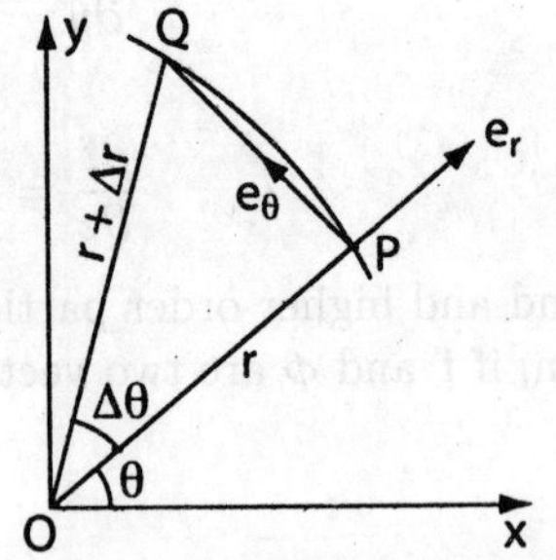

Fig 2.3

Let $\mathbf{e}_r$ and $\mathbf{e}_\theta$ be unit vectors along radial and cross-radial directions respectively. Then $\mathbf{e}_r$ and $\mathbf{e}_\theta$ can be expressed as

$$\mathbf{e}_r = \lim_{\Delta r \to 0} \frac{\Delta \mathbf{r}}{\Delta r} = \frac{\partial \mathbf{r}}{\partial r} \quad (\theta = \text{constant}) \tag{2.31}$$

$$\text{and} \quad \mathbf{e}_\theta = \lim_{\Delta \theta \to 0} \frac{\Delta \mathbf{r}}{r\Delta \theta} = \frac{\partial \mathbf{r}}{r\partial \theta} \quad (r = \text{constant}). \tag{2.32}$$

These unit vectors are not constant since the directions of these vectors continuously change due to the motion of the particle along the curve.

The position vector of the particle can be expressed in Cartesian coordinates as

$$\mathbf{r} = \mathbf{i}x + \mathbf{j}y, \quad \text{i.e.,} \quad \mathbf{r} = \mathbf{i}r\cos\theta + \mathbf{j}r\sin\theta. \tag{2.33}$$

$$\left.\begin{aligned} \therefore \quad \mathbf{e}_r &= \frac{\partial \mathbf{r}}{\partial r} = \mathbf{i}\cos\theta + \mathbf{j}\sin\theta \\ \text{and} \quad \mathbf{e}_\theta &= \frac{1}{r}\frac{\partial \mathbf{r}}{\partial \theta} = -\mathbf{i}\sin\theta + \mathbf{j}\cos\theta \end{aligned}\right\}. \tag{2.34}$$

The unit vectors change along with the change in position of the particles. Differentiating equations (2.34) we get

$$\left.\begin{aligned} \frac{d\mathbf{e}_r}{d\theta} &= -\mathbf{i}\sin\theta + \mathbf{j}\cos\theta = \mathbf{e}_\theta \\ \text{and} \quad \frac{d\mathbf{e}_\theta}{d\theta} &= -\mathbf{i}\cos\theta - \mathbf{j}\sin\theta = -\mathbf{e}_r \end{aligned}\right\}. \tag{2.35}$$

Now the position vector $\mathbf{r}$ of the particle in polar coordinates can be written as

$$\mathbf{r} = r\mathbf{e}_r. \tag{2.36}$$

Differentiating this equation with respect to time, we get

$$\mathbf{v} = \dot{\mathbf{r}} = \frac{d\mathbf{r}}{dt} = \frac{d}{dt}(r\mathbf{e}_r),$$

$$\text{i.e.,}\quad \mathbf{v} = \frac{dr}{dt}\mathbf{e}_r + r\frac{d\mathbf{e}_r}{dt} = \frac{dr}{dt}\mathbf{e}_r + r\frac{d\mathbf{e}_r}{d\theta}\cdot\frac{d\theta}{dt},$$

$$\text{i.e.,}\quad \mathbf{v} = \dot{r}\mathbf{e}_r + r\dot{\theta}\mathbf{e}_\theta \quad \text{(using equation 2.34)..} \tag{2.37}$$

Thus, the radial component of the velocity is

$$\mathbf{v_r} = \dot{\mathbf{r}} \tag{2.38}$$

and transverse or cross-radial component of velocity is

$$\mathbf{v}_\theta = \mathbf{r}\dot{\theta}. \tag{2.39}$$

Differentiating equation (2.37) with respect to time, we get acceleration of the particle as

$$\mathbf{a} = \dot{\mathbf{v}} = \frac{d^2\mathbf{r}}{dt^2} = \ddot{r}\mathbf{e}_r + \dot{r}\frac{d\mathbf{e}_r}{dt} + \dot{r}\dot{\theta}\mathbf{e}_\theta + r\ddot{\theta}\mathbf{e}_\theta + r\dot{\theta}\frac{d\mathbf{e}_\theta}{dt}$$

$$\text{or,}\quad \mathbf{a} = \ddot{r}\mathbf{e}_r + \dot{r}\frac{d\mathbf{e}_r}{d\theta}\frac{d\theta}{dt} + \dot{r}\dot{\theta}\mathbf{e}_\theta + r\ddot{\theta}\mathbf{e}_\theta + r\dot{\theta}\frac{d\mathbf{e}_\theta}{d\theta}\cdot\frac{d\theta}{dt}$$

$$= \ddot{r}\mathbf{e}_r + \dot{r}\dot{\theta}\mathbf{e}_\theta + \dot{r}\dot{\theta}\mathbf{e}_\theta + r\ddot{\theta}\mathbf{e}_\theta - r(\dot{\theta})^2\mathbf{e}_r \quad \text{(using equation 2.34)}$$

$$\therefore\quad \mathbf{a} = [\ddot{r} - r(\dot{\theta})^2]\mathbf{e}_r + (r\ddot{\theta} + 2\dot{r}\dot{\theta})\mathbf{e}_\theta. \tag{2.40}$$

This shows that radial component of acceleration of the particle is

$$a_r = \ddot{r} - r(\dot{\theta})^2 \tag{2.41}$$

and the transverse or cross-radial component of the acceleration of the particle is

$$a_\theta = r\ddot{\theta} + 2\dot{r}\dot{\theta} = \frac{1}{r}\frac{d}{dt}(r^2\dot{\theta}). \tag{2.42}$$

EXAMPLE 2.1. *A particle moves along the curve $x = 2t^2$, $y = t^2 - 4t$, $z = 3t - 5$, where t is the time.*

Find the component of its velocity and acceleration at time $t = 1$ in the direction of $\mathbf{i} - 3\mathbf{j} + 2\mathbf{k}$.

Solution: Here, $\mathbf{r} = 2t^2\mathbf{i} + (t^2 - 4t)\mathbf{j} + (3t - 5)\mathbf{k}$.

$$\therefore\quad \text{Velocity} = \dot{\mathbf{r}} = 4t\mathbf{i} + (2t - 4)\mathbf{j} + 3\mathbf{k}.$$

For $t = 1$, velocity $= 4\mathbf{i} - 2\mathbf{j} + 3\mathbf{k}$.

Hence, component of velocity vector in the direction of $\mathbf{i} - 3\mathbf{j} + 2\mathbf{k}$ is

$$\frac{\mathbf{i} - 3\mathbf{j} + 2\mathbf{k}}{\sqrt{1 + 9 + 4}}\cdot(4\mathbf{i} - 2\mathbf{j} + 3\mathbf{k}) = \frac{16}{\sqrt{14}}.$$

Again, acceleration.

$$\frac{d^2\mathbf{r}}{dt^2} = 4\mathbf{i} + 2\mathbf{j}.$$

Hence, component of acceleration in the direction of $\mathbf{i} - 3\mathbf{j} + 2\mathbf{k}$ is

$$\frac{\mathbf{i} - 3\mathbf{j} + 2\mathbf{k}}{\sqrt{14}}\cdot(4\mathbf{i} + 2\mathbf{j}) = -\frac{2}{\sqrt{14}}.$$

EXAMPLE 2.2. *Evaluate* $\dfrac{d}{dt}\left[\mathbf{r}\cdot\dfrac{d\mathbf{r}}{dt}\times\dfrac{d^2\mathbf{r}}{dt^2}\right]$.

Solution:

$$\text{Given expression} = \frac{d}{dt}\left[\mathbf{r}\frac{d\mathbf{r}}{dt}\frac{d^2\mathbf{r}}{dt^2}\right] = \left[\frac{d\mathbf{r}}{dt}\frac{d\mathbf{r}}{dt}\frac{d^2\mathbf{r}}{dt^2}\right] + \left[\mathbf{r}\frac{d^2\mathbf{r}}{dt^2}\frac{d^2\mathbf{r}}{dt^2}\right] + \left[\mathbf{r}\frac{d\mathbf{r}}{dt}\frac{d^3\mathbf{r}}{dt^3}\right]$$

$$= \left[\mathbf{r}\frac{d\mathbf{r}}{dt}\frac{d^3\mathbf{r}}{dt^3}\right] \qquad [\because \text{ first two terms vanish}]$$

$$= \mathbf{r}\cdot\left(\frac{d\mathbf{r}}{dt}\times\frac{d^3\mathbf{r}}{dt^3}\right).$$

EXAMPLE 2.3. *If* $\mathbf{e}_r$ *denotes a unit vector, prove that* $\mathbf{e}_r\times d\mathbf{e}_r = (\mathbf{r}\times d\mathbf{r})/r^2$, *where* $\mathbf{r} = r\mathbf{e}_r$.

Solution:

$$e_r = \frac{\mathbf{r}}{r}, \quad \therefore \quad d\mathbf{e}_r = \frac{1}{r}\,d\mathbf{r} + \mathbf{r}\left(-\frac{1}{r^2}\right)dr.$$

$$\therefore \quad \mathbf{e}_r\times d\mathbf{e}_r = \mathbf{e}_r\times\left[\frac{1}{r}\,d\mathbf{r} - \frac{\mathbf{r}}{r^2}\,dr\right] = \frac{\mathbf{r}}{r}\times\left[\frac{1}{r}\,d\mathbf{r} - \frac{\mathbf{r}}{r^2}\,dr\right]$$

$$= \frac{\mathbf{r}\times d\mathbf{r}}{r^2}, \qquad \because \mathbf{r}\times\mathbf{r} = 0.$$

II. DIFFERENTIAL OPERATORS

2.7 Scalar and Vector Fields

Any physical quantity, whether scalar or a vector, can be expressed as a continuous function of the position of the point within a region of space. For example, density at any point of a certain body can be expressed as a continuous function of the position of that point. Similarly, the velocity of a particle at any time occupying a certain region can be expressed as a continuous function of position of that point. Such functions are called *point functions* and the regions in which they specify the physical quantities are called *fields*. The fields may be scalar or vector according to the nature of the quantity specified.

If, corresponding to each point $P(x, y, z)$ of a region R, there exists a scalar defined by $\phi(P)$, then $\phi(P)$ is said to be a scalar point function for region R and is denoted by

$$\phi(P) = \phi(x, y, z) \tag{2.43}$$

The series of surfaces upon each of which the scalar point function has a definite constant value, i.e., $\phi(x, y, z) = \text{constant}$, are called *level surfaces*.

Similarly, if corresponding to each point $P(x, y, z)$ of a region R, there exists a vector defined by

$$\mathbf{f}(P) = f_1(x, y, z)\mathbf{i} + f_2(x, y, z)\mathbf{j} + f_3(x, y, z)\mathbf{k} \tag{2.44}$$

then $\mathbf{f}(P)$ is called a *vector point function* for the region R.

2.8 Directional Derivative

Let $\phi(x, y, z)$ be a scalar point function defined in a region R, where AP is a line. The directional derivative of the scalar function along the line AP is defined as

$$\lim_{P \longrightarrow A} \frac{\phi(P) - \phi(A)}{AP} \tag{2.45}$$

provided the limit exists. This limit is a scalar. Thus, directional derivative of a scalar point function is a scalar.

Similarly, if $\mathbf{f}(x, y, z)$ is a vector point function defined in a region R, where AP is a line, then the

$$\lim_{P \longrightarrow A} \frac{\mathbf{f}(P) - \mathbf{f}(A)}{AP} \tag{2.46}$$

if it exists, is called the *directional derivative* of the vector point function in the direction of the given line.

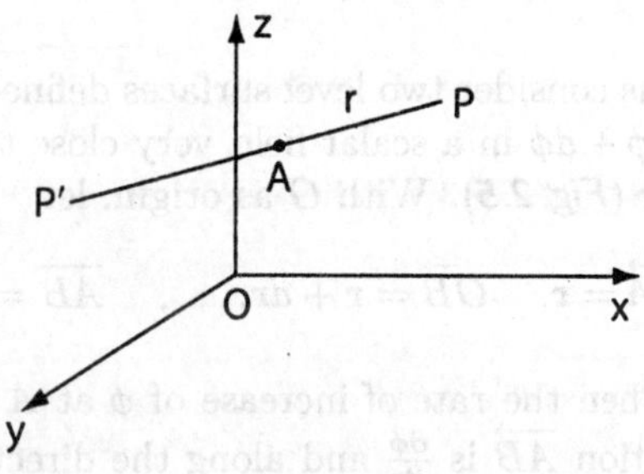

Fig 2.4

Let us now suppose that line AP is of length r having l, m, n as its direction cosines. Taking coordinates of A as (x, y, z), those of P will be $(x + lr, y + mr, z + nr)$. Then, for a scalar point function $\phi(x, y, z)$, the directional derivative along AP is (*Fig 2.4*)

$$\lim_{r \longrightarrow 0} \frac{\phi(x + lr, y + mr, z + nr) - \phi(x, y, z)}{r}$$

$$= \lim_{r \longrightarrow 0} \frac{\phi(x, y, z) + r\left[l\dfrac{\partial \phi}{\partial x} + m\dfrac{\partial \phi}{\partial y} + n\dfrac{\partial \phi}{\partial z}\right] + \cdots - \phi(x, y, z)}{r} \tag{2.47}$$

(expanding by Taylor's theorem)

$$= l\frac{\partial \phi}{\partial x} + m\frac{\partial \phi}{\partial y} + n\frac{\partial \phi}{\partial z}. \tag{2.48}$$

It is to be noted that the direction cosines of the line AP' will be $-l, -m, -n$ and hence, the sign of the directional derivative along AP' will be reversed.

If now we consider a vector point function $\mathbf{f}(x, y, z)$ in the form given by equation (2.44), the directional derivatives along the line AP of each of the scalar functions f_1, f_2, f_3 will be [as per equation (2.48)]

$$\lambda = l\frac{\partial f_1}{\partial x} + m\frac{\partial f_1}{\partial y} + n\frac{\partial f_1}{\partial z},$$

$$\mu = l\frac{\partial f_2}{\partial x} + m\frac{\partial f_2}{\partial y} + n\frac{\partial f_2}{\partial z}$$

$$\text{and} \quad \nu = l\frac{\partial f_3}{\partial x} + m\frac{\partial f_3}{\partial y} + n\frac{\partial f_3}{\partial z}$$

respectively. Hence, the directional derivative of the vector point function $\mathbf{f}(x, y, z)$ along the line AP will be $\lambda\mathbf{i} + \mu\mathbf{j} + \nu\mathbf{k}$ which, after algebraic arrangement becomes

$$l\frac{\partial \mathbf{f}}{\partial x} + m\frac{\partial \mathbf{f}}{\partial y} + n\frac{\partial \mathbf{f}}{\partial z} \qquad (2.49)$$

From equations (2.48) and (2.49), we find that the directional derivative of a scalar point function and a vector point function along any line having direction cosines as l, m, n are of the same form.

2.9 Gradient of a Scalar Function

Let us consider two level surfaces defined by ϕ and $\phi + d\phi$ in a scalar field very close to each other (*Fig 2.5*). With O as origin, let

$$\overrightarrow{OA} = \mathbf{r}, \quad \overrightarrow{OB} = \mathbf{r} + d\mathbf{r}, \quad \therefore \quad \overrightarrow{AB} = d\mathbf{r}. \qquad (2.50)$$

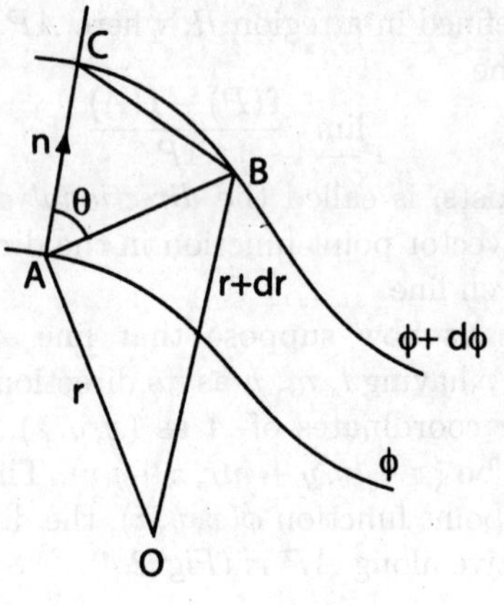

Fig 2.5

Then the rate of increase of ϕ at A in the direction $\overrightarrow{AB}$ is $\frac{\partial \phi}{\partial r}$ and along the direction of unit normal $\vec{n}$ at A which has maximum value is $\frac{\partial \phi}{\partial n}$.

Now, shortest distance between the two surfaces will be

$$AC = d\mathbf{n} = dr\cos\theta. \qquad (2.51)$$

$$\therefore \quad \frac{\partial \phi}{\partial r} = \frac{\partial \phi}{\partial n}\cos\theta. \qquad (2.52)$$

Thus, the maximum rate of change of ϕ is found in the direction of $\vec{n}$ and is a vector denoted by $\frac{\partial \phi}{\partial n}\vec{n}$. This vector is called the *gradient of scalar point function* $\phi(x, y, z)$ at point A and is written as

$$\operatorname{grad}\phi = \frac{\partial \phi}{\partial n}\vec{n}. \qquad (2.53)$$

Hence, *the gradient of a scalar field at any point is a vector,* whose magnitude is equal to the maximum rate of increase of ϕ at that point and direction is along normal to the surface ϕ = constant at that point.

Now, taking dot product of the both sides of equation (2.53) with $d\mathbf{r}$, we get

$$(\operatorname{grad}\phi)\cdot d\mathbf{r} = \frac{\partial \phi}{\partial n}\vec{n}\cdot d\mathbf{r} = \frac{\partial \phi}{\partial n}\,d\mathbf{r}\cos\theta = \frac{\partial \phi}{\partial n}\,d\mathbf{n} \quad \text{[by equation (2.51)]}$$

i.e., $$(\operatorname{grad}\phi)\cdot d\mathbf{r} = d\phi. \qquad (2.54)$$

But

$$\begin{aligned} d\phi &= \frac{\partial \phi}{\partial x}dx + \frac{\partial \phi}{\partial y}dy + \frac{\partial \phi}{\partial z}dz \\ &= \left(\mathbf{i}\frac{\partial \phi}{\partial x} + \mathbf{j}\frac{\partial \phi}{\partial y} + \mathbf{k}\frac{\partial \phi}{\partial z}\right)\cdot(\mathbf{i}\,dx + \mathbf{j}\,dy + \mathbf{k}\,dz) \\ &= (\nabla\phi)\cdot d\mathbf{r}. \end{aligned} \qquad (2.55)$$

Hence, by equations (2.54) and (2.55)

$$(\operatorname{grad}\phi)\cdot d\mathbf{r} = (\nabla\phi)\cdot d\mathbf{r}, \quad \text{i.e.,} \quad \operatorname{grad}\phi = \nabla\phi \tag{2.56}$$

since $d\mathbf{r}$ can be chosen arbitrary. The operator

$$\nabla(\cdots) = \mathbf{i}\frac{\partial\cdots}{\partial x} + \mathbf{j}\frac{\partial\cdots}{\partial y} + \mathbf{k}\frac{\partial\cdots}{\partial z} \tag{2.57}$$

is read *del* or *nabla*.

Now, the directional derivative of a scalar point function ϕ along any line having the direction cosines l, m, n given by equation (2.48) can be written as

$$(l\mathbf{i} + m\mathbf{j} + n\mathbf{k})\cdot\left(\frac{\partial\phi}{\partial x}\mathbf{i} + \frac{\partial\phi}{\partial y}\mathbf{j} + \frac{\partial\phi}{\partial z}\mathbf{k}\right).$$

But $l\mathbf{i} + m\mathbf{j} + n\mathbf{k}$ represents unit vector along the line with direction cosines l, m, n and the second part is $\nabla\phi$. Thus, the *directional derivative of any scalar point function along any direction is obtained by multiplying a unit vector in that direction scalarly by grad* ϕ.

EXAMPLE 2.4. *Find the directional derivative of* $\phi = x^3 + y^3 + z^3$ *at the point* $(1, -1, 2)$ *in the direction of the vector* $\mathbf{i} + 2\mathbf{j} + \mathbf{k}$.

Solution:

$$\frac{\partial\phi}{\partial x} = 3x^2, \quad \frac{\partial\phi}{\partial y} = 3y^2, \quad \frac{\partial\phi}{\partial z} = 3z^2.$$

$\therefore\ \nabla\phi = 3(\mathbf{i}x^2 + \mathbf{j}y^2 + \mathbf{k}z^2) = 3(\mathbf{i} + \mathbf{j} + 4\mathbf{k})$ at the point $(1, -1, 2)$.
Unit vector along

$$\mathbf{i} + 2\mathbf{j} + \mathbf{k} = \frac{1}{\sqrt{6}}(\mathbf{i} + 2\mathbf{j} + \mathbf{k}).$$

$\therefore$ Required directional derivative is

$$\frac{3}{\sqrt{6}}(\mathbf{i} + 2\mathbf{j} + \mathbf{k})\cdot(\mathbf{i} + \mathbf{j} + 4\mathbf{k}) = \frac{21}{\sqrt{6}}.$$

EXAMPLE 2.5. *Find directional derivative of* $\phi = x^2 - 2y^2 + 4z^2$ *at the point* $(1, 1, -1)$ *in the direction* $2\mathbf{i} + \mathbf{j} - \mathbf{k}$. *In what direction the directional derivative at that point is maximum and what is its value?*

Solution: $\nabla\phi = (\mathbf{i}2x - \mathbf{j}4y + \mathbf{k}8z) = 2\mathbf{i} - 4\mathbf{j} - 8\mathbf{k}$ at $(1, 1, -1)$.
Unit vector along

$$2\mathbf{i} + \mathbf{j} - \mathbf{k} = \frac{1}{\sqrt{6}}(2\mathbf{i} + \mathbf{j} - \mathbf{k}).$$

$\therefore$ Required directional derivative is

$$\frac{1}{\sqrt{6}}(2\mathbf{i} + \mathbf{j} - \mathbf{k})\cdot(2\mathbf{i} - 4\mathbf{j} - 8\mathbf{k}) = \frac{4 - 4 + 8}{\sqrt{6}} = \frac{8}{\sqrt{6}}.$$

Directional derivative is maximum along the normal, i.e., along grad ϕ, i.e., along $2\mathbf{i} - 4\mathbf{j} - 8\mathbf{k}$. Hence, its value is

$$|\operatorname{grad}\phi| = \sqrt{4 + 16 + 64} = 2\sqrt{21}.$$

EXAMPLE 2.6. *If ϕ and ψ be two scalar point functions, prove that*

$$grad\,(\phi\psi) = \phi\ grad\,\psi + \psi\ grad\,\phi.$$

Solution:

$$\begin{aligned}\text{grad}\,(\phi\psi) &= \left(\mathbf{i}\frac{\partial}{\partial x} + \mathbf{j}\frac{\partial}{\partial y} + \mathbf{k}\frac{\partial}{\partial z}\right)(\phi\psi)\\ &= \mathbf{i}\left(\phi\frac{\partial\psi}{\partial x} + \psi\frac{\partial\phi}{\partial x}\right) + \mathbf{j}\left(\phi\frac{\partial\psi}{\partial y} + \psi\frac{\partial\psi}{\partial y}\right) + \mathbf{k}\left(\phi\frac{\partial\psi}{\partial z} + \psi\frac{\partial\psi}{\partial z}\right)\\ &= \phi\left(\mathbf{i}\frac{\partial\psi}{\partial x} + \mathbf{j}\frac{\partial\psi}{\partial y} + \mathbf{k}\frac{\partial\psi}{\partial z}\right) + \psi\left(\mathbf{i}\frac{\partial\phi}{\partial x} + \mathbf{j}\frac{\partial\phi}{\partial y} + \mathbf{k}\frac{\partial\phi}{\partial z}\right)\\ &= \phi\ \text{grad}\,\psi + \psi\ \text{grad}\,\phi.\end{aligned}$$

EXAMPLE 2.7. *Prove that $grad\,f(r) \times \mathbf{r} = 0$, where $r^2 = x^2 + y^2 + z^2$.*

Solution: $\phi = f(r) = f\left(\sqrt{x^2 + y^2 + z^2}\right)$.

$$\therefore\quad \frac{\partial\phi}{\partial x} = f'(r)\frac{\partial r}{\partial x} = f'(r)\frac{x}{r}.$$

Similarly

$$\frac{\partial\phi}{\partial y} = f'(r)\frac{y}{r},\quad \frac{\partial\phi}{\partial z} = f'(r)\frac{z}{r}.$$

$$\therefore\quad \text{grad}\,\phi = \frac{1}{r}f'(r)(\mathbf{i}x + \mathbf{j}y + \mathbf{k}z) = \frac{f'(r)}{r}\mathbf{r}.$$

$$\therefore\quad \text{grad}\,f(r) \times \mathbf{r} = \frac{f'(r)}{r}\mathbf{r} \times \mathbf{r} = 0.$$

EXAMPLE 2.8. *Prove that* $\nabla[\mathbf{rab}] = \mathbf{a} \times \mathbf{b}$.

Solution: Let $\mathbf{a} \times \mathbf{b} = \mathbf{c}$. Then, $\nabla[\mathbf{rab}] = \nabla(\mathbf{r} \cdot \mathbf{c})$.
Now, $\mathbf{r} = \mathbf{i}x + \mathbf{j}y + \mathbf{k}z$ and $\mathbf{c} = \mathbf{i}c_1 + \mathbf{j}c_2 + \mathbf{k}c_3$ (say).

$$\therefore\quad \mathbf{r} \cdot \mathbf{c} = c_1x + c_2y + c_3z.$$

$$\therefore\quad \nabla(\mathbf{r} \cdot \mathbf{c}) = c_1\mathbf{i} + c_2\mathbf{j} + c_3\mathbf{k} = \mathbf{c}.$$

Hence proved.

EXAMPLE 2.9. *If* **a** *is a unit vector along a line with direction cosines l, m, n, prove that the directional derivative of a scalar function ϕ along that line is given by* $\mathbf{a} \cdot \boldsymbol{\nabla}\phi$.

Solution: Since **a** is a unit vector along the line with direction cosines l, m, n, we have $\mathbf{a} \cdot \mathbf{i} = 1 \cdot 1 \cdot \cos\alpha = l$; similarly, $\mathbf{a} \cdot \mathbf{j} = m$, $\mathbf{a} \cdot \mathbf{k} = n$.

$$\therefore\quad \mathbf{a} \cdot \boldsymbol{\nabla}\phi = \mathbf{a} \cdot \left(\mathbf{i}\frac{\partial\phi}{\partial x} + \mathbf{j}\frac{\partial\phi}{\partial y} + \mathbf{k}\frac{\partial\phi}{\partial z}\right) = l\frac{\partial\phi}{\partial x} + m\frac{\partial\phi}{\partial y} + n\frac{\partial\phi}{\partial z}$$

which is the directional derivative of ϕ along the line whose direction cosines are l, m, n.

EXAMPLE 2.10. *If* $\phi = \left(3r^2 - 4\sqrt{r} + 6r^{-1/3}\right)$, *find* $\nabla\phi$.

Solution: We know, $r = \left(x^2 + y^2 + z^2\right)^{1/2}$.

$$\therefore \quad \frac{\partial r}{\partial x} = \frac{x}{r}, \quad \frac{\partial r}{\partial y} = \frac{y}{r}, \quad \frac{\partial r}{\partial z} = \frac{z}{r}.$$

$$\therefore \quad \frac{\partial \phi}{\partial x} = \frac{\partial \phi}{\partial r} \cdot \frac{\partial r}{\partial x} = \left(6r - \frac{4}{2\sqrt{r}} - \frac{1}{3} \cdot 6 \cdot r^{-4/3}\right) \cdot \frac{x}{r}$$
$$= 2\left(3 - r^{-3/2} - r^{-7/3}\right) x.$$

Similar expression for $\dfrac{\partial \phi}{\partial y}$ and $\dfrac{\partial \phi}{\partial z}$ will be obtained.

$$\therefore \quad \nabla\phi = \left(\mathbf{i}\frac{\partial \phi}{\partial x} + \mathbf{j}\frac{\partial \phi}{\partial y} + \mathbf{k}\frac{\partial \phi}{\partial z}\right)$$
$$= 2\left(3 - r^{-3/2} - r^{-7/3}\right)(\mathbf{i}x + \mathbf{j}y + \mathbf{k}z)$$
$$= 2\left(3 - r^{-3/2} - r^{-7/3}\right)\mathbf{r}$$

2.10 Properties of $\nabla\phi$

1. grad ϕ is a vector normal to the surface $\phi(x, y, z) =$ constant.

   Proof. Let a point $P(x, y, z)$ be on the surface $\phi(x, y, z) =$ constant $= c$. If $\mathbf{r}$ be the position vector of P, then $\mathbf{r} = \mathbf{i}x + \mathbf{j}y + \mathbf{k}z$.

   $$\therefore \quad d\mathbf{r} = \mathbf{i}\,dx + \mathbf{j}\,dy + \mathbf{k}\,dz.$$

   $$\therefore \quad d\mathbf{r} \cdot \nabla\phi = (\mathbf{i}\,dx + \mathbf{j}\,dy + \mathbf{k}\,dz) \cdot \left(\mathbf{i}\frac{\partial \phi}{\partial x} + \mathbf{j}\frac{\partial \phi}{\partial y} + \mathbf{k}\frac{\partial \phi}{\partial z}\right)$$
   $$= \frac{\partial \phi}{\partial x}dx + \frac{\partial \phi}{\partial y}dy + \frac{\partial \phi}{\partial z}dz$$
   $$= d\phi = 0, \quad \text{since } \phi = \text{constant}.$$

   Hence, $\nabla\phi$ is perpendicular to $d\mathbf{r}$ which lies in the tangent plane. Thus $\nabla\phi$ is normal to the surface. □

2. grad ϕ is invariant, i.e., its value is independent of the choice of rectangular axes.

   Proof. Let $\mathbf{i}, \mathbf{j}, \mathbf{k}$ be any three mutually perpendicular unit vectors and $\phi(x, y, z)$ be any scalar point function. Then

   $$\text{grad}\,\phi = \mathbf{i}\frac{\partial \phi}{\partial x} + \mathbf{j}\frac{\partial \phi}{\partial y} + \mathbf{k}\frac{\partial \phi}{\partial z}, \tag{2.58}$$

   where $\dfrac{\partial \phi}{\partial x}, \dfrac{\partial \phi}{\partial y}, \dfrac{\partial \phi}{\partial z}$ are directional derivatives along the directions $\mathbf{i}$, $\mathbf{j}$, $\mathbf{k}$. Let us now choose another set of three mutually perpendicular unit vectors $\mathbf{i}'$, $\mathbf{j}'$, $\mathbf{k}'$.

We know that, directional derivative of ϕ along a unit vector **a** is given by $\mathbf{a}\cdot\nabla\phi$ (*see* Example 2.9). Hence, directional derivatives of ϕ along unit vectors $\mathbf{i}', \mathbf{j}', \mathbf{k}'$ are $\mathbf{i}'\cdot \text{grad}\,\phi$, $\mathbf{j}'\cdot\text{grad}\,\phi$ and $\mathbf{k}'\cdot\text{grad}\,\phi$ respectively. Hence, $\nabla\phi$ in terms of $\mathbf{i}', \mathbf{j}', \mathbf{k}'$ will be

$$\nabla\phi = \mathbf{i}'\cdot\{\mathbf{i}'\nabla\phi\} + \mathbf{j}'\cdot\{\mathbf{j}'\nabla\phi\} + \mathbf{k}'\cdot\{\mathbf{k}'\nabla\phi\}. \tag{2.59}$$

Coefficient of $\dfrac{\partial\phi}{\partial x}$ in the RHS of equation (2.59) is

$$\mathbf{i}'(\mathbf{i}'\cdot\mathbf{i}) + \mathbf{j}'(\mathbf{j}'\cdot\mathbf{j}) + \mathbf{k}'(\mathbf{k}'\cdot\mathbf{k}) = \mathbf{i}.$$

Similarly, coefficients of $\dfrac{\partial\phi}{\partial y}, \dfrac{\partial\phi}{\partial z}$ in the RHS of equation (2.59) will be **j** and **k** respectively. Thus, equation (2.59) becomes

$$\nabla\phi = \mathbf{i}\frac{\partial\phi}{\partial x} + \mathbf{j}\frac{\partial\phi}{\partial y} + \mathbf{k}\frac{\partial\phi}{\partial z}.$$

Hence proved. □

2.11 Equation of Tangent Plane and Normal

Let us consider a point A with position vector $\mathbf{r_0}$ lying on the surface $\phi(x, y, z) = c$. Let P be any other point on the tangent plane to the surface at A having position vector **r**. Then the vector $\overrightarrow{AP} = \mathbf{r} - \mathbf{r_0}$ lies on the tangent plane. Since $\text{grad}\,\phi$ is along the normal to the surface

$$(\mathbf{r} - \mathbf{r_0})\cdot\text{grad}\,\phi = 0 \tag{2.60}$$

which is the equation of the tangent plane.

Again, let us consider a point P with position vector **r** on the normal to the surface $\phi(x, y, z) = c$ at the point A having position vector $\mathbf{r_0}$. Then the vector $\overrightarrow{AP} = \mathbf{r} - \mathbf{r_0}$ lies along the normal.

$$\therefore \quad (\mathbf{r} - \mathbf{r_0})\times\text{grad}\,\phi = 0 \tag{2.61}$$

which is the equation of normal at the point A.

EXAMPLE 2.11. *Find equation of tangent plane and normal to the surface* $z = x^2 + y^2$ *at the point* $(2, -1, 5)$.

Solution: $\text{grad}\,\phi = \left(\mathbf{i}\dfrac{\partial\phi}{\partial x} + \mathbf{j}\dfrac{\partial\phi}{\partial y} + \mathbf{k}\dfrac{\partial\phi}{\partial z}\right) = (\mathbf{i}2x + \mathbf{j}2y - k) = 4\mathbf{i} - 2\mathbf{j} - \mathbf{k}$ at $(2, -1, 5)$.

Position vector of the point $(2, -1, 5)$ is

$$\mathbf{r_0} = 2\mathbf{i} - \mathbf{j} + 5\mathbf{k}.$$

$\therefore$ Equation of tangent plane is given by $(\mathbf{r} - \mathbf{r_0})\cdot\text{grad}\,\phi = 0$,

$$\text{i.e.,}\quad (\mathbf{i}x + \mathbf{j}y + \mathbf{k}z - 2\mathbf{i} + \mathbf{j} - 5\mathbf{k})\cdot(4\mathbf{i} - 2\mathbf{j} - \mathbf{k}) = 0,$$

$$\text{i.e.,}\quad (x-2)\cdot 4 - 2(y+1) - (z-5) = 0,$$

$$\text{i.e.,}\quad 4x - 2y - z = 5.$$

Again, equation of the normal to the surface is $(\mathbf{r} - \mathbf{r_0})\times\text{grad}\,\phi = 0$,

$$\text{i.e.,}\quad \{(x-2)\mathbf{i} + (y+1)\mathbf{j} + (z-5)\mathbf{k}\}\times(4\mathbf{i} - 2\mathbf{j} - \mathbf{k}),$$

$$\text{i.e.,}\quad \frac{x-2}{4} = \frac{y+1}{-2} = \frac{z-5}{-1}.$$

EXAMPLE 2.12. *Find a unit vector along the normal to the surface $z = x^2 + y^2$ at the point* $(1, 2, 5)$.

Solution: $\text{grad}\,\phi = (\mathbf{i}2x + \mathbf{j}2y - \mathbf{k}) = 2\mathbf{i} + 4\mathbf{j} - \mathbf{k}$ at the point $(1, 2, 5)$.

$\therefore$ Unit vector along the normal is

$$\frac{2\mathbf{i} + 4\mathbf{j} - \mathbf{k}}{\sqrt{4 + 16 + 1}} = \frac{1}{\sqrt{21}}(2\mathbf{i} + 4\mathbf{j} - \mathbf{k}).$$

EXAMPLE 2.13. *Find the angle between the surfaces*

$$\phi_1 \equiv x^2 + y^2 + z^2 - 9 = 0,$$
$$\phi_2 \equiv x^2 + y^2 - z - 3 = 0$$

at the point $(2, -1, 2)$.

Solution: Let $\mathbf{n}_1$, $\mathbf{n}_2$ be the vectors along normals to the surfaces ϕ_1, ϕ_2 at the point $(2, -1, 2)$. Then

$$\mathbf{n}_1 = \text{grad}\,\phi_1 = (\mathbf{i}2x + \mathbf{j}2y + \mathbf{k}2z) = 4\mathbf{i} - 2\mathbf{j} + 4\mathbf{k} \text{ at } (2, -1, 2).$$

Similarly

$$\mathbf{n}_2 = \text{grad}\,\phi_2 = (\mathbf{i}2x + \mathbf{j}2y - \mathbf{k}) = 4\mathbf{i} - 2\mathbf{j} - \mathbf{k} \text{ at } (2, -1, 2).$$

$$\therefore \quad |\mathbf{n}_1| = \sqrt{16 + 4 + 16} = 6$$
$$\text{and} \quad |\mathbf{n}_2| = \sqrt{16 + 4 + 1} = \sqrt{21}.$$

If θ be the angle between $\mathbf{n}_1$ and $\mathbf{n}_2$, then

$$\mathbf{n}_1 \cdot \mathbf{n}_2 = |\mathbf{n}_1|\,|\mathbf{n}_2| \cos\theta$$

$$\text{or,} \quad (4\mathbf{i} - 2\mathbf{j} + 4\mathbf{k}) \cdot (4\mathbf{i} - 2\mathbf{j} - \mathbf{k}) = 6\sqrt{21}\cos\theta, \quad \text{i.e.,} \quad 16 + 4 - 4 = 6\sqrt{21}\cos\theta.$$

$$\therefore \quad \cos\theta = \frac{16}{6\sqrt{21}} = \frac{8}{3\sqrt{21}}.$$

Note. If two surfaces $\phi_1(x, y, z) = 0$ and $\phi_2(x, y, z) = 0$ be orthogonal, then normals at the common point of intersection of the surface will be at right angles to each other. Hence

$$(\text{grad}\,\phi_1) \cdot (\text{grad}\,\phi_2) = 0$$

at the common point of intersection.

2.12 Divergence of a Vector Field

The *divergence of a vector field* is defined to be the net amount of outflow or flux of the vector field per unit volume.

Let us consider a small volume element $dx\,dy\,dz$ in the vector field whose value at the point O is $\mathbf{A}$ (*Fig 2.6*). Let A_x, A_y, A_z be the components of vector function $\mathbf{A}$ and $\frac{\partial A_x}{\partial x}, \frac{\partial A_y}{\partial y}, \frac{\partial A_z}{\partial z}$ be the rates of change of those components along x, y, z axes respectively.

Consider the surfaces marked $OADE$ and $BCFG$. The values of x-component of $\mathbf{A}$ over these surfaces are A_x and $A_x + \frac{\partial A_x}{\partial x}\,dx$ which are assumed to remain constant over the respective surfaces. Hence, the net outflow of the vector $\mathbf{A}$ in the x-direction is

$$\left[\left(A_x + \frac{\partial A_x}{\partial x}dx\right) - A_x\right] dy\,dz = \frac{\partial A_x}{\partial x}\,dx\,dy\,dz. \tag{2.62}$$

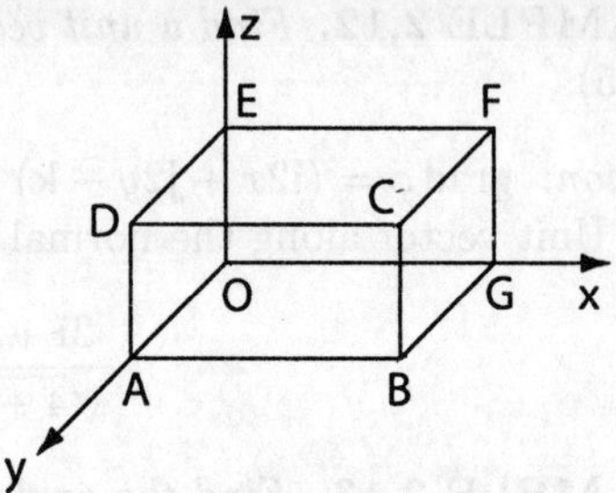

Fig 2.6

Similarly, the net outflow or flux of the vector $\mathbf{A}$ along y and z directions will be given by

$$\frac{\partial A_y}{\partial y}\,dx\,dy\,dz \quad \text{and} \quad \frac{\partial A_z}{\partial z}\,dx\,dy\,dz$$

respectively. Hence the total flux of the vector $\mathbf{A}$ from the infinitesimally small volume element is

$$\left(\frac{\partial A_x}{\partial x} + \frac{\partial A_y}{\partial y} + \frac{\partial A_z}{\partial z}\right) dx\,dy\,dz. \tag{2.63}$$

Thus, the amount of flux of vector $\mathbf{A}$ per unit volume is

$$\frac{\partial A_x}{\partial x} + \frac{\partial A_y}{\partial y} + \frac{\partial A_z}{\partial z}. \tag{2.64}$$

This quantity is called *divergence of vector* $\mathbf{A}$. Thus, we write

$$\operatorname{div}\mathbf{A} = \frac{\partial A_x}{\partial x} + \frac{\partial A_y}{\partial y} + \frac{\partial A_z}{\partial z}. \tag{2.65}$$

Now, we have

$$\mathbf{A} = A_x\mathbf{i} + A_y\mathbf{j} + A_z\mathbf{k} \quad \text{and} \quad \boldsymbol{\nabla} = \frac{\partial}{\partial x}\mathbf{i} + \frac{\partial}{\partial y}\mathbf{j} + \frac{\partial}{\partial z}\mathbf{k}.$$

Hence

$$\boldsymbol{\nabla}\cdot\mathbf{A} = \frac{\partial A_x}{\partial x} + \frac{\partial A_y}{\partial y} + \frac{\partial A_z}{\partial z}.$$

Thus we find

$$\operatorname{div}\mathbf{A} = \boldsymbol{\nabla}\cdot\mathbf{A}. \tag{2.66}$$

If $\operatorname{div}\mathbf{A} > 0$, flux is in the outward direction.
If $\operatorname{div}\mathbf{A} < 0$, flux is in the inward direction.
If $\operatorname{div}\mathbf{A} = 0$, the net flux is zero, i.e., the inflow and outflow of the flux balance each other. A vector which satisfies this condition is called a *solenoidal vector*.

2.13 Integral Form of Divergence

Let us consider a small area $d\sigma$ from the surface of a small volume element of the *Fig 2.6*. If $\mathbf{n}$ be the un[illegible] normal vector to the surface $d\sigma$, then the area vector is $d\boldsymbol{\sigma} = \mathbf{n}\cdot d\sigma$. The outflow of the flux across this area is $\mathbf{A} \cdot \mathbf{n}\, d\sigma$. So the net outflow of the vector $\mathbf{A}$ through the entire surface is

$$\int_\sigma \mathbf{A} \cdot \mathbf{n}\, d\sigma. \tag{2.67}$$

Now, the total flux of the vector $\mathbf{A}$ through the entire surfaces of the small element of volume $dx\, dy\, dz = d\tau$ is $\boldsymbol{\nabla} \cdot \mathbf{A}\, d\tau$.

Hence

$$\boldsymbol{\nabla} \cdot \mathbf{A}\, d\tau = \int_\sigma \mathbf{A} \cdot \mathbf{n}\, d\sigma, \quad \text{i.e.,} \quad \boldsymbol{\nabla} \cdot \mathbf{A} = \lim_{d\tau \longrightarrow 0} \frac{\int_\sigma \mathbf{A} \cdot \mathbf{n}\, d\sigma}{d\tau}. \tag{2.68}$$

2.14 Line Integral of a Vector

Let P, Q be two points in a vector field $\mathbf{A}$. With these two points, two paths marked 1 and 2 are drawn (*Fig 2.7*). If the position vectors of any two adjacent points on the curve are $\mathbf{r}$ and $\mathbf{r}+d\mathbf{r}$, then the length of small element will be $d\mathbf{l} = d\mathbf{r}$. The line integral of the vector from P to Q is given by

$$\int_P^Q \mathbf{A} \cdot d\mathbf{r}. \tag{2.69}$$

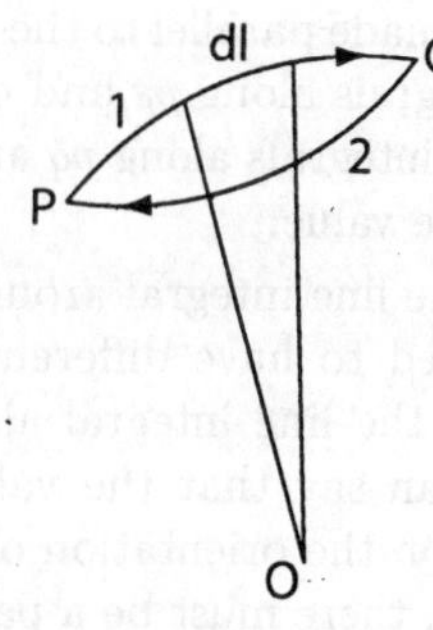

Fig 2.7

If now, the vector $\mathbf{A}$ can be expressed as the gradient of a scalar function ϕ by the equation

$$\mathbf{A} = \operatorname{grad}\phi = \boldsymbol{\nabla}\phi \tag{2.70}$$

the line integral of equation (2.69) becomes

$$\int_P^Q (\operatorname{grad}\phi) \cdot d\mathbf{r} = \int_P^Q d\phi = \phi(Q) - \phi(P). \tag{2.71}$$

Since the values of the scalar function ϕ at points P, Q [i.e., $\phi(P), \phi(Q)$] are fixed, equation (2.71) clarifies that the line integral has same value for different paths. Then, for path 2, the line integral

$$\int_Q^P \mathbf{A} \cdot d\mathbf{r} = \int_Q^P (\operatorname{grad}\phi) \cdot d\mathbf{r} = \phi(Q) - \phi(P). \tag{2.72}$$

Hence, for a closed path PQP, we have

$$\int (\operatorname{grad}\phi) \cdot d\mathbf{r} = 0. \tag{2.73}$$

Thus, when a vector field can be expressed as a gradient of a scalar field,

(i) the line integral of the vector taken between two points does not depend on the path followed;

(ii) the line integral of the vector taken along any closed path is zero. Such a vector field is called *conservative*.

A vector field derived by taking the gradient of a scalar field is called *Scalar Potential field* or *Lamellar Vector field*.

2.15 Curl or Rotation of a Vector Field

We have seen that for lamellar vector field, line integral around a closed path is zero. But for non-lamellar vector fields, i.e., those vector fields which are not derived by taking gradient of a scalar field, the line integral around a closed path is not zero. In such cases curl of a vector plays an important role.

Let us consider a small region of such a vector field where a rectangular path *pqrs* is taken. If the plane of this rectangle is kept perpendicular to the direction of vector **A**, then the line integral along this path is zero, since **A** is perpendicular to every line element **dr** (*Fig 2.8*) of *pqrs*. If however the plane of the rectangle is made parallel to the direction of **A**, the line integrals along *ps* and *qr* will be zero, but the line integrals along *pq* and *rs* will have some definite value.

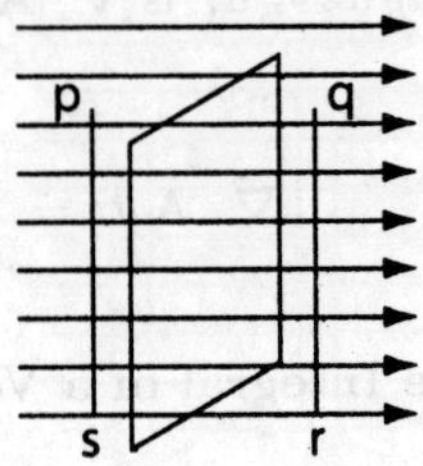

Fig 2.8

Hence the line integral around the closed boundary will have a finite value since the vector **A** is assumed to have different values along *pq* and *rs* edges. For any other intermediate orientation, the line integral also will have a finite value different from that obtained above. Hence we can say that the value of the line integral of the vector **A** around a closed path depends upon the orientation of the small vector area in respect to the direction of the vector field. Hence, there must be a particular orientation of the area for which the line integral along the closed path is maximum. *This maximum value of the line integral computed for unit area is called the* curl *or* rotation *of the vector field.*

The curl of vector **A** is a vector in the direction of the positive normal to the area under consideration which is in the position giving the maximum value of the line integral.

Let, **A** be a vector point function having components A_x, A_y, A_z along x, y and z axes respectively (*Fig 2.9*). Let us consider the rectangular surface marked 1.

The values of the components of vector **A** along sides of rectangle *pqrs* are:

along $ps : A_y$,

along $qr : A_y + \dfrac{\partial A_y}{\partial z} dz$,

along $pq : A_z$

and along $rs : A_z + \dfrac{\partial A_z}{\partial y} dy$.

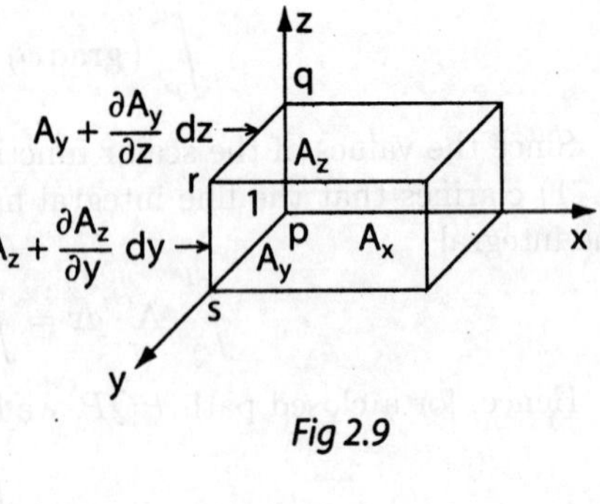

Fig 2.9

Hence, the integral around the closed path *pqrsp* is

$$A_z\, dz + \left(A_y + \frac{\partial A_y}{\partial z} dz\right) dy - \left(A_z + \frac{\partial A_z}{\partial y} dy\right) dz - A_y\, dy$$

[∵ directions of *rs* and *sp* are in opposite sense]

$$= \left(\frac{\partial A_y}{\partial z} - \frac{\partial A_z}{\partial y}\right) dy\, dz.$$

It is obvious that this line integral will have maximum value since the paths of integration are either parallel or anti-parallel to the vector components. Now, area enclosed by the path being $dy\,dz$, the maximum value of the line integral per unit area, i.e.,

$$\text{curl}_x \mathbf{A} = \left(\frac{\partial A_y}{\partial z} - \frac{\partial A_z}{\partial y}\right) \tag{2.74}$$

which is a vector along the positive normal to the area. Similarly, considering rectangular paths whose planes are perpendicular to y and z axes, it can be shown that

$$\left.\begin{aligned} \text{curl}_y \mathbf{A} &= \left(\frac{\partial A_x}{\partial z} - \frac{\partial A_z}{\partial x}\right) \\ \text{and} \quad \text{curl}_z \mathbf{A} &= \left(\frac{\partial A_y}{\partial x} - \frac{\partial A_x}{\partial y}\right) \end{aligned}\right\}. \tag{2.75}$$

$$\begin{aligned} \therefore \quad \text{curl}\,\mathbf{A} &= \mathbf{i}\,\text{curl}_x \mathbf{A} + \mathbf{j}\,\text{curl}_y \mathbf{A} + \mathbf{k}\,\text{curl}_z \mathbf{A} \\ &= \mathbf{i}\left(\frac{\partial A_y}{\partial z} - \frac{\partial A_z}{\partial y}\right) + \mathbf{j}\left(\frac{\partial A_x}{\partial z} - \frac{\partial A_z}{\partial x}\right) + \mathbf{k}\left(\frac{\partial A_y}{\partial x} - \frac{\partial A_x}{\partial y}\right). \end{aligned} \tag{2.76}$$

Equation (2.76) can be written as

$$\text{curl}\,\mathbf{A} = \begin{vmatrix} \mathbf{i} & \mathbf{j} & \mathbf{k} \\ \dfrac{\partial}{\partial x} & \dfrac{\partial}{\partial y} & \dfrac{\partial}{\partial z} \\ A_x & A_y & A_z \end{vmatrix} \tag{2.77}$$

In terms of vector differential operator $\boldsymbol{\nabla}$, we can write

$$\text{curl}\,\mathbf{A} = \boldsymbol{\nabla} \times \mathbf{A}. \tag{2.78}$$

The differential operator $\boldsymbol{\nabla}$ is invariant, and hence the curl of a vector is also invariant. If curl $\mathbf{A} = 0$, the vector $\mathbf{A}$ is called an *irrotational* vector.

2.16 The Laplacian Operator

The operator

$$\nabla^2 = \frac{\partial^2}{\partial x^2} + \frac{\partial^2}{\partial y^2} + \frac{\partial^2}{\partial z^2} \tag{2.79}$$

is known as the *Laplacian operator.*

If ϕ be any scalar function, then

$$\nabla^2\phi = \boldsymbol{\nabla} \cdot (\boldsymbol{\nabla}\phi) = (\boldsymbol{\nabla} \cdot \boldsymbol{\nabla})\phi.$$

Proof.

$$\begin{aligned} \boldsymbol{\nabla} \cdot (\boldsymbol{\nabla}\phi) &= \left(\mathbf{i}\frac{\partial}{\partial x} + \mathbf{j}\frac{\partial}{\partial y} + \mathbf{k}\frac{\partial}{\partial z}\right) \cdot \left(\mathbf{i}\frac{\partial \phi}{\partial x} + \mathbf{j}\frac{\partial \phi}{\partial y} + \mathbf{k}\frac{\partial \phi}{\partial z}\right) \\ &= \frac{\partial^2 \phi}{\partial x^2} + \frac{\partial^2 \phi}{\partial y^2} + \frac{\partial^2 \phi}{\partial z^2}. \end{aligned}$$

Again,
$$(\nabla\cdot\nabla)\phi=\left(\mathbf{i}\frac{\partial}{\partial x}+\mathbf{j}\frac{\partial}{\partial y}+\mathbf{k}\frac{\partial}{\partial z}\right)\cdot\left(\mathbf{i}\frac{\partial}{\partial x}+\mathbf{j}\frac{\partial}{\partial y}+\mathbf{k}\frac{\partial}{\partial z}\right)\phi$$
$$=\left(\frac{\partial^2}{\partial x^2}+\frac{\partial^2}{\partial y^2}+\frac{\partial^2}{\partial z^2}\right)\phi=\frac{\partial^2\phi}{\partial x^2}+\frac{\partial^2\phi}{\partial y^2}+\frac{\partial^2\phi}{\partial z^2}.$$
$$\therefore\quad \nabla^2\phi=\nabla\cdot(\nabla\phi)=(\nabla\cdot\nabla)\phi=\frac{\partial^2\phi}{\partial x^2}+\frac{\partial^2\phi}{\partial y^2}+\frac{\partial^2\phi}{\partial z^2}.$$
□

2.17 Important Vector Relationships

(a) $\operatorname{div}(\operatorname{grad}\phi)=\nabla\cdot\nabla\phi.$

Proof.

$$\operatorname{div}(\operatorname{grad}\phi)=\left(\mathbf{i}\frac{\partial}{\partial x}+\mathbf{j}\frac{\partial}{\partial y}+\mathbf{k}\frac{\partial}{\partial z}\right)\cdot\left(\mathbf{i}\frac{\partial\phi}{\partial x}+\mathbf{j}\frac{\partial\phi}{\partial y}+\mathbf{k}\frac{\partial\phi}{\partial z}\right)$$
$$=\frac{\partial^2\phi}{\partial x^2}+\frac{\partial^2\phi}{\partial y^2}+\frac{\partial^2\phi}{\partial z^2}.$$

Again
$$\nabla\cdot\nabla\phi=\nabla^2\phi=\frac{\partial^2\phi}{\partial x^2}+\frac{\partial^2\phi}{\partial y^2}+\frac{\partial^2\phi}{\partial z^2}.$$

Hence, $\operatorname{div}(\operatorname{grad}\phi)=\nabla\cdot\nabla\phi.$ □

(b) $\operatorname{curl}(\operatorname{grad}\phi)=\mathbf{0}.$

Proof.

$$\operatorname{curl}(\operatorname{grad}\phi)=\nabla\times\nabla\phi=\left(\mathbf{i}\frac{\partial}{\partial x}+\mathbf{j}\frac{\partial}{\partial y}+\mathbf{k}\frac{\partial}{\partial z}\right)\times\left(\mathbf{i}\frac{\partial\phi}{\partial x}+\mathbf{j}\frac{\partial\phi}{\partial y}+\mathbf{k}\frac{\partial\phi}{\partial z}\right).$$

Collecting the coefficients of **i**,
$$\operatorname{curl}(\operatorname{grad}_x\phi)=\left(\frac{\partial^2\phi}{\partial y\partial z}-\frac{\partial^2\phi}{\partial z\partial y}\right)=0.$$

Other coefficients are similarly zero.
Hence, $\operatorname{curl}(\operatorname{grad}\phi)=0.$ □

(c) $\operatorname{div}(\operatorname{curl}\mathbf{A})=\mathbf{0}.$

Proof.

$$\operatorname{div}(\operatorname{curl}\mathbf{A})=\left(\mathbf{i}\frac{\partial}{\partial x}+\mathbf{j}\frac{\partial}{\partial y}+\mathbf{k}\frac{\partial}{\partial z}\right)\cdot\left[\mathbf{i}\left(\frac{\partial A_z}{\partial y}-\frac{\partial A_y}{\partial z}\right)+\mathbf{j}\left(\frac{\partial A_x}{\partial z}-\frac{\partial A_z}{\partial x}\right)+\mathbf{k}\left(\frac{\partial A_y}{\partial x}-\frac{\partial A_x}{\partial y}\right)\right]$$
$$=\frac{\partial}{\partial x}\left(\frac{\partial A_z}{\partial y}-\frac{\partial A_y}{\partial z}\right)+\frac{\partial}{\partial y}\left(\frac{\partial A_x}{\partial z}-\frac{\partial A_z}{\partial x}\right)+\frac{\partial}{\partial z}\left(\frac{\partial A_y}{\partial x}-\frac{\partial A_x}{\partial y}\right)$$
$$=\frac{\partial^2 A_z}{\partial x\partial y}-\frac{\partial^2 A_y}{\partial x\partial z}+\frac{\partial^2 A_x}{\partial y\partial z}-\frac{\partial^2 A_z}{\partial y\partial x}+\frac{\partial^2 A_y}{\partial z\partial x}-\frac{\partial^2 A_x}{\partial y\partial z}=0.$$
□

(d) $\operatorname{curl}(\operatorname{curl}\mathbf{A}) = \boldsymbol{\nabla}\times(\nabla\mathbf{A})$.

Proof.

$$
\begin{aligned}
\operatorname{curl}(\operatorname{curl}\mathbf{A}) &= \boldsymbol{\nabla}\times(\boldsymbol{\nabla}\times\mathbf{A})\\
&= \left(\mathbf{i}\frac{\partial}{\partial x}+\mathbf{j}\frac{\partial}{\partial y}+\mathbf{k}\frac{\partial}{\partial z}\right)\times\begin{vmatrix}\mathbf{i} & \mathbf{j} & \mathbf{k}\\ \frac{\partial}{\partial x} & \frac{\partial}{\partial y} & \frac{\partial}{\partial z}\\ A_x & A_y & A_z\end{vmatrix}\\
&= \left(\mathbf{i}\frac{\partial}{\partial x}+\mathbf{j}\frac{\partial}{\partial y}+\mathbf{k}\frac{\partial}{\partial z}\right)\\
&\quad\times\left[\mathbf{i}\left(\frac{\partial A_z}{\partial y}-\frac{\partial A_y}{\partial z}\right)+\mathbf{j}\left(\frac{\partial A_x}{\partial z}-\frac{\partial A_z}{\partial x}\right)+\mathbf{k}\left(\frac{\partial A_y}{\partial x}-\frac{\partial A_x}{\partial y}\right)\right].
\end{aligned}
$$

Collecting the coefficients for $\mathbf{i}$,

$$\operatorname{curl}_x(\operatorname{curl}\mathbf{A}) = \mathbf{i}\left[\frac{\partial}{\partial y}\left(\frac{\partial A_y}{\partial x}-\frac{\partial A_x}{\partial y}\right)-\frac{\partial}{\partial z}\left(\frac{\partial A_x}{\partial z}-\frac{\partial A_z}{\partial x}\right)\right].$$

Adding and subtracting $\frac{\partial^2 A_x}{\partial x^2}$, we get

$$\operatorname{curl}_x(\operatorname{curl}\mathbf{A}) = \frac{\partial}{\partial x}\left(\frac{\partial A_x}{\partial x}+\frac{\partial A_y}{\partial y}+\frac{\partial A_z}{\partial z}\right)\mathbf{i}-\mathbf{i}\nabla^2 A_x. \tag{2.80}$$

The 1st term on the RHS is x-component of $\operatorname{grad}(\operatorname{div}\mathbf{A})$, because

$$\operatorname{grad}(\operatorname{div}\mathbf{A}) = \left(\mathbf{i}\frac{\partial}{\partial x}+\mathbf{j}\frac{\partial}{\partial y}+\mathbf{k}\frac{\partial}{\partial z}\right)\left(\frac{\partial A_x}{\partial y}+\frac{\partial A_y}{\partial y}+\frac{\partial A_z}{\partial z}\right)$$

so that x-component of

$$\operatorname{grad}(\operatorname{div}\mathbf{A}) = \operatorname{grad}_x(\operatorname{div}\mathbf{A}) = \mathbf{i}\frac{\partial}{\partial x}\left(\frac{\partial A_x}{\partial y}+\frac{\partial A_y}{\partial y}+\frac{\partial A_z}{\partial z}\right).$$

Again, the 2nd term on the RHS of equation (2.80) is x-component of $\nabla^2\mathbf{A}$. Hence

$$\operatorname{curl}_x(\operatorname{curl}\mathbf{A}) = \operatorname{grad}_x(\operatorname{div}\mathbf{A}) - \nabla^2 A_x.$$

Similarly

$$
\begin{aligned}
&& \operatorname{curl}_y(\operatorname{curl}\mathbf{A}) &= \operatorname{grad}_y(\operatorname{div}\mathbf{A}) - \nabla^2 A_y\\
\text{and} && \operatorname{curl}_z(\operatorname{curl}\mathbf{A}) &= \operatorname{grad}_z(\operatorname{div}\mathbf{A}) - \nabla^2 A_z.\\
\therefore && \operatorname{curl}(\operatorname{curl}\mathbf{A}) &= \operatorname{grad}(\operatorname{div}\mathbf{A}) - \nabla^2\mathbf{A}.
\end{aligned}
$$

From here we can also write $\operatorname{grad}(\operatorname{div}\mathbf{A}) = \operatorname{curl}(\operatorname{curl}\mathbf{A}) + \nabla^2\mathbf{A}$. □

(e) $\boldsymbol{\nabla}\cdot(\phi\mathbf{A}) = \phi\boldsymbol{\nabla}\cdot\mathbf{A}+\mathbf{A}\cdot\boldsymbol{\nabla}\phi$.

Proof.

$$
\begin{aligned}
\nabla\cdot(\phi\mathbf{A}) &= \nabla\cdot[\phi A_x\mathbf{i}+\phi A_y\mathbf{j}+\phi A_z\mathbf{k}]\\
&= \left(\mathbf{i}\frac{\partial}{\partial x}+\mathbf{j}\frac{\partial}{\partial y}+\mathbf{k}\frac{\partial}{\partial z}\right)\cdot(\phi A_x\mathbf{i}+\phi A_y\mathbf{j}+\phi A_z\mathbf{k})\\
&= \frac{\partial}{\partial x}(\phi A_x)+\frac{\partial}{\partial y}(\phi A_y)+\frac{\partial}{\partial z}(\phi A_z)\\
&= \phi\left(\frac{\partial A_x}{\partial x}+\frac{\partial A_y}{\partial y}+\frac{\partial A_z}{\partial z}\right)+\left(A_x\frac{\partial \phi}{\partial x}+A_y\frac{\partial \phi}{\partial y}+A_z\frac{\partial \phi}{\partial z}\right)\\
&= \phi\nabla\cdot\mathbf{A}+(\mathbf{i}A_x+\mathbf{j}A_y+\mathbf{k}A_z)\cdot\left(\mathbf{i}\frac{\partial \phi}{\partial x}+\mathbf{j}\frac{\partial \phi}{\partial y}+\mathbf{k}\frac{\partial \phi}{\partial z}\right)\\
&= \phi\nabla\cdot\mathbf{A}+\mathbf{A}\cdot\nabla\phi.
\end{aligned}
$$

□

(f) $\operatorname{grad}(\mathbf{A}\cdot\mathbf{B})=(\mathbf{A}\operatorname{grad})\mathbf{B}+(\mathbf{B}\operatorname{grad})\mathbf{A}+(\mathbf{A}\times\operatorname{curl}\mathbf{B})+(\mathbf{B}\times\operatorname{curl}\mathbf{A})$.

Proof. Considering only the x-component, we have

$$
\begin{aligned}
\operatorname{grad}_x(\mathbf{A}\cdot\mathbf{B}) &= \frac{\partial}{\partial x}(A_xB_x+A_yB_y+A_zB_z)\\
&= A_x\frac{\partial B_x}{\partial x}+B_x\frac{\partial A_x}{\partial x}+A_y\frac{\partial B_y}{\partial x}+B_y\frac{\partial A_y}{\partial x}+A_z\frac{\partial B_z}{\partial x}+B_z\frac{\partial A_z}{\partial x}\\
&= \left(A_x\frac{\partial B_x}{\partial x}+A_y\frac{\partial B_x}{\partial y}+A_z\frac{\partial B_x}{\partial z}\right)+\left(B_x\frac{\partial A_x}{\partial x}+B_y\frac{\partial A_x}{\partial y}+B_z\frac{\partial A_x}{\partial z}\right)\\
&\quad +A_y\frac{\partial B_y}{\partial x}+B_y\frac{\partial A_y}{\partial x}+A_z\frac{\partial B_z}{\partial x}+B_z\frac{\partial A_z}{\partial x}\\
&\quad -A_y\frac{\partial B_x}{\partial y}-A_z\frac{\partial B_x}{\partial z}-B_y\frac{\partial A_x}{\partial y}-B_z\frac{\partial A_x}{\partial z}\\
&= (\mathbf{A}\operatorname{grad})B_x+(\mathbf{B}\operatorname{grad})A_x+A_y\left(\frac{\partial B_y}{\partial x}-\frac{\partial B_x}{\partial y}\right)\\
&\quad +B_y\left(\frac{\partial A_y}{\partial x}-\frac{\partial A_x}{\partial y}\right)-A_z\left(\frac{\partial B_x}{\partial z}-\frac{\partial B_z}{\partial x}\right)-B_z\left(\frac{\partial A_x}{\partial z}-\frac{\partial A_z}{\partial x}\right),
\end{aligned}
$$

i.e.,

$$\operatorname{grad}_x(\mathbf{A}\cdot\mathbf{B})=(\mathbf{A}\operatorname{grad})B_x+(\mathbf{B}\operatorname{grad})A_x+(\mathbf{A}\times\operatorname{curl}\mathbf{B})_x+(\mathbf{B}\times\operatorname{curl}\mathbf{A})_x.$$

Hence, considering y and z components of the LHS and adding them all, we get

$$\operatorname{grad}(\mathbf{A}\cdot\mathbf{B})=(\mathbf{A}\operatorname{grad})\mathbf{B}+(\mathbf{B}\operatorname{grad})\mathbf{A}+(\mathbf{A}\times\operatorname{curl}\mathbf{B})+(\mathbf{B}\times\operatorname{curl}\mathbf{A}).$$

□

(g) $\operatorname{div}(\mathbf{A}\times\mathbf{B})=\mathbf{B}\cdot\operatorname{curl}\mathbf{A}-\mathbf{A}\cdot\operatorname{curl}\mathbf{B}$.

Proof.

$$\begin{aligned}
\operatorname{div}(\mathbf{A}\times\mathbf{B}) &= \left(\mathbf{i}\frac{\partial}{\partial x}+\mathbf{j}\frac{\partial}{\partial y}+\mathbf{k}\frac{\partial}{\partial z}\right)(\mathbf{A}\times\mathbf{B})\\
&= \left(\mathbf{i}\frac{\partial}{\partial x}+\mathbf{j}\frac{\partial}{\partial y}+\mathbf{k}\frac{\partial}{\partial z}\right)\\
&\quad\cdot[\mathbf{i}(A_yB_z-A_zB_y)+\mathbf{j}(A_zB_x-A_xB_z)+\mathbf{k}(A_xB_y-A_yB_x)]\\
&= \frac{\partial}{\partial x}(A_yB_z-A_zB_y)+\frac{\partial}{\partial y}(A_zB_x-A_xB_z)+\frac{\partial}{\partial z}(A_xB_y-A_yB_x)\\
&= B_x\left(\frac{\partial A_z}{\partial y}-\frac{\partial A_y}{\partial z}\right)+B_y\left(\frac{\partial A_x}{\partial z}-\frac{\partial A_z}{\partial x}\right)+B_z\left(\frac{\partial A_y}{\partial x}-\frac{\partial A_x}{\partial y}\right)\\
&\quad-\left[A_x\left(\frac{\partial B_z}{\partial y}-\frac{\partial B_y}{\partial z}\right)+A_y\left(\frac{\partial B_x}{\partial z}-\frac{\partial B_z}{\partial y}\right)+A_z\left(\frac{\partial B_y}{\partial x}-\frac{\partial B_x}{\partial y}\right)\right]\\
&= \mathbf{B}\cdot(\operatorname{curl}\mathbf{A})-\mathbf{A}\cdot(\operatorname{curl}\mathbf{B}).
\end{aligned}$$

□

(h) $\operatorname{curl}(\mathbf{A}\times\mathbf{B})=\mathbf{A}\operatorname{div}\mathbf{B}-\mathbf{B}\operatorname{div}\mathbf{A}+(\mathbf{B}\operatorname{grad})\mathbf{A}-(\mathbf{A}\operatorname{grad})\mathbf{B}$.

Proof.

$$\operatorname{curl}(\mathbf{A}\times\mathbf{B})=\begin{vmatrix}\mathbf{i} & \mathbf{j} & \mathbf{k}\\ \dfrac{\partial}{\partial x} & \dfrac{\partial}{\partial y} & \dfrac{\partial}{\partial z}\\ A_yB_z-B_yA_z & A_zB_x-B_zA_x & A_xB_y-B_xA_y\end{vmatrix}.$$

Considering the x-component only

$$\operatorname{curl}_x(\mathbf{A}\times\mathbf{B})=\frac{\partial}{\partial y}(A_xB_y-B_xA_y)-\frac{\partial}{\partial z}(A_zB_x-B_zA_x).$$

Adding and subtracting $\frac{\partial}{\partial x}(A_xB_x)$, we get

$$\begin{aligned}
\operatorname{curl}_x(\mathbf{A}\times\mathbf{B}) &= \frac{\partial}{\partial x}(A_xB_x)+\frac{\partial}{\partial y}(A_xB_y)+\frac{\partial}{\partial z}(A_xB_z)\\
&\quad-\left[\frac{\partial}{\partial x}(A_xB_x)+\frac{\partial}{\partial y}(A_yB_x)+\frac{\partial}{\partial z}(A_zB_x)\right]\\
&= A_x\left(\frac{\partial B_x}{\partial x}+\frac{\partial B_y}{\partial y}+\frac{\partial B_z}{\partial z}\right)+\left(B_x\frac{\partial A_x}{\partial x}+B_y\frac{\partial A_x}{\partial y}+B_z\frac{\partial A_x}{\partial z}\right)\\
&\quad-B_x\left(\frac{\partial A_x}{\partial x}+\frac{\partial A_y}{\partial y}+\frac{\partial A_z}{\partial z}\right)-\left(A_x\frac{\partial B_x}{\partial x}+A_y\frac{\partial B_x}{\partial y}+A_z\frac{\partial B_x}{\partial z}\right)\\
&= (\mathbf{B}\cdot\nabla)A_x+A_x(\nabla\cdot\mathbf{B})-(\mathbf{A}\cdot\nabla)B_x-B_x(\nabla\cdot\mathbf{A}).
\end{aligned}$$

Considering components in y and z directions and adding, we get

$$\begin{aligned}
\operatorname{curl}(\mathbf{A}\times\mathbf{B}) &= (\mathbf{B}\cdot\nabla)(\mathbf{i}A_x+\mathbf{j}A_y+\mathbf{k}A_z)-(\mathbf{A}\cdot\nabla)(\mathbf{i}B_x+\mathbf{j}B_y+\mathbf{k}B_z)\\
&\quad+(\mathbf{i}A_x+\mathbf{j}A_y+\mathbf{k}A_z)(\nabla\cdot\mathbf{B})-(\mathbf{i}B_x+\mathbf{j}B_y+\mathbf{k}B_z)(\nabla\cdot\mathbf{A})\\
&= (\mathbf{B}\cdot\nabla)\mathbf{A}-(\mathbf{A}\cdot\nabla)\mathbf{B}+\mathbf{A}(\nabla\cdot\mathbf{B})-\mathbf{B}(\nabla\cdot\mathbf{A}).
\end{aligned}$$

$$\therefore\quad \operatorname{curl}(\mathbf{A}\times\mathbf{B})=\mathbf{A}\operatorname{div}\mathbf{B}-\mathbf{B}\operatorname{div}\mathbf{A}+(\mathbf{B}\operatorname{grad})\mathbf{A}-(\mathbf{A}\operatorname{grad})\mathbf{B}.$$

□

(i) $\text{curl}\,(\phi\mathbf{A}) = \phi\,\text{curl}\,\mathbf{A} - \mathbf{A}\times\text{grad}\,\phi$.

Proof.

$$\begin{aligned}\text{curl}\,(\phi\mathbf{A}) &= \nabla\times(\phi\mathbf{A})\\ &= \left(\mathbf{i}\frac{\partial}{\partial x}+\mathbf{j}\frac{\partial}{\partial y}+\mathbf{k}\frac{\partial}{\partial z}\right)\times[\mathbf{i}(\phi A_x)+\mathbf{j}(\phi A_y)+\mathbf{k}(\phi A_z)]\\ &= \left[\frac{\partial}{\partial y}(\phi A_z)-\frac{\partial}{\partial z}(\phi A_y)\right]\mathbf{i}+\left[\frac{\partial}{\partial z}(\phi A_x)-\frac{\partial}{\partial x}(\phi A_z)\right]\mathbf{j}\\ &\quad+\left[\frac{\partial}{\partial x}(\phi A_y)-\frac{\partial}{\partial y}(\phi A_x)\right]\mathbf{k}.\end{aligned}$$

Considering only the x-component

$$\text{curl}_x(\phi\mathbf{A}) = \left[\phi\left(\frac{\partial A_z}{\partial y}-\frac{\partial A_y}{\partial z}\right)+A_z\frac{\partial\phi}{\partial y}-A_y\frac{\partial\phi}{\partial z}\right]\mathbf{i}.$$

Similarly

$$\text{curl}_y(\phi\mathbf{A}) = \left[\phi\left(\frac{\partial A_x}{\partial z}-\frac{\partial A_z}{\partial x}\right)+A_x\frac{\partial\phi}{\partial z}-A_z\frac{\partial\phi}{\partial x}\right]\mathbf{j}$$

and

$$\text{curl}_z(\phi\mathbf{A}) = \left[\phi\left(\frac{\partial A_z}{\partial y}-\frac{\partial A_y}{\partial z}\right)+A_z\frac{\partial\phi}{\partial y}-A_y\frac{\partial\phi}{\partial z}\right]\mathbf{k}.$$

Adding, we get

$$\begin{aligned}\text{curl}\,(\phi\mathbf{A}) &= \phi\,\text{curl}\,\mathbf{A}+\left(\frac{\partial\phi}{\partial x}\mathbf{i}+\frac{\partial\phi}{\partial y}\mathbf{j}+\frac{\partial\phi}{\partial z}\mathbf{k}\right)\times(A_x\mathbf{i}+A_y\mathbf{j}+A_z\mathbf{k})\\ &= \phi\,\text{curl}\,\mathbf{A}+(\text{grad}\,\phi)\times\mathbf{A}\\ &= \phi\,\text{curl}\,\mathbf{A}-\mathbf{A}\times(\text{grad}\,\phi).\end{aligned}$$

□

EXAMPLE 2.14. *If $\phi = x^3+y^3+z^3-3xyz$, then find div (grad ϕ) and curl(grad ϕ).*

Solution:

$$\begin{aligned}\text{grad}\,\phi &= \mathbf{i}\frac{\partial\phi}{\partial x}+\mathbf{j}\frac{\partial\phi}{\partial y}+\mathbf{k}\frac{\partial\phi}{\partial z} = \mathbf{i}(3x^2-3yz)+\mathbf{j}(3y^2-3zx)+\mathbf{k}(3z^2-3xy)\\ &= \phi_1\mathbf{i}+\phi_2\mathbf{j}+\phi_3\mathbf{k}\quad\text{(say)}.\end{aligned}$$

$$\therefore\quad \text{div}\,(\text{grad}\,\phi) = \frac{\partial\phi_1}{\partial x}+\frac{\partial\phi_2}{\partial y}+\frac{\partial\phi_3}{\partial z} = 6x+6y+6z$$

$$\begin{aligned}\text{curl}\,(\text{grad}\,\phi) &= \begin{vmatrix}\mathbf{i} & \mathbf{j} & \mathbf{k}\\ \frac{\partial}{\partial x} & \frac{\partial}{\partial y} & \frac{\partial}{\partial z}\\ \phi_1 & \phi_2 & \phi_3\end{vmatrix}\\ &= \left[\mathbf{i}\left(\frac{\partial\phi_3}{\partial y}-\frac{\partial\phi_2}{\partial z}\right)+\mathbf{j}\left(\frac{\partial\phi_1}{\partial z}-\frac{\partial\phi_3}{\partial x}\right)+\mathbf{k}\left(\frac{\partial\phi_2}{\partial x}-\frac{\partial\phi_1}{\partial y}\right)\right]\\ &= 3[\mathbf{i}(-x+x)+\mathbf{j}(-y+y)+\mathbf{k}(-z+z)] = 0.\end{aligned}$$

EXAMPLE 2.15. *Prove that $div(\,grad\,\mathbf{r}^n) = n(n+1)r^{n-2}$.*

Solution: Let, $\mathbf{r} = ix + jy + kz$. $\therefore\ r = \left(x^2+y^2+z^2\right)^{1/2}$.

$$\therefore \quad \frac{\partial r}{\partial x} = \frac{2x}{2\sqrt{x^2+y^2+z^2}} = \frac{x}{r}.$$

Similarly

$$\frac{\partial r}{\partial y} = \frac{y}{r}, \quad \frac{\partial r}{\partial z} = \frac{z}{r}. \tag{1}$$

Let, $\phi = r^n$,

$$\therefore \quad \frac{\partial \phi}{\partial x} = nr^{n-1}\frac{\partial r}{\partial x} = nr^{n-1}\frac{x}{r}, \quad \text{i.e.,} \quad \frac{\partial \phi}{\partial x} = nxr^{n-2}.$$

Similarly

$$\frac{\partial \phi}{\partial y} = nyr^{n-2}, \quad \frac{\partial \phi}{\partial z} = nzr^{n-2}. \tag{2}$$

$$\therefore \quad \text{grad}\,\phi = i\frac{\partial \phi}{\partial x} + j\frac{\partial \phi}{\partial y} + k\frac{\partial \phi}{\partial z}; \quad \text{div}\,(\,\text{grad}\,\phi) = \frac{\partial^2 \phi}{\partial x^2} + \frac{\partial^2 \phi}{\partial y^2} + \frac{\partial^2 \phi}{\partial z^2}.$$

Now, from equation (2)

$$\frac{\partial^2 \phi}{\partial x^2} = n\left[1\cdot r^{n-2} + x(n-2)r^{n-3}\cdot\frac{x}{r}\right] \quad \text{[by using equation (1)]}$$

$$= nr^{n-2}\left[1 + \frac{n-2}{r^2}x^2\right].$$

Similar expressions will be obtained for $\dfrac{\partial^2 \phi}{\partial y^2}$ and $\dfrac{\partial^2 \phi}{\partial z^2}$.

Hence

$$\text{div}\,(\,\text{grad}\,\phi) = \sum \frac{\partial^2 \phi}{\partial x^2} = nr^{n-2}\left[3 + \frac{n-2}{r^2}\left(x^2+y^2+z^2\right)\right]$$

$$= nr^{n-2}\left(3 + \frac{n-2}{r^2}\cdot r^2\right) = nr^{n-2}(3+n-2)$$

$$= n(n+1)r^{n-2}.$$

EXAMPLE 2.16. *Prove that $div\; r^n\mathbf{r} = (n+3)r^n$.*

Solution:

$$r^n\mathbf{r} = r^n(x\mathbf{i} + y\mathbf{j} + z\mathbf{k}), \quad r = \left(x^2+y^2+z^2\right)^{1/2}$$

$$= \mathbf{i}\phi_1 + \mathbf{j}\phi_2 + \mathbf{k}\phi_3 \quad \text{(say)}.$$

$$\therefore \quad \text{div}\, r^n\mathbf{r} = \frac{\partial \phi_1}{\partial x} + \frac{\partial \phi_2}{\partial y} + \frac{\partial \phi_3}{\partial z}.$$

Now, $\phi_1 = xr^n$,

$$\therefore \quad \frac{\partial \phi_1}{\partial x} = r^n + nxr^{n-1}\frac{\partial r}{\partial x} = r^n + nxr^{n-1}\frac{x}{r} = r^n + nr^{n-2}\cdot x^2.$$

Similar expressions for $\frac{\partial \phi_2}{\partial y}, \frac{\partial \phi_3}{\partial z}$ will be obtained.

Hence

$$\operatorname{div} r^n \mathbf{r} = 3r^n + nr^{n-2}\left(x^2+y^2+z^2\right) = 3r^n + nr^{n-2}\cdot r^2 = (3+n)r^n.$$

EXAMPLE 2.17. *Prove that* $div(\mathbf{r}\, grad\, \mathbf{r}^{-3}) = 3r^{-4}$.

Solution:

$$\begin{aligned}\operatorname{grad}\mathbf{r}^{-3} &= \left[\mathbf{i}\frac{\partial}{\partial x}(r^{-3}) + \mathbf{j}\frac{\partial}{\partial y}(r^{-3}) + \mathbf{k}\frac{\partial}{\partial z}(r^{-3})\right]\\ &= -3r^{-4}\left\{\mathbf{i}\frac{x}{r} + \mathbf{j}\frac{y}{r} + \mathbf{k}\frac{z}{r}\right\} = -3r^{-5}(\mathbf{i}x + \mathbf{j}y + \mathbf{k}z).\end{aligned}$$

$$\therefore\quad \mathbf{r}\operatorname{grad}\mathbf{r}^{-3} = -3r^{-4}(\mathbf{i}x+\mathbf{j}y+\mathbf{k}z) = \mathbf{i}\phi_1 + \mathbf{j}\phi_2 + \mathbf{k}\phi_3 \quad \text{(say)}.$$

$$\therefore\quad \operatorname{div}(\mathbf{r}\operatorname{grad}\mathbf{r}^{-3}) = \frac{\partial\phi_1}{\partial x} + \frac{\partial\phi_2}{\partial y} + \frac{\partial\phi_3}{\partial z}.$$

Now, $\phi_1 = -3r^{-4}x$,

$$\therefore\quad \frac{\partial\phi_1}{\partial x} = -3r^{-4} + 12r^{-5}x\frac{\partial r}{\partial x},$$

$$\text{i.e.,}\quad \frac{\partial\phi_1}{\partial x} = -3r^{-4} + 12r^{-5}\cdot x\cdot\frac{x}{r} = -3r^{-4} + 12r^{-6}x^2.$$

$$\begin{aligned}\therefore\quad \frac{\partial\phi_1}{\partial x} + \frac{\partial\phi_2}{\partial y} + \frac{\partial\phi_3}{\partial z} &= 3(-3r^{-4}) + 12r^{-6}(x^2+y^2+z^2)\\ &= -9r^4 + 12r^{-4} = 3r^{-4}.\end{aligned}$$

EXAMPLE 2.18. *Find the constants a, b, c, so that the vector*

$$\phi = (2x+3y+az)\mathbf{i} + (bx+2y+3z)\mathbf{j} + (2x+cy+3z)\mathbf{k}$$

is irrotational.

Solution: The vector ϕ will be irrotational if $\operatorname{curl}\phi = 0$.

Now

$$\operatorname{curl}\phi = \begin{vmatrix}\mathbf{i} & \mathbf{j} & \mathbf{k}\\ \frac{\partial}{\partial x} & \frac{\partial}{\partial y} & \frac{\partial}{\partial z}\\ \phi_1 & \phi_2 & \phi_3\end{vmatrix}$$

where, $\phi_1 = 2x+3y+az$, $\phi_2 = bx+2y+3z$ and $\phi_3 = 2x+cy+3z$.

$$\begin{aligned}\therefore\quad \operatorname{curl}\phi &= \mathbf{i}\left(\frac{\partial\phi_3}{\partial y} - \frac{\partial\phi_2}{\partial z}\right) + \mathbf{j}\left(\frac{\partial\phi_1}{\partial z} - \frac{\partial\phi_3}{\partial x}\right) + \mathbf{k}\left(\frac{\partial\phi_2}{\partial x} - \frac{\partial\phi_1}{\partial y}\right)\\ &= \mathbf{i}(c+1) + \mathbf{j}(a-4) + \mathbf{k}(b-2). \qquad (1)\end{aligned}$$

If $\operatorname{curl}\phi = 0$, then from equation (1)

$$c+1=0,\quad a-4=0,\quad b-2=0.$$

$$\therefore\quad a=4,\quad b=2,\quad c=-1.$$

EXAMPLE 2.19. *Prove that*

$$\mathbf{b}\cdot\nabla\left(\mathbf{a}\cdot\nabla\frac{1}{r}\right)=\frac{3}{r^5}(\mathbf{a}\cdot\mathbf{r})(\mathbf{b}\cdot\mathbf{r})-\frac{\mathbf{a}\cdot\mathbf{b}}{r^3},$$

where **a**, **b** *are constant vectors.*

Solution:

$$\nabla\left(\frac{1}{r}\right)=\left(\mathbf{i}\frac{\partial}{\partial x}+\mathbf{j}\frac{\partial}{\partial y}+\mathbf{k}\frac{\partial}{\partial z}\right)\frac{1}{r}=\left(-\mathbf{i}r^{-2}\frac{\partial r}{\partial x}-\mathbf{j}r^{-2}\frac{\partial r}{\partial y}-\mathbf{k}r^{-2}\frac{\partial r}{\partial z}\right).$$

Since

$$r=\left(x^2+y^2+z^2\right)^{1/2},\quad \frac{\partial r}{\partial x}=\frac{x}{r},\quad \frac{\partial r}{\partial y}=\frac{y}{r},\quad \frac{\partial r}{\partial z}=\frac{z}{r}.$$

$$\therefore\quad \nabla\left(\frac{1}{r}\right)=-r^{-3}(\mathbf{i}x+\mathbf{j}y+\mathbf{k}z)=-\frac{\mathbf{r}}{r^3}. \tag{1}$$

Let, $\mathbf{a}=\mathbf{i}a_1+\mathbf{j}a_2+\mathbf{k}a_3$.
Then

$$\nabla\left[\mathbf{a}\cdot\nabla\left(\frac{1}{r}\right)\right]=-\left(i\frac{\partial}{\partial x}+j\frac{\partial}{\partial y}+k\frac{\partial}{\partial z}\right)\frac{(\mathbf{a}\cdot\mathbf{r})}{r^3}. \tag{2}$$

Now

$$\begin{aligned}\frac{\partial}{\partial x}\left(\frac{(\mathbf{a}\cdot\mathbf{r})}{r^3}\right)&=\frac{-3}{r^4}(\mathbf{a}\cdot\mathbf{r})\frac{\partial r}{\partial x}+\frac{1}{r^3}\left(\frac{\partial\mathbf{a}}{\partial x}\cdot\mathbf{r}+\mathbf{a}\cdot\frac{\partial\mathbf{r}}{\partial x}\right)\\&=\frac{-3}{r^4}(\mathbf{a}\cdot\mathbf{r})\frac{x}{r}+\frac{1}{r^3}(0+\mathbf{a}\cdot\mathbf{i})=\frac{-3}{r^5}(\mathbf{a}\cdot\mathbf{r})x+\frac{a_1}{r^3}.\end{aligned} \tag{3}$$

Hence, adding similar expressions from equation (3), we get

$$\nabla\left[\mathbf{a}\cdot\nabla\left(\frac{1}{r}\right)\right]=\frac{3}{r^5}(\mathbf{a}\cdot\mathbf{r})\mathbf{r}-\frac{\mathbf{a}}{r^3}.$$

$$\therefore\quad \mathbf{b}\cdot\nabla\left(\mathbf{a}\cdot\nabla\frac{1}{r}\right)=\mathbf{b}\cdot\left[\frac{3}{r^5}(\mathbf{a}\cdot\mathbf{r})\mathbf{r}-\frac{\mathbf{a}}{r^3}\right]=\frac{3}{r^5}(\mathbf{a}\cdot\mathbf{r})(\mathbf{b}\cdot\mathbf{r})-\frac{1}{r^3}(\mathbf{b}\cdot\mathbf{a}).$$

EXAMPLE 2.20. *If* $\boldsymbol{\phi}=(\mathbf{a}\times\mathbf{r})r^n$, *show that*

$$div\,\boldsymbol{\phi}=0 \text{ and } curl\,\boldsymbol{\phi}=(n+2)r^n\mathbf{a}-nr^{n-2}\,(\mathbf{a}\cdot\mathbf{r})\mathbf{r}.$$

Solution: Let, $\mathbf{r}=\mathbf{i}x+\mathbf{j}y+\mathbf{k}z$, $a=\mathbf{i}a_1+\mathbf{j}a_2+\mathbf{k}a_3$.

$$\therefore\quad r=\left(x^2+y^2+z^2\right)^{1/2},\quad \frac{\partial r}{\partial x}=\frac{x}{r},\quad \frac{\partial r}{\partial y}=\frac{y}{r},\quad \frac{\partial r}{\partial z}=\frac{z}{r}.$$

$$\therefore\quad (\mathbf{a}\times\mathbf{r})=\begin{vmatrix}\mathbf{i}&\mathbf{j}&\mathbf{k}\\a_1&a_2&a_3\\x&y&z\end{vmatrix}=\mathbf{i}(a_2z-a_3y)+\mathbf{j}(a_3x-a_1z)+\mathbf{k}(a_1y-a_2x).$$

$$\therefore\ \boldsymbol{\phi}=(\mathbf{a}\times\mathbf{r})r^n=\mathbf{i}(a_2z-a_3y)r^n+\mathbf{j}(a_3x-a_1z)r^n+\mathbf{k}(a_1y-a_2x)r^n,$$

i.e., $\boldsymbol{\phi}=\mathbf{i}\phi_1+\mathbf{j}\phi_2+\mathbf{k}\phi_3,$

where, $\phi_1 = (a_2z - a_3y)r^n$, $\phi_2 = (a_3x - a_1z)r^n$, $\phi_3 = (a_1y - a_2x)r^n$.

$$\therefore \quad \operatorname{div} \boldsymbol{\phi} = \frac{\partial \phi_1}{\partial x} + \frac{\partial \phi_2}{\partial y} + \frac{\partial \phi_3}{\partial z}. \tag{1}$$

Now

$$\frac{\partial \phi_1}{\partial x} = (a_2z - a_3y)nr^{n-1}\frac{\partial r}{\partial x} = nr^{n-2}(a_2zx - a_3xy), \quad \because \frac{\partial r}{\partial x} = \frac{x}{r}.$$

Similar expressions can be obtained for $\dfrac{\partial \phi_2}{\partial y}$ and $\dfrac{\partial \phi_3}{\partial z}$. Putting these expressions in equation (1), we get

$$\operatorname{div} \boldsymbol{\phi} = nr^{n-2}\left[(a_2zx - a_3xy) + (a_3xy - a_1yz) + (a_1yz - a_2zx)\right] = 0.$$

Again

$$\operatorname{curl} \boldsymbol{\phi} = \begin{vmatrix} \mathbf{i} & \mathbf{j} & \mathbf{k} \\ \dfrac{\partial}{\partial x} & \dfrac{\partial}{\partial y} & \dfrac{\partial}{\partial z} \\ \phi_1 & \phi_2 & \phi_3 \end{vmatrix}$$

$$= \mathbf{i}\left(\frac{\partial \phi_3}{\partial y} - \frac{\partial \phi_2}{\partial z}\right) + \mathbf{j}\left(\frac{\partial \phi_1}{\partial z} - \frac{\partial \phi_3}{\partial x}\right) + \mathbf{k}\left(\frac{\partial \phi_2}{\partial x} - \frac{\partial \phi_1}{\partial y}\right). \tag{2}$$

Now

$$\frac{\partial \phi_3}{\partial y} - \frac{\partial \phi_2}{\partial z} = \frac{\partial}{\partial y}(a_1y - a_2x)r^n - \frac{\partial}{\partial z}(a_3x - a_1z)r^n,$$

i.e.,

$$\begin{aligned}
\frac{\partial \phi_3}{\partial y} - \frac{\partial \phi_1}{\partial z} &= a_1r^n + (a_1y - a_2x)nr^{n-1} \cdot \frac{y}{r} - \left[-a_1r^n + (a_3x - a_1z)nr^{n-1} \cdot \frac{z}{r}\right] \\
&= 2a_1r^n + nr^{n-2}(a_1y^2 - a_2xy - a_3xz + a_1z^2) \\
&= 2a_1r^n + nr^{n-2}[a_1(x^2 + y^2 + z^2) - x(a_1x + a_2y + a_3z)] \\
&= 2a_1r^n + nr^{n-2} \cdot a_1r^2 - nr^{n-2}(\mathbf{a} \cdot \mathbf{r})x \\
&= (n+2)r^na_1 - nr^{n-2}(\mathbf{a} \cdot \mathbf{r})x.
\end{aligned}$$

Similar expressions can be obtained for 2nd and 3rd portions of equation (2).

$$\begin{aligned}
\therefore \quad \operatorname{curl} \boldsymbol{\phi} &= (n+2)r^n(\mathbf{i}a_1 + \mathbf{j}a_2 + \mathbf{k}a_3) - nr^{n-2}(\mathbf{a} \cdot \mathbf{r})(\mathbf{i}x + \mathbf{j}y + \mathbf{k}z) \\
&= (n+2)r^n\mathbf{a} - nr^{n-2}(\mathbf{a} \cdot \mathbf{r})\mathbf{r}.
\end{aligned}$$

EXAMPLE 2.21. *Find the values of λ, μ, ν, so that the vector*

$$\mathbf{F} = \mathbf{i}(x + \lambda y + 4z) + \mathbf{j}(2x - 3y + \mu z) + \mathbf{k}(\nu x - y + 2z)$$

is conservative. Find also the scalar function ϕ such that $\boldsymbol{\nabla}\phi = \mathbf{F}$.

Solution:

$$\boldsymbol{\nabla} \times \mathbf{F} = \begin{vmatrix} \mathbf{i} & \mathbf{j} & \mathbf{k} \\ \dfrac{\partial}{\partial x} & \dfrac{\partial}{\partial y} & \dfrac{\partial}{\partial z} \\ F_x & F_y & F_z \end{vmatrix},$$

where, $F_x = x + \lambda y + 4z$, $F_y = 2x - 3y + \mu z$, $F_z = \nu x - y + 2z$.

$$\therefore \quad \nabla \times \mathbf{F} = \mathbf{i}\left(\frac{\partial F_z}{\partial y} - \frac{\partial F_y}{\partial z}\right) + \mathbf{j}\left(\frac{\partial F_x}{\partial z} - \frac{\partial F_z}{\partial x}\right) + \mathbf{k}\left(\frac{\partial F_y}{\partial x} - \frac{\partial F_x}{\partial y}\right)$$

$$= \mathbf{i}(-1-\mu) + \mathbf{j}(4-\nu) + \mathbf{k}(2-\lambda) = 0.$$

$$\therefore \quad \mu = -1, \quad \nu = 4, \quad \lambda = 2.$$

Hence, $\mathbf{F}$ is conservative for $\lambda = 2$, $\mu = -1$, $\nu = 4$ and can be expressed as $\mathbf{F} = \nabla\phi$, where ϕ is a scalar potential function.

Since

$$\nabla\phi = \mathbf{i}\frac{\partial\phi}{\partial x} + \mathbf{j}\frac{\partial\phi}{\partial y} + \mathbf{k}\frac{\partial\phi}{\partial z} \quad \text{and} \quad \mathbf{F} = \mathbf{i}F_x + \mathbf{j}F_y + \mathbf{k}F_z$$

we can write putting the values of λ, μ, ν

$$\left.\begin{aligned} \frac{\partial\phi}{\partial x} &= F_x = x + 2y + 4z, \\ \frac{\partial\phi}{\partial y} &= F_y = 2x - 3y - z \\ \text{and} \quad \frac{\partial\phi}{\partial z} &= F_z = 4x - y + 2z \end{aligned}\right\}. \tag{1}$$

Integrating the equations (1) with respect to x, y, z respectively, we get

$$\left.\begin{aligned} \phi &= \frac{x^2}{2} + 2xy + 4xz + \psi_1(y,z), \\ \phi &= 2xy - \frac{3}{2}y^2 - yz + \psi_2(x,z) \\ \text{and} \quad \phi &= 4xz - yz + z^2 + \psi_3(x,y) \end{aligned}\right\}, \tag{2}$$

where ψ_1, ψ_2, ψ_3 are to be suitably chosen in order that function ϕ were identical in all these three equations. By inspection, we get

$$\psi_1(y,z) = -\frac{3}{2}y^2 - yz + z^2$$

$$\psi_2(x,z) = \frac{x^2}{2} + z^2 + 4xz$$

$$\psi_3(x,y) = \frac{x^2}{2} + 2xy - \frac{3y^2}{2}$$

Hence, required scalar potential function is

$$\phi(x,y,z) = \frac{1}{2}x^2 - \frac{3}{2}y^2 + z^2 + 2xy - yz + 4xz.$$

III. INTEGRATION OF VECTORS

If there is a vector $\mathbf{A}(t)$ defined by the parameter t such that

$$\mathbf{A}(t) = \mathbf{i}A_x(t) + \mathbf{j}A_y(t) + \mathbf{k}A_z(t) \tag{2.81}$$

then the integration of the vector $\mathbf{A}(t)$ is defined as

$$\int \mathbf{A}(t)\,dt = \int [\mathbf{i}A_x(t) + \mathbf{j}A_y(t) + \mathbf{k}A_z(t)]\,dt. \tag{2.82}$$

Vector integrations are of three types—line integration, surface integration and volume integration.

2.18 Line Integration of Vectors

We have already discussed the line integral of a vector along a continuously differentiable curve in the Section 2.16. We have seen that the line integral of a vector $\mathbf{A}$ along a curve between two points P and Q is defined as

$$\int_P^Q \mathbf{A}\cdot d\mathbf{r}. \tag{2.83}$$

We have also seen that value of the line integral (2.83) does not depend on the choice of the path from P to Q when the vector $\mathbf{A}$ can be expressed as a scalar potential function.

Now, if the vector $\mathbf{A}$ represents a force acting on a particle which moves along the curve from P to Q, then the line integral (2.83) represents the work done by the force $\mathbf{A}$. If this line integral along a closed path is zero, the vector $\mathbf{A}$ is called *conservative.*

If the vector $\mathbf{A}$ represents the velocity of a fluid flow, the integral (2.83) is called the *circulation* of $\mathbf{A}$ about the curve C obtained between points P and Q. If this integral around a closed path is zero, then $\mathbf{A}$ is called *irrotational.*

Let us now assume that the vector $\mathbf{A}$ can be expressed in terms of a parameter t as in the equation (2.81). Then, since $\mathbf{r} = \mathbf{i}x + \mathbf{j}y + \mathbf{k}z$, $\therefore$ $d\mathbf{r} = \mathbf{i}\,dx + \mathbf{j}\,dy + \mathbf{k}\,dz.$

$$\therefore \quad \mathbf{A}\cdot d\mathbf{r} = (A_x\mathbf{i} + A_y\mathbf{j} + A_z\mathbf{k})\cdot(\mathbf{i}\,dx + \mathbf{j}\,dy + \mathbf{k}\,dz) = A_x\,dx + A_y\,dy + A_z\,dz.$$

$$\therefore \quad \int \mathbf{A}\cdot d\mathbf{r} = \int (A_x\,dx + A_y\,dy + A_z\,dz) = \int \left(A_x\frac{dx}{dt} + A_y\frac{dy}{dt} + A_z\frac{dz}{dt}\right)dt. \tag{2.84}$$

EXAMPLE 2.22. *Find the work done by the force*

$$\mathbf{F} = (2x - y + 2z)\mathbf{i} + (x + y)\mathbf{j} + (3x - 2y - 5z)\mathbf{k}$$

in moving a particle once around the circle $x^2 + y^2 = 4$.

Solution: Equation of the circle in parametric form is given by $x = 2\cos t$, $y = 2\sin t$.

$$\therefore \quad \mathbf{r} = \mathbf{i}x + \mathbf{j}y + \mathbf{k}z = \mathbf{i}(2\cos t) + \mathbf{j}(2\sin t) \quad (\because\ z = 0),$$

$$\text{and} \quad \mathbf{F} = (2x - y)\mathbf{i} + (x + y)\mathbf{j} + (3x - 2y)\mathbf{k}$$

$$\text{i.e.,} \quad \mathbf{F} = (4\cos t - 2\sin t)\mathbf{i} + (2\cos t + 2\sin t)\mathbf{j} + (6\cos t - 4\sin t)\mathbf{k}.$$

In moving around the circle once, the parameter t will vary from 0 to 2π.

Hence, the work done by the force is

$$\int \mathbf{F} \cdot d\mathbf{r} = \int \mathbf{F} \cdot \frac{d\mathbf{r}}{dt}\, dt. \tag{1}$$

Since

$$\frac{d\mathbf{r}}{dt} = -2\mathbf{i} \sin t + 2\mathbf{j} \cos t,$$

the integral is

$$\begin{aligned} &\int_0^{2\pi} [-2 \sin t(4 \cos t - 2 \sin t) + 2 \cos t(2 \cos t + 2 \sin t)]\, dt \\ &= \int_0^{2\pi} \left[4\left(\sin^2 t + \cos^2 t\right) - 4 \sin t \cos t\right] dt \\ &= \int_0^{2\pi} (4 - 4 \sin t \cos t)\, dt = \left(4t - 2\sin^2 t\right)_0^{2\pi} = 8\pi. \end{aligned}$$

EXAMPLE 2.23. *Find the circulation of velocity vector* $\mathbf{V}$ *given by*

$$\mathbf{V} = \left(2x + y^2\right)\mathbf{i} + (3y - 4x)\mathbf{j}$$

along the curve $y = x^2$ *from* $(0, 0)$ *to* $(1, 1)$ *and along the curve* $y^2 = x$ *from* $(1, 1)$ *to* $(0, 0)$.

Solution: Here $\mathbf{r} = \mathbf{i}x + \mathbf{j}y$, $\therefore\ d\mathbf{r} = \mathbf{i}\, dx + \mathbf{j}\, dy$. Hence, the circulation is

$$\int \mathbf{V} \cdot d\mathbf{r} = \int \left[\left(2x + y^2\right) dx + (3y - 4x)\, dy\right]. \tag{1}$$

Now, for the curve $y = x^2$, $dy = 2x\, dx$ and x varies from 0 to 1. Hence, the integral (1) is

$$\int_0^1 \left[\left(2x + x^4\right) dx + \left(3x^2 - 4x\right)(2x)\, dx\right] = \int_0^1 \left(x^4 + 6x^3 - 8x^2 + 2x\right) dx = \frac{1}{30}.$$

Again, for the curve $y^2 = x$, $dx = 2y\, dy$ and y varies from 1 to 0. Hence, the circulation is

$$\int_1^0 \left[\left(2y^2 + y^2\right)(2y\, dy) + \left(3y - 4y^2\right) dy\right] = \int_1^0 \left(6y^3 - 4y^2 + 3y\right) dy = -\frac{49}{30}.$$

2.19 Two Important Theorems

We now discuss about two important theorems on vector line integrals.

Theorem 2.1. *If* $\mathbf{F} = \nabla\phi$ *is a single-valued force vector field having continuous partial derivatives, then the work done in moving a particle from one point* P *to another point* Q *both remaining within the field, is independent of the path joining the two points.*

Conversely, if $\int \mathbf{F} \cdot d\mathbf{r}$ *is independent of the path joining any two points within the force field, then there exists a scalar function* ϕ *such that* $\mathbf{F} = \nabla\phi$.

Proof. We know that the work done in moving a particle from point P to Q in the force field $\mathbf{F}$ is

$$\int_P^Q \mathbf{F}\cdot d\mathbf{r} = \int_P^Q \nabla\phi\cdot d\mathbf{r} \quad (\because \mathbf{F} = \nabla\phi \text{ given}).$$

Now

$$\nabla\phi = \mathbf{i}\frac{\partial\phi}{\partial x} + \mathbf{j}\frac{\partial\phi}{\partial y} + \mathbf{k}\frac{\partial\phi}{\partial z} \quad \text{and} \quad d\mathbf{r} = \mathbf{i}\,dx + \mathbf{j}\,dy + \mathbf{k}\,dz.$$

$$\therefore \quad \int_P^Q \nabla\phi\cdot d\mathbf{r} = \int_P^Q \left(\frac{\partial\phi}{\partial x}dx + \frac{\partial\phi}{\partial y}dy + \frac{\partial\phi}{\partial z}dz\right) = \int_P^Q d\phi = \phi(Q) - \phi(P).$$

Hence the work done depends only on the values of the function ϕ at the end points P and Q of the path joining them. Clearly, the function ϕ should be single-valued having continuous partial derivatives.

To prove the converse, we accept that the integral $\int \mathbf{F}\cdot d\mathbf{r}$ is independent of the path of integration. Then, for any path joining $P(x_1, y_1, z_1)$ to any point $Q(x, y, z)$, we can write

$$\phi(x, y, z) = \int_P^Q \mathbf{F}\cdot d\mathbf{r} = \int_P^Q \mathbf{F}\frac{d\mathbf{r}}{ds}\,ds.$$

Differentiating both sides, we get

$$\frac{d\phi}{ds} = \mathbf{F}\cdot\frac{d\mathbf{r}}{ds}. \tag{2.85}$$

Since $\dfrac{d\phi}{ds}$ is the directional derivative of ϕ,

$$\therefore \quad \frac{d\phi}{ds} = \operatorname{grad}\phi\cdot\frac{d\mathbf{r}}{ds} = \nabla\phi\cdot\frac{d\mathbf{r}}{ds}. \tag{2.86}$$

$\therefore$ From equation (2.85) and (2.86), we have

$$\nabla\phi\cdot\frac{d\mathbf{r}}{ds} = \mathbf{F}\cdot\frac{d\mathbf{r}}{ds}, \quad \text{or,} \quad (\nabla\phi - \mathbf{F})\cdot\frac{d\mathbf{r}}{ds} = 0.$$

$\therefore$ $\nabla\phi - \mathbf{F} = 0$, since $d\mathbf{r}$ is arbitrary.
$\therefore$ $\mathbf{F} = \nabla\phi$. □

Theorem 2.2. *If $\mathbf{F}$ is a conservative field (i.e., $\mathbf{F} = \nabla\phi$), then curl$\mathbf{F} = 0$; conversely, if curl$\mathbf{F} = 0$, then $\mathbf{F}$ is a conservative field.*

Proof. Since $\mathbf{F}$ is a conservative field, it can be expressed as the gradient of a scalar function ϕ, i.e., $\mathbf{F} = \operatorname{grad}\phi = \nabla\phi$.

$\therefore$ curl $\mathbf{F}$ = curl grad $\phi = \nabla\times\nabla\phi = 0$ [*see* relation (b) of Section 2.17].

To prove the converse theorem, we assume that $\mathbf{F} = \mathbf{i}F_1 + \mathbf{j}F_2 + \mathbf{k}F_3$.

Now, it is given that curl $\mathbf{F} = 0$; then we have

$$\frac{\partial F_3}{\partial y} = \frac{\partial F_2}{\partial z}, \quad \frac{\partial F_3}{\partial x} = \frac{\partial F_1}{\partial z}, \quad \frac{\partial F_2}{\partial x} = \frac{\partial F_1}{\partial y}. \tag{2.87}$$

We are to prove that $\mathbf{F} = \nabla\phi$.

Let us consider a path joining any two points $P(x_1, y_1, z_1)$ and $Q(x, y, z)$.

We select the path consisting of three portions PR, RS and SQ which are parallel to x, y and z axes respectively. Then (*Fig 2.10*)

$$\phi(x,y,z) = \int_P^Q \mathbf{F}\cdot d\mathbf{r}$$
$$= \int_P^R \mathbf{F}\cdot d\mathbf{r} + \int_R^S \mathbf{F}\cdot d\mathbf{r} + \int_S^Q \mathbf{F}\cdot d\mathbf{r}. \tag{2.88}$$

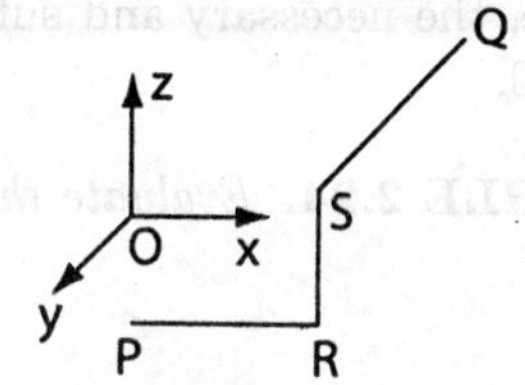

Fig 2.10

Now

$$\mathbf{F}\cdot d\mathbf{r} = (\mathbf{i}F_1 + \mathbf{j}F_2 + \mathbf{k}F_3)\cdot(\mathbf{i}\,dx + \mathbf{j}\,dy + \mathbf{k}\,dz) = F_1 dx + F_2 dy + F_3 dz. \tag{2.89}$$

Along PR, $dy = dz = 0$.

$$\therefore \quad \mathbf{F}\cdot d\mathbf{r} = F_1(x, y_1, z_1)\,dx.$$

Again, along RS, $dx = dz = 0$, and along SQ, $dx = dy = 0$. Hence, along RS, $\mathbf{F}\cdot d\mathbf{r} = F_2(x,y,z_1)\,dy$, and along SQ, $\mathbf{F}\cdot d\mathbf{r} = F_3(x,y,z)\,dz$.

$$\therefore \quad \phi(x,y,z) = \int_{x_1}^{x} F_1(x,y_1,z_1)\,dx + \int_{y_1}^{y} F_2(x,y,z_1)\,dy + \int_{z_1}^{z_2} F_3(x,y,z)\,dz.$$

$$\therefore \quad \frac{\partial \phi}{\partial z} = F_3(x,y,z). \tag{2.90}$$

Now

$$\frac{\partial \phi}{\partial y} = F_2(x,y,z_1) + \int_{z_1}^{z} \frac{\partial F_3}{\partial y}(x,y,z)\,dz.$$

But, $\dfrac{\partial F_3}{\partial y} = \dfrac{\partial F_2}{\partial z}$ [from equation (2.87)]

$$\therefore \quad \frac{\partial \phi}{\partial y} = F_2(x,y,z_1) + \int_{z_1}^{z} \frac{\partial F_2}{\partial z}(x,y,z)\,dz = F_2(x,y,z_1) + [F_2(x,y,z) - F_2(x,y,z_1)].$$

$$\therefore \quad \frac{\partial \phi}{\partial y} = F_2(x,y,z). \tag{2.91}$$

Again

$$\frac{\partial \phi}{\partial x} = F_1(x,y_1,z_1) + \int_{y_1}^{y} \frac{\partial}{\partial x}F_2(x,y,z_1)\,dy + \int_{z_1}^{z} \frac{\partial}{\partial x}F_3(x,y,z)\,dz.$$

Using relations given by equation (2.87)

$$\frac{\partial \phi}{\partial x} = F_1(x,y_1,z_1) + \int_{y_1}^{y} \frac{\partial}{\partial y}F_1(x,y,z_1)\,dy + \int_{z_1}^{z} \frac{\partial}{\partial z}F_1(x,y,z)\,dz$$
$$= F_1(x,y_1,z_1) + [F_1(x,y,z_1) - F_1(x,y_1,z_1)] + [F_1(x,y,z) - F_1(x,y,z_1)].$$

$$\therefore \quad \frac{\partial \phi}{\partial x} = F_1(x,y,z). \tag{2.92}$$

From equations (2.90)–(2.92), we get

$$\mathbf{F} = \mathbf{i}F_1 + \mathbf{j}F_2 + \mathbf{k}F_3 = \mathbf{i}\frac{\partial \phi}{\partial x} + \mathbf{j}\frac{\partial \phi}{\partial y} + \mathbf{k}\frac{\partial \phi}{\partial z}, \quad \text{i.e.,} \quad \mathbf{F} = \nabla\phi.$$

Hence, the necessary and sufficient condition that a vector field $\mathbf{F}$ be conservative is that $\operatorname{curl}\mathbf{F} = 0$. □

EXAMPLE 2.24. *Evaluate the integral*

$$\int \frac{-\mathbf{i}y^3 + \mathbf{j}x^3}{(x^2+y^2)^2} \cdot d\mathbf{r}$$

around the boundary of a square $x = \pm a$, $y = \pm a$ in counter-clockwise direction.

Solution: $d\mathbf{r} = \mathbf{i}dx + \mathbf{j}dy$.

$$\therefore \quad I = \int \frac{-y^3 dx + x^3 dy}{(x^2+y^2)^2}. \tag{1}$$

Now, along AB, $y = -a$, $\therefore\ dy = 0$; and x varies from $-a$ to a (*Fig 2.11*).

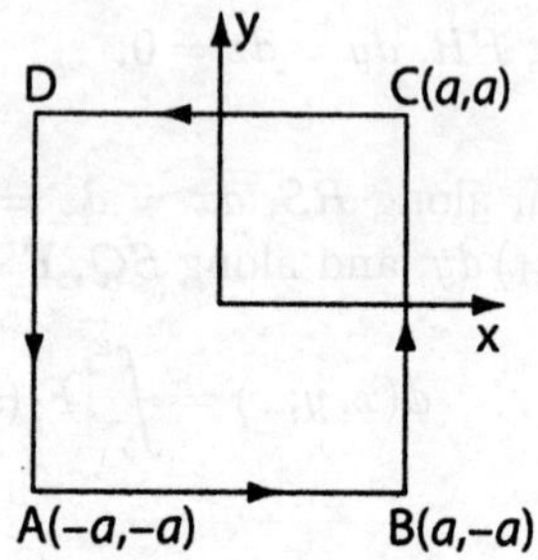

Fig 2.11

$$\therefore \quad I_1 = \int_{-a}^{a} \frac{a^3 dx}{(x^2+y^2)^2} \quad \text{[putting in (1) } y = -a,\ dy = 0]$$

$$= \int_{-\pi/4}^{\pi/4} \frac{a^3(a\sec^2\theta)\,d\theta}{a^4\sec^4\theta} \quad (\text{putting } x = a\tan\theta)$$

$$= \int_{-\pi/4}^{\pi/4} \cos^2\theta d\theta = \int_{-\pi/4}^{\pi/4} \frac{1}{2}(1+\cos 2\theta)\,d\theta$$

$$= \frac{1}{2}\left[\theta + \frac{\sin 2\theta}{2}\right]_{-\pi/4}^{\pi/4} = \frac{\pi}{4} + \frac{1}{2}.$$

Similarly

along BC, $I_2 = -\left(\frac{\pi}{4}+\frac{1}{2}\right)$, along CD, $I_3 = \frac{\pi}{4}+\frac{1}{2}$; and along DA, $I_4 = -\left(\frac{\pi}{4}+\frac{1}{2}\right)$.

Hence, the integral over the boundary of the square $I = I_1 + I_2 + I_3 + I_4 = 0$.

EXAMPLE 2.25. *Show that $\mathbf{F} = \mathbf{i}(\sin y + z) + \mathbf{j}(x\cos y - z) + \mathbf{k}(x - y)$ is conservative. Find a scalar function ϕ such that $\mathbf{F} = \nabla\phi$. Also, find the work done in moving a particle from $A(0,0,1)$ to $B(1,\pi/2,1)$.*

Solution: If $\mathbf{F}$ is to be a conservative field, $\operatorname{curl}\mathbf{F} = 0$.

Now

$$\operatorname{curl}\mathbf{F} = \begin{vmatrix} \mathbf{i} & \mathbf{j} & \mathbf{k} \\ \dfrac{\partial}{\partial x} & \dfrac{\partial}{\partial y} & \dfrac{\partial}{\partial z} \\ \sin y + z & x\cos y - z & x - y \end{vmatrix}$$

$$= \mathbf{i}(-1+1) + \mathbf{j}(1-1) + \mathbf{k}(\cos y - \cos y) = 0.$$

Hence, the vector $\mathbf{F}$ is conservative. Now, let

$$\mathbf{F} = \nabla\phi = \mathbf{i}\frac{\partial\phi}{\partial x} + \mathbf{j}\frac{\partial\phi}{\partial y} + \mathbf{k}\frac{\partial\phi}{\partial z}.$$

Then

$$\mathbf{F}\cdot d\mathbf{r} = \left(\mathbf{i}\frac{\partial\phi}{\partial x} + \mathbf{j}\frac{\partial\phi}{\partial y} + \mathbf{k}\frac{\partial\phi}{\partial z}\right)\cdot(\mathbf{i}\,dx + \mathbf{j}\,dy + \mathbf{k}\,dz),$$

i.e.,

$$F_1dx + F_2dy + F_3dz = \frac{\partial\phi}{\partial x}\,dx + \frac{\partial\phi}{\partial y}\,dy + \frac{\partial\phi}{\partial z}\,dz = d\phi.$$

$$\begin{aligned}\therefore\quad d\phi &= (\sin y + z)\,dx + (x\cos y - z)\,dy + (x-y)\,dz\\ &= (\sin y dx + x\cos y\,dy) + (zdx + xdz) - (zdy + ydz)\\ \text{or,}\quad d\phi &= d(x\sin y) + d(xz) - d(yz).\end{aligned}$$

$\therefore\ \phi = x\sin y + xz - yz +$ constant.

$\therefore$ Work done $= \Big[\phi(x,y,z)\Big]_A^B = \phi(B) - \phi(A) = 2 - \pi/2.$

2.20 Surface Integrals

Let us consider a smooth surface defined by $z = f(x,y)$ having continuous first order partial derivatives. Let dS be a small area of the surface and $\mathbf{n}$ the unit normal vector in the outward direction to this small area. Then the area vector corresponding to this small portion of the surface is $d\sigma = \mathbf{n}\,dS$. The normal surface integral of a continuous vector point function $\mathbf{F}(\mathbf{r})$ is defined as

$$\int_S \mathbf{F}\cdot d\sigma = \int_S \mathbf{F}\cdot\mathbf{n}\,dS. \tag{2.93}$$

Now, the projection of vector area $\mathbf{n}\,dS$ on xy plane whose unit normal is $\mathbf{k}$ is given by

$$\mathbf{n}\,dS\cdot\mathbf{k} = (\mathbf{n}\cdot\mathbf{k})\,dS. \tag{2.94}$$

But the projection of dS on xy plane is $dx\,dy$. Hence

$$(\mathbf{n}\cdot\mathbf{k})\,dS = dx\,dy,\quad\text{i.e.,}\quad dS = \frac{dx\,dy}{\mathbf{n}\cdot\mathbf{k}}. \tag{2.95}$$

Thus, the surface integral (2.93) can also be written as

$$\int_S \mathbf{F}\cdot\mathbf{n}\frac{dx\,dy}{\mathbf{n}\cdot\mathbf{k}}. \tag{2.96}$$

Similarly, considering the projection of vector area on yz and zx plane, the surface integral (2.93) can be expressed in the following forms:

$$\int_S \mathbf{F}\cdot\mathbf{n}\frac{dy\,dz}{\mathbf{n}\cdot\mathbf{i}} \tag{2.97}$$

and

$$\int_S \mathbf{F}\cdot\mathbf{n}\frac{dz\,dx}{\mathbf{n}\cdot\mathbf{j}}. \tag{2.98}$$

EXAMPLE 2.26. *Evaluate* $\int_S \mathbf{F}\cdot\mathbf{n}\,dS$, *where* $\mathbf{F} = y\mathbf{i} + 2x\mathbf{j} + z\mathbf{k}$ *and* S *is the surface of the plane* $2x + y = 6$ *in the first octant cut off by the plane* $z = 4$.

Solution: Normal along the surface

$$\phi = 2x + y - 6 \tag{1}$$

is $\operatorname{grad}\phi = 2\mathbf{i} + \mathbf{j}$.

$\therefore$ Unit normal

$$\mathbf{n} = \frac{1}{\sqrt{5}}(2\mathbf{i}+\mathbf{j}). \quad \therefore \quad \int_S \mathbf{F}\cdot\mathbf{n}\,dS = \int_S \mathbf{F}\cdot\mathbf{n}\frac{dz\,dx}{\mathbf{n}\cdot\mathbf{j}},$$

where $\mathbf{n}\cdot\mathbf{j} = \frac{1}{\sqrt{5}}$ and

$$\mathbf{F}\cdot\mathbf{n} = (y\mathbf{i} + 2x\mathbf{j} + z\mathbf{k})\cdot\frac{1}{\sqrt{5}}(2\mathbf{i}+\mathbf{j}) = \frac{2}{\sqrt{5}}(x+y)$$

$$= \frac{2}{\sqrt{5}}(x+6-2x) = \frac{2}{\sqrt{5}}(6-x)$$

$$\therefore \quad \int \mathbf{F}\cdot\mathbf{n}\,dS = \int_0^4\int_0^3 \frac{2}{\sqrt{5}}(6-x)\sqrt{5}\,dz\,dx = 2\int_0^3 (6-x)\,dx \int_0^4 dz$$

$$= 8\int_0^3 (6-x)\,dx = 108.$$

EXAMPLE 2.27. *Evaluate* $\int_S \mathbf{F}\cdot\mathbf{n}\,dS$, *where S is the sphere* $x^2+y^2+z^2=1$ *in the first octant and* $\mathbf{F} = yz\mathbf{i} + zx\mathbf{j} + xy\mathbf{k}$.

Solution: Here, $\phi = x^2+y^2+z^2-1$, $\therefore$ $\nabla\phi = 2(x\mathbf{i}+y\mathbf{j}+z\mathbf{k})$.

$\therefore$ Unit normal

$$\mathbf{n} = \frac{2(x\mathbf{i}+y\mathbf{j}+z\mathbf{k})}{\sqrt{4x^2+4y^2+4z^2}} = x\mathbf{i}+y\mathbf{j}+z\mathbf{k},$$

$$\mathbf{F}\cdot\mathbf{n} = (yz\mathbf{i}+zx\mathbf{j}+xy\mathbf{k})\cdot(x\mathbf{i}+y\mathbf{j}+z\mathbf{k}) = 3xyz$$

and $\mathbf{n}\cdot\mathbf{k} = (x\mathbf{i}+y\mathbf{j}+z\mathbf{k})\cdot\mathbf{k} = z.$

$$\therefore \quad \int_S \mathbf{F}\cdot\mathbf{n}\,dS = \int_S \mathbf{F}\cdot\mathbf{n}\frac{dx\,dy}{\mathbf{n}\cdot\mathbf{k}} = \int_0^1\int_0^{\sqrt{1-x^2}} 3xyz\frac{dx\,dy}{z}$$ ($\because$ projection of surface on xy plane is $x^2+y^2=1$)

$$= 3\int_0^1 x\left(\frac{y^2}{2}\right)_0^{\sqrt{1-x^2}} dx = \frac{3}{2}\int_0^1 x(1-x^2)\,dx = \frac{3}{8}.$$

EXAMPLE 2.28. *Evaluate* $\int_S \mathbf{r}\cdot d\boldsymbol{\sigma}$, *where S is the part of ellipsoid* $\frac{x^2}{a^2}+\frac{y^2}{b^2}+\frac{z^2}{c^2}=1$ *lying above the plane* $z=0$, *and* $\mathbf{r} = \mathbf{i}x+\mathbf{j}y+\mathbf{k}z$.

Solution: We know

$$\int_S \mathbf{r}\cdot d\boldsymbol{\sigma} = \int \mathbf{r}\cdot\mathbf{n}\,dS = \int \mathbf{r}\cdot\mathbf{n}\frac{dx\,dy}{\mathbf{n}\cdot\mathbf{k}}.$$

Now, any normal to the given ellipsoid is

$$\operatorname{grad}\phi = \frac{2x}{a^2}\mathbf{i} + \frac{2y}{b^2}\mathbf{j} + \frac{2z}{c^2}\mathbf{k}. \tag{1}$$

Hence, unit normal to the surface is

$$\mathbf{n} = \frac{\mathbf{i}x/a^2 + \mathbf{j}y/b^2 + \mathbf{k}z/c^2}{(x^2/a^4 + y^2/b^4 + z^2/c^4)^{1/2}} \tag{2}$$

$$\therefore \quad \mathbf{n}\cdot\mathbf{k} = \frac{z/c^2}{(x^2/a^4 + y^2/b^4 + z^2/c^4)^{1/2}} \tag{3}$$

$$\text{and} \quad \mathbf{r}\cdot\mathbf{n} = \frac{x^2/a^2 + y^2/b^2 + z^2/c^2}{(x^2/a^4 + y^2/b^4 + z^2/c^4)^{1/2}} = \frac{1}{(x^2/a^4 + y^2/b^4 + z^2/c^4)^{1/2}}. \tag{4}$$

$$\therefore \quad \int_S \mathbf{r}\cdot d\boldsymbol{\sigma} = \int_S \mathbf{r}\cdot\mathbf{n}\frac{dx\,dy}{\mathbf{n}\cdot\mathbf{k}} = \int \frac{c^2}{z}\,dx\,dy \quad \text{[using equations (3) and (4)]}$$

$$\therefore \quad I = c^2 \iint \frac{dx\,dy}{c(1 - x^2/a^2 - y^2/b^2)^{1/2}}.$$

Here x varies from $-a$ to a and y varies from $-b(1 - x^2/a^2)^{1/2}$ to $b(1 - x^2/a^2)^{1/2}$. Thus, the integral

$$I = 4c \int_0^a \int_0^{b\sqrt{1-x^2/a^2}} \frac{b}{\sqrt{1 - x^2/a^2 - y^2/b^2}}\,dx \cdot \left(\frac{1}{b}\,dy\right).$$

Substituting

$$\frac{y^2}{b^2} = \left(1 - \frac{x^2}{a^2}\right)\sin^2\theta,$$

the integral will become

$$I = 4bc \int_0^a \left[\sin^{-1}\frac{y}{b\sqrt{1 - x^2/a^2}}\right]_0^{b\sqrt{1-x^2/a^2}} dx = 4bc \int_0^a \frac{\pi}{2}\,dx = 2\pi abc.$$

EXAMPLE 2.29. *Evaluate* $\int_S \phi\mathbf{n}\,dS$, *where* $\phi = \frac{3}{8}xyz$ *and S is the surface of the cylinder* $x^2 + y^2 = 16$ *included in the first octant between* $z = 0$ *and* $z = 5$.

Solution: Normal along the surface of the cylinder is

$$\nabla(x^2 + y^2) = 2x\mathbf{i} + 2y\mathbf{j}. \tag{1}$$

$\therefore$ Unit normal

$$\mathbf{n} = \frac{2x\mathbf{i} + 2y\mathbf{j}}{\sqrt{4x^2 + 4y^2}} = \frac{x\mathbf{i} + y\mathbf{j}}{4}. \quad \therefore \ \mathbf{n}\cdot\mathbf{j} = \frac{y}{4}.$$

$$\therefore \quad \int_S \phi\mathbf{n}\,dS = \iint \phi\mathbf{n}\frac{dx\,dz}{\mathbf{n}\cdot\mathbf{j}} = \iint \frac{3}{8}xyz\left(\frac{x\mathbf{i} + y\mathbf{j}}{4}\right)\frac{4}{y}\,dx\,dz.$$

Here x varies from 0 to 4 and z varies from 0 to 5. Hence, the integral

$$I = \frac{3}{8}\int_0^5\int_0^4 \left[x^2z\mathbf{i} + xz\sqrt{16 - x^2}\mathbf{j}\right] dx\,dz = \frac{3}{8}\int_0^4 \left[x^2\mathbf{i} + x\sqrt{16 - x^2}\mathbf{j}\right]\left[\frac{z^2}{2}\right]_0^5 dx$$

$$= \frac{3}{8}\cdot\frac{25}{2}\left[\frac{x^3}{3}\mathbf{i} - \frac{(16 - x^2)^{3/2}}{3}\mathbf{j}\right]_0^4 = \frac{75}{16}\cdot\frac{64}{3}(\mathbf{i} + \mathbf{j}) = 100(\mathbf{i} + \mathbf{j}).$$

2.21 Volume Integrals

Consider a closed surface in space enclosing a volume V. If $\mathbf{F}(\mathbf{r})$ be a continuous vector point function, then volume integration is defined as

$$\int_V \mathbf{F} \cdot dV. \tag{2.99}$$

In Cartesian coordinates, it is written as

$$\iiint (\mathbf{i}F_1 + \mathbf{j}F_2 + \mathbf{k}F_3)\, dx\, dy\, dz$$

$$= \mathbf{i} \iiint F_1 dx\, dy\, dz + \mathbf{j} \iiint F_2 dx\, dy\, dz + \mathbf{k} \iiint F_3 dx\, dy\, dz. \tag{2.100}$$

EXAMPLE 2.30. *Evaluate* $\int_V (2x + y)\, dV$, *where* V *is the closed region bounded by the cylinder* $z = 4 - x^2$ *and the planes* $x = 0$, $y = 0$, $y = 2$ *and* $z = 0$.

Solution: The given cylinder $z = 4 - x^2$ meets x-axis (i.e., $y = 0$, $z = 0$) at $x^2 = 4$, i.e., $x = 2$, i.e., at the point $(2, 0, 0)$. Similarly, it meets z-axis (i.e., $x = 0$, $y = 0$) at $z = 4$, i.e., at the point $(0, 0, 4)$. Hence, the limits of integration are as under:

$$x = 0 \text{ to } x = 2,$$
$$y = 0 \text{ to } y = 2$$
$$\text{and } z = 0 \text{ to } z = 4 - x^2.$$

$$\therefore \quad \int_V (2x + y)\, dV = \iiint (2x + y)\, dx\, dy\, dz = \iint (2x + y) \Big[z \Big]_0^{4-x^2} dx\, dy$$

$$= \iint (2x + y)(4 - x^2) dx\, dy = \int \left[2x(4 - x^2)y + (4 - x^2)\frac{y^2}{2} \right]_0^2 dx$$

$$= \int_0^2 [4x(4 - x^2) + 2(4 - x^2)] dx = 2\int_0^2 (4 + 8x - x^2 - 2x^3)\, dx = \frac{80}{3}.$$

EXAMPLE 2.31. *Evaluate* $\int_V \nabla \times \mathbf{F}\, dV$, *where* $\mathbf{F} = (x^2 - 2z)\mathbf{i} + 2xy\mathbf{j} - 4xz\mathbf{k}$ *and* V *is a closed region bounded by the planes* $x = 0$, $y = 0$, $z = 0$ *and* $2x + y + z = 6$.

Solution:

$$\nabla \times \mathbf{F} = \begin{vmatrix} \mathbf{i} & \mathbf{j} & \mathbf{k} \\ \dfrac{\partial}{\partial x} & \dfrac{\partial}{\partial y} & \dfrac{\partial}{\partial z} \\ x^2 - 2z & 2xy & -4xz \end{vmatrix} = (4z - 2)\mathbf{j} + 2y\mathbf{k}.$$

The limits of integration are:

$$z = \text{from } 0 \text{ to } (6 - 2x - y),$$
$$y = \text{from } 0 \text{ to } (6 - 2x)\ .$$
$$\text{and} \quad x = \text{from } 0 \text{ to } 3.$$

$$\therefore \quad \int_V \nabla \times \mathbf{F}\, dV = \int_V [(4z - 2)\mathbf{j} + 2y\mathbf{k}]\; dx\, dy\, dz$$

$$= \iint \left[\left(\frac{4z^2}{2} - 2z\right)\mathbf{j} + 2yz\mathbf{k}\right]_0^{6-2x-y} dx\, dy$$

$$= \iint \left[\left\{2(6 - 2x - y)^2 - 2(6 - 2x - y)\right\}\mathbf{j} + 2y(6 - 2x - y)\mathbf{k}\right] dx\, dy.$$

The first integral

$$I_1 = 2\iint (4x^2 + 4xy + y^2 - 22x - 11y + 30)\, dx\, dy$$

$$= 2\int_0^3 (4x^2 - 22x + 30)y - \left(11\frac{y^2}{2} - 4x\frac{y^2}{2} - \frac{y^3}{3}\right)\Bigg|_0^{6-2x}$$

$$= 2\int_0^3 \left(-\frac{8x^3}{3} + 22x^2 - 60x + 54\right) dx$$

$$= 2\left[-\frac{2x^4}{3} + 22\frac{x^3}{3} - 60\frac{x^2}{2} + 54x\right]_0^3 = 92.$$

Similarly, the second integral

$$I_2 = 2\iint (6y - 2xy - y^2)\, dx\, dy = 2\int_0^3 \left(3y^2 - xy^2 - \frac{y^3}{3}\right)_0^{6-2x} dx = \frac{8}{3}\int_0^3 (3 - x)^3\, dx,$$

i.e., $I_2 = 72$. $\therefore\ I = 92\mathbf{i} + 72\mathbf{j}$.

2.22 Gauss's Divergence Theorem

Gauss's divergence theorem relates a volume integral with the surface integral. It states that, the volume integral of the divergence of a vector point function **A** taken over any volume V is equal to the surface integral of the vector **A** taken over the closed surface surrounding the volume V.

Thus

$$\int_V \operatorname{div} \mathbf{A}\, dV = \int_S \mathbf{A} \cdot \mathbf{n}\, dS, \tag{2.101}$$

where **n** is the unit outward drawn normal vector.

Proof. Let $\mathbf{A} = \mathbf{i}A_x + \mathbf{j}A_y + \mathbf{k}A_z$. Then

$$\operatorname{div} \mathbf{A} = \frac{\partial A_x}{\partial x} + \frac{\partial A_y}{\partial y} + \frac{\partial A_z}{\partial z}.$$

$$\therefore \quad \int_V \operatorname{div} \mathbf{A}\, dV = \iiint_V \left(\frac{\partial A_x}{\partial x} + \frac{\partial A_y}{\partial y} + \frac{\partial A_z}{\partial z}\right) dx\, dy\, dz. \tag{2.102}$$

Now, let us consider the 1st integral of the RHS, i.e.,

$$\iiint_V \frac{\partial A_x}{\partial x}\, dx\, dy\, dz.$$

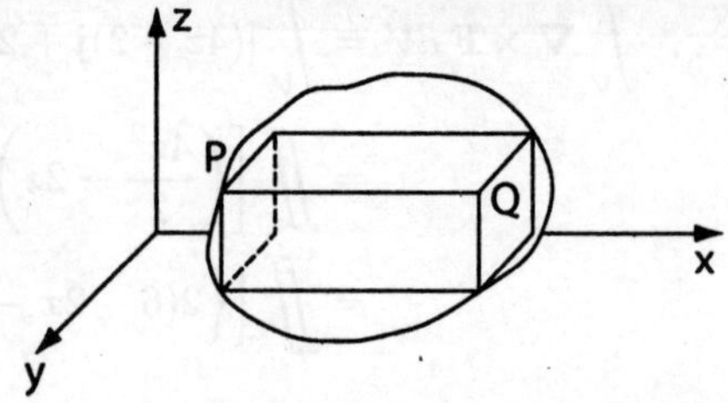

Fig 2.12

We consider a strip of cross-section $dy\,dz$ extending from point P to Q which are on the boundary of the surface (*Fig 2.12*).

Then

$$\iiint_V \frac{\partial A_x}{\partial x}\, dx\, dy\, dz = \iint [A_x(x_2, y, z) - A_x(x_1, y, z)] dy\, dz \tag{2.103}$$

where, x_2, y, z are coordinates of Q, and x_1, y, z are coordinates of P.

Now, at point P, outward drawn normal and A_x are in opposite directions; so flux is negative. But at point Q, outward drawn normal and A_x are in the same direction so that flux is positive.

Thus,

$$\text{at point } P,\ A_x dy\, dz = -A_x dS_x,\ \text{i.e., } dy\, dz = -dS_x$$
$$\text{and at point } Q,\ A_x dy\, dz = A_x dS_x,\quad \text{i.e., } dy\, dz = dS_x.$$

Substituting these values in equation (2.103), we get

$$\iiint_V \frac{\partial A_x}{\partial x}\, dx\, dy\, dz = \iint [A_x(x_2, y, z)\, dS_x + \iint A_x(x_1, y, z)\, dS_x.$$

The 1st integral on the RHS is taken over the right-hand projection part of the surface S, while the 2nd integral is taken over the left-hand projection part of the surface S. Hence, in general

$$\iiint_V \frac{\partial A_x}{\partial x}\, dx\, dy\, dz = \iint_S A_x dS_x.$$

Similarly, we have

$$\iiint_V \frac{\partial A_y}{\partial y}\, dx\, dy\, dz = \iint_S A_y dS_y$$

$$\text{and} \quad \iiint_V \frac{\partial A_z}{\partial z}\, dx\, dy\, dz = \iint_S A_z dS_z.$$

Hence

$$\iiint_V \left(\frac{\partial A_x}{\partial x} + \frac{\partial A_y}{\partial y} + \frac{\partial A_z}{\partial z}\right) dx\, dy\, dz = \iint_S (A_x dS_x + A_y dS_y + A_z dS_z),$$

i.e.,

$$\int_V \operatorname{div} \mathbf{A}\, dV = \int_S \mathbf{A} \cdot \mathbf{n}\, dS.$$

Thus the theorem is proved. □

EXAMPLE 2.32. *Evaluate* $\int \mathbf{F}\cdot\mathbf{n}\,dS$, *where* $\mathbf{F} = 4xz\mathbf{i} - y^2\mathbf{j} + yz\mathbf{k}$ *and* S *is the surface of the cube bounded by* $x = 0$, $x = 1$, $y = 0$, $y = 1$, $z = 0$ *and* $z = 1$ *by using divergence theorem.*

Solution:

$$\int_S \mathbf{F}\cdot\mathbf{n}\,dS = \int_V \operatorname{div}\mathbf{F}\,dV.$$

Now

$$\operatorname{div}\mathbf{F} = \frac{\partial}{\partial x}(4xz) + \frac{\partial}{\partial y}(-y^2) + \frac{\partial}{\partial z}(yz), \quad \text{or,} \quad \operatorname{div}\mathbf{F} = 4z - 2y + y = 4z - y.$$

$$\therefore \quad \int_V \operatorname{div}\mathbf{F}\,dV = \iiint (4z - y)\,dx\,dy\,dz = \int_0^1\int_0^1 \left(2z^2 - yz\right)_0^1 dx\,dy = \int_0^1\int_0^1 (2-y)\,dx\,dy$$

$$= \int_0^1 \left(2y - \frac{y^2}{2}\right)_0^1 dx = \int_0^1 \frac{3}{2}\,dx = \frac{3}{2}.$$

EXAMPLE 2.33. *Prove that* $\int_S \mathbf{F}\cdot\mathbf{n}\,dS = -4\pi\int \rho\,dV$, *where* $\mathbf{F} = \nabla\phi$ *and* $\nabla^2\phi = -4\pi\rho$.

Solution:

$$\int_S \mathbf{F}\cdot\mathbf{n}\,dS = \int_V \operatorname{div}\mathbf{F}\,dV = \int_V \nabla\cdot\mathbf{F}\,dV = \int \nabla\cdot(\nabla\phi)\,dV$$

$$= \int_V \nabla^2\phi\,dV = \int -4\pi\rho\,dV = -4\pi\int_V \rho\,dV.$$

EXAMPLE 2.34. *Show that* $\int_S (ax\mathbf{i}+by\mathbf{j}+cz\mathbf{k})\cdot\mathbf{n}\,dS = \frac{4}{3}\pi(a+b+c)$, *where* S *is the surface of sphere* $x^2 + y^2 + z^2 = 1$.

Solution: $\int_S \mathbf{F}\cdot\mathbf{n}\,dS = \int_V \operatorname{div}\mathbf{F}\cdot dV.$

Now

$$\operatorname{div}\mathbf{F} = \frac{\partial}{\partial x}(ax) + \frac{\partial}{\partial y}(by) + \frac{\partial}{\partial z}(cz) = a + b + c.$$

$\therefore$ Given surface integral is

$$\int_V (a+b+c)\,dV = (a+b+c)V,$$

where V = volume of the sphere $= \frac{4}{3}\pi r^3 = \frac{4}{3}\pi$ ($\because\ r = 1$).

Hence, given surface integral $= \frac{4\pi}{3}(a+b+c)$.

EXAMPLE 2.35. *Prove that* $\int_V \frac{dV}{r^2} = \int \frac{\mathbf{r}\cdot\mathbf{n}}{r^2}\,dS.$

Solution: Let, $\mathbf{F} = \frac{\mathbf{r}}{r^2} = \frac{1}{r^2}(x\mathbf{i} + y\mathbf{j} + z\mathbf{k})$.

Then

$$\begin{aligned}
\operatorname{div}\mathbf{F} &= \frac{\partial}{\partial x}\left(\frac{x}{r^2}\right) + \frac{\partial}{\partial y}\left(\frac{y}{r^2}\right) + \frac{\partial}{\partial z}\left(\frac{z}{r^2}\right) \\
&= \left(\frac{1}{r^2} - \frac{2x^2}{r^4}\right) + \left(\frac{1}{r^2} - \frac{2y^2}{r^4}\right) + \left(\frac{1}{r^2} - \frac{2z^2}{r^4}\right) \\
&= \frac{3}{r^2} - \frac{2}{r^4}(x^2 + y^2 + z^2) = \frac{3}{r^2} - \frac{2}{r^2} \quad (\because\ x^2 + y^2 + z^2 = r^2) \\
&= \frac{1}{r^2}.
\end{aligned}$$

$\therefore$ By divergence theorem

$$\int_S \frac{\mathbf{r}\cdot\mathbf{n}}{r^2}\, dS = \int_V \frac{dV}{r^2}.$$

EXAMPLE 2.36. *Evaluate* $\int \mathbf{F}\cdot\mathbf{n}\, dS$, *where* $F = x^2\mathbf{i} + y^2\mathbf{j} + z^2\mathbf{k}$ *and* S *is the surface of the solid cut off by the plane* $x + y + z = a$ *from the 1st octant.*

Solution:

$$\int \mathbf{F}\cdot\mathbf{n}\, dS = \iiint 2(x + y + z)\, dx\, dy\, dz,$$

since

$$\operatorname{div}\mathbf{F} = \frac{\partial}{\partial x}(x^2) + \frac{\partial}{\partial y}(y^2) + \frac{\partial}{\partial z}(z^2) = 2(x + y + z).$$

Now, the limits of integration are:

$$\begin{aligned}
&\text{for } z,\ z = 0 \text{ to } a - x - y, \\
&\text{for } y,\ y = 0 \text{ to } a - x \\
\text{and}\quad &\text{for } x,\ x = 0 \text{ to } a.
\end{aligned}$$

$$\therefore\ I = 2\iint \left[(x + y)z + \frac{z^2}{2}\right]_0^{a-x-y} dx\, dy,$$

i.e.,

$$\begin{aligned}
I &= \iint (a - x - y)(2x + 2y + a - x - y)\, dx\, dy = \iint (a - x - y)(a + x + y)\, dx\, dy \\
&= \iint (a^2 - x^2 - y^2 - 2xy)\, dx\, dy = \int_0^a \left[a^2 y - x^2 y - \frac{y^3}{3} - xy^2\right]_0^{a-x} dx \\
&= \frac{1}{3}\int_0^a (a - x)^2 (2a + x)\, dx \quad \text{[after simplification]} \\
&= \frac{a^4}{4}.
\end{aligned}$$

EXAMPLE 2.37. *Prove that* $\int_V \nabla \times \mathbf{B}\, dV = \int_S (\mathbf{n} \times \mathbf{B})\, dS$.

Solution: Let us take a constant vector $\mathbf{C}$, and consider the integration of vector $\mathbf{B} \times \mathbf{C}$. Then by divergence theorem

$$\int_V \nabla \cdot (\mathbf{B} \times \mathbf{C}) = \int_S (\mathbf{B} \times \mathbf{C}) \cdot \mathbf{n}\, dS. \tag{1}$$

But, $\boldsymbol{\nabla} \cdot (\mathbf{B} \times \mathbf{C}) = \mathbf{C} \cdot (\boldsymbol{\nabla} \times \mathbf{B})$ and $(\mathbf{B} \times \mathbf{C}) \cdot \mathbf{n} = \mathbf{B} \cdot (\mathbf{C} \times \mathbf{n}) = \mathbf{C} \cdot (\mathbf{n} \times \mathbf{B})$.

$$\therefore \quad \int_V \mathbf{C} \cdot (\boldsymbol{\nabla} \times \mathbf{B})\, dV = \int_S \mathbf{C} \cdot (\mathbf{n} \times \mathbf{B})\, dS \quad \text{[from (1)]},$$

$$\text{i.e.,} \quad \mathbf{C} \int_V (\boldsymbol{\nabla} \times \mathbf{B})\, dV = \mathbf{C} \int_S (\mathbf{n} \times \mathbf{B})\, dS \quad [\because\ \mathbf{C} = \text{constant vector}]$$

$$\text{or,} \quad \int_V (\boldsymbol{\nabla} \times \mathbf{B})\, dV = \int_S (\mathbf{n} \times \mathbf{B})\, dS.$$

2.23 Green's Theorem

It states that, if ϕ and ψ are two scalar functions having continuous derivatives within a region V bounded by closed surface S, then

$$\int_V (\phi \nabla^2 \psi - \psi \nabla^2 \phi)\, dV = \int_S (\phi \nabla \psi - \psi \nabla \phi) \cdot \mathbf{n}\, dS, \tag{2.104}$$

where $\mathbf{n}$ is the outward-drawn unit normal vector to the surface S.

Proof. From Gauss's divergence theorem, we have

$$\int_S \mathbf{F} \cdot \mathbf{n}\, dS = \int_V \operatorname{div} \mathbf{F}\, dV.$$

Now, let $\mathbf{F} = \phi \nabla \psi$.

$$\therefore \quad \operatorname{div} \mathbf{F} = \nabla \cdot (\phi \nabla \psi) = \nabla \phi \cdot \nabla \psi + \phi \nabla^2 \psi.$$

$$\therefore \quad \int_S \phi \nabla \psi \cdot \mathbf{n}\, dS = \int_V \nabla \phi \cdot \nabla \psi\, dV + \int_V \phi \nabla^2 \psi. \tag{2.105}$$

Interchanging ϕ and ψ in equation (2.105), we get

$$\int_S \psi \nabla \phi \cdot \mathbf{n}\, dS = \int_V \nabla \psi \cdot \nabla \phi\, dV + \int_V \psi \nabla^2 \phi. \tag{2.106}$$

Subtracting, we have

$$\int_S (\phi \nabla \psi - \psi \nabla \phi) \cdot \mathbf{n}\, dS = \int_V (\phi \nabla^2 \psi - \psi \nabla^2 \phi)\, dV.$$

Hence, the theorem is proved. □

EXAMPLE 2.38. *Prove that* $\int_V \boldsymbol{\nabla} \phi \cdot \mathit{curl}\, \mathbf{F}\, dV = \int_S (\mathbf{F} \times \boldsymbol{\nabla} \phi) \cdot \mathbf{n} dS.$

Solution: Let

$$\mathbf{F} \times \nabla\phi = \mathbf{A}. \tag{1}$$

Now, by divergence theorem

$$\int_S (\mathbf{F} \times \nabla\phi) \cdot \mathbf{n} dS = \int_S \mathbf{A} \cdot \mathbf{n} dS = \int_V \operatorname{div} \mathbf{A}\, dV.$$

But

$$\operatorname{div} \mathbf{A} = \operatorname{div}(\mathbf{F} \times \nabla\phi) = \nabla\phi \cdot \operatorname{curl} \mathbf{F} - \mathbf{F} \operatorname{curl} \nabla\phi$$

[by relation (g) of Section 2.17]

$$= \nabla\phi \cdot \operatorname{curl} \mathbf{F} \quad (\because \operatorname{curl} \nabla\phi = 0).$$

$$\therefore \int_S (\mathbf{F} \times \nabla\phi) \cdot \mathbf{n} dS = \int_V \nabla\phi \cdot \operatorname{curl} \mathbf{F}\, dV.$$

Hence proved.

EXAMPLE 2.39. *Prove that*

$$\int_V (\phi \, curl\; curl \psi - \psi \, curl\; curl \phi)\, dV = \int_S [(\psi \times \, curl \phi) - (\phi \times \, curl \psi)] \cdot \mathbf{n}\, dS.$$

Solution: By divergence theorem, we have

$$\int_S (\psi \times \operatorname{curl} \phi)\mathbf{n} \cdot dS = \int_V \operatorname{div}[(\psi \times \operatorname{curl} \phi)] dV$$

$$= \int_V (\operatorname{curl} \phi \cdot \operatorname{curl} \psi - \psi \operatorname{curl} \operatorname{curl} \phi)\, dV \tag{1}$$

[by relation (g) of Section 2.17]

Similarly

$$\int_S (\phi \times \operatorname{curl} \psi) \cdot \mathbf{n}\, dS = \int_V (\operatorname{curl} \psi \cdot \operatorname{curl} \phi - \phi \operatorname{curl} \operatorname{curl} \psi)\, dV. \tag{2}$$

Subtracting (2) from (1) we get the result.

2.24 Green's Theorem in the Plane

It states that, if M and N are continuous functions of x and y having continuous derivative in a region S bounded by a closed curve C, then

$$\int_C (M dx + N dy) = \iint_S \left(\frac{\partial N}{\partial x} - \frac{\partial M}{\partial y} \right) dx\, dy. \tag{2.107}$$

Proof. We will assume that any line parallel to either of the axes cuts the curve C bounding the region S in at most two points.

Let the region S be bounded between lines $x = a$ and $x = b$ so that the boundary curve C may be considered to be of two parts AEB and BFA (*Fig 2.13*). Let the equations to curve C_1 represented by portion AEB be $y = f_1(x)$ and that of curve C_2 represented by portion BFA be $y = f_2(x)$.

Then

$$\iint_S \frac{\partial M}{\partial y}\,dx\,dy = \int_{x=a}^{b}\left[\int_{y=f_1(x)}^{y=f_2(x)} \frac{\partial M}{\partial y}\,dy\right]dx$$

$$= \int_{x=a}^{b}\Big[M(x,y)\Big]_{f_1(x)}^{f_2(x)}\,dx$$

$$= \int_{x=a}^{b} \underset{(\text{along } C_2)}{M[x,f_2(x)]dx} - \int_{a}^{b} \underset{(\text{along } C_1)}{M[x,f_1(x)]dx}$$

$$= -\int_{a}^{b} M[x,f_2(x)]dx - \int_{a}^{b} M[x,f_1(x)]dx$$

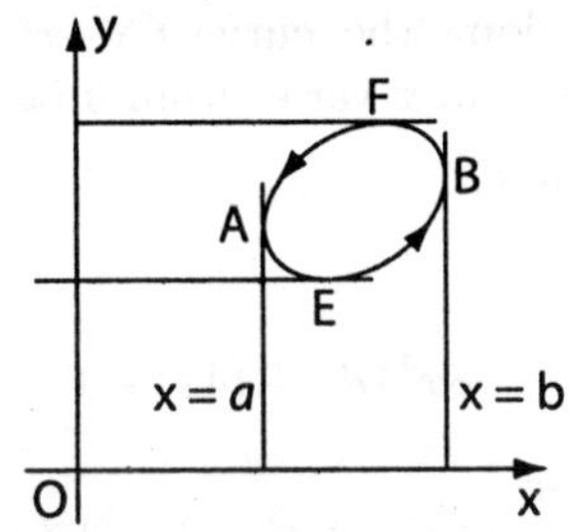

Fig 2.13

[since sense of description of C_2 is opposite to C_1]

Hence

$$\iint_S \frac{\partial M}{\partial y}\,dx\,dy = -\int_C M\,dx. \tag{2.108}$$

Similarly, taking the region to be bounded between the lines $y = c, y = d$ and equation of curves EBF as $x = \phi_1(y)$, and curve FAE as $x = \phi_2(y)$, we can write

$$\iint_S \frac{\partial N}{\partial x}\,dx\,dy = \int_{y=c}^{d}\int_{x=\phi_1(y)}^{x=\phi_2(y)}\left[\frac{\partial N}{\partial x}\,dx\right]dy = \int_{y=c}^{y=d}\Big[N(x,y)\Big]_{x=\phi_1(y)}^{x=\phi_2(y)}dy.$$

$$= \int_{y=c}^{y=d} [N\{\phi_2(y),y\} - N\{\phi_1(y),y\}]dy$$

$$= \int_{y=c}^{d} [N\{\phi_2(y),y\} + N\{\phi_1(y),y\}]dy = \int_C N(x,y)\,dy. \tag{2.109}$$

Hence, from equation (2.108) and (2.109), we get by adding

$$\int_C (M\,dx + N\,dy) = \iint_S \left(\frac{\partial N}{\partial x} - \frac{\partial M}{\partial y}\right)dx\,dy.$$

Thus the theorem is proved. □

In vector notation, the above relation can be written as

$$\int_C \mathbf{F}\cdot d\mathbf{r} = \int_S \operatorname{curl}\mathbf{F}\cdot\mathbf{n}dS \tag{2.110}$$

where $\mathbf{F}$ is expressed as

$$\mathbf{F} = M(x,y)\mathbf{i} + N(x,y)\mathbf{j} \quad \text{and} \quad \mathbf{r} = x\mathbf{i} + y\mathbf{j}. \tag{2.111}$$

EXAMPLE 2.40. *Verify Green's theorem in the plane for*

$$\int_C (3x^2 - 8y^2)\,dx + (4y - 6xy)\,dy,$$

where C is the boundary of the region defined by $y = \sqrt{x}$ and $y = x^2$.

Solution: $y = \sqrt{x}$, i.e., $y^2 = x$ and $y = x^2$ are two parabolas intersecting at $(0,0)$ and $(1,1)$ (*see Fig 2.14*).

Now, along the curve C_1, $x^2 = y$, i.e., $dy = 2x dx$, and x varies from 0 to 1.

$\therefore$ Along C_1,

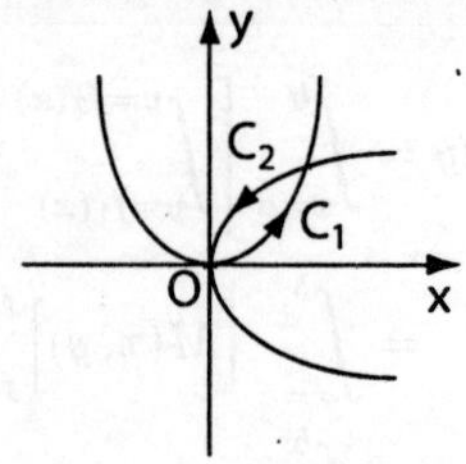

Fig 2.14

$$\int_0^1 [(3x^2 - 8x^4)\,dx + (4x^2 - 6x^3)2x\,dx]$$

$$= \int_0^1 (3x^2 + 8x^3 - 20x^4)\,dx = -1. \quad (1)$$

Along C_2, $x = y^2$, i.e., $dx = 2y\,dy$, and y varies from $y = 1$ to $y = 0$.
$\therefore$ Along C_2,

$$\int_1^0 [(3y^4 - 8y^2)2y\,dy + (4y - 6y^3)\,dy] = \int_1^0 (6y^5 - 22y^3 + 4y)\,dy = \frac{5}{2}. \quad (2)$$

Hence, the line integral along C

$$= -1 + \frac{5}{2} = \frac{3}{2}. \quad (3)$$

Again

$$M = 3x^2 - 8y^2, \quad \therefore \quad \frac{\partial M}{\partial y} = -16y$$

and

$$N = 4y - 6xy, \quad \therefore \quad \frac{\partial N}{\partial x} = -6y.$$

$$\therefore \quad \left(\frac{\partial N}{\partial x} - \frac{\partial M}{\partial y}\right) = +10y.$$

$$\therefore \quad \iint_S \left(\frac{\partial M}{\partial y} - \frac{\partial N}{\partial x}\right) dx\,dy = \int_{x=0}^1 \int_{y=x^2}^{\sqrt{x}} 10y\,dx\,dy = \int_0^1 (5y^2)_{x^2}^{\sqrt{x}}\,dx$$

$$= 5\int_0^1 (x - x^4)\,dx = \frac{3}{2}. \quad (4)$$

Hence, from (3) and (4), Green's theorem is verified.

EXAMPLE 2.41. *Evaluate by Green's theorem*

$$\int_C (e^{-x} \sin y\,dx + e^{-x} \cos y\,dy),$$

where C is the rectangle with vertices $(0,0)$, $(\pi,0)$, $(\pi,\pi/2)$, $(0,\pi/2)$ and hence verify Green's theorem.

Solution: Here

$$M = e^{-x} \sin y, \quad \therefore \quad \frac{\partial M}{\partial y} = e^{-x} \cos y$$

and

$$N = e^{-x} \cos y, \quad \therefore \quad \frac{\partial N}{\partial x} = -e^{-x} \cos y.$$

$$\therefore \int_C (M\,dx + N\,dy) = \int_{x=0}^{\pi}\int_{y=0}^{\pi/2}\left(\frac{\partial N}{\partial x} - \frac{\partial M}{\partial y}\right) dx\,dy = \int_0^{\pi} -2e^{-x}\,dx \int_0^{\pi/2} \cos y\,dy$$

$$= -\int_0^{\pi} 2e^{-x} = 2(e^{-\pi} - 1).$$

By direct method: Along OA, $y = 0$, $\therefore$ $dy = 0$, $\therefore$ line integral $= 0$.

Along AB, $x = \pi$, $\therefore$ $dx = 0$ (*Fig 2.15*).

$\therefore$ Line integral $= \int_0^{\pi/2} e^{-\pi}\cos y\,dy = e^{-\pi}$.

Along BC, $y = \pi/2$, $\therefore$ $dy = 0$.

$\therefore$ Line integral $= \int_{\pi}^{0} e^{-x}dx = -\left(e^{-x}\right)_{\pi}^{0}$

$= e^{-\pi} - 1.$

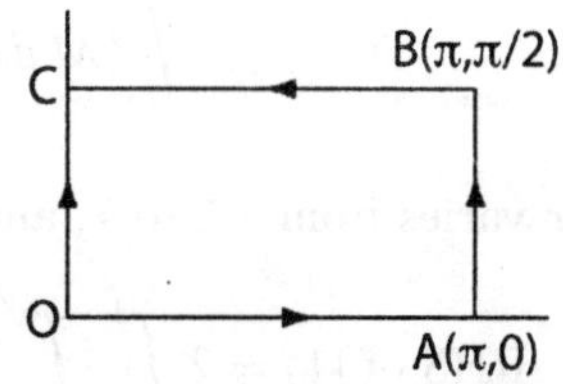

Fig 2.15

Along CO, $x = 0$, $\therefore$ $dx = 0$, and the line integral $= \int_{\pi/2}^{0} \cos y\,dy = \left(\sin y\right)_{\pi/2}^{0} = -1.$

Hence, line integral along rectangle $= 0 + e^{-\pi} + (e^{-\pi} - 1) - 1 = 2(e^{-\pi} - 1)$.

EXAMPLE 2.42. *Prove that* $\int_S \mathbf{n}\times(\mathbf{a}\times\mathbf{r})\,dS = 2\mathbf{a}V$, *where* $\mathbf{a}$ *is a constant vector.*

Solution: Let, $\mathbf{a} = a_1\mathbf{i} + a_2\mathbf{j} + a_3\mathbf{k}$.

Then

$$\int_S \mathbf{n}\times(\mathbf{a}\times\mathbf{r})\,dS = \int_V \nabla\times(\mathbf{a}\times\mathbf{r})\,dV. \quad (1)$$

Now

$$\nabla\times(\mathbf{a}\times\mathbf{r}) = \mathbf{i}\times\frac{\partial}{\partial x}(\mathbf{a}\times\mathbf{r}) + \mathbf{j}\times\frac{\partial}{\partial y}(\mathbf{a}\times\mathbf{r}) + \mathbf{k}\times\frac{\partial}{\partial z}(\mathbf{a}\times\mathbf{r}).$$

But

$$\frac{\partial}{\partial x}(\mathbf{a}\times\mathbf{r}) = \frac{\partial\mathbf{a}}{\partial x}\times\mathbf{r} + \mathbf{a}\times\frac{\partial\mathbf{r}}{\partial x} = 0 + (\mathbf{i}a_1 + \mathbf{j}a_2 + \mathbf{k}a_3)\times\mathbf{i}.$$

$$\therefore \quad \mathbf{i}\times\frac{\partial}{\partial x}(\mathbf{a}\times\mathbf{r}) = \mathbf{i}\times(\mathbf{a}\times\mathbf{i}) = \mathbf{i}\cdot\mathbf{i}\mathbf{a} - \mathbf{i}\cdot\mathbf{a}\mathbf{i} = \mathbf{a} - a_1\mathbf{i}.$$

Similarly

$$\mathbf{j}\times\frac{\partial}{\partial y}(\mathbf{a}\times\mathbf{r}) = \mathbf{a} - a_2\mathbf{j} \quad \text{and} \quad \mathbf{k}\times\frac{\partial}{\partial z}(\mathbf{a}\times\mathbf{r}) = \mathbf{a} - a_3\mathbf{k}.$$

$$\therefore \quad \nabla\times(\mathbf{a}\times\mathbf{r}) = 3\mathbf{a} - (a_1\mathbf{i} + a_2\mathbf{j} + a_3\mathbf{k}) = 2\mathbf{a}.$$

$\therefore$ From (1)

$$\int_S \mathbf{n}\times(\mathbf{a}\times\mathbf{r})\,dS = \int_V \nabla\times(\mathbf{a}\times\mathbf{r})\,dV = \int 2\mathbf{a}\,dV = 2\mathbf{a}V.$$

Hence proved.

EXAMPLE 2.43. *Evaluate* .

$$\int_C (2x^2 - y^2)\,dx + (x^2 + y^2)\,dy,$$

where C *is the boundary of the surface in* xy *plane enclosed by* x*-axis and the semi-circle* $x^2 + y^2 = 1$.

Solution: Here

$$M = 2x^2 - y^2, \quad \therefore \quad \frac{\partial M}{\partial y} = -2y$$

and

$$N = x^2 + y^2, \quad \therefore \quad \frac{\partial N}{\partial x} = 2x.$$

$$\therefore \quad \int_C (M\,dx + N\,dy) = \iint_S \left(\frac{\partial N}{\partial x} - \frac{\partial M}{\partial y}\right) dx\,dy. \tag{1}$$

Here x varies from -1 to 1, and y varies from 0 to $\sqrt{1-x^2}$.

$$\therefore \quad \text{RHS of (1)} = 2\int_{-1}^{1}\int_{0}^{\sqrt{1-x^2}} (x+y)\,dx\,dy = 2\int_{-1}^{1}\left[x\sqrt{1-x^2} + \frac{1}{2}(1-x^2)\right] dx$$

$$= 2\int_0^1 (1-x^2)\,dx \qquad \left[\because \int_{-1}^{1} x\sqrt{1-x^2}\,dx = 0\right]$$

$$= \frac{4}{3}.$$

2.25 Stokes' Theorem

Stokes' theorem establishes a relationship between a surface integral and a line integral. It states that the surface integral of the curl of a vector point function **A** taken over any surface S is equal to the line integral of vector point function **A** around the periphery of the surface. Thus

$$\int_S (\nabla \times \mathbf{A}) \cdot \mathbf{n}\,dS = \int_C \mathbf{A} \cdot d\mathbf{r}, \tag{2.112}$$

where **n** is outward-drawn unit normal to the surface.

Proof. Let the surface S be represented by $z = f(x, y)$, where $f(x, y)$ is a single-valued continuous function having continuous derivatives. The projection of this surface on the xy plane is a region R bounded by a simple closed curve P (*Fig 2.16*).

Now, we take

$$\left.\begin{aligned} \mathbf{A} &= \mathbf{i}A_x + \mathbf{j}A_y + \mathbf{k}A_z \\ \text{and} \quad \mathbf{r} &= \mathbf{i}x + \mathbf{j}y + \mathbf{k}z \end{aligned}\right\} \tag{2.113}$$

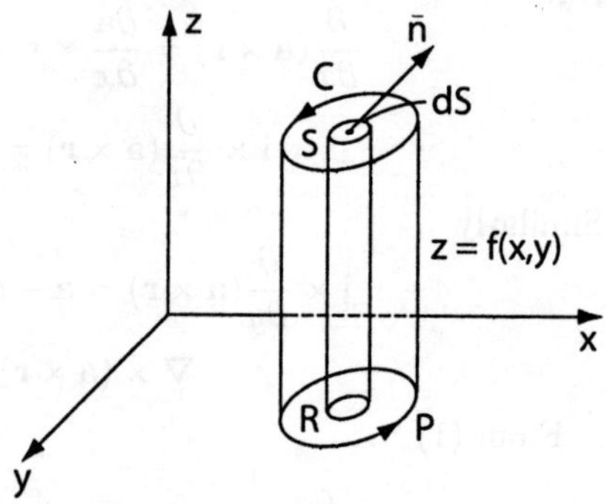

Fig 2.16

where **r** is position vector of any point on S. Then

$$\nabla \times \mathbf{A} = \nabla \times (\mathbf{i}A_x + \mathbf{j}A_y + \mathbf{k}A_z).$$

$$\therefore \quad \nabla \times A_x\mathbf{i} = \begin{vmatrix} \mathbf{i} & \mathbf{j} & \mathbf{k} \\ \dfrac{\partial}{\partial x} & \dfrac{\partial}{\partial y} & \dfrac{\partial}{\partial z} \\ A_x & 0 & 0 \end{vmatrix} = \frac{\partial A_x}{\partial z}\mathbf{j} - \frac{\partial A_x}{\partial y}\mathbf{k}.$$

$$\therefore\ (\nabla \times A_x\mathbf{i}) \cdot \mathbf{n}\, dS = \left(\frac{\partial A_x}{\partial z}\mathbf{n}\cdot\mathbf{j} - \frac{\partial A_x}{\partial y}\mathbf{n}\cdot\mathbf{k}\right) dS. \tag{2.114}$$

Now

$$\mathbf{r} = \mathbf{i}x + \mathbf{j}y + \mathbf{k}z = \mathbf{i}x + \mathbf{j}y + \mathbf{k}f(x, y) \quad [\because\ z = f(x, y)].$$

$$\therefore\quad \frac{\partial \mathbf{r}}{\partial y} = \mathbf{j} + \mathbf{k}\frac{\partial f}{\partial y} = \mathbf{j} + \mathbf{k}\frac{\partial z}{\partial y}.$$

$$\therefore\quad \mathbf{n}\cdot\frac{\partial \mathbf{r}}{\partial y} = \mathbf{n}\cdot\mathbf{j} + \mathbf{n}\cdot\mathbf{k}\frac{\partial z}{\partial y}. \tag{2.115}$$

But $\frac{\partial \mathbf{r}}{\partial y}$ is a vector which is tangent to S, and $\mathbf{n}$ is normal to the surface. Hence, these two vectors are perpendicular to each other. Thus

$$\mathbf{n}\cdot\frac{\partial \mathbf{r}}{\partial y} = 0, \quad \text{i.e.,} \quad \mathbf{n}\cdot\mathbf{j} + \mathbf{n}\cdot\mathbf{k}\frac{\partial z}{\partial y} = 0 \quad [\text{from (2.115)}].$$

$$\therefore\quad \mathbf{n}\cdot\mathbf{j} = -\mathbf{n}\cdot\mathbf{k}\frac{\partial z}{\partial y}. \tag{2.116}$$

From equations (2.114) and (2.116), we get

$$(\nabla \times A_x\mathbf{i}) \cdot \mathbf{n}\, dS = \left(\frac{\partial A_x}{\partial z}\mathbf{n}\cdot\mathbf{j} - \frac{\partial A_x}{\partial y}\mathbf{n}\cdot\mathbf{k}\right) dS = -\left(\frac{\partial A_x}{\partial z}\frac{\partial z}{\partial y} + \frac{\partial A_x}{\partial y}\right)\mathbf{n}\cdot\mathbf{k}\, dS. \tag{2.117}$$

Now, we can write $A_x(x, y, z)$ as a function of x, y only, say $\phi(x, y)$, since z is a function of x, y.

$$\therefore\quad A_x(x, y, z) = \phi(x, y). \tag{2.118}$$

Differentiating partially w.r. to y both sides, we get

$$\frac{\partial A_x}{\partial y} + \frac{\partial A_x}{\partial z}\cdot\frac{\partial z}{\partial y} = \frac{\partial \phi}{\partial y}. \tag{2.119}$$

Again, a small projected area $dx\, dy$ on the xy plane of a small area dS from surface S is given by

$$dx\, dy = \mathbf{n}\cdot\mathbf{k}\, dS. \tag{2.120}$$

Hence, using these relations we get from equation (2.117)

$$(\nabla \times A_x\mathbf{i}) \cdot \mathbf{n}\, dS = -\frac{\partial \phi}{\partial y}\, dx\, dy. \tag{2.121}$$

Now, by Green's theorem for the plane we have

$$\int_C M\, dx + N\, dy = \iint_S \left(\frac{\partial N}{\partial x} - \frac{\partial M}{\partial y}\right) dx\, dy.$$

If we let $M = \phi$ and $N = 0$ in this equation, we can write

$$\int_P \phi\, dx = -\iint_R \frac{\partial \phi}{\partial y}\, dx\, dy. \tag{2.122}$$

Now, at each point (x, y) on the boundary P, the value of $\phi(x, y)$ is equal to the value of A_x at each point (x, y, z) on C.

$$\therefore \quad \int_P \phi\, dx = \int_C A_x\, dx, \quad \text{i.e.,} \quad \int_C A_x\, dx = \int_p \phi\, dx = -\iint_R \frac{\partial \phi}{\partial y}\, dx\, dy. \tag{2.123}$$

Hence, from equations (2.121) and (2.123), we get

$$\int_S (\nabla \times A_x \mathbf{i}) \cdot \mathbf{n}\, dS = \int_C A_x dx. \tag{2.124}$$

Considering the projection of surface S on yz and xz planes and following the same argument, we can get

$$\left.\begin{aligned} \int_S (\nabla \times A_y \mathbf{j}) \cdot \mathbf{n}\, dS &= \int_C A_y\, dy \\ \text{and} \quad \int_S (\nabla \times A_z \mathbf{k}) \cdot \mathbf{n}\, dS &= \int_C A_z\, dz \end{aligned}\right\}. \tag{2.125}$$

Adding these relations

$$\int_S (\nabla \times \mathbf{A}) \cdot \mathbf{n}\, dS = \int_C (A_x\, dx + A_y\, dy + A_z\, dz) = \int_C \mathbf{A} \cdot d\mathbf{r}.$$

Hence the theorem is proved. □

EXAMPLE 2.44. *Prove that*

$$\int_C \phi\, d\mathbf{r} = \iint_S (\mathbf{n} \times \nabla\phi)\, dS.$$

Solution: Let **a** be a constant vector. Then

$$\begin{aligned} \int_C (\mathbf{a}\phi)\, d\mathbf{r} &= \iint_S \operatorname{curl}(\mathbf{a}\phi) \cdot \mathbf{n}\, dS \quad \text{[by Stokes' theorem]} \\ &= \iint_S (\phi \operatorname{curl} \mathbf{a} - \mathbf{a} \times \operatorname{grad} \phi) \cdot \mathbf{n}\, dS \\ &= \iint_S (\nabla\phi \times \mathbf{a}) \cdot \mathbf{n}\, dS = \iint_S \mathbf{a} \cdot (\nabla\phi \times \mathbf{n})\, dS, \end{aligned}$$

i.e.,

$$\mathbf{a} \cdot \int_C \phi\, d\mathbf{r} = \iint_S \mathbf{a} \cdot (\nabla\phi \times \mathbf{n})\, dS, \quad \text{or,} \quad \int_C \phi\, d\mathbf{r} = \iint_S (\nabla\phi \times \mathbf{n})\, dS$$

(since **a** is constant).

Hence proved.

EXAMPLE 2.45. *Evaluate by Stokes' theorem*

$$\int_C (e^x\, dx + 2y\, dy - dz),$$

where C is the curve $x^2 + y^2 = 4$, $z = 2$.

Solution: Here, $\mathbf{A} = \mathbf{i}A_x + \mathbf{j}A_y + \mathbf{k}A_z$, so that, $A_x = e^x$, $A_y = 2y$, $A_z = -1$.

Thus, the given integral is $\int_C \mathbf{A} \cdot d\mathbf{r}$, where $d\mathbf{r} = \mathbf{i}\,dx + \mathbf{j}\,dy + \mathbf{k}\,dz$.

By Stokes' theorem

$$\int_C \mathbf{A} \cdot d\mathbf{r} = \iint_S (\operatorname{curl} \mathbf{A}) \cdot \mathbf{n}\,dS,$$

where the surface S is bounded by the circle $x^2 + y^2 = 4$, $z = 2$; so its centre is at $(0, 0, 2)$ and radius $= 2$ units. Moreover, $\mathbf{n} = \mathbf{k}$.

$$\therefore \quad \operatorname{curl} \mathbf{A} = \begin{vmatrix} \mathbf{i} & \mathbf{j} & \mathbf{k} \\ \dfrac{\partial}{\partial x} & \dfrac{\partial}{\partial y} & \dfrac{\partial}{\partial z} \\ e^x & 2y & -1 \end{vmatrix} = 0. \qquad \therefore \quad \operatorname{curl} \mathbf{A} \cdot \mathbf{n} = 0.$$

Hence the integral $= 0$.

EXAMPLE 2.46. *Verify Stokes' theorem for* $\mathbf{A} = (2x - y)\mathbf{i} - yz^2\mathbf{j} - y^2z\mathbf{k}$, *where* S *is the upper half surface of the sphere* $x^2 + y^2 + z^2 = 1$ *and* C *is its boundary.*

Solution: By Stokes' theorem

$$\int_S \operatorname{curl} \mathbf{A} \cdot \mathbf{n}\,dS = \int_C \mathbf{A} \cdot d\mathbf{r}.$$

Here C is the boundary of upper half of the given sphere in xy plane, i.e., it is the circle $x^2 + y^2 = 1$ whose parametric equation can be written as $x = \cos t$, $y = \sin t$.

$$\therefore \quad \int_C \mathbf{A} \cdot d\mathbf{r} = \int_C [(2x - y)\,dx - yz^2\,dy - y^2\,z\,dz] = \int_0^{2\pi} (2x - y)\frac{dx}{dt}\,dt \quad [\because\ z = 0 \text{ on } C]$$

$$= \int_0^{2\pi} (2\cos t - \sin t)(-\sin t)\,dt = \int_0^{2\pi} \left(\sin^2 t - 2\cos t \sin t\right)\,dt = \pi. \qquad (1)$$

Again

$$\operatorname{curl} \mathbf{A} = \begin{vmatrix} \mathbf{i} & \mathbf{j} & \mathbf{k} \\ \dfrac{\partial}{\partial x} & \dfrac{\partial}{\partial y} & \dfrac{\partial}{\partial z} \\ 2x - y & -yz^2 & -y^2z \end{vmatrix} = \mathbf{k}.$$

$$\therefore \quad \iint_S \operatorname{curl} \mathbf{A} \cdot \mathbf{n}\,dS = \int_S \mathbf{n} \cdot \mathbf{k}\,dS = \iint_R \mathbf{n} \cdot \mathbf{k}\frac{dx\,dy}{\mathbf{n} \cdot \mathbf{k}} = \iint_R dx\,dy$$

[R is the projection of S on xy plane]

$$\therefore \quad \iint_S \operatorname{curl} \mathbf{A} \cdot \mathbf{n}\,dS = \int_{x=-1}^{1} \int_{y=-\sqrt{1-x^2}}^{y=\sqrt{1-x^2}} dx\,dy = 4\int_0^1 \int_0^{\sqrt{1-x^2}} dx\,dy = 4\int_0^1 \sqrt{1 - x^2}\,dx = \pi. \qquad (2)$$

Hence, by (1) and (2), Stokes' theorem is verified.

IV. APPLICATIONS IN PHYSICS

We have already seen in Section 2.6 how components of velocity and acceleration of a particle moving in a plane are determined by using the vector calculus. Here we shall discuss some topics from different corners of physics and see how conveniently vector methods are used.

2.26 Equation of Continuity of Flowing Fluids

We know that divergence of a vector field **A** is the net amount of outflow of the vector **A** per unit volume and is written as div **A**. Let us now suppose that a quantity of fluid of density ρ is flowing with velocity **V**. Then the divergence of the velocity vector is div $\rho\mathbf{V}$. Consider the flow of fluid within a small volume $d\tau$. Due to outflow of the fluid the density of the fluid inside the volume $d\tau$ will decrease; but by the principle of conservation of mass, the total outflow of mass should be equal to the total decrease in mass within the small volume due to change of fluid density. Hence

$$\nabla \cdot (\rho\mathbf{V})\, d\tau = -\frac{d\rho}{dt}\, d\tau, \quad \text{i.e.,} \quad \nabla \cdot (\rho\mathbf{V}) + \frac{d\rho}{dt} = 0. \tag{2.126}$$

This equation is called the *equation of continuity* for flow of fluid.

If the fluid is incompressible, its density will remain constant, i.e., $\dfrac{d\rho}{dt} = 0$. In this case equation (2.126) becomes

$$\nabla \cdot (\rho\mathbf{V}) = 0, \quad \text{i.e.,} \quad \nabla \cdot \mathbf{V} = 0. \tag{2.127}$$

Let us now consider a system of flow of fluid of a fixed amount to which some fluid from outside is added so that increase of mass of fluid per unit volume per unit time is represented by $\phi(x, y, z)$. Then, the equation of mass conservation will be as under

rate of fluid outflow = rate of fluid mass added + rate of decrease of fluid

$$\text{i.e.,} \quad \nabla(\rho\mathbf{V})\, d\tau = -\frac{d\rho}{dt}\, d\tau + \phi\, d\tau, \quad \text{i.e.,} \quad \nabla(\rho\mathbf{V}) + \frac{d\rho}{dt} = \phi. \tag{2.128}$$

2.27 Angular Velocity of a Rotating Body

Let us consider a rigid body rotating with a constant angular velocity $\boldsymbol{\omega}$ about an axis passing through a point O within the body (*Fig 2.17*). Let P be any point within the body, such that $\overrightarrow{OP} = \mathbf{r}$. Then the linear velocity of the point P is given by

$$\mathbf{V} = \boldsymbol{\omega} \times \mathbf{r}. \tag{2.129}$$

If the body, in addition to its rotational motion, moves as a whole with linear velocity $\mathbf{v_0}$, then the total linear velocity of the point P is

$$\mathbf{V} = \mathbf{v_0} + \boldsymbol{\omega} \times \mathbf{r}. \tag{2.130}$$

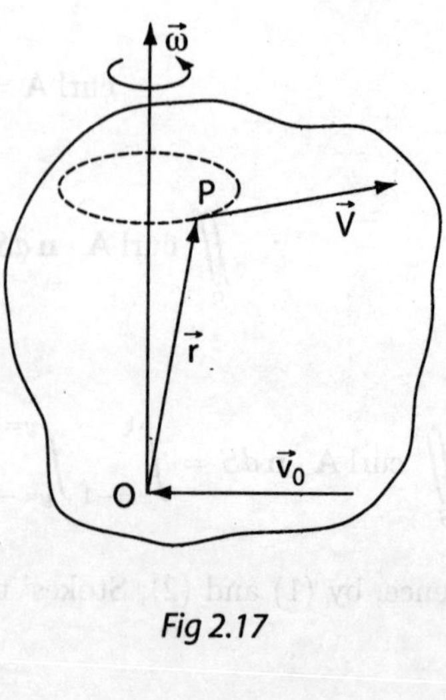

Fig 2.17

Now, $\nabla \times \mathbf{V} = \nabla \times \mathbf{v_0} + \nabla \times (\boldsymbol{\omega} \times \mathbf{r})$.
But $\mathbf{v_0}$ = constant, $\therefore\ \nabla \times \mathbf{v_0} = 0$.
Hence, $\nabla \times \mathbf{V} = \nabla \times (\boldsymbol{\omega} \times \mathbf{r})$.
Let us suppose that $\boldsymbol{\omega} = \mathbf{i}\omega_x + \mathbf{j}\omega_y + \mathbf{k}\omega_z$ and $\mathbf{r} = \mathbf{i}x + \mathbf{j}y + \mathbf{k}z$.
Then

$$\mathbf{V} = \boldsymbol{\omega} \times \mathbf{r} = \begin{vmatrix} \mathbf{i} & \mathbf{j} & \mathbf{k} \\ \omega_x & \omega_y & \omega_z \\ x & y & z \end{vmatrix} = \mathbf{i}(\omega_y z - \omega_z y) + \mathbf{j}(\omega_z x - \omega_x z) + \mathbf{k}(\omega_x y - \omega_y x).$$

$$\therefore\ \operatorname{curl} \mathbf{V} = \begin{vmatrix} \mathbf{i} & \mathbf{j} & \mathbf{k} \\ \dfrac{\partial}{\partial x} & \dfrac{\partial}{\partial y} & \dfrac{\partial}{\partial z} \\ \omega_y z - \omega_z y & \omega_z x - \omega_x z & \omega_x y - \omega_y x \end{vmatrix} = 2(\mathbf{i}\omega_x + \mathbf{j}\omega_y + \mathbf{k}\omega_z) = 2\boldsymbol{\omega}.$$

$$\therefore\ \boldsymbol{\omega} = \tfrac{1}{2}(\nabla \times \mathbf{V}). \tag{2.131}$$

Thus, angular velocity of rotation of a rigid body is equal to half the curl of linear translational velocity of a point within the rigid body. Therefore, curl $\mathbf{V}$ describes rotation of a vector field $\mathbf{V}$ and hence the name curl or rot. If the curl of velocity vector is zero, the motion will be irrotational.

2.28 Gauss's Theorem in Electrostatics

The theorem states that the flux of electric intensity across any closed surface having a unique outward-drawn normal at every point on it is 4π times the total charge enclosed by the surface. If the charge is not enclosed, flux is zero.

Proof. Let S be a closed surface and $P(x, y, z)$ be any point having position vector $\mathbf{r}$ relative to an origin O. Let a charge of amount q be placed at O (*Fig 2.18*). Then electric intensity at P due to q is

$$\mathbf{E} = \frac{q}{r^2} \cdot \frac{\mathbf{r}}{|\mathbf{r}|} = \frac{q}{r^3} \cdot \mathbf{r}. \tag{2.132}$$

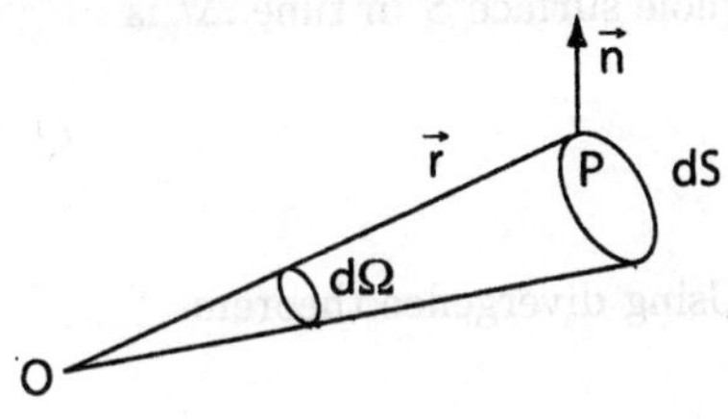

Fig 2.18

Now, if we consider a small area dS of the surface where positive unit normal vector is $\mathbf{n}$, then the flux of electric intensity due to q through elementary area dS is

$$\frac{q}{r^3}\mathbf{r} \cdot d\mathbf{S} = \frac{q}{r^3}\mathbf{r} \cdot \mathbf{n}\, dS. \tag{2.133}$$

Hence, the flux of electric intensity through the whole surface is

$$q \iint\limits_S \frac{\mathbf{r} \cdot \mathbf{n}}{r^3}\, dS = q \iint d\Omega, \tag{2.134}$$

where $d\Omega$ is the solid angle subtended at O by the elementary surface dS.

When the charge q is inside the boundary of the surface, the integral (2.134) represents the total solid angle about the point O and hence its value is $4\pi q$.

Let us now suppose that the point O is outside of the surface, Then, by divergence theorem, integral (2.134) becomes

$$q \int_V \operatorname{div}\left(\frac{\mathbf{r}}{r^3}\right) dV. \tag{2.135}$$

But

$$\nabla \cdot \left(\frac{\mathbf{r}}{r^3}\right) = \nabla \left(\frac{1}{r^3}\right) \cdot \mathbf{r} + \left(\frac{1}{r^3}\right) \cdot \nabla \mathbf{r} = -3r^{-5}\mathbf{r} \cdot \mathbf{r} + 3r^{-3} = 0 \quad (\text{provided } \mathbf{r} \neq 0).$$

Hence, the integral (2.135) vanishes when the charge is placed outside the surface. Thus the theorem is proved. □

2.29 Heat Flow in Solids

Let us consider the problem of heat transfer through a body by means of conduction. We make the following assumptions:

1. The temperature $T(x, y, z, t)$ at any point of the body is a finite and continuous function.
2. Rate of heat flow per unit area is equal to $-K\nabla T$ according to Fourier's law. Here K is called the *thermal conductivity* of the material of the body.
3. Temperature gradient and heat flow within the body are normal to the isothermal surface, i.e., the surface on which temperature is same everywhere.

Now, if S be an arbitrarily closed surface enclosing volume V within the body, then the amount of heat flowing out of a small surface dS in time Δt is

$$dQ = -[(K\nabla T) \cdot \mathbf{n}\, dS]\, \Delta t,$$

where $\mathbf{n}$ is the unit normal vector to the surface S. Hence, the total amount of heat flowing out of whole surface S in time Δt is

$$Q = -\Delta t \iint_S K\nabla T \cdot \mathbf{n} dS.$$

Using divergence theorem

$$Q = -\Delta t \iiint_V \nabla \cdot (K\nabla T)\, dV. \tag{2.136}$$

Again, let $c =$ specific heat of the material of the body, $\rho =$ density and $\dfrac{\partial T}{\partial t}$ is the rate of increase of temperature within the volume V of the body. Then the amount of heat flowing out of the volume V of the body is given by

$$Q = -\Delta t \iiint_V c\rho \frac{\partial T}{\partial t}\, dV. \tag{2.137}$$

From equations (2.136) and (2.137), we get

$$\iiint_V \left\{ c\rho \frac{\partial T}{\partial t} - \nabla \cdot (K\nabla T) \right\} dV = 0.$$

Since the volume dV is arbitrary

$$c\rho\frac{\partial T}{\partial t} = \nabla \cdot (K\nabla T).$$

This is the equation of heat conduction which can be written as

$$\frac{\partial T}{\partial t} = \lambda^2\nabla^2 T, \tag{2.138}$$

where $\lambda^2 = \dfrac{K}{c\rho}$ which is called the *thermal diffusivity* of the material.

2.30 Motion of a Charged Particle in Electromagnetic Field

Let a particle of mass m and charge q be under the influence of an electric field of intensity $\mathbf{E}$ as well as a magnetic field of intensity $\mathbf{B}$. Initially the particle is at rest, i.e., at $t = 0$, velocity $\mathbf{v} = 0$.

Let

$$\left.\begin{aligned} \mathbf{E} &= \mathbf{i}E_x + \mathbf{j}E_y + \mathbf{k}E_z, \\ \mathbf{B} &= \mathbf{i}B_x + \mathbf{j}B_y + \mathbf{k}B_z \\ \text{and}\quad \mathbf{v} &= \mathbf{i}\dot{x} + \mathbf{j}\dot{y} + \mathbf{k}\dot{z}. \end{aligned}\right\} \tag{2.139}$$

Now, the Lorentz force acting on the particle due to electric and magnetic fields is

$$\mathbf{F} = \mathbf{F_e} + \mathbf{F_m} = q\mathbf{E} + q(\mathbf{v}\times\mathbf{B}), \tag{2.140}$$

where force due to electric field only is

$$\mathbf{F_e} = q\mathbf{E} \tag{2.141}$$

and the force due to magnetic field only is

$$\mathbf{F_m} = q(\mathbf{v}\times\mathbf{B}). \tag{2.142}$$

From equation (2.142), we have

$$\frac{\mathbf{F_m}}{q} = \mathbf{v}\times\mathbf{B} = \begin{vmatrix} \mathbf{i} & \mathbf{j} & \mathbf{k} \\ \dot{x} & \dot{y} & \dot{z} \\ B_x & B_y & B_z \end{vmatrix},$$

i.e., $\mathbf{F_m} = q[\mathbf{i}(B_z\dot{y} - B_y\dot{z}) + \mathbf{j}(B_x\dot{z} - B_z\dot{x}) + \mathbf{k}(B_y\dot{x} - B_x\dot{y})]$.

$\therefore$ The net force on the particle is

$$\mathbf{F} = q[\mathbf{i}(E_x + B_z\dot{y} - B_y\dot{z}) + \mathbf{j}(E_y + B_x\dot{z} - B_z\dot{x}) + \mathbf{k}(E_z + B_y\dot{x} - B_x\dot{y})]. \tag{2.143}$$

Again, if acceleration of the particle at any time is $\mathbf{a}$, then

$$\mathbf{F} = m\mathbf{a} = m(\mathbf{i}\ddot{x} + \mathbf{j}\ddot{y} + \mathbf{k}\ddot{z}). \tag{2.144}$$

From equation (2.143) and (2.144), we have

$$m\ddot{x} = q(E_x + B_z\dot{y} - B_y\dot{z}) \tag{2.145}$$
$$m\ddot{y} = q(E_y + B_x\dot{z} - B_z\dot{x}) \tag{2.146}$$
$$m\ddot{z} = q(E_z + B_y\dot{x} - B_x\dot{y}) \tag{2.147}$$

These are the equations of motion of the charged particle under the simultaneous influence of electric and magnetic fields.

Multiplying equation (2.145) by B_x and (2.146) by B_y, and adding the result, we get

$$m(B_x\ddot{x} + B_y\ddot{y}) = q(E_xB_x + E_yB_y) - qB_z(B_y\dot{x} - B_x\dot{y}). \tag{2.148}$$

Again, from equation (2.147)

$$m\ddot{z} - qE_z = q(B_y\dot{x} - B_x\dot{y}). \tag{2.149}$$

Substituting this relation into equation (2.148), we get

$$m(B_x\ddot{x} + B_y\ddot{y}) = q(E_xB_x + E_yB_y) - B_z(m\ddot{z} - qE_z),$$
$$\text{i.e.,}\quad m(B_x\ddot{x} + B_y\ddot{y} + B_z\ddot{z}) = q(E_xB_x + E_yB_y + E_zB_z),$$
$$\text{i.e.,}\quad m(\mathbf{B}\cdot\mathbf{a}) = q(\mathbf{B}\cdot\mathbf{E}),\quad \text{or,}\quad \mathbf{B}\cdot(m\mathbf{a}) = \mathbf{B}\cdot(q\mathbf{E}). \tag{2.150}$$

Equation (2.150) indicates that component of instantaneous acceleration vector along the direction of **B** is always constant.

2.31 Euler's Equations of Motion of Fluids

In Eulerian method of fluid mechanics, a point (x, y, z) is taken within the fluid and the motion is studied. As a result, the velocity components v_x, v_y, v_z along the axes are all functions of (x, y, z, t).

Let us consider an infinitesimally small element of volume $dx\,dy\,dz = dV$ of non-viscous fluid of density ρ under the action of an external force **F** per unit mass of fluid having components F_x, F_y, F_z along the axes. Then the net force acting on the element in x-direction due to internal pressure p is (*Fig 2.19*)

$$p\,dy\,dz - \left(p + \frac{\partial p}{\partial x}dx\right) dy\,dz = -\frac{\partial p}{\partial x}\,dV. \tag{2.151}$$

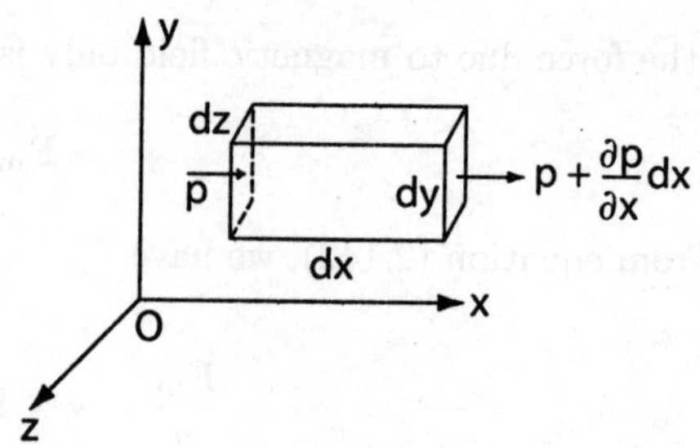

Fig 2.19

The external force acting on the element in x-direction is

$$F_x\rho dV \quad (\because \text{ mass of element} = \rho dV). \tag{2.152}$$

Thus, the resultant force acting on the element in x-direction is

$$F_x\rho dV - \frac{\partial p}{\partial x}\,dV. \tag{2.153}$$

The force impressed on the elementary mass due to acceleration in x-direction is

$$\rho dV \frac{d}{dt}(v_x) = \frac{dv_x}{dt} \rho dV. \tag{2.154}$$

Hence, we have

$$F_x \rho dV - \frac{\partial \rho}{\partial x} dV = \frac{dv_x}{dt} \rho\, dV,$$

i.e.,

$$\frac{dv_x}{dt} = F_x - \frac{1}{\rho}\frac{\partial p}{\partial x}. \tag{2.155}$$

Similarly, accelerations in y- and z-directions are

$$\frac{dv_y}{dt} = F_y - \frac{1}{\rho}\frac{\partial p}{\partial y} \tag{2.156}$$

$$\text{and} \quad \frac{dv_z}{dt} = F_z - \frac{1}{\rho}\frac{\partial p}{\partial z}. \tag{2.157}$$

Since velocity components are functions of x, y, z and t, we have

$$\frac{dv_x}{dt} = \left(\frac{\partial}{\partial t} + v_x\frac{\partial}{\partial x} + v_y\frac{\partial}{\partial y} + v_z\frac{\partial}{\partial z}\right) v_x, \tag{2.158}$$

$$\frac{dv_y}{dt} = \left(\frac{\partial}{\partial t} + v_x\frac{\partial}{\partial x} + v_y\frac{\partial}{\partial y} + v_z\frac{\partial}{\partial z}\right) v_y \tag{2.159}$$

$$\text{and} \quad \frac{dv_z}{dt} = \left(\frac{\partial}{\partial t} + v_x\frac{\partial}{\partial x} + v_y\frac{\partial}{\partial y} + v_z\frac{\partial}{\partial z}\right) v_z. \tag{2.160}$$

Multiplying these equations by **i**, **j**, **k** respectively, and then adding, we get

$$\begin{aligned}\left(\mathbf{i}\frac{dv_x}{dt} + \mathbf{j}\frac{dv_y}{dt} + \mathbf{k}\frac{dv_z}{dt}\right) &= \left(\mathbf{i}\frac{\partial v_x}{\partial t} + \mathbf{j}\frac{\partial v_y}{\partial t} + \mathbf{k}\frac{\partial v_z}{\partial t}\right) \\ &\quad + \left[(\mathbf{i}v_x + \mathbf{j}v_y + \mathbf{k}v_z)\cdot\left(\mathbf{i}\frac{\partial}{\partial x} + \mathbf{j}\frac{\partial}{\partial y} + \mathbf{k}\frac{\partial}{\partial z}\right)\right](\mathbf{i}v_x + \mathbf{j}v_y + \mathbf{k}v_z),\end{aligned}$$

i.e.,

$$(\mathbf{i}F_x + \mathbf{j}F_y + \mathbf{k}F_z) - \frac{1}{\rho}\left(\mathbf{i}\frac{\partial p}{\partial x} + \mathbf{j}\frac{\partial p}{\partial y} + \mathbf{k}\frac{\partial p}{\partial z}\right) = \frac{\partial \mathbf{v}}{\partial t} + (\mathbf{v}\cdot\nabla)\mathbf{v},$$

i.e.,

$$\frac{\partial \mathbf{v}}{\partial t} + (\mathbf{v}\cdot\nabla)\mathbf{v} = \mathbf{F} - \frac{1}{\rho}\nabla p. \tag{2.161}$$

This is the Eulerian equation of motion for fluid flow.

Now, we have a vector relation (*see* relation (h) of Section 2.17)

$$\nabla(\mathbf{a}\cdot\mathbf{b}) = (\mathbf{a}\cdot\nabla)\mathbf{b} + (\mathbf{b}\cdot\nabla)\mathbf{a} + \mathbf{a}\times(\nabla\times\mathbf{b}) + \mathbf{b}\times(\nabla\times\mathbf{a}).$$

If we set here $\mathbf{a}=\mathbf{b}=\mathbf{v}$, we have

$$\begin{aligned}\nabla(\mathbf{v}\cdot\mathbf{v}) &= 2(\mathbf{v}\cdot\nabla)\mathbf{v}+2\mathbf{v}\times(\nabla\times\mathbf{v}),\\ \text{i.e.,}\quad (\mathbf{v}\cdot\nabla)\mathbf{v} &= \tfrac{1}{2}\nabla(\mathbf{v}^2)-\mathbf{v}\times(\nabla\times\mathbf{v}).\end{aligned}\tag{2.162}$$

Using this relation, the equation (2.161) is transformed to

$$\frac{\partial\mathbf{v}}{\partial t}+\frac{1}{2}\nabla(\mathbf{v}^2)-\mathbf{v}\times(\nabla\times\mathbf{v})=\mathbf{F}-\frac{1}{\rho}\nabla p.\tag{2.163}$$

If the external field is conservative, i.e., $\mathbf{F}=-\nabla\phi$, where ϕ is the scalar potential of the field, then equation (2.163) reduces to

$$\begin{aligned}&\frac{\partial\mathbf{v}}{\partial t}-\mathbf{v}\times(\nabla\times v)+\nabla\left(\phi+\frac{p}{\rho}+\frac{\mathbf{v}^2}{2}\right)=0,\\ \text{i.e.,}\quad &\frac{\partial\mathbf{v}}{\partial t}-2\mathbf{v}\times\boldsymbol{\omega}+\nabla\left(\phi+\frac{p}{\rho}+\frac{\mathbf{v}^2}{2}\right)=0,\end{aligned}\tag{2.164}$$

where $\boldsymbol{\omega}$ is the vorticity vector defined by the equation

$$(\nabla\times\mathbf{v})=\tfrac{1}{2}\boldsymbol{\omega},\quad \text{i.e.,}\quad \operatorname{curl}\mathbf{v}=\tfrac{1}{2}\boldsymbol{\omega}.\tag{2.165}$$

If the fluid motion is irrotational, $\boldsymbol{\omega}=0$ and then the velocity vector can be expressed in terms of velocity potential function ψ in the form $\mathbf{v}=-\nabla\psi$. Then equation (2.164) becomes

$$\nabla\left(\phi+\frac{p}{\rho}+\frac{\mathbf{v}^2}{2}-\frac{\partial\psi}{\partial t}\right)=0.\tag{2.166}$$

In case the motion is steady, $\dfrac{\partial\psi}{\partial t}=0$. Then the equation (2.166) reduces to

$$\phi+\frac{p}{\rho}+\frac{\mathbf{v}^2}{2}=\text{constant}.\tag{2.167}$$

This is known as *Bernoulli's equation.* It states that, the sum of kinetic energy $(\frac{1}{2}\mathbf{v}^2)$, potential energy (ϕ) and pressure, each for unit mass of fluid has a constant value for every point in the fluid.

2.32 Steady-state Magnetic Phenomena

Steady-state magnetic phenomena occur where there is no change in the net charge density anywhere. Here we shall discuss the basic equations of magnetostatics under steady-state condition.

Let us consider a small element of length $d\mathbf{l}$ (pointing towards current flow) of a filamentary wire carrying current I. With respect to an origin, let the coordinate vector of the middle point of the element be $\mathbf{x}'$ and that of a point of observation P be $\mathbf{x}$. Then, according to Biot-Savart law, the elemental magnetic flux density at the point P in magnitude and direction is given by

$$d\mathbf{B}=\frac{1}{c}Id\mathbf{l}\times\frac{(\mathbf{x}-\mathbf{x}')}{|\mathbf{x}-\mathbf{x}'|^3}\tag{2.168}$$

where c is the velocity of light in vacuum.

If the current density is J, then, for elemental area da

$$Id\mathbf{l} = \mathbf{J} \cdot da d\mathbf{l} = \mathbf{J}\, d^3x'$$

so that equation (2.168) becomes

$$d\mathbf{B} = \frac{1}{c}\mathbf{J}(\mathbf{x}') \times \frac{(\mathbf{x}-\mathbf{x}')}{|\mathbf{x}-\mathbf{x}'|^3}\, d^3x'.$$

Hence, the total magnetic induction in general form can be written as

$$\mathbf{B} = \frac{1}{c}\int \mathbf{J}(\mathbf{x}') \times \frac{(\mathbf{x}-\mathbf{x}')}{|\mathbf{x}-\mathbf{x}'|^3}\, d^3x'. \tag{2.169}$$

Now, the vector factor in the integrand is

$$\frac{(\mathbf{x}-\mathbf{x}')}{|\mathbf{x}-\mathbf{x}'|^3} = -\boldsymbol{\nabla}\left(\frac{1}{|\mathbf{x}-\mathbf{x}'|}\right).$$

With the use of this relation, equation (2.169) reduces to

$$\mathbf{B} = -\frac{1}{c}\int \mathbf{J}(\mathbf{x}') \times \boldsymbol{\nabla}\left(\frac{1}{|\mathbf{x}-\mathbf{x}'|}\right) d^3x'. \tag{2.170}$$

Since the gradient operation involves $\mathbf{x}$ but not the integration variable $\mathbf{x}'$, the equation (2.170) can be written as

$$\mathbf{B}(\mathbf{x}) = \frac{1}{c}\boldsymbol{\nabla} \times \int \mathbf{J}(\mathbf{x}') \cdot \frac{d^3x'}{|\mathbf{x}-\mathbf{x}'|}. \tag{2.171}$$

From equation (2.171), it follows that divergence of $\mathbf{B}$ vanishes.
Hence

$$\boldsymbol{\nabla} \cdot \mathbf{B} = 0. \tag{2.172}$$

This is the first equation of magnetostatics. This implies that $\mathbf{B}$ is solenoidal.
Let us now calculate curl of $\mathbf{B}$.

$$\boldsymbol{\nabla} \times \mathbf{B} = \frac{1}{c}\boldsymbol{\nabla} \times \boldsymbol{\nabla} \times \int \frac{\mathbf{J}(\mathbf{x}')}{|\mathbf{x}-\mathbf{x}'|} d^3x'. \tag{2.173}$$

Now, for any vector field $\mathbf{a}$, we have the identity

$$\boldsymbol{\nabla} \times (\boldsymbol{\nabla} \times \mathbf{a}) = \boldsymbol{\nabla}(\boldsymbol{\nabla} \cdot \mathbf{a}) - \boldsymbol{\nabla}^2\mathbf{a}.$$

With the help of this identity, we write equation (2.173) in the form

$$\boldsymbol{\nabla} \times \boldsymbol{B} = \frac{1}{c}\boldsymbol{\nabla}\int \mathbf{J}(\mathbf{x}') \cdot \boldsymbol{\nabla}\left(\frac{1}{|\mathbf{x}-\mathbf{x}'|}\right) d^3x' - \frac{1}{c}\int \mathbf{J}(\mathbf{x}')\boldsymbol{\nabla}^2\left(\frac{1}{|\mathbf{x}-\mathbf{x}'|}\right) d^3x'. \tag{2.174}$$

But, we know that

$$\boldsymbol{\nabla}\left(\frac{1}{|\mathbf{x}-\mathbf{x}'|}\right) = -\boldsymbol{\nabla}'\left(\frac{1}{|\mathbf{x}-\mathbf{x}'|}\right) \quad \text{and} \quad \boldsymbol{\nabla}^2\left(\frac{1}{|\mathbf{x}-\mathbf{x}'|}\right) = -4\pi\delta(x-x').$$

Using these in equation (2.174), we get

$$\nabla \times B = -\frac{1}{c}\nabla \int \mathbf{J}(\mathbf{x}') \cdot \nabla' \left(\frac{1}{|\mathbf{x}-\mathbf{x}'|}\right) d^3x' + \frac{4\pi}{c}\int \mathbf{J}(\mathbf{x}')\delta(\mathbf{x}-\mathbf{x}')\, d^3x'$$

$$= \frac{4\pi}{c}\mathbf{J} + \frac{1}{c}\nabla \int \frac{\nabla' \cdot \mathbf{J}(\mathbf{x}')}{|\mathbf{x}-\mathbf{x}'|}\, d^3x'$$

But, for steady-state magnetic phenomena, $\nabla \cdot J = 0$. Hence, we get

$$\nabla \times \mathbf{B} = \frac{4\pi}{c}\mathbf{J}. \tag{2.175}$$

This is the second equation of magnetostatics. Thus, the basic differential laws of magnetostatics are governed by

$$\left.\begin{aligned} \nabla \times \mathbf{B} &= \frac{4\pi}{c}\mathbf{J} \\ \nabla \cdot \mathbf{B} &= 0 \end{aligned}\right\}. \tag{2.176}$$

We now intend to solve these equations. If $\nabla \cdot \mathbf{B} = 0$ everywhere, then there must be a vector field $\mathbf{A}(\mathbf{x})$, such that

$$\mathbf{B}(\mathbf{x}) = \nabla \times \mathbf{A}(\mathbf{x}). \tag{2.177}$$

The vector field $\mathbf{A}(\mathbf{x})$ is called *vector potential.* Then, comparing with equation (2.171), we can write the general form of $\mathbf{A}(\mathbf{x})$ as

$$\mathbf{A}(\mathbf{x}) = \frac{1}{c}\int \frac{\mathbf{J}(\mathbf{x}')}{|\mathbf{x}-\mathbf{x}'|}\, d^3x' + \nabla\psi(x), \tag{2.178}$$

where $\psi(x)$ is an arbitrary scalar function giving freedom to transform the vector potential from $\mathbf{A} \longrightarrow \mathbf{A} + \nabla\psi$. This transformation is called *gauge transformation.* This transformation allows us to make $\nabla \cdot \mathbf{A}$ to have any convenient functional form to our wish.

Now, substituting (2.177) into the 1st equation of (2.176), we get

$$\nabla \times (\nabla \times A) = \frac{4\pi}{c}\mathbf{J}, \quad \text{i.e.,} \quad \nabla(\nabla \cdot A) - \nabla^2 \mathbf{A} = \frac{4\pi}{c}\mathbf{J}. \tag{2.179}$$

If we now take advantage of the freedom of transformation and select vector potential $\mathbf{A}$, such that $\nabla \cdot \mathbf{A} = 0$ (this choice being known as *Coulomb gauge*), then from equation (2.179), we find that

$$\nabla^2 \mathbf{A} = -\frac{4\pi}{c}\mathbf{J}. \tag{2.180}$$

Thus, each rectangular component of the vector potential $\mathbf{A}$ satisfies Poisson's equation.

If, again, the current density J vanishes in some region of space, then the first equation in (2.176) gives $\nabla \times \mathbf{B} = 0$. This implies that we can introduce a magnetic scalar potential ϕ such that

$$\mathbf{B} = -\nabla\phi. \tag{2.181}$$

Then from second equation in (2.176), we have

$$\nabla \cdot (-\nabla\phi) = 0, \quad \text{i.e.,} \quad \nabla^2\phi = 0. \tag{2.182}$$

Thus, the magnetic scalar potential satisfies Laplace equation.

EXERCISE

1. A particle moves along the curve $x = t^3 + 1$, $y = t^2$, $z = 2t + 5$, where t is the time. Find components of its velocity and acceleration at $t = 1$ in the direction $\mathbf{i} + 2\mathbf{j} + 3\mathbf{k}$.

2. Prove, from definition of differentiation, that

$$\frac{d}{dt}(\mathbf{a} \times \mathbf{b}) = \mathbf{b} \times \frac{d\mathbf{b}}{dt} + \frac{d\mathbf{a}}{dt} \times \mathbf{b}.$$

3. For a particle moving in a circular orbit given by $\mathbf{r} = \mathbf{i}r\cos\theta t + \mathbf{j}r\sin\theta t$, evaluate (a) $\mathbf{r} \times \dot{\mathbf{r}}$, (b) $\ddot{\mathbf{r}} + \theta^2\mathbf{r}$.

4. If $r = \mathbf{a}e^{nt} + \mathbf{b}e^{-nt}$, where $\mathbf{a}$, $\mathbf{b}$ are constant vectors, show that

$$\frac{d^2\mathbf{r}}{dt^2} = n^2\mathbf{r}.$$

5. If $\phi(x, y, z) = xy^2z$ and $\mathbf{a} = \mathbf{i}xz - \mathbf{j}xy + \mathbf{k}yz^2$, find $\dfrac{\partial^3}{\partial x^2 \partial z}(\phi\mathbf{a})$ at $(2, 1, 2)$.

6. If $\phi = \dfrac{1}{r}$, show that $\operatorname{grad}\phi = \dfrac{-\mathbf{r}}{r^3}$, where $r^2 = x^2 + y^2 + z^2$.

7. If ϕ is a function of x, y, z and $\mathbf{r} = \mathbf{i}x + \mathbf{j}y + \mathbf{k}z$, show that

$$\frac{d\phi}{ds} = \nabla\phi \cdot \frac{d\mathbf{r}}{ds}.$$

8. Find the directional derivative of $\phi = xy + yz + zx$ at the point $(1, 2, 2)$ in the direction of $2\mathbf{i} + \mathbf{j} + 3\mathbf{k}$.

9. Find the unit vector normal to the surface $x^4 + 5xyz + 3z^2 = 0$ at the point $(1, 1, 1)$.

10. A curve is defined by intersection of two surfaces $x^2 + y^2 + z^2 = 1$ and $x + y + z = 1$. Find the equation of tangent line to that curve at the point $(1, 1, 0)$.

11. Find directional derivative of the function $\phi = xy^2 + yz^2 + zx^2$ along the tangent to the curve $x = t$, $y = 2t^2$, $z = t^3$ at the point $(1, 2, 1)$.

12. Prove that the acute angle between the surfaces $xy^2z - 3x - z^2 = 0$ and $3x^2 - y^2 + 2z = 1$ at the point $(1, -2, 1)$ is $\cos^{-1}(3/7\sqrt{6})$.

13. Find the values of the constants a, b, c, so that the directional derivative of $\phi = ax^2 + by^2 + cz^2$ at $(1, 1, 2)$ has a maximum value of 4 in the direction parallel to y-axis.

14. If $\mathbf{A} = (x + y + 1)\mathbf{i} + (2x + 3)\mathbf{j} - (x + y)\mathbf{k}$, find $\operatorname{curl}\mathbf{A}$ and $\mathbf{A} \cdot \operatorname{curl}\mathbf{A}$.

15. Prove that $\operatorname{curl}\operatorname{grad} r^n = 0$, where $r = \left(x^2 + y^2 + z^2\right)^{1/2}$.

16. Prove that $r^n\mathbf{r}$ is irrotational. Find n when it is solenoidal.

17. Prove that $\nabla \cdot \left[r\nabla\left(\dfrac{1}{r^2}\right)\right] = \dfrac{3}{r^4}$.

18. Find the constants a, b, c, so that the vector $\mathbf{A}$ given by

$$\mathbf{A} = (2x + 3y + az)\mathbf{i} + (bx + 2y + 3z)\mathbf{j} + (2x + cy + 3z)\mathbf{k}$$

will be irrotational.

19. If $\mathbf{a}$ is a constant vector, prove that $\operatorname{div}(\mathbf{a} \times \mathbf{f}) = -\mathbf{a}\operatorname{curl}\mathbf{f}$.

20. If $\phi(x, y, z)$ is a solution of Laplacian equation, show that $\operatorname{grad}\phi$ is a vector which is both solenoidal and irrotational.

21. Prove that $\operatorname{div}\left[\frac{\phi(r)}{r}\mathbf{r}\right] = \frac{1}{r^2}\frac{d}{dr}[r^2\phi(r)]$.

22. If ϕ and ψ are scalar functions, such that a vector function $\mathbf{A}$ is given by $\mathbf{A} = \phi\operatorname{grad}\psi$, prove that $\mathbf{A} \cdot \operatorname{curl}\mathbf{A} = 0$.

23. Find the total work done in moving a particle in a force field given by

$$F = 3xy\mathbf{i} - 5xz\mathbf{j} + 10x\mathbf{k}$$

along the curve $x = t^2 + 1$, $y = 2t^2$, $z = t^3$ from $t = 0$ to $t = 1$.

24. Evaluate the integral $\int_c \mathbf{F} \cdot d\mathbf{r}$, where $\mathbf{F} = (x^2 + y^2)\mathbf{i} - 2xy\mathbf{j}$ and the curve C is the boundary of the rectangle $y = 0$, $x = a$ and $y = b$, $x = 0$.

25. If $\mathbf{F} = xy\mathbf{i} + yz\mathbf{j} + x^2\mathbf{k}$, evaluate $\int_C \mathbf{F} \times d\mathbf{r}$, where C is the curve $x = t^2$, $y = 2t$, $z = t^3$ from $t = 0$ to $t = 1$.

26. Evaluate $\int_C (x^2 + y^2 + z^2)\, ds$, where C is the arc of circular helix $\mathbf{r}(t) = \mathbf{i}\cos t + \mathbf{j}\sin t + \mathbf{k}3t$ from $A(1, 0, 0)$ to $B(1, 0, 6\pi)$.

27. Show that the vector function

$$\mathbf{F} = \mathbf{i}(y\sin z - \sin x) + \mathbf{j}(x\sin z + 2yz) + \mathbf{k}(xy\cos z + y^2)$$

is irrotational. Also, find a scalar function ϕ, such that $F = \nabla\phi$.

28. Evaluate $\int_C \{yz\,dx + (xz + 1)\,dy + xy\,dz\}$, where C is any path from $(1, 0, 0)$ to $(2, 1, 4)$.

29. Evaluate $\int_S \mathbf{F} \cdot \mathbf{n}\,dS$, where S is the surface of the cylinder $x^2 + y^2 = 9$ included in the first octant between $z = 0$ and $z = 4$, and $\mathbf{F} = \mathbf{i}z + \mathbf{j}x - \mathbf{k}yz$.

30. Evaluate $\int_S \mathbf{F} \cdot \mathbf{n}\,dS$, where $\mathbf{F} = 2y\mathbf{i} - 3\mathbf{j} + x^2\mathbf{k}$ and S is the surface of the parabolic cylinder $y^2 = 8x$ in the first octant bound by the planes $y = 4$ and $z = 6$.

31. Verify divergence theorem by evaluating the integral $\int_S (4xz\mathbf{i} - y^2\mathbf{j} + yz\mathbf{k}) \cdot \mathbf{n}\,dS$, where S is the surface bounded by $x = 0$, $x = 1$, $y = 0$, $y = 1$, $z = 0$, and $z = 1$.

32. Prove that $\int_S \phi \cdot \mathbf{n}\,dS = \int_V \nabla\phi\,dV$.

33. Show that $\int_S \nabla\phi \times \nabla\psi \cdot \mathbf{n}\,dS = 0$.

34. Verify Stokes' theorem for $\mathbf{F} = (2x-y)\mathbf{i} - yz^2\mathbf{j} - y^2z\mathbf{k}$, where S is the upper half of the sphere $x^2+y^2+z^2=1$ and C is its boundary.

35. Verify Stokes' theorem for $\mathbf{F} = y\mathbf{i} + (x-2xz)\mathbf{j} - xy\mathbf{k}$, where the surface S is the part of the sphere $x^2+y^2+z^2=a^2$ above the xy plane.

36. Verify Green's theorem in the plane for $\int_C (xy+y^2)\,dx + x^2dy$, where C is the closed curve of the region bounded by $y=x$ and $y=x^2$.

37. Show that $\oint_C r^2 d\mathbf{r} = -2\int_S \mathbf{r} \times d\mathbf{s}$, where $\mathbf{r} = \mathbf{i}x + \mathbf{j}y + \mathbf{k}z$.

38. Show that the magnitude of $\oint dr \times r$ taken round a curve in xy plane is twice the area enclosed by the curve.

39. Show that the necessary and sufficient condition that $F_x dx + F_y\,dy + F_z\,dz$ is a perfect differential is that, $\nabla \times \mathbf{F} = 0$, where $\mathbf{F} = \mathbf{i}F_x + \mathbf{j}F_y + \mathbf{k}F_z$.

40. Prove that $\int_V \nabla \times \mathbf{B}\,dV = \int_S \mathbf{n} \times \mathbf{B}\,dS$, where $\mathbf{n}$ is unit outward-drawn normal to the surface S enclosing the volume V.

ANSWERS

1. $\frac{10}{\sqrt{40}}$.
3. $\theta r^2\mathbf{k}, 0$.
5. $8\mathbf{i} - 2\mathbf{j}$.
8. $\frac{20}{\sqrt{14}}$.
9. $\frac{1}{\sqrt{227}}(9\mathbf{i} + 5\mathbf{j} + 11\mathbf{k})$.
10. $x+y=2, z=0$.
11. $\frac{41}{\sqrt{26}}$.
13. $a=0, b=2, c=0$.
14. $-\mathbf{i}+\mathbf{j}+\mathbf{k}, 2-2y$.
18. $a=2, b=3, c=3$.
23. $14\frac{1}{7}$.
24. $2\sqrt{10}\pi\,(1+12\pi^2)$.
25. $\frac{16}{35}\mathbf{i} - \frac{2}{3}\mathbf{j} + \frac{1}{3}\mathbf{k}$.
26. $2\sqrt{10}\pi(1+12\pi^2)$.
28. 9.
29. 42.
30. 132.

Chapter 3

Ordinary Differential Equations

3.1 Introduction

A differential equation is an equation in terms of derivatives of an unknown function to be determined. The derivatives occurring in the equation may be total derivatives when the unknown function is of single independent variable. In such a case, we call the differential equation *an ordinary differential equation.* However, if the unknown function happens to depend on more than one variable, then the derivatives occurring in the differential equation should be partial. Such types of differential equations containing partial derivatives of the unknown functions are called *partial differential equations.*

The order of a differential equation is the order of the highest derivative occurring in the differential equation. Thus, the differential equations

$$\left.\begin{aligned} \frac{dx}{dy} + xy &= 2 \\ m\frac{dv}{dt} + kt &= 0 \end{aligned}\right\} \tag{3.1}$$

are of first order, while differential equations

$$\left.\begin{aligned} \frac{d^2y}{dx^2} + p(x)\frac{dy}{dx} + q(x) &= 0 \\ \frac{d^2x}{dt^2} + k\frac{dx}{dt} + r &= f(x) \end{aligned}\right\} \tag{3.2}$$

are of second order.

The degree of a differential equation is the power of the highest derivative of the equation. A differential equation is said to be linear if it does not contain any higher power of dependent variable or of derivatives or of a product of the two. Thus, the first equation in (3.1) is not linear, though it is of first order. A linear differential equation is said to be homogeneous if all of the terms in the equation contain the unknown function or its derivatives. Thus the first equation in (3.2) is homogeneous, while the second equation is non-homogeneous.

A solution of a differential equation is a function which, when substituted in the differential equation produces an identity. For example, the differential equation

$$\left(\frac{dy}{dx}\right)^2 + x\frac{dy}{dx} - y = 0 \tag{3.3}$$

has two solutions like $y_1(x) = 1 - x$ and $y_2(x) = 2(x + 2)$.

This can be verified easily by finding the derivatives of the functions y_1 and y_2, and putting them in equation (3.3). The serious questions on existence and uniqueness of solutions of differential equations will not be discussed in this book.

Differential equations act as pivot to many physical problems universally. In this chapter, we shall limit our discussions to techniques of solving differential equations. Physical applications of differential equations of first order only will be treated in this chapter while those of second order will be discussed in detail in the next chapter.

I. FIRST ORDER DIFFERENTIAL EQUATIONS

3.2 Method of Separation of Variables

If the equation can be put in the form

$$\phi(y)\,dy = \psi(x)\,dx \tag{3.4}$$

such that the left-hand side of equation (3.4) is a function of y only, and the right-hand side is a function of x only, we call the differential equation *separable*. In such cases, the solution can be obtained by direct integration. Here, we write

$$\int \phi(y)\,dy = \int \psi(x)\,dx + C, \tag{3.5}$$

where C is an arbitrary constant of integration.

EXAMPLE 3.1. *Solve the differential equation* $y\dfrac{dy}{dx} + xy^2 = x$.

Solution: The equation can be written as

$$y\,dy + x\left(y^2 - 1\right) dx = 0, \quad \text{i.e.,} \quad \frac{y\,dy}{y^2 - 1} + x\,dx = 0$$

Upon integration, we get

$$\frac{1}{2}\ln\left(y^2 - 1\right) + \frac{1}{2}x^2 = C, \quad \text{i.e.,} \quad \left(y^2 - 1\right)e^{x^2} = \text{constant}.$$

EXAMPLE 3.2. *Solve the differential equation* $\left(1 + y^2\right)x + \left(1 + x^2\right)\dfrac{dy}{dx} = 0$.

Solution: The equation can be written as

$$\frac{x\,dx}{1 + x^2} + \frac{dy}{1 + y^2} = 0.$$

Integrating, we get

$$\frac{1}{2}\ln\left(1 + x^2\right) + \tan^{-1} y = C_1$$

$$\text{or,} \quad \ln\sqrt{1 + x^2} + \tan^{-1} y = C_1,$$

$$\text{i.e.,} \quad \tan^{-1} y = C_1 - \ln\sqrt{1 + x^2} = \ln\frac{C}{\sqrt{1 + x^2}}$$

$$\text{or,} \quad y = \tan\left(\ln\frac{C}{\sqrt{1 + x^2}}\right).$$

3.3 Equations of General Form

Linear differential equations of first order have the general form

$$\frac{dy}{dx} + Q(x)\, y = R(x). \tag{3.6}$$

This is a non-homogeneous equation. If we put $R(x) = 0$, it becomes homogeneous and at the same time separable. Thus, we write

$$\frac{dy}{y} = -Q(x)\, dx.$$

We get the solution by integration, $\ln y = -\int Q(x)\, dx + C$ which can be written as

$$y = A \exp\left[-\int Q(x)\, dx\right], \quad \text{where } A = e^c.$$

The quantity $\exp\left[\int Q(x)\, dx\right]$ is called *integrating factor.*

Let us now go back to equation (3.6). Multiplying the equation by the integrating factor, we get

$$\left[\frac{dy}{dx} + Q(x)\, y\right] e^{\int Q(x)\, dx} = R(x)e^{\int Q(x)\, dx},$$

$$\text{i.e.,} \quad \frac{d}{dx}\left[ye^{\int Q(x)\, dx}\right] = R(x)e^{\int Q(x)\, dx}.$$

Integrating, we get

$$ye^{\int Q(x)\, dx} = \int \left[R(x)e^{\int Q(x)\, dx} dx\right] + C.$$

$$\therefore \quad y = e^{-\int Q(x)\, dx} \int \left[R(x)e^{\int Q(x)\, dx} dx\right] + Ce^{-\int Q(x)\, dx}. \tag{3.7}$$

This is the general solution of the equation (3.6).

EXAMPLE 3.3. *Solve the differential equation* $x\dfrac{dy}{dx} + 2y = 12x.$

Solution: The equation can be written as $\dfrac{dy}{dx} + \dfrac{2}{x}y = 12.$

Thus, here $Q(x) = \dfrac{2}{x}$, $R(x) = 12.$

Hence, the integrating factor is $e^{\int Q(x)\, dx} = e^{\int \frac{2}{x} dx} = e^{\ln x^2} = x^2.$

Thus, we can write

$$y \cdot x^2 = \int 12x^2 + C = 4x^3 + C. \quad \therefore \quad y = 4x + \frac{C}{x^2}.$$

3.4 Conversion to Separable Form

Sometimes it is possible to convert linear differential equations to separable form. A class of such types of differential equations are written in the form

$$\frac{dy}{dx} = f\left(\frac{y}{x}\right). \tag{3.8}$$

In such cases, we make the substitution $y = zx$, where z is the new variable. Then

$$\frac{dy}{dx} = z + x\frac{dz}{dx}, \quad \frac{y}{x} = z$$

so that equation (3.8) is converted to

$$x\frac{dz}{dx} + z = f(z), \quad \text{i.e.,} \quad \frac{dz}{f(z) - z} = \frac{dx}{x},$$

which is now in separable form and can be integrated.

EXAMPLE 3.4. *Solve the differential equation* $x^3 + y^3 - xy^2\dfrac{dy}{dx} = 0.$

Solution: Dividing the given equation by x^3, we get

$$1 + \left(\frac{y}{x}\right)^3 - \left(\frac{y}{x}\right)^2\frac{dy}{dx} = 0. \tag{1}$$

We define a new variable z, such that $z = \dfrac{y}{x}$, i.e., $y = zx$. Then

$$\frac{dy}{dx} = z + x\frac{dz}{dx}.$$

Substituting these relations into the equation (1), we have

$$1 + z^3 - z^2\left(z + x\frac{dz}{dx}\right) = 0, \quad \text{i.e.,} \quad 1 - xz^2\frac{dz}{dx} = 0, \quad \text{i.e.,} \quad \frac{dx}{x} - z^2 dz = 0.$$

Integrating

$$\ln x - \frac{z^3}{3} = C_1, \quad \text{i.e.,} \quad \frac{z^3}{3} = \ln Cx, \quad C = \text{constant}$$

$$\text{i.e.,} \quad \left(\frac{y}{x}\right)^3 = 3\ln Cx, \quad \text{(putting back original variable)}$$

$$\text{i.e.,} \quad y = x\ln Ax^3, \quad A = C^3 = \text{constant}.$$

EXAMPLE 3.5. *Solve the differential equation* $(x + 2y + 1)\dfrac{dy}{dx} = x + 2y + 4.$

Solution: Let, $x + 2y = z$, so that

$$1 + 2\frac{dy}{dx} = \frac{dz}{dx}, \quad \text{i.e.,} \quad \frac{dy}{dx} = \frac{1}{2}\left(\frac{dz}{dx} - 1\right). \tag{1}$$

Using equation (1), we convert the given differential equation to

$$\frac{1}{2}(z+1)\left(\frac{dz}{dx}-1\right)=z+4, \quad \text{i.e.,} \quad z\frac{dz}{dx}-z+\frac{dz}{dx}-1=2z+8,$$

$$\text{i.e.,} \quad (z+1)\frac{dz}{dx}=3(z+3), \quad \text{i.e.,} \quad \frac{z+1}{z+3}dz=3dx,$$

$$\text{i.e.,} \quad \left(1-\frac{2}{z+3}\right)dz=3dx.$$

Integrating

$$z-2\ln(z+3)=3x+C.$$

Putting expression for z, we get

$$(x+2y)-2\ln(x+2y+3)=3x+C, \quad \text{i.e.,} \quad y=x+\ln(x+2y+3)+C'.$$

3.5 Exact Differential

Let us consider a function $\phi(x,y)$ = constant which is continuous along with its first-order partial derivatives. Then, the total differential of $\phi(x,y)$ is

$$d\phi=\frac{\partial\phi}{\partial x}dx+\frac{\partial\phi}{\partial y}dy=0.$$

This can be expressed as

$$d\phi=M(x,y)\,dx+N(x,y)\,dy=0, \tag{3.9}$$

where

$$M(x,y)=\frac{\partial\phi}{\partial x} \quad \text{and} \quad N(x,y)=\frac{\partial\phi}{\partial y}.$$

Since the function $\phi(x,y)$ has continuous first order derivatives, we can write

$$\frac{\partial^2\phi}{\partial y\partial x}=\frac{\partial^2\phi}{\partial x\partial y}, \quad \text{i.e.,} \quad \frac{\partial M}{\partial y}=\frac{\partial N}{\partial x}.$$

This is the necessary and sufficient condition that expression (3.9) be an exact differential equation. Under this condition, the solution of the differential equation will be of the form $\phi(x,y)$ = constant.

EXAMPLE 3.6. *Solve the differential equation* $x\left(x^2+2y^2\right)dx+y\left(2x^2+y^2\right)dy=0$.

Solution: Here, $M(x,y)=x\left(x^2+2y^2\right)$.

$$\therefore \quad \frac{\partial M}{\partial y}=4xy.$$

Again

$$N(x,y)=y\left(2x^2+y^2\right). \quad \therefore \quad \frac{\partial N}{\partial x}=4xy.$$

Hence

$$\frac{\partial M}{\partial y}=\frac{\partial N}{\partial x}.$$

Thus the given differential equation is exact. So its solution is $\phi(x,y) = C$, where

$$\frac{\partial \phi}{\partial x} = M = x\left(x^2 + 2y^2\right)$$

so that

$$\phi(x,y) = \frac{x^4}{4} + x^2y^2 + f(y). \tag{1}$$

Again

$$\frac{\partial \phi}{\partial y} = N = y\left(2x^2 + y^2\right)$$

so that

$$\phi(x,y) = \frac{y^4}{4} + x^2y^2 + g(x), \tag{2}$$

where, $f(y)$ and $g(x)$ are functions of y and x respectively which appear due to integration of partial derivatives of $\phi(x,y)$ with respect to x and y respectively. For the consistence of two solutions (1) and (2), it is necessary that $f(y) = y^4/4$ and $g(x) = x^4/4$.

Hence, the solution is

$$\phi(x,y) \equiv \frac{x^4}{4} + x^2y^2 + \frac{y^4}{4} = \text{constant}, \quad \text{i.e.,} \quad x^4 + 4x^2y^2 + y^4 = \text{constant}.$$

EXAMPLE 3.7. *Solve the differential equation $xy^2dx + \left(2 + x^2y\right)dy = 0$ with the condition that, $y(1) = 2$.*

Solution: Here, $M(x,y) = xy^2$, $\therefore \dfrac{\partial M}{\partial y} = 2xy$ and $N(x,y) = 2 + x^2y$, $\therefore \dfrac{\partial N}{\partial x} = 2xy.$

Hence

$$\frac{\partial M}{\partial y} = \frac{\partial N}{\partial x}.$$

Thus, the given differential equation is exact.

Hence, its solution is $\phi(x,y) = \text{constant}$, where

$$\frac{\partial \phi}{\partial x} = M(x,y) = xy^2, \quad \text{i.e.,} \quad \phi(x,y) = \frac{x^2y^2}{2} + f(y) \tag{1}$$

and

$$\frac{\partial \phi}{\partial y} = N(x,y) = 2 + x^2y, \quad \text{i.e.,} \quad \phi(x,y) = 2y + \frac{x^2y^2}{2} + g(x). \tag{2}$$

For consistency of equations (1) and (2), $f(y) = 2y$ and $g(x) = 0$.

$$\therefore \quad \phi(x,y) = \frac{x^2y^2}{2} + 2y.$$

Thus, the solution of the differential equation is

$$\frac{x^2y^2}{2} + 2y = C, \quad C = \text{constant}.$$

By the condition given, for $x = 1$, $y = 2$. Thus, $C = 6$. The solution is given by

$$\frac{x^2y^2}{2} + 2y = 6.$$

3.6 Bernoulli's Equation

The non-linear differential equation of the form

$$\frac{dy}{dx} + P(x)\, y = Q(x)\, y^n, \quad (n \neq 1) \tag{3.10}$$

is known as *Bernoulli's equation*. The non-linear equation can be reduced to a linear equation with the substitution

$$z = y^{1-n}. \tag{3.11}$$

Then

$$\frac{dz}{dx} = (1-n)\, y^{-n} \frac{dy}{dx}.$$

Multiplying equation (3.10) by $(1-n)\, y^{-n}$, we get

$$(1-n)\, y^{-n} \frac{dy}{dx} + (1-n)\, y^{1-n} P(x) = (1-n) Q(x).$$

In terms of the variable z, the equation takes the form

$$\frac{dz}{dx} + (1-n) P(x) z = (1-n) Q(x). \tag{3.12}$$

Following the method outlined in Section 3.3, the integrating factor is found to be

$$I(x) = e^{\int (1-n) P(x)\, dx} \tag{3.13}$$

so that the solution of equation (3.12) is given by

$$z = \frac{1}{I(x)} \int (1-n) Q(x) I(x)\, dx + \frac{C}{I(x)}. \tag{3.14}$$

Finally, we are to go back to the original variable y with the help of equation (3.11).

EXAMPLE 3.8. *Solve the differential equation* $y^4 \frac{dy}{dx} + \frac{y^5}{x} = x$.

Solution: The given equation can be written as

$$\frac{dy}{dx} + \frac{y}{x} = x y^{-4}.$$

Thus, here $P(x) = 1/x$, $Q(x) = x$, $n = -4$.
$\therefore$ Integrating factor is

$$I(x) = \exp\left[\int 5 \frac{dx}{x}\right] = e^{5 \ln x} = x^5.$$

The solution of the differential equation is

$$z = \frac{1}{x^5} \int 5x \cdot x^5\, dx + \frac{C}{x^5} = \frac{5}{7} x^2 + \frac{C}{x^5}.$$

Again, $z = y^{1-n} = y^5$, $\quad \therefore\ y = z^{1/5}$.

$$\therefore \quad y = \left(\frac{5}{7} x^2 + \frac{C}{x^5}\right)^{1/5}$$

which is the final solution.

EXAMPLE 3.9. *Solve the differential equation* $\dfrac{dy}{dx} - xy = x^3y^2$.

Solution: Here, $P(x) = -x$, $Q(x) = x^3$, $n = 2$.

$\therefore$ The integrating factor is

$$I(x) = \exp\left[\int (1-2)(-x)\,dx\right] = e^{x^2/2}.$$

$\therefore$ The solution of the differential equation is

$$z = \frac{1}{e^{x^2/2}} \int (1-2)\, x^3 e^{x^2/2} dx + \frac{C}{e^{x^2/2}}. \tag{1}$$

For the integration, put $v = \dfrac{x^2}{2}$, $\quad \therefore\ dv = x\,dx$.

$$I_1 = \int x^3 e^{x^2/2} dx = \int 2ve^v dv = 2\left(ve^v - e^v\right).$$

$$\therefore I_1 = 2e^{x^2/2}\left(\frac{x^2}{2} - 1\right)$$

Hence, from equation (1)

$$z = -2\left(\frac{x^2}{2} - 1\right) + \frac{C}{e^{x^2/2}}.$$

$$\therefore \quad ze^{x^2/2} = -\left(x^2 - 2\right) e^{x^2/2} + C.$$

Since $z = y^{1-n} = y^{-1}$, $\quad \therefore\ y = \dfrac{1}{z}$.

$$\therefore \quad y = \frac{e^{x^2/2}}{(2-x^2)\, e^{x^2/2} + C} = \frac{1}{(2-x^2) + Ce^{-x^2/2}}.$$

3.7 Clairaut's Equation

Equations of the type

$$y = xp + \phi(p), \quad \text{where } p = \frac{dy}{dx} \tag{3.15}$$

are known as *Clairaut's equation*. These are in general non-linear equations but can be reduced to linear form.

Differentiating the equation (3.15) once with respect to x, we get

$$p = x\frac{dp}{dx} + p + \frac{d\phi}{dp} \cdot \frac{dp}{dx}, \quad \text{i.e.,} \quad \left(x + \frac{d\phi}{dp}\right) \frac{dp}{dx} = 0.$$

Thus, if $\dfrac{dp}{dx} = 0$, i.e., $p = \text{constant} = C_1$, we get, $y = C_1 x + C_2$.

This is the general solution of the equation which represents a system of straight lines. When

$$x + \frac{d\phi}{dp} = 0, \quad \text{i.e.,} \quad x = -\frac{d\phi}{dp},$$

the equation (3.15) becomes

$$y = -p\frac{d\phi}{dp} + \phi(p)$$

which gives so-called special solution of the equation.

EXAMPLE 3.10. *Solve the differential equation* $y = xp - \frac{p^2}{4}$, $p = \frac{dy}{dx}$.

Solution: Differentiating the equation with respect to x, we have

$$p = x\frac{dp}{dx} + p - \frac{p}{2}\frac{dp}{dx}, \quad \text{i.e.,} \quad \left(x - \frac{p}{2}\right)\frac{dp}{dx} = 0.$$

Thus, either $\frac{dp}{dx} = 0$, when $p = \text{constant} = C_1$, so that

$$y = C_1 x + C_2, \tag{1}$$

$$\text{or,} \quad x - \frac{p}{2} = 0, \quad \text{i.e.,} \quad 2x = \frac{dy}{dx}, \quad \text{i.e.,} \quad y = x^2 + C_3. \tag{2}$$

Thus, equation (1) gives the general solution and equation (2) gives the special solution.

3.8 Lagrange's Equation

Differential equation of the form

$$y = x\phi(p) + \psi(p), \tag{3.16}$$

where $p = \frac{dy}{dx}$ and $\phi(p)$, $\psi(p)$ are functions of p is known as *Lagrange's equation.* If $\phi(p) = p$, we get Clairaut's equation. Let us suppose that $\phi(p) \neq p$.

Differentiating equation (3.16) with respect to x, we get

$$p = \phi(p) + \{x\phi'(p) + \psi'(p)\}\frac{dp}{dx},$$

$$\text{i.e.,} \quad \{\phi(p) - p\}\frac{dx}{dp} + x\phi'(p) + \psi'(p) = 0,$$

$$\text{i.e.,} \quad \frac{dx}{dp} + \frac{\phi'(p)}{\phi(p) - p}x = -\frac{\psi'(p)}{\phi(p) - p}$$

which can be easily integrated by the previous methods. This will give x in terms of p in the form

$$x = cf(p) + g(p). \tag{3.17}$$

EXAMPLE 3.11. *Solve the differential equation* $y = x(p + a) - \frac{1}{4}(p + a)^2$, $p = \frac{dy}{dx}$.

Solution: Differentiating the given equation with respect to x, we get

$$p = p + a + x\frac{dp}{dx} - \frac{1}{2}(p + a)\frac{dp}{dx}, \quad \text{i.e.,} \quad a + \left\{x - \frac{1}{2}(p + a)\right\}\frac{dp}{dx} = 0,$$

$$\text{i.e.,} \quad \frac{dx}{dp} + \frac{x}{a} = \frac{p + a}{2a}.$$

By the method of Section 3.3, the integrating factor is $I(p) = e^{\int \frac{1}{a} dp} = e^{p/a}$. Hence, the solution is given by

$$x = e^{-p/a}\left[\int \frac{p+a}{2a} e^{p/a} dp\right] + Ce^{-p/a},$$

$$\text{i.e.,}\quad x = \frac{e^{-p/a}}{2a}\cdot a\left[(p+a)e^{p/a} - ae^{p/a}\right] + Ce^{-p/a},$$

$$\text{i.e.,}\quad x = Ce^{-p/a} + \frac{p}{2}. \tag{1}$$

Substituting equation (1) into the given differential equation, we have

$$y = C(p+a)e^{-p/a} + \tfrac{1}{2}p(p+a) - \tfrac{1}{4}(p+a)^2.$$

If, in particular case $C = 0$, then

$$y = \frac{1}{2}p(p+a) - \frac{1}{4}(p+a)^2 = \frac{1}{4}\left(p^2 - a^2\right), \quad \text{i.e.,}\quad p^2 = 4y + a^2,$$

$$\text{i.e.,}\quad \frac{dy}{dx} = \pm\sqrt{4y+a^2}, \quad \text{i.e.,}\quad \int \frac{dy}{\pm\sqrt{4y+a^2}} = \int dx,$$

$$\text{i.e.,}\quad \pm 2\sqrt{4y+a^2} = x, \quad \text{i.e.,}\quad \frac{x^2}{4} - 4y - a^2 = 0.$$

3.9 Homogeneous Equations

A function $f(x, y)$ of variables x and y is said to be homogeneous of degree n, if, for any λ, the identity $f(\lambda x, \lambda y) = \lambda^n f(x, y)$ is true.

A differential equation of the form

$$\frac{dy}{dx} = \frac{\phi(x,y)}{\psi(x,y)},$$

where, $\phi(x, y)$ and $\psi(x, y)$ are homogeneous functions of first degree is called *first-order homogeneous equation*. Such equations can be converted to separable variables form with the substitution $y = zx$, where z is the new variable.

EXAMPLE 3.12. *Solve the differential equation* $\dfrac{dy}{dx} = \dfrac{x^2+y^2}{2xy}$.

Solution: Let, $y = zx$, where z is a new variable. Then

$$\frac{dy}{dx} = z + x\frac{dz}{dx}.$$

Thus, the given equation takes the form

$$z + x\frac{dz}{dx} = \frac{x^2 + (xz)^2}{2x(xz)} = \frac{1+z^2}{2z}, \quad \text{or,}\quad x\frac{dz}{dx} = \frac{1+z^2}{2z} - z = \frac{1-z^2}{2z},$$

$$\text{i.e.,}\quad \frac{2z\,dz}{1-z^2} = \frac{dx}{x}.$$

Integrating, we get

$$\int \frac{2z\,dz}{1-z^2} = \int \frac{dx}{x}, \quad \text{or,} \quad -\ln(1-z^2) = \ln x - \ln C,$$

$$\text{or,} \quad \ln x(1-z^2) = \ln C, \quad \text{or,} \quad x\left(1-\frac{y^2}{x^2}\right) = C.$$

3.10 Non-homogeneous Equations

A differential equation of the form

$$\frac{dy}{dx} = \frac{ax+by+c}{a_1x+b_1y+c_1}, \quad (c, c_1 \neq 0)$$

is a non-homogeneous equation of first order. The equation can be converted into homogeneous form by a change of variable $x = x_1 + h$ and $y = y_1 + k$. Then the differential equation takes the form

$$\frac{dy}{dx} = \frac{ax_1+by_1+ah+bk+c}{a_1x_1+b_1y_1+a_1h+b_1k+c_1}.$$

The values of h and k are selected in such a way that $ah+bk+c=0$ and $a_1h+b_1k+c_1=0$.

Thus, the differential equation is changed to homogeneous form and can be solved by the method of Section 3.9.

EXAMPLE 3.13. *Solve* $\dfrac{dy}{dx} = \dfrac{x+y+1}{x-y-7}$.

Solution: Let, $x = x_1 + h$ and $y = y_1 + k$. Then equation for h and k are $h+k+1=0$ and $h-k-7=0$.

Solving these, we get $h = 3$, $k = -4$.

Thus, the differential equation in new variables becomes

$$\frac{dy_1}{dx_1} = \frac{x_1+y_1}{x_1-y_1}. \tag{1}$$

We put, $y_1 = zx_1$, so that

$$\frac{dy_1}{dx_1} = z + x_1\frac{dz}{dx_1}.$$

Thus, equation (1) becomes

$$z + x_1\frac{dz}{dx_1} = \frac{1+z}{1-z}, \quad \text{or,} \quad x_1\frac{dz}{dx_1} = \frac{1+z}{1-z} - z = \frac{1+z^2}{1-z}.$$

$$\therefore \quad \frac{(1-z)\,dz}{1+z^2} = \frac{dx_1}{x_1}.$$

Integrating

$$\tan^{-1} z - \tfrac{1}{2}\ln(1+z^2) = \ln x_1 + \ln C, \quad \text{or,} \quad \tan^{-1} z = \ln Cx_1\sqrt{1+z^2}$$

$$\text{or,} \quad Cx_1\sqrt{1+z^2} = e^{\tan^{-1} z}.$$

Now, going back to old variables, we have $x = x_1 + 3$, $y = y_1 - 4$, $y_1 = zx_1$, i.e., $x_1 = x - 3$, $y_1 = y + 4$ and $z = y_1/x_1$.

$$\therefore \quad C(x-3)\left\{1+\left(\frac{y+4}{x-3}\right)^2\right\}^{1/2} = e^{\tan^{-1}\left(\frac{y+4}{x-3}\right)}$$

$$\text{or,} \quad C\left\{(x-3)^2+(y+4)^2\right\}^{1/2} = e^{\tan^{-1}\left(\frac{y+4}{x-3}\right)}.$$

· *Note.* This method fails if it happens that $ax+by = k(a_1x+b_1y)$. In such cases we substitute $ax+by = z$ and solve the equation by the method of separation of variables (*see* Example 3.5).

II. PHYSICAL APPLICATIONS

3.11 Pressure Variation for a Static Compressible Fluid

The hydrostatic pressure of a fluid at rest varies only in the vertical direction and is given by the differential equation

$$\frac{dp}{dz} = -\rho g, \tag{3.18}$$

where p is the pressure at height z, ρ is the density of the fluid and g is the acceleration due to gravity.

· For a compressible fluid like air, density is a function of pressure and temperature, and is given by

$$p = \rho RT, \tag{3.19}$$

where R is a constant which is a characteristics of a gas.

Eliminating ρ in the above two equations we obtain the equation of static equilibrium of the air medium in the form

$$\frac{dp}{p} = -\frac{g}{RT}dz. \tag{3.20}$$

Let us consider the temperature-altitude variation in air in the form

$$T = T_0 - \lambda z, \tag{3.21}$$

where T_0 is the temperature at sea level and λ is a constant called *lapse rate.* The relation given in (3.21) is found to be valid for the layer of air about 11 km height from earth's surface known as *troposphere.*

Now, from equation (3.21) by differentiation, we get

$$dT = -\lambda dz. \tag{3.22}$$

Substituting for dz from equation (3.22) into equation (3.20), we get

$$\frac{dp}{p} = \frac{g}{\lambda R}\frac{dT}{T}. \tag{3.23}$$

Integrating

$$\ln p = \frac{g}{\lambda R}\ln T + C, \quad C = \text{constant}. \tag{3.24}$$

If we assume that at sea level $p = p_0$ and $T = T_0$, then using these values in equation (3.24), we have

$$\ln p_0 - \frac{g}{\lambda R} \ln T_0 = C. \tag{3.25}$$

Thus, equation (3.24) becomes

$$\ln \frac{p}{p_0} = \frac{g}{\lambda R} \ln \frac{T}{T_0}, \quad \text{i.e.,} \quad \ln \frac{p}{p_0} = \frac{g}{\lambda R} \ln \left(1 - \frac{\lambda z}{T_0}\right) \quad \text{[using equation (3.21)]}$$

$$\text{or,} \quad \frac{p}{p_0} = \left(1 - \frac{\lambda z}{T_0}\right)^{g/\lambda R} \tag{3.26}$$

From equations (3.19) and (3.21), we can write

$$\frac{p}{p_0} = \frac{\rho T}{\rho_0 T_0}, \quad \rho_0 = \text{density at sea level}$$

so that

$$\frac{\rho}{\rho_0} = \frac{p}{p_0}\frac{T_0}{T} = \frac{p}{p_0}\left(1 - \frac{\lambda z}{T_0}\right)^{-1}.$$

Thus

$$\frac{\rho}{\rho_0} = \left(1 - \frac{\lambda z}{T_0}\right)^{g/\lambda R - 1}. \tag{3.27}$$

3.12 Fluid Flow through an Aperture

Consider a cylindrical tank of diameter D containing non-viscous liquid having a small aperture of diameter d near the bottom. The velocity of fluid flow through the aperture is given by according to Torricelli

$$v = \sqrt{2gh}, \tag{3.28}$$

where h is the height of free surface of the liquid from the aperture, and g is the acceleration due to gravity. We intend to develop a differential equation giving a relationship of the height of the free surface of fluid above the aperture and time. We assume that at time $t = 0$, the height of the fluid surface above the aperture is h_0. Now, during a small period of time dt the volume of fluid escaping from the aperture will be

$$dV = v\left(\frac{\pi d^2}{4}\right) dt = \sqrt{2gh}\left(\frac{\pi d^2}{4}\right) dt. \tag{3.29}$$

If dh be the decrease in fluid level in the tank due to escape of fluid, then the small change in volume of the fluid is

$$dV = -\frac{\pi D^2}{4}\, dh. \tag{3.30}$$

From equations (3.29) and (3.30), we get

$$\frac{dh}{dt} = -\sqrt{2gh}\,\frac{d^2}{D^2}. \tag{3.31}$$

This is an ordinary differential equation of first order and is separable in the form

$$\frac{dh}{\sqrt{h}} = -\sqrt{2g}\,\frac{d^2}{D^2} dt. \tag{3.32}$$

Integrating and using initial condition that $h = h_0$ at $t = 0$, we get

$$h(t) = \left[-\sqrt{\frac{g}{2}} \frac{d^2}{D^2} t + \sqrt{h_0} \right]^2. \tag{3.33}$$

Equation (3.33) gives the height of the fluid surface at any instant of time t.

The time t required to completely empty the tank, (i.e., when $h = 0$) is given by

$$t = \frac{D^2}{d^2} \sqrt{\frac{2h_0}{g}}. \tag{3.34}$$

3.13 Time of Cooling

Let us consider a body of mass m and specific heat s the temperature around which is $\theta_0\,°$C. We assume that the temperature of the body falls to $\theta_2\,°$C from $\theta_1\,°$C in a period of time t due to radiation.

If the temperature of the body changes by a small amount $d\theta$ during a small period of time dt, then by Newton's law of cooling

$$\frac{dQ}{dt} = k(\theta - \theta_0), \tag{3.35}$$

where, Q is the total heat content of the body and k is a constant the value of which depends on the surface area of the body.

Now, $Q = ms\theta$, so that

$$\frac{dQ}{dt} = -ms\frac{d\theta}{dt} \tag{3.36}$$

negative sign has been taken, since θ decreases with increase of time t.

From equations (3.35) and (3.36), we write

$$-ms\frac{d\theta}{dt} = k(\theta - \theta_0), \quad \text{or,} \quad \frac{d\theta}{\theta - \theta_0} = -\frac{k}{ms}dt.$$

Integrating within proper limits

$$-\int_{\theta_1}^{\theta_2} \frac{d\theta}{\theta - \theta_0} = \frac{k}{ms}\int_0^t dt, \quad \text{i.e.,} \quad t = \frac{ms}{k} \ln \frac{\theta_1 - \theta_0}{\theta_2 - \theta_0}. \tag{3.37}$$

3.14 Freezing of Pond

Let us consider the freezing of a pond where the temperature of air is below zero degree centigrade, say $-\theta\,°$C. We assume that after time t, an ice-block of thickness x has been formed over the water surface. Let, after a small time dt, the thickness of ice-block increases by a small amount dx. Then, the increase in mass of the ice-block during time dt is

$$dm = A\rho\, dx, \tag{3.38}$$

where, A = bottom surface area of ice-block and ρ = density of ice at $0\,°$C.

If the latent heat of ice is L, then the quantity of heat taken by ice from water of the pond is

$$dQ = A\rho L\,dx. \tag{3.39}$$

This amount of heat has been conducted through ice-block in a small period of time dt. The temperature at the bottom surface of ice being 0 °C and that on the top surface (i.e., air) being $-\theta$ °C, we write

$$A\rho L\,dx = \frac{kA\theta}{x}dt, \quad \text{i.e.,} \quad \frac{k\theta}{\rho L}dt = x\,dx. \tag{3.40}$$

This is the differential equation governing the rate of ice-formation in the pond.

Integrating equation (3.40)

$$\frac{k\theta}{\rho L}t = \frac{x^2}{2} + C. \tag{3.41}$$

From initial condition, $x = 0$ when $t = 0$.

Hence, from equation (3.41) we find $C = 0$. Thus, the solution becomes

$$\frac{k\theta}{\rho L}t = \frac{x^2}{2}$$

which gives

$$t = \frac{\rho L}{2k\theta}x^2. \tag{3.42}$$

Equation (3.42) determines the time taken for the formation of ice of thickness x.

If, however, the initial conditions are such that at time $t = 0$, $x = x_1$ and at $t = t$, $x = x_2$, then from equation (3.41) we get

$$\left.\begin{aligned} 0 &= \frac{x_1^2}{2} + C \\ \text{and} \quad \frac{k\theta}{\rho L}t &= \frac{x_2^2}{2} + C \end{aligned}\right\}. \tag{3.43}$$

Eliminating C, we find

$$t = \frac{\rho L}{2k\theta}\left(x_2^2 - x_1^2\right). \tag{3.44}$$

3.15 Rate of Chemical Reactions

The rate of a chemical reaction is the rate at which the concentrations of reacting substances vary with time. According to the law of mass action, the rate of any uni-molecular reaction

$$A \longrightarrow \text{products}$$

at any instant of time t is directly proportional to the concentration of A present in the system at that instant of time. Thus

$$-\frac{dC}{dt} = kC, \tag{3.45}$$

where C = concentration of A and the proportionality factor k is called *specific rate* or *specific velocity constant* of the reaction.

Let a = initial concentration of A and x = decrease in concentration of A due to reaction up to time t. Then, equation (3.45) becomes

$$-\frac{d}{dt}(a-x) = k(a-x), \quad \text{i.e.,} \quad \frac{dx}{dt} = k(a-x). \tag{3.46}$$

This is the differential equation giving the rate of a unimolecular reaction in terms of initial concentration and the amount of substance reacted. Initial conditions of the reaction are:

$$\text{at } t=0,\ x=0, \quad \text{and} \quad \text{at } t=t,\ x=x. \tag{3.47}$$

Thus, integrating the equation (3.46), we get

$$\int_0^x \frac{dx}{a-x} = \int_0^t kdt, \quad \text{i.e.,} \quad \ln\frac{a}{a-x} = kt. \tag{3.48}$$

For any bimolecular reaction like

$$A + B \longrightarrow \text{Products}$$

the rate equation can be written as

$$\frac{dx}{dt} = k_1(a-x)(b-x) \tag{3.49}$$

where, a = initial concentration of A,
b = initial concentration of B,
x = decrease in concentration of each A and B due to reaction up to time t
and k_1 = specific rate constant for such a reaction and is equal to the rate when both A and B are of unit concentrations.

Now, from equation (3.49) we get

$$\int \frac{dx}{(a-x)(b-x)} = \int k_1 dt,$$

$$\text{or,} \quad \int \frac{-1}{(a-b)}\left[\frac{1}{a-x} - \frac{1}{b-x}\right] dx = \int k_1 dt,$$

$$\text{i.e.,} \quad \frac{1}{a-b}\left[\ln(a-x) - \ln(b-x)\right] = k_1 t + C,$$

$$\text{or,} \quad \frac{1}{a-b}\ln\frac{a-x}{b-x} = k_1 t + C. \tag{3.50}$$

Since at $t=0$, $x=0$, we get

$$\frac{1}{a-b}\ln\frac{a}{b} = C. \tag{3.51}$$

Thus, equation (3.50) becomes

$$\frac{1}{a-b}\ln\frac{a-x}{b-x} = k_1 t + \frac{1}{a-b}\ln\frac{a}{b},$$

$$\text{or,} \quad \frac{1}{a-b}\ln\frac{(a-x)b}{(b-x)a} = k_1 t, \quad \text{or,} \quad \ln\frac{(a-x)b}{(b-x)a} = (a-b)k_1 t.$$

$$\therefore \quad \frac{(a-x)b}{(b-x)a} = e^{(a-b)k_1 t},$$

$$\text{or,} \quad ab - bx = (ab - ax)\cdot e^{\xi t}, \quad \text{where } \xi = (a-b)k_1,$$

$$\text{or,} \quad x\left(ae^{\xi t} - b\right) = ab\left(e^{\xi t} - 1\right).$$

$$\therefore \quad x = \frac{ab\left(e^{\xi t}-1\right)}{ae^{\xi t} - b}, \quad \text{or,} \quad x = \frac{a\left(1-e^{\xi t}\right)}{\left(1-\frac{a}{b}e^{\xi t}\right)}. \tag{3.52}$$

Equation (3.52) gives the concentration of the product of reactants A and B at any instant of time t.

3.16 Falling Body under Gravity

Let us consider a body of mass m falling freely under the influence of gravity through a medium whose resistance varies as its velocity. We shall discuss its motion. The equation of motion of the body is

$$m\frac{dv}{dt} = mg - \mu v, \quad \mu = \text{constant}, \tag{3.53}$$

$$\text{i.e.,} \quad \frac{dv}{g - kv} = dt, \quad \text{where } k = \frac{\mu}{m}. \tag{3.54}$$

Integrating, we get

$$\ln(g - kv) = -kt + C_1. \tag{3.55}$$

We assume that the body was released from rest, i.e., at $t = 0$, $v = 0$. Then from equation (3.55), we get

$$\ln g = C_1. \tag{3.56}$$

Hence

$$\ln\left(1 - \frac{k}{g}v\right) = -kt, \quad \text{or,} \quad v = \frac{g}{k}\left(1 - e^{-kt}\right). \tag{3.57}$$

The expression (3.57) gives the velocity of the particle at any instant of time. It is clear that if

$$t \longrightarrow \infty, \quad v \longrightarrow \frac{g}{k} = \frac{mg}{\mu},$$

a constant value. This is called *terminal velocity.*

Again, if x be distance fallen during time t, then

$$\frac{dx}{dt} = v = \frac{g}{k}\left(1 - e^{-kt}\right), \quad \text{or,} \quad dx = \frac{g}{k}\left(1 - e^{-kt}\right)dt. \tag{3.58}$$

Integrating, we get

$$x = \frac{g}{k}\left(t + \frac{e^{-kt}}{k}\right) + C_2. \tag{3.59}$$

At $t = 0$, $x = 0$, and hence

$$C_2 = -\frac{g}{k^2}. \tag{3.60}$$

Thus, from equations (3.59) and (3.60), we get

$$x = \frac{g}{k^2}\left(e^{-kt} - 1 + kt\right), \quad \text{or,} \quad x = g\frac{m^2}{\mu^2}\left(e^{\frac{-\mu}{m}t} - 1 + \frac{\mu}{m}t\right). \tag{3.61}$$

This equation (3.61) determines the distance traversed by the body after time t.

3.17 Motion of a Body with Varying Mass

Let us consider a spherical rain-drop falling freely under gravity and receiving in each instant an increase of volume equal to λ times its surface area at that instant. We shall discuss its motion.

Let, at any instant of time, radius of the rain-drop be r and its mass M.

Then, $M = \frac{4}{3}\pi\rho r^3$, $\rho =$ density.

$$\therefore \quad \frac{dM}{dt} = 4\pi\rho r^2 \frac{dr}{dt}. \tag{3.62}$$

By the condition given

$$\frac{dM}{dt} = \lambda \cdot 4\pi r^2. \tag{3.63}$$

From equations (3.62) and (3.63), we get

$$\frac{dr}{dt} = \frac{\lambda}{\rho} = k \text{ (say)}. \tag{3.64}$$

Integrating

$$r = a + kt, \tag{3.65}$$

where $a =$ initial radius of rain-drop.

Thus

$$M(t) = \text{mass at time } t = \frac{4}{3}\pi\rho(a+kt)^3. \tag{3.66}$$

Now, the equation of motion is

$$\frac{d}{dt}\left(M\frac{dx}{dt}\right) = Mg, \tag{3.67}$$

where $x =$ distance traversed in time t.

Thus, from equations (3.66) and (3.67), we can write

$$\frac{d}{dt}\left[(a+kt)^3\frac{dx}{dt}\right] = (a+kt)^3 g.$$

Integrating

$$(a+kt)^3\frac{dx}{dt} = \frac{(a+kt)^4}{4k}g + C_1. \tag{3.68}$$

Now, at $t = 0$, velocity $= 0$. This gives

$$C_1 = -\frac{a^4}{4k}g. \tag{3.69}$$

Hence, from equations (3.68) and (3.69)

$$(a+kt)^3\frac{dx}{dt} = \frac{(a+kt)^4}{4k}g - \frac{a^4}{4k}g, \quad \text{i.e.,} \quad \frac{dx}{dt} = \frac{g}{4k}\left[(a+kt) - \frac{a^4}{(a+kt)^3}\right].$$

Integrating

$$x = \frac{g}{4k^2}\left[\frac{(a+kt)^2}{2} + \frac{a^4}{2(a+kt)^2}\right] + C_2.$$

Again, at $t = 0$, distance fallen = 0.
Hence

$$C_2 = -\frac{g}{4k^2}a^2. \tag{3.70}$$

Hence

$$x = \frac{g}{8k^2}\left[(a+kt)^2 - 2a^2 + \frac{a^4}{(a+kt)^2}\right] = \frac{g}{8k^2}\left[(a+kt) - \frac{a^2}{(a+kt)}\right]^2$$

$$= \frac{gt^2}{8k^2}\left[\frac{2a+kt}{a+kt}\right]^2. \tag{3.71}$$

3.18 Growth of Current with Inductance and Resistance in Series

Let us consider a circuit consisting of resistance R and inductance L connected in series (*Fig 3.1*).

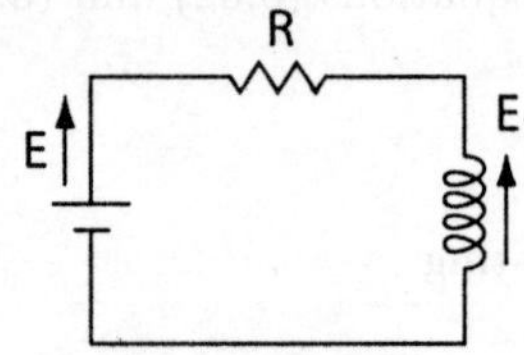

Fig 3.1

Let, at any instant of time t, the amount of current established in the circuit be i. Then induced e.m.f. E_1 would be

$$E_1 = L\frac{di}{dt}. \tag{3.72}$$

The instantaneous e.m.f. equation is

$$E - E_1 = Ri, \quad \text{i.e.,} \quad E - L\frac{di}{dt} = Ri, \quad \text{or,} \quad \frac{di}{dt} + \frac{R}{L}i = \frac{E}{L}. \tag{3.73}$$

Equation (3.73), when solved, will give a relationship between current and time. Applying the method of Section 3.6, so that here

$$Q(x) = \frac{R}{L} \quad \text{and} \quad R(x) = \frac{E}{L}$$

we get the solution

$$i(t) = e^{-Rt/L}\int \frac{E}{L}e^{Rt/L}dt + Ce^{-Rt/L} = \frac{E}{R} + Ce^{-Rt/L}. \tag{3.74}$$

Now, when $t = 0$, $i = 0$. Hence

$$0 = \frac{E}{R} + C, \quad \text{i.e.,} \quad C = -\frac{E}{R}. \tag{3.75}$$

Thus, the solution (3.74) is written in the form

$$i(t) = \frac{E}{R}\left(1 - e^{-Rt/L}\right). \tag{3.76}$$

If i_0 be the steady value of the current, then

$$i_0 = \frac{E}{R}. \tag{3.77}$$

The expression (3.76) with use of relation (3.77) becomes

$$i(t) = i_0\left(1 - e^{-Rt/L}\right). \tag{3.78}$$

This shows that the steady value of the current is reached only when $t \longrightarrow \infty$.
Let us denote

$$\lambda = \frac{L}{R}. \tag{3.79}$$

This λ is called *time constant* of the circuit and determines the rate of current growth in a circuit. Then the expression (3.78) with use of (3.79) can be written as

$$i(t) = i_0\left(1 - e^{-t/\lambda}\right). \tag{3.80}$$

It is obvious from equation (3.80) that the current in the circuit at $t = \lambda$ is given by

$$i(\lambda) = i_0\left(1 - \frac{1}{2.72}\right) = 0.63 i_0. \tag{3.81}$$

Thus, we define the time constant λ as the time when the current in the circuit grows to 63% of final steady value.

Again, from equation (3.80), rate of current growth is given by

$$\frac{di}{dt} = \frac{i_0}{\lambda}e^{-t/\lambda} = \frac{1}{\lambda}(i_0 - i). \tag{3.82}$$

This shows that the growth rate of current depends on the value of λ—greater it is, smaller is the value of current growth rate and vice versa.

III. SECOND ORDER DIFFERENTIAL EQUATIONS

3.19 Incomplete Differential Equations

The second-order differential equations have, in general, the following form:

$$F(x, y, y', y'') = 0. \tag{3.83}$$

If, in this equation, any of the quantities x, y, y' are absent, the differential equation is called *incomplete*. In such cases, the second-order differential equations can be transformed into first-order differential equations by the following methods:

(a) If in the differential equation the variable x is not present, the order of the equation can be lowered by the substitution $y' = p(y)$. Then

$$y'' = \frac{d}{dx}(y') = \frac{dp}{dy}\cdot\frac{dy}{dx} = p'p \tag{3.84}$$

so that the equation (3.83) takes the form

$$F(y, p, p') = 0 \tag{3.85}$$

which is of the first order.

EXAMPLE 3.14. *Solve* $2yy'' = y'^2 + 1$.
Solution: We substitute $y' = p(y)$. Then, $y'' = p'p$.
Thus, the given equation is transformed into a first-order equation

$$2ypp' = p^2 + 1, \tag{1}$$

$$\text{i.e.,} \quad \frac{d}{dy}(p^2) - \frac{p^2}{y} = \frac{1}{y}. \tag{2}$$

Equation is now solved by the method of Section 3.3 and we get

$$p = \pm\sqrt{Cy - 1}. \tag{3}$$

$$\therefore \quad y' = \pm\sqrt{Cy - 1}.$$

Solving we get

$$4(Cy - 1) = C^2(x + C_1).$$

EXAMPLE 3.15. *Solve* $yy'' + (y')^2 = 1$.
Solution: Let, $y' = p(y)$.
Then

$$y'' = \frac{dp}{dx} = \frac{dp}{dy} \cdot \frac{dy}{dx} = p'p.$$

The given differential equation is transformed into

$$yp'p + p^2 = 1, \quad \text{i.e.,} \quad yp\frac{dp}{dy} = 1 - p^2, \quad \text{i.e.,} \quad \frac{p\,dp}{1 - p^2} = \frac{dy}{y}. \tag{1}$$

Integrating, we get

$$-\frac{1}{2}\ln|1 - p^2| = \ln|y| - \ln|C_1|, \quad \text{i.e.,} \quad y = \frac{C_1}{\sqrt{|1 - p^2|}}.$$

This gives

$$\text{either,} \quad 1 - p^2 = \frac{C_1^2}{y^2} \tag{2}$$

$$\text{or,} \quad p^2 - 1 = \frac{C_1^2}{y^2}. \tag{3}$$

Considering equation (2)

$$p = \pm\frac{\sqrt{y^2 - C_1^2}}{y}, \quad \text{or,} \quad \frac{dy}{dx} = \pm\frac{\sqrt{y^2 - C_1^2}}{y}, \quad \text{i.e.,} \quad \pm\frac{y\,dy}{\sqrt{y^2 - C_1^2}} = dx.$$

Integrating, $\pm\sqrt{y^2 - C_1^2} = x + C_2$

$$\therefore \quad y^2 - C_1^2 = (x + C_2)^2. \tag{4}$$

If we now consider equation (3), we get

$$p = \pm\frac{\sqrt{y^2 + C_1^2}}{y}, \quad \text{i.e.,} \quad \pm\frac{y\,dy}{\sqrt{y^2 + C_1^2}} = dx.$$

Integrating, $\pm\sqrt{y^2 + C_1^2} = x + C_3$,

$$\text{i.e.,} \quad y^2 + C_1^2 = (x + C_3)^2 \tag{5}$$

Combining the solutions (4) and (5), we can write

$$y^2 = x^2 + cx + d$$

which is the general solution of the given equation.

(b) If, in the differential equation, the variable 'y' is not present, the order of the equation can be reduced by the substitution

$$\frac{dy}{dx} = p(x) \tag{3.86}$$

so that the equation (3.83) takes the form

$$F(x, p, p') = 0 \tag{3.87}$$

EXAMPLE 3.16. *Solve* $y''(e^x + 1) + y' = 0$.

Solution: With the substitution $\dfrac{dy}{dx} = p(x)$, the given equation changes to

$$\frac{dp}{dx}(e^x + 1) + p = 0, \quad \text{i.e.,} \quad \frac{dp}{p} + \frac{dx}{e^x + 1} = 0.$$

Integrating

$$\ln p - \ln\left(1 + e^{-x}\right) = \ln C_1, \quad \text{i.e.,} \quad p = C_1\left(1 + e^{-x}\right), \quad \text{i.e.,} \quad \frac{dy}{dx} = C_1\left(1 + e^{-x}\right).$$

Integrating

$$y = C_1\left(x - e^{-x}\right) + C_2$$

which is the general solution required.

EXAMPLE 3.17. *Solve* $\left(1 + x^2\right) y'' + 2xy' = 0$.

Solution: With the substitution, $\dfrac{dy}{dx} = p(x)$, the given equation changes to

$$\frac{dp}{dx}\left(1 + x^2\right) + 2xp = 0, \quad \text{i.e.,} \quad \frac{dp}{p} = -\frac{2x\,dx}{1 + x^2}.$$

Integrating

$$\ln p = -\ln\left(1 + x^2\right) + \ln C_1, \qquad \text{i.e.,} \quad p = \frac{C_1}{1 + x^2}.$$

$$\therefore \quad \frac{dy}{dx} = \frac{C_1}{1 + x^2}, \quad \text{i.e.,} \quad dy = \frac{C_1 dx}{1 + x^2}.$$

Integrating

$$y = C_1 \tan^{-1} x + C_2$$

which is the required general solution.

3.20 Linear Differential Equations

Linear differential equations of second order has the general form

$$\frac{d^2y}{dx^2} + P(x)\frac{dy}{dx} + Q(x)\,y = R(x), \tag{3.88}$$

where $P(x)$, $Q(x)$ are coefficient functions and $R(x)$ is the force function. If the function $R(x)$ is zero, the equation is called *homogeneous.*

If y_1 and y_2 are solutions of the homogeneous equation

$$\frac{d^2y}{dx^2} + P(x)\frac{dy}{dx} + Q(x)\,y = 0 \tag{3.89}$$

then

$$y(x) = C_1y_1(x) + C_2y_2(x) \tag{3.90}$$

is also a solution. This is known as *theorem of Superposition Principle.*

The Wronskian, $W(x, y_1, y_2)$ of the two solutions y_1, y_2 is defined as follows:

$$W(x) = \begin{vmatrix} y_1 & y_2 \\ y_1' & y_2' \end{vmatrix} = y_1y_2' - y_2y_1'. \tag{3.91}$$

Now

$$\begin{aligned} W'(x) &= y_1y_2'' + y_1'y_2' - y_2y_1'' - y_1'y_2' = y_1y_2'' - y_2y_1'' \\ &= y_1\left\{-P(x)\,y_2' - Q(x)\,y_2\right\} - y_2\left\{-P(x)\,y_1' - Q(x)\,y_1'\right\} \end{aligned}$$

$[\because\ y_1, y_2$ are solutions of equation (3.89)]

i.e.,

$$W'(x) = -P(x)\left\{y_1y_2' - y_2y_1'\right\} = -P(x)W(x),$$

i.e.,

$$W'(x) + P(x)W(x) = 0. \tag{3.92}$$

This is a first-order differential equation for $W(x)$. Its solution can be written as

$$W(x) = Ce^{-\int P(x)\,dx}. \tag{3.93}$$

We assume that the function $P(x)$ is continuous. Then

$$\int P(x)\,dx$$

will also be continuous, so that

$$e^{\int P(x)\,dx} > 0.$$

Hence, from equation (3.93), we find that

$$\left.\begin{aligned} W &= 0, \quad \text{if and only if} \quad C = 0, \\ W &> 0, \quad \text{if and only if} \quad C > 0 \\ \text{and} \quad W &< 0, \quad \text{if and only if} \quad C < 0 \end{aligned}\right\}. \tag{3.94}$$

If $W(x) \neq 0$, we say that the solutions y_1, y_2 of the equation (3.89) are linearly independent. In such a case, the pair of solutions y_1, y_2 is called a *basic solution set.*

3.21 Homogeneous Linear Equations with Constant Coefficients

The homogeneous linear second-order differential equations with constant coefficients have the following form:

$$\frac{d^2y}{dx^2} + p\frac{dy}{dx} + qy = 0. \tag{3.95}$$

We seek the solutions in the form

$$y = e^{\lambda x}. \tag{3.96}$$

Substituting this solution function into equation (3.95), we find that

$$e^{\lambda x}\left(\lambda^2 + p\lambda + q\right) = 0 \tag{3.97}$$

which shows that $e^{\lambda x}$ is a solution of equation (3.95) only when

$$\lambda^2 + p\lambda + q = 0. \tag{3.98}$$

This is called *characteristic equation.* It has two roots given by

$$\left.\begin{aligned} \lambda_1 &= -\frac{p}{2} + \frac{1}{2}\sqrt{p^2 - 4q} \\ \text{and} \quad \lambda_2 &= -\frac{p}{2} - \frac{1}{2}\sqrt{p^2 - 4q} \end{aligned}\right\}. \tag{3.99}$$

Thus, solutions of equation (3.95) are given by

$$y_1 = e^{\lambda_1 x}, \quad y_2 = e^{\lambda_2 x}. \tag{3.100}$$

Now, the Wronskian of these solutions is

$$W(x) = \begin{vmatrix} e^{\lambda_1 x} & e^{\lambda_2 x} \\ \lambda_1 e^{\lambda_1 x} & \lambda_2 e^{\lambda_2 x} \end{vmatrix} = (\lambda_2 - \lambda_1)e^{(\lambda_1 + \lambda_2)\,x}. \tag{3.101}$$

Thus, the solutions y_1 and y_2 given by equations (3.100) will be linearly independent only when $\lambda_1 \neq \lambda_2$. In this case, the general solution is

$$y(x) = C_1 e^{\lambda_1 x} + C_2 e^{\lambda_2 x}. \tag{3.102}$$

Now, let us consider the quantity

$$\xi = p^2 - 4q. \tag{3.103}$$

If $\xi > 0$, the two roots λ_1, λ_2 are real. In such a ease, the solution (3.102) takes the form

$$y(x) = e^{-px/2}\left\{C_1 e^{\sqrt{\xi}x/2} + C_2 e^{-\sqrt{\xi}x/2}\right\} \tag{3.104}$$

$$\text{or,} \quad y(x) = e^{-px/2}\left\{C \sinh\left(\frac{1}{2}\sqrt{\xi}x\right) + D\cosh\left(\frac{1}{2}\sqrt{\xi}x\right)\right\}. \tag{3.105}$$

If $\xi < 0$, the two roots λ_1, λ_2 are imaginary. Then the general solution (3.102) takes the form

$$y(x) = e^{-px/2}\left\{C_1 e^{\frac{i\sqrt{\xi}x}{2}} + C_2 e^{\frac{-i\sqrt{\xi}x}{2}}\right\} \tag{3.106}$$

$$\text{or,}\quad y(x) = e^{-px/2}\left\{C\sin\frac{\sqrt{\xi}x}{2} + D\cos\frac{\sqrt{\xi}x}{2}\right\} \tag{3.107}$$

If $\xi = 0$, the two roots λ_1, λ_2 are equal, and hence the Wronskian of solutions (3.102) is zero. Thus, two solutions given by equation (3.100) are not linearly independent so that the solution is not acceptable as a general solution. Let us seek the second solution in the form

$$y_2(x) = z(x)e^{\lambda x}. \tag{3.108}$$

Substituting this into equation (3.95), we have

$$e^{\lambda x}\left[\frac{d^2z}{dx^2} + (2\lambda + p)\frac{dz}{dx} + (\lambda^2 + p\lambda + q)\,z\right] = 0. \tag{3.109}$$

Now, coefficient of $z = 0$ by equation (3.98) and coefficient of $\dfrac{dz}{dx} = 0$, since we assumed equal roots of (3.98) which are $\lambda = -p/2$. Thus, from equation (3.109), we get

$$\frac{d^2z}{dx^2} = 0 \tag{3.110}$$

whose solution is $z = x$. Thus, the second solution is given by

$$y_2(x) = xe^{\lambda x} = xe^{-px/2}. \tag{3.111}$$

Hence, the general solution is

$$y(x) = C_1 e^{-px/2} + C_2 x e^{-px/2}. \tag{3.112}$$

EXAMPLE 3.18. *Find a general solution of* $\dfrac{d^2y}{dx^2} - \dfrac{dy}{dx} - 6y = 0$ *in terms of exponentials.*

Solution: We take the solution as $y = e^{\lambda x}$. Then, the characteristic equation is

$$\lambda^2 - \lambda - 6 = 0,$$
$$\text{i.e.,}\quad (\lambda - 3)(\lambda + 2) = 0.$$

The roots of the characteristic equation are $\lambda = 3$ and $\lambda = -2$. Hence, two independent solutions are

$$y_1(x) = e^{3x},\quad y_2(x) = e^{-2x}.$$

Thus, the general solution is

$$y(x) = C_1 e^{3x} + C_2 e^{-2x}.$$

EXAMPLE 3.19. *Find a general solution of* $\dfrac{d^2y}{dx^2} - 4\dfrac{dy}{dx} + 4y = 0.$

Solution: If we assume the solution in the form $y = e^{\lambda x}$, the characteristic equation is found to be

$$\lambda^2 - 4\lambda + 4 = 0,$$
$$\text{i.e.,} \quad (\lambda - 2)^2 = 0.$$

Thus, the roots of characteristic equations are equal and each is equal to $\lambda = 2$. Hence, two independent solutions are

$$y_1(x) = e^{2x}, \quad y_2(x) = xe^{2x}.$$

Thus, the general solution is

$$y(x) = C_1 e^{2x} + C_2 x e^{2x}.$$

EXAMPLE 3.20. *Find a general solution of* $\dfrac{d^2y}{dx^2} + 2\dfrac{dy}{dx} + 10y = 0.$

Solution: Assuming the solution in the form $y = e^{\lambda x}$, we find the characteristic equation as

$$\lambda^2 + 2\lambda + 10 = 0,$$
$$\text{i.e.,} \quad (\lambda + 1)^2 = -9,$$
$$\text{i.e.,} \quad \lambda_1 = -1 + 3i, \ \lambda_2 = -1 - 3i.$$

Hence, two independent solutions are

$$y_1(x) = e^{(-1+3i)\,x}, \quad y_2(x) = e^{(-1-3i)\,x}.$$

Thus, the general solution of the equation is

$$y(x) = e^{-x}\left\{C_1 e^{3ix} + C_2 e^{-3ix}\right\} = e^{-x}\left\{C \sin 3x + D \cos 3x\right\}.$$

3.22 Non-homogeneous Linear Equations with Constant Coefficients

Non-homogeneous linear equations with constant coefficients are of the type

$$\frac{d^2y}{dx^2} + p\frac{dy}{dx} + qy = r(x). \tag{3.113}$$

The general solution of these equations is

$$y = y_c + y_p, \tag{3.114}$$

where y_c = complementary solution of the corresponding homogeneous equation, i.e., when $r(x) = 0$ and y_p = particular solution of equation (3.113).

We have already discussed how to find the complementary solution and so here we shall see how to find the particular solution, y_p.

There are many methods available for finding y_p, but here we shall discuss the method of *undetermined coefficients.* The nature of particular solution y_p will obviously depend on the nature of the function $r(x)$ in equation (3.113). We consider below different cases.

Case I. *If $r(x)$ is a polynomial of degree n and zero is not a root of the characteristic equation, then y_p is taken in the form*

$$y_p = a_0 + a_1 x + \cdots + a_n x^n. \tag{3.115}$$

If zero is a single root of the characteristic equation, then

$$y_p = x(a_0 + a_1 x + \cdots + a_n x^n). \tag{3.116}$$

Case II. *If $r(x)$ is an exponential function of the form $Ce^{\lambda x}$, and λ is not a solution of the characteristic equation, then*

$$y_p = Ae^{\lambda x}. \tag{3.117}$$

If λ is a single root of the characteristic equation, then

$$y_p = Axe^{\lambda x}. \tag{3.118}$$

If λ is a double root of the characteristic equation, then

$$y_p = Ax^2 e^{\lambda x}. \tag{3.119}$$

Case III. *If $r(x)$ is a sine or cosine function in the form $C \sin \lambda x$ or $D \cos \lambda x$, and $i\lambda$ is not a root of the characteristic equation, then*

$$y_p = A \cos \lambda x + B \sin \lambda x. \tag{3.120}$$

If $i\lambda$ is a root of the characteristic equation, then

$$y_p = x(A \cos \lambda x + B \sin \lambda x). \tag{3.121}$$

EXAMPLE 3.21. *Find a general solution of* $\dfrac{d^2y}{dx^2} + 4\dfrac{dy}{dx} + 4y = x^2 + x + 4.$

Solution: The characteristic equation corresponding to the homogeneous equation

$$\frac{d^2y}{dx^2} + 4\frac{dy}{dx} + 4y = 0$$

is $\lambda^2 + 4\lambda + 4 = 0$, i.e., $\lambda = -2, -2$.

Hence, the complementary solution is

$$y_c = C_1 e^{-2x} + C_2 x e^{-2x}. \tag{1}$$

The particular integral may be taken in the form

$$y_p = a_0 + a_1 x + a_2 x^2. \tag{2}$$

Substituting expression (2) into the original equation, we get

$$2a_2 + 4(a_1 + 2a_2 x) + 4(a_0 + a_1 x + a_2 x^2) = x^2 + x + 4.$$

Comparing coefficients of various powers of x, we have

$$x^0 \longrightarrow (2a_2 + 4a_1 + 4a_0) = 4$$
$$x \longrightarrow (8a_2 + 4a_1) = 1$$
$$x^2 \longrightarrow 4a_2 = 1$$

Solving these we get $a_2 = \frac{1}{4}$, $a_1 = -\frac{1}{4}$, $a_0 = \frac{9}{8}$.
Hence

$$y_p = \frac{x^2}{4} - \frac{x}{4} + \frac{9}{8}.$$

Thus, the general solution of the given equation is

$$y = C_1 e^{-2x} + C_2 x e^{-2x} + \frac{x^2}{4} - \frac{x}{4} + \frac{9}{8}.$$

EXAMPLE 3.22. *Find a general solution of* $\dfrac{d^2y}{dx^2} - y = e^x$.

Solution: For homogeneous equation, the characteristic equation is

$$\lambda^2 - 1 = 0, \quad \text{i.e.,} \quad \lambda = 1, -1.$$

Hence, the complementary solution is

$$y_c = C_1 e^x + C_2 e^{-x}.$$

Since $\lambda = 1$ is a root of the characteristic equation, we seek the particular integral in the form

$$y_p = Axe^x.$$

Substituting this into the original differential equation, we get

$$A\,(xe^x + 2e^x - xe^x) = e^x. \quad \therefore \quad A = \tfrac{1}{2}.$$

Hence, particular integral is

$$y_p = \tfrac{1}{2} x e^x.$$

Thus, the general solution of the given equation is

$$y = C_1 e^x + C_2 e^{-x} + \tfrac{1}{2} x e^x.$$

EXAMPLE 3.23. *Solve the initial value problem* $\dfrac{d^2y}{dx^2} + 4y = 2\sin x$, *given* $y(0) = 1$, $\dfrac{dy}{dx}(0) = 0$.

Solution: The characteristic equation for the solution $y = e^{\lambda x}$ is

$$\lambda^2 + 4 = 0, \quad \text{i.e.,} \quad \lambda = 2i, -2i.$$

Hence, the solution of the homogeneous equation is

$$y_c = C_1 \cos 2x + C_2 \sin 2x. \tag{1}$$

We assume the particular solution

$$y_p = A \sin x + B \cos x. \tag{2}$$

Substituting this into original differential equation, we get

$$(-A \sin x - B \cos x) + 4(A \sin x + B \cos x) = 2 \sin x.$$

Comparing coefficients of $\sin x$, $\cos x$ on both sides, we find

$$A = \tfrac{2}{3}, \quad B = 0.$$

Hence, the particular solution is

$$y_p = \tfrac{2}{3} \sin x.$$

Thus, the general solution of the given differential equation is

$$y = C_1 \cos 2x + C_2 \sin 2x + \frac{2}{3} \sin x.$$

$$\therefore \quad \frac{dy}{dx} = -2C_1 \sin 2x + 2C_2 \cos 2x + \frac{2}{3} \cos x.$$

Using the given initial conditions

$$y(0) = C_1 = 1, \qquad \text{i.e.,} \quad C_1 = 1$$

$$\text{and} \quad \frac{dy}{dx}(0) = 2C_2 + \tfrac{2}{3} = 0, \qquad \text{i.e.,} \quad C_2 = -\tfrac{1}{3}.$$

Hence, the specific solution of the initial value problem is

$$y = \cos 2x - \tfrac{1}{3} \sin 2x + \tfrac{2}{3} \sin x.$$

3.23 Second-order Homogeneous Equations with Variable Coefficients

Second-order homogeneous linear differential equations with variable coefficients are of the general form

$$\frac{d^2y}{dx^2} + p(x)\frac{dy}{dx} + q(x)\, y = 0. \tag{3.122}$$

The most powerful method of solving these types of equations is the Power Series Method which will be discussed later in this chapter. Here, we shall explore other methods of solving the linear second-order homogeneous differential equations.

(a) Transforming to Normal Form

If the coefficient $p(x)$ of $\frac{dy}{dx}$ is absent, the differential equation (3.122) is said to be of normal form. The original equation (3.122) can be transformed to normal form with the substitution

$$y(x) = z(x) \exp\left\{-\tfrac{1}{2} \int p(x)\, dx\right\}. \tag{3.123}$$

This can be easily verified by setting

$$y(x) = z(x) f(x) \tag{3.124}$$

and substituting into equation (3.122) when we shall get

$$f z'' + (2f' + pf)\, z' + (f'' + pf' + qf)\, z = 0. \tag{3.125}$$

In this equation, the coefficient of $z' = 0$, if

$$2f' + pf = 0, \quad \text{i.e.,} \quad f' = -\tfrac{1}{2} pf. \tag{3.126}$$

Equation (3.126) when solved determines the function $f(x)$ in the form

$$f(x) = \exp\left\{-\tfrac{1}{2}\int p(x)\,dx\right\}. \tag{3.127}$$

Thus, the transformation (3.123) is established. Now, equation (3.125), when coefficient of z' is zero, is reduced to

$$fz'' + (f'' + pf' + qf)\,z = 0. \tag{3.128}$$

Differentiating equation (3.126), we get

$$\begin{aligned} 2f'' &= -pf' - fp', \\ \text{i.e.,}\quad f'' &= -\tfrac{1}{2}\,(pf' + fp') \\ &= -\tfrac{1}{2}\left\{p\left(-\tfrac{1}{2}pf\right) + fp'\right\} \quad \text{[by using equation (3.126)]} \\ &= \tfrac{1}{4}p^2 f - \tfrac{1}{2}fp'. \end{aligned} \tag{3.129}$$

Substituting expressions for f'' and f' into equation (3.128), we get

$$z'' + R(x)z = 0, \tag{3.130}$$

where

$$R(x) = q(x) - \tfrac{1}{2}p'(x) - \tfrac{1}{4}p^2(x). \tag{3.131}$$

Thus, the differential equation (3.130) is the normal form of the original equation (3.122) in a new variable z. If the function $R(x)$ happens to be a constant, then the equation (3.130) can be easily solved by previous methods. If the function $R(x)$ is not constant, other possible transformations may be sought for to convert the equation (3.130) with constant coefficient.

EXAMPLE 3.24. *Find a general solution of* $xy'' - 2(x-1)\,y' + 2(x-1)\,y = 0$ *by converting to its normal form.*

Solution: Here, $p(x) = -\dfrac{2(x-1)}{x}$, $q(x) = \dfrac{2(x-1)}{x}$.

$$\begin{aligned} \therefore\quad R(x) &= \frac{2(x-1)}{x} - \frac{2}{2}\left(\frac{1}{x^2}\right) - \frac{4}{4}\left(1 - \frac{1}{x}\right)^2 \\ &= 2\left(1 - \frac{1}{x}\right) - \frac{1}{x^2} - \left(1 - \frac{2}{x} + \frac{1}{x^2}\right) = 1. \end{aligned}$$

Hence, the normal form of the given equation is $z'' + z = 0$ whose general solution is

$$z(x) = C_1 \cos x + C_2 \sin x.$$

Now, using equation (3.127)

$$f(x) = \exp\left\{-\frac{1}{2}\int p(x)\,dx\right\} = \exp\left\{-\frac{1}{2}\int -2\left(1 - \frac{1}{x}\right)dx\right\} = e^{x - \ln x} = e^x/x.$$

Thus, the general solution of the original equation is

$$y = \frac{e^x}{x}(C_1 \cos x + C_2 \sin x).$$

(b) Changing Independent Variable

It may sometimes be convenient to change independent variable x to a new variable z for reducing the equation (3.122) into normal form. Then, by differentiation

$$\left.\begin{aligned} \frac{dy}{dx} &= \frac{dy}{dz}\cdot\frac{dz}{dx} \\ \text{and}\quad \frac{d^2y}{dx^2} &= \frac{d^2y}{dz^2}\left(\frac{dz}{dx}\right)^2 + \frac{dy}{dz}\cdot\frac{d^2z}{dx^2} \end{aligned}\right\}. \tag{3.132}$$

Substituting these expressions into the original equation (3.122) we get after some rearrangement

$$\frac{d^2y}{dz^2} + P\frac{dy}{dz} + Qy = 0, \tag{3.133}$$

where

$$P = \frac{\dfrac{d^2z}{dx^2} + p(x)\dfrac{dz}{dx}}{\left(\dfrac{dz}{dx}\right)^2} \tag{3.134}$$

and

$$Q = \frac{q(x)}{\left(\dfrac{dz}{dx}\right)^2}. \tag{3.135}$$

We select z in such a way that

$$\frac{d^2z}{dx^2} + p(x)\frac{dz}{dx} = 0, \tag{3.136}$$

i.e.,

$$z = \int \exp\left[-\int p(x)\,dx\right] dx. \tag{3.137}$$

In that case, equation (3.133) reduces to

$$\frac{d^2y}{dz^2} + Qy = 0 \tag{3.138}$$

which is a normal form.

Now, if Q happens to be a constant, we can solve the equation (3.138) by previous methods. If however Q is not a constant, we can try some other transformations to convert the equation (3.138) with constant Q.

Case I. *As an example, let us suppose that*

$$Q = \frac{A}{z^2}. \tag{3.139}$$

In this case, we make another substitution

$$\lambda = \ln z. \tag{3.140}$$

Then, the equation

$$\frac{d^2y}{dz^2}+\frac{A}{z^2}y=0$$

is transformed into

$$\frac{d^2y}{d\lambda^2}-\frac{dy}{d\lambda}+Ay=0 \tag{3.141}$$

which has only constant coefficients and hence can be solved easily by previous methods.

Case II. *Again, let us suppose that P and Q in equation* (3.133) *have the following forms:*

$$P=\frac{A}{z},\quad Q=\frac{B}{z^2}. \tag{3.142}$$

Then the equation (3.133) becomes

$$\frac{d^2y}{dz^2}+\frac{A}{z}\frac{dy}{dz}+\frac{B}{z^2}y=0. \tag{3.143}$$

We make the same substitution as in equation (3.140) and find

$$\frac{dy}{dz}=\frac{1}{z}\frac{dy}{d\lambda}\quad\text{and}\quad\frac{d^2y}{dz^2}=\frac{1}{z^2}\left(\frac{d^2y}{d\lambda^2}-\frac{dy}{d\lambda}\right). \tag{3.144}$$

Substituting these derivatives into equation (3.143), and making rearrangements, we get

$$\frac{d^2y}{d\lambda^2}+(A-1)\frac{dy}{d\lambda}+By=0. \tag{3.145}$$

This is an ordinary differential equation with constant coefficients and hence can be solved by previous methods.

Let us consider two examples given below.

EXAMPLE 3.25. *Solve:* $\dfrac{d^2y}{dx^2}+\dfrac{2}{x}\dfrac{dy}{dx}+\dfrac{a^2}{x^4}y=0.$

Solution: We substitute, $z=\dfrac{a}{x}$.

Then

$$\frac{dz}{dx}=-\frac{a}{x^2},\quad\frac{d^2z}{dx^2}=\frac{2a}{x^3}.$$

Using equation (3.134), we find $P=0$, $Q=1$.

Thus, the given differential equation becomes

$$\frac{d^2y}{dz^2}+y=0.$$

Its solution is

$$y=C_1\cos z+C_2\sin z.$$

Putting the value of $z=\dfrac{a}{x}$, we get

$$y=C_1\cos\left(\frac{a}{x}\right)+C_2\sin\left(\frac{a}{x}\right).$$

EXAMPLE 3.26. *Solve* $\dfrac{d^2y}{dx^2}+\dfrac{3}{x}\dfrac{dy}{dx}+\dfrac{5}{x^2}y=0.$

Solution: Let us substitute, $z=\ln x$.

Then

$$\frac{dz}{dx}=\frac{1}{x},\quad \frac{d^2z}{dx^2}=-\frac{1}{x^2}.$$

Using equations (3.134) and (3.135), we find

$$P=\frac{-\dfrac{1}{x^2}+\dfrac{3}{x^2}}{\dfrac{1}{x^2}}=2\quad\text{and}\quad Q=\frac{\dfrac{5}{x^2}}{\dfrac{1}{x^2}}=5.$$

Hence, the given equation is transformed into

$$\frac{d^2y}{dz^2}+2\frac{dy}{dz}+5y=0.$$

Its solution is

$$y=e^{-z}(C_1\cos 2z+C_2\sin 2z).$$

Hence, the solution in terms of x is

$$y=\frac{1}{x}\left\{C_1\cos(2\ln x)+C_2\sin(2\ln x)\right\}.$$

Case III. *Let us now consider the case when Q in equation* (3.135) *is unity. Then, we have*

$$\left(\frac{dz}{dx}\right)^2=q(x),\quad i.e.,\quad \frac{dz}{dx}=\sqrt{q(x)}.$$

This gives z in terms of x for the required transformation. Thus

$$z=\int\sqrt{q(x)}. \tag{3.146}$$

In that case, expression for P in equation (3.134) can be found. Here

$$\frac{dz}{dx}=\sqrt{q(x)},\quad\therefore\quad \frac{d^2z}{dx^2}=\frac{1}{2\sqrt{q}}\cdot q'.$$

$$\therefore\quad P=\frac{\frac{q'}{2\sqrt{q}}+p\sqrt{q}}{q},\quad\text{i.e.,}\quad P=\frac{q'+2pq}{2q^{3/2}}. \tag{3.147}$$

Hence, if P is a constant quantity, the original differential equation (3.122) is transformed into a linear second-order differential equation with constant coefficients (with coefficient of y equal to unity).

EXAMPLE 3.27. *Solve* $\dfrac{d^2y}{dx^2}-\dfrac{3}{x}\dfrac{dy}{dx}+16x^6y=0.$

Solution: Here, $p(x) = -\frac{3}{x}$, $q(x) = 16x^6$.
We substitute

$$z = \int \sqrt{q(x)} = \int 4x^3 = x^4.$$

We find P in equation (3.147). Thus

$$P = \frac{96x^5 - 96x^5}{2\left(16x^6\right)^{3/2}} = 0.$$

Hence, the given equation is transformed into

$$\frac{d^2y}{dz^2} + y = 0.$$

Its solution is

$$y = C_1 \cos z + C_2 \sin z.$$

In terms of old variable x, we get

$$y = C_1 \cos\left(x^4\right) + C_2 \sin\left(x^4\right).$$

EXAMPLE 3.28. *Solve* $\frac{d^2y}{dx^2} + \tan x \frac{dy}{dx} + \left(\cos^2 x\right) y = 0.$

Solution: Here, $p(x) = \tan x$, $q(x) = \cos^2 x$.
We substitute

$$z = \int \sqrt{q(x)} = \int \cos x = \sin x.$$

Then, value of P in equation (3.147) is

$$P = \frac{-2\cos x \sin x + 2\tan x \cdot \cos^2 x}{2\left(\cos^2 x\right)^{3/2}} = 0.$$

Hence, the given equation is transformed into

$$\frac{d^2y}{dz^2} + y = 0$$

Its solution is

$$y = C_1 \cos z + C_2 \sin z.$$

Thus the solution of the given equation in terms of old variable x is

$$y = C_1 \cos(\sin x) + C_2 \sin(\sin x).$$

(c) Knowing One of the Complementary Solutions

Let, $y = y_1$ be one of the complementary solutions of the given second-order linear differential equation (3.122). This complementary solution may be obtained by inspection. Here, for the complete solution of the equation, we make the substitution

$$y = z(x)\, y_1(x). \tag{3.148}$$

Then the equation (3.122) takes the form

$$\frac{d^2z}{dx^2} + P(x)\frac{dz}{dx} + Q(x)z = 0, \tag{3.149}$$

where

$$P(x) = p(x) + \frac{2}{y_1}\frac{dy_1}{dx} \tag{3.150}$$

$$\text{and} \quad Q(x) = \frac{1}{y_1}\left(\frac{d^2y_1}{dx^2} + p\frac{dy_1}{dx} + q\right).$$

Since y_1 is a complementary solution of equation (3.122), $Q(x) = 0$. Hence, equation (3.149) is reduced to

$$\frac{d^2z}{dx^2} + P(x)\frac{dz}{dx} = 0. \tag{3.151}$$

In order to solve this equation, we substitute

$$\xi(x) = \frac{dz}{dx}. \tag{3.152}$$

Then equation (3.151) is reduced to a linear first-order differential equation

$$\frac{d\xi}{dx} + P(x)\xi = 0. \tag{3.153}$$

Integrating, we get

$$\xi = e^{-\int P(x)\,dx}. \tag{3.154}$$

Using equation (3.152), we find

$$z = \int \left[e^{-\int P(x)\,dx}\right] dx + C_1.$$

Hence, we can find the complete solution with the use of equation (3.148).

EXAMPLE 3.29. *Solve* $\dfrac{d^2y}{dx^2} - \dfrac{3}{x}\dfrac{dy}{dx} + \dfrac{4}{x^2}y = 0.$

Solution: By inspection, we find that $y_1 = x^2$ is a complementary solution of the given equation. Hence, we substitute

$$y = zx^2.$$

Then

$$\frac{dy}{dx} = 2zx + x^2\frac{dz}{dx}$$

$$\frac{d^2y}{dx^2} = 2z + 2x\frac{dz}{dx} + 2x\frac{dz}{dx} + x^2\frac{d^2z}{dx^2}.$$

Putting these derivatives into the given equation, we get

$$\left(x^2\frac{d^2z}{dx^2} + 4x\frac{dz}{dx} + 2z\right) - \frac{3}{x}\left(2zx + x^2\frac{dz}{dx}\right) + \frac{4}{x^2}\left(zx^2\right) = 0,$$

$$\text{or,} \quad x^2\frac{d^2z}{dx^2} + x\frac{dz}{dx} = 0, \quad \text{i.e.,} \quad \frac{d^2z}{dx^2} + \frac{1}{x}\frac{dz}{dx} = 0. \tag{1}$$

Let, $\xi = \dfrac{dz}{dx}$; then the equation (1) becomes

$$\frac{d\xi}{dx} + \frac{1}{x}\xi = 0.$$

Integrating

$$\ln(\xi x) = \ln C_1, \quad \text{i.e.,} \quad \xi x = C_1, \quad \text{or,} \quad \xi = \frac{dz}{dx} = \frac{C_1}{x}.$$

Integrating, $z = C_1 \ln x + C_2$.
Hence

$$y = zx^2 = x^2 \left(C_1 \ln x + C_2\right).$$

3.24 Riccati Equation

A first-order non-linear differential equation of the form

$$\frac{dy}{dx} + p(x)\, y^2 + q(x)\, y + r(x) = 0 \tag{3.155}$$

is called *Riccati equation.* This equation has got much importance in particle dynamics.

The linear homogeneous second-order differential equation has got an interesting property that it can always be converted into an equation of Riccati type with a substitution of $y' = yz$. Because, in that case we have

$$y'' = yz' + zy' = yz' + z(yz) = yz' + yz^2$$

which, upon putting into the equation (3.122) produces the first-order equation

$$\begin{aligned} &(yz' + yz^2) + p(x)\, yz + q(x)\, y = 0, \\ \text{i.e.,} \quad & z' + z^2 + p(x)z + q(x) = 0. \end{aligned}$$

This is an equation of Riccati type.

We are, however, more interested in reducing Riccati's equation to a homogeneous second-order linear differential equation, so that its solution can be found by one of the previous methods. This can be achieved by the substitution

$$y = \frac{1}{p(x)z(x)}\frac{dz}{dx}. \tag{3.156}$$

Then

$$y' = \frac{z''}{pz} - \frac{z'}{pz}\left(\frac{p'}{p} + \frac{z'}{z}\right). \tag{3.157}$$

Putting these derivatives into Riccati equation (3.155), we find, after simplification, that it is reduced to

$$z'' + \left(q - \frac{p'}{p}\right) z' + prz = 0. \tag{3.158}$$

This is a linear homogeneous second-order differential equation which can be solved by previous method.

EXAMPLE 3.30. *Solve* $\dfrac{dy}{dx} + y^2 - \dfrac{2}{x}y + \dfrac{2}{x^2} = 0$ *by converting it to a second-order equation.*

Solution: Comparing this equation with Riccati equation, we find

$$p(x) = 1, \quad q(x) = -\frac{2}{x}, \quad r(x) = \frac{2}{x^2}.$$

We substitute

$$y = \frac{1}{p(x)z(x)}\frac{dz}{dx} = \frac{1}{z}\frac{dz}{dx}. \tag{1}$$

Then the given equation becomes

$$\frac{d^2z}{dx^2} - \frac{2}{x}\frac{dz}{dx} + \frac{2}{x^2}z = 0. \tag{2}$$

We find, by inspection, that $z = x$ is a complementary solution of this equation. Hence, we substitute, $z = vx$. Then equation (2) is reduced to

$$x\frac{d^2v}{dx^2} = 0, \quad \text{i.e.,} \quad \frac{d^2v}{dx^2} = 0.$$

Its solution is

$$v(x) = C_1x + C_2.$$

Hence

$$z = v(x) \cdot x = C_1x^2 + C_2x.$$

Thus, by the substitution in equation (1), we have

$$y = \frac{1}{z}\frac{dz}{dx} = \frac{2C_1x + C_2}{(C_1x + C_2)x}$$

which is the required solution.

3.25 Cauchy-Euler Equation

The class of second-order homogeneous linear differential equations with variable coefficients denoted by

$$x^2\frac{d^2y}{dx^2} + ax\frac{dy}{dx} + by = 0, \tag{3.159}$$

where a, b are constants, is called *Cauchy-Euler equation* of second order. We accept the solutions of this equation in the form

$$y(x) = x^\lambda. \tag{3.160}$$

Substituting this function into the equation (3.159), we obtain a characteristic equation for λ in the form

$$\lambda(\lambda - 1) + a\lambda + b = 0,$$

$$\text{i.e.,} \quad \lambda^2 + (a-1)\lambda + b = 0. \tag{3.161}$$

If this equation has got two distinct roots λ_1, λ_2, then the two linearly independent solutions are

$$y_1 = x^{\lambda_1} \quad \text{and} \quad y_2 = x^{\lambda_2}. \tag{3.162}$$

Hence, the general solution in this case is

$$y = C_1 x^{\lambda_1} + C_2 x^{\lambda_2} \tag{3.163}$$

provided $x = 0$ does not lie within interval of solution.

If, however

$$(a-1)^2 = 4b, \quad \text{i.e.,} \quad b = \left(\frac{a-1}{2}\right)^2 \tag{3.164}$$

roots of equation (3.161) will be equal. In such a case, we assume that $y_1 = x^\lambda$ is one of the solutions, and the other solution is taken in the form

$$y_2 = z(x)\, y_1 = x^\lambda \cdot z(x). \tag{3.165}$$

Substituting (3.165) into equation (3.159) and making rearrangements, we get

$$x^2 z'' + x(2\lambda + a)z' + [\lambda(\lambda-1) + a\lambda + b]z = 0.$$

Now, coefficient of z is zero by virtue of equation (3.161) and coefficient of z' is unity, since for equal roots $\lambda = -(a-1)/2$, so we find

$$x^2 z'' + xz' = 0,$$

$$\text{i.e.,} \quad xz'' + z' = 0. \tag{3.166}$$

Assuming $p = z'$, we get

$$xp' + p = 0$$

which gives, $p = \dfrac{1}{x}$, so that

$$z = |\ln(x)|. \tag{3.167}$$

Thus, the general solution for the double root is

$$y(x) = [C_1 + C_2 |\ln(x)|]\, |x|^\lambda \tag{3.168}$$

provided $x = 0$ does not remain within the region of the solution.

EXAMPLE 3.31. *Solve* $x^2 \dfrac{d^2y}{dx^2} + 9x\dfrac{dy}{dx} + 12y = 0.$

Solution: Assuming a solution, $y = x^\lambda$, we find the characteristic equation

$$\lambda^2 + 8\lambda + 12 = 0.$$

Two roots are $\lambda_1 = -2$, $\lambda_2 = -6$.

Hence, the general solution is

$$y(x) = C_1 e^{-2x} + C_2 e^{-6x}.$$

Note. The Cauchy-Euler equation can be transformed into a linear second-order equation with constant coefficient by changing the variable in the form

$$x = e^z, \quad \text{i.e.,} \quad z = \ln x.$$

The transformed equation in this case becomes

$$\frac{d^2y}{dz^2} + (a-1)\frac{dy}{dz} + by = 0.$$

3.26 Method of Variation of Parameters

Let us consider second-order non-homogeneous linear differential equations of the type

$$\frac{d^2y}{dx^2} + p(x)\frac{dy}{dx} + q(x)\,y = r(x), \tag{3.169}$$

where the forcing function $r(x)$ may be a continuous function of any type.

We shall discuss the method of variation of parameters due to Lagrange to find the particular solution of equation (3.169).

Let, $y_1(x)$ and $y_2(x)$ are two linearly independent solutions of the homogeneous equation

$$\frac{d^2y}{dx^2} + p(x)\frac{dy}{dx} + q(x)\,y = 0 \tag{3.170}$$

so that, the general solution of homogeneous equation is

$$y = C_1y_1(x) + C_2y_2(x). \tag{3.171}$$

Let, $y_p(x)$ be the particular solution of (3.169) corresponding to forcing function $r(x)$. Then, the general solution of equation is given by

$$y = C_1y_1(x) + C_2y_2(x) + y_p(x). \tag{3.172}$$

We now assume that

$$y_p(x) = y_1(x)z_1(x) + y_2(x)z_2(x), \tag{3.173}$$

where $z_1(x)$, $z_2(x)$ are unknown functions to be determined.

Differentiating equation (3.173), we get

$$y_p' = (z_1y_1' + z_2y_2') + (y_1z_1' + y_2z_2'). \tag{3.174}$$

We seek the functions z_1 and z_2 in such a way that

$$y_1z_1' + y_2z_2' = 0. \tag{3.175}$$

Then from equation (3.174), we have

$$\frac{dy_p}{dx} = z_1\frac{dy_1}{dx} + z_2\frac{dy_2}{dx}. \tag{3.176}$$

Differentiating this equation once more

$$\frac{d^2y_p}{dx^2} = z_1\frac{d^2y_1}{dx^2} + z_2\frac{d^2y_2}{dx^2} + \frac{dz_1}{dx}\cdot\frac{dy_1}{dx} + \frac{dz_2}{dx}\cdot\frac{dy_2}{dx}. \tag{3.177}$$

We now substitute this into equation (3.169) and find that

$$z_1\left(\frac{d^2y_1}{dx^2} + p\frac{dy_1}{dx} + qy_1\right) + z_2\left(\frac{d^2y_2}{dx^2} + p\frac{dy_2}{dx} + qy_2\right) + \frac{dz_1}{dx}\cdot\frac{dy_1}{dx} + \frac{dz_2}{dx}\cdot\frac{dy_2}{dx} = r(x). \tag{3.178}$$

Expressions within the parentheses are both zero, since y_1 and y_2 are solutions of the homogeneous equation (3.170). Thus, we have

$$\frac{dz_1}{dx}\cdot\frac{dy_1}{dx} + \frac{dz_2}{dx}\cdot\frac{dy_2}{dx} = r(x). \tag{3.179}$$

Equations (3.175) and (3.179) are now solved simultaneously. We get

$$\left.\begin{aligned} \frac{dz_1}{dx} &= -\frac{y_2 r(x)}{W(x)} \\ \text{and} \quad \frac{dz_2}{dx} &= \frac{y_1 r(x)}{W(x)} \end{aligned}\right\}, \tag{3.180}$$

where, $W(x)$ is the Wronskian of solution functions $y_1(x)$ and $y_2(x)$ and is equal to, as we know

$$W(x) = y_1 \frac{dy_2}{dx} - y_2 \frac{dy_1}{dx}. \tag{3.181}$$

Integrating equations (3.180), we find

$$\left.\begin{aligned} z_1(x) &= -\int \frac{y_2(x) r(x)}{W(x)} dx \\ \text{and} \quad z_2(x) &= \int \frac{y_1(x) r(x)}{W(x)} dx \end{aligned}\right\} \tag{3.182}$$

Knowing the functions $z_1(x)$ and $z_2(x)$ by equation (3.182), we can find the particular solution using equation (3.173).

EXAMPLE 3.32. *Solve* $\dfrac{d^2y}{dx^2} + 5\dfrac{dy}{dx} + 4y = xe^x$.

Solution: For homogeneous equation, the characteristic equation is

$$\lambda^2 + 5\lambda + 4 = 0, \quad \text{i.e.,} \quad \lambda = -1, -4.$$

Hence, the two linearly independent solutions of homogeneous equation are

$$y_1 = e^{-x} \text{ and } y_2 = e^{-4x}.$$

Here, $r(x) = xe^x$, and $W(x) = -3e^{-5x}$.
Using equations (3.182), we find $z_1(x)$ and $z_2(x)$

$$z_1(x) = -\int \frac{e^{-4x}(xe^x)}{-3e^{-5x}} dx = \frac{1}{3}\int xe^{2x} dx = \frac{1}{3}\left[\frac{x}{2}e^{2x} - \frac{1}{2}e^{2x}\right] = \frac{1}{6}e^{2x}(x-1)$$

$$z_2(x) = \int \frac{e^{-x}\cdot(xe^x)}{-3e^{-5x}} dx = -\frac{1}{3}\int xe^{5x} dx = -\frac{1}{3}\left[\frac{x}{5}e^{5x} - \frac{1}{5}e^{5x}\right] = -\frac{1}{15}e^{5x}(x-1).$$

Hence, the general solution of the given equation is

$$y = C_1 e^{-x} + C_2 e^{-4x} + e^{-x}\cdot\frac{1}{6}e^{2x}(x-1) - e^{-4x}\cdot\frac{1}{15}e^{5x}(x-1)$$

$$\text{i.e.,} \quad y = C_1 e^{-x} + C_2 e^{-4x} + e^x(x-1)\left(\frac{1}{6} - \frac{1}{15}\right) = C_1 e^{-x} + C_2 e^{-4x} + \frac{e^x}{10}(x-1).$$

IV. SERIES METHOD

Series method is a very common and powerful method of solving differential equations in terms of power series. We, therefore, intend to discuss on power series before proceeding to the method.

3.27 Power Series

An infinite series in ascending integral powers of x in the form

$$S = a_0 + a_1(x - x_0) + \cdots + a_n(x - x_0)^n + \cdots, \tag{3.183}$$

where x_0, a_0, a_1, …, a_n are all constants, is called a *power series*. Thus, in a power series, there should not remain a single term with negative or fractional power. The series

$$\frac{1}{x} + x + x^{3/2} + x^2 + \cdots$$

is therefore not a power series.

The power series (3.183) is said to converge at a point $x = a$, if the partial sum up to m terms of the series

$$S_m = a_0 + a_1(x - x_0) + \cdots a_m(x - x_0)^m \tag{3.184}$$

tends to a definite limit as $m \longrightarrow \infty$. Otherwise, the power series is said to diverge. If we consider the point $x = x_0$, the limit

$$\underset{m \longrightarrow \infty}{S_m} = a_0.$$

Thus, the power series about a point always converges at that point. We consider the point $x = x_0$ at the middle of an interval and draw a circle of radius R, such that

$$|\, x - x_0 \,| < R. \tag{3.185}$$

If the power series converges at any point within that circle, we say that R is the radius of convergence of the power series. This radius of convergence is given by

$$R = \lim_{n \to \infty} \left| \frac{a_n}{a_{n+1}} \right|. \tag{3.186}$$

If a function $f(x)$ can be expanded in a power series like

$$f(x) = \sum_{n=0}^{\infty} a_n(x - x_0)^n \tag{3.187}$$

with radius of convergence $R > 0$, we say that the function $f(x)$ is analytic (or regular) at the point $x = x_0$.

Now, if the function $f(x)$ given by equation (3.187) is analytic at $x = x_0$, then it can be differentiated as well as integrated term by term and the radius of convergence does not change in the process. Thus, differentiating equation (3.187) on both sides, we get

$$f'(x) = \sum_{n=0}^{\infty} n a_n (x - x_0)^{n-1} \tag{3.188}$$

$$f''(x) = \sum_{n=0}^{\infty} n(n-1) a_n (x - x_0)^{n-2} \tag{3.189}$$

$$\cdots\cdots\cdots\cdots\cdots\cdots\cdots\cdots\cdots\cdots$$

and so on. Putting $x = x_0$, we find the coefficients of the power series in terms of its derivatives at $x = x_0$ in the form

$$a_n = \frac{1}{n!} f^{(n)}(x_0), \quad n = 1, 2, \cdots \infty. \tag{3.190}$$

Thus, if $f(x)$ is analytic at $x = x_0$, it can be expanded about $x = x_0$ in the form

$$f(x) = \sum_{n=0}^{\infty} \frac{1}{n!} f^{(n)}(x_0)(x - x_0)^n. \tag{3.191}$$

The power series in (3.191) is called *Taylor's series* expanded about the mid-point $x = x_0$. In a special case, if the power series is expanded about origin $x = 0$, the series is called *Maclaurin's series*. As an example, Taylor's series expansion of e^x about origin can be easily derived using equation (3.191) in the form

$$e^x = 1 + x + \frac{x^2}{2!} + \cdots$$

3.28 Power Series Solutions of Differential Equations

Let us consider the differential equation

$$\frac{d^2y}{dx^2} + p(x)\frac{dy}{dx} + q(x) = 0. \tag{3.192}$$

If the functions $p(x)$ and $q(x)$ are analytic at $x = x_0$, then this point $x = x_0$ is an ordinary point of the equation (3.192). If either $p(x)$ or $q(x)$ is not analytic at $x = x_0$, then $x = x_0$ is a singular point of the equation (3.192).

Let us suppose that $x = 0$ is an ordinary point of the equation (3.192). Then, there exists a pair of basic solutions

$$y_1(x) = \sum_{n=0}^{\infty} a_n x^n, \quad y_2(x) = \sum_{n=0}^{\infty} b_n x^n \tag{3.193}$$

for the above differential equation. Moreover, both the series in (3.193) converge for $|x| < R$, the radius of convergence. Hence, if $p(x)$ and $q(x)$ are simple polynomials, the differential equation can be solved by series method, and solution will converge for all x. We shall consider few examples in order to understand the method.

EXAMPLE 3.33. *Find power series solution of* $\dfrac{d^2y}{dx^2} + x\dfrac{dy}{dx} + y = 0.$

Solution: Let us assume the solution in the form

$$y = \sum_{n=0}^{\infty} a_n x^n. \tag{1}$$

Then

$$\frac{dy}{dx} = \sum_{n=1}^{\infty} n a_n x^{n-1} \quad \text{and} \quad \frac{d^2y}{dx^2} = \sum_{n=2}^{\infty} n(n-1) a_n x^{n-2}.$$

Substituting these derivatives into the given equation, we find

$$\sum_{n=2}^{\infty} n(n-1)a_n x^{n-2} + x\sum_{n=1}^{\infty} n a_n x^{n-1} + \sum_{n=0}^{\infty} a_n x^n = 0.$$

Let us collect all the coefficients for x^n. In doing so, we note that, the first series starts from $n = 2$, hence coefficient of x^n will be given for $n \longrightarrow n + 2$. Similarly, the second series starts from $n=1$, but since it is getting multiplied by x, the coefficient of x^n will be given for $n \longrightarrow n$. Same logic is applied for the third series. Thus, the coefficient of x^n is

$$(n+2)(n+1)a_{n+2} + na_n + a_n = 0, \quad \text{i.e.,} \quad a_{n+2} = -\frac{a_n}{n+2}.$$

Replacing $n+2$ by n, we get

$$a_n = -\tfrac{1}{n}a_{n-2}. \tag{2}$$

This is a recursion formula for obtaining coefficients of the series (1). Taking $n=2$ onwards, we find,

$$a_2 = -\tfrac{1}{2}a_0 \qquad a_5 = -\tfrac{1}{5}a_3 = \frac{1}{(3)(5)}\cdot a_1$$

$$a_3 = -\tfrac{1}{3}a_1 \qquad a_6 = -\tfrac{1}{6}a_4 = -\frac{1}{2\cdot 4\cdot 6}\,a_0$$

$$a_4 = -\tfrac{1}{4}a_2 = \frac{1}{(2)(4)}\cdot a_0 \qquad a_7 = -\tfrac{1}{7}a_5 = -\frac{1}{3\cdot 5\cdot 7}\cdot a_1$$

Thus, all the even coefficients are related to a_0 and all the odd coefficients are related to a_1. Hence, the solution of the differential equation is

$$y(x) = a_0\left(1 - \frac{x^2}{2} + \frac{x^4}{2\cdot 4} - \frac{x^6}{2\cdot 4\cdot 6} + \cdots\right) + a_1\left(x - \frac{x^3}{3} + \frac{x^5}{3\cdot 5} - \frac{x^7}{3\cdot 5\cdot 7} + \cdots\right).$$

This solution can be represented as

$$y(x) = a_0 y_1(x) + a_1 y_2(x).$$

Now, the Wronskian of the solutions at $x=0$ is

$$W = \begin{vmatrix} y_1(0) & y_2(0) \\ y_1'(0) & y_2'(0) \end{vmatrix} = \begin{vmatrix} 1 & 0 \\ 0 & 1 \end{vmatrix} = 1.$$

Thus, $W \neq 0$, which confirms that the solutions are the pair of basic solutions.

EXAMPLE 3.34. *Find a power series solution to the differential equation*

$$\frac{d^2y}{dx^2} + \lambda^2 y = 0, \quad \lambda = \textit{constant}.$$

Solution: Let us assume the solution in the form

$$y(x) = \sum_{n=0}^{\infty} a_n x^n. \tag{1}$$

Then

$$\frac{d^2y}{dx^2} = \sum_{n=2}^{\infty} n(n-1)a_n x^{n-2}.$$

Substituting these into the given differential equation, we get

$$\sum_{n=2}^{\infty} n(n-1)a_n x^{n-2} + \lambda^2 \sum_{n=0}^{\infty} a_n x^n = 0.$$

Collecting all the coefficients for x^n and equating them to zero, we have

$$(n+2)(n+1)a_{n+2} + \lambda^2 a_n = 0$$

$$\text{i.e.,} \quad a_{n+2} = -\lambda^2 \frac{a_n}{(n+1)(n+2)}. \tag{2}$$

Putting $n = 0, 1, 2, \ldots$, etc., we find

$$a_2 = -\frac{\lambda^2}{1 \cdot 2} a_0 \qquad a_4 = -\frac{\lambda^2}{3 \cdot 4} a_2 = \frac{\lambda^4}{1 \cdot 2 \cdot 3 \cdot 4} a_0$$

$$a_3 = -\frac{\lambda^2}{2 \cdot 3} a_1 \qquad a_5 = -\frac{\lambda^2}{4 \cdot 5} a_3 = \frac{\lambda^2}{2 \cdot 3 \cdot 4 \cdot 5} a_1$$

Thus we find, even coefficients are related to a_0, and odd coefficients are related to a_1. Two independent solutions are

$$y_1(x) = 1 - \frac{\lambda^2}{1 \cdot 2} x^2 + \frac{\lambda^4 x^4}{1 \cdot 2 \cdot 3 \cdot 4} - \cdots \tag{3}$$

$$\text{and} \quad y_2(x) = x - \frac{\lambda^2}{2 \cdot 3} x^3 + \frac{\lambda^4}{2 \cdot 3 \cdot 4 \cdot 5} x^5 - \cdots \tag{4}$$

Solutions (3) and (4) can be simply written as

$$y_1(x) = \cos \lambda x$$

$$\text{and} \quad y_2(x) = \sin \lambda x.$$

Thus, the general solution of the differential equation is

$$y(x) = a_0 \cos \lambda x + a_1 \sin \lambda x.$$

It is easy to verify that the Wronskian of pair of basic solution is not zero for $x = 0$.

3.29 Method of Frobenius

Let us consider the differential equation

$$x^2 \frac{d^2 y}{dx^2} + x p(x) \frac{dy}{dx} + q(x)\, y = 0, \tag{3.194}$$

where, functions $p(x)$ and $q(x)$ are analytic at $x = 0$, but $x = 0$ is not an ordinary point but a regular singular point of the equation (3.194). By virtue of Fuchs theorem, if the differential equation

$$y'' + P(x)\, y' + Q(x)\, y = 0$$

has a singular point at $x = x_0$, then the solution of the equation by series method will converge if and only if $(x - x_0)P(x)$ and $(x - x_0)^2 Q(x)$ have finite values at $x = x_0$. It is however to be noted that the series solution may not be necessarily a power series. In this case, we cannot have a power series solution of the equation but, at least one solution of the equation can be obtained in the form

$$y(x) = \sum_{n=0}^{\infty} a_n x^{n+r} \quad (a_0 \neq 0). \tag{3.195}$$

Such a series is called *Frobenius series.*

Since $p(x)$ and $q(x)$ are analytic at $x = 0$, these can be expanded in power series in the form

$$p(x) = \sum_{n=0}^{\infty} b_n x^n \tag{3.196}$$

$$q(x) = \sum_{n=0}^{\infty} c_n x^n \tag{3.197}$$

Differentiating (3.195) twice, we have

$$\left.\begin{aligned} y' &= \sum_{n=0}^{\infty} (n+r) a_n x^{n+r-1} \\ y'' &= \sum_{n=0}^{\infty} (n+r)(n+r-1) a_n x^{n+r-2} \end{aligned}\right\}. \tag{3.198}$$

Substituting equations (3.195)–(3.198) into (3.194), we get

$$\sum_{n=0}^{\infty} (n+r)(n+r-1) a_n x^{n+r} + \left(\sum_{n=0}^{\infty} b_n x^{n+1}\right)\left(\sum_{n=0}^{\infty} (n+r) a_n x^{n+r-1}\right) + \left(\sum_{n=0}^{\infty} c_n x^n\right)\left(\sum_{n=0}^{\infty} a_n x^{n+r}\right) = 0. \tag{3.199}$$

Collecting terms containing lowest power of x, i.e., x^r, and equating to zero, we get

$$\{r(r-1) + b_0 r + c_0\}\, a_0 x^r = 0$$

$$\text{i.e.,}\quad f(r) \equiv r(r-1) + b_0 r + c_0 = 0. \tag{3.200}$$

This is called *indicial equation* which provides two values of r. Let r_1 and r_2 are two roots of the indicial equation.

Then three cases may arise.

Case I. *When r_1 and r_2 are two distinct roots not differing by an integer.*

In this case, for $r = r_1$, one solution is

$$y_1(x) = \sum_{n=0}^{\infty} a_n x^{n+r_1} \tag{3.201}$$

and the other solution for $r = r_2$, is

$$y_2(x) = \sum_{n=0}^{\infty} a_n x^{n+r_2}. \tag{3.202}$$

The coefficients $a_0, a_1, \ldots, a_n$ are to be found from the relation (3.199). It can easily be verified that the Wronskian of the solutions $y_1(x)$ and $y_2(x)$ at $x = 0$ is not zero. Hence the two solutions are linearly independent.

Now, let

$$r_1 > r_2 \quad \text{and} \quad \lambda = r_1 - r_2. \tag{3.203}$$

Collecting coefficients of x^{n+r} from equation (3.199) and equating them to zero, we find

$$\{(n+r)(n+r-1) + b_0(n+r) + c_0\}\, a_n + \sum_{m=1}^{n} \{(n+r-m)b_m + c_m\}\, a_{n-m} = 0$$

$$\text{i.e.,} \quad f(n,r)a_n = -\sum_{m=1}^{n} g(n,m)a_{n-m} \tag{3.204}$$

where

$$f(n,r) = (n+r)(n+r-1) + b_0(n+r) + c_0 \tag{3.205}$$

$$\text{and} \quad g(n,m) = (n+r-m)b_m + c_m. \tag{3.206}$$

Thus, for $r = r_1$

$$f(n,r_1) = (n+r_1)(n+r_1-1) + b_0(n+r_1) + c_0. \tag{3.207}$$

Also, from equation (3.200)

$$f(r_1) = f(0,r_1) = r_1(r_1-1) + b_0 r_1 + c_0 = 0. \tag{3.208}$$

Using equations (3.208) and (3.203), we can write $f(r) = (r-r_1)(r-r_2)$, so that

$$f(n,r_1) = n(n+\lambda) \tag{3.209}$$

$$\text{and} \quad f(n,r_2) = n(n-\lambda). \tag{3.210}$$

Hence, from the recursion formula (3.204), we have

$$\text{for } r = r_1, \quad n(n+\lambda)a_n = -\sum_{m=1}^{n} g(n,m)a_{n-m} \tag{3.211}$$

$$\text{and} \quad \text{for } r = r_2, \quad n(n-\lambda)a_n = -\sum_{m=1}^{n} g(n,m)a_{n-m}. \tag{3.212}$$

It is obvious from (3.211) that we can always find coefficients $a_0, a_1, \ldots$, etc., since $n(n+\lambda)$ is never zero. Thus, the Frobenius method guarantees at least one solution of the equation (3.194). But, for $r = r_2$, equation (3.212) for determining the coefficients $a_0, a_1, \ldots$, etc., fails if $\lambda = r_1 - r_2$ happens to be an integer. The recursion formula can be useful to find coefficients $a_0, a_1, \ldots$, etc., only when λ is not an integer. In such a case, we get a pair of basic solution of the equation (3.194).

Again, if $\lambda = 0$, i.e., $r_1 = r_2$, i.e., two roots of the indicial equation are equal, then equation (3.211), (3.212) degenerates into one and the same. Hence, here also, we get only one solution of the differential equation (3.194).

Let us now consider examples when r_1 and r_2 are distinct not differing by an integer. This will help to realise how coefficients are found using the recursion formulae.

EXAMPLE 3.35. *Find a general solution of the equation valid in the vicinity of origin*

$$2x\frac{d^2y}{dx^2} + (1-x)\frac{dy}{dx} + y = 0.$$

Solution: Let us assume the solution in the form

$$y = \sum_{n=0}^{\infty} a_n x^{n+r}. \tag{1}$$

Then

$$\frac{dy}{dx} = \sum_{n=0}^{\infty} (n+r) a_n x^{n+r-1} \tag{2}$$

$$\frac{d^2y}{dx^2} = \sum_{n=0}^{\infty} (n+r)(n+r-1) a_n x^{n+r-2}. \tag{3}$$

Substituting these in the given equation

$$\sum_{n=0}^{\infty} 2(n+r)(n+r-1) a_n x^{n+r-1} + (1-x) \sum_{n=0}^{\infty} (n+r) a_n x^{n+r-1} + \sum_{n=0}^{\infty} a_n x^{n+r} = 0. \tag{4}$$

Equating to zero the lowest power of x, i.e., x^{r-1}, we get

$$\begin{aligned} &[2r(r-1)+r]a_0 = 0, \\ \text{i.e.,}\quad & 2r(r-1)+r = 0, \quad \text{since } a_0 \neq 0, \\ \text{i.e.,}\quad & r = 0,\ 1/2. \end{aligned} \tag{5}$$

Equating to zero the coefficients of x^r, x^{r+1}, ..., we get

$$\begin{aligned} x^r &: [2r(r+1)+r+1]a_1 + (1-r)a_0 = 0 \\ x^{r+1} &: [2(r+1)(r+2)+r+2]a_2 + \{1-(r+1)\}\, a_1 = 0 \\ x^{r+2} &: [2(r+2)(r+3)+r+3]a_3 + \{1-(r+2)\}\, a_2 = 0 \\ &\cdots\cdots\cdots\cdots\cdots\cdots\cdots\cdots\cdots\cdots \\ x^{n+r} &: [2(r+n)(r+n+1)+(r+n+1)]a_{n+1} + \{1-(r+n)\}\, a_n = 0. \end{aligned}$$

Thus, we find a general recursion formula

$$a_{n+1} = -\frac{a_n(1-r-n)}{f(n,r)} = \frac{r+n-1}{f(n,r)} a_n, \tag{6}$$

where

$$f(n,r) = 2(r+n)(r+n+1) + (r+n+1). \tag{7}$$

For $r = 0$,

$$a_{n+1} = \frac{-(1-n)a_n}{f(n,0)} = \frac{n-1}{f(n,0)} a_n, \tag{8}$$

where

$$f(n,0) = 2n(n+1) + (n+1) = (n+1)(2n+1). \tag{9}$$

For $r = \frac{1}{2}$,

$$a_{n+1} = \frac{-\left(\frac{1}{2} - n\right) a_n}{f\left(n, \frac{1}{2}\right)} = \frac{n - \frac{1}{2}}{f\left(n, \frac{1}{2}\right)} a_n, \tag{10}$$

where

$$f\left(n, \tfrac{1}{2}\right) = \left(n + \tfrac{3}{2}\right)(2n+2) = (2n+3)(n+1). \tag{11}$$

From equation (8), we find the coefficients of the solution in the case of $r = 0$ as below:

$$a_1 = -a_0, \quad a_2 = a_3 = \cdots a_n = 0.$$

Thus, the solution is

$$y_1 = a_0(1 - x).$$

From equation (10), the coefficients for the case $r = \frac{1}{2}$ are given by:

$$\begin{aligned} a_1 &= -\frac{1}{2} \cdot \frac{a_0}{3 \cdot 1} = \frac{-a_0}{3!} \\ a_2 &= \frac{a_1}{2 \cdot 5 \cdot 2} = -\frac{1}{2^2} \cdot \frac{a_0}{2 \cdot 3 \cdot 5} = -\frac{a_0}{5!} \\ a_3 &= \frac{3}{2} \cdot \frac{a_2}{7 \cdot 3} = -\frac{3}{2^3} \cdot \frac{a_0}{2 \cdot 3 \cdot 5 \cdot 7 \cdot 3} = -\frac{3a_0}{7!}. \end{aligned}$$

So, the second solution is

$$y_2 = a_0 x^{1/2} \left\{ 1 - \frac{x}{3!} - \frac{x^2}{5!} - \frac{3x^3}{7!} - \cdots \right\}.$$

Thus, the general solution of the given equation is

$$y = y_1(x) + y_2(x) = a_0(1 - x) + b_0 x^{1/2} \left(1 - \frac{x}{3!} - \frac{x^2}{5!} - \frac{3x^3}{7!} - \cdots \right),$$

where we have replaced arbitrary constant a_0 by b_0 in the second solution.

EXAMPLE 3.36. *Solve* $\left(2x + x^3\right) \dfrac{d^2y}{dx^2} + 3\dfrac{dy}{dx} - 2xy = 0.$

Solution: The given equation has regular point at $x = 0$, and so we can find a solution in series developed about $x = 0$. Let the solution of the equation be

$$y = \sum_{n=0}^{\infty} a_n x^{n+r}. \tag{1}$$

Then

$$\frac{dy}{dx} = \sum_{n=0}^{\infty} a_n (n+r)\, x^{n+r-1} \tag{2}$$

$$\frac{d^2y}{dx^2} = \sum_{n=0}^{\infty} a_n (n+r)(n+r-1)\, x^{n+r-2}. \tag{3}$$

Substituting these in the given equation, we get

$$\sum_{n=0}^{\infty} a_n \{(n+r)(n+r-1)-2\} x^{n+r+1}$$
$$+\sum_{n=0}^{\infty} a_n \{2(n+r)(n+r-1)+3(n+r)\} x^{n+r-1}=0.$$

Equating to zero the coefficient of lowest power of x, i.e., x^{r-1}, we have

$$2r(r-1)+3r=0, \quad \text{i.e.,} \quad r=0,-\tfrac{1}{2}. \tag{4}$$

Equating to zero the coefficient of x^{n+r+1}, we get a recursion formula for determining the unknown coefficients in the form

$$a_n \{(n+r)(n+r-1)-2\} = -a_{n+2} \{2(n+r+2)(n+r+1)+3(n+r+2)\}$$

$$\text{or,} \quad a_{n+2} = -\frac{(n+r)(n+r-1)-2}{(n+r+2)(2n+2r+5)} a_n. \tag{5}$$

For $r=0$,

$$a_{n+2} = -\frac{n(n-1)-2}{(n+2)(2n+5)} a_n. \tag{6}$$

$$\therefore \quad a_2 = +\frac{2}{10} a_0 = \frac{1}{5} a_0$$
$$a_3 = \frac{2}{3\cdot 7} a_1$$
$$a_4 = 0, a_6 = 0, a_8 = 0, \cdots$$
$$a_5 = -\frac{4}{5\cdot 11} a_3 = -\frac{2(4)}{(3)(5)(7)(11)} a_1$$
$$a_7 = \frac{18}{7\cdot 15} a_5 = \frac{(2)(4)(18)}{(3)(5)(7)(7)(11)(5)} a_1$$

....

Thus, the solution for $r=0$ is

$$y_1(x) = a_0\left(1+\frac{x^2}{5}\right) + a_1\left(x + \frac{2x^3}{21} - \frac{8}{(55)(21)} x^5 + \frac{144}{(225)(539)} x^7 + \cdots\right).$$

For $r = -1/2$,

$$a_{n+2} = -\frac{\left(n-\frac{1}{2}\right)\left(n-\frac{3}{2}\right)-2}{\left(n+\frac{3}{2}\right)(2n+4)} a_n. \tag{7}$$

$$
\begin{aligned}
a_2 &= \frac{5}{4\cdot 6}a_0 \\
a_3 &= \frac{9}{4\cdot 15}a_1 = \frac{3}{20}a_1 \\
a_4 &= \frac{5}{4\cdot 28}a_2 = \frac{25}{(96)(28)}a_0 \\
a_5 &= -\frac{7}{4\cdot 45}a_3 = -\frac{7\cdot 9}{(4\cdot 15)(4\cdot 45)}a_1 = -\frac{7}{1200}a_1 \\
a_6 &= -\frac{27}{4\cdot 66}a_4 = \frac{27\cdot 25}{4(66)(96)(28)}a_0 \\
&\cdots\cdots\cdots\cdots\cdots\cdots
\end{aligned}
$$

Thus, the solution for $r = -\frac{1}{2}$ is

$$y_2(x) = a_0 x^{-1/2}\left(1 + \frac{5}{24}x^2 + \frac{25}{96\cdot 28}x^4 + \cdots\right) + a_1 x^{-1/2}\left(x + \frac{3}{20}x^3 - \frac{7}{1200}x^5 + \cdots\right).$$

Case II. *Roots of indicial equation differ by an integer.*

As before, we consider the equation

$$x^2\frac{d^2y}{dx^2} + xp(x)\frac{dy}{dx} + q(x)\,y = 0 \tag{3.213}$$

Now, let

$$L(\cdots) = x^2\frac{d^2\cdots}{dx^2} + xp(x)\frac{d\cdots}{dx} + q(x)(\cdots)$$

be a differential operator so that equation (3.213) can be expressed as

$$L(y) \equiv 0. \tag{3.214}$$

We consider $L(y)$ and a function $y(x)$ given by

$$y(x) = \sum_{n=0}^{\infty} a_n x^{n+r}. \tag{3.215}$$

For this function $y(x)$, we find that

$$L(y) = f(r)a_0x^r + \sum_{n=1}^{\infty}\left[f(n,r)a_n + \sum_{m=1}^{n} g(n,m)a_{n-m}\right]x^{n+r}, \tag{3.216}$$

where $f(r)$, $f(n,r)$, $f(n,m)$ are as defined in equations (3.200), (3.204) and (3.207) respectively.

Now, we select the coefficients such that the infinite series portion of equation (3.216) becomes identically zero, so that

$$L(y) = f(r)a_0x^r. \tag{3.217}$$

In such a case, the recursion formula for obtaining coefficients a_1, a_2, ... is given by

$$f(n,r)a_n = -\sum_{m=1}^{n} g(n,m)a_{n-m}, \tag{3.218}$$

where $n = 1, 2, 3, \ldots$. It is to be noted that we do not choose any value of r in the process of calculating coefficients by equation (3.218).

Now, if the function $y(x)$ given by equation (3.215) has to be a solution of equation (3.213), we must have

$$L(y) = 0, \quad \text{i.e.,} \quad L\left[f(r)a_0x^r\right] = 0, \quad \text{i.e.,} \quad f(r) = 0.$$

This provides for two values of r. We now consider an example.

EXAMPLE 3.37. *Find general solutions of* $\dfrac{d^2y}{dx^2} + x\dfrac{dy}{dx} + 2y = 0.$

Solution: Here, $p(x) = x^2$, $q(x) = 2x^2$.

Hence, all b_n's are zero, excepting $b_2 = 1$, and all c_n's are zero excepting $c_2 = 2$ [equations (3.196) and (3.197)].

Using equation (3.200)

$$f(r) = r(r-1), \quad \text{so that } r = 0, 1. \tag{1}$$

Using equations (3.206) and (3.207), we get

$$\begin{aligned} f(n,r) &= (n+r)(n+r-1) = (n+r)^2 - (n+r) \\ g(n,m) &= (n+r-m)b_m + c_m \end{aligned} \tag{2}$$

so that

$$g(n,2) = (n+r-2)\cdot 1 + 2 = n + r \tag{3}$$

and for every other m,

$$g(n,m) = 0. \tag{4}$$

Therefore, by equation (3.216)

$$\begin{aligned} L(y) = \left(r^2 - r\right)a_0x^r &+ r(r+1)a_1x^{r+1} \\ &+ \left[\{(r+2)^2 - (r+2)\}a_2 + (2+r)a_0\right]x^{r+2} \\ &- \left[\{(n+r)^2 - n + r\}a_n + (n+r)a_{n-2}\right]x^{n+r}. \end{aligned}$$

We set all coefficients of $x^{r+1}, \ldots$ to zero. This gives us

$$r(r+1)a_1 = 0, \text{ i.e., } a_1 = 0$$

$$\begin{aligned} \{(r+2)^2 - (r+2)\}a_2 &= -(r+2)a_0, \quad \text{i.e.,} \quad a_2 = \frac{-1}{r+1}a_0 \\ \{(r+3)^2 - (r+3)\}a_3 &= -(r+3)a_1, \quad \text{i.e.,} \quad a_3 = 0 \\ \{(r+4)^2 - (r+4)\}a_4 &= -(r+4)a_2, \quad \text{i.e.,} \quad a_4 = \frac{a_0}{(r+1)(r+3)} \end{aligned}$$

....

Thus, we get $a_1 = a_3 = a_5 = \ldots = 0$ and

$$a_{2k} = \frac{(-1)^k a_0}{(r+1)(r+3)\cdots\{r+(2k-1)\}}.$$

Setting $a_0 = 1$ arbitrarily, we write the solution as

$$y(x) = x^r + \sum_{k=1}^{\infty} \frac{(-1)^k x^{2k+r}}{(r+1)(r+3)\cdots(r+\overline{2k-1})}.$$

Hence, for $r = 0$, we have the solution

$$y_1(x) = 1 - x^2 + \frac{x^4}{3} - \frac{x^6}{15} + \cdots$$

and for $r = 1$, we have the solution

$$y_2(x) = x - \frac{x^3}{2} + \frac{x^5}{8} - \frac{x^7}{48} + \cdots$$

Case III. *Roots of indicial equations are equal.*

Let the two roots of indicial equation

$$f(r) \equiv r(r-1) + b_0 r + c_0 = 0$$

be equal. Then

$$\lambda = r_1 - r_2 = 0$$

In this case, equation (3.217) becomes

$$L[(y)_r] = (r - r_1)^2 a_0 x^r. \tag{3.219}$$

We set $a_0 = 1$ without any loss of generality. Equation (3.219) indicates that $y(x)$ is a solution of the differential equation for $r = r_1$.

Let us now consider

$$\frac{\partial}{\partial r} L[y(x)] = L\left[\frac{\partial}{\partial r} y(x)\right] = \frac{\partial}{\partial r}\left[(r-r_1)^2 x^r\right]$$
$$= 2(r - r_1)\, x^r + (r - r_1)^2 x^r \ln x. \tag{3.220}$$

For $r = r_1$, expression on the right-hand side becomes zero. Hence

$$L\left[\frac{\partial}{\partial r} y(x)\right] = 0$$

which shows that

$$\frac{\partial}{\partial r} y(x) \quad \text{for } r = r_1$$

is also a solution of the equation. Thus

$$y(x) \quad \text{and} \quad \frac{\partial}{\partial r} y(x) \quad \text{for } r = r_1$$

are basic solution set. We consider an example.

EXAMPLE 3.38. *Solve* $x\dfrac{d^2y}{dx^2} + \dfrac{dy}{dx} + xy = 0.$

Solution: Let the solution be

$$y = \sum_{n=0}^{\infty} a_n x^{n+r}. \tag{1}$$

Then

$$\frac{dy}{dx} = \sum_{n=0}^{\infty} a_n(n+r)\,x^{n+r-1} \tag{2}$$

$$\frac{d^2y}{dx^2} = \sum_{n=0}^{\infty} a_n(n+r)(n+r-1)\,x^{n+r-2}. \tag{3}$$

Substituting these in the given equation

$$\sum_{n=0}^{\infty} a_n(n+r)(n+r-1)\,x^{n+r-1} + \sum_{n=0}^{\infty} a_n(n+r)\,x^{n+r-1} + \sum_{n=0}^{\infty} a_n x^{n+r+1} = 0. \tag{4}$$

Equating to zero the coefficient of x^{r-1}, we get the indicial equation

$$[r(r-1)+r]a_0 = 0, \quad \text{i.e.,} \quad r = 0, 0.$$

Thus, two roots of indicial equation are equal. Equating to zero the coefficients of x^r, we get

$$[(r+1)r + r + 1]a_1 = 0, \quad \text{i.e.,} \quad a_1 = 0. \tag{5}$$

General recursion formula here is

$$[(n+2+r)(n+r+1) + (n+r+2)]a_{n+2} = -a_n$$

i.e.,

$$a_{n+2} = -\frac{1}{(n+r+2)^2} \cdot a_n. \tag{6}$$

Thus

$$a_2 = -\frac{a_0}{(r+2)^2}$$

$$a_1 = a_3 = a_5 = \cdots = 0$$

$$a_4 = -\frac{1}{(r+4)^2}a_2 = \frac{1}{(r+2)^2(r+4)^2}a_0$$

$$a_6 = -\frac{1}{(r+6)^2}a_4 = -\frac{1}{(r+2)^2(r+4)^2(r+6)^2}a_0.$$

Hence, one solution is

$$y_1(x)_r = a_0 x^r \left[1 - \frac{x^2}{(r+2)^2} + \frac{x^4}{(r+2)^2(r+4)^2} - \frac{x^6}{(r+2)^2(r+4)^2(r+6)^2} \cdots \right]. \tag{7}$$

For $r = 0$, this solution becomes, letting $a_0 = 1$,

$$y_1(x) = 1 - \frac{x^2}{2^2} + \frac{x^4}{2^2 \cdot 4^2} - \frac{x^6}{2^2 \cdot 4^2 \cdot 6^2} + \cdots \tag{8}$$

The other solution would be given by

$$y_2(x) = \frac{\partial}{\partial r}[y_1(x)_r] \quad \text{for } r = 0. \tag{9}$$

In order to differentiate (7) with respect to r, we let

$$y_1(x)_r = x^r F(r), \tag{10}$$

where

$$F(r) = \left[1 - \frac{x^2}{(r+2)^2} + \frac{x^4}{(r+2)^2(r+4)^2} - \cdots\right].$$

Then

$$\log y_1 = r \log x + \log F.$$

Differentiating partially with respect to r, we get

$$\frac{1}{y_1}\frac{\partial y_1}{\partial r} = \log x + \frac{1}{F}\frac{\partial F}{\partial r}.$$

$$\therefore \quad \frac{\partial y_1}{\partial r} = y_1 \log x + \frac{y_1}{F}\frac{\partial F}{\partial r}, \quad \text{i.e.,} \quad \frac{\partial y_1}{\partial r} = y_1 \log x + x^r \frac{\partial F}{\partial r}. \tag{11}$$

Now

$$\frac{\partial F}{\partial r} = \left[\frac{x^2}{(r+2)^2}\cdot\frac{2}{(r+2)} - \frac{x^4}{(r+2)^2(r+4)^2}\cdot\left(\frac{2}{r+2}+\frac{2}{r+4}\right)+\cdots\right].$$

Thus

$$\left.\frac{\partial F}{\partial r}\right|_{r=0} = \left[\frac{x^2}{2^2} - \frac{x^4}{2^2 \cdot 4^2}\left(1+\frac{1}{2}\right) + \frac{x^6}{2^2 \cdot 4^2 \cdot 6^2}\left(1+\frac{1}{2}+\frac{1}{3}\right) - \cdots\right]$$
$$= y_2(x) \text{ (say)}.$$

$\therefore$ From equation (11), we get

$$\left(\frac{\partial y_1}{\partial r}\right)_{r=0} = y_1 \log x + y_2(x),$$

where, $y_2(x)$ and $y_1(x)$ are already defined.

Therefore, complete solution of the equation is

$$y(x) = Ay_1(x) + B[y_1(x)\log x + y_2(x)] = (A + B\log x)\,y_1(x) + By_2(x)$$
$$= (A + B\log x)\left[1 - \frac{x^2}{2^2} + \frac{x^4}{2^2 \cdot 4^2} - \frac{x^6}{2^2 \cdot 4^2 \cdot 6^2} + \cdots\right]$$
$$+ B\left[\frac{x^2}{2^2} - \frac{x^4}{2^2 \cdot 4^2}\left(1+\frac{1}{2}\right) + \frac{x^6}{2^2 \cdot 4^2 \cdot 6^2}\left(1+\frac{1}{2}+\frac{1}{3}\right) - \cdots\right],$$

where A, B are two unknown constants which can be determined from boundary conditions.

EXAMPLE 3.39. *Solve* $x\dfrac{d^2y}{dx^2} + \dfrac{dy}{dx} + 2y = 0.$

Solution: Let, $y = \sum_{n=0}^{\infty} a_n x^{n+r}$ be a solution.

Differentiating, we get

$$\frac{dy}{dx} = \sum_{n=0}^{\infty} a_n (n+r)\, x^{n+r-1}$$

$$\frac{d^2y}{dx^2} = \sum_{n=0}^{\infty} a_n (n+r)(n+r-1)\, x^{n+r-2}.$$

Substituting these into the differential equation

$$\sum_{n=0}^{\infty} a_n (n+r)(n+r-1)\, x^{n+r-1} + \sum_{n=0}^{\infty} a_n (n+r)\, x^{n+r-1} + \sum_{n=0}^{\infty} 2a_n x^{n+r} = 0. \tag{1}$$

Equating to zero the coefficient of x^{r-1},

$$r(r-1) + r = 0, \quad \text{i.e.,} \quad r = 0, 0.$$

Thus, the roots of indicial equations are equal. The general recursion formula for finding the coefficients $a_1, a_2, \ldots$ is obtained from (1).

$$[(n+r)(n+r-1) + (n+r)]a_n = -2a_{n-1},$$

i.e.,

$$(n+r)^2 a_n = -2a_{n-1} \tag{2}$$

Putting $n = 1, 2, \ldots$, etc., we find

$$(r+1)^2 a_1 = -2a_0, \quad \text{i.e.,} \quad a_1 = -\frac{2a_0}{(r+1)^2}$$

$$(r+2)^2 a_2 = -2a_1, \quad \text{i.e.,} \quad a_2 = \frac{2^2 a_0}{(r+2)^2 (r+1)^2}$$

$$(r+3)^2 a_3 = -2a_2, \quad \text{i.e.,} \quad a_3 = -\frac{2^3 a_0}{(r+3)^2 (r+2)^2 (r+1)^2}$$

....

and so on.

Thus, taking $a_0 = 1$, we get one solution of the equation in the form

$$y_1(x) = x^r \left[1 - \frac{2x}{(r+1)^2} + \frac{2^2 x^2}{(r+2)^2 (r+1)^2} - \frac{2^3 x^3}{(r+3)^2 (r+2)^2 (r+1)^2} + \cdots\right]. \tag{3}$$

For $r = 0$, this solution becomes

$$y_1(x) = \left[1 - \frac{2x}{1^2} + \frac{2^2 x^2}{2^2 \cdot 1^2} - \frac{2^3 x^3}{3^2 \cdot 2^2 \cdot 1^2} + \cdots\right]. \tag{4}$$

The second independent solution would be

$$y_2(x) = \frac{\partial}{\partial r} [y_1(x)_r] \quad \text{for } r = 0.$$

Now, let, $y_1(x)_r = x^r F(r)$, where

$$F(r) = 1 - \frac{2x}{(r+1)^2} + \frac{2^2x^2}{(r+2)^2(r+1)^2} - \frac{2^3x^3}{(r+3)^2(r+2)^2(r+1)^2} - \cdots$$

Then

$$\log y_1 = r\log x + \log F.$$

Differentiating partially with respect to r, we get

$$\frac{1}{y_1}\frac{\partial y_1}{\partial r} = \log x + \frac{1}{F}\frac{\partial F}{\partial r}.$$

$$\therefore \quad \frac{\partial y_1}{\partial r} = y_1 \log x + \frac{y_1}{F}\frac{\partial F}{\partial r} = y_1\log x + x^r\frac{\partial F}{\partial r}. \tag{5}$$

Now

$$\frac{\partial F}{\partial r} = \left[\frac{2x}{(r+1)^2}\cdot\frac{2}{(r+1)} - \frac{2^2x^2}{(r+2)^2(r+1)^2}\left(\frac{2}{r+1}+\frac{2}{r+2}\right)\right.$$
$$\left. + \frac{2^3x^3}{(r+1)^2(r+2)^2(r+3)^2}\left(\frac{2}{r+1}+\frac{2}{r+2}+\frac{2}{r+3}\right) - \cdots\right].$$

$$\therefore \left.\frac{\partial F}{\partial r}\right|_{r=0} = 2\left[\frac{2x}{1^2} - \frac{2^2x^2}{1^2\cdot 2^2}\left(1+\frac{1}{2}\right) + \frac{2^3x^3}{1^2\cdot 2^2\cdot 3^2}\left(1+\frac{1}{2}+\frac{1}{3}\right) - \cdots\right]$$
$$= p(x), \text{ say.} \tag{6}$$

Thus, from (5), we get

$$\left(\frac{\partial y_1}{\partial r}\right)_{r=0} = y_1\log x + p(x).$$

Therefore, the complete solution of the equation is

$$y(x) = Ay_1(x) + B\{y_1(x)\log x + p(x)\}$$
$$= (A + B\log x)\left[1 - \frac{2x}{1^2} + \frac{2^2x^2}{1^2\cdot 2^2} - \frac{2^3x^3}{1^2\cdot 2^2\cdot 3^2} + \cdots\right]$$
$$+ B\left[\frac{2x}{1^2} - \frac{2^2x^2}{1^2\cdot 2^2}\left(1+\frac{1}{2}\right) + \frac{2^3x^3}{1^2\cdot 2^2\cdot 3^2}\left(1+\frac{1}{2}+\frac{1}{3}\right) - \cdots\right].$$

3.30 Roots of Indicial Equation Differing by Integer

We have seen in Example 3.37 that although the roots $r_1 = 1$, $r_2 = 0$ differed by an integer, we could find out two independent solutions of the differential equation. But this does not happen in general. Let us see an example.

Suppose, we are to solve the differential equation

$$x(1-2x)\frac{d^2y}{dx^2} - (6x+1)\frac{dy}{dx} - 2y = 0.$$

Assuming the solution in the form

$$y(x) = \sum_{n=0}^{\infty} a_n x^{n+r}$$

we find, as usual, its derivatives. Thus

$$\frac{dy}{dx} = \sum_{n=0}^{\infty} a_n (n+r)\, x^{n+r-1}$$

$$\frac{d^2y}{dx^2} = \sum_{n=0}^{\infty} a_n (n+r)(n+r-1)\, x^{n+r-2}.$$

Substituting y, y' and y'' in the given differential equation, we get

$$\sum_{n=0}^{\infty} a_n (n+r)(n+r-2)\, x^{n+r-1} - 2\sum_{n=0}^{\infty} [(n+r)(n+r-1) + 3(n+r) + 1]\, a_n x^{n+r} = 0,$$

i.e., $$\sum_{n=0}^{\infty} a_n (n+r)(n+r-2)\, x^{n+r-1} - 2\sum_{n=0}^{\infty} a_n (n+r+1)^2 x^{n+r} = 0.$$

Equating to zero the coefficient of x^{r-1}, we get

$$r(r-2) = 0, \quad \text{i.e.,} \quad r = 0, 2.$$

Thus, the roots differ by an integer 2.

Now, equating to zero the coefficient of x^{n+r}, we get the recursion formula in general:

$$(n+r+1)(n+r-1)a_{n+1} = 2(n+r+1)^2 a_n,$$

$$\text{i.e.,} \quad a_{n+1} = \frac{2(n+r+1)}{n+r-1} \cdot a_n. \tag{3.221}$$

For $r = 0$, this relation reduces to

$$a_{n+1} = \frac{2(n+1)}{n-1},$$

which becomes indeterminate for $n = 1$. Hence, the solution series as assumed before cannot give two independent solutions.

Therefore, we intend here to give a general method for solving equations whose indicial roots differ by an integer.

If r_1, r_2 are two roots, then the indicial equation is

$$f(r) = (r - r_1)(r - r_2). \tag{3.222}$$

In this case equation (3.217) becomes

$$L\,[y(x)_r] = (r - r_1)(r - r_2)a_0 x^r. \tag{3.223}$$

Let us find

$$L\left[\frac{\partial}{\partial r} y(x)_r\right]_{r=r_1}$$

for equation (3.223).

We have

$$\frac{\partial}{\partial r}\,[(r - r_1)(r - r_2)\, x^r] = (r - r_1)(r - r_2)\, x^r \ln x + x^r\,[2r - (r_1 + r_2)]. \tag{3.224}$$

Expression (3.224) does not vanish for $r = r_1$ (or, for $r = r_2$). Thus, here we cannot seek the second solution in the form

$$\left[\frac{\partial}{\partial r} y(x)_r\right]_{r=r_1}$$

as in the case of equal roots. We, however, can select a_0 in the form

$$a_0 = (r - r_2)\xi, \quad \xi = \text{constant} \tag{3.225}$$

so that a situation of double root case is forcibly created. Then

$$\begin{aligned}\frac{\partial}{\partial r}\left[(r-r_1)(r-r_2)^2\xi x^r\right] &= (r-r_1)(r-r_2)^2\xi x^r \ln x + (r-r_2)^2\xi x^r \\ &\quad + 2(r-r_1)(r-r_2)\xi x^r.\end{aligned}$$

This vanishes identically for $r = r_2$. Hence, we can seek for the solution of the differential equation in the form

$$y_2(x) = \left[\frac{\partial}{\partial r} y_1(x)_r\right]_{r=2},$$

where coefficient a_0 of $y_1(x)$ has been changed by using (3.225).

EXAMPLE 3.40. *Solve:* $x(1-2x)\,y'' - (6x+1)\,y' - 2y = 0$.

Solution: This is the same equation discussed before. From equation (3.221), we find the coefficients $a_1, a_2, \ldots$ in the form

$$\left.\begin{aligned} a_1 &= \frac{2(r+1)}{r-1} a_0 \\ a_2 &= \frac{2(r+2)}{r} a_1 = \frac{2^2(r+1)(r+2)}{r(r-1)} a_0 \\ a_3 &= \frac{2(r+3)}{r+1} a_2 = \frac{2^3(r+1)(r+2)(r+3)}{r(r-1)(r+1)} a_0 \\ &\cdots\cdots\cdots\cdots\cdots\cdots \end{aligned}\right\}. \tag{1}$$

Hence, the solution of the differential equation is

$$y(x) = a_0 x^r\left[1 + \frac{r+1}{r-1}(2x) + \frac{(r+1)(r+2)}{r(r-1)}(2x)^2 + \frac{(r+1)(r+2)(r+3)}{r(r-1)(r+1)}(2x)^3 + \cdots\right]. \tag{2}$$

We now use equation (3.225) to change a_0 to

$$a_0 = (r - r_2)\xi = r\xi \quad (\because\ r_2 = 0). \tag{3}$$

Then the solution is

$$\begin{aligned} y(x) &= \xi x^r\left[r + \frac{r(r+1)}{r-1}(2x) + \frac{(r+1)(r+2)}{r-1}(2x)^2 + \frac{(r+1)(r+2)(r+3)}{(r-1)(r+1)}(2x)^3 + \cdots\right] \\ &= \xi x^r F(r), \text{ say.} \end{aligned} \tag{4}$$

For $r = 0$, one solution is

$$y(x) = \xi\left[-8x^2 - 2\cdot 3\cdot 8x^3 - 2\cdot 3\cdot 4\cdot 16x^4 - \cdots\right]$$
$$= -8\xi x^2\left[1 + \frac{2\cdot 3}{2}(2x) + \frac{3\cdot 4}{2}(2x)^2 + \cdots\right]. \quad (5)$$

Now, for differentiating equation (4) partially with respect to r, we write

$$\log y = [r\log x + \log F]\xi.$$

Differentiating

$$\frac{1}{y}\frac{\partial y}{\partial r} = \left[\log x + \frac{1}{F}\frac{\partial F}{\partial r}\right]\xi.$$

$$\therefore \quad \frac{\partial y}{\partial r} = \left[y\log x + \frac{y}{F}\frac{\partial F}{\partial r}\right]\xi = \left[y\log x + x^r\frac{\partial F}{\partial r}\right]\xi.$$

Now

$$\left.\frac{\partial F}{\partial r}\right|_{r=0} = \left[1 - 2x - 5(2x)^2 - 11(2x)^3 - \cdots\right] = P(x), \text{ say.}$$

$$\therefore \quad \left.\frac{\partial y}{\partial r}\right|_{r=0} = -8\xi x^2\log x\left[1 + \frac{2\cdot 3}{2}(2x) + \frac{3\cdot 4}{2}(2x)^2 + \cdots\right]$$
$$+ \xi\left[1 - (2x) - 5(2x)^2 - 11(2x)^3 - \cdots\right]. \quad (6)$$

Thus, one solution is given by equation (5) and the other by equation (6).

3.31 Wronskian Method

Let us consider the differential equation

$$\frac{d^2y}{dx^2} + p(x)\frac{dy}{dx} + q(x) = 0. \quad (3.226)$$

We assume that by series method or otherwise, one solution can be found. Let it be

$$y = y_1(x). \quad (3.227)$$

The other solution, $y = y_2(x)$ should be linearly independent so that the Wronskian of the solutions $y_1(x)$, $y_2(x)$ given by

$$W(x) = y_1(x)\,y_2'(x) - y_2(x)\,y_1'(x) \quad (3.228)$$

should not vanish at the point of observation.

We have also seen that the Wronskian is given by (*see* equation 3.93)

$$W(x) = Ce^{-\int p(x)\,dx} \quad (3.229)$$

provided $p(x)$ is a continuous function.

Equating (3.228) and (3.229), we write

$$y_1(x)\,y_2'(x) - y_2(x)\,y_1'(x) = Ce^{\int -p(x)\,dx},$$

$$\text{i.e.,}\quad \frac{y_1(x)\,y_2'(x) - y_2(x)\,y_1'(x)}{y_1^2(x)} = \frac{Ce^{\int -p(x)\,dx}}{y_1^2(x)},$$

$$\text{i.e.,}\quad \frac{d}{dx}\left[\frac{y_2(x)}{y_1(x)}\right] = \frac{Ce^{\int -p(x)\,dx}}{y_1^2(x)}.$$

$$\therefore \quad y_2(x) = y_1(x)\int \frac{Ce^{-\int p(x)\,dx}}{y_1^2(x)}. \tag{3.230}$$

Thus, if one solution $y_1(x)$ of the differential equation is known, the other independent solution can be found by using equation (3.230). However, in practice, this method rarely makes us fortunate.

EXAMPLE 3.41. *Find the second solution of* $\dfrac{d^2y}{dx^2} + \xi^2 y = 0$. *Given,* $\cos \xi y$ *is a solution.*

Solution: Here, $p(x) = 0$. Hence, $\displaystyle\int -p dx = 0$.

$$\therefore \quad e^{-\int p(x)\,dx} = 1.$$

$$\therefore \quad y_2(x) = \cos \xi x \int \frac{C}{\cos^2 \xi x} dx = C \cos \xi x \int \sec^2 \xi x\,dx = \frac{C}{\xi} \sin \xi x,$$

which is the second solution.

EXAMPLE 3.42. *Find the solution of* $\dfrac{d^2y}{dx^2} + \dfrac{1}{x}\dfrac{dy}{dx} - \dfrac{y}{x^2} = 0$.

Solution: If we seek for the solution in the form

$$y(x) = \sum_{n=0}^{\infty} a_n x^{n+r}$$

and substitute its derivatives into the equation, we get a recursion formula

$$a_n\left[(n+r)^2 - 1\right] = 0. \tag{1}$$

Putting $n = 0$, $r^2 - 1 = 0$, i.e., $r = \pm 1$.
Taking positive value of r, we find from (1) that

$$a_n\left[(n+1)^2 - 1\right] = 0.$$

This shows that all a_n's are zero, excepting for $n = 0$. Therefore, one solution of the equation is

$$y_1(x) = a_0 x. \tag{2}$$

Now, here $p(x) = \dfrac{1}{x}$; hence

$$e^{-\int p(x)\,dx} = e^{-\int \frac{dx}{x}} = \frac{1}{x}.$$

Therefore, the other solution of the equation is given by equation (3.230)

$$y_2(x) = y_1(x) \int \frac{A}{x} \cdot \frac{1}{(a_0 x)^2} dx = \frac{Ax}{a_0} \int \frac{dx}{x^3} = -\frac{A}{2a_0} \frac{1}{x} = b_0 \frac{1}{x}.$$

Therefore, the solution of the equation is

$$y(x) = a_0 x + b_0/x.$$

Summary

We now summarise the results of different methods for solving homogeneous second-order differential equations.

1. If the coefficient functions $p(x)$ and $q(x)$ of equation (3.194) are both analytic at $x = 0$, i.e., $x = 0$ is a regular point of the differential equation, we can always find two independent solutions.
2. If the roots of the indicial equation are distinct and do not differ by an integer, the series method will provide two independent solutions.
3. If the roots of the indicial equation are equal, the series method provides one solution and the other solution will be given by

$$\frac{\partial}{\partial r}[y_1(x)_r] \quad \text{for } r = r_1$$

and will always contain a $\ln x$ term.

4. If the roots of the indicial equation differ by an integer, one solution is obtained by series method, and the other is given by

$$\frac{\partial}{\partial r}[y_1(x)_r] \quad \text{for } r = r_2,$$

where coefficient a_0 of $y_1(x)$ obtained is to be multiplied by $(r - r_2)$ before taking the partial derivative. Here, the solution is of the form

$$y_2(x) = x^{r_2} \sum_{n=0}^{\infty} b_n x^n \quad \text{or,} \quad y_2(x) = Ay_1 \ln x + x^{r_2} \sum_{n=0}^{\infty} b_n x^n.$$

3.32 Non-homogeneous Equations

Let us consider the differential equation

$$\frac{d^2y}{dx^2} + p(x)\frac{dy}{dx} + q(x)\,y = r(x), \tag{3.231}$$

where $r(x)$ is a continuous function not equal to zero. This is a non-homogeneous equation and its general solution will be the sum of general solution of the homogeneous equation

$$\frac{d^2y}{dx^2} + p(x)\frac{dy}{dx} + q(x)\,y = 0 \tag{3.232}$$

and a particular solution of equation (3.231).

We have seen how to find general solution of homogeneous equation (3.232). Here we shall discuss how to find particular solution of non-homogeneous equation (3.231).

In Section 3.23, we have discussed the method of variation of parameters and solved as an example, a differential equation with constant coefficients. The same technique is applicable for non-homogeneous equations with coefficients as functions of x. Thus, if $y_1(x)$ and $y_2(x)$ are two independent solutions of homogeneous equation (3.232), the particular solution of equation (3.231) will be of the form

$$y_p(x) = y_1(x)u_1(x) + y_2(x)u_2(x), \tag{3.233}$$

where

$$u_1(x) = -\int \frac{y_2(x)r(x)}{W} dx \tag{3.234}$$

$$\text{and} \quad u_2(x) = \int \frac{y_1(x)r(x)}{W} dx \tag{3.235}$$

$W(x)$ being the Wronskian of solutions $y_1(x)$ and $y_2(x)$, and is given by

$$W(x) = Ae^{-\int p(x)\,dx}. \tag{3.236}$$

An alternative method of finding the particular solution of (3.231) is to assume it in a power series. Then, substituting into the equation (3.231) and comparing coefficients of like terms, we find the unknown coefficients of particular solution.

Let us assume that

$$y_p(x) = \sum_{n=0}^{\infty} C_n x^n. \tag{3.237}$$

If we substitute this along with its derivatives into equation (3.231), we will get

$$\sum_{n=0}^{\infty} C_n n(n-1)\, x^{n-2} + p(x) \sum_{n=0}^{\infty} C_n n x^{n-1} + q(x) \sum_{n=0}^{\infty} C_n x^n = r(x).$$

We can compare like terms only if $r(x)$ is replaced by its power series. Moreover, the functions $p(x)$ and $q(x)$ should either be constants or polynomials of x. Otherwise, these functions also should be replaced by series expansion.

EXAMPLE 3.43. *Find the general solution of* $\dfrac{d^2y}{dx^2} + \dfrac{1}{x}\dfrac{dy}{dx} - \dfrac{y}{x^2} = x^2$ *using the method of variation of parameters.*

Solution: Here, $p(x) = {}^1/_x$. Hence, the Wronskian

$$W = e^{-\int p(x)\,dx} = e^{-\int \frac{dx}{x}} = \frac{1}{x} \quad [A = \text{taken } 1].$$

Moreover, we know (*see* Example 3.42), that the two independent solutions of the given homogeneous equation are

$$y_1(x) = a_0 x, \quad y_2(x) = \frac{b_0}{x}.$$

$$\therefore \quad u_1(x) = -\int \frac{b_0}{x} \cdot x^2 \cdot x\,dx = -b_0 \frac{x^3}{3}$$

$$\text{and} \quad u_2(x) = \int a_0 x \cdot x^2 \cdot x\,dx = a_0 \frac{x^5}{5}.$$

Hence, the particular solution is

$$y_p(x) = y_1(x)u_1(x) + y_2(x)u_2(x) = (a_0 x)\left(-b_0 \frac{x^3}{3}\right) + \frac{b_0}{x}\left(a_0 \frac{x^5}{5}\right) = Ax^4.$$

EXAMPLE 3.44. *Find the general solution of* $x\dfrac{d^2y}{dx^2} + \dfrac{dy}{dx} + 2y = x^2$.

Solution: We have found the two independent solutions of this homogeneous equation (*see* Example 3.39). Here we shall find the particular solution only.

Let us assume the particular solution of the equation in the form

$$y_p(x) = \sum_{n=0}^{\infty} b_n x^n \tag{1}$$

Then

$$y_p' = \sum_{n=0}^{\infty} b_n n x^{n-1} \tag{2}$$

$$y_p'' = \sum_{n=0}^{\infty} b_n n(n-1)\, x^{n-2}. \tag{3}$$

Substituting (1), (2), (3) into the given equation, we find

$$\sum_{n=1}^{\infty} b_n n(n-1)\, x^{n-1} + \sum_{n=1}^{\infty} b_n n x^{n-1} + 2\sum_{n=1}^{\infty} b_n x^n = x^2,$$

$$\text{i.e.,} \quad \sum_{n=1}^{\infty} b_n n^2 x^{n-1} + 2\sum_{n=0}^{\infty} b_n x^n = x^2.$$

Equating the coefficients of similar power of x on both sides, we get

$$x^0 \longrightarrow b_1 + 2b_0 = 0, \qquad \text{i.e.,} \quad b_1 = -2b_0$$

$$x^1 \longrightarrow 4b_2 + 2b_1 = 0, \quad \text{i.e.,} \quad b_2 = -\frac{1}{2}b_1 = b_0$$

$$x^2 \longrightarrow 9b_3 + 2b_2 = 1, \quad \text{i.e.,} \quad b_3 = \frac{(1-2b_0)}{9}$$

$$x^3 \longrightarrow 16b_4 + 2b_3 = 0, \quad \text{i.e.,} \quad b_4 = -\frac{b_3}{8} = -\frac{1-2b_0}{72}$$

$$\cdots\cdots\cdots\cdots \qquad \cdots\cdots\cdots\cdots$$

Hence, we can write

$$y_p(x) = b_0 - 2b_0 x + b_0 x^2 + \frac{1-2b_0}{9}x^3 - \frac{1-2b_0}{72}x^4 + \cdots$$

The value of b_0 is generally obtained from initial conditions.

Thus, the general solution is

$$y(x) = Ay_1(x) + By_2(x) + y_p(x).$$

3.33 Simultaneous Differential Equations

Let us consider the following system of linear differential equations

$$\left.\begin{aligned}\frac{dx}{dt} &= a_1x + b_1y + c_1z \\ \frac{dy}{dt} &= a_2x + b_2y + c_2z \\ \frac{dz}{dt} &= a_3x + b_3y + c_3z\end{aligned}\right\}, \tag{3.238}$$

where a_1, b_1, ..., etc., are constants, t is the argument and $x(t)$, $y(t)$, $z(t)$ are functions of t. Such a system of equations is called a *system of linear homogeneous differential equations* with constant coefficients. The system may consist of any number of equations.

We shall see to how to solve such a system of equation with the help of examples.

EXAMPLE 3.45. *Solve:*

$$\frac{dx}{dt} - 3x - 2y = 0$$
$$\frac{dy}{dt} + 5x + 3y = 0.$$

Solution: Differentiating the first equation, we get

$$\frac{d^2x}{dt^2} = 3\frac{dx}{dt} + 2\frac{dy}{dt}. \tag{1}$$

From second equation, we have

$$\frac{dy}{dt} = -5x - 3y. \tag{2}$$

Substituting equation (2) into (1)

$$\frac{d^2x}{dt^2} = 3\frac{dx}{dt} + 2(-5x - 3y) = 3\frac{dx}{dt} - 10x - 6y. \tag{3}$$

From first equation, we have

$$2y = \frac{dx}{dt} - 3x. \tag{4}$$

Substituting equation (4) into (3), we get

$$\frac{d^2x}{dt^2} = 3\frac{dx}{dt} - 10x - 3\left(\frac{dx}{dt} - 3x\right), \quad \text{i.e.,} \quad \frac{d^2x}{dt^2} = -x, \quad \text{i.e.,} \quad \frac{d^2x}{dt^2} + x = 0.$$

Its solution is

$$x = A\cos t + B\sin t. \tag{5}$$

Hence, from equation (4), we get

$$y = \tfrac{1}{2}[-A\sin t + B\cos t] - \tfrac{3}{2}(A\cos t + B\sin t) = \tfrac{1}{2}(B - 3A)\cos t - \tfrac{1}{2}(A + 3B)\sin t.$$

EXAMPLE 3.46. *Solve:*

$$\frac{dx}{dt} - \frac{dy}{dt} - y = -e^t$$
$$\frac{dy}{dt} + x - y = e^{2t}.$$

Solution: We denote the differential operator

$$D = \frac{d\cdots}{dt}.$$

Then given equations can be written as

$$Dx - (D+1)\,y = -e^t \qquad (1)$$
$$(D-1)\,y + x = e^{2t}. \qquad (2)$$

Eliminating x from equations (1) and (2)

$$D(D-.1)\,y + (D+1)\,y = D\left(e^{2t}\right) + e^t$$
$$\text{or,}\quad \left(D^2+1\right)\,y = D\left(e^{2t}\right) + e^t,$$
$$\text{i.e.,}\quad \frac{d^2y}{dt^2} + 1 = \frac{d}{dt}\left(e^{2t}\right) + e^t = 2e^{2t} + e^t.$$

Its complementary solution is

$$y = A\cos t + B\sin t. \qquad (3)$$

Particular integral is

$$y = \frac{2}{5}e^{2t} + \frac{1}{2}e^t. \qquad (4)$$

Thus, general solution for y is

$$y = A\cos t + B\sin t + \frac{2}{5}e^{2t} + \frac{1}{2}e^t. \qquad (5)$$

From equation (2), we get

$$\begin{aligned} x &= e^{2t} - (D-1)\,y = e^{2t} - Dy + y \\ &= e^{2t} - \left[-A\sin t + B\cos t + \frac{4}{5}e^{2t} + \frac{1}{2}e^t\right] + \left(A\cos t + B\sin t + \frac{2}{5}e^{2t} + \frac{1}{2}e^t\right) \\ &= (A-B)\cos t + (A+B)\sin t + \frac{3}{5}e^{2t}. \end{aligned}$$

Note. The system of differential equations may be of any higher order. If the coefficients are constants, the method of solving such a system is actually the same as above. If, however, the coefficients are functions, these methods cannot be used. We shall not discuss about systems of differential equations with variable coefficients. This is beyond the scope of this book.

EXERCISE

1. Solve the following equations by separating the variables:
   (a) $(x^2-1)\dfrac{dy}{dx}+2xy^2=0,\ y(0)=1;$
   (b) $x\dfrac{dy}{dx}+y=y^2,\ y(1)=0.5;$
   (c) $2x^2y\dfrac{dy}{dx}+y^2=2;$
   (d) $\sqrt{y^2+1}\,dx=xy\,dy.$

2. Solve the following homogeneous equations:
   (a) $(y^2-2xy)\,dx+x^2dy=0;$
   (b) $2x^3\dfrac{dy}{dx}=y\left(2x^2-y^2\right);$
   (c) $x\dfrac{dy}{dx}-y=x\tan\left(\dfrac{y}{x}\right);$
   (d) $(y+\sqrt{xy})\,dx=x\,dy;$
   (e) $(x+4y)\dfrac{dy}{dx}=2x+3y-5;$
   (f) $y\,dx+x(2xy+1)\,dy=0.$

3. Solve the following equations by integrating factor method:
   (a) $(2x+1)\dfrac{dy}{dx}=4x+2y;$
   (b) $\dfrac{dy}{dx}=2x\left(x^2+y\right);$
   (c) $(x+y^2)\,dy=y\,dx;$
   (d) $(1-2xy)\dfrac{dy}{dx}=y(y-1).$

4. Check for exactness of the equations and solve them:
   (a) $2xy\,dx+(x^2-y^2)\,dy=0;$
   (b) $(2-9xy^2)\,x\,dx+(4y^2-6x^3)\,y\,dy=0;$
   (c) $e^{-y}dx-(2y+xe^{-y})\,dy=0;$
   (d) $2x\left(1+\sqrt{x^2-y}\right)dx-\sqrt{x^2-y}\,dy=0.$

5. Solve the following equations of Clairaut and Lagrange:
   (a) $y=x\dfrac{dy}{dx}-\left(\dfrac{dy}{dx}\right)^2;$
   (b) $y+x\dfrac{dy}{dx}=4\sqrt{\dfrac{dy}{dx}};$
   (c) $y=x\left(\dfrac{dy}{dx}\right)^2-2\left(\dfrac{dy}{dx}\right)^3;$
   (d) $x\dfrac{dy}{dx}-y=\ln\dfrac{dy}{dx}.$

6. Temperature at sea level is 283 K and the lapse rate is assigned a value of 6.5 K/km. Find the ratio of air pressure at a height of 10 km and that at sea level.

7. Water is flowing out through an aperture at the bottom of a rectangular tank of base 2 m×3 m. Initially, water level is 4.5 m. If the diameter of the aperture is 10 cm, find the time when the tank will be empty. Derive the formula you use.

8. In a pond, there is already an ice-block of 5 cm thickness. If the temperature of air is $-20\,°$C, how long will it take for the thickness of ice to become double? Coefficient of thermal conductivity of ice = 1.045(SI), density of ice = 910 kg/m^3 and latent heat of ice = 3.36×10^5 J/kg.

9. A body is falling freely under gravity through the medium where resistance to the body is proportional to its velocity. Find the time when its velocity will be half of the terminal velocity.

10. A raindrop, while falling freely under gravity, accumulates an increase in volume in each instant which is 0.001 times its surface area. If its initial radius is 0.6 cm, find the time when its radius will be doubled. Also, find the distance covered during that time.

11. With substitution $p = y'$, solve the following equations:
(a) $y'' + (y')^2 + 1 = 0$;
(b) $y'' = 2yy'$;
(c) $2xy'y'' = (y')^2 - 1$;
(d) $(1 - x^2)\, y'' + xy' = 2$.

12. Find general solutions of:
(a) $y'' + 4y' + 3y = 0$;
(b) $4y'' + 4y' + y = 0$;
(c) $y'' - 3y' + 3y = 0$;
(d) $y'' - 2y' + y = 0$;
(e) $2y'' - 5y' + 2y = 0$;
(f) $y'' + 4y = 0$.

13. Find solutions of:
(a) $\dfrac{d^2y}{dx^2} + 9y = x^2 + \sin 3x$;
(b) $\dfrac{d^2y}{dx^2} - 16y = e^{4x}$;
(c) $\dfrac{d^2y}{dx^2} + \dfrac{dy}{dx} + 2y = 4e^x + 2x^2$;
(d) $\dfrac{d^2y}{dx^2} + 4\dfrac{dy}{dx} + 5y = x^2 + 5$, $y(0) = 0$ and $\dfrac{dy}{dx}(0) = 0$;
(e) $\dfrac{d^2y}{dx^2} + 6\dfrac{dy}{dx} + 10y = \cos 2x$, $y(0) = 0$ and $\dfrac{dy}{dx}(0) = 0$.

14. Find normal forms of the equations and solve them:
(a) $x^2\dfrac{d^2y}{dx^2} - 2x\dfrac{dy}{dx} + (a^2x^2 + 2)\, y = 0$;
(b) $x\dfrac{d^2y}{dx^2} - 2(x-1)\dfrac{dy}{dx} + 2(x-1)\, y = 0$;
(c) $x^2\dfrac{d^2y}{dx^2} + x\dfrac{dy}{dx} + (x^2 - 1)\, y = 0$;
(d) $\dfrac{d^2y}{dx^2} - 2x\dfrac{dy}{dx} + 2y = 0$.

15. Solve by changing independent variable the following equations:
(a) $x^2\dfrac{d^2y}{dx^2} + 3x\dfrac{dy}{dx} + 5y = 0$;
(b) $x\dfrac{d^2y}{dx^2} + (8x^2 - 1)\dfrac{dy}{dx} + 20x^3y = 0$;
(c) $2x\dfrac{d^2y}{dx^2} + (5x^2 - 2)\dfrac{dy}{dx} + 2x^3y = 0$.

16. Solve the following Riccati equations:
(a) $y' = x^2y^2 - x^4 + 1$;
(b) $y' = y^2 + 2x - x^4$.

17. Solve the following equations:
(a) $y'' - 2y' + y = 6xe^x$;
(b) $y'' - 4y' + 8y = e^{2x} + \sin 2x$;
(c) $y'' - 3y' + 2y = x\cos x$;
(d) $x^2y'' + xy' - y = 2x^2$;
(e) $xy'' - y' = (1 + x)\, x$;

18. Solve the following equations by series method:

(a) $y'' + x^2y = 0$;
(b) $(4 - x^2)\, y'' + 2y = 0$, with $y(0) = 0$, $y'(0) = 1$;
(c) $(1 - x)\, y'' - 2yy' + y = 0$;

(d) $xy'' + y\ln(1-x) = 0$;

(e) $y'' - x^2y' + y\sin x = 0$.

19. One solution of $y'' + 2\xi y' + \xi^2 y = 0$ (ξ = constant) is $e^{-\xi x}$. Find a second independent solution by Wronskian method.

20. Find the general solution of $y'' - \tan x y' + 2y = 0$ if $y = \sin x$ is a solution.

21. Show that, the equation $xy'' - 2(x-1)\,y' + 2(x-1)\,y = 0$ can be reduced to normal form with constant coefficients. Then solve it.

22. Show that the general solution of $(1+x)\,y'' + xy' - y = 0$ has the form $y = Ax + Be^{-x}$.

23. By Frobenius method, one solution of $xy'' - (1+2x)\,y' + (1+x)\,y = 0$ is obtained as $y = Ae^x$. Find the second solution.

24. Find the solutions of the following initial value problems:

(a) $y'' + x^2y = x$, $y(0) = y'(0) = 0$;

(b) $y'' + (1-x)\,y = 4x$, $y(0) = 1$, $y'(0) = 0$.

25. Solve the equation $(1-x)\,y' - y = 2x$ in the interval $x = 1$ to $x = 2$. Expanding about $x = 2$, determine the value of $y(x)$ at $x = 1.9$ if $y(2) = 1$.

26. Solve the following systems of equations:

(a) $\dfrac{dx}{dt} + 5x + y = e^t$,

$\dfrac{dy}{dt} - x + 3y = e^{2t}$;

(b) $\dfrac{dx}{dt} + \dfrac{dy}{dt} - 2y = \cos t$,

$\dfrac{dx}{dt} - \dfrac{dy}{dt} + 2x = -\frac{1}{2}\sin t$;

(c) $2\dfrac{dx}{dt} + \dfrac{dy}{dt} - 4x - y = e^t$,

$\dfrac{dx}{dt} + 3x + y = 0$;

(d) $\dfrac{dx}{dt} + 2x - 3y = t$,

$\dfrac{dy}{dt} - 3x + 2y = e^{2t}$;

(e) $\dfrac{d^2x}{dt^2} - 3x - 4y = 0$,

$\dfrac{d^2y}{dt^2} + x + y = 0$.

ANSWERS

1. (a) $y\ln\{|x^2-1|+C\} = 1$;

(b) $y(1-Cx) = 1, y = 0, y(1+x) = 1$;

(c) $y^2 - 2 = Ce^{1/x}$;

(d) $\ln|x| = C + \sqrt{y^2+1}$.

2. (a) $x(y-x) = Cy, y = 0;$
(b) $x = \pm y\sqrt{\ln Cx}, y = 0;$
(c) $\sin\left(\frac{y}{x}\right) = Cx;$
(d) $x \ln Cx = 2\sqrt{xy}, y = 0;$
(e) $(y-x+5)^5(x+2y-2) = C;$
(f) $y^2 \exp\left(-\frac{1}{xy}\right) = C, y = 0, x = 0.$

3. (a) $y = (2x+1)(C + \ln|2x+1|) + 1;$
(b) $y = Ce^{x^2} - x^2 - 1;$
(c) $x = y^2 + Cy;$
(d) $(y-1)^2 x = y - \ln Cy, y = 0, y = 1.$

4. (a) $3x^2y - y^3 = C;$
(b) $x^3 - 3x^3y^2 + y^4 = C;$
(c) $xe^{-y} - y^2 = C;$
(d) $x^2 + \frac{2}{3}(x^3 - y)^{3/2} = C.$

5. (a) $y = Cx - C^2, 4y = x^2;$
(b) $x\sqrt{p} = \ln p + C, y = \sqrt{p}(4 - \ln p - C);$
(c) $x = C(p-1)^{-2} + 2p + 1,$
$y = Cp^2(p-1)^{-2} + p^2, y = x - 2, y = 0;$
(d) $y = Cx - \ln C, y = \ln x + 1.$

6. 0.934.

7. $t = \frac{4ab}{\pi d^2}\sqrt{\frac{2h_0}{g}}, t = 732$ sec.

8. 15.24 hour.

9. 0.693 m/μ, m = mass of the body, μ = constant.

10. 600ρ secs; $99.3 \times 10^{10}\rho^2$, ρ = density.

11. (a) $x + C_2 = \pm \sin^{-1}\left(\frac{e^y}{C}\right);$
(b) $y = C_1 \tan(C_1 x + C_2);$
(c) $9C_1^2(y - C_2)^2 = 4(C_1 x + 1)^3;$
(d) $y = C_1\left[x\sqrt{x^2-1} - \ln\left|x + \sqrt{x^2-1}\right|\right] + x^2 + C_2.$

12. (a) $y = C_1 e^{-x} + C_2 e^{-3x};$
(b) $y = e^{-x/2}(C_1 + C_2 x);$
(c) $y = e^{3x/2}\left(C_1 \sin\frac{\sqrt{3}x}{2} + C_2 \cos\frac{\sqrt{3}x}{2}\right);$
(d) $y = e^x(C_1 + C_2 x);$
(e) $y = C_1 e^{2x} + C_2 e^{x/2};$
(f) $y = C_1 \cos 2x + C_2 \sin 2x.$

13. (a) $y = Aby + B\sin 3x + \frac{x^2}{2} - \frac{2}{81} - \frac{x}{6}\cos 3x;$
(b) $y = Ae^{4x} + Be^{-4x} + \frac{1}{8}xe^{4x};$
(c) $y = e^{-x/2}\left(A\cos\frac{\sqrt{7}x}{2} + B\sin\frac{\sqrt{7}x}{2}\right) + e^x + x^2 - x - \frac{1}{2};$
(d) $y = e^{-2x}\left(-\frac{147}{125}\cos x - \frac{254}{125}\sin x\right) + \frac{x^2}{5} - \frac{8x}{25} + \frac{147}{125};$
(e) $y = \frac{e^{-3x}}{18}(\cos x - \sin x) - \frac{1}{18}\cos 2x + \frac{1}{9}\sin 2x.$

14. (a) $y = x(A\cos ax + B\sin ax)$;
(b) $y = \dfrac{e^x}{x}(A\cos x + B\sin x)$;
(c) $y = 1 - \dfrac{3}{4x^2}$;
(d) $y = 3 - x^2$.

15. (a) $y = e^{-x}[A\cos(2\ln x) + B\sin(2\ln x)]$;
(b) $y = e^{x^2}\left[A\cos(x^2) + B\sin(x^2)\right]$;
(c) $y = e^{-x^2/2}\left[A\cos\left(\dfrac{x^2}{4}\right) + B\sin\left(\dfrac{x^2}{4}\right)\right]$.

16. (a) $y = x - \dfrac{e^{x^4/2}}{A + \int_0^x x^2 e^{x^4/2}dx}$;
(b) $y = x^2 - \dfrac{e^{2x^3/3}}{A + \int e^{2x^3/3}dx}$.

17. (a) $y = (A + Bx + x^3)e^x$;
(b) $y = e^{2x}(A\cos 2x + B\sin 2x) + \dfrac{1}{4}e^{2x} + \dfrac{1}{10}\cos 2x$;
(c) $y = Ae^x + Be^{2x} + \left(\dfrac{x}{10} - \dfrac{3}{25}\right)\cos x - \left(\dfrac{3x}{10} - \dfrac{17}{50}\right)\sin x$;
(d) $y = Ax + \dfrac{B}{x} + \dfrac{2x^2}{3}$;
(e) $y = Ax^2 + B + \dfrac{x^3}{3} - \dfrac{x^2}{4} + \dfrac{x^2}{2}\ln x$.

18. (a) $y = A\left(1 - \dfrac{x^4}{12} + \dfrac{x^8}{672} + \cdots\right) + B\left(x - \dfrac{x^5}{20} + \dfrac{x^9}{1440} + \cdots\right)$;
(b) $y = x - 2\sum \dfrac{1}{4k^2 - 1}\left(\dfrac{x}{2}\right)^{2k+1}$;
(c) $y = Ay_1(x) + By_2(x)$

$$y_1(x) = 1 - \frac{x^2}{2} - \frac{x^3}{2} - \frac{11x^4}{24} - \cdots$$
$$y_2(x) = x + x^2 + \frac{5x^3}{6} + \frac{3x^4}{4} + \cdots;$$

(d) $y_1(x) = 1 + \dfrac{x^2}{2} + \dfrac{x^3}{12} + \dfrac{5x^4}{72} + \cdots$

$$y_2(x) = x + \frac{x^3}{6} + \frac{x^4}{24} + \cdots;$$

(e) $y = x + \dfrac{x^6}{180} + \dfrac{x^8}{6720} + \cdots$.

19. $y_2 = xe^{-\xi x}$.

20. $y = A\sin x + B\left[2 - \sin x \cdot \ln\dfrac{1 + \sin x}{1 - \sin x}\right]$.

23. $y_2 = xe^x$.

24. (a) $y(x) = \dfrac{1}{6}\left[x^3 + \sum \dfrac{(-1)^k x^{4k+1}}{6.7\cdots(4k+2)(4k+3)}\right]$;

(b) $y(x) = 1 - \frac{x^2}{2} + \frac{5x^3}{6} + \frac{x^4}{24} + \cdots$.

26. (a) $x = (A + Bt)e^{-4t} + \frac{4}{25}e^{t} - \frac{1}{36}e^{2t}$

$y = -[A + (1 + t)B]e^{-4t} + \frac{1}{25}e^{t} + \frac{7}{36}e^{2t};$

(b) $x = Ae^{\sqrt{2}t} + Be^{-\sqrt{2}t} + \frac{1}{6}\cos t$

$y = A\left(1 + \sqrt{2}\right)e^{\sqrt{2}t} + B\left(1 - \sqrt{2}\right)e^{\sqrt{2}t} - \frac{5}{12}\cos t - \frac{1}{3}\sin t;$

(c) $x = A\cos t + B\sin t - \frac{1}{2}e^{t}$

$y = (A - 3B)\sin t - (B + 3A)t + 2e^{t};$

(d) $x = Ae^{-5t} + Be^{t} + \frac{3}{7}e^{2t} - \frac{2}{5}t - \frac{23}{25}$

$y = -Ae^{-5t} + Be^{t} + \frac{4}{7}e^{2t} - \frac{3}{5}t - \frac{12}{25};$

(e) $x = (A + Bt)e^{t} + (C + Dt)e^{-t}$

$y = [B(1 - t) - A]\,e^{t} - [C + D(1 + t)]\,e^{-t/2}.$

Chapter 4

Differential Equations in Physical Processes

We have, in the previous chapter, studied some physical processes leading to ordinary differential equations of first order. Here in this chapter, we intend to peep into a few corners of physics, where extensive use of differential equations of second order has been made.

I. ELECTRICAL CURRENTS VARYING WITH TIME

4.1 Charging and Discharging through Inductance

Let us consider an electrical circuit having a condenser and an inductive coil connected in series. Let (*Fig 4.1*),

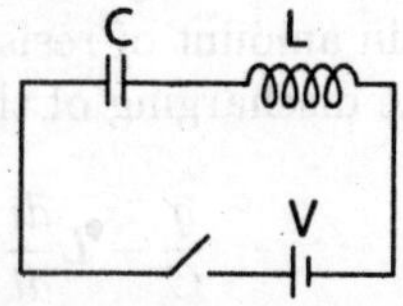

Fig 4.1

L = inductance of the coil;
C = capacitance of the condenser;
q = instantaneous charge on the positive plate of the condenser;
i = instantaneous current in the circuit;
V = steady voltage of the cell.

Then, the e.m.f. equation is written as

$$L\frac{di}{dt} + \frac{q}{C} = V, \quad \text{or,} \quad L\frac{d^2q}{dt^2} + \frac{q}{C} = V, \quad \text{i.e.,} \quad \frac{d^2q}{dt^2} + \beta^2 q = \gamma, \tag{4.1}$$

where

$$\beta^2 = \frac{1}{LC}, \quad \gamma = \frac{V}{L}. \tag{4.2}$$

Thus, the charge on the condenser plate at any instant of time is determined by solving the differential equation (4.1).

For the solution of equation (4.1), we assume

$$q = Ae^{\lambda t}. \tag{4.3}$$

Substituting the expression (4.3) in homogeneous equation (4.1), we have

$$\lambda^2 + \beta^2 = 0, \quad \text{i.e.,} \quad \lambda = \pm j\beta, \quad j = \sqrt{-1}. \tag{4.4}$$

Particular solution is

$$\frac{\gamma}{\beta^2} = \frac{V}{L}\Big/\frac{1}{LC} = CV. \tag{4.5}$$

Hence, the complete solution is

$$q = Ae^{j\beta t} + Be^{-j\beta t} + CV. \tag{4.6}$$

Initially at $t = 0$, we have

$$q = 0, \quad \frac{dq}{dt} = 0. \tag{4.7}$$

These conditions give

$$\begin{aligned} A + B + CV &= 0 \\ A - B &= 0. \end{aligned}$$

Solving these, we get

$$A = B = -\tfrac{1}{2}CV = -\tfrac{1}{2}q_0, \tag{4.8}$$

where q_0 is the final charge developed in the steady state.

Hence, the solution given by equation (4.6) can be written as

$$q = q_0(1 - \cos\beta t), \quad \beta = \sqrt{\frac{1}{LC}}. \tag{4.9}$$

This shows that the charge on the condenser oscillates around q_0. But in practice, the inductive coil has certain amount of resistance which dampens the charging process.

During the discharging of the condenser, we can write the e.m.f. equation as

$$\frac{q}{C} - L\frac{di}{dt} = 0, \quad i = -\frac{dq}{dt}. \quad \therefore \quad \frac{d^2q}{dt^2} + \beta^2 q = 0. \tag{4.10}$$

Its solution is

$$q = Ae^{j\beta t} + Be^{-j\beta t}. \tag{4.11}$$

Now, at $t = 0$,

$$q = q_0 \quad \text{and} \quad \frac{dq}{dt} = 0.$$

Therefore, we have from equation (4.11)

$$\begin{aligned} A + B &= q_0 \\ A - B &= 0. \\ \therefore \quad A = B &= \tfrac{1}{2}q_0. \end{aligned}$$

Solution function (4.11) becomes

$$q = q_0 \cos\beta t, \quad \beta = \sqrt{\frac{1}{LC}}. \tag{4.12}$$

This shows that discharging of the condenser goes on indefinitely. This also does not happen in practice because of the presence of certain resistance in the circuit.

4.2 Charging in L-C-R Circuit

Let us consider the most general circuit consisting of inductance (L), capacitance (C) and resistance (R) (*Fig 4.2*).

The charge q on the condenser at any instant is determined by the equation

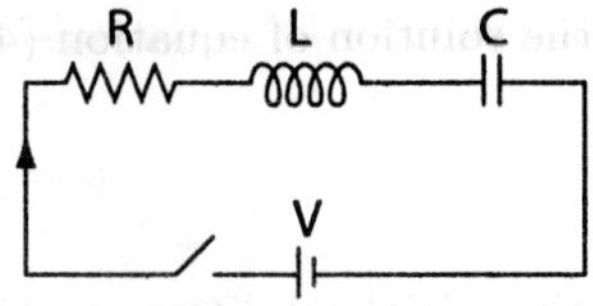

Fig 4.2

$$Ri + L\frac{di}{dt} + \frac{q}{C} = V,$$

$$\text{i.e.,} \quad \frac{d^2q}{dt^2} + \frac{R}{L}\frac{dq}{dt} + \frac{q}{LC} = \frac{V}{L},$$

$$\text{i.e.,} \quad \frac{d^2q}{dt^2} + 2\alpha\frac{dq}{dt} + \beta^2 q = \gamma, \qquad (4.13)$$

where

$$2\alpha = \frac{R}{L}, \quad \beta^2 = \frac{1}{LC}, \quad \gamma = \frac{V}{L}.$$

Assuming the solution of homogeneous equation (4.13) in the form $q = Ae^{\lambda t}$ we find the characteristic equation as

$$\lambda^2 + 2\alpha\lambda + \beta^2 = 0. \qquad (4.14)$$

We now consider the following situations:

Case I. *When* $\alpha^2 > \beta^2$, *i.e.,* $\frac{R^2}{4L^2} > \frac{1}{LC}$.

In this case, the roots of the equation (4.14) are real. Substituting $\sqrt{\alpha^2 - \beta^2} = \xi$, we can write the solution of homogeneous equation as $q_1 = e^{-\alpha t}\left[Ae^{\xi t} + Be^{-\xi t}\right]$.

The particular integral, as before, is $CV = q_0$.

Thus, the general solution is given by

$$q = q_0 + e^{-\alpha t}\left[Ae^{\xi t} + Be^{-\xi t}\right]. \qquad (4.15)$$

Initially, at $t = 0$, we have

$$q = 0 \quad \text{and} \quad \frac{dq}{dt} = 0. \qquad (4.16)$$

Using these conditions in equation (4.15), we get $A+B = -q_0$ and $\xi(A-B) - \alpha(A+B) = 0$.
Solving we get

$$\left.\begin{aligned} A &= -\frac{1}{2}q_0\left(1 + \frac{\alpha}{\xi}\right) \\ B &= -q_0 - A = -\frac{1}{2}q_0\left(1 - \frac{\alpha}{\xi}\right) \end{aligned}\right\}. \qquad (4.17)$$

Thus, the specific solution of equation (4.13) is

$$q = q_0 - \tfrac{1}{2}q_0\, e^{-\alpha t}\left[\left(1 + \frac{\alpha}{\xi}\right)e^{\xi t} + \left(1 - \frac{\alpha}{\xi}\right)e^{-\xi t}\right]. \qquad (4.18)$$

This shows that the charge on the condenser grows exponentially, and as $t \longrightarrow \infty$, $q \longrightarrow q_0$.

Case II. *When* $\alpha^2 < \beta^2$, *i.e.,* $\frac{R^2}{4L^2} < \frac{1}{LC}$, *i.e.,* $R^2 < \frac{4L}{C}$.

In this case, we substitute $\xi = \sqrt{\beta^2 - \alpha^2}$, so that

$$\sqrt{\alpha^2 - \beta^2} = j\xi, \quad j = \sqrt{-1}. \qquad (4.19)$$

Then the solution of equation (4.13) becomes

$$q = q_0 + e^{-\alpha t}\left[Ae^{j\xi t} + Be^{-j\xi t}\right]. \tag{4.20}$$

Using the initial conditions (4.16), we get

$$\left.\begin{aligned} A &= -\frac{1}{2}q_0\left(1+\frac{\alpha}{\xi}\right) \\ B &= -\frac{1}{2}q_0\left(1-\frac{\alpha}{\xi}\right) \end{aligned}\right\}. \tag{4.21}$$

Hence, the specific solution of equation (4.13) is

$$\begin{aligned} q &= q_0 - \tfrac{1}{2}q_0\, e^{-\alpha t}\left[\left(1+\frac{\alpha}{\xi}\right)e^{j\xi t} + \left(1-\frac{\alpha}{\xi}\right)e^{-j\xi t}\right] \\ &= q_0 - q_0\, e^{-\alpha t}\left[\cos\xi t + \frac{\alpha}{\xi}\sin\xi t\right] \\ &= q_0 - \frac{D}{\xi}q_0\, e^{-\alpha t}\,\sin(\xi t + \phi), \end{aligned} \tag{4.22}$$

where $\xi = D\sin\phi$ and $\alpha = D\cos\phi$, so that

$$\tan\phi = \frac{\xi}{\alpha} = \frac{\sqrt{\beta^2-\alpha^2}}{\alpha} \quad\text{and}\quad D^2 = \alpha^2 + \xi^2 = \alpha^2 + \beta^2 - \alpha^2 = \beta^2.$$

The final solution given in explicit form is

$$q = q_0\left[1 - \frac{e^{-Rt/2L}\sqrt{\dfrac{1}{LC}}}{\sqrt{\dfrac{1}{LC} - \dfrac{R^2}{4L^2}}}\left\{\sin\sqrt{\frac{1}{LC} - \frac{R^2}{4L^2}}\,t + \phi\right\}\right], \tag{4.23}$$

where

$$\phi = \tan^{-1}\frac{\xi}{\alpha} = \tan^{-1}\frac{\sqrt{\dfrac{1}{LC} - \dfrac{R^2}{4L^2}}}{\dfrac{R}{2L}}. \tag{4.24}$$

The current in the circuit calculated from equation (4.23) will be obtained in the form

$$i = \frac{dq}{dt} = \frac{V}{L}\cdot\frac{e^{-Rt/2L}}{\sqrt{\dfrac{1}{LC} - \dfrac{R^2}{4L^2}}}\sin\sqrt{\frac{1}{LC} - \frac{R^2}{4L^2}}\,t. \tag{4.25}$$

Equations (4.23) and (4.25) show that both charge and current reach final value after some oscillations about the initial values. *Fig 4.3* shows how the charge on the condenser during the process of charging varies with time. From the figure, it is seen that the maximum charge is much greater than the final steady charge on the condenser.

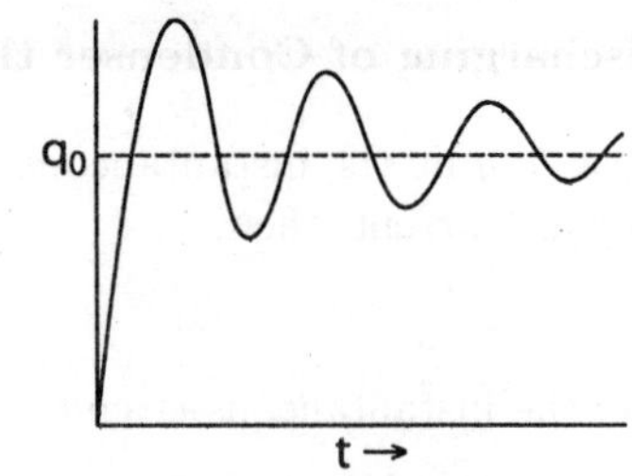

Fig 4.3

If T be the period of oscillation of charge or current, then

$$T = \frac{2\pi}{\xi} = \frac{2\pi}{\sqrt{\dfrac{1}{LC} - \dfrac{R^2}{4L^2}}}. \tag{4.26}$$

The frequency of oscillation of charge or current is given by

$$f = \frac{1}{T} = \frac{1}{2\pi}\sqrt{\frac{1}{LC} - \frac{R^2}{4L^2}}. \tag{4.27}$$

Case III. *When* $\alpha^2 = \beta^2$, *i.e.*, $\frac{R^2}{4L^2} = \frac{1}{LC}$, *i.e.*, $R^2 = \frac{4L}{C}$.

In this case, $\xi = 0$ and roots of the characteristic equation are equal. Hence, the solution of equation (4.13) is

$$q = q_0 + (A + Bt)\, e^{-\alpha t}. \tag{4.28}$$

Using the initial conditions (4.16), we find

$$A = -q_0, \quad B = -q_0\alpha. \tag{4.29}$$

Hence, the specific solution of equation (4.13) is

$$q = q_0\left[1 - (1 + \alpha t)\, e^{-\alpha t}\right] = q_0\left[1 - \left(1 + \frac{R}{2L}t\right) e^{-Rt/2L}\right]. \tag{4.30}$$

The instantaneous current is

$$i = \frac{dq}{dt} = q_0 \frac{R^2 t}{4L^2}\, e^{-Rt/2L}. \tag{4.31}$$

From equation (4.27), we find that the frequency of oscillation is zero in this case. Thus, when

$$R = \frac{4L}{C} \tag{4.32}$$

the oscillation just begins or just stops. This phenomenon is known as *critical damping* and the resistance given by (4.32) is called *critical damping resistance*.

4.3 Discharging of Condenser through Inductive Coil

In this case, if q be the instantaneous charge on the positive plate of condenser, and if i be the instantaneous current, then

$$i = -\frac{dq}{dt}. \tag{4.33}$$

Hence, the instantaneous circuit equation is

$$\frac{q}{C} - Ri - L\frac{di}{dt} = 0 \quad \text{or,} \quad \frac{d^2q}{dt^2} + \frac{R}{L}\frac{dq}{dt} + \frac{q}{LC} = 0 \quad \text{or,} \quad \frac{d^2q}{dt^2} + 2\alpha\frac{dq}{dt} + \beta^2 q = 0, \tag{4.34}$$

where, as before $2\alpha = R/L$, $\beta^2 = 1/LC$.

The factor R/L is known as *attenuation constant.*

Case I. *When* $\alpha^2 > \beta^2$, *i.e.,* $R^2 > \frac{4L}{C}$.

Proceeding in the same way as before, we write the solution of equation (4.34) as

$$q = e^{-\alpha t}\left[Ae^{\xi t} + Be^{-\xi t}\right], \tag{4.35}$$

where

$$\xi = \sqrt{\alpha^2 - \beta^2}. \tag{4.36}$$

Initial conditions are

$$\left.\begin{aligned} &\text{(i)} \quad \text{at } t = 0, \quad q = q_0 \\ \text{and} \quad &\text{(ii)} \quad \text{at } t = 0, \quad i = 0 \end{aligned}\right\}. \tag{4.37}$$

Using these conditions in equation (4.35), we find

$$\left.\begin{aligned} A &= \frac{q_0}{2}\left(1 + \frac{\alpha}{\xi}\right) \\ B &= \frac{q_0}{2}\left(1 - \frac{\alpha}{\xi}\right) \end{aligned}\right\}. \tag{4.38}$$

Thus, the specific solution is

$$q = \tfrac{1}{2} q_0 e^{-\alpha t}\left[\left(1 + \frac{\alpha}{\xi}\right)e^{\xi t} + \left(1 - \frac{\alpha}{\xi}\right)e^{-\xi t}\right]. \tag{4.39}$$

The solution in explicit form is

$$\begin{aligned} q = \frac{q_0}{2}\Bigg[&\left(1 + \frac{R/2L}{\sqrt{\frac{R^2}{4L^2} - \frac{1}{LC}}}\right)\exp\left(-\frac{R}{2L} + \sqrt{\frac{R^2}{4L^2} - \frac{1}{LC}}\right)t \\ &+ \left(1 - \frac{R/2L}{\sqrt{\frac{R^2}{4L^2} - \frac{1}{LC}}}\right)\exp\left(-\frac{R}{2L} - \sqrt{\frac{R^2}{4L^2} - \frac{1}{LC}}\right)t\Bigg]. \end{aligned} \tag{4.40}$$

Case II. *When* $\alpha^2 < \beta^2$, *i.e.,* $\frac{R^2}{4L^2} < \frac{1}{LC}$, *i.e.,* $R^2 < \frac{4L}{C}$.

In this case, as before, we get the solution

$$q = e^{-\alpha t}\left[Ae^{j\xi t} + Be^{-j\xi t}\right], \quad j = \sqrt{-1}. \tag{4.41}$$

Using the initial conditions and proceeding exactly as before, we find

$$q = \frac{D}{\xi} q_0 \, e^{-\alpha t} \sin(\xi t + \psi), \tag{4.42}$$

where

$$\tan\psi = \frac{\xi}{\alpha} = \frac{\sqrt{\beta^2 - \alpha^2}}{\alpha}, \quad D^2 = \alpha^2 + \xi^2. \tag{4.43}$$

Thus, the final solution in explicit form is

$$q = \frac{q_0 \cdot \sqrt{\dfrac{1}{LC}} \cdot e^{-Rt/2L}}{\sqrt{\dfrac{1}{LC} - \dfrac{R^2}{4L^2}}} \sin\left[\left(\sqrt{\frac{1}{LC} - \frac{R^2}{4L^2}}\right) t + \psi\right]. \tag{4.44}$$

The instantaneous current will be given by

$$i = -\frac{dq}{dt} = \frac{V}{L} \frac{e^{-Rt/2L}}{\sqrt{\dfrac{1}{LC} - \dfrac{R^2}{4L^2}}} \sin\left\{\sqrt{\frac{1}{LC} - \frac{R^2}{4L^2}} t\right\}. \tag{4.45}$$

Thus, both the charge and current reach zero value after a few oscillations.
The natural frequency of the circuit is

$$f = \frac{\xi}{2\pi} = \frac{1}{2\pi}\sqrt{\frac{1}{LC} - \frac{R^2}{4L^2}}. \tag{4.46}$$

If R is small compared with $2\sqrt{L/C}$, then

$$f = \frac{1}{2\pi\sqrt{LC}} = f_0, \tag{4.47}$$

where f_0 is the frequency of free oscillations.

Fig 4.4 illustrates the nature of discharge occurring in a circuit. It is obvious that this nature depends on the circuit resistance and hence will be different for different resistance values.

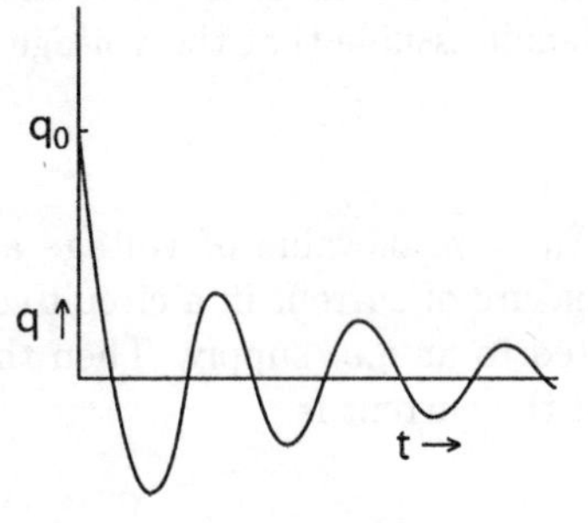

Fig 4.4

Case III. *When* $\alpha^2 = \beta^2$, *i.e.,* $R^2 = \frac{4L}{C}$.

In this case, the roots of the characteristic equation are equal. Hence, the solution of equation (4.34) is

$$\left.\begin{aligned} q &= (A + Bt)\, e^{-\alpha t} \\ \text{and so,} \quad i &= -Be^{-\alpha t} + (A + Bt)\, \alpha e^{-\alpha t} \end{aligned}\right\}. \tag{4.48}$$

Using the initial conditions (4.38) we find

$$A = q_0 \quad \text{and} \quad B = \alpha q_0. \tag{4.49}$$

Thus

$$q = q_0(1+\alpha t)\, e^{-\alpha t} = q_0\left(1+\frac{Rt}{2L}\right) e^{-Rt/2L} \tag{4.50}$$

$$\text{and} \quad i = q_0 \frac{R^2 t}{4L^2} \cdot e^{-Rt/2L}. \tag{4.51}$$

Discussion. If T be the period and q_1, q_2 are two successive maxima of the charge q given by equation (4.44), then

$$\frac{q_1}{q_2} = \frac{e^{-\alpha t}}{e^{-\alpha(t+T)}} = e^{\alpha T}.$$

The quantity δ defined by

$$\delta = \ln(q_1/q_2) = \alpha T = 2\pi \frac{R/2L}{\sqrt{\dfrac{1}{LC} - \dfrac{R^2}{4L^2}}} \tag{4.52}$$

is known as the *logarithmic decrement per cycle of the circuit.* When the resistance of the circuit is very small

$$\delta = \frac{\pi R}{L}\sqrt{LC} = \pi R\sqrt{\frac{C}{L}}. \tag{4.53}$$

4.4 Theory of Alternating Currents

An electric current is said to be alternating if it passes through a complete cycle of changes at regular intervals. Each cycle consists of two half-cycles. The current is entirely positive in one half-cycle and is entirely negative in the other.

We shall assume that the voltage applied to the circuit is sinusoidal and is expressed as

$$V = V_0 \cos \omega t, \tag{4.54}$$

where V_0 = peak value of voltage and $\omega = 2\pi f$, where f = frequency. We now intend to study nature of current in a circuit containing resistance, inductance and capacitance in series connected to an a.c. supply. Then the differential equation for determining the instantaneous value of the current is

$$\frac{q}{C} + Ri = V - L\frac{di}{dt}. \tag{4.55}$$

Here all the notations have usual meaning. Since $i = dq/dt$, we write the equation (4.55) in the form

$$\frac{d^2 i}{dt^2} + \frac{R}{L}\frac{di}{dt} + \frac{1}{LC} i = -\frac{\omega V_0}{L} \sin \omega t. \tag{4.56}$$

We shall consider the case when

$$\frac{R^2}{L^2} < \frac{4}{LC}, \quad \text{i.e.,} \quad \frac{R}{L} < \frac{2}{\sqrt{LC}}, \tag{4.57}$$

which is the only important instance in practice. Then the complementary solution may be written from equation (4.42) in the form

$$i = Ae^{-\alpha t}\sin(\xi t + \psi), \tag{4.58}$$

where α and ξ are already defined there and A and ψ are integration constants.

We seek the particular solution in the form

$$i_p = P\cos\omega t + Q\sin\omega t. \tag{4.59}$$

Substituting into equation (4.56), we get

$$\left[P\left(\frac{1}{LC}-\omega^2\right)+Q\cdot\frac{R\omega}{L}\right]\cos\omega t+\left[Q\left(\frac{1}{LC}-\omega^2\right)-P\cdot\frac{R\omega}{L}\right]\sin\omega t = -\frac{V_0\omega}{L}\sin\omega t. \tag{4.60}$$

Comparing coefficients of $\cos\omega t$ and $\sin\omega t$,

$$\left.\begin{aligned} P\frac{R}{L}\omega - Q\left(\frac{1}{LC}-\omega^2\right) &= \frac{V_0\omega}{L} \\ P\left(\frac{1}{LC}-\omega^2\right)+Q\frac{R}{L}\omega &= 0 \end{aligned}\right\}. \tag{4.61}$$

Solving for P and Q, we get

$$\left.\begin{aligned} P &= \frac{V_0R\omega^2}{L^2\left[\frac{R^2}{L^2}\omega^2+\left(\frac{1}{LC}-\omega^2\right)^2\right]} \\ Q &= -\frac{\frac{V_0\omega}{L}\left(\frac{1}{LC}-\omega^2\right)}{\left[\frac{R^2}{L^2}\omega^2+\left(\frac{1}{LC}-\omega^2\right)^2\right]} \end{aligned}\right\}. \tag{4.62}$$

The particular solution (4.59) can be expressed as

$$i_p = \frac{\frac{V_0\omega}{L}}{\sqrt{\left(\frac{1}{LC}-\omega^2\right)^2+\frac{R^2}{L^2}\omega^2}}\cos(\omega t+\phi), \tag{4.63}$$

where

$$\cot\phi = \frac{R}{\left(\omega L-\frac{1}{\omega C}\right)}. \tag{4.64}$$

Thus, the final solution of the equation is

$$i = Ae^{-Rt/2L}\sin\left(\sqrt{\frac{1}{LC}-\frac{4R^2}{L^2}}\,t+\psi\right)+\frac{V_0\omega}{L}\frac{\cos(\omega t+\phi)}{\sqrt{\left(\frac{1}{LC}-\omega^2\right)^2+\frac{R^2}{L^2}\omega^2}}. \tag{4.65}$$

After the lapse of a finite period of time, the exponential part of the solution (4.65) vanishes, so that the expression of current becomes

$$i = \frac{V_0 \omega}{L} \frac{\cos(\omega t + \phi)}{\sqrt{\left(\frac{1}{LC} - \omega^2\right)^2 + \frac{R^2 \omega^2}{L^2}}} = \frac{V_0 \cos(\omega t + \phi)}{\sqrt{R^2 + \left(\omega L - \frac{1}{\omega C}\right)^2}}. \tag{4.66}$$

From equation (4.65), it is clear that at the beginning, there are two types of oscillations—one is natural oscillation with frequency

$$f_1 = \frac{1}{2\pi} \sqrt{\frac{1}{LC} - \frac{R^2}{4L^2}} \tag{4.67}$$

and the other a forced oscillation with frequency

$$f_2 = \frac{\omega}{2\pi}. \tag{4.68}$$

The amplitude of the forced oscillation is

$$\frac{V_0}{\sqrt{R^2 + \left(\omega L - \frac{1}{\omega C}\right)^2}}. \tag{4.69}$$

In circuits where resistance is very small so that R can be neglected, the frequency of natural oscillation is given by

$$f_1 = \frac{1}{2\pi\sqrt{LC}} \tag{4.70}$$

and the amplitude of the forced vibration is

$$\frac{V_0}{\omega L - \frac{1}{\omega C}}. \tag{4.71}$$

After the extinction of natural oscillation, there remains only forced vibration of amplitude given by equation (4.69). Its value is maximum when

$$\omega L - \frac{1}{\omega C} = 0, \quad \text{i.e.,} \quad \omega^2 = \frac{1}{LC}. \tag{4.72}$$

In that case, from equations (4.63) and (4.64), we get

$$i = i_p = \frac{V_0}{R} \cos \omega t, \quad \text{i.e.,} \quad \phi = 0. \tag{4.73}$$

The impedance of the circuit is given by

$$Z = \sqrt{R^2 + \left(\omega L - \frac{1}{\omega C}\right)^2}. \tag{4.74}$$

The current in terms of impedance is expressed as [from equation (4.66)]

$$i = I_0 \cos(\omega t + \phi), \quad I_0 = \frac{V_0}{Z}. \tag{4.75}$$

The r.m.s. current is given by $I = I_0/\sqrt{2}$, so that

$$\sqrt{2}I = \frac{V_0}{Z} = \frac{V_0}{\sqrt{R^2 + \left(\omega L - \frac{1}{\omega C}\right)^2}}. \tag{4.76}$$

Differentiating with respect to L, we get

$$\sqrt{2}\frac{dI}{dL} = -\frac{V_0\omega}{\left[R^2 + \left(\omega L - \frac{1}{\omega C}\right)^2\right]^{3/2}}\left(\omega L - \frac{1}{\omega C}\right).$$

For maximum value of I,

$$\frac{dI}{dL} = 0, \quad \text{i.e., either,} \quad \omega = 0 \quad \text{or,} \quad \omega L - \frac{1}{\omega C} = 0.$$

But $\omega = 0$ corresponds to d.c. circuit and hence is ignored. From second condition, we get, for maximum value of r.m.s. current

$$\omega L - \frac{1}{\omega C} = 0, \quad \text{i.e.,} \quad L = \frac{1}{\omega^2 C}. \tag{4.77}$$

Thus, if we consider a circuit with condenser and resistance alone, and then go on adding inductance, the r.m.s. current also will go on increasing till a maximum value is reached. At this point, the inductance will be given by equation (4.77).

4.5 Growth of Current in Coupled Circuits

Let us consider a mutually coupled circuit consisting of a primary and a secondary circuit (*Fig 4.5*). Let,

L_1 = self-inductance of primary circuit;
R_1 = resistance in primary circuit;
L_2 = self-inductance of secondary circuit;
R_2 = resistance in the secondary circuit;
M = mutual inductance between circuits;
i_1 = instantaneous current in primary circuit;
i_2 = instantaneous current in secondary circuit;
and V = e.m.f. of the cell.

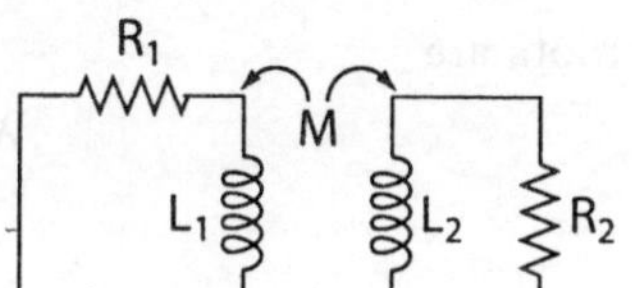

Fig 4.5

Then circuit equations at any instant t are written as

For primary circuit:

$$R_1 i_1 + L_1\frac{di_1}{dt} + M\frac{di_2}{dt} = V. \tag{4.78}$$

For secondary circuit:

$$R_2 i_2 + L_2\frac{di_2}{dt} + M\frac{di_1}{dt} = 0. \tag{4.79}$$

We change the variables as

$$x = i_1 - \frac{V}{R_1} \quad \text{and} \quad y = i_2. \tag{4.80}$$

Then equations (4.78) and (4.79) become

$$\left.\begin{aligned} L_1\frac{dx}{dt}+M\frac{dy}{dt}+R_1x=0 \\ \text{and}\quad L_2\frac{dy}{dt}+M\frac{dx}{dt}+R_2y=0 \end{aligned}\right\}. \tag{4.81}$$

Using the differential operator

$$D(\cdots)=\frac{d(\cdots)}{dt} \tag{4.82}$$

equations (4.81) are written in the form

$$\left.\begin{aligned} (L_1D+R_1)\,x+M\,Dy=0 \\ \text{and}\quad (L_2D+R_2)\,y+M\,Dx=0 \end{aligned}\right\}. \tag{4.83}$$

Eliminating x from equations (4.83), we get

$$aD^2y+bDy+cy=0, \tag{4.84}$$

where

$$\left.\begin{aligned} a&=L_1L_2-M^2>0, \\ b&=R_1L_2+R_2L_1 \\ \text{and}\quad c&=R_1R_2. \end{aligned}\right\} \tag{4.85}$$

We assume the solution of equation (4.84) in the form

$$y=e^{\lambda t}. \tag{4.86}$$

Substituting (4.86) into equation (4.84), we get the characteristic equation as

$$a\lambda^2+b\lambda+c=0 \tag{4.87}$$

whose roots are

$$\left.\begin{aligned} \lambda_1&=-\frac{b}{2a}+\frac{1}{2a}\sqrt{b^2-4ac} \\ \lambda_2&=-\frac{b}{2a}-\frac{1}{2a}\sqrt{b^2-4ac} \end{aligned}\right\}. \tag{4.88}$$

Now

$$\begin{aligned} b^2-4ac&=(R_1L_2+R_2L_1)^2-4(L_1L_2-M^2)\,R_1R_2 \\ &=(R_1L_2-R_2L_1)^2+4R_1R_2L_1L_2M^2>0. \end{aligned}$$

Hence, roots λ_1, λ_2 are real. Moreover, λ_1 is negative since $b>\sqrt{b^2-4ac}$ and λ_2 is negative which is obvious. Hence, the solution of equation is written as

$$y(t)=Ae^{-\lambda_1 t}+Be^{-\lambda_2 t}. \tag{4.89}$$

From second equation of (4.81), we have

$$M\frac{dx}{dt}=-L_2\frac{dy}{dt}-R_2y. \tag{4.90}$$

Substituting $y(t)$ from equation (4.89) into equation (4.90), we get

$$\frac{dx}{dt} = \frac{A(\lambda_1 L_2 - R_2)}{M}\, e^{-\lambda_1 t} + \frac{B(\lambda_2 L_2 - R_2)}{M}\, e^{-\lambda_2 t}. \tag{4.91}$$

Integrating, we get

$$x = A\xi e^{-\lambda_1 t} + B\eta e^{-\lambda_2 t}, \tag{4.92}$$

where

$$\xi = \frac{R_2 - \lambda_1 L_2}{\lambda_1 M}, \quad \eta = \frac{R_2 - \lambda_2 L_2}{\lambda_2 M}. \tag{4.93}$$

Initially, at $t = 0$,

$$\left.\begin{aligned} i_1 = 0, \quad \text{i.e.,} \quad x = -\frac{V}{R_1} \\ \text{and} \quad i_2 = 0, \quad \text{i.e.,} \quad y = 0 \end{aligned}\right\}. \tag{4.94}$$

Using these initial conditions to the solution functions (4.89) and (4.92), we get

$$\left.\begin{aligned} A + B = 0 \\ \text{and} \quad A\xi + B\eta = -\frac{V}{R_1} \end{aligned}\right\}. \tag{4.95}$$

Solving these, we find

$$\left.\begin{aligned} A = -\frac{V}{R_1}\cdot\frac{1}{\xi - \eta} \\ B = \frac{V}{R_1}\cdot\frac{1}{\xi - \eta} \end{aligned}\right\}. \tag{4.96}$$

Using equation (4.93), we find

$$\xi - \eta = \frac{R_2}{M}\cdot\frac{\lambda_1 - \lambda_2}{\lambda_1\lambda_2}. \tag{4.97}$$

Thus, we write the final solutions

$$i_1 = x + \frac{V}{R_1} = \frac{V}{R_1}\left[1 - \frac{(R_2 - \lambda_1 L_2)\lambda_2}{R_2(\lambda_1 - \lambda_2)}\, e^{-\lambda_1 t} + \frac{(R_2 - \lambda_2 L_2)\lambda_1}{R_2(\lambda_1 - \lambda_2)}\, e^{-\lambda_2 t}\right] \tag{4.98}$$

$$\text{and} \quad i_2 = y = \frac{MV\lambda_1\lambda_2}{R_1 R_2(\lambda_1 - \lambda_2)}\left[e^{-\lambda_2 t} - e^{-\lambda_1 t}\right]. \tag{4.99}$$

From these solutions, we find that, with lapse of time i_1 attains a steady value V/R_1. But i_2 gradually increases from its zero value at $t = 0$ to some maximum at $t = t_1$, and then reaches to zero value. If we differentiate equation (4.99) and equate to zero for maximum value at $t = t_1$, then we have

$$\lambda_2 e^{-\lambda_2 t_1} - \lambda_1 e^{-\lambda_1 t_1} = 0, \quad \text{i.e.,} \quad e^{(\lambda_1 - \lambda_2)t_1} = \frac{\lambda_1}{\lambda_2}, \quad \text{i.e.,} \quad t_1 = \frac{1}{\lambda_1 - \lambda_2}\ln\left(\frac{\lambda_1}{\lambda_2}\right). \tag{4.100}$$

A graphical presentation of the growth of currents in primary and secondary circuits is shown in the *Fig 4.6*.

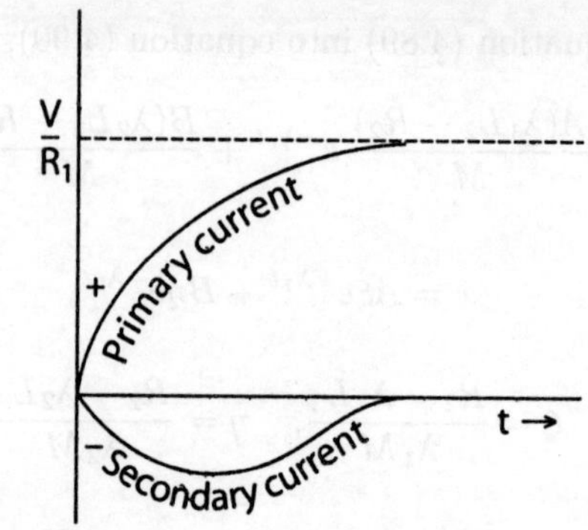

Fig 4.6

4.6 Decay of Current in Coupled Circuit

We have seen that, the current in the primary circuit attains a steady value of V/R_1, after some time. Let us now cut off the source e.m.f. Then the current in the primary and secondary circuit will decay with time. The circuit equations at any instant t are as follows:

For primary circuit:

$$L_1\frac{di_1}{dt} + R_1 i_1 + M\frac{di_2}{dt} = 0. \tag{4.101}$$

For secondary circuit:

$$L_2\frac{di_2}{dt} + R_2 i_2 + M\frac{di_1}{dt} = 0. \tag{4.102}$$

The initial conditions are

$$\text{at} \quad t = 0, \quad i_1 = \frac{V}{R_1} \quad \text{and} \quad i_2 = 0. \tag{4.103}$$

Proceeding exactly as before, we find the solutions of the above equations as

$$i_1 = \frac{V}{R_1}\left[\frac{(R_2 - \lambda_1 L_2)\lambda_2}{R_2(\lambda_1 - \lambda_2)}\, e^{-\lambda_1 t} - \frac{(R_2 - \lambda_2 L_2)}{R_2(\lambda_1 - \lambda_2)}\, e^{-\lambda_2 t}\right] \tag{4.104}$$

$$\text{and} \quad i_2 = \frac{MV\lambda_1\lambda_2}{R_1 R_2(\lambda_1 - \lambda_2)}\left(e^{-\lambda_1 t} - e^{-\lambda_2 t}\right). \tag{4.105}$$

Equations (4.104) and (4.105) clarify that the primary current i_1 decays exponentially from its value of V/R_1 to zero, and the secondary current increases to a maximum value as before, and then reaches to zero value. But, in this case, direction of flow of i_2 will be opposite to that of previous case which is evident from equations (4.99) and (4.105).

A graphical representation of decay of primary and secondary currents is shown in the *Fig 4.7.*

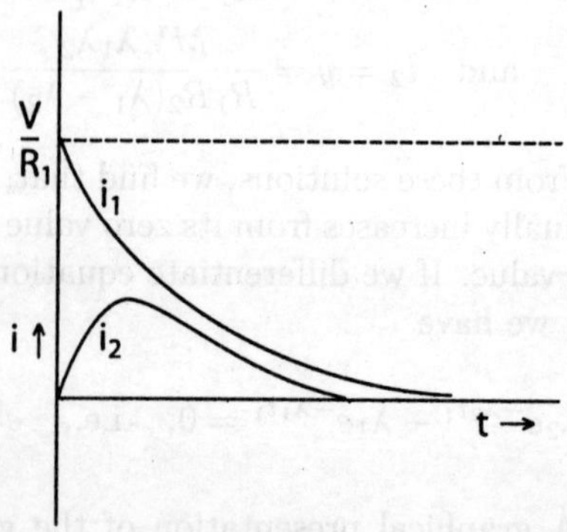

Fig 4.7

II. BENDING OF BEAMS

4.7 Differential Equation of Equilibrium

Let us consider a beam of weight per unit length q and Young's modulus of its material E. We assume that a plane cross-section of the beam remains plane after bending.

We isolate an element of length dx from the beam. This will be under the actions of shearing force Q and moment M at the left end, and shearing force $Q + dQ$ and moment $M + dM$ at the right end (*Fig 4.8*). Considering the equilibrium of this element, sum of all forces along vertical axis and sum of moments along a transverse axis should be zero. Hence

$$Q + dQ - Q - q\,dx = 0$$

and

$$M + dM + q\,dx \cdot \frac{dx}{2} - M - Q\,dx = 0.$$

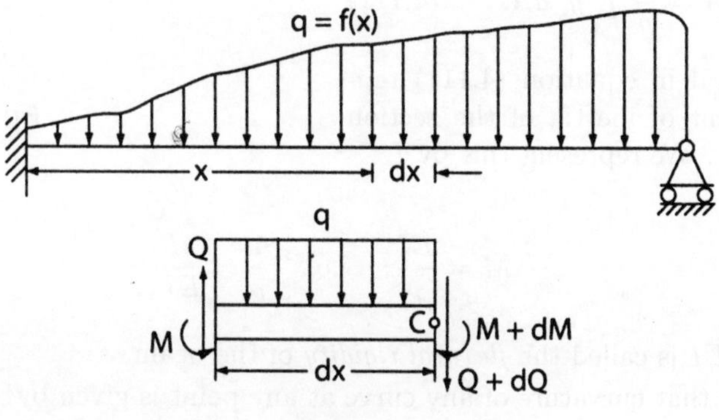

Fig 4.8

From these equations, we get

$$\frac{dQ}{dx} = q \tag{4.106}$$

$$\text{and} \quad \frac{dM}{dx} = Q. \tag{4.107}$$

Now, deformation in pure bending causes rotation of plane cross-sections with respect to one another. Let us consider two adjacent sections of the beam at distance dx apart. Assuming the left end to be fixed, rotation of the right section due to bending is, say $d\theta$ (*Fig 4.9*). Due to this rotation, top layers elongate and bottom layers shorten. Thus, there exists a layer in between which remains unaltered. This layer is called *neutral.* The change in curvature due to rotations of sections will be

$$\frac{1}{\rho} = \frac{d\theta}{dx}. \tag{4.108}$$

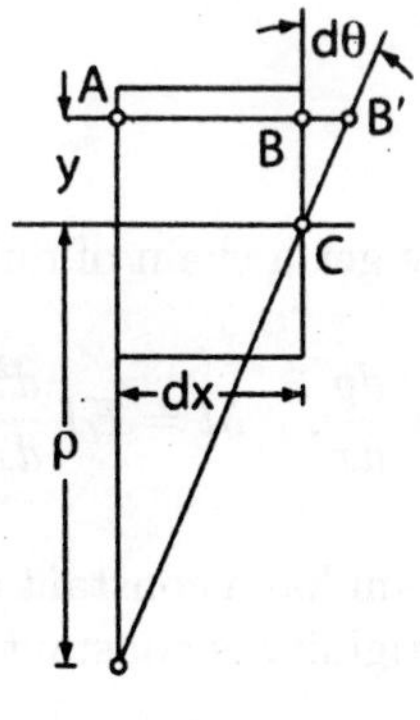

Fig 4.9

An arbitrary segment AB of original length dx increases by an amount $BB' = y\,d\theta$, where $y =$ distance of segment AB from neutral layer. Thus, unit elongation of the layer is

$$\epsilon = \frac{y\,d\theta}{dx} = \frac{y}{\rho}. \tag{4.109}$$

By Hooke's law, the stress in the layer is

$$\sigma = E\epsilon = E\frac{y}{\rho}. \tag{4.110}$$

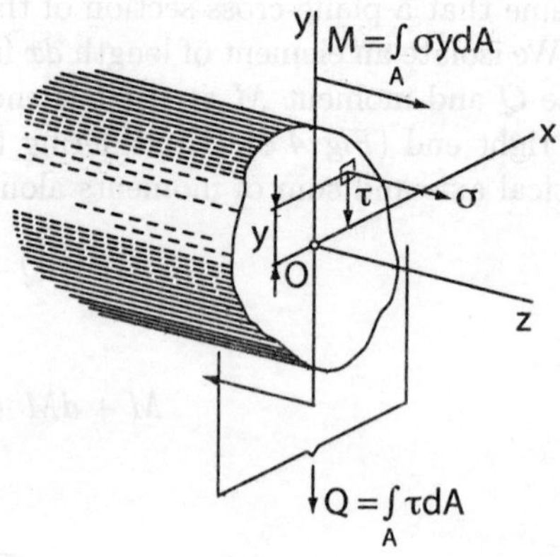

Fig 4.10

Let us now consider a small area dA at distance y from neutral axis (i.e., where stress $\sigma = 0$) on a plane section (*Fig 4.10*).

Taking moments of all the forces about neutral axis, we get

$$M = \int_A \sigma y\,dA = \frac{E}{\rho}\int_A y^2\,dA. \tag{4.111}$$

But the integral in equation (4.111) represents the moment of inertia of the section about neutral axis. We represent this by I.

Thus, we get

$$M = \frac{EI}{\rho}, \quad \text{i.e.,} \quad \frac{1}{\rho} = \frac{M}{EI}. \tag{4.112}$$

The quantity EI is called the *flexural rigidity* of the beam.

Now, we know that curvature of any curve at any point is given by

$$\frac{1}{\rho} = \frac{y''}{\left[1+(y')^2\right]^{3/2}}. \tag{4.113}$$

We shall limit our discussion to small displacements. Then we can neglect $(y')^2$ in comparison to unity and write $1/\rho \simeq y''$ whence, using equation (4.112), we get

$$y'' = \frac{M}{EI}. \tag{4.114}$$

Thus, we get a chain of differential relations

$$\theta = \frac{dy}{dx}, \quad M = EI\frac{d^2y}{dx^2}, \quad Q = \frac{d}{dx}\left(EI\frac{d^2y}{dx^2}\right), \quad q = \frac{d^2}{dx^2}\left(EI\frac{d^2y}{dx^2}\right). \tag{4.115}$$

If the beam has a constant cross-section throughout and is made of a homogeneous material, its flexural rigidity is constant. In such a case, equations (4.115) are as follows:

$$\theta = \frac{dy}{dx}, \quad M = EI\frac{d^2y}{dx^2}, \quad Q = EI\frac{d^3y}{dx^3}, \quad q = EI\frac{d^4y}{dx^4} \tag{4.116}$$

4.8 Uniformly Loaded Beam

For a uniformly loaded beam of length l, we have

$$EI\,\frac{d^4y}{dx^4} = q.$$

Integrating four times in succession, we get

$$EI\,\frac{d^3y}{dx^3} = qx + C_1 \tag{4.117}$$

$$EI\,\frac{d^2y}{dx^2} = \frac{qx^2}{2} + C_1x + C_2 \tag{4.118}$$

$$EI\,\frac{dy}{dx} = \frac{qx^3}{6} + C_1\frac{x^2}{2} + C_2x + C_3 \tag{4.119}$$

$$EI\,y = \frac{qx^4}{24} + C_1\frac{x^3}{6} + C_2\frac{x^2}{2} + C_3x + C_4, \tag{4.120}$$

where C_1, C_2, C_3, C_4 are constants of integration which are to be found from boundary conditions.

Case I. *Beam simply supported.*

If the beam is supported on knife-edges at its ends $x = 0$ and $x = l$, then we have the following boundary conditions:

(i) At $x = 0$, deflection $y = 0$, moment $EI\,\dfrac{d^2y}{dx^2} = 0$;

(ii) At $x = l$, deflection $y = 0$, moment $EI\,\dfrac{d^2y}{dx^2} = 0$; and

(iii) At $x = l/2$, $\theta = 0$ due to symmetry.

Using condition at $x = 0$, we get

$$C_4 = C_2 = 0. \tag{4.121}$$

Using condition at $x = l$, we get

$$0 = \frac{ql^2}{2} + C_1l, \qquad \therefore\quad C_1 = -\frac{ql}{2}. \tag{4.122}$$

Using condition at $x = l/2$, we get

$$0 = \frac{1}{6}\cdot\frac{ql^3}{8} + \tfrac{1}{2}C_1\frac{l^2}{4} + C_3 = \frac{ql^3}{48} - \frac{ql^3}{16} + C_3, \qquad \therefore\quad C_3 = \frac{ql^3}{24}. \tag{4.123}$$

Thus, the values of the constants are given by

$$C_1 = -\frac{ql}{2}, \quad C_2 = 0, \quad C_3 = \frac{ql^3}{24}, \quad C_4 = 0. \tag{4.124}$$

Hence, the solution of the differential equation is

$$EI\,y = \frac{qx^4}{24} - \frac{qlx^3}{12} + \frac{ql^3x}{24} = \frac{qx}{24}\left(x^3 - 2x^2l + l^3\right) = \frac{qx}{24}(l-x)\left(l^2 + lx - x^2\right).$$

$$\therefore\quad y = \frac{qx}{24EI}(l-x)(l^2 + lx - x^2). \tag{4.125}$$

The maximum deflection will be at the middle point where $x = l/2$. Hence

$$y_{\max} = \frac{q}{24EI} \cdot \frac{l}{2}\left(l - \frac{l}{2}\right)\left(l^2 + \frac{l^2}{2} - \frac{l^2}{4}\right) = \frac{5}{384} \cdot \frac{ql^4}{EI}. \tag{4.126}$$

Case II. *Beams clamped at both ends.*

If the beam is clamped at ends $x = 0$ and $x = l$, then the boundary conditions are as follows:
(i) At $x = 0$, deflection $y = 0$, and rotation $\theta = 0$;
(ii) At $x = l$, deflection $y = 0$, and rotation $\theta = 0$.
Using conditions (i), we get

$$C_3 = C_4 = 0. \tag{4.127}$$

Using conditions (ii), we get

$$\left.\begin{aligned} \frac{1}{6}ql^3 + \frac{1}{2}C_1l^2 + C_2l &= 0 \\ \text{and} \quad \frac{1}{24}ql^4 + \frac{1}{6}C_1l^3 + \frac{1}{2}C_2l^2 &= 0 \end{aligned}\right\} \tag{4.128}$$

$$\left.\begin{aligned} \text{or,} \quad \frac{1}{6}ql^2 + \frac{1}{2}C_1l + C_2 &= 0 \\ \text{and} \quad \frac{1}{12}ql^2 + \frac{1}{3}C_1l + C_2 &= 0 \end{aligned}\right\}$$

Solving these, we get

$$C_1 = -\frac{ql}{2}, \quad C_2 = \frac{ql^2}{12}. \tag{4.129}$$

Hence, the deflection of the beam is given by

$$EI\,y = \frac{qx^4}{24} - \frac{qlx^3}{12} + \frac{ql^2x^2}{24} = \frac{qx^2}{24}(l - x)^2. \tag{4.130}$$

Therefore, the deflection of the beam at the middle point where $x = l/2$, is

$$EI\,y_{\max} = \frac{q}{24}\left(\frac{l}{2}\right)^2\left(l - \frac{l}{2}\right)^2 = \frac{1}{384} \cdot ql^4.$$

$$\therefore \quad y_{\max} = \frac{1}{384} \cdot \frac{ql^4}{EI}. \tag{4.131}$$

From equations (4.126) and (4.131), it is clear that the maximum deflection at the centre of the beam under uniformly distributed load with clamped edges is only one-fifth of the maximum deflection of the same beam but supported on knife-edges at its ends.

4.9 Point Load at Any Position on the Beam

Let a point load P be acting vertically on the beam at a distance of $x = a$ from the left end which is taken as the origin. Neglecting the self-load of the beam, we consider the following cases:

Case I. *Cantilever beam.*

The bending moment at a section distant x from origin is (*Fig 4.11*)

$$M = -P(x-a) = P(a-x).$$

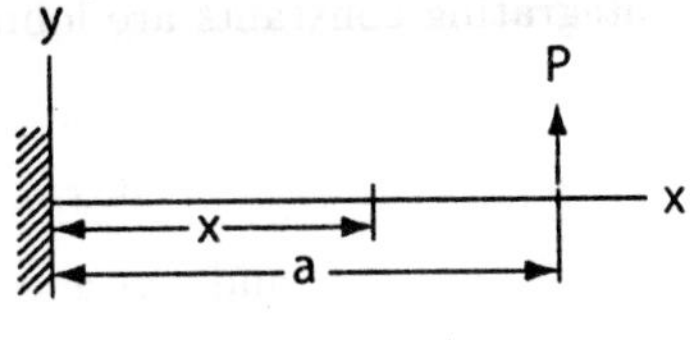

Fig 4.11

Integrating twice, we get

$$EI\,y = P\left(a\frac{x^2}{2} - \frac{x^3}{6} + C_1x + C_2\right).$$

In the present case, the boundary conditions are:

$$\text{at} \quad x = 0, \quad y = 0 \quad \text{and} \quad \frac{dy}{dx} = 0.$$

Using these conditions, we find $C_1 = 0$, $C_2 = 0$.
Thus, deflection function of the beam is

$$y = \frac{P}{EI}\left(a\frac{x^2}{2} - \frac{x^3}{6}\right). \tag{4.132}$$

Maximum deflection will occur at the point of application of the force, i.e., at $x = a$. Hence

$$y_{\max} = \frac{Pa^3}{3EI}. \tag{4.133}$$

Case II. *Simply supported beam.*

Let the distance of the point load P from left end be a, so that its distance from right end is $(l-a)$ (*Fig 4.12*).

Then, reactions at two supports are

$$R_1 = P\left(\frac{l-a}{l}\right) \quad \text{and} \quad R_2 = \frac{Pa}{l} \tag{4.134}$$

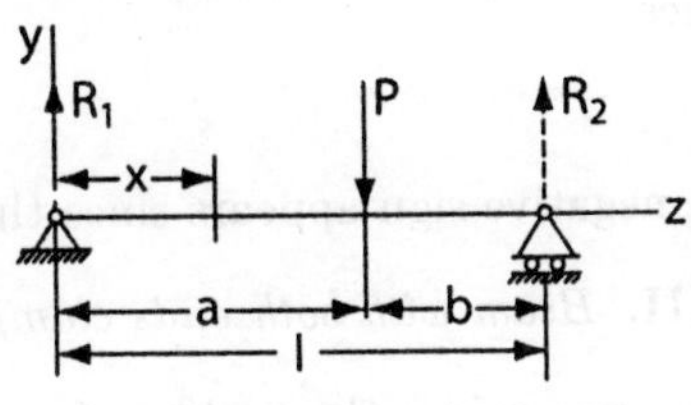

Fig 4.12

both acting in vertically upward direction. We consider two portions of the beam as under:

$$\left.\begin{array}{ll}\text{portion I} & : 0 \le x \le a \\ \text{portion II} & : a \le x \le l\end{array}\right\}. \tag{4.135}$$

The bending moment in these portions are

$$M_1 = P\left(\frac{l-a}{l}\right)x \quad \text{and} \quad M_2 = P\left(\frac{l-a}{l}\right)x - P(x-a) \tag{4.136}$$

Integrating twice, we find

$$y_1 = \frac{P}{EI}\left[\frac{l-a}{l}\cdot\frac{x^3}{6} + C_1x + C_2\right]$$

$$\text{and} \quad y_2 = \frac{P}{EI}\left[\frac{l-a}{l}\cdot\frac{x^3}{6} - \frac{x^3}{6} + \frac{ax^2}{2} + C_3x + C_4\right]. \tag{4.137}$$

The integrating constants are found from the boundary conditions which are as follows:

$$\text{at } x = 0, \quad y_1 = 0;$$
$$\text{at } x = l, \quad y_2 = 0$$
$$\text{and} \quad \text{at } x = a, \quad y_1 = y_2 \quad \text{and} \quad \frac{dy_1}{dx} = \frac{dy_2}{dx}.$$

The last condition is called the *condition of continuity.*
Using these conditions, we find

$$C_1 = \frac{a}{6l}(3al - 2l^2 - a^2), \quad C_2 = 0,$$

$$C_3 = -\frac{a}{6l}(2l^2 + a^2), \qquad C_4 = \frac{a^3}{6}.$$

Hence, for $0 \leq x \leq a$,

$$y_1 = \frac{P}{EI}\left[\frac{l-a}{l} \cdot \frac{x^3}{6} + \frac{ax}{6l}(3al - 2l^2 - a^2)\right] \tag{4.138}$$

and for $a \leq x \leq l$,

$$y_2 = \frac{P}{EI}\left[-\frac{ax^3}{6l} + \frac{ax^2}{2} - \frac{ax}{6l}(2l^2 + a^2) + \frac{a^3}{6}\right] \tag{4.139}$$

At the point of application of the force

$$x = a \quad \text{and} \quad y_1 = y_2 = -\frac{Pa^2}{3EI\,l}(l-a)^2. \tag{4.140}$$

If the force is applied at the centre of the beam, then the deflection will be maximum there. Hence

$$y_{\max} = -\frac{Pl^3}{48EI}. \tag{4.141}$$

The negative sign appears since the direction of y-axis is taken upwards.

Case III. *Beam with both ends clamped.*

Here we consider the problem from basic differential equation. We have

$$EI\frac{d^4y}{dx^4} = 0, \quad \because\ q = 0. \tag{4.142}$$

For the region $0 \leq x \leq a$, we get by integrating four times the above equation

$$Q_1 \longrightarrow EI\frac{d^3y_1}{dx^3} = A_1 \tag{4.143}$$

$$M_1 \longrightarrow EI\frac{d^2y_1}{dx^2} = A_1x + B_1 \tag{4.144}$$

$$\theta_1 \longrightarrow EI\frac{dy_1}{dx} = A_1\frac{x^2}{2} + B_1x + C_1 \tag{4.145}$$

$$EI\,y_1 = \frac{A_1x^3}{6} + \frac{B_1x^2}{2} + C_1x + D_1. \tag{4.146}$$

Similar equations are obtained for the region $a \leq x \leq l$. These are:

$$Q_2 \longrightarrow EI\frac{d^3y_2}{dx^3} = A_2 \tag{4.147}$$

$$M_2 \longrightarrow EI\frac{d^2y_2}{dx^2} = A_2x + B_2 \tag{4.148}$$

$$\theta_2 \longrightarrow EI\frac{dy_2}{dx} = A_2\frac{x^2}{2} + B_2x + C_2 \tag{4.149}$$

$$EI\,y_2 = \frac{A_2x^3}{6} + \frac{B_2x^2}{2} + C_2x + D_2. \tag{4.150}$$

The constants of integration are to be found from boundary conditions and from conditions of continuity.

The boundary conditions for $0 \leq x \leq a$ are:

$$\text{at} \quad x = 0, \quad y_1 = 0 \quad \text{and} \quad \theta_1 = 0. \tag{4.151}$$

Using these conditions, we get

$$C_1 = 0, \quad D_1 = 0. \tag{4.152}$$

The continuity equations at $x = a$ are:

$$\left.\begin{aligned} Q_2 - Q_1 &= P \\ M_1 &= M_2 \\ \theta_1 &= \theta_2 \\ y_1 &= y_2 \end{aligned}\right\}. \tag{4.153}$$

Using these conditions, we have

$$A_1 - A_2 = -P \tag{4.154}$$

$$B_1 - B_2 = Pa \tag{4.155}$$

$$C_1 - C_2 = -\frac{Pa^2}{2} \tag{4.156}$$

$$D_1 - D_2 = \frac{Pa^3}{6} \tag{4.157}$$

Using equations (4.147)–(4.152), we have

$$C_2 = \frac{Pa^2}{2}, \quad D_2 = -\frac{Pa^3}{6}. \tag{4.158}$$

Now

$$\text{at} \quad x = l, \quad y_2 = 0 \quad \text{and} \quad \theta_2 = 0. \tag{4.159}$$

Using these conditions in equations (4.149) and (4.150)

$$\tfrac{1}{2}A_2l^2 + B_2l + \tfrac{1}{2}Pa^2 = 0 \tag{4.160}$$

$$\text{and} \quad \tfrac{1}{6}A_2l^3 + \tfrac{1}{2}B_2l^2 = \frac{Pa^3}{6} - \frac{1}{2}Pa^2l. \tag{4.161}$$

These equations are recast as

$$\left.\begin{aligned} A_2 l^2 + 2B_2 l &= -Pa^2 \\ \text{and} \quad A_2 l^3 + 3B_2 l^2 &= Pa^3 - 3Pa^2 l \end{aligned}\right\}. \tag{4.162}$$

Solving these, we find

$$B_2 = \frac{Pa^2}{l^2}(a - 2l) \tag{4.163}$$

$$\text{and} \quad A_2 = \frac{Pa^2}{l^3}(3l - 2a). \tag{4.164}$$

From equation (4.154)

$$A_1 = A_2 - P = -\frac{P}{l^3}(l^3 + 2a^3 - 3a^2 l) = -\frac{P}{l^3}(l - a)^2 (l + 2a). \tag{4.165}$$

From equation (4.155)

$$B_1 = B_2 + Pa = Pa + \frac{Pa^2}{l^2}(a - 2l) = \frac{Pa}{l^2}(l - a)^2. \tag{4.166}$$

Thus, all the constants of integration are found out. Expressions of the deflection functions are:

for $0 \le x \le a$,

$$y_1(x) = \frac{Px^2}{EI}\left[-\frac{(l-a)^2}{6l^3}(l + 2a)\,x + \frac{a(l-a)^2}{2l^2}\right] \tag{4.167}$$

and for $0 \le x \le l$,

$$y_2(x) = \frac{P}{EI}\left[\frac{a^2}{6l^3}(3l - 2a)\,x^3 - \frac{a^2}{2l^2}(2l - a)\,x^2 + \frac{a^2 x}{2} - \frac{a^3}{6}\right]. \tag{4.168}$$

Maximum deflection will be at the point of loading when $x = a$. Thus, from equation (4.167), we get

$$y_1\Big|_{x=a} = \frac{Pa^2 (l-a)^2}{EI}\left[-\frac{(l+2a)\,a}{6l^3} + \frac{a}{2l^2}\right] = \frac{Pa^3 (l-a)^3}{3EI\,l^3}. \tag{4.169}$$

If the load is at the centre of the beam, i.e., $a = l/2$, the maximum deflection of the beam is

$$y_{\max} = \frac{Pl^3}{192EI}. \tag{4.170}$$

From equations (4.141) and (4.170), we find that the maximum deflection of a beam under a point load at the centre with clamped edges is one-fourth of the maximum deflection of the same beam but supported on knife-edges at its ends.

4.10 Beams on Elastic Foundation

Let us consider a beam subjected to external forces and resting upon a set of closely spaced springs whose spring constant is k. Under the action of external forces, the beam will deflect and hence the springs will be compressed by the same amount. The springs will exert reaction forces on the bottom surface of the beam which will be proportional to the deflection of the beam and directed upwards. The springs being very closely spaced, the reaction forces may be taken to be uniformly distributed having magnitude equal to

$$q_r = ky. \tag{4.171}$$

Any continuous medium that exerts reaction forces in accordance with equation (4.171) is called an *elastic foundation* (*Fig 4.13*).

Soil foundation has been modelled as an elastic foundation having the above property. This is known as *Winkler's foundation.*

Let us now suppose that a beam carrying uniformly distributed load q is resting upon elastic foundation. Then the differential equation for bending of the beam will be obtained from last equation of (4.116) by replacing q with $q - q_r$.

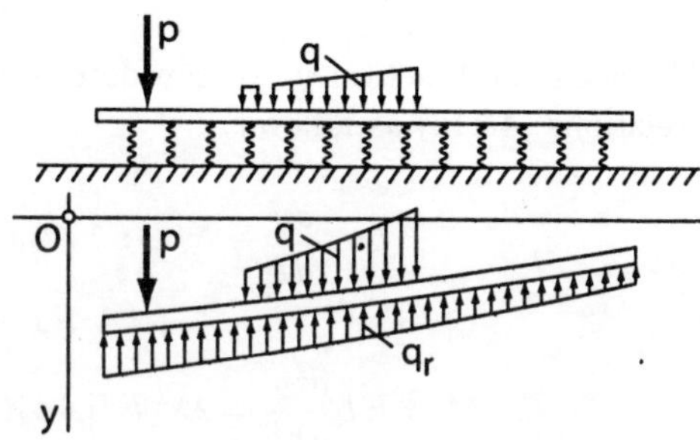

Fig 4.13

Thus, we have, for a beam of constant rigidity

$$EI\,\frac{d^4y}{dx^4} = q - q_r, \quad \text{i.e.,} \quad EI\,\frac{d^4y}{dx^4} + ky = q, \quad \text{or,} \quad \frac{d^4y}{dx^4} + 4\lambda^4 y = \frac{q}{EI}, \tag{4.172}$$

where

$$\lambda^4 = \frac{k}{4EI}. \tag{4.173}$$

Assuming the solution in the form $y = e^{mx}$ we get the characteristic equation as

$$m^4 + 4\lambda^4 = 0, \quad \text{i.e.,} \quad (m^2 + 2\lambda^2)^2 = 4m^2\lambda^2, \quad \text{i.e.,} \quad m^2 + 2\lambda^2 = \pm 2m\lambda.$$

Thus, we have two equations

$$(m + \lambda)^2 = -\lambda^2, \quad \text{and} \quad (m - \lambda)^2 = -\lambda^2.$$

Hence, the roots of the characteristic equation are

$$m_{1,2} = -\lambda \pm i\lambda, \quad m_{3,4} = \lambda \pm i\lambda.$$

Thus, the solution of the differential equation is

$$y = e^{-\lambda x}(A\sin\lambda x + B\cos\lambda x) + e^{\lambda x}(C\cos\lambda x + D\sin\lambda x) + y_p.$$

The above solution can be written as

$$y = A\phi_1(x) + B\phi_2(x) + C\phi_3(x) + D\phi_4(x) + y_p, \tag{4.174}$$

where

$$\left.\begin{aligned}\phi_1(x) &= \sin\lambda x \sinh\lambda x\\ \phi_2(x) &= \sin\lambda x \cosh\lambda x\\ \phi_3(x) &= \cos\lambda x \sinh\lambda x\\ \phi_4(x) &= \cos\lambda x \cosh\lambda x\end{aligned}\right\} \tag{4.175}$$

and A, B, C, D are integration constants and y_p is the particular solution, which in the case of uniformly distributed load, is

$$y_p = \frac{q}{4\lambda^4 EI} = \frac{q}{k}. \tag{4.176}$$

Knowing the function y, we can determine bending moments and shearing forces by using the relations (4.116) as follows:

$$\theta = \frac{dy}{dx} = \lambda\,[(B-C)\phi_1(x) + (A-D)\phi_2(x) + (A+D)\phi_3(x) + (B+C)\phi_4(x)\,] \tag{4.177}$$

$$M = EI\,\frac{d^2y}{dx^2} = 2\lambda^2\,EI[A\phi_4(x) + B\phi_3(x) - C\phi_2(x) - D\phi_1(x)\,] \tag{4.178}$$

$$M = EI\,\frac{d^3y}{dx^3} = 2\lambda^3\,EI[\,(B-C)\phi_4(x) + (A-D)\phi_3(x) - (A+D)\phi_2(x) - (B+C)\phi_1(x)\,]. \tag{4.179}$$

The constants of integration are to be found out from boundary conditions. If the end of the beam is free, then $Q = M = 0$ at that end.

If the beam is subjected to a point load P at the centre of the beam (i.e., say at $x = 0$), then the shearing force to the right of that section will be $-P/2$, so that we have the condition

$$EI\,\frac{d^3y}{dx^3}\bigg|_{x=0} = -\frac{P}{2}. \tag{4.180}$$

If the beam is subjected to either a point load at the centre or an uniformly distributed load, then from symmetry, rotation at the centre will be zero. Hence

$$\frac{dy}{dx}\bigg|_{x=0} = 0. \tag{4.181}$$

III. STABILITY OF STRUTS

A strut is a slender rod subjected primarily to a thrust. If the thrust upon a strut is gradually increased, a state of displacement occurs when a small increase of load will prompt a transition to a new state. In such a case, we say that the strut has lost stability and the load under which this occurs is known as *critical load.*

4.11 The Euler Problem

Let us consider a strut subjected to two equal forces P at its ends along the direction of its axis (*Fig 4.14*). The strut will be slightly bent now. Considering any point with coordinates (x, y) on the elastic curve, we have, for small deflections

$$M = EI\frac{d^2y}{dx^2}.$$

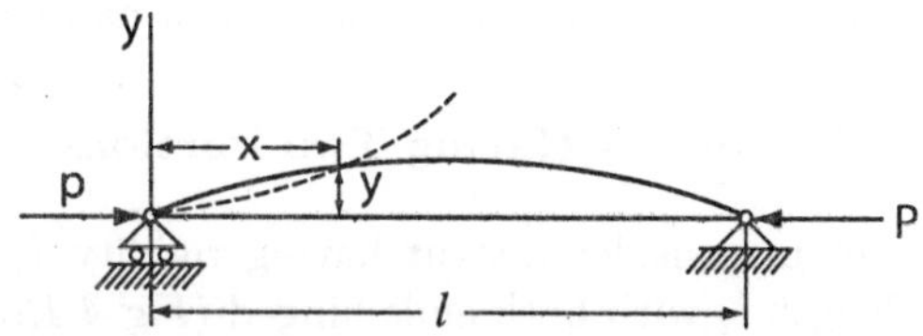

Fig 4.14

Bending of the strut occurs in the plane of minimum rigidity so I denotes the minimum moment of inertia of the cross-section.

Now, the external bending moment due to the load P at the point (x, y) is $-Py$.

Negative sign has been taken since the curvature of the elastic curve becomes more and more negative for gradual increase of P.

Thus

$$EI\frac{d^2y}{dx^2} = -Py, \quad \text{or,} \quad \frac{d^2y}{dx^2} + \xi^2 y = 0, \tag{4.182}$$

where

$$\xi^2 = \frac{P}{EI}. \tag{4.183}$$

Solution of equation (4.182) can be written as

$$y = A\cos\xi x + B\sin\xi x, \tag{4.184}$$

where A, B are constants of integration which are to be determined from boundary conditions. Let the rod having length l is not allowed to move along y-direction at its ends. Thus at $x = 0$ and $x = l$, $y = 0$.

These conditions give us $A = 0$ and $B\sin\xi l = 0$.

But B cannot be zero, since in that case the rod maintains its straight-line shape. Hence, we accept

$$\sin\xi l = 0, \quad \text{i.e.,} \quad \xi l = n\pi, \tag{4.185}$$

where n is any integer.

From equations (4.183) and (4.185), we get

$$\frac{P}{EI} = \left(\frac{n\pi}{l}\right)^2, \quad \text{i.e.,} \quad P = \frac{n^2\pi^2}{l^2}EI. \tag{4.186}$$

Thus, P given by equation (4.186) is the force required to keep the rod in a curvilinear position. Its minimum value is obtained for $n = 1$, and equals to

$$P_{\text{cr}} = \frac{\pi^2}{l^2}EI. \tag{4.187}$$

This problem was first formulated and solved by Euler and the load given by equation (4.187) is termed in his name as *Euler's critical load.*

Now, we find from equation (4.184) that the shape of the rod is given by

$$y = B\sin\xi x \quad (\because\ A = 0)$$

$$\text{i.e.,} \quad y = B\sin\left(\frac{n\pi}{l}x\right).$$

This shows that the shape of the rod after bending for $n = 1$ is a half-wave of a sine curve with maximum deflection $= B$ at the centre. For any other value of n, the shape of the rod will be represented by n-number of half-waves of sine curve.

4.12 Struts Having Two Portions with Different Rigidities

Let us consider a strut having rigidity I_1 for a portion of length l_1, and I_2 for a portion of length l_2 with both ends hinged (*Fig 4.15*).

Differential equation (4.182) for two portions are

$$EI\frac{d^2y_1}{dx^2}+\xi_1^2 y_1 = 0 \quad \text{and} \quad EI\frac{d^2y_2}{dx^2}+\xi_2^2 y_2 = 0 \tag{4.188}$$

where

$$\xi_1^2 = \frac{P}{EI_1} \quad \text{and} \quad \xi_2^2 = \frac{P}{EI_2}. \tag{4.189}$$

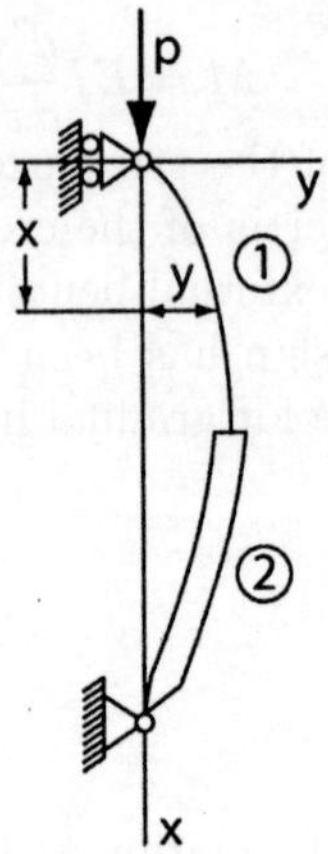

Fig 4.15

Solutions of the differential equations (4.188) are

$$\left.\begin{aligned} y_1 &= A_1\cos\xi_1 x + B_1\sin\xi_1 x \\ \text{and} \quad y_2 &= A_2\cos\xi_2 x + B_2\sin\xi_2 x \end{aligned}\right\}. \tag{4.190}$$

We have the following boundary conditions:

$$\left.\begin{aligned} &\text{at} \quad x = 0, \quad y_1 = 0 \\ &\text{at} \quad x = l, \quad y_2 = 0 \qquad (l = l_1 + l_2) \\ &\text{at} \quad x = l_1, \quad y_1 = y_2 \\ \text{and} \quad &\text{at} \quad x = l_1, \quad \frac{dy_1}{dx} = \frac{dy_2}{dx} \end{aligned}\right\}. \tag{4.191}$$

From these boundary conditions, we get $A_1 = 0$ and

$$B_1\sin\xi_1 l_1 = A_2\cos\xi_2 l_1 + B_2\sin\xi_2 l_1$$

$$B_1\xi_1\cos\xi_1 l_1 = -A_2\xi_2\sin\xi_2 l_1 + B_2\xi_2\cos\xi_2 l_1$$

$$A_2\cos\xi_2 l + B_2\sin\xi_2 l = 0.$$

Eliminating B_1, A_2, B_2 from these equations, we get

$$\begin{vmatrix} \sin\xi_1 l_1 & -\cos\xi_2 l_1 & -\sin\xi_2 l_1 \\ \xi_1\cos\xi_1 l_1 & \xi_2\sin\xi_2 l_1 & -\xi_2\cos\xi_2 l_1 \\ 0 & \cos\xi_2 l & \sin\xi_2 l \end{vmatrix} = 0. \tag{4.192}$$

The smallest non-zero root obtained from this equation will give the critical load. In particular, if we assume that $EI_2 = 4EI_1$, (i.e., $\xi_1 = 2\xi_2$) and $l_1 = l_2 = \frac{1}{2}l$, then we shall get the transcendental equation $\tan\xi_1 l = \sqrt{2}$, i.e., $\xi_1 l = 0.955$.

Thus

$$P_{\text{cr}} = \frac{14.6\,EI}{2} \tag{4.193}$$

4.13 Struts Having Different End Conditions

We differentiate equation (4.182) twice with respect to x, and get

$$\frac{d^4y}{dx^4} + \xi^2 \frac{d^2y}{dx^2} = 0. \tag{4.194}$$

The solution of this equation is

$$y = A \sin \xi x + B \cos \xi x + Cx + D. \tag{4.195}$$

Let us now consider different end conditions.

Case I. *One end fixed and the other free.*

Let the end at $x = 0$ be fixed and load P acting at free end. Then the boundary conditions are:

$$\text{at } x = 0, \quad y = 0 \quad \text{and} \quad \frac{dy}{dx} = 0$$

$$\text{at } x = l, \quad M = EI \frac{d^2y}{dx^2} = 0$$

$$\text{and} \quad \text{at } x = l, \quad Q = -EI \frac{d^3y}{dx^3} = P\frac{dy}{dx}.$$

Using these boundary conditions, we have

$$B + D = 0$$
$$A\xi + C = 0$$

$$A\xi^2 \sin \xi l + B\xi^2 \cos \xi l = 0, \quad \text{and} \quad A = 0.$$

Hence, we get $A = C = 0$, $B = -D$, so that

$$\cos \xi l = 0, \quad \because \; B \neq 0$$

$$\text{or,} \quad \xi l = (2n+1)\frac{\pi}{2}, \quad n = 0, 1, 2, 3. \tag{4.196}$$

The critical load is obtained for $n = 0$, and is given by

$$P_{\text{cr}} = \frac{\pi^2 EI}{4l^2}. \tag{4.197}$$

The equation of the bent-up rod will be

$$y = B(\cos x - 1). \tag{4.198}$$

Case II. *Both ends clamped.*

In this case, the boundary conditions are:

$$\text{at } x = 0, \quad y = 0 \quad \text{and} \quad \frac{dy}{dx} = 0$$

$$\text{at } x = l, \quad y = 0 \quad \text{and} \quad \frac{dy}{dx} = 0.$$

Using these boundary conditions, we get

$$B + D = 0$$
$$A\xi + C = 0$$

$$A \sin \xi l + B \cos \xi l + Cl + D = 0$$

and $$A\xi \cos \xi l - B\xi \sin \xi l + C = 0.$$

From these equations, we get two equations for A and B in the form

$$A(\sin \xi l - \xi l) + B(\cos \xi l - 1) = 0$$

and $$A(\cos \xi l - 1) - B \sin \xi l = 0.$$

Eliminating A and B ($\because A \neq 0, B \neq 0$)

$$\begin{vmatrix} \sin \xi l - \xi l & \cos \xi l - 1 \\ \cos \xi l - 1 & -\sin \xi l \end{vmatrix} = 0, \quad \text{i.e.,} \quad \sin \frac{\xi l}{2} \left[\sin \frac{\xi l}{2} - \frac{\xi l}{2} \cos \frac{\xi l}{2} \right] = 0.$$

If

$$\sin \frac{\xi l}{2} = 0, \xi l = 2n\pi \quad (n = 1, 2, 3, \ldots)$$

so that minimum value of $\xi l = 2\pi$.

If

$$\sin \frac{\xi l}{2} - \frac{\xi l}{2} \cos \frac{\xi l}{2} = 0, \quad \text{then} \quad \tan \frac{\xi l}{2} = \frac{\xi l}{2}.$$

In this case, the minimum value of ξl is approximately equal to 9. Hence we take $\xi l = 2\pi$, so that the critical load is

$$P_{\text{cr}} = \frac{4\pi^2}{l^2} EI. \tag{4.199}$$

Case III. *One end hinged and the other clamped.*

Let the rod be hinged at $x = l$ and clamped at $x = 0$. Then the boundary conditions are:

$$\text{at } x = 0, \quad y = 0 \quad \text{and} \quad \frac{dy}{dx} = 0$$

$$\text{at } x = l, \quad y = 0 \quad \text{and} \quad \frac{d^2y}{dx^2} = 0.$$

Using these boundary conditions, we get the following equations for unknown constants:

$$B + D = 0$$
$$A\xi + C = 0$$

$$A \sin \xi l + B \cos \xi l + Cl + D = 0$$
$$A \sin \xi l + B \cos \xi l = 0.$$

The solutions are $B = -A\xi l$, $C = -A\xi$, $D = A\xi l$, and

$$A(\sin \xi l - \xi l \cos \xi l) = 0,$$

whence, for $A \neq 0$,

$$\sin \xi l - \xi l \cos \xi l = 0,$$

$$\text{i.e.,} \quad \tan \xi l = \xi l.$$

The minimum value of $\xi l = 4.493$ (approx.).
Hence, the critical load is

$$P_{\text{cr}} = \frac{20.19}{l^2} EI \quad \text{or,} \quad P_{\text{cr}} \simeq \frac{2\pi^2}{l^2} EI. \tag{4.200}$$

Thus, we have obtained different values of critical load depending upon the end conditions given in equations (4.187), (4.197), (4.199) and (4.200). All these expressions can be presented in the form

$$P_{\text{cr}} = \frac{\pi^2}{l_0^2} EI, \tag{4.201}$$

where $l_0 = \eta l$ is called *effective length* of the rod and η is called the *coefficient of effectiveness* for length of the rod.

Equation (4.200) is thus a more general form of Euler's formula for critical load. The value of η for different cases are:

$\eta = 1$, for both ends hinged
$\eta = 2$, for one end fixed and the other free
$\eta = 1/2$, for both ends clamped
and $\eta = 1/\sqrt{2}$, for one end hinged and the other clamped.

4.14 Struts under Special Loadings

We have so far considered that the thrust P on the rod was acting in the direction of its axis. However, in practice, the loading conditions may be different from what we have considered. We shall look into following three cases:

Case I. *Strut having initial curvature.*

Let us consider a strut with both ends hinged having some initial curvature so that for any point on the strut, there exists a relation (*Fig 4.16*)

$$y_0 = y_0\,(x).$$

Now, under a load P in the direction of two hinges, the strut bends by an amount $y(x)$ then, the total displacement is

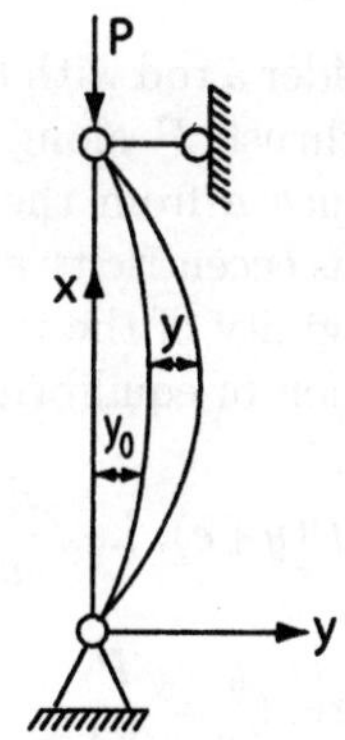

Fig 4.16

$$y_1 = y(x) + y_0(x).$$

Therefore, the equation of equilibrium of the rod can be written as

$$EI \frac{d^2y}{dx^2} = -P\{y(x) + y_0(x)\}, \quad \text{i.e.,} \quad \frac{d^2y}{dx^2} + \xi^2 y = -\xi^2 y_0(x). \tag{4.202}$$

Let us assume that the initial bent-up curve of the rod is given by

$$y_0\,(x) = f_0 \sin \frac{\pi x}{l}, \tag{4.203}$$

which represents a half-wave of sine curve with maximum amplitude f_0 at the centre. In this case, we accept the solution of equation (4.202) in the form

$$y(x) = A \sin \frac{\pi x}{l}, \tag{4.204}$$

which satisfies the boundary conditions at $x = 0$ and $x = l$, $y = 0$.

Substituting equations (4.203) and (4.204) into (4.202), we find

$$A = \frac{f_0 \xi^2}{\frac{\pi^2}{l^2} - \xi^2}. \tag{4.205}$$

Thus, the equation of deflected curve is

$$y = \frac{f_0 \xi^2}{\left(\frac{\pi}{l}\right)^2 - \xi^2} \sin \frac{\pi x}{l}. \tag{4.206}$$

The additional deflection at the centre due to load P is

$$f = \frac{f_0 \xi^2}{\left(\frac{\pi}{l}\right)^2 - \xi^2}, \quad \text{i.e.,} \quad f = \frac{f_0}{\theta - 1}, \tag{4.207}$$

where

$$\theta = \frac{P_{cr}}{P}, \quad \text{and} \quad P_{cr} = \frac{\pi^2}{l^2} EI.$$

From equation (4.207) it is clear that when $P \longrightarrow P_{cr}$, $\theta \longrightarrow 1$ and hence $f \longrightarrow \infty$, i.e., the rod looses its stability.

Case II. *Load position having an eccentricity.*

Let us consider a rod with both ends hinged subject to a thrust P along the line having constant distance e from the axis of the rod (*Fig 4.17*). This eccentricity e dies in the plane of minimum rigidity of the rod. Then the differential equation of equilibrium of the rod is

$$EI \frac{d^2 y}{dx^2} == -P(y+e), \text{ i.e., } \frac{d^2 y}{dx^2} + \xi^2 y = -\xi^2 e, \tag{4.208}$$

where, as before, $\xi^2 = \dfrac{P}{EI}$.

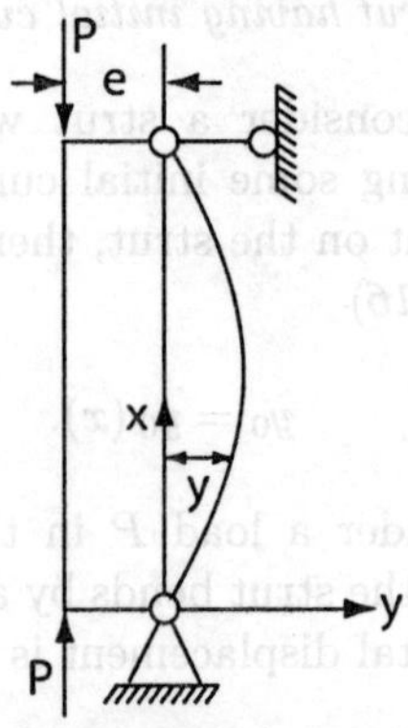

Fig 4.17

General solution of equation (4.208) is

$$y = A \sin \xi x + B \cos \xi x - e.$$

The boundary conditions are, as before at $x = 0$ and $x = l$, $y = 0$.

Using these conditions, we find $A = e \tan{}^{\xi l}/_2$, $B = e$.

Thus, the specific solution is

$$y = e \left[\tan \frac{\xi l}{2} \sin \xi x + \cos x - 1 \right]. \tag{4.209}$$

Maximum deflection at $x = l/2$ is

$$f = e\left[\tan\frac{\xi l}{2}\sin\frac{\xi l}{2} + \cos\frac{\xi l}{2} - 1\right] = e\left(\frac{1}{\cos\frac{\xi l}{2}} - 1\right) = e\left(\frac{1}{\cos\frac{\pi}{2}\sqrt{\frac{P}{P_{cr}}}} - 1\right), \quad (4.210)$$

where P_{cr} is as given by (4.187).

Here also, we see that as $P \longrightarrow P_{cr}$, $f \longrightarrow \infty$, that is, the rod looses its stability.

Case III. *Combination of axial and lateral load.*

Let us consider a rod of length l with both ends hinged and subject to an axial thrust P along with a lateral uniformly distributed load of intensity q (*Fig 4.18*).

Then the equation of equilibrium of the rod is given by

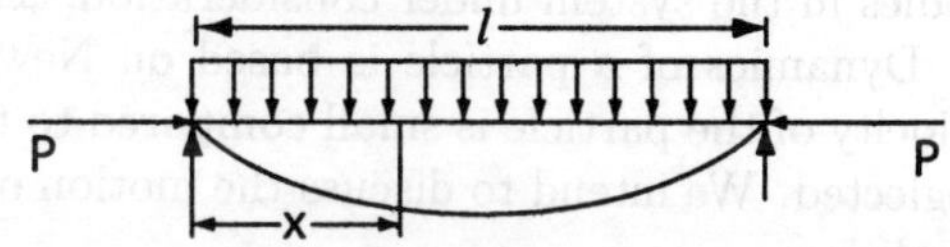

Fig 4.18

$$EI\frac{d^2y}{dx^2} = -Py - \frac{ql}{2}x + \frac{qx^2}{2}$$

$$\text{or,} \quad \frac{d^2y}{dx^2} + \xi^2 y = -\frac{q\xi^2}{P}\left(\frac{lx}{2} - \frac{x^2}{2}\right). \quad (4.211)$$

The complementary solution of this equation can be written as $y = A_1\cos\xi x + B_1\sin\xi x$. The particular integral is assumed as $y_p = \alpha x^2 + \beta x + \gamma$.

Substituting this into equation (4.211)

$$2\alpha + \xi^2(\alpha x^2 + \beta x + \gamma) = -\frac{q\xi^2}{P}\left(\frac{lx}{2} - \frac{x^2}{2}\right).$$

Comparing coefficients of x^2, x and constant quantity, we get

$$\alpha = \frac{q}{2P}, \quad \beta = -\frac{ql}{2P}, \quad \gamma = -\frac{q}{P\xi^2}.$$

Hence

$$y_p = -\frac{q}{P}\left(\frac{lx}{2} - \frac{x^2}{2}\right) - \frac{q}{P\xi^2}.$$

Hence, the general solution is

$$\xi^2 y = A\cos\xi x + B\sin\xi x - \frac{q\xi^2}{P}\left(\frac{lx}{2} - \frac{x^2}{2}\right) - \frac{q}{P}.$$

The boundary conditions, as before, are at $x = 0$ and $x = l$, $y = 0$.

Using these conditions, we get

$$A = \frac{q}{P}, \quad B = \frac{q}{P}\tan\frac{\xi l}{2}.$$

Hence, the final solution is

$$\xi^2 y = \frac{q}{P}\left[\tan\frac{\xi l}{2}\sin\xi x + \cos\xi x - 1\right] - \frac{\xi^2 q}{P}\left(\frac{lx}{2} - \frac{x^2}{2}\right). \quad (4.212)$$

The maximum deflection at the centre of the rod is

$$f = \frac{q}{P\xi^2}\left(\frac{1}{\cos\frac{\xi l}{2}} - 1\right) - \frac{1}{P}\frac{ql^2}{8} \quad \text{or,} \quad f = \frac{q}{P\xi^2}\left(\frac{1}{\cos\frac{\pi}{2}\sqrt{P/P_{\text{cr}}}} - 1\right) - \frac{1}{P}\frac{ql^2}{8}, \tag{4.213}$$

where P_{cr} is, as before, given by (4.187).

IV. DYNAMICS OF A PARTICLE

A particle is a point body having some mass but negligible physical dimensions relative to other bodies in the system under consideration. This is just a mathematical idealisation.

Dynamics of a particle is based on Newton's laws of motion. We shall assume that the velocity of the particle is small compared to that of light so that relativistic variation of mass is neglected. We intend to discuss the motion of a particle in various situations and under various conditions.

4.15 Basic Principles

If a particle of mass m is moving with velocity $\mathbf{v}$, then its momentum is $\mathbf{p} = m\mathbf{v}$.

According to Newton's second law of motion, the external force acting on the particle is

$$\mathbf{F} = \frac{d\mathbf{p}}{dt}.$$

Therefore, if there is no external force acting on the particle, then

$$\mathbf{F} = \frac{d\mathbf{p}}{dt} = 0, \quad \text{i.e.,} \quad \mathbf{p} = \text{constant}.$$

Thus, momentum of a particle under the action of no external forces remains constant.

The angular momentum $\mathbf{L}$ of the moving particle about a point O is defined as $\mathbf{L} = \mathbf{r} \times \mathbf{p}$, where $\mathbf{r}$ is the position vector of the particle with respect to origin at O.

The torque $\mathbf{N}$ acting on the particle is equal to the rate of change of its angular momentum. Thus

$$\mathbf{N} = \frac{d\mathbf{L}}{dt} = \frac{d}{dt}(\mathbf{r} \times \mathbf{p}) = \dot{\mathbf{r}} \times \mathbf{p} + \mathbf{r} \times \dot{\mathbf{p}}.$$

Now

$$\dot{\mathbf{r}} \times \mathbf{p} = \mathbf{v} \times m\mathbf{v} = 0 \quad \text{and} \quad \mathbf{r} \times \dot{\mathbf{p}} = \mathbf{r} \times \frac{d\mathbf{p}}{dt} = \mathbf{r} \times \mathbf{F}.$$

Hence $\mathbf{N} = \mathbf{r} \times \mathbf{F}$.

If $\mathbf{N} = 0$, then $\frac{d\mathbf{L}}{dt} = 0$, i.e., $\mathbf{L}$ = constant.

Therefore, the angular momentum of a moving particle will be constant if no torque is acting on it.

Work done by the external force in moving a particle from position P to position Q is given by

$$W = \int_P^Q \mathbf{F} \cdot d\mathbf{r}.$$

Now

$$\mathbf{F}\cdot d\mathbf{r} = m\frac{d\mathbf{v}}{dt}\cdot d\mathbf{r} = m\frac{d\mathbf{v}}{dt}\cdot\frac{d\mathbf{r}}{dt}\cdot dt,$$

i.e., $$\mathbf{F}\cdot d\mathbf{r} = m\left(\frac{d\mathbf{v}}{dt}\cdot\mathbf{v}\right)dt = \tfrac{1}{2}m\,d(\mathbf{v}^2).$$

$$\therefore\quad W = \int_P^Q \tfrac{1}{2}m\,d(v^2) = \tfrac{1}{2}m(v_2^2 - v_1^2) = T_2 - T_1,$$

where T_1 and T_2 are kinetic energies of the particles at positions P and Q respectively.

If $W > 0$, work is said to be done by the force on the particle when its kinetic energy is increased.

If $W < 0$, work is said to be done by the particle against the force due to which its kinetic energy is decreased. This occurs in the case of dissipative forces like the frictional force.

We have already seen that the work done by a conservative force in moving the particle along a closed curve is zero (Section 2.13). Moreover, the force vector $\mathbf{F}$ can be expressed as gradient of a scalar function V, so that we write $\mathbf{F} = -\operatorname{grad} V$, where V is the potential energy of the particle.

In this case, work done is

$$W = \int_P^Q \mathbf{F}\cdot d\mathbf{r} = -\int_P^Q \nabla V\cdot d\mathbf{r} = -\int_P^Q\left(\frac{\partial V}{\partial x}dx + \frac{\partial V}{\partial y}dy + \frac{\partial V}{\partial z}dz\right)$$

$$= -\int_P^Q dV = V_1 - V_2.$$

Hence, the work done by the force in displacing a particle from position P to position Q is equal to the difference of potential energies of the particle at those positions.

Thus $W = T_2 - T_1 = V_1 - V_2$ so that $T_1 + V_1 = T_2 + V_2 = \text{constant}$.

Hence, the sum of kinetic and potential energies of the particle during its motion in a conservative force always remains a constant.

The basic problem in dynamics is to solve the equation of motion given by $m\ddot{\mathbf{r}} = \mathbf{F}(\mathbf{r}, \dot{\mathbf{r}}, t)$, where $\ddot{\mathbf{r}}$ is the acceleration of the particle and $\mathbf{F}(\mathbf{r}, \dot{\mathbf{r}}, t)$ is the external force acting on the particle, which in general depends on position and velocity of the particle apart from time.

4.16 Motion of a Charged Particle in Electromagnetic Field

Let us consider a particle of mass m carrying a charge q, moving in an electromagnetic field with velocity $\mathbf{v}$. Then Lorentz force acting on the particle is $\mathbf{F} = q(\mathbf{E} + \mathbf{v}\times\mathbf{B})$, where $\mathbf{E}$ is the intensity of electric field and $\mathbf{B}$ is the magnetic induction.

Hence, the equation of motion of the particle is

$$m\frac{d\mathbf{v}}{dt} = q(\mathbf{E} + \mathbf{v}\times\mathbf{B}). \tag{4.214}$$

We shall consider some special cases.

Case I. *Motion in a constant electric field.*

We assume that the particle initially is at rest and is located at the point (x_0, y_0, z_0). If we take the constant electric field along x-axis, i.e., $\mathbf{E} = (E, 0, 0)$, then the equations of motion in Cartesian coordinate system are $m\ddot{x} = qE$, $m\ddot{y} = 0$, $m\ddot{z} = 0$.

Solutions of these equations are

$$\left.\begin{aligned} x &= x_0 + \frac{1}{2}\frac{qE}{m}t^2 \\ y &= y_0, \quad z = z_0 \end{aligned}\right\}. \tag{4.215}$$

Thus, acceleration of the particle at any instant of time is

$$a = \frac{qE}{m}. \tag{4.216}$$

If the particle falls through a potential difference of V volts, its initial potential energy qV is converted into kinetic energy. Hence, if v is the velocity of the particle acquired, then $\frac{1}{2}mv^2 = qV$.

$$\therefore \quad v = \sqrt{\frac{2qV}{m}}. \tag{4.217}$$

Case II. *Motion in a constant magnetic field.*

In this case, the equation of motion of the particle is

$$m\frac{d\mathbf{v}}{dt} = q\mathbf{v} \times \mathbf{B}.$$

We take dot product with velocity vector $\mathbf{v}$ on both sides. Then we get

$$m\frac{d\mathbf{v}}{dt} \cdot \mathbf{v} = q\mathbf{v} \cdot (\mathbf{v} \times \mathbf{B}), \quad \text{i.e.,} \quad \frac{d}{dt}\left(\tfrac{1}{2}m\mathbf{v}^2\right) = q(\mathbf{v} \times \mathbf{v}) \cdot \mathbf{B} = 0.$$

$$\therefore \quad \tfrac{1}{2}mv^2 = \text{constant}. \tag{4.218}$$

Thus, the kinetic energy and hence the speed of particle remain constant. Only its direction is getting changed due to the action of the magnetic field.

Let us assume that the magnetic induction $\mathbf{B}$ is constant and is directed along x-axis. If v_1 and v_2 be component velocities of the particle perpendicular and parallel to the direction of $\mathbf{B}$ respectively, then the corresponding components of forces acting on the particle are:

$$\mathbf{F}_1 = q\mathbf{v}_1 \times \mathbf{B} \tag{4.219}$$

$$\text{and} \quad \mathbf{F}_2 = q\mathbf{v}_2 \times \mathbf{B} = 0,$$

$$\text{i.e.,} \quad \mathbf{F}_2 = m\frac{d\mathbf{v}_2}{dt} = 0,$$

$$\text{i.e.,} \quad v_2 = \text{constant}. \tag{4.220}$$

But, $v^2 = v_1^2 + v_2^2$; hence, from equation (4.218) we find that v_1 is also constant.

Now, from equation (4.214), we get

$$m\frac{d\mathbf{v}_1}{dt} = q\mathbf{v}_1 \times \mathbf{B}.$$

This shows that the magnitude of v_1 is constant but its direction continuously changes.

We shall determine the trajectory of the particle. Taking $\mathbf{B} = \mathbf{i}B$, equation of motion in Cartesian coordinates is

$$m\frac{d\mathbf{v}}{dt} = q\mathbf{v} \times \mathbf{i}B,$$

$$\left.\begin{aligned} \text{i.e.,} \quad \frac{d\mathbf{v}}{dt} &= \xi \mathbf{v} \times \mathbf{i}, \\ \text{where} \quad \xi &= \frac{qB}{m} \end{aligned}\right\}. \tag{4.221}$$

Equation (4.221) in component form becomes

$$\frac{dv_x}{dt} = 0, \quad \frac{dv_y}{dt} = \xi v_z, \quad \frac{dv_z}{dt} = -\xi v_y.$$

From the first equations, we have

$$v_x = \text{constant} = v_2 \text{ and so, } x = x_0 + v_2 t. \tag{4.222}$$

Eliminating v_y from second and third equation, we get

$$\frac{d}{dt}\left(-\frac{1}{\xi}\frac{dv_z}{dt}\right) = \xi v_z, \quad \text{i.e.,} \quad \frac{d^2 v_z}{dt^2} + \xi^2 v_z = 0. \tag{4.223}$$

Its solution is

$$v_z = A\cos\xi t + B\sin\xi t. \tag{4.224}$$

$$\therefore \quad v_y = -\frac{1}{\xi}\frac{dv_z}{dt} = -A\sin\xi t + B\cos\xi t. \tag{4.225}$$

Integrating the equations (4.224) and (4.225), we get

$$\left.\begin{aligned} \xi(y - y_0) &= A\cos\xi t + B\sin\xi t \\ \text{and} \quad \xi(z - z_0) &= A\sin\xi t - B\cos\xi t \end{aligned}\right\}, \tag{4.226}$$

where y_0 and z_0 are integration constants. It may be noted that y_0, z_0 do not correspond to initial positions at $t = 0$.

Now, squaring and adding both sides of equation (4.226), we get

$$(y - y_0)^2 + (z - z_0)^2 = \frac{1}{\xi^2}(A^2 + B^2) = \text{constant}. \tag{4.227}$$

This represents a circle with centre at (y_0, z_0). The trajectory of the particle is given by equations (4.222) and (4.226). Hence, it will be moving in circular orbit about x-axis and at the same time, it will have a translational motion with constant velocity along x-axis. Thus, the trajectory will be a helix (*Fig 4.19*).

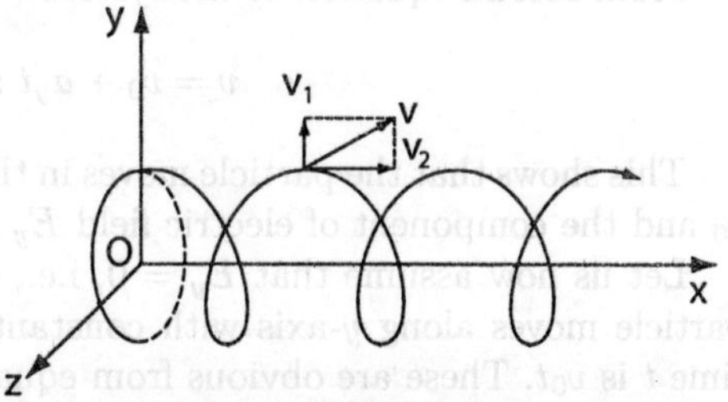

Fig 4.19

Case III. *Motion in combined field.*

Let the particle at time $t = 0$ be at the origin of coordinates with velocity components (u_0, v_0, w_0). Let us assume that **B** is along y-axis and **E** is in the xy-plane making some angle with x-axis. That is

$$\mathbf{B} = (0, B, 0)$$

$$\text{and} \quad \mathbf{E} = (E_x, E_y, 0).$$

Then the equations of motions (4.214) in the component form becomes

$$\frac{du}{dt} = a_x - \xi w, \quad \frac{dv}{dt} = a_y, \quad \frac{dw}{dt} = \xi u, \tag{4.228}$$

where

$$a_x = \frac{qE_x}{m}, \quad a_y = \frac{qE_y}{m}, \quad \text{and} \quad \xi = \frac{qB}{m}. \tag{4.229}$$

From first and third equations of the system (4.228) we have

$$\frac{d^2u}{dt^2} = -\xi\frac{dw}{dt} = -\xi^2 u, \quad \text{i.e.,} \quad \frac{d^2u}{dt^2} + \xi^2 u = 0.$$

Its solution is $u = C\cos\xi t + A\sin\xi t$.
Let us assume that at $t = 0$, $u_0 = 0$. Then, $C = 0$.
Hence

$$u = A\sin\xi t. \tag{4.230}$$

From third equation of (4.228)

$$\frac{dw}{dt} = \xi A\sin\xi t.$$

Integrating

$$w - w_0 = \xi A\int_0^t \sin\xi t\,dt, \quad \text{i.e.,} \quad w - w_0 = A(1 - \cos\xi t). \tag{4.231}$$

Integrating equations (4.230) and (4.231), we get

$$x = \int_0^t A\sin\xi t = \frac{A}{\xi}(1 - \cos\xi t) \tag{4.232}$$

$$\text{and} \quad z = w_0 t + \frac{A}{\xi}(\xi t - \sin\xi t). \tag{4.233}$$

Equations (4.232) and (4.233) give the parametric representation of a cycloid.
From second equation of the system (4.228)

$$v = v_0 + a_y t \text{ so that } y = v_0 t + \tfrac{1}{2}a_y t^2. \tag{4.234}$$

This shows that the particle moves in the direction of y-axis due to initial velocity component v_0 and the component of electric field E_y along the y-axis.

Let us now assume that $E_y = 0$, i.e., the fields are crossed. Then, $a_y = 0$, and hence the particle moves along y-axis with constant velocity v_0 so that its distance from origin at any time t is $v_0 t$. These are obvious from equation (4.234).

Now, from equations (4.228) and (4.230), we write

$$\frac{du}{dt} = a_x - \xi w, \quad \text{i.e.,} \quad A\xi\cos\xi t = a_x - \xi w.$$

Thus, for $t = 0$, $A\xi = a_x - \xi w_0$,

$$\therefore\ A = \frac{1}{\xi}(a_x - \xi w_0).$$

Putting back the value of ξ from equation (4.229), we get

$$A = \left(\frac{E_x}{B} - w_0\right).$$

Hence, we can write equation (4.230) in the form

$$u = \left(\frac{E_x}{B} - w_0\right) \sin \xi t. \tag{4.235}$$

Equation (4.235) shows that, if $w_0 = E_x/B$, then $u = 0$, i.e., there is no deviation of the particle along x-axis. Such particles are called to be filtered out. Thus, the crossed combination of electric and magnetic fields has got a practical importance in that, by suitably choosing **E** and **B**, particles of desired velocity may be obtained and particles of undesired velocity may be filtered out. Such a combination of electromagnetic field works as a velocity filter for charged particles.

4.17 Motion of a Particle in a Resisting Medium

Let us consider a particle of mass m moving under gravity in a medium the resistance of which varies as the square of its velocity.

Case I. *Particle falling down freely.*

In this case, the equation of motion of the particle is

$$m\frac{d^2x}{dt^2} = mg - \mu m v^2, \tag{4.236}$$

where μ = a constant, v = velocity of the particle.
Writing the above equation as

$$\frac{d^2x}{dt^2} = g\left(1 - \frac{\mu}{g}v^2\right) \tag{4.237}$$

we find that the acceleration of the particle becomes zero when its velocity is given by

$$g\left(1 - \frac{\mu}{g}v^2\right) = 0, \quad \text{i.e.,} \quad V_0 = \sqrt{\frac{g}{\mu}}. \tag{4.238}$$

This velocity is called the *terminal velocity*.
Now, from equation (4.237), we get

$$\frac{d^2x}{dt^2} = g\left(1 - \frac{v^2}{V_0^2}\right), \tag{4.239}$$

$$\text{or,} \quad v\frac{dv}{dx} = \frac{g}{V_0^2}(V_0^2 - v^2), \quad \text{i.e.,} \quad \frac{g\,dx}{V_0^2} = \frac{v\,dv}{V_0^2 - v^2}.$$

Integrating, we get

$$\frac{2g}{V_0^2}x = -\log(V_0^2 - v^2) + C.$$

Initially, for $x = 0$, $v = 0$; $\therefore\ C = \log V_0^2$.

$$\therefore\quad \frac{2g}{V_0^2}x = \log V_0^2 - \log(V_0^2 - v^2), \quad \text{i.e.,}\quad \frac{2g}{V_0^2}x = \log\frac{V_0^2}{V_0^2 - v^2},$$

$$\text{or,}\quad \frac{V_0^2}{V_0^2 - v^2} = e^{2gx/V_0^2} \quad \text{or,}\ v^2 = V_0^2\left(1 - e^{-2gx/V_0^2}\right). \tag{4.240}$$

This gives a relation between velocity v and the distance x traversed.

Now, we write equation (4.239) in the form

$$\frac{dv}{dt} = \frac{g}{V_0^2}(V_0^2 - v^2), \quad \text{or,}\quad \frac{dv}{V_0^2 - v^2} = \frac{g}{V_0^2}dt.$$

Integrating, we get

$$\frac{1}{2V_0}\log\frac{V_0 + v}{V_0 - v} = \frac{g}{V_0^2}t + C_1.$$

Initially, at $t = 0$, $v = 0$; $\therefore\ C_1 = 0$ ($\because\ \log 1 = 0$).

$$\therefore\quad \frac{1}{2}\log\frac{V_0 + v}{V_0 - v} = \frac{gt}{V_0}, \quad \text{i.e.,}\quad \tanh^{-1}\frac{v}{V_0} = \frac{gt}{V_0}.$$

$$\therefore\quad v = V_0\tanh\left(\frac{gt}{V_0}\right). \tag{4.241}$$

This gives the velocity of the particle at any instant of time t.

Again, from equations (4.240) and (4.241), eliminating v, we get

$$V_0^2\left(1 - e^{-2gx/V_0^2}\right) = V_0^2\tanh^2\left(\frac{gt}{V_0}\right),$$

$$\text{i.e.,}\quad e^{-2gx/V_0^2} = 1 - \tanh^2\left(\frac{gt}{V_0}\right) = \operatorname{sech}^2\left(\frac{gt}{V_0}\right).$$

$$\therefore\quad e^{2gx/V_0^2} = \cosh^2\left(\frac{gt}{V_0}\right) \quad \text{or,}\quad e^{gx/V_0^2} = \cosh\frac{gt}{V_0}.$$

$$\therefore\quad \frac{gx}{V_0^2} = \log\left(\cosh\frac{gt}{V_0}\right), \quad \text{i.e.,}\quad x = \frac{V_0^2}{g}\log\left(\cosh\frac{gt}{V_0}\right). \tag{4.242}$$

This gives the distance fallen after time t by the particle.

EXAMPLE 4.1. *A particle falls from rest under gravity in a medium whose resistance varies as the square of the velocity. If V_0 be the terminal velocity and v_0 the velocity which it would have acquired if there were no resistance, show that*

$$\left(\frac{v}{v_0}\right)^2 = 1 - \frac{1}{2}\left(\frac{v_0}{V_0}\right)^2 + \frac{1}{2\cdot 3}\left(\frac{v_0}{V_0}\right)^4 - \cdots$$

Solution: We have, from equation (4.240), the velocity at any distance x as

$$v^2 = V_0^2\left(1 - e^{-2gx/V_0^2}\right). \tag{1}$$

Now, if there were no resistance, the velocity v_0 at any distance x will be given by

$$v_0^2 = 2gx. \tag{2}$$

Eliminating x from equations (1) and (2)

$$v^2 = V_0^2\left(1 - e^{-v_0^2/V_0^2}\right),$$

$$\text{or,}\quad v^2 = V_0^2\left[1 - \left\{1 - \left(\frac{v_0}{V_0}\right)^2 + \frac{1}{2!}\left(\frac{v_0}{V_0}\right)^4 - \cdots\right\}\right],$$

$$\text{or,}\quad v^2 = v_0^2 - \frac{1}{2}\frac{v_0^4}{V_0^2} + \frac{1}{2\cdot 3}\frac{v_0^6}{V_0^4} - \cdots,$$

$$\text{or,}\quad \frac{v^2}{v_0^2} = 1 - \frac{1}{2}\frac{v_0^2}{V_0^2} + \frac{1}{2\cdot 3}\frac{v_0^4}{V_0^4} - \cdots$$

EXAMPLE 4.2. *If the resistance varies as the fourth power of the velocity, show that the kinetic energy of a particle of mass m falling freely under gravity at a distance x is*

$$E\tanh\left(\frac{mgx}{E}\right),$$

where E = terminal energy.

Solution: Equation of motion is

$$m\frac{d^2x}{dt^2} = mg - \mu m v^4, \quad \text{i.e.,}\quad \frac{d^2x}{dt^2} = g\left(1 - \frac{\mu}{g}v^4\right).$$

If V_0 be the terminal velocity, then acceleration is zero. Hence

$$g\left(1 - \frac{\mu}{g}V_0^4\right) = 0, \quad \therefore\ V_0^4 = \frac{g}{\mu}. \tag{1}$$

Equation of motion can now be written as

$$v\frac{dv}{dx} = g\left(1 - \frac{v^4}{V_0^4}\right) = \frac{g}{V_0^4}\left(V_0^4 - v^4\right).$$

$$\therefore\quad \frac{2v\,dv}{(V_0^2)^2 - (v^2)^2} = \frac{2g}{V_0^4}\,dx.$$

Integrating, we get

$$\frac{1}{2V_0^2}\log\frac{V_0^2 + v^2}{V_0^2 - v^2} = \frac{2gx}{V_0^4} + C.$$

Initially, at $x = 0$, $v = 0$; $\therefore\ C = 0$.
Hence

$$\frac{1}{2V_0^2}\log\frac{V_0^2 + v^2}{V_0^2 - v^2} = \frac{2gx}{V_0^4},$$

$$\text{or,}\quad \tanh^{-1}\frac{v^2}{V_0^2} = \frac{2gx}{V_0^2} \quad \left[\text{remember: } \frac{1}{2}\log\frac{a+x}{a-x} = \tanh^{-1}\frac{x}{a}\right].$$

$$\therefore \quad v^2 = V_0^2 \tanh\left(\frac{2gx}{V_0^2}\right).$$

$\therefore$ Kinetic energy at position x is

$$\tfrac{1}{2}mv^2 = \tfrac{1}{2}mV_0^2 \tanh\left(\frac{2gx}{V_0^2}\right). \tag{2}$$

Now, terminal energy is

$$E = \tfrac{1}{2}mV_0^2. \tag{3}$$

Thus, from equations (2) and (3), we get

$$\text{Kinetic energy} = E \tanh\left(\frac{2gx}{V_0^2}\right).$$

Case II. *Particle moving upward.*

Let the particle be projected upwards with an initial velocity of V_0. Then its equation of motion is

$$m\frac{d^2x}{dt^2} = -mg - \mu mv^2, \quad \text{i.e.,} \quad \frac{d^2x}{dt^2} = -g\left(1 + \frac{\mu}{g}v^2\right),$$

$$\text{i.e.,} \quad v\frac{dv}{dx} = -g\left(1 + \frac{v^2}{k^2}\right), \quad k^2 = \frac{g}{\mu}. \tag{4.243}$$

$$\therefore \quad -\int \frac{2v\,dv}{k^2 + v^2} = \frac{2g}{k^2}x + A, \quad \text{or,} \quad -\log(k^2 + v^2) = \frac{2g}{k^2}x + A.$$

Initially, at $x = 0$, $v = V_0$;

$$\therefore \quad A = -\log\left(k^2 + V_0^2\right).$$

$$\therefore \quad \frac{2gx}{k^2} = \log\frac{V_0^2 + k^2}{v^2 + k^2}. \tag{4.244}$$

Again, from equation (4.243), we have

$$\frac{dv}{dt} = -g\left(1 + \frac{v^2}{k^2}\right). \quad \therefore \quad \frac{dv}{k^2 + v^2} = -\frac{g}{k^2}\,dt.$$

Integrating

$$\frac{1}{k}\tan^{-1}\frac{v}{k} = -\frac{gt}{k^2} + B.$$

Initially, at $t = 0$, $v = V_0$;

$$\therefore \quad B = \frac{1}{k}\tan^{-1}\frac{V_0}{k}. \quad \therefore \quad \frac{gt}{k} = \tan^{-1}\frac{V_0}{k} - \tan^{-1}\frac{v}{k}. \tag{4.245}$$

Equation (4.244) gives the velocity when the particle has gone up a distance x, and equation (4.245) gives the velocity at any time t.

EXAMPLE 4.3. *A particle of mass m is projected vertically upward under gravity, the resistance of air being mk times the velocity. Show that, the greatest height attained by the particle is*

$$\frac{V^2}{g}\left[\lambda - \log(1 + \lambda)\right],$$

where V = terminal velocity and λV is the initial velocity.

Solution: If the particle falls freely under gravity, then its equation of motion is

$$\frac{d^2x}{dt^2} = g - kv.$$

Hence, if V = terminal velocity, then

$$g - kV = 0, \quad \text{i.e.,} \quad V = \frac{g}{k}. \tag{4.246}$$

The particle has been projected upwards. Hence, its equation of motion is

$$\frac{d^2x}{dt^2} = -g - kv \quad \text{or,} \quad v\frac{dv}{dx} = -g\left(1 + \frac{v}{V}\right), \quad \text{[using equation (2.46)]}$$

$$\text{or,} \quad \frac{v\,dv}{V+v} = -\frac{g}{V}\,dx \quad \text{or,} \quad \left(1 - \frac{V}{V+v}\right)dv = -\frac{g}{V}\,dx.$$

Integrating, we get

$$v - V\log(V+v) = -\frac{gx}{V} + A. \tag{4.247}$$

Initially, at $x = 0$, $v = \lambda V$; hence

$$\lambda V - V\log(V + \lambda V) = A.$$

Thus, equation (4.247) becomes

$$v - V\log(V+v) - \lambda V + V\log(V+\lambda V) = -\frac{gx}{V}. \tag{4.248}$$

If H be the greatest height, then $v = 0$ at $x = H$.
Hence we get

$$-V\log V - \lambda V + V\log(V+\lambda V) = -\frac{g}{V}H,$$

$$\text{or,} \quad H = \frac{V}{g}\left\{\lambda V - V\log\frac{V+\lambda V}{V}\right\} = \frac{V^2}{g}\{\lambda - \log(1+\lambda)\}.$$

EXAMPLE 4.4. *If the resistance varies as fourth power of the velocity, show that the kinetic energy of a particle of mass m projected upward against gravity at a depth x below highest point is,*

$$E\tan\left(\frac{mgx}{E}\right),$$

where E = terminal energy.

Solution: Equation of motion of the particle is

$$m\frac{d^2x}{dt^2} = -mg - \mu m v^4, \quad \text{i.e.,} \quad \frac{d^2x}{dt^2} = -g\left(1 + \frac{\mu}{g}v^4\right) = -g\left(1 + \frac{v^4}{V_0^4}\right), \tag{1}$$

where $V_0^4 = g/\mu$.

$\therefore$ From equation (1), we get

$$v\frac{dv}{dx} = -\frac{g}{V_0^4}\left(V_0^4 + v^4\right), \quad \text{or,} \quad \frac{2v\,dv}{(V_0^2)^2 + (v^2)^2} = -\frac{2g}{V_0^4}\,dx.$$

Integrating

$$\frac{1}{V_0^2}\tan^{-1}\frac{v^2}{V_0^2} = -\frac{2gx}{V_0^4} + C. \tag{2}$$

If, at $x = 0$, velocity of projection $= u$, then

$$\frac{1}{V_0^2}\tan^{-1}\frac{u^2}{V_0^2} = C. \tag{3}$$

Hence, from equations (2) and (3)

$$\frac{2gx}{V_0^2} = \tan^{-1}\frac{u^2}{V_0^2} - \tan^{-1}\frac{v^2}{V_0^2}. \tag{4}$$

If H be the greatest height reached, then at $x = H$, $v = 0$. Thus, from equation (4)

$$\frac{2gH}{V_0^2} = \tan^{-1}\frac{u^2}{V_0^2}. \tag{5}$$

Subtracting (4) from (5)

$$\frac{2g}{V_0^2}(H - x) = \tan^{-1}\frac{v^2}{V_0^2}.$$

$$\therefore \quad v^2 = V_0^2\tan\frac{2g}{V_0^2}z, \quad z = H - x.$$

$$\therefore \text{ Kinetic energy} = \tfrac{1}{2}mv^2 = \tfrac{1}{2}mV_0^2\tan\frac{2g}{V_0^2}z.$$

But, terminal energy

$$E = \tfrac{1}{2}mV_0^2.$$

Therefore, kinetic energy $= E\tan\left(\dfrac{mgz}{E}\right)$.

Case III. *Motion of a projectile.*

Let, a particle of mass m be projected with a velocity u at an angle α to the horizon. The resistance of air is equal to μm times the velocity.

Let, $P(x, y)$ be any point on the trajectory of the particle at time t when tangent at P makes an angle ϕ with x-axis (*Fig 4.20*).

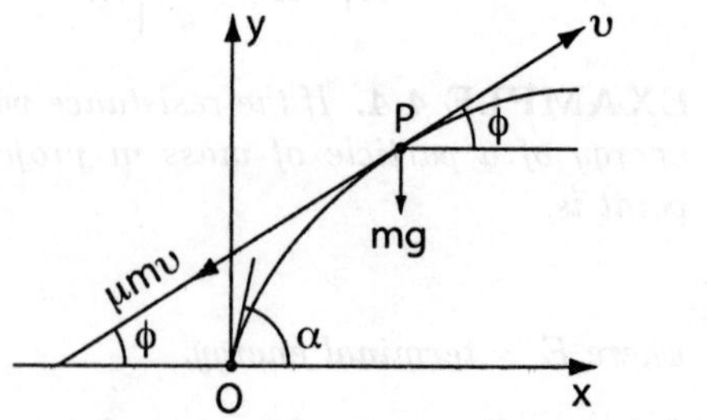

Fig 4.20

Then

$$\frac{dy}{dx} = \tan\phi, \quad \frac{dy}{ds} = \sin\phi \quad \text{and} \quad \frac{dx}{dx} = \cos\phi.$$

The equation of motion along x-axis is

$$m\frac{d^2x}{dt^2} = -\mu mv\cos\phi, \quad \text{i.e.,} \quad \frac{d^2x}{dt^2} = -\mu\frac{ds}{dt}\cdot\frac{dx}{ds} = -\mu\frac{dx}{dt}. \tag{4.249}$$

$$\therefore \quad \frac{\ddot{x}}{\dot{x}} = -\mu.$$

Integrating

$$\log(\dot{x}) = -\mu t + A.$$

Initially, at $t = 0$, $\dot{x} = u \cos \alpha = U$ (say).

$$\therefore \quad \log U = A.$$

Hence

$$\log \frac{\dot{x}}{U} = -\mu t, \quad \text{or,} \quad \dot{x} = \frac{dx}{dt} = Ue^{-\mu t}. \tag{4.250}$$

Integrating again

$$x = -\frac{U}{\mu} e^{-\mu t} + B.$$

Initially, at $t = 0$, $x = 0$; $\therefore$ $B = U/\mu$.
Hence

$$x = \frac{U}{\mu} \left(1 - e^{-\mu t}\right). \tag{4.251}$$

Again, equation of motion along y-axis is

$$\frac{d^2y}{dt^2} = -g - \mu v \sin\phi = -g - \mu \frac{ds}{dt} \cdot \frac{dy}{ds} = -\left(g + \mu \frac{dy}{dt}\right), \quad \text{or,} \quad \frac{\mu \ddot{y}}{g + \mu \dot{y}} = -\mu. \tag{4.252}$$

Integrating

$$\log(g + \mu \dot{y}) = -\mu t + C.$$

Initially, at $t = 0$, $\dot{y} = u \sin \alpha = V$ (say).

$$\therefore \quad \log(g + \mu V) = C.$$

Hence

$$\log \frac{g + \mu \dot{y}}{g + \mu V} = -\mu t, \quad \text{or,} \quad g + \mu \frac{dy}{dt} = (g + \mu V)\, e^{-\mu t}. \tag{4.253}$$

Equations (4.250) and (4.253), give horizontal and vertical components of velocity at any time.

Integrating equation (4.253) we get

$$gt + \mu y = -\frac{1}{\mu}(g + \mu V)\, e^{-\mu t} + D. \tag{4.254}$$

Initially, at $t = 0$, $y = 0$; hence

$$D = \frac{1}{\mu}(g + \mu V). \tag{4.255}$$

Thus, we get

$$gt + \mu y = \frac{1}{\mu}\,(g + \mu V)\left(1 - e^{-\mu t}\right). \tag{4.256}$$

Equations (4.251) and (4.256) give horizontal and vertical displacements of the projectile after a time t. Hence, these equations represent the parametric equation of the trajectory.

Now, from equation (4.251), we have

$$\frac{\mu x}{U} = 1 - e^{-\mu t}, \quad \text{i.e.,} \quad e^{-\mu t} = 1 - \frac{\mu x}{U}.$$

$$\therefore \quad t = -\frac{1}{\mu} \log\left(1 - \frac{\mu x}{U}\right). \tag{4.257}$$

Substituting t into equation (4.256)

$$-\frac{g}{\mu} \log\left(1 - \frac{\mu x}{U}\right) + \mu y = \frac{1}{\mu}(g + \mu V)\frac{\mu x}{U},$$

$$\text{or,} \quad y = \frac{g}{\mu^2} \log\left(1 - \frac{\mu x}{U}\right) + \frac{x}{\mu U}(g + \mu V). \tag{4.258}$$

Equation (4.258) gives the trajectory in Cartesian coordinates.
Now, we introduce new variables

$$X = 1 - \frac{\mu x}{U} \tag{4.259}$$

$$\text{and} \quad Y = y - \frac{g + \mu V}{\mu^2}. \tag{4.260}$$

Then

$$\frac{x}{U} = \frac{1 - X}{\mu}. \tag{4.261}$$

Using this in equation (4.258), we get

$$y = \frac{g}{\mu^2} \log X + \frac{1 - X}{\mu^2}(g + \mu V),$$

$$\text{or,} \quad y - \frac{1}{\mu^2}(g + \mu V) = \frac{g}{\mu^2} \log X - \frac{1}{\mu^2}(g + \mu V)X,$$

$$\text{or,} \quad Y = \frac{g}{\mu^2} \log X - \frac{1}{\mu^2}(g + \mu V)X,$$

which is of the form

$$Y + aX = b \log X. \tag{4.262}$$

This gives another form of representation of the trajectory.

Let us now determine the range of the projectile. If R be the horizontal range, then at that point $x = R$ and $y = 0$.

Hence, from equation (4.258), we get

$$0 = \frac{g}{\mu^2} \log\left(1 - \frac{\mu R}{U}\right) + \frac{R}{\mu U}(g + \mu V),$$

$$\text{i.e.,} \quad gU \log\left(1 - \frac{\mu R}{U}\right) + \mu R(g + \mu V) = 0,$$

$$\text{or,} \quad gU\left[-\frac{\mu R}{U} - \frac{\mu^2 R^2}{2U^2} - \frac{\mu^3 R^3}{3U^3} - \cdots\right] + \mu R(g + \mu V) = 0,$$

$$\text{or,} \quad \left[-g - \frac{\mu R g}{2U} - \frac{\mu^2 R^2 g}{3U^2} - \cdots\right] + g + \mu V = 0,$$

$$\text{or,} \quad \frac{Rg}{2U} = V - \frac{1}{3}\frac{\mu R^2 g}{U^2} \quad \text{(approximately).}$$

$$\therefore \quad R = \frac{2UV}{g} - \frac{2}{3}\frac{\mu R^2}{U}. \tag{4.263}$$

Thus, in the first approximation

$$R = \frac{2UV}{g} = \frac{2u^2 \sin\alpha \cos\alpha}{g}. \tag{4.264}$$

Putting this value of R into equation (4.263), we get

$$R = \frac{2UV}{g} - \frac{2\mu}{3U} \cdot \frac{4U^2V^2}{g^2},$$

$$\text{or,} \quad R = \frac{2UV}{g} - \frac{8\mu}{3g^2}UV^2.$$

$$\therefore \quad R = \frac{2}{g}u^2 \sin\alpha \cos\alpha - \frac{8\mu}{3g^2}u^2 \cos\alpha \sin^2\alpha. \tag{4.265}$$

We shall now determine the time of flight. If T be the time of flight after which the particle comes back to the plane containing point of projection, then at $t = T$, $y = 0$.

Hence, from equation (4.256), we get

$$gT = \frac{1}{\mu}(g + \mu V)(1 - e^{-\mu T}),$$

$$\text{or,} \quad \mu gT = (g + \mu V)\left[1 - \left(1 - \mu T + \frac{\mu^2T^2}{2!} - \frac{\mu^3T^3}{3!} + \cdots\right)\right],$$

$$\text{or,} \quad \mu gT = (g + \mu V)\left(\mu T - \frac{\mu^2T^2}{2!} + \frac{\mu^3T^3}{3!} - \cdots\right),$$

$$\text{or,} \quad \mu gT = \mu gT + \mu^2\left(VT - \frac{gT^2}{2}\right) + \mu^3\left(\frac{gT^3}{6} - \frac{VT^2}{2}\right) + \cdots,$$

$$\text{or,} \quad 0 = \mu^2 T\left[\left(V - \frac{gT}{2}\right) + \mu\left(\frac{gT^2}{6} - \frac{VT}{2}\right)\right] + \cdots$$

$$\therefore \quad T = \frac{2V}{g} + \frac{2\mu}{g}\left(\frac{gT^2}{6} - \frac{VT}{2}\right) + \cdots \tag{4.266}$$

Hence, in the first approximation

$$T = \frac{2V}{g}. \tag{4.267}$$

Putting this value in RHS of equation (4.266), we get, in the second approximation

$$T = \frac{2V}{g} + \frac{2\mu}{g}\left[\frac{g}{6} \cdot \frac{4V^2}{g^2} - \frac{V}{2} \cdot \frac{2V}{g}\right],$$

$$\text{or,} \quad T = \frac{2V}{g} - \frac{2\mu}{3} \cdot \frac{V^2}{g^2}. \tag{4.268}$$

Let us now determine the greatest height reached by the particle and the corresponding time.

At the greatest height, vertical component of the velocity of the particle is zero. If T_0 be the time to reach this height, then from equation (4.253)

$$g = (g + \mu V)\, e^{-\mu T_0}.$$

$$\therefore e^{\mu T_0} = \frac{g + \mu V}{g} = 1 + \frac{\mu V}{g}.$$

$$\therefore \quad T_0 = \frac{1}{\mu} \log\left(1 + \frac{\mu V}{g}\right). \tag{4.269}$$

Hence, if the greatest height attained is H, then, from equation (4.256)

$$gT_0 + \mu H = \left(1 + \frac{\mu V}{g}\right) \frac{g}{\mu} \left(1 - e^{-\mu T_0}\right),$$

$$\text{or,} \quad \frac{g}{\mu} \log\left(1 + \frac{\mu V}{g}\right) + \mu H = \frac{g}{\mu}\left(1 + \frac{\mu V}{g}\right) \frac{\mu V}{g + \mu V}.$$

$$\therefore \quad H = \frac{V}{\mu} - \frac{g}{\mu^2} \log\left(1 + \frac{\mu V}{g}\right). \tag{4.270}$$

EXAMPLE 4.5. *A particle of unit mass is projected with velocity u at an angle α with the horizon in a medium whose resistance is μ times the velocity. Show that, its direction will again make an angle α with the horizon after a time*

$$\frac{1}{\mu} \log\left\{1 + \frac{2\mu}{g} V\right\},$$

where V is the initial velocity in vertical direction.

Solution: At any instant of t, if θ be the angle made with horizon by the direction of motion, then

$$\tan\theta = \frac{dy}{dx} = \frac{dy}{dt} \Big/ \frac{dx}{dt}. \tag{1}$$

Hence, by given condition

$$\tan(\pi - \alpha) = \frac{\dot{y}}{\dot{x}}. \tag{2}$$

From equations (4.250) and (4.253), we substitute expressions for $\dot{x}$ and $\dot{y}$. Thus

$$-\tan\alpha = \frac{(g + \mu V)e^{-\mu t} - g}{\mu U e^{-\mu t}},$$

$$\text{or,} \quad -\mu U e^{-\mu t} \cdot \tan\alpha = (g + \mu V)\, e^{-\mu t} - g. \tag{3}$$

Now

$$\left.\begin{aligned} U &= u\cos\alpha \\ \text{and} \quad V &= u\sin\alpha \end{aligned}\right\}. \tag{4}$$

Therefore, we get

$$-\mu u \sin\alpha e^{-\mu t} = (g + \mu u \sin\alpha)\, e^{-\mu t} - g,$$

$$\text{or,} \quad g e^{\mu t} = g + 2\mu V,$$

$$\text{or,} \quad e^{\mu t} = 1 + \frac{2\mu V}{g}.$$

$$\therefore \quad t = \frac{1}{\mu} \log\left(1 + \frac{2\mu V}{g}\right).$$

EXAMPLE 4.6. *Show that, in the motion of a heavy particle in a medium where the resistance varies as the velocity, the time to reach the greatest height is less than half the total time of flight.*

Solution: From equation (4.268), we have the time of flight

$$T = \frac{2V}{g} - \frac{2\mu}{3}\frac{V^2}{g^2}$$

and, from equation (4.269), we have the time to reach the greatest height

$$T_0 = \frac{1}{\mu}\log\left(1 + \frac{\mu V}{g}\right).$$

We have to prove that

$$2T_0 - T < 0.$$

Now

$$\begin{aligned} 2T_0 - T &= \frac{2}{\mu}\log\left(1 + \frac{\mu V}{g}\right) - \left(\frac{2V}{g} - \frac{2\mu}{3}\frac{V^2}{g^2}\right) \\ &= \frac{2}{\mu}\left[\frac{\mu V}{g} - \frac{1}{2}\frac{\mu^2 V^2}{g^2}\cdots\right] - \frac{2V}{g} + \frac{2\mu}{3}\frac{V^2}{g^2} \\ &= -\frac{\mu V^2}{g^2} + \frac{2\mu}{3}\frac{V^2}{g^2}\cdots \\ &= -\frac{1}{3}\frac{\mu V^2}{g^2} < 0. \end{aligned}$$

Hence proved.

4.18 Constrained Motion

We shall now consider motion of a particle along a definite curve. In these cases, the equations of motion are preferably considered along the tangent and normal to the curve.

Let us now determine tangential and normal accelerations from those along the Cartesian coordinates.

We know

$$\frac{dx}{dt} = \frac{dx}{ds}\cdot\frac{ds}{dt}.$$

$$\therefore \quad \frac{d^2x}{dt^2} = \frac{d^2x}{ds^2}\left(\frac{ds}{dt}\right)^2 + \frac{dx}{ds}\cdot\frac{d^2s}{dt^2}. \tag{4.271}$$

Similarly

$$\frac{d^2y}{dt^2} = \frac{d^2y}{ds^2}\left(\frac{ds}{dt}\right)^2 + \frac{dy}{ds}\cdot\frac{d^2s}{dt^2}. \tag{4.272}$$

From differential calculus, we know that curvature at any point of a curve is

$$\frac{1}{\rho} = -\frac{d^2x}{ds^2}\bigg/\frac{dy}{ds} = \frac{d^2y}{ds^2}\bigg/\frac{dx}{ds}. \tag{4.273}$$

From these relations, we have

$$\frac{d^2x}{dt^2} = -\frac{1}{\rho}\frac{dy}{ds}\left(\frac{ds}{dt}\right)^2 + \frac{dx}{ds}\cdot\frac{d^2s}{dt^2}$$
$$= -\frac{\sin\phi}{\rho}\left(\frac{ds}{dt}\right)^2 + \cos\phi\cdot\frac{d^2s}{dt^2} \tag{4.274}$$

and
$$\frac{d^2y}{dt^2} = \frac{1}{\rho}\frac{dx}{ds}\left(\frac{ds}{dt}\right)^2 + \frac{dy}{ds}\cdot\frac{d^2s}{dt^2}$$
$$= \frac{\cos\phi}{\rho}\cdot\left(\frac{ds}{dt}\right)^2 + \sin\phi\cdot\frac{d^2s}{dt^2}, \tag{4.275}$$

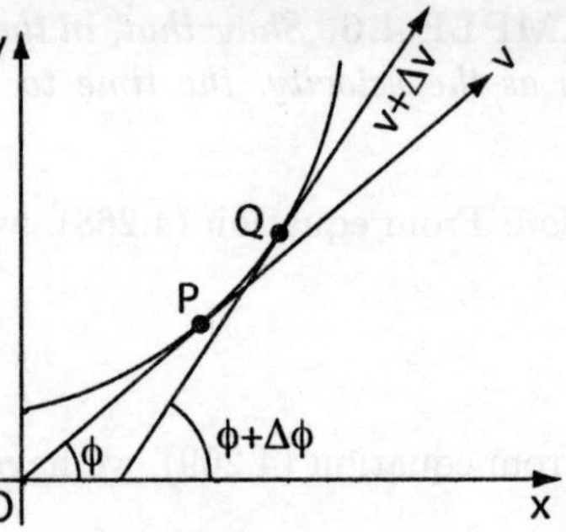

Fig 4.21

where ϕ is the angle which the tangent at a point to the curve makes with x-axis (*Fig 4.21*).

Therefore, the acceleration along the tangent is

$$f_t = \frac{d^2x}{dt^2}\cos\phi + \frac{d^2y}{dt^2}\sin\phi = \frac{d^2s}{dt^2} = \frac{dv}{ds}\cdot\frac{ds}{dt} = v\frac{dv}{ds} \tag{4.276}$$

and, the acceleration along the normal is

$$f_n = -\frac{d^2x}{dt^2}\sin\phi + \frac{d^2y}{dt^2}\cos\phi = \frac{v^2}{\rho}, \tag{4.277}$$

where v is the velocity of the particle.

We shall consider the constrained motion of a particle in the following cases:

Case I. *Motion along a smooth curve.*

Let a particle of mass m move along a smooth curve under the action of external forces having components X and Y along Cartesian coordinates. Equation of motion along the tangent (*Fig 4.21*) is

$$mv\frac{dv}{ds} = X\cos\phi + Y\sin\phi = X\frac{dx}{ds} + Y\frac{dy}{ds}. \tag{4.278}$$

Equation of motion along the normal is

$$m\frac{v^2}{\rho} = -X\sin\phi + Y\cos\phi + R = -X\frac{dy}{ds} + Y\frac{dx}{ds} + R, \tag{4.279}$$

where R is the reaction force along normal. Now, from equation (4.278), we get on integration

$$\tfrac{1}{2}mv^2 = \int (X\,dx + Y\,dy). \tag{4.280}$$

Let us now suppose that the external force is such that

$$X\,dx + Y\,dy$$

is the complete differential of a function $\psi(x,y)$, so that

$$X = \frac{\partial \psi}{\partial x} \quad \text{and} \quad Y = \frac{\partial \psi}{\partial y}. \tag{4.281}$$

Then, we get

$$\tfrac{1}{2}mv^2 = \int \left(\frac{\partial \psi}{\partial x}\,dx + \frac{\partial \psi}{\partial y}\,dy \right) = \psi(x,y) + C. \tag{4.282}$$

If the particle started initially with a velocity V from a point (x_0, y_0), then

$$\tfrac{1}{2}mV^2 = \psi(x_0, y_0) + C. \tag{4.283}$$

Hence, from equations (4.282) and (4.283)

$$\tfrac{1}{2}m(v^2 - V^2) = \psi(x,y) - \psi(x_0, y_0). \tag{4.284}$$

Thus, the change in kinetic energy of the particle is independent of the path followed between initial and final position. Therefore, the change in kinetic energy will be same for all types of restraining curves.

Again, from the definition of work, $X\,dx + Y\,dy$ represents the work done by forces X and Y during a small displacement ds along the curve. Hence, from equation (4.284), we conclude that, change in kinetic energy of the particle is equal to the work done by the external forces. The function $\psi(x,y)$ is known as the *work function* of the system of forces. The kind of forces considered are called, as we already know, conservative forces.

Potential energy of the particle of a position $P(x_1, y_1)$ will be equal to the work done by the forces in moving the particle to P. Thus, potential energy is

$$\int_{(x,y)}^{(x_1,y_1)} (X\,dx + Y\,dy) = \Big[\psi(x,y)\Big]_{(x,y)}^{(x_1,y_1)} = \psi(x_1, y_1) - \psi(x,y). \tag{4.285}$$

Hence, from equations (4.282) and (4.285), we find that the sum of kinetic energy and potential energy of the particle at P is

$$\psi(x,y) + C + \psi(x_1,y_1) - \psi(x,y) = \psi(x_1,y_1) + C = \text{a constant}. \tag{4.286}$$

Hence, when a particle moves under the action of a conservative system of forces, the sum of its kinetic energy and potential energy remains constant throughout its motion.

EXAMPLE 4.7. *A smooth parabolic tube is placed vertex downwards in a vertical plane. A particle slides down the tube from rest under gravity. Prove that, the reaction of the tube in any position is $2mg(h+a)/\rho$, where m = mass of the particle, h = initial vertical height of the particle from vertex, ρ = radius of curvature and $4a$ = latus rectum.*

Solution: Tangential equation of motion is (*Fig 4.22*)

$$mv\frac{dv}{ds} = -mg\sin\phi, \quad \text{i.e.,} \quad v\frac{dv}{ds} = -g\frac{dy}{ds}, \quad \text{or,} \quad v\,dv = -g\,dy.$$

Integrating

$$v^2 = -2gy + C.$$

Now, when $y = h$, $v = 0$; $\therefore\ C = 2gh$.

$$\therefore\quad v^2 = 2g(h-y). \qquad (1)$$

Equation of motion along the normal is

$$m\frac{v^2}{\rho} = R - mg\cos\phi.$$

$$\therefore\quad R = mg\cos\phi + \frac{m}{\rho}\cdot 2g(h-y) \quad \text{[by using (1)]}.$$

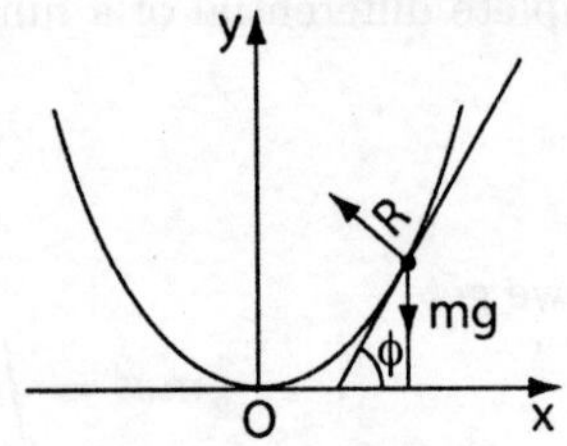

Fig 4.22

Now, equation of the parabola is $x^2 = 4ay$.

$$\therefore\quad \tan\phi = \frac{dy}{dx} = \frac{x}{2a} = \sqrt{\frac{y}{a}}.$$

$$\therefore\quad \cos\phi = \sqrt{\frac{a}{a+y}}. \qquad (2)$$

Again, the radius of curvature is

$$\rho = \frac{\left[1+(y')^2\right]^{3/2}}{y''} = 2a\left(\frac{a+y}{a}\right)^{3/2}. \qquad (3)$$

Using equations (2) and (3), we get

$$R = mg\sqrt{\frac{a}{a+y}} + \frac{m}{\rho}\cdot 2g(h-y) = \frac{mg}{\rho}\cdot 2a\left(\frac{y+a}{a}\right) + \frac{m}{\rho}\cdot 2g(h-y)$$

$$= \frac{mg}{\rho}[2(y+a)+2(h-y)] = \frac{2mg}{\rho}(h+a).$$

EXAMPLE 4.8. *From the lowest point of a smooth hollow elliptic cylinder of axes $2a, 2b$ with minor axis vertical, a particle is projected in a vertical plane perpendicular to the axis of the cylinder. Show that, it will leave the cylinder, if the velocity of projection V satisfies the relation*

$$\sqrt{2gb} < V < \left[g\frac{a^2+4b^2}{b}\right]^{1/2}.$$

Solution: Let V_1 and V_2 be the velocities of the particle at points A and B (*Fig 4.23*). Then, from the principle of work and energy, we have

$$\tfrac{1}{2}m(V_1^2 - V^2) = -mgb,$$

i.e., $\quad V_1^2 = V^2 - 2gb \qquad (1)$

and $\quad \tfrac{1}{2}m(V_2^2 - V^2) = -2mgb,$

i.e., $\quad V_2^2 = V^2 - 4gb. \qquad (2)$

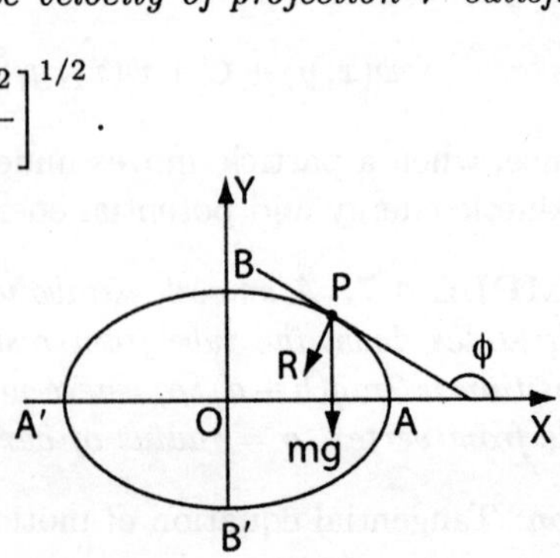

Fig 4.23

The particle may leave the surface at any point between A and B. Hence, in order that it crosses the point A, velocity V_1 at A should be positive. Thus, from (1), we get

$$V^2 - 2gb > 0, \quad \text{i.e.,} \quad V > \sqrt{2gb}. \qquad (3)$$

Now, equation of motion along normal at B is

$$\frac{mv^2}{\rho} = mg + R, \quad \text{i.e.,} \quad R = \frac{m}{\rho}(V^2 - 4gb) - mg, \tag{4}$$

where R is the normal reaction at B. Now, radius of curvature at $B = a^2/b$ (this is left as an exercise). Hence, from equation (4)

$$R = m\frac{b}{a^2}\,(V^2 - 4gb) - mg. \tag{5}$$

The particle will leave the surface if $R = 0$. That is

$$\frac{b}{a^2}(V^2 - 4gb) = g, \quad \text{or,} \quad V^2 = \frac{ga^2}{b} + 4gb = \frac{g}{b}(a^2 + 4b^2).$$

$$\therefore \quad V = \sqrt{\frac{g}{b}\,(a^2 + 4b^2)}. \tag{6}$$

Hence, the particle will leave the surface if

$$\sqrt{2gb} < V < \sqrt{\frac{g}{b}\,(a^2 + 4b^2)}.$$

Case II. *Motion along a rough curve.*

Let a particle of mass m be sliding down a rough curve under gravity (*Fig 4.24*). The equations of motion along tangent and normal to the curve are

$$mv\frac{dv}{ds} = mg\sin\phi - \mu R \tag{4.287}$$

$$\text{and} \quad m\frac{v^2}{\rho} = mg\cos\phi - R, \tag{4.288}$$

where $R =$ normal reaction and $\mu =$ coefficient of friction.

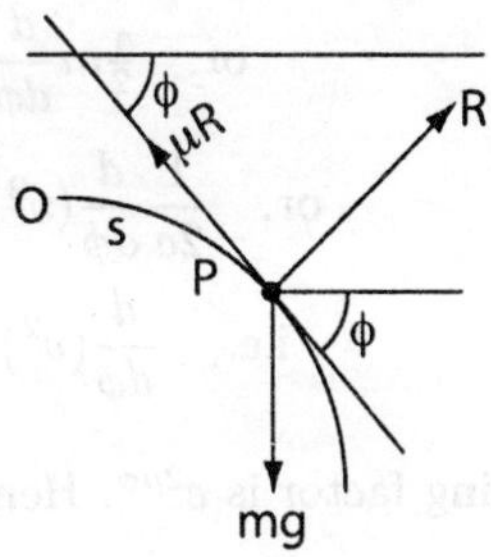

Fig 4.24

Eliminating R from these equations, we get

$$v\frac{dv}{ds} - \mu\frac{v^2}{\rho} = g\,(\sin\phi - \mu\cos\phi),$$

$$\text{or,} \quad \frac{\rho}{2}\frac{d}{ds}(v^2) - \mu v^2 = g\rho\,(\sin\phi - \mu\cos\phi),$$

$$\text{or,} \quad \frac{d}{ds}(v^2)\cdot\frac{ds}{d\phi} - 2\mu v^2 = 2g\rho\,(\sin\phi - \mu\cos\phi),$$

$$\text{or,} \quad \frac{d}{d\phi}(v^2) - 2\mu v^2 = 2g\rho\,(\sin\phi - \mu\cos\phi). \tag{4.289}$$

This is a linear first-order differential equation in v^2. The integrating factor is

$$e^{-\int 2\mu\, d\phi} = e^{-2\mu\phi}.$$

Hence

$$v^2 \cdot e^{-2\mu\phi} = 2g \int \rho e^{-2\mu\phi} (\sin\phi - \mu\cos\phi) + C. \tag{4.290}$$

Equation (4.290) determines v^2 when the curve is given, i.e., ρ is known. Then, from equation (4.288), the reaction force R can be found out.

EXAMPLE 4.9. *A particle is projected along the inner surface of a rough sphere and is acted on by no forces. Show that, it will return to the point of projection at the end of time*

$$\frac{a}{\mu V}\,(e^{2\mu\pi} - 1).$$

where a = radius of the sphere, V = velocity of projection and μ = coefficient of friction.

Solution: Equations of motion are (there is no force acting)

$$mv\frac{dv}{ds} = -\mu R \tag{1}$$

$$\text{and} \quad m\frac{v^2}{\rho} = R. \tag{2}$$

From equation (1), we have

$$\tfrac{1}{2}m\frac{d}{ds}(v^2) = -\mu R,$$

$$\text{or,} \quad \tfrac{1}{2}m\frac{d}{d\phi}(v^2)\frac{d\phi}{ds} = -\mu\frac{mv^2}{\rho} \quad \text{[by using equation (2)]},$$

$$\text{or,} \quad \frac{1}{2a}\frac{d}{d\phi}(v^2) + \frac{\mu v^2}{a} = 0, \quad (\because\ \rho = a),$$

$$\text{i.e.,} \quad \frac{d}{d\phi}(v^2) + 2\mu v^2 = 0. \tag{3}$$

Integrating factor is $e^{2\mu\phi}$. Hence, after integration, we get

$$v^2\, e^{2\mu\phi} = C. \tag{4}$$

Initially, $v = V$ when $\phi = 0$; $\therefore\ C = V^2$.
Thus, from equation (4), we have

$$v^2\, e^{2\mu\phi} = V^2, \quad \text{or,} \quad v\, e^{\mu\phi} = V. \tag{5}$$

Now

$$v = \frac{ds}{dt} = a\frac{d\phi}{dt}.$$

Hence, from equation (5)

$$a\, e^{\mu\phi}\, d\phi = V\, dt.$$

Integrating

$$\frac{a}{\mu}\, e^{\mu\phi} = Vt + A.$$

Initially, at $t = 0$, $\phi = 0$; $\therefore\ A = a/\mu$.

Hence

$$\frac{a}{\mu}\left(e^{\mu\phi}-1\right)=Vt. \tag{6}$$

If T is the time taken by the particle to return to the point of projection, then, at $t = T$, $\phi = 2\pi$. Hence, from equation (6)

$$VT=\frac{a}{\mu}\left(e^{2\mu\pi}-1\right).$$

$$\therefore \quad T=\frac{a}{\mu V}\left(e^{2\mu\pi}-1\right).$$

EXAMPLE 4.10. *A particle is projected horizontally with velocity V_0 along the inside of a rough vertical circle from the lowest point. Prove that, if it completes the circle, it will return to lowest point with a velocity V given by*

$$V^2=V_0^2\,e^{-4\pi\mu}+\frac{2ga}{1+4\mu^2}\left(1-2\mu^2\right)\left(1-e^{-4\pi\mu}\right).$$

Solution: Equations of motion are

$$mv\frac{dv}{ds}=-\mu R-mg\sin\phi \tag{1}$$

$$m\frac{v^2}{\rho}=R-mg\cos\phi. \tag{2}$$

Eliminating R from equations (1) and (2), we get

$$\tfrac{1}{2}m\frac{d}{ds}(v^2)=-\mu m\frac{v^2}{\rho}-mg\,(\sin\phi+\mu\cos\phi),$$

$$\text{or,}\quad \tfrac{1}{2}m\frac{d}{d\phi}(v^2)\frac{d\phi}{ds}=-\mu\frac{mv^2}{a}-mg\,(\sin\phi+\mu\cos\phi),$$

$$\text{or,}\quad \frac{d}{d\phi}(v^2)+2\mu v^2=-2ga\,(\sin\phi+\mu\cos\phi). \tag{3}$$

Integrating factor is

$$e^{\int 2\mu\,d\phi}=e^{2\mu\phi}. \tag{4}$$

Hence, the solution of differential equation (3) is

$$v^2\cdot e^{2\mu\phi}=-2ga\int\left[e^{2\mu\phi}\,(\sin\phi+\mu\cos\phi)\right]+C. \tag{5}$$

Now

$$I_1=\int e^{2\mu\phi}\sin\phi\,d\phi=\frac{e^{2\mu\phi}}{1+4\mu^2}\,(2\mu\sin\phi-\cos\phi)$$

$$\text{and}\quad I_2=\mu\int e^{2\mu\phi}\cos\phi\,d\phi=\frac{\mu e^{2\mu\phi}}{1+4\mu^2}\,(2\mu\cos\phi+\sin\phi).$$

Using these integrals, expression (5) becomes

$$v^2\cdot e^{2\mu\phi}=-2ga\frac{e^{2\mu\phi}}{1+4\mu^2}\left[3\mu\sin\phi-(1-2\mu^2)\cos\phi\right]+C. \tag{6}$$

Initially, $v = V_0$, when $\phi = 0$. This gives

$$C = V_0^2 - \frac{2ga}{1+4\mu^2}(1-2\mu^2). \tag{7}$$

From equations (6) and (7), we get

$$v^2 \cdot e^{2\mu\phi} = -2ga\frac{e^{2\mu\phi}}{1+4\mu^2}\left[3\mu\sin\phi - (1-2\mu^2)\cos\phi\right] + V_0^2 - \frac{2ga}{1+4\mu^2}(1-2\mu^2). \tag{8}$$

If the particle returns to the point of projection with velocity V, then $\phi = 2\pi$. Putting $\phi = 2\pi$ in equation (8) and dividing throughout by $e^{2\mu\phi}$, we get

$$V^2 = V_0^2\, e^{-4\pi\mu} + \frac{2ga}{1+4\mu^2}\left(1-2\mu^2\right)\left(1-e^{-4\pi\mu}\right).$$

4.19 Oscillatory Motion

Let us imagine a body of mass m attached to one end of a spring, the other end of which is fixed at roof-ceiling, so that the system hangs vertically. If now, the body is pushed a little downward and released immediately, the body will oscillate vertically. Such a system is called a *single-degree system.* In order to study the motion of such a system we idealize the body as a particle.

We shall consider oscillatory motion of a particle in the following cases:

Case I. *Free vibration (harmonic oscillator).*

Let the spring constant be k, i.e., the force acting on the particle will be k times the displacement and opposite to direction of motion.

Hence, the equation of motion is

$$m\frac{d^2x}{dt^2} = -kx, \quad \text{or,} \quad \frac{d^2x}{dt^2} + \xi^2 x = 0, \tag{4.291}$$

where

$$\xi^2 = k/m. \tag{4.292}$$

This equation is the standard form for simple harmonic motion and the system is called *simple harmonic oscillator.* The general solution of this equation is

$$x = A\sin(\xi t + \psi), \tag{4.293}$$

where A and ψ are integration constants, which are to be determined from initial conditions. If initially, at $t = 0$, $x = x_0$ and $\dot{x} = v = v_0$, then

$$\left.\begin{aligned} A &= \sqrt{x_0^2 + \frac{v_0^2}{\xi^2}} \\ \text{and}\quad \psi &= \tan^{-1}\left(\frac{\xi x_0}{v_0}\right) \end{aligned}\right\}. \tag{4.294}$$

If T be the period of motion, then

$$\xi t + \psi + 2\pi = \xi (t + T) + \psi, \quad \text{i.e.,} \quad T = \frac{2\pi}{\xi}. \tag{4.295}$$

Frequency of vibration is

$$n = \frac{1}{T} = \frac{\xi}{2\pi}, \tag{4.296}$$

so that

$$\xi = 2n\pi. \tag{4.297}$$

From equation (4.297), we get the physical significance of the constant ξ, which is actually number of oscillations in 2π seconds.

The velocity of the particle at any instant of time t is

$$v = \frac{dx}{dt} = A\xi \cos(\xi t + \psi), \quad \text{i.e.,} \quad v = \xi \sqrt{A^2 - x^2}. \tag{4.298}$$

Hence, the kinetic energy of the particle at time t is

$$E_k = \tfrac{1}{2}mv^2 = \tfrac{1}{2}m\xi^2 (A^2 - x^2). \tag{4.299}$$

The potential energy of the particle at any instant is

$$E_p = \int_0^x m\xi^2 \, x \, dx = \tfrac{1}{2}m\xi^2 x^2. \tag{4.300}$$

From equations (4.299) and (4.300), we find that the total energy of the particle at any instant is

$$E = E_k + E_p = \tfrac{1}{2}m\xi^2 A^2 = \text{constant}. \tag{4.301}$$

Case II. *Free vibration under non-linear force function.*

In the previous case, we assumed that the restoring force of the spring was proportional to the displacement. But, in reality, this type of linear force function may not occur. We shall therefore assume that the restoring force function is non-linear and discuss a general method for the solution of the equation of motion. If $F(x)$ is the given restoring force function, then the equation of motion is written in the form

$$m\frac{d^2x}{dt^2} + F(x) = 0, \quad \text{or,} \quad \frac{d^2x}{dt^2} + f(x) = 0, \tag{4.302}$$

where $f(x) = \dfrac{F(x)}{m}$.

In such cases, we are generally interested to know the time period of vibration.

Now

$$\frac{d^2x}{dt^2} = \frac{d}{dt}\left(\frac{dx}{dt}\right) = \frac{dv}{dt} = \frac{dv}{dx} \cdot \frac{dx}{dt} = v\frac{dv}{dx}.$$

Hence, equation (4.302) can be written as

$$v\frac{dv}{dx} = -f(x), \quad \text{i.e.,} \quad v\,dv = -f(x)\,dx. \tag{4.303}$$

Let the particle during its oscillatory motion attains maximum amplitude $x = \alpha$ when the velocity momentarily becomes zero.

Then, integrating (2), we get

$$\int_0^v v\,dv = -\int_\alpha^x f(x)\,dx = \int_x^\alpha f(x)\,dx. \qquad \therefore \quad v^2 = 2P(x), \tag{4.304}$$

where

$$P(x) = \int_x^\alpha f(x)\,dx.$$

Thus, from equation (4.304), we get

$$v = -\sqrt{2P(x)} \tag{4.305}$$

negative sign being taken, since the direction of motion in the interval (x, α) is in opposite direction.

Hence, we get

$$v = \frac{dx}{dt} = -\sqrt{2P(x)}.$$

Integrating

$$t = -\int_\alpha^x \frac{dx}{\sqrt{2P(x)}} = \int_x^\alpha \frac{dx}{\sqrt{2P(x)}}. \tag{4.306}$$

If we now assume that the restoring force function is symmetric, so that time for its motion from $x = 0$ to $x = \alpha$ denotes one-forth of the time period, then

$$\frac{T}{4} = \int_0^\alpha \frac{dx}{\sqrt{2P(x)}}. \tag{4.307}$$

This equation gives the relation between the period of vibration with its amplitude. The exactness of the value will depend on the integration results. It may happen that the integral is fairly complicated when we are to resort to approximate method. We shall, as an example, discuss the motion of a simple pendulum.

Simple Pendulum

A simple pendulum consists of a heavy particle attached at the end of a light inextensible rod which is suspended from a rigid support at the other end so that the system is free to oscillate under gravity. When the displacement of the particle from its equilibrium position is small, it oscillates as a simple harmonic oscillator (*Fig 4.25*).

At any instant of time, if θ be the angle made by the pendulum with the vertical, its equation of motion is $m\ddot{x} = -mg\sin\theta$, i.e., $m\ddot{x} = -mg\,\theta$, if θ is small

$$\text{i.e.,} \quad \ddot{x} + \frac{g}{l}x = 0. \tag{4.308}$$

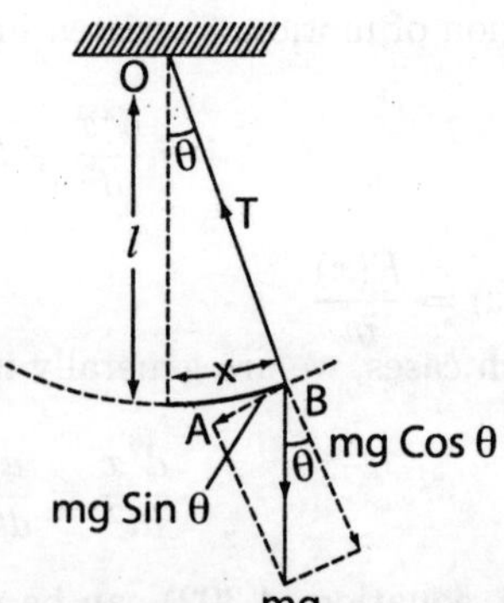

Fig 4.25

This equation is similar to (4.302). Hence, the time period is

$$T = 2\pi\sqrt{l/g}. \tag{4.309}$$

If, however, θ is not assumed to be small, we cannot write $\sin\theta = \theta$. In such a case, the equation becomes

$$\ddot{\theta} + \frac{g}{l}\sin\theta = 0 \quad [\because\ x = l\theta] \tag{4.310}$$

Multiplying both sides of this equation by

$$\frac{d\theta}{dt}\cdot dt, \quad \text{i.e.,} \quad d\theta$$

we get

$$\ddot{\theta}\cdot\dot{\theta}\,dt = -\frac{g}{l}\sin\theta\,d\theta, \quad \text{i.e.,} \quad \tfrac{1}{2}d\left(\dot{\theta}^2\right) = \frac{g}{l}d(\cos\theta).$$

Integrating

$$\left(\dot{\theta}\right)^2 = \frac{2g}{l}\cos\theta + C.$$

If we assume that the maximum angular displacement is α, then at $\theta = \alpha$, $\dot{\theta} = 0$. Hence

$$C = -\frac{2g}{l}\cos\alpha.$$

Therefore

$$\left(\dot{\theta}\right)^2 = \frac{2g}{l}(\cos\theta - \cos\alpha), \quad \text{or,} \quad \left(\dot{\theta}\right)^2 = \frac{4g}{l}\left(\sin^2\frac{\alpha}{2} - \sin^2\frac{\theta}{2}\right).$$

$$\therefore \quad \frac{d\theta}{\sqrt{\sin^2\frac{\alpha}{2} - \sin^2\frac{\theta}{2}}} = 2\sqrt{\frac{g}{l}}\,dt.$$

Integrating between the limits $\theta = \pm\alpha$,

$$\int_{-\alpha}^{\alpha}\frac{d\theta}{\sqrt{\sin^2\frac{\alpha}{2} - \sin^2\frac{\theta}{2}}} = \int_0^{T_\alpha/2} 2\sqrt{\frac{g}{l}}\,dt = \sqrt{\frac{g}{l}}\cdot T_\alpha, \tag{4.311}$$

where T_α = the period of oscillation of the pendulum. The integral in (4.311) is an elliptic integral and cannot be integrated by usual methods. We make the substitution

$$\sin\frac{\theta}{2} = \sin\frac{\alpha}{2}\sin\phi$$

so that

$$\frac{1}{2}\cos\frac{\theta}{2}d\theta = \sin\frac{\alpha}{2}\cos\phi\,d\phi, \quad \text{i.e.,} \quad d\theta = \frac{2\sin\frac{\alpha}{2}\cos\phi\,d\phi}{\sqrt{1 - \sin^2\frac{\alpha}{2}\sin^2\phi}}.$$

The limits of integration become

$$\text{at } \theta = -\alpha, \quad \phi = -\frac{\pi}{2}; \quad \text{and at } \theta = \alpha, \quad \phi = \frac{\pi}{2}.$$

Hence, from equation (4.311), we get

$$\sqrt{\frac{g}{l}} \cdot T_\alpha = \int_{-\pi/2}^{\pi/2} \frac{2d\phi}{\sqrt{1-\sin^2\frac{\alpha}{2}\sin^2\phi}} = \int_{-\pi/2}^{\pi/2} 2\left[1-\sin^2\frac{\alpha}{2}\sin^2\phi\right]^{-1/2} d\phi$$

$$= \int_{-\pi/2}^{\pi/2} 2\left[1+\frac{1}{2}\sin^2\frac{\alpha}{2}\sin^2\phi+\frac{3}{8}\sin^4\frac{\alpha}{2}\sin^4\phi+\cdots\right] d\phi,$$

i.e., $$\sqrt{\frac{g}{l}} \cdot T_\alpha = 2\left[\pi+\frac{1}{2}\left(\frac{\pi}{2}\right)\sin^2\frac{\alpha}{2}+\frac{3}{8}\left(\frac{3\pi}{8}\right)\sin^4\frac{\alpha}{2}+\cdots\right].$$

Hence

$$T_\alpha = T\left[1+\frac{1}{4}\sin^2\frac{\alpha}{2}+\frac{9}{64}\sin^4\frac{\alpha}{2}+\cdots\right], \tag{4.312}$$

where, as we know, $T = 2\pi\sqrt{l/g}$, which is the period of the pendulum for small oscillations. Equation (4.312) gives the time period in terms of maximum angular amplitude α.

Thus

$$\text{for } \alpha = \pi/2, \quad T_\alpha = T\left(1+\frac{1}{8}+\frac{9}{256}+\cdots\right) = 1.16T$$

$$\text{for } \alpha = \pi/3, \quad T_\alpha = T\left(1+\frac{1}{16}+\frac{9}{1024}+\cdots\right) = 1.063T.$$

With decrease of α, T_α will approach T.

Case III. *Free vibration with viscous damping.*

Let there be a viscous damping c with units kg/sec, which is proportional to the velocity. Then the equation of motion of the particle considered in Case I becomes

$$m\frac{d^2x}{dt^2} = -c\frac{dx}{dt} - kx, \quad \text{i.e.,} \quad \frac{d^2x}{dt^2}+2\alpha\frac{dx}{dt}+\xi^2 x = 0, \tag{4.313}$$

where $\alpha = \dfrac{c}{2m}$ and $\xi^2 = \dfrac{k}{m}$.

Considering a solution in the form $x = e^{\lambda t}$ we get the characteristic equation as

$$\lambda^2 + 2\alpha\lambda + \xi^2 = 0.$$

Roots of this equation are

$$\left.\begin{aligned} \lambda_1 &= -\alpha + \sqrt{\alpha^2-\xi^2} \\ \lambda_2 &= -\alpha - \sqrt{\alpha^2-\xi^2} \end{aligned}\right\}.$$

Thus, the solution of the equation can be written as

$$x(t) = e^{-\alpha t}\left[Ae^{\psi t}+Be^{-\psi t}\right], \tag{4.314}$$

where $\psi = \sqrt{\alpha^2-\xi^2}$.

It is obvious that the nature of the solution depends on the value of ψ.

If $\psi > 0$, i.e., $\alpha^2 - \xi^2 > 0$, the damping is very large. The roots are real and the oscillation is performed according to equation (4.314).

If $\psi = 0$, i.e., $\alpha^2 - \xi^2 = 0$, the damping is said to be *critical.* In this case, the roots are real and equal, and the solution of equation (4.313) is given by

$$x(t) = e^{-\alpha t}(A + Bt). \tag{4.315}$$

If $\psi < 0$, i.e., $\alpha^2 - \xi^2 < 0$, the damping is very small and the motion is said to be *underdamped.* This is the most common and interesting case. In this case, the solution of the differential equation is written as

$$x(t) = Ae^{-\alpha t} \sin\left(\sqrt{\xi^2 - \alpha^2}\, t + \phi\right). \tag{4.316}$$

The constants of integration A and ϕ are to be determined from initial conditions. If initially at $t = 0$, $x = x_0$ and $\dot{x} = v_0$, then we get

$$A = \sqrt{x_0^2 + \frac{(v_0 + \alpha x_0)^2}{\xi^2 - \alpha^2}} \tag{4.317}$$

$$\text{and} \quad \phi = \tan^{-1}\left(\frac{x_0\sqrt{\xi^2 - \alpha^2}}{v_0 + \alpha x_0}\right). \tag{4.318}$$

The motion represented by equation (4.316) will be oscillating in nature with a gradual decrease of the amplitude (*Fig 4.26*). The maximum amplitude will occur at the points where

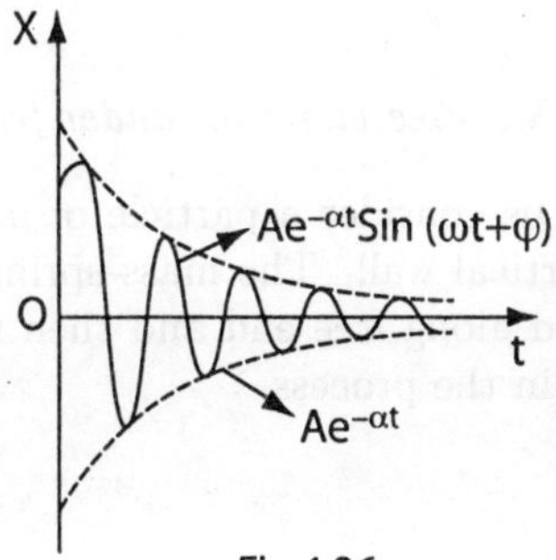

Fig 4.26

$$\frac{dx}{dt} = e^{-\alpha t}\,\omega \cos(\omega t + \phi) - \alpha e^{-\alpha t}\sin(\omega t + \phi) = 0,$$

i.e., $\tan(\omega t + \phi) = \omega/\alpha, \quad \omega = \sqrt{\xi^2 - \alpha^2}$.

Hence, in general

$$\omega t + \phi = \tan^{-1}\left(\frac{\omega}{\alpha}\right) + n\pi. \tag{4.319}$$

Hence we get the corresponding times for maximum amplitudes as

$$t = \frac{1}{\omega}\left[\tan^{-1}\left(\frac{\omega}{\alpha}\right) - \phi + n\pi\right]. \tag{4.320}$$

If t_1 and t_2 be the time for two successive maximum amplitudes, then

$$\frac{x_n}{x_{n+2}} = e^{-\alpha(t_1 - t_2)} \cdot \frac{\sin(\omega t_1 + \phi)}{\sin(\omega t_2 + \phi)},$$

$$\text{or,} \quad \frac{x_n}{x_{n+2}} = e^{2\pi\alpha/\omega} \cdot \frac{\sin\left[\tan^{-1}\left(\frac{\omega}{\alpha}\right) + n\pi\right]}{\sin\left[\tan^{-1}\left(\frac{\omega}{\alpha}\right) + (n+2)\pi\right]} \quad \text{[by using (4.320)}$$

$$= e^{2\pi\alpha/\omega} \cdot \frac{\sin\left\{\tan^{-1}\left(\frac{\omega}{\alpha}\right)\right\}\cos n\pi}{\sin\left\{\tan^{-1}\left(\frac{\omega}{\alpha}\right)\right\}\cos(n+2)\pi},$$

$$\text{i.e.,} \quad \frac{x_n}{x_{n+2}} = e^{2\pi\alpha/\omega}, \tag{4.321}$$

which is independent of time.

The logarithm of the ratio (4.321) is called the *logarithmic decrement*, δ. Thus

$$\delta = \ln \frac{x_n}{x_{n+2}} = 2\pi \frac{\alpha}{\omega},$$

$$\text{i.e.,} \quad \delta = 2\pi \left(\frac{c}{2m}\right) \cdot \frac{1}{\sqrt{\left(\frac{k}{m}\right) - \left(\frac{c}{2m}\right)^2}},$$

$$\text{i.e.,} \quad \delta = \frac{2\pi c}{\sqrt{4km - c^2}}. \tag{4.322}$$

Now, for critical damping, we have

$$\alpha^2 - \xi^2 = 0, \quad \text{i.e.,} \quad \frac{c^2}{4m^2} = \frac{k}{m}. \quad \therefore \quad c_{\text{cr}}^2 = 4mk.$$

Equation (4.322) can thus be written as

$$\delta = \frac{2\pi c}{\sqrt{c_{\text{cr}}^2 - c^2}}. \tag{4.323}$$

Case IV. *Free vibration under friction force.*

Let us consider a particle of mass m attached to a spring the other end of which is fixed to a vertical wall. The mass-spring system is lying on a rough horizontal plane. The particle is pulled along free end and then released immediately. We shall study the vibratory motion caused in the process.

Let

$$m = \text{mass of the particle,}$$
$$k = \text{spring constant}$$
$$\text{and} \quad \mu = \text{coefficient of friction.}$$

Then, the equation of motion is

$$m\frac{d^2x}{dt^2} + kx \pm \mu mg = 0 \quad \text{or,} \quad \frac{d^2x}{dt^2} + \xi^2 x \pm \xi^2 \alpha = 0, \tag{4.324}$$

where

$$\xi^2 = \frac{k}{m}, \quad \alpha = \frac{\mu mg}{k}. \tag{4.325}$$

The sign of the last term of the equation (4.324) will depend on the direction of motion.

Let, at the instant of time $t = 0$, the particle begins its motion when it is at $x = x_0$. Then, the frictional force will be negative, so that equation of motion becomes

$$\frac{d^2x}{dt^2} + \xi^2 (x - \alpha) = 0. \tag{4.326}$$

The solution of this equation with the initial conditions as mentioned is

$$x = (x_0 - \alpha) \cos \xi t + \alpha. \tag{4.327}$$

Hence, the velocity at any instant is

$$\frac{dx}{dt} = -\xi\,(x_0 - \alpha)\,\sin\xi t. \tag{4.328}$$

This shows that, the velocity of the particle becomes zero instantaneously when $\xi t = \pi$. For this value of ξt, the displacement of the particle will be on the other side of the origin whose magnitude is

$$x_1 = (x_0 - \alpha)\,\cos\pi + \alpha = -x_0 + 2\alpha. \tag{4.329}$$

At this moment, if the restoring force of the spring is stronger than the friction force, i.e., $|\,x\,| > \alpha$, then the particle will move in the positive direction towards the origin. In that case, the equation of motion becomes

$$\frac{d^2x}{dt^2} + \xi^2 x + \xi^2\alpha = 0. \tag{4.330}$$

If we count the time from the moment of beginning this motion, we have the initial conditions as

$$\text{at} \quad t = 0, x = x_1 \quad \text{and} \quad v = 0.$$

Thus, the solution of equation (4.330) is

$$x = (x_1 + \alpha)\,\cos\xi t - \alpha.$$

$$\therefore \quad \frac{dx}{dt} = -\xi\,(x_1 + \alpha)\,\sin\xi t.$$

Hence, the velocity as before becomes instantaneously zero for $\xi t = \pi$. At this moment, the displacement from the origin is

$$x_2 = -(x_1 + \alpha) - \alpha = -x_1 - 2\alpha,$$

$$\text{i.e.,} \quad x_2 = x_0 - 4\alpha. \tag{4.331}$$

If we now consider the motion from its initial position at $x = x_2$, we will find that during every period the amplitude is getting diminished by an amount equal to

$$4\alpha = \frac{4\mu mg}{k}. \tag{4.332}$$

The motion will continue as long as the amplitude $|\,x\,| > \alpha$. As soon as the amplitude becomes so small that the restoring force of the spring is less than the friction force, the motion will stop. If this happens after n number of periods, when amplitude of the particle is x_{2n}, then

$$|\,x_{2n}\,| = x_0 - 4n\alpha, \text{ i.e., } n = \frac{1}{4\alpha}\,\{x_0 - |\,x_{2n}\,|\}\,.$$

This will determine the number of cycles of the motion performed by the particle. A graphical presentation of the motion is given in the *Fig 4.27*.

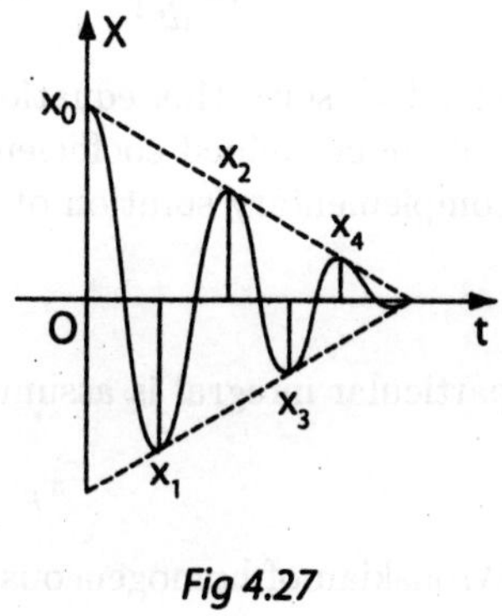

Fig 4.27

4.20 Energy Dissipation in Damped Oscillations

In the damped oscillation there will be a gradual loss of energy. This loss of energy at any instant is numerically equal to

$$E_l = \text{damping force} \times \text{velocity} = (c\dot{x}) \times \dot{x} = c(\dot{x})^2.$$

Now, from the equation of motion (4.313), we have

$$c\left(\frac{dx}{dt}\right)^2 = -\left[m\frac{dx}{dt}\cdot\frac{d^2x}{dt^2} + kx\frac{dx}{dt}\right] = -\left[\frac{1}{2}m\frac{d}{dt}(\dot{x})^2 + \frac{k}{2}\frac{d}{dt}(x^2)\right]$$
$$= -\frac{d}{dt}(T+V) = -\frac{dE}{dt},$$

$$\text{or,} \quad \frac{dE}{dt} = -c\left(\frac{dx}{dt}\right)^2, \tag{4.333}$$

where E is the total energy of motion.

Thus, the total energy of the particle goes on decreasing due to damping. When the damping force is proportional to velocity, the rate of decrease in energy is proportional to square of the velocity. In particular case, if $c = 0$, i.e., there is no damping, from equation (4.333), we have

$$\frac{dE}{dt} = 0, \quad \text{i.e.,} \quad E = T + V = \text{constant}.$$

4.21 Forced Oscillations

We have seen that in a damped system, the amplitude of oscillation goes on decreasing due to energy dissipation which causes the oscillations discontinued. Hence, in order to maintain the oscillations, an external energy has to be supplied to the system to compensate the energy loss. This can be done by providing an external force, which depends on time. The oscillations are then called *forced oscillations*. Forced oscillations can also occur when there is no damping.

Let us consider a simple system at any instant of time, there are two forces working on it—(i) elastic force of the spring which is proportional to displacement of the particle, and (ii) the external time-dependent force $P(t)$. Hence, the equation of motion is

$$m\frac{d^2x}{dt^2} = -kx + P(t), \quad \text{i.e.,} \quad \frac{d^2x}{dt^2} + \xi^2 x = \frac{P(t)}{m}. \tag{4.334}$$

We intend to solve this equation in general case of external force $P(t)$ by the method of variation of undetermined coefficient (*see* Section 3.23).

The complementary solution of the equation is

$$x = A\sin\xi t + B\cos\xi t. \tag{4.335}$$

The particular integral is assumed as

$$x_p = z_1(t)\,\sin\xi t + z_2(t)\,\cos\xi t. \tag{4.336}$$

The Wronskian of homogeneous independent solutions is

$$W = \sin\xi t\,(-\xi\sin\xi t) - \cos\xi t\,(\xi\cos\xi t) = -\xi. \tag{4.337}$$

Then

$$z_1(t) = \frac{1}{m\xi}\int_0^t P(\tau)\cos\xi\tau\,d\tau \tag{4.338}$$

and $$z_2(t) = -\frac{1}{m\xi}\int_0^t P(\tau)\sin\xi\tau\,d\tau. \tag{4.339}$$

Hence, the particular integral can be written as

$$x_p = \frac{1}{m\xi}\left[\sin\xi t\int_0^t P(\tau)\cos\xi\tau\,d\tau - \cos\xi t\int_0^t P(\tau)\sin\xi\tau\,d\tau\right]$$

$$= \frac{1}{m\xi}\int_0^t P(\tau)\sin\xi(t-\tau)\,d\tau. \tag{4.340}$$

Hence, the general solution of the equation (4.334) is

$$x = A\sin\xi t + B\cos\xi t + \frac{1}{m\xi}\int_0^t P(\tau)\sin\xi(t-\tau)\,d\tau. \tag{4.341}$$

The constants of integration A and B are to be determined from initial conditions. We assume that, initially at $t = 0$, $x = x_0$ and velocity $\dot{x} = v_0$. Now, expression for velocity is obtained from equation (4.341) as

$$v = \frac{dx}{dt} = A\xi\cos\xi t - B\xi\sin\xi t + \frac{1}{m}\int_0^t P(\tau)\cos\xi(t-\tau)\,d\tau. \tag{4.342}$$

Now, using the initial conditions, we find

$$A = \frac{v_0}{\xi} \quad\text{and}\quad B = x_0. \tag{4.343}$$

Hence, the final solution is

$$x(t) = x_0\cos\xi t + \frac{v_0}{\xi}\sin\xi t + \frac{1}{m\xi}\int_0^t P(\tau)\sin\xi(t-\tau)\,d\tau. \tag{4.344}$$

If the initial conditions are such that at $t = 0$, $x_0 = 0$ and $v_0 = 0$, then the solution (4.344) becomes

$$x(t) = \frac{1}{m\xi}\int_0^t P(\tau)\sin\xi(t-\tau)\,d\tau. \tag{4.345}$$

Integrating by parts and assuming that at $t = 0$, $P(t) = 0$, the expression (4.345) takes the form

$$x(t) = \frac{P(t)}{m\xi^2} - \frac{1}{m\xi^2}\int_0^t \frac{dP(\tau)}{d\tau}\cos\xi(t-\tau)\,d\tau. \tag{4.346}$$

We now consider the following cases:

Case I. *Force $P(t)$ is a linear function.*

Let, $P(t) = at$, then $\dfrac{dP}{dt} = a$.

From equation (4.346), we get

$$x(t) = \frac{at}{m\xi^2} - \frac{a}{m\xi^2}\sin\xi t = \frac{a}{m\xi^2}(t - \sin\xi t). \tag{4.347}$$

Case II. *Force $P(t)$ is a harmonic function.*

Let, $P(t) = P_0 \sin \omega t$.

This is a harmonic function with P_0 as amplitude and ω as frequency.

Then, using equation (4.345), we get, for $\xi \neq \omega$

$$x(t) = \frac{P_0}{m\xi}\int_0^t \sin \omega t \,\sin \xi\,(t-\tau)\,d\tau = \frac{P_0}{m(\xi^2-\omega^2)}\left\{\sin \omega t - \frac{\omega}{\xi}\sin \xi t\right\}. \tag{4.348}$$

It is obvious from equation (4.348) that, when initial conditions are such that $x_0 = v_0 = 0$ at $t = 0$, then the oscillation consists of two parts—the first part describes oscillation of frequency ω, and the second part describes oscillation with natural frequency ξ. The first oscillation is known as *forced*, and the second as *free*. The combination of these two oscillations with different frequencies ω and ξ shows that the effect of free oscillation becomes negligible after some time so that the oscillation becomes stationary. Then, it is sufficient to consider only the forced oscillation part, so that equation (4.348) is taken as

$$x(t) = \frac{P_0}{m(\xi^2-\omega^2)}\sin \omega t. \tag{4.349}$$

Hence, amplitude of the forced oscillation is

$$A = \frac{P_0}{m\,(\xi^2-\omega^2)}, \quad \text{or,} \quad A = \frac{P_0/k}{1-\dfrac{\omega^2}{\xi^2}} = \frac{x_s}{1-\dfrac{\omega^2}{\xi^2}}, \tag{4.350}$$

where $x_s = P_0/k$ is the deflection under static load.

Hence, the factor

$$\frac{A}{x_s} = \frac{1}{1-\dfrac{\omega^2}{\xi^2}} \tag{4.351}$$

may be called *dynamic coefficient.*

4.22 Forced Oscillation with Damping

Let us now consider the presence of damping in the forced oscillation under a sinusoidal force function. The equation of motion in this case is

$$m\frac{d^2x}{dt^2} + c\frac{dx}{dt} + kx = P_0 \sin \omega t,$$

$$\left.\begin{aligned} \text{or,}\quad & \frac{d^2x}{dt^2} + 2\alpha\frac{dx}{dt} + \xi^2 x = \frac{P_0}{m}\sin \omega t, \\ \text{where, as before,}\quad & \alpha = \frac{c}{2m}, \quad \xi^2 = \frac{k}{m}. \end{aligned}\right\} \tag{4.352}$$

This is a linear non-homogeneous differential equation having complementary solution and the particular integral. The complementary solution, as we have seen, describes the damped oscillation. We are interested in a steady-state oscillation given by the particular integral of equation (4.352) which does not die out due to external driving force. Thus, for the study for forced vibration, we ignore the complementary solution and seek the particular integral in the form

$$x = A\sin(\omega t - \phi). \tag{4.353}$$

Substituting this into equation (4.352), writing

$$\frac{P_0}{m}\sin\omega t = \frac{P_0}{m}\sin(\omega t - \phi + \phi)$$

and comparing coefficients of sine and cosine terms, we find

$$\left.\begin{aligned} -A\omega^2 + \xi^2 A &= \frac{P_0}{m}\cos\phi \\ \text{and} \quad 2\alpha A\omega &= \frac{P_0}{m}\sin\phi \end{aligned}\right\}. \tag{4.354}$$

Solving these equations, we get

$$A = \frac{P_0/m}{\sqrt{(\xi^2-\omega^2)^2 + 4\alpha^2\omega^2}} \tag{4.355}$$

$$\text{and} \quad \tan\phi = \frac{2\alpha\omega}{\xi^2 - \omega^2}. \tag{4.356}$$

Thus, the solution (4.353) is written as

$$x = \frac{P_0/m}{\sqrt{(\xi^2-\omega^2)^2 + 4\alpha^2\omega^2}}\sin(\omega t - \phi). \tag{4.357}$$

From this equation (4.355), we find that the amplitude A of the oscillation depends on the frequency of the applied load. Hence, for some particular value of this frequency, the amplitude will be maximum. This is called *amplitude resonance*. Let us now determine this maximum value for which

$$\frac{dA}{d\omega} = 0 \quad \text{and} \quad \frac{d^2A}{d\omega^2} < 0.$$

This gives

$$-2(\xi^2 - \omega^2)2\omega + 8\alpha^2\omega = 0,$$

$$\text{i.e.,} \quad \omega^2 = \xi^2 - 2\alpha^2.$$

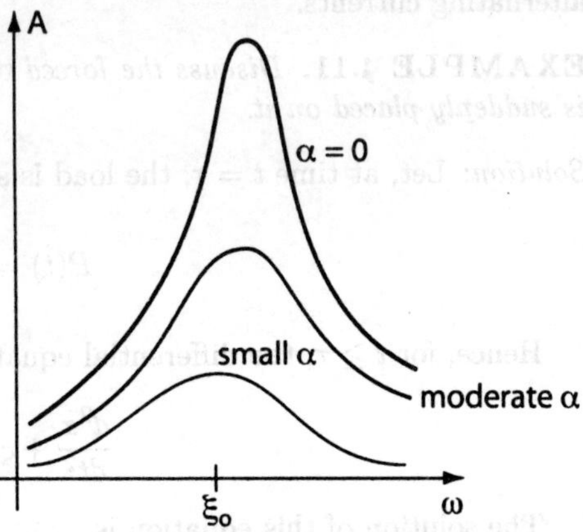

Fig 4.28

Hence, the resonant frequency is

$$\omega' = \sqrt{\xi^2 - 2\alpha^2}, \quad \xi^2 > 2\alpha^2. \tag{4.358}$$

With this frequency, the maximum amplitude is

$$A_{\max} = \frac{P_0/m}{\sqrt{4\alpha^4 + 4\alpha^2(\xi^2 - 2\alpha^2)}} = \frac{P_0/m}{2\alpha\sqrt{\xi^2 - \alpha^2}}. \tag{4.359}$$

Thus, A_{max} depends upon damping factor α^2. This is shown in the *Fig 4.28.*

Again, the velocity at any instant of the particle is given by

$$\frac{dx}{dt} = A\omega \cos(\omega t - \phi)$$

so that the velocity amplitude is

$$A\omega = \frac{P_0/m}{\sqrt{f(\omega)}}, \tag{4.360}$$

where

$$f(\omega) = \left(\frac{\xi^2 - \omega^2}{\omega}\right)^2 + 4\alpha^2. \tag{4.361}$$

Thus, the velocity amplitude will be maximum when $f(\omega)$ is minimum. Now, $f(\omega)$ will be minimum when

$$\left(\frac{\xi^2 - \omega^2}{\omega}\right)^2 = 0, \quad \text{i.e.,} \quad \omega = \xi. \tag{4.362}$$

Hence, the velocity amplitude is maximum when frequency of the driving force is equal to the natural frequency of the oscillator. At this moment resonance occurs and we call it *velocity resonance.* Now, the maximum value of velocity amplitude is obtained from (4.360) and (4.362). This is equal to

$$(A\omega)_{max} = \frac{P_0}{2m\alpha} = \frac{P_0}{c}, \tag{4.363}$$

which shows that it also depends on the damping coefficient. In the absence of damping, i.e., when $c = 0$, the maximum value of velocity amplitude becomes infinite.

Electric current being analogous to velocity, velocity resonance has got much importance in alternating currents.

EXAMPLE 4.11. *Discuss the forced oscillation of the spring-mass system when a unit load is suddenly placed on it.*

Solution: Let, at time $t = \tau$, the load is suddenly placed, so that

$$P(t) = \begin{cases} 0, & \text{for } t < \tau; \\ 1, & \text{for } t \geq \tau. \end{cases}$$

Hence, for $t \geq \tau$, the differential equation of motion is

$$\frac{d^2x}{dt^2} + \xi^2 x = \frac{1}{m}, \quad \xi^2 = \frac{k}{m}.$$

The solution of this equation is

$$x = A \sin \xi t + B \cos \xi t + \frac{1}{m\xi^2}.$$

Initial conditions are

$$\text{at} \quad t = \tau, \quad x = 0 \quad \text{and} \quad \frac{dx}{dt} = 0.$$

Hence, we get

$$A = -\frac{\sin \xi\tau}{k}, \quad B = -\frac{\cos \xi\tau}{k}.$$

Thus, the solution is

$$x = \tfrac{1}{k}\,[1 - \cos\xi(t-\tau)].$$

Maximum amplitude occurs at $\xi\,(t-\tau) = \pi$.

Its value is $x_{\max} = 2/k$ which is double the static deflection for placing unit load on the mass.

EXAMPLE 4.12. *In a spring-mass system, $m = 5$ kg and $k = 125$ N/m. It is found that the logarithmic decrement is 0.2. Find the damping coefficient. If now, a forcing function $P = 8\sin 3t$ is imposed on the system, find the maximum amplitude of steady-state motion.*

Solution: From equation (4.323)

$$\delta = \frac{2\pi c}{\sqrt{4km - c^2}}.$$

From here, we find

$$c = 2\sqrt{km}\,\frac{\delta}{\sqrt{\delta^2 + 4\pi^2}}.$$

Hence, in this case

$$c = 2\sqrt{125\times 5}\,\frac{0.2}{\sqrt{(0.2)^2 + 4\pi^2}} \simeq 1.60 \text{ kg/s}.$$

Maximum amplitude is given by equation (4.359), when

$$\xi^2 = \frac{k}{m} = \frac{125}{5} = 25, \quad P_0 = 8 \text{ kg}, \quad \alpha = \frac{c}{2m} = \frac{1.60}{2\times 5} = 0.16.$$

$$\therefore \quad A_{\max} = \frac{P_0}{c\sqrt{\xi^2 - \alpha^2}} = \frac{8}{1.6\times\sqrt{25 - (0.16)^2}} \simeq 1.0 \text{ m}.$$

EXERCISE

1. Two condensers of capacities C_1 and C_2 each has one plate earthed. Initially, the first condenser is charged with q_0 and the second one is uncharged. Two insulated plates are connected at time $t = 0$ by a wire of resistance R and inductance L. Show that the current i flowing in the wire satisfies the equation

$$L\frac{d^2 i}{dt^2} + R\frac{di}{dt} + \left(\frac{1}{C_1} + \frac{1}{C_2}\right) i = 0.$$

2. The insulated plate of a condenser C is given a charge q_0. This plate is then connected through a resistance R to a battery of e.m.f. V. Show that, after a time t, the charge on the condenser is given by

$$Q = VC + (q_0 - VC)\,e^{-t/CR}.$$

3. Determine the maximum deflection of a cantilever beam of flexural rigidity EI under a uniformly distributed load.

4. A rod of flexural rigidity EI is resting on elastic foundation with a point load P at the centre. Determine the maximum bending moment.
5. A strut with two ends hinged has two equal portions, the rigidity of one portion being four times the other. Determine its critical load.
6. A particle is projected upwards with a velocity u in a medium whose resistance is kv^2 per unit mass. Show that at any instant, its distance from the highest point is

$$\frac{1}{k} \log \sec\left(t\sqrt{gk}\right),$$

where t is the time to reach the highest point.
7. In the motion of a projectile, resistance is constant and equal to kg. Show that the velocity at any point is given by

$$v\,(1-\sin\phi)^k = u_0\,(\cos\phi)^{k-1}$$

ϕ being the slope and u_0 the velocity at the highest point.
8. From the lowest point of a smooth hollow cylinder whose cross-section is one-half of the lemniscate $r^2 = a^2\cos 2\theta$ with axis vertical and node downwards, a particle is projected with velocity V along the inner surface. Show that, it will make a complete revolution if $3V^2 > 7ag$.
9. A bead slides down a rough circular wire in a vertical plane starting from rest at the end of a horizontal diameter. When it has described an angle θ about the centre, show that the square of its angular velocity is

$$\frac{2ga}{a\,(1+4\mu^2)}\left[(1-2\mu^2)\sin\theta + 3\mu\left(\cos\theta - e^{-2\mu\theta}\right)\right],$$

where a = radius of the circle and μ = the coefficient of friction.
10. In a spring-mass system, the upper end of the spring is given a downward velocity v_0 when harmonic motion at $t = 0$ given by

$$x_1 = \frac{v_0}{\omega}\sin\omega t$$

begins. Determine the expression for the motion of the mass.

ANSWERS

3. $\dfrac{ql^4}{8EI}$, where l = length of beam and q = load distribution.

4. $M_{\max} = \dfrac{P}{4k}\dfrac{\sinh^2(kl)+\sin^2(kl)}{\sinh(kl)\cosh(kl)+\sin(kl)\cos(kl)}$.

5. $P_{cr} = \dfrac{14.6\,EI}{l^2}$.

10. $x = \dfrac{v_0/\omega}{1-\frac{\omega^2}{\xi^2}}\left(\sin\omega t - \dfrac{\omega^3}{\xi^3}\sin\xi t\right)$.

Chapter 5

Vector Spaces and Matrices

I. VECTOR SPACE

5.1 Definition of a Vector Space

Mathematically, by *space* is meant a set of elements of which certain well-defined mathematical operations remain valid. These elements may be vectors, functions or any other quantities. Let V and F are two non-empty sets, such that elements of V are vectors and are denoted by $\boldsymbol{\alpha}$, $\boldsymbol{\beta}$, $\boldsymbol{\gamma}$, etc., and those of F are scalars denoted by a, b, c, etc. Then, a vector space (also called *linear space*) V_n is a set over field F, such that elements of V_n obey certain mathematical operations like addition and multiplication by a scalar which are as follows:

Let $\boldsymbol{\alpha}$, $\boldsymbol{\beta}$, $\boldsymbol{\gamma}$ be elements of V and a, b are scalars in F. Then

(i) $\boldsymbol{\alpha}+\boldsymbol{\beta}$ will also be in V;

(ii) $\boldsymbol{\alpha}+\boldsymbol{\beta}=\boldsymbol{\beta}+\boldsymbol{\alpha}$;

(iii) $\boldsymbol{\alpha}+(\boldsymbol{\beta}+\boldsymbol{\gamma})=(\boldsymbol{\alpha}+\boldsymbol{\beta})+\boldsymbol{\gamma}$;

(iv) $a(\boldsymbol{\alpha}+\boldsymbol{\beta})=a\boldsymbol{\alpha}+a\boldsymbol{\beta}$;

(v) $\boldsymbol{\alpha}+(-\boldsymbol{\alpha})=\mathbf{0}$, where $\mathbf{0}$ is the null vector;

(vi) $(ab)\boldsymbol{\alpha}=a(b\boldsymbol{\alpha})$;

(vii) $1\cdot\boldsymbol{\alpha}=\boldsymbol{\alpha}$, where 1 is unity in field F.

5.2 Vector Space of Ordered n-Tuples

Let $\boldsymbol{\alpha}, \boldsymbol{\beta}$ and $\boldsymbol{\gamma}$ be three elements of V, so that

$$\begin{aligned}\boldsymbol{\alpha}&=(a_1,a_2,\ldots,a_n),\\ \boldsymbol{\beta}&=(b_1,b_2,\ldots,b_n)\\ \text{and}\quad \boldsymbol{\gamma}&=(c_1,c_2,\ldots,c_n),\end{aligned}$$

where a_i, b_i, c_i are all in F.

Let a be an element of F. Then, we define addition of $\boldsymbol{\alpha}$ and $\boldsymbol{\beta}$ as

$$\begin{aligned}\boldsymbol{\alpha}+\boldsymbol{\beta}&=(a_1,a_2,\ldots,a_n)+(b_1,b_2,\ldots,b_n)\\ &=(a_1+b_1,a_2+b_2,\ldots,a_n+b_n)\quad\text{(vector addition)}\end{aligned}\tag{5.1}$$

and multiplication of $\boldsymbol{\alpha}$ by a as

$$\begin{aligned}a\boldsymbol{\alpha}&=a(a_1,a_2,\ldots,a_n)\\ &=(aa_1,aa_2,\ldots,aa_n)\quad\text{(multiplication of a vector by a scalar)}\end{aligned}\tag{5.2}$$

With the use of these two definitions, we can verify all the postulates of vector space V. Let us see how.

(i) $\boldsymbol{\alpha}+\boldsymbol{\beta}=(a_1,a_2,\ldots,a_n)+(b_1,b_2,\ldots,b_n)$

$=(a_1+b_1,a_2+b_2,\ldots,a_n+b_n)$ (since a_i, b_i are commutative in F)

$=(b_1+a_1,b_2+a_2,\ldots,b_n+a_n)$

$=(b_1,b_2,\ldots,b_n)+(a_1,a_2,\ldots,a_n).$

$$\therefore \quad \boldsymbol{\alpha}+\boldsymbol{\beta}=\boldsymbol{\beta}+\boldsymbol{\alpha}. \tag{5.3}$$

(ii) $\boldsymbol{\alpha}+(\boldsymbol{\beta}+\boldsymbol{\gamma})=(a_1,a_2,\ldots,a_n)+\{(b_1,b_2,\ldots,b_n)+(c_1,c_2,\ldots,c_n)\}$

$=(a_1,a_2,\ldots,a_n)+(b_1+c_1,b_2+c_2,\ldots,b_n+c_n)$

$=[a_1+(b_1+c_1),a_2+(b_2+c_2),\ldots,a_n+(b_n+c_n)]$

$=[(a_1+b_1)+c_1,(a_2+b_2)+c_2,(a_n+b_n)+c_n]$

(since a_i, b_i, c_i are associative in F)

$=[(a_1+b_1,a_2+b_2,\ldots,a_n+b_n)+(c_1,c_2,\ldots,c_n)$

$=[(a_1,a_2,\ldots,a_n)+(b_1,b_2,\ldots,b_n)]+(c_1,c_2,\ldots,c_n).$

$$\therefore \quad \boldsymbol{\alpha}+(\boldsymbol{\beta}+\boldsymbol{\gamma})=(\boldsymbol{\alpha}+\boldsymbol{\beta})+\boldsymbol{\gamma}. \tag{5.4}$$

(iii) $a(\boldsymbol{\alpha}+\boldsymbol{\beta})=a[(a_1,a_2,\ldots,a_n)+(b_1,b_2,\ldots,b_n)]$

$=a(a_1+b_1,a_2+b_2,\ldots,a_n+b_n)$

$=[a(a_1+b_1),a(a_2+b_2),\ldots,a(a_n+b_n)]$

$=[aa_1+ab_1,aa_2+ab_2,\ldots,aa_n+ab_n]$

$=a(a_1,a_2,\ldots,a_n)+a(b_1,b_2,\ldots,b_n).$

$$\therefore \quad a(\boldsymbol{\alpha}+\boldsymbol{\beta})=a\boldsymbol{\alpha}+a\boldsymbol{\beta}. \tag{5.5}$$

(iv) $\boldsymbol{\alpha}+(-\boldsymbol{\alpha})=(a_1,a_2,\ldots,a_n)+(-a_1,-a_2,\ldots,-a_n)$

$=(a_1-a_1,a_2-a_2,\ldots,a_n-a_n)=(0,0,\ldots,0).$

$$\therefore \quad \boldsymbol{\alpha}+(-\boldsymbol{\alpha})=\mathbf{0}. \tag{5.6}$$

Thus, for each $\boldsymbol{\alpha}$ in V, there exists a $(-\boldsymbol{\alpha})$ also in V, such that $\boldsymbol{\alpha}+(-\boldsymbol{\alpha})$ is a null vector.

(v) If a, b are in F, then, $(ab)\boldsymbol{\alpha}=(ab)(a_1,a_2,\ldots,a_n)=[a(ba_1),a(ba_2),\ldots,a(ba_n)]$

$=a[ba_1,ba_2,\ldots,ba_n]=a[b(a_1,a_2,\ldots,a_n)].$

$$\therefore \quad (ab)\boldsymbol{\alpha}=a(b\boldsymbol{\alpha}). \tag{5.7}$$

(vi) If unity is in F, then, $1\cdot\boldsymbol{\alpha}=1(a_1,a_2,\ldots,a_n)=(1\cdot a_1,1\cdot a_2,\ldots,1\cdot a_n)$

$=(a_1,a_2,a_n).$

$$\therefore \quad 1\cdot\boldsymbol{\alpha}=\boldsymbol{\alpha}. \tag{5.8}$$

Thus, V defined as $V=(a_1,a_2,\ldots,a_n)$ with all a_i in F, is a vector space.

We now mention some properties of a vector space.

Property 1. For each element, say $\boldsymbol{\alpha}$ in V, there exists a null vector $\mathbf{0}$ in V, such that

$$\boldsymbol{\alpha}+\mathbf{0}=\boldsymbol{\alpha} \quad \text{and} \quad \mathbf{0}+\boldsymbol{\alpha}=\boldsymbol{\alpha}.$$

Now

$$\begin{aligned}\boldsymbol{\alpha}+\mathbf{0} &= (a_1, a_2, \ldots, a_n) + (0, 0, \ldots, 0)\\ &= (a_1+0, a_2+0, \ldots, a_n+0)\\ &= (a_1, a_2, \ldots, a_n) = \boldsymbol{\alpha}.\end{aligned}$$

Similarly, $\mathbf{0}+\boldsymbol{\alpha}=\boldsymbol{\alpha}$.

$$\therefore \quad \mathbf{0}+\boldsymbol{\alpha}=\boldsymbol{\alpha}+\mathbf{0}=\boldsymbol{\alpha}. \tag{5.9}$$

Property 2. If $a \neq 0$, then $a\boldsymbol{\alpha}=\mathbf{0}$ implies $\boldsymbol{\alpha}=\mathbf{0}$.

Since $a \neq 0$, a^{-1} should exist. Hence

$$(a^{-1}a)\boldsymbol{\alpha}=\mathbf{0}, \quad \text{i.e.,} \quad 1\cdot\boldsymbol{\alpha}=\mathbf{0}, \quad \text{i.e.,} \quad \boldsymbol{\alpha}=\mathbf{0}.$$

Property 3. If a, b are scalars in F and $\boldsymbol{\alpha}$ is a non-zero vector element in V, then $a\boldsymbol{\alpha}=b\boldsymbol{\alpha}$ implies $a=b$.

Let, $a\boldsymbol{\alpha}=b\boldsymbol{\alpha}$; then, $a\boldsymbol{\alpha}-b\boldsymbol{\alpha}=\mathbf{0}$, i.e., $(a-b)\boldsymbol{\alpha}=\mathbf{0}$.

$\therefore \quad a-b=0$, since $\boldsymbol{\alpha}\neq\mathbf{0}$.

$\therefore \quad a=b$.

5.3 Linear Independence

A set of vectors $\{\boldsymbol{\alpha}_i\}$, $i=1,2,\ldots,n$ in a vector space is said to be linearly dependent, if there exists a corresponding set of scalars a_i, $i=1,2,\ldots,n$, not all zero, such that

$$\sum_{i=1}^{n} a_i\boldsymbol{\alpha}_i = 0. \tag{5.10}$$

If

$$\sum_{i=1}^{n} a_i\boldsymbol{\alpha}_i = 0$$

implies that $a_i=0$ for all i, then the vectors $\{\boldsymbol{\alpha}_i\}$ are said to be linearly independent.

Let us consider the vectors

$$\boldsymbol{\alpha}_1=(1,0,1), \quad \boldsymbol{\alpha}_2=(0,1,1), \quad \boldsymbol{\alpha}_3=(1,1,1).$$

With a_1, a_2, a_3 as three scalars, we find that

$$a_1\boldsymbol{\alpha}_1+a_2\boldsymbol{\alpha}_2+a_3\boldsymbol{\alpha}_3=(0,0,0)$$

if

$$a_1(1,0,1)+a_2(0,1,1)+a_3(1,1,1)=(0,0,0).$$

That is, if

$$\begin{aligned} a_1 \qquad + a_3 &= 0\\ a_2 + a_3 &= 0\\ a_1 + a_2 + a_3 &= 0,\end{aligned}$$

which is possible only if $a_1=a_2=a_3=0$. Thus, the given vectors are linearly independent.

Dimension of a Vector Space

A vector space is said to be n-dimensional if it contains exactly n linearly independent vectors. If a large number of linearly independent vectors exist in a vector space, it is called *infinite-dimensional.*

From the definition of linear independence of a set of vectors, the following deductions can easily be made:

1. If two vectors $\boldsymbol{\alpha}$ and $\boldsymbol{\beta}$ be linearly dependent, then one of them is a scalar multiple of the other.

   Proof. Let $\boldsymbol{\alpha}$ and $\boldsymbol{\beta}$ be two linearly dependent vectors in a vector space V. Then, for two scalars a and b, $a \neq 0$, $b \neq 0$, we must have the relation

   $$a\boldsymbol{\alpha} + b\boldsymbol{\beta} = \mathbf{0},$$

   i.e., $a\boldsymbol{\alpha} = -b\boldsymbol{\beta}$, i.e., $\boldsymbol{\alpha} = \left(-\frac{b}{a}\right)\boldsymbol{\beta}, \because a \neq 0.$

   Thus, $\boldsymbol{\alpha}$ is a scalar multiple of $\boldsymbol{\beta}$. Similarly

   $$b\boldsymbol{\beta} = -a\boldsymbol{\alpha}, \quad \text{i.e.,} \quad \boldsymbol{\beta} = \left(-\frac{a}{b}\right)\boldsymbol{\alpha}, \quad \because b \neq 0.$$

   Thus, $\boldsymbol{\beta}$ is a scalar multiple of $\boldsymbol{\alpha}$. □

2. The system of the set of n vectors $\mathbf{e}_1 = (1, 0, 0, \ldots, 0)$, $\mathbf{e}_2 = (0, 1, 0, \ldots, 0)$, $\ldots$, $\mathbf{e}_n = (0, 0, 0, \ldots, 1)$ defined over the field F is linearly independent. Here, 1 denotes the unity in the field F.

   Proof. Let $a_i (i = 1, 2, \ldots, n)$ be scalars in F. Then,

   $$\sum_{i=1}^{n} a_i \mathbf{e}_i = \mathbf{0}$$

   implies that $(a_1, a_2, \ldots, a_n) = (0, 0, \ldots, 0)$.

   $$\therefore \quad a_1 = a_2 = \cdots = a_n = 0.$$

   Thus, the given set of vectors is linearly independent. □

3. Any subset of a linearly independent set of vectors in a vector space is also linearly independent.

   Proof. Let $\{\boldsymbol{\alpha}_i\}$, $i = 1, 2, \ldots, n$ be a set of linearly independent vectors. Then there exist scalars $a_i (i = 1, 2, \ldots, n)$, such that

   $$\sum_{i=1}^{n} a_i \boldsymbol{\alpha}_i = \mathbf{0}$$

   implies $a_i = 0$, $i = 1, 2, \ldots, n$.
   Now, let us consider another set of vectors $\{\boldsymbol{\alpha}_i\}$, $i = 1, 2, \ldots, r$ where $r < n$. Then, obviously, this set will be a subset of the former.
   It is clear that

   $$\sum_{i=1}^{r} a_i \boldsymbol{\alpha}_i = \mathbf{0}$$

also implies that $a_i (i = 1, 2, \ldots, r)$ is zero.

This shows that the subset of a set of linearly independent vectors is also linearly independent.

We now have the following two important theorems: □

Theorem 5.1. *A set of non-zero vectors in a vector field is linearly dependent, if and only if some elements of that set is a linear combination of others.*

Proof. Let V be a vector space over the field F. Let $\{\boldsymbol{\alpha}_i\}$, $i = 1, 2, \ldots, n$ be a set of vectors in V which are linearly dependent. Then there exist scalars $a_i (i = 1, 2, \ldots, n)$ in F, such that

$$\sum_{i=1}^{n} a_i \boldsymbol{\alpha}_i = \mathbf{0}. \tag{5.11}$$

Since the set is linearly dependent, at least one of the scalars a_i is not equal to zero. Let us assume that $a_k \neq 0$ $(k \leq n)$. Then, from equation (5.11), we get

$$a_k \boldsymbol{\alpha}_k = -\sum_{i=1}^{n} a_i \boldsymbol{\alpha}_i \quad (i \neq k). \tag{5.12}$$

Since $a_k \neq 0$, a_k^{-1} should exist. Hence we can write

$$\boldsymbol{\alpha}_k = -\sum_{i=1}^{n} \left(a_i a_k^{-1}\right) \boldsymbol{\alpha}_i \quad (i \neq k).$$

Thus, the vector $\boldsymbol{\alpha}_k$ can be expressed as a linear combination of others of the set $\{\boldsymbol{\alpha}_i\}$.

Conversely, let $\boldsymbol{\alpha}_m$ of the set of vectors $\{\boldsymbol{\alpha}_i\}$, $i = 1, 2, \ldots, n$ can be expressed as a linear combination of other elements of the set, that is

$$\boldsymbol{\alpha}_m = \sum_{i=1}^{n} b_i \boldsymbol{\alpha}_i \quad (i \neq m), \tag{5.13}$$

where b_i's are scalars in F.

Then, from equation (5.13), we write from postulate of vector space

$$\sum b_i \boldsymbol{\alpha}_i + (-1)\boldsymbol{\alpha}_m = 0.$$

This shows that the set of vectors $\{\boldsymbol{\alpha}_i\}$, $i = 1, 2, \ldots, n$ is linearly dependent. □

Theorem 5.2. *A set of non-zero vectors in a vector space is linearly dependent, if and only if one of them be a linear combination of the preceding vectors.*

Proof. Let V be a vector space over a field F. Let the non-zero vectors $\{\boldsymbol{\alpha}_i\}$, $i = 1, 2, \ldots, n$ in V are linearly dependent. Then, there exist scalars $a_i (i = 1, 2, \ldots, n)$ in F, not all of which are zero, such that the relation

$$\sum_{i=1}^{n} a_i \boldsymbol{\alpha}_i = 0 \tag{5.14}$$

holds good.

Now, without any loss of generality, we can assume that the scalars

$$a_{k+1}, a_{k+2}, a_{k+3}, \ldots, a_n$$

are all zero. Then we write

$$\sum_{i=1}^{k} a_i\boldsymbol{\alpha}_i + \sum_{i=k+1}^{n} a_i\boldsymbol{\alpha}_i = 0. \tag{5.15}$$

Since $a_k \neq 0$, its inverse should exist.
Hence, from equation (5.15), we get

$$\boldsymbol{\alpha}_k = -\sum_{i=1}^{k-1} a_k^{-1} a_i\boldsymbol{\alpha}_i.$$

Thus, $\boldsymbol{\alpha}_k$ is a linear combination of the vectors preceding to it.

Conversely, let $\boldsymbol{\alpha}_m$ of the set of vectors $\{\boldsymbol{\alpha}_i\}$, $i = 1, 2, \ldots, n$ can be expressed as a linear combination of its preceding vectors.

Then

$$\boldsymbol{\alpha}_m = \sum_{i=1}^{m-1} b_i\boldsymbol{\alpha}_i \quad \text{for all } b_i \text{ in } F.$$

Hence, by the postulate of vector space

$$\sum_{i=1}^{m-1} b_i\boldsymbol{\alpha}_i + (-1)\boldsymbol{\alpha}_m = \mathbf{0}. \tag{5.16}$$

Thus

$$\sum_{i=1}^{m-1} b_i\boldsymbol{\alpha}_i + (-1)\boldsymbol{\alpha}_m + \sum_{i=m+1}^{n} a_i\boldsymbol{\alpha}_i = \mathbf{0}.$$

Hence, the vectors $\{\boldsymbol{\alpha}_i\}$, $i = 1, 2, \ldots, n$ are linearly dependent.

EXAMPLE 5.1. *Examine whether the set of vectors* $\boldsymbol{\alpha}_1 = (2, -3, 1)$, $\boldsymbol{\alpha}_2 = (3, -1, 5)$, $\boldsymbol{\alpha}_3 = (1, -4, 3)$ *are linearly dependent or independent.*

Solution: Let there be three scalars a_1, a_2, a_3, such that

$$a_1(2, -3, 1) + a_2(3, -1, 5) + a_3(1, -4, 3) = (0, 0, 0).$$

Then

$$2a_1 + 3a_2 + a_3 = 0 \tag{1}$$

$$-3a_1 - a_2 - 4a_3 = 0 \tag{2}$$

$$a_1 + 5a_2 + 3a_3 = 0. \tag{3}$$

Multiplying (3) by 3 and adding with (2)

$$14a_2 + 5a_3 = 0. \tag{4}$$

Multiplying (3) by 2 and subtracting from (1)

$$-7a_2 - 5a_3 = 0. \tag{5}$$

Thus, the solutions of equations are

$$a_1 = 0, \quad a_2 = 0, \quad a_3 = 0.$$

Hence, the given system is linearly independent.

EXAMPLE 5.2. *Show that the vectors* $\boldsymbol{\alpha}_1 = (-1, 2, 1)$, $\boldsymbol{\alpha}_2 = (3, 0, -1)$, $\boldsymbol{\alpha}_3 = (-5, 4, 3)$ *are linearly dependent.*

Solution: Let a_1, a_2, a_3 be scalars not all zero. Then, suppose

$$a_1\boldsymbol{\alpha}_1 + a_2\boldsymbol{\alpha}_2 + a_3\boldsymbol{\alpha}_3 = (0, 0, 0),$$

i.e.,

$$a_1(-1, 2, 1) + a_2(3, 0, -1) + a_3(-5, 4, 3) = (0, 0, 0).$$

Thus, we have

$$\begin{aligned} -\ a_1 + 3a_2 - 5a_3 &= 0 \\ 2a_1 + 4a_3 &= 0 \\ a_1 - a_2 + 3a_3 &= 0. \end{aligned}$$

The system of equations has a non-zero solution

$$a_1 = -2, \quad a_2 = 1, \quad a_3 = 1.$$

Hence, the given vectors are linearly dependent.

EXAMPLE 5.3. *Show that the vector* $(2, -5, 3)$ *cannot be expressed as a linear combination of vectors*

$$\boldsymbol{\alpha}_1 = (1, -3, 2), \quad \boldsymbol{\alpha}_2 = (-2, -4, -1), \quad \boldsymbol{\alpha}_3 = (1, -5, 7).$$

Solution: Let us suppose that the vector $(2, -5, 3)$ can be expressed as linear combination of $\boldsymbol{\alpha}_1, \boldsymbol{\alpha}_2, \boldsymbol{\alpha}_3$. Then, there should exist scalars a_1, a_2, a_3, not all zero, so that

$$a_1\boldsymbol{\alpha}_1 + a_2\boldsymbol{\alpha}_2 + a_3\boldsymbol{\alpha}_3 = (2, -5, 3),$$

i.e.,

$$a_1(1, -3, 2) + a_2(-2, -4, -1) + a_3(1, -5, 7) = (2, -5, 3). \tag{1}$$

Thus, we get

$$a_1 - 2a_2 + a_3 = 2 \tag{2}$$

$$-3a_1 - 4a_2 - 5a_3 = -5 \tag{3}$$

$$2a_1 - a_2 + 7a_3 = 3. \tag{4}$$

It is seen that the determinant

$$\begin{vmatrix} 1 & -2 & 1 \\ -3 & -4 & -5 \\ 2 & -1 & 7 \end{vmatrix} = 0.$$

Hence, the solution of system of equations does not exist. Therefore, the given vector cannot be expressed as a linear combination of three other given vectors.

5.4 Basis of a Vector Space

Let V be a vector space over the field F. If an arbitrary vector ψ in V can be represented as a linear combination of the vectors $\{\alpha_i\}$ in V so that

$$\psi = \sum_{i=1}^{n} a_1 \alpha_i, \tag{5.17}$$

where a_i are scalars, then $\{\alpha_i\}$ is said to *span* the vector space V.

A set of linearly independent vectors in V that spans the vector space is called a *basis* for V. Thus, if we imagine a subset S of V over F such that S contains linearly independent vectors, and each vector of V is a linear combination of a finite number of elements in S, then S is the basis for V.

As an example, let us consider the vector space V_3 of real numbers of which

$$S = \{(1,0,0), (0,1,0), (0,0,1)\} \tag{5.18}$$

is a subset. Obviously, this set S is linearly independent. Moreover, any vector (ψ_1, ψ_2, ψ_3) of V_3 can be expressed as a linear combination of S as

$$(\psi_1, \psi_2, \psi_3) = \psi_1(1,0,0) + \psi_2(0,1,0) + \psi_3(0,0,1).$$

Hence, the set S is the basis for V_3 over real numbers.

The number of elements in any basis set of a vector space with finite dimension is called *dimension of the vector space.*

If the basis contains n elements, then the vector space V is n-dimensional. The vector space is then written as V_n. The vector space of real numbers is 3-dimensional and it is denoted by R^3. The set S given by equation (5.18) is the fundamental basis in R^3.

EXAMPLE 5.4. *Show that the vectors* $\{(1,-2,3), (2,3,1), (-1,3,2)\}$ *form a basis of the vector space* V_3 *over real numbers.*

Solution: We know that the set $\{(1,0,0), (0,1,0), (0,0,1)\}$ forms a basis for V_3 over real numbers, and hence its dimension is 3. Thus, the given set will form the basis of the vector space V_3 over real numbers if the set is linearly independent.

Now, the set will be linearly independent, if for scalars a_1, a_2, a_3,

$$a_1(1,-2,3) + a_2(2,3,1) + a_3(-1,3,2) = (0,0,0) \tag{1}$$

will produce $a_1 = a_2 = a_3 = 0$. Let us now check it.

From (1), we have

$$\begin{aligned} a_1 + 2a_2 - a_3 &= 0 \\ -2a_1 + 3a_2 + 3a_3 &= 0 \\ 3a_1 + a_2 + 2a_3 &= 0. \end{aligned}$$

Solving these equations, we have

$$a_1 = a_2 = a_3 = 0.$$

Thus, the given set is linearly independent and hence forms a basis for V_3 over real numbers.

EXAMPLE 5.5. *Find a basis of R^3 containing vectors* $(1,1,2)$ *and* $(3,5,2)$.

Solution: The vectors $(1,1,2)$ and $(3,5,2)$ are linearly independent; because, for any two scalars a_1, a_2, the relation

$$a_1(1,1,2) + a_2(3,5,2) = (0,0,0)$$

produces $a_1 = 0$ and $a_2 = 0$.

Let us consider the fundamental basis of R^3

$$\{(1,0,0),(0,1,0),(0,0,1)\}.$$

Now, for any three scalars x, y, z, let us consider the linear relation

$$x(1,1,2) + y(3,5,2) + z(1,0,0) = (0,0,0).$$

This gives

$$\begin{aligned} x + 3y + z &= 0 \\ x + 5y \quad &= 0 \\ 2x + 2y \quad &= 0. \end{aligned}$$

From here, we get, $x = y = z = 0$.

Thus, the vectors $(1,1,2),(3,5,2),(1,0,0)$ are linearly independent and hence forms a basis of R^3.

EXAMPLE 5.6. *Show that the set of vectors* $S = \{(1,0,0,0),(0,1,0,0),(0,0,1,0),(0,0,0,1)\}$ *is a basis of* R^4.

Hence, show that vectors $\boldsymbol{\alpha} = (1,2,-1,-2)$, $\boldsymbol{\beta} = (2,3,0,-1)$, $\boldsymbol{\gamma} = (1,2,1,4)$ *and* $\boldsymbol{\delta} = (1,3,-1,0)$ *form a basis of the vector space V_4 over real numbers.*

Solution: The given set S is a subset of a four-dimensional vector space V_4 of real numbers. It is obvious that the set S is linearly independent. Moreover, any vector $(\alpha_1, \alpha_2, \alpha_3, \alpha_4)$ of V_4 can be expressed as a linear combination of S in the form

$$(\alpha_1,\alpha_2,\alpha_3,\alpha_4) = \alpha_1(1,0,0,0) + \alpha_2(0,1,0,0) + \alpha_3(0,0,1,0) + \alpha_4(0,0,0,1).$$

Hence, the set S is the basis for V_4 over real numbers.

Now, the given vectors $\boldsymbol{\alpha}, \boldsymbol{\beta}, \boldsymbol{\gamma}$ and $\boldsymbol{\delta}$ will form the basis of the vector space V_4 over real numbers, if they are linearly independent. Let, for scalars a_1, a_2, a_3, a_4

$$a_1(1,2,-1,-2) + a_2(2,3,0,-1) + a_3(1,2,1,4) + a_4(1,3,-1,0) = (0,0,0,0).$$

This gives

$$\begin{aligned} a_1 + 2a_2 + a_3 + a_4 &= 0 \\ 2a_1 + 3a_2 + 2a_3 + 3a_4 &= 0 \\ -a_1 + 0 + a_3 - a_4 &= 0 \\ -2a_1 - a_2 + 4a_3 + 0 &= 0. \end{aligned}$$

The solution of this system of equations is $a_1 = a_2 = a_3 = a_4 = 0$.

Therefore, the vectors $\boldsymbol{\alpha}, \boldsymbol{\beta}, \boldsymbol{\gamma}, \boldsymbol{\delta}$ are linearly independent. Hence, it forms a basis for V_4 in the field of real numbers.

We now intend to prove some important theorems concerning the basis of a vector space.

Theorem 5.3. *Every vector in a finite-dimensional vector space V over the field F can be uniquely expressed as a linear combination of the vectors of its basis.*

Proof. Let, $S = \{\alpha_i\}$, $i = 1, 2, \ldots, n$ be a basis of the vector space V over F. Then, any vector α in V can be expressed as a linear combination of the vectors of S. That is

$$\alpha = \sum_{i=1}^{n} a_i \alpha_i, \tag{5.19}$$

where a_i are scalars in F. We are to prove that the above expression is unique.

Now, if possible, let there be another set of scalars b_i $(i = 1, 2, \ldots, n)$ in F, such that

$$\alpha = \sum_{i=1}^{n} b_i \alpha_i. \tag{5.20}$$

Subtracting equation (5.20) from equation (5.19), we get

$$\sum a_i \alpha_i - \sum b_i \alpha_i = \alpha - \alpha = \mathbf{0},$$

i.e.,

$$\sum (a_i - b_i) \alpha_i = \mathbf{0}. \tag{5.21}$$

Since $\{\alpha_i\}$ is a linearly independent set of vectors, equation (5.21) implies that

$$a_i - b_i = 0, \quad \text{i.e.,} \quad a_i = b_i.$$

Hence, the expression for α as a linear combination of vectors $\{\alpha_i\}$ is unique. □

Theorem 5.4. *The set of n vectors* $\mathbf{e}_1(1, 0, 0, \ldots, 0)$, $\mathbf{e}_2(0, 1, 0, \ldots, 0)$, $\ldots$, $\mathbf{e}_n(0, 0, \ldots, 1)$ *is a basis of vector space* V_n *over* F.

Proof. Let $a_i (i = 1, 2, \ldots, n)$ be any scalars in F. Then,

$$\sum a_i \mathbf{e}_i = \mathbf{0}$$

means that $a_1(1, 0, 0, \ldots, 0) + a_2(0, 1, 0, \ldots, 0) + \cdots + a_n(0, 0, \ldots, 1) = (0, 0, \ldots, 0)$.

Therefore, $a_1 = a_2 = \cdots = a_n = 0$.

Hence, the given set of vectors is linearly independent.

Now, let $\beta = \{a_i\}$, $i = 1, 2, \ldots, n$ be any vector of V_n over F. Then we can write

$$(a_1, a_2, \ldots, a_n) = a_1(1, 0, 0, \ldots, 0) + a_2(0, 1, 0, \ldots, 0) + \cdots + a_n(0, 0, \ldots, 1),$$

i.e., $\beta = a_1\mathbf{e}_1 + a_2\mathbf{e}_2 + \cdots + a_n\mathbf{e}_n$.

Thus, any vector in V_n can be expressed as a linear combination of the elements of the given set of vectors.

Hence, the given set of vectors is a basis of V_n over F.

This basis is called the *standard basis* of V_n over real numbers. □

Theorem 5.5. *If* $\{\alpha_i\}$, $i = 1, 2, \ldots, n$ *be a set of n independent vectors and* β *be a non-null vector, such that*

$$\beta = \sum_{i=1}^{n} a_i \alpha_i,$$

where $a_j \neq 0$ *and* $a_i (i = 1, 2, \ldots, n)$ *are scalars, then the vectors*

$$\alpha_1, \alpha_2, \ldots, \alpha_{j-1}, \beta, \alpha_{j+1}, \ldots, \alpha_n$$

are linearly independent.

Proof. Let $b_i (i = 1, 2, \ldots, n)$ be a set of scalars, such that

$$b_1\alpha_1 + b_2\alpha_2 + \cdots b_{j-1}\alpha_{j-1} + b_j\beta + b_{j+1}\alpha_{j+1} + \cdots b_n\alpha_n = 0,$$

$$\text{i.e.,} \quad \sum_{1}^{j-1} b_i\alpha_i + b_j\left(\sum_{1}^{n} a_i\alpha_i\right) + \sum_{j+1}^{n} b_i\alpha_i = 0,$$

$$\text{or,} \quad (b_1 + a_1b_j)\alpha_1 + (b_2 + a_2b_j)\alpha_2 + \cdots$$
$$+ b_ja_j\alpha_j + (b_{j+1} + b_ja_{j+1})\alpha_{j+1} + \cdots + (b_n + b_ja_n)\alpha_n = 0.$$

Since $\{\alpha_i\}$ are linearly independent, we have $b_1 + a_1b_j = 0$, $b_2 + a_2b_j = 0$, ..., $b_ja_j = 0$, $b_{j+1} + a_{j+1}b_j = 0$, ..., $b_n + a_nb_j = 0$.

From these we have $b_ja_j = 0$, i.e., $b_j = 0$ since $a_j \neq 0$. Putting $b_j = 0$ into other equations, we have

$$b_1 = b_2 = \cdots = b_n = 0.$$

This implies that the vectors

$$\alpha_1, \alpha_2, \ldots, \alpha_{j-1}, \beta, \alpha_{j+1}, \ldots, \alpha_n$$

are linearly independent. □

Theorem 5.6. *Let $\{\alpha_i\}$, $i = 1, 2, \ldots, n$ be a basis of vector space V over the field F and a non-null vector β can be expressed as a linear combination of these vectors in the form*

$$\beta = \sum_{i=1}^{n} a_i\alpha_i, \quad (i = 1, 2, \ldots, n),$$

where $a_i (i = 1, 2, \ldots, n)$ are scalars in F. If $a_i \neq 0$, then β can replace α_i in the basis of V. (Replacement theorem)

Proof. We have

$$\beta = \sum_{i=1}^{n} a_i\alpha_i. \quad \therefore \quad a_i\alpha_i = \beta - \sum_{1}^{i-1} a_i\alpha_i - \sum_{i+1}^{n} a_i\alpha_i. \tag{5.22}$$

Since $a_i \neq 0$, there should exist a_i^{-1}. Hence, from equation (5.22)

$$\alpha_i = a_i^{-1}\left[\beta - \sum_{i=1}^{i-1} a_i\alpha_i - \sum_{i+1}^{n} a_i\alpha_i\right]. \tag{5.23}$$

Let, for some scalars $b_i (i = 1, 2, \ldots, n)$ in F

$$\sum_{i=1}^{i-1} b_i\alpha_i + b_i\beta + \sum_{i+1}^{n} b_i\alpha_i = 0, \tag{5.24}$$

i.e.,

$$\sum_{i=1}^{i-1} b_i\alpha_i + b_i\sum_{i=1}^{n} a_i\alpha_i + \sum_{i+1}^{n} b_i\alpha_i = 0. \tag{5.25}$$

Since $\{\alpha_i\}$ is a basis of V, this set of vectors is linearly independent. Thus, all coefficients of α_i are separately zero. Hence, by equating to zero the coefficients of $\alpha_1, \alpha_2, \ldots, \alpha_n$, we get

$$\left.\begin{array}{rr} b_1 + a_1 b_i = 0, & b_2 + a_2 b_i = 0 \\ \ldots, & b_{i-1} + a_{i-1} b_i = 0 \\ b_i a_i = 0, \quad \ldots, & b_n + a_n b_i = 0 \end{array}\right\}. \tag{5.26}$$

These give $b_i = 0$, since $a_i \neq 0$.

Hence, $b_1 = b_2 = \cdots = b_n = 0$.

This implies that the set of vectors $(\alpha_1, \alpha_2, \ldots, \alpha_{i-1}, \beta, \alpha_{i+1}, \ldots, \alpha_n)$ is linearly independent.

Now, let γ be an arbitrary vector in V. Then, for scalars c_i $(i = 1, 2, \ldots, n)$ in F, γ should be expressible as a linear combination of the basis vector $\{\alpha_i\}$. Thus

$$\gamma = \sum c_i \alpha_i, \quad i = 1, 2, \ldots, n. \tag{5.27}$$

Putting here the value of α_i from equation (5.23), we have

$$\gamma = \sum_{i=1}^{i-1} c_i \alpha_i + c_i a_i^{-1}(\beta - a_1 \alpha_1 - \cdots - a_{i-1}\alpha_{i-1} - a_{i+1}\alpha_{i+1} - \cdots - a_n \alpha_n) + \sum_{i+1}^{n} c_i \alpha_i,$$

$$\text{or,} \quad \gamma = \sum_{i=1}^{i-1} d_i \alpha_i + d_i \beta + \sum_{i+1}^{n} d_i \alpha_i, \tag{5.28}$$

where coefficients d_i's are given by

$$\begin{array}{ll} d_1 = c_1 - a_1 c_i a_i^{-1}, & d_2 = c_2 - a_2 c_i a_i^{-1} \\ \ldots & \ldots \\ d_{i-1} = c_{i-1} - a_{i-1} c_i a_i^{-1}, & d_i = c_i a_i^{-1} \\ d_{i+1} = c_{i+1} - a_{i+1} c_i a_i^{-1}, & \ldots, \qquad \text{etc.} \end{array}$$

Equation (5.28) implies that

$$\gamma \in L(\alpha_1, \alpha_2, \ldots, \alpha_{i-1}, \beta, \alpha_{i+1}, \ldots, \alpha_n).$$

Hence

$$V \subset L(\alpha_1, \alpha_2, \ldots, \alpha_{i-1}, \beta, \alpha_{i+1}, \ldots, \alpha_n). \tag{5.29}$$

Now,

$$L(\alpha_1, \alpha_2, \ldots, \alpha_{i-1}, \beta, \alpha_{i+1}, \ldots, \alpha_n)$$

is the smallest sub-space containing the set

$$\{\alpha_1, \alpha_2, \ldots, \alpha_{i-1}, \beta, \alpha_{i+1}, \ldots, \alpha_n).$$

Therefore,

$$L\{\alpha_1, \alpha_2, \ldots, \alpha_{i-1}, \beta, \alpha_{i+1}, \ldots, \alpha_n) \subset V. \tag{5.30}$$

From equations (5.29) and (5.30), we get

$$L\{\alpha_1, \alpha_2, \ldots, \alpha_{i-1}, \beta, \alpha_{i+1}, \ldots, \alpha_n) = V. \tag{5.31}$$

We know, $\alpha_1, \alpha_2, \ldots, \alpha_{i-1}, \beta, \alpha_{i+1}, \ldots, \alpha_n$ are linearly independent. Hence, this set is a basis of V. Thus, β can replace α_i from the basis $\{\alpha_i\}$, $i = 1, 2, \ldots, n$ of V. □

Note: If V be a vector space over the field F and S any non-empty sub-set of V, then the linear span of S is defined as the set of all linear combinations of finite sets of elements of S. It is denoted as $L(S)$.

5.5 Coordinates

Let V be a finite-dimensional vector space over F, and let $B = \{\alpha_i\}$, $i = 1, 2, \ldots, n$ be a basis of V whose vectors are enumerated in some well-defined way, i.e., ordered. Let α be a vector in V. Then there exists a unique n-tuple a_i $(i = 1, 2, \ldots, n)$ of scalars such that

$$\alpha = \sum_{i=1}^{n} a_i \alpha_i. \tag{5.32}$$

This n-tuple is called the *n-tuple of coordinates* of α relative to the ordered basis B. If the order of vectors in B is changed, the coordinates of α will also change.

EXAMPLE 5.7. *Find the coordinates of the vector $(-1, 3, 1)$ relative to the ordered basis*

$$B = \{(2,1,0), (2,1,1), (2,2,1)\}.$$

Solution: Let there be scalars a, b, c, such that

$$a(2,1,0) + b(2,1,1) + c(2,2,1) = (-1,3,1).$$

This gives the following equations:

$$\begin{aligned} 2a + 2b + 2c &= -1 \\ a + b + 2c &= 3 \\ b + c &= 1. \end{aligned}$$

Solving these, we get

$$a = -\tfrac{3}{2}, \quad b = -\tfrac{5}{2}, \quad c = \tfrac{7}{2}.$$

Hence, the coordinates of the given vector are $\left(-\frac{3}{2}, -\frac{5}{2}, \frac{7}{2}\right)$.

EXAMPLE 5.8. *Find the coordinates of the polynomial $(x - 3x^2)$ relative to the ordered basis $\{1 - x, 1 + x, 1 - x^2\}$ in a vector space of polynomials.*

Solution: Let a, b, c be three scalars, such that

$$a(1 - x) + b(1 + x) + c(1 - x^2) = x - 3x^2.$$

This gives the equations

$$\begin{aligned} a + b + c &= 0 \\ -a + b &= 1 \\ -c &= -3. \end{aligned}$$

Solving these, we get

$$a = -2, \quad b = -1, \quad c = 3.$$

Hence, the coordinates of the given polynomial relative to ordered basis are $(-2, -1, 3)$.

5.6 Norms and Inner Products

Let V be a vector space defined over the field F of real or complex numbers. The inner product of the pair of vectors ψ and ϕ in V is denoted by

$$(\psi, \phi) \text{ in } F$$

such that the following properties are valid:

(i) $(\psi, \phi) = \overline{(\phi, \psi)}$, ϕ, ψ in V

$$\left.\begin{aligned} &\text{(ii)}\ (\psi, \phi + \xi) = (\psi, \phi) + (\psi, \xi) \\ &\text{(iii)}\ (\psi + \phi, \xi) = (\psi, \xi) + (\phi, \xi) \end{aligned}\right\} \phi, \psi, \xi \text{ in } V$$

and (iv) $(\psi, \psi) > 0$ unless $\psi = 0$.

In (i), $\overline{(\phi, \psi)}$ denotes complex conjugate of the number (ϕ, ψ).

The vector space V with an inner product defined in it is called the *inner product space.* The inner product space is called *Euclidean space* (E_n) if the field F is of real numbers. If the field F is of complex numbers, the inner product space is called a *unitary space.*

The **norm** or **length** of a vector ψ is defined as the positive square root of the inner product (ψ, ψ) and is denoted as $\| \psi \|$, that is

$$\| \psi \| = (\psi, \psi)^{1/2}. \tag{5.33}$$

In n-dimension, if the vector ψ is represented by

$$\psi = (x_1, x_2, \ldots, x_n),$$

then the norm is

$$\| \psi \| = \left(\sum_{i=1}^{n} x_1^2 \right)^{1/2}. \tag{5.34}$$

5.7 Important Theorems

Theorem 5.7. *If V be an inner product space over a field F, then*

(i) $(a\psi - b\phi, \xi) = a(\psi, \xi) - b(\phi, \xi)$,

(ii) $(\psi, a\phi + b\xi) = \bar{a}(\psi, \phi) + \bar{b}(\psi, \xi)$,

where a, b are in F, $\bar{a}, \bar{b}$, are complex conjugate of a, b and ψ, ϕ, ξ are in V.

Proof. (i) We have

$$\begin{aligned} (a\psi - b\phi, \xi) &= (a\psi + (-b\phi), \xi) = a(\psi, \xi) + (-b)(\phi, \xi) \quad \text{(linearity)} \\ &= a(\psi, \xi) - b(\phi, \xi). \end{aligned} \tag{5.35}$$

(ii) We have

$$\begin{aligned} (\psi, a\phi + b\xi) &= \overline{(a\phi + b\xi, \psi)} \quad \text{property (i)} \\ &= \overline{a\,(\phi, \psi) + b\,(\xi, \psi)} = \bar{a}\,\overline{(\phi, \psi)} + \bar{b}\,\overline{(\xi, \psi)} \\ &= \bar{a}\,(\psi, \phi) + \bar{b}\,(\psi, \xi). \end{aligned} \tag{5.36}$$

□

Theorem 5.8. *If ψ and ϕ be any two vectors in an inner product space V over field F, then*

$$|(\psi,\phi)| \le \|\psi\|\,\|\phi\|, \tag{5.37}$$

where, the left-hand side expression denotes the modulus of the number (ψ,ϕ), real or complex.

This theorem is known as ***Schwarz's inequality.***

Proof. If any of the vectors ψ, ϕ be a null vector, then the theorem holds obviously. So, let us assume that $\psi \ne$ a null vector. Consider the vector

$$\xi = \phi - \frac{(\phi,\psi)}{\|\psi\|^2}\cdot\psi, \quad \frac{1}{\|\psi\|^2} \text{ is a real positive.}$$

Then we have,

$$\begin{aligned}
(\xi,\xi) &= \left(\phi - \frac{(\phi,\psi)}{\|\psi\|^2}\cdot\psi,\ \phi - \frac{(\phi,\psi)}{\|\psi\|^2}\psi\right)\\
&= \left(\phi,\ \phi - \frac{(\phi,\psi)}{\|\psi\|^2}\psi\right) - \frac{(\phi,\psi)}{\|\psi\|^2}\left(\psi,\ \phi - \frac{(\phi,\psi)}{\|\psi\|^2}\psi\right) \quad \text{(by linearity)}\\
&= (\phi,\phi) - \frac{\overline{(\phi,\psi)}}{\|\psi\|^2}(\phi,\psi) - \frac{(\phi,\psi)}{\|\psi\|^2}(\psi,\phi) + \frac{(\phi,\psi)}{\|\psi\|^2}\cdot\frac{\overline{(\phi,\psi)}}{\|\psi\|^2}(\psi,\psi)\\
&\qquad\qquad \text{(by theorem 1)}\\
&= \|\phi\|^2 - \frac{\overline{(\phi,\psi)}\,(\phi,\psi)}{\|\psi\|^2} - \frac{\overline{(\psi,\phi)}\,(\psi,\phi)}{\|\psi\|^2} + \frac{\overline{(\phi,\psi)}\,(\phi,\psi)}{\|\psi\|^2\,\|\psi\|^2}\|\psi\|^2\\
&= \|\phi\|^2 - \frac{(\psi,\phi)\,\overline{(\psi,\phi)}}{\|\psi\|^2} \quad [\because (\psi,\phi) \text{ and } \overline{(\psi,\phi)} \text{ are complex conjugates}]\\
&= \|\phi\|^2 - \frac{|(\psi,\phi)|^2}{\|\psi\|^2}
\end{aligned}$$

Again, we have

$$(\xi,\xi) = \|\xi\|^2 \ge 0.$$

Hence, from above, we have

$$\|\phi\|^2 - \frac{|(\psi,\phi)|^2}{\|\psi\|^2} \ge 0, \quad \text{or,} \quad |(\psi,\phi)|^2 \le \|\phi\|^2\,\|\psi\|^2$$

$$\text{or,} \quad |(\psi,\phi)| \le \|\psi\|\,\|\phi\|.$$

Hence, the theorem is proved. □

Theorem 5.9. *If ψ and ϕ be any two vectors in an inner product space V over F, then*

$$\|\psi + \phi\| \le \|\psi\| + \|\phi\|. \tag{5.38}$$

Proof. From the definition of norm, we have

$$\begin{aligned}
\|\psi+\phi\|^2 &= (\psi+\phi,\ \psi+\phi) = (\psi,\ \psi+\phi) + (\phi,\ \psi+\phi)\\
&= (\psi,\psi) + (\psi,\phi) + (\phi,\psi) + (\phi,\phi)\\
&= \|\psi\|^2 + (\psi,\phi) + \overline{(\psi,\phi)} + \|\phi\|^2 \qquad \left[\because\ (\phi,\psi) = \overline{(\psi,\phi)}\right]\\
&= \|\psi\|^2 + 2\,\mathrm{Re}\,(\psi,\phi) + \|\phi\|^2\\
&\le \|\psi\|^2 + 2\,|(\psi,\phi)| + \|\phi\|^2 \qquad [\because\ \mathrm{Re}\,(\psi,\phi) \le |(\psi,\phi)|]\\
&\le \|\psi\|^2 + 2\|\psi\|\,\|\phi\| + \|\phi\|^2 \quad \text{(by Schwarz's theorem)}\\
&\le (\|\psi\| + \|\phi\|)^2.
\end{aligned}$$

$$\therefore\quad \|\psi+\phi\| \le \|\psi\| + \|\phi\|.$$

Hence, the theorem is proved. This theorem is known as *triangle inequality.* □

EXAMPLE 5.9. *If α, β be vectors in an inner product space and $\|\alpha\| = \|\beta\|$, then show that $(\alpha+\beta, \alpha-\beta) = 0$.*

Solution: We have

$$\begin{aligned}
(\alpha+\beta, \alpha-\beta) &= (\alpha,\alpha) - (\alpha,\beta) + (\beta,\alpha) - (\beta,\beta)\\
&= \|\alpha\|^2 - (\alpha,\beta) + (\beta,\alpha) - \|\beta\|^2\\
&= -(\alpha,\beta) + (\beta,\alpha). \qquad (1)
\end{aligned}$$

Now

$$\text{if } (\alpha,\beta) = (a-ib)(c+id) \quad \text{(say)}$$
$$\text{then } (\beta,\alpha) = (a+ib)(c-id).$$

Thus

$$(\alpha,\beta) - (\beta,\alpha) = 0.$$

Hence, from (1), we get

$$(\alpha+\beta, \alpha-\beta) = 0.$$

EXAMPLE 5.10. *If ψ and ϕ be two vectors in an inner product space, then prove the parallelogram law*

$$\|\psi+\phi\|^2 + \|\psi-\phi\|^2 = 2\|\psi\|^2 + 2\|\phi\|^2.$$

Solution: We have

$$\begin{aligned}
\|\psi+\phi\|^2 &= (\psi+\phi,\ \psi+\phi) = (\psi,\ \psi+\phi) + \phi(\psi+\phi)\\
&= (\psi,\psi) + (\psi,\phi) + (\phi,\psi) + (\phi,\phi)\\
&= \|\psi\|^2 + (\psi,\phi) + (\phi,\psi) + \|\phi\|^2. \qquad (1)
\end{aligned}$$

Again

$$\begin{aligned}
\|\psi-\phi\|^2 &= (\psi-\phi,\ \psi-\phi) = (\psi,\ \psi-\phi) - (\phi,\ \psi-\phi)\\
&= (\psi,\psi) - (\psi,\phi) - (\phi,\psi) + (\phi,\phi)\\
&= \|\psi\|^2 - (\psi,\phi) - (\phi,\psi) + \|\phi\|^2. \qquad (2)
\end{aligned}$$

Adding (1) and (2)

$$\| \psi + \phi \|^2 + \| \psi - \phi \|^2 = 2\| \psi \|^2 + 2\| \phi \|^2.$$

Hence proved.

EXAMPLE 5.11. *If α, β be two vectors in an inner product space V over F, then show that*

$$4(\alpha, \beta) = \| \alpha + \beta \|^2 - \| \alpha - \beta \|^2 + i\| \alpha + i\beta \|^2 - i\| \alpha - i\beta \|^2.$$

Solution: We have, from Example 5.10

$$\| \alpha + \beta \|^2 = \| \alpha \|^2 + (\alpha, \beta) + (\beta, \alpha) + \| \beta \|^2 \quad (1)$$

$$\| \alpha - \beta \|^2 = \| \alpha \|^2 - (\alpha, \beta) - (\beta, \alpha) + \| \beta \|^2. \quad (2)$$

Again

$$\begin{aligned} i\| \alpha + i\beta \|^2 &= i(\alpha + i\beta, \alpha - i\beta) = i\left[\| \alpha \|^2 - i(\alpha, \beta) + i(\beta, \alpha) + \| \beta \|^2 \right] \\ &= i\| \alpha \|^2 + (\alpha, \beta) - (\beta, \alpha) + i\| \beta \|^2. \end{aligned} \quad (3)$$

Also

$$\begin{aligned} i\| \alpha - i\beta \|^2 &= i(\alpha - i\beta, \alpha + i\beta) = i\left[\| \alpha \|^2 + i(\alpha, \beta) - i(\beta, \alpha) + \| \beta \|^2 \right] \\ &= i\| \alpha \|^2 - (\alpha, \beta) + (\beta, \alpha) + i\| \beta \|^2. \end{aligned} \quad (4)$$

Hence, the given RHS expression, by using (1) to (4), is $4(\alpha, \beta)$.

EXAMPLE 5.12. *Using Schwarz's inequality, prove the Cauchy inequality*

$$\left(\sum a_i b_i \right)^2 \le \left(\sum a_i \right)^2 \left(\sum b_i \right)^2, \quad (1)$$

where a_i, b_i $(i = 1, 2, \ldots, n)$ are all real numbers.

Solution: Let α, β be two vectors in V over the field of real numbers, so that

$$\alpha = (a_1, a_2, \ldots, a_n)$$

$$\text{and} \quad \beta = (b_1, b_2, \ldots, b_n).$$

Then, we have the inner product of

$$(\alpha, \beta) = \sum_{i=1}^{n} a_i b_i. \qquad \therefore \quad |(\alpha, \beta)|^2 = \left| \sum a_i b_i \right|^2. \quad (2)$$

Again

$$\| \alpha \|^2 = (\alpha, \alpha) = |a_1|^2 + |a_2|^2 + \cdots + |a_n|^2 \quad (3)$$

$$\text{and} \quad \| \beta \|^2 = (\beta, \beta) = |b_1|^2 + |b_2|^2 + \cdots + |b_n|^2. \quad (4)$$

We have, by Schwarz's inequality

$$|(\alpha, \beta)|^2 \le \| \alpha \|^2 \| \beta \|^2.$$

Hence, from (1), (2), (3), we get

$$(a_1 b_1 + a_2 b_2 + \cdots + a_n b_n)^2 \le \left(a_1^2 + a_2^2 + \cdots + a_n \right)^2 \times \left(b_1^2 + b_2^2 + \cdots + b_n \right)^2.$$

5.8 Orthogonal Set of Vectors

Two vectors ψ, ϕ in an inner product space are said to be orthogonal if

$$(\psi, \phi) = 0. \tag{5.39}$$

Let, S be a set of vectors in an inner product space. If any two distinct vectors of the set are orthogonal, then we say that the set S is orthogonal.

If ψ and ϕ are two orthogonal vectors in a Euclidean space, then

$$\begin{aligned} \| \psi + \phi \|^2 &= (\psi + \phi,\, \psi + \phi) = (\psi, \psi) + 2(\psi, \phi) + (\phi, \phi) \\ &= \| \psi \|^2 + \| \phi \|^2 \quad [\because (\psi, \phi) = 0] \end{aligned} \tag{5.40}$$

This is known as *Pythagoras theorem.*

An important property of orthogonal set of non-null vectors is that they are linearly independent. Let us prove this.

Let, $\{\psi_i\}$, $i = 1, 2, \ldots, m$ be an arbitrary finite subset containing m distinct vectors of an orthogonal set S of non-null vectors in an inner product space V.

Let there be scalars a_i, $i = 1, 2, \ldots, m$ in F, such that

$$\sum_{i=1}^{m} a_i \psi_i = 0. \tag{5.41}$$

Now, we have

$$\begin{aligned} \left(\sum_{i=1}^{m} a_i \psi_i, \psi_j \right) &= \sum_{i=1}^{m} a_i (\psi_i,\, \psi_j) \quad (\text{for } 1 \leq j \leq m) \\ &= a_j\, (\psi_j, \psi_j), \quad \text{since } (\psi_i, \psi_j) = 0 \text{ for } i \neq j \\ &= a_j \| \psi_j \|^2. \end{aligned} \tag{5.42}$$

Again, from equation (5.41), we have

$$\left(\sum_{i=1}^{m} a_i \psi_i, \psi_j \right) = (0, \psi_j) = 0. \tag{5.43}$$

From equations (5.42) and (5.43), we get

$$a_j \| \psi_j \|^2 = 0, \quad j \leq m. \tag{5.44}$$

Since $\| \psi_j \|^2 \neq 0$ as ψ_j is a non-null vector, equation (5.44) implies that

$$a_j = 0. \tag{5.45}$$

Therefore, we conclude that the subset $\{\psi_i\}$, $i = 1, 2, \ldots, m$ is linearly independent. Since the subset is arbitrary, every subset of S should be linearly independent. Hence, the set S is linearly independent.

5.9 Orthonormal Set of Vectors

An orthogonal set of vectors in which each element has norm unity is called *orthonormal set.* Thus, the set $\{\psi_i\}$ of vectors in an inner product space is orthonormal, if (i) $\| \psi_i \|^2 = 1$ and (ii) $(\psi_i, \psi_j) = 0$, $i \neq j$, i.e., $(\psi_i, \psi_j) = \delta_{ij}$, where

$$\delta_{ij} = \begin{cases} 1, & \text{if } i = j; \\ 0, & \text{if } i \neq j. \end{cases} \tag{5.46}$$

δ_{ij} is known as *Kronecker delta.*

Thus, if $\{\phi_i\}$, $i = 1, 2, \ldots, m$ be a set of non-null orthogonal vectors, then the set of vectors $\{\xi_i\}$, $i = 1, 2, \ldots, m$ will be orthonormal, where

$$\xi_i = \frac{\phi_i}{\| \phi_i \|}. \tag{5.47}$$

Therefore, as in the case of orthogonal sets of non-null vectors, any orthonormal set of vectors in an inner product space is linearly independent.

5.10 Gram-Schmidt Process of Orthonormalisation

It is always possible to construct an orthonormal set $\{\alpha_1, \alpha_2, \ldots, \alpha_n\}$ from a linearly independent set $\{\psi_1, \psi_2, \ldots, \psi_n\}$. We shall discuss the method of orthonormalisation known as *Gram-Schmidt process.*

The set $\{\psi_i\}$ being linearly independent none of the vectors is a null vector.

We define

$$\gamma_1 = \psi_1.$$

Then

$$\alpha_1 = \frac{\gamma_1}{\| \gamma_1 \|} = \frac{\psi_1}{\| \psi_1 \|} \tag{5.48}$$

is of unit norm.

Now, we define

$$\gamma_2 = \psi_2 - (\alpha_1, \psi_2)\, \alpha_1. \tag{5.49}$$

Then

$$\alpha_2 = \frac{\gamma_2}{\| \gamma_2 \|}$$

is of unit norm.

Again, we define

$$\gamma_3 = \psi_3 - (\alpha_1, \psi_3)\alpha_1 - (\alpha_2, \psi_3)\alpha_2. \tag{5.50}$$

Then

$$\alpha_3 = \frac{\gamma_3}{\| \gamma_3 \|}$$

is of unit norm.

Thus, in general, we define

$$\gamma_i = \psi_i - \sum_{k=1}^{i-1} (\alpha_k, \psi_i)\alpha_k, \quad i > 1. \tag{5.51}$$

Then

$$\alpha_i = \frac{\gamma_i}{\| \gamma_i \|}$$

is of unit norm.

Thus, as long as $\psi_i \neq 0$, the continued process as above will give vectors of unit norm.

Again, we find that

$$(\alpha_2, \alpha_1) = \frac{1}{\| \gamma_2 \|} (\gamma_2, \alpha_1)$$

$$= \frac{1}{\| \gamma_2 \|} [(\psi_2, \alpha_1) - (\alpha_1, \psi_2) \cdot (\alpha_1, \alpha_1)] = 0.$$

Thus, α_2 is orthogonal to α_1. In the same way, we can prove that

$$(\alpha_3, \alpha_2) = 0 \quad \text{and} \quad (\alpha_3, \alpha_1) = 0$$

so that $(\alpha_1, \alpha_2, \alpha_3)$ is an orthonormal set. The method of proof may be generalized by induction, so that the set of vectors $\{\alpha_i\}$ obtained by the above process will be orthonormal.

EXAMPLE 5.13. *Construct an orthonormal set from*

$$\psi_1 = \begin{bmatrix} 1 \\ 1 \\ 1 \end{bmatrix}, \quad \psi_2 = \begin{bmatrix} 1 \\ 0 \\ 1 \end{bmatrix}, \quad \psi_3 = \begin{bmatrix} -1 \\ -2 \\ 1 \end{bmatrix}.$$

Solution: Following the steps of Section 5.10, we write

$$\| \psi_1 \| = \sqrt{3}, \quad \text{so} \quad \alpha_1 = \frac{1}{\sqrt{3}} \begin{bmatrix} 1 \\ 1 \\ 1 \end{bmatrix}. \tag{1}$$

$$\gamma_2 = \psi_2 - (\alpha_1, \psi_2)\alpha_1 = \begin{bmatrix} 1 \\ 0 \\ 1 \end{bmatrix} - \begin{bmatrix} 1/\sqrt{3} \\ 1/\sqrt{3} \\ 1/\sqrt{3} \end{bmatrix} \begin{bmatrix} 1 \\ 0 \\ 1 \end{bmatrix} \alpha_1$$

$$= \begin{bmatrix} 1 \\ 0 \\ 1 \end{bmatrix} - \frac{2}{\sqrt{3}} \cdot \frac{1}{\sqrt{3}} \begin{bmatrix} 1 \\ 1 \\ 1 \end{bmatrix} = \begin{bmatrix} 1 \\ 0 \\ 1 \end{bmatrix} - \frac{2}{3} \begin{bmatrix} 1 \\ 1 \\ 1 \end{bmatrix} = \frac{1}{3} \begin{bmatrix} 1 \\ 2 \\ 1 \end{bmatrix}.$$

$$\therefore \quad \| \gamma_2 \| = \frac{\sqrt{2}}{\sqrt{3}}. \quad \therefore \quad \alpha_2 = \frac{1}{\sqrt{6}} \begin{bmatrix} 1 \\ 2 \\ 1 \end{bmatrix}. \tag{2}$$

Again

$$\gamma_3 = \psi_3 - (\alpha_1, \psi_3)\, \alpha_1 - (\alpha_2, \psi_3)\, \alpha_2.$$

Now

$$(\alpha_1, \psi_3) = \begin{bmatrix} 1/\sqrt{3} \\ 1/\sqrt{3} \\ 1/\sqrt{3} \end{bmatrix} \cdot \begin{bmatrix} -1 \\ -2 \\ 1 \end{bmatrix} = -\frac{2}{\sqrt{3}}.$$

$$\therefore \quad (\alpha_1, \psi_3)\alpha_1 = -\frac{2}{\sqrt{3}}\begin{bmatrix} 1/\sqrt{3} \\ 1/\sqrt{3} \\ 1/\sqrt{3} \end{bmatrix} = -\frac{2}{3}\begin{bmatrix} 1 \\ 1 \\ 1 \end{bmatrix}.$$

Again

$$(\alpha_2, \psi_3) = \frac{1}{\sqrt{6}}\begin{bmatrix} 1 \\ 2 \\ 1 \end{bmatrix} \cdot \begin{bmatrix} -1 \\ -2 \\ 1 \end{bmatrix} = -\frac{4}{\sqrt{6}}.$$

$$\therefore \quad (\alpha_2, \psi_3)\,\alpha_2 = -\frac{4}{\sqrt{6}} \cdot \frac{1}{\sqrt{6}}\begin{bmatrix} 1 \\ 2 \\ 1 \end{bmatrix} = -\frac{2}{3}\begin{bmatrix} 1 \\ 2 \\ 1 \end{bmatrix}.$$

$$\therefore \quad \gamma_3 = \begin{bmatrix} -1 \\ -2 \\ 1 \end{bmatrix} + \frac{2}{3}\begin{bmatrix} 1 \\ 1 \\ 1 \end{bmatrix} + \frac{2}{3}\begin{bmatrix} 1 \\ 2 \\ 1 \end{bmatrix} = \frac{1}{3}\begin{bmatrix} 1 \\ 0 \\ 7 \end{bmatrix}.$$

$$\therefore \quad \| \gamma_3 \| = \frac{\sqrt{50}}{3}, \qquad \therefore \quad \alpha_3 = \frac{\gamma_3}{\| \gamma_3 \|} = \frac{1}{\sqrt{50}}\begin{bmatrix} 1 \\ 0 \\ 7 \end{bmatrix}.$$

EXAMPLE 5.14. *Construct an orthonormal base for R^3 out of the vectors*

$$\psi_1 = (1, 2, 2), \quad \psi_2 = (1, -1, 2), \quad \psi_3 = (1, 0, 1).$$

Solution:

$$\psi_1 = (1, 2, 2); \quad \therefore \quad \| \psi_1 \| = 3.$$

$$\therefore \quad \alpha_1 = \frac{\psi_1}{\| \psi_1 \|} = \frac{1}{3}(1, 2, 2) = \left(\frac{1}{3}, \frac{2}{3}, \frac{2}{3}\right). \tag{1}$$

$$\gamma_2 = \psi_2 - (\alpha_1, \psi_2)\,\alpha_1 = (1, -1, 2) - \left(\frac{1}{3}, \frac{2}{3}, \frac{2}{3}\right) \cdot (1, -1, 2)\,\alpha_1$$

$$= (1, -1, 2) - \left(\frac{1}{3}, \frac{2}{3}, \frac{2}{3}\right) = \left(\frac{2}{3}, -\frac{5}{3}, \frac{4}{3}\right).$$

$$\therefore \quad \| \gamma_2 \|^2 = \frac{1}{9}(4 + 25 + 16) = 5, \qquad \therefore \quad \| \gamma_2 \| = \sqrt{5}.$$

$$\therefore \quad \alpha_2 = \frac{1}{\sqrt{5}}\left(\frac{2}{3}, -\frac{5}{3}, \frac{4}{3}\right).$$

Again

$$\gamma_3 = \psi_3 - (\alpha_1, \psi_3)\,\alpha_1 - (\alpha_2, \psi_3)\,\alpha_2.$$

Now

$$(\alpha_1, \psi_3)\alpha_1 = \left(\frac{1}{3}, \frac{2}{3}, \frac{2}{3}\right) \cdot (1, 0, 1)\alpha_1 = \alpha_1 = \left(\frac{1}{3}, \frac{2}{3}, \frac{2}{3}\right).$$

Also

$$(\alpha_2, \psi_3)\,\alpha_2 = \frac{1}{\sqrt{5}}\left(\frac{2}{3}, -\frac{5}{3}, \frac{4}{3}\right) \cdot (1, 0, 1)\,\alpha_2 = \frac{2}{\sqrt{5}} \cdot \frac{1}{\sqrt{5}}\left(\frac{2}{3}, -\frac{5}{3}, \frac{4}{3}\right) = \frac{2}{5}\left(\frac{2}{3}, -\frac{5}{3}, \frac{4}{3}\right).$$

$$\therefore \quad \gamma_3 = (1, 0, 1) - \left(\frac{1}{3}, \frac{2}{3}, \frac{2}{3}\right) - \frac{2}{5}\left(\frac{2}{3}, -\frac{5}{3}, \frac{4}{3}\right) = \left(\frac{2}{5}, 0, -\frac{1}{5}\right).$$

$$\therefore \quad \| \gamma_3 \| = \frac{1}{\sqrt{5}}, \quad \therefore \quad \frac{\gamma_3}{\| \gamma_3 \|} = \frac{\sqrt{5}}{5}(2, 0, -1).$$

$$\therefore \quad \alpha_3 = \frac{1}{\sqrt{5}}(2, 0, -1).$$

Hence, the required orthonormal basis for R^3 is

$$\left\{ \left(\frac{1}{3}, \frac{2}{3}, \frac{2}{3} \right), \quad \frac{1}{\sqrt{5}} \left(\frac{2}{3}, -\frac{5}{3}, \frac{4}{3} \right), \quad \frac{1}{\sqrt{5}}(2, 0, -1) \right\}.$$

Note. We have so far denoted the inner product of two vectors ψ and ϕ as (ψ, ϕ). But in quantum physics, it is denoted as $\langle \psi \mid \phi \rangle$, where $\langle \phi \mid$ is called *bra* vector and $\mid \phi \rangle$ is called *ket* vector. This type of notation is due to *Dirac*.

5.11 Linear Operators

An operator is a device for obtaining a unique vector ϕ_i in the vector space V_n for each ψ_i in V_n. Thus, if A be an operator, then

$$\phi_i = A\psi_i. \tag{5.52}$$

In Dirac's notation, we write

$$\mid \phi \rangle = A \mid \psi \rangle. \tag{5.53}$$

An operator A is said to be linear, if

$$A(\alpha\psi + \beta\phi) = \alpha A\psi + \beta A\phi, \tag{5.54}$$

where α, β are scalars.

Let us consider the operator

$$A(x, y) = (x, y).$$

Then, for scalars α and β

$$\begin{aligned} A[\alpha(x_1, y_1) + \beta(x_2, y_2)] &= A[\alpha x_1 + \beta x_2, \alpha y_1 + \beta y_2] = (\alpha x_1 + \beta x_2, \alpha y_1 + \beta y_2) \\ &= \alpha(x_1, y_1) + \beta(x_2, y_2) = \alpha A(x_1, y_1) + \beta A(x_2, y_2). \end{aligned}$$

This shows that the operator A is linear.

5.12 Linear Differential Operators

Operators in terms of derivatives of a function is called a *differential operator.* As an example, the expression

$$L(u) \equiv \frac{d^2u}{dx^2} + p(x)\frac{du}{dx} + q(x)\,u \tag{5.55}$$

always defines a function $R(x)$ for each permissible input function $u(x)$. Thus, we call L a *differential operator.*

Given any two functions $u_1(x)$ and $u_2(x)$, and two scalars a_1 and a_2, we find that

$$\begin{aligned}L\,[a_1u_1(x) + a_2u_2(x)] &= \frac{d^2}{dx^2}\,[a_1u_1(x) + a_2u_2(x)] + p(x)\frac{d}{dx}[a_1u_1(x) \\ &\quad + a_2u_2(x)] + q(x)\,[a_1u_1(x) + a_2u_2(x)] \\ &= a_1\left[\frac{d^2u_1}{dx^2} + p(x)\frac{du_1}{dx} + q(x)u_1\right] \\ &\quad + a_2\left[\frac{d^2u_2}{dx^2} + p(x)\frac{du_2}{dx} + q(x)u_2\right] \\ &= a_1L(u_1) + a_2L(u_2). \end{aligned} \tag{5.56}$$

Thus, the operator L is linear.

From equation (5.56), we get the following consequences: (i) $L[u + v] = L[u] + L[v]$, (ii) $L[\xi u] = \xi L[u]$, where ξ = a constant.

5.13 Arithmetics of Linear Operators

Let us consider two linear operators A and B in the vector space V.

Sum of operators. The sum (or difference) of the operators A and B is denoted by $A \pm B$ and is defined as

$$(A \pm B)\psi = A\psi \pm B\psi, \tag{5.57}$$

where ψ is any vector in V.

Product of operators. The product of the operators A and B is denoted by AB and is defined as

$$AB\psi = A(B\psi). \tag{5.58}$$

The product of operators is generally not commutative, i.e.,

$$AB \neq BA. \tag{5.59}$$

The difference $AB - BA$ is called *commutator* of A and B, and is symbolically written as

$$[A, B] = AB - BA. \tag{5.60}$$

Thus, if the commutator of A and B is zero, the operators A and B are commutative.

If A and B are linear operators, then it can be easily proved that their product is also a linear operator.

Inverse of an operator. If for an operator A

$$A\phi = \psi, \tag{5.61}$$

where ϕ and ψ are two vectors in V, then the inverse operator of A is denoted by A^{-1} and is defined as an operator, such that

$$A^{-1}\psi = \phi. \tag{5.62}$$

EXAMPLE 5.15. *Find the commutators of two operators*

$$A[u(x)] = \frac{d}{dx}[u(x)] \quad \textit{and} \quad B[u(x)] = x^2\,u(x).$$

Solution: We have

$$A[B\,u(x)] = A[x^2\,u(x)] = \frac{d}{dx}[x^2\,u(x)] = x^2\frac{du}{dx} + 2xu. \tag{1}$$

Again

$$B[A\,u(x)] = B\left[\frac{du}{dx}\right] = x^2\,\frac{du}{dx}. \tag{2}$$

Thus, $[A, B]\,u(x) = 2xu(x)$.

EXAMPLE 5.16. *If A is a linear operator, show that its inverse is also linear.*

Solution: Let there be two vectors ϕ_1 and ϕ_2 for which there exist vectors ψ_1 and ψ_2, such that

$$\left.\begin{aligned} A\phi_1 &= \psi_1 \\ \text{and}\quad A\phi_2 &= \psi_2 \end{aligned}\right\}. \tag{1}$$

Now, since A is a linear operator, we can write

$$A(a_1\phi_1 + a_2\phi_2) = a_1 A\phi_1 + a_2 A\phi_2, \tag{2}$$

where a_1, a_2 are scalars.

From (2), we get by definition

$$a_1\phi_1 + a_2\phi_2 = A^{-1}(a_1 A\phi_1 + a_2 A\phi_2), \tag{3}$$

i.e.,

$$a_1 A^{-1}\psi_1 + a_2 A^{-1}\psi_2 = A^{-1}\,(a_1\psi_1 + a_2\psi_2) \quad \text{[from (1)]}. \tag{4}$$

This shows that A^{-1} is also a linear operator.

II. THEORY OF MATRICES

5.14 Introduction

Theory of matrices was born as a result of solving system of linear algebraic equations more than hundred years ago. But the use of matrices by scientists and engineers covering a spectrum of topics like statistics, differential equations, numerical solutions of problems of elasticity, etc., started around the early middle of twentieth century. In this section, we shall develop the theory of matrices covering terminologies and arithmetic operations on matrices.

A matrix is a rectangular array of quantities arranged in m rows and n columns like

$$A = \begin{bmatrix} a_{11} & a_{12} & \cdots & a_{1n} \\ a_{21} & a_{22} & \cdots & a_{2n} \\ \cdots & \cdots & \cdots & \cdots \\ \cdots & \cdots & \cdots & \cdots \\ a_{m1} & a_{m2} & \cdots & a_{mn} \end{bmatrix}, \tag{5.63}$$

where a_{ij} is an element of ith row and jth column. The matrix A above may also be denoted by

$$A = [a_{ij}], \quad \begin{pmatrix} i = 1, 2, \ldots, m \\ j = 1, 2, \ldots, n \end{pmatrix}. \tag{5.64}$$

The order of the matrix is $m \times n$ (we read as m by n).
Thus

$$A_1 = \begin{bmatrix} 1 & 1 & 0 \\ 5 & 7 & 2 \end{bmatrix}, \quad A_2 = \begin{bmatrix} 6 & 7 \\ 4 & 3 \\ 0 & 1 \end{bmatrix}, \quad A_3 = \begin{bmatrix} 1 & 2 & 9 \\ 6 & 0 & 5 \\ 2 & 9 & 1 \end{bmatrix}$$

are matrices of orders 2×3, 3×2 and 3×3 respectively.
The elements of matrices may be real or complex numbers or functions.

5.15 Terminologies

Row matrix. If there be only one row of elements in the matrix, it is called a *row matrix.* Thus

$$A = [3 \quad 0 \quad 2 \quad 9]$$

is a row matrix of order 1×4.
Column matrix. A matrix having elements in one column only is called a *column matrix.* Thus

$$A = \begin{bmatrix} 2 \\ 9 \\ 1 \\ 0 \end{bmatrix}$$

is a column matrix of order 4×1.
Null matrix. If all the elements of a matrix be zero, then it is called a *null matrix.* Thus

$$A = \begin{bmatrix} 0 & 0 & 0 \\ 0 & 0 & 0 \end{bmatrix}$$

is a null matrix of order 2×3.
Negative of matrix. The negative of a matrix A is denoted by $-A$ and is obtained when the signs of all the elements of A are reversed. Thus

$$\text{if} \quad A = \begin{bmatrix} 2 & -3 \\ 4 & 5 \end{bmatrix}, \quad \text{then} \quad -A = \begin{bmatrix} -2 & 3 \\ -4 & -5 \end{bmatrix}.$$

Transpose of a matrix. The transpose of a matrix A is obtained by changing its rows into columns and columns into rows. It is denoted by A' or A^T. Thus

$$\text{if} \quad A = \begin{bmatrix} 1 & 2 & 0 \\ 6 & -3 & 3 \\ 5 & 1 & 3 \end{bmatrix}, \quad \text{then} \quad A^T = \begin{bmatrix} 1 & 6 & 5 \\ 2 & -3 & 1 \\ 0 & 3 & 3 \end{bmatrix}.$$

Square matrix. If the number of rows of a matrix be equal to the number of its columns, then the matrix is called a *square matrix.* Thus

$$A = \begin{bmatrix} 2 & 1 & 1 \\ 3 & 6 & -2 \\ 4 & -9 & 0 \end{bmatrix}$$

is a square matrix of 3×3.

Diagonal matrix. A square matrix having all its non-diagonal elements as zero is called a *diagonal matrix.* Thus

$$A = \begin{bmatrix} 2 & 0 & 0 \\ 0 & 6 & 0 \\ 0 & 0 & -4 \end{bmatrix}$$

is a diagonal matrix of order 3×3.

Scalar matrix. If the elements of a diagonal matrix are all equal, then the matrix is called a *scalar matrix.* Thus

$$A = \begin{bmatrix} 6 & 0 \\ 0 & 6 \end{bmatrix}$$

is a scalar matrix of order 2×2.

Unit matrix. If the elements of a diagonal matrix are all equal to unity, then the matrix is called a *unit* (or *identity*) *matrix.* It is denoted by I. Thus

$$I = \begin{bmatrix} 1 & 0 & 0 \\ 0 & 1 & 0 \\ 0 & 0 & 1 \end{bmatrix}$$

is a unit matrix of order 3×3.

Determinant of a matrix. The determinant whose elements are corresponding elements of a square matrix A is called the *determinant of matrix* A and is denoted by $\det A$ or $|A|$. Thus, if

$$A = \begin{bmatrix} 1 & 7 & 2 \\ 0 & 9 & -6 \\ 2 & -3 & 4 \end{bmatrix}$$

then $\det A$ or $|A| = \begin{vmatrix} 1 & 7 & 2 \\ 0 & 9 & -6 \\ 2 & -3 & 4 \end{vmatrix}$.

Singular matrix. A square matrix A is called a *singular matrix,* if $\det A = 0$. Thus

$$A = \begin{bmatrix} 2 & 5 & 19 \\ 1 & -2 & -4 \\ -3 & 2 & 0 \end{bmatrix}$$

is a singular matrix, since

$$\det A = \begin{vmatrix} 2 & 5 & 19 \\ 1 & -2 & -4 \\ -3 & 2 & 0 \end{vmatrix} = 0.$$

If $\det A \neq 0$, then the matrix is called a *regular* or *non-singular* matrix.

Triangular matrices. Square matrices whose elements below the diagonal are zero are called *upper triangular matrices.* Thus

$$U_1 = \begin{vmatrix} 0 & 1 & 2 \\ 0 & -1 & 3 \\ 0 & 0 & 6 \end{vmatrix}, \quad U_2 = \begin{vmatrix} 0 & 5 \\ 0 & -7 \end{vmatrix}, \quad U_3 = [9]$$

are all upper triangular matrices.

Similarly, square matrices whose elements above the diagonal are zero, are called *lower triangular matrices.*
Thus

$$L_1 = \begin{bmatrix} 0 & 0 & 0 \\ 1 & -1 & 0 \\ 2 & 3 & 6 \end{bmatrix}, \quad L_2 = \begin{bmatrix} 0 & 0 \\ 5 & -7 \end{bmatrix}$$

are all lower triangular matrices.
Banded matrix. A square matrix having a small width of elements both above and below the diagonal and all the remaining elements are zero, is called a *banded matrix.* Thus

$$A = \begin{bmatrix} 2 & 3 & 0 & 0 & 0 \\ 0 & 6 & 7 & 0 & 0 \\ 0 & 0 & 1 & -2 & 0 \\ 0 & 0 & 0 & 4 & 0 \\ 0 & 0 & 0 & 0 & -8 \end{bmatrix}$$

is a banded matrix with bandwidth of 2.
Symmetric matrix. If a square matrix A is such that

$$A^T = A, \tag{5.65}$$

then the matrix A is called a *symmetric matrix.* Thus

$$A = \begin{bmatrix} 2 & -4 & 6 \\ -4 & 3 & 8 \\ 6 & 8 & -7 \end{bmatrix}$$

is a symmetric matrix.
In general, the elements of such a matrix are related by $a_{ij} = a_{ji}$ for all i and j.
Skew-symmetric matrix. A square matrix is said to be skew-symmetric, if its transpose is equal to its negative. That is

$$A^T = -A. \tag{5.66}$$

Hence elements of such a matrix is related by $a_{ij} = -a_{ji}$ for all i and j.
Thus

$$A = \begin{bmatrix} 0 & 2 & 3 \\ -2 & 1 & -4 \\ -3 & 4 & 7 \end{bmatrix}$$

is a skew-symmetric matrix.
Any square matrix can be expressed as the sum of a symmetric and a skew-symmetric matrix in the following manner:

$$A = \tfrac{1}{2}\left(A + A^T\right) + \tfrac{1}{2}\left(A - A^T\right), \tag{5.67}$$

where first part in RHS is a symmetric matrix and the second part is skew-symmetric.

5.16 Arithmetics of Matrices

1. *Equality of matrices.* Two matrices A and B are conformable to be equal, if and only if they are of same order.

Two matrices of same order are equal, if and only if the corresponding elements of the matrices are equal. Thus, matrices $[a_{ij}]$ and $[b_{ij}]$ both of $m \times n$ order are equal, if

$$a_{ij} = b_{ij} \tag{5.68}$$

for each pair of values of i and j. Thus, matrices

$$A = \begin{bmatrix} 1 & 3^2 & 2^2 \\ 2^2 & 7 & 4^2 \end{bmatrix} \quad \text{and} \quad B = \begin{bmatrix} 1 & 9 & 4 \\ 4 & 7 & 16 \end{bmatrix} \quad \text{are equal.}$$

2. *Addition and subtraction of matrices.* Two matrices of same order are only conformable to addition and subtraction.
The sum of two matrices of *same* order is the matrix whose elements are obtained by adding the corresponding elements of the given matrices.
Thus, if $[a_{ij}]$ and $[b_{ij}]$ be two matrices of order $m \times n$, then the sum of these matrices denoted by $[c_{ij}]$ will also be of order $m \times n$, such that

$$c_{ij} = a_{ij} + b_{ij} \tag{5.69}$$

for all values of i and j.

Thus, if $A = \begin{bmatrix} 0 & 2 & 3 \\ 4 & 6 & 9 \\ 1 & 8 & 7 \end{bmatrix}$, $B = \begin{bmatrix} 1 & 0 & 0 \\ 0 & 3 & 2 \\ 0 & 0 & 6 \end{bmatrix}$, then $C = A + B = \begin{bmatrix} 1 & 2 & 3 \\ 4 & 9 & 11 \\ 1 & 8 & 13 \end{bmatrix}$.

The difference of two matrices A and B of same order is denoted by $A - B$ and is defined as the sum of A with negative of B, i.e., $A + (-B)$. Thus, with matrices A and B as above, we get

$$A - B = \begin{bmatrix} -1 & 2 & 3 \\ 4 & 3 & 7 \\ 1 & 8 & 1 \end{bmatrix}.$$

Addition of matrices obeys the following laws:

(a) Matrix addition is commutative. That is, for two matrices A and B of same order

$$A + B = B + A. \tag{5.70}$$

(b) Matrix addition is associative. That is, for three matrices A, B and C

$$A + (B + C) = (A + B) + C. \tag{5.71}$$

(c) The transpose of the sum of two matrices is equal to the sum of their transposes. That is, for two matrices A and B of same order

$$[A + B]^T = A^T + B^T.$$

3. *Multiplication of matrices.* If a matrix $[a_{ij}]$ of order $m \times n$ is multiplied by a scalar λ, then the product is the matrix $[b_{ij}]$ of same order, whose elements are

$$b_{ij} = \lambda a_{ij} \quad \text{for all } i \text{ and } j. \tag{5.72}$$

Thus

$$3 \times \begin{bmatrix} 6 & 0 & 4 & 1 \\ 2 & 9 & 1 & 0 \end{bmatrix} = \begin{bmatrix} 18 & 0 & 12 & 3 \\ 6 & 27 & 3 & 0 \end{bmatrix}.$$

It is obvious that scalar multiplication of matrices is commutative, associative and distributive. Moreover, if a matrix A is multiplied by a scalar λ, then transpose of the product λA is equal to the product of λ and transpose of A. That is

$$[\lambda A]^T = \lambda [A]^T. \tag{5.73}$$

Two matrices are conformable to the multiplication, if and only if the number of rows in one matrix will be equal to the number of columns in another. Thus, if

$$A = [a_{ij}]_{m\times n} \quad \text{and} \quad B = [b_{ij}]_{n\times p},$$

then the product

$$AB = [c_{ik}]_{m\times p}, \tag{5.74}$$

where $c_{ik} = \sum_{j=1}^{n} a_{ij} b_{jk}$.

Thus, the product matrix is of order $m \times p$, and its element at ith row and kth column is the sum of the product of elements of ith row of A and the corresponding elements of the kth column of B. Thus, in expanded form, if

$$A = \begin{bmatrix} a_{11} & a_{12} & \cdots & a_{1n} \\ \vdots & \vdots & & \vdots \\ a_{i1} & a_{i2} & \cdots & a_{in} \\ \vdots & \vdots & & \vdots \\ a_{m1} & a_{m2} & \cdots & a_{mn} \end{bmatrix} \quad \text{and} \quad B = \begin{bmatrix} b_{11} & b_{12} & \cdots & b_{1k} & \cdots & b_{1p} \\ \cdots & \cdots & \cdots & b_{2k} & \cdots & \cdots \\ \cdots & \cdots & \cdots & \vdots & \cdots & \cdots \\ b_{n1} & b_{n2} & \cdots & b_{nk} & \cdots & b_{np} \end{bmatrix},$$

then the element c_{ik} of the product of matrices A and B will be the sum of product of row shown in A and the corresponding elements of column shown in B. Thus

$$c_{ik} = a_{i1}b_{1k} + a_{i2}b_{2k} + \cdots + a_{in}b_{nk}. \tag{5.75}$$

In the product AB of two matrices A and B, A is called the *pre-factor* and B is called the *post-factor*. As an example, if

$$A = \begin{bmatrix} 2 \\ 3 \\ 4 \end{bmatrix} \quad \text{and} \quad B = [2 \;\; 1 \;\; 0],$$

then

$$AB = \begin{bmatrix} 2\cdot 2 & 2\cdot 1 & 2\cdot 0 \\ 3\cdot 2 & 3\cdot 1 & 3\cdot 0 \\ 4\cdot 2 & 4\cdot 1 & 4\cdot 0 \end{bmatrix} = \begin{bmatrix} 4 & 2 & 0 \\ 6 & 3 & 0 \\ 8 & 4 & 0 \end{bmatrix}.$$

Two following peculiarities in matrix multiplication are observed:

1. If $AB = 0$, we cannot say that either A or B is zero. As an example, if

$$A = \begin{bmatrix} 2 & 3 & 0 \\ -7 & 4 & 0 \\ 1 & 9 & 0 \end{bmatrix} \quad \text{and} \quad B = \begin{bmatrix} 0 & 0 & 0 \\ 0 & 0 & 0 \\ -3 & 2 & -7 \end{bmatrix},$$

then

$$AB = \begin{bmatrix} 0 & 0 & 0 \\ 0 & 0 & 0 \\ 0 & 0 & 0 \end{bmatrix}$$

although neither A nor B is zero. Similarly, if

$$A = \begin{bmatrix} 1 & 1 \\ -1 & -1 \end{bmatrix} \quad \text{and} \quad B = \begin{bmatrix} 1 & 1 \\ -1 & -1 \end{bmatrix},$$

then $AB = 0$, but neither A nor B is a zero matrix.

2. If, for three matrices A, B, C, there exists a relation like

$$AB = AC, \tag{5.76}$$

we cannot say that $B = C$, even if $A \neq 0$. As an example, if

$$A = \begin{bmatrix} 2 & 3 & 0 \\ -7 & 4 & 0 \\ 1 & 9 & 0 \end{bmatrix}, \quad B = \begin{bmatrix} 1 & -7 & 2 \\ 2 & -7 & 1 \\ 4 & 5 & 6 \end{bmatrix} \quad \text{and} \quad C = \begin{bmatrix} 1 & -7 & 2 \\ 2 & -7 & 1 \\ 6 & -4 & 2 \end{bmatrix},$$

then,

$$AB = \begin{bmatrix} 8 & -35 & 0 \\ 1 & 21 & 0 \\ 19 & -70 & 0 \end{bmatrix} = AC$$

but, as we can see, $B \neq C$, and $A \neq 0$.

Properties of Matrix Multiplication

1. Matrix multiplication is not commutative in general. That is, if A and B are two matrices conformable to multiplication, then

$$AB \neq BA \qquad \text{in general.} \tag{5.77}$$

In some special cases, $AB = BA$. For example, if

$$A = \begin{bmatrix} 2 & 3 \\ 4 & 5 \end{bmatrix} \quad \text{and} \quad B = \begin{bmatrix} 1 & 0 \\ 0 & 1 \end{bmatrix},$$

then, $AB = BA$.
Again, if

$$A = \begin{bmatrix} 3 & 0 \\ 0 & 4 \end{bmatrix} \quad \text{and} \quad B = \begin{bmatrix} -1 & 0 \\ 0 & 1 \end{bmatrix}$$

then $AB = BA$.
In such cases, matrices A and B are said to commute with each other.

2. Matrix multiplication is associative. That is, if A, B, C are three matrices conformable for the products AB and BC, then

$$(AB) \cdot C = A \cdot (BC). \tag{5.78}$$

A matrix A is said to be involuntary if

$$A^2 = I. \tag{5.79}$$

3. Matrix multiplication is distributive with respect to addition. That is, if A, B, C be three matrices conformable to the sum $B + C$ and to product $A \cdot B$ and $A \cdot C$, then
$$A \cdot (B + C) = AB + AC. \tag{5.80}$$

4. In matrix multiplication, a scalar matrix behaves like a scalar multiplier. That is, if α be a scalar matrix of order $m \times m$, and A be a matrix of $m \times m$, then
$$[\alpha]\,[A] = \alpha\,[A]. \tag{5.81}$$

5. The transpose of the product of two matrices is the product of their transposes taken in the reversed order. That is, if A and B are two matrices conformable to the multiplication, then
$$[AB]^T = B^T A^T. \tag{5.82}$$

EXAMPLE 5.17. *Find A when*
$$\begin{bmatrix} 4 & 3 & 2 \\ -5 & 1 & 3 \end{bmatrix} + 5A = \begin{bmatrix} -1 & 3 & 7 \\ 5 & 6 & 3 \end{bmatrix}.$$

Then show that, $[A^T]^T = A$.

Solution: We have
$$\begin{bmatrix} 4 & 3 & 2 \\ -5 & 1 & 3 \end{bmatrix} + 5A = \begin{bmatrix} -1 & 3 & 7 \\ 5 & 6 & 3 \end{bmatrix}.$$
$$\therefore\quad 5A = \begin{bmatrix} -1 & 3 & 7 \\ 5 & 6 & 3 \end{bmatrix} - \begin{bmatrix} 4 & 3 & 2 \\ -5 & 1 & 3 \end{bmatrix} = \begin{bmatrix} -5 & 0 & 5 \\ 10 & 5 & 0 \end{bmatrix}.$$
$$\therefore\quad A = \begin{bmatrix} -1 & 0 & 1 \\ 2 & 1 & 0 \end{bmatrix};\quad \therefore\quad A^T = \begin{bmatrix} -1 & 2 \\ 0 & 1 \\ 1 & 0 \end{bmatrix}.$$
$$\therefore\quad [A^T]^T = \begin{bmatrix} -1 & 0 & 1 \\ 2 & 1 & 0 \end{bmatrix} = A.$$

Hence proved.

EXAMPLE 5.18. *Find for what values of x, the matrix*
$$A = \begin{bmatrix} 3-x & 2 & 2 \\ 2 & 4-x & 1 \\ -2 & -4 & -1-x \end{bmatrix}$$
is singular.

Solution: The matrix A will be singular when $\det A = 0$.

Now
$$\begin{aligned} \det A &= (3-x)\{(4-x)(-1-x)+4\} - 2\{2(-1-x)+2\} + 2\{-8+2(4-x)\} \\ &= (3-x)(x^2-3x) + 4x - 4x \\ &= -x(x-3)(x-3). \end{aligned}$$

Hence, $\det A = 0$ for $x = 0$ and 3 when the matrix A will be singular.

EXAMPLE 5.19. *Find A and B, when*

$$2A+B=\begin{bmatrix}2&2&5\\5&4&3\\1&1&4\end{bmatrix} \text{ and } A-2B=\begin{bmatrix}1&6&5\\5&2&-1\\-2&-2&2\end{bmatrix}.$$

Solution: We have

$$2A+B=\begin{bmatrix}2&2&5\\5&4&3\\1&1&4\end{bmatrix} \tag{1}$$

$$\text{and} \quad A-2B=\begin{bmatrix}1&6&5\\5&2&-1\\-2&-2&2\end{bmatrix}. \tag{2}$$

Multiplying (1) by 2, and then adding to (2), we have

$$2(2A+B)+(A-2B)=2\begin{bmatrix}2&2&5\\5&4&3\\1&1&4\end{bmatrix}+\begin{bmatrix}1&6&5\\5&2&-1\\-2&-2&2\end{bmatrix}$$

$$=\begin{bmatrix}4&4&10\\10&8&6\\2&2&8\end{bmatrix}+\begin{bmatrix}1&6&5\\5&2&-1\\-2&-2&2\end{bmatrix},$$

$$\text{or,} \quad 5A=\begin{bmatrix}5&10&15\\15&10&5\\0&0&10\end{bmatrix}, \quad \therefore \quad A=\begin{bmatrix}1&2&3\\3&2&1\\0&0&2\end{bmatrix}.$$

From (1), we get

$$B=\begin{bmatrix}2&2&5\\5&4&3\\1&1&4\end{bmatrix}-2\begin{bmatrix}1&2&3\\3&2&1\\0&0&2\end{bmatrix}=\begin{bmatrix}2&2&5\\5&4&3\\1&1&4\end{bmatrix}-\begin{bmatrix}2&4&6\\6&4&2\\0&0&4\end{bmatrix}$$

$$=\begin{bmatrix}0&-2&-1\\-1&0&1\\1&1&0\end{bmatrix}.$$

EXAMPLE 5.20. *If* $\begin{bmatrix}4\\1\\3\end{bmatrix}A=\begin{bmatrix}-4&8&4\\-1&2&1\\-3&6&3\end{bmatrix}$, *then find A.*

Solution: Since $\begin{bmatrix}4\\1\\3\end{bmatrix}$ is a 3×1 matrix and the product with A is a 3×3 matrix, A should be a 1×3 matrix. Let

$$A=[x \quad y \quad z]. \tag{1}$$

Then

$$\begin{bmatrix}4\\1\\3\end{bmatrix}[x \quad y \quad z]=\begin{bmatrix}-4&8&4\\-1&2&1\\-3&6&3\end{bmatrix}.$$

$$\therefore \quad 4x = -4, \quad 4y = 8, \quad 4z = 4$$

$$\therefore \quad x = -1, \quad y = 2, \quad z = 1$$

$\therefore \quad A = [-1 \quad 2 \quad 1].$

EXAMPLE 5.21. *Express the matrix* $A = \begin{bmatrix} 2 & 1 & 0 \\ 1 & -1 & -2 \\ 4 & 2 & 0 \end{bmatrix}$ *as the sum of a symmetric and a skew-symmetric matrix.*

Solution: We have

$$A = \begin{bmatrix} 2 & 1 & 0 \\ 1 & -1 & -2 \\ 4 & 2 & 0 \end{bmatrix}. \quad \therefore \quad A^T = \begin{bmatrix} 2 & 1 & 4 \\ 1 & -1 & 2 \\ 0 & -2 & 0 \end{bmatrix}.$$

$\therefore$ Symmetric part of matrix A is

$$A_s = \frac{1}{2}(A + A^T) = \frac{1}{2}\begin{bmatrix} 2 & 1 & 0 \\ 1 & -1 & -2 \\ 4 & 2 & 0 \end{bmatrix} + \frac{1}{2}\begin{bmatrix} 2 & 1 & 4 \\ 1 & -1 & 2 \\ 0 & -2 & 0 \end{bmatrix} = \begin{bmatrix} 2 & 1 & 2 \\ 1 & -1 & 0 \\ 2 & 0 & 0 \end{bmatrix}.$$

The skew-symmetric part of A is

$$A_a = \frac{1}{2}(A - A^T) = \frac{1}{2}\begin{bmatrix} 2 & 1 & 0 \\ 1 & -1 & -2 \\ 4 & 2 & 0 \end{bmatrix} - \frac{1}{2}\begin{bmatrix} 2 & 1 & 4 \\ 1 & -1 & 2 \\ 0 & -2 & 0 \end{bmatrix} = \begin{bmatrix} 0 & 0 & -2 \\ 0 & 0 & -2 \\ 2 & 2 & 0 \end{bmatrix}.$$

Thus

$$A = A_s + A_a = \begin{bmatrix} 2 & 1 & 2 \\ 1 & -1 & 0 \\ 2 & 0 & 0 \end{bmatrix} + \begin{bmatrix} 0 & 0 & -2 \\ 0 & 0 & -2 \\ 2 & 2 & 0 \end{bmatrix} = \begin{bmatrix} 2 & 1 & 0 \\ 1 & -1 & -2 \\ 4 & 2 & 0 \end{bmatrix}$$

which is checked.

EXAMPLE 5.22. *Given two matrices*

$$A = \begin{bmatrix} 3 & 0 & 1 \\ 2 & -2 & 1 \\ 0 & 2 & 0 \end{bmatrix} \quad \text{and} \quad B = \begin{bmatrix} 2 & -1 & 1 \\ 1 & 0 & 0 \\ 2 & 0 & 1 \end{bmatrix}.$$

Show that, $AB \neq BA$.

Solution: We have

$$AB = \begin{bmatrix} 3 & 0 & 1 \\ 2 & -2 & 1 \\ 0 & 2 & 0 \end{bmatrix}\begin{bmatrix} 2 & -1 & 1 \\ 1 & 0 & 0 \\ 2 & 0 & 1 \end{bmatrix} = \begin{bmatrix} 8 & -3 & 4 \\ 4 & -2 & 3 \\ 2 & 0 & 0 \end{bmatrix} \tag{1}$$

$$\text{and} \quad BA = \begin{bmatrix} 2 & -1 & 1 \\ 1 & 0 & 0 \\ 2 & 0 & 1 \end{bmatrix}\begin{bmatrix} 3 & 0 & 1 \\ 2 & -2 & 1 \\ 0 & 2 & 0 \end{bmatrix} = \begin{bmatrix} 4 & 4 & 1 \\ 3 & 0 & 1 \\ 6 & 2 & 2 \end{bmatrix} \tag{2}$$

$$\therefore \quad AB \neq BA.$$

EXAMPLE 5.23. *If m be a positive integer, and*

$$A = \begin{bmatrix} 3 & -4 \\ 1 & -1 \end{bmatrix},$$

then show that

$$A^m = \begin{bmatrix} 1+2m & -4m \\ m & 1-2m \end{bmatrix}.$$

Solution: We have

$$A = \begin{bmatrix} 3 & -4 \\ 1 & -1 \end{bmatrix}.$$

$$\therefore \quad A^2 = A \cdot A = \begin{bmatrix} 3 & -4 \\ 1 & -1 \end{bmatrix}\begin{bmatrix} 3 & -4 \\ 1 & -1 \end{bmatrix}, \quad \text{i.e.,} \quad A^2 = \begin{bmatrix} 5 & -8 \\ 2 & -3 \end{bmatrix} = \begin{bmatrix} 1+2\cdot 2 & -4\cdot 2 \\ 2 & 1-2\cdot 2 \end{bmatrix}.$$

Again

$$A^3 = A^2 \cdot A = \begin{bmatrix} 5 & -8 \\ 2 & -3 \end{bmatrix}\begin{bmatrix} 3 & -4 \\ 1 & -1 \end{bmatrix},$$

$$\text{i.e.,} \quad A^3 = \begin{bmatrix} 7 & -12 \\ 3 & -5 \end{bmatrix} = \begin{bmatrix} 1+2\cdot 3 & -4\cdot 3 \\ 3 & 1-2\cdot 3 \end{bmatrix}.$$

Thus, by induction, we get

$$A^m = \begin{bmatrix} 1+2m & -4m \\ m & 1-2m \end{bmatrix}.$$

5.17 Properties of Symmetric Matrices

Property 1. The product of a matrix and its transpose is a symmetric matrix.

Proof. Let A be a matrix of order $m \times n$, so that its transpose A^T is of order $n \times m$. Therefore, AA^T will be a square matrix of order $m \times m$.

Now,

$$[AA^T]^T = [A^T]^T A^T = AA^T.$$

Hence, AA^T is symmetric. □

Property 2. The product of two symmetric matrices of same order is symmetric, if the product is commutative.

Proof. Let A and B are two symmetric matrices with their product being commutative. Hence

$$AB = BA. \tag{5.83}$$

Since product of A and B is symmetric,

$$\therefore \quad [AB]^T = BA, \quad \text{or,} \quad B^T A^T = BA = AB.$$

Hence, AB is symmetric. □

Property 3. If A be a square matrix, then $(A+A^T)$ is symmetric and $(A-A^T)$ is skew-symmetric.

Proof.

$$\left(A + A^T\right)^T = A^T + \left[A^T\right]^T = A^T + A.$$

Hence, $A + A^T$ is symmetric.

Again

$$\left(A - A^T\right)^T = A^T - A = -\left(A - A^T\right).$$

Hence, $A - A^T$ is skew-symmetric.

Thus, as we have noted earlier, every square matrix A can be expressed as the sum of a symmetric and a skew-symmetric matrix like

$$A = \tfrac{1}{2}\left(A + A^T\right) + \tfrac{1}{2}\left(A - A^T\right).$$

□

EXAMPLE 5.24. *Verify that* $[AB]^T = B^T A^T$, *where*

$$A = \begin{bmatrix} 1 & 2 \\ 2 & 3 \\ 3 & 4 \end{bmatrix} \quad \textit{and} \quad B = \begin{bmatrix} 2 & 3 & 0 \\ 1 & 2 & 3 \end{bmatrix}.$$

Solution: We have

$$AB = \begin{bmatrix} 1 & 2 \\ 2 & 3 \\ 3 & 4 \end{bmatrix} \begin{bmatrix} 2 & 3 & 0 \\ 1 & 2 & 3 \end{bmatrix} = \begin{bmatrix} 1\cdot 2 + 2\cdot 1 & 1\cdot 3 + 2\cdot 2 & 0 + 2\cdot 3 \\ 2\cdot 2 + 3\cdot 1 & 2\cdot 3 + 3\cdot 2 & 0 + 3\cdot 3 \\ 3\cdot 2 + 4\cdot 1 & 3\cdot 3 + 4\cdot 2 & 0 + 4\cdot 3 \end{bmatrix}$$

$$= \begin{bmatrix} 4 & 7 & 6 \\ 7 & 12 & 9 \\ 10 & 17 & 12 \end{bmatrix}.$$

$$\therefore \quad [AB]^T = \begin{bmatrix} 4 & 7 & 10 \\ 7 & 12 & 17 \\ 6 & 9 & 12 \end{bmatrix}. \tag{1}$$

Again

$$B^T A^T = \begin{bmatrix} 2 & 1 \\ 3 & 2 \\ 0 & 3 \end{bmatrix} \begin{bmatrix} 1 & 2 & 3 \\ 2 & 3 & 4 \end{bmatrix} = \begin{bmatrix} 2\cdot 1 + 1\cdot 2 & 2\cdot 2 + 1\cdot 3 & 2\cdot 3 + 1\cdot 4 \\ 3\cdot 1 + 2\cdot 2 & 3\cdot 2 + 2\cdot 3 & 3\cdot 3 + 2\cdot 4 \\ 0 + 3\cdot 2 & 0 + 3\cdot 3 & 0 + 3\cdot 4 \end{bmatrix}$$

$$= \begin{bmatrix} 4 & 7 & 10 \\ 7 & 12 & 17 \\ 6 & 9 & 12 \end{bmatrix}. \tag{2}$$

Thus, from (1) and (2), we get

$$[AB]^T = B^T A^T.$$

EXAMPLE 5.25. *If* $f(x) = x^2 - 5x + 6$ *and* $A = \begin{bmatrix} 2 & 0 & 1 \\ 2 & 1 & 3 \\ 1 & -1 & 0 \end{bmatrix}$, *then show that*

$$f(A) = \begin{bmatrix} 1 & -1 & -3 \\ -1 & -1 & -10 \\ -5 & 4 & 4 \end{bmatrix}.$$

Solution: We have, $A = \begin{bmatrix} 2 & 0 & 1 \\ 2 & 1 & 3 \\ 1 & -1 & 0 \end{bmatrix}$.

$$\therefore \quad A^2 = A \cdot A = \begin{bmatrix} 2 & 0 & 1 \\ 2 & 1 & 3 \\ 1 & -1 & 0 \end{bmatrix} \begin{bmatrix} 2 & 0 & 1 \\ 2 & 1 & 3 \\ 1 & -1 & 0 \end{bmatrix} = \begin{bmatrix} 5 & -1 & 2 \\ 9 & -2 & 5 \\ 0 & -1 & -2 \end{bmatrix}.$$

$$\therefore \quad f(A) = A^2 - 5A + 6I$$

$$= \begin{bmatrix} 5 & -1 & 2 \\ 9 & -2 & 5 \\ 0 & -1 & -2 \end{bmatrix} - 5 \begin{bmatrix} 2 & 0 & 1 \\ 2 & 1 & 3 \\ 1 & -1 & 0 \end{bmatrix} + 6 \begin{bmatrix} 1 & 0 & 0 \\ 0 & 1 & 0 \\ 0 & 0 & 1 \end{bmatrix}$$

$$= \begin{bmatrix} 5 & -1 & 2 \\ 9 & -2 & 5 \\ 0 & -1 & -2 \end{bmatrix} - \begin{bmatrix} 10 & 0 & 5 \\ 10 & 5 & 15 \\ 5 & -5 & 0 \end{bmatrix} + \begin{bmatrix} 6 & 0 & 0 \\ 0 & 6 & 0 \\ 0 & 0 & 6 \end{bmatrix}$$

$$= \begin{bmatrix} 1 & -1 & -3 \\ -1 & -1 & -10 \\ -5 & 4 & 4 \end{bmatrix}.$$

Hence proved.

EXAMPLE 5.26. *If* $A(\theta) = \begin{bmatrix} \cos\theta & -\sin\theta \\ \sin\theta & \cos\theta \end{bmatrix}$, *then show that*

$$A(\theta)A(\phi) = A(\phi)A(\theta) = A(\theta + \phi).$$

Solution: We have

$$A(\theta) = \begin{bmatrix} \cos\theta & -\sin\theta \\ \sin\theta & \cos\theta \end{bmatrix}. \quad \therefore \quad A(\phi) = \begin{bmatrix} \cos\phi & -\sin\phi \\ \sin\phi & \cos\phi \end{bmatrix}.$$

$$\therefore \quad A(\theta)\,A(\phi) = \begin{bmatrix} \cos\theta & -\sin\theta \\ \sin\theta & \cos\theta \end{bmatrix} \begin{bmatrix} \cos\phi & -\sin\phi \\ \sin\phi & \cos\phi \end{bmatrix}$$

$$= \begin{bmatrix} \cos\theta\cos\phi - \sin\theta\sin\phi & -\cos\theta\sin\phi - \sin\theta\cos\phi \\ \sin\theta\cos\phi + \cos\theta\sin\phi & -\sin\theta\sin\phi + \cos\theta\cos\phi \end{bmatrix}$$

$$= \begin{bmatrix} \cos(\theta+\phi) & -\sin(\theta+\phi) \\ \sin(\theta+\phi) & \cos(\theta+\phi) \end{bmatrix} = A(\theta + \phi). \tag{1}$$

Similarly

$$A(\phi)\,A(\theta) = \begin{bmatrix} \cos\phi & -\sin\phi \\ \sin\phi & \cos\phi \end{bmatrix} \begin{bmatrix} \cos\theta & -\sin\theta \\ \sin\theta & \cos\theta \end{bmatrix}$$

$$= \begin{bmatrix} \cos(\theta+\phi) & -\sin(\theta+\phi) \\ \sin(\theta+\phi) & \cos(\theta+\phi) \end{bmatrix} = A\,(\theta + \phi). \tag{2}$$

Hence, from (1) and (2), we get

$$A(\theta)\,A(\phi) = A(\phi)\,A(\theta) = A(\theta + \phi).$$

Hence proved.

5.18 Calculus of Matrices

Derivative of a matrix. If the elements of a matrix are functions of a variable x, then the derivative of the matrix with respect to that variable is equal to the matrix whose elements are derivatives of each element separately.

Thus

$$\text{if } A = \begin{bmatrix} x^3 & e^x & 3x \\ \log x & x^2 & 0 \end{bmatrix}, \quad \text{then } \frac{dA}{dx} = \begin{bmatrix} 3x^2 & e^x & 3 \\ 1/x & 2x & 0 \end{bmatrix}.$$

Integration of a matrix. The integral of a matrix with respect to a variable x is the matrix whose elements are integral of each element with respect to x separately.

Thus

$$\text{if } A = \begin{bmatrix} x & e^x & x^3 \\ e^{-x} & 0 & x^2 \end{bmatrix}$$

then

$$\int A\,dx = \begin{bmatrix} x^2/2 & e^x & x^4/4 \\ -e^{-x} & c & x^3/3 \end{bmatrix} + \begin{bmatrix} a_1 & a_2 & a_3 \\ a_4 & 0 & a_5 \end{bmatrix},$$

where a_1, a_2, a_3, a_4, a_5 and c are constants of integration.

5.19 Orthogonal Matrices

A square matrix A is said to be orthogonal if its transpose A^T when multiplied by itself produces a unit matrix. That is when

$$A^T \cdot A = I. \tag{5.84}$$

Thus, matrix $A = \begin{bmatrix} 1 & 0 \\ 0 & 1 \end{bmatrix}$ is orthogonal.

Also

$$\text{matrix} \quad A = \begin{bmatrix} \cos\theta & -\sin\theta \\ \sin\theta & \cos\theta \end{bmatrix}.$$

is orthogonal, since

$$A^T \cdot A = \begin{bmatrix} \cos\theta & \sin\theta \\ -\sin\theta & \cos\theta \end{bmatrix} \begin{bmatrix} \cos\theta & -\sin\theta \\ \sin\theta & \cos\theta \end{bmatrix} = \begin{bmatrix} 1 & 0 \\ 0 & 1 \end{bmatrix}.$$

The orthogonal matrices have the following properties:

Property 1. The orthogonal matrices are non-singular.

Proof. Let A be an orthogonal matrix. Then, $A^T \cdot A = I$.

$$\therefore \quad \det |A^T \cdot A| = |A^T||A| = |A|^2 = I = 1.$$

$$\therefore \quad |A| \neq 0.$$

Hence, A is non-singular. □

Property 2. The product of two orthogonal matrices is orthogonal.

Proof. Let A and B be two orthogonal matrices. Then

$$A^T \cdot A = I \quad \text{and} \quad B^T \cdot B = I.$$

Therefore,

$$\begin{aligned}[AB]^T \cdot [AB] &= \left[B^T \cdot A^T\right] \cdot [AB] = B^T \left[A^T \cdot A\right] B \\ &\qquad [\because \text{ matrix multiplication is associative}] \\ &= B^T I B = B^T B = I.\end{aligned}$$

Hence, matrix AB is orthogonal. □

Property 3. The transpose of an orthogonal matrix is orthogonal.

Proof. Let A be an orthogonal matrix. Then

$$A^T \cdot A = I. \tag{5.85}$$

We are to prove that A^T is orthogonal. That is

$$\left(A^T\right)^T \cdot A^T = I. \tag{5.86}$$

Now, the transpose of the transpose of a matrix is the matrix itself, i.e., $\left(A^T\right)^T = A$. Hence, we are to prove $A \cdot A^T = I$, which is true by equation (5.85). Hence, A^T is orthogonal. □

EXAMPLE 5.27. *Show that the matrix*

$$A = \frac{1}{3}\begin{bmatrix} -1 & 2 & -2 \\ -2 & 1 & 2 \\ 2 & 2 & 1 \end{bmatrix}$$

is orthogonal.

Solution: We have

$$A = \frac{1}{3}\begin{bmatrix} -1 & 2 & -2 \\ -2 & 1 & 2 \\ 2 & 2 & 1 \end{bmatrix} \quad \therefore \quad A^T = \frac{1}{3}\begin{bmatrix} -1 & -2 & 2 \\ 2 & 1 & 2 \\ -2 & 2 & 1 \end{bmatrix}.$$

$$\begin{aligned}\therefore \quad A^T \cdot A &= \frac{1}{9}\begin{bmatrix} -1 & -2 & 2 \\ 2 & 1 & 2 \\ -2 & 2 & 1 \end{bmatrix}\begin{bmatrix} -1 & 2 & -2 \\ -2 & 1 & 2 \\ 2 & 2 & 1 \end{bmatrix} \\ &= \frac{1}{9}\begin{bmatrix} 9 & 0 & 0 \\ 0 & 9 & 0 \\ 0 & 0 & 9 \end{bmatrix} = \begin{bmatrix} 1 & 0 & 0 \\ 0 & 1 & 0 \\ 0 & 0 & 1 \end{bmatrix} = I.\end{aligned}$$

Hence, the given matrix is orthogonal.

EXAMPLE 5.28. *Find the value of λ for which the matrix*

$$A = \begin{bmatrix} \cos\psi & -\sin\psi & 0 \\ \sin\psi & \cos\psi & 0 \\ 0 & 0 & \lambda \end{bmatrix}$$

will be orthogonal.

Solution: For the matrix A to be orthonormal

$$A^T \cdot A = I. \tag{1}$$

Now

$$A^T \cdot A = \begin{bmatrix} \cos\psi & \sin\psi & 0 \\ -\sin\psi & \cos\psi & 0 \\ 0 & 0 & \lambda \end{bmatrix} \begin{bmatrix} \cos\psi & -\sin\psi & 0 \\ \sin\psi & \cos\psi & 0 \\ 0 & 0 & \lambda \end{bmatrix}.$$

$$\therefore \quad A^T \cdot A = \begin{bmatrix} 1 & 0 & 0 \\ 0 & 1 & 0 \\ 0 & 0 & \lambda^2 \end{bmatrix}. \tag{2}$$

From (1) and (2), we find that the RHS matrix in (2) should be an identity matrix. Hence, $\lambda^2 = 1$, i.e., $\lambda = \pm 1$.

5.20 Minors and Cofactors

Let A be a matrix of order $m \times n$. If we suppress the ith row and jth column, the determinant thus formed is called the *minor* of a_{ij}. Thus, if

$$A = \begin{bmatrix} 2 & 4 & 9 \\ 0 & -2 & 3 \\ 4 & -5 & 6 \end{bmatrix}$$

the minors of elements of first row are

$$\begin{bmatrix} -2 & 3 \\ -5 & 6 \end{bmatrix}, \quad \begin{bmatrix} 0 & 3 \\ 4 & 6 \end{bmatrix} \quad \text{and} \quad \begin{bmatrix} 0 & -2 \\ 4 & -5 \end{bmatrix} \quad \text{respectively.}$$

The cofactor of a_{ij} is denoted by A_{ij} and is equal to $(-1)^{i+j}$ times its minor. Thus, if

$$A = \begin{bmatrix} a_{11} & a_{12} & a_{13} \\ a_{21} & a_{22} & a_{23} \\ a_{31} & a_{32} & a_{33} \end{bmatrix}$$

then, the cofactor matrix is

$$A_c = \begin{bmatrix} A_{11} & A_{12} & A_{13} \\ A_{21} & A_{22} & A_{23} \\ A_{31} & A_{32} & A_{33} \end{bmatrix}, \tag{5.87}$$

where, the cofactors are

$$A_{11} = (-1)^{1+1} \begin{vmatrix} a_{22} & a_{23} \\ a_{32} & a_{33} \end{vmatrix}, \quad A_{12} = (-1)^{1+2} \begin{vmatrix} a_{21} & a_{23} \\ a_{31} & a_{33} \end{vmatrix}$$

$$\cdots\cdots\cdots\cdots \qquad \cdots\cdots\cdots\cdots$$

$$\cdots\cdots\cdots\cdots \qquad \cdots\cdots\cdots\cdots$$

$$A_{32} = (-1)^{3+2} \begin{vmatrix} a_{11} & a_{13} \\ a_{21} & a_{23} \end{vmatrix}, \quad A_{33} = (-1)^{3+3} \begin{vmatrix} a_{11} & a_{12} \\ a_{21} & a_{22} \end{vmatrix}$$

5.21 Adjoint and Reciprocal Matrices

The adjoint of a matrix A is defined as the transpose of the matrix whose elements are cofactors of corresponding elements in det A. It is denoted by adj A. Thus

$$\operatorname{adj} A = A_c^T. \tag{5.88}$$

If A be a non-singular square matrix, then the reciprocal matrix of A is defined as

$$\frac{\operatorname{adj} A}{|A|}. \tag{5.89}$$

EXAMPLE 5.29. *Compute the adjoint and reciprocal of the matrix*

$$A = \begin{bmatrix} 3 & 2 & 1 \\ 1 & 1 & 1 \\ 5 & 1 & -1 \end{bmatrix}.$$

Solution:

$$A_{11} = (-1)^2 \begin{vmatrix} 1 & 1 \\ 1 & -1 \end{vmatrix} = -2, \quad A_{12} = (-1)^3 \begin{vmatrix} 2 & 1 \\ 1 & -1 \end{vmatrix} = 3,$$

$$A_{13} = (-1)^4 \begin{vmatrix} 2 & 1 \\ 1 & 1 \end{vmatrix} = 1, \quad A_{21} = (-1)^3 \begin{vmatrix} 1 & 1 \\ 5 & -1 \end{vmatrix} = 6,$$

$$A_{22} = (-1)^4 \begin{vmatrix} 3 & 1 \\ 5 & -1 \end{vmatrix} = -8, \quad A_{23} = (-1)^5 \begin{vmatrix} 3 & 1 \\ 1 & 1 \end{vmatrix} = -2,$$

$$A_{31} = (-1)^4 \begin{vmatrix} 1 & 1 \\ 5 & 1 \end{vmatrix} = -4, \quad A_{32} = (-1)^5 \begin{vmatrix} 3 & 2 \\ 5 & 1 \end{vmatrix} = 7,$$

$$A_{33} = (-1)^6 \begin{vmatrix} 3 & 2 \\ 1 & 1 \end{vmatrix} = 1.$$

Hence

$$\operatorname{adj} A = \begin{bmatrix} -2 & 3 & 1 \\ 6 & -8 & -2 \\ -4 & 7 & 1 \end{bmatrix}.$$

Now

$$\det A = \begin{vmatrix} 3 & 2 & 1 \\ 1 & 1 & 1 \\ 5 & 1 & -1 \end{vmatrix} = 3 \begin{vmatrix} 1 & 1 \\ 1 & -1 \end{vmatrix} - 2 \begin{vmatrix} 1 & 1 \\ 5 & -1 \end{vmatrix} + 1 \begin{vmatrix} 1 & 1 \\ 5 & 1 \end{vmatrix}$$

$$= 3(-2) - 2(-6) + (-4) = 2.$$

Therefore, the reciprocal matrix is

$$= \frac{\operatorname{adj} A}{\det A} = \frac{1}{2} \begin{vmatrix} -2 & 3 & 1 \\ 6 & -8 & -2 \\ -4 & 7 & 1 \end{vmatrix}.$$

Theorem 5.10. *If A be a square matrix, then the product of matrices A and $\operatorname{adj} A$ is a scalar matrix each diagonal element of which is $|A|$.*

Proof. Let, $A = [a_{ij}]$ of $m \times n$ order. Then, the element of ith row and jth column of $A \cdot (\operatorname{adj} A)$ will be scalar product of elements of ith row of A with jth column of $\operatorname{adj} A$. It is equal to

$$\sum_{k=1}^{n} a_{ik} \cdot A_{jk} = \begin{cases} |A|, & \text{for } i = j; \\ 0, & \text{for } i \neq j. \end{cases}$$

$$\therefore \quad A \cdot (\operatorname{adj} A) = |A| \cdot I.$$

Similarly

$$(\operatorname{adj} A)\, A = |A|\, I. \tag{5.90}$$

Hence, the theorem follows. From this theorem it is obvious that, if A be a singular matrix, then $|A| = 0$, so that, $A \cdot (\operatorname{adj} A) = 0$ (null matrix). □

5.22 Inverse of a Matrix

If A be a non-singular square matrix and if there exists another square matrix B of the same order, such that

$$AB = BA = I, \tag{5.91}$$

then B is called the *inverse* of A and is denoted by A^{-1}.

Now, we know that

$$A \cdot (\operatorname{adj} A) = |A|\, I, \quad \text{or,} \quad A \cdot \left(\frac{1}{|A|} \operatorname{adj} A \right) = I, \quad \because \; |A| \neq 0. \tag{5.92}$$

Thus, product of matrix A and its reciprocal is an identity matrix. That is, reciprocal of a matrix is its inverse. We shall therefore accept *reciprocal* and *inverse* to be synonymous.

From equation (5.92), we find that inverse of a non-singular matrix is obtained by dividing each element of adjoint matrix A by $\det A$.

Properties of the Inverse Matrix

Property 1. The necessary and sufficient condition for existence of inverse of a square matrix A is

$$|A| \neq 0. \tag{5.93}$$

Proof. To prove that the condition is necessary we assume that inverse of A exists and is equal to B. Then, by definition

$$AB = BA = I.$$

$$\therefore \quad |AB| = I, \quad \text{or,} \quad |A|\,|B| = I,$$

which is possible only when $|A| \neq 0$.

To prove the sufficiency, we consider the matrix

$$B = \frac{\operatorname{adj} A}{|A|}$$

which exists if $|A| \neq 0$.

$$\therefore \quad AB = \frac{A}{|A|}\,\text{adj}\,A = \frac{|A|\,I}{|A|} \quad \text{[by equation (5.92)]}$$
$$= I,$$

which shows that matrix B is the inverse of A. Therefore, condition $|A| \neq 0$ is sufficient for existence of inverse of the matrix A.

The matrix A is called *invertible* if its inverse exists. □

Property 2. The inverse of a matrix, if it exists, is unique.

Proof. Let us assume to the contrary that the matrix A has got B and C as two inverses. Then, by definition, $AB = I$ and $AC = I$.

Now, we have

$$C\,(AB) = (CA)\,B \quad \text{(associativity of multiplication)}$$
$$\text{or,} \quad CI = IB.$$
$$\therefore \quad C = B.$$

Hence, the property follows. □

Property 3. The inverse of product of two matrices of same order is the product of their inverses in reverse order.

Proof. Let A and B be two non-singular square matrices of the same order. Let us consider the products

$$[AB]\,[B^{-1}A^{-1}] = A\,[BB^{-1}]\,A^{-1} \quad \text{(associativity of matrix multiplication)}$$
$$= AI\,A^{-1} = AA^{-1} = I.$$

Again,

$$[B^{-1}A^{-1}]\,[AB] = B^{-1}\,[A^{-1}\,A]\,B = B^{-1}IB = B^{-1}\,B = I.$$
$$\therefore \quad [AB]\,[B^{-1}A^{-1}] = [B^{-1}A^{-1}]\,[AB] = I,$$

which shows that $B^{-1}A^{-1}$ is the inverse of AB, i.e.,

$$[AB]^{-1} = B^{-1}\,A^{-1}. \tag{5.94}$$

□

Property 4. If a matrix A is invertible, then its inverse A^{-1} is also invertible, and $[A^{-1}]^{-1} = A$.

Proof. Let A be a non-singular square matrix so that A^{-1} exists, and $|A| \neq 0$.

Now,

$$|A| \cdot |A^{-1}| = |AA^{-1}| = |I| = 1,$$

which shows that $|A^{-1}| \neq 0$.

Hence, inverse of A^{-1} does exist.

Again, we have, $A^{-1}A = AA^{-1} = I$.

Hence, A is the inverse of A^{-1}, and so

$$[A^{-1}]^{-1} = A. \tag{5.95}$$

□

Property 5. Inverse of the transpose of a square matrix is the transpose of its inverse.

Proof. Let A be a non-singular square matrix having an inverse A^{-1}.

Then, we have, $AA^{-1} = I = A^{-1}A$.

On transposing, we get

$$[AA^{-1}]^T = [I]^T, \quad \text{or,} \quad [A^{-1}]^T[A]^T = I. \tag{5.96}$$

Also

$$[A^{-1}A]^T = [I]^T, \quad \text{or,} \quad [A^T][A^{-1}]^T = I. \tag{5.97}$$

From these relations, we get

$$[A^{-1}]^T = [A^T]^{-1}. \tag{5.98}$$

Hence, the property is proved. □

5.23 Trace of a Matrix

The algebraic sum of the diagonal elements of a square matrix A is called its *trace* and is denoted by $\mathrm{Tr}(A)$.

If $A = [a_{ij}]$ be of order $n \times n$, then its trace

$$\mathrm{Tr}(A) = \sum_{i=1}^{n} a_{ii}.$$

Following are the properties of trace:

Property 1. The trace of sum of two matrices is the sum of their traces.

If A and B are two matrices, so that $A = [a_{ij}]$ and $B = [b_{ij}]$, both of order $n \times n$, then

$$\mathrm{Tr}(A+B) = \sum_{i=1}^{n}(a_{ii}+b_{ii}) = \sum_{i=1}^{n} a_{ii} + \sum_{i=1}^{n} b_{ii},$$

i.e., $$\mathrm{Tr}(A+B) = \mathrm{Tr}(A) + \mathrm{Tr}(B). \tag{5.99}$$

Property 2. If a matrix is multiplied by a scalar matrix, then its trace becomes multiplied by that scalar.

Let A be an $n \times n$ matrix. If it is multiplied by a scalar matrix having λ as its diagonal elements, then

$$\mathrm{Tr}(\lambda A) = \sum_{i=1}^{n} \lambda a_{ii} = \lambda\,\mathrm{Tr}(A). \tag{5.100}$$

Property 3. Trace of a product of two matrices is independent of the order of multiplication.

Let A and B be two matrices of order $n \times n$, so that $A = [a_{ij}]$ and $B = [b_{ij}]$.

Then,

$$\mathrm{Tr}(AB) = \sum_{i=1}^{n}(AB)_{ii} = \sum_{i=1}^{n}\left\{\sum_{j=1}^{n} a_{ij}b_{ji}\right\} = \sum_{j=1}^{n}\sum_{i=1}^{n} a_{ij}b_{ji}$$

$$= \sum_{j=1}^{n}\sum_{i=1}^{n} b_{ji}a_{ij} = \sum_{j=1}^{n}(BA)_{jj} = \mathrm{Tr}(BA).$$

Hence

$$\mathrm{Tr}\,(AB) = \mathrm{Tr}\,(BA). \tag{5.101}$$

EXAMPLE 5.30. *If* $|A| = 2$ *and* $\operatorname{adj} A = \begin{bmatrix} -2 & 3 & 1 \\ 6 & -8 & -2 \\ -4 & 7 & 1 \end{bmatrix}$, *then find A.*

Solution: We know, $A^{-1} = \dfrac{\operatorname{adj} A}{|A|}$.

Hence

$$A^{-1} = \frac{1}{2}\begin{bmatrix} -2 & 3 & 1 \\ 6 & -8 & -2 \\ -4 & 7 & 1 \end{bmatrix}. \tag{1}$$

$$\therefore \quad A = \left(A^{-1}\right)^{-1}.$$

Thus, we are to find inverse of matrix in (1). First, we calculate adjoint of matrix in (1).

$$A_{11} = (-1)^2 \begin{vmatrix} -8 & -2 \\ 7 & 1 \end{vmatrix} = 6, \quad A_{12} = (-1)^3 \begin{vmatrix} 3 & 1 \\ 7 & 1 \end{vmatrix} = 4,$$

$$A_{13} = (-1)^4 \begin{vmatrix} 3 & 1 \\ -8 & -2 \end{vmatrix} = 2, \quad A_{21} = (-1)^3 \begin{vmatrix} 6 & -2 \\ -4 & 1 \end{vmatrix} = 2,$$

$$A_{22} = (-1)^4 \begin{vmatrix} -2 & 1 \\ -4 & 1 \end{vmatrix} = 2, \quad A_{23} = (-1)^5 \begin{vmatrix} -2 & 1 \\ 6 & -2 \end{vmatrix} = 2,$$

$$A_{31} = (-1)^4 \begin{vmatrix} 6 & -8 \\ -4 & 7 \end{vmatrix} = 10, \quad A_{32} = (-1)^5 \begin{vmatrix} -2 & 3 \\ -4 & 7 \end{vmatrix} = 2,$$

$$A_{33} = (-1)^6 \begin{vmatrix} -2 & 3 \\ 6 & -8 \end{vmatrix} = -2.$$

Hence, adjoint of matrix (1) is

$$\frac{1}{4}\begin{bmatrix} 6 & 4 & 2 \\ 2 & 2 & 2 \\ 10 & 2 & -2 \end{bmatrix}. \tag{2}$$

The determinant of matrix in (1) is

$$\frac{1}{8}\left\{-2\begin{vmatrix} -8 & -2 \\ 7 & 1 \end{vmatrix} - 3\begin{vmatrix} 6 & -2 \\ -4 & 1 \end{vmatrix} + 1\begin{vmatrix} 6 & -8 \\ -4 & 7 \end{vmatrix}\right\} = \frac{1}{8}\{-2(6) - 3(-2) + 1(10)\}$$

$$= \frac{1}{8}(-12 + 6 + 10) = \frac{1}{2}.$$

Hence

$$A = \left(A^{-1}\right)^{-1} = \left(\frac{1}{4} \div \frac{1}{2}\right)\begin{bmatrix} 6 & 4 & 2 \\ 2 & 2 & 2 \\ 10 & 2 & -2 \end{bmatrix} = \begin{bmatrix} 3 & 2 & 1 \\ 1 & 1 & 1 \\ 5 & 1 & -1 \end{bmatrix}.$$

5.24 Complex Matrices

If some or all elements of a matrix are complex numbers, then the matrix is called a *complex matrix*. Any complex matrix A can be written as $A = U + iV$, where U and V are real matrices.

Conjugate matrix: Conjugate of a complex matrix A is obtained by replacing each element of A with its conjugate. Conjugate of A is denoted by $\bar{A}$. Thus, if

$$A = U + iV, \quad \text{then} \quad \bar{A} = U - iV. \tag{5.102}$$

Properties of complex conjugate matrices: If A and B are two complex matrices conformable to addition and multiplication, then

(a) $\overline{A+B} = \bar{A} + \bar{B}$;

(b) $\overline{\lambda A} = \bar{\lambda}\bar{A}$, λ = complex number;

(c) $\overline{AB} = \bar{A}\bar{B}$.

Transposed conjugate matrix: Transposed conjugate of a complex matrix A is obtained by transposing the conjugate $\bar{A}$ and is denoted by A^*. Thus, if

$$A = \begin{bmatrix} 2+3i & -1+2i \\ i & 5-6i \end{bmatrix},$$

then

$$\bar{A} = \begin{bmatrix} 2-3i & -1-2i \\ -i & 5+6i \end{bmatrix} \quad \text{and} \quad A^* = \begin{bmatrix} 2-3i & -i \\ -1-2i & 5+6i \end{bmatrix} = [\bar{A}]^T.$$

Properties of transposed conjugate matrices: If A and B are two complex matrices conformable to addition and multiplication, then

(a) $[A^*]^* = A$;

(b) $[A+B]^* = A^* + B^*$;

(c) $[\lambda A]^* = \bar{\lambda}A^*$, λ is any complex number;

(d) $[AB]^* = B^*A^*$.

5.25 Hermitian and Skew-Hermitian Matrices

A square complex matrix whose transposed conjugate coincides with itself is called *Hermitian matrix*. Thus, if A be a square complex matrix, then it is called *Hermitian matrix* if

$$A^* = A. \tag{5.103}$$

Taking transpose of both sides

$$[A^*]^T = A^T, \quad \text{i.e.,} \quad \left[[\bar{A}]^T\right]^T = A^T, \quad \text{i.e.,} \quad \bar{A} = A^T. \tag{5.104}$$

Thus, transpose of a Hermitian matrix coincides with its conjugate.

If A is a matrix $[a_{ij}]$, equation (5.104) gives

$$\bar{a}_{ij} = a_{ji}. \tag{5.105}$$

This shows that (ij)th element of conjugate matrix coincides with (ji)th element of the Hermitian matrix. Setting $i = j$ in this equation, we have

$$a_{ii} = \bar{a}_{ii}, \tag{5.106}$$

which shows that every diagonal element of Hermitian matrix is the same as the diagonal element of its conjugate. This is possible only if the diagonal elements of the Hermitian matrix are real. Thus, the matrix

$$A = \begin{bmatrix} 2 & 1+i & 3-2i \\ 2-i & 4 & i \\ 3 & 6+3i & -7 \end{bmatrix}$$

is a Hermitian matrix.

A square complex matrix whose conjugate transpose coincides with itself preceded with a negative sign is called *Skew-Hermitian matrix*. Thus, A is a skew-Hermitian matrix if

$$A^* = -A, \quad \text{i.e.,} \quad A^T = -\bar{A}. \tag{5.107}$$

This yields

$$a_{ii} = -\bar{a}_{ii}. \tag{5.108}$$

Thus, every diagonal element of a skew-Hermitian matrix is either zero or purely imaginary number. Hence, the matrix

$$A = \begin{bmatrix} 0 & 2+i & -i \\ 4+3i & -5i & -3+2i \\ 7-6i & 1+3i & 2i \end{bmatrix}$$

is a skew-Hermitian matrix.

EXAMPLE 5.31. *If A be a square complex matrix, then show that $(A + A^*)$ is a Hermitian matrix and $(A - A^*)$ is a skew-Hermitian matrix.*

Solution: We have

$$\begin{aligned} (A + A^*)^* &= A^* + (A^*)^* \quad \text{[by property (b)]} \\ &= A^* + A \quad \text{[by property (a)]}. \end{aligned}$$

Hence, by definition, $A + A^*$ is Hermitian.

Again

$$(A - A^*)^* = A^* - (A^*)^* = A^* - A = -(A - A^*).$$

Hence, by definition, $A - A^*$ is skew-Hermitian.

EXAMPLE 5.32. *Prove that, the determinant of a Hermitian matrix is real.*

Solution: Let A be a Hermitian matrix.

Then

$$A^* = A, \quad \text{or,} \quad A^T = \bar{A} \quad \text{[by equation (5.73)]}$$

$$\therefore \quad |A^T| = |\bar{A}|. \tag{1}$$

But, $|A^T| = |A|$, and $|\bar{A}| = \overline{|A|}$.

Hence, from equation (1), we get $|A| = \overline{|A|}$ which shows that the determinant is real.

EXAMPLE 5.33. *Prove that the determinant of a skew-Hermitian matrix is either zero or purely imaginary.*

Solution: If A be a skew-Hermitian matrix, then

$$A^* = -A, \qquad \text{or,} \quad A^T = -\bar{A}.$$

$$\therefore \quad |A^T| = -|\bar{A}|, \quad \text{or,} \quad -|A| = -\overline{|A|},$$

which shows that the determinant of a skew-Hermitian matrix is either zero or purely imaginary.

Unitary matrix: A square matrix A is said to be unitary if its inverse coincides with its conjugate transpose, i.e.,

$$A^* = A^{-1}, \quad \text{or,} \quad \left[\bar{A}\right]^T = A^{-1}, \quad \text{i.e.,} \quad A^T = \overline{[A^{-1}]}.$$

Thus, for a matrix A to be unitary

$$AA^* = A^*A = A^{-1}A = I.$$

$$\therefore \quad \det |AA^*| = \det |I| = 1.$$

$$\therefore \quad |A|\,|\bar{A}| = 1.$$

Thus, product of moduli of determinants of a unitary matrix and its conjugate is unity.

For example, if

$$A = \begin{bmatrix} 1/\sqrt{2} & i/\sqrt{2} \\ -i/\sqrt{2} & 1/\sqrt{2} \end{bmatrix},$$

then

$$\bar{A} = \begin{bmatrix} 1/\sqrt{2} & -i/\sqrt{2} \\ i/\sqrt{2} & 1/\sqrt{2} \end{bmatrix}, \quad A^* = \begin{bmatrix} 1/\sqrt{2} & i/\sqrt{2} \\ -i/\sqrt{2} & 1/\sqrt{2} \end{bmatrix}$$

and hence, $AA^* = I$. Therefore, A is a unitary matrix.

A unitary matrix with elements as real numbers is said to be orthogonal. In this case, $A^* = A^T$, so that $A^T \cdot A = I$, which shows that the matrix A is orthogonal [*see* Equation (5.84)].

5.26 Elementary Row Operations

For any matrix A, it is possible to make the following row operations:

(a) *Interchange of any two rows.* If ith row is interchanged with jth row, the operation is denoted by $R_i \longleftrightarrow R_j$.

For example,

$$\begin{bmatrix} 2 & 1 & 3 \\ 0 & 4 & 9 \\ 5 & 6 & 7 \end{bmatrix} R_1 \longleftrightarrow R_3 \begin{bmatrix} 5 & 6 & 7 \\ 0 & 4 & 9 \\ 2 & 1 & 3 \end{bmatrix}.$$

(b) *Multiplication of a row by a constant* λ. If ith row is multiplied by λ, the operation is denoted by $R_i \longleftrightarrow \lambda R_i$.

For example,

$$\begin{bmatrix} 2 & 3 \\ 5 & 7 \end{bmatrix} R_1 \longleftrightarrow 3R_1 \begin{bmatrix} 6 & 9 \\ 5 & 7 \end{bmatrix}.$$

(c) *Addition* (or *subtraction*) *of λ-times of a row to another row.* The operation is denoted by $R_i \longleftrightarrow R_i \pm \lambda R_j$.
For example,

$$\begin{bmatrix} 2 & 3 \\ 6 & 7 \end{bmatrix} R_2 \longleftrightarrow R_2 - 2R_1 \begin{bmatrix} 2 & 3 \\ 2 & 1 \end{bmatrix}.$$

Such elementary operations may also be applied to columns. These are known as *elementary column operations.*

5.27 Row-Echelon Normal Form

The elementary row operations discussed before may be applied to any matrix A in such a manner that the ultimate transformed matrix has the following properties:

(i) All the zero rows are its last rows.
(ii) The first non-zero element in a non-zero row is 1. Each column containing this 1 will have all other elements zero.
(iii) The leading 1 in row i will be to the left of the leading 1 in row j, if $i < j$.

Any matrix having above properties is said to be in row-echelon normal form, in short, RENF. As an example, the matrix

$$A = \begin{bmatrix} 1 & 3 & 4 \\ 0 & 1 & 0 \\ 0 & 0 & 0 \end{bmatrix}$$

is in RENF.

It is important to note that every matrix can be transformed to RENF *uniquely* by a finite sequence of row operations.

5.28 Matrix Inversion by Row Operation

If a square matrix A is reduced to the unit matrix by row operations, then the same sequence of row operation applied to unit matrix will give A^{-1}. We shall take an example.

EXAMPLE 5.34. *Find A^{-1}, if* $A = \begin{bmatrix} 2 & 3 & 1 \\ 1 & 2 & 3 \\ 3 & 1 & 2 \end{bmatrix}$.

Solution: Let us assume that

$$\begin{bmatrix} 2 & 3 & 1 \\ 1 & 2 & 3 \\ 3 & 1 & 2 \end{bmatrix} = \begin{bmatrix} 1 & 0 & 0 \\ 0 & 1 & 0 \\ 0 & 0 & 1 \end{bmatrix} A.$$

We undertake the following operations:
(1) operation: $R_3 \longrightarrow R_3 - 3R_2$

$$\begin{bmatrix} 2 & 3 & 1 \\ 1 & 2 & 3 \\ 0 & -5 & -7 \end{bmatrix} = \begin{bmatrix} 1 & 0 & 0 \\ 0 & 1 & 0 \\ 0 & -3 & 1 \end{bmatrix} A$$

(2) operation: $R_2 \longrightarrow R_2 - \frac{1}{2}R_1$

$$\begin{bmatrix} 2 & 3 & 1 \\ 0 & 1/2 & 5/2 \\ 0 & -5 & -7 \end{bmatrix} = \begin{bmatrix} 1 & 0 & 0 \\ -1/2 & 1 & 0 \\ 0 & -3 & 1 \end{bmatrix} A$$

(3) operation: $R_3 \longrightarrow R_3 + 10R_2$

$$\begin{bmatrix} 2 & 3 & 1 \\ 0 & 1/2 & 5/2 \\ 0 & 0 & 18 \end{bmatrix} = \begin{bmatrix} 1 & 0 & 0 \\ -1/2 & 1 & 0 \\ -5 & 7 & 1 \end{bmatrix} A$$

(4) operation: $R_3 \longrightarrow \frac{1}{18}R_3$, then $R_1 \longrightarrow R_1 - R_3$

$$\begin{bmatrix} 2 & 3 & 0 \\ 0 & 1/2 & 5/2 \\ 0 & 0 & 1 \end{bmatrix} = \begin{bmatrix} 23/18 & -7/18 & -1/18 \\ -1/2 & 1 & 0 \\ -5/18 & 7/18 & 1/18 \end{bmatrix} A$$

(5) operation: $R_1 \longrightarrow \frac{1}{2}R_1, R_2 \longrightarrow 2R_2$

$$\begin{bmatrix} 1 & 3/2 & 0 \\ 0 & 1 & 5 \\ 0 & 0 & 1 \end{bmatrix} = \begin{bmatrix} 23/36 & -7/36 & -1/36 \\ -1 & 2 & 0 \\ -5/18 & 7/18 & 1/18 \end{bmatrix} A$$

(6) operation: $R_1 \longrightarrow R_1 - \frac{3}{2}R_2$

$$\begin{bmatrix} 1 & 0 & -15/2 \\ 0 & 1 & 5 \\ 0 & 0 & 1 \end{bmatrix} = \begin{bmatrix} 77/36 & -115/36 & -1/36 \\ -1 & 2 & 0 \\ -5/18 & 7/18 & 1/18 \end{bmatrix} A$$

(7) operation: $R_1 \longrightarrow R_1 + \frac{15}{2}R_3, R_2 \longrightarrow R_2 - 5R_3$

$$\begin{bmatrix} 1 & 0 & 0 \\ 0 & 1 & 0 \\ 0 & 0 & 1 \end{bmatrix} = \begin{bmatrix} 2/36 & -10/36 & 14/36 \\ 7/18 & 1/18 & -5/18 \\ -5/18 & 7/18 & 1/18 \end{bmatrix} A = \frac{1}{18}\begin{bmatrix} 1 & -5 & 7 \\ 7 & 1 & -5 \\ -5 & 7 & 1 \end{bmatrix} A.$$

Thus

$$A^{-1} = \frac{1}{18}\begin{bmatrix} 1 & -5 & 7 \\ 7 & 1 & -5 \\ -5 & 7 & 1 \end{bmatrix}$$

since, $I = A^{-1}A$.

5.29 Rank of a Matrix

Rank of a matrix is an integral number r such that, at least one minor of order r of that matrix is non-zero, while all the minors of order $(r+1)$ of that matrix are zero. *For example*, let

$$A = \begin{bmatrix} 1 & 2 & 3 \\ 3 & 4 & 5 \\ 4 & 6 & 8 \end{bmatrix}.$$

Now, the determinant of 3rd order is

$$\begin{vmatrix} 1 & 2 & 3 \\ 3 & 4 & 5 \\ 4 & 6 & 8 \end{vmatrix} = (32-30) - 2(24-20) + 3(18-16) = 0.$$

Hence, rank of the matrix is not 3.
Now, let us consider minors of order 2:

$$\begin{vmatrix} 1 & 2 \\ 3 & 4 \end{vmatrix}, \quad \begin{vmatrix} 2 & 3 \\ 4 & 5 \end{vmatrix}, \quad \begin{vmatrix} 1 & 2 \\ 4 & 6 \end{vmatrix}, \quad \begin{vmatrix} 2 & 3 \\ 6 & 8 \end{vmatrix},$$

$$\begin{vmatrix} 3 & 4 \\ 4 & 6 \end{vmatrix}, \quad \begin{vmatrix} 4 & 5 \\ 6 & 8 \end{vmatrix}, \quad \begin{vmatrix} 3 & 5 \\ 4 & 8 \end{vmatrix}, \quad \begin{vmatrix} 1 & 3 \\ 3 & 5 \end{vmatrix}$$

$$\text{and} \begin{vmatrix} 1 & 3 \\ 4 & 8 \end{vmatrix}.$$

None of these minors are zero. Hence, the rank of the matrix is 2.
The rank of a matrix has the following properties depicted in the form of theorems:

Theorem 5.11. *The rank of a matrix is not altered due to interchange of rows.*

Proof. Let A be a matrix of rank r. Let its ith and jth rows are interchanged forming the matrix B. Thus, any square sub-matrix A_0 of order $(r+1)$ of A is transformed into a square sub-matrix B_0 of order $(r+1)$ of B. Every $(r+1)$-rowed minor A_0 of matrix A is zero, since its rank is r.

Now, if ith and jth rows lie outside the sub-matrix A_0, then

$$|B_0| = 0, \quad \text{since } |A_0| = 0. \tag{5.109}$$

If both these rows lie inside A_0, then

$$|B_0| = \pm |A_0| = 0, \tag{5.110}$$

since interchange in rows in a determinant only changes in sign of its value.

If one row in A_0 is interchanged with a row outside it, then

$$|B_0| = 0, \quad \text{since } |A_0| = 0. \tag{5.111}$$

Thus, rank B_0 is $\leq r$.

Similarly, if we form A from the matrix B by interchange of rows, we shall have rank of $B_0 \geq r$.

Hence, rank of $B_0 = r$, which proves the theorem. □

Theorem 5.12. *The rank of a matrix is not altered if the elements of any row are multiplied by a non-zero number.*

Proof. Let the ith row of a matrix A of rank r be multiplied by a constant λ. Due to this, let any square sub-matrix A_0 of order $(r+1)$ of A is transformed into a square sub-matrix B_0 of order $(r+1)$ of B.

Now, if the ith row lies outside the sub-matrix A_0, then

$$|B_0| = 0 \quad [\text{since } |A_0| = 0]. \tag{5.112}$$

If the ith row lies inside the sub-matrix A_0, then

$$|B_0| = \lambda\,|A_0| = 0 \quad [\text{since } |A_0| = 0] \tag{5.113}$$

since, value of a determinant gets multiplied by λ due to multiplication of a row by λ.

Hence, the rank of B_0 is $\leq r$.

Similarly, if we form A from the matrix B by a similar transformation, we shall have the rank of B_0 is $\geq r$.

Hence, rank of $B_0 = r$.

Thus, the theorem is proved. □

Theorem 5.13. *The rank of a matrix is not altered if λ times the elements of a row is added to corresponding elements of another row, where λ is any given number.*

Proof. Let the ith row of a matrix A of rank r be multiplied by λ and added to the corresponding elements of jth row. Let, any square sub-matrix A_0 of order $(r+1)$ from A is transformed into a square sub-matrix B_0 of order $(r+1)$ of matrix B.

Now, if jth row lies outside the sub-matrix A_0, then

$$|B_0| = 0 \quad [\because\ |A_0| = 0]. \tag{5.114}$$

If jth row lies inside the sub-matrix A_0, then

$$|B_0| = |A_0| + \lambda\,|C_0|, \tag{5.115}$$

where elements of sub-matrix C_0 are the same as those of A_0 excepting the jth row which is λ-times the ith row. Since rank of A is r, therefore, $|A_0| = 0$ and $|C_0| = 0$ by virtue of Theorem 5.11.

Hence, from equation (5.115)

$$|B_0| = 0. \tag{5.116}$$

Hence, the rank of B_0 is $\leq r$.

With similar arguments as before, the rank of $B_0 \geq r$.

Hence, the rank of $B_0 = r$.

This proves the theorem. □

Theorem 5.14. *The rank of a matrix is not altered if the matrix is pre-multiplied or post-multiplied by a non-singular matrix.*

Proof. Let A be a non-singular matrix of order $m \times n$ and of rank r. Let a_{ij} be its non-zero element. Then, interchanging the ith row with the first row, and jth column with the first column, we get a new matrix B such that $b_{11} = a_{ij} = \lambda$ (say). We divide the first row of B by λ and get a new matrix C such that $C_{11} = 1$. Now, by elementary operations, we can transform the matrix A ultimately to the form

$$A \longrightarrow B \longrightarrow C \longrightarrow\longrightarrow\longrightarrow U = \begin{bmatrix} I_r & 0 \\ 0 & 0 \end{bmatrix},$$

where I_r is identity matrix of order r which is the rank of matrix A. Since each elementary transformation has been affected by pre-multiplication or post-multiplication by elementary matrices, we can say, that the rank of a matrix does not change due to such operations. Thus, we can say, that corresponding to every $m \times n$ matrix of rank r, there exist non-singular $m \times m$ matrix P and a non-singular $n \times n$ matrix Q such that

$$PAQ = \begin{bmatrix} I_r & 0 \\ 0 & 0 \end{bmatrix}. \tag{5.117}$$

Hence, rank of a matrix does not change if it is pre-multiplied or post-multiplied with a non-singular matrix. □

Note. 1. Elementary matrices are obtained from unit matrices by elementary operations.

2. If A be a non-singular square matrix of order n, then its rank will be n, so that

$$PAQ = I_n. \tag{5.118}$$

3. This method of diagonalisation is very efficient in finding the rank of a matrix.

Theorem 5.15. *The rank of product of two matrices never exceeds the rank of either matrix, i.e.,*

$$r(AB) \le \min[r(A), r(B)]. \tag{5.119}$$

Proof. Let r_1, r_2 and r be ranks of the matrices A, B and C respectively, where $C = AB$. Let A, B and C have orders of $m \times n, n \times p$ and $m \times p$ respectively. We now consider a non-singular matrix P such that

$$PA = \begin{bmatrix} H \\ 0 \end{bmatrix}, \tag{5.120}$$

where H is a matrix of order $r_1 \times n$ and of rank r_1 and 0 is a null matrix of order $(m - r_1) \times n$.

Then

$$PAB = \begin{bmatrix} H \\ 0 \end{bmatrix} B. \tag{5.121}$$

Since rank of a matrix remains unaltered after pre-multiplication with a non-singular matrix P, hence

$$\text{rank of } PAB = \text{rank of } AB = r. \tag{5.122}$$

Since matrix on the RHS in equation (5.121) has at most r_1 non-zero rows, its rank is $\le r_1$

$$\text{i.e., rank of } \begin{bmatrix} H \\ 0 \end{bmatrix} B \le r_1. \tag{5.123}$$

Thus, from equations (5.122) and (5.123), we get $r \le r_1$

$$\text{i.e., rank of } AB \le \text{rank of } A. \tag{5.124}$$

Again,

$$\begin{aligned} \text{rank of } AB &\le \text{rank of } (AB)^T \\ &\le \text{rank of } B^T A^T \\ &\le \text{rank of } B^T \quad \text{[by equation (5.124) above]} \\ &\le \text{rank of } B. \end{aligned}$$

Therefore

$$\text{rank of } AB \leq \text{rank of } B. \tag{5.125}$$

From equations (5.124) and (5.125), we have rank of $(AB) \leq \min[\text{rank}\ (A),\ \text{rank}\ (B)]$. Hence, the theorem is proved.

EXAMPLE 5.35. *Find the rank of*

$$A = \begin{bmatrix} 0 & 1 & -3 & -1 \\ 1 & 0 & 1 & 1 \\ 3 & 1 & 0 & 2 \\ 1 & 1 & -2 & 0 \end{bmatrix}$$

by reducing it to RENF *form.*

Solution: We have

$$A = \begin{bmatrix} 0 & 1 & -3 & -1 \\ 1 & 0 & 1 & 1 \\ 3 & 1 & 0 & 2 \\ 1 & 1 & -2 & 0 \end{bmatrix} = \begin{bmatrix} 0 & 1 & -3 & -1 \\ 1 & 0 & 1 & 1 \\ 0 & 1 & -3 & -1 \\ 0 & 1 & -3 & -1 \end{bmatrix} \quad \begin{matrix} \text{by } R_3 \longrightarrow R_3 - 3R_2 \\ \text{and } R_4 \longrightarrow R_4 - R_2 \end{matrix}$$

$$= \begin{bmatrix} 1 & 0 & 1 & 1 \\ 0 & 1 & -3 & -1 \\ 0 & 1 & -3 & -1 \\ 0 & 1 & -3 & -1 \end{bmatrix} \quad \text{by } R_1 \longleftrightarrow R_2$$

$$= \begin{bmatrix} 1 & 0 & 1 & 1 \\ 0 & 1 & -3 & 1 \\ 0 & 0 & 0 & 0 \\ 0 & 0 & 0 & 0 \end{bmatrix} \quad \begin{matrix} \text{by } R_3 \longrightarrow R_3 - R_2 \\ \text{and } R_4 \longrightarrow R_4 - R_2 \end{matrix}$$

This is in row-echelon normal form. It is obvious that its rank is 2.

EXAMPLE 5.36. *Find the rank of*

$$A = \begin{bmatrix} 1 & 3 & 4 & 5 \\ 1 & 2 & 6 & 7 \\ 1 & 5 & 0 & 1 \end{bmatrix}.$$

Solution: We have

$$A = \begin{bmatrix} 1 & 3 & 4 & 5 \\ 1 & 2 & 6 & 7 \\ 1 & 5 & 0 & 1 \end{bmatrix} = \begin{bmatrix} 1 & 3 & 4 & 5 \\ 0 & -1 & 2 & 2 \\ 0 & 2 & -4 & -4 \end{bmatrix} \quad \begin{matrix} \text{by } R_2 \longrightarrow R_2 - R_1 \\ \text{and } R_3 \longrightarrow R_3 - R_1 \end{matrix}$$

$$= \begin{bmatrix} 1 & 3 & 4 & 5 \\ 0 & -1 & 2 & 2 \\ 0 & 0 & 0 & 0 \end{bmatrix} \quad \text{by } R_3 \longrightarrow R_3 + 2R_2$$

$$= \begin{bmatrix} 1 & 3 & 10 & 11 \\ 0 & -1 & 0 & 0 \\ 0 & 0 & 0 & 0 \end{bmatrix} \quad \begin{matrix} \text{by } C_3 \longrightarrow C_3 + 2C_2 \\ \text{and } C_4 \longrightarrow C_4 + 2C_2 \end{matrix}$$

$$= \begin{bmatrix} 1 & 3 & 0 & 0 \\ 0 & -1 & 0 & 0 \\ 0 & 0 & 0 & 0 \end{bmatrix} \quad \begin{matrix} \text{by } C_3 \longrightarrow C_3 - 10C_1 \\ \text{and } C_4 \longrightarrow C_4 - 11C_1 \end{matrix}$$

The matrix is thus reduced to normal form. Its rank is 2, the non-zero minor is of order 2×2.

EXAMPLE 5.37. *Reduce the matrix to normal form and find its rank:*

$$A = \begin{bmatrix} 2 & 6 & 5 \\ 2 & 7 & 6 \\ 5 & 16 & 13 \end{bmatrix}.$$

Solution: We have

$$A = \begin{bmatrix} 2 & 6 & 5 \\ 2 & 7 & 6 \\ 5 & 16 & 13 \end{bmatrix} = \begin{bmatrix} 2 & 6 & 5 \\ 0 & 1 & 1 \\ 1 & 4 & 3 \end{bmatrix} \quad \text{by } R_3 \longrightarrow R_3 - 2R_1 \text{ and } R_2 \longrightarrow R_2 - R_1$$

$$= \begin{bmatrix} 1 & 4 & 3 \\ 0 & 1 & 1 \\ 2 & 6 & 5 \end{bmatrix} \quad \text{by } R_1 \longleftrightarrow R_3$$

$$= \begin{bmatrix} 1 & 4 & 3 \\ 0 & 1 & 1 \\ 0 & -2 & -1 \end{bmatrix} \quad \text{by } R_3 \longrightarrow R_3 - 2R_1$$

$$= \begin{bmatrix} 1 & 4 & 3 \\ 0 & 1 & 1 \\ 0 & 0 & 1 \end{bmatrix} \quad \text{by } R_3 \longrightarrow R_3 + 2R_2$$

$$= \begin{bmatrix} 1 & 4 & 0 \\ 0 & 1 & 0 \\ 0 & 0 & 1 \end{bmatrix} \quad \text{by } R_1 \longrightarrow R_1 - 3R_3 \text{ and } R_2 \longrightarrow R_2 - R_3$$

$$= \begin{bmatrix} 1 & 0 & 0 \\ 0 & 1 & 0 \\ 0 & 0 & 1 \end{bmatrix} \quad \text{by } R_1 \longrightarrow R_1 - 4R_2$$

which is of normal form. Its rank is 3.

III. APPLICATIONS OF MATRICES

5.30 Solution of Linear Algebraic Equations

Matrices have got an immediate application in solving the systems of linear algebraic equations.

Let there be a system of m linear equations in n unknowns as below:

$$\left.\begin{array}{l} a_{11}x_1 + a_{12}x_2 + \cdots + a_{1n}x_n = b_1 \\ a_{21}x_1 + a_{22}x_2 + \cdots + a_{2n}x_n = b_2 \\ \cdots\cdots\cdots\cdots\cdots\cdots\cdots\cdots \\ \cdots\cdots\cdots\cdots\cdots\cdots\cdots\cdots \\ a_{m1}x_1 + a_{m2}x_2 + \cdots + a_{mn}x_n = b_n \end{array}\right\}. \tag{5.126}$$

If all of b_i's are zero, then the system of equations is said to be homogeneous, and if at least one of b_i's is not zero, the system of equations is said non-homogeneous.

The system of equations (5.126) can be written in the matrix form

$$Ax = B, \tag{5.127}$$

where

$$A = \begin{bmatrix} a_{11} & a_{12} & \cdots & a_{1n} \\ a_{21} & a_{22} & \cdots & a_{2n} \\ \cdots & \cdots & \cdots & \cdots \\ a_{m1} & a_{m2} & \cdots & a_{mn} \end{bmatrix} \tag{5.128}$$

is an $m \times n$ matrix the elements of which are coefficients of linear equations (5.126) and is called *coefficient matrix* of the system,

$$X = \begin{bmatrix} x_1 \\ x_2 \\ \vdots \\ x_n \end{bmatrix} \quad \text{and} \quad B = \begin{bmatrix} b_1 \\ b_2 \\ \vdots \\ b_m \end{bmatrix}. \tag{5.129}$$

are column matrices of orders $n \times 1$ and $m \times 1$ respectively.

The matrix given by

$$A_b = \begin{bmatrix} a_{11} & a_{12} & \cdots & a_{1n} & b_1 \\ a_{21} & a_{22} & \cdots & a_{2n} & b_2 \\ \cdots & \cdots & \cdots & \cdots & \cdots \\ \cdots & \cdots & \cdots & \cdots & \cdots \\ a_{m1} & a_{m2} & \cdots & a_{mn} & b_m \end{bmatrix} \tag{5.130}$$

is called *augmented matrix* which is of order $n \times (m+1)$ and is denoted by A_b.

We now intend to see under what conditions the solutions of the system (5.126) are consistent.

Theorem 5.16. *The set of linear equations given by* (5.126) *is consistent if and only if matrices A and A_b have same rank, i.e.,*

$$\textit{rank}\,(A) = \textit{rank}\,(A_b). \tag{5.131}$$

Proof. Let the matrix A be denoted by column vectors ψ_i, $i = 1, 2, \ldots, n$. Then the system of equations may be written as

$$[\psi_1, \psi_2, \ldots, \psi_n] \begin{bmatrix} x_1 \\ x_2 \\ \vdots \\ x_n \end{bmatrix} = \mathbf{B} \quad \text{or,} \quad \psi_1 x_1 + \psi_2 x_2 + \cdots + \psi_n x_n = \mathbf{B}. \tag{5.132}$$

If rank of matrix $\mathbf{A}$ is r, then at least r columns are linearly independent. We assume that first r columns are linearly independent, i.e., ψ_i, $i = 1, 2, \ldots, r$ are linearly independent. Therefore, each of the remaining $(n - r)$ columns may be expressed as a linear combination of these r columns.

Let the system be consistent, i.e., there is at least one set of solution. Then, for scalars λ_i, $i = 1, 2, \ldots, n$, we have

$$\sum_{i=1}^{r} \lambda_i \psi_i + \sum_{i=r+1}^{n} \lambda_i \psi_i = \mathbf{B}. \tag{5.133}$$

Since $\{\psi_j\}$, $j = i+1$ to n, can be expressed as linear combinations of $\{\psi_i\}$, $i = 1, \ldots, r$, then the second part in the LHS of equation (5.133) can be expressed in terms of $\{\psi_i\}$, $i = 1, \ldots, r$, so that **B** can be put in terms of $\{\psi_i\}$, $i = 1, \ldots, r$. Thus, the maximum number of linearly independent columns in augmented matrix is also r. Thus, **A** and $\mathbf{A_b}$ will have same rank.

This proves the necessity of the condition.

To prove that the condition is sufficient, we assume that both the matrices have rank equal to r.

Then, maximum number of columns of $\mathbf{A_b}$ which are linearly independent is r. Hence, **B** can be expressed as a linear combination of $\{\psi_i\}$, $i = 1, 2, \ldots, r$, i.e.,

$$\sum_{i=1}^{r} \xi_i \psi_i = \mathbf{B}, \tag{5.134}$$

where ξ_i, $i = 1, 2, \ldots, r$ are scalars.

Equation (5.134) may be written as

$$\sum_{i=1}^{r} \xi_i \psi_i + \sum_{j=i+1}^{n} 0 \cdot \psi_j = \mathbf{B}. \tag{5.135}$$

Comparing this equation with (5.132), we get

$$\left.\begin{aligned} x_1 = \xi_1,\ x_2 = \xi_2,\ \ldots, x_r = \xi_r \\ x_{r+1} = x_{r+2} = \cdots = x_n = 0 \end{aligned}\right\}. \tag{5.136}$$

Thus, the given system has a solution, i.e., the system is consistent.

This proves the sufficiency of the condition. □

EXAMPLE 5.38. *Test the consistency of the following equations:*

$$\begin{aligned} 2x_1 + \ x_2 + 4x_3 &= 4 \\ x_1 - 3x_2 - \ x_3 &= 5 \\ -3x_1 + 2x_2 - 2x_3 &= 1 \\ 8x_1 - 3x_2 + 8x_3 &= 2. \end{aligned}$$

Solution: Here

$$A = \begin{bmatrix} 2 & 1 & 4 \\ 1 & -3 & -1 \\ -3 & 2 & -2 \\ 8 & -3 & 8 \end{bmatrix} \quad \text{and} \quad A_b = \begin{bmatrix} 2 & 1 & 4 & 4 \\ 1 & -3 & -1 & 5 \\ -3 & 2 & -2 & 1 \\ 8 & -3 & 8 & 2 \end{bmatrix}.$$

Now, fourth order determinant of A vanishes, so that rank of $A < 4$. But none of third order determinants of A vanishes. Hence, the rank of $A = 3$. Now, determinant of A_b is

$$\begin{vmatrix} 2 & 1 & 4 & 4 \\ 1 & -3 & -1 & 5 \\ -3 & 2 & -2 & 1 \\ 8 & -3 & 8 & 2 \end{vmatrix} = \begin{vmatrix} 2 & 1 & 0 & 0 \\ 1 & -3 & -3 & 3 \\ -3 & 2 & 4 & 7 \\ 8 & -3 & -8 & -14 \end{vmatrix} \quad \text{by subtracting 2 times col. 1 from col. 3 and 4}$$

$$= \begin{vmatrix} 2 & 0 & 0 & 0 \\ 1 & -7/2 & -3 & 3 \\ -3 & 7/2 & 4 & 7 \\ 8 & -7 & -8 & -14 \end{vmatrix} \quad \text{by subtracting 1/2 times col. 1 from col. 2}$$

$$= 2\begin{vmatrix} -7/2 & -3 & 3 \\ 7/2 & 4 & 7 \\ -7 & -8 & -14 \end{vmatrix}$$

$$= 2\begin{vmatrix} -7/2 & -3 & 0 \\ 7/2 & 4 & 11 \\ -7 & -8 & -22 \end{vmatrix} \quad \text{by adding col. 2 to col. 3}$$

$$\neq 0.$$

Hence, rank of the augmented matrix is also 3. Hence, the system of equations is consistent.

5.31 Homogeneous System

In the case of homogeneous system, the equation (5.127) is written as

$$AX = 0. \tag{5.137}$$

Now, if $|A| \neq 0$, then A^{-1} exists. Pre-multiplying both sides of (5.137), we have $A^{-1}AX = A^{-1}0$, i.e., $IX = 0$ which shows that $X = 0$, i.e., $x_1 = x_2 = \cdots = x_n = 0$. This is called *trivial solution.* If, however, the matrix A is singular, i.e., $|A| = 0$, then the equations will have infinite solutions, some of which may be non-zero solutions.

As an example, let us consider the system

$$\begin{aligned} x + 2y + 3z &= 0 \\ 2x + 3y + 4z &= 0 \\ 3x + 4y + 5z &= 0. \end{aligned}$$

Thus, matrix A here is

$$\begin{bmatrix} 1 & 2 & 3 \\ 2 & 3 & 4 \\ 3 & 4 & 5 \end{bmatrix}$$

whose determinant is zero. That is the matrix is singular.

Making elementary row operations like $R_2 \longrightarrow R_2 - 2R_1$ and $R_3 \longrightarrow R_3 - 3R_1$, we can write the equations as

$$\begin{bmatrix} 1 & 2 & 3 \\ 0 & -1 & -2 \\ 0 & -2 & -4 \end{bmatrix} \begin{bmatrix} x \\ y \\ z \end{bmatrix} = \begin{bmatrix} 0 \\ 0 \\ 0 \end{bmatrix}.$$

This gives three equations like

$$\left.\begin{aligned} x+2y+3z&=0\\ y+2z&=0\\ 2y+4z&=0 \end{aligned}\right\}$$

of which the last two equations are same, giving $y=-2z$.

Putting this into the first equation, we get, $x-z=0$, i.e., $x=z$.

Thus, there will be an infinite number of solutions.

We now simply state the following two important properties of homogeneous system of linear equations:

1. If the matrix A of order $m\times n$ has rank r, then the number of linearly independent solutions of equations (5.137) is $n=r$.
2. If the matrix A is of order $n\times n$, then system of equations (5.137) has a non-zero solution, if and only if rank $(A)<n$.

5.32 Solution of a Non-homogeneous System

Let us consider the system of equations

$$AX=B. \tag{5.138}$$

We assume, that the matrix A is non-singular so that A^{-1} exists. Pre-multiplying equation (5.138) by A^{-1} on both sides, we get

$$A^{-1}AX=A^{-1}B,\quad \text{i.e.,}\quad IX=A^{-1}B;\quad \therefore\ X=A^{-1}B. \tag{5.139}$$

Thus, if we can find A^{-1}, then we get the solution by using equation (5.139). Obviously, this method fails if matrix A is singular.

The solution given by (5.139) will also be unique. Assume, to the contrary, that the system (5.138) has two solutions X_1 and X_2. Then

$$AX_1=B\quad\text{and}\quad AX_2=B,\quad\text{so that}\quad AX_1=AX_2.$$

Pre-multiplying by A^{-1} both sides, we get

$$A^{-1}AX_1=A^{-1}AX_2,\quad\text{or,}\quad IX_1=IX_2;\quad \therefore\ X_1=X_2.$$

Thus, the solution is unique.

EXAMPLE 5.39. *Solve:*

$$\begin{aligned} x\quad\ \ +\ z&=4\\ 2x+y+\ z&=7\\ x+y+2z&=9. \end{aligned}$$

Solution: Here

$$A=\begin{bmatrix}1&0&1\\2&1&1\\1&1&2\end{bmatrix}.$$

First we find its inverse.

We have

$$\det A = 1(2-1) + 1(2-1) = 2.$$

Cofactors are:

$$\begin{array}{lll} A_{11} = \;\;1, & A_{12} = \;\;1, & A_{13} = -1, \\ A_{21} = -3, & A_{22} = \;\;1, & A_{23} = \;\;1, \\ A_{31} = \;\;1, & A_{32} = -1, & A_{33} = \;\;1. \end{array}$$

Thus

$$A^{-1} = \frac{\text{adj}\, A}{|A|} = \frac{1}{2}\begin{bmatrix} 1 & 1 & -1 \\ -3 & 1 & 1 \\ 1 & -1 & 1 \end{bmatrix}.$$

Hence, solutions of equations are given by:

$$\begin{bmatrix} x \\ y \\ z \end{bmatrix} = \frac{1}{2}\begin{bmatrix} 1 & 1 & -1 \\ -3 & 1 & 1 \\ 1 & -1 & 1 \end{bmatrix}\begin{bmatrix} 4 \\ 7 \\ 9 \end{bmatrix}.$$

Therefore, $x = 1$, $y = 2$, $z = 3$.

EXAMPLE 5.40. *Solve the system of equations by reducing the matrix to normal form:*

$$\begin{aligned} x \quad\;\; + \;\; z &= 0 \\ 2x - 3y + 4z &= -4 \\ - \;\; y + 4z &= 2. \end{aligned}$$

Solution: We have

$$\begin{bmatrix} 1 & 0 & 1 \\ 2 & -3 & 4 \\ 0 & -1 & 4 \end{bmatrix}\begin{bmatrix} x \\ y \\ z \end{bmatrix} = \begin{bmatrix} 0 \\ -4 \\ 2 \end{bmatrix}$$

or, $$\begin{bmatrix} 1 & 0 & 1 \\ 0 & -3 & 2 \\ 0 & -1 & 4 \end{bmatrix}\begin{bmatrix} x \\ y \\ z \end{bmatrix} = \begin{bmatrix} 0 \\ -4 \\ 2 \end{bmatrix} \qquad \text{by } R_2 \longrightarrow R_2 - 2R_1$$

or, $$\begin{bmatrix} 1 & 0 & 1 \\ 0 & 1 & 6 \\ 0 & -1 & 4 \end{bmatrix}\begin{bmatrix} x \\ y \\ z \end{bmatrix} = \begin{bmatrix} 0 \\ 8 \\ 2 \end{bmatrix} \qquad \begin{array}{l} \text{by } R_2 \longrightarrow -R_2 \\ \text{and } R_2 \longrightarrow R_2 + 2R_3 \end{array}$$

or, $$\begin{bmatrix} 1 & 0 & 1 \\ 0 & 1 & 6 \\ 0 & 0 & 1 \end{bmatrix}\begin{bmatrix} x \\ y \\ z \end{bmatrix} = \begin{bmatrix} 0 \\ 8 \\ 1 \end{bmatrix} \qquad \begin{array}{l} \text{by } R_3 \longrightarrow R_3 + R_2 \\ \text{then } R_3 \longrightarrow \frac{1}{10}R_3 \end{array}$$

or, $$\begin{bmatrix} 1 & 0 & 0 \\ 0 & 1 & 0 \\ 0 & 0 & 1 \end{bmatrix}\begin{bmatrix} x \\ y \\ z \end{bmatrix} = \begin{bmatrix} -1 \\ 2 \\ 1 \end{bmatrix} \qquad \begin{array}{l} \text{by } R_1 \longrightarrow R_1 - R_3 \\ \text{and } R_2 \longrightarrow R_2 - 6R_3 \end{array}$$

Hence, the solution is $x = -1$, $y = 2$ and $z = 1$.

EXAMPLE 5.41. *Solve the system of equations*

$$\begin{aligned} x + 2y - \;\; z &= 6 \\ 2x + 5y + 3z &= 9 \\ 3x + \;\; y + 2z &= 3. \end{aligned}$$

Solution: We have

$$\begin{bmatrix} 1 & 2 & -1 \\ 2 & 5 & 3 \\ 3 & 1 & 2 \end{bmatrix}\begin{bmatrix} x \\ y \\ z \end{bmatrix} = \begin{bmatrix} 6 \\ 9 \\ 3 \end{bmatrix}$$

$$\text{or,} \quad \begin{bmatrix} 1 & 2 & -1 \\ 0 & 1 & 5 \\ 0 & -5 & 5 \end{bmatrix}\begin{bmatrix} x \\ y \\ z \end{bmatrix} = \begin{bmatrix} 6 \\ -3 \\ -15 \end{bmatrix} \quad \begin{array}{l} \text{by } R_2 \longrightarrow R_2 - 2R_1 \\ R_3 \longrightarrow R_3 - 3R_1 \end{array}$$

$$\text{or,} \quad \begin{bmatrix} 1 & 2 & -1 \\ 0 & 1 & 5 \\ 0 & 0 & 6 \end{bmatrix}\begin{bmatrix} x \\ y \\ z \end{bmatrix} = \begin{bmatrix} 6 \\ -3 \\ -6 \end{bmatrix} \quad \begin{array}{l} \text{by } R_3 \longrightarrow \frac{1}{5}R_3 \\ \text{then } R_3 \longrightarrow R_3 + R_2 \end{array}$$

$$\text{or,} \quad \begin{bmatrix} 1 & 0 & -11 \\ 0 & 1 & 5 \\ 0 & 0 & 1 \end{bmatrix}\begin{bmatrix} x \\ y \\ z \end{bmatrix} = \begin{bmatrix} 12 \\ -3 \\ -1 \end{bmatrix} \quad \begin{array}{l} \text{by } R_3 \longrightarrow \frac{1}{6}R_3 \\ \text{then } R_1 \longrightarrow R_1 - 2R_2 \end{array}$$

$$\text{or,} \quad \begin{bmatrix} 1 & 0 & 0 \\ 0 & 1 & 0 \\ 0 & 0 & 1 \end{bmatrix}\begin{bmatrix} x \\ y \\ z \end{bmatrix} = \begin{bmatrix} 1 \\ 2 \\ -1 \end{bmatrix} \quad \begin{array}{l} \text{by } R_1 \longrightarrow R_1 + 11R_3 \\ \text{and } R_2 \longrightarrow R_2 - 5R_3 \end{array}$$

Hence, the solution is $x = 1$, $y = 2$ and $z = -1$.

5.33 Eigenvalues and Eigenvectors

Let A be a square matrix of order n and $\mathbf{X}$ a non-zero vector. Now, if there exists a scalar λ such that

$$A\mathbf{X} = \lambda\mathbf{X}, \tag{5.140}$$

then the vector $\mathbf{X}$ is defined as an eigenvector and λ is defined as an eigenvalue corresponding to the eigenvector $\mathbf{X}$.

Equation (5.140) can also be written as

$$A\mathbf{X} = \lambda I\mathbf{X}, \quad \text{or,} \quad (A - \lambda I)\,\mathbf{X} = 0, \tag{5.141}$$

where I is identity matrix.

Equation (5.141) is a homogeneous matrix equation which can have non-zero solution if and only if the matrix $(A - \lambda I)$ is singular, i.e.,

$$|\, A - \lambda I \,| = 0. \tag{5.142}$$

This is known as *characteristic equation of matrix* and determines the eigenvalues of matrix A.

Equation (5.142) in expanded form is

$$\begin{bmatrix} a_{11} - \lambda & a_{12} & \cdots & a_{1n} \\ a_{21} & a_{22} - \lambda & \cdots & a_{2n} \\ \cdots & \cdots & \cdots & \cdots \\ \cdots & \cdots & \cdots & \cdots \\ a_{n1} & a_{n2} & \cdots & a_{nn} - \lambda \end{bmatrix} = 0 \tag{5.143}$$

$$\text{or,} \quad (-1)^n \left(\lambda^n + \alpha_1 \lambda^{n-1} + \cdots + \alpha_n\right) = 0, \tag{5.144}$$

where $\alpha_1, \alpha_2, \ldots, \alpha_n$ are constants expressed in terms of the coefficients a_{ij} of the matrix A.

It is obvious from equation (5.143) that if the matrix A is having elements in the leading diagonal only, then the characteristic equation is

$$(a_{11} - \lambda)(a_{22} - \lambda) \cdots (a_{nn} - \lambda) = 0, \tag{5.145}$$

which has $a_{11}, a_{22}, \ldots, a_{nn}$ as its roots. That is, the characteristic roots of a diagonal matrix are the elements in its leading diagonal.

After knowing the eigenvalues from equation (5.144), the corresponding eigenvectors $\mathbf{X}$ can be found by solving the homogeneous equations (5.141).

EXAMPLE 5.42. *Find eigenvalues and eigenvectors of the matrix*

$$A = \begin{bmatrix} 2 & -2 & 0 \\ -2 & 1 & -2 \\ 0 & -2 & 0 \end{bmatrix}.$$

Solution: The characteristic equation is

$$\begin{bmatrix} 2-\lambda & -2 & 0 \\ -2 & 1-\lambda & -2 \\ 0 & -2 & -\lambda \end{bmatrix} = 0,$$

$$\text{or,}\quad (2-\lambda)\{-\lambda(1-\lambda) - 4\} + 2\{2\lambda\} = 0,$$

$$\text{i.e.,}\quad \lambda^3 - 3\lambda^2 - 6\lambda + 8 = 0,$$

$$\text{or,}\quad (\lambda - 1)(\lambda + 2)(\lambda - 4) = 0. \quad \therefore \quad \lambda = 1, -2, 4.$$

These are the eigenvalues of the matrix.

For $\lambda = 1$, the eigenvectors are given by

$$\begin{bmatrix} 2-1 & -2 & 0 \\ -2 & 1-1 & -2 \\ 0 & -2 & -1 \end{bmatrix} \begin{bmatrix} \mathbf{x}_1 \\ \mathbf{x}_2 \\ \mathbf{x}_3 \end{bmatrix} = \begin{bmatrix} 0 \\ 0 \\ 0 \end{bmatrix} \quad \text{or,} \quad \begin{bmatrix} 1 & -2 & 0 \\ -2 & 0 & -2 \\ 0 & -2 & -1 \end{bmatrix} \begin{bmatrix} \mathbf{x}_1 \\ \mathbf{x}_2 \\ \mathbf{x}_3 \end{bmatrix} = \begin{bmatrix} 0 \\ 0 \\ 0 \end{bmatrix},$$

which gives the system $\mathbf{x}_1 = 2\mathbf{x}_2$, $2\mathbf{x}_1 = -2\mathbf{x}_3$, $2\mathbf{x}_2 = -\mathbf{x}_3$,

$$\text{i.e.,}\quad \frac{\mathbf{x}_1}{2} = \frac{\mathbf{x}_2}{1} = \frac{\mathbf{x}_3}{-2} = \mathbf{k}_1$$

so that the infinite set of solutions is

$$\begin{bmatrix} \mathbf{x}_1 \\ \mathbf{x}_2 \\ \mathbf{x}_3 \end{bmatrix} = \mathbf{k}_1 \begin{bmatrix} 2 \\ 1 \\ -2 \end{bmatrix}.$$

For $\lambda = -2$, the eigenvectors are given by

$$\begin{bmatrix} 2-(-2) & -2 & 0 \\ -2 & 1-(-2) & -2 \\ 0 & -2 & -(-2) \end{bmatrix} \begin{bmatrix} \mathbf{x}_1 \\ \mathbf{x}_2 \\ \mathbf{x}_3 \end{bmatrix} = \begin{bmatrix} 0 \\ 0 \\ 0 \end{bmatrix},$$

$$\text{or,}\quad \begin{bmatrix} 4 & -2 & 0 \\ -2 & 3 & -2 \\ 0 & -2 & 2 \end{bmatrix} \begin{bmatrix} \mathbf{x}_1 \\ \mathbf{x}_2 \\ \mathbf{x}_3 \end{bmatrix} = \begin{bmatrix} 0 \\ 0 \\ 0 \end{bmatrix},$$

$$\text{or,}\quad \begin{bmatrix} 4 & -2 & 0 \\ -2 & 1 & 0 \\ 0 & -2 & 2 \end{bmatrix} \begin{bmatrix} \mathbf{x}_1 \\ \mathbf{x}_2 \\ \mathbf{x}_3 \end{bmatrix} = \begin{bmatrix} 0 \\ 0 \\ 0 \end{bmatrix} \quad \text{by } R_2 \longrightarrow R_2 + R_3,$$

which gives the system $4\mathbf{x}_1 = 2\mathbf{x}_2$, $2\mathbf{x}_1 = \mathbf{x}_2$, $2\mathbf{x}_2 = 2\mathbf{x}_3$.

The first two equations are same, and from second and third equations, we get

$$\frac{x_1}{1} = \frac{x_2}{2} = \frac{x_3}{2} = k_2$$

so that the eigenvectors are

$$\begin{bmatrix} x_1 \\ x_2 \\ x_3 \end{bmatrix} = k_2 \begin{bmatrix} 1 \\ 2 \\ 2 \end{bmatrix}.$$

For $\lambda = 4$, the eigenvectors are given by

$$\begin{bmatrix} 2-4 & -2 & 0 \\ -2 & 1-4 & -2 \\ 0 & -2 & -4 \end{bmatrix} \begin{bmatrix} x_1 \\ x_2 \\ x_3 \end{bmatrix} = \begin{bmatrix} 0 \\ 0 \\ 0 \end{bmatrix},$$

$$\text{or,} \quad \begin{bmatrix} -2 & -2 & 0 \\ -2 & -3 & -2 \\ 0 & -2 & -4 \end{bmatrix} \begin{bmatrix} x_1 \\ x_2 \\ x_3 \end{bmatrix} = \begin{bmatrix} 0 \\ 0 \\ 0 \end{bmatrix},$$

$$\text{or,} \quad \begin{bmatrix} -2 & -2 & 0 \\ 1 & 0 & -2 \\ 0 & -2 & -4 \end{bmatrix} \begin{bmatrix} x_1 \\ x_2 \\ x_3 \end{bmatrix} = \begin{bmatrix} 0 \\ 0 \\ 0 \end{bmatrix} \quad \text{by } R_2 \longrightarrow R_2 - \tfrac{3}{2}R_1,$$

which gives the system $2x_1 = -2x_2$, $x_1 = 2x_3$, $2x_2 = -4x_3$.

The above system is equivalent to

$$\frac{x_1}{2} = \frac{x_2}{-2} = \frac{x_3}{1} = k_3.$$

Hence, the eigenvectors are

$$\begin{bmatrix} x_1 \\ x_2 \\ x_3 \end{bmatrix} = k_3 \begin{bmatrix} 2 \\ -2 \\ 1 \end{bmatrix}.$$

We shall now elucidate some interesting theoretical aspects of eigenvalues and eigenvectors of matrices.

Theorem 5.17. *The determinant of matrix A is the product of its eigenvalues.*

Proof. From equations (5.142) and (5.144), we have the identity

$$|A - \lambda I| = (-1)^n \left\{\lambda^n + \alpha_1 \lambda^{n-1} + \cdots + \alpha_n\right\}. \tag{5.146}$$

If $\lambda_1, \lambda_2, \ldots, \lambda_n$ are the eigenvalues of A, the roots of the right-hand side polynomial of (5.146) will be $\lambda_1, \lambda_2, \ldots, \lambda_n$. Then this polynomial should be equivalent to

$$(\lambda_1 - \lambda)(\lambda_2 - \lambda) \cdots (\lambda_n - \lambda). \tag{5.147}$$

Therefore, we can write from equations (5.146) and (5.147)

$$|A - \lambda I| = (\lambda_1 - \lambda)(\lambda_2 - \lambda) \cdots (\lambda_n - \lambda). \tag{5.148}$$

Putting $\lambda = 0$ on both sides

$$|A| = \lambda_1 \lambda_2 \cdots \lambda_n. \tag{5.149}$$

Hence, the theorem is proved. □

We also note from equation (5.149) that if the matrix has a zero eigenvalue, then $|A|$ becomes zero, i.e., the matrix is singular.

Theorem 5.18. *If A be a square matrix, then its trace is the sum of its eigenvalues.*

Proof. If A be a square matrix of order n, then

$$\mathrm{Tr}\,(A) = \sum_{i=1}^{n} a_{ii}.$$

Now, if $\lambda_1, \lambda_2, \ldots, \lambda_n$ be eigenvalues of A, then we have the identity

$$|A - \lambda I| = (-1)^n(\lambda - \lambda_1)(\lambda - \lambda_2)\cdots(\lambda - \lambda_n). \tag{5.150}$$

The coefficient of λ^{n-1} in LHS is

$$(-1)^{n-1} \sum_{i=1}^{n} a_{ii} \tag{5.151}$$

and the coefficient of λ^{n-1} in RHS is

$$(-1)^{n+1} \sum_{i=1}^{n} \lambda_i. \tag{5.152}$$

Equating these quantities, we have

$$(-1)^{n-1} \sum_{i=1}^{n} a_{ii} = (-1)^{n+1} \sum_{i=1}^{n} \lambda_i. \quad \therefore \quad \sum_{i=1}^{n} a_{ii} = \sum_{i=1}^{n} \lambda_i. \tag{5.153}$$

Hence, the theorem is proved. □

Both these theorems can be verified in the Example 5.42.

Theorem 5.19. *If λ_1 be an eigenvalue of A of order $n \times n$, and if the rank of the matrix $[A - \lambda_1 I]$ is $(n-k)$, then λ_1 is repeated k times as a root of the characteristics equation of A.*

Proof. If rank of $[A - \lambda_1 I]$ is $n-k$, then any minor of this determinant of order $(n-k)$ is non-zero. Since matrix $[A - \lambda_1 I]$ is of order $n \times n$, k rows of this matrix must be linearly dependent on each other. Hence, λ_1 is repeated k times. □

EXAMPLE 5.43. *Find the eigenvalues of*

$$A = \begin{bmatrix} 1 & 2 & 0 \\ 2 & 4 & 0 \\ 4 & 8 & 0 \end{bmatrix}.$$

Solution: We find that, $|A| = 0$. Hence, the matrix is singular and one of the eigenvalues is $\lambda_1 = 0$.

Again, any minor of second order is zero so that rank of the matrix is < 2. Since matrix A is not a null matrix, its rank is 1. Hence, $\lambda = 0$ will be repeated $(n-k)$, i.e., $(3-1) = 2$ times. Again, sum of eigenvalues is equal to trace of A.

$$\therefore \quad 1 + 4 = 0 + 0 + \lambda_3, \quad \therefore \ \lambda_3 = 5.$$

Thus, eigenvalues are $(0, 0, 5)$.

EXAMPLE 5.44. *Show that the rank of a matrix, every element of which is unity, is 1. Hence, find its eigenvalues.*

Solution: Let the matrix be

$$A = \begin{bmatrix} 1 & 1 & \cdots & 1 \\ 1 & 1 & \cdots & 1 \\ \cdots & \cdots & \cdots & \cdots \\ \cdots & \cdots & \cdots & \cdots \\ 1 & 1 & \cdots & 1 \end{bmatrix} \text{ of order } n \times n.$$

The matrix A is singular, since $|A| = 0$.

Any minor of order 2 and above is zero. Since A is a non-zero matrix, its rank = 1.

Again, since A is singular, $\lambda = 0$ is one of the characteristic root. Since rank = 1, the root $\lambda = 0$ is repeated $(n-1)$ times.

Thus, sum of eigenvalues $= 0 + 0 + \cdots + \lambda_n = \lambda_n$.

But, sum of eigenvalues $= \operatorname{Tr}(A) = n$.

Hence, $\lambda_n = n$.

Thus, all the eigenvalues are zero, excepting one whose value is n.

Theorem 5.20. *Two eigenvectors of a square matrix corresponding to its two distinct eigenvalues are linearly independent.*

Proof. Let A be a square matrix having two distinct eigenvalues λ_1 and λ_2. Let $\mathbf{X}_1$ and $\mathbf{X}_2$ be two eigenvectors corresponding to these λ_1, λ_2 respectively. Then

$$\left.\begin{aligned} A\mathbf{X}_1 &= \lambda_1 \mathbf{X}_1 \\ \text{and} \quad A\mathbf{X}_2 &= \lambda_2 \mathbf{X}_2 \end{aligned}\right\}. \tag{5.154}$$

Consider the relation

$$\xi_1 \mathbf{X}_1 + \xi_2 \mathbf{X}_2 = 0, \tag{5.155}$$

$$\text{i.e.,} \quad \xi_1 A\mathbf{X}_1 + \xi_2 A\mathbf{X}_2 = 0, \tag{5.156}$$

where ξ_1, ξ_2 are two scalars.

In view of equation (5.154), the equation (5.156) becomes

$$\xi_1 \lambda_1 \mathbf{X}_1 + \xi_2 \lambda_2 \mathbf{X}_2 = 0. \tag{5.157}$$

Again, from equation (5.155)

$$\xi_1 \lambda_1 \mathbf{X}_1 + \xi_2 \lambda_1 \mathbf{X}_2 = 0. \tag{5.158}$$

Subtracting

$$\xi_2 (\lambda_1 - \lambda_2) \mathbf{X}_2 = \mathbf{0}.$$

Since $\mathbf{X}_2 \neq 0$, $\lambda_1 - \lambda_2 \neq 0$, this gives $\xi_2 = 0$.

Similarly, we can prove that $\xi_1 = 0$. Hence, relation (5.155) is possible for $\xi_1 = \xi_2 = 0$, i.e., when $\mathbf{X}_1$ and $\mathbf{X}_2$ are independent.

In general, eigenvectors corresponding to distinct eigenvalues are linearly independent. □

Theorem 5.21. *Eigenvalues of a Hermitian matrix are all real.*

Proof. Let A be a Hermitian matrix. Then

$$A^* = A, \tag{5.159}$$

where A^* is its conjugate transpose. Let λ be an eigenvalue corresponding to which eigenvector is $\mathbf{X}$. Then

$$A\mathbf{X} = \lambda\mathbf{X}. \tag{5.160}$$

Pre-multiplying both sides by conjugate transpose of $\mathbf{X}$, i.e., $\mathbf{X}^*$,

$$\begin{aligned} \mathbf{X}^*(A\mathbf{X}) &= \mathbf{X}^*(\lambda\mathbf{X}), \\ \text{i.e.,}\quad \mathbf{X}^*A\mathbf{X} &= \lambda\mathbf{X}^*\mathbf{X}. \end{aligned} \tag{5.161}$$

Taking conjugate transpose on both sides, we get

$$\begin{aligned} (\mathbf{X}^*A\mathbf{X})^* &= (\lambda\mathbf{X}^*\mathbf{X})^* \\ \text{or,}\quad \mathbf{X}^*A^*(\mathbf{X}^*)^* &= \bar{\lambda}\mathbf{X}^*(\mathbf{X}^*)^*. \end{aligned} \tag{5.162}$$

Now, by the property of complex conjugates

$$(\mathbf{X}^*)^* = \mathbf{X} \quad (\textit{see } \text{Section 5.24}). \tag{5.163}$$

Using equations (5.159) and (5.163), equation (5.162) is written as

$$\begin{aligned} \mathbf{X}^*A\mathbf{X} &= \bar{\lambda}\mathbf{X}^*\mathbf{X} \\ \mathbf{X}^*\lambda\mathbf{X} &= \bar{\lambda}\mathbf{X}^*\mathbf{X} \quad \text{i.e.,}\quad (\lambda - \bar{\lambda})\mathbf{X}^*\mathbf{X} = 0. \end{aligned}$$

Thus, $\lambda - \bar{\lambda} = 0$, i.e., $\lambda = \bar{\lambda}$, since $\mathbf{X}^*\mathbf{X} \neq 0$.
Hence, λ is real. □

Theorem 5.22. *Eigenvalues of a skew-Hermitian matrix are purely imaginary or zero.*

Proof. Let A be a skew-Hermitian matrix having λ as an eigenvalue corresponding to which $\mathbf{X}$ is an eigenvector. Then

$$\begin{aligned} A\mathbf{X} &= \lambda\mathbf{X}, \\ \text{or,}\quad i(A\mathbf{X}) &= i(\lambda\mathbf{X}), \\ \text{i.e.,}\quad (iA)\mathbf{X} &= (i\lambda)\mathbf{X}. \end{aligned}$$

This shows that $i\lambda$ is an eigenvalue of matrix (iA) which is a Hermitian. Hence, eigenvalues of iA are all real, i.e., $i\lambda$ should be real. Hence, λ should be either zero or purely imaginary. □

Theorem 5.23. *Eigenvalues of an orthogonal matrix are of unit modulus.*

Proof. Let A be an orthogonal matrix. Then

$$A^T A = I. \tag{5.164}$$

Let X be an eigenvector corresponding to an eigenvalue λ of A. Then

$$A\mathbf{X} = \lambda\mathbf{X}. \tag{5.165}$$

Taking transpose on both sides

$$[A\mathbf{X}]^T = [\lambda\mathbf{X}]^T, \quad \text{or,} \quad \mathbf{X}^T A^T = \lambda\mathbf{X}^T. \tag{5.166}$$

Post-multiplying LHS with $A\mathbf{X}$ and RHS with $\lambda\mathbf{X}$ [since they are equal]

$$\begin{aligned} &\mathbf{X}^T A^T A\mathbf{X} &&= \lambda\mathbf{X}^T\lambda\mathbf{X} \\ \text{or,}\quad &\mathbf{X}^T(A^T A)\mathbf{X} &&= \lambda^2\mathbf{X}^T\mathbf{X}, \\ \text{or,}\quad &\mathbf{X}^T I\mathbf{X} &&= \lambda^2\mathbf{X}^T\mathbf{X} \quad \text{[by equation (5.164)]} \\ \text{or,}\quad &(1-\lambda^2)\mathbf{X}^T\mathbf{X} = 0. \end{aligned}$$

Now, $\mathbf{X}$ and $\mathbf{X}^T$ are non-zero vectors since these are eigenvectors. Hence

$$1 - \lambda^2 = 0, \quad \text{i.e.,} \quad |\lambda| = 1.$$

This proves the theorem. □

Theorem 5.24. *The eigenvalues of unitary matrices are of unit modulus.*

Proof. Let A be a unitary matrix. Then

$$A^* A = I. \tag{5.167}$$

Let λ be an eigenvalue of A and corresponding to this, let $\mathbf{x}$ be an eigenvector. Then

$$A\mathbf{x} = \lambda\mathbf{x}. \tag{5.168}$$

Taking conjugate transpose

$$(A\mathbf{x})^* = (\lambda\mathbf{x})^*, \quad \text{or,} \quad \mathbf{x}^* A^* = \bar{\lambda}\mathbf{x}^*. \tag{5.169}$$

Post-multiplying by $A\mathbf{x}$ in RHS and $\lambda\mathbf{x}$ in LHS, [since they are equal]

$$\begin{aligned} \mathbf{x}^* A^* A\mathbf{x} &= \bar{\lambda}\mathbf{x}^*\lambda\mathbf{x} \\ \text{or,}\quad \mathbf{x}^* I\mathbf{x} &= \bar{\lambda}\lambda\mathbf{x}^*\mathbf{x} \quad \text{[by equation (5.167)]} \\ \text{or,}\quad (1-\bar{\lambda}\lambda)\,\mathbf{x}^*\mathbf{x} &= 0, \end{aligned}$$

which shows that $\bar{\lambda}\lambda = 1$, or, $|\lambda| = 1$, since $\mathbf{x}^*\mathbf{x} \neq 0$.

This proves the theorem. □

Theorem 5.25. *Every matrix satisfies its own characteristic equation. That is, if the characteristic equation of the matrix be*

$$\lambda^n + \alpha_1\lambda^{n-1} + \cdots + \alpha_n = 0,$$

then

$$A^n + \alpha_1 A^{n-1} + \cdots + A_n = 0. \tag{5.170}$$

This theorem is known as Cayley-Hamilton theorem. We here forgo the proof of this theorem.

5.34 Diagonalisation of Matrix

In solving many physical problems using matrices, it becomes desirable to reduce the matrix to a diagonal form having all the non-diagonal elements equal to zero. This process of reducing a matrix to diagonal form is known as *diagonalisation*.

We know that eigenvalues of diagonal matrix are just its diagonal elements. Thus, if a square matrix A is reduced to a diagonal matrix D with the use of some transformation matrix P which is non-singular, such that

$$D = P^{-1}AP, \tag{5.171}$$

then, matrices A and D will have same eigenvalues. In such case, we say that the matrices A and D are *similar*.

Theorem 5.26. *A necessary and sufficient condition for a matrix A to be diagonalisable is that A has all linearly independent eigenvectors.*

Proof. To prove that the condition is necessary, we accept that there exists a non-singular matrix P, such that

$$P^{-1}AP = D. \tag{5.172}$$

Pre-multiplying this equation by P,

$$PP^{-1}AP = PD, \quad \text{i.e.,} \quad AP = PD. \tag{5.173}$$

Let column vectors of P are $\{\mathbf{X}_i\}$, $i = 1, 2, \ldots, n$.
Then from equation (5.173)

$$\begin{aligned} A[\mathbf{X}_1, \mathbf{X}_2, \ldots, \mathbf{X}_n] &= [\mathbf{X}_1, \mathbf{X}_2, \ldots, \mathbf{X}_n]\, D \\ \text{or,}\quad [A\mathbf{X}_1, A\mathbf{X}_2, \ldots, A\mathbf{X}_n] &= [\mathbf{X}_1, \mathbf{X}_2, \ldots, \mathbf{X}_n] \quad \text{diag}\,(\lambda_1, \lambda_2, \ldots, \lambda_n) \\ &= [\lambda_1\mathbf{X}_1, \lambda_2\mathbf{X}_2, \ldots, \lambda_n\mathbf{X}_n], \end{aligned}$$

which gives $A\mathbf{X}_1 = \lambda_1\mathbf{X}_1$, $A\mathbf{X}_2 = \lambda_2\mathbf{X}_2$, $\ldots$, $A\mathbf{X}_n = \lambda_n\mathbf{X}_n$.

Hence, $\mathbf{X}_1$, $\mathbf{X}_2$, $\ldots$, $\mathbf{X}_n$ are eigenvectors of A corresponding to eigenvalues λ_1, λ_2, $\ldots$, λ_n. Therefore, the condition is necessary.

Now, to prove that the condition is sufficient, we assume that the set of vectors $\{\mathbf{X}_i\}$, $i = 1, 2, \ldots, n$ are linearly independent set of eigenvectors of A. If λ_i $(i = 1, 2, \ldots, n)$ be the corresponding eigenvalues, then, by definition

$$\begin{aligned} A\mathbf{X}_i &= \lambda_i\mathbf{X}_i, \quad i = 1, 2, \ldots, n \\ \text{or,}\quad A[\mathbf{X}_1, \mathbf{X}_2, \ldots, \mathbf{X}_n] &= [\lambda_1\mathbf{X}_1, \lambda_2\mathbf{X}_2, \ldots, \lambda_n\mathbf{X}_n] \\ \text{or,}\quad AP &= [\mathbf{X}_1\mathbf{X}_2 \cdots \mathbf{X}_n] \quad \text{diag}\,(\lambda_1\lambda_2 \cdots \lambda_n) \\ \text{or,}\quad AP &= PD, \end{aligned} \tag{5.174}$$

where P is the same column matrix as defined earlier with vectors $\{\mathbf{X}_i\}$.

Since all the columns of P are linearly independent, the matrix P is non-singular and hence its inverse P^{-1} exists.

Therefore, by pre-multiplying with P^{-1}, we get

$$P^{-1}AP = D = \text{diag}\,(\lambda_1, \lambda_2, \ldots, \lambda_n). \tag{5.175}$$

Thus, we get the matrix A diagonalised. Therefore, linear independence of matrix A is a sufficient condition for its diagonalisation. □

Simple matrix. A matrix with all linearly independent eigenvectors is called *simple*. Thus, the identity matrix of order n is simple. A matrix which is not simple is called *defective*. *Every real and symmetric matrix is simple, and every simple matrix is diagonalisable.*

Thus, in view of Theorem 5.27, we can say: *For every simple matrix A, there exists a non-singular matrix P such that $P^{-1}AP = D$.*

Now, by Theorem 20 we know that the eigenvectors corresponding to distinct eigenvalues of a matrix are linearly independent. Conversely, it can also be proved (it is left for the readers) that if a set of eigenvectors of a matrix are linearly independent, then the corresponding eigenvalues should be distinct. In view of this, we can say that, a simple matrix should have all its eigenvalues distinct, since all its eigenvectors are linearly independent. Thus, we have a theorem:

Theorem 5.27. *If a matrix A has all distinct eigenvalues, then A is simple.*

EXAMPLE 5.45. *Test whether the matrix*

$$A = \begin{bmatrix} \alpha & 1 & 0 \\ 0 & \beta & 1 \\ 0 & 0 & \gamma \end{bmatrix}, \quad \alpha \neq \beta \neq \gamma$$

is simple or not.

Solution: Characteristic equation of the matrix is

$$\begin{bmatrix} \alpha-\lambda & 1 & 0 \\ 0 & \beta-\lambda & 1 \\ 0 & 0 & \gamma-\lambda \end{bmatrix} = 0, \quad \text{i.e.,} \quad (\alpha-\lambda)(\beta-\lambda)(\gamma-\lambda) = 0.$$

Hence, roots are $\lambda_1 = \alpha$, $\lambda_2 = \beta$, $\lambda_3 = \gamma$. Since $\alpha \neq \beta \neq \gamma$, the eigenvalues λ_1, λ_2, λ_3 are distinct. Hence, by Theorem 27, the matrix is simple.

EXAMPLE 5.46. *Reduce the matrix*

$$A = \begin{bmatrix} 1 & 2 & -2 \\ 2 & 1 & 2 \\ -2 & 2 & 1 \end{bmatrix}$$

to a diagonal form.

Solution: The characteristic equation of the matrix A is

$$\begin{bmatrix} 1-\lambda & 2 & -2 \\ 2 & 1-\lambda & 2 \\ -2 & 2 & 1-\lambda \end{bmatrix} = 0$$

$$\text{or,} \quad (\lambda-3)(\lambda^2-9) = 0,$$

which gives $\lambda_1 = 3$, $\lambda_2 = 3$, $\lambda_3 = -3$.

For $\lambda = 3$, the eigenvectors are given by

$$\begin{bmatrix} -2 & 2 & -2 \\ 2 & -2 & 2 \\ -2 & 2 & -2 \end{bmatrix} \begin{bmatrix} x_1 \\ x_2 \\ x_3 \end{bmatrix} = \begin{bmatrix} 0 \\ 0 \\ 0 \end{bmatrix},$$

$$\text{or,} \quad \begin{bmatrix} 0 & 0 & 0 \\ 0 & 0 & 0 \\ -2 & 2 & -2 \end{bmatrix} \begin{bmatrix} x_1 \\ x_2 \\ x_3 \end{bmatrix} = \begin{bmatrix} 0 \\ 0 \\ 0 \end{bmatrix}, \quad \begin{matrix} \text{by } R_1 \longrightarrow R_1 - R_3 \\ \text{and } R_2 \longrightarrow R_2 + R_3 \end{matrix}$$

which is equivalent to $-2x_1 + 2x_2 - 2x_3 = 0$.

Thus, the eigenvectors are $k_1(1,1,0)$ and $k_2(1,2,1)$; k_1, k_2 = constants.

For $\lambda = -3$, the eigenvector is given by

$$\begin{bmatrix} 4 & 2 & -2 \\ 2 & 4 & 2 \\ -2 & 2 & 4 \end{bmatrix} \begin{bmatrix} \mathbf{x}_1 \\ \mathbf{x}_2 \\ \mathbf{x}_3 \end{bmatrix} = \begin{bmatrix} 0 \\ 0 \\ 0 \end{bmatrix},$$

$$\text{or,} \quad \begin{bmatrix} 6 & 6 & 0 \\ 0 & 6 & 6 \\ -2 & 2 & 4 \end{bmatrix} \begin{bmatrix} \mathbf{x}_1 \\ \mathbf{x}_2 \\ \mathbf{x}_3 \end{bmatrix} = \begin{bmatrix} 0 \\ 0 \\ 0 \end{bmatrix}, \quad \begin{array}{l} \text{by } R_1 \longrightarrow R_1 + R_2 \\ \text{and } R_2 \longrightarrow R_2 + R_3 \end{array}$$

which is equivalent to the system $\mathbf{x}_1 + \mathbf{x}_2 = 0, \mathbf{x}_2 + \mathbf{x}_3 = 0, -\mathbf{x}_1 + \mathbf{x}_2 + 2\mathbf{x}_3 = 0$, i.e., $\frac{\mathbf{x}_1}{1} = -\frac{\mathbf{x}_2}{1} = \frac{\mathbf{x}_3}{1}$.

Hence, the eigenvector is $k_3(1,-1,1)$.

The three eigenvectors are linearly independent which can be verified easily. Hence by Theorem 26, the matrix A is diagonalisable.

Now, let

$$P = \begin{bmatrix} 1 & 1 & 1 \\ 1 & 2 & -1 \\ 0 & 1 & 1 \end{bmatrix}.$$

Then, $|P| = 1 \cdot (2+1) - 1 + 1 = 3$.

The cofactors are

$$\begin{array}{lll} A_{11} = \ \ 3, & A_{12} = \ \ 0, & A_{13} = -3, \\ A_{21} = -1, & A_{22} = \ \ 1, & A_{23} = \ \ 2, \\ A_{31} = \ \ 1, & A_{32} = -1, & A_{33} = \ \ 1. \end{array}$$

Thus

$$P^{-1} = \frac{1}{3} \begin{bmatrix} 3 & 0 & -3 \\ -1 & 1 & 2 \\ 1 & -1 & 1 \end{bmatrix}.$$

Now

$$AP = \begin{bmatrix} 1 & 2 & -2 \\ 2 & 1 & 2 \\ -2 & 2 & 1 \end{bmatrix} \begin{bmatrix} 1 & 1 & 1 \\ 1 & 2 & -1 \\ 0 & 1 & 1 \end{bmatrix} = \begin{bmatrix} 3 & 3 & -3 \\ 3 & 6 & 3 \\ 0 & 3 & -3 \end{bmatrix}.$$

$$\therefore \quad P^{-1}AP = \frac{1}{3} \begin{bmatrix} 3 & 0 & -3 \\ -1 & 1 & 2 \\ 1 & -1 & 1 \end{bmatrix} \begin{bmatrix} 3 & 3 & -3 \\ 3 & 6 & 3 \\ 0 & 3 & -3 \end{bmatrix}$$

$$= \frac{1}{3} \begin{bmatrix} 9 & 0 & 0 \\ 0 & 9 & 0 \\ 0 & 0 & -9 \end{bmatrix} = \begin{bmatrix} 3 & 0 & 0 \\ 0 & 3 & 0 \\ 0 & 0 & -3 \end{bmatrix},$$

which is the required diagonal form.

5.35 Power and Exponential of a Square Matrix

Let A be a square matrix which is diagonalisable. If P be the diagonalising matrix, then

$$P^{-1}AP = D, \quad \text{or,} \quad A = PDP^{-1}. \tag{5.176}$$

Thus, any power of A, say A^n, is

$$A^n = PD^nP^{-1}. \tag{5.177}$$

The exponential of a square matrix A is defined as

$$e^A = I + A + \frac{A^2}{2!} + \frac{A^3}{3!} + \cdots + \frac{A^k}{k!} + \cdots$$

If P be the diagonalising matrix for A, then $P^{-1}AP = D$.
Then, we have

$$P^{-1}e^AP = \sum_{i=0}^{n} P^{-1}\left(\frac{A^k}{k!}\right)P$$

$$\text{i.e.,}\quad P^{-1}e^AP = I + D + \frac{D^2}{2!} + \cdots = e^D.$$

$$\therefore\quad e^A = Pe^DP^{-1}. \tag{5.178}$$

EXAMPLE 5.47. *If* $A = \begin{bmatrix} 1 & 1 & 1 \\ 1 & 1 & 1 \\ 1 & 1 & 1 \end{bmatrix}$, *find* A^{10} *and* e^A.

Solution: The characteristic equation of matrix A is

$$\begin{bmatrix} 1-\lambda & 1 & 1 \\ 1 & 1-\lambda & 1 \\ 1 & 1 & 1-\lambda \end{bmatrix} = 0, \quad \text{or,} \quad \lambda^3 - 3\lambda^2 = 0$$

which gives, $\lambda = 0, 0, 3$.

For $\lambda = 0$, the eigenvector is given by

$$\begin{bmatrix} 1 & 1 & 1 \\ 1 & 1 & 1 \\ 1 & 1 & 1 \end{bmatrix}\begin{bmatrix} \mathbf{x}_1 \\ \mathbf{x}_2 \\ \mathbf{x}_3 \end{bmatrix} = \begin{bmatrix} 0 \\ 0 \\ 0 \end{bmatrix},$$

which gives only one equation, i.e., $\mathbf{x} + \mathbf{y} + \mathbf{z} = 0$.

We accept the solutions $(-1, 1, 0)$ and $(-1, 0, 1)$ for two cases of $\lambda = 0$.

For $\lambda = 3$, the eigenvectors are given by

$$\begin{bmatrix} -2 & 1 & 1 \\ 1 & -2 & 1 \\ 1 & 1 & -2 \end{bmatrix}\begin{bmatrix} \mathbf{x}_1 \\ \mathbf{y}_1 \\ \mathbf{z}_1 \end{bmatrix} = \begin{bmatrix} 0 \\ 0 \\ 0 \end{bmatrix},$$

$$\text{or,}\quad \begin{bmatrix} -3 & +3 & 0 \\ 0 & -3 & +3 \\ 1 & 1 & -2 \end{bmatrix}\begin{bmatrix} \mathbf{x}_1 \\ \mathbf{y}_1 \\ \mathbf{z}_1 \end{bmatrix} = \begin{bmatrix} 0 \\ 0 \\ 0 \end{bmatrix} \quad \begin{array}{l}\text{by } R_1 \longrightarrow R_1 - R_2 \\ \text{and } R_2 \longrightarrow R_2 - R_3\end{array}$$

This gives, $\mathbf{x}_1 - \mathbf{y}_1 = 0$, $\mathbf{y}_1 - \mathbf{z}_1 = 0$, $\mathbf{x}_1 + \mathbf{y}_1 - 2\mathbf{z}_1 = 0$, i.e., $\mathbf{x}_1 = \mathbf{y}_1 = \mathbf{z}_1$.

We accept the solution $(1, 1, 1)$.

Therefore

$$P = \begin{bmatrix} -1 & -1 & 1 \\ 1 & 0 & 1 \\ 0 & 1 & 1 \end{bmatrix} \tag{5.179}$$

$$\det|P| = -1(-1) + 1(+1) + 1(1) = 3.$$

Cofactors are

$$\begin{aligned} &A_{11} = -1, \quad && A_{12} = 2, \quad && A_{13} = -1, \\ &A_{21} = -1, && A_{22} = -1, && A_{23} = 2, \\ &A_{31} = 1, && A_{32} = 1, && A_{33} = 1. \end{aligned}$$

Hence

$$P^{-1} = \frac{1}{3}\begin{bmatrix} -1 & 2 & -1 \\ -1 & -1 & 2 \\ 1 & 1 & 1 \end{bmatrix}. \tag{5.180}$$

$$\therefore \quad AP = \begin{bmatrix} 1 & 1 & 1 \\ 1 & 1 & 1 \\ 1 & 1 & 1 \end{bmatrix}\begin{bmatrix} -1 & -1 & 1 \\ 1 & 0 & 1 \\ 0 & 1 & 1 \end{bmatrix} = \begin{bmatrix} 0 & 0 & 3 \\ 0 & 0 & 3 \\ 0 & 0 & 3 \end{bmatrix}.$$

$$\therefore \quad D = P^{-1}AP = \frac{1}{3}\begin{bmatrix} -1 & 2 & -1 \\ -1 & -1 & 2 \\ 1 & 1 & 1 \end{bmatrix}\begin{bmatrix} 0 & 0 & 3 \\ 0 & 0 & 3 \\ 0 & 0 & 3 \end{bmatrix}$$

$$= \frac{1}{3}\begin{bmatrix} 0 & 0 & 0 \\ 0 & 0 & 0 \\ 0 & 0 & 9 \end{bmatrix} = \begin{bmatrix} 0 & 0 & 0 \\ 0 & 0 & 0 \\ 0 & 0 & 3 \end{bmatrix}.$$

$$\therefore \quad A^{10} = PD^{10}P^{-1}$$

$$= \frac{1}{3}\begin{bmatrix} -1 & -1 & 1 \\ 1 & 0 & 1 \\ 0 & 1 & 1 \end{bmatrix}\begin{bmatrix} 0 & 0 & 0 \\ 0 & 0 & 0 \\ 0 & 0 & 3^{10} \end{bmatrix}\begin{bmatrix} -1 & 2 & -1 \\ -1 & -1 & 2 \\ 1 & 1 & 1 \end{bmatrix}$$

$$= \frac{1}{3}\begin{bmatrix} -1 & -1 & 1 \\ 1 & 0 & 1 \\ 0 & 1 & 1 \end{bmatrix}\begin{bmatrix} 0 & 0 & 0 \\ 0 & 0 & 0 \\ 3^{10} & 3^{10} & 3^{10} \end{bmatrix}$$

$$= \frac{1}{3}\begin{bmatrix} 3^{10} & 3^{10} & 3^{10} \\ 3^{10} & 3^{10} & 3^{10} \\ 3^{10} & 3^{10} & 3^{10} \end{bmatrix} = 3^{29}\begin{bmatrix} 1 & 1 & 1 \\ 1 & 1 & 1 \\ 1 & 1 & 1 \end{bmatrix}.$$

Again

$$e^A = Pe^DP^{-1}$$

$$= \frac{1}{3}\begin{bmatrix} -1 & -1 & 1 \\ 1 & 0 & 1 \\ 0 & 1 & 1 \end{bmatrix}\begin{bmatrix} 0 & 0 & 0 \\ 0 & 0 & 0 \\ 0 & 0 & e^3 \end{bmatrix}\begin{bmatrix} -1 & 2 & -1 \\ -1 & -1 & 2 \\ 1 & 1 & 1 \end{bmatrix}$$

$$= \frac{1}{3}e^9\begin{bmatrix} 1 & 1 & 1 \\ 1 & 1 & 1 \\ 1 & 1 & 1 \end{bmatrix}.$$

5.36 Transformation Matrices

Let there be two sets of n variables x_i, y_i; $i = 1, 2, \ldots, n$ connected by the linear equations

$$y_i = \sum a_{ij} x_j. \tag{5.181}$$

The set of equations (5.181) clarifies that for any given set of values for x_i, there corresponds a definite value for y_i. This set of equations is defined as a linear transformation connecting the variables y_i with x_i. The equation (5.181) in expanded form looks like

$$\begin{aligned} y_1 &= a_{11}x_1 + a_{12}x_2 + \cdots + a_{1n}x^n \\ y_2 &= a_{21}x_1 + a_{22}x_2 + \cdots + a_{2n}x^n \\ &\cdots\cdots\cdots\cdots\cdots\cdots \\ &\cdots\cdots\cdots\cdots\cdots\cdots \\ y_n &= a_{n1}x_1 + a_{n2}x_2 + \cdots + a_{nn}x^n \end{aligned}$$

The matrix formed by the coefficients of the set of above equations is said to be a transformation matrix which is

$$A = \begin{bmatrix} a_{11} & a_{12} & \cdots & a_{1n} \\ a_{21} & a_{22} & \cdots & a_{2n} \\ \cdots & \cdots & \cdots & \cdots \\ \cdots & \cdots & \cdots & \cdots \\ a_{n1} & a_{n2} & \cdots & a_{nn} \end{bmatrix}. \tag{5.182}$$

If now we define the following matrices with the sets of two variables as elements like

$$\mathbf{X} = \begin{bmatrix} x_1 \\ x_2 \\ \vdots \\ x_n \end{bmatrix} \quad \text{and} \quad \mathbf{Y} = \begin{bmatrix} y_1 \\ y_2 \\ \vdots \\ y_n \end{bmatrix} \tag{5.183}$$

then the equations (5.181) can be written as

$$\mathbf{Y} = A\mathbf{X}. \tag{5.184}$$

The matrix A may or may not be a non-singular. If A is non-singular, then the transformation is called *non-singular.* In this case A^{-1} exists, so that equation (5.184) can be written as

$$\mathbf{X} = A^{-1}\mathbf{Y}. \tag{5.185}$$

This shows that the vectors $\mathbf{X}$ and $\mathbf{Y}$ are in the same vector space V and they maintain a one-to-one correspondence between them.

Consider the linear transformation defined by equations (5.184) or (5.185) when the vectors $\mathbf{X}$ and $\mathbf{Y}$ are orthogonal. In this case, the inner product of $\mathbf{X}$ and $\mathbf{Y}$ (*see* Section 5.8) should vanish, i.e ,

$$(\mathbf{X}, \mathbf{Y}) = 0. \tag{5.186}$$

Moreover, the norms of lengths of vectors $\mathbf{X}$ and $\mathbf{Y}$ are

$$\| \mathbf{X} \| = \left(\sum_{i=1}^{n} x_i^2 \right)^{1/2} \tag{5.187}$$

$$\text{and} \quad \| \mathbf{Y} \| = \left(\sum_{i=1}^{n} y_i^2 \right)^{1/2}, \quad \text{respectively.} \tag{5.188}$$

The transformation is said to be orthogonal if the norms or lengths of vectors **X** and **Y** are preserved. That is

$$\sum_{i=1}^{n} x_i^2 = \sum_{i=1}^{n} y_i^2,$$

$$\text{or,} \quad \sum_{i=1}^{n} x_i^2 = \sum_i \left[\sum_j a_{ij} x_j \sum_k a_{ik} x_k \right] \quad \text{[using (5.181)]}$$

$$= \sum_{j,k} x_j x_k \sum_i a_{ij} a_{ik}.$$

Comparing coefficients of x_i^2 on both sides, we get

$$\sum a_{ij} a_{ik} = \delta_{jk} \tag{5.189}$$

so that

$$\left.\begin{aligned} \delta_{jk} &= 1, \quad \text{for } j = k \\ \text{and} \quad \delta_{jk} &= 0, \quad \text{for } j \neq k \end{aligned}\right\}. \tag{5.190}$$

The conditions (5.189) can be expressed in terms of the matrix A in the form

$$A^T A = I, \tag{5.191}$$

which is exactly the condition for the matrix A to be orthogonal (*see* Section 5.19). In fact, the condition is necessary as well as sufficient for the transformation to be orthogonal.

Let us now consider another set of variables z_i $(i = 1, 2, \ldots, n)$ such that two linear transformations given by

$$\mathbf{Y} = P\mathbf{X} \quad \text{and} \quad \mathbf{Z} = Q\mathbf{Y} \tag{5.192}$$

exist. This immediately suggests that the vector **Z** is also associated with the vector **X** by the matrix equation

$$\mathbf{Z} = Q\mathbf{Y} = QP\mathbf{X}. \tag{5.193}$$

Thus, here QP is the transformation matrix corresponding to the simultaneous effect of two linear transformations given by (5.192). Again, if the two linear transformations are orthogonal, then the resultant transformation given by (5.193) will also be orthogonal. Because

$$\begin{aligned} [PQ][PQ]^T &= PQQ^T P^T \\ &= PIP^T \qquad (\because Q \text{ is orthogonal}) \\ &= PP^T = I \quad (\because P \text{ is orthogonal}). \end{aligned}$$

We shall use these ideas in the next chapter for the transformation of coordinate systems.

5.37 Systems of Linear Differential Equations

Let us consider a system of linear first-order differential equations

$$\left.\begin{aligned} x' &= a_1 x + b_1 y + c_1 z \\ y' &= a_2 x + b_2 y + c_2 z \\ z' &= a_3 x + b_3 y + c_3 z \end{aligned}\right\}, \tag{5.194}$$

where a_i, b_i, c_i are constants and primes denote differentiation with respect to variable t.

The system of equations can be written in matrix form

$$\mathbf{X}' = A\mathbf{X}, \tag{5.195}$$

where

$$\mathbf{A} = \begin{bmatrix} a_1 & b_1 & c_1 \\ a_2 & b_2 & c_2 \\ a_3 & b_3 & c_3 \end{bmatrix}, \text{ and } \mathbf{X} = \begin{bmatrix} x \\ y \\ z \end{bmatrix}. \tag{5.196}$$

Such a system is known as *homogeneous.* We assume the solution of (5.195) in the form

$$\mathbf{X}(t) = \boldsymbol{\xi} e^{\lambda t}, \quad \boldsymbol{\xi} = \text{constant}. \tag{5.197}$$

Then from equation (5.195) we get, by using (5.197)

$$\mathbf{A}\xi = \boldsymbol{\lambda\xi}, \quad \text{since } e^{\lambda t} \neq 0. \tag{5.198}$$

Thus, the problem of solving the system of linear differential equations with constant coefficients reduces to solving the corresponding algebraic eigenvalue problem.

We assume that the matrix A is simple having linearly independent eigenvectors $\boldsymbol{\xi}_1$, $\boldsymbol{\xi}_2$, $\boldsymbol{\xi}_3$ corresponding to the three eigenvalues λ_1, λ_2, λ_3, so that the solution of the system (5.195) is given by

$$\mathbf{X}(t) = \boldsymbol{\xi}_i e^{\lambda_i t}, \quad i = 1, 2, 3. \tag{5.199}$$

The general solution of the equation is

$$\begin{aligned} \mathbf{X}(t) &= c_1 x + c_2 y + c_3 z \\ &= c_1 \boldsymbol{\xi}_1 e^{\lambda_1 t} + c_2 \boldsymbol{\xi}_2 e^{\lambda_2 t} + c_3 \boldsymbol{\xi}_3 e^{\lambda_3 t}, \end{aligned} \tag{5.200}$$

where c_1, c_2, c_3 are constants to be determined from initial conditions. If we assume that, at $t = 0$, $\mathbf{X}(0) = \mathbf{x}_0$, then

$$\begin{bmatrix} x_0 \\ y_0 \\ z_0 \end{bmatrix} = c_1 \xi_1 + c_2 \xi_2 + c_3 \xi_3,$$

$$\text{or,} \quad [\mathbf{x}_0] = [\boldsymbol{\xi}]\,[\mathbf{c}], \quad \text{so that } [\mathbf{c}] = \boldsymbol{\xi}^{-1}\mathbf{x}_0, \tag{5.201}$$

where

$$[\boldsymbol{\xi}] = [\boldsymbol{\xi}_1, \ \boldsymbol{\xi}_2, \ \boldsymbol{\xi}_3], \quad \text{and} \quad [\mathbf{c}] = \begin{bmatrix} c_1 \\ c_2 \\ c_3 \end{bmatrix}. \tag{5.202}$$

This discussion can easily be extended to a system of n-equations.

EXAMPLE 5.48. *Find solution of the system*

$$\left.\begin{aligned} x' &= x + 2y - 2z \\ y' &= 2x + y + 2z \\ z' &= -2x + 2y + z \end{aligned}\right\} \text{ with condition, } \begin{bmatrix} x(0) \\ y(0) \\ z(0) \end{bmatrix} = \begin{bmatrix} 1 \\ 1 \\ 1 \end{bmatrix}.$$

Solution: Here, $A = \begin{bmatrix} 1 & 2 & -2 \\ 2 & 1 & 2 \\ -2 & 2 & 1 \end{bmatrix}$.

Characteristic roots are $\lambda_1 = 3$, $\lambda_2 = 3$, $\lambda_3 = -3$ and corresponding eigenvectors are (*see* Example 5.46)

$$\xi_1 = \begin{bmatrix} 1 \\ 1 \\ 0 \end{bmatrix}, \quad \xi_2 = \begin{bmatrix} 1 \\ 2 \\ 1 \end{bmatrix}, \quad \xi_3 = \begin{bmatrix} 1 \\ -1 \\ 1 \end{bmatrix}.$$

Thus, the general solution is

$$\mathbf{X}(t) = c_1 \begin{bmatrix} 1 \\ 1 \\ 0 \end{bmatrix} e^{3t} + c_2 \begin{bmatrix} 1 \\ 2 \\ 1 \end{bmatrix} e^{3t} + c_3 \begin{bmatrix} 1 \\ -1 \\ 1 \end{bmatrix} e^{-3t}.$$

Using the initial conditions

$$\begin{bmatrix} 1 & 1 & 1 \\ 1 & 2 & -1 \\ 0 & 1 & 1 \end{bmatrix} \begin{bmatrix} c_1 \\ c_2 \\ c_3 \end{bmatrix} = \begin{bmatrix} 1 \\ 1 \\ 1 \end{bmatrix}$$

we find

$$\begin{bmatrix} c_1 \\ c_2 \\ c_3 \end{bmatrix} = \frac{1}{3} \begin{bmatrix} 3 & 0 & -3 \\ -1 & 1 & 2 \\ 1 & -1 & 1 \end{bmatrix} \begin{bmatrix} 1 \\ 1 \\ 1 \end{bmatrix} = \frac{1}{3} \begin{bmatrix} 0 \\ 2 \\ 1 \end{bmatrix}.$$

Hence, the solution is

$$\mathbf{X}(t) = \frac{2}{3} \begin{bmatrix} 1 \\ 2 \\ 1 \end{bmatrix} e^{3t} + \frac{1}{3} \begin{bmatrix} 1 \\ -1 \\ 1 \end{bmatrix} e^{-3t}.$$

5.38 Non-homogeneous Systems

The non-homogeneous systems of linear differential equations are given by

$$\mathbf{X}' = \mathbf{AX} + \mathbf{F}, \tag{5.203}$$

where

$$\mathbf{F} = \begin{bmatrix} F_1(t) \\ F_2(t) \\ F_3(t) \end{bmatrix}. \tag{5.204}$$

As before, we assume that the matrix A is simple so that it has linearly independent set of eigenvectors. We have found the solution for homogeneous system in the form of equation (5.200) which can be written as

$$\mathbf{X}(t) = \psi(t)c, \quad c = \{c_1, c_2, c_3\}, \tag{5.205}$$

where

$$\psi(t) = \left[\xi_1 e^{\lambda_1 t}, \xi_2 e^{\lambda_2 t}, \xi_3 e^{\lambda_3 t} \right]. \tag{5.206}$$

Obviously

$$\psi(\mathbf{0}) = [\xi_1, \xi_2, \xi_3]$$

which is non-singular.

We now find the particular solution of non-homogenuous system by the method of variation of parameters. Let us assume that

$$\mathbf{X}_p(t) = \boldsymbol{\psi}(t)\boldsymbol{\xi}(t) \tag{5.207}$$

is a solution of equation (5.203). Then

$$\mathbf{X}'_p(t) = \boldsymbol{\psi}'(t)\boldsymbol{\xi}(t) + \boldsymbol{\xi}'(t)\boldsymbol{\psi}(t). \tag{5.208}$$

Now

$$\begin{aligned}\boldsymbol{\psi}'(t) &= \left[\lambda_1\,\xi_1\,e^{\lambda_1 t},\ \lambda_2\,\xi_2\,e^{\lambda_2 t},\ \lambda_3\,\xi_3\,e^{\lambda_3 t}\right] \\ &= \left[A\xi_1\,e^{\lambda_1 t},\ A\xi_2\,e^{\lambda_2 t},\ A\xi_3\,e^{\lambda_3 t}\right]\\ \text{or,}\quad \boldsymbol{\psi}'(t) &= \mathbf{A}\boldsymbol{\psi}(t).\end{aligned} \tag{5.209}$$

Hence, equation (5.208) becomes

$$\mathbf{X}'_p(t) = A\boldsymbol{\psi}(t)\boldsymbol{\xi}(t) + \boldsymbol{\psi}(t)\boldsymbol{\xi}'(t). \tag{5.210}$$

Again, since $\mathbf{X}_p(t)$ is assumed to be a solution of the equation (5.203), then

$$\mathbf{X}'_p(t) = \mathbf{A}\,\boldsymbol{\psi}(t)\,\boldsymbol{\xi}(t) + \mathbf{F}(t). \tag{5.211}$$

Hence, from equations (5.210) and (5.211), we find

$$\boldsymbol{\psi}(t)\boldsymbol{\xi}'(t) = \mathbf{F}(t). \tag{5.212}$$

This is the condition that the assumed solution (5.207) actually does satisfy the system of differential equation (5.203). From this condition, we can find $\boldsymbol{\xi}(t)$ by integration, and then putting into (5.207), we get the final solution.

EXAMPLE 5.49. *Find a particular solution of*

$$\begin{aligned}x' &= x + 2y + e^t\\ y' &= 3x + 2y + e^{4t}.\end{aligned}$$

Solution: Here, $A = \begin{bmatrix} 1 & 2 \\ 3 & 2 \end{bmatrix}$, which has eigenvalues $\lambda_1 = 4$, $\lambda_2 = -1$ and the corresponding eigenvectors are $(2, 3)$ and $(1, -1)$.

$$\therefore\quad \boldsymbol{\psi}(t) = \begin{bmatrix} 2e^{4t} & e^{-t} \\ 3e^{4t} & -e^{-t} \end{bmatrix}.$$

$$\therefore\quad \boldsymbol{\psi}'(t) = \begin{bmatrix} 8e^{4t} & -e^{-t} \\ 12e^{4t} & e^{t} \end{bmatrix} = \begin{bmatrix} 1 & 2 \\ 3 & 2 \end{bmatrix}\boldsymbol{\psi}(t),$$

which verifies equation (5.209).

The condition given by equation (5.212) is

$$\begin{bmatrix} 2e^{4t} & e^{-t} \\ 3e^{4t} & -e^{-t} \end{bmatrix}\boldsymbol{\xi}' = \begin{bmatrix} e^{t} \\ e^{4t} \end{bmatrix},$$

$$\text{or,}\quad \begin{bmatrix} 5e^{4t} & 0 \\ 0 & e^{-t} \end{bmatrix}\boldsymbol{\xi}' = \begin{bmatrix} e^{t} + e^{4t} \\ \frac{2}{5}e^{4t} - \frac{3}{5}e^{t} \end{bmatrix} \qquad \begin{array}{l}\text{by } R_1 \longrightarrow R_1 + R_2 \\ \text{then } R_2 \longrightarrow R_2 - \frac{3}{5}R_1\end{array}$$

$$\therefore \quad \boldsymbol{\xi}'(t) = \begin{bmatrix} \frac{e^{-3t}}{5} + \frac{1}{5} \\ \frac{2}{5}e^{5t} - \frac{3}{5}e^{2t} \end{bmatrix}. \qquad \therefore \quad \boldsymbol{\xi}(t) = \begin{bmatrix} -\frac{e^{-3t}}{15} + \frac{1}{5}t \\ \frac{2}{25}e^{5t} - \frac{3}{10}e^{2t} \end{bmatrix}.$$

$$\therefore \quad \mathbf{X}_p(t) = \psi(t)\boldsymbol{\xi}(t) = \begin{bmatrix} 2e^{4t} & e^{-t} \\ 3e^{4t} & -e^{-t} \end{bmatrix} \begin{bmatrix} -\frac{1}{15}e^{-3t} + \frac{1}{5}t \\ \frac{2}{25}e^{5t} - \frac{3}{10}e^{2t} \end{bmatrix},$$

$$\text{i.e.,} \quad \mathbf{X}_p(t) = \begin{bmatrix} \frac{2}{5}te^{4t} + \frac{2}{25}e^{4t} - \frac{13}{30}e^{t} \\ \frac{3}{5}te^{4t} - \frac{2}{25}e^{4t} + \frac{1}{10}e^{t} \end{bmatrix}.$$

5.39 Particular Forms for F(t)

The particular integral can be found easily if the so-called forcing function $F(t)$ has particular forms like (i) $\mathbf{F}(t) = \mathbf{a}$, a constant, (ii) $\mathbf{F}(t) = \mathbf{a}e^{\omega t}$, and (iii) $\mathbf{F}(t) = \mathbf{a}e^{i\omega t}$, where ω is not an eigenvalue of the matrix A.

In the case of (i), we assume the particular solution as $\mathbf{X}_p(t) = \mathbf{k}$, a constant. Then, substituting into the differential equation

$$\mathbf{X}' = \mathbf{AX} - \mathbf{F}$$

we find

$$\mathbf{0} = \mathbf{Ak} - \mathbf{a}$$

so that

$$\mathbf{k} = \mathbf{A}^{-1}\mathbf{a}. \tag{5.213}$$

In the case of (ii), we assume the particular solution in the form

$$\mathbf{X}_p(t) = \mathbf{k}e^{\omega t} \tag{5.214}$$

so that, after substituting into the differential equation, we get

$$(\mathbf{A} - \omega\mathbf{I})\mathbf{k} = \mathbf{a}. \tag{5.215}$$

Since ω is not an eigenvalue of A, equation (5.213) will provide a solution for k. If ω happens to be an eigenvalue of A, then the matrix $(\mathbf{A} - \omega\mathbf{I})$ will be a singular one due to which it cannot be inverted. Thus, the solution is acceptable only when ω is not an eigenvalue of the matrix A.

In the case of (iii), we write the solution

$$\mathbf{X}_p(t) = \operatorname{Re}\{\mathbf{X}_p(t)\} + i\operatorname{Im}\{\mathbf{X}_p(t)\}.$$

Since, $e^{i\omega t} = \cos\omega t + i\sin\omega t$, so that

$$\operatorname{Re}(e^{i\omega t}) = \cos\omega t$$
$$\text{and} \quad \operatorname{Im}(e^{i\omega t}) = \sin\omega t$$

we can break the differential equation into two as

$$[\operatorname{Re}\mathbf{X}_p(t)]' = A[\operatorname{Re}\mathbf{X}_p(t)] - \mathbf{a}\cos\omega t \tag{5.216}$$
$$\text{and} \quad [\operatorname{Im}\mathbf{X}_p(t)]' = A[\operatorname{Im}\mathbf{X}_p(t)] - \mathbf{a}\sin\omega t. \tag{5.217}$$

Thus, if the forcing function $F(t)$ is of the form $\sin\omega t$ or $\cos\omega t$ (I repeat, ω is not an eigenvalue of the matrix A), we assume the particular solution as

$$\mathbf{X}_p(t) = \mathbf{k}e^{i\omega t}$$

and find the value of $\mathbf{k}$. Then, we take the real part of the solution if

$$\mathbf{F}(t) = \mathbf{a}\cos\omega t \tag{5.218}$$

and the imaginary part, if

$$\mathbf{F}(t) = \mathbf{a}\sin\omega t. \tag{5.219}$$

Now, it might happen that in a particular problem $F(t)$ occurs as a combination of all these forms like

$$\mathbf{F}(t) = \mathbf{a} - \mathbf{b}e^{\omega t} + \mathbf{c}e^{i\omega t}. \tag{5.220}$$

In this case, we find the particular solution for each of the forcing functions separately and accept the final solution as a linear combination of them. That is, if

$$X_{1p}(t), \quad X_{2p}(t), \quad X_{3p}(t) \tag{5.221}$$

are particular solutions corresponding to forcing functions

$$\mathbf{a}, \quad \mathbf{b}e^{\omega t}, \quad \mathbf{c}e^{i\omega t} \tag{5.222}$$

respectively, then the final particular solution will be

$$X_p(t) = \alpha X_{1p}(t) + \beta X_{2p}(t) + \gamma X_{3p}(t), \tag{5.223}$$

where α, β, γ are scalar quantities.

This is true by the principle of superposition and can be easily proved.

For initial value problems, the value of $\mathbf{X}(t)$ at $t = t_0$ (say) may be given. Now, the general solution of non-homogeneous equation being the solution of corresponding homogeneous equation plus the particular solution, we write

$$\mathbf{X}(t) = \psi(t)\mathbf{c} + \mathbf{X}_p(t) \tag{5.224}$$

Using the initial conditions

$$\mathbf{X}(t_0) = \psi(t_0)\mathbf{c} + \mathbf{X}_p(t_0) \tag{5.225}$$

so that

$$\mathbf{c} = \psi^{-1}(t_0)\ \{\mathbf{X}(t_0) - \mathbf{X}_p(t_0)\}, \tag{5.226}$$

which determines constants of integration.

EXAMPLE 5.50. *Find a particular solution of*

$$\mathbf{X}' = \begin{bmatrix} -2 & 4 \\ 4 & 4 \end{bmatrix}\mathbf{X} + e^{3t}\begin{bmatrix} 1 \\ -5 \end{bmatrix}.$$

Solution: We find that the eigenvalues of A here are 6, −4. Hence, we can assume the particular solution in the form $\mathbf{X}_p(t) = \mathbf{k}e^{3t}$.

Substituting this in the given differential equation

$$3\mathbf{k}e^{3t} = \begin{bmatrix} -2 & 4 \\ 4 & 4 \end{bmatrix} \mathbf{k}e^{3t} + e^{3t}\begin{bmatrix} 1 \\ -5 \end{bmatrix},$$

$$\text{or,}\quad 3\mathbf{k}\begin{bmatrix} 1 & 0 \\ 0 & 1 \end{bmatrix} = \begin{bmatrix} -2 & 4 \\ 4 & 4 \end{bmatrix}\mathbf{k} + \begin{bmatrix} 1 \\ -5 \end{bmatrix},$$

$$\text{or,}\quad \begin{bmatrix} 5 & -4 \\ -4 & -1 \end{bmatrix}\mathbf{k} = \begin{bmatrix} 1 \\ -5 \end{bmatrix}.$$

This gives, $\mathbf{k} = \begin{bmatrix} 1 \\ 1 \end{bmatrix}$.

Hence, $\mathbf{X}_p(t) = e^{3t}\begin{bmatrix} 1 \\ 1 \end{bmatrix}$.

EXAMPLE 5.51. *Find the particular solution of* $\mathbf{X}' = \begin{bmatrix} 1 & 2 \\ 3 & 2 \end{bmatrix}\mathbf{X} + \cos 2t\begin{bmatrix} 0 \\ 1 \end{bmatrix}$.

Solution: We consider the equation

$$\mathbf{X}' = \begin{bmatrix} 1 & 2 \\ 3 & 2 \end{bmatrix}\mathbf{X} + e^{i2t}\begin{bmatrix} 0 \\ 1 \end{bmatrix}. \tag{1}$$

We assume the particular solution as

$$\mathbf{X}_p(t) = \mathbf{k}e^{i2t}. \tag{2}$$

Then, from (1) by using (2), we get

$$2i\mathbf{k}e^{i2t} = \mathbf{k}e^{i2t}\begin{bmatrix} 1 & 2 \\ 3 & 2 \end{bmatrix} + e^{i2t}\begin{bmatrix} 0 \\ 1 \end{bmatrix} \quad \text{or,}\quad \begin{bmatrix} 2i-1 & -2 \\ -3 & 2i-2 \end{bmatrix}\mathbf{k} = \begin{bmatrix} 0 \\ 1 \end{bmatrix} \tag{3}$$

which gives, $\mathbf{k} = \dfrac{1}{25}\begin{bmatrix} -4+3i \\ 2+11i \end{bmatrix}$.

$$\therefore\quad \mathbf{X}_p(t) = \frac{1}{25}\, e^{i2t}\begin{bmatrix} -4+3i \\ 2+11i \end{bmatrix} \quad \text{[from (2)]}$$

$$\text{or,}\quad \mathbf{X}_p(t) = \frac{1}{25}\,(\cos 2t + i\sin 2t)\begin{bmatrix} -4+3i \\ 2+11i \end{bmatrix}$$

$$= \frac{1}{25}\begin{bmatrix} -4\cos 2t - 3\sin 2t \\ 2\cos 2t - 11\sin 2t \end{bmatrix} + \frac{i}{25}\begin{bmatrix} 3\cos 2t - 4\sin 2t \\ 11\cos 2t + 2\sin 2t \end{bmatrix}.$$

We have to take the real part of the solution. Hence

$$\mathbf{X}_p(t) = \frac{1}{25}\begin{bmatrix} -4\cos 2t - 3\sin 2t \\ 2\cos 2t - 11\sin 2t \end{bmatrix}.$$

EXERCISE

1. Show that the vectors $(2, -3, 1)$, $(3, -1, 5)$ and $(1, -4, 3)$ are linearly independent.
2. Show that the vectors $(1, 0, 0)$, $(0, 1, 0)$, $(0, 0, 1)$ and $(1, 1, 1)$ form a linearly dependent set, but any three of them are linearly independent.
3. Test whether the vectors

$$\boldsymbol{\alpha} = \begin{bmatrix} 1 & -2 & 4 \\ 3 & 0 & -1 \end{bmatrix}, \quad \boldsymbol{\beta} = \begin{bmatrix} 2 & -4 & 8 \\ 6 & 0 & -2 \end{bmatrix}$$

are linearly dependent or not.
4. Find the basis of R^3 containing the vectors $(1, 1, 0)$ and $(1, 1, 1)$.
5. Show that the set of vectors $(1, 2, -3)$, $(2, -3, 1)$, $(3, -1, 1)$ does not form a basis for the vector space V over real numbers.
6. Show that the set

$$S = \{ (1,0,0),\ (1,1,0),\ (1,1,1),\ (0,1,0) \}$$

spans the vector space but is not the basis set.
7. Find the coordinate vector of $(2, 3, 1)$ relative to the basis

$$B = \{ (1,1,1),\ (1,1,0),\ (0,1,0) \}.$$

8. Show that the coordinates of the vector (α, β, γ) with respect to the basis set

$$S = \{ (1,0,0),\ (1,1,0),\ (1,1,1) \}$$

are $(\alpha - \beta, \beta - \gamma, \gamma)$.
9. If in an inner product space

$$\| \alpha + \beta \| = \| \alpha \| + \| \beta \|,$$

then show that α and β are linearly dependent.
10. Show that vectors $(2, -2, 1)$, $(1, 2, -3)$, $(2, 1, 2)$ form orthogonal basis of Euclidean space R^3 with standard inner product.
11. Apply the Gram-Schmidt process to the following sets of vectors to find an orthonormal basis for V_3 over real numbers:
    (a) $S = \{ (1,0,1), (1,0,-1), (1,3,4) \}$;
    (b) $S = \{ (1,0,0), (1,1,0), (1,1,1) \}$.
12. Examine whether the following operators are linear or not:
    (a) $L(\ldots) = x^2 \dfrac{d^2(\ldots)}{dx^2} + x\dfrac{d(\ldots)}{dx}$;
    (b) $\mathbf{Ar} = \boldsymbol{\alpha} \times \mathbf{r}$, $\boldsymbol{\alpha}$ = constant vector.
13. If $A = \begin{bmatrix} 1 & 2 & -2 \\ 4 & 1 & 5 \\ -3 & 2 & 4 \end{bmatrix}$, find B, when $A^T + 2B = 3I$.
14. Show that $AB \neq BA$, where

$$A = \begin{bmatrix} 1 & 3 & 0 \\ -1 & 2 & 1 \\ 0 & 0 & 1 \end{bmatrix} \quad \text{and} \quad B = \begin{bmatrix} 2 & 3 & 4 \\ 1 & 2 & 3 \\ -1 & 1 & 2 \end{bmatrix}.$$

15. If $A = \begin{bmatrix} 1 & 1 \\ x & y \\ x^2 & y^2 \end{bmatrix}$, find AA^T.

16. If n be a positive integer, and

$$A = \begin{bmatrix} \cos\theta & \sin\theta \\ -\sin\theta & \cos\theta \end{bmatrix},$$

then show that

$$A^n = \begin{bmatrix} \cos n\theta & \sin n\theta \\ -\sin n\theta & \cos n\theta \end{bmatrix}.$$

17. Verify that $[AB]^T = B^T A^T$, where

$$A = \begin{bmatrix} 1 & 2 & 3 \\ 3 & -2 & 1 \end{bmatrix} \quad \text{and} \quad B = \begin{bmatrix} 1 & 2 \\ 2 & 0 \\ -1 & 1 \end{bmatrix}.$$

18. If the matrices A, B are orthogonal, then show that $\begin{bmatrix} A & 0 \\ 0 & B \end{bmatrix}$ is also orthogonal.

19. Find the inverse of the matrix

$$\begin{bmatrix} \cos\theta & -\sin\theta & 0 \\ \sin\theta & \cos\theta & 0 \\ 0 & 0 & 1 \end{bmatrix}.$$

20. Find inverse of the matrices by elementary row operations:

(a) $A = \begin{bmatrix} 1 & -1 & 1 \\ 2 & -1 & 0 \\ 1 & 0 & 0 \end{bmatrix}$; (b) $A = \begin{bmatrix} 1 & 2 & 2 \\ 2 & 1 & 2 \\ 2 & 2 & 1 \end{bmatrix}$.

21. Find, by reducing to normal form, the rank of the following matrices:

(a) $\begin{bmatrix} 1 & 2 & 3 \\ 3 & 4 & 5 \\ 4 & 6 & 8 \end{bmatrix}$; (c) $\begin{bmatrix} 8 & 1 & 3 & 6 \\ 0 & 3 & 2 & 2 \\ -8 & -1 & -3 & 4 \end{bmatrix}$.

(b) $\begin{bmatrix} 1 & 0 & 5 \\ -2 & 1 & 0 \\ 4 & 0 & 1 \end{bmatrix}$;

22. Find the value of λ for which the system of linear equations

$$\left.\begin{aligned} 2x_1 + x_2 - x_3 &= 12 \\ x_1 - x_2 - 2x_3 &= -3 \\ 3x_2 + 3x_3 &= \lambda \end{aligned}\right\} \text{ is consistent.}$$

23. Solve, by matrix method the following systems of equations:

(a) $\begin{aligned} x + y + z &= 8 \\ x - y + 2z &= 6 \\ 3x + 5y - 7z &= 14; \end{aligned}$ (b) $\begin{aligned} x + y - z &= 6 \\ 2x - 3y + z &= 1 \\ 2x - 4y + 2z &= 1. \end{aligned}$

24. Solve, by reducing the matrix to normal form, the following systems of equations:

(a) $\begin{aligned} 2x - 3y + z &= 1 \\ x + 2y - 3z &= 4 \\ 4x - y - 2z &= 8, \end{aligned}$ (b) $\begin{aligned} 2x + 3y + z &= 11 \\ x + y + z &= 6 \\ 5x - y + 10z &= 34. \end{aligned}$

25. Find eigenvalues and eigenvectors of matrices:

(a) $\begin{bmatrix} -1 & 2 & 2 \\ 2 & 2 & 2 \\ -3 & -6 & -6 \end{bmatrix}$; (b) $\begin{bmatrix} 6 & -2 & 2 \\ -2 & 3 & -1 \\ 2 & -1 & 3 \end{bmatrix}$.

26. Diagonalise the matrices:

(a) $A = \begin{bmatrix} 5 & 1 & 1 \\ 1 & 5 & -1 \\ 1 & -1 & 5 \end{bmatrix}$; (b) $A = \begin{bmatrix} 1 & -1 & 1 \\ -1 & 1 & -1 \\ 1 & -1 & 1 \end{bmatrix}$.

27. If $A = \begin{bmatrix} 1 & 0 & 0 \\ 1 & 0 & 1 \\ 0 & 1 & 0 \end{bmatrix}$, then find A^{50}.

28. Show that the matrix in the Question No. 19 is unitary.

29. Find the solutions of the following systems of linear differential equations:

(a) $$\left.\begin{aligned} \frac{dx}{dt} &= 3x + 5y + z \\ \frac{dy}{dt} &= -2x - 2y \\ \frac{dz}{dt} &= 2x + 7y + 3z \end{aligned}\right\}, \text{ with } \begin{bmatrix} x(0) \\ y(0) \\ z(0) \end{bmatrix} = \begin{bmatrix} 1 \\ 1 \\ 0 \end{bmatrix};$$

(b) $$\left.\begin{aligned} \frac{dx}{dt} &= y + z \\ \frac{dy}{dt} &= x + z \\ \frac{dz}{dt} &= x + y \end{aligned}\right\}, \text{ with } \begin{bmatrix} x(0) \\ y(0) \\ z(0) \end{bmatrix} = \begin{bmatrix} 1 \\ 1 \\ 0 \end{bmatrix}.$$

30. Show that the matrices of $n \times n$ order (a) J_n and (b) $J_n - I$ are simple. All elements of J_n are unity.

31. Find the particular solutions of the following systems of linear differential equations

(a) $$\left.\begin{aligned} \frac{dx}{dt} &= x + y + 2e^t \\ \frac{dy}{dt} &= 4x + y + 3e^{-t} \end{aligned}\right\};$$

(b) $$\left.\begin{aligned} \frac{dx}{dt} &= \frac{1}{2}(3x + y) + \sin\theta \\ \frac{dy}{dt} &= \frac{1}{2}(x + 3y) + \cos\theta \end{aligned}\right\}.$$

32. Find a solution of the initial value problem

$$\left.\begin{aligned} \frac{dx}{dt} &= x \\ \frac{dy}{dt} &= -x + 3y \end{aligned}\right\} \text{ with } \begin{bmatrix} x(0) \\ y(0) \end{bmatrix} = \begin{bmatrix} 2 \\ 1 \end{bmatrix}.$$

33. Show that the rank of the matrix $\begin{bmatrix} I & 0 \\ 0 & 0 \end{bmatrix}$ is r, where I is of order $r \times r$.

34. Show that the matrix

$$\begin{bmatrix} 3i & 2+4i & 4-5i \\ -2+4i & -4i & 5+6i \\ -4-5i & -5+6i & 0 \end{bmatrix}$$

is skew-Hermitian.

ANSWERS

3. Linearly dependent.
4. $(1,1,0)$, $(1,1,1)$, $(1,0,0)$.
7. $(1,1,1)$.
11. (a) $\left\{\frac{1}{\sqrt{2}}(1,0,1), \frac{1}{\sqrt{2}}(1,0,-1), (0,1,0)\right\}$; (b) $\{(1,0,0), (0,1,0), (0,0,1)\}$.
12. (a) Linear, (b) Linear.
13. $\frac{1}{2}\begin{bmatrix} 2 & -4 & 3 \\ -2 & 2 & -2 \\ -2 & -5 & -1 \end{bmatrix}$.
15. $\begin{bmatrix} 2 & x+y & x^2+y^2 \\ x+y & x^2+y^2 & x^3+y^3 \\ x^2+y^2 & x^3+y^3 & x^4+y^4 \end{bmatrix}$.
19. $\begin{bmatrix} \cos\theta & \sin\theta & 0 \\ -\sin\theta & \cos\theta & 0 \\ 0 & 0 & 1 \end{bmatrix}$.
20. (a) $\begin{bmatrix} 0 & 0 & 1 \\ 0 & -1 & 2 \\ 1 & -1 & 1 \end{bmatrix}$;

    (b) $\frac{1}{5}\begin{bmatrix} -3 & 2 & 2 \\ 2 & -3 & 2 \\ 2 & 2 & -3 \end{bmatrix}$.
21. (a) 2; (b) 3; (c) 3.
22. $\lambda = 18$.
23. (a) $x = 5, y = 5/3, z = 4/3$; (b) $x = 6, y = 11/2, z = 11/2$.
24. (a) $x = 3, y = 2, z = 1$; (b) $x = 3, y = 1, z = 2$.
25. (a) $\lambda_1 = 0, \lambda_2 = -2, \lambda_3 = -3$; $\mathbf{x}_1 = \begin{bmatrix} 0 \\ -1 \\ 1 \end{bmatrix}$, $\mathbf{x}_2 = \begin{bmatrix} -2 \\ 1 \\ 0 \end{bmatrix}$, $\mathbf{x}_3 = \begin{bmatrix} -1 \\ 0 \\ 1 \end{bmatrix}$;

    (b) $\lambda_1 = 2, \lambda_2 = 2, \lambda_3 = 8$; $\mathbf{x}_1 = \begin{bmatrix} 1 \\ 2 \\ 0 \end{bmatrix}$, $\mathbf{x}_2 = \begin{bmatrix} -1 \\ 0 \\ 2 \end{bmatrix}$, $\mathbf{x}_3 = \begin{bmatrix} 2 \\ -1 \\ 1 \end{bmatrix}$.
26. (a) $\begin{bmatrix} 6 & 0 & 0 \\ 0 & 6 & 0 \\ 0 & 0 & 3 \end{bmatrix}$; (b) $\begin{bmatrix} 3 & 0 & 0 \\ 0 & 0 & 0 \\ 0 & 0 & 0 \end{bmatrix}$.
27. $\begin{bmatrix} 1 & 0 & 0 \\ 25 & 1 & 0 \\ 25 & 0 & 1 \end{bmatrix}$.

Chapter 6

Coordinate Systems and Their Transformation

6.1 Curvilinear Coordinates

Let us consider three curved surfaces defined by q_1 = constant, q_2 = constant and q_3 = constant, such that any two surfaces always intersect to produce a curve and all the three surfaces intersect at a point. We can take the curves of intersections as reference axes to construct a coordinate system. We call such a coordinate system *a curvilinear coordinate system* (*Fig 6.1*).

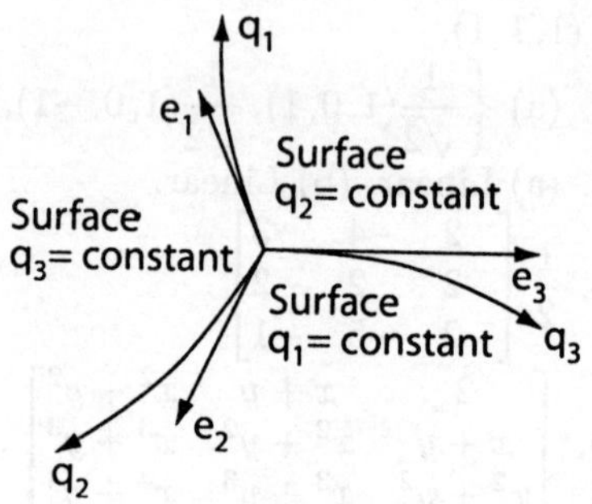

Fig 6.1

Let the curve of intersection by surfaces q_2 = constant and q_3 = constant be denoted as q_1-axis. Similarly, q_2 and q_3 axes are defined. Let the Cartesian coordinates of any point $P(x, y, z)$ be related to the curvilinear coordinates as

$$\left.\begin{array}{l} x = x\,(q_1,\ q_2,\ q_3) \\ y = y\,(q_1,\ q_2,\ q_3) \\ z = z\,(q_1,\ q_2,\ q_3) \end{array}\right\}. \tag{6.1}$$

This gives transformation of Cartesian coordinate system to curvilinear system. Since there is one-to-one correspondence between the two systems of coordinates, it is possible to express the curvilinear coordinates of $P\,(q_1,\ q_2,\ q_3)$ in terms of Cartesian coordinates in the manner

$$\left.\begin{array}{l} q_1 = q_1\,(x,\ y,\ z) \\ q_2 = q_2\,(x,\ y,\ z) \\ q_3 = q_3\,(x,\ y,\ z) \end{array}\right\}. \tag{6.2}$$

The functions defined in (6.2) are continuous having first-order continuous derivatives. Now, if $\mathbf{r}$ be the position vector of the point P, then the vectors along tangents to q_1, q_2, q_3 axes will be

$$\frac{\partial \mathbf{r}}{\partial q_1},\quad \frac{\partial \mathbf{r}}{\partial q_2},\quad \frac{\partial \mathbf{r}}{\partial q_3}, \tag{6.3}$$

respectively. Then unit vectors along these tangents are

$$\mathbf{e}_1 = \frac{1}{h_1}\frac{\partial \mathbf{r}}{\partial q_1},\quad \mathbf{e}_2 = \frac{1}{h_2}\frac{\partial \mathbf{r}}{\partial q_2},\quad \mathbf{e}_3 = \frac{1}{h_3}\frac{\partial \mathbf{r}}{\partial q_3}, \tag{6.4}$$

where

$$h_i = \left| \frac{\partial \mathbf{r}}{\partial q_i} \right|. \tag{6.5}$$

In Cartesian coordinate system

$$\mathbf{r} = \mathbf{i}x + \mathbf{j}y + \mathbf{k}z \tag{6.6}$$

so that

$$\left.\begin{aligned} \frac{\partial \mathbf{r}}{\partial q_1} &= \mathbf{i}\frac{\partial x}{\partial q_1} + \mathbf{j}\frac{\partial y}{\partial q_1} + \mathbf{k}\frac{\partial z}{\partial q_1} \\ \frac{\partial \mathbf{r}}{\partial q_2} &= \mathbf{i}\frac{\partial x}{\partial q_2} + \mathbf{j}\frac{\partial y}{\partial q_2} + \mathbf{k}\frac{\partial z}{\partial q_2} \\ \frac{\partial \mathbf{r}}{\partial q_3} &= \mathbf{i}\frac{\partial x}{\partial q_3} + \mathbf{j}\frac{\partial y}{\partial q_3} + \mathbf{k}\frac{\partial z}{\partial q_3} \end{aligned}\right\}. \tag{6.7}$$

In order that the coordinate system defined by three surfaces $q_1 =$ constant, $q_2 =$ constant and $q_3 =$ constant uniquely defines a point of intersection, the vectors in (6.7) should be non-coplanar; that is the scalar triple products of these vectors should not vanish, the condition for which is

$$\begin{vmatrix} \dfrac{\partial x}{\partial q_1} & \dfrac{\partial y}{\partial q_1} & \dfrac{\partial z}{\partial q_1} \\ \dfrac{\partial x}{\partial q_2} & \dfrac{\partial y}{\partial q_2} & \dfrac{\partial z}{\partial q_2} \\ \dfrac{\partial x}{\partial q_3} & \dfrac{\partial y}{\partial q_3} & \dfrac{\partial z}{\partial q_3} \end{vmatrix} \neq 0. \tag{6.8}$$

This determinant is known as *Jacobian of transformation* and the condition (6.8) is written in compact form as

$$\frac{\partial\,(x,\ y,\ z)}{\partial\,(q_1,\ q_2,\ q_3)} \neq 0. \tag{6.9}$$

Rudiments of Jacobians are provided in the Appendix B.

If $P(\mathbf{r})$ and $Q(\mathbf{r} + d\mathbf{r})$ be two neighbouring points, then

$$dr^2 = dx^2 + dy^2 + dz^2. \tag{6.10}$$

Since x, y, z are functions of q_1, q_2, q_3 defined by equation (6.1), we have

$$\left.\begin{aligned} dx &= \frac{\partial x}{\partial q_1}\,dq_1 + \frac{\partial x}{\partial q_2}\,dq_2 + \frac{\partial x}{\partial q_3}\,dq_3 \\ dy &= \frac{\partial y}{\partial q_1}\,dq_1 + \frac{\partial y}{\partial q_2}\,dq_2 + \frac{\partial y}{\partial q_3}\,dq_3 \\ dz &= \frac{\partial z}{\partial q_1}\,dq_1 + \frac{\partial z}{\partial q_2}\,dq_2 + \frac{\partial z}{\partial q_3}\,dq_3 \end{aligned}\right\}. \tag{6.11}$$

Thus, from equation (6.10) using (6.11), we write

$$dr^2 = \sum a_{ij}\,dq_i\,dq_j, \tag{6.12}$$

where

$$a_{ij} = \frac{\partial x}{\partial q_i}\frac{\partial x}{\partial q_j} + \frac{\partial y}{\partial q_i}\frac{\partial y}{\partial q_j} + \frac{\partial z}{\partial q_i}\frac{\partial z}{\partial q_j}, \tag{6.13}$$

which depends on the nature of the coordinate system.

6.2 Orthogonal Curvilinear Coordinates

The curvilinear coordinate system would be orthogonal if the unit vectors given by equation (6.4) along the tangents to the axes are orthogonal. This gives

$$\mathbf{e}_i \cdot \mathbf{e}_j = \begin{cases} 0, & i \neq j \\ 1, & i = j \end{cases} \tag{6.14}$$

Moreover, in the expression of dr^2 in equation (6.12), cross-product terms will vanish. Hence

$$a_{ij} = 0, \quad \text{for } i \neq j. \tag{6.15}$$

Thus

$$dr^2 = (h_1\, dq_1)^2 + (h_2\, dq_2)^2 + (h_3\, dq_3)^2, \tag{6.16}$$

where

$$h_1 = \sqrt{a_{11}}, \quad h_2 = \sqrt{a_{22}}, \quad h_3 = \sqrt{a_{33}}. \tag{6.17}$$

The quantities h_1, h_2, h_3 are called *scale factors* which may have dimensions.

The orthogonal curvilinear coordinates are convenient to study physical problems due to which we shall restrict our discussion on them only.

6.3 Elements of Length, Area and Volume

We have

$$\mathbf{r} = \mathbf{i}x + \mathbf{j}y + \mathbf{k}z$$

$$\therefore \quad d\mathbf{r} = \mathbf{i}dx + \mathbf{j}dy + \mathbf{k}dz.$$

In curvilinear coordinates

$$d\mathbf{r} = \frac{\partial \mathbf{r}}{\partial q_1}\, dq_1 + \frac{\partial \mathbf{r}}{\partial q_2}\, dq_2 + \frac{\partial \mathbf{r}}{\partial q_3}\, dq_3. \tag{6.18}$$

Using equation (6.4), we get

$$d\mathbf{r} = \mathbf{e}_1\, h_1\, dq_1 + \mathbf{e}_2\, h_2\, dq_2 + \mathbf{e}_3\, h_3\, dq_3. \tag{6.19}$$

Equation (6.19) gives the element of length. Now, plane area is a vector given by the cross-product of two vectors. Thus, element of area in this case is

$$d\mathbf{r}_1 \times d\mathbf{r}_2 = |\,\mathbf{e}_1 \times \mathbf{e}_2\,|\; h_1\, h_2\, dq_1\, dq_2$$

$$\text{or,} \quad d\mathbf{r}_1 \times d\mathbf{r}_2 = h_1\, h_2\, dq_1\, dq_2. \tag{6.20}$$

Similarly

$$d\mathbf{r}_2 \times d\mathbf{r}_3 = h_2\, h_3\, dq_2\, dq_3$$

$$\text{and} \quad d\mathbf{r}_3 \times d\mathbf{r}_1 = h_3\, h_1\, dq_3\, dq_1.$$

Again, volume of a parallelopiped is given by the scalar triple product of the three vectors forming the sides of the parallelopiped. Hence, volume element having sides dr_1, dr_2, dr_3 is

$$dV = d\mathbf{r}_1 \cdot d\mathbf{r}_2 \times d\mathbf{r}_3 = (\mathbf{e}_1 \cdot \mathbf{e}_2 \times \mathbf{e}_3)\, h_1\, h_2\, h_3\, dq_1\, dq_2\, dq_3$$

$$\text{or,} \quad dV = h_1\, h_2\, h_3\, dq_1\, dq_2\, dq_3. \tag{6.21}$$

Thus, equation (6.21) gives volume element in curvilinear coordinates. Now, we intend to obtain general expressions for differential operators in curvilinear coordinates. But before that we shall establish an interesting relation between the vectors $\frac{\partial \mathbf{r}}{\partial q_i}$ and ∇q_i.

We have

$$d\mathbf{r} = \mathbf{i}dx + \mathbf{j}dy + \mathbf{k}dz$$

$$\text{and} \quad \nabla q_1 = \mathbf{i}\,\frac{\partial q_1}{\partial x} + \mathbf{j}\,\frac{\partial q_1}{\partial y} + \mathbf{k}\,\frac{\partial q_1}{\partial z}.$$

$$\therefore \quad \nabla q_1 \cdot d\mathbf{r} = \frac{\partial q_1}{\partial x}\, dx + \frac{\partial q_1}{\partial y}\, dy + \frac{\partial q_1}{\partial z}\, dz,$$

$$\text{or,} \quad \nabla q_1 \cdot d\mathbf{r} = dq_1. \tag{6.22}$$

Again, if we multiply both sides of equation (6.18) scalarly by ∇q_1, we get

$$\nabla q_1 \cdot d\mathbf{r} = \left(\nabla q_1\,\frac{\partial \mathbf{r}}{\partial q_1}\right) dq_1 + \left(\nabla q_1\,\frac{\partial \mathbf{r}}{\partial q_2}\right) dq_2 + \left(\nabla q_1\,\frac{\partial \mathbf{r}}{\partial q_3}\right) dq_3. \tag{6.23}$$

Comparing equations (6.22) and (6.23), we get

$$\nabla q_1 \cdot \frac{\partial \mathbf{r}}{\partial q_1} = 1, \quad \nabla q_1 \cdot \frac{\partial \mathbf{r}}{\partial q_2} = 0, \quad \nabla q_1 \cdot \frac{\partial \mathbf{r}}{\partial q_3} = 0.$$

Similarly, by multiplying equation (6.18) scalarly by ∇q_2 and ∇q_3, we can get

$$\nabla q_2 \cdot \frac{\partial \mathbf{r}}{\partial q_1} = 0, \quad \nabla q_2 \cdot \frac{\partial \mathbf{r}}{\partial q_2} = 1, \quad \nabla q_2 \cdot \frac{\partial \mathbf{r}}{\partial q_3} = 0$$

$$\text{and} \quad \nabla q_3 \cdot \frac{\partial \mathbf{r}}{\partial q_1} = 0, \quad \nabla q_3 \cdot \frac{\partial \mathbf{r}}{\partial q_2} = 0, \quad \nabla q_3 \cdot\cdot \frac{\partial \mathbf{r}}{\partial q_3} = 1.$$

From these relations, we can say that vectors

$$\frac{\partial \mathbf{r}}{\partial q_i} \quad \text{and} \quad \nabla q_i \quad (i = 1, 2, 3)$$

form a reciprocal system of triads (*see* Sections 1.17 and 1.18). Thus, we can find vectors ∇q_i in the following manner:

$$\nabla q_1 = \frac{\frac{\partial \mathbf{r}}{\partial q_2} \times \frac{\partial \mathbf{r}}{\partial q_3}}{\left[\frac{\partial \mathbf{r}}{\partial q_1}\ \frac{\partial \mathbf{r}}{\partial q_2}\ \frac{\partial \mathbf{r}}{\partial q_3}\right]} = \frac{(\mathbf{e}_2\, h_2) \times (\mathbf{e}_3\, h_3)}{(\mathbf{e}_1\, h_1) \cdot (\mathbf{e}_2\, h_2) \times (\mathbf{e}_3\, h_3)} = \frac{h_2\, h_3\, (\mathbf{e}_2 \times \mathbf{e}_3)}{h_1\, h_2\, h_3\, (\mathbf{e}_1 \cdot \mathbf{e}_2 \times \mathbf{e}_3)}),$$

$$\text{or,} \quad \nabla q_1 = \frac{\mathbf{e}_1}{h_1} \quad (\because\ \mathbf{e}_1 \cdot \mathbf{e}_2 \times \mathbf{e}_3 = 1).$$

We can get other two similar relations. Hence we write

$$\left.\begin{aligned} \mathbf{e}_1 &= h_1 \nabla q_1 \\ \mathbf{e}_2 &= h_2 \nabla q_2 \\ \mathbf{e}_3 &= h_3 \nabla q_3 \end{aligned}\right\}. \tag{6.24}$$

6.4 Gradient in Orthogonal Curvilinear Coordinates

We consider a scalar function $\psi(q_1, q_2, q_3)$, where q_1, q_2, q_3 are functions of x, y, z expressed by equation (6.2).

Then, we have

$$\left.\begin{aligned} \frac{\partial \psi}{\partial x} &= \frac{\partial \psi}{\partial q_1}\frac{\partial q_1}{\partial x} + \frac{\partial \psi}{\partial q_2}\frac{\partial q_2}{\partial x} + \frac{\partial \psi}{\partial q_3}\frac{\partial q_3}{\partial x} \\ \frac{\partial \psi}{\partial y} &= \frac{\partial \psi}{\partial q_1}\frac{\partial q_1}{\partial y} + \frac{\partial \psi}{\partial q_2}\frac{\partial q_2}{\partial y} + \frac{\partial \psi}{\partial q_3}\frac{\partial q_3}{\partial y} \\ \frac{\partial \psi}{\partial z} &= \frac{\partial \psi}{\partial q_1}\frac{\partial q_1}{\partial z} + \frac{\partial \psi}{\partial q_2}\frac{\partial q_2}{\partial z} + \frac{\partial \psi}{\partial q_3}\frac{\partial q_3}{\partial z} \end{aligned}\right\}. \tag{6.25}$$

Multiplying the equations (6.25) by **i**, **j**, **k** respectively and adding, we get

$$\nabla\psi = \mathbf{i}\frac{\partial \psi}{\partial x} + \mathbf{j}\frac{\partial \psi}{\partial y} + \mathbf{k}\frac{\partial \psi}{\partial z} = \frac{\partial \psi}{\partial q_1}\nabla q_1 + \frac{\partial \psi}{\partial q_2}\nabla q_2 + \frac{\partial \psi}{\partial q_3}\nabla q_3.$$

By using equations (6.24)

$$\nabla\psi = \frac{\mathbf{e}_1}{h_1}\frac{\partial \psi}{\partial q_1} + \frac{\mathbf{e}_2}{h_2}\frac{\partial \psi}{\partial q_2} + \frac{\mathbf{e}_3}{h_3}\frac{\partial \psi}{\partial q_3}. \tag{6.26}$$

6.5 Divergence in Orthogonal Curvilinear Coordinates

Let us consider a vector point function $\mathbf{V}(q_1, q_2, q_3)$ having components $\mathbf{V}_1$, $\mathbf{V}_2$, $\mathbf{V}_3$ along the unit vectors $\mathbf{e}_1$, $\mathbf{e}_2$, $\mathbf{e}_3$. Then

$$\mathbf{V} = V_1\,\mathbf{e}_1 + V_2\,\mathbf{e}_2 + V_3\,\mathbf{e}_3. \tag{6.27}$$

Since the unit vectors $\mathbf{e}_1$, $\mathbf{e}_2$, $\mathbf{e}_3$ are orthogonal

$$\mathbf{e}_1 = \mathbf{e}_2 \times \mathbf{e}_3, \quad \mathbf{e}_2 = \mathbf{e}_3 \times \mathbf{e}_1, \quad \mathbf{e}_3 = \mathbf{e}_1 \times \mathbf{e}_2. \tag{6.28}$$

Let us consider the component $\mathbf{V}_1$ only of the vector point function. We have

$$\begin{aligned} \mathbf{V}_1 &= V_1\,\mathbf{e}_1 = V_1\,\mathbf{e}_2 \times \mathbf{e}_3 \\ &= V_1\,h_2\,h_3\,\nabla q_2 \times \nabla q_3 \quad \text{[by using (6.24)]}. \end{aligned}$$

$$\therefore \quad \operatorname{div}\mathbf{V}_1 = \nabla(V_1\,h_2\,h_3)\cdot(\nabla q_2 \times \nabla q_3) + V_1\,h_2\,h_3\nabla\cdot(\nabla q_2 \times \nabla q_3).$$

But

$$\nabla\cdot(\nabla q_2 \times \nabla q_3) = \nabla q_3\cdot\nabla\times\nabla q_2 - \nabla q_2\cdot\nabla\times\nabla q_3 = 0.$$

Thus

$$\operatorname{div}\mathbf{V}_1 = \nabla(V_1\,h_2\,h_3)\cdot(\nabla q_2 \times \nabla q_3). \tag{6.29}$$

Now, for any function $f(q_1)$, we have

$$\nabla f(q_1) = \boldsymbol{i}\,\frac{\partial f}{\partial x} + \boldsymbol{j}\,\frac{\partial f}{\partial y} + \boldsymbol{k}\,\frac{\partial f}{\partial z} = \boldsymbol{i}\,\frac{\partial f}{\partial q_1}\,\frac{\partial q_1}{\partial x} + \boldsymbol{j}\,\frac{\partial f}{\partial q_1}\,\frac{\partial q_1}{\partial y} + \boldsymbol{k}\,\frac{\partial f}{\partial q_1}\,\frac{\partial q_1}{\partial z} = \frac{\partial f}{\partial q_1}\nabla q_1.$$

Using this identity, we can write equation (6.29) as

$$\begin{aligned}
\operatorname{div}\mathbf{V}_1 &= \frac{\partial}{\partial q_1}\,(V_1\,h_2\,h_3)\;\nabla q_1\cdot\nabla q_2\times\nabla q_3\\
&= \frac{\boldsymbol{e}_1\cdot\boldsymbol{e}_2\times\boldsymbol{e}_3}{h_1\,h_2\,h_3}\,\frac{\partial}{\partial q_1}\,(V_1\,h_2\,h_3)\quad\text{[by using (6.24)]}\\
&= \frac{1}{h_1\,h_2\,h_3}\,\frac{\partial}{\partial q_1}\,(V_1\,h_2\,h_3)\quad(\because\ \boldsymbol{e}_1\cdot\boldsymbol{e}_2\times\boldsymbol{e}_3 = 1).
\end{aligned}$$

Similarly, we can find $\operatorname{div}\mathbf{V}_2$ and $\operatorname{div}\mathbf{V}_3$. Thus, we get

$$\operatorname{div}\mathbf{V} = \frac{1}{h_1\,h_2\,h_3}\left[\frac{\partial}{\partial q_1}\,(V_1\,h_2\,h_3) + \frac{\partial}{\partial q_2}\,(V_2\,h_3\,h_1) + \frac{\partial}{\partial q_3}\,(V_3\,h_1\,h_2)\right]. \tag{6.30}$$

Equation (6.30) gives divergence of the vector point function $\mathbf{V}$ in orthogonal curvilinear coordinates.

6.6 Curl in Orthogonal Curvilinear Coordinates

Here, we have, from equation (6.27)

$$\mathbf{V} = V_1\,\mathbf{e}_1 + V_2\,\mathbf{e}_2 + V_3\,\mathbf{e}_3.$$

By using equation (6.24), we get

$$\mathbf{V} = V_1\,h_1\nabla q_1 + V_2\,h_2\nabla q_2 + V_3\,h_3\nabla q_3.$$

$$\therefore\quad \nabla\times\mathbf{V} = \nabla\times(V_1\,h_1\nabla q_1) + \nabla\times(V_2\,h_2\nabla q_2) + \nabla\times(V_3\,h_3\nabla q_3). \tag{6.31}$$

Let us consider the first term in equation (6.31).
Thus

$$\nabla\times(V_1\,h_1\nabla q_1) = \frac{\partial}{\partial q_1}(V_1\,h_1)\,\nabla q_1\times\nabla q_1 + \frac{\partial}{\partial q_2}(V_1\,h_1)\,\nabla q_2\times\nabla q_1 + \frac{\partial}{\partial q_3}(V_1\,h_1)\,\nabla q_3\times\nabla q_1. \tag{6.32}$$

Now, from equation (6.24) we have

$$\nabla q_1\times\nabla q_2 = \frac{\mathbf{e}_1\times\mathbf{e}_2}{h_1\,h_2} = \frac{\mathbf{e}_3}{h_1\,h_2}.$$

$$\therefore\quad \nabla q_2\times\nabla q_1 = \frac{-\boldsymbol{e}_3}{h_1 h_2},\quad \nabla q_3\times\nabla q_1 = \frac{\boldsymbol{e}_2}{h_3 h_1}.$$

Thus, equation (6.32) becomes

$$\nabla\times(V_1\,h_1\nabla q_1) = \frac{\mathbf{e}_2}{h_3\,h_1}\,\frac{\partial}{\partial q_3}\,(V_1\,h_1) - \frac{\mathbf{e}_3}{h_1\,h_2}\,\frac{\partial}{\partial q_2}\,(V_1\,h_1).$$

Similarly, we can get

$$\nabla \times (V_2\, h_2 \nabla q_2) = \frac{\mathbf{e}_3}{h_1\, h_2} \frac{\partial}{\partial q_1} (V_2\, h_2) - \frac{\mathbf{e}_1}{h_2\, h_3} \frac{\partial}{\partial q_3} (V_2\, h_2)$$

$$\nabla \times (V_3\, h_3 \nabla q_3) = \frac{\mathbf{e}_1}{h_2\, h_3} \frac{\partial}{\partial q_2} (V_3\, h_3) - \frac{\mathbf{e}_2}{h_3\, h_1} \frac{\partial}{\partial q_3} (V_3\, h_3).$$

Thus

$$\begin{aligned} \operatorname{curl} \mathbf{V} &= \nabla \times \mathbf{V} \\ &= \frac{\mathbf{e}_1}{h_2\, h_3} \left[\frac{\partial}{\partial q_2} (V_3\, h_3) - \frac{\partial}{\partial q_3} (V_2\, h_2) \right] + \frac{\mathbf{e}_2}{h_3\, h_1} \left[\frac{\partial}{\partial q_3} (V_1\, h_1) - \frac{\partial}{\partial q_1} (V_3\, h_3) \right] \\ &\quad + \frac{\mathbf{e}_3}{h_1\, h_2} \left[\frac{\partial}{\partial q_1} (V_2\, h_2) - \frac{\partial}{\partial q_2} (V_1\, h_1) \right], \end{aligned} \tag{6.33}$$

which can also be written as

$$\frac{1}{h_1\, h_2\, h_3} \begin{vmatrix} \mathbf{e}_1 h_1 & \mathbf{e}_2 h_2 & \mathbf{e}_3 h_3 \\ \dfrac{\partial}{\partial q_1} & \dfrac{\partial}{\partial q_2} & \dfrac{\partial}{\partial q_3} \\ V_1\, h_1 & V_2\, h_2 & V_3\, h_3 \end{vmatrix}. \tag{6.34}$$

6.7 Laplacian Operator

We have, from equation (6.30)

$$\operatorname{div} \mathbf{V} = \nabla \cdot \mathbf{V} = \frac{1}{h_1\, h_2\, h_3} \left[\frac{\partial}{\partial q_1} (V_1\, h_2\, h_3) + \frac{\partial}{\partial q_2} (V_2\, h_3\, h_1) + \frac{\partial}{\partial q_3} (V_3\, h_1\, h_2) \right], \tag{6.35}$$

where $\mathbf{V} = \mathbf{e}_1 V_1 + \mathbf{e}_2 V_2 + \mathbf{e}_3 V_3$.

Let there be a scalar point function ϕ, such that

$$\nabla \phi = \left[\frac{\mathbf{e}_1}{h_1} \frac{\partial \phi}{\partial q_1} + \frac{\mathbf{e}_2}{h_2} \frac{\partial \phi}{\partial q_2} + \frac{\mathbf{e}_3}{h_3} \frac{\partial \phi}{\partial q_3} \right].$$

If we substitute in equation (6.35) the relations

$$V_1 = \frac{1}{h_1} \frac{\partial \phi}{\partial q_1}, \quad V_2 = \frac{1}{h_2} \frac{\partial \phi}{\partial q_2}, \quad V_3 = \frac{1}{h_3} \frac{\partial \phi}{\partial q_3}$$

so that $\mathbf{V} = \nabla \phi$, then we get

$$\nabla^2 \phi = \frac{1}{h_1\, h_2\, h_3} \left[\frac{\partial}{\partial q_1} \left(\frac{h_2\, h_3}{h_1} \frac{\partial \phi}{\partial q_1} \right) + \frac{\partial}{\partial q_2} \left(\frac{h_3\, h_1}{h_2} \frac{\partial \phi}{\partial q_2} \right) + \frac{\partial}{\partial q_3} \left(\frac{h_1\, h_2}{h_3} \frac{\partial \phi}{\partial q_3} \right) \right]. \tag{6.36}$$

Thus, equation (6.36) gives the Laplacian of a scalar function ϕ in orthogonal curvilinear coordinates. We shall now discuss some special cases of orthogonal curvilinear coordinates.

6.8 Circular Cylindrical Coordinates

In this coordinate system, the three curvilinear coordinates are

$$\begin{aligned} q_1 &= r, \qquad 0 \le r \le \infty, \\ q_2 &= \theta, \qquad 0 \le \theta \le 2\pi \\ \text{and} \quad q_3 &= z, \qquad -\infty \le z \le +\infty. \end{aligned}$$

The Cartesian coordinates corresponding to these cylindrical coordinates are

$$\left.\begin{aligned} x &= r\cos\theta \\ y &= r\sin\theta \\ z &= z \end{aligned}\right\}. \tag{6.37}$$

Thus, $x^2 + y^2 = r^2 =$ constant describes a family of right circular cylinder with z-axis as its own axis (*Fig 6.2*).

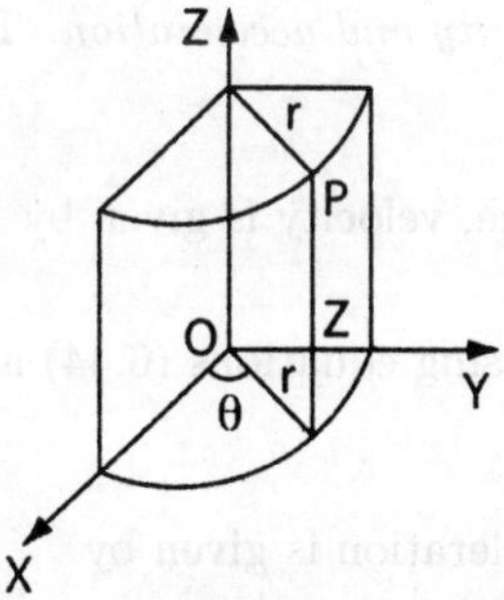

Fig 6.2

The scale factors are

$$\left.\begin{aligned} h_1 &= h_r = 1 \\ h_2 &= h_\theta = r \\ h_3 &= h_z = 1 \end{aligned}\right\}. \tag{6.38}$$

The position vector of any point is

$$\mathbf{r} = \mathbf{i}r\cos\theta + \mathbf{j}r\sin\theta + \mathbf{k}. \tag{6.39}$$

(i) *Volume element*

$$dV = r\,dr\,d\theta\,dz \quad [\text{by } (6.21)]. \tag{6.40}$$

(ii) *Unit vectors*

$$\mathbf{e}_r = \frac{\partial\mathbf{r}/\partial r}{|\,\partial\mathbf{r}/\partial r\,|} = \mathbf{i}\cos\theta + \mathbf{j}\sin\theta, \tag{6.41}$$

$$\mathbf{e}_\theta = \frac{\partial\mathbf{r}/\partial\theta}{|\,\partial\mathbf{r}/\partial\theta\,|} = -\mathbf{i}\sin\theta + \mathbf{j}\cos\theta; \tag{6.42}$$

$\mathbf{e}_z = \mathbf{k}$, unit vector in positive z-direction.
It is obvious that

$$\left.\begin{aligned} \mathbf{e}_r \times \mathbf{e}_\theta &= \mathbf{k} \\ \mathbf{e}_\theta \times \mathbf{k} &= \mathbf{e}_r \\ \mathbf{k} \times \mathbf{e}_r &= \mathbf{e}_\theta \end{aligned}\right\}. \tag{6.43}$$

Hence, the unit vectors are mutually orthogonal. The unit vectors are not constant, but change continuously along with time. The time derivatives of unit vectors are

$$\dot{\mathbf{e}}_r = (-\mathbf{i}\sin\theta + \mathbf{j}\cos\theta)\,\dot\theta = \dot\theta\mathbf{e}_\theta \tag{6.44}$$

$$\text{and} \quad \dot{\mathbf{e}}_\theta = (-\mathbf{i}\cos\theta - \mathbf{j}\sin\theta)\,\dot\theta = -\dot\theta\mathbf{e}_r. \tag{6.45}$$

(iii) *Differential operators*

$$\nabla\psi = \mathbf{e}_r \frac{\partial \psi}{\partial r} + \mathbf{e}_\theta \frac{1}{r}\frac{\partial \psi}{\partial \theta} + \mathbf{k}\frac{\partial \psi}{\partial z} \tag{6.46}$$

$$\text{div}\,\mathbf{V} = \frac{1}{r}\frac{\partial}{\partial r}(rV_r) + \frac{1}{r}\frac{\partial V_\theta}{\partial \theta} + \frac{\partial V_z}{\partial z} \tag{6.47}$$

$$\nabla^2\phi = \frac{1}{r}\frac{\partial}{\partial r}\left(r\frac{\partial \phi}{\partial r}\right) + \frac{1}{r^2}\frac{\partial^2 \phi}{\partial \theta^2} + \frac{\partial^2 \phi}{\partial z^2}. \tag{6.48}$$

$$\nabla \times \mathbf{V} = \frac{1}{r}\begin{vmatrix} \mathbf{e}_r & r\mathbf{e}_\theta & \mathbf{k} \\ \dfrac{\partial}{\partial r} & \dfrac{\partial}{\partial \theta} & \dfrac{\partial}{\partial z} \\ V_r & rV_\theta & V_z \end{vmatrix}. \tag{6.49}$$

(iv) *Velocity and acceleration.* The position vector of a point in (6.39) can be written as

$$\mathbf{r} = r\mathbf{e}_r + \mathbf{k}z.$$

Hence, velocity is given by

$$\mathbf{v} = \dot{\mathbf{r}} = \dot{r}\mathbf{e}_r + r\dot{\mathbf{e}}_r + \mathbf{k}\dot{z}.$$

By using equations (6.44) and (6.45)

$$\mathbf{v} = \dot{r}\mathbf{e}_r + r\dot{\theta}\mathbf{e}_\theta + \mathbf{k}\dot{z}. \tag{6.50}$$

Acceleration is given by

$$\mathbf{a} = \dot{\mathbf{v}} = \ddot{r}\mathbf{e}_r + \dot{r}\dot{\mathbf{e}}_r + r\ddot{\theta}\mathbf{e}_\theta + \dot{r}\dot{\theta}\mathbf{e}_\theta + r\dot{\theta}\dot{\mathbf{e}}_\theta + \mathbf{k}\ddot{z}.$$

By using equations (6.44) and (6.45)

$$\mathbf{a} = \left(\ddot{r} - r\dot{\theta}^2\right)\mathbf{e}_r + \left(r\ddot{\theta} + 2\dot{r}\dot{\theta}\right)\mathbf{e}_\theta + \mathbf{k}\ddot{z}. \tag{6.51}$$

6.9 Spherical Polar Coordinates

In this coordinate system

$$q_1 = r, \quad q_2 = \theta, \quad q_3 = \phi. \tag{6.52}$$

The Cartesian coordinates of any point P are related to the spherical polar coordinates in the following manner (*Fig 6.3*):

$$\left.\begin{aligned} x &= r\sin\theta\cos\phi \\ y &= r\sin\theta\sin\phi \\ z &= r\cos\theta \end{aligned}\right\}. \tag{6.53}$$

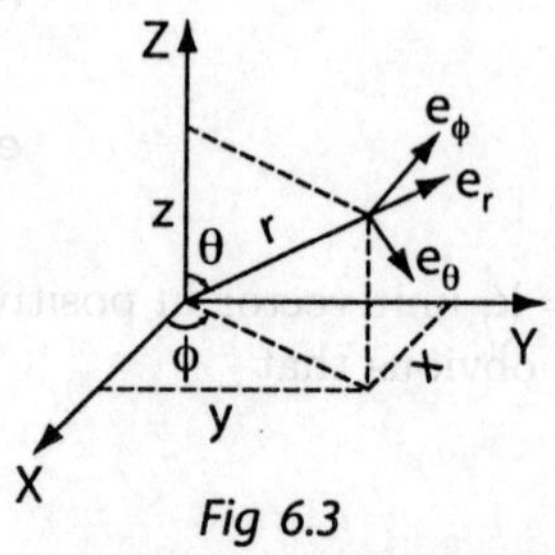

Fig 6.3

Thus

$$r = \sqrt{x^2 + y^2 + z^2},$$

$$\theta = \cos^{-1} z/r,$$

$$\text{and} \quad \phi = \tan^{-1} y/x. \tag{6.54}$$

We observe that,

(a) $r =$ constant describes a family of concentric spheres with centre at the origin.
(b) $\theta =$ constant describes a family of right circular cones with vertex at the origin.
(c) $\phi =$ constant gives the half-planes through z-axis.

From equation (6.53)

$$\begin{aligned} dx &= \sin\theta\cos\phi\, dr + r\cos\theta\cos\phi\, d\theta - r\sin\theta\sin\phi\, d\phi \\ dy &= \sin\theta\sin\phi\, dr + r\cos\theta\sin\phi\, d\theta + r\sin\theta\cos\phi\, d\phi \\ dz &= \cos\theta dr - r\sin\theta d\theta. \end{aligned}$$

Thus

$$|\, dr^2\,| = dr^2 + r^2 d\theta^2 + r^2\sin^2\theta d\phi^2.$$

Hence, the scale factors are

$$\left.\begin{aligned} h_r &= h_1 = 1 \\ h_\theta &= h_2 = r \\ h_\phi &= h_3 = r\sin\theta \end{aligned}\right\}. \tag{6.55}$$

(i) *Volume element*

$$dV = r^2\sin\theta dr\, d\theta\, d\phi. \tag{6.56}$$

(ii) *Unit vectors.* The position vector of any point is given by

$$\mathbf{r} = \mathbf{i}x + \mathbf{j}y + \mathbf{k}z = \mathbf{i}r\sin\theta\cos\phi + \mathbf{j}r\sin\theta\sin\phi + \mathbf{k}r\cos\theta. \tag{6.57}$$

Thus, the unit vectors along the curvilinear coordinate axes are:

$$\left.\begin{aligned} \mathbf{e}_r &= \frac{\partial\mathbf{r}/\partial r}{|\,\partial\mathbf{r}/\partial r\,|} = \mathbf{i}\sin\theta\cos\phi + \mathbf{j}\sin\theta\sin\phi + \mathbf{k}\cos\theta \\ \mathbf{e}_\theta &= \frac{\partial\mathbf{r}/\partial\theta}{|\,\partial\mathbf{r}/\partial\theta\,|} = \mathbf{i}\cos\theta\cos\phi + \mathbf{j}\cos\theta\sin\phi - \mathbf{k}\sin\theta \\ \mathbf{e}_\phi &= \frac{\partial\mathbf{r}/\partial\phi}{|\,\partial\mathbf{r}/\partial\phi\,|} = -\mathbf{i}\sin\phi + \mathbf{j}\cos\phi \end{aligned}\right\}. \tag{6.58}$$

It can be verified that

$$\mathbf{e}_r\cdot\mathbf{e}_\theta = \mathbf{e}_\theta\cdot\mathbf{e}_\phi = \mathbf{e}_\phi\cdot\mathbf{e}_r = 0, \tag{6.59}$$

which shows that the unit vectors are mutually orthogonal. These vectors also satisfy the relations

$$\left.\begin{aligned} \mathbf{e}_r\times\mathbf{e}_\theta &= \mathbf{e}_\phi \\ \mathbf{e}_\theta\times\mathbf{e}_\phi &= \mathbf{e}_r \\ \mathbf{e}_\phi\times\mathbf{e}_r &= \mathbf{e}_\theta \end{aligned}\right\}. \tag{6.60}$$

(iii) *Time-derivatives of unit vectors*

$$\begin{aligned}\dot{\mathbf{e}}_r &= \mathbf{i}\left(\dot{\theta}\cos\theta\cos\phi - \dot{\phi}\sin\theta\sin\phi\right) + \mathbf{j}\left(\dot{\theta}\cos\theta\sin\phi + \dot{\phi}\sin\theta\cos\phi\right) - k\dot{\theta}\sin\theta \\ &= \dot{\theta}\,(\mathbf{i}\cos\theta\cos\phi + \mathbf{j}\cos\theta\sin\phi - \mathbf{k}\sin\theta) + \dot{\phi}\sin\theta\,(-\mathbf{i}\sin\phi + \mathbf{j}\cos\phi) \\ &= \dot{\theta}\mathbf{e}_\theta + \dot{\phi}\sin\theta\mathbf{e}_\phi. \end{aligned} \tag{6.61}$$

$$\begin{aligned}\dot{\mathbf{e}}_\theta &= \mathbf{i}\left(-\dot{\theta}\sin\theta\cos\phi - \dot{\phi}\cos\theta\sin\phi\right) + \mathbf{j}\left(-\dot{\theta}\sin\theta\sin\phi + \dot{\phi}\cos\theta\cos\phi\right) - \mathbf{k}\dot{\theta}\cos\theta \\ &= -\dot{\theta}\,(\mathbf{i}\sin\theta\cos\phi + \mathbf{j}\sin\theta\sin\phi + \mathbf{k}\cos\theta) + \dot{\phi}\cos\theta\,(-\mathbf{i}\sin\phi + \mathbf{j}\cos\phi) \\ &= -\dot{\theta}\mathbf{e}_r + \dot{\phi}\cos\theta\mathbf{e}_\phi. \end{aligned} \tag{6.62}$$

$$\dot{\mathbf{e}}_\phi = -\mathbf{i}\dot{\phi}\cos\phi - \mathbf{j}\dot{\phi}\sin\phi.$$

Now, from equation (6.58), we find that

$$\sin\theta\mathbf{e}_r + \cos\theta\mathbf{e}_\theta = \mathbf{i}\cos\phi + j\sin\phi.$$

Hence

$$\dot{\mathbf{e}}_\phi = -\dot{\phi}\,(\sin\theta\mathbf{e}_r + \cos\theta\mathbf{e}_\theta). \tag{6.63}$$

(iv) *Differential operators*

$$\nabla\psi = \mathbf{e}_r\frac{\partial\psi}{\partial r} + \mathbf{e}_\theta\frac{\partial\psi}{r\partial\theta} + \mathbf{e}_\phi\frac{1}{r\sin\theta}\frac{\partial\psi}{\partial\phi} \tag{6.64}$$

$$\nabla\cdot\mathbf{V} = \frac{1}{r^2\sin\theta}\left[\sin\theta\frac{\partial}{\partial r}(r^2V_r) + r\frac{\partial}{\partial\theta}(\sin\theta V_\theta) + r\frac{\partial V_\phi}{\partial\phi}\right] \tag{6.65}$$

$$\nabla^2\Phi = \frac{1}{r^2\sin\theta}\left[\sin\theta\frac{\partial}{\partial r}\left(r^2\frac{\partial\Phi}{\partial r}\right) + \frac{\partial}{\partial\theta}\left(\sin\theta\frac{\partial\Phi}{\partial\theta}\right) + \frac{1}{\sin\theta}\frac{\partial^2\Phi}{\partial\phi^2}\right] \tag{6.66}$$

$$\nabla\times\mathbf{V} = \frac{1}{r^2\sin\theta}\begin{vmatrix}\mathbf{e}_r & r\mathbf{e}_\theta & r\sin\theta\mathbf{e}_\phi \\ \dfrac{\partial}{\partial r} & \dfrac{\partial}{\partial\theta} & \dfrac{\partial}{\partial\phi} \\ V_r & rV_\theta & r\sin\theta V_\phi\end{vmatrix}. \tag{6.67}$$

(v) *Velocity and acceleration.* From equations (6.57) and (6.58), we find that the position vector of any point is given by

$$\mathbf{r} = r\mathbf{e}_r. \tag{6.68}$$

Hence, velocity is

$$\mathbf{v} = \dot{\mathbf{r}} = \dot{r}\mathbf{e}_r + r\dot{\mathbf{e}}_r.$$

By using equation (6.61)

$$\mathbf{v} = \dot{r}\mathbf{e}_r + r\dot{\theta}\mathbf{e}_\theta + r\sin\theta\dot{\phi}\mathbf{e}_\phi. \tag{6.69}$$

Again, acceleration is given by

$$\begin{aligned}\mathbf{a} = \dot{\mathbf{v}} &= (\ddot{r}\mathbf{e}_r + \dot{r}\dot{\mathbf{e}}_r) + (\dot{r}\dot{\theta}\mathbf{e}_\theta + r\ddot{\theta}\mathbf{e}_\theta + r\dot{\theta}\dot{\mathbf{e}}_\theta) \\ &\quad + (\dot{r}\sin\theta\dot{\phi}\mathbf{e}_\phi + r\dot{\theta}\cos\theta\dot{\phi}\mathbf{e}_\phi + r\sin\theta\ddot{\phi}\mathbf{e}_\phi + r\sin\theta\dot{\phi}\dot{\mathbf{e}}_\phi).\end{aligned}$$

Putting here the expressions for $\dot{\mathbf{e}}_r$ and $\dot{\mathbf{e}}_\theta$ from equations (6.61) and (6.62), and simplifying, we get

$$\mathbf{a} = (\ddot{r} - r\dot{\theta}^2 - r\sin^2\theta\dot{\phi}^2)\,\mathbf{e}_r + (r\ddot{\theta} + 2\dot{r}\dot{\theta} - r\sin\theta\cos\theta\dot{\phi}^2)\,\mathbf{e}_\theta + (r\sin\theta\ddot{\phi} + 2\dot{r}\dot{\phi}\sin\theta + 2r\cos\theta\dot{\theta}\dot{\phi})\,\mathbf{e}_\phi. \quad (6.70)$$

6.10 Unit Vectors of Cartesian Coordinates

Unit vectors **i**, **j**, **k** of Cartesian coordinate system can be expressed in terms of unit vectors of orthogonal curvilinear coordinate system.

In circular cylindrical coordinate system, the unit vectors have been expressed in equations (6.41) and (6.42). Solving these equations simultaneously, we get

$$\left.\begin{aligned} \mathbf{i} &= \mathbf{e}_r\cos\theta - \mathbf{e}_\theta\sin\theta \\ \mathbf{j} &= \mathbf{e}_r\sin\theta + \mathbf{e}_\theta\cos\theta \\ \mathbf{k} &= \mathbf{e}_z \end{aligned}\right\}. \quad (6.71)$$

Similarly, in spherical coordinates, the unit vectors along the curvilinear coordinate axes are given by equations (6.58). Solving these equations simultaneously, we get

$$\left.\begin{aligned} \mathbf{i} &= \mathbf{e}_r\sin\theta\cos\phi + \mathbf{e}_\theta\cos\theta\cos\phi - \mathbf{e}_\phi\sin\phi \\ \mathbf{j} &= \mathbf{e}_r\sin\theta\sin\phi + \mathbf{e}_\theta\cos\theta\sin\phi + \mathbf{e}_\phi\cos\phi \\ \mathbf{k} &= \mathbf{e}_r\cos\theta - \mathbf{e}_\theta\sin\theta \end{aligned}\right\}. \quad (6.72)$$

6.11 Elliptic Cylindrical Coordinates

In this coordinates system, the curvilinear coordinates are:

$$\left.\begin{aligned} q_1 &= \xi, & 0 &\le \xi < \infty \\ q_2 &= \eta, & 0 &\le \eta \le 2\pi \\ q_3 &= z, & -\infty &\le z \le +\infty \end{aligned}\right\}. \quad (6.73)$$

The Cartesian coordinates corresponding to these elliptic cylindrical coordinates are given by (*Fig 6.4*)

$$\left.\begin{aligned} x &= a\cosh\xi\cos\eta \\ y &= a\sinh\xi\sin\eta \\ z &= z \end{aligned}\right\}. \quad (6.74)$$

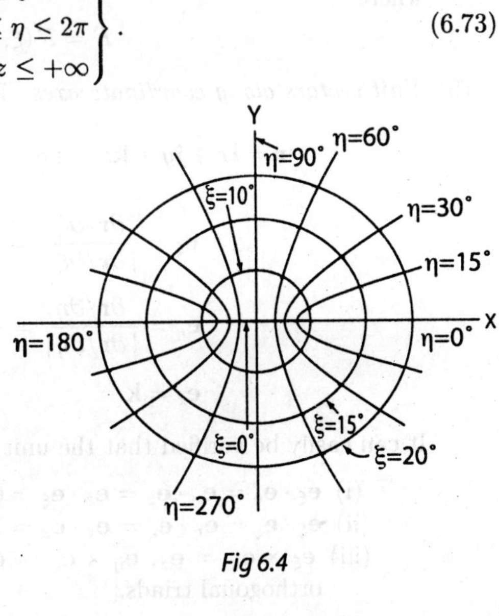

Fig 6.4

From equations (6.74), we find

$$\frac{x^2}{a^2\cosh^2\xi} + \frac{y^2}{a^2\sinh^2\xi} = 1 \quad (6.75)$$

and $$\frac{x^2}{a^2\cos^2\eta} - \frac{y^2}{a^2\sin^2\eta} = 1. \quad (6.76)$$

Equation (6.75) represents a family of ellipses with x-axis as the major axis when $\xi =$ constant. Similarly, equation (6.76) represents a family of hyperbolas with focal points on x-axis when $\eta =$ constant. Thus, the family of coordinate surfaces in this system are:

(i) Elliptic cylinder for $\xi =$ constant;
(ii) Hyperbolic cylinder for $\eta =$ constant;
(iii) Planes parallel to xy-plane for $z =$ constant.

(a) *Scale factors.* From equations (6.74), we have

$$\begin{aligned} dx &= a\sinh\xi\cos\eta\,d\xi - a\cosh\xi\sin\eta\,d\eta \\ dy &= a\cosh\xi\sin\eta\,d\xi + a\sinh\xi\cos\eta\,d\eta \\ \text{and}\quad dz &= dz. \end{aligned}$$

Therefore

$$\begin{aligned} ds^2 &= dx^2 + dy^2 + dz^2 \\ &= a^2\sinh^2\xi\cos^2\eta\,(d\xi)^2 + a^2\cosh^2\xi\sin^2\eta\,(d\eta)^2 \\ &\quad + a^2\cosh^2\xi\sin^2\eta\,(d\xi)^2 + a^2\sinh^2\xi\cos^2\eta\,(d\eta)^2 + dz^2 \\ &= a^2(\sinh^2\xi + \sin^2\eta)\,(d\xi)^2 + a^2(\sinh^2\xi + \sin^2\eta)\,(d\eta)^2 + dz^2. \end{aligned}$$

This gives the scale factors as

$$\left.\begin{aligned} h_1 &= h_\xi = a\sqrt{F} \\ h_2 &= h_\eta = a\sqrt{F} \\ h_3 &= h_z = 1 \end{aligned}\right\}, \tag{6.77}$$

where

$$F = F(\xi, \eta) = (\sinh^2\xi + \sin^2\eta)\,. \tag{6.78}$$

(b) *Unit vectors along coordinate axes.* The position vector of any point P is given by

$$\mathbf{r} = \mathbf{i}x + \mathbf{j}y + \mathbf{k}z, \quad \text{i.e.,} \quad \mathbf{r} = \mathbf{i}a\cosh\xi\cos\eta + \mathbf{j}a\sinh\xi\sin\eta + \mathbf{k}z.$$

$$\left.\begin{aligned} \therefore\quad \mathbf{e}_\xi &= \frac{\partial\mathbf{r}/\partial\xi}{|\partial\mathbf{r}/\partial\xi|} = \frac{1}{\sqrt{F}}\,(\mathbf{i}\sinh\xi\cos\eta + \mathbf{j}\cosh\xi\sin\eta) \\ \mathbf{e}_\eta &= \frac{\partial\mathbf{r}/\partial\eta}{|\partial\mathbf{r}/\partial\eta|} = \frac{1}{\sqrt{F}}\,(-\mathbf{i}\cosh\xi\sin\eta + \mathbf{j}\sinh\xi\cos\eta) \\ \mathbf{e}_z &= \mathbf{k} \end{aligned}\right\}. \tag{6.79}$$

It can easily be verified that the unit vectors satisfy the following relations:

(i) $\mathbf{e}_\xi\cdot\mathbf{e}_\eta = \mathbf{e}_\eta\cdot\mathbf{e}_z = \mathbf{e}_z\cdot\mathbf{e}_\xi = 0$, which shows that they are orthogonal;
(ii) $\mathbf{e}_\xi\cdot\mathbf{e}_\xi = \mathbf{e}_\eta\cdot\mathbf{e}_\eta = \mathbf{e}_z\cdot\mathbf{e}_z = 1$, which shows that they are unit vectors; and
(iii) $\mathbf{e}_\xi\times\mathbf{e}_\eta = \mathbf{e}_z$, $\mathbf{e}_\eta\times\mathbf{e}_z = \mathbf{e}_\xi$, $\mathbf{e}_z\times\mathbf{e}_\xi = \mathbf{e}_\eta$, which shows that they are orthogonal triads.

These unit vectors are not constants but continuously change the direction as ξ, η change. Solving simultaneously the equations (6.79), we find the unit vectors $\mathbf{i}, \mathbf{j}, \mathbf{k}$ of the Cartesian system in terms of unit vectors in elliptic coordinate system as

$$\left.\begin{aligned} \mathbf{i} &= \frac{1}{\sqrt{F}}\left(\mathbf{e}_\xi \sinh\xi\cos\eta - \mathbf{e}_\eta \cosh\xi\sin\eta\right) \\ \mathbf{j} &= \frac{1}{\sqrt{F}}\left(\mathbf{e}_\xi \cosh\xi\sin\eta + \mathbf{e}_\eta \sinh\xi\cos\eta\right) \\ \mathbf{k} &= \mathbf{e}_z \end{aligned}\right\}. \tag{6.80}$$

(c) *Differential operators.*

$$\operatorname{div}\mathbf{V} = \frac{1}{a^2F}\left[\frac{\partial}{\partial\xi}\left(V_\xi\, a^2F\right) + \frac{\partial}{\partial\eta}\left(V_\eta\, a\sqrt{F}\right) + \frac{\partial}{\partial z}\left(V_z\, a\sqrt{F}\right)\right]. \tag{6.81}$$

$$\nabla\psi = \frac{\mathbf{e}_\xi}{a\sqrt{F}}\frac{\partial\psi}{\partial\xi} + \frac{\mathbf{e}_\eta}{a\sqrt{F}}\frac{\partial\psi}{\partial\eta} + \frac{\mathbf{e}_z}{1}\frac{\partial\psi}{\partial z}. \tag{6.82}$$

$$\nabla\times\mathbf{V} = \frac{1}{a^2F}\begin{bmatrix} \mathbf{e}_\xi\, a\sqrt{F} & \mathbf{e}_\eta\, a\sqrt{F} & \mathbf{e}_z \\ \dfrac{\partial}{\partial\xi} & \dfrac{\partial}{\partial\eta} & \dfrac{\partial}{\partial z} \\ V_\xi\, a\sqrt{F} & V_\eta\, a\sqrt{F} & V_z \end{bmatrix}. \tag{6.83}$$

$$\nabla^2\phi = \frac{1}{a^2F}\left[\frac{\partial}{\partial\xi}\left(\frac{\partial\phi}{\partial\xi}\right) + \frac{\partial}{\partial\eta}\left(\frac{\partial\phi}{\partial\eta}\right) + \frac{\partial}{\partial z}\left(a^2F\frac{\partial\phi}{\partial z}\right)\right]. \tag{6.84}$$

6.12 Parabolic Cylindrical Coordinates

In this coordinate system, the curvilinear coordinates are:

$$q_1 = \lambda, \quad -\infty < \lambda < +\infty$$
$$q_2 = \mu, \quad 0 \le \mu < +\infty$$
$$q_3 = z, \quad -\infty \le z \le +\infty$$

The Cartesian coordinates corresponding to these parabolic cylindrical coordinates are given by

$$\left.\begin{aligned} x &= \lambda\mu \\ y &= \tfrac{1}{2}\left(\mu^2 - \lambda^2\right) \\ z &= z \end{aligned}\right\}. \tag{6.85}$$

From these equations we find that for $\lambda =$ constant and $\mu =$ constant, we get two sets of orthogonal parabolas (*Fig 6.5*).

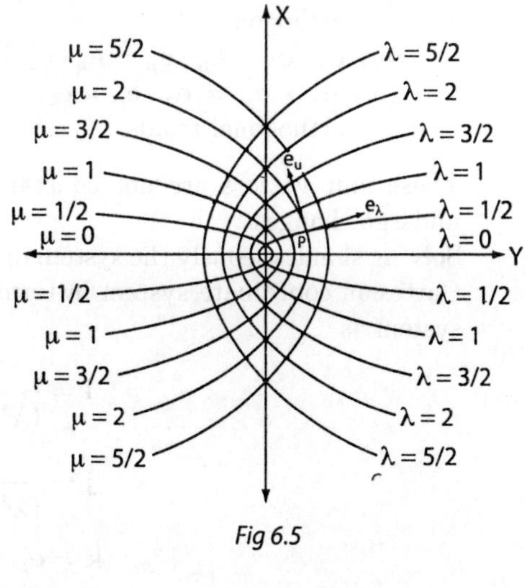

Fig 6.5

(a) *Scale factor.* From equations (6.85), we have

$$dx = \lambda\, d\mu + \mu\, d\lambda$$
$$dy = \mu\, d\mu - \lambda\, d\lambda$$
$$dz = dz.$$

Hence

$$ds^2 = dx^2 + dy^2 + dz^2 = (\lambda^2 + \mu^2)\, d\lambda^2 + (\lambda^2 + \mu^2)\, d\mu^2 + dz^2.$$

This gives the scale factors as

$$\left.\begin{aligned} h_1 &= h_\lambda = (\lambda^2 + \mu^2)^{1/2} \\ h_2 &= h_\mu = (\lambda^2 + \mu^2)^{1/2} \\ h_3 &= h_z = 1 \end{aligned}\right\}. \tag{6.86}$$

(b) *Unit vectors along coordinate axes.* The position vector of any point P is given by

$$\mathbf{r} = \mathbf{i}x + \mathbf{j}y + \mathbf{k}z = \mathbf{i}\lambda\mu + \mathbf{j}\tfrac{1}{2}\left(\mu^2 - \lambda^2\right) + \mathbf{k}z.$$

$$\left.\begin{aligned} \therefore\quad \mathbf{e}_\lambda &= \frac{\partial \mathbf{r}/\partial\lambda}{|\partial \mathbf{r}/\partial\lambda|} = \frac{1}{(\lambda^2 + \mu^2)^{1/2}}\, [\mathbf{i}\mu - \mathbf{j}\lambda] \\ \mathbf{e}_\mu &= \frac{\partial \mathbf{r}/\partial\mu}{|\partial \mathbf{r}/\partial\mu|} = \frac{1}{(\lambda^2 + \mu^2)^{1/2}}\, [\mathbf{i}\lambda - \mathbf{j}\mu] \\ \mathbf{e}_z &= \mathbf{k} \end{aligned}\right\}. \tag{6.87}$$

These unit vectors satisfy the following relations:

(i) $\mathbf{e}_\lambda \cdot \mathbf{e}_\mu = \mathbf{e}_\mu \cdot \mathbf{e}_z = \mathbf{e}_z \cdot \mathbf{e}_\lambda = 0$, which shows that the unit vectors are orthogonal;
(ii) $\mathbf{e}_\lambda \cdot \mathbf{e}_\lambda = \mathbf{e}_\mu \cdot \mathbf{e}_\mu = \mathbf{e}_z \cdot \mathbf{e}_z = 1$, which shows that the unit vectors; and
(iii) $\mathbf{e}_\lambda \times \mathbf{e}_\mu = \mathbf{e}_z$, $\mathbf{e}_\mu \times \mathbf{e}_z = \mathbf{e}_\lambda$, $\mathbf{e}_z \times \mathbf{e}_\lambda = \mathbf{e}_\mu$, which shows that they are orthogonal triads.

These unit vectors are not constants, they continuously change the direction as λ, μ undergo changes.
Solving simultaneously the system of equations (6.87), we find the unit vectors **i**, **j**, **k** of the Cartesian coordinate system in terms of unit vectors of parabolic cylindrical coordinate system as

$$\left.\begin{aligned} \mathbf{i} &= \frac{1}{(\lambda^2 + \mu^2)^{1/2}}\, [\mathbf{e}_\lambda\mu + \mathbf{e}_\mu\lambda] \\ \mathbf{j} &= \frac{1}{(\lambda^2 + \mu^2)^{1/2}}\, [\mathbf{e}_\mu\mu + \mathbf{e}_\lambda\lambda] \\ \mathbf{k} &= \mathbf{e}_z \end{aligned}\right\}. \tag{6.88}$$

(c) *Differential operators.*

$$\nabla\psi = \frac{1}{(\lambda^2+\mu^2)^{1/2}}\left[e_\lambda\frac{\partial\psi}{\partial\lambda} + e_\mu\frac{\partial\psi}{\partial\mu}\right] + e_z\frac{\partial\psi}{\partial z}$$

$$\text{div}\,\mathbf{V} = \frac{1}{f(\lambda,\mu)}\left[\frac{\partial}{\partial\lambda}\left(V_\lambda\sqrt{f}\right) + \frac{\partial}{\partial\mu}\left(V_\mu\sqrt{f}\right) + \frac{\partial}{\partial z}(V_z f)\right],$$

where $f = f(\lambda,\mu) = \lambda^2+\mu^2$

$$\nabla\times\mathbf{V} = \frac{1}{f(\lambda,\mu)}\begin{bmatrix} \mathbf{e}_\lambda\sqrt{f} & \mathbf{e}_\mu\sqrt{f} & \mathbf{e}_z \\ \dfrac{\partial}{\partial\lambda} & \dfrac{\partial}{\partial\mu} & \dfrac{\partial}{\partial z} \\ V_\lambda\sqrt{f} & V_\mu\sqrt{f} & V_z \end{bmatrix}$$

$$\nabla^2\phi = \frac{1}{f(\lambda,\mu)}\left[\frac{\partial}{\partial\lambda}\left(\frac{\partial\phi}{\partial\lambda}\right) + \frac{\partial}{\partial\mu}\left(\frac{\partial\phi}{\partial\mu}\right) + \frac{\partial}{\partial z}\left(f\frac{\partial\phi}{\partial z}\right)\right].$$

EXAMPLE 6.1. *Express the vector* $\mathbf{V} = \mathbf{i}x + \mathbf{j}2y + \mathbf{k}z$ *in circular cylindrical coordinates.*

Solution: $\mathbf{V} = \mathbf{i}x + \mathbf{j}2y + \mathbf{k}z$.

Substituting the values of unit vectors **i**, **j**, **k** from equation (6.71), we get

$$\begin{aligned} \mathbf{V} &= \mathbf{i}r\cos\theta + \mathbf{j}2r\sin\theta + \mathbf{k}z \\ &= r\cos\theta\,(\mathbf{e}_r\cos\theta - \mathbf{e}_\theta\sin\theta) + 2r\sin\theta\,(\mathbf{e}_r\sin\theta + \mathbf{e}_\theta\cos\theta) + \mathbf{e}_z z, \\ \text{or,}\quad \mathbf{V} &= \mathbf{e}_r\,(r\cos^2\theta + 2r\sin^2\theta) + \mathbf{e}_\theta\,(-r\sin\theta\cos\theta + 2r\sin\theta\cos\theta) + \mathbf{e}_z z \\ &= r\mathbf{e}_r\,(1+\sin^2\theta) + r\mathbf{e}_\theta\sin\theta\cos\theta + \mathbf{e}_z z. \end{aligned}$$

EXAMPLE 6.2. *Express the vector* $\mathbf{V} = \mathbf{i}2x - \mathbf{j}y + 3\mathbf{k}z$ *in spherical coordinates.*

Solution: $\mathbf{V} = \mathbf{i}2x - \mathbf{j}y + 3\mathbf{k}z = 2\mathbf{i}r\sin\theta\cos\phi - \mathbf{j}r\sin\theta\sin\phi + 3\mathbf{k}r\cos\theta.$

Substituting the values of unit vectors **i**, **j**, **k** from equation (6.72), we get

$$\begin{aligned} \mathbf{V} &= 2r\sin\theta\cos\phi\,(\mathbf{e}_r\sin\theta\cos\phi + \mathbf{e}_\theta\cos\theta\cos\phi - \mathbf{e}_\phi\sin\phi) \\ &\quad - r\sin\theta\sin\phi\,(\mathbf{e}_r\sin\theta\sin\phi + \mathbf{e}_\theta\cos\theta\sin\phi + \mathbf{e}_\phi\cos\phi) \\ &\quad + 3r\cos\theta\,(\mathbf{e}_r\cos\theta - \mathbf{e}_\theta\sin\theta) \\ &= r\mathbf{e}_r\,(2\sin^2\theta\cos^2\phi - \sin^2\theta\sin^2\phi + 3\cos^2\theta) \\ &\quad + r\mathbf{e}_\theta\,(2\sin\theta\cos\theta\cos^2\phi - \sin\theta\cos\theta\sin^2\phi - 3\sin\theta\cos\theta) \\ &\quad + r\mathbf{e}_\phi\,(-2\sin\theta\sin\phi\cos\phi - \sin\theta\sin\phi\cos\phi) \\ &= r\mathbf{e}_r\,(3\sin^2\theta\cos^2\phi + 4\cos^2\theta - 1) \\ &\quad + r\mathbf{e}_\theta\sin\theta\cos\theta\,(2\cos^2\phi - \sin^2\phi - 3) \\ &\quad - 3r\mathbf{e}_\phi\sin\theta\sin\phi\cos\phi. \end{aligned}$$

EXAMPLE 6.3. *Express partial differential operators*

$$\frac{\partial}{\partial x}, \quad \frac{\partial}{\partial y}, \quad \frac{\partial}{\partial z}$$

in circular cylindrical coordinates.

Solution: In circular cylindrical coordinates, we have, $x = r\cos\theta$, $y = r\sin\theta$, $z = z$.

Thus, $r = \sqrt{x^2 + y^2}$, $\theta = \tan^{-1} y/x$.

Therefore

$$\frac{\partial r}{\partial x} = \frac{x}{r} = \cos\theta, \qquad \frac{\partial \theta}{\partial x} = -\frac{y}{x^2 + y^2} = -\frac{y}{r^2} = -\frac{\sin\theta}{r}$$

and

$$\frac{\partial r}{\partial y} = \frac{y}{r} = \sin\theta, \qquad \frac{\partial \theta}{\partial y} = \frac{x}{x^2 + y^2} = \frac{x}{r^2} = \frac{\cos\theta}{r}.$$

Therefore

$$\frac{\partial}{\partial x} = \frac{\partial}{\partial r}\frac{\partial r}{\partial x} + \frac{\partial}{\partial \theta}\frac{\partial \theta}{\partial x} = \cos\theta\frac{\partial}{\partial r} - \frac{\sin\theta}{r}\frac{\partial}{\partial \theta}$$

and $$\frac{\partial}{\partial y} = \frac{\partial}{\partial r}\frac{\partial r}{\partial y} + \frac{\partial}{\partial \theta}\frac{\partial \theta}{\partial y} = \sin\theta\,\frac{\partial}{\partial r} + \frac{\cos\theta}{r}\frac{\partial}{\partial \theta}.$$

EXAMPLE 6.4. *Express partial differential operators*

$$\frac{\partial}{\partial x}, \quad \frac{\partial}{\partial y}, \quad \frac{\partial}{\partial z}$$

in spherical coordinates.

Solution: In spherical coordinates, we have

$$\left.\begin{aligned} x &= r\sin\theta\cos\phi \\ y &= r\sin\theta\sin\phi \\ z &= r\cos\theta \end{aligned}\right\}. \tag{1}$$

From here we get, $r = \sqrt{x^2 + y^2 + z^2}$

$$\tan\phi = \frac{y}{x}, \quad \tan\theta = \frac{\sqrt{x^2 + y^2}}{z}. \tag{2}$$

Hence, we get

$$\left.\begin{aligned} \frac{\partial r}{\partial x} &= \frac{x}{r} = \sin\theta\cos\phi \\ \frac{\partial r}{\partial y} &= \frac{y}{r} = \sin\theta\sin\phi \\ \frac{\partial r}{\partial z} &= \frac{z}{r} = \cos\theta \end{aligned}\right\}. \tag{3}$$

$$\left.\begin{aligned} \frac{\partial \phi}{\partial x} &= -\frac{y}{x^2 + y^2} = -\frac{\sin^2\phi}{y} = -\frac{\sin\phi}{r\sin\theta} \\ \frac{\partial \phi}{\partial y} &= \frac{x}{x^2 + y^2} = \frac{\cos^2\phi}{x} = \frac{\cos\phi}{r\sin\theta} \end{aligned}\right\} \tag{4}$$

From (2), we have

$$\left.\begin{aligned} \sec^2\theta &= 1+\tan^2\theta = \frac{r^2}{z^2} \\ \sin\theta &= \frac{\sqrt{x^2+y^2}}{r} \\ \cos\theta &= \frac{z}{r} \end{aligned}\right\}. \tag{5}$$

$$\therefore \quad \frac{\partial\theta}{\partial x} = \frac{x}{\sqrt{x^2+y^2}}\cdot\frac{1}{z}\cdot\frac{z^2}{r^2} = \frac{x}{r\sin\theta}\cdot\frac{z}{r^2} \quad \text{[by using (5)]},$$

$$\text{i.e.,} \quad \frac{\partial\theta}{\partial x} = \frac{1}{r}\cos\theta\cos\phi \quad \text{[by using (1)]}.$$

Similarly

$$\frac{\partial\theta}{\partial y} = \frac{1}{r}\cos\theta\sin\phi; \quad \text{also,} \quad \frac{\partial\theta}{\partial z} = -\frac{\sin\theta}{r}.$$

Therefore

$$\begin{aligned} \frac{\partial}{\partial x} &= \frac{\partial}{\partial x}\cdot\frac{\partial r}{\partial x} + \frac{\partial}{\partial\theta}\cdot\frac{\partial\theta}{\partial x} + \frac{\partial}{\partial\phi}\cdot\frac{\partial\phi}{\partial x} \\ &= \sin\theta\cos\phi\frac{\partial}{\partial r} + \frac{1}{r}\cos\theta\cos\phi\frac{\partial}{\partial\theta} - \frac{\sin\phi}{r\sin\theta}\cdot\frac{\partial}{\partial\phi} \end{aligned} \tag{6}$$

$$\begin{aligned} \frac{\partial}{\partial y} &= \frac{\partial}{\partial r}\cdot\frac{\partial r}{\partial y} + \frac{\partial}{\partial\theta}\cdot\frac{\partial\theta}{\partial y} + \frac{\partial}{\partial\phi}\cdot\frac{\partial\phi}{\partial y} \\ &= \sin\theta\sin\phi\frac{\partial}{\partial r} + \frac{1}{r}\cos\theta\sin\phi\frac{\partial}{\partial\theta} + \frac{\cos\phi}{r\sin\theta}\cdot\frac{\partial}{\partial\phi}. \end{aligned} \tag{7}$$

$$\begin{aligned} \frac{\partial}{\partial z} &= \frac{\partial}{\partial r}\cdot\frac{\partial r}{\partial z} + \frac{\partial}{\partial\theta}\cdot\frac{\partial\theta}{\partial z} \\ &= \cos\theta\frac{\partial}{\partial r} - \frac{\sin\theta}{r}\cdot\frac{\partial}{\partial\theta}. \end{aligned} \tag{8}$$

TRANSFORMATION OF COORDINATES

The coordinate system may change due to a rotatory and translation motion. The physical quantities expressed in the original coordinate system need to be expressed differently due to change in coordinate systems. We shall discuss here how this transformation can be made.

6.13 Rotation of Two-Dimensional Coordinates

Let us consider a point $P(x, y)$ in a two-dimensional coordinate system OX and OY. Let now the coordinate frame rotate in anti-clockwise direction by an angle θ so that OX', OY' are the new positions of the axes (*Fig 6.6*). If the coordinates of the point P in the new coordinate frame be x', y', then, from the *Fig 6.6*, we find

$$x' = OA + AB + BC = x\cos\theta + a\sin\theta + b\sin\theta = x\cos\theta + y\sin\theta \quad [\because\ a+b=y]$$

$$y' = y\cos\theta - x\sin\theta \quad \text{or,} \quad y' = -x\sin\theta + y\cos\theta.$$

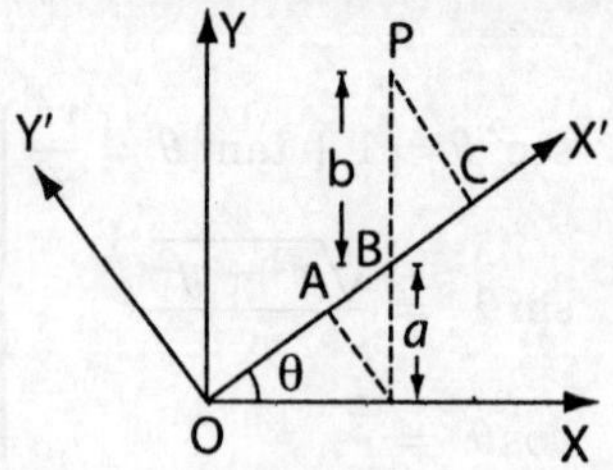

Fig 6.6

In matrix notation, we write

$$\begin{bmatrix} x' \\ y' \end{bmatrix} = \begin{bmatrix} \cos\theta & \sin\theta \\ -\sin\theta & \cos\theta \end{bmatrix} \begin{bmatrix} x \\ y \end{bmatrix}, \quad \text{i.e.,} \quad \begin{bmatrix} x' \\ y' \end{bmatrix} = [R] \begin{bmatrix} x \\ y \end{bmatrix}, \tag{6.89}$$

where

$$[R] = \begin{bmatrix} \cos\theta & \sin\theta \\ -\sin\theta & \cos\theta \end{bmatrix} \tag{6.90}$$

is called the *rotation matrix* or *transformation matrix*.

Thus, equation (6.89) determines the coordinates of the point P in the new coordinate frame in terms of the original coordinates x, y. Again, from equation (6.89), we get

$$\begin{bmatrix} x \\ y \end{bmatrix} = [R]^{-1} \begin{bmatrix} x' \\ y' \end{bmatrix}, \tag{6.91}$$

where

$$[R]^{-1} = \begin{bmatrix} \cos\theta & -\sin\theta \\ \sin\theta & \cos\theta \end{bmatrix}. \tag{6.92}$$

Thus, equation (6.91) determines the coordinates of the point $P(x, y)$ in terms of the new coordinates x', y'.

Now, the coefficients of the rotation matrix may be written as

$$\left.\begin{aligned} a_{11} &= \quad \cos\theta = \quad \cos(x', x) \\ a_{12} &= \quad \sin\theta = \cos(90 - \theta) = \cos(x', y) \\ a_{21} &= -\sin\theta = \cos(90 + \theta) = \cos(y', x) \\ a_{22} &= \quad \cos\theta = \quad \cos(y', y) \end{aligned}\right\}. \tag{6.93}$$

In equations (6.93), it is clear that, by $\cos(x', y)$ is meant cosine of the angle between the axes OX' and OY.

6.14 Rotation of Axes in Three-Dimensions

Let the coordinates of a point P in the original frame of reference be denoted by x_1, x_2, x_3. Let there be a rotation of the reference frame about the origin O so that the coordinates in the new frame of reference are x_1', x_2', x_3'. Then, this transformation will be related by

$$\begin{bmatrix} x_1' \\ x_2' \\ x_3' \end{bmatrix} = [R] \begin{bmatrix} x_1 \\ x_2 \\ x_3 \end{bmatrix}, \tag{6.94}$$

where the rotation matrix is written as

$$[R] = \begin{bmatrix} a_{11} & a_{12} & a_{13} \\ a_{21} & a_{22} & a_{23} \\ a_{31} & a_{32} & a_{33} \end{bmatrix}. \tag{6.95}$$

The coefficients of rotation matrix are given by

$$a_{11} = \cos(x_1', x_1), \quad a_{12} = \cos(x_1', x_2), \quad a_{13} = \cos(x_1', x_3);$$
$$a_{21} = \cos(x_2', x_1), \quad a_{22} = \cos(x_2', x_2), \quad a_{23} = \cos(x_2', x_3);$$
$$a_{31} = \cos(x_3', x_1), \quad a_{32} = \cos(x_3', x_2), \quad a_{33} = \cos(x_3', x_3).$$

From symmetry of the above expressions, we can write, in general

$$a_{ij} = \cos(x_i', x_j). \tag{6.96}$$

Let, for example, the frame of axes has rotated about x_3-axis by an angle θ in anti-clockwise direction (*Fig 6.7*). Then

$$a_{11} = \cos\theta, \quad a_{12} = \cos(90 - \theta) = \sin\theta, \quad a_{13} = 0;$$
$$a_{21} = \cos(90 + \theta) = -\sin\theta, \quad a_{22} = \cos\theta, \quad a_{23} = 0;$$
$$a_{31} = 0, \quad a_{32} = 0, \quad a_{33} = 0.$$

Thus, the rotation matrix is

$$[R] = \begin{bmatrix} \cos\theta & \sin\theta & 0 \\ -\sin\theta & \cos\theta & 0 \\ 0 & 0 & 1 \end{bmatrix}. \tag{6.97}$$

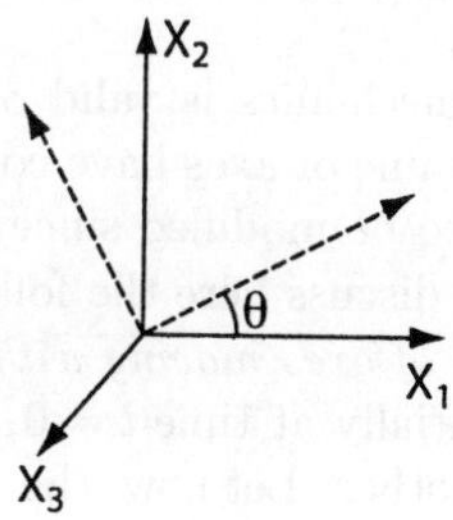

Fig 6.7

Therefore, the two sets of coordinates are related by

$$\begin{bmatrix} x_1' \\ x_2' \\ x_3' \end{bmatrix} = \begin{bmatrix} \cos\theta & \sin\theta & 0 \\ -\sin\theta & \cos\theta & 0 \\ 0 & 0 & 1 \end{bmatrix} \begin{bmatrix} x_1 \\ x_2 \\ x_3 \end{bmatrix}. \tag{6.98}$$

Since the inverse of the matrix $[R]$ is

$$[R]^{-1} = \begin{bmatrix} \cos\theta & -\sin\theta & 0 \\ \sin\theta & \cos\theta & 0 \\ 0 & 0 & 1 \end{bmatrix}, \tag{6.99}$$

we also have the relation

$$\begin{bmatrix} x_1 \\ x_2 \\ x_3 \end{bmatrix} = \begin{bmatrix} \cos\theta & -\sin\theta & 0 \\ \sin\theta & \cos\theta & 0 \\ 0 & 0 & 1 \end{bmatrix} \begin{bmatrix} x_1' \\ x_2' \\ x_3' \end{bmatrix}. \tag{6.100}$$

In the most general case of rotation of the axes with the origin remaining fixed, we assume that the unit vectors along the old and new axes are respectively **i**, **j**, **k** and **i**′, **j**′, **k**′. Then the position vector of any point P will be equal in both the coordinate systems. That is

$$\mathbf{r} \equiv \mathbf{i}x_1 + \mathbf{j}x_2 + \mathbf{k}x_3 = \mathbf{i}'x_1' + \mathbf{j}'x_2' + \mathbf{k}'x_3'. \tag{6.101}$$

Taking dot product on both sides of equation (6.101) by **i**, **j**, **k** respectively, we get

$$\left.\begin{aligned} x_1 &= (\mathbf{i}'\cdot\mathbf{i})\,x_1' + (\mathbf{j}'\cdot\mathbf{i})\,x_2' + (\mathbf{k}'\cdot\mathbf{i})\,x_3' \\ x_2 &= (\mathbf{i}'\cdot\mathbf{j})\,x_1' + (\mathbf{j}'\cdot\mathbf{j})\,x_2' + (\mathbf{k}'\cdot\mathbf{j})\,x_3' \\ x_3 &= (\mathbf{i}'\cdot\mathbf{k})\,x_1' + (\mathbf{j}'\cdot\mathbf{k})\,x_2' + (\mathbf{k}'\cdot\mathbf{k})\,x_3' \end{aligned}\right\}. \tag{6.102}$$

In matrix notation, this is written as

$$\begin{bmatrix} x_1 \\ x_2 \\ x_3 \end{bmatrix} = [R]\begin{bmatrix} x_1' \\ x_2' \\ x_3' \end{bmatrix}, \tag{6.103}$$

where the rotation matrix $[R]$ is given by

$$[R] = \begin{bmatrix} \mathbf{i}'\cdot\mathbf{i} & \mathbf{j}'\cdot\mathbf{i} & \mathbf{k}'\cdot\mathbf{i} \\ \mathbf{i}'\cdot\mathbf{j} & \mathbf{j}'\cdot\mathbf{j} & \mathbf{k}'\cdot\mathbf{j} \\ \mathbf{i}'\cdot\mathbf{k} & \mathbf{j}'\cdot\mathbf{k} & \mathbf{k}'\cdot\mathbf{k} \end{bmatrix}. \tag{6.104}$$

6.15 Moving Frame of Axes

Newtonian mechanics is valid only in absolute frame of axes, that is frame of axes at rest. But, if the frame of axes have continuous motion, translational or rotational, the second law of motion has to be modified since some other forces, called *inertial forces,* come into play.

We shall discuss here the following cases of moving frame of axes:

1. *Frame of axes moving with constant velocity.*
   Let initially at time $t = 0$, there be two frames of Cartesian axes A_1, A_2 coincident with one another. Let now, the frame of axes A_2 move with constant velocity **v** in the direction of positive x-axis. If the position vector of any point P with respect to the frame of axes A_1, A_2 at time t be $\mathbf{r}_1$ and $\mathbf{r}_2$ respectively, then

$$\mathbf{r}_1 = \mathbf{r}_2 + \mathbf{v}t \tag{6.105}$$

$$\therefore \quad \dot{\mathbf{r}}_1 = \dot{\mathbf{r}}_2 + \mathbf{v} \tag{6.106}$$

$$\text{and} \quad \ddot{\mathbf{r}}_1 = \ddot{\mathbf{r}}_2. \tag{6.107}$$

   Thus, absolute velocity is equal to the vector sum of the relative velocity and the velocity of the frame A_2 of axes, but accelerations in the two frames are same. Hence, the Newtonian mechanics are equally applicable in both the frames of axes.

2. *Frame of axes moving with constant acceleration.*
   Let, as before, there be two frames of axes A_1 and A_2 which were coincident initially at $t = 0$. Now, let the frame of axes A_2 move with uniform acceleration **a** in the negative direction of z-axis. If the position vector of any point with respect to the frames of axes A_1, A_2 at any time t be $\mathbf{r}_1$, $\mathbf{r}_2$ respectively, then

$$\mathbf{r}_1 = \mathbf{r}_2 - \tfrac{1}{2}\mathbf{a}t^2 \tag{6.108}$$

$$\therefore \quad \dot{\mathbf{r}}_1 = \dot{\mathbf{r}}_2 - \mathbf{a}t \tag{6.109}$$

$$\text{and} \quad \ddot{\mathbf{r}}_1 = \ddot{\mathbf{r}}_2 - \mathbf{a}. \tag{6.110}$$

Thus, for the motion of a particle of mass m in the frame of axes A_1, the impressed force is

$$\mathbf{F} = m\ddot{\mathbf{r}}_1 = m(\ddot{\mathbf{r}}_2 - \mathbf{a}), \tag{6.111}$$

whereas, the same impressed force in the frame of axes A_2 is

$$\mathbf{F} = m\ddot{\mathbf{r}}_2 = m\ddot{\mathbf{r}}_1 + m\mathbf{a}. \tag{6.112}$$

Thus, there is an inertial force $m\mathbf{a}$ impressed on the particle in frame of axes A_2 in addition to force by Newton's second law.

If the acceleration $\mathbf{a}$ of the frame of axes A_2 be assumed to be identical with the acceleration due to gravity $\mathbf{g}$, the inertial force exactly balances the gravity force. Thus, by an accelerated frame of axes, gravitational field may be produced or neutralised, that is, accelerated frames of axes are equivalent to gravitational field.

3. *Uniformly rotating frame of axes.*
Let there be a frame of axes A at rest having $\mathbf{i}$, $\mathbf{j}$, $\mathbf{k}$ as unit vectors along its axes. Let another frame A_1 having $\mathbf{i}'$, $\mathbf{j}'$, $\mathbf{k}'$ as unit vectors along the axes, rotate uniformly with respect to the frame of axes A about an axis passing through the common origin. Let the position vectors of a point P with respect to the frames of axes be $\mathbf{r}$. Then

$$\left.\begin{aligned} \mathbf{r} &= \mathbf{i}x + \mathbf{j}y + \mathbf{k}z \\ \text{and also,}\quad \mathbf{r} &= \mathbf{i}'x_1 + \mathbf{j}'y_1 + \mathbf{k}'z_1 \end{aligned}\right\}. \tag{6.113}$$

The unit vectors $\mathbf{i}'$, $\mathbf{j}'$ and $\mathbf{k}'$ are not constant since they are rotating and continuously changing the direction. Hence, in rotating frame

$$\frac{d\mathbf{r}}{dt} = \left(\mathbf{i}'\frac{dx_1}{dt} + \mathbf{j}'\frac{dy_1}{dt} + \mathbf{k}'\frac{dz_1}{dt}\right) + \left(x_1\frac{d\mathbf{i}'}{dt} + y_1\frac{d\mathbf{j}'}{dt} + z_1\frac{d\mathbf{k}'}{dt}\right). \tag{6.114}$$

Now, since $\mathbf{i}'$, $\mathbf{j}'$, $\mathbf{k}'$ are rotating vectors

$$\frac{d\mathbf{i}'}{dt} = \boldsymbol{\omega}\times\mathbf{i}', \quad \frac{d\mathbf{j}'}{dt} = \boldsymbol{\omega}\times\mathbf{j}', \quad \frac{d\mathbf{k}'}{dt} = \boldsymbol{\omega}\times\mathbf{k}', \tag{6.115}$$

where $\boldsymbol{\omega}$ is the uniform angular velocity of rotation.
Substituting expressions in (6.115) into second part of (6.114), we get

$$x_1\frac{d\mathbf{i}'}{dt} + y_1\frac{d\mathbf{j}'}{dt} + z_1\frac{d\mathbf{k}'}{dt} = x_1(\boldsymbol{\omega}\times\mathbf{i}') + y_1(\boldsymbol{\omega}\times\mathbf{j}') + z_1(\boldsymbol{\omega}\times\mathbf{k}'),$$

$$\text{or,}\quad x_1\frac{d\mathbf{i}'}{dt} + y_1\frac{d\mathbf{j}'}{dt} + z_1\frac{d\mathbf{k}'}{dt} = \boldsymbol{\omega}\times(x_1\mathbf{i}' + y_1\mathbf{i}' + z_1\mathbf{i}') = \boldsymbol{\omega}\times\mathbf{r}. \tag{6.116}$$

Now, the first three terms in equation (6.114) represent the time derivative of position vector when unit vectors $\mathbf{i}'$, $\mathbf{j}'$, $\mathbf{k}'$ are constants. Thus, in rotating frame

$$\mathbf{i}'\frac{dx_1}{dt} + \mathbf{j}'\frac{dy_1}{dt} + \mathbf{k}'\frac{dz_1}{dt} = \frac{d'\mathbf{r}}{dt}. \tag{6.117}$$

Using equations (6.116) and (6.117), we write

$$\frac{d\mathbf{r}}{dt} = \frac{d'\mathbf{r}}{dt} + \boldsymbol{\omega}\times\mathbf{r}, \tag{6.118}$$

where, we have denoted the operator d/dt in rotating frame by d'/dt.

Equation (6.118) is an operator equation, the operator being

$$\frac{d(\cdots)}{dt} = \frac{d'(\cdots)}{dt} + \boldsymbol{\omega} \times (\cdots). \tag{6.119}$$

This operator equation gives relations between time derivatives of any vector in the fixed and rotating coordinate systems. If we operate it on ω, we get

$$\frac{d\boldsymbol{\omega}}{dt} = \frac{d'\boldsymbol{\omega}}{dt} \quad [\because \boldsymbol{\omega} \times \boldsymbol{\omega} = 0], \tag{6.120}$$

which shows that angular acceleration is the same in both the fixed and rotating coordinate systems.

Now, differentiating the equation (6.118) once with respect to time t, we get

$$\frac{d^2\mathbf{r}}{dt^2} = \left(\frac{d'}{dt} + \boldsymbol{\omega}\times\right)\left(\frac{d'\mathbf{r}}{dt} + \boldsymbol{\omega} \times \mathbf{r}\right) = \frac{d'^2\mathbf{r}}{dt^2} + 2\left(\boldsymbol{\omega} \times \frac{d'\mathbf{r}}{dt}\right) + \boldsymbol{\omega} \times (\boldsymbol{\omega} \times \mathbf{r}). \tag{6.121}$$

The first term on the RHS of equation (6.121) represents the acceleration in the rotating frame while the third term represents the acceleration in the fixed frame of a point fixed in rotating system. The second term, which is equal to twice the vector product of angular velocity, and relative velocity in rotating frame, is called the *Coriolis acceleration.*

From equation (6.121), we find the impressed force on a particle of mass m as

$$\mathbf{F} = m\frac{d^2\mathbf{r}}{dt^2} = m\frac{d'^2\mathbf{r}}{dt^2} + 2m\left(\boldsymbol{\omega} \times \frac{d'\mathbf{r}}{dt}\right) + m\left[\boldsymbol{\omega} \times (\boldsymbol{\omega} \times \mathbf{r})\right]$$

so that the equation of motion of the particle in rotating coordinate system becomes

$$m\frac{d'^2\mathbf{r}}{dt^2} = \mathbf{F} - 2m\left(\boldsymbol{\omega} \times \frac{d'\mathbf{r}}{dt}\right) - m\{\boldsymbol{\omega} \times (\boldsymbol{\omega} \times \mathbf{r})\}. \tag{6.122}$$

The inertial force $-m\{\boldsymbol{\omega} \times (\boldsymbol{\omega} \times \mathbf{r})\}$ is the centrifugal force acting due to rotation of the coordinate axes.

Thus, the forces acting on a particle of mass m in the rotating frame are:

(i) the real force $\mathbf{F} = m\dfrac{d^2\mathbf{r}}{dt^2}$ in fixed frame,

(ii) the centrifugal force $-m\boldsymbol{\omega} \times (\boldsymbol{\omega} \times \mathbf{r})$ arising out of the rotation of coordinate axes,

(iii) the Coriolis force $-2m\left(\boldsymbol{\omega} \times \dfrac{d'\mathbf{r}}{dt}\right)$ due to the motion of the particle in the rotating system.

If the angular velocity ω is not uniform, the equation (6.121) will get modified. In this case

$$\begin{aligned}\frac{d^2\mathbf{r}}{dt^2} &= \left(\frac{d'}{dt} + \boldsymbol{\omega}\times\right)\left(\frac{d'\mathbf{r}}{dt} + \boldsymbol{\omega} \times \mathbf{r}\right)\\ &= \frac{d'^2\mathbf{r}}{dt^2} + 2\left(\boldsymbol{\omega} \times \frac{d'\mathbf{r}}{dt}\right) + \boldsymbol{\omega} \times (\boldsymbol{\omega} \times \mathbf{r}) + \frac{d'\boldsymbol{\omega}}{dt} \times \mathbf{r}.\end{aligned}$$

Thus, equation (6.122) becomes

$$m\frac{d'^2\mathbf{r}}{dt^2} = \mathbf{F} - 2m\left(\boldsymbol{\omega} \times \frac{d'\mathbf{r}}{dt}\right) - m\frac{d'\boldsymbol{\omega}}{dt} \times \mathbf{r} - m \times \boldsymbol{\omega} \times (\boldsymbol{\omega} \times \mathbf{r}). \tag{6.123}$$

6.16 Rotation of Rigid Bodies

The rotation of rigid bodies are very conveniently studied by introducing three angular coordinates. These coordinates are achieved by transforming a given Cartesian coordinate system to another by means of three successive rotations in a given sequence. These three successive angles of rotation are called *Eulerian angles.*

Let there be a Cartesian coordinate system OX, OY, OZ. The first rotation is given to this initial system by angle θ in anti-clockwise direction about z-axis (*Fig 6.8*). Let the unit vectors $\mathbf{i}$, $\mathbf{j}$, $\mathbf{k}$ along the initial axes of coordinates be transformed to unit vectors $\mathbf{i}_1$, $\mathbf{j}_1$, $\mathbf{k}_1$ due to rotation. Then, the transformation matrix is given by, according to equation (6.92)

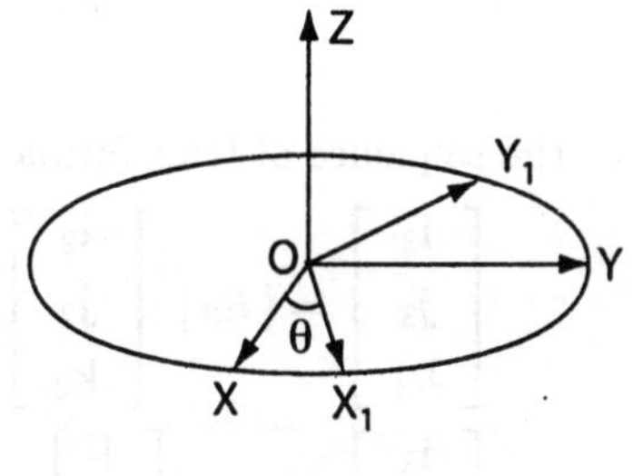

Fig 6.8

$$[R_1] = \begin{bmatrix} \cos\theta & \sin\theta & 0 \\ -\sin\theta & \cos\theta & 0 \\ 0 & 0 & 1 \end{bmatrix}. \quad (6.124)$$

Let, this new coordinate axes be OX_1, OY_1, OZ_1. The x-axis, OX_1 is called the *line of nodes*.

The second rotation is given about OX_1 axis to the xy-plane by an angle ϕ in anti-clockwise direction. This causes OZ_1 axis to turn through the angle ϕ (*Fig 6.9*). Let this new coordinate axes be OX_2, OY_2, OZ_2. If $\mathbf{i}_2$, $\mathbf{j}_2$, $\mathbf{k}_2$ be unit vectors along these coordinate axes, then

$$\begin{aligned} \mathbf{i}_2 &= \mathbf{i}_1 \\ \mathbf{j}_2 &= \mathbf{j}_1 \cos\phi + \mathbf{k}_1 \sin\phi \\ \mathbf{k}_2 &= -\mathbf{j}_1 \sin\phi + \mathbf{k}_1 \cos\phi. \end{aligned}$$

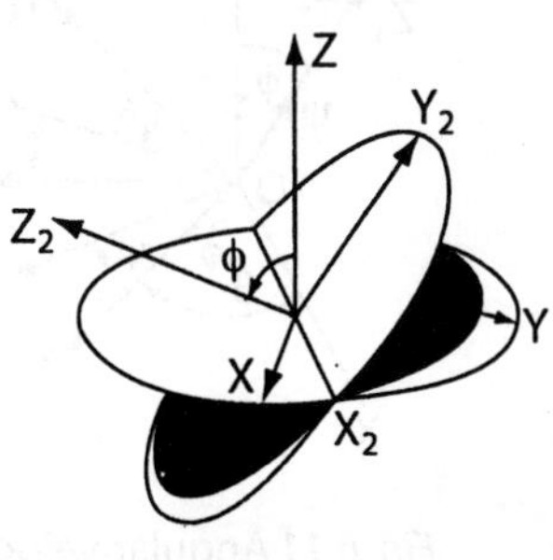

Fig 6.9

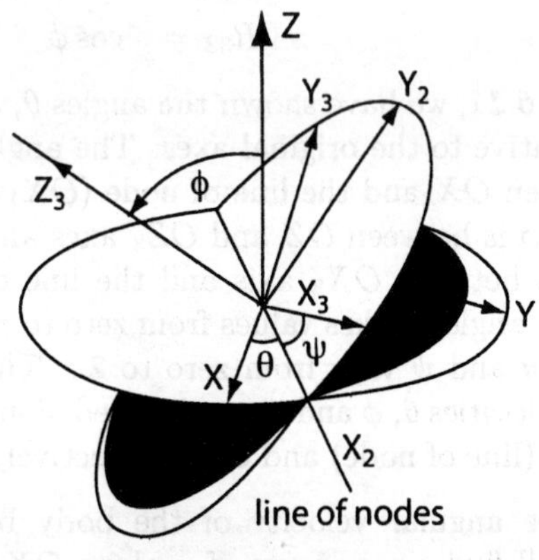

Fig 6.10

Hence, the rotation matrix is

$$[R_2] = \begin{bmatrix} 1 & 0 & 0 \\ 0 & \cos\phi & \sin\phi \\ 0 & -\sin\phi & \cos\phi \end{bmatrix} \quad (6.125)$$

The third rotation is given about OZ_2 axis by an angle ψ in anti-clockwise direction. This causes OX_2 axis to turn through the angle ψ (*Fig 6.10*). Let OX_3, OY_3, OZ_3 be the new axes of coordinates along which let the unit vectors be $\mathbf{i}_3$, $\mathbf{j}_3$, $\mathbf{k}_3$. Then the rotation matrix is given by (similar to the first rotation)

$$[R_3] = \begin{bmatrix} \cos\psi & \sin\psi & 0 \\ -\sin\psi & \cos\psi & 0 \\ 0 & 0 & 1 \end{bmatrix}. \tag{6.126}$$

Thus, the sequence of transformation performed is represented in matrix form as

$$\begin{bmatrix} \mathbf{i}_3 \\ \mathbf{j}_3 \\ \mathbf{k}_3 \end{bmatrix} = [R_3] \begin{bmatrix} \mathbf{i}_2 \\ \mathbf{j}_2 \\ \mathbf{k}_2 \end{bmatrix} = [R_3][R_2] \begin{bmatrix} \mathbf{i}_1 \\ \mathbf{j}_1 \\ \mathbf{k}_1 \end{bmatrix} = [R_3][R_2][R_1] \begin{bmatrix} \mathbf{i} \\ \mathbf{j} \\ \mathbf{k} \end{bmatrix},$$

$$\text{i.e.,} \quad \begin{bmatrix} \mathbf{i}_3 \\ \mathbf{j}_3 \\ \mathbf{k}_3 \end{bmatrix} = [R] \begin{bmatrix} \mathbf{i} \\ \mathbf{j} \\ \mathbf{k} \end{bmatrix}, \tag{6.127}$$

where

$$[R] = [R_3][R_2][R_1]. \tag{6.128}$$

Multiplication of matrices on RHS of equation (6.128) produces the elements of the matrix $[R]$ as follows:

$$\left.\begin{aligned} R_{11} &= \cos\psi\cos\theta - \sin\psi\sin\theta\cos\phi \\ R_{12} &= \cos\psi\sin\theta + \sin\psi\cos\theta\cos\phi \\ R_{13} &= \sin\psi\sin\phi \\ R_{21} &= -\sin\psi\cos\theta - \cos\psi\sin\theta\cos\phi \\ R_{22} &= -\sin\psi\sin\theta + \cos\psi\cos\theta\cos\phi \\ R_{23} &= \cos\psi\sin\phi \\ R_{33} &= \cos\phi \end{aligned}\right\}. \tag{6.129}$$

In *Fig 6.11*, we have shown the angles θ, ϕ and ψ relative to the original axes. The angle θ is between OX and the line of node (OX_1), the angle ϕ is between OZ and OZ_3 axes and angle ψ is between OX_3 axis and the line of node. The angle ϕ takes values from zero to π, and both θ and ψ vary from zero to 2π. The angular velocities $\dot\theta$, $\dot\phi$ and $\dot\phi$ are directed along OZ, OX_1 (line of node) and OZ_3 respectively.

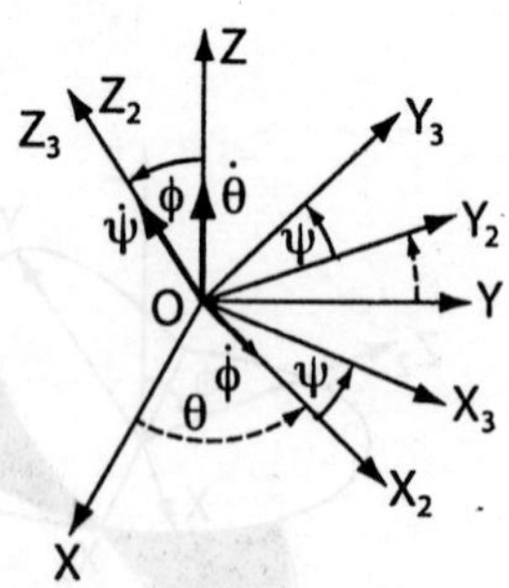

Fig 6.11 Angular velocities $\dot\varphi, \dot\psi$ and $\dot\theta$

Let the angular velocity of the body be ω. We shall find components of ω along OX_3, OY_3, OZ_3 axes.

Components of $\dot\theta$ along OX_3, OY_3 and OZ_3 are:

$$\left.\begin{aligned} \dot\theta_x &= \dot\theta\sin\phi\sin\psi \\ \dot\theta_y &= \dot\theta\sin\phi\cos\psi \\ \dot\theta_z &= \dot\theta\cos\phi \end{aligned}\right\}. \tag{6.130}$$

Components of $\dot{\phi}$ along OX_3, OY_3, and OZ_3 are:

$$\left.\begin{aligned} \dot{\phi}_x &= \dot{\phi}\cos\psi \\ \dot{\phi}_y &= -\dot{\phi}\sin\psi \\ \dot{\phi}_z &= 0 \end{aligned}\right\}. \tag{6.131}$$

Components of $\dot{\psi}$ along OX_3, OY_3 and OZ_3 are

$$\left.\begin{aligned} \dot{\psi}_x &= \dot{\psi}_y = 0 \\ \dot{\psi}_z &= \dot{\psi}. \end{aligned}\right\}. \tag{6.132}$$

Hence, the component of ω along OX_3, OY_3 and OZ_3 are:

$$\left.\begin{aligned} \omega_x &= \dot{\theta}\sin\phi\sin\psi + \dot{\phi}\cos\psi \\ \omega_y &= \dot{\theta}\sin\phi\cos\psi - \dot{\phi}\sin\psi \\ \omega_z &= \dot{\theta}\cos\phi + \dot{\psi} \end{aligned}\right\}. \tag{6.133}$$

EXAMPLE 6.5. *A three-dimensional frame of reference is rotating with angular velocity ω about an axis passing through origin. If* $\mathbf{i}, \mathbf{j}, \mathbf{k}$ *and* $\mathbf{i}_1, \mathbf{j}_1, \mathbf{k}_1$ *be the unit vectors along the axes in old and new coordinate system, then find the rotation matrix R and its derivative with respect to time.*

Solution: If $\mathbf{r}$ be the position vector of any point, then

$$\mathbf{r} = \mathbf{i}x + \mathbf{j}y + \mathbf{k}z$$

$$\text{and also,}\quad \mathbf{r} = \mathbf{i}_1 x_1 + \mathbf{j}_1 y_1 + \mathbf{k}_1 z_1.$$

Therefore

$$\mathbf{i}x + \mathbf{j}y + \mathbf{k}z = \mathbf{i}_1 x_1 + \mathbf{j}_1 y_1 + \mathbf{k}_1 z_1. \tag{1}$$

Taking dot product on both sides of equation (1) by $\mathbf{i}$, $\mathbf{j}$ and $\mathbf{k}$ in sequence, we get

$$\left.\begin{aligned} x &= (\mathbf{i}_1\cdot\mathbf{i})\,x_1 + (\mathbf{j}_1\cdot\mathbf{i})\,y_1 + (\mathbf{k}_1\cdot\mathbf{i})\,z_1 \\ y &= (\mathbf{i}_1\cdot\mathbf{j})\,x_1 + (\mathbf{j}_1\cdot\mathbf{j})\,y_1 + (\mathbf{k}_1\cdot\mathbf{j})\,z_1 \\ z &= (\mathbf{i}_1\cdot\mathbf{k})\,x_1 + (\mathbf{j}_1\cdot\mathbf{k})\,y_1 + (\mathbf{k}_1\cdot\mathbf{k})\,z_1 \end{aligned}\right\}. \tag{2}$$

The system of equations (2) can be written as

$$\begin{bmatrix} x \\ y \\ z \end{bmatrix} = [R]\begin{bmatrix} x_1 \\ y_1 \\ z_1 \end{bmatrix}, \tag{3}$$

where

$$[R] = \begin{bmatrix} \mathbf{i}_1\cdot\mathbf{i} & \mathbf{j}_1\cdot\mathbf{i} & \mathbf{k}_1\cdot\mathbf{i} \\ \mathbf{i}_1\cdot\mathbf{j} & \mathbf{j}_1\cdot\mathbf{j} & \mathbf{k}_1\cdot\mathbf{j} \\ \mathbf{i}_1\cdot\mathbf{k} & \mathbf{j}_1\cdot\mathbf{k} & \mathbf{k}_1\cdot\mathbf{k} \end{bmatrix}, \tag{4}$$

which is the required rotation matrix.

Hence

$$\frac{d}{dt}[R] = \begin{bmatrix} \frac{d\mathbf{i}_1}{dt}\cdot\mathbf{i} & \frac{d\mathbf{j}_1}{dt}\cdot\mathbf{i} & \frac{d\mathbf{k}_1}{dt}\cdot\mathbf{i} \\ \frac{d\mathbf{i}_1}{dt}\cdot\mathbf{j} & \frac{d\mathbf{j}_1}{dt}\cdot\mathbf{j} & \frac{d\mathbf{k}_1}{dt}\cdot\mathbf{j} \\ \frac{d\mathbf{i}_1}{dt}\cdot\mathbf{k} & \frac{d\mathbf{j}_1}{dt}\cdot\mathbf{k} & \frac{d\mathbf{k}_1}{dt}\cdot\mathbf{k} \end{bmatrix}. \tag{5}$$

Since $\mathbf{i}_1$, $\mathbf{j}_1$, $\mathbf{k}_1$ are rotating vectors with angular velocity ω,

$$\frac{d\mathbf{i}_1}{dt} = \boldsymbol{\omega} \times i_1, \quad \frac{d\mathbf{j}_1}{dt} = \boldsymbol{\omega} \times j_1, \quad \frac{d\mathbf{k}_1}{dt} = \boldsymbol{\omega} \times k_1.$$

Hence, the time derivative of rotation matrix can be written as

$$\frac{d}{dt}[R] = \begin{bmatrix} (\omega\times\mathbf{i}_1)\cdot\mathbf{i} & (\omega\times\mathbf{j}_1)\cdot\mathbf{i} & (\omega\times\mathbf{k}_1)\cdot\mathbf{i} \\ (\omega\times\mathbf{i}_1)\cdot\mathbf{j} & (\omega\times\mathbf{j}_1)\cdot\mathbf{j} & (\omega\times\mathbf{k}_1)\cdot\mathbf{j} \\ (\omega\times\mathbf{i}_1)\cdot\mathbf{k} & (\omega\times\mathbf{j}_1)\cdot\mathbf{k} & (\omega\times\mathbf{k}_1)\cdot\mathbf{k} \end{bmatrix}. \tag{6}$$

6.17 Effect on Acceleration Due to Gravity

Let us consider a coordinate system on the surface of the earth. Due to the continous rotation of the earth about its own axis with an angular velocity ω, the coordinate frame also will be rotating with the same angular velocity. Let there be a particle of mass m acted upon by some external force $\mathbf{F}$ apart from the gravitational force $m\mathbf{g}$. Then, the equation of motion of the particle in rotating frame of axes will be [*see* equation (6.122)]

$$m\frac{d^2\mathbf{r}}{dt^2} = \mathbf{F} + m\left[\mathbf{g} - \boldsymbol{\omega} \times (\boldsymbol{\omega} \times \boldsymbol{r})\right] - 2m\left(\boldsymbol{\omega} \times \frac{d\mathbf{r}}{dt}\right). \tag{6.134}$$

The centrifugal force $-\boldsymbol{\omega} \times (\boldsymbol{\omega} \times \boldsymbol{r})$ always acts radially away from the centre. Thus, the term

$$\mathbf{g}_e = \mathbf{g} - \boldsymbol{\omega} \times (\boldsymbol{\omega} \times \mathbf{r}) \tag{6.135}$$

indicates the effective value of the acceleration due to gravity which arises out of rotation of the earth.

Now, at the poles, vectors $\boldsymbol{\omega}$ and $\mathbf{r}$ are parallel; hence, $\boldsymbol{\omega} \times \boldsymbol{r} = 0$, so that there is no reduction in the value of the acceleration due to gravity. Similarly, at equator, the vectors $\boldsymbol{\omega}$ and $\mathbf{r}$ are at right-angles, so that $\boldsymbol{\omega} \times \boldsymbol{r}$ is maximum. Thus, the value of $\mathbf{g}_e$ at the equator is minimum.

6.18 Effect of Coriolis Force

Let us consider a particle of mass m situated at a point P on the surface of the earth having latitude λ. Let the particle be moving with a velocity $\mathbf{v}$. We take the rotating frame of axes at P, such that: (i) x-axis is directed towards east; (ii) y-axis is directed towards north; and (iii) z-axis is in the direction of the plumb line at the geographical latitude (*Fig 6.12*).

Then, the components of angular velocity ω of the earth in the northern hemisphere along the coordinate axes are

$$\left.\begin{aligned} \omega_x &= 0 \\ \omega_y &= \omega\cos\lambda \\ \omega_z &= \omega\sin\lambda \end{aligned}\right\}. \qquad (6.136)$$

The component ω_z will be directed upward in the northern hemisphere and downward in the southern hemisphere.

The Coriolis force acting on the particle due to the rotation of the frame of axes is

$$\mathbf{F}_c = -2m\boldsymbol{\omega}\times\frac{d\mathbf{r}}{dt} = -2m\boldsymbol{\omega}\times\mathbf{v}, \qquad (6.137)$$

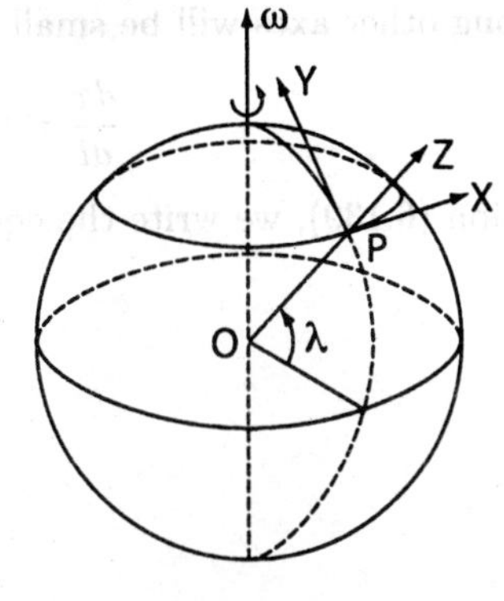

Fig 6.12

which will be directed at right angles to the plane containing $\boldsymbol{\omega}$ and $\mathbf{v}$. Hence, if we assume that the particle is moving in a plane perpendicular to the axis of rotation of the earth, then, due to Coriolis force, the particle will be deflected eastward in northern hemisphere, and westward in the southern hemisphere.

The maximum value of this force will be at poles and equal to $2m\omega v$. At equator, its value is zero, so that particles moving horizontally will not be affected.

The effect of Coriolis force plays an important role in guided missiles due to their large velocity and time of flight. It also plays an important role in the formation of cyclones. When a mass of air rushes to any low-pressure region, it gets deflected due to Coriolis force to the right in the northern hemisphere, and to the left in the southern hemisphere. As a result, in a cyclone, the wind whirls in the anti-clockwise direction in northern hemisphere, and in clockwise sense in the southern hemisphere.

We now intend to find the effect of Coriolis force on the motion of a particle on earth. Let $\mathbf{i}$, $\mathbf{j}$, $\mathbf{k}$ be the unit vectors along x, y and z axes of the rotating frame which we have already defined. We assume that there is no resistance of air, viscosity, etc., in the motion of the particle. Then, the equation of motion can be written as

$$\begin{aligned} m\{\mathbf{i}\ddot{x}+\mathbf{j}\ddot{y}+\mathbf{k}\ddot{z}\} &= -mg_e\mathbf{k} - 2m\,(\mathbf{i}\omega_x+\mathbf{j}\omega_y+\mathbf{k}\omega_z)\times(\mathbf{i}\dot{x}+\mathbf{j}\dot{y}+\mathbf{k}\dot{z}) \\ &= -mg_e\mathbf{k} - 2m\,(\mathbf{j}\omega\cos\lambda+\mathbf{k}\omega\sin\lambda)\times(\mathbf{i}\dot{x}+\mathbf{j}\dot{y}+\mathbf{k}\dot{z}) \qquad (6.138) \end{aligned}$$

[using equation (6.136)]

$$\text{i.e.,}\quad \left.\begin{aligned} m\frac{d^2x}{dt^2} &= -2m\omega\cos\lambda\frac{dz}{dt} + 2m\omega\sin\lambda\frac{dy}{dt} \\ m\frac{d^2y}{dt^2} &= -2m\omega\sin\lambda\frac{dx}{dt} \\ m\frac{d^2z}{dt^2} &= -mg_e + 2m\omega\cos\lambda\frac{dx}{dt} \end{aligned}\right\}. \qquad (6.139)$$

We consider the following cases:

Case I. *A particle falling freely.*

Let a particle fall freely towards the earth through a height h which is very small. We can, therefore, neglect the variation of g. Moreover, for particle falling along z-axis, the Coriolis force along other axes will be small, so that we can neglect them. Thus, putting

$$\frac{dx}{dt}=0,\quad \frac{dy}{dt}=0\quad \text{and}\quad \frac{dz}{dt}=-gt \tag{6.140}$$

in equation (6.139), we write the equations of motion of the particle as

$$\frac{d^2x}{dt^2}=2\omega gt\cos\lambda \tag{6.141}$$

$$\frac{d^2y}{dt^2}=0 \tag{6.142}$$

$$\frac{d^2z}{dt^2}=-g. \tag{6.143}$$

Integrating equations (6.141) and (6.143) we get, assuming that, at $t=0$, $x=0$, $\dot{x}=0$, $z=h$, $\dot{z}=0$,

$$x=\tfrac{1}{3}\omega gt^3\cos\lambda \tag{6.144}$$

$$z=h-\tfrac{1}{2}gt^2. \tag{6.145}$$

From equation (6.145), we get the time of fall (when $z=0$)

$$t=\sqrt{\frac{2h}{g}}. \tag{6.146}$$

The particle during this time will move along x-direction by an amount

$$\frac{2}{3}\omega\cos\lambda\left(\frac{2h^3}{g}\right)^{1/2}. \tag{6.147}$$

This shows the effect of Coriolis force on a freely falling body.

Case II. *The particle thrown vertically downwards.*

Let the particle be thrown vertically downwards with a velocity v_0. Then, the initial conditions are, at $t=0$

$$\left.\begin{aligned} x&=0, & \frac{dx}{dt}&=0\\ z&=h, & \frac{dz}{dt}&=-v_0 \end{aligned}\right\} \tag{6.148}$$

Integrating equation (6.143) with initial conditions in (6.148), we get

$$\frac{dz}{dt}=-gt-v_0. \tag{6.149}$$

Putting this into first equation of (6.139), we write

$$\frac{d^2x}{dt^2}=2\omega\cos\lambda\,(gt+v_0). \tag{6.150}$$

Integrating twice, with due care for initial conditions in (6.148), we get

$$x=\omega v_0t^2\cos\lambda+\tfrac{1}{3}\omega gt^3\cos\lambda. \tag{6.151}$$

This gives the amount of shift towards east at any time t.

Case III. *The particle thrown vertically upwards.*

Let the particle be thrown vertically upwards with an initial velocity v_0. then the initial conditions are, at $t = 0$

$$\left.\begin{aligned} x = 0, \quad \frac{dx}{dt} = 0 \\ z = 0, \quad \frac{dz}{dt} = v_0 \end{aligned}\right\} \tag{6.152}$$

Integrating equation (6.143) with initial conditions in (6.152), we get

$$\frac{dz}{dt} = -gt + v_0. \tag{6.153}$$

Then, the first equation of motion in (6.139) becomes

$$\frac{d^2x}{dt^2} = 2\omega \cos\lambda \, (gt - v_0). \tag{6.154}$$

Integrating this twice with due care for the initial conditions in (6.152), we get

$$x = 2\omega \cos\lambda \left(\frac{1}{6}gt^3 - \frac{v_0 t^2}{2}\right). \tag{6.155}$$

Now, the time to reach the greatest height will be given by equation (6.153) when vertical velocity is zero. Thus

$$0 = -gt + v_0, \quad \text{i.e.,} \quad t = v_0/g. \tag{6.156}$$

Equation (6.156) gives the time to reach the point, where the particle comes to rest. Same time will be taken by the particle to come down to the point of projection. Hence, the total time of flight will be

$$2t = \frac{2v_0}{g}. \tag{6.157}$$

Putting this into equation (6.155), we get the shift of the particle eastward when it returns as

$$x = 2\omega \cos\lambda \left[\frac{1}{6}g\left(\frac{2v_0}{g}\right)^3 - \frac{1}{2}v_0\left(\frac{2v_0}{g}\right)^2\right]$$

$$\text{or,} \quad x = -\frac{4v_0^3}{3g^2}\omega \cos\lambda. \tag{6.158}$$

Again, from energy consideration

$$\tfrac{1}{2}mv_0^2 = mgh, \quad \text{i.e.,} \quad v_0 = \sqrt{2gh}, \tag{6.159}$$

where h is the greatest height attained. Putting this into equation (6.157), the total time of flight is also given by

$$\frac{2v_0}{g} = \frac{2}{g}\sqrt{2gh} = 2\sqrt{\frac{2h}{g}}, \tag{6.160}$$

which is same as in equation (6.146). Since the particle falls freely during its return journey, we can use equation (6.144) to find its shift. Hence

$$x = \frac{1}{3}\omega\cos\lambda\left[g\cdot\left(\frac{8h}{g}\right)^{3/2}\right]$$

$$= \frac{16}{3}\omega\sqrt{\frac{2h^3}{g}}\cos\lambda. \tag{6.161}$$

Case IV. *Particle thrown at some angle with horizon.*

Let a particle of mass m be thrown upwards with velocity V_0 at an angle α with the horizon towards the east. Then the initial conditions are: at $t=0$,

$$\left.\begin{aligned} x=0,\ \frac{dx}{dt}&=V_0\cos\alpha\\ y=0,\ \frac{dy}{dt}&=0\\ z=0,\ \frac{dz}{dt}&=V_0\sin\alpha,\end{aligned}\right\} \tag{6.162}$$

where V_0 is the initial velocity of throw. We assume that due to the motion of the particle in xy-plane the deflection by Coriolis force along y-direction is negligible. Hence

$$\left.\begin{aligned} \dot{x}&=V_0\cos\alpha\\ \dot{y}&=0\\ \dot{z}&=V_0\sin\alpha-gt\end{aligned}\right\}. \tag{6.163}$$

The equations of motion following the set of equations (6.139) are:

$$\frac{d^2x}{dt^2} = -2\omega\cos\lambda\,(V_0\sin\alpha - gt) \tag{6.164}$$

$$\frac{d^2y}{dt^2} = -2\omega\sin\lambda\,V_0\cos\alpha \tag{6.165}$$

$$\frac{d^2z}{dt^2} = -g + 2\omega\cos\lambda\,V_0\cos\alpha. \tag{6.166}$$

Integrating equation (6.166), we get, by using the initial conditions in (6.162)

$$\frac{dz}{dt} = -gt + 2\omega V_0\cos\lambda\cos\alpha\,t + V_0\sin\alpha. \tag{6.167}$$

Maxinum height would be reached when

$$\frac{dz}{dt}=0,\quad \text{i.e.,}\quad t=\frac{V_0\sin\alpha}{g-2\omega V_0\cos\lambda\cos\alpha}. \tag{6.168}$$

The magnitude of earth's angular velocity is

$$\omega = \frac{2\pi}{24\times 60\times 60} \simeq 7.3\times 10^{-5}\ \text{rad/sec}.$$

Thus, the denominator in (6.168) may be approximated to g, so that we can write

$$t = \frac{V_0 \sin\alpha}{g}. \tag{6.169}$$

Hence, the time of flight, i.e., the time taken to reach the ground is

$$\frac{2V_0 \sin\alpha}{g}. \tag{6.170}$$

Integrating equation (6.165) twice with respect to time t and using the boundary conditions in (6.162), we get

$$y = -V_0\omega \sin\lambda \cos\alpha\, t^2. \tag{6.171}$$

Substituting here the value of t from equation (6.169), we get

$$y = -V_0\omega \sin\lambda \cos\alpha \left(\frac{2V_0 \sin\alpha}{g}\right)^2 = -\frac{4V_0^3\omega}{g^2} \sin\lambda \sin^2\alpha \cos\alpha. \tag{6.172}$$

Equation (6.172) gives the amount of lateral deflection when the particle reaches the ground. Integrating equation (6.164) and using the boundary conditions in (6.162), we get

$$\frac{dx}{dt} = -2\omega\cos\lambda \left(V_0 \sin\alpha\, t - \tfrac{1}{2}\, gt^2\right) + V_0 \cos\alpha \tag{6.173}$$

$$x = -2\omega\cos\lambda \left(V_0 \sin\alpha \cdot \tfrac{t^2}{2} - \tfrac{1}{3}\, gt^3\right) + V_0 \cos\alpha \cdot t. \tag{6.174}$$

The horizontal range of the particle will be obtained by taking the value of t from equation (6.169) and substituting into equation (6.174). Thus

$$R' = \frac{4V_0^3\omega}{3g^2} \cos\lambda \sin^3\alpha + \frac{2V_0^2}{g} \sin\alpha\cos\alpha. \tag{6.175}$$

If the earth did not rotate, the horizontal range of the projectile would have been

$$R = \frac{2V_0^2}{g} \sin\alpha\cos\alpha. \tag{6.176}$$

Using the equation (6.176), equation (6.175) can be written as

$$R' = \frac{1}{3}\sqrt{\frac{2R^3}{g}}\, \omega\cos\lambda \sqrt{\tan^3\alpha} + R. \tag{6.177}$$

Hence, the change in the horizontal range of the particle is given by

$$R' - R = \frac{1}{3}\sqrt{\frac{2R^3}{g}}\, \omega\cos\lambda \sqrt{\tan^3\alpha}. \tag{6.178}$$

Again, integrating equation (6.167) and using the boundary condition that $z(0) = 0$, we get

$$z = V_0 \sin\alpha \cdot t - \tfrac{1}{2}\, gt^2 + \omega V_0 \cos\lambda \cos\alpha \cdot t^2. \tag{6.179}$$

The maximum height reached by the particle would be obtained by putting the value of t from equation (6.169) into the equation (6.179).

Thus

$$H' = \frac{V_0^2 \sin^2 \alpha}{2g} + \frac{V_0^3 \omega}{g^2} \cos \lambda \sin^2 \alpha \cos \alpha. \tag{6.180}$$

If the earth did not rotate, the maximum height attained would have been

$$H = \frac{V_0^2 \sin^2 \alpha}{2g}. \tag{6.181}$$

Hence, from equations (6.180) and (6.181), we can write

$$H' = H + \frac{V_0^3 \omega}{g^2} \cos \lambda \sin^2 \alpha \cos \alpha. \tag{6.182}$$

Thus, the difference in maximum height attained due to the rotation of the earth is

$$H' - H = \frac{V_0^3 \omega}{g^2} \cos \lambda \sin^2 \alpha \cos \alpha. \tag{6.183}$$

EXERCISE

1. Prove that if q_1, q_1, q_3 are orthogonal curvilinear coordinates, then the Jacobian of transformation from x, y, z coordinates to q_1, q_2, q_3 coordinate system defined by
$$J\left(\frac{x, y, z}{q_1, q_2, q_3}\right)$$
is $h_1 h_2 h_3$.
2. Show that, in spherical polar coordinates
$$-i\left(x\frac{\partial}{\partial y} - y\frac{\partial}{\partial x}\right) = -i\frac{\partial}{\partial \phi}, \quad i = \sqrt{-1}.$$
3. Express the position vector $\mathbf{i}3x - \mathbf{j}2y + \mathbf{k}z$ in
   (a) spherical polar coordinates, and
   (b) elliptic cylindrical coordinates.
4. The velocity of a particle in Cartesian coordinates is given by
$$V = \mathbf{i}\frac{x}{x^2+y^2} + \mathbf{j}\frac{y}{x^2+y^2}.$$
Express V in circular cylindrical coordinates. Also, examine the existence of a velocity potential.
5. In an equilateral triangle, origin is taken at the centroid, and three mutually perpendicular lines from the centroid are taken as axes. The transformation of coordinates are given by
$$\left.\begin{aligned} p_1 &= r + \tfrac{x}{2} - \tfrac{\sqrt{3}}{2}y \\ p_2 &= r + \tfrac{x}{2} + \tfrac{\sqrt{3}}{2}y \\ p_3 &= r - x \end{aligned}\right\},$$
where r is the in-radius of the triangle. Find expression for Laplacian operator.

6. A three-dimensional Cartesian frame is rotated through angles (a) 45°, (b) 90°, (c) 180° about the z-axis. Find the transformation matrix in each case.
7. In a rotating frame of reference, deduce the operator

$$\frac{d}{dt} = \frac{d'}{dt} + \omega\times$$

by matrix method.
8. A bead slides along a smooth circular wire in a vertical plane. The wire is rotating about its vertical diameter with uniform angular velocity. Discuss its motion.
9. A projectile is fired horizontally from a point at height h from the ground with initial velocity V_0. Find its lateral shift when it reaches the ground. The latitude of the point of projection is λ.
10. A projectile is fired at an angle α to the horizon from a place having latitude λ with initial velocity V_0. Show that, its time of flight is

$$\frac{2V_0 \sin\alpha}{g}\left(1 + \frac{2\omega V_0}{g}\cos\lambda\cos\alpha\right).$$

ANSWERS

3. (a) $(5\cos^2\theta - 2)\, r\mathbf{e}_r - 5\sin\theta\cos\theta r\mathbf{e}_\theta + z\mathbf{e}_z$;

(b) $\dfrac{a}{\sqrt{F}}\sinh\xi\cosh\xi\,(5\cos^2\eta - 2)\,\mathbf{e}_\xi - \dfrac{a}{\sqrt{F}}\sin\eta\cos\eta.(5\cosh^2\xi - 2)\,\mathbf{e}_\eta + z\mathbf{e}_z$.

4. $\dfrac{1}{r}\mathbf{e}_r$.

5. $\nabla = \dfrac{\partial^2}{\partial p_1^2} + \dfrac{\partial^2}{\partial p_2^2} + \dfrac{\partial^2}{\partial p_3^2} - \dfrac{\partial^2}{\partial p_1 \partial p_3} - \dfrac{\partial^2}{\partial p_2 \partial p_3} - \dfrac{\partial^2}{\partial p_3 \partial p_1}$.

6. (a) $\begin{bmatrix} 1/\sqrt{2} & 1/\sqrt{2} & 0 \\ -1/\sqrt{2} & 1/\sqrt{2} & 0 \\ 0 & 0 & 1 \end{bmatrix}$; (b) $\begin{bmatrix} 0 & 1 & 0 \\ -1 & 0 & 0 \\ 0 & 0 & 1 \end{bmatrix}$; (c) $\begin{bmatrix} -1 & 0 & 0 \\ 0 & -1 & 0 \\ 0 & 0 & 1 \end{bmatrix}$.

9. $y = -\dfrac{2hV_0\omega\sin\lambda}{g - 2\omega V_0\cos\lambda}$.

Chapter 7

Beta and Gamma Functions

7.1 Beta Function

The Beta function is defined as a definite Euler's integral of first kind

$$B(m,n) = \int_0^1 x^{m-1} \cdot (1-x)^{n-1}\, dx, \tag{7.1}$$

where m, n are positive numbers, i.e., $m > 0$, $n > 0$. For m and n negative real numbers, the integral in equation (7.1) does not converge. This we can prove with the help of theorem called *μ-test for convergence.* It states that:

If $f(x)$ be an integrable function in the arbitrary interval $(a+\epsilon, b)$, where $0 < \epsilon < (b-a)$, then the integral

$$\int_a^b f(x)\, dx \tag{7.2}$$

converges, if

$$\lim_{x \longrightarrow a+} (x-a)^{\mu} f(x) = \xi, \quad \text{for } 0 < \mu < 1 \tag{7.3}$$

and diverges, if

$$\lim_{x \longrightarrow a+} (x-a)^{\mu} f(x) = \xi (\neq 0), \quad \text{or,} \quad \pm\infty, \text{ for } \mu \geq 1. \tag{7.4}$$

We shall not go to prove this theorem, but use this μ-test to establish convergence and divergence of the Beta function.

We find that the integral in equation (7.1) is proper when $m, n \geq 1$, but is improper at the lower limit when $m < 1$, and at the upper limit when $n < 1$. Hence, we split the integral into two parts as

$$\int_0^1 x^{m-1} (1-x)^{n-1}\, dx = I_1 + I_2, \tag{7.5}$$

where

$$I_1 = \int_0^{1/2} x^{m-1} (1-x)^{n-1}\, dx \tag{7.6}$$

$$\text{and} \quad I_2 = \int_{1/2}^1 x^{m-1} (1-x)^{n-1}\, dx. \tag{7.7}$$

We take here

$$f(x) = x^{m-1} (1-x)^{n-1}. \tag{7.8}$$

Thus

$$x^{1-m} f(x) = x^{1-m} \cdot x^{m-1} \cdot (1-x)^{n-1} = (1-x)^{n-1},$$

which tends to 1 as $x \longrightarrow 0+$.

Thus, by condition (7.3), the integral I_1 converges for $0 < m < 1$ at the lower limit.

Again, as $x \longrightarrow 0+$

$$xf(x) = x \cdot x^{1-m} \cdot (1-x)^{n-1} = x^m (1-x)^{n-1}$$

which tends to 1 for $m = 0$, and to ∞ when $m < 0$. Thus, the first integral I_1 converges when $m > 0$ and diverges for $m < 0$ at the lower limit.

In order to test the second integral I_2, we make the substitution

$$x = 1 - y. \tag{7.9}$$

Then the integral I_2 is converted into the integral I_1 with m and n interchanged. Hence, following the above discussion, we can conclude that the integral I_2 converges when $n > 0$ and diverges when $n < 0$. Thus, the Beta function defined by the integral in equation (7.1) is convergent for $m > 0$, $n > 0$, and divergent for $m < 0$, $n < 0$.

7.2 Properties of Beta Function

Property 1. The Beta function is symmetrical in m and n, i.e.

$$B(m,n) = B(n,m). \tag{7.10}$$

Proof. If we make a change of variables given by equation (7.9), we have

$$B(m,n) = \int_1^0 (1-y)^{m-1} [1-(1-y)]^{n-1} (-dy) = \int_0^1 y^{n-1} (1-y)^{m-1} \, dy = B(n,m).$$

□

Property 2. $B(m,n) = \dfrac{m-1}{m+n-1} B(m-1, n) = \dfrac{n-1}{m+n-1} B(m, n-1)$.

Proof. We have

$$B(m,n) = \int_0^1 x^{m-1} (1-x)^{n-1} \, dx.$$

Integrating by parts

$$B(m,n) = (1-x)^{n-1} \frac{x^m}{m}\bigg|_0^1 + \int_0^1 \frac{x^m}{m} (n-1)(1-x)^{n-2} \, dx = \frac{n-1}{m} \int_0^1 x^m (1-x)^{n-2} \, dx.$$

Now

$$x^m = x^{m-1} - x^{m-1}(1-x). \tag{7.11}$$

Substituting this identity (7.11) into the above integral, we have

$$B(m,n) = \frac{n-1}{m}\int_0^1 x^{m-1}(1-x)^{n-2}\,dx - \frac{n-1}{m}\int_0^1 x^{m-1}(1-x)^{n-1}\,dx$$
$$= \frac{n-1}{m}B(m,n-1) - \frac{n-1}{m}B(m,n).$$

$$\therefore \quad B(m,n) = \frac{n-1}{m+n-1}B(m,n-1). \tag{7.12}$$

Interchanging m, n, we get

$$B(n,m) = \frac{m-1}{m+n-1}B(n,\, m-1). \tag{7.13}$$

□

Property 3. $B(m,n) = B(n,m) = \dfrac{(n-1)!\,(m-1)!}{(m+n-1)!}$, for m, n integers.

Proof. Let n be an integer. Then, using the relation (7.12) successively, we get

$$B(m,n) = \frac{n-1}{m+n-1}\cdot\frac{n-2}{m+n-2}\cdots\frac{1}{m+1}B(m,1). \tag{7.14}$$

But

$$B(m,1) = \int_0^1 x^{m-1}\,dx = \frac{1}{m}.$$

Hence, equation (7.14) becomes

$$B(m,n) = \frac{1\cdot 2\cdots(n-1)}{m(m+1)\cdots(m+n-1)}. \tag{7.15}$$

If m and n are both integers, then, from equation (7.14), we can write by successive use of equation (7.13)

$$B(m,n) = B(n,m) = \frac{1\cdot 2\cdots(n-1)}{(m+1)(m+2)\cdots(m+n-1)}B(m,1)$$
$$= \frac{1\cdot 2\cdots(n-1)}{(m+1)\cdots(m+n-1)}\cdot\frac{(m-1)(m-2)\cdots 2\cdot 1}{m(m-1)\cdots 2}B(1,1).$$

But $B(1,1) = 1$.
Hence

$$B(m,n) = B(n,m) = \frac{(n-1)!\,(m-1)!}{(m+n-1)!}. \tag{7.16}$$

□

Property 4. $B(m,m) = \dfrac{1}{2^{2m-1}}B\left(\dfrac{1}{2}, m\right).$

Proof. Substituting $n = m$ in the integral in equation (7.1), we get

$$B(m,m) = \int_0^1 x^{m-1}(1-x)^{m-1}\,dx,$$

$$\text{or,} \quad B(m,m) = 2\int_0^{1/2}\left[\frac{1}{4} - \left(\frac{1}{2} - x\right)^2\right]^{m-1} dx \tag{7.17}$$

(since the integrand is symmetrical about the line $x = 1/2$).

We substitute

$$\frac{1}{2} - x = \frac{\sqrt{y}}{2}. \tag{7.18}$$

Then equation (7.17) becomes

$$B(m,m) = \frac{1}{2^{2m-1}}\int_0^1 y^{-1/2}(1-y)^{m-1}\,dy,$$

$$\text{i.e.,} \quad B(m,m) = \frac{1}{2^{2m-1}} B\left(\frac{1}{2}, m\right). \tag{7.19}$$

□

7.3 Other Forms of Beta Function

Beta functions can also be represented by other forms of integrals. These are given below:

1. If we substitute

$$x = \frac{y}{1+y}, \quad \text{i.e.,} \quad y = \frac{x}{1-x} \tag{7.20}$$

in the integral in equation (7.1), we get

$$B(m,n) = \int_0^\infty \frac{y^{m-1}}{(1+y)^{m-1}} \cdot \frac{1}{(1+y)^{n-1}} \cdot \frac{dy}{(1+y)^2}.$$

$$\therefore \quad B(m,n) = \int_0^\infty \frac{y^{m-1}}{(1+y)^{m+n}}\,dy. \tag{7.21}$$

2. The integral in equation (7.21) can be written as

$$B(m,n) = \int_0^1 \frac{y^{m-1}}{(1+y)^{m+n}}\,dy + \int_1^\infty \frac{y^{m-1}}{(1+y)^{m+n}}\,dy. \tag{7.22}$$

In the second integral, we substitute

$$y = 1/x. \tag{7.23}$$

Then, the second integral in equation (7.22) becomes

$$I_2 = \int_1^0 \frac{\left(\frac{1}{x}\right)^{m-1}}{\left(1+\frac{1}{x}\right)^{m+n}}\left(-\frac{1}{x^2}\right)dx = \int_0^1 \frac{x^{n-1}}{(1+x)^{m+n}}\,dx = \int_0^1 \frac{y^{n-1}}{(1+y)^{m+n}}\,dy.$$

Substituting this into the integral in equation (7.22), we get

$$B(m,n) = \int_0^1 \frac{y^{m-1} + y^{n-1}}{(1+y)^{m+n}}\,dy. \tag{7.24}$$

3. If we substitute

$$y = \frac{\alpha}{\beta} x, \quad (\alpha, \beta \text{ constants}) \tag{7.25}$$

in the integral in equation (7.21), we get

$$B(m,n) = \int_0^\infty \frac{y^{m-1}}{(1+y)^{m+n}}\, dy = \alpha^m \beta^n \int_0^\infty \frac{x^{m-1}}{(\alpha x + \beta)^{m+n}}\, dx. \tag{7.26}$$

4. If we substitute

$$x = \sin^2 \theta \tag{7.27}$$

in the integral in equation (7.1), we get

$$B(m,n) = 2\int_0^{\pi/2} (\sin\theta)^{2m-2} (\cos\theta)^{2n-2} \sin\theta \cos\theta \, d\theta,$$

$$\text{or,} \quad B(m,n) = 2\int_0^{\pi/2} (\sin\theta)^{2m-1} (\cos\theta)^{2n-1}\, d\theta. \tag{7.28}$$

5. If we substitute

$$\alpha x = y, \quad (\alpha = \text{constant}) \tag{7.29}$$

in the integral in equation (7.1), we get

$$B(m,n) = \int_0^\infty \left(\frac{y}{\alpha}\right)^{m-1} \left(\frac{\alpha - y}{\alpha}\right)^{n-1} \frac{dy}{\alpha},$$

$$\text{i.e.,} \quad B(m,n) = \frac{1}{\alpha^{m+n-1}} \int_0^\infty y^{m-1} (\alpha - y)^{n-1}\, dy. \tag{7.30}$$

Thus, the Beta function can be defined by the integrals in equations (7.20), (7.23), (7.25), (7.27) and (7.29).

7.4 Incomplete Beta Function

The Beta function is called *incomplete* when its upper limit is a variable quantity. Thus, the incomplete Beta function is defined as

$$B_\xi(m,n) = \int_0^\xi x^{m-1} (1-x)^{n-1}\, dx, \quad \text{for } 0 \le \xi \le 1. \tag{7.31}$$

It's series expansion is given by

$$B_\xi(m,n) = \xi^m \left\{ \frac{1}{m} + \frac{1-n}{m+1}\xi + \frac{(1-n)(2-n)}{2!\,(m+2)}\xi^2 + \cdots + \frac{(1-n)(2-n)\cdots(p-n)}{p!\,(m+p)}\xi^p + \cdots \right\}. \tag{7.32}$$

Obviously, for $\xi = 1$, $B_\xi(m,n) = B(m,n)$.
Hence, the series expansion for regular Beta function can be obtained by putting $\xi = 1$ in the equation (7.32). Thus

$$B(m,n) = \frac{1}{m} + \frac{1-n}{m+1} + \frac{(1-n)(2-n)}{2!\,(m+2)} + \cdots + \frac{(1-n)(2-n)\cdots(p-n)}{p!\,(m+p)} + \cdots \tag{7.33}$$

7.5 Gamma Function

The Gamma function is defined in terms of Euler's definite integral of second kind as

$$\Gamma(m) = \int_0^\infty x^{m-1}\, e^{-x}\, dx, \quad \text{where } m > 0. \tag{7.34}$$

Let us discuss the convergence of Gamma function. We split the integral in equation (7.34) into two parts, since it becomes an improper integral at lower limit for $0 < m < 1$. Thus

$$\int_0^\infty e^{-x}\, x^{m-1}\, dx = I_1 + I_2,$$

where

$$I_1 = \int_0^1 e^{-x}\, x^{m-1}\, dx \tag{7.35}$$

$$\text{and} \quad I_2 = \int_1^\infty e^{-x}\, x^{m-1}\, dx. \tag{7.36}$$

We take

$$f(x) = e^{-x}\, x^{m-1}. \tag{7.37}$$

Then

$$x^{1-m}\, f(x) = x^{1-m}\, e^{-x}\, x^{m-1} = e^{-x},$$

which tends to 1 when $x \longrightarrow 0+$.

Thus, by μ-test, I_1 is absolutely convergent for $0 < m < 1$.

Again

$$x^2 f(x) = x^2 \cdot e^{-x} \cdot x^{m-1} = e^{-x}\, x^{m+1}, \tag{7.38}$$

which tends to zero when $x \longrightarrow \infty$.

Thus, by μ-test, I_2 converges absolutely for $m > 0$.

Hence, we conclude, that the Gamma function defined by the integral in equation (7.33) converges for $m > 0$.

7.6 Recursion Formula for Gamma Function

We have

$$\Gamma(m) = \int_0^\infty x^{m-1}\, e^{-x}\, dx.$$

Hence

$$\Gamma(m+1) = \int_0^\infty x^m\, e^{-x}\, dx. \tag{7.39}$$

Integrating by parts the integral in equation (7.39), we get

$$\Gamma(m+1) = \left(-x^m\, e^{-x}\right)\Big|_0^\infty + m\int_0^\infty x^{m-1}\, e^{-x}\, dx, \quad \text{i.e.,} \quad \Gamma(m+1) = m\Gamma(m). \tag{7.40}$$

This is the recursion formula for Gamma function.

For integral value of m, we get from equation (7.40)

$$\begin{aligned}\Gamma(m+1) &= m\Gamma(m) = m(m-1)\Gamma(m-1) = \cdots \\ &= m\,(m-1)\cdots 2\cdot 1\cdot \Gamma(1) = m!\,\Gamma(1).\end{aligned} \tag{7.41}$$

But

$$\Gamma(1) = \int_0^{\infty} e^{-x}\,dx = 1.$$

Hence

$$\Gamma(m+1) = m! \tag{7.42}$$

If $0 < m < 1$, then $m+1 > 0$; hence, $\Gamma(m+1)$ is defined and so is defined $\Gamma(m)$ which follows from equation (7.40).

Again

$$\Gamma(m) = \frac{1}{m}\,\Gamma(m+1), \quad \text{which} \longrightarrow \infty \text{ as } m \longrightarrow 0.$$

Thus, with successive use of (7.40), we find $\Gamma(m)$ is infinite for $m = 0$ and negative integers.

7.7 Relationship with Beta Function

We have

$$\Gamma(m) = \int_0^{\infty} x^{m-1}\,e^{-x}\,dx.$$

Substituting $x = ty$, where $t > 0$, we get

$$\Gamma(m) = t^m \int_0^{\infty} y^{m-1}\,e^{-ty}\,dy.$$

Changing $m \longrightarrow m+n$, and $t \longrightarrow 1+t$, we get

$$\frac{\Gamma(m+n)}{(1+t)^{m+n}} = \int_0^{\infty} y^{m+n-1}\,e^{-(1+t)\,y}\,dy. \tag{7.43}$$

Multiplying both sides of equation (7.43) by t^{m-1} and integrating the result on t in the interval $(0, \infty)$, we get

$$\Gamma(m+n)\int_0^{\infty} \frac{t^{m-1}}{(1+t)^{m+n}}\,dt = \int_0^{\infty}\left[\int_0^{\infty} y^{m+n-1} e^{-(1+t)\,y}\,dy\right] t^{m-1}\,dt. \tag{7.44}$$

Now, the integral in the RHS of (7.44) is

$$\begin{aligned} &\int_0^{\infty}\left[\int_0^{\infty} (ty)^{m-1}\,e^{-ty}\,d(ty)\right] y^{n-1}\,e^{-y}\,dy \\ &= \int_0^{\infty}\left[\int_0^{\infty} x^{m-1}\,e^{-x}\,dx\right] y^{n-1}\,e^{-y}\,dy \qquad [\because\ x = ty] \\ &= \Gamma(m)\int_0^{\infty} y^{n-1}\,e^{-y}\,dy = \Gamma(m)\cdot\Gamma(n). \end{aligned} \tag{7.45}$$

The LHS of equation (7.44) can be written, by using relation (7.21), as equal to

$$\Gamma\,(m+n)\,B\,(m,n).$$

Hence, we get

$$\Gamma(m+n)\,B(m,n) = \Gamma(m)\,\Gamma(n), \quad \text{i.e.,} \quad B(m,n) = \frac{\Gamma(m)\,\Gamma(n)}{\Gamma(m+n)}. \tag{7.46}$$

Equation (7.46) gives the relationship between the Beta and Gamma functions.

7.8 Infinite Products

Let x_1, x_2, ..., x_n be a sequence of real numbers. The continuous infinite product of these numbers is denoted by

$$x_1\,x_2\,\cdots\,x_n\,\cdots = \prod_{n=1}^{\infty} x_n. \tag{7.47}$$

Let us take the continuous product up to n numbers defined as

$$P_n = x_1\,x_2\,\cdots\,x_n. \tag{7.48}$$

If

$$\lim_{n\longrightarrow\infty} P_n = Q \quad \text{(finite or infinite)},$$

then this limit-value Q is said to be the value of infinite product in (7.47). Then, we write

$$Q = x_1\,x_2\,\cdots\,x_n\,\cdots = \prod_{n=1}^{\infty} x_n. \tag{7.49}$$

If the limiting value Q is non-zero finite, we say that the infinite product in (7.49) is convergent; otherwise, it is divergent. If we consider that

$$x_n\,(n=1,2,\ldots) > 0 \tag{7.50}$$

then from equation (7.48), we get

$$\ln P_n = \ln x_1 + \ln x_2 + \cdots + \ln x_n. \tag{7.51}$$

Assuming that limit Q exists as a non-zero value, then, from (7.51), we can write

$$\lim_{n\longrightarrow\infty} (\ln x_1 + \ln x_2 + \cdots + \ln x_n) = \ln Q. \tag{7.52}$$

Equation (7.52) clarifies that, if the limit of the infinite product (7.47) exists as a positive quantity Q, then the series

$$\sum_{n=1}^{\infty} \ln x_n \tag{7.53}$$

converges to the sum $\ln Q$. Conversely, if the series in (7.53) converges, then we have

$$\begin{aligned}\lim_{n\longrightarrow\infty} (\ln x_1 + \ln x_2 + \cdots + \ln x_n) &= \lim_{n\longrightarrow\infty} \ln\,(x_1 \cdot x_n \cdots x_n) = \lim_{n\longrightarrow\infty} \ln P_n \\ &= \ln \lim_{n\longrightarrow\infty} P_n = \ln Q,\end{aligned} \tag{7.54}$$

which shows that the limit of (7.48) exists as $n \longrightarrow \infty$. Moreover, it is positive.

Thus, if $x_n > 0$ $(n = 1, 2, \ldots)$, the necessary and sufficient condition for the convergence of the infinite product in (7.47) is that the infinite series in (7.53) should be convergent.

Functions like $\sin x$, $\cos x$ can be expressed as infinite products. These formulae are given without proof:

$$\sin x = x \prod_{n=1}^{\infty} \left(1 - \frac{x^2}{n^2 \pi^2}\right), \tag{7.55}$$

$$\cos x = \prod_{n=1}^{\infty} \left(1 - \frac{x^2}{\xi^2 \pi^2}\right), \quad \xi = \frac{2n-1}{2}, \tag{7.56}$$

$$\sinh x = x \prod_{n=1}^{\infty} \left(1 + \frac{x^2}{n^2 \pi^2}\right) \tag{7.57}$$

$$\text{and} \quad \cosh x = \prod_{n=1}^{\infty} \left(1 + \frac{x^2}{\xi^2 \pi^2}\right), \quad \xi = \frac{2n-1}{2}. \tag{7.58}$$

7.9 Euler-Gauss's Formula

The Euler-Gauss's formula defines Gamma function in terms of an infinite limit like

$$\Gamma(n) = \lim_{m \longrightarrow \infty} \frac{1 \cdot 2 \cdots m}{n(n+1) \cdots (n+m)} m^n \tag{7.59}$$

for n real or complex and m positive integers. To prove this formula, we consider the limit

$$\lim_{m \longrightarrow \infty} \left\{ m(1 - z^{1/m}) \right\}, \quad \text{which is equal to} \quad \lim_{m \longrightarrow \infty} \frac{1 - z^{1/m}}{1/m}$$

$$= \lim_{m \longrightarrow \infty} \frac{-z^{1/m} \ln z \, (-1/m^2)}{-1/m^2} = -\ln z = \ln 1/z. \tag{7.60}$$

Now, we substitute in equation (7.34)

$$z = e^{-x}, \quad \text{i.e.,} \quad x = \ln 1/z. \tag{7.61}$$

Then, the Gamma function can be written as

$$\Gamma(n) = \int_0^1 \left(\ln \frac{1}{z}\right)^{n-1} dz.$$

Using relation (7.60)

$$\Gamma(n) = \lim_{m \longrightarrow \infty} \int_0^1 \left\{ m\left(1 - z^{1/m}\right) \right\}^{n-1} dz. \tag{7.62}$$

We substitute, $z = t^m$. Then equation (7.62) becomes

$$\Gamma(n) = \lim_{m \longrightarrow \infty} \left[m^n \int_0^1 (1-t)^{n-1} t^{m-1} \, dt \right],$$

$$\text{or,} \quad \Gamma(n) = \lim_{m \longrightarrow \infty} \{ m^n B(m, n) \}.$$

Using the relation (7.15), we get

$$\Gamma(n) = \lim_{m \longrightarrow \infty} \frac{1 \cdot 2 \cdots m}{n\,(n+1)\cdots(n+m)}\, m^n, \tag{7.63}$$

which proves the result.

Putting $n = 1$ in equation (7.63), we find

$$\Gamma(1) = \lim_{m \longrightarrow \infty} \frac{1 \cdot 2 \cdots m}{1 \cdot 2 \cdots (m+1)} \cdot m = \lim_{m \longrightarrow \infty} \frac{m}{m+1} = \lim_{m \longrightarrow \infty} \frac{1}{1+\frac{1}{m}},$$

i.e., $\Gamma(1) = 1$, which we got earlier.

From equation (7.63), it is also clear that $\Gamma(n) \longrightarrow \infty$, for $m = 0$, and negative integers.

7.10 Gamma Function as an Infinite Product

Gamma function can be expressed as an infinite product in the form

$$\Gamma(m) = \frac{1}{m} \prod_{k=1}^{\infty} \frac{(1+1/k)^m}{1+m/k}, \tag{7.64}$$

where $m \neq 0, -1, -2, \ldots$

To prove this, let us consider the partial product

$$\begin{aligned} P_{k-1} &= \frac{\left(1+\frac{1}{1}\right)^m \left(1+\frac{1}{2}\right)^m \cdots \left(1+\frac{1}{k-1}\right)^m}{m\left(1+\frac{m}{1}\right)\left(1+\frac{m}{2}\right)\cdots\left(1+\frac{m}{k-1}\right)} \\ &= \frac{2^m \cdot 3^m \cdots k^m \cdot 1 \cdot 2 \cdots (k-1)}{1^m \cdot 2^m \cdots (k-1)^m \cdot m(m+1)\cdots(m+k-1)} \\ &= \frac{k^m\,(k-1)!}{m(m+1)\cdots(m+k-1)}. \end{aligned}$$

Hence

$$\begin{aligned} \lim_{k \longrightarrow \infty} P_{k-1} &= \lim_{k \longrightarrow \infty} \frac{k^m\,(k-1)!}{m\,(m+1)\cdots(m+k-1)} \\ &= \Gamma(m), \quad \text{[by definition of equation (7.59)]}. \end{aligned}$$

Thus the equivalence of relation (7.64) is established.

Again, we have from equation (7.40)

$$\Gamma(m+1) = m\Gamma(m) = \prod_{k=1}^{\infty} \frac{\left(1+\frac{1}{k}\right)^m}{1+\frac{m}{k}} \quad \text{[using (7.64)]} \tag{7.65}$$

We make the change $m \longrightarrow -m$, then

$$\Gamma(1-m) = \prod_{k=1}^{\infty} \frac{\left(1+\frac{1}{k}\right)^{-m}}{1-\frac{m}{k}}. \tag{7.66}$$

We multiply equations (7.64) and (7.66), and get

$$\Gamma(m)\,\Gamma(1-m) = \frac{1}{m\prod_{k=1}^{\infty}\left(1-\frac{m^2}{k^2}\right)}. \tag{7.67}$$

If we put $x = m\pi$ in equation (7.55), we get

$$\sin m\pi = m\pi \prod_{k=1}^{\infty}\left(1-\frac{m^2}{k^2}\right). \tag{7.68}$$

From equations (7.67) and (7.68), we can write

$$\Gamma(m)\,\Gamma(1-m) = \frac{\pi}{\sin m\pi}. \tag{7.69}$$

This is an important relation.

7.11 Value of $\Gamma\left(\frac{1}{2}\right)$

If we put $m = 1/2$ in equation (7.69), we get

$$\left\{\Gamma\left(\tfrac{1}{2}\right)\right\}^2 = \pi, \quad \text{or,} \quad \Gamma\left(\tfrac{1}{2}\right) = \sqrt{\pi}. \tag{7.70}$$

Again, we know

$$\Gamma(m+1) = m\Gamma(m).$$

Putting $m = -\frac{1}{2}$,

$$\Gamma\left(\tfrac{1}{2}\right) = -\tfrac{1}{2}\Gamma\left(-\tfrac{1}{2}\right)$$

which gives

$$\Gamma\left(-\tfrac{1}{2}\right) = -2\sqrt{\pi}. \tag{7.71}$$

Similarly, we find

$$\Gamma\left(-\tfrac{3}{2}\right) = -\tfrac{2}{3}\Gamma\left(-\tfrac{1}{2}\right) = \tfrac{4}{5}\sqrt{\pi}. \tag{7.72}$$

In the same way, we can find the values of Gamma function for $m = -5/2, -7/2, \ldots$, etc.

The value of $\Gamma\left(\frac{1}{2}\right)$ as in equation (7.70) can also be obtained directly from the definition of Gamma function. We have, from definition

$$\Gamma\left(\tfrac{1}{2}\right) = \int_0^\infty x^{-1/2}\, e^{-x}\, dx.$$

We substitute, $x = y^2$, then

$$\Gamma\left(\tfrac{1}{2}\right) = 2\int_0^\infty e^{-y^2}\, dy \tag{7.73}$$

$$\text{or,} \quad \Gamma\left(\tfrac{1}{2}\right) = 2\int_0^\infty e^{-z^2}\, dz \quad \text{(in terms of another variable } z\text{)}.$$

Therefore

$$\left\{\Gamma\left(\tfrac{1}{2}\right)\right\}^2 = 4\int_0^\infty e^{-y^2}\,dy \int_0^\infty e^{-z^2}\,dz = 4\int_0^\infty\int_0^\infty e^{-(y^2+z^2)}\,dy\,dz.$$

Here, we substitute

$$y = r\cos\phi$$
$$\text{and}\quad z = r\sin\phi.$$

Then

$$\left\{\Gamma\left(\tfrac{1}{2}\right)\right\}^2 = 4\int_{\phi=0}^{\pi/2}\left[\int_{r=0}^{\infty} r\,e^{-r^2}\,dr\right]d\phi = 4\int_0^{\pi/2}\left[-\tfrac{1}{2}\,e^{-r^2}\right]_0^\infty d\phi = 2\int_0^{\pi/2} d\phi = \pi.$$

$$\therefore\quad \Gamma\left(\tfrac{1}{2}\right) = \sqrt{\pi}.$$

7.12 Euler's Constant

The Euler's constant γ is defined by

$$\gamma = \lim_{m\longrightarrow\infty}\left(1 + \frac{1}{2} + \frac{1}{3} + \cdots + \frac{1}{m} - \ln m\right). \tag{7.74}$$

This constant γ has the following integral representations which we simply mention without proof:

(1) $$\gamma = \int_0^1 \left(1 - e^{-x}\right)\frac{dx}{x} - \int_1^\infty e^{-x}\,\frac{dx}{x}; \tag{7.75}$$

(2) $$\gamma = \int_0^1 \frac{1-\cos x}{x}\,dx - \int_1^\infty \frac{\cos x}{x}\,dx. \tag{7.76}$$

7.13 Weierstrass's Formula

Let us consider the infinite product

$$P = \prod_{n=1}^{\infty}\frac{e^{1/n}}{1+\frac{1}{n}}. \tag{7.77}$$

Then, the partial product

$$P_n = \frac{e^{1+\frac{1}{2}+\cdots+\frac{1}{n}}}{\left(1+\frac{1}{1}\right)\left(1+\frac{1}{2}\right)\cdots\left(1+\frac{1}{n}\right)}. \tag{7.78}$$

Now

$$1 + \tfrac{1}{2} + \cdots + \tfrac{1}{n} = \gamma + \ln n + \epsilon, \tag{7.79}$$

where ϵ is infinitely small. Then, equation (7.78) can be written as

$$P_n = \frac{1}{n+1}\,e^{\gamma+\ln n+\epsilon} = e^{\gamma}\,\frac{n}{n+1}\,e^{\epsilon}.$$

Hence

$$\lim_{n\longrightarrow\infty} P_n = \lim_{n\longrightarrow\infty} e^{\gamma}\frac{n}{n+1}e^{\epsilon} = e^{\gamma}. \tag{7.80}$$

Thus, from equations (7.77) and (7.80), we get

$$e^{\gamma} = \prod_{n=1}^{\infty}\frac{e^{1/n}}{1+\frac{1}{n}}. \tag{7.81}$$

Taking the expression (7.81) to the power of m on both sides, we get

$$e^{m\gamma} = \prod_{n=1}^{\infty}\frac{e^{m/n}}{(1+\frac{1}{n})^m}. \tag{7.82}$$

From equation (7.64), we have

$$\Gamma(m) = \frac{1}{m}\prod_{n=1}^{\infty}\frac{(1+\frac{1}{n})^m}{1+\frac{m}{n}}. \tag{7.83}$$

We multiply equation (7.82) and (7.83), and we get

$$e^{m\gamma}\Gamma(m) = \frac{1}{m}\prod_{n=1}^{\infty}\frac{e^{m/n}}{1+m/n} \tag{7.84}$$

$$\text{or,}\quad \frac{1}{\Gamma(m)} = me^{m\gamma}\prod_{n=1}^{\infty}\left(1+\frac{m}{n}\right)e^{-m/n}. \tag{7.85}$$

This is known as Weierstrass's formula which defines the Gamma function in terms of an infinite product. A graph of the function $1/\Gamma(m)$ is provided in *Fig 7.1.*

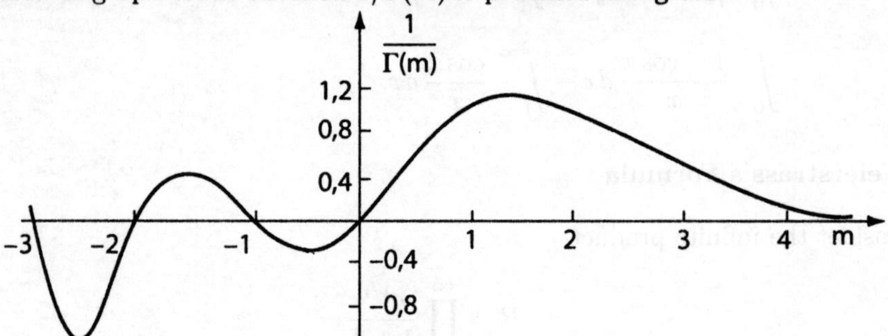

Fig 7.1

7.14 Relations of Gamma Function with Certain Integrals

We have seen that Beta and Gamma functions are related with one another by the formula

$$B(m,n) = \frac{\Gamma(m)\,\Gamma(n)}{\Gamma(m+n)}. \tag{7.86}$$

We have also seen in Section 7.3 that Beta functions are also defined by certain integrals. It is, therefore, quite natural that those integrals will represent the RHS of equation (7.86) expressed in gamma functions.

Thus, we get the following integrals in terms of Gamma functions:

1. From equations (7.21) and (7.86), we write

$$\int_0^\infty \frac{y^{m-1}}{(1+y)^{m+n}}\,dy = \frac{\Gamma(m)\,\Gamma(n)}{\Gamma(m+n)}. \tag{7.87}$$

2. From equations (7.24) and (7.86), we write

$$\int_0^1 \frac{y^{m-1}+y^{n-1}}{(1+y)^{m+n}}\,dy = \frac{\Gamma(m)\,\Gamma(n)}{\Gamma(m+n)}. \tag{7.88}$$

3. From equations (7.26) and (7.86), we write

$$\alpha^m\,\beta^n \int_0^\infty \frac{x^{m-1}}{(\alpha x+\beta)^{m+n}}\,dx = \frac{\Gamma(m)\,\Gamma(n)}{\Gamma(m+n)}. \tag{7.89}$$

4. From equations (7.30) and (7.86), we write

$$\frac{1}{\alpha^{m+n-1}} \int_0^\infty y^{m-1}\,(\alpha-y)^{n-1}\,dy = \frac{\Gamma(m)\,\Gamma(n)}{\Gamma(m+n)}. \tag{7.90}$$

5. From equations (7.28) and (7.86), we write

$$\int_0^{\pi/2} (\sin\theta)^{2m-1}\,(\cos\theta)^{2n-1}\,d\theta = \frac{\Gamma(m)\,\Gamma(n)}{2\Gamma(m+n)}. \tag{7.91}$$

The relation (7.91) gives opportunity to get further important information depicted below:

1. If in equation (7.91) we put

$$\left.\begin{aligned} 2m-1 = p,\ \text{i.e., } m = \tfrac{1}{2}(p+1) \\ \text{and}\quad 2n-1 = q,\ \text{i.e., } n = \tfrac{1}{2}(q+1) \end{aligned}\right\} \tag{7.92}$$

then we get

$$\int_0^{\pi/2} (\sin\theta)^p\,(\cos\theta)^q\,d\theta = \frac{\Gamma\left(\dfrac{p+1}{2}\right)\Gamma\left(\dfrac{q+1}{2}\right)}{2\Gamma\left(\dfrac{p+q}{2}+1\right)}, \quad p,\ q > -1. \tag{7.93}$$

2. Special cases of relation (7.93) are:

(a) If $p = 0$, then

$$\int_0^{\pi/2} (\cos\theta)^q\,d\theta = \frac{\Gamma\left(\dfrac{1}{2}\right)\Gamma\left(\dfrac{q+1}{2}\right)}{2\Gamma\left(\dfrac{q}{2}+1\right)}.$$

By using relation (7.70)

$$\int_0^{\pi/2} (\cos\theta)^q\,d\theta = \frac{\Gamma\left(\dfrac{q+1}{2}\right)\sqrt{\pi}}{2\Gamma\left(\dfrac{q}{2}+1\right)}. \tag{7.94}$$

(b) If $q = 0$, then

$$\int_0^{\pi/2} (\sin\theta)^p\, d\theta = \frac{\Gamma\left(\frac{1}{2}\right)\Gamma\left(\frac{p+1}{2}\right)}{2\Gamma\left(\frac{p}{2}+1\right)} = \frac{\Gamma\left(\frac{p+1}{2}\right)\sqrt{\pi}}{2\Gamma\left(\frac{p}{2}+1\right)}. \tag{7.95}$$

3. *Legendre's duplication formula.* Legendre's duplication formula can be derived from the relation (7.91) in the following manner:
We put $m = n$ in equation (7.91) and get

$$\frac{\{\Gamma(m)\}^2}{2\Gamma(2m)} = \int_0^{\pi/2} (\sin\theta)^{2m-1}(\cos\theta)^{2m-1}\, d\theta = \frac{1}{2^{2m-1}} \int_0^{\pi/2} (\sin 2\theta)^{2m-1}\, d\theta.$$

Let, $2\theta = \phi$, then, we get

$$\frac{\{\Gamma(m)\}^2}{2\Gamma(2m)} = \frac{1}{2^m}\int_0^{\pi} (\sin\phi)^{2m-1}\, d\phi,$$

$$\text{i.e.,}\quad \frac{\{\Gamma(m)\}^2}{2\Gamma(2m)} = \frac{1}{2^{2m-1}}\int_0^{\pi/2} (\sin\theta)^{2m-1}\, d\theta.$$

Hence

$$\int_0^{\pi/2} (\sin\theta)^{2m-1}\, d\theta = \frac{2^{2m-2}\{\Gamma(m)\}^2}{\Gamma(2m)}. \tag{7.96}$$

Again, if we put $2n - 1 = 0$, i.e., $n = 1/2$ in equation (7.91), we get

$$\int_0^{\pi/2} (\sin\theta)^{2m-1}\, d\theta = \frac{\Gamma(m)\sqrt{\pi}}{2\Gamma\left(m+\frac{1}{2}\right)}. \tag{7.97}$$

From equations (7.96) and (7.97), we write

$$\frac{\Gamma(m)\sqrt{\pi}}{2\Gamma\left(m+\frac{1}{2}\right)} = \frac{2^{2m-2}\{\Gamma(m)\}^2}{\Gamma(2m)},$$

$$\text{or,}\quad \sqrt{\pi}\,\Gamma(2m) = 2^{2m-1}\Gamma(m)\Gamma\left(m+\frac{1}{2}\right). \tag{7.98}$$

This is known as *Legendre's duplication formula.*

7.15 Evaluation of Certain Integrals

1. Let us consider the integral

$$I = \int_0^\infty x^k e^{-mx^n}\, dx, \quad k > -1,\ m > 0, n > 0.$$

Here we substitute, $mx^n = \xi$. Then

$$I = \frac{1}{\alpha}\int_0^\infty \xi^{\frac{(k+1)}{n}-1} e^{-\xi}\, d\xi.$$

Thus

$$\int_0^\infty x^k\, e^{-x^n}\, dx = \frac{1}{\alpha}\Gamma\left(\frac{k+1}{n}\right), \tag{7.99}$$

where

$$\alpha = nm^{(k+1)/n}. \tag{7.100}$$

2. Consider the integral

$$I = \int_0^1 \frac{dx}{\sqrt{1-x^n}}.$$

Here we substitute, $x^n = \sin^2\theta$.
Then

$$I = \frac{2}{n}\int_0^{\pi/2} \frac{(\sin\theta)^{(2-n)/n}\cos\theta}{\cos\theta}\, d\theta = \frac{2}{n}\int_0^{\pi/2} (\sin\theta)^{\frac{2}{n}-1}\, d\theta$$

$$= \frac{2\Gamma\left(\frac{1}{2}\right)\Gamma\left(\frac{1}{n}\right)}{2n\,\Gamma\left(\frac{1}{n}+\frac{1}{2}\right)} \quad \text{[by using equation (7.95)]}.$$

Thus

$$\int_0^1 \frac{dx}{\sqrt{1-x^n}} = \frac{\sqrt{\pi}\,\Gamma\left(\frac{1}{n}\right)}{n\,\Gamma\left(\frac{1}{n}+\frac{1}{2}\right)}. \tag{7.101}$$

3. Let us consider the integral

$$I = \int_0^\infty e^{-ax^2}\, dx.$$

Here we substitute, $ax^2 = t^2$.
Then

$$I = \frac{1}{\sqrt{a}}\int_0^\infty e^{-t^2}\, dt = \frac{1}{2\sqrt{a}}\Gamma\left(\frac{1}{2}\right) \quad \text{[by using equations (7.70) and (7.73)]}.$$

Thus

$$\int_0^\infty e^{-ax^2}\, dx = \frac{1}{2}\frac{\sqrt{\pi}}{\sqrt{a}} = \frac{1}{2}\sqrt{\frac{\pi}{a}}. \tag{7.102}$$

4. Consider the integral

$$I = \int_0^1 (1-x^n)^{1/n}\, dx.$$

Here we substitute, $x^n = \sin^2\phi$, then

$$dx = \frac{2}{n}(\sin\phi)^{(2-n)/n}\cos\phi\, d\phi.$$

Thus

$$I = \frac{2}{n}\int_0^{\pi/2} (\cos\phi)^{\frac{2}{n}+1}\,(\sin\phi)^{\frac{2}{n}-1}\,d\phi$$

$$= \frac{2}{n}\cdot\frac{\Gamma\left(\dfrac{1}{n}+1\right)\Gamma\left(\dfrac{1}{n}\right)}{2\Gamma\left(\dfrac{2}{n}+1\right)} \quad \text{[by using equation (7.91)]}$$

$$= \frac{1}{n}\cdot\frac{\dfrac{1}{n}\Gamma\left(\dfrac{1}{n}\right)\Gamma\left(\dfrac{1}{n}\right)}{\dfrac{2}{n}\Gamma\left(\dfrac{2}{n}\right)} \quad \text{[by using equation (7.40)].}$$

Thus

$$\int_0^1 (1-x^n)^{1/n}\,dx = \frac{1}{n}\cdot\frac{\left\{\Gamma\left(\dfrac{1}{n}\right)\right\}^2}{2\Gamma\left(\dfrac{2}{n}\right)}. \tag{7.103}$$

5. Let us consider the integral

$$I = \int_0^\infty \frac{x^{n-1}}{1+x}\,dx.$$

If we put $m+n=1$ in the equation (7.21), we get

$$I = \int_0^\infty \frac{x^{n-1}}{1+x}\,dx = B(n,\,1-n) = \frac{\Gamma(n)\,\Gamma(1-n)}{\Gamma(1)},$$

$$\text{i.e.,}\quad I = \Gamma(n)\,\Gamma(1-n) = \frac{\pi}{\sin n\pi} \quad \text{[by using equation (7.69)].}$$

Hence

$$\int_0^1 \frac{x^{n-1}}{1+x}\,dx = \frac{\pi}{\sin n\pi}. \tag{7.104}$$

6. Let us consider the integral

$$I = \int_{-\infty}^\infty \exp\left(-\lambda x^2 + \mu x\right)dx, \quad \lambda, \mu = \text{constants.}$$

Here

$$I = \int_{-\infty}^\infty \exp\left\{-\lambda\left[x^2 - \frac{\mu}{\lambda}x + \left(\frac{\mu}{2\lambda}\right)^2 - \left(\frac{\mu}{2\lambda}\right)^2\right]\right\}dx = \exp\left(\frac{\mu^2}{4\lambda}\right)\int_{-\infty}^\infty e^{-\lambda t^2}\,dt,$$

where $t = x - (\mu/2\lambda)$.
Using the value of the integral in (7.102), we get

$$\int_{-\infty}^\infty \exp\left(-\lambda x^2 + \mu x\right)dx = e^{\mu^2/4\lambda}\,\sqrt{\pi/\lambda}. \tag{7.105}$$

7. Let us consider the integral

$$I = \int_0^\infty \frac{x^{m-1}}{e^x - 1}\, dx, \quad \text{where } m > 0.$$

Then

$$\begin{aligned} I &= \int_0^\infty x^{m-1} \left(\sum_{n=1}^{\infty} e^{-nx} \right) dx = \sum_{n=1}^{\infty} \int_0^\infty x^{m-1}\, e^{-nx}\, dx \\ &= \sum_{n=1}^{\infty} \frac{1}{n^m} \int_0^\infty t^{m-1} e^{-t} dt \quad (\text{putting } t = nx) \\ &= \Gamma(m) \sum_{n=1}^{\infty} \frac{1}{n^m} = \Gamma(m)\, \zeta(m), \end{aligned}$$

where $\zeta(m) = \sum_{n=1}^{\infty} \frac{1}{n^m}$ is the *Riemann's zeta function*.

Hence

$$\int_0^\infty \frac{x^{m-1}}{e^x - 1}\, dx = \Gamma(m)\, \zeta(m). \tag{7.106}$$

7.16 Dirichlet's Integral

The volume integral

$$I = \iiint\limits_{\Omega} x^{a-1}\, y^{b-1}\, z^{c-1} dx\, dy\, dz \tag{7.107}$$

defined in the region given by

$$x \geq 0, \quad y \geq 0, \quad z \geq 0, \quad x + y + z \leq 1 \tag{7.108}$$

is known as *Dirichlet's integral.*

Its value can be expressed in terms of Gamma function.

The integral (7.107) represents the volume enclosed within the coordinate planes and the plane

$$\left.\begin{aligned} x + y + z &= 1 \\ \text{i.e.,} \quad z &= 1 - x - y \end{aligned}\right\}. \tag{7.109}$$

Let us evaluate the integral (7.107)

$$\begin{aligned} I &= \int_{x=0}^{1} \int_{y=0}^{1-x} \int_{z=0}^{1-x-y} x^{a-1}\, y^{b-1}\, z^{c-1}\, dx\, dy\, dz \\ &= \int_{x=0}^{1} \int_{y=0}^{1-x} x^{a-1}\, y^{b-1} \left[\frac{z^c}{c} \right]_0^{1-x-y} dx\, dy \\ &= \int_{x=0}^{1} \frac{1}{c} \int_{y=0}^{1-x} x^{a-1}\, y^{b-1}\, (1 - x - y)^c\, dx\, dy. \end{aligned}$$

Now, let $y = (1 - x)\, \xi$, i.e., $dy = (1 - x)\, d\xi$ and, for $y = 0$, $\xi = 0$; for $y = 1 - x$, $\xi = 1$.

Hence

$$I = \int_{x=0}^{1} \tfrac{1}{c} \int_{\xi=0}^{1} x^{a-1} (1-x)^{b-1} \xi^{b-1} (1-x)^c (1-\xi)^c (1-x)\, dx\, d\xi,$$

$$\text{i.e.,} \quad I = \frac{1}{c} \int_{x=0}^{1} x^{a-1} (1-x)^{b+c} dx \int_{\xi=0}^{1} \xi^{b-1} (1-\xi)^c d\xi$$

$$= \frac{1}{c} B(a, b+c+1)\, B(b, c+1)$$

$$= \frac{1}{c} \frac{\Gamma(a)\,\Gamma(b+c+1)}{\Gamma(a+b+c+1)} \cdot \frac{\Gamma(b)\,\Gamma(c+1)}{\Gamma(b+c+1)}$$

$$= \frac{\Gamma(a)\,\Gamma(b)\,\Gamma(c)}{\Gamma(a+b+c+1)} \quad [\because\ \Gamma(c+1) = c\Gamma(c)].$$

Hence

$$\iiint\limits_{\Omega} x^{a-1}\, y^{b-1}\, z^{c-1}\, dx\, dy\, dz = \frac{\Gamma(a)\,\Gamma(b)\,\Gamma(c)}{\Gamma(a+b+c+1)}. \tag{7.110}$$

7.17 Logarithmic Derivatives of Gamma Functions

Let us take the natural logarithm of the equation (7.84). Then, we get

$$\ln \Gamma(m) = -m\gamma - \ln m + \sum_{n=1}^{\infty} \left[\frac{m}{n} - \ln\left(1 + \frac{m}{n}\right) \right]. \tag{7.111}$$

Differentiating equation (7.111) with respect to m, we get

$$\frac{\Gamma'(m)}{\Gamma(m)} = -\gamma - \frac{1}{m} + \sum_{n=1}^{\infty} \left[\frac{1}{n} - \frac{1}{m+n} \right] = -\gamma - \frac{1}{m} + \sum_{n=1}^{\infty} \frac{m}{n(m+n)} \tag{7.112}$$

Thus, the logarithmic derivative of Gamma function is

$$\frac{\Gamma'(m)}{\Gamma(m)} = \psi_1(m), \tag{7.113}$$

where

$$\psi_1(m) = -\gamma - \frac{1}{m} + m \sum_{n=1}^{\infty} \frac{1}{n(m+n)}. \tag{7.114}$$

The function $\psi_1(m)$ is called *Diagamma function.*
For $m = 1$, we get from equation (7.114)

$$\psi_1(1) = -\gamma - 1 + \sum_{n=1}^{\infty} \frac{1}{n(n+1)}$$

$$\text{or,} \quad \psi_1(1) = -\gamma, \text{ since } \sum_{n=1}^{\infty} \frac{1}{n(n+1)} = 1. \tag{7.115}$$

Thus

$$\frac{\Gamma'(1)}{\Gamma(1)} = -\gamma, \quad \text{i.e.,} \quad \Gamma'(1) = -\gamma. \tag{7.116}$$

Again, we have the recursion formula

$$\Gamma(m+1) = m\Gamma(m).$$

Taking natural logarithm on both sides, we get

$$\ln\Gamma(m+1) = \ln m + \ln\Gamma(m).$$

Differentiating with respect to m,

$$\frac{\Gamma'(m+1)}{\Gamma(m+1)} = \frac{1}{m} + \frac{\Gamma'(m)}{\Gamma(m)}. \tag{7.117}$$

Substituting $m = 1, 2, \ldots, k$, successively

$$\frac{\Gamma'(2)}{\Gamma(2)} = \frac{1}{1} + \frac{\Gamma'(1)}{\Gamma(1)} = -\gamma + 1$$

$$\frac{\Gamma'(3)}{\Gamma(3)} = \frac{\Gamma'(2)}{\Gamma(2)} + \frac{1}{2} = -\gamma + 1 + \frac{1}{2}$$

$$\cdots\cdots\cdots\cdots\cdots\cdots\cdots\cdots\cdots\cdots$$

$$\cdots\cdots\cdots\cdots\cdots\cdots\cdots\cdots\cdots\cdots$$

$$\frac{\Gamma'(k+1)}{\Gamma(k+1)} = \frac{\Gamma'(k)}{\Gamma(k)} + \frac{1}{k} = -\gamma + 1 + \frac{1}{2} + \cdots + \frac{1}{k}.$$

Thus

$$\frac{\Gamma'(k+1)}{\Gamma(k+1)} = -\gamma + \sum_{n=1}^{k} \frac{1}{n}. \tag{7.118}$$

This gives a recursion formula for Diagamma functions.
Derivatives of Diagamma functions are called *Polygamma functions.*
Differentiating equation (7.114) with respect to m, we get

$$\psi_2(m) = \frac{1}{m^2} + \sum_{n=1}^{\infty} \frac{1}{(m+n)^2}, \quad \text{i.e.,} \quad \psi_2(m) = \sum_{n=0}^{\infty} \frac{1}{(m+n)^2}. \tag{7.119}$$

Differentiating once more, we have

$$\psi_3(m) = \frac{d}{dm}\,\psi_2(m) = (-1)^3\, 2! \sum_{n=0}^{\infty} \frac{1}{(m+n)^3}. \tag{7.120}$$

Thus, we can write, in general

$$\psi_k(m) = (-1)^k\,\Gamma(k) \sum_{n=0}^{\infty} \frac{1}{(m+n)^k}, \quad k \geq 2. \tag{7.121}$$

Again, from equation (7.121), we have

$$\psi_k(m+1) = (-1)^k\,\Gamma(k) \sum_{n=0}^{\infty} \frac{1}{(m+n+1)^k} = (-1)^k\,\Gamma(k) \sum_{n=1}^{\infty} \frac{1}{(m+n)^k}$$

$$= (-1)^k\,\Gamma(k) \sum_{n=0}^{\infty} \left[\frac{1}{(m+n)^k} - \frac{1}{m^k}\right],$$

$$\text{i.e.,} \quad \psi_k(m+1) = \psi_k(m) - (-1)^k\,\Gamma(k)\,\frac{1}{m^k}. \tag{7.122}$$

Equation (7.122) gives a recursion formula for Polygamma functions.

7.18 Integral Representation of Diagamma Functions

We know, that the Diagamma function is given by

$$\psi_1(m) = -\gamma - \frac{1}{m} + \sum_{n=1}^{\infty}\left(\frac{1}{n} - \frac{1}{m+n}\right) \quad \text{[from (7.114)]}$$

$$= -\gamma + \sum_{n=1}^{\infty}\frac{1}{n} - \sum_{n=0}^{\infty}\frac{1}{m+n}. \tag{7.123}$$

Let us consider the right-hand side terms one by one:

(i) $$\gamma = \lim_{m \longrightarrow \infty}\left(1 + \frac{1}{2} + \frac{1}{3} + \cdots + \frac{1}{m} - \ln m\right) \quad \text{[by (7.74)]}$$

$$= \sum_{n=1}^{\infty}\frac{1}{n} - \lim_{m \longrightarrow \infty}\int_0^{\infty}\frac{e^{-x} - e^{-mx}}{x}\,dx = \sum_{n=1}^{\infty}\frac{1}{n} - \int_0^{\infty}\frac{e^{-x}}{x}\,dx. \tag{7.124}$$

(ii) $$\sum_{n=0}^{\infty}\frac{1}{m+n} = \lim_{n \longrightarrow \infty}\sum_{k=0}^{n}\int_0^{\infty} e^{-(m+k)x}\,dx = \lim_{n \longrightarrow \infty}\int_0^{\infty}\sum_{k=0}^{n} e^{-(m+k)x}\,dx.$$

Now

$$\sum_{k=0}^{n} e^{-(m+k)x}$$

is a geometric series with its first term $= e^{-mx}$ and common ratio $= e^{-x}$ whose sum is

$$\frac{e^{-mx}}{1 - e^{-x}}.$$

Hence

$$\sum_{n=0}^{\infty}\frac{1}{m+n} = \int_0^{\infty}\frac{e^{-mx}}{1 - e^{-x}}. \tag{7.125}$$

Hence, using relations (7.124) and (7.125) in equation (7.123), we get

$$\psi_1(m) = \int_0^{\infty}\left\{\frac{e^{-x}}{x} - \frac{e^{-mx}}{1 - e^{-x}}\right\}dx, \tag{7.126}$$

which gives the integral representation of the Diagamma function.

7.19 Stirling's Formula

Stirling's formula determines an approximate value for Gamma function $\Gamma(m+1)$ or $m!$ when m is very large.

We have, from definition

$$\Gamma(m+1) = \int_0^{\infty} x^m\, e^{-x}\,dx = \int_0^{\infty} e^{\ln x^m}\, e^{-x}\,dx = \int_0^{\infty} e^{(\ln x^m - x)}\,dx. \tag{7.127}$$

Now, we substitute

$$x = m + \sqrt{m}\, y \tag{7.128}$$

so that

$$dx = \sqrt{m}\, dy, \quad \text{and} \quad \text{for } x = 0,\ y = -\sqrt{m}. \tag{7.129}$$

Moreover, the expression

$$\ln x^m - x = m \ln x - x$$

$$\text{and} \quad \ln x = \ln(m + \sqrt{m}\, y) = \ln m + \ln\left(1 + \frac{y}{\sqrt{m}}\right) = \ln m + \frac{y}{\sqrt{m}} - \frac{y^2}{2m} + \cdots$$

$$\text{i.e.,} \quad \ln x \simeq \ln m + \frac{y}{\sqrt{m}} - \frac{y^2}{2m} \quad [\text{for } m \text{ large}]. \tag{7.130}$$

Substituting relations (7.128)–(7.130) in the integral in (7.127), we get

$$\begin{aligned}
\Gamma(m+1) &= \int_{-\sqrt{m}}^{\infty} e^{\left\{m \ln m + y\sqrt{m} - \frac{y^2}{2} - m - \sqrt{m}\, y\right\}} \sqrt{m}\, dy = e^{(m \ln m - m)} \sqrt{m} \int_{-\sqrt{m}}^{\infty} e^{-y^2/2}\, dy \\
&= \sqrt{m}\, m^m\, e^{-m} \left[\int_{-\infty}^{\infty} e^{-y^2/2}\, dy - \int_{-\infty}^{-\sqrt{m}} e^{-y^2/2}\, dy\right] \\
&= \sqrt{m}\, m^m\, e^{-m} \int_{-\infty}^{\infty} e^{-y^2/2}\, dy,
\end{aligned} \tag{7.131}$$

since the integral

$$\int_{-\infty}^{-\sqrt{m}} e^{-y^2/2}\, dy \longrightarrow 0 \quad \text{for large values of } m.$$

Thus, from equation (7.131), we get

$$\begin{aligned}
\Gamma(m+1) &= 2\sqrt{m}\, m^m\, e^{-m} \int_0^{\infty} e^{-y^2/2}\, dy \\
&= 2\sqrt{m}\, m^m\, e^{-m}\, \tfrac{1}{2} \sqrt{2\pi} \quad [\text{using equation (7.102)}] \\
\text{or,} \quad m! &= \sqrt{2m\pi}\, m^m\, e^{-m} \quad [\because\ \Gamma(m+1) = m!].
\end{aligned} \tag{7.132}$$

Equation (7.132) is the Stirling's formula giving an approximate value of $m!$ for large m.

7.20 Incomplete Gamma Functions

The Gamma functions with variable limit are defined as *Incomplete Gamma functions.* There are two types of incomplete gamma functions which are defined as

$$\gamma(m, x) = \int_0^x e^{-\xi}\, \xi^{m-1}\, d\xi \tag{7.133}$$

$$\text{and} \quad \Gamma(m, x) = \int_x^{\infty} e^{-\xi}\, \xi^{m-1}\, d\xi, \tag{7.134}$$

where $m > 0$.

Obviously, we find that

$$\gamma(m,x)+\Gamma(m,x)=\int_0^\infty e^{-\xi}\,\xi^{m-1}\,d\xi,$$

$$\text{i.e.,}\quad \gamma(m,x)+\Gamma(m,x)=\Gamma(m). \tag{7.135}$$

Both the incomplete gamma functions can be expanded in series for integral as well as non-integral values of m. These are given below:

1. *For integral values of m*

$$\gamma(m,x)=(m-1)!\left(1-e^{-x}\sum_{k=0}^{m-1}\frac{x^k}{k!}\right) \tag{7.136}$$

$$\text{and}\quad \Gamma(m,x)=(m-1)!\,e^{-x}\sum_{k=0}^{m-1}\frac{x^k}{k!}. \tag{7.137}$$

2. *For non-integral values of m*

$$\gamma(m,x)=x^m\sum_{k=0}^{\infty}(-1)^k\frac{x^k}{k!\,(m+k)} \tag{7.138}$$

$$\text{and}\quad \Gamma(m,x)=x^{m-1}\,e^{-x}\sum_{k=1}^{\infty}(-1)^k\frac{(k-m)!}{\Gamma(1-m)\,x^k}. \tag{7.139}$$

EXAMPLE 7.1. *Evaluate:* $\int_0^\infty \frac{x^a}{a^x}\,dx.$

Solution: Here, $a^x=e^{x\log a}$.

Thus

$$I=\int_0^\infty\frac{x^a}{a^x}=\int_0^\infty x^a\,e^{-x\log a}\,dx.$$

Now, put $x\log a=z$, so that $dx=\frac{dz}{\log a}$.

$$\therefore\quad I=\int_0^\infty\left(\frac{z}{\log a}\right)^a e^{-z}\frac{dz}{\log a}=\frac{1}{(\log a)^{a+1}}\int_0^\infty z^a\,e^{-z}\,dz=\frac{\Gamma(a+1)}{(\log a)^{a+1}}.$$

EXAMPLE 7.2. *Evaluate:* $\int_0^{\pi/2}\sqrt{\tan\theta}\,d\theta.$

Solution:

$$I=\int_0^{\pi/2}\sqrt{\tan\theta}\,d\theta=\int_0^{\pi/2}\sin^{1/2}\theta\cos^{-1/2}\theta\,d\theta=\frac{1}{2}\Gamma\left(\frac{3}{4}\right)\Gamma\left(\frac{1}{4}\right)\quad\text{[by equation (7.91)]}.$$

Now, putting $m=1/4$ in duplication formula (7.98), we get

$$\Gamma\left(\frac{3}{4}\right)\Gamma\left(\frac{1}{4}\right)=\sqrt{2}\cdot\sqrt{\pi}\,\Gamma\left(\frac{1}{2}\right)=\sqrt{2}\,\pi.$$

Hence

$$I=\frac{1}{2}\left(\sqrt{2}\,\pi\right)=\frac{\pi}{\sqrt{2}}.$$

EXAMPLE 7.3. *Prove that, for n a positive integer*

$$B(m,n) = \frac{(n-1)!}{m(m+1)\cdots(m+n-1)}.$$

Solution: We have

$$B(m,n) = \int_0^1 x^{m-1}(1-x)^{n-1}\,dx.$$

Integrating by parts, we get

$$B(m,n) = (1-x)^{n-1}\frac{x^m}{m}\bigg|_0^1 + \frac{n-1}{m}\int_0^1 x^m(1-x)^{n-2}\,dx = \frac{n-1}{m}\int_0^1 x^m(1-x)^{n-2}\,dx. \quad (1)$$

Continuing the integration by parts

$$B(m,n) = \frac{(n-1)(n-2)}{m(m+1)}\int_0^1 x^{m+1}(1-x)^{n-3}\,dx$$

$$\cdots\cdots\cdots\cdots\cdots\cdots\cdots\cdots\cdots\cdots$$

$$\cdots\cdots\cdots\cdots\cdots\cdots\cdots\cdots\cdots\cdots$$

$$= \frac{(n-1)(n-2)\cdots 2\cdot 1}{m(m+1)\cdots(m+n-2)}\cdot\int_0^1 x^{m+n-2}\,dx,$$

$$\text{i.e.,}\quad B(m,n) = \frac{(n-1)!}{m(m+1)\cdots(m+n-1)}. \quad (2)$$

EXAMPLE 7.4. *Prove that*

$$1\cdot 3\cdot 5\cdots(2n-1) = \frac{2^n}{\sqrt{\pi}}\,\Gamma\left(n+\frac{1}{2}\right)$$

and hence show that

$$\Gamma\left(n+\frac{1}{2}\right) = \frac{(2n)!}{2^{2n}\,n!}\sqrt{\pi}.$$

Solution: For n a positive integer, we have from equation (2) of above example.

$$B(n,m) = B(m,n) = \frac{(n-1)!}{m(m+1)\cdots(m+n-1)}.$$

Putting $m = 1/2$, we get

$$B\left(n,\tfrac{1}{2}\right) = \frac{2^n(n-1)!}{1\cdot 3\cdots(2n-1)},$$

$$\text{or,}\quad \frac{\Gamma(n)\,\Gamma\left(\frac{1}{2}\right)}{\Gamma\left(n+\frac{1}{2}\right)} = \frac{2^n(n-1)!}{1\cdot 3\cdots(2n-1)},$$

$$\text{or,}\quad \frac{(n-1)!\sqrt{\pi}}{\Gamma\left(n+\frac{1}{2}\right)} = \frac{2^n(n-1)!}{1\cdot 3\cdots(2n-1)}.$$

$$\therefore\quad 1\cdot 3\cdots(2n-1) = \frac{2^n}{\sqrt{\pi}}\,\Gamma\left(n+\frac{1}{2}\right).$$

Again

$$1 \cdot 3 \cdots (2n-1) = 1 \cdot 3 \cdots (2n-1)\,\frac{2 \cdot 4 \cdot 6 \cdots 2n}{2 \cdot 4 \cdot 6 \cdots 2n} = \frac{(2n)!}{2^n\, n!}. \tag{2}$$

Hence, from equations (1) and (2), we get

$$\frac{2^n}{\sqrt{\pi}}\,\Gamma\left(n+\frac{1}{2}\right) = \frac{(2n)!}{2^n\, n!}.$$

Therefore

$$\Gamma\left(n+\frac{1}{2}\right) = \frac{(2n)!}{2^{2n}\, n!}\,\sqrt{\pi}. \tag{3}$$

EXAMPLE 7.5. *Evaluate:*

$$\int_0^\infty e^{-x^2}\, x^{2n-1}\, dx, \quad n > 0.$$

Solution:

$$I = \int_0^\infty e^{-x^2}\, x^{2n-1}\, dx.$$

Put, $x^2 = t$; then, $2x\, dx = dt$.

$$\therefore \quad I = \frac{1}{2}\int_0^\infty e^{-t}\, t^{n-1}\, dt = \frac{1}{2}\,\Gamma(n).$$

EXAMPLE 7.6. *Evaluate:*

$$I = \int_0^1 \frac{x^{m-1}\,(1-x)^{n-1}}{(a+x)^{m+n}}\, dx.$$

Solution: Put

$$\frac{x(1+a)}{a+x} = z. \tag{1}$$

Then

$$x = \frac{az}{1+a-z} \quad \text{and} \quad dx = \frac{(a+x)^2}{a(1+a)}\, dz. \tag{2}$$

Therefore

$$(1-x) = 1 - \frac{az}{1+a-z} = \frac{(1+a)(1-z)}{1+a-z} \tag{3}$$

$$\text{and} \quad (a+x) = a + \frac{az}{1+a-z} = \frac{a(a+1)}{1+a-z}. \tag{4}$$

From (2) and (4), we have

$$dx = \frac{a(a+1)}{(1+a-z)^2}\, dz. \tag{5}$$

Thus

$$I = \int_0^1 \left(\frac{az}{1+a-z}\right)^{m-1} \frac{(1+a)^{n-1}\,(1-z)^{n-1}}{(1+a-z)^{n-1}} \cdot \frac{(1+a-z)^{m+n}}{[a(a+1)]^{m+n}}\, \frac{a(a+1)}{(1+a-z)^2}\, dz$$

$$= \frac{1}{a^n(1+a)^m} \int_0^1 z^{m-1}(1-z)^{n-1}\, dz = \frac{1}{a^n(1+a)^m}\, B(m,n).$$

EXAMPLE 7.7. *Show that*

$$\Gamma\left(\frac{1}{3}\right)\Gamma\left(\frac{2}{3}\right)=\frac{2\pi}{\sqrt{3}}.$$

Solution: Let

$$P=\Gamma\left(\frac{1}{3}\right)\Gamma\left(\frac{2}{3}\right). \tag{1}$$

Then

$$P=\Gamma\left(\frac{2}{3}\right)\Gamma\left(\frac{1}{3}\right). \tag{2}$$

Multiplying, we get

$$\begin{aligned} P^2 &= \Gamma\left(\frac{1}{3}\right)\Gamma\left(\frac{2}{3}\right)\Gamma\left(\frac{2}{3}\right)\Gamma\left(\frac{1}{3}\right) \\ &= \Gamma\left(\frac{1}{3}\right)\Gamma\left(1-\frac{1}{3}\right)\Gamma\left(\frac{2}{3}\right)\Gamma\left(1-\frac{2}{3}\right), \\ \text{or,}\quad P^2 &= \frac{\pi}{\sin(\pi/3)}\cdot\frac{\pi}{\sin(2\pi/3)} \quad \text{[by using equation (7.69)]} \\ &= \frac{4\pi^2}{3}. \\ \therefore\quad P &= \frac{2\pi}{\sqrt{3}}. \end{aligned}$$

EXAMPLE 7.8. *Prove that*

$$\int_a^b (x-a)^m(b-x)^n\,dx=(b-a)^{m+n+1}\,\frac{\Gamma(m+1)\,\Gamma(n+1)}{\Gamma(m+n+2)}.$$

Solution: Put, $x-a=(b-a)\,y$, i.e., $x=a+(b-a)\,y$.
Then, $dx=(b-a)\,dy$; $b-x=(b-a)\,(1-y)$.

$$\begin{aligned} \therefore\quad I &= \int_0^1 (b-a)^m\,y^m\,(b-a)^n\,(1-y)^n\,(b-a)\,dy=(b-a)^{m+n+1}\int_0^1 y^m(1-y)^n\,dy \\ &= (b-a)^{m+n+1}\,B(m+1,n+1) \quad \text{[by definition]} \\ &= (b-a)^{m+n+1}\,\frac{\Gamma(m+1)\Gamma(n+1)}{\Gamma(m+n+2)} \quad \text{[using (7.46)].} \end{aligned}$$

EXAMPLE 7.9. *Prove that*

$$\int_{-1}^{+1}(1+x)^p\,(1-x)^q\,dx=2^{p+q+1}\,\frac{\Gamma(p+1)\,\Gamma(q+1)}{\Gamma(p+q+2)},$$

where $p,\ q>-1$.

Solution: Put $1+x=2y$, so that $x=2y-1$, $dx=2dy$.
Then

$$\begin{aligned} I &= \int_0^1 (2y)^p\,2^q\,(1-y)^q\cdot 2dy=2^{p+q+1}\int_0^1 y^p\,(1-y)^q\,dy \\ &= 2^{p+q+1}\cdot B(p+1,q+1) \quad \text{[by definition]} \\ &= 2^{p+q+1}\cdot\frac{\Gamma(p+1)\,\Gamma(q+1)}{\Gamma(p+q+2)} \quad \text{[using (7.46)].} \end{aligned}$$

EXAMPLE 7.10. *Show that, for n positive integers*

$$\Gamma\left(-n+\frac{1}{2}\right)=\frac{(-1)^n\,2^n\,\sqrt{\pi}}{(2n-1)\,(2n-3)\cdots 3\cdot 1}.$$

Solution: We have

$$\Gamma(x)\,\Gamma(1-x)=\frac{\pi}{\sin\pi x}\quad\text{[equation (7.69)].}$$

We also have

$$\Gamma(1-x)=-x\Gamma\,(-x)\quad\text{[by putting } m=-x \text{ in equation (7.40)]}$$

From these two relations, we get

$$\Gamma(x)\{-x\Gamma(-x)\}=\frac{\pi}{\sin\pi x}.$$

$$\therefore\quad \Gamma(-x)=-\frac{\pi}{x\sin\pi x}\cdot\frac{1}{\Gamma(x)}. \tag{1}$$

Putting $x=\left(n-\frac{1}{2}\right)$ in equation (1), we get

$$\Gamma\left(-n+\frac{1}{2}\right)=-\frac{\pi}{\left(n-\frac{1}{2}\right)\sin\left(n-\frac{1}{2}\right)\pi}\cdot\frac{1}{\Gamma\left(n-\frac{1}{2}\right)},$$

$$\text{or,}\quad \Gamma\left(-n+\frac{1}{2}\right)=-\frac{\pi}{\sin\left(n-\frac{1}{2}\right)\pi}\cdot\frac{1}{\Gamma\left(n+\frac{1}{2}\right)}. \tag{2}$$

Now

$$\Gamma\left(n+\frac{1}{2}\right)=\left(n-\frac{1}{2}\right)\left(n-\frac{3}{2}\right)\cdots\frac{3}{2}\cdot\frac{1}{2}\Gamma\left(\frac{1}{2}\right)$$
$$=\frac{(2n-1)\,(2n-3)\cdots 3\cdot 1}{2^n}\sqrt{\pi}. \tag{3}$$

Using equation (3) in the equation (2), we get

$$\Gamma\left(-n+\frac{1}{2}\right)=\frac{(-1)^n\,2^n\,\sqrt{\pi}}{(2n-1)\,(2n-3)\cdots 3\cdot 1}. \tag{4}$$

In particular

for $n=1$, $\Gamma\left(-\frac{1}{2}\right)=-2\sqrt{\pi}$;

for $n=2$, $\Gamma\left(-\frac{3}{2}\right)=\frac{4}{3}\sqrt{\pi}$;

for $n=3$, $\Gamma\left(-\frac{5}{2}\right)=-\frac{8}{15}\sqrt{\pi};\cdots$, etc.

Thus, equation (4) gives the values of Gamma functions for any negative-half value of the argument.

EXERCISE

1. Show that, $B(m,n) = B(m+1,n) + B(m,n+1)$.

2. Prove that:

(a) $\displaystyle\int_0^1 \frac{dx}{\sqrt{1-x^4}} = \frac{1}{4}\frac{\Gamma\left(\frac{1}{2}\right)\Gamma\left(\frac{1}{4}\right)}{\Gamma\left(\frac{3}{4}\right)}$;

(b) $\displaystyle\int_0^{\pi/2} \frac{d\theta}{\sqrt{\sin\theta}} \times \int_0^{\pi/2} \sqrt{\sin\theta}\, d\theta = \pi$;

(c) $\displaystyle\int_0^1 \log\Gamma(x)\, dx = \frac{1}{2}\log(2\pi)$;

(d) $\displaystyle\int_0^\infty e^{-ax^2}\cos bx\, dx = \frac{e^{-b^2/4a}}{2}\sqrt{\frac{\pi}{a}}$;

(e) $\displaystyle\int_0^1 x^{2p}(1-x^2)^q\, dx = \frac{1}{2}\frac{\Gamma\left(p+\frac{1}{2}\right)\Gamma(q+1)}{\Gamma\left(p+q+\frac{3}{2}\right)}$.

3. Show that:

(a) $\displaystyle\Gamma\left(\frac{1}{m}\right)\Gamma\left(\frac{2}{m}\right)\cdots\Gamma\left(\frac{m-1}{m}\right) = \left\{\frac{(2\pi)^{m-1}}{m}\right\}^{1/2}$;

(b) $\displaystyle\frac{d}{d\xi}\left[\frac{1}{\Gamma(\xi)}\right] = \frac{\psi(\xi)}{\Gamma(\xi)}$, where $\displaystyle\psi(\xi) = \frac{d}{d\xi}\ln\Gamma(\xi)$.

4. For incomplete gamma functions, prove that

(a) $\displaystyle\gamma(a,x) = e^{-x}\sum_{m=0}^{\infty}\frac{(a-1)!}{(a+m)!}\,x^{a+m}$;

(b) $\Gamma(a+1,x) = a\Gamma(a,x) + x^a\, e^{-x}$.

Chapter 8

Complex Variables

I. COMPLEX NUMBERS

8.1 Definition of a Complex Number

A complex number z is defined as an ordered pair (x, y) where both x and y are real numbers. It is written as

$$z = x + iy, \quad \text{where } i = \sqrt{-1}. \tag{8.1}$$

The numbers x and y are called *real* and *imaginary parts* of the complex number z and are denoted as

$$\left.\begin{aligned} x &= \operatorname{Re}(z) \\ y &= \operatorname{Im}(z) \end{aligned}\right\} \tag{8.2}$$

The symbol $i = \sqrt{-1}$ was, for the first time, introduced by Euler and the era of complex number began.

It is to be noted that the order of the pair (x, y) cannot be interchanged, i.e.

$$(x, y) \neq (y, x). \tag{8.3}$$

If we introduced polar coordinates, so that

$$\left.\begin{aligned} x &= r\cos\theta \\ y &= r\sin\theta \end{aligned}\right\}, \tag{8.4}$$

then

$$r^2 = x^2 + y^2, \quad \text{i.e.,} \quad r = \left(x^2 + y^2\right)^{1/2} \quad \text{and} \quad \theta = \tan^{-1}\frac{y}{x}. \tag{8.5}$$

The quantity r is called the *modulus* of the complex number z, and the quantity θ is called the *amplitude* of the complex number z. It is obvious that the modulus of a complex number is zero, if and only if both x and y are zero, i.e., when $z = x + iy = 0$.

If the real part of a complex number is zero, i.e., $x = 0$, then $z = iy$; we say that the number z is purely imaginary. Thus, the symbol i can be taken as unit of imaginary numbers.

If the amplitude θ of a complex number is $\pi/2$, then $x = 0$ and $y = r$. Therefore, it becomes a purely imaginary number. Again, if the amplitude θ of a complex number is zero or π, then $x = r$ or $-r$ and $y = 0$. Hence, the number becomes purely real.

8.2 Algebra of Complex Numbers

1. *Equality of complex numbers*: Two complex numbers (x_1, y_1) and (x_2, y_2) are said to be equal if and only if their real and imaginary parts are separately equal, i.e.,

$$x_1 = x_2 \quad \text{and} \quad y_1 = y_2. \tag{8.6}$$

2. *Addition (Subtraction)*: The addition or subtraction of two complex numbers is the sum or difference of real and imaginary parts of the numbers separately, i.e., if $z_1 = x_1 + iy_1$ and $z_2 = x_2 + iy_2$, then

$$z_1 \pm z_2 = (x_1 \pm x_2) + i(y_1 \pm y_2). \tag{8.7}$$

The addition of complex numbers obeys commutative and associative laws. Thus

$$z_1 + z_2 = (x_1 + x_2) + i(y_1 + y_2) = (x_2 + x_1) + i(y_2 + y_1),$$

$$\text{i.e.,} \quad z_1 + z_2 = z_2 + z_1. \tag{8.8}$$

Again, if $z_3 = x_3 + iy_3$ be another complex number, then

$$\begin{aligned} z_1 + (z_2 + z_3) &= (x_1 + iy_1) + (x_2 + iy_2 + x_3 + iy_3) \\ &= (x_1 + iy_1 + x_2 + iy_2) + (x_3 + iy_3), \end{aligned}$$

$$\text{i.e.,} \quad z_1 + (z_2 + z_3) = (z_1 + z_2) + z_3. \tag{8.9}$$

For a complex number z, $-z$ is called its *additive inverse* and $z + (-z) = 0$.

3. *Multiplication*: Multiplication of a complex number z by a scalar λ is equal to λ times z, i.e.,

$$\lambda z = \lambda(x + iy) = (\lambda x, \lambda y). \tag{8.10}$$

Multiplication of two complex numbers z_1 and z_2 is given by

$$\begin{aligned} z_1 z_2 &= (x_1 + iy_1)(x_2 + iy_2) \\ &= (x_1 x_2 + y_1 y_2) + i(x_1 y_2 + x_2 y_1) \\ &= \{(x_1 x_2 + y_1 y_2), (x_1 y_2 + x_2 y_1)\}. \end{aligned} \tag{8.11}$$

The multiplication of complex numbers obey the *distributive* law, i.e., for three numbers z_1, z_2, z_3,

$$(z_1 + z_2) z_3 = z_1 z_3 + z_2 z_3. \tag{8.12}$$

Let us show this. We have

$$\begin{aligned} (z_1 + z_2) z_3 &= (x_1 + iy_1 + x_2 + iy_2)(x_3 + iy_3) \\ &= \{(x_1 + x_2) + i(y_1 + y_2)\}(x_3 + iy_3) \\ &= \{(x_1 + x_2)x_3 - (y_1 + y_2)y_3\} \\ &\qquad + i\{(x_1 + x_2)y_3 - (y_1 + y_2)x_3\} \\ &= \{(x_1 x_3 - y_1 y_3) + i(x_1 y_3 + x_3 y_1)\} \\ &\qquad + \{(x_2 x_3 - y_2 y_3) + i(x_2 y_3 + x_3 y_2)\} \\ &= (x_1 + iy_1)(x_3 + iy_3) + (x_2 + iy_2)(x_3 + iy_3), \end{aligned}$$

$$\text{i.e.,} \quad (z_1 + z_2) z_3 = z_1 z_3 + z_2 z_3.$$

The complex number $(1, 0)$ is known as *unity* of the system of complex numbers. The complex number $(0, 1)$ is denoted as i, so that $i^2 = (0, 1).\,(0, 1) = -1$, i.e., $i = \sqrt{-1}$.

4. *Inverse of a complex number*: The complex number z_2 is called the *inverse of the complex number* z_1 if

$$z_1 \cdot z_2 = (1, 0), \quad \text{i.e., unity.} \tag{8.13}$$

Thus, if $z_1 = x_1 + iy_1$ and $z_2 = x_2 + iy_2$, then the equation (8.13) gives

$$\{(x_1x_2 - y_1y_2) + i(x_1y_2 + x_2y_1)\} = 1 + i \cdot 0,$$

i.e.,

$$\left.\begin{aligned} x_1x_2 - y_1y_2 &= 1 \\ \text{and} \quad x_1y_2 - x_2y_1 &= 0 \end{aligned}\right\}. \tag{8.14}$$

Solving this system of equation, we get

$$x_2 = \frac{x_1}{x_1^2 + y_1^2}, \quad y_2 = \frac{-y_1}{x_1^2 + y_1^2}. \tag{8.15}$$

It is obvious that the solutions (8.15) exists if and only if $x_1^2 + y_1^2 \neq 0$, i.e., $|z_1| \neq 0$, i.e., $z \neq 0$. Thus, every non-zero complex number has a unique inverse.

5. *Division*: Division of complex numbers is defined as the inverse of the operation of multiplication. If $z_1 = (x_1 + iy_1)$ and $z_2 = (x_2 + iy_2)$ be two complex numbers such that $z_2 \neq 0$, then

$$\frac{z_1}{z_2} = z_1 z_2^{-1} = (x_1 + iy_1)\left(\frac{x_2}{x_2^2 + y_2^2} - i\frac{y_2}{x_2^2 + y_2^2}\right) \quad \text{[by equation (8.15)]}$$

$$= \left[x_1\left(\frac{x_2}{x_2^2 + y_2^2}\right) + y_1\left(\frac{y_2}{x_2^2 + y_2^2}\right),\ x_1\left(\frac{-y_2}{x_2^2 + y_2^2}\right) + y_1\left(\frac{x_2}{x_2^2 + y_2^2}\right)\right],$$

$$\text{i.e.,} \quad \frac{z_1}{z_2} = \left[\frac{x_1x_2 + y_1y_2}{x_2^2 + y_2^2}, \frac{x_2y_1 - x_1y_2}{x_2^2 + y_2^2}\right] \tag{8.16}$$

8.3 Complex Conjugate

The complex number obtained by changing the sign of the imaginary part of $z = x + iy$, i.e., $x - iy$ is called the *complex conjugate* or simply conjugate of z. It is denoted by $\bar{z}$. Thus

$$\text{if } z = x + iy, \text{ then } \bar{z} = x - iy. \tag{8.17}$$

Obviously, conjugate of $\bar{z}$ is z. We have the following properties:

1. $\overline{z_1 + z_2} = \bar{z} + \bar{z}_2$.

   If $z_1 = x_1 + iy_1$ and $z_2 = x_2 + iy_2$, then $z_1 + z_2 = (x_1 + x_2) + i(y_1 + y_2)$. Hence

$$\overline{z_1 + z_2} = (x_1 + x_2) - i(y_1 + y_2) = (x_1 - iy_1) + (x_2 - iy_2),$$

$$\text{i.e.,} \quad \overline{z_1 + z_2} = \bar{z}_1 + \bar{z}_2. \tag{8.18}$$

2. $\overline{z_1 \cdot z_2} = \bar{z}_1\bar{z}_2$.

   We have, from equation (8.11)

$$z_1z_2 = (x_1x_2 - y_1y_2) + i(x_1y_2 + x_2y_1). \tag{8.19}$$

Hence

$$\overline{z_1z_2} = (x_1x_2 - y_1y_2) - i(x_1y_2 + x_2y_1) = (x_1 - iy_1)(x_2 - iy_2),$$

$$\text{i.e.,} \quad \overline{z_1z_2} = \bar{z}_1 \cdot \bar{z}_2. \tag{8.20}$$

3. $$\overline{\left(\frac{z_1}{z_2}\right)} = \frac{\bar{z}_1}{\bar{z}_2} \text{ (proof is left to readers).} \tag{8.21}$$

4. $$z \cdot \bar{z} = (x+iy)(x-iy) = x^2+y^2 = |\,z\,|^2 = |\,\bar{z}\,|^2. \tag{8.22}$$

5. $$z+\bar{z} = (x+iy)+(x-iy) = 2x = 2\,\mathrm{Re}\,(z). \tag{8.23}$$

6. $$z-\bar{z} = (x+iy)-(x-iy) = 2iy = 2\,\mathrm{Im}\,(z). \tag{8.24}$$

8.4 Geometrical Representation of Complex Numbers

We take rectangular axes OX and OY, where OX is called *real axis* and OY is called *imaginary axis*. Let $z = (x+iy) = (x,y)$ be an ordered pair. The point P whose coordinates are (x,y) in this rectangular system will represent the complex number $z = (x+iy)$. The plane xy is known as *Argand's plane* or *Argand's diagram*. Each point of this plane uniquely represents a complex number.

The complex conjugate of z is $\bar{z} = x - iy$. This number will, therefore, be represented by the point P' having coordinates $(x, -y)$ in the Argand's diagram (*Fig 8.1*).

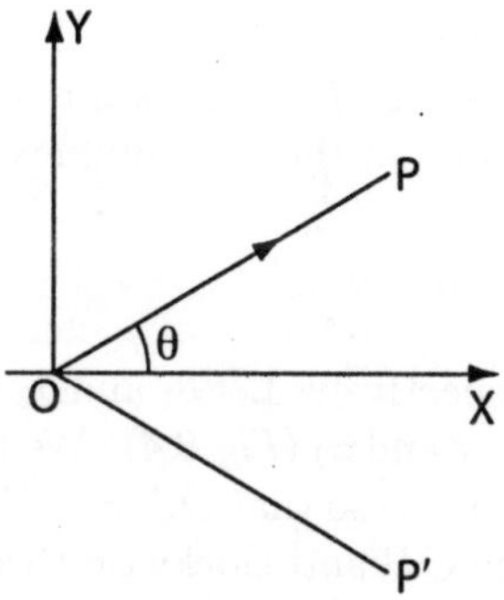

Fig 8.1

8.5 Vector Representation of Complex Numbers

If $P(x,y)$ be the point representing the complex number $z = x+iy$ in the complex plane, then $|\,z\,| = \sqrt{x^2+y^2} = r$, and amplitude is, $\theta = \tan^{-1}\frac{y}{x}$. Thus, the line OP has a definite magnitude (r) and direction (θ). Hence, OP can be treated as a vector. Its end point P will represent the number z.

With this concept, we shall see how the sum, difference, product and quotient of two complex numbers can be represented vectorially.

1. *Sum of two complex numbers*: Let the complex numbers

$$z_1 = x_1 + iy_1$$
$$\text{and} \quad z_2 = x_2 + iy_2$$

be represented by vectors $\overrightarrow{OP}$ and $\overrightarrow{OQ}$ in the complex plane. We construct the parallelogram $OPQR$. Then, by vector addition, the diagonal $\overrightarrow{OR}$ will represent the sum of the vectors $\overrightarrow{OP}$ and $\overrightarrow{OQ}$, i.e., the sum of complex numbers z_1 and z_2 (*Fig 8.2*) will be represented by the point R in complex plane.

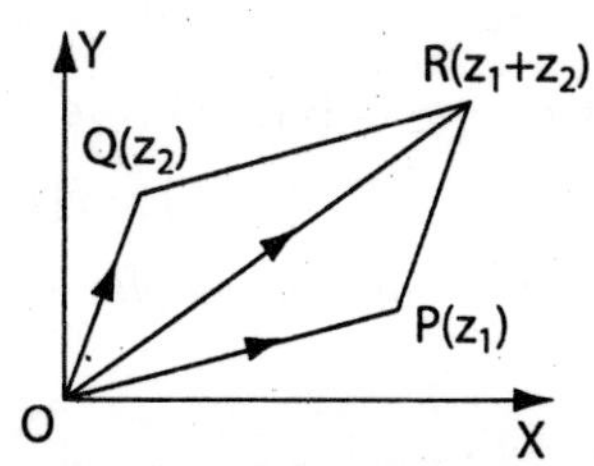

Fig 8.2

2. *Difference of two complex numbers*: As before, let the points P and Q represent the two complex numbers

$$z_1 = x_1 + iy_1$$
$$\text{and} \quad z_2 = x_2 + iy_2.$$

We produce OQ backwards up to Q' such that $OQ = OQ'$ (*Fig 8.3*). Then Q' represents the number $-z_2$. Now, we construct the parallelogram $OPRQ'$. Then the point R will represent the complex number $z = z_1 + (-z_2) = z_1 - z_2$.

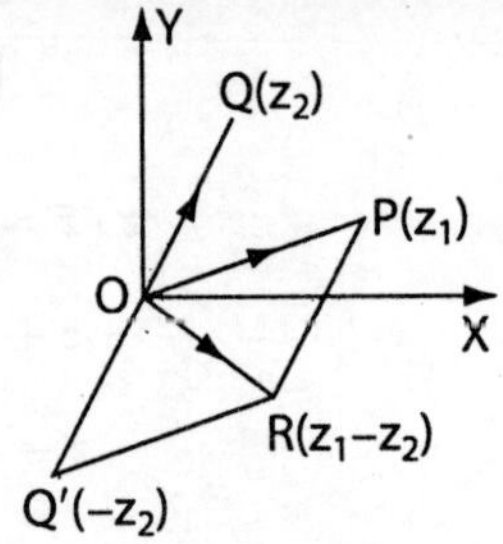

Fig 8.3

3. *Product of two complex numbers*: Let the end-points P and Q of vectors $\overrightarrow{OP}$ and $\overrightarrow{OQ}$ represent the two complex numbers

$$z_1 = x_1 + iy_1$$
$$\text{and} \quad z_2 = x_2 + iy_2$$

respectively. Let θ_1 and θ_2 be the amplitude of z_1 and z_2 (*Fig 8.4*). We take a point A on real axis so that $OA = 1$. We now rotate the line OP anti-clockwise through an angle θ_2 (which is amplitude of z_2) and draw another line through Q so that $\angle OQR = \angle OAP$. These two lines meet at R. Then, amplitude of $\overline{OR} = \theta_1 + \theta_2$. From similar triangles OPA and ORQ, we have

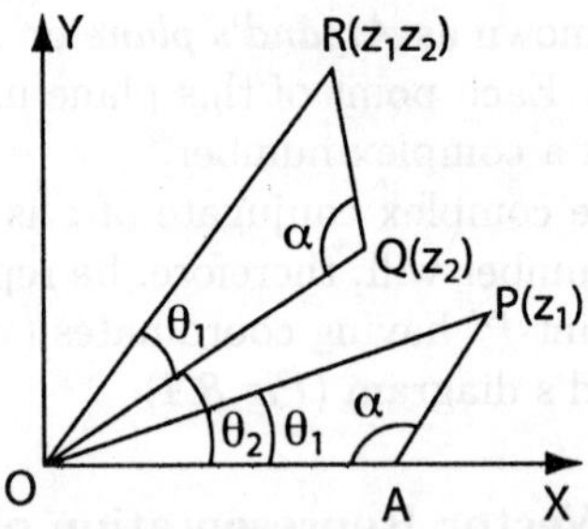

Fig 8.4

$$\frac{OR}{OP} = \frac{OQ}{OA}, \quad \text{i.e.,} \quad OR = OP \cdot OQ = r_1 r_2 \quad (\because OA = 1).$$

Thus, the vector OR has modulus equal to $r_1 r_2$ and amplitude equal to $\theta_1 + \theta_2$. Therefore, the point R represents the complex number $z = z_1 z_2$. Analytically, if

$$\left.\begin{aligned} z_1 &= r_1(\cos\theta_1 + i\sin\theta_1) = r_1 e^{i\theta_1}, \ |z_1| = r_1 \\ \text{and} \quad z_2 &= r_2(\cos\theta_2 + i\sin\theta_2) = r_2 e^{i\theta_2}, \ |z_2| = r_2 \end{aligned}\right\} \tag{8.25}$$

then, $z_1 z_2 = \left(r_1 e^{i\theta_1}\right)\left(r_2 e^{i\theta_2}\right) = r_1 r_2 e^{i(\theta_1+\theta_2)}$, so that

$$|z_1 z_2| = r_1 r_2, \text{ and amp of } z = \theta_1 + \theta_2 = \text{amp } z_1 + \text{amp } z_2. \tag{8.26}$$

4. *Quotient of two complex numbers*: Let the points P, Q represent the complex numbers

$$z_1 = x_1 + iy_1$$
$$\text{and} \quad z_2 = x_1 + iy_2.$$

We take a point A on the real axis so that $OA = 1$. We now rotate the line OP in clockwise direction through an angle θ_2 (which is the amplitude of z_2), and draw a line through A so that $\angle OAR = \angle OQP$. These two lines meet at the point R which will represent the complex number $z = z_1/z_2$. From similar triangles OPQ and OAR (*Fig 8.5*) we have

$$\frac{OR}{OA} = \frac{OP}{OQ}, \quad \text{i.e.,} \quad OR = \frac{r_1}{r_2} \quad [\because OA = 1]$$

Moreover, OR makes an angle $\theta_2 - \theta_1 = -(\theta_1 - \theta_2)$ with real axis, so that amplitude of R is $-(\theta_1 - \theta_2)$.

Analytically, if $z_1 = r_1 e^{i\theta_1}$ and $z_2 = r_2 e^{i\theta_2}$, then

$$\frac{z_1}{z_2} = \frac{r_1}{r_2} e^{i(\theta_1 - \theta_2)}. \qquad (8.27)$$

Therefore

$$\left|\frac{z_1}{z_2}\right| = \frac{r_1}{r_2} = \frac{|z_1|}{|z_2|} \qquad (8.28)$$

and, amplitude of

$$\frac{z_1}{z_2} = \theta_1 - \theta_2 = \text{amp}\, z_1 - \text{amp}\, z_2. \quad (8.29)$$

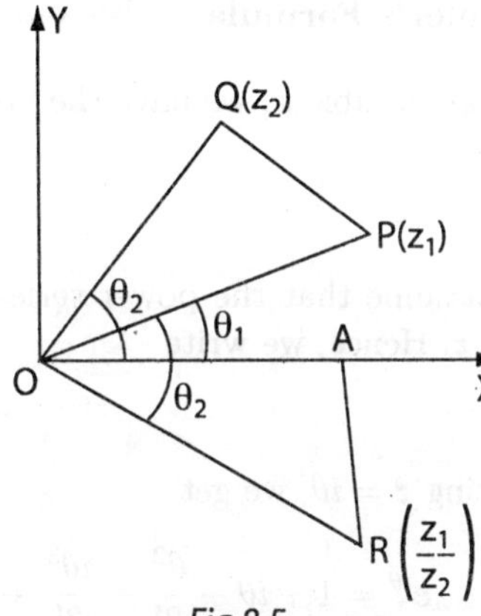

Fig 8.5

8.6 Angle between Lines

Let, z, z_1, z_2 be the points A, B, C in the complex plane. The vectors $\overrightarrow{AB}$ and $\overrightarrow{AC}$ are represented by $(z_1 - z)$ and $(z_2 - z)$ respectively. Then, angle between the lines AB and AC is given by

$$\theta = \text{amp}\,\frac{z_1 - z}{z_2 - z}, \qquad -\pi \le \theta \le \pi. \qquad (8.30)$$

Hence, if this angle $\theta = \pm\pi/2$, we conclude that the number

$$\frac{z_1 - z}{z_2 - z}$$

is purely imaginary. Again, if the points A, B, C are collinear, then the angle θ between them is zero or π. Hence, the number $(z_1-z)/(z_2-z)$ would be purely real.

8.7 Cube Roots of Unity (Imaginary)

Let ω be an imaginary cube root of unity.

Then, $\omega^3 = 1$ or, $(\omega - 1)(\omega^2 + \omega + 1) = 0$.

Since ω is imaginary, $\omega - 1 \ne 0$. Hence, $\omega^2 + \omega + 1 = 0$.

Solving, we get

$$\omega = 1/2\left(-1 \pm i\sqrt{3}\right). \qquad (8.31)$$

If we take, $\omega = 1/2\left(-1 + i\sqrt{3}\right)$, then

$$\omega^2 = 1/4\left(1 - 3 - 2i\sqrt{3}\right) = 1/2\left(-1 - i\sqrt{3}\right),$$

which is the other root. Similarly, if we take $\omega = 1/2\left(-1 - i\sqrt{3}\right)$, then

$$\omega^2 = 1/2\left(-1 + i\sqrt{3}\right).$$

Thus, the imaginary cube roots of unity are ω and ω^2.

8.8 Euler's Formula

For a real variable x, we have the power series expansion of e^x as

$$e^x = 1 + \frac{x}{1!} + \frac{x^2}{2!} + \cdots . \tag{8.32}$$

We assume that the power series expansion in equation (8.32) is also valid for a complex number z. Hence, we write

$$e^z = 1 + \frac{z}{1!} + \frac{z^2}{2!} + \cdots \tag{8.33}$$

Putting $z = i\theta$, we get

$$e^{i\theta} = 1 + i\theta - \frac{\theta^2}{2!} - \frac{i\theta^3}{3!} + \cdots = \left(1 - \frac{\theta^2}{2!} + \frac{\theta^4}{4!} - \cdots\right) + i\left(\theta - \frac{\theta^3}{3!} + \frac{\theta^5}{5!} - \cdots\right)$$

$$\text{i.e.,} \quad e^{i\theta} = \cos\theta + i\sin\theta. \tag{8.34}$$

Equation (8.34) is the well-known Euler's formula. It may be noted that we have already used this formula in equations (8.25) and (8.26) while using the polar coordinate representation of a complex number.

8.9 DeMoivre's Theorem

The theorem states that, for all positive and negative integral values as well as fractional values of n,

$$(\cos\theta + i\sin\theta)^n = \cos n\theta + i\sin\theta.$$

Let us consider complex numbers

$$\begin{aligned} z_1 &= r_1(\cos\theta_1 + i\sin\theta_1) \\ z_2 &= r_2(\cos\theta_2 + i\sin\theta_2) \\ &\cdots\cdots\cdots\cdots\cdots\cdots \\ &\cdots\cdots\cdots\cdots\cdots\cdots \\ z_n &= r_n(\cos\theta_n + i\sin\theta_n) \end{aligned}$$

Then, their product will be given by

$$z_1 z_2 \cdots z_n = r_1 r_2 \cdots r_n \left[\cos(\theta_1 + \theta_2 + \cdots + \theta_n) + i\sin(\theta_1 + \theta_2 + \cdots + \theta_n)\right].$$

If now we assume $z_1 = z_2 = \cdots = z_n = z$, then the product becomes

$$z^n = r^n(\cos n\theta + i\sin n\theta),$$

i.e., $r^n(\cos\theta + i\sin\theta)^n = r^n(\cos n\theta + i\sin n\theta)$.

Thus

$$(\cos\theta + i\sin\theta)^n = \cos n\theta + i\sin n\theta. \tag{8.35}$$

By using Euler's formula (8.34), we also can write

$$e^{in\theta} = \cos n\theta + i\sin n\theta. \tag{8.36}$$

We have proved the theorem for n, a positive integer. But it can also be proved for n, a negative integer and a fraction.

As a direct consequence of the theorem, we get the following formulae:

1. $$(\cos\theta - i\sin\theta)^n = \cos n\theta - i\sin n\theta. \tag{8.37}$$

2. $$(\cos m\theta + i\sin m\theta)^n = (\cos n\theta + i\sin n\theta)^m. \tag{8.38}$$

The DeMoivre's theorem can be used in extracting any assigned root of a complex number. Let z be a complex number which is given in polar coordinates as

$$z = r(\cos\theta + i\sin\theta). \tag{8.39}$$

Expression (8.39) can be written as

$$z = r\{\cos(\theta + 2k\pi) + i\sin(\theta + 2k\pi)\}, \tag{8.40}$$

where $k = 0, 1, 2, \ldots$.

If we are to find the nth root of z, we get by DeMoivre's theorem

$$z^{1/n} = r^{1/n}\left\{\cos\xi_k + i\sin\xi_k\right\}, \tag{8.41}$$

where

$$\xi_k = \frac{\theta + 2k\pi}{n}, \quad \text{for } k = 0, 1, \ldots, (n-1). \tag{8.42}$$

For values greater than $(n-1)$, i.e., for $k = n$, $(n+1), \ldots$, we will get the same values already obtained for $k = 0, 1, 2, \ldots, (n-1)$ repeated over and over again. Moreover, no two roots will be obtained same for $k = 0, 1, 2, \ldots, (n-1)$. Thus, $z^{1/n}$ has n distinct roots. If $z = x + iy$, the roots are given by

$$\left(x^2 + y^2\right)^{1/2n}\left\{\cos\eta_k + i\sin\eta_k\right\}, \tag{8.43}$$

where

$$\eta_k = \frac{1}{n}\left\{2k\pi + \tan^{-1}\frac{y}{x}\right\} \tag{8.44}$$

for $k = 0, 1, 2, \ldots, n-1$.

EXAMPLE 8.1. *Find the conjugate of the complex number* $\dfrac{2-i}{(1+2i)^2}$.

Solution:

$$z = \frac{2-i}{(1+2i)^2} = \frac{2-i}{1+4i+4i^2} = \frac{2-i}{-3+4i}$$

$$\text{or,}\quad z = \frac{-(2-i)(3+4i)}{(3-4i)(3+4i)} = \frac{-(6+8i-3i-4i^2)}{9-16i^2} = \frac{-5(2+i)}{9+16} = -\frac{1}{5}(2+i).$$

Hence, the conjugate of $z = -\frac{1}{5}(2-i)$.

EXAMPLE 8.2. *If z_1 and z_2 be two complex numbers, then show that*

$$|z_1 \pm z_2| \le |z_1| + |z_2|.$$

Solution:

$$\begin{aligned}
|z_1+z_2|^2 &= |z_1+z_2|\,|\overline{z_1+z_2}| \\
&= (z_1+z_2)(\bar{z}_1+\bar{z}_2) \quad \text{(by equation 8.18)} \\
&= z_1\bar{z}_1 + z_1\bar{z}_2 + z_2\bar{z}_1 + z_2\bar{z}_2 \\
&= |z_1|^2 + |z_2|^2 + z_1\bar{z}_2 + \overline{z_1\bar{z}_2} \quad [\text{since } \overline{z_1\bar{z}_2} = \bar{z}_1\bar{\bar{z}}_2 = \bar{z}_1 z_2].
\end{aligned}$$

$$\begin{aligned}
\therefore \quad |z_1+z_2|^2 &= |z_1|^2 + |z_2|^2 + 2\,\mathrm{Re}\,(z_1\bar{z}_2) \quad \text{(by equation 8.23)} \\
&\le |z_1|^2 + |z_2|^2 + 2\,|z_1\bar{z}_2| \\
&\le |z_1|^2 + |z_2|^2 + 2\,|z_1|\,|z_2| \quad (\because |z_2| = |\bar{z}_2|).
\end{aligned}$$

$$\therefore \quad |z_1+z_2|^2 \le \{|z_1| + |z_2|\}^2.$$

$$\therefore \quad |z_1+z_2| \le |z_1| + |z_2|.$$

If we replace z_2 by $-z_2$, then $|z_1 - z_2| \le |z_1| + |-z_2| \le |z_1| + |z_2|$,

$$\text{since } |z_2| = |-z_2|.$$

EXAMPLE 8.3. *For any complex number z, show that*

$$|z| \ge 1/\sqrt{2}\,\{|Re(z)| + |Im(z)|\}.$$

Solution: Let $z = x + iy$, then $\mathrm{Re}\,(z) = x$, $\mathrm{Im}\,(z) = y$ and $|z| = \sqrt{x^2+y^2}$.

Hence

$$2\,|z|^2 = 2\,(x^2+y^2). \tag{1}$$

$$|\mathrm{Re}\,(z)| + |\mathrm{Im}\,(z)| = |x| + |y|.$$

Now

$$2\,|z|^2 - \{|\mathrm{Re}\,(z)| + |\mathrm{Im}\,(z)|\}^2 = 2\,(x^2+y^2) - (x+y)^2 = (x-y)^2 \ge 0$$

for all real values of x and y.

Hence

$$|z|^2 \ge 1/2\,\{|\mathrm{Re}\,(z)| + |\mathrm{Im}\,(z)|\}^2,$$

$$\text{i.e.,} \quad |z| \ge 1/\sqrt{2}\{|\mathrm{Re}\,(z)| + |\mathrm{Im}\,(z)|\}.$$

EXAMPLE 8.4. *Find equation to a circle having centre at the point a and radius r.*

Solution: If $P(z)$ be any point on the circumference of the circle, then, $|z-a| = r$.

Squaring, we get

$$\begin{aligned}
|z-a|^2 &= r^2 \\
\text{or,} \quad (z-a)(\overline{z-a}) &= r^2 \\
\text{or,} \quad (z-a)(\bar{z}-\bar{a}) &= r^2 \\
\text{or,} \quad z\bar{z} - a\bar{z} - \bar{a}z &= (r^2 - a\bar{a}),
\end{aligned}$$

which is the required equation to the circle.

EXAMPLE 8.5. *Let z be a variable complex number such that* (i) *amp* $\frac{(z-i)}{(z+i)} = \frac{\pi}{4}$ *and* (ii) $|z^2+z| < 1$. *Find the loci of the point z in both cases.*

Solution: Let $z = x + iy$ be the complex number.

(i) $\text{amp}\,\dfrac{z-i}{z+i} = \dfrac{\pi}{4}$, i.e., $\text{amp}\,\{x+iy-i\} - \text{amp}\,\{x+iy+i\} = \dfrac{\pi}{4}$,

i.e., $\text{amp}\,\{x+i(y-1)\} - \text{amp}\,\{x+i(y+1)\} = \dfrac{\pi}{4}$,

i.e., $\tan^{-1}\dfrac{y-1}{x} - \tan^{-1}\dfrac{y+1}{x} = \dfrac{\pi}{4}$,

i.e., $\tan^{-1}\left[\dfrac{\dfrac{y-1}{x} - \dfrac{y+1}{x}}{1+\dfrac{y^2-1}{x^2}}\right] = \dfrac{\pi}{4}$.

$$\therefore \qquad \frac{-\dfrac{2}{x}}{1+\dfrac{y^2-1}{x^2}} = \tan\frac{\pi}{4} = 1.$$

Therefore

$$x^2+y^2-1 = -2x \quad \text{or,} \quad x^2+y^2+2x-1 = 0,$$

which is the equation of a circle.

(ii) $|z^2+z| = |(x+iy)^2+(x+iy)| = |(x^2-y^2+x)+i(2xy+y)|$.

Hence, $|z^2+z| < 1$ means

$$\begin{aligned} &(x^2-y^2+x)^2+(2xy+y)^2 < 1\\ \text{or,}\quad &x^4+y^4+x^2-2x^2y^2+2x^3-2y^2x+4x^2y^2+y^2+4y^2x < 1\\ \text{or,}\quad &x^4+y^4+2x^2y^2+2x^3+x^2+y^2+2y^2x < 1\\ \text{or,}\quad &(x^2+y^2)(x^2+y^2+2x+1) < 1. \end{aligned}$$

Hence, the locus of the point is interior of the curve

$$(x^2+y^2)(x^2+y^2+2x+1) = 1.$$

EXAMPLE 8.6. *If the ratio $(z-i)/(z-1)$ be purely imaginary, then show that the point z lies on a circle.*

Solution: Let, $z = x+iy$.

Then, $z - i = x + i(y-1)$, $z - 1 = (x-1) + iy$.

$$\begin{aligned} \therefore \qquad \frac{z-i}{z-1} = \frac{x+i(y-1)}{(x-1)+iy} &= \frac{\{x+i(y-1)\}\{(x-1)-iy\}}{\{(x-1)+iy\}\{(x-1)-iy\}}\\ &= \frac{\{x(x-1)+y(y-1)\}+i\{(x-1)(y-1)-xy\}}{\{(x-1)^2+y^2\}}. \end{aligned} \qquad (1)$$

Since the given ratio is purely imaginary, the real part in equation (1) should vanish. Hence

$$\begin{aligned} &x(x-1)+y(y-1) = 0\\ \text{or,}\quad &\left(x-\frac{1}{2}\right)^2+\left(y-\frac{1}{2}\right)^2 = \frac{1}{2} = \left(\frac{1}{\sqrt{2}}\right)^2. \end{aligned}$$

This represents a circle with radius $1/\sqrt{2}$ and centre at $(1/2, 1/2)$.

Thus, the point z lies on the circle whose radius is $1/\sqrt{2}$ and centre at $1/2\,(1+i)$.

EXAMPLE 8.7. *If z be a complex number, find iz $\left(i=\sqrt{-1}\right)$ and interpret the result.*

Solution: Let, $z = r(\cos\theta + i\sin\theta)$.

We can write, $i = 0 + i\cdot 1 = \cos \pi/2 + i \sin \pi/2$.

$$\begin{aligned}\therefore \quad iz &= r\,(\cos \pi/2 + i\sin \pi/2)\,(\cos\theta + i\sin\theta)\\ &= r\left\{\cos(\theta + \pi/2) + i\sin(\theta+\pi/2)\right\}.\end{aligned}$$

Thus, iz represents a vector obtained by rotating the vector z through an angle $\pi/2$ in the positive sense.

EXAMPLE 8.8. *If ω be the cube root of unity, show that*

$$(1-\omega)(1-\omega^2)(1-\omega^4)(1-\omega^8) = 9.$$

Solution: Since ω is the cube root of unity, we have

$$\omega^3 = 1 \quad \text{and} \quad 1+\omega+\omega^2 = 0. \tag{1}$$

Now

$$1-\omega^4 = 1-\omega\cdot\omega^3 = 1-\omega \tag{2}$$

$$\begin{aligned}\text{and} \quad 1-\omega^8 &= (1-\omega^4)(1+\omega^4) = (1-\omega\cdot\omega^3)(1+\omega\cdot\omega^3)\\ &= (1-\omega)(1+\omega) = 1-\omega^2. \end{aligned}\tag{3}$$

Hence

$$\begin{aligned}(1-\omega)(1-\omega^2)(1-\omega^4)(1-\omega^8) &= (1-\omega)(1-\omega^2)(1-\omega)(1-\omega^2)\\ &\qquad\qquad [\text{using (2) and (3)}]\\ &= (1-\omega^2-\omega+\omega^3)^2\\ &= (1+1+1)^2 \quad [\text{using (1)}]\\ &= 9.\end{aligned}$$

EXAMPLE 8.9. *The four points z_1, z_2, z_3, z_4 representing complex numbers are concyclic. Find the necessary condition for this.*

Solution: Let $A(z_1)$, $B(z_2)$, $C(z_3)$ and $D(z_4)$ be the points on a circle in the complex plane taken in cyclic order. Then the angles $\angle ABD$ and $\angle ACD$ will be on the same segment $ABCD$ of the circle. Hence

$$\angle ABD = \angle ACD, \quad \text{i.e.,} \quad \angle ABD - \angle ACD = 0,$$

$$\text{i.e.,} \quad \operatorname{amp}\frac{z_2-z_1}{z_4-z_2} - \operatorname{amp}\frac{z_3-z_1}{z_4-z_3} = 0,$$

$$\text{i.e.,} \quad \operatorname{amp}\left[\frac{z_2-z_1}{z_4-z_2} \div \frac{z_3-z_1}{z_4-z_3}\right] = 0.$$

Hence, the number

$$\frac{z_2-z_1}{z_4-z_2} \div \frac{z_3-z_1}{z_4-z_3}, \quad \text{i.e.,} \quad \frac{(z_2-z_1)}{(z_4-z_2)}\frac{(z_4-z_3)}{(z_3-z_1)}$$

is purely real. This is the required condition.

EXAMPLE 8.10. (i) *Find the values of* $\left(\sqrt{3}-i\right)^{1/7}$; (ii) *Solve:* $x^5 = 1$.

Solution: (i) We have

$$\left(\sqrt{3}-i\right) = 2\left(\frac{\sqrt{3}}{2} - \frac{1}{2}i\right) = 2\left(\cos\frac{\pi}{6} - i\sin\frac{\pi}{6}\right)$$

$$= 2\left\{\cos\left(2m\pi + \frac{\pi}{6}\right) - i\sin\left(2m\pi + \frac{\pi}{6}\right)\right\},$$

where $m = 0$, or any integer.

Hence

$$\left(\sqrt{3}-i\right)^{1/7} = 2^{1/7}\left\{\cos\left(2m\pi + \frac{\pi}{6}\right) - i\sin\left(2m\pi + \frac{\pi}{6}\right)\right\}^{1/7}$$

$$= 2^{1/7}\left\{\cos\left(2m + \frac{1}{6}\right)\frac{\pi}{7} - i\sin\left(2m + \frac{1}{6}\right)\frac{\pi}{7}\right\},$$

where $m = 0, 1, 2, 3, 4, 5, 6$.

(ii) We have, $x^5 = 1$, i.e., $x^5 = \cos 0 + i\sin 0 = \cos 2k\pi + i\sin 2k\pi$, where $k = 0$, or any integer.

$$\therefore \quad x = (\cos 2k\pi + i\sin 2k\pi)^{1/5} = \cos\frac{2k\pi}{5} + i\sin\frac{2k\pi}{5},$$

where $k = 0, 1, 2, 3, 4$.

II. COMPLEX FUNCTIONS

8.10 Introduction

Before entering into the theory of functions of complex variables, we like to define some useful terminologies.

(a) *Set of points*: A collection of points in the complex plane is said to be a *set of points.* The number of points may be finite or infinite.

A set of points is said to be *bounded,* if there exists a constant M, such that $|z| \leq M$ for every point z of the set.

(b) *Neighbourhood of a point*: The neighbourhood of a point z_0 in the complex plane is the set of all points z, such that $|z - z_0| < \epsilon$, where ϵ is an arbitrary small positive number. We call ϵ the *radius* of this neighbourhood.

If, from the neighbourhood of a point z_0, the point z_0 is excluded, we call this a *deleted neighbourhood.*

The neighbourhood of a point at infinity is defined to be the region outside the circle $|z| = R$, where R is a positive real large number.

(c) *Limiting point*: A point z_0 is called a *limiting point* of the set of points S, if every neighbourhood of z_0 contains a point of S distinct from z_0. For example, each point on the circumference of the circle $|z| = R$ is a limit point of the set $|z| < R$. The limit point may or may not belong to the set. The points on the circumference of the circle do not belong to the set $|z| < R$, but each point inside the circle is also a limit point that belongs to the set.

The limiting points are of two types—interior and boundary points.

A limiting point z_0 of a set S is called an *interior point* if there exists a neighbourhood of z_0 containing all the points of S. This means that z_0 belongs to S. If, however, every neighbourhood of z_0 contains, apart from all points of S, points not belonging to S, then z_0 is called a *boundary point.* Thus, all points inside the circle $|z| = R$ are interior points of the set defined by $|z| < R$, whereas all points on the circumference of the circle are boundary points of the set $|z| < R$.

A set which contains entirely the interior points is called *open.* Thus, the set of points defined by $|z| < R$ is open.

A set is called *closed* if every limit point of the set belongs to it. Thus, the set of points defined by $|z| \le R$ is closed.

(d) *Domain*: A set of points in the complex plane is said to be *connected,* if every pair of points of the set can be joined by a polygonal arc which consists only of the points of the set. An open connected set of points is called a *Domain.* If the limiting points of the set are added to a domain, the resulting set is called a *closed region.* Two domains which have no common point are said to be separated.

Let us now define a function of a complex variable.

8.11 Definition of a Function

Let (x, y) be an ordered pair of real variables, such that $z = x + iy$ represents points in the Argand plane. Let $u(x, y)$, $v(x, y)$ be two functions of real variables x, y, such that $w = u(x, y) + iv(x, y)$. Now, if for every point z in a domain D there corresponds one or more values of w, we say w is a function of z, and write $w = f(z)$.

Thus

$$w = f(z) = u(x, y) + iv(x, y). \tag{8.44}$$

If, for each value of z in D, the function w has only one value, we say the function w is *single-valued;* otherwise, it is called *multiple-valued.*

8.12 Continuity of Functions

The function $f(z)$, defined in a bounded closed region D, is said to be continuous at the point z_0, if for any positive number ϵ, however small, there exists a number δ depending on ϵ, such that

$$|f(z) - f(z_0)| < \epsilon \tag{8.45}$$

for all z satisfying the condition $|z - z_0| < \delta$.

If a function $f(z)$ is continuous at every point of a bounded closed region D, it is said to be continuous in D.

A function $f(z)$ which is continuous in a closed bounded region D is *uniformly* continuous there.

A function $f(z)$, defined in a domain D, is said to be *bounded* in D, if there exists a positive constant M, such that

$$|f(z)| < M \text{ for each } z \text{ in } D. \tag{8.46}$$

If $f(z)$ is continuous in a bounded closed region D, then it is bounded in D.

8.13 Differentiability of Functions

If $f(z)$ is single-valued function defined in a domain D of the complex plane, we say that $f(z)$ is differentiable at a point z_0 of D, if the increment ratio

$$\frac{f(z)-f(z_0)}{z-z_0} \tag{8.47}$$

tends to a unique finite limit as $z \longrightarrow z_0$ *along any path* of the domain D. This unique limit is then called *differential coefficient* or *derivative of the function* $f(z)$ at $z = z_0$.

In order that a function $f(z)$ should be differentiable, it is necessary that it should be continuous. Let $f(z)$ be differentiable at $z = z_0$, i.e., $f'(z_0)$ exists. Then, for non-zero value of h,

$$f(z_0+h)-f(z_0) = \frac{f(z_0+h)-f(z_0)}{h}\cdot h$$

so that

$$\lim_{h\longrightarrow 0}\{f(z_0+h)-f(z_0)\} = \lim_{h\longrightarrow 0}\frac{f(z_0+h)-f(z_0)}{h}\cdot h,$$
$$= f'(z_0)\cdot \lim_{h\longrightarrow 0} h = 0$$

which shows that $f(z)$ is continuous. But a function which is continuous, may not be differentiable.

As an example, let us consider the function $f(z) = |\,z\,|^2$, i.e., $f(z) = x^2+y^2$. Obviously, the function $f(z)$ is continuous everywhere. Let us consider its differentiability at z_0. Now

$$f'(z_0) = \lim_{h\longrightarrow 0}\frac{f(z_0+h)-f(z_0)}{h} = \lim_{h\longrightarrow 0}\frac{|\,z_0+h\,|^2-|\,z_0^2\,|}{h}$$
$$= \lim_{h\longrightarrow 0}\frac{(z_0+h)(\bar{z}_0+\bar{h})-z_0\bar{z}_0}{h} = \lim_{h\longrightarrow 0}\left(\bar{z}_0+\bar{h}+z_0\frac{\bar{h}}{h}\right)$$
$$= \lim_{h\longrightarrow 0}\ \bar{z}_0+z_0\frac{\bar{h}}{h}\quad (\text{since } \bar{h}\longrightarrow 0 \text{ as } h\longrightarrow 0).$$

At $z_0 = 0$, above limit is clearly zero, so that $f'(0)$ exists. But, if $z_0 \neq 0$, then let $h = re^{i\theta}$, so that $\bar{h} = re^{-i\theta}$.

$$\therefore\ \frac{\bar{h}}{h} = \frac{re^{-i\theta}}{re^{i\theta}} = e^{-2i\theta} = \cos 2\theta - i\sin 2\theta,$$

which does not tend to a unique limit as $h \longrightarrow 0$ in any manner. Thus, we find that the function $f(z) = |\,z\,|^2$ is continuous everywhere but nowhere it is differentiable except at the origin.

EXAMPLE 8.11. *If*

$$f(z) = \frac{x^3y(y-ix)}{x^6+y^2}\quad (z\neq 0),\ f(0)=0$$

prove that the increment ratio $\{f(z)-f(0)\}/z$ *tends to zero as* $z \longrightarrow 0$ *along any radius vector, but not as* $z \longrightarrow 0$ *in any other manner.*

Solution: We have

$$f(z) = \frac{x^3y(y-ix)}{x^6+y^2} = \frac{x^3y(-i^2y-ix)}{x^6+y^2},$$

$$\text{i.e.,}\quad f(z) = -\frac{x^3yiz}{x^6+y^2},\quad f(0)=0 \text{ (given)}.$$

Hence

$$\lim_{z \to 0} \frac{f(z) - f(0)}{z} = \lim_{z \to 0} -\frac{x^3 yi}{x^6 + y^2}.$$

Let $z \longrightarrow 0$ along a radius vector $y = mx$.

Then

$$\lim_{z \to 0} -\frac{x^3(mx)i}{x^6 + (mx)^2} = \lim_{x \to 0} -\frac{mx^4 i}{x^6 + m^2 x^2} = 0.$$

If now $z \longrightarrow 0$ along the curve $y = x^3$, then

$$\lim_{z \to 0} \frac{f(z) - f(0)}{z} = \lim_{x \to 0} -\frac{x^6 i}{x^6 + x^6} = -\frac{1}{2}.$$

Hence proved.

8.14 Analytic Function

If a function $f(z)$ be single-valued and differentiable at every point of a domain D, save possibly at a finite number of exceptional points, then the function $f(z)$ is said to be analytic in the domain D. The exceptional points are called *singular points* or *singularities*.

If the function $f(z)$ has no singularity in the domain D, then we say that the function $f(z)$ is *regular* in D. For analytic or regular function, we prefer to use the notation $df(z)/dz$ for its derivatives.

We now give some examples of analytic functions.

(a) *Polynomials*: Simplest analytic function is $f(z) = z^n$, where n is positive integer. Now

$$\frac{f(z) - f(z_0)}{z - z_0} = \frac{z^n - z_0^n}{z - z_0} = z^{n-1} + z_0 z^{n-2} + \cdots + z_0^{n-1},$$

which tends to nz_0^{n-1} as $z \longrightarrow z_0$. Thus, the function is regular in every bounded domain for n positive integer. Similarly, $f(z) = z^{-n}$ where n is positive integer, is analytic in every bounded domain. Its only singularity is the origin. The function

$$f(z) = a_0 + a_1 z + a_2 z^2 + \cdots + a_n z^n, \tag{8.48}$$

where n is positive integer and a_k's are constants, is called a *polynomial.* Since, by above discussion, every term of the polynomial is regular, the polynomial $f(z)$ is analytic and regular in every bounded domain.

(b) *Power series*: An infinite series having terms in ascending integral powers of z in the form

$$f(z) = a_0 + a_1 z + a_2 z^2 + \cdots + a_n z^n + \cdots, \tag{8.49}$$

where the coefficients a_k's are all constants, is called a *power series.* Let

$$\left| \frac{a_n}{a_{n+1}} \right| \longrightarrow R \text{ as } n \longrightarrow \infty. \tag{8.50}$$

This R is called the *radius of convergence* of the power series. We state the following theorem without proof:

Theorem 8.1. *If a power series has a non-zero radius of convergence, its sum is an analytic function, regular within its circle of convergence given by* $|z| = R$.

(c) *Exponential function*: The exponential function is defined by the power series

$$\exp z = 1 + \sum_{n=1}^{\infty} \frac{z^n}{n!}. \tag{8.51}$$

We find that

$$\lim_{n \longrightarrow \infty} \left| \frac{a_n}{a_{n+1}} \right| = \lim_{n \longrightarrow \infty} \left| \frac{(n+1)!}{n!} \right| = \infty.$$

Hence, by theorem 8.1, $\exp z$ is an analytic function having no singularity in any bounded region of the complex plane. Term-by-term differentiation will lead to the fact that

$$\frac{d}{dz}(\exp z) = \exp z. \tag{8.52}$$

(d) *Trigonometrical functions*: The trigonometrical functions of a complex variable z are defined by equations

$$\sin z = \sum_{n=0}^{\infty} (-1)^n \frac{z^{2n+1}}{(2n+1)!} \tag{8.53}$$

$$\cos z = \sum_{n=0}^{\infty} (-1)^n \frac{z^{2n}}{(2n)!}. \tag{8.54}$$

It is obvious that the radius of convergence of each of these series is infinite. Hence, the functions $\sin z$ and $\cos z$ are analytic and regular in every bounded domain in z-plane.

Term-by-term differentiations of the functions $\sin z$ and $\cos z$ give us

$$\frac{d}{dz}(\sin z) = \cos z; \quad \frac{d}{dz}(\cos z) = -\sin z. \tag{8.55}$$

Now, if $z = x + iy$, where x and y are real, then

$$\sin z = \sin x \cosh y + i \cos x \sinh y.$$

Hence, $\sin z$ vanishes if and only if

$$\sin x \cosh y = 0, \cos x \sinh y = 0.$$

But for y real, $\cosh y \geq 0$. Hence, the first equation implies that

$$\sin x = 0, \quad \text{i.e.,} \quad x = n\pi. \tag{8.56}$$

For this value of x, the second equation gives

$$\sinh y = 0, \quad \text{i.e.,} \quad y = 0. \tag{8.57}$$

Thus, $\sin z$ vanishes if and only if $z = n\pi$. Similarly, it can be shown that $\cos z$ vanishes if and only if $z = (z + 1/2)\,\pi$, where n is an integer or zero.

(e) *Hyperbolic sine and cosine functions*: These functions are defined as

$$\sinh z = \frac{1}{2}\left(e^z - e^{-z}\right), \quad \cosh z = \frac{1}{2}\left(e^z + e^{-z}\right). \tag{8.58}$$

Equations defining $\sin z$ and $\cos z$ in terms of $\exp(iz)$ are

$$\sin z = \frac{1}{2i}\left(e^{iz} - e^{-iz}\right), \quad \cos z = \frac{1}{2}\left(e^{iz} + e^{-iz}\right). \tag{8.59}$$

From these, we can get, by putting $z = iz$,

$$\sin iz = i \sinh z, \quad \cos iz = \cosh z. \tag{8.60}$$

Similarly

$$\sinh iz = i \sin z, \quad \cosh iz = \cos z. \tag{8.61}$$

These equations help to deduce properties of hyperbolic functions from corresponding trigonometric functions.

(f) *Natural logarithm*: Natural logarithm of z should be defined such that $e^{\ln z} = z$. Now, if $z = re^{i\theta}$, then

$$\ln z = \ln(re^{i\theta}) = \ln r + i\theta. \tag{8.62}$$

Thus

$$e^{\ln z} = e^{\ln r + i\theta} = e^{\ln r} e^{i\theta} = re^{i\theta} = z.$$

Now, $\ln z$ defined by equation (8.62) gives infinite number of values since θ is multi-valued function. For each particular value of θ, we get what is called a *branch of logarithm.* The most important branch is the *principal value of* $\ln z$ obtained for $0 \le \theta \le 2\pi$.

(g) *The function* z^α: The function z^α, for any real or complex value of α, is defined by the equation

$$z^\alpha = e^{\alpha \ln z}. \tag{8.63}$$

With this definition, the law of indices holds good.
Because

$$z^\alpha \cdot z^\beta = e^{\alpha \ln z} \cdot e^{\beta \ln z} = e^{(\alpha+\beta)\ln z} = z^{\alpha+\beta}.$$

The function z^α is one-valued, although we found earlier that $\ln z$ is many-valued.

(h) *Inverse trigonometric and hyperbolic functions*: Inverse trigonometric functions are $w = \sin^{-1} z$, $w = \cos^{-1} z$, etc.

Similarly, inverse hyperbolic functions are $w = \sinh^{-1} z$, $w = \cosh^{-1} z$, etc.

But, we are to get explicit expressions for the above functions in terms of z. Let us find $\cos^{-1} z$. Let

$$w = \cos^{-1} z, \quad \text{i.e.,} \quad \cos w = z$$

$$\text{i.e.,} \quad 1/2\left(e^{iw} + e^{-iw}\right) = z$$

$$\text{or,} \quad e^{2iw} - 2ze^{iw} + 1 = 0 \quad \text{(multiplying both sides by } e^{iw}\text{)}$$

$$\text{or,} \quad \lambda^2 - 2z\lambda + 1 = 0, \text{ where } \lambda = e^{iw}.$$

$$\therefore \quad \lambda = z + (z^2 - 1)^{1/2}$$

$$\text{i.e.,} \quad e^{iw} = z + (z^2 - 1)^{1/2}$$

$$\text{or,} \quad w = -i \ln\left\{z + (z^2 - 1)^{1/2}\right\}.$$

Thus, $\cos^{-1} z = -\ln\left\{z + (z^2 - 1)^{1/2}\right\}$.

Similarly, we get other functions like

$$\left.\begin{aligned}\sin^{-1} z &= -i\ln\left\{iz + \left(1 - z^2\right)^{1/2}\right\}\\ \sinh^{-1} z &= \ln\left\{z + \left(1 + z^2\right)^{1/2}\right\}\\ \cosh^{-1} z &= \ln\left\{z + \left(z^2 - 1\right)^{1/2}\right\}\end{aligned}\right\}. \tag{8.64}$$

8.15 The Cauchy-Riemann Equations

Let there be two functions $u(x, y)$ and $v(x, y)$ of real variables x, y such that

$$f(z) = u(x, y) + iv(x, y) \tag{8.65}$$

is a function of complex variable $z = x + iy$.

Let us assume that $f(z)$ is analytic at any point z of its domain of definition. Then

$$f'(z) = \lim_{h \longrightarrow 0} \frac{f(z + h) - f(z)}{h}. \tag{8.66}$$

If we take h to be real, then equation (8.66) implies that

$$\lim_{h \longrightarrow 0} \frac{u(x + h, y) - u(x, y)}{h} + i\frac{v(x + h, y) - v(x, y)}{h} \tag{8.67}$$

must exist and equal to $f'(z)$.

Equation (8.67) gives

$$f'(z) = \frac{\partial u}{\partial x} + i\frac{\partial v}{\partial x}. \tag{8.68}$$

Similarly, by taking h to be purely imaginary, we find that

$$f'(z) = \frac{\partial v}{\partial y} - i\frac{\partial u}{\partial y}. \tag{8.69}$$

The two expressions (8.68) and (8.69) must be, by definition, identical. Hence

$$\frac{\partial u}{\partial x} + i\frac{\partial v}{\partial x} = \frac{\partial v}{\partial y} - i\frac{\partial u}{\partial y}.$$

Equating the real and imaginary parts, we get

$$\frac{\partial u}{\partial x} = \frac{\partial v}{\partial y}, \quad \frac{\partial v}{\partial x} = -\frac{\partial u}{\partial y}. \tag{8.70}$$

These are known as *Cauchy-Riemann equations.* Hence, we have the theorem.

Theorem 8.2. *The necessary condition for the function $f(z)$ to be analytic at any point $z = x + iy$ of its domain D is that the partial derivatives u_x, u_y, v_x, v_y should exist and satisfy the Cauchy-Riemann differential equations* (8.70).

The conditions of this theorem are not sufficient for a function $f(z)$ to be analytic. As an example, we consider the function

$$f(z) = u(x, y) + iv(x, y), \tag{8.71}$$

where

$$u(x, y) = \frac{x^3 - y^3}{x^2 + y^2} \quad \text{and} \quad v(x, y) = \frac{x^3 + y^3}{x^2 + y^2}$$

and $f(0) = 0$, i.e., $u(0, 0) = 0$ and $v(0, 0) = 0$.

We can find that the four partial derivatives exist and they have the values

$$\frac{\partial u}{\partial x} = 1, \quad \frac{\partial v}{\partial y} = 1, \quad \frac{\partial u}{\partial y} = 1, \quad \frac{\partial v}{\partial x} = -1.$$

Hence, they satisfy the Cauchy-Riemann conditions. But

$$f'(0) = \lim_{z \to 0} \frac{f(z) - f(0)}{z} = \lim_{\substack{x \to 0 \\ y \to 0}} \frac{(x^3 - y^3) + i(x^3 + y^3)}{x^2 + y^2}$$

takes different values if we approach to the point zero along the line $y = mx$. Hence, $f'(0)$ does not exist.

We, therefore, intend to find the sufficient conditions for differentiability of a function $f(z)$.

Let $z = x + iy$ and $z' = (x + h) + i(y + k)$ be two points in D. We assume that u_x, u_y exist and are continuous. Then, by the mean value theorem for functions of two variables, we have

$$\delta u = \{u_x(x, y) + \epsilon\}h + \{u_y(x, y) + \eta\}k, \tag{8.72}$$

where ϵ and η tend to zero as $z' \longrightarrow z$, i.e., as $h, k \longrightarrow 0$. Similarly

$$\delta v = \{v_x(x, y) + \epsilon_1\}h + \{v_y(x, y) + \eta_1\}k, \tag{8.73}$$

where ϵ_1 and η_1 tend to zero as $z' \longrightarrow z$.

Now

$$\begin{aligned} f(z') - f(z) = \delta u + i\delta v &= [(u_x + \epsilon)h + (u_y + \eta)k] + i[(v_x + \epsilon_1)h + (v_y + \eta_1)k] \\ &= (u_x + iv_x)h + (u_y + iv_y)k + \epsilon_2 h + \eta_2 k, \end{aligned} \tag{8.74}$$

where $\epsilon_2 = \epsilon + i\epsilon_1, \eta_2 = \eta + i\eta_1$.

Again, by Cauchy-Riemann conditions

$$u_x = v_y \quad \text{and} \quad u_y = -v_x = i^2 v_x. \tag{8.75}$$

Using equations (8.75), equation (8.74) becomes

$$\begin{aligned} f(z') - f(z) &= (u_x + iv_x)h + i(iv_x + u_x)k + \epsilon_2 h + \eta_2 k \\ &= (u_x + iv_x)(h + ik) + \epsilon_2 h + \eta_2 k. \end{aligned}$$

Hence

$$\frac{f(z') - f(z)}{z' - z} = (u_x + iv_x) + \frac{\epsilon_2 h + \eta_2 k}{h + ik}. \tag{8.76}$$

Therefore, if $z' \longrightarrow 0$, i.e., $(h, k) \longrightarrow 0$, then we get from (8.76), $f'(z) = u_x + iv_x$, since ϵ_2 and $\eta_2 \longrightarrow 0$ as $(h, k) \longrightarrow 0$. Thus, $f(z)$ is differentiable at each point of D and hence, it is analytic. Hence, we have the following theorem:

Theorem 8.3. *The continuous single-valued function $f(z) = u(x, y) + iv(x, y)$ is analytic, regular in D, if the four partial derivatives u_x, u_y, v_x, v_y exist, are continuous, and satisfy Cauchy-Riemann equations at each point of the domain D.*

8.16 Laplace's Differential Equation

Let

$$f(z) = u(x, y) + iv(x, y) \tag{8.77}$$

be an analytic function in D, so that we have from Cauchy-Riemann equations

$$\frac{\partial u}{\partial x} = \frac{\partial v}{\partial y}, \quad \frac{\partial u}{\partial y} = -\frac{\partial v}{\partial x}. \tag{8.78}$$

We assume that the second order partial derivatives of the functions $u(x, y)$ and $v(x, y)$ exist and are continuous.

Differentiating equations (8.78) with respect to x and y respectively and adding, we get

$$\frac{\partial^2 u}{\partial x^2} + \frac{\partial^2 u}{\partial y^2} = \frac{\partial^2 v}{\partial x \partial y} - \frac{\partial^2 v}{\partial x \partial y} = 0, \quad \text{i.e.,} \quad \nabla^2 u = 0. \tag{8.79}$$

Again, differentiating the equations (8.78) with respect to y and x respectively and subtracting, we get

$$\frac{\partial^2 u}{\partial x \partial y} - \frac{\partial^2 u}{\partial x \partial y} = \frac{\partial^2 v}{\partial y^2} + \frac{\partial^2 v}{\partial x^2} = 0, \quad \text{i.e.,} \quad \nabla^2 v = 0. \tag{8.80}$$

Thus, we have shown that the real and imaginary parts $u(x, y)$ and $v(x, y)$ of the analytic function $f(z)$ satisfy Laplace's equation. Such functions are called *harmonic functions.*

8.17 Cauchy-Riemann's Equations in Polar Coordinates

The Cauchy-Riemann's equations are

$$\frac{\partial u}{\partial x} = \frac{\partial v}{\partial y} \quad \text{and} \quad \frac{\partial u}{\partial y} = -\frac{\partial v}{\partial x}. \tag{8.81}$$

In polar coordinates

$$\left.\begin{aligned} x = r\cos\theta \text{ and } y = r\sin\theta \\ \therefore \quad r = \left(x^2 + y^2\right)^{1/2}, \ \theta = \tan^{-1} y/x \end{aligned}\right\} \tag{8.82}$$

Therefore, we have

$$\left.\begin{aligned} &\frac{\partial r}{\partial x} = \frac{1}{2\sqrt{x^2 + y^2}} \cdot 2x = \frac{x}{r} = \cos\theta; \quad \frac{\partial r}{\partial y} = \frac{y}{r} = \sin\theta \\ &\frac{\partial \theta}{\partial x} = \frac{1}{1 + y^2/x^2}\left(-\frac{y}{x^2}\right) = \frac{-y}{x^2 + y^2} = -\frac{\sin\theta}{r} \\ &\frac{\partial \theta}{\partial y} = \frac{1}{1 + y^2/x^2}\left(\frac{1}{x}\right) = \frac{x}{x^2 + y^2} = \frac{\cos\theta}{r} \end{aligned}\right\} \tag{8.83}$$

Now

$$\frac{\partial u}{\partial x} = \frac{\partial u}{\partial r} \cdot \frac{\partial r}{\partial x} + \frac{\partial u}{\partial \theta} \cdot \frac{\partial \theta}{\partial x} = \cos\theta \frac{\partial u}{\partial r} - \frac{\sin\theta}{r}\frac{\partial u}{\partial \theta} \quad \text{[using equation (8.83)]} \tag{8.84}$$

and

$$\frac{\partial v}{\partial y} = \frac{\partial v}{\partial r} \cdot \frac{\partial r}{\partial y} + \frac{\partial v}{\partial \theta} \cdot \frac{\partial \theta}{\partial y} = \sin\theta \frac{\partial v}{\partial r} + \frac{\cos\theta}{r}\frac{\partial v}{\partial \theta}. \tag{8.85}$$

But, we have

$$\frac{\partial u}{\partial x} = \frac{\partial v}{\partial y}.$$

Hence, from equations (8.84) and (8.85), we get

$$\cos\theta\frac{\partial u}{\partial r} - \frac{\sin\theta}{r}\frac{\partial u}{\partial \theta} = \sin\theta\frac{\partial v}{\partial r} + \frac{\cos\theta}{r}\frac{\partial v}{\partial \theta}. \tag{8.86}$$

Again

$$\frac{\partial u}{\partial y} = \frac{\partial u}{\partial r}\cdot\frac{\partial r}{\partial y} + \frac{\partial u}{\partial \theta}\cdot\frac{\partial \theta}{\partial y} = \sin\theta\frac{\partial u}{\partial r} + \frac{\cos\theta}{r}\frac{\partial u}{\partial \theta} \tag{8.87}$$

and

$$\frac{\partial v}{\partial x} = \frac{\partial v}{\partial r}\cdot\frac{\partial r}{\partial x} + \frac{\partial v}{\partial \theta}\cdot\frac{\partial \theta}{\partial x} = \cos\theta\frac{\partial v}{\partial r} - \frac{\sin\theta}{r}\frac{\partial v}{\partial \theta}. \tag{8.88}$$

By Cauchy-Riemann's equation

$$\frac{\partial u}{\partial y} = -\frac{\partial v}{\partial x}.$$

Hence, from equations (8.87) and (8.88), we have

$$\sin\theta\frac{\partial u}{\partial r} + \frac{\cos\theta}{r}\frac{\partial u}{\partial \theta} = -\cos\theta\frac{\partial v}{\partial r} + \frac{\sin\theta}{r}\frac{\partial v}{\partial \theta}. \tag{8.89}$$

Multiplying equation (8.86) by $\cos\theta$ and (8.89) by $\sin\theta$, and adding, we get

$$\frac{\partial u}{\partial r} = \frac{1}{r}\frac{\partial v}{\partial \theta}. \tag{8.90}$$

Again, multiplying equation (8.86) by $\sin\theta$ and (8.89) by $\cos\theta$, and subtracting, we get

$$\frac{\partial v}{\partial r} = -\frac{1}{r}\frac{\partial u}{\partial \theta}. \tag{8.91}$$

Equations (8.90) and (8.91) are the Cauchy-Riemann's equations in polar coordinates.

EXAMPLE 8.12. *If $w = f(z) = u(x, y) + iv(x, y)$ is analytic in D, show that*

$$\frac{dw}{dz} = \frac{\partial w}{\partial x} = \frac{1}{i}\frac{\partial w}{\partial y}.$$

Also, show that, in polar coordinates

$$\frac{dw}{dz} = (\cos\theta - i\sin\theta)\frac{\partial w}{\partial r} = \frac{-i}{r}(\cos\theta - i\sin\theta)\frac{\partial w}{\partial \theta}.$$

Solution: From equation (8.68), we have

$$\frac{dw}{dz} = f'(z) = \frac{\partial u}{\partial x} + i\frac{\partial v}{\partial x} = \frac{\partial}{\partial x}(u + iv) = \frac{\partial w}{\partial x}. \tag{1}$$

Again, from equation (8.69), we have

$$\frac{dw}{dz} = f'(z) = \frac{\partial v}{\partial y} - i\frac{\partial u}{\partial y} = \frac{1}{i}\left(\frac{\partial u}{\partial y} + i\frac{\partial v}{\partial y}\right) = \frac{1}{i}\frac{\partial}{\partial y}(u + iv) = \frac{1}{i}\frac{\partial w}{\partial y}. \tag{2}$$

Hence, from equations (1) and (2)

$$\frac{dw}{dz} = \frac{\partial w}{\partial x} = \frac{1}{i}\frac{\partial w}{\partial y}. \tag{3}$$

Now

$$\frac{\partial w}{\partial x} = \frac{\partial w}{\partial r}\cdot\frac{\partial r}{\partial x} + \frac{\partial w}{\partial \theta}\cdot\frac{\partial \theta}{\partial x} = \frac{\partial w}{\partial r}\cdot\cos\theta - \frac{\sin\theta}{r}\left(\frac{\partial u}{\partial \theta} + i\frac{\partial v}{\partial \theta}\right) \quad \text{[using (8.83)]}$$

$$= \cos\theta\frac{\partial w}{\partial r} - \frac{\sin\theta}{r}\left(-r\frac{\partial v}{\partial r} + ir\frac{\partial u}{\partial r}\right) \quad \text{[using (8.90) and (8.91)]}$$

$$= \cos\theta\frac{\partial w}{\partial r} - i\sin\theta\frac{\partial}{\partial r}(u+iv) = \cos\theta\frac{\partial w}{\partial r} - i\sin\theta\frac{\partial w}{\partial r}$$

i.e., $$\frac{\partial w}{\partial x} = (\cos\theta - i\sin\theta)\frac{\partial w}{\partial r}. \tag{4}$$

Again

$$\frac{\partial w}{\partial y} = \frac{\partial w}{\partial r}\cdot\frac{\partial r}{\partial y} + \frac{\partial w}{\partial \theta}\cdot\frac{\partial \theta}{\partial y} = \sin\theta\left(\frac{\partial u}{\partial r} + i\frac{\partial v}{\partial r}\right) + \frac{\cos\theta}{r}\frac{\partial w}{\partial \theta}$$

$$= \sin\theta\left(\frac{1}{r}\frac{\partial v}{\partial \theta} - \frac{i}{r}\frac{\partial u}{\partial \theta}\right) + \frac{\cos\theta}{r}\frac{\partial w}{\partial \theta} \quad \text{[using equations (8.90) and (8.91)]}$$

$$= \frac{-i\sin\theta}{r}\frac{\partial}{\partial \theta}(u+iv) + \frac{\cos\theta}{r}\frac{\partial w}{\partial \theta} = \frac{-i\sin\theta}{r}\frac{\partial w}{\partial \theta} + \frac{\cos\theta}{r}\frac{\partial w}{\partial \theta}$$

$$\therefore \quad \frac{1}{i}\frac{\partial w}{\partial y} = \frac{-\sin\theta}{r}\frac{\partial w}{\partial \theta} - i\frac{\cos\theta}{r}\frac{\partial w}{\partial \theta} = \frac{-i}{r}(\cos\theta - i\sin\theta)\frac{\partial w}{\partial \theta}. \tag{5}$$

But, we have shown in equation (3)

$$\frac{\partial w}{\partial x} = \frac{1}{i}\frac{\partial w}{\partial y}.$$

Hence, from equations (4) and (5), we get

$$(\cos\theta - i\sin\theta)\frac{\partial w}{\partial r} = \frac{-i}{r}(\cos\theta - i\sin\theta)\frac{\partial w}{\partial \theta}. \tag{6}$$

Hence proved.

EXAMPLE 8.13. *Examine whether the function*

$$f(z) = \begin{cases} \dfrac{x^2y^3(x+iy)}{x^4+y^6}, & \textit{for } z \neq 0 \\ 0, & \textit{for } z = 0 \end{cases}$$

is analytic or not.

Solution: Here

$$u(x,y) = \frac{x^3y^3}{x^4+y^6}, \quad v(x,y) = \frac{x^2y^4}{x^4+y^6}$$

with conditions $u(0,0) = 0$ and $v(0,0) = 0$. It is obvious that all the four partial derivatives u_x, u_y, v_x, v_y are equal to zero at the origin. Hence, Cauchy-Riemann's equations are satisfied at the origin.

Now, let us find $f'(z)$ at the origin.

$$f'(0) = \lim_{z \longrightarrow 0} \frac{f(z) - f(0)}{z} = \lim_{\substack{x \longrightarrow 0 \\ y \longrightarrow 0}} \frac{x^2 y^3 (x + iy)}{(x^4 + y^6)(x + iy)} = \lim_{\substack{x \longrightarrow 0 \\ y \longrightarrow 0}} \frac{x^2 y^3}{x^4 + y^6}.$$

Let us assume that $z \longrightarrow 0$ along the curve $y^3 = mx^2$.

Then

$$f'(0) = \lim_{x \longrightarrow 0} \frac{mx^4}{x^4 + m^2 x^4}, \quad \text{i.e.,} \quad f'(0) = \lim_{x \longrightarrow 0} \frac{mx^4}{(1 + m^2)x^4} = \frac{m}{1 + m^2},$$

which are different for different values of m. Hence, $f'(0)$ does not exist. Thus the function is not analytic at the origin.

EXAMPLE 8.14. *If $f(z) = u + iv$ is an analytic function of $z = x + iy$, show that*

(i) $L\,|\,Rf(z)\,|^2 = 2\,|\,f'(z)\,|^2$;

(ii) $L\,|\,f(z)\,|^2 = 4\,|\,f'(z)\,|^2$;

(iii) $L\,|\,f(z)\,|^p = p^2\,|\,f(z)\,|^{p-2}\,|\,f'(z)\,|^2$,

where, L is the differential operator $= \dfrac{\partial^2}{\partial x^2} + \dfrac{\partial^2}{\partial y^2}$.

Solution: (i) We have, $f(z) = u + iv$.

$$\therefore \quad |\,f(z)\,|^2 = u^2 + v^2. \tag{1}$$

Now

$$\frac{\partial}{\partial x}(u^2) = 2u \frac{\partial u}{\partial x}.$$

$$\therefore \quad \frac{\partial^2}{\partial x^2}(u^2) = 2\left\{ u \frac{\partial^2 u}{\partial x^2} + \left(\frac{\partial u}{\partial x} \right)^2 \right\}. \tag{2}$$

Similarly

$$\frac{\partial^2}{\partial y^2}(u^2) = 2\left\{ u \frac{\partial^2 u}{\partial y^2} + \left(\frac{\partial u}{\partial y} \right)^2 \right\}. \tag{3}$$

Hence

$$\left(\frac{\partial^2}{\partial x^2} + \frac{\partial^2}{\partial y^2} \right) u^2 = 2u \left(\frac{\partial^2 u}{\partial x^2} + \frac{\partial^2 u}{\partial y^2} \right) + 2\left\{ \left(\frac{\partial u}{\partial x} \right)^2 + \left(\frac{\partial u}{\partial y} \right)^2 \right\}.$$

Since u should be harmonic, $L(u) = 0$.

$$\therefore \quad L(u^2) = 2\left\{ \left(\frac{\partial u}{\partial x} \right)^2 + \left(\frac{\partial u}{\partial y} \right)^2 \right\}. \tag{4}$$

Now

$$f'(z) = \frac{dw}{dz} = \frac{\partial w}{\partial x} \quad \text{[see Example 8.12]}$$

$$= \frac{\partial u}{\partial x} + i \frac{\partial v}{\partial x} = \frac{\partial u}{\partial x} - i \frac{\partial u}{\partial y}.$$

$$\therefore \quad |f'(z)|^2 = \left(\frac{\partial u}{\partial x}\right)^2 + \left(\frac{\partial u}{\partial y}\right)^2. \tag{5}$$

From equations (4) and (5), we get

$$L\,|Rf(z)|^2 = 2\,|f'(z)|^2. \tag{6}$$

Hence proved.

(ii) Proceeding as before, we can get

$$L\,|\operatorname{Im} f(z)|^2 = 2\,|f'(z)|^2. \tag{7}$$

Adding equations (6) and (7), we get

$$L(u^2+v^2) = 4\,|f'(z)|^2$$

$$\text{i.e.,}\quad L\,|f(z)|^2 = 4\,|f'(z)|^2 \quad \text{[by using (1)]}$$

Hence proved.

(iii) We have, $z = x+iy$, $\therefore\ \bar{z} = x - iy$.

$$\therefore \quad x = \frac{1}{2}(z+\bar{z}), \quad y = \frac{1}{2i}(z-\bar{z}).$$

$$\therefore \quad \frac{\partial}{\partial z} = \frac{\partial}{\partial x}\frac{\partial x}{\partial z} + \frac{\partial}{\partial y}\frac{\partial y}{\partial z} = \frac{1}{2}\left(\frac{\partial}{\partial x} - i\frac{\partial}{\partial y}\right) \quad [\because \frac{1}{i} = -i].$$

Similarly

$$\frac{\partial}{\partial \bar{z}} = \frac{1}{2}\left(\frac{\partial}{\partial x} + i\frac{\partial}{\partial y}\right).$$

$$\therefore \quad 4\frac{\partial^2}{\partial z \partial \bar{z}} = \left(\frac{\partial}{\partial x} - i\frac{\partial}{\partial y}\right)\left(\frac{\partial}{\partial x} + i\frac{\partial}{\partial y}\right), \quad \text{i.e.,}\quad 4\frac{\partial^2}{\partial z\partial \bar{z}} = \frac{\partial^2}{\partial x^2} + \frac{\partial^2}{\partial y^2}. \tag{8}$$

Now, $|f(z)|^2 = f(z)\cdot f(\bar{z})$.

$$\therefore \quad L\,|f(z)|^p = \left(\frac{\partial^2}{\partial x^2} + \frac{\partial^2}{\partial y^2}\right)\{f(z)\cdot f(\bar{z})\}^{p/2}$$

$$= 4\frac{\partial^2}{\partial z \partial \bar{z}}\left\{[f(z)]^{p/2}[f(\bar{z})]^{p/2}\right\} \quad \text{[by using (8)].}$$

$$\therefore \quad L\,|f(z)|^p = 4\frac{\partial}{\partial z}\left[\{f(z)\}^{p/2}\frac{p}{2}\{f(\bar{z})\}^{p/2-1}f'(\bar{z})\right]$$

$$= 4\cdot\frac{p^2}{4}\left[\{f(z)\cdot f(\bar{z})\}^{p/2-1}f'(z)f'(\bar{z})\right]$$

$$= p^2\,|f(z)|^{p-2}\,|f'(z)|^2 = \text{RHS}.$$

EXAMPLE 8.15. *If $w = f(z) = u+iv$ be analytic, then show that families of curves defined by $u = c_1$ and $v = c_2$ constitute an orthogonal system.*

Hence, verify this in the case when $f(z) = e^x(\cos y + i\sin y)$.

Solution: We have, $u(x,y) = c_1$.

Differentiating with respect to x, we get

$$\frac{\partial u}{\partial x} + \frac{\partial u}{\partial y} \cdot \frac{dy}{dx} = 0. \therefore \quad \frac{dy}{dx} = -\frac{u_x}{u_y} = m_1 \text{ (say)}. \tag{1}$$

Similarly, from the equation of curve $v(x,y) = c_2$, we get

$$\frac{dy}{dx} = -\frac{v_x}{v_y} = m_2 \text{ (say)}. \tag{2}$$

Since $f(z)$ is analytic, we have, from the Cauchy-Riemann's equation

$$u_x = v_y \quad \text{and} \quad u_y = -v_x. \tag{3}$$

Equations (1) and (2) by using (3) become

$$m_1 = -\frac{u_x}{u_y} = \frac{u_x}{v_x}; \quad m_2 = -\frac{v_x}{v_y} = -\frac{v_x}{u_x}.$$

Hence

$$m_1 m_2 = \frac{u_x}{v_x} \cdot \left(-\frac{v_x}{u_x}\right) = -1. \tag{4}$$

This shows that the families of curves $u = c_1$ and $v = c_2$ intersect at right angles. Hence, the families of curves form an orthogonal system.

If $f(z) = e^x(\cos y + i \sin y)$, we have

$$u(x,y) = e^x \cos y$$
$$\text{and} \quad v(x,y) = e^x \sin y.$$

$$\therefore \quad u_x = e^x \cos y$$
$$u_y = -e^x \sin y.$$

From equation (1)

$$m_1 = -\frac{u_x}{u_y} = \cot y.$$

Again

$$v_x = e^x \sin y$$
$$v_y = -e^x \cos y.$$

From equation (2)

$$m_2 = -\frac{v_x}{v_y} = -\tan y.$$

$$\therefore \quad m_1 \cdot m_2 = \cot y(-\tan y) = -1.$$

Hence, the families of curves $u = c_1$ and $v = c_2$ form an orthogonal system.

8.18 Conjugate Functions and Their Constructions

If $f(z) = u(x,y)+iv(x,y)$ be analytic, then the two functions $u(x,y)$ and $v(x,y)$ of real variables x, y are called *conjugate functions*. If one of these two functions, say $u(x,y)$, be given, the other function $v(x,y)$ can be determined by the following methods:

(a) *Direct method*: Since $v(x,y)$ is a function of two variables x and y, we have

$$dv = \frac{\partial v}{\partial x}dx + \frac{\partial v}{\partial y}dy$$

$$\text{or,} \quad dv = -\frac{\partial u}{\partial y}dx + \frac{\partial u}{\partial x}dy \quad \text{[using C-R conditions]}$$

$$= M\,dx + N\,dy, \tag{8.93}$$

where

$$M = -\frac{\partial u}{\partial y}, \quad N = \frac{\partial u}{\partial x}. \tag{8.94}$$

Hence

$$\frac{\partial M}{\partial y} = -\frac{\partial^2 u}{\partial y^2}; \quad \frac{\partial N}{\partial x} = \frac{\partial^2 u}{\partial x^2}.$$

Thus

$$\frac{\partial M}{\partial y} - \frac{\partial N}{\partial x} = -\left(\frac{\partial^2 u}{\partial x^2} + \frac{\partial^2 u}{\partial y^2}\right) = 0,$$

since $u(x,y)$ is harmonic.

Hence

$$\frac{\partial N}{\partial x} = \frac{\partial M}{\partial y},$$

which shows that the differential in equation (8.93) is exact. Therefore, we can integrate it to get $v(x,y)$, i.e.,

$$v(x,y) = \int \left\{\frac{\partial u}{\partial x}dy - \frac{\partial u}{\partial y}dx\right\} + C. \tag{8.95}$$

(b) *Thomson's method*: Since $z = x + iy$, $\bar{z} = x - iy$, so that

$$\left.\begin{aligned} x &= \tfrac{1}{2}\,(z+\bar{z}) \\ y &= \tfrac{1}{2i}\,(z-\bar{z}) \end{aligned}\right\} \tag{8.96}$$

$$\therefore \quad f(z) = u(x,y) + iv(x,y) = u\left(\frac{z+\bar{z}}{2}, \frac{z-\bar{z}}{2i}\right) + iv\left(\frac{z+\bar{z}}{2}, \frac{z-\bar{z}}{2i}\right). \tag{8.97}$$

If we set $z = \bar{z}$, we get from equation (8.96) $x = z$ and $y = 0$. Then, we get

$$f(z) = u(z,0) + iv(z,0). \tag{8.98}$$

Now, we have

$$f'(z) = \frac{\partial u}{\partial x} + i\frac{\partial v}{\partial x} \quad \text{[from equation (8.68)]}$$
$$= \frac{\partial u}{\partial x} - i\frac{\partial u}{\partial y} \quad \text{[using C-R condition]}$$
$$\text{i.e.,}\quad f'(z) = \phi_1(x,y) - i\phi_2(x,y) \text{ (say)}$$
$$= \phi_1(z,0) - i\phi_2(z,0).$$
$$\therefore\quad f(z) = \int \{\phi_1(z,0) - i\phi_2(z,0)\}\, dz + C. \tag{8.99}$$

EXAMPLE 8.16. *Prove that the function* $u(x,y) = x^3 - 3xy^2 + 3x^2 - 3y^2 + 1$ *satisfies Laplace's equation. Determine the corresponding analytic function* $f(z)$ *by* (i) *Direct method and* (ii) *Thomson's method.*

Solution: We have, $u(x,y) = x^3 - 3xy^2 + 3x^2 - 3y^2 + 1$.

$$\therefore\quad \frac{\partial u}{\partial x} = 3x^2 - 3y^2 + 6x \tag{1}$$
$$\frac{\partial u}{\partial y} = -6xy - 6y \tag{2}$$
$$\frac{\partial^2 u}{\partial x^2} = 6x + 6, \quad \frac{\partial^2 u}{\partial y^2} = -6x - 6. \tag{3}$$

From (3), we find that

$$\frac{\partial^2 u}{\partial x^2} + \frac{\partial^2 u}{\partial y^2} = 6x + 6 - 6x - 6 = 0.$$

Hence, $u(x,y)$ satisfies Laplace's equation. Let us now determine $v(x,y)$.
(i) *Direct method*:

$$v(x,y) = \int \left(\frac{\partial u}{\partial x}dy - \frac{\partial u}{\partial y}dx\right)$$
$$= \int \{(3x^2 - 3y^2 + 6x)\, dy + (6xy + 6y)dx\} \quad \text{[using (1) and (2)]}$$
$$= 3x^2y - y^3 + 6xy + C. \tag{4}$$

Repeated terms have been ignored.
(ii) *Thomson's method*: Putting $x = z$ and $y = 0$ in (1) and (2), we get

$$\phi_1(z,0) = 3z^2 + 6z, \quad \phi_2(z,0) = 0.$$

Hence

$$f(z) = \int \phi_1(z,0)dz + C = \int (3z^2 + 6z)dz + C, \quad \text{i.e.,}\quad f(z) = z^3 + 3z^2 + C.$$

EXAMPLE 8.17. *Find the analytic function* $f(z) = u + iv$, *if* $u = e^x(x\cos y - y\sin y)$ *using both the methods.*

Solution: We have, $u = e^x(x\cos y - y\sin y)$.

$$\therefore \quad \frac{\partial u}{\partial x} = e^x(x\cos y + \cos y - y\sin y) \tag{1}$$

$$\text{and} \quad \frac{\partial u}{\partial y} = e^x(-x\sin y - \sin y - y\cos y). \tag{2}$$

(i) *By direct method:*

$$v(x,y) = \int \left\{ \frac{\partial u}{\partial x} dy - \frac{\partial u}{\partial y} dx \right\} + C.$$

Now

$$\int \frac{\partial u}{\partial x} dy = \int e^x(x\cos y + \cos y - y\sin y)\, dy$$
$$= e^x(x\sin y + \sin y + y\cos y - \sin y)$$
$$= e^x(x\sin y + y\cos y) \tag{3}$$

$$\text{and} \quad -\int \frac{\partial u}{\partial y} dx = \int e^x(x\sin y + \sin y + y\cos y)\, dx$$
$$= e^x[(x-1)\sin y + \sin y + y\cos y]. \tag{4}$$

Hence, from (3) and (4), we get by ignoring repeated terms

$$v(x,y) = e^x(x\sin y + y\cos y) + C.$$

(ii) *By Thomson's method*: Putting $x = z$, $y = 0$ in (1) and (2), we get

$$\phi_1(z,0) = e^z(z+1),$$
$$\phi_2(z,0) = 0.$$

Hence

$$f(z) = \int \{\phi_1(z,0) - i\phi_2(z,0)\} dz + C = \int e^z(z+1) dz + C,$$

$$\text{i.e.,} \quad f(z) = ze^z + C.$$

8.19 Integration of Analytic Functions

We begin with definitions of certain useful terms before discussing the theorems on the integration of analytic functions.

Jordan arc: Let $x(t)$ and $y(t)$ be continuous functions of a real variable t. The set of points in the complex plane given by

$$z = x(t) + iy(t) \tag{8.100}$$

is called a *continuous arc*. If for a particular value of t, there corresponds a single point z in the complex plane, then the continuous arc is called a *Jordan arc*. An arc of a Jordan curve is said to be regular, if the derivatives $x'(t)$, $y'(t)$ are also continuous within the range of definition.

Contour: A Jordan curve consisting of a continuous chain of finite number of regular arcs is called a *contour*. If the starting point of the first arc and end point of the last arc coincide, then the contour is said to be *closed.* Anti-clockwise direction along a closed contour is taken to be positive.

Domain: The domain of definition of a Jordan curve is a set of points in Argand plane, such that any two points belonging to the set can be joined by a continuous curve. A domain in which every closed curve can be shrunk to a point without passing out of the region is called a *simply connected domain.* Otherwise, it is called *multiply-connected.*

Rectifiable arc: Let a Jordan arc be defined by equation (8.100). Let $z_0, z_1, \ldots, z_n$ be points on this arc corresponding to the values $t_0, t_1, \ldots, t_n$ of the parameter t, such that

$$t_0 < t_1 < t_2 \cdots < t_n = T. \tag{8.101}$$

We join the points $z_0, z_1, \ldots, z_n$ successively by straight lines to get a polygon. The length of this polygonal arc would be

$$S = \sum_{r=1}^{n} |z_r - z_{r-1}|. \tag{8.102}$$

If S tends to a limit l as $n \longrightarrow \infty$ and the greatest of numbers $(t_r - t_{r-1})$ (called *norms*) tends to zero, we say that the arc is *rectifiable* and that its length is l.

Every regular arc is rectifiable. Moreover, it can be shown that

$$l = \int_{t_0}^{T} (\dot{x}^2 + \dot{y}^2)^{1/2}\, dt. \tag{8.103}$$

8.20 Riemann's Definition of Integration

Let there be a function $f(z)$ of complex variable z. The function may not necessarily be analytic, but it has a definite finite value on a rectifiable arc L given by $z = x(t) + iy(t)$, for $t_0 \leq t \leq T$.

We sub-divide this arc into n smaller arcs by points $z_0, z_1, \ldots, z_n$ corresponding to the values of the parameter t as $t_0, t_1, \ldots, t_n = T$ satisfying equation (8.101). Let ξ_r be a point between z_r and z_{r-1} (*Fig 8.6*). Now, if the sum

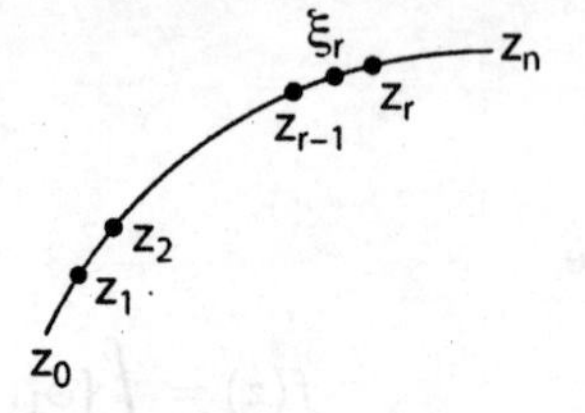

Fig 8.6

$$S = \sum_{r=1}^{n} f(\xi_r)(z_r - z_{r-1}) \tag{8.104}$$

tends to a unique limit J as $n \longrightarrow \infty$, and the greatest of the numbers $(t_r - t_{r-1})$ tends to zero, we say that $f(z)$ is integrable from z_0 to z_n of the arc L. In this case, we write

$$J = \int_L f(z)\, dz. \tag{8.105}$$

The complex integral defined by equation (8.105) depends not only on initial and final points on the arc L, but also on the actual form of L. For the existence of the integral J, it is sufficient that the function $f(z)$ is continuous on the arc L.

The following properties of the integration can easily be verified:

1. If P, Q are the starting and end points of the arc L and A is a point in between them, then

$$\int_{PQ} f(z) = \int_{PA} f(z) + \int_{AQ} f(z). \tag{8.106}$$

2. If the arc L is traversed in opposite direction defined by arc L_1, then

$$\int_{L_1} f(z)\,dz = -\int_L f(z)\,dz. \tag{8.107}$$

3. If λ be a constant, then

$$\int_L \lambda f(z)\,dz = \lambda \int_L f(z)dz. \tag{8.108}$$

As an example, let us consider the integral

$$\int_L f(z)\,dz = \int_L z\,dz,$$

the arc L being defined between $z_0 = a$ and $z_n = b$.

Since the integrand is continuous, the integral exists. Now

$$\int_L z\,dz = \lim_{n\longrightarrow\infty} \sum_{r=1}^{n} z_r(z_r - z_{r-1}).$$

Also

$$\int_L z\,dz = \lim_{n\longrightarrow\infty} \sum_{r=1}^{n} z_{r-1}(z_r - z_{r-1}).$$

Hence, adding these, we get

$$\begin{aligned}\int_L z\,dz &= \frac{1}{2}\lim_{n\longrightarrow\infty} \sum_{r=1}^{n}(z_r - z_{r-1})(z_r + z_{r-1}) = \frac{1}{2}\lim_{n\longrightarrow\infty} \sum_{r=1}^{n}(z_r^2 - z_{r-1}^2)\\ &= \frac{1}{2}\lim_{n\longrightarrow\infty} \left\{(z_1^2 - z_0^2) + (z_2^2 - z_1^2) + \cdots + (z_n^2 - z_{n-1}^2)\right\}\\ &= \frac{1}{2}\left(z_1^2 - z_0^2\right) = \frac{1}{2}\left(b^2 - a^2\right).\end{aligned}$$

If the curve L happens to be closed, then z_0 and z_n coincide, i.e., $a = b$. Hence

$$\oint f(z)\,dz = \oint z\,dz = 0.$$

Let us consider another integral

$$\int_L |\,z\,|\,dz.$$

By definition

$$\int_L |\,z\,|\,dz = \lim_{n\longrightarrow\infty} \sum_{r=1}^{n} |\,z_r - z_{r-1}\,| = \lim_{n\longrightarrow\infty} \left[|\,z_1 - z_0\,|\right] + |\,z_2 - z_1\,| + \cdots.$$

Since each term represents the length of the chord between corresponding points, we find, in the limit when $n \longrightarrow \infty$, the sum of all chords equal to length of the arc L. Hence

$$\int_L |\,z\,|\,dz = l, \tag{8.109}$$

where l = length of the curve L.

8.21 Absolute Value of a Complex Integral

Theorem 8.4. *If a function $f(z)$ is continuous on a contour C of length l, such that $|f(z)| \leq M$, then*

$$\left| \int_C f(z)\, dz \right| \leq Ml. \tag{8.110}$$

Proof. We have, by definition

$$\int_C f(z)\, dz = \lim_{n \to \infty} \sum_{r=1}^{n} f(\xi_r)(z_r - z_{r-1}).$$

Hence

$$\left| \int_C f(z)\, dz \right| = \lim_{n \to \infty} \left| \sum_{r=1}^{n} f(\xi_r)(z_r - z_{r-1}) \right|.$$

Now, modulus of sum of n complex numbers is less than or equal to sum of their moduli (*see* Example 8.2).

Hence, we can write

$$\begin{aligned} \left| \int_C f(z)\, dz \right| &\leq \lim_{n \to \infty} \sum_{r=1}^{n} |f(\xi_r)(z_r - z_{r-1})| \\ &\leq \lim_{n \to \infty} \sum_{r=1}^{n} |f(\xi_r)|\,|z_r - z_{r-1}| \quad [\because |z_1 z_2| = |z_1|\,|z_2|] \\ &\leq M \lim_{n \to \infty} \sum_{r=1}^{n} |z_r - z_{r-1}|. \end{aligned}$$

$$\therefore \quad \left| \int_C f(z)\, dz \right| \leq M \int_C |z|\, dz \leq Ml \quad \text{[using equation (8.109)]},$$

which proves the theorem. □

Green's Theorem

The theorem establishes a relation between a line integral and a surface integral.

Theorem 8.5. *If $M(x, y)$ and $N(x, y)$ are two functions of x and y having continuous first-order derivatives $\frac{\partial M}{\partial y}$ and $\frac{\partial N}{\partial x}$ at every point within and on the closed contour C, then*

$$\int_C (M\, dx + N\, dy) = \int_S \left(\frac{\partial N}{\partial x} - \frac{\partial M}{\partial y} \right) dx\, dy, \tag{8.111}$$

where S denotes the area enclosed by the curve C.

8.22 Cauchy's Theorem (Original)

Theorem 8.6. *If $f(z)$ is an analytic function of z, and $f'(z)$ exists and is continuous at each point within and on the closed contour C, then*

$$\int_C f(z)\, dz = 0. \tag{8.112}$$

Proof. (Elementary) Let $z = x + iy$, and $f(z) = u(x, y) + iv(x, y)$.
Then

$$\int_C f(z)dz = \int_C (u + iv)(dx + i\,dy) = \int_C (u\,dx - v\,dy) + i\int_C (v\,dx + u\,dy). \tag{8.113}$$

Let S be the closed region which consists of all points within and on C.

Since $f(z)$ is analytic, the partial derivatives u_x, u_y, v_x, v_y are continuous and hence, Green's theorem can be applied to both the integrals in (8.113). Thus, we get

$$\begin{aligned}\int_C f(z)\,dz &= \int_S \left(-\frac{\partial v}{\partial x} - \frac{\partial u}{\partial y}\right) dx\,dy + i\int_S \left(\frac{\partial u}{\partial x} - \frac{\partial v}{\partial y}\right) dx\,dy \\ &= \int_S \left(\frac{\partial u}{\partial y} - \frac{\partial u}{\partial y}\right) dx\,dy + i\int_S \left(\frac{\partial v}{\partial y} - \frac{\partial v}{\partial y}\right) dx\,dy\end{aligned}$$

[since, by C-R condition, $u_x = v_y$ and $v_x = -u_y$]

Thus

$$\int_C f(z)\,dz = 0,$$

which proves the theorem. □

8.23 Cauchy's Theorem (General Form)

In the original Cauchy's theorem, it was assumed that $f'(z)$ exists and is continuous at each point within and on C. But Goursat pointed out that the continuity of $f'(z)$ is not necessary, and that Cauchy's theorem is true if it is only assumed that $f'(z)$ exists at each point within and on C. Thus, in general form, the Cauchy's theorem states that:

Theorem 8.7. *If $f(z)$ is an analytic function, continuous within and on a closed rectifiable curve C, and if $f'(z)$ exists at each point within C, then*

$$\int_C f(z)\,dz = 0.$$

This theorem is also known as *Cauchy-Goursat theorem.*

We make an important deduction from this theorem.

If $f(z)$ is analytic in a simply connected domain, then the value of the integral

$$\int_{z_0}^{z_1} f(z)\,dz$$

is independent of the path joining the points z_0 and z_1. If we make two paths C_1 and C_2 joining these points, then by Cauchy's theorem (*Fig 8.7*)

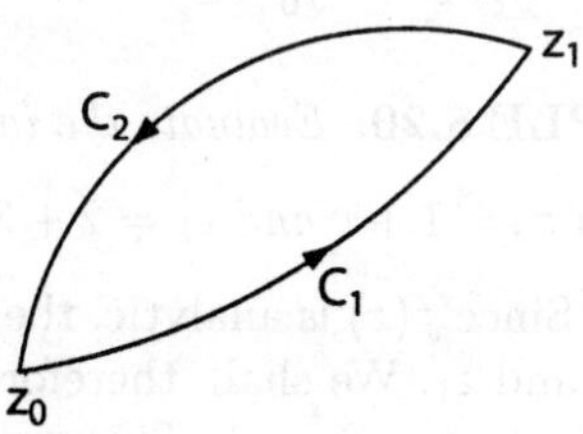

Fig 8.7

$$\int_{C_1+C_2} f(z)\,dz = 0, \quad \text{i.e.,} \quad \int_{C_1} f(z)\,dz + \int_{C_2} f(z)\,dz = 0$$

or, $\displaystyle\int_{C_1} f(z)dz = -\int_{C_2} f(z)dz = \int_{-C_2} f(z)dz,$

where, C_1 is the curve from z_0 to z_1, C_2 is the curve from z_1 to z_0 so that $-C_2$ is the curve from z_0 to z_1.

EXAMPLE 8.18. *Find the value of* $\int_0^{2+i} z^2\,dz$ *along the following contours:* (i) *the straight line from* 0 *to* $2+i$; (ii) *polygonal line from* 0 *to* 2 *and from* 2 *to* $2+i$.

Solution:

$$\int_0^{2+i} z^2 dz = \int_0^{2+i} (x^2-y^2+2xy\,i)(dx+i\,dy)$$
$$= \int_0^{2+i} \{(x^2-y^2)\,dx - 2xy\,dy\} + i\int_0^{2+i} \{2xy\,dx + (x^2-y^2)\,dy\}.$$

(i) For the straight line from 0 to $2+i$, it has parametric equation $x=2t$, $y=t$, $0\le t\le 1$. Thus, $dx = 2dt$, $dy = dt$. Hence, the integral becomes

$$\int_0^{2+i} z^2 dz = \int_0^1 \{(4t^2-t^2)(2dt) - 2(2t)t\,dt\} + i\int_0^1 \{4t^2(2dt) + (4t^2-t^2)\,dt\}$$
$$= \int_0^1 2t^2\,dt + i\int_0^1 11t^2\,dt = \frac{2}{3} + i\frac{11}{3}.$$

(ii) In this case, we have two paths—one along real axis when $z=x$ and $y=0$, and the other along imaginary axis when $z = 2+iy$ and $dx = 0$. Thus

$$\int_0^{2+i} z^2\,dz = \int_0^2 x^2\,dx + i\int_0^1 (2+iy)^2\,dy = \frac{8}{3} + i\int_0^1 (4-y^2+4iy)\,dy$$
$$= \frac{8}{3} + i\left(4 - \frac{1}{3} + 2i\right) = \frac{2}{3} + \frac{11i}{3}.$$

EXAMPLE 8.19. *Evaluate* $\int_C \frac{dz}{z^n}$, *n is positive integer* > 1, *where C is unit circle with centre at origin.*

Solution: Parametric representation of unit circle is $z = e^{i\theta}$, so that $dz = ie^{i\theta}d\theta$.

$$\therefore \quad \int_C \frac{dz}{z^n} = \int_0^{2\pi} \frac{ie^{i\theta}}{e^{ni\theta}}d\theta = \int_0^{2\pi} ie^{(1-n)i\theta}\,d\theta = \left.\frac{ie^{(1-n)i\theta}}{i(1-n)}\right|_0^{2\pi} = \frac{1}{1-n}(1-1) = 0.$$

EXAMPLE 8.20. *Evaluate the integral* $\int_{C_1} (3z+2)\,dz$, *where* C_1 *represents any curve between the points* $z_0 = 1+i$ *and* $z_1 = 2+3i$.

Solution: Since $f(z)$ is analytic, the value of the integral will be independent of the path joining points z_0 and z_1. We shall, therefore, take the curve C_1 to be a straight line joining z_0 and z_1. Its equation is $y = 2x-1$. This gives $dy = 2dx$, also x changes from 1 to 2.

Hence

$$\int_{C_1} (3z+2)\,dz = \int_{C_1} (3x+i3y+2)(dx+idy)$$
$$= \int_{C_1} (3x+2)\,dx - 3y\,dy + i\int_{C_1} 3y\,dx + (3x+2)\,dy.$$

Changing y to x, we get the integral

$$\int_1^2 (3x+2)\,dx - 3(2x-1)\cdot 2dx + i\int_1^2 3(2x-1)\,dx + (3x+2)\cdot 2dx$$

$$= \int_1^2 (-9x+8)\,dx + i\int_1^2 (12x+1)\,dx = \left(-\frac{9x^2}{2}+8x\right)\Big|_1^2 + i(6x^2+x)\Big|_1^2$$

$$= -\frac{11}{2} + 19i.$$

Theorem 8.8. *If C_1 and C_2 are two closed contours, C_2 lying completely within C_1, then*

$$\int_{C_1} f(z)dz = \int_{C_2} f(z)dz \tag{8.114}$$

provided that $f(z)$ is continuous in the closed annulus and regular in the open annulus bounded by C_1 and C_2.

Proof. The contours C_1 and C_2 are shown in the *Fig 8.8.* We take points P, Q on C_1 and R, S on C_2. Then we join P to S and Q to R by two polygonal arc having no common point and which do not cross C_1 or C_2. This forms two closed contours like L_1RQPSL_1 and L_2SPQRL_2 which we represent by γ_1 and γ_2 respectively.

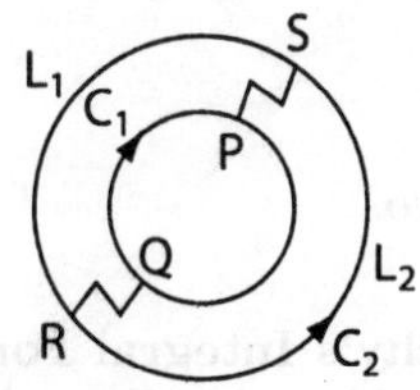

Fig 8.8

Then, by Cauchy's theorem

$$\int_{\gamma_1} f(z)\,dz + \int_{\gamma_2} f(z)\,dz = 0. \tag{8.115}$$

Now, γ_1 contains the arc PS from P to S, while γ_2 contains PS in opposite sense. Thus, two integrals along PS cancel each other. Similarly, two integrals along QR also cancel each other. Hence, equation (8.115) gives

$$\int_{C_1} f(z)\,dz - \int_{C_2} f(z)\,dz = 0,$$

where the sense of description of contours C_1 and C_2 are taken anti-clockwise.

Hence, the theorem is proved.

This theorem can be easily extended. □

Theorem 8.9. *If C is a closed curve, and $C_1, C_2, \ldots$ are other closed curves lying completely within C, and if a function $f(z)$ is analytic in the region between these curves and continuous on C, then*

$$\int_C f(z)\,dz = \int_{C_1} f(z)\,dz + \int_{C_2} f(z)\,dz + \cdots, \tag{8.116}$$

where the integrals along each curve is taken in the anti-clockwise direction.

EXAMPLE 8.21. *Evaluate the integral $\int_C \frac{dz}{z-a}$, where C is a cardioid given by the equation $r = \alpha(1+\cos\theta)$, and $z = a$ is a point inside C.*

Solution: Let us imagine a circle with centre at the point a and radius equal to ρ. We call this curve C_1. Then, the function

$$f(z) = \frac{1}{z-a}$$

is continuous in the closed annulus and regular in the open annulus bounded by C and C_1. Hence

$$\int_C \frac{dz}{z-a} = \int_{C_1} \frac{dz}{z-a}.$$

Now, on C_1, $z = a + \rho(\cos\theta + i\sin\theta) = a + \rho e^{i\theta}$, $0 \le \theta \le 2\pi$, so that $dz = \rho i e^{i\theta}\, d\theta$. Hence

$$\int_{C_1} \frac{dz}{z-a} = \int_0^{2\pi} \frac{\rho i e^{i\theta}}{\rho e^{i\theta}}\, d\theta = \int_0^{2\pi} i\, d\theta = 2\pi i.$$

It may be noted that, if the point $z = a$ happens to be outside the boundary of C, then the function $f(z) = 1/(z-a)$ becomes analytic at any point within and on C. In that case the integral

$$\int_C \frac{dz}{z-a}$$

should be zero.

8.24 Cauchy's Integral Formula

Cauchy's integral formula gives the value of an analytic function at a point within a closed contour in terms of its values on the contour.

Theorem 8.10. *If $f(z)$ is an analytic function, regular within a closed contour C, and continuous within and on C, and if $z = a$ be any point within C, then*

$$f(a) = \frac{1}{2\pi i} \int_C \frac{f(z)}{z-a}\, dz. \tag{8.117}$$

Proof. We enclose the point $z = a$ by a small circle of radius ρ and call this circle C_1.

Then, by Theorem 8.8, we have

$$\int_C \frac{f(z)\, dz}{z-a} = \int_{C_1} \frac{f(z)\, dz}{z-a}. \tag{8.118}$$

Now, on C_1, $z - a = \rho e^{i\theta}$; hence, $dz = \rho i e^{i\theta}\, d\theta$.
Thus

$$\int_{C_1} \frac{f(z)\, dz}{z-a} = \int_0^{2\pi} \frac{f(a+\rho e^{i\theta})}{\rho e^{i\theta}} \rho i e^{i\theta}\, d\theta = i\int_0^{2\pi} f(a+\rho e^{i\theta})\, d\theta. \tag{8.119}$$

Since $f(z)$ is continuous within and on C, we have

$$|f(a+\rho e^{i\theta}) - f(a)| < \eta,$$

where η is very small.

Hence, in the limit

$$\lim_{\rho \to 0} f(a + \rho e^{i\theta}) = f(a).$$

Hence, from equation (8.119), we get

$$\int_{C_1} \frac{f(z)}{z-a}\,dz = i\int_0^{2\pi} f(a)\,d\theta = 2\pi i\, f(a). \tag{8.120}$$

From equations (8.118) and (8.120), we have

$$\int_C \frac{f(z)}{z-a}dz = 2\pi i\, f(a).$$

Hence

$$f(a) = \frac{1}{2\pi i}\int_C \frac{f(z)\,dz}{z-a}.$$

Thus, the theorem is proved. □

8.25 Derivatives of Analytic Functions

Let $f(z)$ be an analytic function in D; C is a closed contour in D surrounding the point $z = a$. We take another point $z = a + h$ in the neighbourhood of $z = a$. Then by Theorem 8.10, we have

$$f(a) = \frac{1}{2\pi i}\int_C \frac{f(z)\,dz}{z-a} \tag{8.121}$$

$$\text{and}\quad f(a+h) = \frac{1}{2\pi i}\int_C \frac{f(z)\,dz}{z-a-h}. \tag{8.122}$$

Hence

$$\begin{aligned}\frac{f(a+h)-f(a)}{h} &= \frac{1}{2\pi ih}\int_C \left[\frac{1}{z-a-h} - \frac{1}{z-a}\right] f(z)\,dz\\ &= \frac{1}{2\pi i}\int_C \left[\frac{f(z)}{(z-a-h)(z-a)}\right] dz.\end{aligned}$$

Thus, in the limit

$$\lim_{h\longrightarrow 0}\frac{f(a+h)-f(a)}{h} = \lim_{h\longrightarrow 0}\frac{1}{2\pi i}\int_C \frac{f(z)\,dz}{(z-a-h)(z-a)}$$

$$\text{or,}\quad f'(a) = \frac{1}{2\pi i}\int_C \frac{f(z)}{(z-a)^2}\,dz. \tag{8.123}$$

Equation (8.123) is known as *Cauchy's Integral formula.* Thus, we have:

Theorem 8.11. *If a function $f(z)$ is analytic in D, then its derivative at any point $z = a$ inside D is given by*

$$f'(a) = \frac{1}{2\pi i}\int_C \frac{f(z)\,dz}{(z-a)^2}.$$

EXAMPLE 8.22. *Evaluate the integral* $\displaystyle\int_C \frac{e^z(z^2+1)}{(z-1)^2}\,dz$, *where C is the circle $|z| = 2$.*

Solution: Since C is given by $|z| = 2$, the point $z = 1$ lies within C. Hence, by Theorem 8.11, we have

$$\int_C \frac{e^z(z^2+1)}{(z-1)^2}\,dz = \int_C \frac{f(z)}{(z-1)^2}dz = 2\pi i f'(1). \tag{1}$$

Now, $f(z) - e^z(z^2+1)$, $\therefore\ f'(z) = e^z(z^2+2z+1)$.
Hence, $f'(1) = 4e$.
Thus, from (1), we get

$$\int_C \frac{f(z)}{(z-1)^2}\,dz = 2\pi i(4e) = 8\pi e i.$$

EXAMPLE 8.23. *Evaluate the integral*

$$\int_C \frac{4z^2+16z+14}{z^3+6z^2+11z+6}\,dz,$$

where C is the circle $|z| = 5/2$.

Solution: The integrand is

$$\frac{4z^2+16z+14}{z^3+6z^2+11z+6} = \frac{4z^2+16z+14}{(z+1)(z+2)(z+3)} = \frac{1}{z+1} + \frac{2}{z+2} + \frac{3}{z+3}.$$

Hence, the integral is

$$J = \int_C \left(\frac{1}{z+1} + \frac{2}{z+2} + \frac{3}{z+3}\right) dz. \tag{1}$$

Now, the points $z = -1$ and $z = -2$ lie within the circle $|z| = 5/2$; but the point $z = -3$ lies outside C. Hence, by Theorem 8.10

$$\int_C \frac{dz}{z+1} = 2\pi i f(-1) = 2\pi i \quad [\because\ f(z) = 1],$$

$$\int_C \frac{dz}{z+2} = 2\pi i f(-2) = 4\pi i$$

and $$\int_C \frac{dz}{z+3} = 0$$

Hence, from (1), we get, $J = 2\pi i + 4\pi i = 6\pi i$.

8.26 Higher Order Derivatives of Analytic Functions

The integral formula (8.123) leads us to conclude that $f'(a)$ is analytic in D, i.e., the derivative of an analytic function is also an analytic function. Let us prove that $f''(z)$ exists at each point within D.

Now

$$\frac{f'(a+h) - f'(a)}{h} = \frac{1}{2\pi i h}\int_C \left[\frac{1}{(z-a-h)^2} - \frac{1}{(z-a)^2}\right] f(z)\,dz$$

$$= \frac{1}{2\pi i h}\int_C \left[\frac{2h}{(z-a-h)^2(z-a)} - \frac{h^2}{(z-a-h)^2(z-a)^2}\right] f(z)\,dz.$$

Taking limit $h \longrightarrow 0$,

$$f''(a) = \frac{2}{2\pi i} \int_C \frac{f(z)\,dz}{(z-a)^3}. \tag{8.125}$$

Proceeding in a similar manner, we can show that $f''(z)$ is regular in D and its derivative $f'''(z)$ is given by

$$f'''(a) = \frac{3!}{2\pi i} \int_C \frac{f(z)}{(z-a)^4}. \tag{8.126}$$

In general, we will get

$$f^{(n)}(a) = \frac{n!}{2\pi i} \int_C \frac{f(z)}{(z-a)^{n+1}}. \tag{8.127}$$

Thus, we have the theorem:

Theorem 8.12. *If $f(z)$ is analytic function regular within a closed contour C, and continuous within and on C, it possesses derivatives of all orders which are regular within C. The nth derivative of $f(z)$ is given by*

$$f^{(n)}(a) = \frac{n!}{2\pi i} \int_C \frac{f(z)}{(z-a)^{n+1}}. \tag{8.128}$$

EXAMPLE 8.24. *Evaluate the integral*

$$\int_C \frac{\sin z}{(z-\pi/4)^3}\,dz,$$

where C is the circle given by $|z| = 2$.

Solution: We use the formula (8.127). Here, $n = 2$, $f(z) = \sin z$ and $a = \pi/4$. The point $z = \pi/4$ lies within the circle $|z| = 2$. Hence, we have

$$f''\left(\frac{\pi}{4}\right) = \frac{2!}{2\pi i} \int_C \frac{\sin z}{(z-\pi/4)^3}.$$

$$\therefore \quad \int_C \frac{\sin z}{(z-\frac{\pi}{4})^3}\,dz = \pi i f''\left(\frac{\pi}{4}\right) = \pi i\left(-\sin\frac{\pi}{4}\right) = -\frac{\pi i}{\sqrt{2}}.$$

8.27 Cauchy's Inequality

Let $f(z)$ be an analytic function regular within a circle of radius R. If $|f(z)| \le M$ for every z on C, then

$$|f^n(a)| \le n!\frac{M}{R^n}. \tag{8.129}$$

Equation (8.129) is known as *Cauchy's inequality.* Let us now prove this inequality. We have, from equation (8.128)

$$f^n(a) = \frac{n!}{2\pi i} \int_C \frac{f(z)}{(z-a)^{n+1}}\,dz.$$

$$\therefore \quad |f^n(a)| = \frac{n!}{2\pi i} \int_C \left| \frac{f(z)}{(z-a)^{n+1}} \right| dz \le \frac{n!}{|2\pi i|} \int_C \frac{|f(z)|\,|dz|}{|(z-a)^{n+1}|}.$$

Now, $z - a = Re^{i\theta}$, so that $dz = Rie^{i\theta}\, d\theta$.

$$\therefore \quad |f^n(a)| \le \frac{n!}{2\pi}\frac{M}{R^{n+1}}\int_C |Ri\, e^{i\theta}\, d\theta| \quad [\because |f(z)| \le M]$$

$$\le \frac{n!}{2\pi}\frac{M}{R^{n+1}}\int_0^{2\pi} R\, d\theta \le \frac{n!}{2\pi}\frac{M}{R^{n+1}} \cdot 2\pi R.$$

$$\therefore \quad |f^n(a)| \le n!\frac{M}{R^n}.$$

Hence proved.

8.28 Liouville's Theorem

A function which is analytic in every finite region of the z-plane is called an *integral function.*

Now, let there be an integral function $f(z)$, such that, for all values of z, $|f(z)| \le M$, where M is constant. Then, if a be any point in the z-plane, $f(z)$ is regular when $|z - a| \le R$, however large R may be. Moreover, $f(z)$ satisfies there the inequality $|f(z)| \le M$. Hence, from Cauchy's inequality, we have

$$|f'(a)| \le M/R. \tag{8.130}$$

By making $R \longrightarrow \infty$, we get, $f'(a) = 0$. Since a is arbitrary, we can say that the derivative of $f(z)$ vanishes everywhere in the z-plane. Hence, $f(z)$ is constant. Thus, we have Liouville's theorem:

Theorem 8.13. *If $f(z)$ is an integral function satisfying $|f(z)| \le M$ for all values of z, M being a constant, then $f(z)$ is a constant.*

8.29 Morera's Theorem

This theorem is the converse of the Cauchy's integral theorem. It states that:

Theorem 8.14. *If a function $f(z)$ is continuos in a simply-connected region D, and if for every closed contour C in D,*

$$\int_C f(z)\, dz = 0,$$

then $f(z)$ is analytic in D.

Proof. Let $z = a$ be a fixed point and z a variable point in the domain D.

Since, along a closed contour C

$$\int_C f(z)\, dz = 0,$$

then the integral

$$\int_a^z f(z)\, dz$$

will depend only on end points (*see* Section 8.24).

Let us denote

$$F(z) = \int_a^z f(t)\, dt, \tag{8.131}$$

where t is another complex variable of integration.

For another variable point $z + h$ adjacent to z, we can write

$$F(z+h) = \int_a^{z+h} f(t)\,dt. \tag{8.132}$$

Hence

$$F(z+h) - F(z) = \int_a^{z+h} f(t)\,dt - \int_a^{z} f(t)\,dt = \int_a^{z+h} f(t)\,dt + \int_z^{a} f(t)\,dt$$
$$= \int_z^{z+h} f(t)\,dt. \tag{8.133}$$

Therefore

$$\frac{F(z+h) - F(z)}{h} - f(z) = \frac{1}{h}\int_z^{z+h} f(t)dt - \frac{f(z)}{h}h = \frac{1}{h}\int_z^{z+h} \{f(t) - f(z)\}\,dt. \tag{8.134}$$

Since $f(t)$ is continuous in the domain D, the right-hand side of equation (8.134) vanishes as $h \longrightarrow 0$. Hence, we get

$$\lim_{h\longrightarrow 0} \frac{1}{h}\{F(z+h) - F(z)\} = f(z),$$

which shows that $F'(z)$ exists for all values of z and is equal to $f(z)$. Hence, $F(z)$ is analytic in D. But we know that derivative of an analytic function is also analytic. Hence, $f(z)$ is also analytic. This proves the theorem. □

8.30 Taylor's Theorem

A function analytic at the points inside a circle C can be represented by an infinite series in powers of $(z - a)$, where a is a point within C. We shall see how.

Let $f(z)$ be an analytic function within and on C given by $|z - a| = R$ with centre at a; and z is a point inside C, such that $|z - a| = r < R$ (*Fig 8.9*). Then, by Cauchy's integral formula, we have

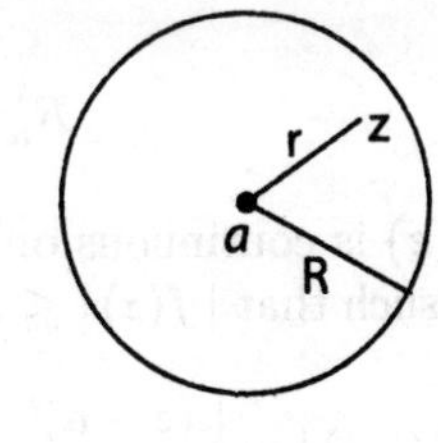

Fig 8.9

$$f(z) = \frac{1}{2\pi i}\int_C \frac{f(t)\,dt}{t - z}. \tag{8.135}$$

Now

$$t - z = (t-a) - (z-a) = (t-a)\left\{1 - \frac{z-a}{t-a}\right\}.$$

$$\therefore \quad \frac{1}{t-z} = \frac{1}{t-a}\left\{1 - \frac{z-a}{t-a}\right\}^{-1}. \tag{8.136}$$

We have the algebraic identity

$$\frac{1}{1-x} = 1 + x + x^2 + \cdots + x^{n-1} + \frac{x^n}{1-x}. \tag{8.137}$$

Putting $x = \frac{z-a}{t-a}$ in equation (8.136), we get

$$\frac{1}{t-z} = \frac{1}{t-a}\left[1 + \frac{z-a}{t-a} + \left(\frac{z-a}{t-a}\right)^2 + \cdots + \left(\frac{z-a}{t-a}\right)^{n-1} + R_n(t,z)\right], \tag{8.138}$$

where

$$R_n(t,z) = \frac{1}{1 - \dfrac{z-a}{t-a}}\left(\frac{z-a}{t-a}\right)^n = \frac{t-a}{t-z}\cdot\left(\frac{z-a}{t-a}\right)^n. \tag{8.139}$$

Using the expansion (8.138) in Cauchy's integral (8.135), we have

$$f(z) = \frac{1}{2\pi i}\int_C \frac{f(t)}{t-a}\left[1 + \frac{z-a}{t-a} + \left(\frac{z-a}{t-a}\right)^2 + \cdots + \left(\frac{z-a}{t-a}\right)^{n-1} + R_n(t,z)\right] dt. \tag{8.140}$$

We take pth term in equation (8.140). This is equal to

$$\frac{1}{2\pi i}\int_C \frac{(z-a)^{p-1}}{(t-a)^p} f(t)\,dt = \frac{(z-a)^{p-1}}{2\pi i}\int_C \frac{f(t)\,dt}{(t-a)^p}$$
$$= \frac{(z-a)^{p-1}}{(p-1)!} f^{(p-1)}(a) \quad \text{[by formula (8.128)]}.$$

Hence, from equation (8.140), we get

$$f(z) = f(a) + \frac{z-a}{1!}f'(a) + \frac{(z-a)^2}{2!}f''(a) + \cdots + \frac{(z-a)^{n-1}}{(n-1)!} + R_n^*(t,z), \tag{8.141}$$

where

$$R_n^*(t,z) = \frac{(z-a)^n}{2\pi i}\int_C \frac{f(t)\,dt}{(t-z)(t-a)^n}. \tag{8.142}$$

Since $f(z)$ is continuous on the circle C, it is bounded there and so there exists a positive number M such that $|f(z)| \le M$ on C. Since $|t-z| \ge R-r$, then, by Theorem (8.4), we have

$$|R_n^*(t,z)| = \left|\frac{(z-a)^n}{2\pi i}\int_C \frac{f(t)\,dt}{(t-z)(t-a)^n}\right| \le \frac{r^n}{2\pi}\frac{M}{R-r}\cdot\frac{2\pi R}{R^n} = \frac{MR}{R-r}\left(\frac{r}{R}\right)^n.$$

Since $r/R < 1$, $(r/R)^n \longrightarrow 0$ as $n \longrightarrow \infty$.
Thus, the series expansion (8.141) becomes

$$f(z) = f(a) + \frac{(z-a)}{1!}f'(a) + \frac{(z-a)^2}{2!}f''(a) + \cdots + \frac{(z-a)^n}{n!}f^{(n)}(a) + \cdots. \tag{8.143}$$

The result is known as *Taylor's theorem.* Thus, we have:

Theorem 8.15. *If $f(z)$ is an analytic function, regular in the neighbourhood of $|z-a| < R$ of the point $z = a$, then $f(z)$ can be expressed in that neighbourhood as a convergent power series in the form*

$$f(z) = f(a) + \sum_{n=1}^{\infty} f^{(n)}(a)\frac{(z-a)^n}{n!}. \tag{8.144}$$

If in a particular case, the centre of the circle is taken at the origin, i.e., $a = 0$, the series in equation (8.144) becomes

$$f(z) = f(0) + \sum_{n=1}^{\infty} f^{(n)}(0)\frac{z^n}{n!}. \tag{8.145}$$

This is known as *Maclaurin's series.*

EXAMPLE 8.25. *Find a series expansion for* (i) $f(z) = \dfrac{1}{(1+z)^m}$, (ii) $f(z) = \ln(1+z)$ *about the origin.*

Solution: (i) We have, $f(z) = (1+z)^{-m}$.

$$\begin{aligned} \therefore \quad f'(z) &= -m(1+z)^{-m-1} \\ f''(z) &= -m(-m-1)(1+z)^{-m-2} \\ f'''(z) &= -m(-m-1)(-m-2)(1+z)^{-m-3}, \ldots, \text{ etc.} \end{aligned}$$

Thus, for $z = 0$, we get

$$\begin{aligned} f'(0) &= -m \\ f''(0) &= m(m+1) \\ f'''(0) &= -m(m+1)(m+2), \text{ and so on.} \end{aligned}$$

Hence, the Taylor's series becomes

$$\frac{1}{(1+z)^m} = 1 - mz + m(m+1)\frac{z^2}{2!} - m(m+1)(m+2)\frac{z^3}{3!} + \cdots$$

Since there is a singular point at $z = -1$, the series converges for $|z| < 1$ and diverges for $|z| \geq 1$. For $m = 1$, we get

$$\frac{1}{1+z} = 1 - z + z^2 - z^3 + \cdots \tag{1}$$

(ii) The series expansion in (1) becomes useful to get Taylor series representation of $\ln(1+z)$. We have

$$\ln(1+z) = \int \frac{dz}{1+z}, \quad \text{i.e.,} \quad \ln(1+z) = z - \frac{z^2}{2} + \frac{z^3}{3} - \frac{z^4}{4} + \cdots + A,$$

which is obtained by term-by-term integration of the series. The constant of integration $A = 0$, since for $z = 0$, $\ln(1+z) = 0$. Hence

$$\ln(1+z) = z - \frac{z^2}{2} + \frac{z^3}{3} - \frac{z^4}{4} + \cdots$$

The point $z = -1$ is a singularity for the function and so the series expansion is convergent for $|z| < 1$.

EXAMPLE 8.26. *Expand in series the function*

$$f(z) = \frac{1}{(z-1)(z-3)}$$

in the regions (i) $0 < |z| < 1$, (ii) $1 < |z| < 3$, (iii) $|z| > 3$.

Solution:

$$f(z) = \frac{1}{(z-1)(z-3)} = \frac{1}{2}\left(\frac{1}{z-3} - \frac{1}{z-1}\right). \tag{1}$$

The function has singularities at $z = 3$ and $z = 1$.

(i) For $0 < |z| < 1$, we write

$$f(z) = -\frac{1}{6}\left(1 - \frac{z}{3}\right)^{-1} + \frac{1}{2}(1-z)^{-1}. \tag{2}$$

Now, we have the series expansion

$$(1-z)^{-1} = 1 + z + z^2 + \cdots \quad (\text{for } 0 < |z| < 1). \tag{3}$$

We use this well-known series for (2), and get

$$f(z) = -\frac{1}{6}\sum_{n=0}^{\infty}\left(\frac{z}{3}\right)^n + \frac{1}{2}\sum_{n=0}^{\infty} z^n = \sum_{n=0}^{\infty}\frac{1}{2}\left(1 - \frac{1}{3^{n+1}}\right)z^n.$$

(ii) For $1 < |z| < 3$, we write equation (1) in the form

$$f(z) = -\frac{1}{6}\left(1 - \frac{z}{3}\right)^{-1} - \frac{1}{2z}\left(1 - \frac{1}{z}\right)^{-1}.$$

Using the series (3), we get

$$f(z) = -\frac{1}{6}\sum_{n=0}^{\infty}\left(\frac{z}{3}\right)^n - \frac{1}{2z}\sum_{n=0}^{\infty}\left(\frac{1}{z}\right)^n.$$

(iii) For $|z| > 3$, we write equation (1) in the form

$$f(z) = \frac{1}{2z}\left(1 - \frac{3}{z}\right)^{-1} - \frac{1}{2z}\left(1 - \frac{1}{z}\right)^{-1}.$$

$\therefore$ Series expansion for $f(z)$ is

$$f(z) = \frac{1}{2z}\sum_{n=0}^{\infty}\left(\frac{3}{z}\right)^n - \frac{1}{2z}\sum_{n=0}^{\infty}\left(\frac{1}{z}\right)^n = \frac{1}{2}\sum_{n=0}^{\infty}(3^n - 1)\frac{1}{z^{n+1}}.$$

EXAMPLE 8.27. *Find a Taylor's series expansion of the function*

$$f(z) = \frac{2z^3 + 1}{z^2 + z}$$

in the neighbourhood of the point $z = i$.

Solution:

$$f(z) = \frac{2z^3 + 1}{z^2 + z} = 2(z-1) + \frac{1}{z} + \frac{1}{z+1} = \phi_1(z) + \phi_2(z) + \phi_3(z),$$

where, $\phi_1(z) = 2(z-1)$, $\phi_2(z) = \dfrac{1}{z}$ and $\phi_3(z) = \dfrac{1}{z+1}$.

(i) $\phi_1(z) = 2(z-1), \therefore\ \phi_1'(z) = 2,\ \phi_1'' = \phi_1''' = \cdots = 0.$

$$\therefore \quad \phi_1(z) = \sum_{n=0}^{\infty} a_n (z-i)^n,$$

where

$$a_0 = \frac{\phi_1(i)}{1!} = 2(i-1), \quad a_1 = \frac{\phi_1'(i)}{1!} = 2, \quad a_3 = \cdots = 0.$$

$$\therefore \quad \phi_1(z) = 2(z-1) = 2(i-1) + 2(z-i). \tag{1}$$

(ii) $\phi_2(z) = \dfrac{1}{z} = \displaystyle\sum_{n=0}^{\infty} a_n (z-i)^n$, where $a_n = \dfrac{1}{n!}\phi_2^{(n)}(i)$.

Now

$$\phi_2'(z) = -\frac{1}{z^2}, \quad \phi_2''(z) = (-1)^2 2! \frac{1}{z^3}, \cdots$$

so that

$$\phi_2^{(n)}(z) = (-1)^n n! \frac{1}{z^{n+1}}.$$

$$\therefore \quad \phi_2(z) = \frac{1}{z} = \sum_{n=0}^{\infty} (-1)^n \frac{(z-i)^n}{(i)^{n+1}}. \tag{2}$$

(iii) $\phi_3(z) = \dfrac{1}{z+1}, \quad \therefore\ \phi_3^{(n)}(z) = \dfrac{(-1)^n n!}{(z+1)^{n+1}}.$

$$\therefore \quad \text{Coefficients } a_n = \frac{\phi_3^{(n)}(i)}{n!} = \frac{(-1)^n}{(1+i)^{n+1}}. \tag{3}$$

$$\therefore \quad \phi_3(z) = \sum_{n=0}^{\infty} (-1)^n \cdot \frac{(z-i)^n}{(1+i)^{n+1}}.$$

Thus, adding (1), (2) and (3), we get

$$f(z) = 2(i-1) + 2(z-i) + \sum_{n=0}^{\infty} (-1)^n (z-i)^n \left\{ \frac{1}{(1+i)^{n+1}} + \frac{1}{(i)^{n+1}} \right\}.$$

8.31 Laurent's Series

Let a function $f(z)$ has a singular point at $z = a$. We wish to expand the function $f(z)$ in a series about the point $z = a$. For this purpose, we draw two circles C_1 and C_2 with centres at $z = a$ and radii $R_1, R_2 (R_2 > R_1)$. Since $z = a$ is a singular point, $f(z)$ cannot be expanded about this point in Taylor's series. We, therefore, make a cutting between the circles, so that the region becomes simply-connected bounded by the curve D (*Fig 8.10*).

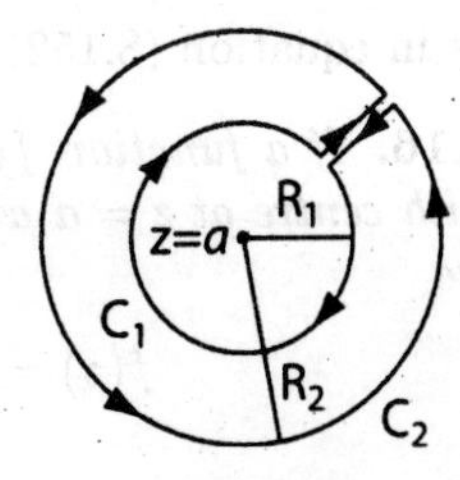

Fig 8.10

Then, the Cauchy's integral formula becomes

$$f(z) = \frac{1}{2\pi i}\int_D \frac{f(t)}{t-z}\,dt = \frac{1}{2\pi i}\int_{C_2} \frac{f(t)}{t-z}\,dt - \int_{C_1} \frac{f(t)}{t-z}\,dt. \tag{8.146}$$

The sign of the second integral has been taken negative in order to keep its sense of direction positive.

Since the singular point does not lie inside or on C_2, the 1st integral on the right-hand side in equation (8.146) holds good for all t on it. But, the second integral does not hold good for all t on C_1. In order to get the corresponding result on C_1, we interchange t and z in the second integral. Hence

$$f(z) = \frac{1}{2\pi i}\int_{C_2} \frac{f(t)}{t-z}\,dt + \int_{C_1} \frac{f(t)}{z-t}\,dt. \tag{8.147}$$

Now, for the 1st integral, we follow exactly the arguments of Section 8.30, and find

$$\frac{1}{2\pi i}\int_{C_2} \frac{f(t)\,dt}{t-z} = a_0 + a_1(z-a) + a_2(z-a)^2 + \cdots, \tag{8.148}$$

where

$$a_n = \frac{1}{2\pi i}\int \frac{f(t)\,dt}{(t-a)^{n+1}}. \tag{8.149}$$

For the second integral, we write

$$\frac{1}{z-t} = \frac{1}{z-a}\left(1 - \frac{t-a}{z-a}\right)^{-1}$$

and, proceeding to analogous arguments as before, we get

$$\frac{1}{2\pi i}\int_{C_1} \frac{f(t)}{z-t}\,dt = b_0 + b_1(z-a)^{-1} + b_2(z-a)^{-2} + \cdots, \tag{8.150}$$

where

$$b_n = \frac{1}{2\pi i}\int_{C_1} f(t)(t-a)^{n-1}\,dt. \tag{8.151}$$

Thus, using equations (8.144) and (8.146), we write

$$f(z) = \sum_{n=0}^{\infty} a_n(z-a)^n + \sum_{n=0}^{\infty} b_n(z-a)^{-n}. \tag{8.152}$$

The series in equation (8.152) is known as *Laurent's series.* Thus, we have the theorem:

Theorem 8.16. *If a function $f(z)$ is analytic in the annulus between two concentric circles C_1 and C_2 with centre at $z = a$ and radii R_1 and R_2, then $f(z)$ can be expanded as the sum of two series like*

$$f(z) = \sum_{n=0}^{\infty} a_n(z-a)^n + \sum_{n=0}^{\infty} b_n(z-a)^{-n}, \tag{8.153}$$

where coefficients a_n, b_n are given by

$$a_n = \frac{1}{2\pi i}\int_{C_2} \frac{f(t)\,dt}{(t-a)^{n+1}}; \quad b_n = \frac{1}{2\pi i}\int_{C_1} \frac{f(t)\,dt}{(t-a)^{-n+1}}. \tag{8.154}$$

EXAMPLE 8.28. *Prove that*

$$\cosh\left(z+\frac{1}{z}\right)=a_0+\sum_{n=1}^{\infty}a_n\left(z^n+\frac{1}{z^n}\right),$$

where

$$a_n=\frac{1}{2\pi}\int_0^{2\pi}\cos n\theta\cosh(2\cos\theta)\,d\theta.$$

Solution: The function

$$f(z)=\cosh\left(z+\frac{1}{z}\right)$$

is analytic everywhere except at origin. Hence, $f(z)$ can be expanded in Laurent's series around origin.

$$\therefore\quad \cosh\left(z+\frac{1}{z}\right)=a_0+\sum_{n=1}^{\infty}a_nz^n+\sum_{n=1}^{\infty}b_nz^{-n}, \tag{1}$$

where, by equations (8.149) and (8.151)

$$a_n=\frac{1}{2\pi i}\int_C\cosh\left(z+\frac{1}{z}\right)\frac{dz}{z^{n+1}} \tag{2}$$

$$\text{and}\quad b_n=\frac{1}{2\pi i}\int_C\cosh\left(z+\frac{1}{z}\right)z^{n-1}\,dz. \tag{3}$$

By taking C as a circle of unit radius, we get

$$z=e^{i\theta},\quad \frac{1}{z}=e^{-i\theta},\quad z+\frac{1}{z}=2\cos\theta,\quad dz=ie^{i\theta}\,d\theta.$$

$$\begin{aligned}\therefore\quad a_n&=\frac{1}{2\pi i}\int_0^{2\pi}\cosh(2\cos\theta)\cdot\frac{ie^{i\theta}\,d\theta}{e^{i(n+1)\,\theta}}=\frac{1}{2\pi}\int_0^{2\pi}\cosh(2\cos\theta)\,e^{-in\theta}\,d\theta\\&=\frac{1}{2\pi}\int_0^{2\pi}\cosh(2\cos\theta)(\cos n\theta-i\sin n\theta)\,d\theta=\frac{1}{2\pi}\int_0^{2\pi}\cosh(2\cos\theta)\cos n\theta\,d\theta.\end{aligned} \tag{4}$$

Here the integral

$$\int_0^{2\pi}\cosh(2\cos\theta)\sin n\theta\,d\theta=0,$$

because the integrand $\psi(\theta)$ (say) has the property, $\psi(2\pi-\theta)=-\psi(\theta)$ and so

$$\int_0^{2\pi}\psi(\theta)\,d\theta=0.$$

Similarly, we will get

$$b_n=a_n. \tag{5}$$

Hence, the result is proved.

EXAMPLE 8.29. *Show that, if $c > 0$, then*

$$f(z) = e^{z+c^3/2z^2} = \sum_{-\infty}^{\infty} a_n z^n,$$

where

$$a_n = \frac{1}{2\pi c^n} \int_0^{2\pi} e^{f(\theta)} F(\theta)\, d\theta,$$

where

$$f(\theta) = c\left(\cos\theta + \tfrac{1}{2}\cos 2\theta\right)$$
$$\text{and} \quad F(\theta) = \cos\{c\sin\theta(1-\cos\theta) - n\theta\}.$$

Solution: The function $f(z)$ is regular except at $z = 0$. Hence, $f(z)$ can be expanded in Laurent's series. Thus

$$f(z) = \sum_{n=0}^{\infty} a_n z^n + \sum_{n=1}^{\infty} b_n z^{-n} = \sum_{-\infty}^{\infty} a_n z^n, \tag{1}$$

where

$$a_n = \frac{1}{2\pi i} \int_C \frac{f(z)}{z^{n+1}}\, dz. \tag{2}$$

We take a circle with centre at $z = 0$ and radius equal to c. Then

$$z = ce^{i\theta}, \quad dz = cie^{i\theta}\, d\theta. \tag{3}$$

Hence

$$z + \frac{c^3}{2z^2} = ce^{i\theta} + \frac{c^3}{2c^2} e^{-2i\theta} = \left(c\cos\theta + \frac{c}{2}\cos 2\theta\right) + i\left(c\sin\theta - \frac{c}{2}\sin 2\theta\right)$$
$$= \alpha + i\beta \text{ (say)}, \tag{4}$$

where

$$\left.\begin{aligned} \alpha &= c\cos\theta + \tfrac{c}{2}\cos 2\theta \\ \beta &= c\sin\theta - \tfrac{c}{2}\sin 2\theta \end{aligned}\right\}. \tag{5}$$

Hence, from (2), we get

$$a_n = \frac{1}{2\pi i} \int_0^{2\pi} \frac{e^{\alpha+i\beta} cie^{i\theta}}{c^{n+1} e^{i(n+1)\theta}}\, d\theta = \frac{1}{2\pi c^n} \int_0^{2\pi} e^{\alpha} \cdot e^{i\beta} \cdot e^{-in\theta}\, d\theta$$
$$= \frac{1}{2\pi c^n} \int_0^{2\pi} e^{\alpha} \cdot e^{i(\beta - n\theta)}\, d\theta = \frac{1}{2\pi c^n} \int_0^{2\pi} e^{\alpha} \{\cos(\beta - n\theta) + i\sin(\beta - n\theta)\}\, d\theta$$
$$= \frac{1}{2\pi c^n} \int_0^{2\pi} e^{\alpha} \cos(\beta - n\theta)\, d\theta \quad \text{[since the second integral vanishes as before]}$$

Putting the values of α, β, we get the required result.

EXAMPLE 8.30. *By using the integral representation of $f^{(n)}(c)$, prove that*

$$\left(\frac{x^n}{n!}\right)^2 = \frac{1}{2\pi i} \int_C \frac{x^n e^{xz}}{n!\, z^{n+1}}\, dz,$$

where C is any closed curve surrounding origin.

Hence, prove that

$$\sum_{n=0}^{\infty}\left(\frac{x^n}{n!}\right)^2 = \frac{1}{2\pi}\int_0^{2\pi} e^{2x\cos\theta}\,d\theta.$$

Solution: From equation (8.127), we get by putting $a = 0$,

$$f^{(n)}(0) = \frac{n!}{2\pi i}\int_C \frac{f(z)}{z^{n+1}}\,dz. \tag{1}$$

We take, $f(z) = e^{xz}$, so that $f^{(n)}(z) = x^n e^{xz}$, $f^{(n)}(0) = x^n$.
Hence, from (1), we get

$$\frac{x^n}{n!} = \frac{.1}{2\pi i}\int_C \frac{e^{xz}\,dz}{z^{n+1}}. \tag{2}$$

Multiplying both sides by $x^n/n!$, we get the 1st part of the result

$$\left(\frac{x^n}{n!}\right)^2 = \frac{1}{2\pi i}\int_C \frac{x^n e^{xz}\,dz}{n!\,z^{n+1}}. \tag{3}$$

Now, taking sum on both sides of (3), we get

$$\begin{aligned}\sum_{n=0}^{\infty}\left(\frac{x^n}{n!}\right)^2 &= \frac{1}{2\pi i}\int_C \sum_{n=0}^{\infty}\frac{x^n e^{xz}}{n!\,z^{n+1}}\,dz = \frac{1}{2\pi i}\int_C dz\frac{e^{xz}}{z}\sum_{n=0}^{\infty}\frac{(x/z)^n}{n!}\\ &= \frac{1}{2\pi i}\int_C \frac{e^{xz}}{z}\cdot e^{x/z}dz = \frac{1}{2\pi i}\int_C \frac{e^{xz+\frac{x}{z}}}{z}\,dz.\end{aligned} \tag{4}$$

We choose C to be a unit circle with centre at origin. Then

$$z = e^{i\theta},\quad z + \frac{1}{z} = 2\cos\theta \quad\text{and}\quad dz = ie^{i\theta}\,d\theta.$$

Hence, the integral in (4) becomes

$$\sum_{n=0}^{\infty}\left(\frac{x^n}{n!}\right)^2 = \frac{1}{2\pi i}\int_0^{2\pi} e^{2x\cos\theta}\cdot\frac{ie^{i\theta}}{e^{i\theta}}\,d\theta = \frac{1}{2\pi}\int_0^{2\pi} e^{2x\cos\theta}\,d\theta.$$

Hence proved.

8.32 Conformal Representation

Let us consider a complex function

$$w = f(z) = u(x,y) + iv(x,y), \tag{8.158}$$

where

$$z = x + iy. \tag{8.159}$$

This means that, for a point in z-plane defined by (8.159), there will correspond at least one point in w-plane defined by (8.158). Thus, the set of equations maps a domain or curve in z-plane into corresponding domain or curve in w-plane. If, for each point z in the z-plane, there corresponds one only one point in the w-plane, and conversely, we say that the correspondence between the two planes is one-to-one.

Let us now assume that in the z-plane, there are two curves C_1 and C_2 intersecting at z_0 at an angle θ between them. Further, let these curves be mapped into two curves Γ_1 and Γ_2 in w-plane intersecting at w_0 at an angle ϕ. If $\theta = \phi$, i.e., angles between two sets of intersecting curves are equal, we say that the mapping (or transformation) is *isogonal.* If, $\theta = \phi$, and also the sense of rotation is preserved (i.e., if we consider rotation from C_1 curve to C_2 in z-plane, it should correspond to rotation from Γ_1 curve to Γ_2 in w-plane), then the mapping is said to be *conformal.* In the *Fig 8.11,* the curve Γ_1' in dotted line and Γ_2 are not conformal, although angle between them is same.

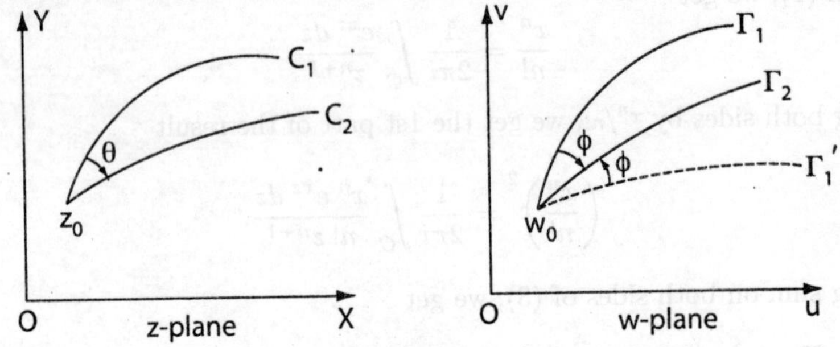

Fig 8.11

It is, therefore, obvious that every function $w = f(z)$ cannot represent a conformal mapping. Let us now find a necessary condition that the function should satisfy in order to represent a conformal mapping.

Let ds and $d\sigma$ denote small line elements in z-plane and w-plane respectively. Then

$$\left.\begin{aligned} ds^2 &= dx^2 + dy^2 \\ \text{and} \quad d\sigma^2 &= du^2 + dv^2 \end{aligned}\right\}. \tag{8.160}$$

But

$$\left.\begin{aligned} du &= u_x\, dx + u_y\, dy \\ \text{and} \quad dv &= v_x\, dx + v_y\, dy \end{aligned}\right\}.$$

Making these substitutions in equation (8.160), we have

$$d\sigma^2 = \left(u_x^2 + v_x^2\right) dx^2 + \left(u_y^2 + v_y^2\right) dy^2 + 2\left(u_x u_y + v_x v_y\right) dx\, dy.$$

For conformal mapping, $d\sigma^2 = h^2\left(dx^2 + dy^2\right)$, where $h(\neq 0)$ depends only on x and y. Hence, we get the following differential equations

$$\left.\begin{aligned} u_x^2 + v_x^2 &= h^2 \\ u_y^2 + v_y^2 &= h^2 \\ u_x u_y + v_x v_y &= 0 \end{aligned}\right\}. \tag{8.161}$$

The first two equations give $u_x^2 + v_x^2 = u_y^2 + v_y^2$, which is satisfied, if $u_x = v_y$ and $v_x = -u_y$. But these are Cauchy-Riemann's condition for a function to be analytic. Hence, we have the theorem:

Theorem 8.17. *If the function $w = f(z)$ represents a conformal mapping, then the function $f(z)$ is necessarily an analytic function.*

The third of equation (8.154) is satisfied if $u_x = -v_y$ and $u_y = -v_x$, which are nothing but Cauchy-Riemann's conditions for a function

$$\bar{w} = f(z) = u(x, y) - iv(x, y). \tag{8.162}$$

The function in equation (8.162) is equivalent to that in equation (8.158) followed by a reflection in the real axis of the w-plane. Therefore, this transformation conserves only magnitude of the angle of intersection of curves but reverses its sense. That is why this transformation is isogonal and not conformal.

We now intend to see the sufficient condition for $w = f(z)$ to represent a conformal mapping. Let us suppose that $f(z)$ is an analytic function. Let C be a regular arc in z-plane given by parametric equation

$$z = x(t) + iy(t). \tag{8.163}$$

Then, its map in w-plane is a regular arc Γ given by

$$w = f\{x(t) + iy(t)\}. \tag{8.164}$$

We take a point z_0 on C and draw a tangent to C which makes an angle θ with x-axis. Then

$$\theta = \tan^{-1}\left(\frac{\dot{y}}{\dot{x}}\right) = \text{amp}\ \{\dot{x}(t) + i\dot{y}(t)\}. \tag{8.165}$$

We now take the point w_0 on Γ corresponding to z_0 on C and draw a tangent to Γ which makes an angle ϕ with x-axis in w-plane. Then

$$\begin{aligned}\phi &= \text{amp}\left[\frac{d}{dt} f\{x(t) + iy(t)\}\right] \\ &= \text{amp}\,[f'(z)\{\dot{x}(t) + i\dot{y}(t)\}] \\ &= \text{amp}\, f'(z) + \text{amp}\,\{\dot{x}(t) + i\dot{y}(t)\} \\ &= \text{amp}\, f'(z) + \theta \quad \text{[by equation (8.165)]}.\end{aligned}$$

Thus

$$\phi - \theta = \text{amp}\, f'(z). \tag{8.166}$$

Equation (8.166) shows that the angle $\phi - \theta$ does not depend on the particular curve through the point $z_0(w_0)$. Thus, the transformation conserves both the magnitude and sense of rotation of the angle of intersection of two curves. Thus follows the theorem:

Theorem 8.18. *At each point z of a domain where $f(z)$ is analytic and $f'(z) \neq 0$, the mapping $w = f(z)$ is conformal.*

8.33 Homographic Transformation

The transformation defined by

$$w = \frac{az + b}{cz + d}, \tag{8.167}$$

where a, b, c, d are complex constants, such that

$$ad - bc \neq 0 \tag{8.168}$$

is called a *homographic transformation.*

Equation (8.167) can be written as

$$w = \frac{a}{c} + \frac{bc - ad}{c^2} \cdot \frac{1}{z + \frac{d}{c}}. \tag{8.169}$$

We shall show that the transformation (8.169) actually is a combination of three types of transformations:

1. If we set, $w_1 = z + \dfrac{d}{c}$, then this transformation will translate every point in z-plane through a distance equal to $\left|\dfrac{d}{c}\right|$. This type of transformation is called *translation.*

2. Now we set, $w_2 = \dfrac{1}{z_1}$, where $z_1 = z + \dfrac{d}{c}$. Let, $z_1 = r_1 e^{i\theta_1}$, so that $\bar{z}_1 = r_1 e^{-i\theta_1}$.

   $\therefore \dfrac{1}{z_1} = \dfrac{1}{r_1} e^{-i\theta_1}$ and $\dfrac{1}{\bar{z}_1} = \dfrac{1}{r_1} e^{i\theta_1}$.

   Hence, $|w_2| = \dfrac{1}{r_1}$, and $\operatorname{amp} w_2 = -\theta_1 = \operatorname{amp} \bar{z}_1$.

   Let ζ be the transformation given by

$$\zeta = \frac{1}{\bar{z}_1} = \frac{1}{r_1} e^{i\theta_1}, \quad \text{so that } |\zeta||\bar{z}_1| = 1. \tag{8.170}$$

   Then, the point ζ lies on the radius drawn through the point z. Its distance from the centre of the circle is such that $|\zeta||z| = 1$. This transformation is called *inversion* with respect to the circle $|\bar{z}_1| = 1$. The final transformation will be given by $w_2 = \bar{\zeta} = \dfrac{1}{z_1}$ which is reflection on real axis. Thus, the transformation given by $w_2 = \dfrac{1}{z_1}$ is the resultant of two transformations.

3. Now we set, $w_3 = \beta z_2$, $\beta = \dfrac{bc - ad}{c^2}$.

   Let

$$\left.\begin{aligned} z_2 &= r_2\, e^{i\theta_2} \\ \text{and} \quad \beta &= \lambda e^{i\phi} \end{aligned}\right\} \tag{8.171}$$

   Then

$$w_3 = \lambda r_2 e^{i(\theta_2 + \phi)}. \tag{8.172}$$

   Equation (8.172) shows that $|w_3| = |\lambda r_2|$ is magnified in the ratio of $|\lambda|$.

   Moreover, the $\operatorname{amp} w_3$ is increased by ϕ which is the $\operatorname{amp} \beta$. Thus, this transformation causes rotation as well as magnification. If $|\beta| = 1$, there will be no magnification.

   Lastly, $w = \dfrac{a}{c} + \beta z_2$ is again a transformation causing translation.

   Thus, the homographic transformation (8.167) consists of three types of transformations—translation, inversion and rotation with magnification.

EXAMPLE 8.31. *Show that both the transformations* (i) $w = \dfrac{1+z}{1-z}$; (ii) $w = \dfrac{z-i}{z+i}$ *transform unit circle* $|w| \leq 1$ *into planes. Also, identify the planes.*

Solution: (i) $w = \dfrac{1+z}{1-z}$, $z = x + iy$, $\bar{z} = x - iy$.

$$\therefore \quad |w|^2 - 1 = w\bar{w} - 1 = \frac{1+z}{1-z} \cdot \frac{1+\bar{z}}{1-\bar{z}} - 1 = \frac{2(z+\bar{z})}{|1-z|^2} = \frac{4x}{|1-z|^2}.$$

Hence, $|w| = 1$ corresponds to $x = 0$, which is the imaginary axis in z-plane. Thus, $|w| \leq 1$ transforms $\text{Re}\,(z) \leq 0$, i.e., left-hand of the plane.

(ii) $w = \dfrac{z-i}{z+i}$.

$$\therefore \quad |w|^2 - 1 = w \cdot \bar{w} - 1 = \frac{z-i}{z+i} \cdot \frac{\bar{z}+i}{\bar{z}-i} - 1 = \frac{2i(z-\bar{z})}{|z+i|^2} = \frac{2i(2iy)}{|z+i|^2} = -\frac{4y}{|z+i|^2}.$$

Hence, $|w| = 1$ corresponds to $y = 0$, which is the real axis in z-plane. Also, $|w| < 1$ corresponds to $y > 0$. Thus, $|w| \leq 1$ is transformed into upper half-plane given by $\text{Im}\,(z) > 0$.

EXAMPLE 8.32. *Show that the transformation* $w = \dfrac{iz+3}{5z+2i}$ *maps the real axis in the z-plane into a circle in the w-plane. Find radius and centre of the circle*

Solution: We have

$$w = \frac{iz+3}{5z+2i}.$$

Hence

$$z = \frac{3-2iw}{5w-i}, \quad \therefore \quad \bar{z} = \frac{3+2i\bar{w}}{5\bar{w}+i}. \tag{1}$$

Now, real axis in z-plane is given by $y = 0$; hence, $z - \bar{z} = 2iy = 0$.

Thus

$$\frac{3-2iw}{5w-i} - \frac{3+2i\bar{w}}{5\bar{w}+i} = 0.$$

This gives

$$20iw\bar{w} + 13(w-\bar{w}) - 6i = 0$$
$$\text{or,} \quad 20i(u^2+v^2) + 13(2iv) - 6i = 0$$
$$\text{or,} \quad u^2 + v^2 + \frac{13}{10}v - \frac{3}{10} = 0,$$

which represents a circle in w-plane with centre at $\left(0, -\frac{13}{20}\right)$.

The radius of the circle is $\sqrt{\left(\frac{13}{20}\right)^2 + \frac{3}{10}} = \frac{17}{20}$.

Now, the centre of the circle in w-plane is $w = 0 - i\frac{13}{20}$. Hence, the corresponding point in z-plane is, from (1)

$$z = \frac{3 - 2i\left(-i\frac{13}{20}\right)}{5\left(-i\frac{13}{20}\right) - i} = \frac{3 - \frac{13}{10}}{-i\left(\frac{13}{4}+1\right)} = \frac{2}{5}i.$$

Thus, the point is, $x = 0$, $y = \frac{2}{5}i$.

8.34 Special Homographic Transformations

The following homographic transformations are very useful in solving problems in physics. These are given without any deductions:

(A) The homographic transformation

$$w = e^{i\lambda}\left(\frac{z-a}{z-\bar{a}}\right) \tag{8.173}$$

maps the half-plane Im $(z) = y \geq 0$ onto the circle $|w| \leq 1$ in the w-plane, subject to the condition that In $(a) \geq 0$.

(B) The homographic transformation

$$w = e^{i\lambda}\left(\frac{z-a}{z+\bar{a}}\right) \tag{8.174}$$

maps the half-plane Re $(z) = x \geq 0$ onto the circle $|w| \leq 1$ in the w-plane, subject to the condition that Re $(a) > 0$.

(C) The homographic transformation

$$w = e^{i\lambda}\left(\frac{z-a}{\bar{a}z-1}\right) \tag{8.175}$$

maps the unit circle $|z| \leq 1$ in z-plane onto the unit circle $|w| \leq 1$ in w-plane, subject to the condition that $|a| < 1$.

(D) The homographic transformation

$$w = \rho_1\rho_2 e^{i\lambda}\left(\frac{z-a}{\bar{a}z-\rho_1^2}\right) \tag{8.176}$$

maps the circle $|z| \leq \rho_1$ in the z-plane onto the circle $|w| \leq \rho_2$ in w-plane, subject to the condition that $|a| < \rho_1$.

Let us verify one of the above transformations, say the transformation in (C).

Here, we have

$$w = e^{i\lambda}\left(\frac{z-a}{\bar{a}z-1}\right), \quad \therefore \quad \bar{w} = e^{-i\lambda}\left(\frac{\bar{z}-\bar{a}}{a\bar{z}-1}\right).$$

Hence

$$|w|^2 - 1 = w\bar{w} - 1 = e^{i\lambda}\left(\frac{z-a}{\bar{a}z-1}\right)\cdot e^{-i\lambda}\left(\frac{\bar{z}-\bar{a}}{a\bar{z}-1}\right) - 1.$$

The numerator is

$$(z\bar{z} - \bar{a}z - a\bar{z} + a\bar{a}) - (a\bar{a}z\bar{z} - \bar{a}z - a\bar{z} + 1) = z\bar{z}(1 - a\bar{a}) - 1(1 - a\bar{a})$$

$$= \left\{|z|^2 - 1\right\}\left\{1 - |a|^2\right\}. \tag{8.177}$$

From equation (8.177), we find that, for $|z| = 1$, $|w| = 1$; for $|z| < 1$ and $|a| < 1$, $|w|^2 - 1$ is negative, i.e., $|w| < 1$. Hence, $|z| \leq 1$ is mapped onto $|w| \leq 1$.

In the same way, we can verify all the other transformations.

EXAMPLE 8.33. *Determine the region in w-plane into which the rectangular region bounded by lines $x = 0$, $y = 0$, $x = 2$, $y = 2$ in the z-plane is mapped under the transformation $w = (3+i)z + (4-i)$.*

Solution: We have, $w = u + iv$ and $z = x + iy$.

Hence, the given transformation is

$$w = (3+i)z + (4-i)$$

i.e., $$u + iv = (3+i)(x+iy) + (4-i) = (3x - y + 4) + i(x + 3y - 1).$$

Thus

$$\left.\begin{aligned} u &= 3x - y + 4 \\ v &= x + 3y - 1 \end{aligned}\right\}. \tag{1}$$

Thus, when $x = 0$, $u = -y + 4$ and $v = 3y - 1$.

$$\therefore \quad 3u + v = 11. \tag{2}$$

When $y = 0$, $u = 3x + 4$ and $v = x - 1$.

$$\therefore \quad u - 3v = 7. \tag{3}$$

When $x = 2$, $u = 10 - y$ and $v = 1 + 3y$.

$$\therefore \quad 3u + v = 31. \tag{4}$$

When $y = 2$, $u = 3x + 2$ and $v = x + 5$.

$$\therefore \quad u - 3v = -13. \tag{5}$$

The region mapped in w-plane is bounded by the straight lines (2)–(5). The origin in z-plane is mapped into the plane $w = 4 - i$.

EXAMPLE 8.34. *Discuss about the region into which the interior of the circle* $|z| = 1$ *is mapped by the transformation* $w = i\dfrac{1-z}{1+z}$.

Solution: We have, $w = u + iv$ and $z = x + iy$. Hence, the given transformation is

$$u + iv = i\frac{1-(x+iy)}{1+(x+iy)} = \frac{y + i(1-x)}{(1+x)+iy} = \frac{\{y + i(1-x)\}\{(1+x) - iy\}}{\{(1+x)+iy\}\{(1+x)-iy\}}.$$

This gives

$$u = \frac{2y}{(1+x)^2 + y^2} \quad \text{and} \quad v = -\frac{x^2 + y^2 - 1}{(1+x)^2 + y^2}. \tag{1}$$

Hence, for $|z| = 1$, $v = 0$, i.e., real axis in w-plane. For interior of the circle $|z| < 1$, we have $x^2 + y^2 < 1$. Thus, from (1), we get $v =$ positive. Again, for lower semi-circle in z-plane, y is negative, i.e., $u =$ negative. Thus, interior of $|z| = 1$ is mapped into the plane above real axis, upper half of $|z| = 1$ into positive half of real axis and lower half of $|z| = 1$ into negative half of real axis.

EXERCISE

1. Find modulus and amplitude of the complex numbers:
(a) $1 + i\tan\frac{3}{5}\pi$; (b) $\frac{(2+3i)^2}{2+i}$; (c) $\frac{1+2i}{1-(1-i)^2}$.
2. If z_2 be complex conjugate of z_1 and z_4 be complex conjugate of z_3, then show that $\text{amp}\,\frac{z_1}{z_4} = \text{amp}\,\frac{z_3}{z_2}$.
3. If z_1, z_2, z_3 be vertices of an isosceles triangle, right-angled at z_2, prove that
$$z_1^2 + 2z_2^2 + z_3^2 = 2z_2(z_1 + z_3).$$
4. For any two complex numbers z_1 and z_2, prove that
$$|z_1 + z_2|^2 + |z_1 - z_2|^2 = 2\left\{|z_1|^2 + |z_2|^2\right\}.$$
5. Find the locus of the point z, if
(a) $\text{amp}\left(\frac{z-1}{z+1}\right) = \lambda$ (constant); (b) $\frac{z+1}{z-1}$ is purely imaginary.
6. Show that the points $(2+2i)$, $(-2-2i)$, $(-2\sqrt{3}+2\sqrt{3}i)$ are vertices of an equilateral triangle whose centroid is $2(-1+i)/\sqrt{3}$ and area is $8\sqrt{3}$ sq units.
7. If ω be the cube root of unity, show that $(1-\omega+\omega^2)(1-\omega^2+\omega^4)(1-\omega^4+\omega^8)\ldots$ to $2n$ factors $= 2^{2n}$.
8. Find the values of:
(a) $(1+i\sqrt{3})^{1/5}$; (b) $\frac{3}{(2+i)^3}$.
9. Solve the equation $z^n = (z+1)^n$, and show that the roots in Argand diagram are collinear.
10. Show that the equation $(a\bar{z} + \bar{a}z)^2 = 2(b\bar{z} + \bar{b}z) + c$, where c is real, represents the most general parabola in the complex plane.
11. Evaluate
$$\int_C (z^2 + 3z + 2)\,dz,$$
where C is the arc of the cycloid, $x = a(\theta + \sin\theta)$, $y = a(1 - \cos\theta)$ between the points $(0,0)$ and $(a\pi, 2a)$.
12. Expand in series, the following functions:
(a) $\frac{1}{(1+z^2)(z+2)}$; (b) $\frac{(z-2)(z+2)}{(z+1)(z+4)}$; (c) $\frac{z+3}{z(z^2-z-2)}$.
for the regions, (i) $|z| < 1$, (ii) $1 < |z| < 4$.
13. If $f(z) = \sum_{n=0}^{\infty} a_n z^n$, $(|z| < R)$ and $|f(z)| < M$ on $|z| = r < R$, then prove that
$$|a_n| r^n \le M.$$
14. Show that $\sin\{c(z + 1/z)\}$ can be expanded in a series of the form
$$\sum_{n=1}^{\infty} a_n z^n + \sum_{n=1}^{\infty} b_n z^{-n},$$
where
$$a_n = b_n = \frac{1}{2\pi}\int_0^{2\pi} \sin(2c\cos\theta)\cos n\theta\,d\theta.$$

15. Evaluate

$$\int_C \frac{4z-3}{z^2+2z+3}\, dz,$$

where C is the circle, $|z+2+i|=3$.

16. A function $f(z)$ is analytic within and on a unit circle. The function has the value on the circle

$$[a-(\cos\theta - i\sin\theta)]\,\frac{1}{F(\theta)},$$

where, $F(\theta) = a^2 - 2a\cos\theta + 1$, $a > 1$, and θ is the vectorial angle at points on the circumference. Determine the function $f(z)$.

17. Determine the region of the w-plane into which the region in the z-plane defined by $x=0$, $y=0$, $x=2$ and $y=3$ is mapped under the transformation $w = \sqrt{2}\, e^{i\pi/4}\, z$.

18. Show that the equation

$$\text{amp}\left(\frac{z-z_1}{z-z_2}\right) = \lambda \quad (\lambda = \text{real constant})$$

represents a family of circles every member of which passes through two fixed points.

19. Show that the transformation $w = \frac{z+i}{z-i}$ maps $|w| \leq 1$ onto lower half-plane $\text{Im}\,(z) > 0$.

20. Show that the transformation $w = e^{i\lambda}\frac{z-c}{cz-1}$ maps the circle $|z|=1$ onto $|w|=1$, and makes the points $z=1$ and $z=-1$ correspond to $w=1$ and $w=-1$ respectively.

21. Show that the region $|z-c| \leq R$ is mapped conformably on $|w| \leq 1$ by the transformation

$$w = \frac{R(z-a)}{R^2-(z-c)(\bar{a}-\bar{c})}\, e^{i\lambda},$$

where λ is real, and $z=a$ is the point which is transformed into origin.

ANSWERS

1. (a) $-\sec\dfrac{3\pi}{5}, -\dfrac{2\pi}{5}$; (b) $\dfrac{13}{\sqrt{5}}, \tan^{-1}\dfrac{29}{2}$; (c) 1, 0.

5. (a) A family of circles: $x^2+y^2-\dfrac{2y}{\tan\lambda}-1=0$;

   (b) A circle with centre at $\dfrac{1}{2}(-1+i)$ and radius $\dfrac{1}{\sqrt{2}}$.

8. (a) $2^{1/5}\left[\cos\left(2m+\dfrac{1}{3}\right)\dfrac{\pi}{5} + i\sin\left(2m+\dfrac{1}{3}\right)\dfrac{\pi}{5}\right]$, $m = 0, 1, 2, \ldots$

   (b) $\dfrac{3}{5^{3/2}}(\cos 3\theta - i\sin 3\theta)$, where $\theta = \tan^{-1}\dfrac{1}{2}$.

11. $I = 2\pi a + \dfrac{1}{3}(\pi a + 2ai)^3 + 4ai + \dfrac{3}{2}(\pi a + 2ai)^2$.

12. (a) i. $f(z) = \dfrac{1}{10}\displaystyle\sum_{n=0}^{\infty}(-1)^n\left(\frac{z}{2}\right)^n + \frac{2-z}{5}\sum_{n=0}^{\infty}(-1)^n z^{2n}$;

ii. $f(z) = \frac{1}{10}\sum_{n=0}^{\infty}(-1)^n\left(\frac{z}{2}\right)^n + \frac{2-z}{5z^2}\sum_{n=0}^{\infty}(-1)^n\left(\frac{1}{z^2}\right)^n.$

(b) i. $f(z) = -1 + \sum_{n=1}^{\infty}(-1)^{n+1}\left(1+4^{-n}\right)z^n;$

ii. $f(z) = 1 - \frac{1}{z}\sum_{n=0}^{\infty}(-1)^n\left(\frac{1}{z}\right)^n - \sum_{n=0}^{\infty}(-1)^n\left(\frac{z}{4}\right)^n.$

(c) i. $f(z) = -\frac{3}{2z} + \frac{2}{3}\sum_{n=0}^{\infty}(-1)^n z^n - \frac{5}{12}\sum_{n=0}^{\infty}\left(\frac{z}{2}\right)^n;$

ii. $f(z) = -\frac{3}{2z} + \frac{2}{3z}\sum_{n=0}^{\infty}(-1)^n\left(\frac{1}{z}\right)^n + \frac{5}{6z}\sum_{n=0}^{\infty}\left(\frac{2}{z}\right)^n.$

16. $f(z) = \frac{1}{a-z}.$

Chapter 9

Calculus of Residues

9.1 Singularities of Complex Functions

Let $f(z)$ be an analytic function in a domain D and a is any point within it. Then, by Taylor's theorem, we can expand $f(z)$ about $z = a$, i.e.,

$$f(z) = \sum_{n=0}^{\infty} a_n (z-a)^n, \tag{9.1}$$

where

$$a_n = \frac{f^{(n)}(a)}{n!}. \tag{9.2}$$

Let

$$\left.\begin{aligned} a_1 = a_2 &= \cdots a_{m-1} = 0, \text{ but } a_m \neq 0 \\ \text{i.e., } f(a) = f'(a) &= \cdots f^{(m-1)}(a) = 0, \text{ but } f^{(m)} \neq 0 \end{aligned}\right\}. \tag{9.3}$$

In such a case, we say that the analytic function $f(z)$ has a zero of order m. Then, we can write

$$f(z) = (z-a)^m \phi(z), \tag{9.4}$$

where $\phi(z)$ is analytic having non-zero value at $z = a$.

If, in particular, $m = 1$, i.e., $f(a) = 0$, but $f'(a) \neq 0$, we say that $f(z)$ has a simple zero at $z = a$.

Singularities: The point at which a function $f(z)$ ceases to be analytic is called a *singular point* of $f(z)$. The function $f(z)$ is said to have a singularity at that point.

For example, the function $f(z) = \dfrac{1}{z - ia}$ is analytic everywhere except at $z = ia$. Hence, $z = ia$ is a singularity of $f(z)$.

Singularities of an analytic function may be of two types—isolated and non-isolated singularities.

A singular point $z = a$ of the function $f(z)$ is said to be isolated if there is no other singularity in its neighbourhood. For example, the function

$$f(z) = \frac{z+1}{z^2+4}$$

has two isolated singularities at $z = 2i$ and $z = -2i$.

We can draw a small circle with centre at an isolated singularity such that the function $f(z)$ becomes analytic in the annulus of that circle and any other concentric larger circle. Hence, $f(z)$ can be expanded in Laurent's series as

$$f(z) = \sum_{n=0}^{\infty} a_n(z-a)^n + \sum_{n=1}^{\infty} b_n(z-a)^{-n}, \tag{9.5}$$

where the second part

$$\sum_{n=1}^{\infty} b_n(z-a)^{-n}$$

is called the *principal part* of $f(z)$ at its isolated singularity $z = a$. We consider three cases:

1. If the coefficients b_n in equation (9.5) are all zero, then the singularity $z = a$ is called *non-isolated* or *removable* singularity. In this case

$$\lim_{z \to a} f(z) = l \neq 0. \tag{9.6}$$

For example, the function

$$f(z) = \frac{\sinh z}{z}$$

has a singularity at $z = 0$ which is removable, because

$$f(z) = \frac{1}{z}\left(z + \frac{z^3}{3!} + \frac{z^5}{5!} + \cdots\right) = 1 + \frac{z^2}{3!} + \frac{z^4}{5!} + \cdots$$

does not contain any term containing negative powers of z.

2. If the principal part of $f(z)$ contains a finite number of terms, say m terms, then the singularity at $z = a$ is called a *pole of order m*. A pole of order one is called a *simple pole*. Thus, for a function $f(z)$ having a pole of order m at $z = a$

$$f(z) = \sum_{n=0}^{\infty} a_n(z-a)^n + \sum_{n=1}^{m} b_n(z-a)^{-n}. \tag{9.7}$$

For example, the function

$$f(z) = \frac{\sinh z}{z^5} = \frac{1}{z^5}\left(z + \frac{z^3}{3!} + \frac{z^5}{5!} + \frac{z^7}{7!} + \cdots\right) = \frac{1}{z^4} + \frac{1}{3!z^2} + \frac{1}{5!} + \frac{z^2}{7!} + \cdots$$

has a pole of order 2 at $z = 0$, since there are only two terms containing negative powers of z. In the case of a pole of order m at $z = a$, we have

$$\lim_{z \to a} (z-a)^m f(z) = l \neq 0. \tag{9.8}$$

3. If the principal part of $f(z)$ has an infinite number of terms, then the singularity at $z = a$ is called an essential singularity. For example

$$f(z) = \cos\left(\frac{1}{z}\right) = 1 - \frac{1}{2!z^2} + \frac{1}{4!z^4} - \cdots$$

has essential singularity at $z = 0$, since there are an infinite number of terms containing negative powers of z.

We now shall prove the following theorems:

Theorem 9.1. *If $f(z)$ has pole at $z = a$, then*

$$\lim_{z \to a} |f(z)| = \infty. \tag{9.9}$$

Proof. Let, $f(z)$ has a pole of order m at $z = a$. Then, Laurent's expansion gives

$$\begin{aligned} f(z) &= \sum_{n=0}^{\infty} a_n (z-a)^n + \sum_{n=1}^{m} b_n (z-a)^{-n} \\ &= \sum_{n=0}^{\infty} a_n (z-a)^n + \frac{1}{(z-a)^m} \sum_{n=1}^{m} b_n (z-a)^{m-n}. \end{aligned} \tag{9.10}$$

Now, as $z \longrightarrow a$, $\sum_{n=1}^{m} b_n (z-a)^{m-n} \longrightarrow b_m$.

Hence, $f(z) \longrightarrow \infty$ as $z \longrightarrow a$.

Theorem 9.2. *If an analytic function $f(z)$ has a pole of order m at $z = a$, then $\frac{1}{f(z)}$ has a zero of order m at $z = a$, and conversely.*

Proof. If $f(z)$ has a pole of order m at $z = a$, then from (9.10), we have

$$(z-a)^m f(z) = \sum_{n=0}^{\infty} a_n (z-a)^{m+n} + \sum_{n=1}^{m} b_n (z-a)^{m-n} = \psi(z), \text{ (say)} \tag{9.11}$$

such that $\psi(z)$ is analytic and $\psi(a) \neq 0$.

Then, from (9.11), we have

$$\frac{1}{f(z)} = \frac{(z-a)^m}{\psi(z)},$$

which $\longrightarrow 0$ as $z \longrightarrow a$, since $\psi(a) \neq 0$. This shows that $\frac{1}{f(z)}$ has zero of order m.

Similarly, if $\frac{1}{f(z)}$ has a zero of order m at $z = a$, we can write

$$\frac{1}{f(z)} = (z-a)^m \cdot \psi(z),$$

where $\psi(z)$ is analytic and $\psi(a) \neq 0$. Therefore

$$(z-a)^m f(z) = \frac{1}{\psi(z)},$$

where $\frac{1}{\psi(z)}$ is analytic.

This shows that $f(z)$ has a pole of order m at $z = a$.

9.2 The Point at Infinity

The relation $t = \frac{1}{z}$ gives a continuous one-to-one correspondence between points on t and z planes. Thus, the point at infinity in z-plane corresponds to the origin in t-plane. In order to study the behaviour of the function $f(z)$ at $z = \infty$, we therefore substitute $t = \frac{1}{z}$ in $f(z)$, and study the resulting function $f(\frac{1}{t})$ at $t = 0$. Thus, if a function $f(z)$ has a zero, a pole, or an essential singularity at infinity, the function $f(\frac{1}{z})$ has a zero, a pole, or an essential singularity at the origin. For example, the function $\sin z$ has a simple zero at the origin, and the function $\sin(\frac{1}{z})$ has a simple zero at infinity.

Polynomials have got an interesting property which we state below without proof.

Theorem 9.3. *A function $f(z)$ is a polynomial of degree m, if and only if, $f(z)$ has no singularities in the finite part of the z-plane but has a pole of order m at infinity.*

EXAMPLE 9.1. *Examine the singularities of*

$$f(z) = \frac{\cot \pi z}{(z-a)^2} \text{ at } z = 0 \text{ and } z = \infty.$$

Solution: We have

$$f(z) = \frac{\cot \pi z}{(z-a)^2} = \frac{\cos \pi z}{(\sin \pi z)(z-a)^2}.$$

Hence, the poles are given by $(\sin \pi z)(z-a)^2 = 0$, i.e., $\sin \pi z = 0$ and $(z-a)^2 = 0$.

Hence, the poles are given by $z = n$, $(n = 0, \pm 1, \ldots)$ and $z = a$ (of order 2).

At $z = \infty$, we get the limit point of poles, and hence, $z = \infty$ is an isolated essential singularity.

EXAMPLE 9.2. *Discuss the singularities of the function*

$$f(z) = \frac{z^2-2}{z^3} \cos\left(\frac{1}{1+z}\right).$$

Solution: The zeros of the function $f(z)$ are given by

$$(z^2-2)\cos\left(\frac{1}{1+z}\right) = 0,$$

i.e., by $z^2 - 2 = 0$, and $\cos\left(\frac{1}{1+z}\right) = 0$

i.e., by $z = \pm\sqrt{2}$, and $\frac{1}{1+z} = 2n\pi + \frac{\pi}{2}$ $(n = 0, \pm 1, \ldots)$

i.e., by $z = \pm\sqrt{2}$, and $z = 1 - \dfrac{1}{2n\pi + \frac{\pi}{2}}$ $(n = 0, \pm 1, \pm 2, \ldots)$.

This shows that the function has two zeros at $z = \sqrt{2}$ and $z = -\sqrt{2}$, and the limit point of zeros given by

$$z = 1 - \frac{1}{2n\pi + \frac{\pi}{2}}$$

is 1. Therefore, $z = 1$ is an isolated essential singularity.

Poles of the function $f(z)$ are given by $z^3 = 0$. Thus, $z = 0$ is a pole of $f(z)$ of order three.

Note 1. Limit point of zeros is an isolated essential singularity.

Note 2. Limit point of poles is a non-essential singularity.

9.3 Residue at a Pole

Let us consider a single-valued function $f(z)$ having pole of order m at $z = a$. We draw a circle C with centre at $z = a$ and radius r, so that this circle does not enclose any other singularity of $f(z)$. Then, $f(z)$ is analytic within and on C except at $z = a$, and can therefore be expanded in Laurent's series as

$$f(z) = \sum_{n=0}^{\infty} a_n (z-a)^n + \sum_{n=1}^{m} b_n (z-a)^{-n}. \tag{9.12}$$

The coefficient b_1 is called the *residue* of $f(z)$ at the pole $z = a$.
Now, the coefficients b_n are in general given by

$$b_n = \frac{1}{2\pi i} \int_C f(z)(z-a)^{n-1}\, dz \tag{9.13}$$

so that, the residue of $f(z)$ at $z = a$ is (i.e., for $n = 1$)

$$b_1 = \frac{1}{2\pi i} \int_C f(z)\, dz. \tag{9.14}$$

Thus, the value of residue b_1 given by equation (9.14) does not depend on the order of the pole. Hence, we accept this equation as the general definition of residue at a pole. We shall see later that this definition of residue is a powerful tool to evaluate certain integrals around closed contours.

EXAMPLE 9.3. *The only singularities of a single-valued function $f(z)$ are poles of order 1 and 2 at $z = 1$ and $z = -2$. The residues at those points are 2 and 3 respectively. If $f(0) = 8$ and $f(-1) = \frac{7}{2}$, determine the function.*

Solution: Since the function $f(z)$ has a pole of order one at $z = 1$, where residue is 2, the function contains a term like $\frac{2}{(z-1)}$. Similarly, the function contains terms like $\frac{3}{(z+2)}$ and $\frac{\lambda}{(z+2)^2}$, so that we can write

$$f(z) = \sum_{n=0}^{\infty} a_n z^n + \frac{2}{z-1} + \frac{3}{z+2} + \frac{\lambda}{(z+2)^2}, \quad \lambda = \text{a real constant}.$$

Since $f(z)$ has no other singularity, $z = \infty$ is not a pole of $f(z)$, i.e., $t = 0$ is not a pole of $f(\frac{1}{t})$. The principal part of $f(\frac{1}{t})$ is

$$a_0 + \sum_{n=1}^{\infty} a_n t^{-n},$$

which should not contain any terms having negative powers of t. Hence, $a_1 = a_2 = \cdots$ are all zero.

Thus

$$f(z) = a_0 + \frac{2}{z-1} + \frac{3}{z+2} + \frac{\lambda}{(z+2)^2}.$$

Now, $f(0) = 8$ and $f(-1) = \frac{7}{2}$. Hence

$$8 = a_0 - 2 + \tfrac{3}{2} + \tfrac{\lambda}{4}, \quad \text{i.e.,} \quad a_0 + \tfrac{\lambda}{4} = \tfrac{17}{2}$$

$$\text{and} \quad \tfrac{7}{2} = a_0 - 1 + 3 + \lambda, \quad \text{i.e.,} \quad a_0 + \lambda = \tfrac{3}{2}.$$

Solving these, we get

$$a_0 = \frac{65}{6} \quad \text{and} \quad \lambda = -\frac{28}{3}.$$

Hence, the required function is

$$f(z) = \frac{65}{6} + \frac{2}{z-1} + \frac{3}{z+2} - \frac{28}{3(z+2)^2}.$$

9.4 Calculation of Residues

Residues can be calculated in different cases by the following methods:

(a) *Residue at a simple pole.* If $f(z)$ has a pole of order one, i.e., a simple pole at $z = a$, then Laurent's expansion (9.5) becomes

$$f(z) = \sum_{n=0}^{\infty} a_n (z-a)^n + \frac{b_1}{z-a}.$$

Hence, residue at $z = a$ is given by

$$b_1 = \lim_{z \to a} (z-a) f(z). \tag{9.15}$$

If $f(z)$ is of the form

$$f(z) = \frac{\phi(z)}{\psi(z)},$$

where $\psi(a) = 0$, but $\phi(a) \neq 0$, then

$$b_1 = \lim_{z \to a} (z-a) f(z) = \lim_{z \to a} (z-a) \frac{\phi(z)}{\psi(z)}. \tag{9.16}$$

Expanding both $\phi(z)$ and $\psi(z)$ by Taylor's series, we get from equation (9.16)

$$b_1 = \lim_{z \to a} (z-a) \frac{\phi(a) + (z-a)\phi'(a) + \cdots}{\psi(a) + (z-a)\psi'(a) + \cdots}$$

$$\text{i.e.,} \quad b_1 = \frac{\phi(a)}{\psi'(a)} \quad [\because \psi(a) = 0] \tag{9.17}$$

(b) *Residue at a pole of order m.* If $f(z)$ has a pole of order m, we can write $f(z)$ in the form

$$f(z) = \frac{\phi(z)}{(z-a)^m}, \tag{9.18}$$

where $\phi(z)$ is analytic, and $\phi(a) \neq 0$.

Hence, the residue of $f(z)$ at $z = a$ is given by equation (9.14)

$$b_1 = \frac{1}{2\pi i} \int_C f(z)\, dz = \frac{1}{2\pi i} \int_C \frac{\phi(z)}{(z-a)^m} dz,$$

$$\text{i.e.,} \quad b_1 = \frac{\phi^{m-1}(a)}{(m-1)!} \quad \text{[by Cauchy's integral formula].}$$

Hence, residue of $f(z)$ at $z = a$, is

$$b_1 = \frac{1}{(m-1)!}\left(\frac{d}{dz}\right)^{m-1} \{(z-a)^m f(z)\}_{z=a}. \tag{9.19}$$

(c) *Residue at an essential singularity.* If $z = a$ is an essential singularity, i.e., pole of any order, we put $z - a = t$ in Laurent's expansion of $f(z)$, which becomes

$$f(a+t) = \sum_{n=0}^{\infty} a_n t^n + \frac{b_1}{t} + \frac{b_2}{t^2} + \cdots + \frac{b_m}{t^m} + \cdots \tag{9.20}$$

Then the residue of $f(z)$ at $z = a$ is given by the coefficient of $\frac{1}{t}$, i.e., b_1 in the series (9.20).

(d) *Residue at infinity.* The residue at $z = \infty$ of a function $f(z)$ is defined as

$$\text{Res}\,(z = \infty) = -\frac{1}{2\pi i}\int_C f(z)\,dz, \tag{9.21}$$

where C is any closed contour enclosing all the singularities of $f(z)$ except $z = \infty$. Now, if we put $z = \frac{1}{t}$, the integral in equation (9.21) becomes

$$\frac{1}{2\pi i}\int_\gamma \left[-f\left(\frac{1}{t}\right)\frac{1}{t^2}\,dt\right], \tag{9.22}$$

where integration is in positive sense about a small circle γ whose centre is at the origin.

Hence, residue of $f(z)$ at $z = \infty$ will be same as that of the function

$$F(t) = \left\{-f\left(\frac{1}{t}\right)\right\}\frac{1}{t^2}$$

at $t = 0$. This is given by

$$\lim_{t\longrightarrow 0} t\,F(t) = \lim_{t\longrightarrow 0} t\left\{-f\left(\frac{1}{t}\right)\right\}\frac{1}{t^2} = \lim_{t\longrightarrow 0}\left\{-\frac{1}{t}f\left(\frac{1}{t}\right)\right\}$$

$$\text{i.e.,}\quad \text{Res}\,(z=\infty) = \lim_{z\longrightarrow\infty}\{-z\,f(z)\}. \tag{9.23}$$

Thus, we have proved:

Theorem 9.4. *If a function $f(z)$ is analytic at $z = \infty$, then the residue of $f(z)$ at $z = \infty$ is* $\lim_{z\longrightarrow\infty}\{-z\,f(z)\}$.

We now prove another interesting theorem on the residue of a function at infinity:

Theorem 9.5. *The residue of a function $f(z)$ at $z = \infty$ is the negative of the coefficient of $\frac{1}{z}$ in the expansion of $f(z)$ for the value of z in the neighbourhood of $z = \infty$.*

Proof. Let $f(z)$ have a pole of order m at $z = \infty$, so that $f(\frac{1}{z})$ has a zero of order m at $z = 0$. Hence, the function $f(\frac{1}{z})$ can be expanded in a Laurent's series about $z = 0$ in the form

$$f\left(\frac{1}{z}\right) = \sum_{n=0}^{\infty} a_n z^n + \sum_{n=1}^{m} b_n z^{-n}. \tag{9.24}$$

Replacing $\frac{1}{z}$ by z, we get expansion of $f(z)$ in the neighbourhood of $z=\infty$ in the form

$$f(z)=\sum_{n=1}^{m} b_n z^n+\sum_{n=0}^{\infty} a_n z^{-n}.$$

Hence, residue of $f(z)$ at $z=\infty$ is, by definition

$$\begin{aligned}
&-\frac{1}{2\pi i}\int_C f(z)\,dz, \quad C \text{ taken anti-clockwise}\\
&=-\frac{1}{2\pi i}\int_C\left[\sum_{n=1}^{m} b_n z^n+\sum_{n=0}^{\infty} a_n z^{-n}\right]\\
&=-\frac{1}{2\pi i}\int \frac{a_1}{z}\,dz \quad \left[\text{other integrals are of the form } \int_C \frac{dz}{z^k},\ k\neq 1 \text{ and so vanish}\right]\\
&=-\frac{1}{2\pi i}a_1\cdot 2\pi i=-a_1.
\end{aligned}$$

Thus, residue of $f(z)$ at $z=\infty$ is negative of coefficient of $\frac{1}{z}$ in the expansion of $f(z)$ around $z=\infty$. □

EXAMPLE 9.4. *Find the residue of*

$$f(z)=\frac{e^z}{z^2+a^2} \quad \text{at } z=ia.$$

Solution: Here, $z=ia$ is a simple pole.

Putting $\phi(z)=e^z$ and $\psi(z)=z^2+a^2$, so that $\psi'(z)=2z$, we get the residue at $z=ia$ by using equation (9.17) as

$$b_1=\left.\frac{\phi(z)}{\psi'(z)}\right|_{z=ia}=\frac{e^{ia}}{2ia}.$$

EXAMPLE 9.5. *Find the residue of*

$$f(z)=\frac{z^2}{(z-a)(z-b)(z-c)} \quad \text{at } z=\infty.$$

Solution: The residue of $f(z)$ at $z=\infty$ is, by equation (9.23)

$$\begin{aligned}
\lim_{z\to\infty}\{-z f(z)\}&=\lim_{z\to\infty}\left\{-\frac{z^3}{(z-a)(z-b)(z-c)}\right\}\\
&=\lim_{z\to\infty}\left\{-\frac{1}{(1-\frac{a}{z})(1-\frac{b}{z})(1-\frac{c}{z})}\right\}=-1.
\end{aligned}$$

EXAMPLE 9.6. *Find the residue of*

$$f(z)=\frac{z^2}{(z-1)^4(z-2)(z-3)} \quad \text{at } z=1.$$

Solution: The point $z = 1$ is a pole of order 4. We write $f(z)$ as

$$f(z) = \frac{\phi(z)}{(z-1)^4},$$

where

$$\phi(z) = \frac{z^2}{(z-2)(z-3)},$$

$$\text{i.e.,}\quad \phi(z) = 1 - \frac{4}{z-2} + \frac{9}{z-3} \qquad \text{(by expanding in partial fractions)}$$

Hence

$$\phi'''(z) = \left\{ \frac{24}{(z-2)^4} - \frac{30}{(z-3)^4} \right\}$$

$$\therefore\quad \phi'''(z)\Big|_{z=1} = 24 - \frac{27}{4} = \frac{69}{4}.$$

Thus, residue of $f(z)$ at $z = 1$ is [by equation (9.19)]

$$\frac{\phi'''(1)}{3!} = \frac{69}{4 \cdot 3!} = \frac{69}{24}.$$

EXAMPLE 9.7. *Find the residue of*

$$f(z) = \frac{z \cos z}{(z-\pi)^3} \quad at\ z = \pi.$$

Solution: For the given function $f(z)$, $z = \pi$ is a pole of order 3. Hence we take

$$\phi(z) = z \cos z,$$

$$\therefore\quad \phi'(z) = \cos z - z \sin z$$

$$\phi''(z) = -\sin z - (\sin z + z \cos z) = -2 \sin z - z \cos z.$$

The residue at $z = \pi$ is

$$\frac{\phi''(z)}{2!}\Big|_{z=\pi} = \frac{1}{2!}(-2 \sin z - z \cos z)_{z=\pi} = \frac{\pi}{2}. \tag{1}$$

9.5 Cauchy's Residue Theorem

Theorem 9.6. *If a function $f(z)$ is analytic within and on a closed contour C, except at a finite number of poles within C, then*

$$\int_C f(z)\, dz = 2\pi i \Sigma R, \tag{9.25}$$

where ΣR denotes the sum of residues at poles within C, the integration being taken in positive sense.

Proof. Let $z_1, z_2, \ldots, z_n$ be a finite number of poles within C. We draw circles $C_1, C_2, \ldots, C_n$ with centres at $z_1, z_2, \ldots, z_n$ respectively, such that all of them lie entirely within C without overlapping (*Fig 9.1*). Then the function $f(z)$ would be analytic within the region bounded by these circles. Hence, by Cauchy's theorem on multiple-connected region (Theorem 8.9), we have

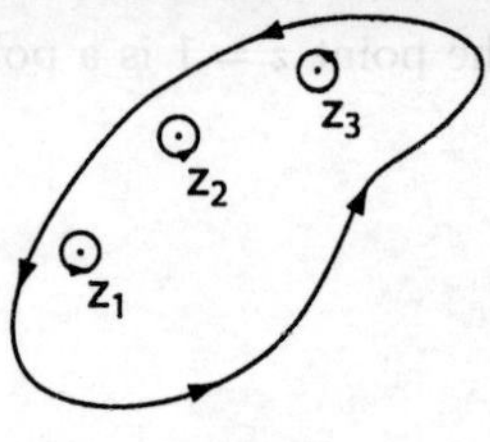

Fig 9.1

$$\int_C f(z)\,dz = \int_{C_1} f(z)\,dz + \int_{C_2} f(z)\,dz + \cdots + \int_{C_n} f(z)\,dz \tag{9.26}$$

all the integrations being taken in anti-clockwise direction.

But, from definition, we have

$$\frac{1}{2\pi i}\int_{C_1} f(z)\,dz = \text{Residue at } z = z_1 \text{ which is } = R_1 \text{ (say) and so on.}$$

Hence, from equation (9.26), we get

$$\int_C f(z)\,dz = 2\pi i\,(R_1 + R_2 + \cdots + R_n) = 2\pi i \Sigma R, \tag{9.27}$$

where $R_1, R_2, \ldots, R_n$ are residues of the function $f(z)$ at $z_1, z_2, \ldots, z_n$ respectively.

Hence, the theorem is proved. □

We have an interesting consequence of this theorem. Let an analytic function $f(z)$ have a finite number of singularities within a closed curve C. Then, by above theorem

$$\int_C f(z)\,dz = 2\pi i \Sigma R, \quad \text{i.e.,} \quad \Sigma R = \frac{1}{2\pi i}\int_C f(z)\,dz. \tag{9.28}$$

Let the function have a singularity at $z = \infty$ also. Then, from definition

$$\text{Residue at } z = \infty \text{ is } -\frac{1}{2\pi i}\int_C f(z)\,dz. \tag{9.29}$$

Hence, from equations (9.28) and (9.29), we find that the sum of the residues at all the singularities of the function $f(z)$ is zero. Thus, we have proved:

Theorem 9.7. *If a function $f(z)$ is analytic except at a finite number of singularities (including that at infinity), then the sum of residues at these singularities is zero.*

There are two useful theorems related to Cauchy's residue theorem which we like to establish now.

Theorem 9.8. *If C be an arc of a circle $|z - a| = r$ between angles θ_1 and $\theta_2 (\theta_1 \le \theta \le \theta_2)$, and if*

$$\lim_{z \to a} (z - a) f(z) = R,$$

then

$$\lim_{r \to 0} \int_C f(z)\,dz = iR(\theta_2 - \theta_1). \tag{9.30}$$

Proof. Since

$$\lim_{z \longrightarrow a} (z-a)\, f(z) = R$$

for a given ϵ, we can always find a small quantity δ depending on ϵ, such that

$$|\,(z-a)\, f(z) - R\,| < \epsilon, \quad \text{for } |\, z-a\,| < \delta. \tag{9.31}$$

But, it is given that $|\,z-a\,| < r$; hence we choose $r < \delta$ so that inequality (9.31) holds. Then we write, $(z-a)f(z) = R + \eta$, where $|\,\eta\,| < \epsilon$,

$$\text{i.e.,} \quad f(z) = \frac{R+\eta}{z-a}. \quad \therefore \quad \int_C f(z)\, dz = \int_C \frac{R+\eta}{z-a}\, dz. \tag{9.32}$$

Now, $z - a = re^{i\theta}$, so that $dz = re^{i\theta} \cdot i\, d\theta$.
Hence, integral (9.32) becomes

$$\int_C f(z)\, dz = \int_{\theta_1}^{\theta_2} \frac{R+\eta}{re^{i\theta}} \cdot r\, e^{i\theta} \cdot i\, d\theta$$

$$\text{or,} \quad \int_C f(z)\, dz = (R+\eta) i \int_{\theta_2}^{\theta_2} d\theta = iR(\theta_2 - \theta_1) + i\eta(\theta_2 - \theta_1). \tag{9.33}$$

Now

$$|\, i\eta\, (\theta_2 - \theta_1)\,| = (\theta_2 - \theta_1)\, |\,\eta\,| < (\theta_2 - \theta_1)\, \epsilon,$$

which tends to zero as $\epsilon \longrightarrow 0$ (consequently $r \longrightarrow 0$).
Hence, from equation (9.33), we get

$$\lim_{r \longrightarrow 0} \int_C f(z)\, dz = iR\, (\theta_2 - \theta_1)\,.$$

Hence, the theorem is proved. □

Theorem 9.9. *If C be an arc $\theta_1 \le \theta \le \theta_2$ of the circle $|\,z\,| = R$ and if*

$$\lim_{R \longrightarrow \infty} z\, f(z) = B,$$

then

$$\lim_{R \longrightarrow \infty} \int_C f(z)\, dz = iB\, (\theta_2 - \theta_1)\,. \tag{9.34}$$

Proof. Since $\lim\limits_{R \longrightarrow \infty} zf(z) = B$, for a given small quantity ϵ, we have, $|\, zf(z) - B\,| < \epsilon$,

$$\text{i.e.,} \quad zf(z) = B + \eta, \text{ where } |\,\eta\,| < \epsilon$$

$$\text{or,} \quad f(z) = \frac{B+\eta}{z}. \tag{9.35}$$

Hence

$$\int_C f(z)\, dz = \int_C \frac{B+\eta}{z}\, dz. \tag{9.36}$$

Putting $z = Re^{i\theta}$, so that $dz = Re^{i\theta} \cdot i\,d\theta$, we get from equation (9.36)

$$\int_C f(z)\,dz = \int_{\theta_1}^{\theta_2} \frac{B+\eta}{Re^{i\theta}} \cdot Re^{i\theta} \cdot i\,d\theta$$

$$= i(B+\eta)(\theta_2-\theta_1) = iB(\theta_2-\theta_1) + i\eta(\theta_2-\theta_1). \tag{9.37}$$

Now

$$|i\eta(\theta_2-\theta_1)| = (\theta_2-\theta_1)|\eta| < (\theta_2-\theta_1)\epsilon. \tag{9.38}$$

Hence, as $\epsilon \longrightarrow 0$ (consequently $R \longrightarrow \infty$), we get, $|i\eta(\theta_2-\theta_1)| \longrightarrow 0$. Thus, taking limit on both sides of equation (9.37), we get

$$\lim_{R\longrightarrow\infty} \int_C f(z)\,dz = iB(\theta_2-\theta_1).$$

Hence, the theorem is proved. □

9.6 Evaluation of Definite Integrals

Theory of residues provide a very useful technique for evaluating a wide variety of definite integrals encountered in many problems of physics. Those integrals can also be evaluated by other techniques, but the use of residue theory generally makes the evaluation easier.

We shall discuss some methods of evaluating definite integrals using different types of contours.

(A) *Integration round a unit circle*: Integrals of the type

$$I = \int_0^{2\pi} f(\sin\theta, \cos\theta)\,d\theta, \tag{9.39}$$

where, the function $f(\sin\theta, \cos\theta)$ is finite for all values of θ, are evaluated by using circle as a closed contour.

In this case, we substitute, $z = e^{i\theta}$, so that

$$\cos\theta = \frac{1}{2}\left(z + \frac{1}{z}\right), \quad \sin\theta = \frac{1}{2i}\left(z - \frac{1}{z}\right), \; dz = ie^{i\theta}\,d\theta.$$

Thus, the integral (9.39) is transformed into the form

$$\int_C F(z)\,dz,$$

where C is the closed unit circle.

Then, we find the poles of $F(z)$, and residues of $F(z)$ at all the poles, and using Cauchy's residue theorem, we get the value of the integral as

$$\int_C f(z)\,dz = 2\pi i \Sigma R. \tag{9.40}$$

Let us consider some examples.

EXAMPLE 9.8. *Evaluate*

$$\int_0^{2\pi} \frac{d\theta}{(1+a^2)-2a\cos\theta}, \quad 0<a<1.$$

Solution: We put $z=e^{i\theta}$, $\therefore\ \cos\theta=\frac{1}{2}\left(z+\frac{1}{z}\right)$, $dz=ie^{i\theta}\,d\theta$.

$$\therefore\quad I=\int_0^{2\pi}\frac{d\theta}{(1+a^2)-2a\cos\theta}=\int_C\frac{1}{(1+a^2)-a\left(z+\frac{1}{z}\right)}\cdot\frac{dz}{iz}=-\frac{1}{ia}\int_C f(z)\,dz, \tag{1}$$

where, $\frac{1}{f(z)}=z^2-z(a+\frac{1}{a})+1=(z-a)(z-\frac{1}{a})$.

Thus, poles of $f(z)$ are $z=a$ and $z=\frac{1}{a}$ out of which only the point $z=a$ lies within the circle $|z|<1$, since $a<1$.

The residue of $f(z)$ at $z=a$ is

$$\begin{aligned}R_1&=\lim_{z\to a}(z-a)f(z)=\lim_{z\to a}(z-a)\frac{1}{(z-a)(z-\frac{1}{a})}\\&=\frac{1}{a-\frac{1}{a}}=\frac{a}{a^2-1}.\end{aligned}$$

Hence, by Cauchy's residue theorem

$$\int_C f(z)\,dz=2\pi i\left(\frac{a}{a^2-1}\right). \tag{2}$$

From equations (1) and (2), we get

$$I=-\frac{1}{ai}\int_C f(z)\,dz=-\frac{1}{ai}2\pi i\left(\frac{a}{a^2-1}\right),\quad \text{i.e.,}\quad I=\frac{2\pi}{1-a^2}.$$

EXAMPLE 9.9. *Evaluate*

$$I=\int_0^{2\pi}\frac{d\theta}{(a+b\cos\theta)^2},\quad a>b>0.$$

Solution: With the substitution $z=e^{i\theta}$, we have

$$\cos\theta=z+\frac{1}{z}\quad\text{and}\quad d\theta=\frac{dz}{iz}.$$

Hence, the given integral is

$$I=\int_0^{2\pi}\frac{d\theta}{(a+b\cos\theta)^2}=\frac{1}{i}\int_C\frac{dz}{z\left\{a+\frac{b}{2}\left(z+\frac{1}{z}\right)\right\}^2},$$

where C = unit circle,

$$\text{i.e.,}\quad I=\frac{4}{i}\int_C F(z)\,dz, \tag{1}$$

where

$$F(z)=\frac{z}{(bz^2+2az+b^2)^2}=\frac{z}{b^2(z-\lambda)^2(z-\mu)^2} \tag{2}$$

with

$$\lambda = \frac{1}{b}\left(-a + \sqrt{a^2 - b^2}\right) \quad \text{and} \quad \mu = \frac{1}{b}\left(-a - \sqrt{a^2 - b^2}\right). \tag{3}$$

Thus, $z = \lambda$ is a pole of order 2 which lies within the circle C.
The residue at $z = \lambda$ is given by

$$R = \lim_{z \to \lambda} \frac{1}{1!}\frac{d}{dz}\left[(z-\lambda)^2 F(z)\right] \qquad \text{[putting } m = 2 \text{ in equation (9.19)]}$$

$$= \frac{1}{b^2}\frac{d}{dz}\left[\frac{z}{(z-\mu)^2}\right]_{z=\lambda} \qquad \text{[putting } F(z) \text{ from (2)]}$$

$$= \frac{1}{b^2}\left[\frac{z-\mu-2z}{(z-\mu)^3}\right]_{z=\lambda}$$

$$= -\frac{1}{b^2}\frac{\lambda+\mu}{(\lambda-\mu)^3}$$

$$\text{or,} \quad R = \frac{a}{4(a^2-b^2)^{\frac{3}{2}}} \qquad \text{[using equation (3)]}$$

Hence, by Cauchy's residue theorem

$$\int_C F(z)\,dz = 2\pi i \Sigma R = 2\pi i\left\{\frac{a}{4(a^2-b^2)^{\frac{3}{2}}}\right\}. \tag{4}$$

Therefore, from (1), the value of the given integral is

$$I = \frac{4}{i}2\pi i\left\{\frac{a}{4(a^2-b^2)^{\frac{3}{2}}}\right\} \qquad \text{[using equation (4)]}$$

$$\text{i.e.,} \quad I = \frac{2\pi a}{(a^2-b^2)^{\frac{3}{2}}}.$$

EXAMPLE 9.10. *Using calculus of residues, show that*

$$\int_0^{2\pi} \frac{\sin^2\theta}{a + b\cos\theta}d\theta = \frac{2\pi}{b^2}\left\{a - \sqrt{a^2-b^2}\right\},$$

where $a > b > 0$.

Solution: The given integral is

$$I = \int_0^{2\pi} \frac{\sin^2\theta\,d\theta}{a+b\cos\theta} = \frac{1}{2}\int_0^{2\pi}\frac{1-\cos 2\theta}{a+b\cos\theta}d\theta = \text{Real part of} \int_0^{2\pi}\frac{1-e^{2i\theta}}{2a+b\left(e^{i\theta}+e^{-i\theta}\right)}d\theta.$$

Now, we put $z = e^{i\theta}$, so that $d\theta = \dfrac{dz}{iz}$.
Hence

$$I = \text{Real part of} \int_C \frac{1-z^2}{2a+b\left(z+\frac{1}{z}\right)}\cdot\frac{dz}{iz} = \text{Real part of } \frac{1}{i}\int_C F(z)\,dz, \tag{1}$$

where

$$F(z) = \frac{1-z^2}{bz^2+2az+b} = \frac{1-z^2}{b(z-\lambda)(z-\mu)}. \tag{2}$$

The poles of $F(z)$ are $z=\lambda$ and $z=\mu$, where

$$\lambda = \frac{1}{b}\left(-a+\sqrt{a^2-b^2}\right) \quad \text{and} \quad \mu = \frac{1}{b}\left(-a-\sqrt{a^2-b^2}\right). \tag{3}$$

But, the only pole lying within unit circle is $z=\lambda$. The residue at this point is

$$\begin{aligned} R &= \lim_{z\longrightarrow\lambda}(z-\lambda)F(z) = \lim_{z\longrightarrow\lambda}(z-\lambda)\frac{1-z^2}{b(z-\lambda)(z-\mu)} \\ &= \frac{1-\lambda^2}{b(\lambda-\mu)} = \frac{\lambda\mu-\lambda^2}{b(\lambda-\mu)} \quad [\because \lambda\mu=1] \\ &= -\frac{\lambda}{b} = -\frac{1}{b^2}\left(-a+\sqrt{a^2-b^2}\right). \end{aligned} \tag{4}$$

Hence, by Cauchy's theorem of residue

$$\int_C F(z)\,dz = 2\pi i\left(-\frac{\lambda}{b}\right) = -\frac{2\pi i}{b^2}\left(-a+\sqrt{a^2-b^2}\right). \tag{5}$$

Hence, from (1), we get

$$I = \text{Real part of } \frac{1}{i}\int_C F(z)\,dz = \frac{2\pi}{b^2}\left(a-\sqrt{a^2-b^2}\right).$$

EXAMPLE 9.11. *Using calculus of residue, prove that*

$$\int_0^{2\pi} e^{\cos\theta}\cdot\cos(\sin\theta - n\theta)\,d\theta = \frac{2\pi}{n!},$$

where n is a positive integer.

Solution: The given integrand is $e^{\cos\theta}\cdot\cos(\sin\theta-n\theta)$

$$\begin{aligned} &= \text{Real part of } e^{\cos\theta}\cdot e^{i(\sin\theta-n\theta)} \\ &= \text{Real part of } e^{(\cos\theta+i\sin\theta)}\cdot e^{-in\theta} \\ &= \text{Real part of } e^{e^{i\theta}}\cdot e^{-in\theta}. \end{aligned}$$

Hence, the given integral is

$$I = \text{Real part of } \int_0^{2\pi} e^{e^{i\theta}}\cdot e^{-in\theta}\,d\theta. \tag{1}$$

We put $z=e^{i\theta}$, so that $d\theta = \frac{dz}{iz}$. Then the integral (1) becomes

$$\begin{aligned} I &= \text{Real part of } \int_C e^z\cdot\frac{1}{z^n}\cdot\frac{dz}{iz}, \quad C \text{ is circle } |z|=1 \\ &= \text{Real part of } \frac{1}{i}\int_C F(z)\,dz, \end{aligned} \tag{2}$$

where $F(z) = \frac{e^z}{z^{n+1}}$.

Obviously, $z = 0$ is a pole of order $(n+1)$ of $F(z)$.
Hence, residue at $z = 0$ is

$$\lim_{z \to 0} \frac{1}{n!} \frac{d^n}{dz^n} \left(z^{n+1} \cdot \frac{e^z}{z^{n+1}} \right) = \frac{1}{n!}. \tag{3}$$

Thus, by Cauchy's residue theorem

$$\int_C F(z)\, dz = 2\pi i \frac{1}{n!}. \tag{4}$$

Hence, from (2) and (4), $I = \dfrac{2\pi}{n!}$.

EXAMPLE 9.12. *Show that*

$$\int_0^{2\pi} \frac{\cos^2 3\theta\, d\theta}{1 - 2a\cos 2\theta + a^2} = \frac{\pi(1 - a + a^2)}{1 - a}, \quad 0 < a < 1.$$

Solution: The given integral is

$$I = \frac{1}{2} \int_0^{2\pi} \frac{1 + \cos 6\theta}{1 - 2a\cos 2\theta + a^2} d\theta$$

$$= \text{Re of } \frac{1}{2} \int_0^{2\pi} \frac{1 + e^{i6\theta}}{1 - 2a\cos 2\theta + a^2} \qquad \text{['Re' means real part]}$$

Putting $z = e^{i\theta}$, so that $\frac{dz}{iz} = d\theta$, and C is unit circle, we get

$$I = \text{Re of } -\frac{1}{2ia} \int_C F(z)\, dz, \tag{1}$$

where

$$F(z) = \frac{z\,(1 + z^6)}{(z^2 - a)\left(z^2 - \frac{1}{a}\right)}. \tag{2}$$

Poles of $F(z)$ are $z = \pm\sqrt{a}$ and $z = \pm\frac{1}{\sqrt{a}}$.
Since $a < 1$, only poles within C are $z = \pm\sqrt{a}$.
Now, residue at $z = +\sqrt{a}$ is

$$R_1 = \lim_{z \to \sqrt{a}} (z - \sqrt{a})\, F(z) = \lim_{z \to \sqrt{a}} (z - \sqrt{a}) \frac{z(1 + z^6)}{(z^2 - a)\left(z^2 - \frac{1}{a}\right)}$$

$$= \lim_{z \to \sqrt{a}} \frac{z(1 + z^6)}{(z + \sqrt{a})\left(z^2 - \frac{1}{a}\right)} = \frac{a(1 + a^3)}{2(a^2 - 1)}.$$

Similarly, residue at $z = -\sqrt{a}$ is

$$R_2 = \frac{a(1 + a^3)}{2(a^2 - 1)}.$$

Hence, sum of the residues

$$R_1 + R_2 = \frac{a(1 + a^3)}{a^2 - 1} = \frac{a(a + 1)(a^2 - a + 1)}{(a - 1)(a + 1)} = \frac{a(a^2 - a + 1)}{a - 1}.$$

Hence, from (1), the given integral is real part of

$$-\frac{1}{2ia}\cdot 2\pi i\frac{a(a^2-a+1)}{a-1}=\pi\frac{1-a+a^2}{1-a}.$$

EXAMPLE 9.13. *Show that*

$$\int_0^{\pi}\tan(\theta+i\alpha)d\theta=i\pi,\quad \textit{where Re }(\alpha)>0.$$

Solution: The integrand is $\dfrac{\sin(\theta+i\alpha)}{\cos(\theta+i\alpha)}=\dfrac{\sin(\theta+i\alpha)\cdot 2\cos(\theta-i\alpha)}{\cos(\theta+i\alpha)\cdot 2\cos(\theta-i\alpha)}$

$$=\frac{\sin 2\theta+\sin 2i\alpha}{\cos 2\theta+\cos 2i\alpha}.$$

Hence, the integral is (putting $\xi=2\theta$)

$$I=\int_0^{\pi}\frac{\sin 2\theta+\sin 2i\alpha}{\cos 2\theta+\cos 2i\alpha}d\theta=\frac{1}{2}\int_0^{2\pi}\frac{\sin\xi+i\sinh 2\alpha}{\cos\xi+\cosh 2\alpha}d\xi.$$

We put $z=e^{i\xi}$, so that $\frac{dz}{iz}=d\xi$, and C is the circle $|z|=1$.
Hence, the integral becomes

$$I=-\frac{1}{2}\int_C F(z)\,dz, \tag{1}$$

where

$$F(z)=\frac{z^2-2z\sinh 2\alpha-1}{z(z^2+1+2z\cosh 2\alpha)}. \tag{2}$$

The poles of $F(z)$ are $z=0$, $z=\lambda$ and $z=\mu$, where

$$\left.\begin{aligned}\lambda&=-\cosh 2\alpha+\sinh 2\alpha\\ \mu&=-\cosh 2\alpha-\sinh 2\alpha\end{aligned}\right\}. \tag{3}$$

Since $\lambda\mu=1$ and $|\lambda|<1$, so $|\mu|>1$. Hence, the simple poles are $z=0$ and $z=\lambda$.
Residue at $(z=0)=\lim\limits_{z\longrightarrow 0} z\,F(z)=-1$.

Residue at $z=\lambda=\lim\limits_{z\longrightarrow\lambda}(z-\lambda)F(z)=\lim\limits_{z\longrightarrow\lambda}(z-\lambda)\dfrac{z^2-2z\sinh 2\alpha-1}{z(z-\lambda)(z-\mu)}$

$$=\frac{\lambda^2-2\lambda\sinh 2\alpha-1}{\lambda(\lambda-\mu)}=\frac{\lambda^2-2\lambda\sinh 2\alpha-\lambda\mu}{\lambda(\lambda-\mu)}$$

$$=\frac{(\lambda-\mu)-2\sinh 2\alpha}{\lambda-\mu}=0\quad\text{[using (3)]}$$

Hence, by Cauchy's theorem

$$\int_C F(z)\,dz=2\pi i(-1+0)=-2\pi i.$$

Thus, from (1)

$$I=-\frac{1}{2}\int_C F(z)\,dz=-\frac{1}{2}(-2\pi i)=\pi i.$$

(B) *Integration round an infinite semi-circle*: Integrals of the form

$$\int_{-\infty}^{\infty} f(x)\,dx \tag{9.41}$$

are evaluated by considering a contour consisting of a semi-circle of large radius R in the upper half-plane bounded by its diameter along the real axis. The value of the above integral, under certain conditions, is

$$\int_{-\infty}^{\infty} f(x)\,dx = 2\pi i \Sigma R^*,$$

where ΣR^* = sum of residues of $f(z)$ at poles within the contour. The conditions are:

(i) The function $f(z)$ is analytic in the upper half of z-plane except at a finite number of poles.

(ii) The function $f(z)$ has no poles on real axis.

(iii) $zf(z) \longrightarrow 0$ as $|z| \longrightarrow \infty$.

We now prove it.

Since $f(z)$ has only singularities in the upper half of z-plane, by Cauchy's residue theorem we have

$$\int_C f(z)\,dz = 2\pi i \Sigma R^*, \tag{9.42}$$

where C is the closed contour consisting of real axis from $-R$ to R, and semi-circle Γ (*Fig 9.2*).

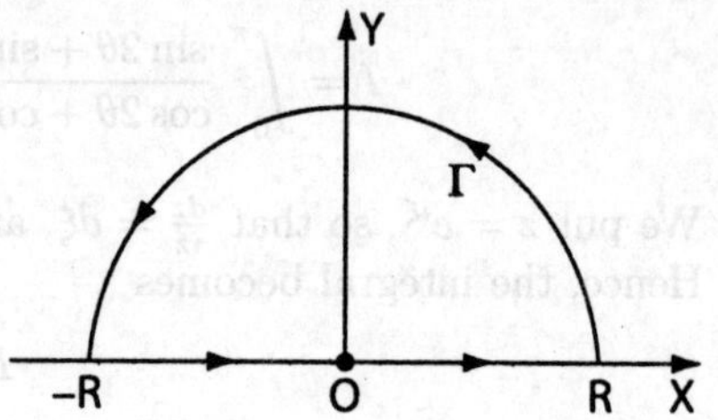

Fig 9.2

Hence, from equation (9.42), we have

$$\int_{-R}^{R} f(x)\,dx + \int_{\Gamma} f(z)\,dz = 2\pi i \Sigma R^* \tag{9.43}$$

Here, ΣR^* = the sum of residues of $f(z)$ at poles in the upper half of the z-plane only, since there are no poles on real axis. Now, since $|z| = R$ on Γ, and

$$\lim_{R \longrightarrow \infty} z\,f(z) = 0, \quad \text{[condition (iii)]}$$

where Γ is the arc $0 \leq \theta \leq \pi$, then by Theorem 9.9, we have, (putting $B = 0$)

$$\int_{\Gamma} f(z)\,dz = 0. \tag{9.44}$$

Moreover, in the limit as $R \longrightarrow \infty$,

$$\lim_{R \longrightarrow \infty} \int_{-R}^{R} f(x)\,dx = \int_{-\infty}^{\infty} f(x)\,dx. \tag{9.45}$$

Thus, proceeding to the limit as $R \longrightarrow \infty$ in equation (9.43) and using (9.44) and (9.45), we get

$$\int_{-\infty}^{\infty} f(x)\,dx = 2\pi i \Sigma R^*.$$

Hence, we have proved the following theorem:

Theorem 9.10. *If $f(z)$ is analytic in the upper half of the z-plane except at a finite number of poles in it, but having no poles on real axis, and if further $zf(z) \longrightarrow 0$ as $|z| \longrightarrow \infty$, then*

$$\int_{-\infty}^{\infty} f(x)\,dx = 2\pi i \Sigma R^*,$$

where ΣR^* represents the sum of residues of the function $f(z)$ at its poles in the upper half of z-plane.

We shall now see some examples.

EXAMPLE 9.14. *Show that*

$$\int_0^{\infty} \frac{dx}{x^4 + a^4} = \frac{\pi\sqrt{2}}{4a^3}, \quad a > 0.$$

Solution: The integrand being an even function, we can write

$$I = \int_0^{\infty} \frac{dx}{x^4 + a^4} = \frac{1}{2}\int_{-\infty}^{\infty} \frac{dx}{x^4 + a^4}. \tag{1}$$

The integral in (1) represents integration of

$$f(z) = \frac{1}{z^4 + a^4} \tag{2}$$

along the entire real axis.

Now, the poles of the function $f(z)$ are given by

$$z^4 + a^4 = 0, \quad \text{i.e.,} \quad z = a(-1)^{\frac{1}{4}}, \tag{3}$$

where $(-1)^{\frac{1}{4}} = \Big\{\cos(2n\pi + \pi) + i\sin(2n\pi + \pi)\Big\}^{\frac{1}{4}}$

$$= \cos\left(\frac{2n\pi + \pi}{4}\right) + i\sin\left(\frac{2n\pi + \pi}{4}\right), \text{ with } n = 0, 1, 2, 3.$$

Hence, the poles are $ae^{i\frac{\pi}{4}}$, $ae^{i\frac{3\pi}{4}}$, $ae^{i\frac{5\pi}{4}}$ and $ae^{i\frac{7\pi}{4}}$.

Out of these, only first two poles remain within the semi-circle, the other two having amplitude $> \pi$ remain outside.

Thus, $f(z)$ is analytic in the upper half of z-plane except at these poles, $f(z)$ has no pole on real axis, and

$$\lim_{z \longrightarrow \infty} z f(z) = \lim_{z \longrightarrow \infty} \frac{z}{z^4 + a^4} = 0.$$

Thus, the function $f(z)$ satisfies all the conditions of Theorem 9.10, and hence

$$\int_{-\infty}^{\infty} \frac{dx}{x^4 + a^4} = 2\pi i \Sigma R^*, \tag{4}$$

where ΣR^* is the sum of residues of $f(z)$ at its poles which are

$$\lambda = ae^{i\frac{\pi}{4}} \quad \text{and} \quad \mu = ae^{i\frac{3\pi}{4}}. \tag{5}$$

Now, residue at $z = \lambda$ is

$$\begin{aligned}\lim_{z\to\lambda}(z-\lambda)f(z) &= \lim_{z\to\lambda}(z-\lambda)\frac{1}{z^4+a^4}\\ &= \lim_{z\to\lambda}(z-\lambda)\frac{1}{z^4-\lambda^4} \quad \text{[by equation (3)]}\\ &= \lim_{z\to\lambda}\frac{1}{4z^3} \quad \text{[using L'Hospital's rule]}\\ &= \frac{1}{4\lambda^3} = -\frac{\lambda}{4a^4} = -\frac{1}{4a^4}ae^{i\frac{\pi}{4}}. \qquad (6)\end{aligned}$$

Similarly, the residue at $z = \mu$ is

$$-\frac{1}{4a^4}ae^{i\frac{3\pi}{4}}. \qquad (7)$$

Hence, sum of residues is

$$\Sigma R^* = -\frac{1}{4a^3}\left(e^{i\frac{\pi}{4}} + e^{i\frac{3\pi}{4}}\right) = -\frac{1}{4a^3}\left(e^{i\frac{\pi}{4}} - e^{\frac{-i\pi}{4}}\right) = -\frac{1}{4a^3}\left(2i\sin\frac{\pi}{4}\right) = -\frac{\sqrt{2}i}{4a^3}. \qquad (8)$$

Hence, from (1), (2) and (8), we get

$$I = \int_0^\infty \frac{dx}{x^4+a^4} = \frac{1}{2}\int_{-\infty}^\infty \frac{dx}{x^4+a^4} = \frac{1}{2}(2\pi i)\left(\frac{-\sqrt{2}i}{4a^3}\right) = \frac{\sqrt{2}\pi}{4a^3}.$$

Hence proved.

EXAMPLE 9.15. *Evaluate the integral*

$$\int_{-\infty}^\infty \frac{x^2-x+2}{x^4+10x^2+9}\,dx.$$

Solution: The given integral represents the integration of the function

$$f(z) = \frac{z^2-z+2}{z^4+10z^2+9} \qquad (1)$$

along the entire length of the real axis.

Now, the poles of $f(z)$ are given by

$$\begin{aligned} z^4+10z^2+9=0, \quad &\text{i.e.,} \quad (z^2+9)(z^2+1)=0,\\ &\text{i.e.,} \quad z=\pm i \quad \text{and} \quad z=\pm 3i. \qquad (2)\end{aligned}$$

Out of the four poles given by equation (2), only two poles $z = i$ and $z = 3i$ remain in the upper half of z-plane.

Thus, $f(z)$ is analytic in the upper half of the z-plane except at those two poles, and $f(z)$ has no pole on real axis. Moreover

$$\lim_{z\to\infty} zf(z) = \lim_{z\to\infty}\frac{z(z^2-z+2)}{z^4+10z^2+9} = 0.$$

Thus, the function $f(z)$ satisfies all the conditions of the Theorem 9.10. Hence

$$\int_{-\infty}^\infty \frac{x^2-x+2}{x^4+10x^2+9}\,dx = 2\pi i\Sigma R^*, \qquad (3)$$

where ΣR^* = sum of residues at poles $z = i$ and $z = 3i$.

Now, residue of $f(z)$ at $z = i$ is

$$\lim_{z \longrightarrow i} (z - i) f(z) = \lim_{z \longrightarrow i} (z - i) \frac{z^2 - z + 2}{(z^2 + 9)(z + i)(z - i)} = \lim_{z \longrightarrow i} \frac{z^2 - z + 2}{(z^2 + 9)(z + i)}$$

$$= \frac{i^2 - i + 2}{(i^2 + 9)(i + i)} = \frac{1 - i}{16i}. \tag{4}$$

Similarly, residue of $f(z)$ at $z = 3i$ is

$$\frac{7 + 3i}{48i}. \tag{5}$$

From (4) and (5), sum of residues is

$$\Sigma R^* = \frac{1 - i}{16i} + \frac{7 + 3i}{48i} = \frac{5}{24i}. \tag{6}$$

Hence, from (3) and (6), we get

$$\int_{-\infty}^{\infty} \frac{x^2 - x + 2}{x^4 + 10x^2 + 9}\, dx = 2\pi i \left(\frac{5}{24i}\right) = \frac{5\pi}{12}.$$

EXAMPLE 9.16. *Evaluate the integral*

$$\int_{-\infty}^{\infty} \frac{dx}{(x^2 + b^2)(x^2 + a^2)^2}, \quad (a > 0,\ b > 0).$$

Solution: The given integral represents the integration of the function

$$f(z) = \frac{1}{(z^2 + a^2)^2 (z^2 + b^2)} \tag{1}$$

along the entire length of the real axis.

The poles of $f(z)$ are given by $(z^2 + a^2)^2 (z^2 + b^2) = 0$, i.e., $z = \pm ib$ and $z = \pm ia$ (twice). Out of these four poles, only two lie in the upper half-plane. These are:

$$z = ib \text{ (simple)} \quad \text{and} \quad z = ia \text{ (order two)}. \tag{2}$$

Thus, $f(z)$ is analytic in the upper half of the z-plane except at those two poles, $f(z)$ has no pole on real axis, and

$$\lim_{z \longrightarrow \infty} z f(z) = \lim_{z \longrightarrow \infty} \frac{z}{(z^2 + a^2)(z^2 + b^2)} = 0.$$

Thus the function $f(z)$ satisfies all the conditions of the theorem 9.10. Hence

$$\int_{-\infty}^{\infty} \frac{dx}{(x^2 + a^2)^2 (x^2 + b^2)} = 2\pi i \Sigma R^*, \tag{3}$$

where ΣR^* is the sum of residues at poles of $f(z)$.

Now, the residue of $f(z)$ at $z = ib$ is

$$R_1 = \lim_{z \longrightarrow ib} (z - ib)\, f(z) = \lim_{z \longrightarrow ib} (z - ib) \frac{1}{(z^2 + a^2)^2 (z - ib)(z + ib)}$$

$$= \frac{1}{2ib(a^2 - b^2)^2} = -\frac{i}{2b(a^2 - b^2)^2}. \tag{4}$$

The residue of $f(z)$ at $z = ia$ of order two, is obtained by using equation (9.19) as

$$R_2 = \frac{d}{dz}\left[\frac{(z - ia)^2}{(z^2 + b^2)(z + ia)^2 (z - ia)^2}\right]_{z=ia}$$

$$= \frac{2(z + ia)\left\{-(z^2 + b^2) - z(z + ia)\right\}}{\left\{(z^2 + b^2)(z + ia)^2\right\}^2} \text{ for } z = ia,$$

$$\text{i.e.,}\quad R_2 = \frac{(3a^2 - b^2)\, i}{4a^3 (a^2 - b^2)^2}. \tag{5}$$

Hence, from (4) and (5), the sum of the residues is

$$\Sigma R^* = R_1 + R_2 = -\frac{i}{2b\,(a^2 - b^2)^2} + \frac{(3a^2 - b^2)\, i}{4a^3\,(a^2 - b^2)^2}$$

$$= \frac{i}{4a^3 b\,(a^2 - b^2)^2}\left\{-2a^3 + b\,(3a^2 - b^2)\right\}$$

$$= \frac{-i}{4a^3 b\,(a^2 - b^2)^2}(b + 2a)(b - a)^2$$

$$= -\frac{i\,(2a + b)}{4a^3 b (a + b)^2}. \tag{6}$$

Hence, from (3) and (6), we get

$$\int_{-\infty}^{\infty} \frac{dx}{(x^2 + b^2)(x^2 + a^2)^2} = 2\pi i\left[\frac{-i(2a + b)}{4a^3 b (a + b)^2}\right] = \frac{2\pi(2a + b)}{4a^3 b (a + b)^2}.$$

We now stop for a while dealing with integrals and look into an useful inequality as well as a lemma due to Jordan.

9.7 Jordan's Inequality

We know that the value of $\cos\theta$ steadily decreases as θ increases from 0 to $\frac{\pi}{2}$. The mean value of $\cos x$ in a range $0 \leq x \leq \theta$, which is given by

$$\frac{1}{\theta}\int_0^{\theta} \cos x\, dx = \frac{\sin\theta}{\theta}$$

also decreases steadily. Hence, for $0 \leq \theta \leq \frac{\pi}{2}$

$$\frac{2}{\pi} \leq \frac{\sin\theta}{\theta} \leq 1, \quad \text{i.e.,}\quad \frac{2\theta}{\pi} \leq \sin\theta \leq \theta. \tag{9.46}$$

This is known as *Jordan's inequality*.

9.8 Jordan's Lemma

Theorem 9.11. *If a function $f(z)$ is analytic in the upper half-plane except at a finite number of poles, and if $f(z) \longrightarrow 0$ uniformly as $z \longrightarrow \infty$, then*

$$\lim_{R \longrightarrow \infty} \int_\Gamma e^{imz} f(z) = 0, \tag{9.47}$$

where $m > 0$ and Γ is the semi-circle $|z| = R$ in the upper half-plane.

Proof. Since $f(z) \longrightarrow 0$ uniformly as $|z| \longrightarrow \infty$ for $0 \le \operatorname{amp} z \le \pi$, we choose R sufficiently large, such that

$$|f(z)| < \epsilon \text{ for } \epsilon > 0. \tag{9.48}$$

Now

$$\left| \int_\Gamma e^{imz} f(z)\, dz \right| = \left| \int_0^\pi e^{imz} f(z)\, Re^{i\theta}\, i\, d\theta \right|, \quad [z = Re^{i\theta}]$$

$$\le \int_0^\pi |f(z)|\, |e^{imz}|\, |Re^{i\theta}\, d\theta|. \tag{9.49}$$

But

$$imz = imRe^{i\theta} = imR(\cos\theta + i\sin\theta).$$

Hence

$$|e^{imz}| = \left| e^{imR(\cos\theta + i\sin\theta)} \right| \le |e^{imR\cos\theta}| \cdot |e^{-mR\sin\theta}| = e^{-mR\sin\theta}. \tag{9.50}$$

Thus, using equations (9.48) and (9.50), equation (9.49) becomes

$$\left| \int_\Gamma e^{imz} f(z)\, dz \right| \le R\epsilon \int_0^\pi |e^{-mR\sin\theta}|\, d\theta$$

$$\le R\epsilon \int_0^\pi e^{-mR\left(\frac{2\theta}{\pi}\right)} d\theta \quad \text{[using Jordan's inequality]}$$

$$= R\epsilon \left(\frac{-\pi}{2mR} \right) \left[e^{-2mR\frac{\theta}{\pi}} \right]_0^\pi$$

$$= \frac{\pi\epsilon}{2m} \left(1 - e^{-2mR}\right),$$

which tends to zero as $R \longrightarrow \infty$ and $\epsilon \longrightarrow 0$.

Hence

$$\lim_{R \longrightarrow \infty} \int_\Gamma e^{imz} f(z)\, dz = 0.$$

Thus the theorem is proved. □

We shall now use this theorem in evaluating a class of integrals given by

$$I = \int_{-\infty}^{\infty} e^{imx} f(x)\, dx. \tag{9.51}$$

(C) *Evaluation of integrals using Jordan's Lemma*: Integrals of the type (9.51) are evaluated round a contour C consisting of a semi-circle Γ of radius R large enough to include all the poles of the integrand in the upper half-plane and also the portion of real axis from $x = -R$ to $x = R$. Therefore, by Cauchy's residue theorem

$$\int_C e^{imz} f(z)\, dz = 2\pi i \Sigma R^*, \tag{9.52}$$

where ΣR^* = sum of residues of all poles of $f(z)$ within the contour C. We assume that $f(z) \longrightarrow 0$ as $z \longrightarrow \infty$.

From equation (9.52), we get

$$\int_{-R}^{R} e^{imx} f(x)\, dx + \int_\Gamma e^{imz} f(z)\, dz = 2\pi i \Sigma R^*. \tag{9.53}$$

Proceeding to the limit as $R \longrightarrow \infty$, and noting that, by virtue of Jordan's lemma (Theorem 9.11), the second integral in LHS vanishes, we get

$$\int_{-\infty}^{\infty} e^{imx} f(x)\, dx = 2\pi i \Sigma R^*$$

$$\text{or,} \quad \int_{-\infty}^{\infty} f(x) \{\cos mx + i \sin mx\}\, dx = 2\pi i \Sigma R^*.$$

Equating real and imaginary parts, we will get the values of the integrals

$$\left.\begin{aligned} &\int_{-\infty}^{\infty} f(x) \cos mx\, dx \\ \text{and} \quad &\int_{-\infty}^{\infty} f(x) \sin mx\, dx \end{aligned}\right\} \text{for } m > 0. \tag{9.54}$$

Let us now evaluate some integrals of these types.

EXAMPLE 9.17. *Evaluate*

$$\int_0^\infty \frac{x \sin mx}{a^2 + x^2}\, dx, \quad m > 0.$$

Solution: We consider the integral

$$\int_C e^{imz} \frac{z}{a^2 + z^2}\, dz, \tag{1}$$

where C is a contour consisting of a semi-circle Γ in the upper half of z-plane of radius R large enough to include all the poles of the integrand in (1), and also the part of the real axis from $x = -R$ to $x = R$.

Then, by Cauchy's residue theorem

$$\int_C e^{imz} \frac{z}{a^2 + z^2}\, dz = 2\pi i \Sigma R^*, \tag{2}$$

where ΣR^* = sum of residues at poles of the integrand in equation (2).

Now

$$\lim_{z \longrightarrow \infty} f(z) = \lim_{z \longrightarrow \infty} \frac{z}{a^2 + z^2} = 0.$$

Hence, by Jordan's lemma

$$\lim_{R \longrightarrow \infty} \int_{\Gamma} e^{imz} f(z)\, dz = 0. \tag{3}$$

Thus, proceeding to limit $R \longrightarrow \infty$, we get from equation (2), using equation (3)

$$\int_{-\infty}^{\infty} e^{imx} \frac{x}{a^2 + x^2}\, dx = 2\pi i \Sigma R^*. \tag{4}$$

Now, the poles of

$$\phi(z) = e^{imz} \frac{z}{a^2 + z^2} \tag{5}$$

are given by $a^2 + z^2 = 0$, i.e., $z = \pm ai$.
But the only pole remaining within C is $z = ai$.
Now, residue of $\phi(z)$ at $z = ai$ is

$$\begin{aligned} R_1 &= \lim_{z \longrightarrow ai} (z - ai)\, \phi(z) = \lim_{z \longrightarrow ai} (z - ai) \frac{z \cdot e^{imz}}{(z + ai)\,(z - ai)} \\ &= \frac{(ai)\, e^{-ma}}{2ai} = \frac{1}{2} e^{-ma}. \end{aligned} \tag{6}$$

Hence, from equations (4) and (6), we get

$$\int_{-\infty}^{\infty} e^{imx} \cdot \frac{x}{a^2 + x^2}\, dx = 2\pi i \left(\frac{1}{2} e^{-ma} \right) = \pi i e^{-ma}.$$

Equating imaginary parts, we get

$$\int_{-\infty}^{\infty} \frac{x \sin mx}{a^2 + x^2}\, dx = \pi e^{-ma} \quad \text{or,} \quad 2 \int_0^{\infty} \frac{x \sin mx}{a^2 + x^2}\, dx = \pi e^{-ma}.$$

$$\therefore \quad \int_0^{\infty} \frac{x \sin mx}{a^2 + x^2}\, dx = \frac{\pi}{2} e^{-ma}.$$

EXAMPLE 9.18. *Prove that*

$$\int_{-\infty}^{\infty} \frac{\cos x}{(x^2 + a^2)\,(x^2 + b^2)}\, dx = \frac{\pi}{a^2 - b^2} \left(\frac{e^{-b}}{b} - \frac{e^{-a}}{a} \right),$$

where $a > b > 0$.

Solution: We consider the integral

$$\int_C \phi(z)\, dz = \int_C e^{iz} f(z)\, dz,$$

where

$$f(z) = \frac{1}{(z^2 + a^2)\,(z^2 + b^2)} \tag{1}$$

and C is a contour consisting of a semi-circle Γ in the upper half of z-plane of radius R large enough to include all the poles of the function $\phi(z)$, and also the part of the real axis from $x = -R$ to $x = R$.

Then, by Cauchy's residue theorem

$$\int_C e^{iz} f(z)\,dz = 2\pi i \Sigma R^*,$$

$$\text{i.e.,} \quad \int_{-R}^{R} e^{ix} f(x)\,dx + \int_\Gamma e^{iz} f(z)\,dz = 2\pi i \Sigma R^*, \tag{2}$$

where ΣR^* = sum of residues of $\phi(z)$ at its poles within C.

Now

$$\lim_{z\longrightarrow\infty} f(z) = \lim_{z\longrightarrow\infty} \frac{1}{(z^2+a^2)(z^2+b^2)} = 0. \tag{3}$$

Hence, by Jordan's lemma, we have

$$\lim_{R\longrightarrow\infty} \int_\Gamma e^{iz} f(z)\,dz = 0. \tag{4}$$

Thus, proceeding to the limit $R \longrightarrow \infty$, we get from equation (2) using equation (4)

$$\int_{-\infty}^{\infty} e^{ix} f(x)\,dx = 2\pi i \Sigma R^*,$$

$$\text{i.e.,} \quad \int_{-\infty}^{\infty} \frac{e^{ix}}{(x^2+a^2)(x^2+b^2)} = 2\pi i \Sigma R^*. \tag{5}$$

Now, the poles of $\phi(z)$ are given by $(z^2+a^2)(z^2+b^2) = 0$, which gives $z = \pm ai$ and $z = \pm bi$.

Out of these four poles, only two poles $z = ai$ and $z = bi$ are within C.

Now, residue at $z = ai$ is

$$R_1 = \lim_{z\longrightarrow ai} (z-ai)\phi(z) = \lim_{z\longrightarrow ai} (z-ai) \frac{e^{iz}}{(z+ai)(z-ai)(z^2+b^2)},$$

$$\text{i.e.,} \quad R_1 = \frac{e^{-a}}{2ai\,(b^2-a^2)}. \tag{6}$$

Similarly, the residue at $z = bi$ is

$$R_2 = \frac{e^{-b}}{2bi\,(a^2-b^2)}. \tag{7}$$

Hence, the sum of the residues is

$$\Sigma R^* = \frac{1}{2i\,(a^2-b^2)} \left\{ \frac{e^{-b}}{b} - \frac{e^{-a}}{a} \right\}. \tag{8}$$

Hence, from equations (5) and (8), we get

$$\int_{-\infty}^{\infty} \frac{e^{ix}}{(x^2+a^2)(x^2+b^2)}\,dx = (2\pi i) \frac{1}{2i\,(a^2-b^2)} \left\{ \frac{e^{-b}}{b} - \frac{e^{-a}}{a} \right\}.$$

Equating the real parts on both sides

$$\int_{-\infty}^{\infty} \frac{\cos x}{(x^2+a^2)(x^2+b^2)}\,dx = \frac{\pi}{a^2-b^2} \left\{ \frac{e^{-b}}{b} - \frac{e^{-a}}{a} \right\}.$$

EXAMPLE 9.19. *Prove that*

$$\int_0^\infty \frac{\sin x}{x^2+4x+5}\,dx = -\frac{\pi}{e}\sin 2 \quad \textit{and} \quad \int_0^\infty \frac{\cos x}{x^2+4x+5}\,dx = \frac{\pi}{e}\cos 2.$$

Solution: We consider the integral

$$\int_C \phi(z)\,dz = \int_C e^{iz} f(z)\,dz,$$

where

$$f(z) = \frac{1}{z^2+4z+5} \tag{1}$$

and C is the contour as explained in previous example. Then, by Cauchy's residue theorem

$$\int_C e^{iz} f(z)\,dz = 2\pi i \Sigma R^*$$

$$\text{or,} \quad \int_{-R}^{R} e^{ix} f(x)\,dx + \int_\Gamma e^{iz} f(z)\,dz = 2\pi i \Sigma R^*, \tag{2}$$

where ΣR^* = sum of residues of $\phi(z)$ at its poles within C.

Now

$$\lim_{z\longrightarrow\infty} f(z) = \lim_{z\longrightarrow\infty} \frac{1}{z^2+4z+5} = 0.$$

Hence, by Jordan's lemma, we have

$$\lim_{R\longrightarrow\infty} \int_\Gamma e^{iz} f(z)\,dz = 0. \tag{3}$$

Thus, proceeding to the limits $R \longrightarrow \infty$, we get from equation (2) using equation (3)

$$\int_{-\infty}^{\infty} e^{ix} f(x)\,dx = 2\pi i \Sigma R^*. \tag{4}$$

Now, the poles of $\phi(z)$ are given by $z^2+4z+5=0$, i.e., $z = -2+i = (\alpha)$ and $-2-i = (\beta)$. Out of these poles, $z = -2+i$ is the only pole within C.

Now, the residue of $\phi(z)$ at $z = \alpha$ is

$$\lim_{z\longrightarrow\alpha} (z-\alpha)\frac{e^{iz}}{(z-\alpha)(z-\beta)} = \frac{e^{i\alpha}}{(\alpha-\beta)} = \frac{e^{i(-2+i)}}{2i} = \frac{e^{-1}\cdot e^{-2i}}{2i}$$

$$= \frac{1}{2ei}(\cos 2 - i\sin 2). \tag{5}$$

Hence, from equations (4) and (5), we get

$$\int_{-\infty}^{\infty} \frac{e^{ix}}{x^2+4x+5}\,dx = 2\pi i \cdot \frac{1}{2ei}(\cos 2 - i\sin 2) = \frac{\pi}{e}(\cos 2 - i\sin 2).$$

Equating real and imaginary parts, we get the required results.

(D) *Integration around indented semi-circle*: We have so far discussed about integrations, where the integrands do not have any pole on real axis. But the method can also be extended when the integrands have a simple pole on the real axis. In such cases, we exclude the poles on real axis by enclosing them with small semi-circles drawn in the upper half of z plane and with centres at the poles. This process is known as *indenting*. After indenting, the integrand becomes regular in the modified contour, so that we can apply Cauchy's residue theorem.

We shall take some examples to demonstrate this technique for evaluating integrals.

EXAMPLE 9.20. *Show that*

$$\int_{-\infty}^{\infty} \frac{\cos px - \cos qx}{x^2}\, dx = -\pi(p-q),$$

when p and q are positive.

Solution: We consider the integral

$$\int_C f(z)\, dz,$$

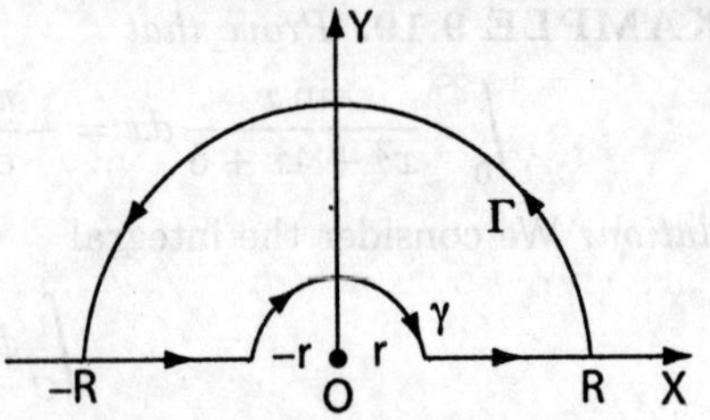

Fig 9.3

where

$$f(z) = \frac{e^{ipz} - e^{iqz}}{z^2} \tag{1}$$

and the contour C is a large semi-circle in the upper half of radius R, indented at origin with a small semi-circle of radius r (*Fig 9.3*). Thus, there are no pole within C. Hence, by Cauchy's residue theorem

$$\int_C f(z)\, dz = 2\pi i \Sigma R^* = 0,$$

$$\text{i.e.,} \quad \int_{-R}^{-r} f(x)\, dx + \int_\gamma f(z)\, dz + \int_r^R f(x)\, dx + \int_\Gamma f(z)\, dz = 0. \tag{2}$$

Now

$$I_1 = \int_\Gamma f(z)\, dz = \int_\Gamma \left(e^{ipz} - e^{iqz}\right) \frac{1}{z^2}\, dz.$$

Here, as $z \longrightarrow \infty$, $\frac{1}{z^2} \longrightarrow 0$. Hence, by Jordan's lemma

$$\lim_{R \longrightarrow \infty} \int_\Gamma \left(e^{ipz} - e^{iqz}\right) \frac{1}{z^2}\, dz = 0. \tag{3}$$

Again

$$I_2 = \int_\gamma f(z)\, dz,$$

where γ is the circle $|z| = r$ and

$$\begin{aligned}
\lim_{z \longrightarrow 0} z\, f(z) &= \lim_{z \longrightarrow 0} z \frac{e^{ipz} - e^{iqz}}{z^2} \\
&= \lim_{z \longrightarrow 0} \frac{1}{z} \left[\left(1 + ipz + \frac{i^2p^2z^2}{2} + \cdots \right) - \left(1 + iqz + i^2q^2z^2 + \cdots \right) \right] \\
&= \lim_{z \longrightarrow 0} i(p-q) - \frac{1}{2}\left(p^2 - q^2\right) z + \cdots \\
&= i(p-q).
\end{aligned} \tag{4}$$

Hence, by Theorem 9.8

$$\lim_{r \longrightarrow 0} \int_\gamma f(z)\, dz = -i(\pi - 0)(p-q)i \tag{5}$$

$$= \pi(p-q). \tag{6}$$

Negative sign has been taken in (5) since the circle γ is described in clockwise direction.

Thus, making $R \longrightarrow \infty$ and $r \longrightarrow 0$, equation (2) with the use of equations (3) and (6) becomes

$$\int_{-\infty}^{0} f(x)\,dx + \pi(p-q) + \int_{0}^{\infty} f(x)\,dx + 0 = 0$$

$$\text{or,} \quad \int_{-\infty}^{\infty} \frac{e^{ipx} - e^{iqx}}{x^2}\,dx = -\pi(p-q).$$

Equating real parts on both sides, we get

$$\int_{-\infty}^{\infty} \frac{\cos px - \cos qx}{x^2} dx = -\pi(p-q).$$

EXAMPLE 9.21. *Prove that*

$$\int_{0}^{\infty} \frac{\sin \pi x}{x(1-x^2)}\,dx = \pi.$$

Solution: We consider the integral

$$\int_C f(z)\,dz,$$

where

$$f(z) = \frac{e^{i\pi z}}{z(1-z^2)}. \tag{1}$$

The function $f(z)$ has poles at $z = 0$, $z = 1$, $z = -1$, all on the real axis. Hence, we choose the contour C to be a large semi-circle $|z| = R$ in the upper half of z-plane, indented at $z = 0$, $z = 1$, $z = -1$ by small semi-circles of radii r_1, r_2 and r_3 (*Fig 9.4*).

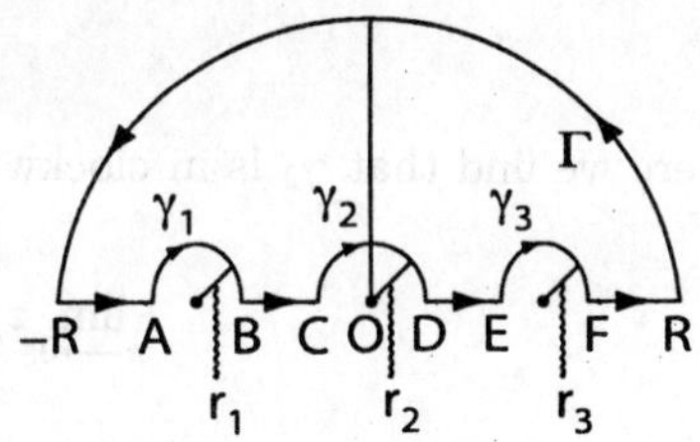

Fig 9.4

Then the given function $f(z)$ has no poles within C, and hence, by Cauchy's residue theorem

$$\int_C f(z)\,dz = 0. \tag{2}$$

Let us split the integral along different paths:

(i) from $-R$ to point A : $I_1 = \int_{-R}^{-(1+r_1)} f(x)\,dx$;

(ii) along semi-circle γ_1 : $I_2 = \int_{\gamma_1} f(z)\,dz$;

(iii) along B to point C : $I_3 = \int_{-(1-r_1)}^{-r_2} f(x)\,dx$;

(iv) along semi-circle γ_2 : $I_4 = \int_{\gamma_2} f(z)\,dz$;

(v) along point D to E: $\quad I_5 = \int_{r_2}^{1-r_3} f(x)\,dx;$

(vi) along semi-circle γ_3: $\quad I_6 = \int_{\gamma_3} f(z)\,dz;$

(vii) from point F to point R: $I_7 = \int_{1+r_3}^{R} f(x)\,dx;$ and

(viii) along large semi-circle: $\quad I_8 = \int_{\Gamma} f(z)\,dz.$

Now, for

$$I_2 = \int_{\gamma_1} f(z)\,dz,$$

where γ_1 is described in clockwise direction, we have

$$\lim_{z \longrightarrow -1} (z+1)f(z) = \lim_{z \longrightarrow -1} (z+1)\frac{e^{i\pi z}}{z(1+z)(1-z)} = \frac{e^{-i\pi}}{-2} = \frac{1}{2}. \tag{3}$$

Hence, by Theorem 9.8, we have

$$I_2 = \lim_{r_1 \longrightarrow 0} \int_{\gamma_1} f(z)\,dz = -i(\pi - 0)\frac{1}{2} = \frac{-i\pi}{2}. \tag{4}$$

Let us now consider

$$I_4 = \int_{\gamma_2} f(z)\,dz.$$

Here we find that γ_2 is in clockwise direction and

$$\lim_{z \longrightarrow 0} zf(z) = \lim_{z \longrightarrow 0} z\frac{e^{i\pi z}}{z(1-z^2)} = 1. \tag{5}$$

Hence, by Theorem 9.8, we have

$$I_4 = \lim_{r_2 \longrightarrow 0} \int_{\gamma_2} f(z)\,dz = -i(\pi - 0)\cdot 1 = -i\pi. \tag{6}$$

Again

$$I_6 = \int_{\gamma_3} f(z)dz,$$

and γ_3 is in clockwise sense.

Moreover

$$\lim_{z \longrightarrow 1} (z-1)f(z) = \lim_{z \longrightarrow 1} (z-1)\frac{e^{i\pi z}}{z(1+z)(1-z)} = \frac{e^{i\pi}}{-2} = \frac{1}{2}. \tag{7}$$

Hence, by Theorem 9.8, we have

$$I_6 = \lim_{r_3 \longrightarrow 0} \int_{\gamma} f(z)\,dz = -i(\pi - 0)\,\frac{1}{2} = -\frac{i\pi}{2}. \tag{8}$$

Lastly

$$I_8 = \int_\Gamma f(z)\,dz,$$

Γ is in anti-clockwise sense.

Moreover

$$\lim_{z\longrightarrow\infty} \frac{1}{z(1-z^2)} = 0.$$

Hence, by Jordan's lemma

$$\lim_{R\longrightarrow\infty} \int_\Gamma e^{i\pi z} \frac{dz}{z(1-z^2)} = 0. \tag{9}$$

Thus, by making r_1, r_2, $r_3 \longrightarrow 0$ and $R \longrightarrow \infty$, and using the results (4), (6), (8) and (9), we get from equation (2)

$$\int_{-\infty}^{-1} f(x)\,dx + \int_{-1}^{0} f(x)\,dx + \int_0^1 f(x)\,dx + \int_1^\infty f(x)\,dx + \left(-\frac{i\pi}{2} - i\pi - \frac{i\pi}{2}\right) = 0$$

$$\text{or,} \int_{-\infty}^{\infty} f(x)\,dx = 2i\pi, \quad \text{i.e.,} \int_{-\infty}^{\infty} \frac{e^{i\pi x}}{x(1-x^2)} = 2i\pi.$$

Equating imaginary parts on both sides, we get

$$\int_{-\infty}^{\infty} \frac{\sin \pi x}{x(1-x^2} \, dx = 2\pi.$$

$$\therefore \quad \int_0^\infty \frac{\sin \pi x}{x(1-x^2)}\, dx = \pi.$$

EXAMPLE 9.22. *Prove that*

$$\int_0^\infty \frac{x^{p-1}}{1+x^2}\,dx = \frac{\pi}{2}\operatorname{cosec}\left(\frac{\pi p}{2}\right),$$

when $0 < p < 2$.

Solution: We consider the integral

$$\int_C f(z)\,dz,$$

where

$$f(z) = \frac{z^{p-1}}{1+z^2} \tag{1}$$

and C is a contour consisting of the upper half of a circle Γ of large radius $|z| = R$, real axis from $-R$ to R, and a small semi-circle γ of radius r with centre at $z = 0$. Since $0 < p < 1$, $z = 0$ is a singularity of $f(z)$, so the contour is indented at $z = 0$.

The poles of $f(z)$ are, $z = \pm i$, where only $z = i$ remains within C.

Residue of $f(z)$ at $z = i$ is

$$\lim_{z\longrightarrow i} (z-i)\frac{z^{p-1}}{(z-i)(z+i)} = \frac{1}{2i}(i)^{p-1} = -\frac{1}{2}(i)^p = -\frac{1}{2}e^{i\pi\frac{p}{2}} \left[\because i = e^{i\pi/2}\right]. \tag{2}$$

Hence, by Cauchy's residue theorem

$$\int_C f(z)\,dz = 2\pi i \Sigma R^* = 2\pi i\left(-\frac{1}{2}e^{i\pi \frac{p}{2}}\right) = -\pi i e^{i\pi \frac{p}{2}}. \tag{3}$$

By splitting the integral for different paths, we get

$$\int_{-R}^{-r} f(x)\,dx + \int_\gamma f(z)\,dz + \int_r^R f(x)\,dx + \int_\Gamma f(z)\,dz = -\pi i e^{i\pi \frac{p}{2}}. \tag{4}$$

Now

$$\lim_{z\longrightarrow 0} z f(z) = \lim_{z\longrightarrow 0} \frac{z^p}{1+z^2} = 0 \quad (\because p > 0).$$

Hence, by Theorem 9.8

$$\int_\gamma f(z)\,dz = -i(\pi - 0)\cdot 0 = 0. \tag{5}$$

Again

$$\lim_{z\longrightarrow\infty} z f(z) = \lim_{z\longrightarrow\infty} \frac{z^p}{1+z^2} = \lim_{z\longrightarrow\infty} \frac{z^{p-2}}{1+\frac{1}{z^2}} = \lim_{z\longrightarrow\infty} \frac{1}{z^{2-p}\left(1+\frac{1}{z^2}\right)} = 0 \tag{6}$$

[since $p < 2$, $2-p$ is positive]

Hence

$$\lim_{R\longrightarrow\infty} \int_\Gamma f(z)\,dz = 0.. \tag{7}$$

Thus, proceeding to the limit $R \longrightarrow \infty$ and $r \longrightarrow 0$, we get from (4)

$$\int_{-\infty}^0 f(x)\,dx + 0 + \int_0^\infty f(x)\,dx + 0 = -\pi i e^{i\pi \frac{p}{2}}$$

$$\text{or,} \quad \int_{-\infty}^0 \frac{x^{p-1}}{1+x^2}\,dx + \int_0^\infty \frac{x^{p-1}}{1+x^2}\,dx = -\pi i e^{i\pi \frac{p}{2}}. \tag{8}$$

Now

$$\int_{-\infty}^0 \frac{x^{p-1}}{1+x^2}\,dx = \int_\infty^0 \frac{(-t)^{p-1}}{1+t^2}(-dt) = \int_0^\infty (-1)(-1)^p \frac{t^{p-1}}{1+t^2}\,dt. \tag{9}$$

Since $e^{i\pi} = (\cos\pi + i\sin\pi) = -1$

$$\therefore \quad e^{i\pi p} = (-1)^p. \tag{10}$$

Using equation (10), we get from (9)

$$\int_{-\infty}^0 \frac{x^{p-1}}{1+x^2}\,dx = -e^{i\pi p}\int_0^\infty \frac{x^{p-1}}{1+x^2}\,dx. \tag{11}$$

Hence, from equations (8) and (11), we get

$$(1 - e^{i\pi p})\int_0^\infty \frac{x^{p-1}}{1+x^2}\,dx = -\pi i e^{i\pi \frac{p}{2}}.$$

Equating real parts on both sides, we get

$$(1 - \cos\pi p)\int_0^\infty \frac{x^{p-1}}{1+x^2}\,dx = -\pi(-1)\sin\pi\frac{p}{2}.$$

$$\therefore \quad \int_0^\infty \frac{x^{p-1}}{1+x^2}\,dx = \frac{\pi\sin\left(\pi\frac{p}{2}\right)}{1-\cos\pi p} = \frac{\pi\sin\left(\pi\frac{p}{2}\right)}{2\sin^2\left(\pi\frac{p}{2}\right)} = \frac{\pi}{2}\operatorname{cosec}\left(\pi\frac{p}{2}\right).$$

(E) *Integration along double circles*: Let us consider integrals of the type

$$\int_0^\infty x^{\alpha-1} f(x)\, dx, \qquad (9.55)$$

where α is not an integer. Hence, $x^{\alpha-1}$ is a many-valued function. For integration of such types of integrals, we choose contour consisting of a large circle $|z| = R$, a small circle $|z| = r$ enclosing the branch point, and a cut along real axis joining these two circles (*Fig 9.5*).

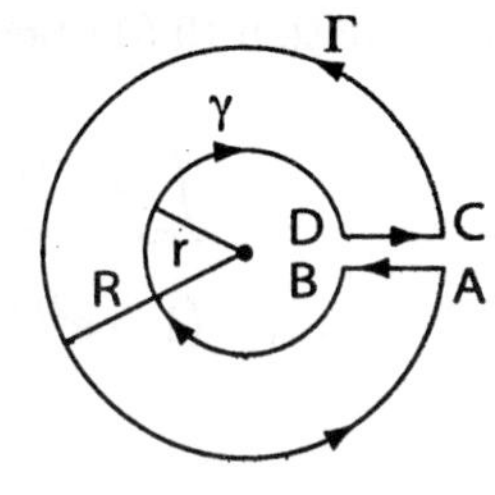

Fig 9.5

Let the integrand in (9.55) be

$$\phi(z) = z^{\alpha-1} f(z) \qquad (9.56)$$

and

$$\left.\begin{aligned} \lim_{z \longrightarrow \infty} z\phi(z) &= 0 \\ \text{and} \quad \lim_{z \longrightarrow 0} z\phi(z) &= 0 \end{aligned}\right\}. \qquad (9.57)$$

We get by Cauchy's residue theorem

$$\int_C \phi(z)\, dz = \int_\Gamma \phi(z)\, dz + \int_A^B \phi(x)\, dx + \int_\gamma \phi(z)\, dz + \int_C^D \phi(x)\, dx = 2\pi i \Sigma R^*, \qquad (9.58)$$

where, ΣR^* = sum of residues of $\phi(z)$ at the poles within the contour C.

Now

$$\left| \int_\Gamma \phi(z)\, dz \right| = \left| \int_\Gamma \phi(z) \cdot z \cdot \frac{dz}{z} \right| = \int_\Gamma |\, z\phi(z)\,|\, i\, d\theta \quad [\text{since on } \Gamma,\ z = Re^{i\theta}]$$

$$\leq z\phi(z) \cdot 2\pi$$

which tends to zero, as $R \longrightarrow \infty$ [by equation (9.57)].

This means

$$\int_\Gamma \phi(z)\, dz = 0. \qquad (9.59)$$

Similarly, we can show

$$\int_\gamma \phi(z)\, dz = 0. \qquad (9.60)$$

Equation (9.58) after using (9.59) and (9.60), becomes

$$\int_A^B \phi(x)\, dx + \int_C^D \phi(x)\, dx = 2\pi i \Sigma R^*. \qquad (9.61)$$

Now, for the integral from C to D, $|z| = x$ and $\theta = 0$, but, for the integral from A to B, $|z| = x$ and $\theta = 2\pi$.

Therefore

$$\left.\begin{aligned} z^{\alpha-1} &= x^{\alpha-1}, && \text{along } CD \\ z^{\alpha-1} &= x^{\alpha-1} e^{2\pi i(\alpha-1)}, && \text{along } AB \\ &= x^{\alpha-1} e^{2\pi i \alpha}, && \text{along } AB \end{aligned}\right\}, \qquad (9.62)$$

since $e^{2\pi i(\alpha-1)} = e^{2\pi i \alpha}(\cos 2\pi - i \sin 2\pi) = e^{2\pi i \alpha}$.

Thus, the equation (9.61) becomes, in the limit $R \longrightarrow \infty$ and $r \longrightarrow 0$

$$\int_{\infty}^{0} e^{2\pi i\alpha} x^{\alpha-1} f(x)\,dx + \int_{0}^{\infty} x^{\alpha-1} f(x)\,dx = 2\pi i \Sigma R^*$$

$$\text{or,} \quad \left(1 - e^{2\pi i\alpha}\right) \int_{0}^{\infty} x^{\alpha-1} f(x)\,dx = 2\pi i \Sigma R^*.$$

Hence

$$\int_{0}^{\infty} x^{\alpha-1} f(x)dx = \frac{2\pi i \Sigma R^*}{1 - e^{2\pi i\alpha}}. \tag{9.63}$$

Thus, we have proved the following theorem:

Theorem 9.12. *If $f(z)$ is a rational function having no pole on real positive axis, and if $\phi(z) = z^{\alpha-1} f(z)$, such that $\lim_{|z| \longrightarrow \infty} z\phi(z) = 0$ and $\lim_{|z| \longrightarrow 0} z\phi(z) = 0$, then*

$$\int_{0}^{\infty} x^{\alpha-1} f(x)\,dx = \frac{2\pi i \Sigma R^*}{1 - e^{2\pi i\alpha}}, \tag{9.64}$$

where, ΣR^* = sum of resides of $\phi(z)$ at all the poles enclosed within the contour as detailed above.

We shall see a few examples now. It may be noted that some problems solved in previous examples can also be solved by this method. Let us take Example 9.22.

EXAMPLE 9.23. *Evaluate, by contour integration*

$$\int_{0}^{\infty} \frac{x^{p-1}}{1+x^2}\,dx, \quad 0 < p < 2.$$

Solution: We consider the function

$$\phi(z) = \frac{z^{p-1}}{1+z^2} \tag{1}$$

and integrate it round the contour C shown in the *Fig 9.5.*

We have already seen, in Example 9.22, that

$$\lim_{z \longrightarrow \infty} z\phi(z) = 0 \quad \text{and} \quad \lim_{z \longrightarrow 0} z\phi(z) = 0.$$

Thus, conditions of the Theorem (9.12) are satisfied.

Hence

$$\int_{0}^{\infty} \frac{x^{p-1}}{1+x^2}\,dx = \frac{2\pi i \Sigma R^*}{1 - e^{2\pi i p}}, \tag{2}$$

where, ΣR^* = sum of residues of $\phi(z)$ at its poles within C. Now, the function $\phi(z)$ in (1) has poles at $z = i$ and $z = -i$.

Residue of $\phi(z)$ at $z = i$ is

$$R_1 = -\frac{1}{2}(i)^p \quad [\textit{see } \text{Example 9.22}] \tag{3}$$

Similarly, residue of $\phi(z)$ at $z = -i$ is

$$R_2 = -\frac{1}{2}(-i)^p. \tag{4}$$

Now

$$i = (-1)(-i) = e^{i\pi} \cdot e^{-i\pi/2}.$$

Hence

$$R_1 = -\frac{1}{2}(i)^p = -\frac{1}{2} e^{i\pi p} \cdot e^{-i\pi \frac{p}{2}}. \tag{5}$$

Similarly

$$-i = (-1)(i) = e^{i\pi} \cdot e^{i\pi/2}.$$

Hence

$$R_2 = -\frac{1}{2}(-i)^p = -\frac{1}{2} e^{i\pi p} \cdot e^{i\pi \frac{p}{2}}. \tag{6}$$

From (5) and (6), we get

$$\begin{aligned} 2\pi i \Sigma R^* &= (2\pi i)\left(-\frac{1}{2}\right) e^{i\pi p}\left[e^{-i\pi \frac{p}{2}} + e^{i\pi \frac{p}{2}}\right] \\ &= -2\pi i e^{i\pi p} \cdot \cos \pi \frac{p}{2}. \end{aligned} \tag{7}$$

Hence, by equation (9.64), we get

$$\begin{aligned} \int_0^\infty \frac{x^{p-1}}{1+x^2}\,dx &= \frac{2\pi i \Sigma R^*}{1-e^{2\pi i p}} = \frac{-2\pi i \cos \pi \frac{p}{2}}{e^{-i\pi p} - e^{i\pi p}} = \frac{-2\pi i \cos \pi \frac{p}{2}}{-2i \sin \pi p} \\ &= \frac{\pi \cos \pi \frac{p}{2}}{2 \sin \pi \frac{p}{2} \cos \pi \frac{p}{2}} = \frac{\pi}{2} \operatorname{cosec}\left(\pi \frac{p}{2}\right). \end{aligned}$$

EXAMPLE 9.24. *Evaluate, by contour integration*

$$\int_0^\infty \frac{x^{\alpha-1} \log x}{(1+x)^2}\,dx, \quad 0 < \alpha < 2.$$

Solution: We consider the function

$$\phi(z) = \frac{z^{\alpha-1} \log z}{(1+z)^2} \tag{1}$$

and integrate it round the contour C shown in the *Fig 9.5*.

The function $\phi(z)$ has poles given by $(1+z)^2 = 0$, i.e., at $z = -1 = e^{\pi i}$ of order two. Hence, residue of $\phi(z)$ at $z = -1 = e^{\pi i}$ is

$$R^* = \frac{1}{(2-1)!}\frac{d}{dz}\left[(1+z)^2\phi(z)\right] \qquad \text{at } z = e^{\pi i},$$

$$\begin{aligned} \text{i.e.,}\quad R^* &= (\alpha-1)z^{\alpha-2}\log z + z^{\alpha-1}\cdot\left(\frac{1}{z}\right) \quad \text{for } z = e^{\pi i} \\ &= e^{\pi i \alpha}\left[1 + (\alpha-1)\log e^{\pi i}\right] \qquad [\because e^{2\pi i} = 1] \\ &= e^{\pi i \alpha}[1 + (\alpha-1)\pi i]. \end{aligned} \tag{2}$$

Hence, by Cauchy's residue theorem, we have

$$I_1 + I_2 + I_3 + I_4 = 2\pi i \Sigma R^*, \tag{3}$$

where

$$I_1 = \int_\Gamma \phi(z)\, dz, \qquad I_2 = \int_R^r \phi\left(xe^{2\pi i}\right) e^{2\pi i} dx,$$

$$I_3 = \int_\gamma \phi(z)\, dz, \qquad I_4 = \int_r^R \phi(x)\, dx \tag{4}$$

and, ΣR^* = sum of the residues of the function $\phi(z)$ at its poles which is given by equation (2). It is to be noted that, in I_2, the path of integration being from R to r, the amplitude of $z = 2\pi$, so that, $z = xe^{2\pi i}$ and $dz = e^{2\pi i}\, dx$.

Now

$$\begin{aligned}\lim_{z\to\infty} z\phi(z) &= \lim_{z\to\infty} \frac{z^\alpha \log z}{(1+z)^2} = \lim_{z\to\infty} \frac{\log z}{z^{-\alpha}(1+z)^2} \\ &= \lim_{z\to\infty} \frac{\frac{1}{z}}{-\alpha z^{-\alpha-1}(1+z)^2 + 2z^{-\alpha}(1+z)} \\ &= \lim_{z\to\infty} \frac{1}{-\alpha z^{(2-\alpha)}(1+\frac{1}{z})^2 + 2z^{(2-\alpha)}(1+\frac{1}{z})} = 0.\end{aligned} \tag{5}$$

Again

$$\lim_{z\to 0} z\phi(z) = \lim_{z\to 0} \frac{z^\alpha \log z}{(1+z)^2} = \lim_{z\to 0} \frac{\alpha z^{\alpha-1} \log z + z^{\alpha-1}}{2(1+z)} = 0. \tag{6}$$

From equation (5), we can write

$$I_1 = \lim_{R\to\infty} \int_\Gamma \phi(z)\, dz = i(2\pi - 0)\cdot 0 = 0. \tag{7}$$

Again, from equation (6), we can write

$$I_3 = \lim_{r\to 0} \int_\gamma \phi(z)\, dz = i(2\pi - 0)\cdot 0 = 0. \tag{8}$$

Thus, proceeding to the limit $R \longrightarrow \infty$ and $r \longrightarrow 0$, we have the relation (3) as

$$\int_\infty^0 \phi\left(xe^{2\pi i}\right) e^{2\pi i}\, dx + \int_0^\infty \phi(x)\, dx = 2\pi i \Sigma R^*. \tag{9}$$

Now

$$\begin{aligned}\int_\infty^0 \phi\left(xe^{2\pi i}\right) e^{2\pi i}\, dx &= -\int_0^\infty \frac{x^{\alpha-1}\left(e^{2\pi i}\right)^{\alpha-1} \log\left(x\cdot e^{2\pi i}\right) e^{2\pi i}}{(1+x)^2}\, dx \\ &= -\int_0^\infty \frac{x^{\alpha-1}}{(1+x)^2}\left[e^{2\pi i\alpha}(\log x + 2\pi i)\right]\, dx.\end{aligned} \tag{10}$$

From equations (2), (9) and (10), equating real and imaginary parts, we get

$$-\int_0^\infty \frac{x^{\alpha-1}}{(1+x)^2}[\cos 2\pi\alpha \cdot \log x - 2\pi \sin 2\pi\alpha - \log x]\, dx$$
$$= -2\pi[\sin \pi\alpha + \pi(\alpha-1)\cos \pi\alpha] \tag{11}$$

and $$-\int_0^\infty \frac{x^{\alpha-1}}{(1+x)^2}[\sin 2\pi\alpha \cdot \log x + 2\pi \cos 2\pi\alpha]\, dx$$
$$= 2\pi[\cos \pi\alpha - \pi(\alpha-1)\sin \pi\alpha]. \tag{12}$$

We multiply equation (11) by $\cos 2\pi\alpha$, and (12) by $\sin 2\pi\alpha$, and add. Then, we get

$$(\cos 2\pi\alpha - 1)\int_0^\infty \frac{x^{\alpha-1}\log x}{(1+x)^2}\,dx = 2\pi[\sin\pi\alpha - \pi(\alpha-1)\cos\pi\alpha]$$

or,

$$\int_0^\infty \frac{x^{\alpha-1}\log x}{(1+x)^2}\,dx = \pi\left[\pi(\alpha-1)\frac{\cos\pi\alpha}{\sin^2\pi\alpha} - \frac{1}{\sin\pi\alpha}\right].$$

In particular, if $\alpha = \frac{7}{6}$, i.e., $\alpha - 1 = \frac{1}{6}$, then

$$\int_0^\infty \frac{x^{\frac{1}{6}}\log x}{(1+x)^2}dx = \frac{\pi^2}{6}\left(-\frac{\sqrt{3}}{2}\right)(2)^2 - \pi(-2) = 2\pi - \frac{\pi^2}{\sqrt{3}}.$$

EXAMPLE 9.25. *Prove that*

$$\int_0^\infty \frac{x^{-\alpha}}{x^2 + 2x\cos\lambda + 1}\,dx = \frac{\pi}{\sin\alpha\pi}\cdot\frac{\sin\alpha\lambda}{\sin\lambda},$$

where $-1 < \alpha < 1$ *and* $-\pi < \lambda < \pi$.

Solution: We consider the function

$$\phi(z) = \frac{z^{-\alpha}}{z^2 + 2z\cos\lambda + 1} \tag{1}$$

and integrate it along the contour C as shown in the *Fig 9.5*.

Now

$$\lim_{z\to\infty} z\phi(z) = \lim_{z\to\infty}\frac{z^{1-\alpha}}{z^2+2z\cos\lambda+1} = \lim_{z\to\infty}\frac{z^{-1-\alpha}}{1+\frac{2}{z}\cos\lambda+\frac{1}{z^2}}$$
$$= 0, \text{ if } \alpha + 1 > 0. \tag{2}$$

Again

$$\lim_{z\to 0} z\phi(z) = \lim_{z\to 0}\frac{z^{1-\alpha}}{z^2+2z\cos\lambda+1}$$
$$= 0, \text{ if } 1 - \alpha > 0. \tag{3}$$

Thus, from equations (2) and (3), we find that $\alpha + 1 > 0$ and $1 - \alpha > 0$,

$$\text{i.e., } -1 < \alpha < 1, \text{ which is necessary.} \tag{4}$$

Now, the poles of $\phi(z)$ are given by

$$z^2 + 2z\cos\lambda + 1 = 0$$
$$\text{i.e.,. } z^2 + z\left(e^{i\lambda} + e^{-i\lambda}\right) + 1 = 0$$
$$\text{i.e., } \left(z + e^{i\lambda}\right)\left(z + e^{-i\lambda}\right) = 0.$$

Thus

$$z_1 = -e^{i\lambda} = e^{i\pi}\cdot e^{i\lambda} = e^{i(\pi+\lambda)} \tag{5}$$

$$\text{and } z_2 = -e^{-i\lambda} = e^{i\pi}\cdot e^{-i\lambda} = e^{i(\pi-\lambda)}. \tag{6}$$

Hence, amplitudes of poles are

$$\left.\begin{aligned} z_1 &= \pi + \lambda \\ \text{and} \quad z_2 &= \pi - \lambda \end{aligned}\right\}. \tag{7}$$

Since $-\pi < \lambda < \pi$, both the poles lie within the contour C.
The residue at pole $z_1 = -e^{i\lambda}$ is

$$\begin{aligned} R_1 &= \lim_{z \longrightarrow z_1} \left(z + e^{\lambda i}\right) \phi(z) = \lim_{z \longrightarrow -e^{i\lambda}} \left(z + e^{i\lambda}\right) \frac{z^{-\alpha}}{\left(z + e^{i\lambda}\right)\left(z + e^{-i\lambda}\right)} \\ &= \frac{\left\{e^{i(\pi+\lambda)}\right\}^{-\alpha}}{-e^{i\lambda} + e^{-i\lambda}} \quad \text{[using equation (5)]} \\ &= \frac{e^{-i\pi\alpha} \cdot e^{-i\lambda\alpha}}{-(2i \sin\lambda)}. \end{aligned} \tag{8}$$

Similarly, residue at $z_2 = -e^{-i\lambda}$ is

$$R_2 = \frac{e^{-i\pi\alpha} \cdot e^{i\lambda\alpha}}{2i \sin\lambda}. \tag{9}$$

Thus, the sum of the residues at poles is

$$\Sigma R^* = R_1 + R_2 = \frac{e^{-i\pi\alpha}}{2i \sin\lambda} \left\{e^{i\lambda\alpha} - e^{-i\lambda\alpha}\right\} = e^{-i\pi\alpha} \cdot \frac{\sin\lambda\alpha}{\sin\lambda}. \tag{10}$$

Now, by Cauchy's residue theorem

$$\int_\Gamma \phi(z)\, dz + \int_R^r \phi(x)\, dx + \int_\gamma \phi(z)\, dz + \int_r^R \phi(x)\, dx = 2\pi i \Sigma R^*. \tag{11}$$

From results obtained in equations (2) and (3), we have

$$\left.\begin{aligned} &\int_\Gamma \phi(z)\, dz = 0 \\ \text{and} \quad &\int_\gamma \phi(z)\, dz = 0 \end{aligned}\right\}. \tag{12}$$

Moreover, in taking integration from R to r, the amplitude of $z = 2\pi$, so that

$$z = xe^{2\pi i}, \quad \therefore \quad dz = e^{2\pi i}\, dx. \tag{13}$$

Hence, proceeding to the limit $R \longrightarrow \infty$ and $r \longrightarrow 0$, and using equations (12) and (13), we get from (11)

$$\int_\infty^0 \phi\left(xe^{2\pi i}\right) dx + \int_0^\infty \phi(x)\, dx = 2\pi i \cdot e^{-i\pi\alpha} \cdot \frac{\sin\lambda\alpha}{\sin\lambda}$$

$$\text{or,} \quad -\int_0^\infty \frac{x^{-\alpha} e^{-2\pi\alpha i}}{x^2 + 2x\cos\lambda + 1}\, dx + \int_0^\infty \frac{x^{-\alpha}}{x^2 + 2x\cos\lambda + 1}\, dx = 2\pi i \cdot e^{-i\pi\alpha} \cdot \frac{\sin\lambda\alpha}{\sin\lambda}$$

$$\text{or,} \quad \left(1 - e^{-2\pi\alpha i}\right) \int_0^\infty \frac{x^{-\alpha}}{x^2 + 2x\cos\lambda + 1}\, dx = 2\pi i \cdot e^{-i\pi\alpha} \cdot \frac{\sin\lambda\alpha}{\sin\lambda}.$$

$$\therefore \quad \int_0^\infty \frac{x^{-\alpha}}{x^2 + 2x\cos\lambda + 1}\, dx = \frac{2\pi i}{\left(e^{i\pi\alpha} - e^{-i\pi\alpha}\right)} \cdot \frac{\sin\lambda\alpha}{\sin\lambda} = \frac{\pi}{\sin\alpha\pi} \cdot \frac{\sin\lambda\alpha}{\sin\lambda}.$$

(F) *Integration around rectangular contours*: The rectangular contours are generally suitable when the integrand contains hyperbolic or periodic functions. Let us take some examples.

EXAMPLE 9.26. *Prove that*

$$\int_{-\infty}^{\infty} \frac{e^{\alpha x}}{1+e^x}\, dx = \pi \operatorname{cosec}(\alpha\pi), \quad 0 < \alpha < 1.$$

Solution: We take the function

$$\phi(z) = \frac{e^{\alpha z}}{1+e^z} \tag{1}$$

and integrate it around a rectangle with two vertices on real axis at $(\pm R, 0)$, and two others at $(\pm R, 2\pi i)$ as shown in the *Fig 9.6*.

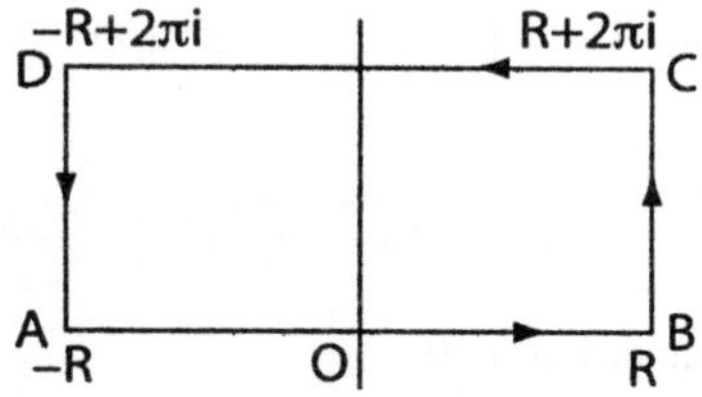

Fig 9.6

By Cauchy's residue theorem

$$\int_C \phi(z)\, dz = 2\pi i \Sigma R^*, \tag{2}$$

where ΣR^* = sum of residues at poles of $\phi(z)$. Now, the poles of $\phi(z)$ are given by $1 + e^z = 0$, i.e., $e^z = -1 = e^{i\pi}$.

$$\therefore \quad z = (2n+1)\pi i, \quad n = 0, \pm 1, \pm 2, \dots. \tag{3}$$

The pole at $z = \pi i$ only lies within C.
Let

$$\phi(z) = \frac{f(z)}{\psi(z)},$$

where $f(z) = e^{\alpha z}$, $\psi(z) = 1 + e^z$, so that $\psi(\pi i) = 0$, but $f(\pi i) \neq 0$.
Then, by equation (9.17), the residue of $\phi(z)$ at $z = \pi i$ is

$$\Sigma R^* = \frac{f(\pi i)}{\psi'(\pi i)} = \frac{e^{i\alpha\pi}}{e^{i\pi}} = -e^{i\alpha\pi}. \tag{4}$$

Thus, from equations (2) and (4), we get

$$I_1 + I_2 + I_3 + I_4 = -2\pi i e^{i\alpha\pi}, \tag{5}$$

where

$$I_1 = \int_{-R}^{R} \phi(x)\, dx \qquad I_2 = \int_{R}^{-R} \phi(x + 2\pi i)\, dx, \tag{6}$$

$$I_3 = \int_{BC} \phi(z)\, dz \qquad I_4 = \int_{DA} \phi(z)\, dz. \tag{7}$$

It is to be noted that

$$\left.\begin{array}{lll} \text{along } AB, & z = x, & dz = dx \\ \text{along } CD, & z = x + 2\pi i, & dz = dx \\ \text{along } BC, & z = R + iy, & dz = i\, dy \\ \text{along } DA, & z = -R + iy, & dz = i\, dy \end{array}\right\}. \tag{8}$$

Thus

$$I_3 = \int_0^{2\pi} \phi(R+iy)\, d(iy) = \int_0^{2\pi} \frac{e^{\alpha(R+iy)}}{1+e^{R+iy}}\, i\, dy.$$

Hence

$$|I_3| \le \int_0^{2\pi} \frac{|e^{\alpha R}|\,|e^{\alpha iy}|\,|i|\, dy}{|1+e^R \cdot e^{iy}|}. \tag{9}$$

Now, $|e^{\alpha iy}| = 1$, $|i| = 1$, and

$$\frac{1}{|1+e^R \cdot e^{iy}|} \le \frac{1}{e^R - 1} \quad \left[\text{Since } \frac{1}{|z_1+z_2|} \le \frac{1}{|z|-|z_2|}\right]$$

Hence, from (9)

$$|I_3| \le \int_0^{2\pi} \frac{e^{\alpha R}}{e^R-1}\, dy, \quad \text{i.e.,} \quad |I_3| \le \frac{2\pi e^{\alpha R}}{e^R - 1} = \frac{2\pi}{(e^R-1)\, e^{-\alpha R}}. \tag{10}$$

Since $0 < \alpha < 1$, the expression (10) tends to zero as $R \longrightarrow \infty$.
Similarly, it can be shown that

$$I_4 = \int_{2\pi}^0 \phi(-R+iy)\, d(iy) \longrightarrow 0 \text{ as } R \longrightarrow \infty. \tag{11}$$

Hence, proceeding to the limit $R \longrightarrow \infty$, we get from equation (5) using (10) and (11)

$$\int_{-\infty}^{\infty} \frac{e^{\alpha x}}{1+e^x}\, dx - \int_{-\infty}^{\infty} \frac{e^{\alpha x} \cdot e^{2\pi i\alpha}}{1+e^x}\, dx = -2\pi i e^{i\alpha\pi}$$

$$\text{or,} \quad \left(1 - e^{2\pi i\alpha}\right) \int_{-\infty}^{\infty} \frac{e^{\alpha x}}{1+e^x}\, dx = -2\pi i e^{i\alpha\pi}.$$

$$\therefore \quad \int_{-\infty}^{\infty} \frac{e^{\alpha x}}{1+e^x}\, dx = -\frac{2\pi i e^{i\alpha\pi}}{1-e^{2\pi i\alpha}} = \frac{-2\pi i}{e^{-i\alpha\pi} - e^{i\alpha\pi}}$$

$$= \frac{2\pi i}{2i \sin \pi\alpha} = \pi \operatorname{cosec} \pi\alpha.$$

EXAMPLE 9.27. *Evaluate, by contour integration*

$$\int_0^{\infty} \frac{\cosh \alpha x}{\cosh x}\, dx, \quad |\alpha| < 1.$$

Solution: We consider the function

$$\phi(z) = \frac{e^{\alpha z}}{\cosh z} \tag{1}$$

and integrate it along the rectangle, whose vertices are $-R$, R, $R+\pi i$, $-R+i\pi$ similar to *Fig 9.6.*

The poles of $\phi(z)$ are given by, $\cosh z = 0$, i.e., $e^z + e^{-z} = 0$, i.e., $e^{2z} = -1 = e^{\pi i}$.

$\therefore\ 2z = (2n+1)\pi i$, or, $z = \left(n + \frac{1}{2}\right)\pi i$, $n = 0, \pm 1, \ldots$

Simple pole $z = \frac{1}{2}\pi i$ lies within C.

Residue of $\phi(z)$ at $z = \frac{1}{2}\pi i$ is

$$R^* = \lim_{z \longrightarrow (\frac{\pi i}{2})} \frac{e^{\alpha z}}{\frac{d}{dz}(\cosh z)} = \frac{e^{\alpha z}}{\sinh z} \quad \text{for } z = \frac{\pi i}{2}$$
$$= \frac{2e^{i\pi \frac{\alpha}{2}}}{\left(e^{i\frac{\pi}{2}} - e^{-i\frac{\pi}{2}}\right)} = \frac{2e^{i\pi\frac{\alpha}{2}}}{i+i} = \frac{1}{i}e^{i\pi\frac{\alpha}{2}}. \tag{2}$$

By Cauchy's residue theorem

$$\int_C \phi(z)\, dz = 2\pi i \Sigma R^* = 2\pi e^{i\pi\frac{\alpha}{2}} \quad \text{or,} \quad I_1 + I_2 + I_3 + I_4 = 2\pi e^{i\pi\alpha/2}, \tag{3}$$

where

$$I_1 = \int_{-R}^{R} \frac{e^{\alpha z}}{\cosh z}\, dz, \quad \text{with } z = x, dz = dx$$
$$I_2 = \int_{BC} \frac{e^{\alpha z}}{\cosh z}\, dz, \quad \text{with } z = R + iy,\ dz = i\, dy, \quad y \text{ varies from 0 to } \pi$$
$$I_3 = \int_{CD} \frac{e^{\alpha z}}{\cosh z}\, dz, \quad \text{with } z = x + i\pi, \qquad x \text{ varies from } -R \text{ to } R$$
$$I_4 = \int_{DA} \frac{e^{\alpha z}}{\cosh z}\, dz, \quad \text{with } z = -R + iy, \qquad y \text{ varies from } \pi \text{ to zero}$$

Now

$$I_2 = \int_0^{\pi} \frac{2e^{\alpha(R+iy)} i\, dy}{e^{R+iy} + e^{-(R+iy)}} = \int_0^{\pi} \frac{2e^{\alpha R} \cdot e^{i\alpha y} \cdot i\, dy}{e^{R} \cdot e^{iy} + e^{-R} \cdot e^{-iy}}. \tag{4}$$

Since, $|\, e^{iy}\, | = |\, e^{-iy}\, | = 1$, $|\, e^{i\alpha y}\, | = 1$, $|\, i\, | = 1$ and for two complex numbers z_1, z_2

$$\frac{1}{|\, z_1 + z_2\, |} \leq \frac{1}{|\, z_1\, | - |\, z_2\, |},$$

we get from (4)

$$|\, I_2\, | \leq \int_0^{\pi} \frac{2e^{\alpha R}\, dy}{e^R - e^{-R}} = 2\int_0^{\pi} \frac{dy}{e^{R(1-\alpha)} - e^{-R(1+\alpha)}}$$

which tends to zero as $R \longrightarrow \infty$ since $|\, \alpha\, | < 1$.

Thus

$$I_2 = \int_{BC} \phi(z)\, dz = 0. \tag{5}$$

Similarly, we can prove that

$$I_4 = \int_{DA} \phi(z)\, dz = 0. \tag{6}$$

Hence, proceeding to the limit $R \longrightarrow \infty$, we get from equation (3) with the use of (5), (6) and (2)

$$\int_{-\infty}^{\infty} \frac{e^{\alpha x}}{\cosh x} dx + \int_{\infty}^{-\infty} \frac{e^{\alpha x} \cdot e^{i\alpha\pi}}{\cosh(x + i\pi)} dx = 2\pi e^{i\pi\frac{\alpha}{2}}. \tag{7}$$

The second integral in equation (7) is

$$\int_{-\infty}^{\infty} \frac{-2e^{\alpha x}\cdot e^{i\alpha\pi}dx}{e^{x}\cdot e^{i\pi}+e^{-x}\cdot e^{-i\pi}} = \int_{-\infty}^{\infty} \frac{2e^{\alpha x}\cdot e^{i\alpha\pi}}{(e^{x}+e^{-x})}\,dx$$

$$= \int_{-\infty}^{\infty} \frac{e^{\alpha x}\cdot e^{i\alpha\pi}}{\cosh x}\,dx \quad [\because e^{i\pi}=e^{-i\pi}=-1] \tag{8}$$

Hence, from (7) and (8), we get

$$(1+e^{i\alpha\pi})\int_{-\infty}^{\infty} \frac{e^{\alpha x}}{\cosh x}\,dx = 2\pi e^{i\alpha\frac{\pi}{2}}$$

$$\text{or,} \quad \int_{-\infty}^{\infty} \frac{e^{\alpha x}}{\cosh x}\,dx = \frac{2\pi e^{i\alpha\frac{\pi}{2}}}{1+e^{i\alpha\pi}} = \frac{\pi\cdot 1}{e^{i\alpha\frac{\pi}{2}}+e^{-i\alpha\frac{\pi}{2}}} = \pi\sec\left(\alpha\frac{\pi}{2}\right)$$

$$\text{i.e.,} \quad \int_{-\infty}^{0} \frac{e^{\alpha x}}{\cosh x}\,dx + \int_{0}^{\infty} \frac{e^{\alpha x}}{\cosh x}\,dx = \pi\sec\left(\alpha\frac{\pi}{2}\right). \tag{9}$$

We replace x by $-x$ in the 1st integral. Then, we get

$$\int_{-\infty}^{0} \frac{e^{\alpha x}}{\cosh x}\,dx = \int_{\infty}^{0} \frac{e^{-\alpha x}(-dx)}{\cosh x}\,dx = \int_{0}^{\infty} \frac{e^{-\alpha x}}{\cosh x}\,dx. \tag{10}$$

From (9) and (10), we get

$$\int_{0}^{\infty} \frac{e^{\alpha x}+e^{-\alpha x}}{\cosh x}\,dx = \pi\sec\left(\alpha\frac{\pi}{2}\right)$$

$$\text{or,} \quad 2\int_{0}^{\infty} \frac{\cosh\alpha x}{\cosh x}\,dx = \pi\sec\left(\alpha\frac{\pi}{2}\right).$$

$$\therefore \quad \int_{0}^{\infty} \frac{\cosh\alpha x}{\cosh x}\,dx = \frac{\pi}{2}\sec\left(\alpha\frac{\pi}{2}\right).$$

EXAMPLE 9.28. *Prove that*

$$\int_{-\infty}^{\infty} e^{-x^2}\cos 2\alpha x\,dx = \sqrt{\pi}\,e^{-\alpha^2}.$$

Solution: We consider the function

$$\phi(z) = e^{-z^2} \tag{1}$$

and integrate it round a rectangular contour C whose vertices are $\pm R$ and $\pm R + i\alpha$. The function $\phi(z)$ is regular within and on C, so that there is no pole of $\phi(z)$. Hence, by Cauchy's theorem

$$\int_C \phi(z)\,dz = \int_C e^{-z^2}\,dz = 0. \tag{2}$$

Equation (2) gives

$$\int_{-R}^{R} e^{-x^2}\,dx + \int_{0}^{\alpha} e^{-(R+iy)^2}\,(i\,dy) + \int_{R}^{-R} e^{-(x+i\alpha)^2}\,dx + \int_{\alpha}^{0} e^{-(R-iy)^2}\,(i\,dy) = 0. \tag{3}$$

Now

$$\left|\int_{0}^{\alpha} e^{-(R+iy)^2}\,(i\,dy)\right| \le \int_{0}^{\alpha}\left|e^{-(R+iy)^2}\right|dy \le \int_{0}^{\alpha} e^{-R^2+y^2}\,dy \le e^{-R^2}\int_{0}^{\alpha} e^{y^2}\,dy$$

which $\longrightarrow 0$ as $R \longrightarrow \infty$.

Similarly, the fourth integral in (3) can be shown to $\longrightarrow 0$ as $R \longrightarrow \infty$.
Hence, from (3) we get, in the limit $R \longrightarrow \infty$,

$$\int_{-\infty}^{\infty} e^{-x^2}\,dx + \int_{\infty}^{-\infty} e^{-(x+i\alpha)^2}\,dx = 0$$

$$\text{i.e.,} \quad \int_{-\infty}^{\infty} e^{-x^2+\alpha^2}(\cos 2\alpha x - i\sin 2\alpha x)\,dx = \int_{-\infty}^{\infty} e^{-x^2}\,dx = \sqrt{\pi}. \tag{4}$$

Equating real part, we get

$$\int_{-\infty}^{\infty} e^{-x^2}\cos 2\alpha x\,dx = \sqrt{\pi}\,e^{-\alpha^2}.$$

Incidentally, we find that, equating the imaginary parts in equation (4)

$$\int_{-\infty}^{\infty} e^{-x^2}\sin 2\alpha x\,dx = 0.$$

(G) *Integration around a sector of a circle*: A contour in the form of a sector of a circle is also used conveniently to evaluate certain types of integrals. We shall consider some examples.

EXAMPLE 9.29. *Prove that*

$$\int_0^{\infty} \sin(x^2)\,dx = \int_0^{\infty} \cos(x^2)\,dx = \frac{1}{2}\sqrt{\frac{\pi}{2}}.$$

Solution: We consider the function

$$\phi(z) = e^{iz^2} \tag{1}$$

and integrate it along the contour C in the form of a sector of a circle shown in *Fig 9.7*.

There is no singularity of the function $\phi(z)$ within the contour C. Hence, by Cauchy's residue theorem

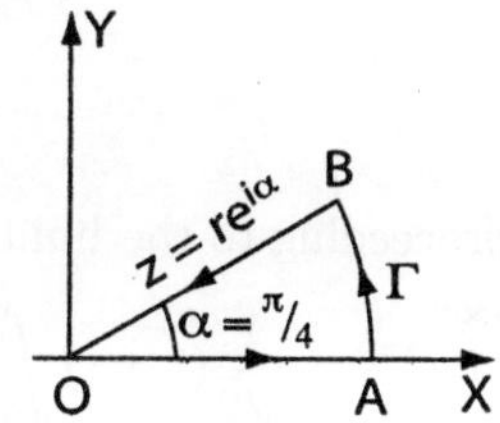

Fig. 9.7

$$\int_C \phi(z)\,dz = 2\pi i \Sigma R^* = 0. \tag{2}$$

This gives

$$\int_{OA} \phi(z)\,dz + \int_{AB} \phi(z)\,dz + \int_{BO} \phi(z)\,dz = 0 \tag{3}$$

Now (i) along $OA : z = x$, x varies from zero to R;

(ii) along $AB : z = Re^{i\theta}$, θ varies from zero to $\frac{\pi}{4}$;

(iii) along $BO : z = re^{i\frac{\pi}{4}}$, r varies from R to zero.

Thus, from (3), we get

$$I_1 + I_2 + I_3 = 0, \tag{4}$$

where

$$\left.\begin{aligned} I_1 &= \int_0^R e^{ix^2}\,dx \\ I_2 &= \int_\Gamma e^{iz^2}\,dz \\ I_3 &= \int_R^0 e^{i(re^{i\frac{\pi}{4}})^2}\,d\left(re^{i\frac{\pi}{4}}\right) \end{aligned}\right\}. \tag{5}$$

Now

$$\begin{aligned}|I_2| &\le \int_\Gamma \left| e^{iz^2} \right| |dz| \\ &\le \int_0^{\frac{\pi}{4}} \left| e^{iR^2 e^{2i\theta}} \right| |iRe^{i\theta}\, d\theta| \quad \text{putting } z = Re^{i\theta} \\ &= \int_0^{\frac{\pi}{4}} \left| e^{-R^2 \sin 2\theta} \cdot e^{iR^2 \cos 2\theta} \right| R\, d\theta. \end{aligned} \tag{6}$$

We put $2\theta = \xi$, so that $d\theta = \frac{1}{2}\, d\xi$.
Then, equation (6) becomes

$$|I_2| \le \frac{R}{2} \int_0^{\frac{\pi}{2}} e^{-R^2 \sin\xi} d\xi \le \frac{R}{2} \int_0^{\frac{\pi}{2}} e^{-R^2 \frac{2\xi}{\pi}} d\xi \quad \text{[by Jordan's inequality]}$$

$$= \frac{R}{2}\left(-\frac{\pi}{2R^2}\right)\left[e^{-2R^2 \frac{\xi}{\pi}}\right]_0^{\frac{\pi}{2}} = -\frac{\pi}{4R}\left[e^{-R^2} - 1\right] \text{ which } \longrightarrow 0 \text{ as } R \longrightarrow \infty. \tag{7}$$

Again

$$i\left(re^{i\frac{\pi}{4}}\right)^2 = ir^2 e^{i\frac{\pi}{2}} = ir^2\left(\cos\frac{\pi}{2} + i\sin\frac{\pi}{2}\right) = i^2 r^2 = -r^2. \tag{8}$$

Hence

$$\begin{aligned} I_3 &= \int_R^0 e^{-r^2} \cdot e^{i\frac{\pi}{4}}\, dr = -\int_0^R e^{-r^2} \cdot e^{i\frac{\pi}{4}}\, dr \\ &= -\frac{1}{\sqrt{2}}(1+i)\int_0^R e^{-x^2}\, dx. \end{aligned} \tag{9}$$

Thus, proceeding to the limit $R \longrightarrow \infty$, we get from equations (4), (7) and (9)

$$\int_0^\infty e^{ix^2} = \frac{1}{\sqrt{2}}(1+i)\int_0^\infty e^{-x^2}\, dx = \frac{1+i}{2\sqrt{2}}\sqrt{\pi} \quad \left(\text{since} \int_0^\infty e^{-x^2}\, dx = \frac{\sqrt{\pi}}{2}\right).$$

Equating real and imaginary parts, we get

$$\int_0^\infty \cos x^2\, dx = \frac{1}{2}\sqrt{\frac{\pi}{2}} \quad \text{and} \quad \int_0^\infty \sin x^2\, dx = \frac{1}{2}\sqrt{\frac{\pi}{2}}.$$

EXAMPLE 9.30. *Prove that*

$$\int_0^\infty e^{-x^2 \cos 2\alpha} \cdot \cos\left(x^2 \sin 2\alpha\right) dx = \frac{\sqrt{\pi}}{2}\cos\alpha$$

$$\textit{and} \quad \int_0^\infty e^{-x^2 \cos 2\alpha} \cdot \sin\left(x^2 \sin 2\alpha\right) dx = \frac{\sqrt{\pi}}{2}\sin\alpha.$$

Solution: We take the same functon $\phi(z)$ as in the previous example, and the same contour C with $\angle AOB = \alpha$. Proceeding exactly in the same manner, we get, in the limit $R \longrightarrow \infty$,

$$\int_0^\infty e^{-x^2}\, dx + \int_\infty^0 e^{-(re^{i\alpha})^2}\, d(re^{i\alpha}) = 0 \tag{1}$$

$$\text{or,} \quad \int_0^\infty e^{-x^2}\, dx - \int_0^\infty e^{-x^2 \cos 2\alpha} \cdot e^{-ix^2 \sin 2\alpha} \cdot e^{i\alpha}\, dx = 0.$$

$$\therefore \quad \int_0^\infty e^{-x^2 \cos 2\alpha} \cdot e^{-ix^2 \sin 2\alpha}\, dx = e^{-i\alpha}\int_0^\infty e^{-x^2}\, dx = e^{-i\alpha} \cdot \frac{\sqrt{\pi}}{2}.$$

Equating real and imaginary parts, we get

$$\int_0^\infty e^{-x^2\cos 2\alpha}\cdot\cos(x^2\sin 2\alpha)\,dx=\frac{\sqrt{\pi}}{2}\cos\alpha$$

and $$\int_0^\infty e^{-x^2\cos 2\alpha}\cdot\sin(x^2\sin 2\alpha)\,dx=\frac{\sqrt{\pi}}{2}\sin\alpha.$$

This is to be noted that, in the last three examples, we have used a known definite integral and evaluated the other integrals by means of a complex change of variables. This is known as *complex transformations* of definite integrals.

EXERCISE

1. The only singularities of a single-valued function $f(z)$ are poles of order 1 and 2 at $z=-1$ and $z=2$ with residues at these poles 1 and 2 respectively. If $f(0)=\frac{7}{4}$, $f(1)=\frac{5}{2}$, determine the function.

2. Discuss the singularities of the following functions:

   (a) $\dfrac{e^{\frac{c}{(z-a)}}}{e^{\frac{z}{a}}-1}$; (b) $\displaystyle\sum_{n=0}^{\infty}\frac{1}{n!}\frac{1}{1+2^{2n}z^2}$.

3. Find the residue of the following functions at the point mentioned:

   (a) $\dfrac{z^4}{z^2+a^2}$ at $z=-ia$; (c) $\dfrac{z^4}{(z-1)^4(z-2)(z-3)}$ at $z=1$;

   (b) $\dfrac{z}{(z-a)(z-b)}$ at infinity; (d) $\dfrac{z^2}{(z-a)(z-b)(z-c)}$ at infinity.

4. Show that:

   (a) $\displaystyle\int_0^{2\pi}\frac{d\theta}{a+b\cos\theta}=\frac{2\pi}{\sqrt{a^2-b^2}}$, $a>b>0$;

   (b) $\displaystyle\int_0^{2\pi}\frac{\cos 2\theta\,d\theta}{5+4\cos\theta}=\frac{\pi}{6}$; (c) $\displaystyle\int_0^{\pi}\frac{a\,d\theta}{a^2+\sin^2\theta}=\frac{\pi}{\sqrt{1+a^2}}$, $a>0$.

5. Prove that:

   (a) $\displaystyle\int_0^\infty\frac{dx}{(1+x^2)^2}=\frac{\pi}{2}$; (b) $\displaystyle\int_0^\infty\frac{x^2\,dx}{x^6+1}=\frac{\pi}{6}$.

6. Prove, by contour integration, that

$$\int_0^\infty\frac{dx}{(a+bx^2)^n}=\frac{\pi}{2^n\sqrt{b}}\cdot\frac{1\cdot3\cdot5\cdots(2n-3)}{1\cdot2\cdot3\cdots(n-1)}\frac{1}{a^{\frac{(2n-1)}{2}}}.$$

7. By contour integration, prove that:

   (a) $\displaystyle\int_0^\infty\frac{\cos mx}{x^2+a^2}\,dx=\frac{\pi}{2a}e^{-ma}$, $(m\geq 0)$;

   (b) $\displaystyle\int_{-\infty}^\infty\frac{a\cos x+x\sin x}{x^2+a^2}\,dx=2\pi e^{-a}\cdot\left[\text{Take }\phi(z)=\frac{e^{iz}}{z-ai},\ (a>0)\right]$;

(c) $\displaystyle\int_0^\infty \frac{\cos mx}{x^4+x^2+1}\,dx = \frac{\pi}{\sqrt{3}}e^{-m\frac{\sqrt{3}}{2}}\sin\left(\frac{m}{2}+\frac{\pi}{6}\right).$

8. Taking semi-circular contours with indents, prove that:

(a) $\displaystyle\int_0^\infty \frac{\sin x\,dx}{x(x^2+a^2)} = \frac{\pi}{2a^2}(1-e^{-a}),\ a>0;$

(b) $\displaystyle\int_0^\infty \frac{\cos x\,dx}{a^2-x^2} = \frac{\pi}{2a}\sin a;$ (c) $\displaystyle\int_0^\infty \frac{x^4\,dx}{x^6-1} = \frac{\pi}{6}\sqrt{3};$

(d) $\displaystyle\int_0^\infty \frac{(\log x)^2}{1+x^2}\,dx = \frac{\pi^3}{8};$ (e) $\displaystyle\int_0^\infty \frac{\log x}{(1+x^2)^2} = -\frac{\pi}{4}.$

9. Using suitable contours, show that

(a) $\displaystyle\int_0^\infty \frac{(\log x)^3}{x^2+x+1}\,dx = \frac{16\pi^3}{81\sqrt{3}};$

(b) $\displaystyle\int_0^\infty \frac{\log(1+x^2)}{1+x^2}\,dx = \pi\log 2.$

10. Using rectangular contour, prove that:

(a) $\displaystyle\int_0^\infty \frac{\sinh \alpha x}{\cosh \pi x}\,dx = \frac{1}{2}\tan\left(\frac{\alpha}{2}\right),\ 0<\alpha<\pi;$

(b) $$\int_0^\pi \frac{x\sin x}{1+a^2-2a\cos x}\,dx = \begin{cases} \dfrac{\pi}{a}\log(1+a), & \text{if } 0<a<1 \\ \dfrac{\pi}{a}\log(1+\dfrac{1}{a}), & \text{if } a>1. \end{cases}$$

[*Hints*: Take $\phi(z) = \frac{z}{a-e^{-iz}}$]

ANSWERS

1. $f(z) = 1 + \frac{1}{z+1} + \frac{1}{z-2} + \frac{3}{(z-2)^2}.$

2. (a) $z=a$ is an isolated essential singularity.
   (b) Simple poles at $z = \pm\frac{i}{2^n},\ n=0,1,2,\ldots$

3. (a) $\frac{ia^3}{2}$; (b) -1; (c) $\frac{175}{16}$; (d) -1.

Chapter 10

Fourier Series

10.1 Periodic Functions

A function $f(x)$ is said to be periodic, if for a constant number $T > 0$,

$$f(x+T) = f(x) \tag{10.1}$$

for any value of x. The number T is called the *period of the function.* It is obvious that, for a function $f(x)$ with period T

$$\left.\begin{aligned} f(x) &= f(x+T) = f(x+2T) = \dots \\ \text{and} \quad f(x) &= f(x-T) = f(x-2T) = \dots \end{aligned}\right\} \tag{10.2}$$

which show that $\pm 2T$, $\pm 3T$, ... are also the periods of the function $f(x)$.

The integral of a periodic function does not change in value, if the range of integration is equal to its period, i.e.,

$$\int_a^b f(x)\,dx = \int_{a+T}^{b+T} f(x)\,dx \tag{10.3}$$

for arbitrary values of a and b. This is also obvious, since, with a change of variable $x = \xi - T$, i.e., $dx = d\xi$,

$$\int_a^b f(x)\,dx = \int_{a+T}^{b+T} f(\xi - T)\,d\xi = \int_{a+T}^{b+T} f(\xi)\,d\xi.$$

Thus, the value of an integral of a periodic function taken between two points at a distance equal to its period remains same, whatever be the initial point of interval.

The simplest periodic function is represented by

$$y = A\sin(\omega x + \phi), \tag{10.4}$$

where A, ω and ϕ are constants. This function is called *harmonic* with amplitude A, frequency ω and initial phase ϕ, and describes sinusoidal vibrations in physical processes.

10.2 Orthogonal System of Functions

The system of finite functions

$$\phi_0(x), \phi_1(x), \phi_2(x), \dots, \phi_n(x), \tag{10.5}$$

is called *orthogonal* in the interval (a, b), if

$$\left.\begin{aligned}\int_a^b \phi_m(x)\phi_n(x)\,dx &= 0, \ m \neq n\\ \text{and} \quad \int_a^b \{\phi_n(x)\}^2\,dx &= M \neq 0\end{aligned}\right\} \tag{10.6}$$

for $n = 0, 1, 2, \ldots$.

If the value of the second integral in (10.6) is 1, i.e., if $M = 1$, the system of functions in (10.5) is called *normalised.*

We give below some examples of trigonometrical orthogonal functions:

(A) The system of functions

$$1, \cos x, \cos 2x, \ldots, \cos nx, \tag{10.7}$$

is orthogonal in the interval $(0, \pi)$.

(B) The system of functions

$$\sin x, \sin 2x, \ldots, \sin nx, \tag{10.8}$$

is orthogonal in the interval $(0, \pi)$.

(C) The system of functions

$$\sin x, \sin 3x, \sin 5x, \ldots, \sin(2n+1)x \tag{10.9}$$

is orthogonal in the interval $(0, \frac{\pi}{2})$.

(D) The systems of functions

$$1, \cos\frac{\pi x}{l}, \cos\frac{2\pi x}{l}, \ldots, \cos\frac{n\pi x}{l} \tag{10.10}$$

and

$$\sin\frac{\pi x}{l}, \quad \sin\frac{2\pi x}{l}, \ldots, \sin\frac{n\pi x}{l} \tag{10.11}$$

are orthogonal in the interval $(0, l)$.

(E) The system of functions

$$\sin\frac{\pi x}{2l}, \sin\frac{3\pi x}{2l}, \ldots, \sin\frac{(2n+1)\pi x}{2l} \tag{10.12}$$

is orthogonal in the interval $(0, l)$.

The orthogonality of all the above systems can easily be verified.

10.3 Right and Left Limits of a Function

Consider a function $f(x)$ with $x = \alpha$ a point within its region of definition. We denote

$$\lim_{x \longrightarrow \alpha - 0} f(x) = f(\alpha - 0) \tag{10.13}$$

$$\text{and} \quad \lim_{x \longrightarrow \alpha + 0} f(x) = f(\alpha + 0) \tag{10.14}$$

where both the limits exist.

The first limit in equation (10.13) is called the *limit of the function* $f(x)$ at $x = \alpha$ *from left*, and the second limit in equation (10.14) is called the *limit of the function* $f(x)$ at $x = \alpha$ *from right.* If the function $f(x)$ is continuous at $x = \alpha$, then by definition, both the above limits exist and are equal, i.e.,

$$f(\alpha - 0) = f(\alpha) = f(\alpha + 0). \tag{10.15}$$

If, however, the function $f(x)$ has a discontinuity at $x = \alpha$, then either or both of them do not exist. If, both the limits at a point of discontinuity $x = \alpha$ exist, we say that this point is a *point of discontinuity of first order*, and call the function $f(x)$ to be piecewise continuous at $x = \alpha$. The quantity

$$\delta = f(\alpha + 0) - f(\alpha - 0) \tag{10.16}$$

is known as the *jump of the function* $f(x)$ at $x = \alpha$.

As an example, we consider the function

$$f(x) = \begin{cases} -x^3, & \text{for } x < 1 \\ 0, & \text{for } x = 1 \\ \sqrt{x}, & \text{for } x > 1 \end{cases} \tag{10.17}$$

The function $f(x)$ is graphically represented in *Fig 10.1*. Its limit at $x = 1$ from left and from right are $f(1 - 0) = -1$ and $f(1 + 0) = 1$, so that its jump at $x = 1$ is

$$\delta = f(1 + 0) - f(1 - 0) = 2.$$

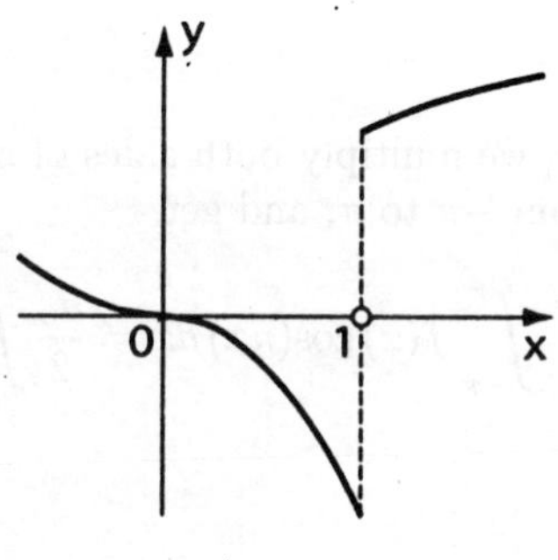

Fig 10.1

A function $f(x)$ is said to be *smooth* in an interval (a, b) if its first derivative is continuous within (a, b). It means that tangents to the curve $y = f(x)$ continuously change direction without break in the interval. A function is said to be *piecewise smooth* in the interval (a, b), if the function and its first derivative are either continuous or have got a point of discontinuity of first order in that interval.

10.4 Fourier Series

The French mathematician Fourier found that any single-valued function $f(x)$ defined in the interval $(-\pi, \pi)$ can be expanded in a trigonometric series in the form

$$f(x) = \frac{a_0}{2} + \sum_{n=1}^{\infty} \{a_n \cos(nx) + b_n \sin(nx)\}, \tag{10.18}$$

where the coefficients a_n and b_n are determined by the use of Euler's formulas:

$$a_n = \frac{1}{\pi} \int_{-\pi}^{\pi} f(x) \cos(nx)\, dx \quad [n = 0, 1, 2, \ldots] \tag{10.19}$$

$$\text{and} \quad b_n = \frac{1}{\pi} \int_{-\pi}^{\pi} f(x) \sin(nx)\, dx \quad [n = 1, 2, \ldots]. \tag{10.20}$$

The trigonometric series associated with a function $f(x)$ in the form given in equation (10.18) is known a *Fourier series,* and the coefficients given in equations (10.19) and (10.20) are known as *Fourier coefficients.*

Fourier, however, could not develop any specific criteria for the convergence of the series in equation (10.18). It was Dirichlet, who could establish the sufficient conditions, called *Dirichlet's conditions,* for the convergence of the Fourier series. We shall discuss about it later. For the time being, we assume that the series in equation (10.18) permits term-by-term integration. Then the Fourier coefficients are obtained in the following manner.

We integrate both sides of equation (10.18) within the limit from $-\pi$ to π, and get

$$\int_{-\pi}^{\pi} f(x)\,dx = \frac{a_0}{2}\int_{-\pi}^{\pi} dx + \sum_{k=1}^{\infty}\int_{-\pi}^{\pi} (a_k \cos kx + b_k \sin kx)\,dx$$

$$= \pi a_0 \quad \text{[since all the other integrals vanish]}$$

Hence

$$a_0 = \frac{1}{\pi}\int_{-\pi}^{\pi} f(x)\,dx. \tag{10.21}$$

Now, we multiply both sides of equation (10.18) by $\cos(nx)$ and integrate the result within limit from $-\pi$ to π, and get

$$\int_{-\pi}^{\pi} f(x)\cos(nx)\,dx = \frac{a_0}{2}\int_{-\pi}^{\pi} \cos(nx)\,dx + \sum_{k=1}^{\infty}\int_{-\pi}^{\pi} a_k \cos kx \cos nx\,dx + \sum_{k=1}^{\infty}\int_{-\pi}^{\pi} b_k \cos kx \sin nx\,dx.$$

$$= \int_{-\pi}^{\pi} \cos^2 nx\,dx \quad \text{[since all other integrals vanish due to orthogonality property]}$$

$$= a_n \pi.$$

Hence

$$a_n = \frac{1}{\pi}\int_{-\pi}^{\pi} f(x)\cos(nx)\,dx. \tag{10.22}$$

Similarly, by multiplying both sides of equation (10.18) by $\sin nx$ and integrating the result as before, we get

$$b_n = \frac{1}{\pi}\int_{-\pi}^{\pi} f(x)\sin(nx)\,dx. \tag{10.23}$$

10.5 Even and Odd Functions

A function $f(x)$ is said to be an *Even function* if, for all x

$$f(-x) = f(x). \tag{10.24}$$

Graphically, an even function $y = f(x)$ will be symmetrical about y-axis (*Fig 10.2*). Moreover

$$\int_{-\alpha}^{\alpha} f(x)\,dx = 2\int_{0}^{\alpha} f(x)\,dx, \tag{10.25}$$

when the function $f(x)$ is defined in the interval $[-\alpha, \alpha]$.

A function $f(x)$ is said to be an *odd function,* if for all x within its interval of definition $[-\alpha, \alpha]$,

$$f(-x) = -f(x). \tag{10.26}$$

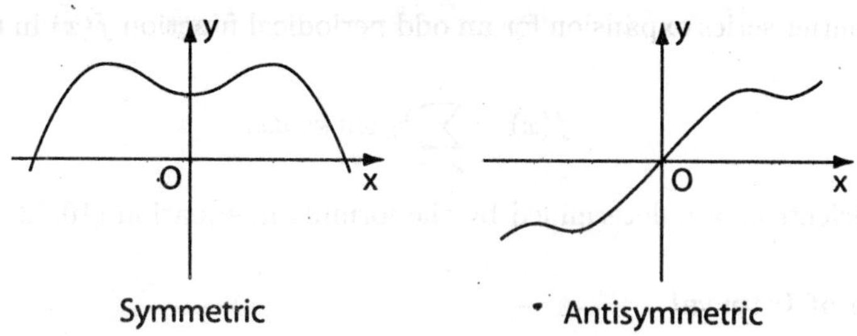

Fig 10.2

Graphically, an odd function $y = f(x)$ will be symmetrical about the origin (*Fig 10.2*). Moreover

$$\int_{-\alpha}^{\alpha} f(x)\,dx = 0. \tag{10.27}$$

It is evident that

(a) product of two odd or two even functions will be an even function,
(b) product of an odd and an even function will be an odd function.

10.6 Fourier Cosine and Sine Series

Let $f(x)$ be an even periodical function defined in the interval $[-\pi, \pi]$.

Since $\cos(nx)$, $(n = 0, 1, 2, \dots)$ is an even function, the function $f(x)\cos nx$ will be an even function. Hence, equation (10.22) following equation (10.25) becomes, (for $n = 0, 1, 2, \dots$)

$$a_n = \frac{1}{\pi}\int_{-\pi}^{\pi} f(x)\cos nx\,dx = \frac{2}{\pi}\int_{0}^{\pi} f(x)\cos nx\,dx. \tag{10.28}$$

Similarly, since the function $\sin(nx)$ is an odd function, the function $f(x)\sin(nx)$ will be an odd function. Hence, equation (10.23) following equation (10.27) becomes (for $n = 1, 2, 3, \dots$)

$$b_n = \frac{1}{\pi}\int_{-\pi}^{\pi} f(x)\sin(nx)\,dx = 0. \tag{10.29}$$

Hence, for an even function $f(x)$, the Fourier series representation is

$$f(x) = \frac{a_0}{2} + \sum_{k=1}^{\infty} a_k \cos kx, \tag{10.30}$$

where the coefficients a_k are determined by the formula in equation (10.28).

Let now $f(x)$ be an odd periodical function in the interval $[-\pi, \pi]$. Then, $f(x)\cos nx$ will be an odd function, as such

$$a_n = \frac{1}{\pi}\int_{-\pi}^{\pi} f(x)\cos(nx)\,dx = 0 \qquad (n = 0, 1, 2, \dots). \tag{10.31}$$

Again, since $f(x)\sin x$ will be an even function, we have

$$b_n = \frac{1}{\pi}\int_{-\pi}^{\pi} f(x)\sin(nx)\,dx = \frac{2}{\pi}\int_{0}^{\pi} f(x)\sin(nx)\,dx \quad (n = 1, 2, \dots). \tag{10.32}$$

Thus, the Fourier series expansion for an odd periodical function $f(x)$ in the interval $[-\pi, \pi]$ becomes

$$f(x) = \sum_{k=1}^{\infty} b_k \sin kx\, dx, \tag{10.33}$$

where the coefficients b_k are determined by the formula in equation (10.32).

10.7 Change of Interval

We have discussed so far about the Fourier expansion in the interval $[-\pi, \pi]$. But in many physical processes, it becomes necessary to develop a Fourier series in an interval, say $[-l, l]$. To do so, we use the transformation

$$y = \frac{\pi x}{l}, \quad \text{i.e.,} \quad x = \frac{ly}{\pi}, \tag{10.34}$$

so that we get the function

$$\phi(y) = f\left(\frac{ly}{\pi}\right).$$

We find the Fourier expansion for $\phi(y)$ as

$$\phi(y) = \frac{a_0}{2} + \sum_{n=1}^{\infty} (a_n \cos ny + b_n \sin ny), \tag{10.35}$$

where

$$a_n = \frac{1}{\pi}\int_{-\pi}^{\pi} \phi(y)\cos(ny)\, dy = \frac{1}{\pi}\int_{-\pi}^{\pi} f\left(\frac{ly}{\pi}\right)\cos ny\, dy, \quad n = 0, 1, 2, \ldots$$

$$\text{and} \quad b_n = \frac{1}{\pi}\int_{-\pi}^{\pi} \phi(y)\sin(ny)\, dy = \frac{1}{\pi}\int_{-\pi}^{\pi} f\left(\frac{ly}{\pi}\right)\sin ny\, dy, \quad n = 0, 1, 2, \ldots.$$

Thus, going back to old variable x, i.e., putting $y = \dfrac{\pi x}{l}$, we get

$$f(x) = \frac{a_0}{2} + \sum_{n=1}^{\infty}\left(a_n \cos\frac{n\pi x}{l} + b_n \sin\frac{n\pi x}{l}\right), \tag{10.36}$$

where

$$a_n = \frac{1}{l}\int_{-l}^{l} f(x)\cos\frac{n\pi x}{l}\, dx \quad (n = 0, 1, 2, \ldots) \tag{10.37}$$

$$\text{and} \quad b_n = \frac{1}{l}\int_{-l}^{l} f(x)\sin\frac{n\pi x}{l}\, dx \quad (n = 0, 1, 2, 3, \ldots). \tag{10.38}$$

10.8 Half-range Expansion

In many physical processes, it becomes necessary to get the Fourier expansion of a function $f(x)$ in the interval $[0, \pi]$. In such cases, we extend the function $f(x)$ periodically in the interval $[-\pi, 0]$. The extension of the function $f(x)$ is done in even manner if we require expansion in cosine series only. If, however, we require expansion in sine series only, the extension of function $f(x)$ is done in odd manner. Then, for the extended function $f(x)$ in even manner, we get following our discussions in Section 10.6

$$f(x) = \frac{a_0}{2} + \sum_{k=1}^{\infty} a_k \cos kx\, dx, \tag{10.39}$$

where the coefficients a_k are given

$$a_k = \frac{2}{\pi}\int_0^{\pi} f(x) \cos kx\, dx, \quad k = 0, 1, 2, \ldots. \tag{10.40}$$

Similarly, for the extended function $f(x)$ in odd manner, we get

$$f(x) = \sum_{k=1}^{\infty} b_k \sin kx\, dx, \tag{10.41}$$

where the coefficients b_k are determined from

$$b_k = \frac{2}{\pi}\int_0^{\pi} f(x) \sin kx\, dx, \quad k = 1, 2, \ldots. \tag{10.42}$$

Such series expansions are known as *Half-range expansions.*

We shall now discuss some examples assuming uniform convergence of corresponding Fourier expansions.

EXAMPLE 10.1. *Expand the function $f(x) = |\,x\,|$, for $-\pi \le x \le \pi$ in Fourier series.*

Solution: The function $f(x)$ is an even function. Hence we expand it only in cosine series. Since $|\,x\,| = x$, for $x \ge 0$, we find the coefficients using equation (10.28).

$$a_0 = \frac{2}{\pi}\int_0^{\pi} x\, dx = \frac{2}{\pi}\left(\frac{x^2}{2}\right)_0^{\pi} = \pi \tag{1}$$

$$\begin{aligned} a_n &= \frac{2}{\pi}\int_0^{\pi} x \cos nx\, dx = -\frac{2}{n\pi}\int_0^{\pi} \sin(nx)\, dx \\ &= \frac{2}{\pi n^2}\Big(\cos nx\Big)_0^{\pi} = \frac{2}{\pi n^2}(\cos n\pi - 1) = \frac{2}{\pi n^2}\Big\{(-1)^n - 1\Big\}. \end{aligned} \tag{2}$$

From (2), it is clear, that

$$\left.\begin{aligned} a_n &= 0, && \text{for } n = \text{even number} \\ a_n &= -\frac{4}{\pi n^2}, && \text{for } n = \text{odd number} \end{aligned}\right\} \tag{3}$$

Hence, for $-\pi \le x \le \pi$

$$|\,x\,| = \frac{\pi}{2} - \frac{4}{\pi}\left(\cos x + \frac{\cos 3x}{3^2} + \frac{\cos 5x}{5^2} + \cdots\right). \tag{4}$$

EXAMPLE 10.2. *Expand in Fourier series, the function $f(x) = x^2$ for $0 < x < 2\pi$.*

Solution: A graph of the function along with its periodic extension is given in *Fig 10.3.*

It is obvious, that the function $f(x)$ is neither odd nor even. The coefficients of expansion are

$$a_0 = \frac{1}{\pi}\int_0^{2\pi} x^2\,dx = \frac{1}{\pi}\left(\frac{x^3}{3}\right)_0^{2\pi} = \frac{8\pi^2}{3} \tag{1}$$

$$\begin{aligned} a_n &= \frac{1}{\pi}\int_0^{2\pi} x^2 \cos nx\,dx \\ &= -\frac{2}{n\pi}\int_0^{2\pi} x\sin nx\,dx \\ &= \frac{2}{\pi n^2}\left(x\cos nx\right)_0^{2\pi} - \frac{2}{\pi n^2}\int_0^{2\pi}\cos nx\,dx, \end{aligned}$$

i.e., $a_n = \dfrac{4}{n^2}$. (2)

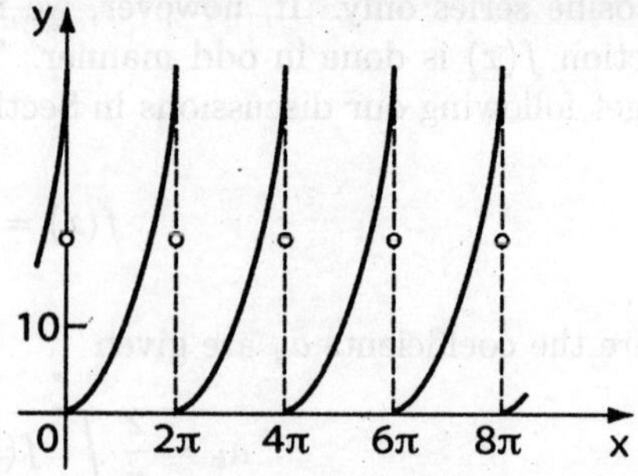

Fig 10.3

$$b_n = \frac{1}{\pi}\int_0^{2\pi} x^2 \sin(nx)\,dx = -\frac{1}{n\pi}\left(x^2\cos nx\right)_0^{2\pi} + \frac{2}{n\pi}\int_0^{2\pi} x\cos nx\,dx$$

$$= -\frac{4\pi}{n} - \frac{2}{\pi n^2}\int_0^{2\pi}\sin(nx)\,dx = -\frac{4\pi}{n}. \tag{3}$$

Hence, for $0 < x < 2\pi$,

$$\begin{aligned} x^2 &= \frac{4\pi^2}{3} + 4\left(\cos x - \pi\sin x + \frac{\cos 2x}{2^2} - \frac{\pi\sin 2x}{2} + \ldots\right) \\ &= \frac{4\pi^2}{3} + 4\sum_{n=1}^{\infty}\frac{\cos nx}{n^2} - 4\pi\sum_{n=1}^{\infty}\frac{\sin nx}{n}. \end{aligned} \tag{4}$$

EXAMPLE 10.3. *Expand in Fourier series the function $f(x) = x$, for $0 < x < 2\pi$.*

Solution: The function is neither odd, nor even. The coefficients of expansion are

$$a_0 = \frac{1}{\pi}\int_0^{2\pi} x\,dx = \frac{1}{\pi}\left(\frac{x^2}{2}\right)_0^{2\pi} = 2\pi \tag{1}$$

$$a_n = \frac{1}{\pi}\int_0^{2\pi} x\cos nx\,dx = -\frac{1}{n\pi}\int_0^{2\pi}\sin nx\,dx + \left(\frac{x}{n\pi}\sin nx\right)_0^{2\pi} = 0 \tag{2}$$

$$b_n = \frac{1}{\pi}\int_0^{2\pi} x\sin(nx)dx = -\frac{1}{n\pi}(x\cos nx)_0^{2\pi} + \frac{1}{n\pi}\int_0^{2\pi}\cos nx\,dx = -\frac{2}{n}. \tag{3}$$

Hence, for $0 < x < 2\pi$,

$$x = \pi - 2\left(\sin x + \frac{\sin 2x}{2} + \frac{\sin 3x}{3} + \cdots\right). \tag{4}$$

EXAMPLE 10.4. *Expand in Fourier series, the function $f(x) = ax^2 + bx + c$, for $0 < x < 2\pi$.*

Solution: Using results (4) of Example 10.2 and (4) of Example 10.3 we get, for $0 < x < 2\pi$.

$$ax^2 + bx + c = \frac{4}{3}a\pi^2 + b\pi + c + 4a\sum_{n=1}^{\infty}\frac{\cos nx}{n^2} - (4\pi a + 2b)\sum_{n=1}^{\infty}\frac{\sin nx}{n}. \quad (1)$$

EXAMPLE 10.5. *Find the Fourier series for the function $f(x) = x^2$ for $-\pi \le x \le \pi$.*

Solution: The function is an even one. The graph of the function along with its periodic extension is given in *Fig 10.4*. The coefficients of expansion are

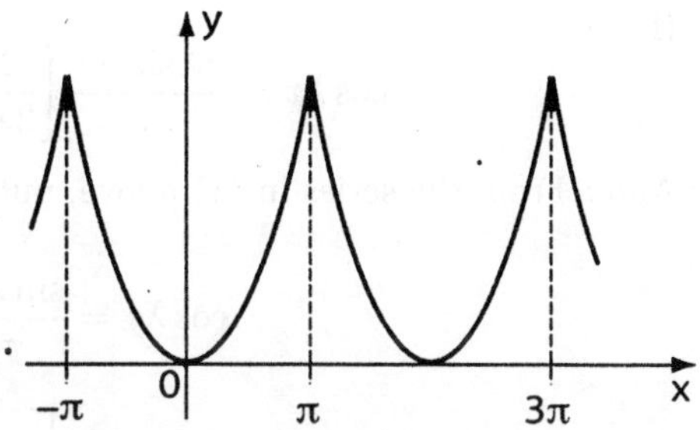

Fig 10.4

$$a_0 = \frac{2}{\pi}\int_0^{\pi} x^2\,dx = \frac{2}{\pi}\left(\frac{x^3}{3}\right)_0^{\pi} = \frac{2\pi^2}{3}. \quad (1)$$

$$a_n = \frac{2}{\pi}\int_0^{\pi} x^2\cos nx\,dx = -\frac{4}{n\pi}\int_0^{\pi} x\sin(nx)\,dx$$

$$= \frac{4}{\pi n^2}\Big(x\cos nx\Big)_0^{\pi} - \frac{4}{\pi n^2}\int_0^{\pi}\cos nx\,dx$$

$$= \frac{4}{n^2}\cos n\pi = (-1)^n\frac{4}{n^2}.$$

Since the function $f(x)$ is even, $b_n = 0$.
Hence, for $-\pi \le x \le \pi$,

$$x^2 = \frac{\pi^2}{3} - 4\left(\cos x - \frac{\cos 2x}{2^2} + \frac{\cos 3x}{3^2} - \cdots\right). \quad (2)$$

EXAMPLE 10.6. *Find the Fourier series for the function*

$$f(x) = \begin{cases} -1, & \textit{for } -\pi < x < 0 \\ 0, & \textit{for } x = 0 \\ 1, & \textit{for } 0 < x < \pi. \end{cases}$$

Solution: The given function is odd. Its graph is shown along with its periodic extension in *Fig 10.5*. The coefficients of expansion are

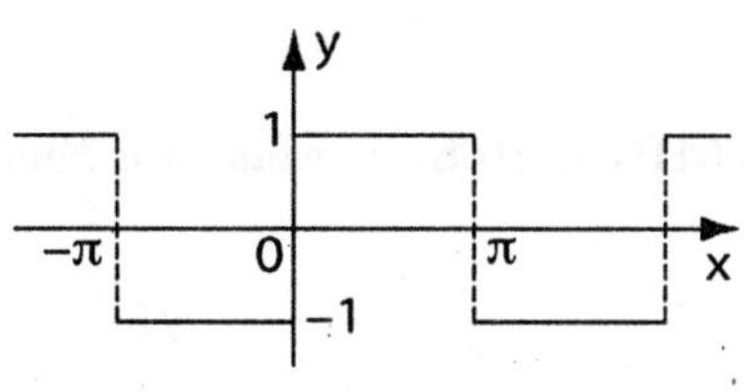

Fig 10.5

$a_n = 0,$

$$b_n = \frac{2}{\pi}\int_0^{\pi}\sin(nx)\,dx = \begin{cases} 0, & \text{for } n = \text{even number} \\ \dfrac{4}{n\pi}, & \text{for } n = \text{odd number.} \end{cases}$$

Hence, the Fourier series is

$$f(x) = \frac{4}{\pi}\left(\sin x + \frac{\sin 3x}{3} + \frac{\sin 5x}{5} + \cdots\right). \quad (1)$$

EXAMPLE 10.7. *Expand in Fourier series, the function $f(x) = \cos\lambda x$, $\lambda \ne$ integer, and $-\pi < x < \pi$.*

Solution: The function is an even one. Hence, we have

$$b_n = 0$$

$$a_n = \frac{2}{\pi}\int_0^{\pi} \cos\lambda x \cos nx\, dx = \frac{1}{\pi}\int_0^{\pi} \{\cos(\lambda+n)\,x + \cos(\lambda-n)\,x\}\, dx$$

$$= \frac{1}{\pi}\left[\frac{\sin(\lambda+n)\pi}{\lambda+n} + \frac{\sin(\lambda-n)\pi}{\lambda-n}\right] = \frac{2}{\pi}\frac{\lambda(-1)^n}{(\lambda^2-n^2)}\cdot \sin\lambda\pi.$$

Hence

$$\cos\lambda x = \frac{2\lambda\sin\lambda\pi}{\pi}\left[\frac{1}{2\lambda^2} - \frac{\cos x}{\lambda^2-1^2} + \frac{\cos 2x}{\lambda^2-2^2}\cdots\right]. \quad (1)$$

Note: From the series in (1) above, putting $x=\pi$, we get

$$\cos\lambda\pi = \frac{\sin\lambda\pi}{\pi}\left[\frac{1}{\lambda} + \sum_{n=1}^{\infty}\frac{2\lambda(-1)^n}{\lambda^2-n^2}\cos n\pi\right]$$

$$\text{i.e.,}\quad \pi\cot\lambda\pi = \frac{1}{\lambda} + \sum_{n=1}^{\infty}\frac{2\lambda}{\lambda^2-n^2}. \quad (2)$$

Again, if we put $x=0$ in equation (1), we get

$$\frac{1}{\sin\lambda\pi} = \frac{1}{\lambda\pi} + 2\sum_{k=1}^{\infty}(-1)^k\frac{\lambda\pi}{(\lambda\pi)^2-(k\pi)^2}.$$

Now, we put $\theta = \lambda\pi$.
Then we get

$$\frac{1}{\sin\theta} = \frac{1}{\theta} + \sum_{k=1}^{\infty}(-1)^k\frac{2\theta}{\theta^2-(k\pi)^2} = \frac{1}{\theta} + \sum_{k=1}^{\infty}(-1)^k\left[\frac{1}{\theta+k\pi} + \frac{1}{\theta-k\pi}\right].$$

Thus we obtain an expansion for the function

$$f(\theta) = \frac{1}{\sin\theta}.$$

EXAMPLE 10.8. *Expand in a Fourier series, the function*

$$f(x) = \begin{cases} \sin x, & \text{for } 0 \le x \le \pi \\ -\sin x, & \text{for } \pi \le x \le 0. \end{cases}$$

Solution: The graphical representation of the function and its periodic extension is given in *Fig 10.6*. The expansion coefficients are

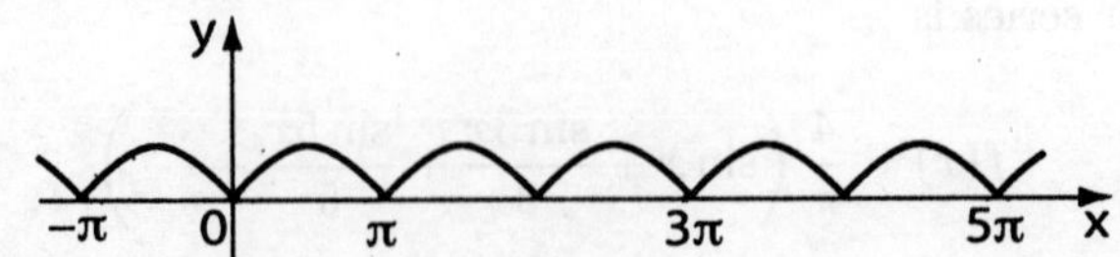

Fig 10.6

$$a_0 = \frac{1}{\pi}\int_{-\pi}^{0}(-\sin x)\,dx + \frac{1}{\pi}\int_{0}^{\pi}\sin x\,dx = \frac{2}{\pi}\int_{0}^{\pi}\sin x\,dx = \frac{4}{\pi}.$$

$$a_n = \frac{1}{\pi}\int_{-\pi}^{0}(-\sin x)\cos nx\,dx + \int_{0}^{\pi}\sin x\cos nx\,dx$$

$$= \frac{2}{\pi}\int_{0}^{\pi}\sin x\cos nx\,dx = \begin{cases} -\dfrac{4}{\pi}\cdot\dfrac{1}{n^2-1}, & (\text{for } n = \text{even number}) \\ 0, & (\text{for } n = \text{odd number}) \end{cases}$$

$b_n = 0$ (the function is even).

Hence

$$f(x) = \frac{2}{\pi} - \frac{4}{\pi}\sum_{n=2}^{\infty}\frac{\cos nx}{n^2-1} \quad (n = \text{even}). \tag{1}$$

It may be noted that the given function represents a full-wave rectifier, and the series in (1) above shows that the rectifier approaches pure direct current very fast.

EXAMPLE 10.9. *Find the Fourier series of the function* $y(t) = a\left(1 - \frac{t}{T}\right)$.

Solution: The function here represents a saw-tooth wave where the displacement is linear with time such that

$$\left.\begin{aligned} y &= a, \quad \text{for } t = 0 \\ y &= 0, \quad \text{for } t = T \end{aligned}\right\} \tag{1}$$

T represents time period such that $T = \dfrac{2\pi}{\omega}$.

Coefficients of Fourier expansion are

$$a_0 = \frac{1}{T}\int_0^T y\,dt = \frac{1}{T}\int_0^T a\left(1 - \frac{t}{T}\right)dt = \frac{a}{2} \tag{2}$$

$$a_n = \frac{2}{T}\int_0^T y\cos(n\omega t)\,dt = \frac{2a}{T}\int_0^T\left(1 - \frac{t}{T}\right)\cos(n\omega t)\,dt = 0$$

$$b_n = \frac{2}{T}\int_0^T y\sin(n\omega t)\,dt = \frac{2a}{T}\int_0^T\left(1 - \frac{t}{T}\right)\sin(n\omega t)\,dt = \frac{a}{n\pi}.$$

Hence

$$y(t) = \frac{a}{2} + \frac{a}{\pi}\sum_{n=1}^{\infty}\frac{\sin(n\omega t)}{n}.$$

EXAMPLE 10.10. *Expand in a Fourier series:*

$$f(x) = \begin{cases} 0, & \text{for } -\pi \le x \le 0 \\ l, & \text{for } 0 \le x \le \pi. \end{cases}$$

and $f(x) = f(x + 2\pi)$.

Solution: The function is periodic in $[-\pi, \pi]$ and represents a square wave. The expansion coefficients are

$$a_0 = \frac{1}{\pi}\int_0^{\pi} l\,dx = l \tag{1}$$

$$a_n = \frac{1}{\pi}\int_0^{\pi} l\cos(nx)\,dx = 0 \tag{2}$$

$$b_n = \frac{1}{\pi}\int_0^{\pi} l\sin(nx)\,dx = \frac{l}{\pi}\Big[-\frac{\cos nx}{n}\Big]_0^{\pi},$$

$$\text{i.e.,}\quad b_n = \begin{cases} 0, & \text{for } n \text{ even number} \\ \dfrac{2l}{n\pi}, & \text{for } n \text{ odd number.} \end{cases}$$

Hence

$$f(x) = \frac{a_0}{2} + \sum_{n=1}^{\infty}\{a_n\cos nx + b_n \sin nx\}$$
$$= \frac{l}{2} + \frac{2l}{\pi}\sum_{n=1}^{\infty}\frac{\sin nx}{n} \quad [n = \text{odd numbers only}].$$

EXAMPLE 10.11. *Find the Fourier expansion of*

$$f(t) = \begin{cases} A\sin\omega t, & \textit{for } 0 < t < \frac{T}{2} \\ 0, & \textit{for } -\frac{T}{2} \le t < 0. \end{cases}$$

where T is the period.

Solution: The function actually represents a half-wave rectifier with a period T. The Fourier coefficients are

$$a_0 = \frac{1}{T}\int_0^{\frac{T}{2}} A\sin\omega t = \frac{A}{\pi} \quad [\omega T = 2\pi] \tag{1}$$

$$a_n = \frac{2}{T}\int_0^{\frac{T}{2}} A\sin\omega t\cos(n\omega t)\,dt = \frac{A}{2\pi}\left[\frac{1-\cos(1+n)\pi}{1+n} + \frac{1-\cos(1-n)\pi}{1-n}\right],$$

$$\text{i.e.,}\quad a_n = \begin{cases} \dfrac{2A}{\pi(1-n^2)}, & \text{for } n = \text{even number} \\ 0, & \text{for } n = \text{odd number} \end{cases}$$

$$a_1 = \frac{2A}{T}\int_0^{\frac{T}{2}} \sin\omega t\cos\omega t\,dt = 0$$

$$b_n = \frac{2A}{T}\int_0^{\frac{T}{2}} \sin\omega t\cdot\sin(n\omega t)\,dt = \frac{A}{T}\int_0^{\frac{T}{2}}[\cos(1-n)\omega t - \cos(1+n)\omega t]\,dt$$

$$= \frac{A}{T}\left[\frac{\sin(1-n)\omega t}{(1-n)\omega} - \frac{\sin(1+n)\omega}{(1+n)\omega}\right]_0^{\frac{T}{2}} = 0, \text{ for } n = 2, 3, 4, \ldots$$

$$b_1 = \frac{2A}{T}\int_0^{\frac{T}{2}} \sin\omega t\cdot\sin\omega t\,dt = \frac{A}{T}\int_0^{\frac{T}{2}}(1-\cos 2\omega t)\,dt = \frac{A}{2}.$$

Hence

$$f(t) = \frac{A}{\pi} - \frac{2A}{\pi}\left[\sum_{n=2}^{\infty} \frac{\cos n\omega t}{n^2-1}\right] + \frac{A}{2}\sin \omega t, \quad n = 2, 4, 6, \ldots$$

10.9 Some Useful Results

(a) From equation (2) of Example 10.5, we have

$$\frac{\pi^2 - 3x^2}{12} = \sum_{n=1}^{\infty} (-1)^{n+1} \frac{\cos nx}{n^2} \quad (-\pi \le x \le \pi). \tag{10.43}$$

Putting $x = 0$, we get

$$\frac{\pi^2}{12} = 1 - \frac{1}{2^2} + \frac{1}{3^2} - \frac{1}{4^2} + \cdots \tag{10.44}$$

(b) From equations (4) of Example 10.2 and (4) of Example 10.3, we have

$$\sum_{n=1}^{\infty} \frac{\cos nx}{n^2} = \frac{1}{12}\left(3x^2 - 6\pi x + 2\pi^2\right), \text{ for } 0 < x < 2\pi. \tag{10.45}$$

Putting $x = 0$, we get

$$\frac{\pi^2}{6} = 1 + \frac{1}{2^2} + \frac{1}{3^2} + \cdots$$

(c) Putting $x = 0$ in equation (4) of Example 10.1, we get

$$\frac{\pi^2}{8} = 1 + \frac{1}{3^2} + \frac{1}{5^2} + \cdots \tag{10.46}$$

(d) Putting $x = \frac{\pi}{2}$ in equation (3) of Example 10.6, we get

$$\frac{\pi}{4} = 1 - \frac{1}{3} + \frac{1}{5} - \frac{1}{7} + \cdots \tag{10.47}$$

EXAMPLE 10.12. *Find the Fourier series for the function*

$$f(x) = \begin{cases} x, & \text{for } 0 \le x \le \frac{l}{2} \\ l - x, & \text{for } \frac{l}{2} < x \le l \end{cases}$$

Solution: The graph of the function and its periodic extension is shown in the *Fig 10.7.* The function is an odd one. The expansion coefficients are $a_n = 0$, $n = 0, 1, 2, \ldots$.

$$b_n = \frac{2}{l}\int_0^l f(x)\sin\frac{n\pi x}{l}\,dx = \frac{2}{l}\int_0^{\frac{l}{2}} x\sin\frac{n\pi x}{l}\,dx + \frac{2}{l}\int_{\frac{l}{2}}^{l} (l-x)\sin\frac{n\pi x}{l}\,dx.$$

We put $z = \frac{\pi x}{l}$. Then

$$b_n = \frac{2l}{\pi^2}\int_0^{\frac{\pi}{2}} z\sin(nz)\,dz + \frac{2l}{\pi^2}\int_{\frac{\pi}{2}}^{\pi} (\pi - z)\sin(nz)\,dz = \frac{4l}{n^2\pi^2}\sin\frac{n\pi}{2}.$$

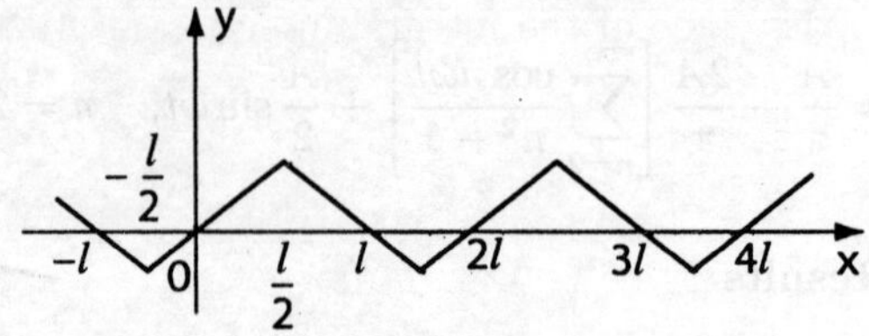

Fig 10.7

Hence

$$f(x) = \frac{4l}{\pi^2}\left(\sin\frac{\pi x}{l} - \frac{1}{3^2}\sin\frac{3\pi x}{l} + \frac{1}{5^2}\sin\frac{5\pi x}{l} - \cdots\right).$$

10.10 Two Important Formulae

We intend to prove the following formulae which would be very useful in discussing the convergence of Fourier series.

(a)
$$\frac{1}{2} + \sum_{k=1}^{n}\cos kx = \frac{\sin(n+\frac{1}{2})x}{2\sin\frac{x}{2}}. \tag{10.48}$$

(b)
$$\frac{1}{\pi}\int_0^{\pi}\frac{\sin(k+\frac{1}{2})x}{2\sin\frac{x}{2}} = \frac{1}{2}. \tag{10.49}$$

Proof. (a) Let

$$S = \frac{1}{2} + \sum_{k=1}^{n}\cos kx.$$

Then

$$\begin{aligned}2S\sin\frac{x}{2} &= \sin\frac{x}{2} + 2\sin\frac{x}{2}\sum_{k=1}^{n}\cos kx = \sin\frac{x}{2} + 2\sin\frac{x}{2}\cdot\cos x + 2\sin\frac{x}{2}\cdot\cos 2x + \cdots\\ &= \sin\frac{x}{2} + \left(\sin\frac{3x}{2} - \sin\frac{x}{2}\right) + \left(\sin\frac{5x}{2} - \sin\frac{3x}{2}\right) + \cdots\\ &\quad + \left\{\sin\left(n+\frac{1}{2}\right)x - \sin\left(n-\frac{1}{2}\right)x\right\}\\ &= \sin\left(n+\frac{1}{2}\right)x.\end{aligned}$$

Hence

$$S = \frac{\sin\left(n+\frac{1}{2}\right)x}{2\sin\frac{x}{2}}.$$ □

(b) We integrate the identity (10.48) in the interval $[-\pi, \pi]$ and divide the result by π. Then we get

$$\frac{1}{\pi}\int_{-\pi}^{\pi}\frac{\sin\left(n+\frac{1}{2}\right)x}{2\sin\frac{x}{2}}\,dx = 1. \tag{10.50}$$

It is to be noted that the value of the integral in equation (10.50) is not dependent on n. Moreover, the integrand in equation (10.50) is an even function. Hence, we can write

$$\frac{1}{\pi}\int_0^{\pi}\frac{\sin\left(n+\frac{1}{2}\right)x}{2\sin\frac{x}{2}}\,dx=\frac{1}{2}. \tag{10.51}$$

10.11 Partial Sum of Fourier Series

Let a periodic function $f(x)$ of period 2π be represented by Fourier series

$$f(x)=\frac{a_0}{2}+\sum_{k=1}^{\infty}(a_k\cos kx+b_k\sin kx). \tag{10.52}$$

Its partial sum considering n-number of terms is denoted as

$$S_n(x)=\frac{a_0}{2}+\sum_{k=1}^{n}(a_k\cos kx+b_k\sin kx). \tag{10.53}$$

We insert here the expression for the Fourier coefficients given by equations (10.19) and (10.20). Then we get

$$\begin{aligned}
S_n(x)&=\frac{1}{2\pi}\int_{-\pi}^{\pi}f(t)\,dt+\frac{1}{\pi}\sum_{k=1}^{n}\int_{-\pi}^{\pi}[f(t)\cos kt\cdot\cos kx\,dt+f(t)\sin kt\cdot\sin kx\,dt]\\
&=\frac{1}{\pi}\int_{-\pi}^{\pi}\left[\frac{1}{2}+\sum_{k=1}^{n}(\cos kt\cos kx+\sin kt\sin kx)\right]f(t)\,dt\\
&=\frac{1}{\pi}\int_{-\pi}^{\pi}f(t)\left[\frac{1}{2}+\sum_{k=1}^{n}\cos k(t-x)\right]dt\\
&=\frac{1}{\pi}\int_{-\pi}^{\pi}f(t)\cdot\frac{\sin\left[(n+\frac{1}{2})(t-x)\right]}{2\sin\frac{(t-x)}{2}}\,dt\quad\text{[by equation (10.48)]}.
\end{aligned}$$

Here we substitute $\xi=t-x$. Then we get

$$S_n(x)=\frac{1}{\pi}\int_{-\pi-x}^{\pi-x}f(x+\xi)\frac{\sin(n+\frac{1}{2})\xi}{2\sin\frac{\xi}{2}}\,d\xi.$$

Since the integrand is a periodic function with respect to the variable ξ having a period 2π, and the interval $[-\pi-x,\pi-x]$ is of length 2π, we can write

$$S_n(x)=\frac{1}{\pi}\int_{-\pi}^{\pi}f(x+\xi)\frac{\sin(n+\frac{1}{2})\xi}{2\sin\frac{\xi}{2}}\,d\xi. \tag{10.54}$$

Equation (10.54) is known as *Dirichlet's Integral formula* for the partial sum of a Fourier series. We now intend to investigate the convergence of the Fourier series.

10.12 Dirichlet's Theorem

If the function $f(x)$ is periodic in $[-\pi, \pi]$ with a period of 2π, such that (i) it is single-valued and bounded, (ii) it has a finite number of maxima and minima and (iii) it has only a finite number of discontinuities in the interval $[-\pi, \pi]$, then

$$S_n(x) = \frac{a_0}{2} + \sum_{n=1}^{\infty}(a_n \cos nx + b_n \sin nx)$$

converges to $f(x)$ as $n \longrightarrow \infty$ at values of x for which $f(x)$ is continuous, and to $\frac{1}{2}[f(x+0) + f(x-0)]$ at points of discontinuity.

Proof. Our aim is to prove that

$$\lim_{n\longrightarrow\infty} S_n(x) = \lim_{n\longrightarrow\infty} \frac{1}{2\pi}\int_{-\pi}^{\pi} f(x+\xi)\frac{\sin(n+\frac{1}{2})\xi}{\sin\frac{\xi}{2}}\,d\xi = f(x).$$

We break the interval $[-\pi, \pi]$ into two parts (i) $[-\pi, 0]$ and (ii) $[0, \pi]$, and consider the above integration in the interval $[0, \pi]$.

Let

$$\phi(\xi) = \frac{f(x+\xi) - f(x+0)}{2\sin\frac{\xi}{2}}, \tag{10.55}$$

where $f(x+0)$ represents the limit of $f(x)$ from right of points x. The function $\phi(\xi)$ is piecewise continuous and bounded, since $f(x)$ is such and $\left|\sin\frac{\xi}{2}\right| \le 1$.

Hence

$$\begin{aligned}\left|\frac{1}{\pi}\int_0^{\pi}\phi(\xi)\sin\lambda\xi\,d\xi\right| &\le \frac{1}{\pi}\,|\phi(\xi)|\,\left|\int_0^{\pi}\sin\lambda\xi\,d\xi\right| \\ &= \frac{M}{\pi}\left|\frac{1-\cos\lambda\pi}{\lambda}\right|, \quad [M = \text{upper bound of } \phi(\xi)]\end{aligned}$$

which $\longrightarrow 0$, as $\lambda \longrightarrow \infty$.

Thus

$$\lim_{\lambda\longrightarrow\infty}\frac{1}{\pi}\int_0^{\pi}\phi(\xi)\sin\lambda\xi\,d\xi = 0. \tag{10.56}$$

Taking $\lambda = (n+\frac{1}{2})$ and proceeding to the limit $n \longrightarrow \infty$, we get from equations (10.55) and (10.56)

$$\lim_{n\longrightarrow\infty}\frac{1}{2\pi}\int_0^{\pi} f(x+\xi)\frac{\sin(n+\frac{1}{2})\xi}{\sin\frac{\xi}{2}}\,d\xi = \lim_{n\longrightarrow\infty}\frac{1}{2\pi}\int_0^{\pi} f(x+0)\frac{\sin(n+\frac{1}{2})\xi}{\sin\frac{\xi}{2}}\,d\xi$$

which is

$$\begin{aligned}&= \frac{f(x+0)}{\pi}\lim_{n\longrightarrow\infty}\int_0^{\pi}\frac{\sin(n+\frac{1}{2})\xi}{2\sin\frac{\xi}{2}}\,d\xi \\ &= \frac{f(x+0)}{\pi}\cdot\frac{\pi}{2} \quad [\text{using equation (10.51)}] \\ &= \frac{1}{2}f(x+0).\end{aligned} \tag{10.57}$$

Similarly, considering integration in the interval $[-\pi, 0]$, we can show that

$$\lim_{n\longrightarrow\infty} \frac{1}{2\pi}\int_{-\pi}^{0} f(x+\xi)\frac{\sin(n+\frac{1}{2})\xi}{\sin\frac{\xi}{2}}\,d\xi = \frac{1}{2}f(x-0), \tag{10.58}$$

where $f(x-0)$ represents the limit of $f(x)$ from the left of the point x.

Adding equations (10.57) and (10.58), we get

$$\lim_{n\longrightarrow\infty} S_n(x) = \lim_{n\longrightarrow\infty} \frac{1}{2\pi}\int_{-\pi}^{\pi} f(x+\xi)\frac{\sin(n+\frac{1}{2})\xi}{\sin\frac{\xi}{2}}\,d\xi = \frac{1}{2}[f(x-0)+f(x+0)]. \tag{10.59}$$

If $f(x)$ is continuous at the point x, then

$$\lim_{n\longrightarrow\infty} S_n(x) = f(x).$$

This proves the theorem. □

10.13 Complex Form of the Fourier Series

Let the Fourier expansion of a single-valued function defined in the interval $[-\pi, \pi]$ be

$$f(x) = \frac{a_0}{2} + \sum_{k=1}^{\infty}(a_k\cos kx + b_k\sin kx) \tag{10.60}$$

so that

$$a_k = \frac{1}{\pi}\int_{-\pi}^{\pi} f(x)\cos kx\,dx, \quad (k=0,1,2,\ldots) \tag{10.61}$$

$$\text{and} \quad b_k = \frac{1}{\pi}\int_{-\pi}^{\pi} f(x)\sin kx\,dx, \quad (k=1,2,3,\ldots). \tag{10.62}$$

Now by Euler's well-known identity, we have

$$e^{i\theta} = \cos\theta + i\sin\theta$$

$$\text{and} \quad e^{-i\theta} = \cos\theta - i\sin\theta.$$

From these identities, we get

$$\cos\theta = \frac{1}{2}\left(e^{i\theta}+e^{-i\theta}\right), \quad \sin\theta = \frac{1}{2i}\left(e^{i\theta}-e^{-i\theta}\right).$$

Hence

$$\cos kx = \frac{1}{2}\left(e^{ikx}+e^{-ikx}\right), \quad \sin kx = \frac{i}{2}\left(e^{-ikx}-e^{ikx}\right). \tag{10.63}$$

We substitute expressions (10.63) into equation (10.60) and get

$$f(x) = \frac{a_0}{2} + \sum_{k=1}^{\infty}\left[\frac{a_k - ib_k}{2}e^{ikx} + \frac{a_k+ib_k}{2}e^{-ikx}\right] \quad \text{or,} \quad f(x) = \sum_{k=-\infty}^{\infty} C_k e^{ikx}, \tag{10.64}$$

where

$$\left.\begin{aligned} C_0 &= \tfrac{1}{2}a_0 \\ C_k &= \tfrac{1}{2}(a_k - ib_k) \\ C_{-k} &= \tfrac{1}{2}(a_k + ib_k) \end{aligned}\right\}. \tag{10.65}$$

Equation (10.64) is the *complex form of Fourier series.* The coefficients given by equation (10.65) are called *complex coefficients of Fourier series* for the function $f(x)$. If we multiply both sides of equation (10.64) by e^{-ikx} and integrate the result with respect to x from $-\pi$ to π, we get

$$\int_{-\pi}^{\pi} f(x)\, e^{-ikx}\, dx = \sum_{k=-\infty}^{\infty} C_k \int_{-\pi}^{\pi} e^{ix(k-m)}\, dx = 2\pi C_k.$$

since, for $k \neq m$,

$$C_k \int_{-\pi}^{\pi} e^{ix(k-m)}\, dx = C_k \int_{-\pi}^{\pi} [\cos(k-m)x + i\sin(k-m)x]\, dx = 0.$$

Hence

$$C_k = \frac{1}{2\pi} \int_{-\pi}^{\pi} f(x)\, e^{-ikx}\, dx. \tag{10.66}$$

Thus, equation (10.66) is the general formula for obtaining the coefficients C_k.

10.14 Complex Conjugate Series

Let us consider the series

$$\sum_{k=1}^{\infty} (a_k - ib_k)\, e^{ixk}, \tag{10.67}$$

where

$$e^{ixk} = \cos kx + i \sin kx \tag{10.68}$$

and a_k, b_k are constants.

Then substituting expression (10.68) into equation (10.67), we get

$$\sum_{k=1}^{\infty} (a_k - ib_k)(\cos kx + i\sin kx)$$
$$= \sum_{k=1}^{\infty} \Big[(a_k \cos kx + b_k \sin kx) + i(-b_k \cos kx + a_k \sin kx)\Big].$$

Thus, the real part of this series plus $\frac{a_0}{2}$ corresponds to the Fourier expansion for the function $f(x)$ as given in equation (10.60). The imaginary part formally produces the series

$$\sum_{k=1}^{\infty} (-b_k \cos kx + a_k \sin kx), \tag{10.69}$$

which is called *conjugate* to the series in equation (10.60). The partial sum up to n-terms of this series is obtained by substituting a_k, b_k from equations (10.61) and (10.62) in the following manner.

$$\tilde{S}_n(x) = \sum_{k=1}^{n} \frac{1}{\pi} \int_{-\pi}^{\pi} [-\sin k\xi \cos kx + \cos k\xi \sin kx]\, f(\xi)\, d\xi$$
$$= -\frac{1}{\pi} \int_{-\pi}^{\pi} f(\xi) \sum_{k=1}^{n} \sin k(\xi - x)\, d\xi. \tag{10.70}$$

Now, it can be proved that (in the same way as in Section 10.10)

$$\sum_{k=1}^{n} \sin kz = \frac{\cos\frac{z}{2} - \cos(n+\frac{1}{2})z}{2\sin\frac{z}{2}}. \tag{10.71}$$

With the use of equation (10.71), equation (10.70) becomes

$$\tilde{S}_n(x) = -\frac{1}{2\pi}\int_{-\pi}^{\pi} f(\xi)\frac{\cos\theta - \cos(2n+1)\theta}{\sin\theta}d\xi, \tag{10.72}$$

where

$$\theta = \frac{1}{2}(\xi - x). \tag{10.73}$$

Integral in equation (10.73) is analogus to Dirichlet's integral.

10.15 Integration of Fourier Series

Let $f(x)$ be a piecewise continuous function in the interval $[-\pi, \pi]$, periodic with period 2π, and has a Fourier expansion

$$f(x) = \frac{a_0}{2} + \sum_{k=1}^{\infty}(a_k \cos kx + b_k \sin kx). \tag{10.74}$$

We consider the function

$$F(x) = \int_0^x \left[f(x) - \frac{a_0}{2}\right] dx, \quad (-\pi \le x \le \pi). \tag{10.75}$$

Now

$$F(\pi) - F(-\pi) = \int_{-\pi}^{\pi} f(x)\,dx - \pi a_0 = 0,$$

which shows that $F(x)$ is also periodic, with period 2π. Hence, $F(x)$ can also be expanded in a Fourier series like

$$F(x) = \frac{A_0}{2} + \sum_{k=1}^{\infty}(A_k \cos kx + B_k \sin kx). \tag{10.76}$$

The coefficients A_k, B_k are determined in the usual manner (excepting A_0). Thus

$$A_k = \frac{1}{\pi}\int_{-\pi}^{\pi} F(x)\cos kx\,dx = \frac{1}{\pi}F(x)\frac{\sin kx}{k}\bigg|_{-\pi}^{\pi} - \frac{1}{k\pi}\int_{-\pi}^{\pi}\left[f(x) - \frac{a_0}{2}\right]\sin kx\,dx$$

$$\text{i.e.,}\quad A_k = -\frac{b_k}{k};\ \text{similarly, } B_k = \frac{a_k}{k}. \tag{10.77}$$

To determine A_0, we put $x = 0$ in equations (10.75) and (10.76). This gives

$$\frac{A_0}{2} = -\sum_{k=1}^{\infty} A_k = \sum_{k=1}^{\infty}\frac{b_k}{k}. \tag{10.78}$$

Hence, the Fourier series in equation (10.76) becomes

$$F(x) = \sum_{k=1}^{\infty} \frac{1}{k}\left[a_k \sin kx + b_k(1 - \cos kx)\right]. \tag{10.79}$$

Thus from equations (10.75) and (10.79), we get

$$\int_0^x f(x)\,dx = \int_0^x \frac{a_0}{2}\,dx + F(x) = \frac{a_0 x}{2} + \sum_{k=1}^{\infty} \frac{1}{k}\left[a_k \sin kx + b_k(1 - \cos kx)\right]$$

$$= \int_0^x \frac{a_0}{2}\,dx + \sum_{k=1}^{\infty} \int_0^x (a_k \cos kx + b_k \sin kx)\,dx. \tag{10.80}$$

Equation (10.80) will be valid for any other interval remaining within $[-\pi, \pi]$. Thus, we have proved the following theorem.

Theorem 10.1. *If a function $f(x)$ is piecewise continuous and periodic in the interval $[-\pi, \pi]$ with a period 2π, then the function $f(x)$ can be integrated term-by-term in any interval remaining within $[-\pi, \pi]$.*

By integrating term-by-term the known Fourier series, we can get Fourier expansion of other functions. For example, we know that, for $0 < x < 2\pi$,

$$\sum_{k=1}^{\infty} \frac{\sin kx}{k} = \frac{\pi - x}{2} \qquad [\textit{see } \text{Example 10.3}]$$

Integrating both sides from 0 to x, we get

$$\sum_{k=1}^{\infty} \frac{1 - \cos kx}{k^2} = \frac{\pi}{2}x - \frac{x^2}{4}. \tag{10.81}$$

This gives

$$\sum_{k=1}^{\infty} \frac{\cos kx}{k^2} = \frac{1}{4}x^2 - \frac{\pi}{2}x + C,$$

where

$$C = \frac{1}{2\pi}\int_0^{2\pi} \left(\frac{\pi}{2}x - \frac{x^2}{4}\right) dx = \frac{\pi^2}{6}.$$

Thus, we get, Fourier expansion of the function

$$\frac{x^2}{4} - \frac{\pi}{2}x + \frac{\pi^2}{6} = \sum_{k=1}^{\infty} \frac{\cos kx}{k^2} \quad (0 < x < 2\pi). \tag{10.82}$$

Let us take another example. We have, from equation (10.43)

$$\frac{\pi^2 - 3x^2}{12} = \sum_{n=1}^{\infty} (-1)^{n+1} \frac{\cos nx}{n^2}, \quad \text{for } -\pi \le x \le \pi.$$

Integrating both sides from 0 to x, we get

$$\frac{\pi^2}{12}x - \frac{x^3}{12} = \sum_{n=1}^{\infty}(-1)^{n+1}\frac{\sin nx}{n^3}, \quad \text{i.e.,} \quad x^3 = \pi^2 x - 12\sum_{n=1}^{\infty}(-1)^{n+1}\frac{\sin nx}{n^3}. \tag{10.83}$$

Again, Fourier expansion for

$$f(x) = x \quad (-\pi \le x \le \pi)$$

may be obtained as (readers may verify)

$$x = 2\sum_{n=1}^{\infty}(-1)^{n+1}\frac{\sin nx}{n}. \tag{10.84}$$

Hence, from equations (10.83) and (10.84), we have

$$x^3 = \sum_{n=1}^{\infty}\left(\frac{2\pi^2}{n} - \frac{12}{n^3}\right)(-1)^{n+1}\sin nx. \tag{10.85}$$

Thus, equation (10.85) is the Fourier expansion of function

$$f(x) = x^3, \quad \text{for } -\pi \le x \le \pi.$$

10.16 Differentiation of Fourier Series

Theorem 10.2. *If $f(x)$ defined in $[-\pi, \pi]$ is a continuous function of period 2π, having derivative $f'(x)$ which is integrable, then, the Fourier series for the function $f'(x)$ can be obtained by term-by-term differentiation of the Fourier series for $f(x)$.*

Proof. Let the Fourier expansion for the function $f(x)$ be

$$f(x) = \frac{a_0}{2} + \sum_{k=1}^{\infty}(a_k \cos kx + b_k \sin kx). \tag{10.86}$$

We assume the Fourier expansion for $f'(x)$ as

$$f'(x) = \frac{A_0}{2} + \sum_{k=1}^{\infty}(A_k \cos kx + B_k \sin kx). \tag{10.87}$$

Then the coefficients A_k, B_k are given by

$$A_0 = \frac{1}{\pi}\int_{-\pi}^{\pi} f'(x)\,dx = f(-\pi) - f(\pi) = 0$$

$$A_k = \frac{1}{\pi}\int_{-\pi}^{\pi} f'(x)\cos kx\,dx = \frac{k}{\pi}\int_{-\pi}^{\pi} f(x)\sin kx\,dx - \frac{1}{\pi}\Big[f(x)\cos kx\Big]_{-\pi}^{\pi} = kb_k.$$

Similarly, $B_k = -ka_k$.
Hence

$$f'(x) = \sum_{k=1}^{\infty} k(b_k \cos kx - a_k \sin kx). \tag{10.88}$$

The term-by-term differentiation of the series in equation (10.86) would also result the series in equation (10.88). Hence the theorem is proved. The importance of the function $f(x)$ to be periodic is realised when we found $A_0 = 0$. Otherwise, term-by-term differentiation will not produce the Fourier series in equation (10.88). □

EXAMPLE 10.13. *Using the Fourier expansion for the function $f(x) = \cos \lambda x$, $\lambda \neq$ integer and $-\pi < x < \pi$, find Fourier series for $\sin \lambda x$, and hence show that*

$$\frac{1}{4}\pi \sec \pi x = \sum_{n=1}^{\infty} \frac{(-1)^n (2n-1)}{4x^2 - (2n-1)^2}.$$

Solution: We have (*see* Example 10.7)

$$\cos \lambda x = \frac{2\lambda \sin \lambda\pi}{\pi}\left[\frac{1}{2\lambda^2} - \frac{\cos x}{\lambda^2 - 1^2} + \frac{\cos 2x}{\lambda^2 - 2^2} + \frac{\cos 3x}{\lambda^2 - 3^2} + \cdots\right].$$

Differentiating both sides with respect to x, we get

$$\sin \lambda x = -\frac{2\sin \lambda\pi}{\pi}\left[\frac{\sin x}{\lambda^2 - 1^2} - \frac{2\sin 2x}{\lambda^2 - 2^2} + \frac{3\sin 3x}{\lambda^2 - 3^2} - \cdots\right]. \tag{1}$$

This gives the Fourier expansion for $\sin \lambda x$, $\lambda \neq$ integer, when $-\pi < x < \pi$.
We put $x = \frac{\pi}{2}$ in equation (1). Then we get

$$\sin\left(\lambda\frac{\pi}{2}\right) = \frac{4\sin(\lambda\frac{\pi}{2})}{\pi}\cos\left(\lambda\frac{\pi}{2}\right)\sum_{k=1}^{n}\frac{(-1)^k(2k-1)}{\lambda^2 - (2k-1)^2}$$

$$\text{or,} \quad \pi \sec\left(\lambda\frac{\pi}{2}\right) = 4\sum_{k=1}^{n}\frac{(-1)^k(2k-1)}{\lambda^2 - (2k-1)^2}. \tag{2}$$

We put $\lambda = 2x$ in equation (2). Then we get

$$\frac{\pi}{4}\sec(\pi x) = \sum_{k=1}^{\infty}\frac{(-1)^k(2k-1)}{4x^2 - (2k-1)^2}.$$

Theorem 10.3. *If the function $f(x)$ is continuous in $[0, \pi]$, has its derivative $f'(x)$ which is integrable, and can be expanded only in Fourier sine series in the form*

$$f(x) = \sum_{k=1}^{\infty} b_k \sin kx, \quad 0 < x < \pi, \tag{10.89}$$

then

$$f'(x) = \sum_{k=1}^{\infty} k b_k \cos kx. \tag{10.90}$$

Proof. We extend the function $f(x)$ in $[-\pi, 0]$. Since the series for $f(x)$ contains only sine-terms, its graph will be an odd function. However, the function becomes continuous in $[-\pi, \pi]$ such that, $f(\pi) = f(-\pi)$. Thus, due to periodic extension throughout the x-axis, the functon $f(x)$ becomes continuous with period 2π. It also has derivative which is integrable. Thus, the function $f(x)$ satisfies all conditions of theorem 10.2. Hence, its Fourier series is differentiable term-by-term. Thus

$$f'(x) = \sum_{k=1}^{\infty} kb_k \cos kx.$$

Hence the theorem is proved.

Similarly, we can prove the following theorem. □

Theorem 10.4. *If the function is continuous in $[0, \pi]$, has its derivative $f'(x)$ which is integrable and can be expanded only in Fourier cosine series in the form*

$$f(x) = \frac{a_0}{2} + \sum_{k=1}^{\infty} a_k \cos kx, \tag{10.91}$$

then

$$f'(x) = \sum_{k=1}^{\infty} -ka_k \sin kx. \tag{10.92}$$

10.17 Generalised Fourier Series

Let a function $f(x)$ defined in $[a, b]$ be represented as a sum of functions which constitute an orthogonal system. That is

$$f(x) = \sum_{n=0}^{\infty} C_n \phi_n(x), \tag{10.93}$$

where the system of functions $\phi_n(x)$ satisfies the following orthogonality conditions

$$\int_a^b \phi_m(x)\phi_n(x)\, dx = 0, \quad m \neq n \tag{10.94}$$

and

$$\int_a^b \{\phi_n(x)\}^2\, dx = M, \quad \text{(constant)}. \tag{10.95}$$

Now, we assume that, if we multiply both sides of equation (10.93) by any $\phi_n(x)$ of the functions of orthogonal system, the result thus obtained is integrable term-by-term, i.e.,

$$f(x)\phi_n(x) = \left[\sum_{n=0}^{\infty} C_n \phi_n(x)\right] \phi_n(x) \tag{10.96}$$

is integrable in $[a, b]$. Then, performing the integration, and using the orthogonality conditions (10.94) and (10.95), we get

$$\int_a^b f(x)\,\phi_n(x)\, dx = C_n \int_a^b [\phi_n(x)]^2\, dx, \quad n = 0, 1, 2, \ldots.$$

Hence

$$C_n = \frac{\displaystyle\int_a^b f(x)\phi_n(x)\, dx}{\displaystyle\int_a^b [\phi_n(x)]^2\, dx}. \tag{10.97}$$

Thus, if the function $f(x)$ can be expanded in a series of the form (10.93), then the coefficients of the series are determined by the formula (10.97).

If the system of function $\phi_0(x), \phi_1(x), \ldots$ be normalised, then

$$\int_a^b [\phi_n(x)]^2\,dx = 1. \tag{10.98}$$

Hence, from equation (10.97), we get

$$C_n = \int_a^b f(x)\phi_n(x)\,dx. \tag{10.99}$$

The series in equation (10.93) is called *generalised Fourier series,* and the coefficients C_n are called *Generalised Fourier coefficients.* The generalised Fourier expansion is permissible if the functions $\phi_0, \phi_1, \ldots$ are continuous and the sum $\Sigma C_n\phi_n(x)$ is convergent.

10.18 Use of Complex Variable in Finding Fourier Series

Let us consider the function

$$f(z) = e^z = 1 + z + \frac{z^2}{2!} + \cdots + \frac{z^n}{n!} + \cdots. \tag{10.100}$$

We substitute, on unit circle $|z| = 1$, $z = e^{i\phi}$,

$$\text{i.e.,}\quad z = \cos\phi + i\sin\phi. \tag{10.101}$$

Then

$$e^z = e^{\cos\phi + i\sin\phi} = e^{\cos\phi}\cdot e^{i\sin\phi} = e^{\cos\phi}\left[\cos(\sin\phi) + i\sin(\sin\phi)\right]. \tag{10.102}$$

The RHS of equation (10.100) is

$$= 1 + (\cos\phi + i\sin\phi) + \frac{1}{2!}(\cos\phi + i\sin\phi)^2 + \cdots$$

$$= \left(1 + \cos\phi + \frac{\cos 2\phi}{2!} + \cdots\right) + i\left(\sin\phi + \frac{\sin 2\phi}{2!} + \cdots\right). \tag{10.103}$$

Hence, equating real and imaginary parts between equation (10.102) and (10.103), we get

$$e^{\cos\phi}\cos(\sin\phi) = 1 + \cos\phi + \frac{\cos 2\phi}{2!} + \cdots \tag{10.104}$$

$$\text{and}\quad e^{\cos\phi}\sin(\sin\phi) = \sin\phi + \frac{\sin 2\phi}{2!}\cdots \tag{10.105}$$

It is obvious that, both the series are convergent in $-\pi < \phi < \pi$.

Thus, we have obtained Fourier series expansions for the functions

$$\left.\begin{aligned} F_1(\phi) &= e^{\cos\phi}\cdot\cos(\sin\phi) \\ \text{and}\quad F_2(\phi) &= e^{\cos\phi}\cdot\sin(\sin\phi) \end{aligned}\right\}\ \text{for } -\pi < \phi < \pi. \tag{10.106}$$

Let us consider another function $f(z) = \sin z$.
We know

$$\sin z = z - \frac{z^3}{3!} + \frac{z^5}{5!} - \cdots \tag{10.107}$$

We substitute, on unit circle $|z| = 1$,

$$z = \cos\phi + i\sin\phi$$

so that

$$z^n = (\cos\phi + i\sin\phi)^n = \cos n\phi + i\sin n\phi.$$

Then, from equation (10.107), we get

$$\sin(\cos\phi + i\sin\phi) = (\cos\phi + i\sin\phi) - \frac{1}{3!}(\cos 3\phi + i\sin 3\phi) + \cdots. \tag{10.108}$$

Remembering that

$$\sin(\alpha + i\beta) = \sin\alpha\cosh\beta + i\cos\alpha\sinh\beta$$

we get from equation (10.108), equating real and imaginary parts

$$\sin(\cos\phi)\cdot\cosh(\sin\phi) = \cos\phi - \frac{\cos 3\phi}{3!} + \frac{\cos 5\phi}{5!} - \cdots \tag{10.109}$$

$$\text{and} \quad \cos(\cos\phi)\cdot\sinh(\sin\phi) = \sin\phi - \frac{\sin 3\phi}{3!} + \frac{\sin 5\phi}{5!} - \cdots \tag{10.110}$$

Both the series in right-hand side of equations (10.109) and (10.110) are fast convergent in $-\pi < \phi < \pi$. Thus, we have obtained Fourier series expansions for the functions

$$\left.\begin{aligned} F_1(\phi) &= \sin(\cos\phi)\cdot\cosh(\sin\phi) \\ \text{and} \quad F_2(\phi) &= \cos(\cos\phi)\cdot\sinh(\sin\phi) \end{aligned}\right\} \text{ for } -\pi < \phi < \pi.$$

EXAMPLE 10.14. *Considering the function*

$$f(z) = \frac{\alpha}{\alpha - z}, \quad \alpha = \textit{real constant},$$

find the Fourier expansion of

$$f(\phi) = \frac{1}{\alpha^2 - 2\alpha\cos\phi + 1}.$$

Solution: We have

$$f(z) = \frac{\alpha}{\alpha - z} = \frac{1}{1 - \frac{z}{\alpha}} = \left(1 - \frac{z}{\alpha}\right)^{-1},$$

$$\text{i.e.,} \quad f(z) = 1 + \frac{z}{\alpha} + \frac{z^2}{\alpha^2} + \cdots. \tag{1}$$

On unit circle $|z| = 1$, we take $z = \cos\phi + i\sin\phi$, so that

$$z^p = \cos p\phi + i\sin p\phi. \tag{2}$$

Then, the function $f(z)$ is

$$f(z)=\frac{\alpha}{\alpha-z}=\frac{\alpha}{(\alpha-\cos\phi)-i\sin\phi}=\frac{\alpha[(\alpha-\cos\phi)+i\sin\phi]}{(\alpha-\cos\phi)^2+\sin^2\phi}$$

$$=\frac{\alpha[(\alpha-\cos\phi)+i\sin\phi]}{\alpha^2-2\alpha\cos\phi+1} \tag{3}$$

Again, RHS of equation (1) is

$$f(z)=1+\frac{1}{\alpha}(\cos\phi+i\sin\phi)+\frac{1}{\alpha^2}(\cos 2\phi+i\sin 2\phi)+\cdots \tag{4}$$

Equating real and imaginary parts in equations (3) and (4), we get

$$\frac{\alpha(\alpha-\cos\phi)}{\alpha^2-2\alpha\cos\phi+1}=\sum_{k=0}^{\infty}\frac{1}{\alpha^k}\cos k\phi \tag{5}$$

and
$$\frac{\alpha\sin\phi}{\alpha^2-2\alpha\cos\phi+1}=\sum_{k=1}^{\infty}\frac{1}{\alpha^k}\sin k\phi. \tag{6}$$

Both the series will converge, if $\alpha>1$.
Now, we subtract $\frac{1}{2}$ from both sides of equation (5), and get

$$\frac{\alpha(\alpha-\cos\phi)}{\alpha^2-2\alpha\cos\phi+1}-\frac{1}{2}=-\frac{1}{2}+\sum_{k=0}^{\infty}\frac{\cos k\phi}{\alpha^k}$$

or,
$$\frac{\alpha^2-1}{2(\alpha^2-2\alpha\cos\phi+1)}=-\frac{1}{2}+\sum_{k=0}^{\infty}\frac{\cos k\phi}{\alpha^k}.$$

Hence

$$\frac{1}{\alpha^2-2\alpha\cos\phi+1}=\frac{-1}{\alpha^2-1}+\frac{2}{\alpha^2-1}\sum_{k=0}^{\infty}\frac{\cos k\phi}{\alpha^k}. \tag{7}$$

Equation (7) is the required Fourier expansion.

10.19 Parseval's Relations

The relation between Fourier coefficients of a function $f(x)$ and the average of the square of that function in its interval of definition is known as *Parseval's relation.* We will discuss three such relations.

(i) Let a function $f(x)$ be defined in $[-\pi,\pi]$ so that its Fourier expansion is

$$f(x)=\frac{a_0}{2}+\sum_{k=1}^{n}(a_k\cos kx+b_k\sin kx). \tag{10.111}$$

Then, average of the square of $f(x)$ in $[-\pi, \pi]$ is

$$\frac{1}{2\pi}\int_{-\pi}^{\pi}[f(x)]^2\,dx = \frac{1}{2\pi}\int_{-\pi}^{\pi}\left[\frac{a_0}{2}+\sum_{k=1}^{\infty}(a_k\cos kx+b_k\sin kx)\right]^2 dx$$

$$= \frac{1}{2\pi}\int_{-\pi}^{\pi}\left[\left(\frac{a_0}{2}\right)^2 + a_k^2\cos^2 kx + b_k^2\sin^2 kx\right] dx \qquad (10.112)$$

(since all other terms vanish due to orthogonality property)

Now

$$\int_{-\pi}^{\pi}\cos^2 kx\,dx = \frac{1}{2}\int_{-\pi}^{\pi}(1+\cos 2kx)\,dx = \frac{1}{2}\left[x+\frac{\sin 2kx}{2k}\right]_{-\pi}^{\pi} = \pi.$$

Similarly

$$\int_{-\pi}^{\pi}\sin^2 kx\,dx = \frac{1}{2}\int_{-\pi}^{\pi}(1-\cos 2kx)\,dx = \pi.$$

Hence, from equation (10.112) we get

$$\frac{1}{\pi}\int_{-\pi}^{\pi}[f(x)]^2\,dx = \frac{1}{2}a_0^2 + \sum_{k=1}^{\infty}\left(a_k^2+b_k^2\right). \qquad (10.113)$$

Equation (10.113) is one of the relations of Parseval.
(ii) If the function $f(x)$ is defined in an arbitrary interval $(-l, l)$, then

$$f(x) = \frac{a_0}{2}+\sum_{k=1}^{\infty}\left(a_k\cos\frac{k\pi x}{l}+b_k\sin\frac{k\pi x}{l}\right).$$

Then, proceeding as before, it can be proved that

$$\frac{1}{l}\int_{-l}^{l}[f(x)]^2\,dx = \frac{1}{2}a_0^2 + \sum_{k=1}^{\infty}\left(a_k^2+b_k^2\right). \qquad (10.114)$$

(iii) If $f(x)$ is defined in $[0, l]$ and expanded in a half-range sine-series like

$$f(x) = \sum_{k=1}^{\infty} b_n \sin\left(\frac{k\pi x}{l}\right)$$

then equation (10.114) becomes

$$\frac{2}{l}\int_{0}^{l}[f(x)]^2\,dx = \sum_{k=1}^{\infty} b_k^2. \qquad (10.115)$$

Similarly, if $f(x)$ is defined in $[0, l]$ and expanded in a half-range cosine-series like

$$f(x) = \frac{a_0}{2}+\sum_{k=1}^{\infty} a_k\cos\frac{k\pi x}{l},$$

then equation (10.114) becomes

$$\frac{2}{l}\int_{0}^{l}[f(x)]^2\,dx = \frac{1}{2}a_0^2 + \sum_{k=1}^{\infty} a_k^2. \qquad (10.116)$$

EXAMPLE 10.15. *Expand in Fourier series, the function $f(x) = e^{ax}$, (a = non-zero constant), in $[-\pi, \pi]$ and show that*

$$\frac{\pi \cosh ax}{2 \sinh a\pi} = \frac{1}{2a} + \sum_{k=1}^{\infty} (-1)^k \frac{a}{a^2 + k^2} \cos kx, \quad (-\pi \leq x \leq \pi)$$

and $$\frac{\pi \sinh ax}{2 \sinh a\pi} = \sum_{k=1}^{\infty} (-1)^{k-1} \frac{k}{a^2 + k^2} \sin kx, \quad (-\pi < x < \pi).$$

Solution: We determine the coefficients of expansion of the function e^{ax} as follows

$$a_0 = \frac{1}{\pi} \int_{-\pi}^{\pi} e^{ax}\, dx = \frac{e^{a\pi} - e^{-a\pi}}{a\pi} = \frac{2 \sinh a\pi}{a\pi}.$$

$$a_k = \frac{1}{\pi} \int_{-\pi}^{\pi} e^{ax} \cos kx\, dx = \frac{1}{\pi} \left[\frac{a \cos kx + k \sin kx}{a^2 + k^2} e^{ax} \right]_{-\pi}^{\pi}$$

$$= \frac{(-1)^k}{\pi} \frac{2a}{a^2 + k^2} \sinh a\pi$$

$$b_k = \frac{1}{\pi} \int_{-\pi}^{\pi} e^{ax} \sin kx\, dx = \frac{1}{\pi} \left[\frac{a \sin kx - k \cos kx}{a^2 + k^2} e^{ax} \right]_{-\pi}^{\pi}$$

$$= \frac{(-1)^{k-1}}{\pi} \cdot \frac{2k}{a^2 + k^2} \sinh a\pi.$$

Hence, we have for $-\pi < x < \pi$,

$$e^{ax} = \frac{2}{\pi} \sinh a\pi \left[\frac{1}{2a} + \sum_{k=1}^{\infty} \frac{(-1)^k}{a^2 + k^2} (a \cos kx - k \sin kx) \right]. \tag{10.117}$$

Replacing x by $-x$, we get from equation (10.117)

$$e^{-ax} = \frac{2}{\pi} \sinh a\pi \left[\frac{1}{2a} + \sum_{k=1}^{\infty} \frac{(-1)^k}{a^2 + k^2} (a \cos kx + k \sin kx) \right]. \tag{10.118}$$

Adding equations (10.117) and (10.118), we get

$$\frac{\pi \cosh ax}{2 \sinh a\pi} = \frac{1}{2a} + \sum_{k=1}^{\infty} \frac{(-1)^k a}{a^2 + k^2} \cos kx.$$

Again, subtracting equation (10.118) from (10.117), we get

$$\frac{\pi \sinh ax}{2 \sinh a\pi} = \sum_{k=0}^{\infty} (-1)^{k-1} \frac{k}{a^2 + k^2} \sin kx.$$

Note: If we put $x = 0$ in equation (10.117), we get

$$\frac{1}{\sinh a\pi} = \frac{1}{\pi a} + \frac{2}{\pi} \sum_{k=1}^{\infty} (-1)^k \cdot \frac{a}{a^2 + k^2}$$

or, $$\frac{1}{\sinh a\pi} = \frac{1}{\pi a} + 2 \sum_{k=1}^{\infty} (-1)^k \frac{a\pi}{(a\pi)^2 + (k\pi)^2}.$$

Let $a\pi = z$. Then

$$\frac{1}{\sinh z} = \frac{1}{z} + \sum_{k=1}^{\infty} (-1)^k \frac{2z}{z^2 + (k\pi)^2} = \frac{1}{z} + \sum_{k=1}^{\infty} (-1)^k \left[\frac{1}{z + ik\pi} + \frac{1}{z - ik\pi}\right],$$

which gives an expansion of $\dfrac{1}{\sinh z}$ in a simple form in terms of a complex function.

EXAMPLE 10.16. *Find Fourier cosine series for the function* $f(x) = \log\left(2\cos\dfrac{x}{2}\right)$ *in* $[-\pi, \pi]$.

Solution: The coefficients for cosine series are

$$a_0 = \frac{2}{\pi}\int_0^{\pi} \log\left(2\cos\frac{x}{2}\right) dx = 2\log 2 + \frac{4}{\pi}\int_0^{\frac{\pi}{2}} \log\cos\theta \, d\theta \quad \left[\theta = \frac{x}{2}\right]$$

$$= 2\log 2 + \frac{4}{\pi}\cdot\frac{\pi}{2}\log\frac{1}{2} = 0 \quad \left[\int_0^{\frac{\pi}{2}} \log\cos\theta\, d\theta \text{ is a standard improper integral whose value is } \tfrac{\pi}{2}\log\tfrac{1}{2}\right]$$

$$a_n = \frac{2}{\pi}\int_0^{\pi} \log\left(2\cos\frac{x}{2}\right)\cos nx \, dx = \frac{2}{\pi}\log\left(2\cos\frac{x}{2}\right)\frac{\sin nx}{n}\Bigg|_0^{\pi} + \frac{1}{n\pi}\int_0^{\pi} \frac{\sin nx \cdot \sin\frac{x}{2}}{\cos\frac{x}{2}} dx$$

$$= (-1)^{n-1}\frac{1}{n\pi}\int_0^{\pi} \frac{\sin nx \cos\frac{x}{2}}{\sin\frac{x}{2}} dx \quad [\text{changing } x \text{ to } (\pi - x)]$$

$$= (-1)^{n-1}\frac{1}{n\pi}\int_0^{\pi} \left[\frac{\sin(n+\frac{1}{2})x}{2\sin\frac{x}{2}} + \frac{\sin(n-\frac{1}{2})x}{2\sin\frac{x}{2}}\right] dx$$

$$= (-1)^{n-1}\frac{1}{n}\left(\frac{1}{2} + \frac{1}{2}\right) = (-1)^{n-1}\cdot\frac{1}{n} \quad [\text{using equation (10.49)}].$$

Hence, the required expansion is

$$\log\left(2\cos\frac{x}{2}\right) = \sum_{n=1}^{\infty} (-1)^{n-1}\frac{\cos nx}{n} \quad (-\pi < x < \pi).$$

If we put $x = \pi - x$ in this expansion, we get

$$-\log\left(2\sin\frac{x}{2}\right) = \sum_{n=1}^{\infty} \frac{\cos nx}{n}.$$

EXAMPLE 10.17. *Find the Fourier series for the function* $f(x) = e^{i\lambda x}$, λ *is not an integer. Hence prove that*

$$\frac{\pi\sin\lambda x}{2\sin\lambda\pi} = \frac{\sin x}{1^2 - \lambda^2} - \frac{2\sin 2x}{2^2 - \lambda^2} + \cdots$$

and

$$\frac{\pi\cos\lambda x}{2\sin\lambda\pi} = \frac{1}{2\pi} + \frac{\lambda\cos x}{1^2 - \lambda^2} - \frac{\lambda\cos 2x}{2^2 - \lambda^2} + \cdots.$$

Solution: Let

$$e^{i\lambda x} = \frac{a_0}{2} + \sum_{k=1}^{\infty} (a_k\cos kx + b_k\sin kx).$$

Then, the coefficients are

$$a_0 = \frac{1}{\pi}\int_{-\pi}^{\pi} e^{i\lambda x}\,dx = \frac{2}{\lambda\pi}\sin\lambda\pi$$

$$a_k = \frac{1}{\pi}\int_{-\pi}^{\pi} e^{i\lambda x}\cos kx\,dx \tag{1}$$

$$\pi a_k = e^{i\lambda x}\frac{\sin kx}{k}\Bigg|_{-\pi}^{\pi} - \int \frac{i\lambda}{k} e^{i\lambda x}\sin kx\,dx$$

$$= i\lambda e^{i\lambda x}\cdot\frac{\cos kx}{k^2}\Bigg|_{-\pi}^{\pi} + \lambda^2\int_{-\pi}^{\pi} e^{i\lambda x}\frac{\cos kx}{k^2}\,dx$$

$$= -\frac{2\lambda}{k^2}\sin\lambda\pi\cos k\pi + \frac{\lambda^2}{k^2}(\pi a_k) \quad \text{[using equation (1)].}$$

Hence

$$\pi a_k\left(1-\frac{\lambda^2}{k^2}\right) = -\frac{2\lambda}{k^2}(-1)^k\sin\lambda\pi.$$

Thus

$$a_k = -(-1)^k\frac{2\lambda}{\pi}\frac{\sin\lambda\pi}{k^2-\lambda^2}. \tag{2}$$

Again

$$b_k = \frac{1}{\pi}\int_{-\pi}^{\pi} e^{i\lambda x}\sin kx\,dx \tag{3}$$

i.e.,
$$\pi b_k = -\left[e^{i\lambda x}\frac{\cos kx}{k}\right]_{-\pi}^{\pi} + \frac{i\lambda}{k}\int_{-\pi}^{\pi} e^{i\lambda x}\cos kx\,dx$$

$$= -\frac{2i}{k}\cos k\pi\sin\lambda\pi + \frac{\lambda^2}{k^2}\int_{-\pi}^{\pi} e^{i\lambda x}\sin kx\,dx$$

$$= -\frac{2i}{k}(-1)^k\sin\lambda\pi + \frac{\lambda^2}{k^2}(\pi b_k) \quad \text{[using equation (3)].}$$

Hence

$$\pi b_k\left(1-\frac{\lambda^2}{k^2}\right) = -\frac{2i(-1)^k}{k}\sin\lambda\pi.$$

Thus

$$b_k = \frac{2i}{\pi}(-1)^k\frac{k\sin\lambda\pi}{k^2-\lambda^2}. \tag{4}$$

Hence, we have

$$e^{i\lambda x} = \frac{2}{\pi}\sin\lambda\pi\left[\frac{1}{2\lambda} + \sum_{k=1}^{\infty}(-1)^{k+1}\left(\frac{\lambda\cos kx}{k^2-\lambda^2} + \frac{ik\sin kx}{k^2-\lambda^2}\right)\right]. \tag{5}$$

Equating real and imaginary parts in equation (5), we get

$$\frac{\pi\sin\lambda x}{2\sin\lambda\pi} = \sum_{k=1}^{\infty}(-1)^{k+1}\frac{k\sin kx}{k^2-\lambda^2}$$

and
$$\frac{\pi\cos\lambda x}{2\sin\lambda\pi} = \frac{1}{2\lambda} + \sum_{k=1}^{\infty}(-1)^{k+1}\frac{\lambda\cos kx}{k^2-\lambda^2}.$$

EXAMPLE 10.18. *Find the Fourier series expansion for the function*

$$f(x) = x^4 - 2\pi x^2, \quad (-\pi \le x \le \pi)$$

and hence show that

$$\sum_{n=1}^{\infty} \frac{1}{n^4} = \frac{\pi^4}{90}.$$

Solution: The function is even. Hence Fourier expansion will be a cosine series.

Let

$$x^4 - 2\pi x^2 = \frac{a_0}{2} + \sum_{k=1}^{\infty} a_k \cos kx. \tag{1}$$

Then the coefficients are:

$$a_0 = \frac{1}{\pi}\int_{-\pi}^{\pi} (x^4 - 2\pi x^2)\, dx = 2\left(\frac{\pi^4}{5} - \frac{2\pi^3}{3}\right)$$

$$a_k = \frac{1}{\pi}\int_{-\pi}^{\pi} (x^4 - 2\pi x^2) \cos kx\, dx. \tag{2}$$

Now

$$I = \int_{-\pi}^{\pi} x^2 \cos kx = x^2 \frac{\sin kx}{k}\Big|_{-\pi}^{\pi} - \int_{-\pi}^{\pi} 2x \frac{\sin kx}{k}\, dx$$

$$= -\frac{2}{k}\left[-x\frac{\cos kx}{k}\Big|_{-\pi}^{\pi} + \int_{-\pi}^{\pi} \frac{\cos kx}{k}\, dx\right] = (-1)^k \frac{4\pi}{k^2} \tag{3}$$

and

$$\int_{-\pi}^{\pi} x^4 \cos kx\, dx = x^4 \frac{\sin kx}{k}\Big|_{-\pi}^{\pi} - \int_{-\pi}^{\pi} 4x^3 \frac{\sin kx}{k}\, dx = \frac{4}{k^2}\left(x^3 \cos kx\right)_{-\pi}^{\pi} - \frac{12}{k^2} I$$

$$= \frac{8\pi^3}{k^2}(-1)^k - \frac{12}{k^2}\left(\frac{4\pi}{k^2}\right)(-1)^k \quad \text{[using equation (3)]}$$

$$= (-1)^k \left(\frac{8\pi^3}{k^2} - \frac{48\pi}{k^4}\right). \tag{4}$$

Hence, from equations (2), (3) and (4), we get

$$a_k = \frac{(-1)^k}{\pi}\left[\frac{8\pi^3}{k^2} - \frac{48\pi}{k^4} - 2\pi\left(\frac{4\pi}{k^2}\right)\right] = (-1)^k\left[\frac{8\pi^2 - 8\pi}{k^2} - \frac{48}{k^4}\right]. \tag{5}$$

Hence, from equation (1), we have

$$x^4 - 2\pi x^2 = \left(\frac{\pi^4}{5} - \frac{2\pi^3}{3}\right) + (8\pi^2 - 8\pi)\sum_{k=1}^{\infty}(-1)^k \frac{\cos kx}{k^2} - 48\sum_{k=1}^{\infty}(-1)^k \frac{\cos kx}{k^4}. \tag{6}$$

Putting $x = 0$ on both sides of equation (6)

$$0 = \frac{\pi^4}{5} - \frac{2\pi^3}{3} - (8\pi^2 - 8\pi)\left[\frac{1}{1^2} - \frac{1}{2^2} + \frac{1}{3^2} - \cdots\right] + 48\left[\frac{1}{1^4} - \frac{1}{2^4} + \frac{1}{3^4} - \cdots\right]$$

$$= \frac{\pi^4}{5} - \frac{2\pi^3}{3} - (8\pi^2 - 8\pi)\left[\left(\frac{1}{1^2} + \frac{1}{2^2} + \frac{1}{3^2}\cdots\right) - \frac{2}{2^2}\left(\frac{1}{1^2} + \frac{1}{2^2} + \frac{1}{3^2}\cdots\right)\right]$$

$$+ 48\left[\left(\frac{1}{1^4} + \frac{1}{2^4} + \cdots\right) - \frac{2}{2^4}\left(\frac{1}{1^4} + \frac{1}{2^4} + \cdots\right)\right],$$

i.e., $$0 = \frac{\pi^4}{5} - \frac{2\pi^3}{3} - (8\pi^2 - 8\pi)\cdot\frac{1}{2}\sum_{n=1}^{\infty}\frac{1}{n^2} + 48\left(\frac{7}{8}\right)\sum_{n=1}^{\infty}\frac{1}{n^4}$$

$$= \frac{\pi^4}{5} - \frac{2\pi^3}{3} - (8\pi^2 - 8\pi)\cdot\frac{1}{2}\left(\frac{\pi^2}{6}\right) + 42\sum_{k=1}^{\infty}\frac{1}{n^4} \tag{7}$$

[using the result in (10.45)]

Hence

$$42\sum_{n=1}^{\infty}\frac{1}{n^4} = \frac{2}{3}\pi^4 - \frac{1}{5}\pi^4 = \frac{7}{15}\pi^4.$$

Hence

$$\sum_{n=1}^{\infty}\frac{1}{n^4} = \frac{\pi^4}{90}.$$

10.20 Fourier Series in Two Dimensions

The system of trigonometric functions

$$\sin mx \cos ny, \quad \sin mx \sin ny$$
$$\cos mx \cos ny, \quad \cos mx \sin ny$$

for $m = 0, 1, 2, \ldots$; and $n = 0, 1, 2, \ldots$ constitute an orthogonal system in a region $(-\pi \leq x \leq \pi, -\pi \leq y \leq \pi)$. Each of the above functions is periodical with a period 2π with respect to both x and y.

Hence, if a function $f(x, y)$ of two variables x, y is defined in the square region $R(-\pi \leq x \leq \pi, -\pi \leq y \leq \pi)$, it can be expanded in a Fourier series, provided the function $f(x, y)$ is continuous and integrable in that region.

Let

$$f(x, y) = A_{0,0} + \sum_{m=0}^{\infty}\sum_{n=0}^{\infty}[A_{mn}\sin mx \sin ny + B_{mn}\cos mx \cos ny]. \tag{10.119}$$

Then, the coefficients are determined by

$$A_{0,0} = \frac{\iint_R f(x, y)\,dx\,dy}{\iint_R dx\,dy} = \frac{1}{4\pi^2}\iint_R f(x, y)\,dx\,dy. \tag{10.120}$$

Similarly

$$A_{m,n} = \frac{1}{\pi^2} \iint_R f(x,y) \sin mx \sin ny \, dx \, dy \tag{10.121}$$

$$B_{m,n} = \frac{1}{\pi^2} \iint_R f(x,y) \cos mx \cos ny \, dx \, dy \tag{10.122}$$

$$B_{m,o} = \frac{1}{2\pi^2} \iint_R f(x,y) \cos mx \, dx \, dy \tag{10.123}$$

$$B_{o,m} = \frac{1}{2\pi^2} \iint_R f(x,y) \cos ny \, dx \, dy. \tag{10.124}$$

Thus, from equation (10.119) we write

$$f(x,y) = \sum_{m=0}^{\infty} \sum_{n=0}^{\infty} (A_{mn} \sin mx \sin y + B_{mn} \cos mx \cos ny) \tag{10.125}$$

We have taken only two functions of the system mentioned above. But, in most general cases, all the functions may appear in the Fourier series (10.119). The other coefficients are determined in the same manner.

In complex form, the Fourier series is written as

$$f(x,y) = \sum_{m=-\infty}^{\infty} \sum_{n=-\infty}^{\infty} C_{mn} e^{i(mx+ny)}, \tag{10.126}$$

where the coefficients are given by

$$C_{mn} = \frac{1}{4\pi^2} \iint f(x,y) e^{-i(mx+ny)} \, dx \, dy \tag{10.127}$$

$$[m = 0, \pm 1, \pm 2, \ldots; n = 0, \pm 1, \pm 2, \ldots]$$

We now shall discuss some physical problems in the light of Fourier series.

EXAMPLE 10.19. *An elastic beam of length l is resting on elastic foundation with two ends hinged. Discuss the deflection of the beam under different loading conditions.*

Solution: If k be the spring constant of the elastic foundation, and EI the flexural rigidity of the beam, then the differential equation for the deflection of the beam is

$$\frac{d^4 w}{dx^4} + 4\lambda^4 w = \frac{q(x)}{\mathrm{EI}}, \tag{1}$$

where $4\lambda^4 = \frac{k}{EI}$, and $q(x)$ is the loading function.

Since the ends of the beam are hinged, we have the boundary conditions as

$$\left. \begin{aligned} &\text{for } x = 0, w = 0 \text{ and (Moment) } \frac{d^2 w}{dx^2} = 0 \\ \text{and} \quad &\text{for } x = l, w = 0 \text{ and (Moment) } \frac{d^2 w}{dx^2} = 0 \end{aligned} \right\}. \tag{2}$$

We assume the solution of equation (1) in the form

$$w = \sum_{k=1}^{\infty} A_k \sin\frac{k\pi x}{l}, \tag{3}$$

where coefficients A_k are so far unknown.

Obviously, the solution function (3) satisfies all the boundary conditions in (2).

Now, we expand the load $q(x)$ in Fourier sine series in $[0, l]$. Then

$$q(x) = \sum_{k=1}^{\infty} Q_k \sin\frac{k\pi x}{l}, \tag{4}$$

where the coefficients Q_k are given by

$$Q_k = \frac{2}{l}\int_0^l q(x)\sin\frac{k\pi x}{l}\,dx. \tag{5}$$

We now substitute the solution function (3) and load function (4) in the differential equation (1). Then we get

$$\sum_{k=1}^{\infty} A_k\left[\left(\frac{k\pi}{l}\right)^4 + 4\lambda^4\right]\sin\frac{k\pi x}{l} = \frac{1}{\text{EI}}\sum_{k=1}^{\infty} Q_k \sin\frac{k\pi x}{l}.$$

This gives

$$A_k = \frac{1}{\text{EI}}\frac{Q_k}{\left(\frac{k\pi}{l}\right)^4 + 4\lambda^4}. \tag{6}$$

Thus, we have the unknown constants A_k. Hence, the deflection function of the beam is

$$w = \frac{1}{\text{EI}}\sum_{k=1}^{\infty}\frac{Q_k}{\xi_k}\sin\frac{k\pi x}{l}, \tag{7}$$

where we have denoted

$$\xi_k = \left(\frac{k\pi}{l}\right)^4 + 4\lambda^4. \tag{8}$$

The coefficient Q_k is obtained from equation (5) when the loading function $q(x)$ is given.

Case I. *Let $q(x) = q = constant$, i.e., the beam is under the action of an uniformly distributed load of intensity q. Then, from equation* (5)

$$Q_k = \frac{2q}{l}\int_0^l \sin\frac{k\pi x}{l}\,dx = \frac{4q}{k\pi}. \tag{9}$$

Hence, equation (7) *becomes*

$$w = \frac{4q}{\pi\text{EI}}\sum_{k=1}^{\infty}\frac{1}{k\xi_k}\sin\frac{k\pi x}{l}. \tag{10}$$

Case II. *Let there be a point load P at a distance of a from the left end of the beam.*

We assume that the point load P is distributed in a very small length of the beam from $(a-\epsilon)$ to $(a+\epsilon)$, so that intensity of load within that region is $\frac{P}{2\epsilon}$. Then, from equation (5), we have

$$Q_k = \frac{2}{l}\int_0^l \frac{P}{2\epsilon}\sin\frac{k\pi x}{l}\,dx = \frac{P}{\epsilon l}\int_{a-\epsilon}^{a+\epsilon}\sin\frac{k\pi x}{l}\,dx = \frac{P}{l}\cdot\frac{l}{k\pi\epsilon}\left[-\cos\frac{k\pi x}{l}\right]_{a-\epsilon}^{a+\epsilon}$$
$$= -\frac{P}{l}\cdot\frac{l}{k\pi\epsilon}\left[\cos\frac{k\pi}{l}(a+\epsilon)-\cos\frac{k\pi}{l}(a-\epsilon)\right] = \frac{2P}{l}\cdot\sin\frac{k\pi a}{l}\cdot\frac{\sin\frac{k\pi\epsilon}{l}}{\frac{k\pi\epsilon}{l}}. \tag{11}$$

Now, as $\epsilon \longrightarrow 0$, $\frac{k\pi\epsilon}{l}$ also $\longrightarrow 0$, and so

$$\lim_{\epsilon\longrightarrow 0}\frac{\sin(\frac{k\pi\epsilon}{l})}{\frac{k\pi\epsilon}{l}} = 1.$$

Hence, from equation (11), we get in the limit as $\epsilon \longrightarrow 0$,

$$Q_k = \frac{2P}{l}\sin\frac{k\pi a}{l}. \tag{12}$$

Thus, the deflection function of the beam is

$$w = \frac{2P}{\text{EI}l}\sum_{k=1}^{\infty}\frac{1}{\xi_k}\sin\frac{k\pi a}{l}\cdot\sin\frac{k\pi x}{l}. \tag{13}$$

If the load P is at the centre of the beam, then $a=\frac{l}{2}$, so that

$$w = \frac{2P}{\text{EI}l}\sum_{k=1}^{\infty}\frac{1}{\xi_k}\sin\frac{k\pi}{2}\cdot\sin\frac{k\pi x}{l}. \tag{14}$$

EXAMPLE 10.20. *Discuss the forced oscillation of a spring-mass system when the force-function is given by*

$$F(t) = \begin{cases} -\frac{2}{\pi}(t-\pi), & \textit{for } -\pi \le t \le -\frac{\pi}{2} \\ \frac{2}{\pi}t, & \textit{for } -\frac{\pi}{2} \le t \le \frac{\pi}{2} \\ -\frac{2}{\pi}(t-\pi), & \textit{for } \frac{\pi}{2} \le t \le \pi \end{cases}$$

and $F(t+2\pi) = f(t)$.

Solution: The forced vibration of an oscillator with damping is represented by the equation

$$m\frac{d^2x}{dt^2} + c\frac{dx}{dt} + kx = F(t), \tag{1}$$

where $m=$ mass of the oscillating particle
$c=$ damping constant
$k=$ spring modulus, assumed constant
and $F(t)=$ external forcing function.

We have discussed this problem earlier. We have found that, if the forcing function $F(t)$ is a sine or cosine function, the steady state solution is a harmonic function having same frequency as that of forcing function. In this case, the forcing function is such that its Fourier expansion is necessary.

Thus, we assume

$$F(t) = \frac{a_0}{2} + \sum_{k=1}^{\infty}(a_k \cos kt + b_k \sin kt), \tag{2}$$

where coefficients are:

$$a_0 = \frac{1}{\pi}\int_{-\pi}^{\pi} F(t)\,dt = \frac{1}{\pi}\left[\int_{-\pi}^{-\frac{\pi}{2}} -\frac{2}{\pi}(t+\pi)\,dt + \int_{-\frac{\pi}{2}}^{\frac{\pi}{2}} \frac{2t}{\pi}\,dt + \int_{\frac{\pi}{2}}^{\pi} -\frac{2}{\pi}(t-\pi)\,dt\right]$$

$$= \frac{1}{\pi}\left[-\frac{2}{\pi}\left(\frac{t^2}{2}+\pi t\right)\Big|_{-\pi}^{-\frac{\pi}{2}} + \frac{t^2}{\pi}\Big|_{-\frac{\pi}{2}}^{\frac{\pi}{2}} - \frac{2}{\pi}\left(\frac{t^2}{2}-\pi t\right)\Big|_{\frac{\pi}{2}}^{\pi}\right]$$

$$= \frac{1}{\pi}\left[-\frac{2}{\pi}\left\{\frac{\pi^2}{8}-\frac{\pi^2}{2}-\frac{\pi^2}{2}+\pi^2\right\}+0-\frac{2}{\pi}\left\{\frac{\pi^2}{2}-\pi^2-\frac{\pi^2}{8}+\frac{\pi^2}{2}\right\}\right] = 0.$$

$$a_k = \frac{1}{\pi}\int_{-\pi}^{\pi} f(t)\cos kt\,dt = 0$$

$$b_k = \frac{1}{\pi}\int_{-\pi}^{\pi} F(t)\sin kt\,dt = \frac{1}{\pi}\{I_1+I_2+I_3\},$$

where

$$I_1 = -\frac{2}{\pi}\int_{-\pi}^{-\frac{\pi}{2}}(t+\pi)\sin kt\,dt = -\frac{2}{\pi}\int_{-\pi}^{-\frac{\pi}{2}}\frac{\cos kt}{k}\,dt = \frac{2}{\pi k^2}\sin\frac{k\pi}{2}$$

$$I_2 = \frac{2}{\pi}\int_{-\frac{\pi}{2}}^{\frac{\pi}{2}} t\sin kt\,dt = \frac{2}{\pi}\int_{-\frac{\pi}{2}}^{\frac{\pi}{2}}\frac{\cos kt}{k}\,dt = \frac{4}{\pi k^2}\sin\frac{k\pi}{2}$$

$$I_3 = -\frac{2}{\pi}\int_{\frac{\pi}{2}}^{\pi}(t-\pi)\sin kt\,dt = -\frac{2}{\pi}\int_{\frac{\pi}{2}}^{\pi}\frac{\cos kt}{k}\,dt = \frac{2}{\pi k^2}\sin\frac{k\pi}{2}.$$

Hence

$$b_k = \frac{1}{\pi}\left(\frac{2}{\pi k^2}+\frac{4}{\pi k^2}+\frac{2}{\pi k^2}\right)\sin\frac{k\pi}{2} = \frac{8}{\pi^2 k^2}\sin\frac{k\pi}{2}, \quad k = 1,3,5,\ldots.$$

Thus

$$F(t) = \frac{8}{\pi^2}\left[\frac{\sin t}{1^2}-\frac{\sin 3t}{3^2}+\frac{\sin 5t}{5^2}-\cdots\right]. \tag{3}$$

Now, we write equation of motion (1) in the form

$$\frac{d^2x}{dt^2}+2\alpha\frac{dx}{dt}+\xi^2 x = \frac{8}{m\pi^2}\left[\frac{\sin t}{1^2}-\frac{\sin 3t}{3^2}+\frac{\sin 5t}{5^2}-\cdots\right], \tag{4}$$

where $2\alpha = \frac{c}{m}$; $\xi^2 = \frac{k}{m}$.

The particular integral of this equation may be written as

$$x(t) = \sum_{k=1}^{\infty}(A_k \sin kt + B_k \cos kt). \tag{5}$$

Substituting equation (5) into equation (4), and comparing coefficients of $\cos kt$ and $\sin kt$, we get

$$\left.\begin{aligned}(\xi^2 - k^2)\, A_k - 2\alpha k B_k &= C_k \\ \text{and} \quad 2\alpha k A_k + (\xi^2 - k^2)\, B_k &= 0\end{aligned}\right\}, \tag{6}$$

where

$$C_k = \frac{8}{m\pi^2}\frac{1}{k^2}\sin\frac{k\pi}{2}. \tag{7}$$

Solving the system of equations in (6), we get

$$A_k = \frac{\xi^2 - k^2}{(\xi^2 - k^2)^2 + 4\alpha^2 k^2} C_k \tag{8}$$

$$\text{and} \quad B_k = \frac{-2\alpha k}{(\xi^2 - k^2)^2 + 4\alpha^2 k^2} C_k. \tag{9}$$

Thus, we have found the particular solution of the equation (1). Since $F(t)$ is a periodic function with period 2π, this solution will be valid for any value of t, so that the solution obtained also represents the general solution of the forced vibration. This is an advantage of the Fourier series expansion.

EXAMPLE 10.21. *The laminar flow of fluid in a rectangular conduit is governed by the differential equation*

$$\frac{\partial^2 u}{\partial y^2} + \frac{\partial^2 u}{\partial z^2} = \frac{k}{\mu}\frac{dP}{dx},$$

where u = velocity of fluid parallel to horizontal x-axis; P = static pressure; k = constant and μ = Newtonian coefficient of viscosity.

Find a general solution of the equation, assuming that the velocity is zero at the walls of the conduit, and maximum at the centre.

Solution: The cross-section of the rectangular conduit is shown in the *Fig 10.8*. Since the velocity of the fluid is zero at the walls, we have

$$u = 0, \text{ for } y = \pm\frac{b}{2} \text{ and for } z = \pm\frac{a}{2}. \quad (1)$$

Again, since the velocity is maximum at the centre of the conduit, we have

$$\frac{\partial u}{\partial y} = \frac{\partial u}{\partial z} = 0, \text{ for } y = 0 \text{ and for } z = 0. \quad (2)$$

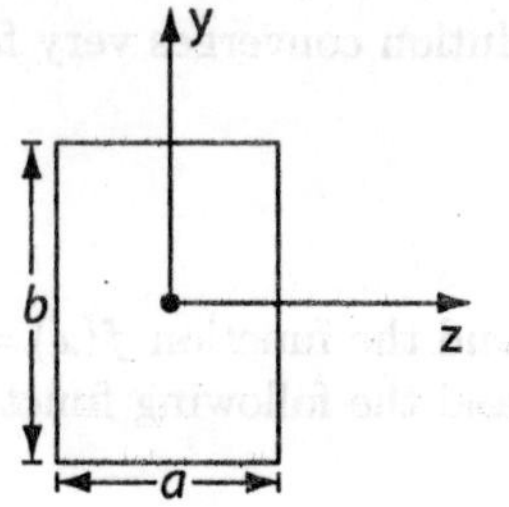

Fig 10.8

Thus, we are to solve the given equation under the boundary conditions given by equations (1) and (2). We write the given equation in the form

$$\frac{\partial^2 u}{\partial y^2} + \frac{\partial^2 u}{\partial z^2} = f(x), \tag{3}$$

where

$$f(x) = \frac{k}{\mu}\frac{dP}{dx}. \tag{4}$$

The solution of the equation (3) is assumed in the form

$$u = \sum_{m=1,3,5}^{\infty} \sum_{n=1,3,5}^{\infty} A_{mm} \cos \frac{m\pi z}{a} \cos \frac{n\pi y}{b}. \tag{5}$$

It is obvious that the solution function (5) satisfies all the boundary conditions in (1) and (2).

The function $f(x)$ in (4) is expanded in Fourier series in two dimension. We write

$$f(x) = \sum_{m=1,3,5,\ldots}^{\infty} \sum_{n=1,3,5,\ldots}^{\infty} B_{mn} \cos \frac{m\pi z}{a} \cos \frac{n\pi y}{b}, \tag{6}$$

where coefficients B_{mn} are given by

$$\begin{aligned} B_{mn} &= \frac{1}{ab} \int_{-\frac{a}{2}}^{\frac{a}{2}} \int_{-\frac{b}{2}}^{\frac{b}{2}} f(x) \cos \frac{m\pi z}{a} \cos \frac{n\pi y}{b}\, dz\, dy \\ &= \frac{4f(x)}{mn\pi^2} \sin \frac{m\pi}{2} \sin \frac{n\pi}{2}, \qquad \left.\begin{matrix} m = 1, 3, 5, \ldots \\ n = 1, 3, 5, \ldots \end{matrix}\right\}. \end{aligned} \tag{7}$$

Substituting the expressions (5) and (6) into the differential equations (3), we find the unknown coefficients A_{mn}

$$A_{mn} = -\frac{B_{mn}}{\left(\frac{m\pi}{a}\right)^2 + \left(\frac{n\pi}{b}\right)^2}. \tag{8}$$

Thus, using equations (5), (7) and (8), we get the solution of the differential equation (3) as

$$u(y, z) = -\frac{4k}{\mu\pi^2} \frac{dP}{dx} \sum_{m=1,3,5,\ldots}^{\infty} \sum_{n=1,3,5,\ldots}^{\infty} \frac{\sin \frac{m\pi}{2} \sin \frac{n\pi}{2}}{mn \left\{ \left(\frac{m\pi}{a}\right)^2 + \left(\frac{n\pi}{b}\right)^2 \right\}} \cdot \cos \frac{m\pi z}{a} \cos \frac{n\pi y}{b}.$$

The solution converges very fast.

EXERCISE

1. Expand the function $f(x) = 1$ for $0 < x < \pi$ in Fourier series.
2. Expand the following function in Fourier cosine series:

$$f(x) = \begin{cases} \cos \frac{\pi x}{l}, & \text{for } 0 \le x \le \frac{l}{2} \\ 0, & \text{for } \frac{l}{2} \le x \le l. \end{cases}$$

3. Find Fourier expansion of $f(t) = \sin t$, for $0 < t < \pi$.
4. Expand in Fourier cosine series $f(x) = \sin ax$, for $0 \le x < \pi$ (a is not an integer). What happens if a becomes an integer?
5. Expand in Fourier series $f(x) = x \sin x$, for $-\pi < x < \pi$.
6. Expand in Fourier series the function

$$f(x) = \begin{cases} \cos x, & \text{for } 0 \le x \le \pi \\ -\cos x, & \text{for } -\pi \le x \le 0. \end{cases}$$

7. Find Fourier expansions for the function

$$f(x) = \begin{cases} -\pi, & \text{for } -\pi < x < 0 \\ x, & \text{for } 0 < x < \pi. \end{cases}$$

Hence prove that

$$\frac{\pi^2}{8} = \frac{1}{1^2} + \frac{1}{3^2} + \frac{1}{5^2} + \cdots$$

8. Expand in Fourier series the function

$$f(x) = \begin{cases} 1 - \frac{x}{2}h, & \text{for } 0 \le x < 2h \\ 0, & \text{for } 2h < x \le \pi. \end{cases}$$

9. Expand the function $f(x) = e^{ax}$, $0 \le x < \pi$ in Fourier sine and cosine series.

10. Find Fourier sine series for the function

$$f(x) = \begin{cases} \sin \frac{\pi x}{l}, & \text{for } 0 \le x \le \frac{l}{2} \\ 0, & \text{for } \frac{l}{2} < x \le l. \end{cases}$$

11. Expand $f(t) = t^2$ for $-\pi < t < \pi$ in Fourier series, and therefrom find Fourier series for $f(t) = t^3$ for $-\pi < t < \pi$.

12. Show that, for $-\pi < x < \pi$, $x \neq 0$

$$\frac{d}{dx} |\sin x| = \begin{cases} -\cos x, & \text{for } -\pi < x < 0 \\ \cos x, & \text{for } 0 < x < \pi. \end{cases}$$

13. Consider the following functions and find in each case Fourier series expansions which arise from them:

(a) $\log(1 + z)$; (b) $\cos z$; (c) $\cosh z$.

14. Starting with power series of $\log(1 + z)$, show that

$$\log \tan(\frac{\theta}{2}) = -2 \sum_{n=1}^{\infty} \frac{1}{(2n-1)} \cos(2n-1)\theta.$$

15. Show that, for x in $[0, \pi]$

$$\sum_{k=1}^{\infty} \frac{\cos(2k-1)x}{(2k-1)^4} = \frac{1}{96}\pi(\pi - 2x)\left(\pi^2 + 2\pi x - 2x^2\right).$$

16. Expand $f(x) = e^x$ for $-\pi < x < \pi$, and deduce that

$$\coth \pi = \frac{1}{\pi} + \frac{2}{\pi}\left[\frac{1}{1+1^2} + \frac{1}{1+2^2} + \frac{1}{1+3^2} + \cdots\right]$$

ANSWERS

1. $1 = \frac{4}{\pi}\left(\sin x + \frac{1}{3}\sin 3x + \frac{1}{5}\sin 5x + \cdots\right).$

2. $f(x) = \frac{1}{\pi} + \frac{1}{2}\cos\frac{\pi x}{l} - \frac{2}{\pi}\sum_{n=1}^{\infty}\frac{(-1)^n}{4n^2-1}\cos\frac{2n\pi x}{l}.$

3. $\sin t = \frac{2}{\pi} - \frac{4}{\pi}\sum_{k=1}^{\infty}\frac{\cos 2kx}{4k^2-1}.$

4. $\sin ax = \frac{1-\cos a\pi}{\pi}\left[1 + 2a\sum_{n=1}^{\infty}\frac{\cos 2nx}{a^2-4n^2}\right] + \frac{2a}{\pi}(1+\cos a\pi)\sum_{n=0}^{\infty}\frac{\cos(2n+1)x}{a^2-(2n+1)^2}.$

5. $x\sin x = 1 - 2\left[\frac{\cos x}{4} + \frac{\cos 2x}{1\cdot 3} - \frac{\cos 3x}{2\cdot 4} + \frac{\cos 4x}{3\cdot 5} \cdots\right]$

6. $f(x) = \frac{8}{\pi}\left[\frac{2}{1\cdot 3}\sin 2x + \frac{4}{3\cdot 5}\sin 4x + \cdots\right].$

7. $f(x) = -\frac{\pi}{4} - \frac{2}{\pi}\sum_{n=1,3,\ldots}^{\infty}\frac{\cos(nx)}{n^2} + 3\sum_{n=1,3,5,\ldots}^{\infty}\frac{\sin(nx)}{n} - \sum_{n=2,4,\ldots}^{\infty}\frac{\sin(nx)}{n}.$

8. $f(x) = \frac{2h}{\pi}\left[\frac{1}{2} + \sum_{k=1}^{\infty}\left(\frac{\sin kh}{kh}\right)^2\cos kx\right].$

9. (a) $e^{ax} = \frac{e^{a\pi}-1}{a\pi} + \frac{2a}{\pi}\sum_{k=1}^{\infty}\frac{(-1)^k e^{a\pi}-1}{a^2+k^2}\cos kx,\ 0 \le x < \pi;$

(b) $e^{ax} = \frac{2}{\pi}\sum_{k=1}^{\infty}\left[1-(-1)^k e^{a\pi}\right]\frac{k}{a^2+k^2}\sin kx,\ 0 < x < \pi.$

10. $f(x) = \frac{1}{2}\sin\frac{\pi x}{l} - \frac{4}{\pi}\sum_{k=1}^{\infty}\frac{(-1)^k k}{4k^2-1}\sin\frac{2\pi kx}{l}.$

11. $x^3 = \pi^2 x + 12\sum_{k=1}^{\infty}(-1)^k\frac{\sin kx}{k^3}.$

13. (a) $\log\left(2\cos\frac{x}{2}\right) = \cos x - \frac{1}{2}\cos 2x + \frac{1}{3}\cos 3x - \cdots$

$\frac{x}{2} = \sin x - \frac{1}{2}\sin 2x + \frac{1}{3}\sin 3x - \cdots$

(b) $\cos(\cos x)\cosh(\sin x) = 1 - \frac{\cos 2x}{2!} + \frac{\cos 4x}{4!} - \cdots$

$\sin(\cos x)\sinh(\sin x) = \frac{\sin 2x}{2!} - \frac{\sin 4x}{4!} + \frac{\sin 6x}{6!} - \cdots$

(c) $\cosh x \cos(\sin x) = 1 + \dfrac{\cos 2x}{2!} + \dfrac{\cos 4x}{4!} + \cdots$

$\cosh x \sin(\sin x) = \sin x + \dfrac{\sin 3x}{3!} + \dfrac{\sin 5x}{5!} + \cdots$

16. $e^x = \dfrac{2}{\pi}\sinh \pi \left[\dfrac{1}{2} + \sum_{k=1}^{\infty} \dfrac{(-1)^k}{k^2+1}(\cos kx - k \sin kx)\right].$

Chapter 11

Equations of Mathematical Physics

11.1 Introduction

In the study of a physical process, we generally assume the field of the physical process to be continuous, in stead of discrete. The study of such processes leads to differential equations which we call equations of mathematical physics. For a process which is described by a function of two or more variables, the study leads to a partial differential equations.

Equations of mathematical physics are supplemented by additional information without which their unique solution is not possible to be determined. This information may be related to boundary conditions, when we call it a *boundary value problem.* If the information relates to the condition at some instant of time, we call it an *initial value problem.* Physical processes are called *mixed-value problems* when part of the information relates to boundary conditions, and other part to the initial conditions.

We have already encountered with some equations of mathematical physics in earlier chapters. We intend to mention two of them.

1. The elastic bending of a rod of arbitrary cross-section is governed by the equation

$$\frac{d^2}{dx^2}\left(\mathrm{EI}\frac{d^2y}{dx^2}\right) = q(x), \tag{11.1}$$

where $q(x)$ is the load/unit length and EI is the flexural rigidity of the rod.

2. The heat conduction through a solid is described by the equation

$$\nabla^2\phi = \frac{1}{h^2}\frac{\partial\phi}{\partial t}, \tag{11.2}$$

where $\phi(x,t)$ is the non-steady state temperature function and h^2 is a constant called *diffusivity.*

In this chapter, we intend to discuss on the equations of mechanics, theory of electric and magnetic phenomena, elasticity and Schröedinger equation. We begin with the Laplace's and Poisson's equations which constitute the pillars of mathematical physics.

POTENTIAL THEORY

11.2 Laplace's Equation

We consider Newtonian attraction of a particle of mass m placed at the point $Q(a,b,c)$ on a unit mass at $P(x,y,z)$. The force of attraction of magnitude

$$F = -\frac{\mu m}{r^2} \tag{11.3}$$

is directed along the line joining the two points and towards the mass m, where μ = a positive constant. The distance r between the two points is given by

$$r = \left[(a-x)^2 + (b-y)^2 + (c-z)^2\right]^{\frac{1}{2}}.$$

The components of the force F in equation (11.3) parallel to the axes are

$$\left.\begin{aligned} X &= \frac{\mu m(a-x)}{r^3} \\ Y &= \frac{\mu m(b-y)}{r^3} \\ \text{and} \quad Z &= \frac{\mu m(c-z)}{r^3} \end{aligned}\right\} \tag{11.4}$$

We now define a function V by the formula

$$V = \frac{\mu m}{r}. \tag{11.5}$$

Thus, if we keep the position of the mass m fixed, V is a function of the coordinates of the point P. Moreover, it is a single-valued function which does not depend on a system of axes of reference.

Now

$$\frac{\partial V}{\partial x} = \frac{\partial V}{\partial r} \cdot \frac{\partial r}{\partial x} = -\frac{\mu m}{r^2} \cdot -\frac{(a-x)}{r} = \mu m \frac{(a-x)}{r^3} = X \text{ [from equation (11.4)]}$$

We have similar expressions for other two directions. Hence we write

$$\frac{\partial V}{\partial x} = X, \quad \frac{\partial V}{\partial y} = Y \quad \text{and} \quad \frac{\partial V}{\partial z} = Z. \tag{11.6}$$

The function V defined by equation (11.5) is called the *gravitational potential.* Thus, we have proved that the partial derivatives of V with respect to x, y, z give the components of attraction force at P due to the mass m at Q in the directions of the axes. The above discussion is clearly valid also for a system of particles.

Let us now extend our discussion for the case of a distributed mass occupying a volume τ and having density ρ. We consider an element of mass dm, so that potential dV due to this mass is

$$dV = \mu \frac{dm}{r} = \frac{\mu \rho d\tau}{r}. \tag{11.7}$$

Hence

$$V = \iiint_{\tau} \frac{\mu \rho \, d\tau}{r} = \mu \iiint_{\tau} \frac{\rho \, dx \, dy \, dz}{\left[(a-x)^2 + (b-y)^2 + (c-z)^2\right]^{\frac{1}{2}}}. \tag{11.8}$$

We differentiate equation (11.8) (which is permissible under integral sign since r is measured from an external point and the results of differentiation with respect to x, y, z are continuous and yield convergent integrals), and get

$$X = \frac{\partial V}{\partial x} = \mu \iiint_{\tau} \frac{\rho(a-x)}{r^3} \, dx \, dy \, dz \tag{11.9}$$

Similarly

$$Y = \frac{\partial V}{\partial y} = \mu \iiint_{\tau} \frac{\rho(b-y)}{r^3} \, dx \, dy \, dz. \tag{11.10}$$

$$Z = \frac{\partial V}{\partial z} = \mu \iiint_{\tau} \frac{\rho(c-z)}{r^3} \, dx \, dy \, dz. \tag{11.11}$$

Differentiating equations (11.6) with respect to x, y, z respectively, we get

$$\frac{\partial^2 V}{\partial x^2} = -\frac{\mu m}{r^3} + 3\,\frac{\mu m\,(a-x)^2}{r^5}$$

$$\frac{\partial^2 V}{\partial y^2} = -\frac{\mu m}{r^3} + 3\,\frac{\mu m\,(b-y)^2}{r^5}$$

and $$\frac{\partial^2 V}{\partial z^2} = -\frac{\mu m}{r^3} + 3\,\frac{\mu m\,(c-z)^2}{r^5}.$$

Thus, by adding, we get

$$\frac{\partial^2 V}{\partial x^2} + \frac{\partial^2 V}{\partial y^2} + \frac{\partial^2 V}{\partial z^2} = 0. \tag{11.12}$$

Again, by differentiating equations (11.9), (11.10) and (11.11) with respect to x, y, z respectively, we get

$$\frac{\partial^2 V}{\partial x^2} = -\mu \iiint_{\tau} \left[\frac{1}{r^3} - \frac{3\,(a-x)^2}{r^5}\right] \rho \, dx \, dy \, dz \tag{11.13}$$

$$\frac{\partial^2 V}{\partial y^2} = -\mu \iiint_{\tau} \left[\frac{1}{r^3} - \frac{3\,(b-y)^2}{r^5}\right] \rho \, dx \, dy \, dz \tag{11.14}$$

$$\frac{\partial^2 V}{\partial z^2} = -\mu \iiint_{\tau} \left[\frac{1}{r^3} - \frac{3\,(c-z)^2}{r^5}\right] \rho \, dx \, dy \, dz \tag{11.15}$$

Adding all these we get

$$\frac{\partial^2 V}{\partial x^2} + \frac{\partial^2 V}{\partial y^2} + \frac{\partial^2 V}{\partial z^2} = 0 \tag{11.16}$$

$$\text{i.e.,} \quad \nabla^2 V = 0. \tag{11.17}$$

Equation (11.16) or (11.17) is known as *Laplace's equation* in three dimension. Thus, we have seen that the gravitational potential satisfies Laplace's equation at every point, where there is no matter.

It is interesting to note that magnetic potential as well as electric potential functions also satisfy Laplace's equation at every point, where there is no magnetic pole or electric charges. This we shall see later.

11.3 Poisson's Equation

We recapitulate the Gauss's theorem. The theorem states that, the outward flux of the force of attraction over any closed surface in a gravitational field of force is equal to -4π times the mass enclosed by the surface.

We are acquainted with this theorem, so we do not offer any proof here. We rather intend to get Poisson's equation for potential.

Let a small volume $d\tau$ of mass of density ρ is enclosed by a surface S. Then, the outward normal component of force acting on a small area dS along normal direction n of the closed surface S is given by

$$\begin{aligned}\frac{\partial V}{\partial n} &= \frac{\partial V}{\partial x}\cdot\frac{\partial x}{\partial n} + \frac{\partial V}{\partial y}\cdot\frac{\partial y}{\partial n} + \frac{\partial V}{\partial z}\cdot\frac{\partial z}{\partial n} \\ &= l'X + m'Y + n'Z, \end{aligned} \tag{11.18}$$

where l', m', n' are direction cosines of the normal to the surface. Then by Gauss theorem

$$-4\pi \iiint \rho d\tau = \iint (l'X + m'Y + n'Z)\, dS.$$

Using Green's theorem, we get

$$\begin{aligned} -4\pi \iiint \rho d\tau &= \iiint \left(\frac{\partial X}{\partial x} + \frac{\partial Y}{\partial y} + \frac{\partial Z}{\partial z}\right) d\tau \\ &= \iiint \left(\frac{\partial^2 V}{\partial x^2} + \frac{\partial^2 V}{\partial y^2} + \frac{\partial^2 V}{\partial z^2}\right) d\tau. \end{aligned}$$

Therefore

$$\iiint \left(\frac{\partial^2 V}{\partial x^2} + \frac{\partial^2 V}{\partial y^2} + \frac{\partial^2 V}{\partial z^2} + 4\pi\rho\right) d\tau = 0.$$

Since the integration is valid for all ranges, the integrand must vanish. Hence

$$\begin{aligned} \frac{\partial^2 V}{\partial x^2} + \frac{\partial^2 V}{\partial y^2} + \frac{\partial^2 V}{\partial z^2} &= -4\pi\rho \\ \text{i.e.,}\quad \nabla^2 V &= -4\pi\rho. \end{aligned} \tag{11.19}$$

Equation (11.19) is known as *Poisson's equation*. Thus, the gravitational potential satisfies Poisson's equation at every point, where there exists matter. Poisson's equation is also satisfied by electrostatic potential and magnetostatic potential at every point, where there exists electric charge or magnetic pole. This also we shall find later.

From equation (11.19), we get

$$\rho = -\frac{1}{4\pi}\nabla^2 V. \tag{11.20}$$

Thus, we can find the volume density of matter when the potential is given at all points of space. If, however, the potential is given by different functions V_1, V_2 on two opposite sides of a surface S, the surface density of matter on S is given by

$$\rho = -\frac{1}{4\pi}\left(\frac{\partial V_2}{\partial n} - \frac{\partial V_1}{\partial n}\right), \tag{11.21}$$

where n is the normal to the surface directed from face 1 to face 2.

EXAMPLE 11.1. *If the potential within an attracting medium having spherical symmetry is given by* $V = (c^2+r^2)^{-\frac{1}{2}}$, *where r is the distance from the centre and c is a constant, prove that the density is given by*

$$\rho = \frac{3c^2}{4\pi(c^2+r^2)^{\frac{5}{2}}}.$$

Solution: We have

$$V = (c^2+r^2)^{-\frac{1}{2}} \tag{1}$$

$$\text{and} \quad \rho = -\frac{1}{4\pi}\nabla^2 V. \tag{2}$$

Due to spherical symmetry, we use the spherical coordinate system when the Laplace operator becomes

$$\nabla^2(\cdots) = \frac{1}{r^2}\frac{\partial}{\partial r}\left(r^2\frac{\partial\cdots}{\partial r}\right). \tag{3}$$

Hence, equation (2) becomes

$$\rho = -\frac{1}{4\pi}\left[\frac{\partial^2 V}{\partial r^2} + \frac{2}{r}\frac{\partial V}{\partial r}\right]. \tag{4}$$

Now

$$\frac{\partial V}{\partial r} = -r(c^2+r^2)^{-\frac{3}{2}} \tag{5}$$

$$\text{and} \quad \frac{\partial^2 V}{\partial r^2} = -(c^2+r^2)^{-\frac{3}{2}} + 3r^2(c^2+r^2)^{-\frac{5}{2}}. \tag{6}$$

Using equations (5) and (6), equation (4) becomes

$$\begin{aligned}\rho &= -\frac{1}{4\pi}\left[-3(c^2+r^2)^{-\frac{3}{2}} + 3r^2(c^2+r^2)^{-\frac{5}{2}}\right]\\ &= -\frac{3}{4\pi}\left[\frac{r^2-(c^2+r^2)}{(c^2+r^2)^{\frac{5}{2}}}\right]\\ &= \frac{3c^2}{4\pi}\cdot\frac{1}{(c^2+r^2)^{\frac{5}{2}}} \quad \text{(proved)}.\end{aligned}$$

11.4 Equipotential Surface

A family of surfaces given by $\phi(x,y,z)$ = constant is called *equipotential surfaces* when the potential function remains constant on these surfaces. This happens under certain conditions which we like to determine.

Since the potential function V is constant whenever the function $\phi(x,y,z)$ is constant, we can write

$$V = f\{\phi(x,y,z)\}. \tag{11.21}$$

Then

$$\frac{\partial V}{\partial x} = f'(\phi)\frac{\partial \phi}{\partial x}$$

$$\frac{\partial^2 V}{\partial x^2} = f'(\phi)\frac{\partial^2\phi}{\partial x^2} + f''(\phi)\left(\frac{\partial\phi}{\partial x}\right)^2.$$

We get similar other two expressions. Adding all these we get

$$\therefore \quad \nabla^2 V = f'(\phi)\nabla^2\phi + f''(\phi)\left[\left(\frac{\partial\phi}{\partial x}\right)^2 + \left(\frac{\partial\phi}{\partial y}\right)^2 + \left(\frac{\partial\phi}{\partial z}\right)^2\right]. \tag{11.22}$$

Since, in free space, $\nabla^2 V = 0$, we get from equation (11.22)

$$\frac{\nabla^2\phi}{\left(\frac{\partial\phi}{\partial x}\right)^2 + \left(\frac{\partial\phi}{\partial y}\right)^2 + \left(\frac{\partial\phi}{\partial z}\right)^2} = -\frac{f''(\phi)}{f'(\phi)} = \psi(\phi), \text{ say.} \tag{11.23}$$

Equation (11.23) is the necessary condition for the surface $\phi(x, y, z) = \text{constant}$ to be equipotential.

Hence, in that case

$$\frac{f''(\phi)}{f'(\phi)} + \psi(\phi) = 0. \tag{11.24}$$

Integrating

$$\log f'(\phi) = A - \int \psi(\phi)\, d\phi$$

$$\text{i.e.,} \quad f'(\phi) = Ae^{-\int \psi(\phi)\, d\phi}.$$

Hence

$$V = f(\phi) = A\int e^{-\int \psi(\phi)\, d\phi}\, d\phi + B, \tag{11.25}$$

where A and B are constants of integration

11.5 Properties of Potential Functions

Property 1. The potential function cannot have a maximum or minimum value at any point of space, where matter does not exist.

Proof. Let the potential function V have a maximum at a point O in the space. We enclose the point O by a small sphere S. Then, potential at any point on S would be less than that at O. The outward normal force R_n would be negative at all points of S. Hence total outward normal force through the surface S given by

$$\iint_S R_n\, dS$$

cannot vanish. This is contrary to the fact that it should vanish, since there is no matter at O. Similar reasoning stands for minimum value. □

Property 2. The mean value over a spherical surface of the potential of any attracting system is equal to the potential at the centre of the sphere due to a mass outside the sphere, plus the mass inside the sphere divided by the radius.

This is known as *Gauss's mean value theorem.*

Proof.

We consider a sphere of radius a. Let P be a point on its surface. We take a small area dS surrounding P. If there is a mass M placed at an external point A (*Fig 11.1*), then potential at P is

$$V_1 = \frac{M}{AP}. \tag{11.26}$$

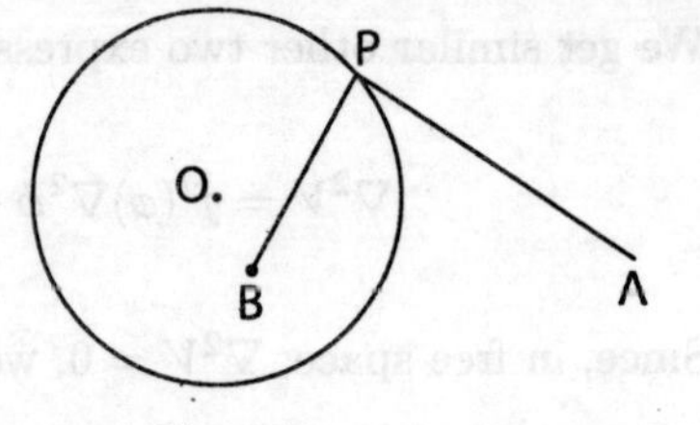

Fig 11.1

Similarly, if there be mass m placed at an internal point B, then the potential at P is

$$V_2 = \frac{m}{BP}. \tag{11.27}$$

Thus, the total potential at P is

$$V = V_1 + V_2 = \frac{M}{AP} + \frac{m}{BP}. \tag{11.28}$$

Now, the mean value of V over the sphere is

$$V_{av} = \frac{1}{4\pi a^2}\iint_S V\,dS = \frac{M}{4\pi a^2}\iint_S \frac{dS}{AP} + \frac{m}{4\pi a^2}\iint_S \frac{dS}{BP} \quad \text{[using equation (11.28)]} \tag{11.29}$$

But

$$\begin{aligned}\iint_S \frac{dS}{AP} &= \text{potential at } A \text{ due to a sphere of unit surface density} \\ &= \frac{4\pi a^2}{OA}\end{aligned} \tag{11.30}$$

and

$$\begin{aligned}\iint \frac{dS}{BP} &= \text{potential at } B \text{ due to same sphere} \\ &= \frac{4\pi a^2}{a} = 4\pi a.\end{aligned} \tag{11.31}$$

Hence, using equations (11.30) and (11.31), we get

$$V_{av} = \frac{M}{OA} + \frac{m}{a}.$$

Hence the theorem is proved. □

Property 3. If two different distributions of matter produce equal potentials over any surface which encloses both the sources, then they produce equal potentials throughout all space external to the surface.

11.6 Solution of Laplace's Equation

Let us consider a region bounded by a closed surface S. We are to find solution of $\nabla^2 V = 0$ inside the region S when values of V and $\frac{\partial V}{\partial n}$ on the boundary of S are given.

We take a point P inside S and draw a small sphere Ω of radius ϵ with centre at P (*Fig 11.2*).

Now, for any two functions ϕ and ψ, we have by Green's theorem

$$\iint_S \left(\phi\frac{\partial\psi}{\partial n} - \psi\frac{\partial\phi}{\partial n}\right) dS = \iiint_\tau (\phi\nabla^2\psi - \psi\nabla^2\phi)\, d\tau. \tag{11.32}$$

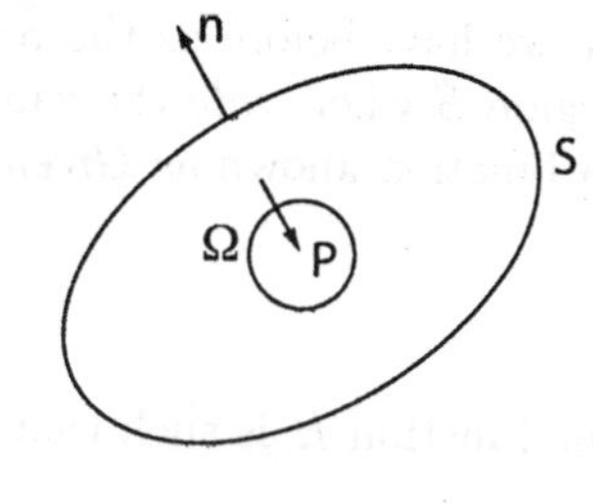

Fig 11.2

We put $\phi = V$ and $\psi = \dfrac{1}{r}$, where r is the radius vector drawn from P inside S.

Since

$$\left.\begin{aligned} \nabla^2\left(\frac{1}{r}\right) &= 0 \\ \text{and} \quad \nabla^2 V &= 0 \end{aligned}\right\} \tag{11.33}$$

throughout the region, the right-hand side of equation (11.32) is zero. Hence, the integral in the left-hand side taken over two surfaces S and Ω is zero. That is

$$\iint_S \left[V\frac{\partial}{\partial n}\left(\frac{1}{r}\right) - \frac{1}{r}\frac{\partial V}{\partial n}\right] dS + \iint_\Omega \left[V\frac{\partial}{\partial n}\left(\frac{1}{r}\right) - \frac{1}{r}\frac{\partial V}{\partial n}\right] d\Omega = 0. \tag{11.34}$$

For the second integral, we have

$$\frac{\partial}{\partial n}\left(\frac{1}{r}\right) = \frac{\partial}{\partial n}\left(\frac{1}{\epsilon}\right) = \frac{1}{\epsilon^2} \quad [\because \text{ direction of } n \text{ is opposite}]$$

and $d\Omega = \epsilon^2 d\omega$, ω = solid angle at P.

Hence

$$\iint_\Omega V\frac{\partial}{\partial n}\left(\frac{1}{r}\right) d\Omega = \iint_\Omega V\, d\omega = 4\pi V_p, \tag{11.35}$$

where V_p = value of V at the point P.

Again

$$\iint_\Omega \frac{1}{r}\frac{\partial V}{\partial n} d\Omega = \iint_\Omega \frac{1}{\epsilon}\frac{\partial V}{\partial n}\left(\epsilon^2\, d\omega\right), \tag{11.36}$$

which $\longrightarrow 0$ as $\epsilon \longrightarrow 0$. Thus with the use of equations (11.35) and (11.36), we get from equation (11.34)

$$V_p = \frac{1}{4\pi}\int_S \left[\frac{1}{r}\frac{\partial V}{\partial n} - V\frac{\partial}{\partial n}\left(\frac{1}{r}\right)\right] dS. \tag{11.37}$$

Equation (11.37) gives the value of V at any point P inside the region S in terms of the values of V and $\frac{\partial V}{\partial n}$ on the surface.

But, according to Dirichlet, *there exists a solution of Laplace's equation which, with its first and second derivatives, is single-valued, finite and continuous throughout a region bounded by a surface S and takes assigned values on the boundary of S.*

Thus, we have before us the problem of finding the solution of Laplace's equation inside a closed region S when *only* the value of V on the boundary of S is given. This we do with the help of a function known as *Green's function.* We define a function G, such that

$$G = H + \frac{1}{r}, \tag{11.38}$$

where the function H is such that

$$\left.\begin{aligned} \nabla^2 H &= 0, \quad \text{within } S \\ \text{and } H &= -\tfrac{1}{r}, \quad \text{on } S. \end{aligned}\right\} \tag{11.39}$$

The function G thus defined is called *Green's function* for the point P and surface S.

From equations (11.38) and (11.39), it is clear that the function G satisfies Laplace's equation inside S, and is zero on S. Now, in equation (11.32), we put $\phi = V$, and $\psi = H$. Then we get the right-hand side of equation (11.32) equal to zero. Hence

$$\iint\limits_S \left(V\frac{\partial H}{\partial n} - H\frac{\partial V}{\partial n} \right) dS = 0. \tag{11.40}$$

We multiply equation (11.40) by $-\dfrac{1}{4\pi}$, and add to the right-hand side of equation (11.37). Then we get

$$\begin{aligned} V_p &= \frac{1}{4\pi}\iint\limits_S \left\{ \left(H + \frac{1}{r} \right)\frac{\partial V}{\partial n} - V\frac{\partial}{\partial n}\left(H + \frac{1}{r} \right) \right\} dS \\ &= \frac{1}{4\pi}\iint\limits_S \left(G\frac{\partial V}{\partial n} - V\frac{\partial G}{\partial n} \right) dS. \end{aligned} \tag{11.41}$$

Since G vanishes on S, we get from equation (11.41)

$$V_p = -\frac{1}{4\pi}\iint\limits_S V\frac{\partial G}{\partial n}\, dS. \tag{11.42}$$

Thus, the problem of finding solution of the Laplace's equation reduces to the determination of Green's function.

There are many other physical problems which lead to Laplace's and Poisson's equation. Steady-state heat-conduction and velocity potential of non-viscous fluid flow satisfy Laplace's equation. Torsion of beams and laminar viscous flow of fluids through conduits satisfy Poisson's equation.

WAVE MOTION

Wave motion occurs in many physical situations like (i) vibrating string, membranes, beams, plates, etc. (ii) acoustic waves, water waves, earthquake waves, etc., (iii) electromagnetic waves, radiation waves, etc. We shall discuss here about derivation of differential equations for wave motion in a few case.

11.7 Vibration of a String

We consider a tightly stretched flexible string undergoing vibration in a vertical plane. The deflection of the string from the horizontal x-axis at any time t is denoted by $u(x, t)$. We make the following assumptions:

1. The mass m per unit length of the string is constant.
2. Weight of the string compared to the tension T in the string is negligible.
3. Slope of the deflection curve is small.
4. Every element of the string moves in the vertical plane during vibration.
5. There is no resistance to motion.

Now we consider a small element of the string (*Fig 11.3*) of length Δx. Forces acting on the element AB are:

(i) Tension T at A at an angle ϕ with horizon.
(ii) Tension $T + \Delta T$ at B at an angle $\phi + \Delta\phi$ with horizon.
(iii) Mass of the element $m\Delta x$ at its centre.

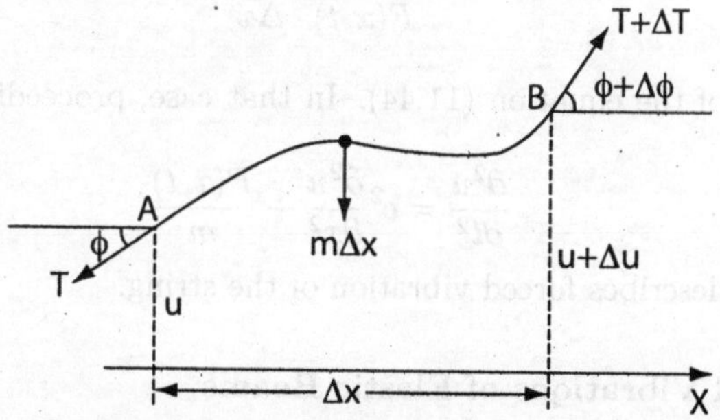

Fig 11.3

By Newton's second law, equation of motion along x-direction is

$$(T + \Delta T)\cos(\phi + \Delta\phi) - T\cos\phi = 0 \quad \text{(since there is no acceleration)} \tag{11.43}$$

Since by assumption (3) ϕ is very small

$$\cos(\phi + \Delta\phi) = \cos\phi \simeq 1.$$

Thus, from equation (11.43) we get $\Delta T = 0$, i.e., tension in the string remains constant. Now, equation of motion along vertical direction is

$$T\sin(\phi + \Delta\phi) - T\sin\phi = m\Delta x \frac{\partial^2}{\partial t^2}\left(u + \frac{\Delta u}{2}\right). \tag{11.44}$$

The left-hand side of this equation is

$$= T\left[\tan(\phi + \Delta\phi) - \tan\phi\right] \quad \text{(since } \phi \text{ is small)}$$

$$= T\left[u_x(x + \Delta x, t) - u_x(x, t)\right], \quad \left(u_x = \frac{\partial u}{\partial x}\right). \tag{11.45}$$

From equations (11.44) and (11.45), dividing both sides by Δx, we get

$$T\left[\frac{u_x(x + \Delta x, t) - u_x(x, t)}{\Delta x}\right] = m\frac{\partial^2}{\partial t^2}\left(u + \frac{\Delta u}{2}\right).$$

Proceeding to the limit $\Delta x \longrightarrow 0$ when Δu also tends to zero, we get

$$T \cdot u_{xx} = m\frac{\partial^2 u}{\partial t^2}, \quad \text{i.e.,} \quad \frac{\partial^2 u}{\partial t^2} = c^2\frac{\partial^2 u}{\partial x^2}, \tag{11.46}$$

where

$$c = \sqrt{\frac{T}{m}}. \tag{11.47}$$

Equation (11.46) is the one-dimensional wave equation. The wave moves normal to the string. The speed c of the wave is given by equation (11.47). It is directly proportional to square root of the tension, and inversely proportional to square root of mass per unit length of the string.

Equation (11.46) describes the free vibration of the string. If on the string acts a force $F(x,t)$ per unit length of the string, there will be an additional force-term

$$F(x,t) \cdot \Delta x \tag{11.48}$$

in the left-hand side of the equation (11.44). In that case, proceeding to the limit $\Delta x \longrightarrow 0$, we get

$$\frac{\partial^2 u}{\partial t^2} = c^2\frac{\partial^2 u}{\partial x^2} + \frac{F(x,t)}{m}. \tag{11.49}$$

Equation (11.49) describes forced vibration of the string.

11.8 Longitudinal Vibrations of Elastic Beams

We consider an elastic bar which vibrates in longitudinal direction. We make the following assumptions:

(i) The bar is of homogeneous material;

(ii) Cross-sectional area of the bar is constant;

(iii) Hook's law of stress and strain is applicable.

We take a small length Δx of the bar whose longitudinal axis coincides with x-axis (*Fig 11.4*).

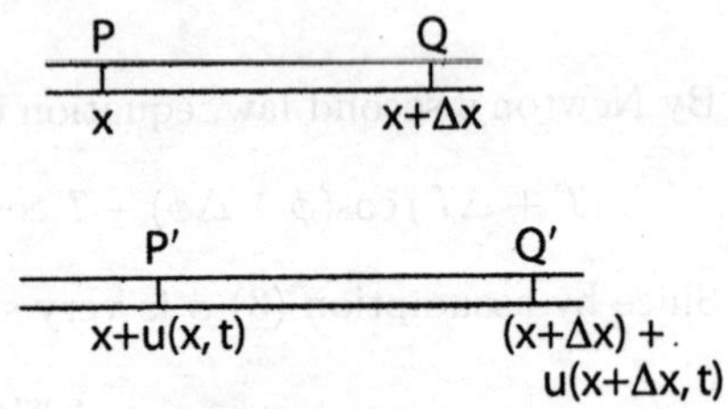

Fig 11.4

Let $u(x,t) =$ position of a beam section at any instant t,

$A =$ cross-sectional area of the beam,

$E =$ Young's modulus of elasticity,

$\rho =$ density of the material of the beam.

Then, total force acting at any cross-section is

$$F = AE\epsilon, \tag{11.50}$$

where

$$\epsilon = \text{strain} = \frac{\text{elongation}}{\text{original length}}.$$

Due to motion, the cross-section of the beam at P moves to P', and that at Q to Q'.

The length of the element

$$P'Q' = \Delta x + u(x + \Delta x, t) - u(x, t).$$

Hence, change in length of the element

$$= u(x + \Delta x, t) - u(x, t).$$

Hence, the strain is

$$\epsilon = \frac{u(x + \Delta x, t) - u(x, t)}{\Delta x}.$$

Proceeding to the limit $\Delta x \longrightarrow 0$, we have

$$\epsilon = \frac{\partial u}{\partial x}, \quad \text{i.e.,} \quad \epsilon = u_x. \tag{11.51}$$

Thus, from equation (11.50), we get

$$F = AE\frac{\partial u}{\partial x} = AEu_x. \tag{11.52}$$

Now, going back to the element PQ, the net force acting on it in x-direction is

$$F_x = AE\left[u_x(x + \Delta x, t) - u_x(x, t)\right]. \tag{11.53}$$

From Newton's second law

$$F_x = \rho A \Delta x \frac{\partial^2 u}{\partial t^2}. \tag{11.54}$$

Hence, from equations (11.53) and (11.54), we get

$$\rho A \frac{\partial^2 u}{\partial t} = AE\left[\frac{u_x(x + \Delta x, t) - u(x, t)}{\Delta x}\right]. \tag{11.55}$$

Proceeding to the limit $\Delta x \longrightarrow 0$,

$$\rho A \frac{\partial^2 u}{\partial t^2} = AE\, u_{xx} = AE\frac{\partial^2 u}{\partial x^2},$$

$$\text{i.e.,} \quad \frac{\partial^2 u}{\partial t^2} = a^2 \frac{\partial^2 u}{\partial x^2}, \tag{11.56}$$

where $a = \sqrt{\dfrac{E}{\rho}}$ = longitudinal wave speed.

Equation (11.56) is an one-dimensional wave equation which describes longitudinal vibrations of elastic beam.

If an additional force $F(x,t)$ per unit volume acts on the beam in longitudinal direction, then the equation (11.55) will contain additional term $F(x,t) \cdot A\Delta x$ in the right-hand side. Thus, in the limit $\Delta x \longrightarrow 0$, we will have

$$\rho A \frac{\partial^2 u}{\partial t^2} = AE\frac{\partial^2 u}{\partial x^2} + F(x,t) \cdot A$$

$$\text{i.e.,} \quad \frac{\partial^2 u}{\partial t^2} = a^2 \frac{\partial^2 u}{\partial x^2} + \frac{F(x,t)}{\rho}. \tag{11.57}$$

Equation (11.57) describes the forced vibration of a longitudinal beam.

11.9 Vibration of Membranes

Membranes are thin two-dimensional elements which offer no resistance to bending due to its deflection. For a vibrating membrane, we make the following assumptions:

1. The mass m per unit area of the membrane is constant.
2. Weight of the membrane compared to the tension T per unit length is negligible.
3. Slope of the deflection surface is small.
4. Every element of the membrane moves in the direction normal to the xy-plane.
5. There is no resistance to motion.

We consider a small element of area $\Delta x \Delta y$ of the membrane (*Fig 11.5*). Forces acting on the element are:

(i) Tensions $T\Delta x$ on edges parallel to x-axis at angles of θ and $\theta + \Delta\theta$ with the horizon.
(ii) Tensions $T\Delta y$ on edges parallel to y-axis at angles of ϕ and $\phi + \Delta\phi$ with the horizon.
(iii) Mass of the element $m\Delta x\Delta y$ at its centre.

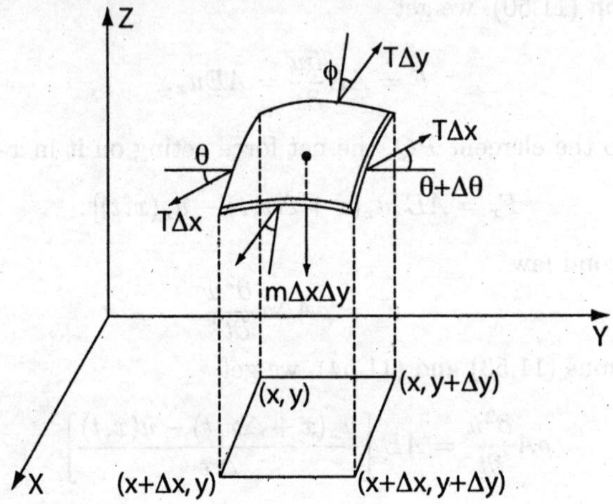

Fig 11.5

Equation of motion along z-axis following Newton's second law is

$$T\Delta x\,[\sin(\theta+\Delta\theta)-\sin\theta]+T\Delta y\,[\sin(\phi+\Delta\phi)-\sin\phi]=m\Delta x\Delta y\frac{\partial^2 u}{\partial t^2}. \tag{11.58}$$

Dividing both sides by $\Delta x\Delta y$, we get

$$T\left[\frac{\sin(\theta+\Delta\theta)-\sin\theta}{\Delta y}+\frac{\sin(\phi+\Delta\phi)-\sin\phi}{\Delta x}\right]=m\frac{\partial^2 u}{\partial t^2}. \tag{11.59}$$

Here $u(x, y, t)$ is the function which describes the surface of the membrane at any instant in its deflected position.

Now, angles θ and ϕ being small, we have

$$\sin(\theta+\Delta\theta) \simeq \tan(\theta+\Delta\theta) = u_y(x, y+\Delta y, t) \tag{11.60}$$
$$\sin\theta \simeq \tan\theta = u_y(x, y, t) \tag{11.61}$$
$$\sin(\phi+\Delta\phi) \simeq \tan(\phi+\Delta\phi) = u_x(x+\Delta x, y, t) \tag{11.62}$$
$$\sin\phi \simeq \tan\phi = u_x(x, y, t). \tag{11.63}$$

From equations (11.60) and (11.61), we have

$$\frac{\sin(\theta+\Delta\theta)-\sin\theta}{\Delta y}=\frac{u_y(x,y+\Delta y,t)-u_y(x,y,t)}{\Delta y}$$

$$\text{which} \longrightarrow u_{yy}=\frac{\partial^2 u}{\partial y^2} \text{ as } \Delta y \longrightarrow 0. \tag{11.64}$$

Again, from equations (11.62) and (11.63), we have

$$\frac{\sin(\phi+\Delta\phi)-\sin\phi}{\Delta x}=\frac{u_x(x+\Delta x,y,t)-u_x(x,y,t)}{\Delta x}$$

$$\text{which} \longrightarrow u_{xx}=\frac{\partial^2 u}{\partial x^2} \text{ as } \Delta x \longrightarrow 0. \tag{11.65}$$

Thus, in the limit $\Delta x \longrightarrow 0$ and $\Delta y \longrightarrow 0$, the equation (11.59) becomes

$$T\left(\frac{\partial^2 u}{\partial x^2}+\frac{\partial^2 u}{\partial y^2}\right)=m\frac{\partial^2 u}{\partial t^2} \quad \text{or,} \quad \frac{\partial^2 u}{\partial t^2}=a^2\left(\frac{\partial^2 u}{\partial x^2}+\frac{\partial^2 u}{\partial y^2}\right), \tag{11.66}$$

where

$$a=\sqrt{\frac{T}{m}}. \tag{11.67}$$

Equation (11.67) is the two-dimensional wave equation which describes free vibration of a membrane.

If an additional force $F(x,t)$ per unit area acts on the surface of membrane vertically downwards, then equation (11.58) will contain an additional term $F(x,t)\Delta x\Delta y$ in the right-hand side. This will give an additional term $F(x,t)$ in the right-hand side of equation (11.59). Thus, equation (11.66) becomes

$$\frac{\partial^2 u}{\partial t^2}=a^2\left(\frac{\partial^2 u}{\partial x^2}+\frac{\partial^2 u}{\partial y^2}\right)+\frac{F(x,t)}{m}, \tag{11.68}$$

which is the equation for forced vibration of membranes.

MAXWELL'S EQUATIONS

11.10 Introduction

Maxwell's equations of electromagnetism consist of a set of four differential equations which summarise all the basic laws of electricity and magnetism established by Gauss, Ampere and Faraday. These equations are:

(i) $\boldsymbol{\Delta}\cdot \boldsymbol{D}=4\pi\rho$, where **D** is the electric displacement and ρ is the charge density;
(ii) $\boldsymbol{\Delta}\cdot \boldsymbol{B}=0$, where **B** is the magnetic induction of the medium;
(iii) $\boldsymbol{\Delta}\times \boldsymbol{E}=-\dfrac{1}{c}\dfrac{\partial \mathbf{B}}{\partial t}$, where **E** is the electric field and c is the velocity of light in vacuum;
(iv) $\boldsymbol{\Delta}\times \boldsymbol{H}=\dfrac{4\pi}{c}\,\mathbf{J}+\dfrac{1}{c}\dfrac{\partial \mathbf{D}}{\partial t}$, where **J** is the current density and **H** is the magnetic field accompanying a current of electricity.

Maxwell's major contribution was to introduce the concept of displacement current and amend the Ampere's law which is valid for steady-state phenomena. The term $\frac{1}{c}\frac{\partial \mathbf{D}}{\partial t}$ in equation (iv) was added by Maxwell to make the system of four equations consistent for time-dependent phenomena. Maxwell showed that these equations can be reduced to wave equation having solutions in vacuum. He thus predicted that light is an electro-magnetic phenomena, and that electromagnetic waves of all frequencies can be produced.

We now intend to discuss briefly the equations of electrodynamics.

11.11 Basic Equations of Electrostatics

The electric field at a point x for a continuous charge distribution $\rho(\mathbf{x}')$ is given by

$$\mathbf{E}(\mathbf{x}) = \int \rho(\mathbf{x}') \frac{(\mathbf{x}-\mathbf{x}')}{|\mathbf{x}-\mathbf{x}'|^3}\, d^3x', \tag{11.69}$$

where $d^3x' = dx_1'\, dx_2'\, dx_3'$, is the volume element in source space. The unprimed x refers to a field point and primed x' to a source point.

Since

$$\frac{(\mathbf{x}-\mathbf{x}')}{|\mathbf{x}-\mathbf{x}'|^3} = -\boldsymbol{\nabla}\left(\frac{1}{|\mathbf{x}-\mathbf{x}'|}\right), \tag{11.70}$$

we write equation (11.69) in the form

$$\mathbf{E}(\mathbf{x}) = -\nabla \int \frac{\rho(\mathbf{x}')}{|\mathbf{x}-\mathbf{x}'|}\, d^3x'. \tag{11.71}$$

Equation (11.71) yields $\boldsymbol{\nabla} \times \mathbf{E}(\mathbf{x}) = 0$, since the curl of the grad of any scalar function is zero.

A static electric field $\mathbf{E}$ is described by the following two equations

$$\boldsymbol{\nabla} \cdot \mathbf{E} = 4\pi\rho \tag{11.72}$$

$$\boldsymbol{\nabla} \times \mathbf{E} = 0. \tag{11.73}$$

The first equation is the result of Gauss's theorem and the second one is proved above.

Equation (11.71) suggests that

$$\mathbf{E}(\mathbf{x}) = -\boldsymbol{\nabla}\phi \tag{11.74}$$

where

$$\phi = \int \rho(\mathbf{x}') \frac{d^3x'}{|\mathbf{x}-\mathbf{x}'|}. \tag{11.75}$$

Here ϕ is a scalar potential function. Putting the value of $\mathbf{E}$ from equation (11.74) into (11.72), we get

$$\nabla^2\phi = -4\pi\rho \tag{11.76}$$

which is a Poisson's equation. In free space, $\rho = 0$, and hence, we have

$$\nabla^2\phi = 0, \tag{11.77}$$

which is a Laplace's equation. Thus, electric potential satisfies Poisson's equation at the point where there is charge distribution, and satisfies Laplace's equation in free space. We had similar results in the case of gravitational potential also.

Let us now seek the modifications of electrostatic equations due to the presence of a dielectric material in the field.

When a dielectric substance is placed in an external electric field, its charged entities like nucleus and electrons are displaced in opposite directions from their mean positions. The dielectric substance is said to be *polarised.* The charges developed due to polarisation form an electric dipole.

We define the *polarisation vector* **P** *as the electric dipole moment per unit volume.*

It can be proved that the potential of a polarised substance consists of two parts: (i) potential produced by surface charge density and (ii) potential produced by a charge of $-\boldsymbol{\nabla}\cdot\mathbf{P}$ per unit volume inside the material. Hence, the total charge density of the polarised material is given by

$$\rho_t = \rho + \rho_p, \tag{11.78}$$

where ρ is the charge density of non-polarised substance called *true charges,* and ρ_p is the density of polarisation charges given by

$$\rho_p = -\boldsymbol{\nabla}\cdot\mathbf{P}(\mathbf{x}). \tag{11.79}$$

Equation (11.72) with the use of equation (11.78) becomes

$$\boldsymbol{\nabla}\cdot\mathbf{E} = 4\pi(\rho+\rho_p) = 4\pi\rho - 4\pi\boldsymbol{\nabla}\cdot\mathbf{P}$$

i.e.,

$$\boldsymbol{\nabla}\cdot(\mathbf{E}+4\pi\mathbf{P}) = 4\pi\rho. \tag{11.80}$$

Maxwell defined electric displacement **D** as

$$\mathbf{D} = \mathbf{E} + 4\pi\mathbf{P}. \tag{11.81}$$

Thus, equation (11.80) is written as

$$\boldsymbol{\nabla}\cdot\mathbf{D} = 4\pi\rho, \tag{11.82}$$

which is Gauss's law for dielectric media.

If field strengths are not strong, experiments have confirmed that

$$\mathbf{P} = \chi_e\mathbf{E},$$

where χ_e is the electrical susceptibility of the material.

In view of this, we write equation (11.81) as

$$\mathbf{D} = \epsilon\mathbf{E}, \tag{11.83}$$

where, $\epsilon = 1 + 4\pi\chi_e$ is defined as dielectric constant of the medium.

11.12 Basic Equation of Magnetostatics

The steady-state phenomena of magnetostatics are governed by the equations

$$\boldsymbol{\nabla}\times\mathbf{B} = \frac{4\pi}{c}\mathbf{J} \tag{11.84}$$

$$\boldsymbol{\nabla}\cdot\mathbf{B} = 0, \tag{11.85}$$

where **B** is the magnetic induction of the medium and **J** is the surface current density. It is also referred to as the flux of charge $\rho\mathbf{v}$, where **v** is the velocity vector of charges.

Equation (11.85) permits to write

$$\mathbf{B} = \nabla \times \mathbf{A}(\mathbf{x}), \tag{11.86}$$

where $\mathbf{A}(\mathbf{x})$ is a vector potential function, called *magnetic vector potential.*

We put equation (11.86) into (11.84), and get

$$\begin{aligned} \nabla \times \nabla \times \mathbf{A} &= \frac{4\pi}{c}\mathbf{J} \\ \text{i.e.,} \quad \nabla(\nabla \cdot \mathbf{A}) - \nabla^2 \mathbf{A} &= \frac{4\pi}{c}\mathbf{J} \\ \text{i.e.,} \quad \nabla^2 \mathbf{A} &= -\frac{4\pi}{c}\mathbf{J}. \end{aligned} \tag{11.87}$$

Thus, the vector potential function $\mathbf{A}(\mathbf{x})$ satisfies Poisson's equation.

Now, we know that matter is a collection of atoms and molecules having positive and negative charges but on the whole electrically neutral. Some of these charges are in continued motion circulating inside the matter. We call it bound current. If the bound current density is $\mathbf{J}_a$ and the true current density due to magnetic induction **B** is **J**, then the total current density becomes

$$\mathbf{J}_t = \mathbf{J}_a + \mathbf{J}. \tag{11.88}$$

The bound current in the matter forms current whirls or magnetic dipoles. When the material is placed in an external magnetic field, these dipoles align themselves in the direction of the induction vector **B**. The material is then said to be magnetised.

If $d\mathbf{m}$ is the dipole moment in a small volume $d\tau$ of the material, then we define the macroscopic vector magnetisation **M** as

$$\mathbf{M} = \frac{d\mathbf{m}}{d\tau}. \tag{11.89}$$

It can be proved that the bound current density $\mathbf{J}_a$ is given by

$$\mathbf{J}_a = c\nabla \times \mathbf{M}. \tag{11.90}$$

Thus, with the use of equations (11.88) and (11.90) the equation (11.84) becomes

$$\begin{aligned} \nabla \times \mathbf{B} &= \frac{4\pi}{c}(J_a + \mathbf{J}) \\ \text{or,} \quad \nabla \times \mathbf{B} &= \frac{4\pi}{c}(c\nabla \times \mathbf{M} + \mathbf{J}) \\ \text{or,} \quad \nabla \times (\mathbf{B} - 4\pi\mathbf{M}) &= \frac{4\pi}{c}\mathbf{J}. \end{aligned} \tag{11.91}$$

We define a new fields **H** called the *magnetic intensity vector* as

$$\mathbf{H} = \mathbf{B} - 4\pi\mathbf{M} \tag{11.92}$$

Then, we get from equation (11.91)

$$\nabla \times \mathbf{H} = \frac{4\pi}{c}\mathbf{J}. \tag{11.93}$$

For paramagnetic and diamagnetic materials, it has been experimentally found that

$$\mathbf{M} = \chi_m \mathbf{H}, \tag{11.94}$$

where χ_m is magnetic susceptibility.

Hence, from equation (11.92), we can write

$$\mathbf{B} = \mathbf{H} + 4\pi\mathbf{M} = \mathbf{H} + 4\pi\chi_m\mathbf{H}, \quad \text{i.e.,} \quad \mathbf{B} = \mu\mathbf{H}, \tag{11.95}$$

where, $\mu = 1 + 4\pi\chi_m$ is the magnetic permeability of the material. For free space, $\mu = 1$.

11.13 Equation of Continuity

It has been established experimentally that the total charge of any closed system remains constant. This is known as the *law of conservation of charge*, and is one of the basic laws of electrodynamics.

We consider a closed surface S surrounding a volume V. If ρ is the charge density at a point within S, then the total quantity of charge inside the surface S is given by

$$\mathbf{Q} = \int_V \rho dV. \tag{11.96}$$

We assume that there is movement of charge through the surface enclosing the volume. The the change in the amount of charge during time dt is

$$d\mathbf{Q} = dt\frac{d\mathbf{Q}}{dt} = -dt\int_V \frac{\partial\rho}{\partial t}dV. \tag{11.97}$$

Again, by definition, the current density is

$$\mathbf{J} = \rho\mathbf{v}, \tag{11.98}$$

where $\mathbf{v}$ is the velocity of charge at a point where the charge density is ρ. Then the charge passing through the surface in time dt is

$$d\mathbf{Q} = dt\int_S \mathbf{J}\cdot\mathbf{n}dS. \tag{11.99}$$

From the law of conservation of charge, the amount of charge flowing out of surface must be equal to the amount of charge flowing in. Hence, from equations (11.97) and (11.99)

$$\begin{aligned}
&\int_S \mathbf{J}\cdot\mathbf{n}dS &&= -\int_V \frac{\partial\rho}{\partial t}dV\\
\text{i.e.,}\quad &\int_V \nabla\cdot\mathbf{J}\,dV &&= -\int_V \frac{\partial\rho}{\partial t}dV \quad \text{(using divergence theorem)}\\
\text{i.e.,}\quad &\int_V \left(\nabla\cdot\mathbf{J} + \frac{\partial\rho}{\partial t}\right) dV = 0.
\end{aligned}$$

Since the volume V is arbitrary, we write

$$\nabla\cdot\mathbf{J} + \frac{\partial\rho}{\partial t} = 0, \tag{11.100}$$

which is the equation of continuity.

11.14 Derivation of Maxwell's Equations

We now shall derive Maxwell's equations.

(a) *Maxwell's first equation*: We consider a continuous distribution of charge of density ρ throughout a volume V. Then, by Gauss's law of electrostatics

$$\int_S \mathbf{D}\cdot\mathbf{n}dS = 4\pi\int_V \rho\, dV$$

$$\text{i.e.,}\quad \int_V \nabla\cdot\mathbf{D}\, dV = 4\pi\int_V \rho\, dV \quad \text{(using divergence theorem).}$$

Since the volume V is arbitrary, we get

$$\nabla\cdot\mathbf{D} = 4\pi\rho. \tag{11.101}$$

(b) *Maxwell's second equation*: The magnetic lines of force are either closed or move to infinity. Mathematically, this means that the surface integral of magnetic induction $\mathbf{B}$ over a closed surface should vanish, i.e.,

$$\int_S \mathbf{B}\cdot\mathbf{n}\, dS = 0$$

$$\text{i.e.,}\quad \int_V \nabla\cdot\mathbf{B}\, dV = 0 \quad \text{(by divergence theorem).}$$

Since the volume dV is arbitrary, we get

$$\nabla\cdot\mathbf{B} = 0. \tag{11.102}$$

(c) *Maxwell's third equation*: We consider a circuit C bounded by an open surface S with unit normal $\mathbf{n}$. The circuit is moving in some direction so that transient current is induced in it. If the magnetic induction in the neighbourhood of the circuit is $\mathbf{B}$, then the magnetic flux linking the circuit is, by definition

$$\phi = \int_S \mathbf{B}\cdot\mathbf{n}\, dS. \tag{11.103}$$

The electromotive force around the circuit is

$$\epsilon = \int_C \left(\mathbf{E} + \frac{\mathbf{v}}{c}\times\mathbf{B}\right) d\mathbf{l}, \tag{11.104}$$

where $\mathbf{v}$ is the velocity of a given point on the circuit $\mathbf{C}$ and $\mathbf{E}$ is the local electric field.

According to Faraday

$$\epsilon = -\frac{1}{c}\frac{\partial\phi}{\partial t}. \tag{11.105}$$

Hence, from equations (11.104) and (11.105), we get

$$\int_C \left(\mathbf{E} + \frac{\mathbf{v}}{c}\times\mathbf{B}\right)\cdot d\mathbf{l} = -\frac{1}{c}\frac{\partial}{\partial t}\int_S \mathbf{B}\cdot\mathbf{n}\, dS$$

$$\text{or,}\quad \int_S (\nabla\times\mathbf{E})\cdot\mathbf{n}\, dS = -\frac{1}{c}\int_S \frac{\partial\mathbf{B}}{\partial t}\cdot\mathbf{n}\, dS \quad \text{(using Stokes' theorem)}$$

$$\text{i.e.,}\quad \int_S \left(\nabla\times\mathbf{E} + \frac{1}{c}\frac{\partial\mathbf{B}}{\partial t}\right)\cdot\mathbf{n}\, dS = 0.$$

Here we have used the identity

$$\mathbf{v} \times \mathbf{B} \cdot d\mathbf{l} = -(\mathbf{v} \times d\mathbf{l}) \cdot \mathbf{B} = 0$$

since $\mathbf{v}$ and $d\mathbf{l}$ are parallel.

Since the circuit C and its bounding surface S are arbitrary, we get

$$\nabla \times \mathbf{E} + \frac{1}{c}\frac{\partial \mathbf{B}}{\partial t} = 0. \tag{11.106}$$

(d) *Maxwell's fourth equation*: From equation (11.101), we have

$$\rho = \frac{1}{4\pi}\nabla \cdot \mathbf{D}.$$

Hence

$$\frac{\partial \rho}{\partial t} = \frac{1}{4\pi}\nabla \cdot \frac{\partial \mathbf{D}}{\partial t}. \tag{11.107}$$

We use equation (11.107) in the equation of continuity (11.100), and get

$$\nabla \cdot \left(\mathbf{J} + \frac{1}{4\pi}\frac{\partial \mathbf{D}}{\partial t}\right) = 0. \tag{11.108}$$

The vector

$$\frac{1}{4\pi}\frac{\partial \mathbf{D}}{\partial t} \tag{11.109}$$

is called the *displacement current density.* Maxwell assumed that this displacement current has magnetic action similar to the conduction current.

Now, as we know, for steady currents, the Ampere's law is [*see* equation (11.93)]

$$\nabla \times \mathbf{H} = \frac{4\pi}{c}\mathbf{J}. \tag{11.110}$$

For time-dependent phenomena, Maxwell assumed that the displacement current density given by equation (11.109) also satisfies equation (11.110). Thus, we replace

$$\mathbf{J} \longrightarrow \mathbf{J} + \frac{1}{4\pi}\frac{\partial \mathbf{D}}{\partial t}$$

in equation (11.110), and get

$$\nabla \times \mathbf{H} = \frac{4\pi}{c}\mathbf{J} + \frac{1}{c}\frac{\partial \mathbf{D}}{\partial t}. \tag{11.111}$$

Thus we get Maxwell's fourth equation. We write the set of Maxwell's equations

$$\nabla \cdot \mathbf{D} = 4\pi\rho \tag{11.112}$$

$$\nabla \cdot \mathbf{B} = 0 \tag{11.113}$$

$$\nabla \times \mathbf{E} + \frac{1}{c}\frac{\partial \mathbf{B}}{\partial t} = 0 \tag{11.114}$$

and

$$\nabla \times \mathbf{H} = \frac{4\pi}{c}\mathbf{J} + \frac{1}{c}\frac{\partial \mathbf{D}}{\partial t} \tag{11.115}$$

These equations are supplemented by other constitutive relations which, for linear, isotropic and homogeneous media, are

$$\left.\begin{aligned} \mathbf{D} &= \epsilon\mathbf{E} = \mathbf{E} + 4\pi\mathbf{P} \\ \mathbf{B} &= \mu\mathbf{H} = \mathbf{H} + 4\pi\mathbf{M} \\ \text{and}\quad \mathbf{J} &= \sigma\mathbf{E}, \end{aligned}\right\} \tag{11.116}$$

where σ = electric conductivity.

On a microscopic level, we consider fields **E** and **B** in place of **D** and **H**, so that Maxwell's microscopic equations are

$$\boldsymbol{\nabla}\cdot\mathbf{E} = 4\pi\rho \tag{11.117}$$

$$\boldsymbol{\nabla}\cdot\mathbf{B} = 0 \tag{11.118}$$

$$\boldsymbol{\nabla}\times\mathbf{E} + \frac{1}{c}\frac{\partial\mathbf{B}}{\partial t} = 0 \tag{11.119}$$

and

$$\boldsymbol{\nabla}\times\mathbf{B} = \frac{4\pi}{c}\mathbf{J} + \frac{1}{c}\frac{\partial\mathbf{E}}{\partial t}. \tag{11.120}$$

11.15 Vector and Scalar Potentials

Equation (11.118) permits us to define **B** in terms of a vector potential **A** :

$$\mathbf{B} = \boldsymbol{\nabla}\times\mathbf{A}. \tag{11.121}$$

Then equation (11.119) can be written as

$$\boldsymbol{\nabla}\times\left(\mathbf{E} + \frac{1}{c}\frac{\partial\mathbf{A}}{\partial t}\right) = 0. \tag{11.122}$$

Equation (11.122) means that we can define a scalar potential ϕ, such that

$$\mathbf{E} + \frac{1}{c}\frac{\partial\mathbf{A}}{\partial t} = -\boldsymbol{\nabla}\phi, \quad \text{i.e.,}\quad \mathbf{E} = -\boldsymbol{\nabla}\phi - \frac{1}{c}\frac{\partial\mathbf{A}}{\partial t}. \tag{11.123}$$

Thus, with the definition of **B** and **E** in terms of **A** and ϕ, two homogeneous equations of Maxwell, i.e., equations (11.118) and (11.119) are automatically satisfied.

Substituting the value of **E** from equation (11.123) into (11.117), we get

$$\boldsymbol{\nabla}\cdot\left(-\boldsymbol{\nabla}\phi - \frac{1}{c}\frac{\partial\mathbf{A}}{\partial t}\right) = 4\pi\rho, \quad \text{i.e.,}\quad \nabla^2\phi + \frac{1}{c}\frac{\partial}{\partial t}(\boldsymbol{\nabla}\cdot\mathbf{A}) = -4\pi\rho. \tag{11.124}$$

Again we substitute the value of **E** from (11.123) and **B** from (11.121) into equation (11.120) and get

$$\boldsymbol{\nabla}\times(\boldsymbol{\nabla}\times\mathbf{A}) = \frac{4\pi}{c}\mathbf{J} + \frac{1}{c}\frac{\partial}{\partial t}\left(-\boldsymbol{\nabla}\phi - \frac{1}{c}\frac{\partial\mathbf{A}}{\partial t}\right)$$

$$\text{i.e.,}\quad -\nabla^2\mathbf{A} + \boldsymbol{\nabla}(\boldsymbol{\nabla}\cdot\mathbf{A}) = \frac{4\pi}{c}\mathbf{J} - \frac{1}{c}\boldsymbol{\nabla}\frac{\partial\phi}{\partial t} - \frac{1}{c^2}\frac{\partial^2\mathbf{A}}{\partial t^2}$$

$$\text{i.e.,}\quad \nabla^2\mathbf{A} - \frac{1}{c^2}\frac{\partial^2\mathbf{A}}{\partial t^2} - \boldsymbol{\nabla}\left(\boldsymbol{\nabla}\cdot\mathbf{A} + \frac{1}{c}\frac{\partial\phi}{\partial t}\right) = -\frac{4\pi}{c}\mathbf{J}. \tag{11.125}$$

Thus, by choosing a vector potential **A** and a scalar potential ϕ, we have reduced the set of four Maxwell's equation to two, namely (11.124) and (11.125). Now, we make the choice of the set of potentials **A** and ϕ in such a manner that

$$\nabla \cdot \mathbf{A} + \frac{1}{c}\frac{\partial \phi}{\partial t} = 0. \tag{11.126}$$

This choice is known as *Lorentz Gauge.*

In view of the equation (11.126), equation (11.125) reduces to

$$\nabla^2 \mathbf{A} - \frac{1}{c^2}\frac{\partial \mathbf{A}}{\partial t^2} = -\frac{4\pi}{c}\mathbf{J}. \tag{11.127}$$

Thus, we finally get the following two uncoupled equations in terms of potentials $(\mathbf{A}, \phi)$ as

$$\left.\begin{aligned} \nabla^2 \phi - \frac{1}{c^2}\frac{\partial^2 \phi}{\partial t^2} &= -4\pi\rho \\ \nabla^2 \mathbf{A} - \frac{1}{c^2}\frac{\partial^2 \mathbf{A}}{\partial t^2} &= -\frac{4\pi}{c}\mathbf{J} \end{aligned}\right\} \tag{11.128}$$

which, along with equation (11.126), form a set of equations equivalent in all respect to Maxwell's equations.

11.16 Wave Equations from Maxwell's Equations

We now like to show that wave equations can be derived from the set of Maxwell's equations. We assume the medium to be homogeneous and isotropic.

We take curl of equation (11.119), and get

$$\begin{aligned} \nabla \times (\nabla \times \mathbf{E}) &= -\frac{1}{c}\nabla \times \frac{\partial \mathbf{B}}{\partial t} = -\frac{1}{c}\frac{\partial}{\partial t}(\nabla \times \mathbf{B}) \\ \text{or,}\quad \nabla \cdot (\nabla \cdot \mathbf{E}) - \nabla^2 \mathbf{E} &= -\frac{1}{c}\frac{\partial}{\partial t}(\nabla \times \mathbf{B}). \end{aligned} \tag{11.129}$$

We have, from equations (11.115) and (11.116)

$$\nabla \times \mathbf{B} = \mu \nabla \times \mathbf{H} = \frac{\mu}{c}\left(4\pi \mathbf{J} + \frac{\partial \mathbf{D}}{\partial t}\right). \tag{11.130}$$

Substituting $(\nabla \times \mathbf{B})$ from equation (11.130) into (11.129), we get

$$\nabla(\nabla \cdot \mathbf{E}) - \nabla^2 \mathbf{E} = -\frac{\mu}{c^2}\frac{\partial}{\partial t}\left(4\pi \mathbf{J} + \frac{\partial \mathbf{D}}{\partial t}\right). \tag{11.131}$$

In the absence of source, i.e., for $\rho = 0$, we have

$$\nabla \cdot \mathbf{D} = 0, \quad \text{i.e.,} \quad \nabla \cdot \mathbf{E} = 0. \tag{11.132}$$

In view of this equation (11.131) becomes

$$\nabla^2 \mathbf{E} - \frac{4\pi\mu\sigma}{c^2}\frac{\partial \mathbf{E}}{\partial t} - \frac{\mu\epsilon}{c^2}\frac{\partial^2 \mathbf{E}}{\partial t^2} = 0, \quad \text{i.e.,} \quad \nabla^2 \mathbf{E} - \frac{\mu\epsilon}{c^2}\frac{\partial^2 \mathbf{E}}{\partial t^2} = \frac{4\pi\mu\sigma}{c^2}\frac{\partial \mathbf{E}}{\partial t}. \tag{11.133}$$

Equation (11.133) is the wave equation for a homogeneous isotropic medium. A similar equation in terms of **B** can be derived by taking curl of equation (11.115). This equation is

$$\nabla^2\mathbf{B} - \frac{\mu\epsilon}{c^2}\frac{\partial^2\mathbf{B}}{\partial t^2} = \frac{4\pi\mu\sigma}{c^2}\frac{\partial\mathbf{B}}{\partial t}. \tag{11.134}$$

If the medium is isotropic dielectric, then $\sigma = 0$. In such a case, the equations (11.133) and (11.134) can be written in the form

$$\left(\nabla^2 - \frac{\mu\epsilon}{c^2}\frac{\partial^2}{\partial t^2}\right)\psi = 0, \tag{11.135}$$

where ψ represents either of **E** and **B**.

In Cartesian coordinate systems, each of the quantities **E** and **B** will have three components like (E_x, E_y, E_z) and (B_x, B_y, B_z). Hence, all these components will satisfy equation (11.135).

We substitute

$$v = \frac{c}{\sqrt{\epsilon\mu}} = \frac{c}{n}, \tag{11.136}$$

where v is called the *phase velocity* of the wave and n is the refractive index of the medium. Then we write equation (11.135) in the form

$$\left(\nabla^2 - \frac{1}{v^2}\frac{\partial^2}{\partial t^2}\right)\psi = 0. \tag{11.137}$$

Equation of the type (11.137) is called *wave equation* with which we got aquainted earlier. This wave equation has a non-zero solution which clarifies that electromagnetic waves can exist when there is no charge.

If the direction of propagation of wave is along x-axis, then the wave function ψ does not depend on other coordinates. In such a case, we get one-dimensional wave equation

$$\frac{\partial^2\psi}{\partial x^2} - \frac{1}{v^2}\frac{\partial^2\psi}{\partial t^2} = 0. \tag{11.138}$$

EXAMPLE 11.2. *Show that, in a homogeneous medium containing charges and currents*

$$\nabla^2\mathbf{E} - \mu\epsilon\frac{\partial^2\mathbf{E}}{\partial t^2} = \frac{1}{\epsilon}\nabla\rho + \mu\frac{\partial}{\partial t}\mathbf{J}$$

$$\textit{and}\quad \nabla^2\mathbf{H} - \mu\epsilon\frac{\partial^2\mathbf{H}}{\partial t^2} = -\nabla\times\mathbf{J}.$$

Solution: Maxwell's equation in MKS system with the use of equations (11.116) becomes

(i) $\nabla\cdot\mathbf{E} = \frac{\rho}{\epsilon}$; (iii) $\nabla\times\mathbf{E} = -\mu\frac{\partial\mathbf{H}}{\partial t}$;

(ii) $\nabla\cdot\mathbf{H} = 0$; (iv) $\nabla\times\mathbf{H} = \mathbf{J} + \epsilon\frac{\partial\mathbf{E}}{\partial t}$.

Taking curl on both sides of equation (iii)

$$\nabla(\nabla\cdot\mathbf{E}) - \nabla^2\mathbf{E} = -\mu\frac{\partial}{\partial t}(\nabla\times\mathbf{H}) = -\mu\frac{\partial}{\partial t}\left[\mathbf{J} + \epsilon\frac{\partial\mathbf{E}}{\partial t}\right] \quad \text{[using equation (iv)]}$$

$$\text{or,}\quad \frac{1}{\epsilon}\nabla\rho - \nabla^2\mathbf{E} = -\mu\frac{\partial\mathbf{J}}{\partial t} - \mu\epsilon\frac{\partial^2\mathbf{E}}{\partial t^2} \quad \text{[using equation (i)].}$$

Hence

$$\nabla^2\mathbf{E} - \mu\epsilon\frac{\partial^2\mathbf{E}}{\partial t^2} = \frac{1}{\epsilon}\nabla\rho + \mu\frac{\partial\mathbf{J}}{\partial t}.$$

Again, taking curl on both sides of equation (iv)

$$\nabla(\mathbf{\nabla}\cdot\mathbf{H}) - \nabla^2\mathbf{H} = \nabla\times\mathbf{J} + \epsilon\frac{\partial}{\partial t}(\mathbf{\nabla}\times\mathbf{E})$$

$$\text{or,} \quad -\nabla^2\mathbf{H} = \nabla\times\mathbf{J} + \epsilon\frac{\partial}{\partial t}\left(-\mu\frac{\partial\mathbf{H}}{\partial t}\right) \quad \text{[using equation (iii)]}$$

$$= \mathbf{\nabla}\times\mathbf{J} - \mu\epsilon\frac{\partial^2\mathbf{H}}{\partial t^2}.$$

Hence

$$\nabla^2\mathbf{H} - \mu\epsilon\frac{\partial^2\mathbf{H}}{\partial t^2} = -\nabla\times\mathbf{J} \quad \text{(proved).}$$

BIHARMONIC EQUATIONS

11.17 Definition

A differential equation of the form

$$\nabla^2\nabla^2\phi = 0, \tag{11.139}$$

where ϕ is a function of x, y, z, is called a *biharmonic equation.* The solution of the biharmonic equation (11.139) is called a *biharmonic function.*

Any harmonic function $\psi(x, y, z)$ satisfies the equation (11.139) and hence is a solution of the biharmonic equation. It can be easily proved that the functions

$$x\psi, y\psi, z\psi \tag{11.140}$$

also satisfy the biharmonic equation. For

$$\nabla^2\nabla^2(x\psi) = 2\nabla^2\left(\frac{\partial\psi}{\partial z}\right) = 0, \tag{11.141}$$

since $\frac{\partial\psi}{\partial z}$ will be a harmonic function.

Similarly, the functions

$$x\frac{\partial\psi}{\partial x}, \quad y\frac{\partial\psi}{\partial y}, \quad z\frac{\partial\psi}{\partial z} \tag{11.142}$$

are also biharmonic.

We now like to show that, a function of the form

$$\phi = r^2\psi, \quad r^2 = x^2 + y^2 + z^2 \tag{11.143}$$

is a solution of biharmonic equation if the function $\psi(x, y, z)$ is harmonic.

Since $r^2 = x^2 + y^2 + z^2$,

$$2r\frac{\partial r}{\partial x} = 2x, \quad 2r\frac{\partial r}{\partial y} = 2y, \quad 2r\frac{\partial r}{\partial z} = 2z. \tag{11.144}$$

Consequently

$$\frac{\partial}{\partial x}\left(r^2\psi\right)=2r\frac{\partial r}{\partial x}\psi+r^2\frac{\partial \psi}{\partial x}=2x\psi+r^2\frac{\partial \psi}{\partial x}$$

$$\begin{aligned}\frac{\partial^2}{\partial x^2}\left(r^2\psi\right)&=2\psi+2x\frac{\partial \psi}{\partial x}+r^2\frac{\partial^2 \psi}{\partial x^2}+2r\frac{\partial r}{\partial x}\frac{\partial \psi}{\partial x}\\&=2\psi+4x\frac{\partial \psi}{\partial x}+r^2\frac{\partial^2 \psi}{\partial x^2}.\end{aligned}$$

Similarly, we get

$$\frac{\partial^2}{\partial y^2}\left(r^2\psi\right)=2\psi+4y\frac{\partial \psi}{\partial y}+r^2\frac{\partial^2 \psi}{\partial y^2}$$

$$\frac{\partial^2}{\partial z^2}\left(r^2\psi\right)=2\psi+4z\frac{\partial \psi}{\partial z}+r^2\frac{\partial^2 \psi}{\partial z^2}.$$

Adding the last three equations

$$\begin{aligned}\nabla^2\left(r^2\psi\right)&=6\psi+4\left(x\frac{\partial \psi}{\partial x}+y\frac{\partial \psi}{\partial y}+z\frac{\partial \psi}{\partial z}\right)+r^2\nabla^2\psi\\&=6\psi+4\left(x\frac{\partial \psi}{\partial x}+y\frac{\partial \psi}{\partial y}+z\frac{\partial \psi}{\partial z}\right)\end{aligned}\tag{11.145}$$

since $\nabla^2\psi=0$.

If we take Laplace operator on both sides of equation (11.145), the right-hand side will vanish, since each of the functions within bracket is biharmonic and the function ψ is harmonic. Thus

$$\nabla^2\nabla^2\left(r^2\psi\right)=0.\tag{11.146}$$

Thus, we conclude that if $\psi(x, y, z)$ is a harmonic function, then the solution of biharmonic equation (11.139) can be taken in any of the following forms:

$$\left.\begin{aligned}&(1)\quad \phi=\psi+(Ax\psi+By\psi+Cz\psi)\\&(2)\quad \phi=\psi+\left(Ax\frac{\partial \psi}{\partial x}+By\frac{\partial \psi}{\partial y}+Cz\frac{\partial \psi}{\partial z}\right)\\&(2)\quad \phi=\psi+\left(Ar^2\frac{\partial \psi}{\partial x}+Br^2\frac{\partial \psi}{\partial y}+Cr^2\frac{\partial \psi}{\partial z}\right)\end{aligned}\right\},\tag{11.147}$$

where A, B, C are arbitrary constants.

The biharmonic equations play an important role in the theory of elasticity. We, however, like to look into the problem of bending of thin elastic plates. Before doing so, we have to know the basic equations of elasticity which are given below without any derivation.

11.18 Equations of Elasticity

Consider an elastic body with density ρ at a given point in a rectangular coordinate system.

Let X, Y, Z—components of body forces applied to unit mass along x, y, z coordinates;
σ_x, σ_y, σ_z—normal stresses on the areas normal to Ox, Oy, Oz axes;
τ_{xy}, τ_{yz}, τ_{zx}—shearing stresses on the same areas as above;
u, v, w—components of displacement of a point in a body along coordinate axes;
ϵ_{xx}, ϵ_{yy}, ϵ_{zz}—normal strains at a point in the body along coordinate axes;
and ϵ_{xy}, ϵ_{yz}, ϵ_{zx}—shearing strains at a point in the body in the planes Oxy, OYz, Ozx respectively.

Then we have the following basic equations of theory of elasticity:

1. *Navier's equations of equilibrium*:

$$\left.\begin{aligned}\frac{\partial \sigma_x}{\partial x}+\frac{\partial \tau_{xy}}{\partial y}+\frac{\partial \tau_{xz}}{\partial z}+\rho X=0\\ \frac{\partial \tau_{yx}}{\partial x}+\frac{\partial \sigma_y}{\partial y}+\frac{\partial \tau_{yz}}{\partial z}+\rho Y=0\\ \frac{\partial \tau_{zx}}{\partial x}+\frac{\partial \tau_{zy}}{\partial y}+\frac{\partial \sigma_z}{\partial z}+\rho Z=0\end{aligned}\right\} \tag{11.148}$$

with the condition that

$$\tau_{xy}=\tau_{yx}, \quad \tau_{xz}=\tau_{zx}, \quad \tau_{yz}=\tau_{zy}. \tag{11.149}$$

2. *Cauchy's geometrical equations*:

$$\left.\begin{aligned}\epsilon_{xx}=\frac{\partial u}{\partial x}, \quad \epsilon_{yy}=\frac{\partial v}{\partial y}, \quad \epsilon_{zz}=\frac{\partial w}{\partial z}\\ \epsilon_{xy}=\frac{\partial u}{\partial y}+\frac{\partial v}{\partial x}, \quad \epsilon_{yz}=\frac{\partial v}{\partial z}+\frac{\partial w}{\partial y}, \quad \epsilon_{zx}=\frac{\partial w}{\partial x}+\frac{\partial u}{\partial z}\end{aligned}\right\} \tag{11.150}$$

3. *Generalised Hook's laws*:

$$\left.\begin{aligned}\epsilon_{xx}=\frac{1}{E}\left[\sigma_x-\mu\left(\sigma_y+\sigma_z\right)\right]\\ \epsilon_{yy}=\frac{1}{E}\left[\sigma_y-\mu\left(\sigma_z+\sigma_x\right)\right]\\ \epsilon_{zz}=\frac{1}{E}\left[\sigma_z-\mu\left(\sigma_x+\sigma_y\right)\right]\\ \epsilon_{xy}=\frac{1}{G}\tau_{xy}, \quad \epsilon_{yz}=\frac{1}{G}\tau_{yz}, \quad \epsilon_{zx}=\frac{1}{G}\tau_{zx}\end{aligned}\right\} \tag{11.151}$$

where E is the Young's modulus and μ is the Poisson's ratio of the material of the body. Here

$$G=\frac{E}{2(1+\mu)}=\text{shear rigidity.}$$

11.19 Bending of Thin Plates

We make the following assumptions:

1. The deflections are small compared with the plate thickness.

2. Normal stresses σ_z and consequently the strains ϵ_{zz} of the plate in the direction of its thickness are neglected, i.e., $\sigma_z = 0$ and $\epsilon_{zz} = 0$.
3. Plane cross-sections of a plate subjected to bending *remain plane* and normal to the deflected *middle plane.*
4. The points on the middle plane of the plate have no displacements along the plane, i.e., for them, at $z = 0$,

$$u = 0, \quad v = 0 \quad \text{and} \quad w = w(x, y). \tag{11.152}$$

In view of assumption (2), the generalized Hook's law for a plate takes the form

$$\left.\begin{aligned} \epsilon_{xx} &= \frac{1}{E}(\sigma_x - \mu\sigma_y) \\ \epsilon_{yy} &= \frac{1}{E}(\sigma_y - \mu\sigma_x) \\ \epsilon_{xy} &= \frac{2(1+\mu)}{E}\tau_{xy} \end{aligned}\right\}. \tag{11.153}$$

The deflected shape of the plate is shown in the *Fig 11.6*, taking x-axis through the middle plane and z-axis vertically downwards.

From the figure we get, the deflection of any point D along x-axis as

$$u = -z\sin\theta \simeq -z\tan\theta = -z\frac{\partial w}{\partial x}, \tag{11.154}$$

where θ is the angle of inclination of the tangent to the deflected middle plane. In equation (11.154), the minus sign appears, since for $z > 0$, $\tan\theta > 0$ but $u < 0$.

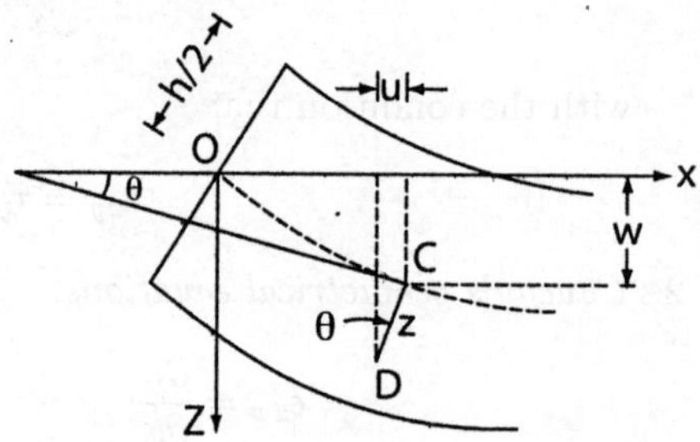

Fig 11.06

Similarly, considering a section of the plate by a plane parallel to Oyz, we can obtain

$$v = -z\frac{\partial w}{\partial y}. \tag{11.155}$$

With the use of equations (11.154) and (11.155), Cauchy's geometrical equations (11.150) assume the forms

$$\left.\begin{aligned} \epsilon_{xx} &= \frac{\partial u}{\partial x} &&= -z\frac{\partial^2 w}{\partial x^2} \\ \epsilon_{yy} &= \frac{\partial v}{\partial y} &&= -z\frac{\partial^2 w}{\partial y^2} \\ \epsilon_{xy} &= \frac{\partial u}{\partial y} + \frac{\partial v}{\partial x} &&= -2z\frac{\partial^2 w}{\partial x \partial y} \end{aligned}\right\}. \tag{11.156}$$

Now, solving equations (11.153) for stresses, we find

$$\left.\begin{aligned} \sigma_x &= \frac{E}{1-\mu^2}(\epsilon_{xx} + \mu\epsilon_{yy}) \\ \sigma_y &= \frac{E}{1-\mu^2}(\epsilon_{yy} + \mu\epsilon_{xx}) \\ \tau_{xy} &= \frac{E}{2(1-\mu^2)}(1-\mu)\epsilon_{xy} \end{aligned}\right\} \tag{11.157}$$

Substituting the values of ϵ_{xx}, ϵ_{yy} and ϵ_{xy} from equations (11.156) into equations (11.157), we express the stresses in terms of deflection function $w(x, y)$ in the following forms:

$$\left.\begin{aligned} \sigma_x &= -\frac{Ez}{1-\mu^2}\left(\frac{\partial^2 w}{\partial x^2} + \mu\frac{\partial^2 w}{\partial y^2}\right) \\ \sigma_y &= -\frac{Ez}{1-\mu^2}\left(\frac{\partial^2 w}{\partial y^2} + \mu\frac{\partial^2 w}{\partial x^2}\right) \\ \tau_{xy} &= -\frac{Ez}{1-\mu^2}(1-\mu)\frac{\partial^2 w}{\partial x \partial y} \end{aligned}\right\}. \tag{11.158}$$

We now find the shearing stresses τ_{xz}, τ_{yz} and also normal stress σ_z from equations of equilibrium. Assuming the load intensity function $q(x, y)$ acting on the upper surface $z = -\frac{h}{2}$ with no body forces, i.e., $X = Y = Z = 0$, we get from the first of equations (11.148)

$$\begin{aligned} \frac{\partial \tau_{xz}}{\partial z} &= -\frac{\partial \sigma_x}{\partial x} - \frac{\partial \tau_{xy}}{\partial y} = \frac{Ez}{1-\mu^2}\left[\frac{\partial^3 w}{\partial x^3} + \mu\frac{\partial^3 w}{\partial x \partial y^2} + (1-\mu)\frac{\partial^3 w}{\partial x \partial y^2}\right] \\ &= \frac{Ez}{1-\mu^2}\frac{\partial}{\partial x}\nabla^2 w. \end{aligned} \tag{11.159}$$

Similarly, from the second of equations (11.148), we find

$$\frac{\partial \tau_{yz}}{\partial z} = \frac{Ez}{1-\mu^2}\frac{\partial}{\partial y}\nabla^2 w. \tag{11.160}$$

Integrating equations (11.159) and (11.160) from $z = +\frac{h}{2}$ to $-\frac{h}{2}$, and remembering that

$$\tau_{xz} = \tau_{yz} = 0 \text{ for } z = \pm\frac{h}{2}, \tag{11.161}$$

we find

$$\left.\begin{aligned} \tau_{xz} &= \phi(z)\frac{\partial}{\partial x}\nabla^2 w \\ \tau_{yz} &= \phi(z)\frac{\partial}{\partial y}\nabla^2 w \end{aligned}\right\}, \tag{11.162}$$

where we have denoted

$$\phi(z) = \frac{E}{2(1-\mu^2)}\left(z^2 - \frac{h^2}{4}\right). \tag{11.163}$$

Now, from third of equations (11.148), we get

$$\frac{\partial \sigma_z}{\partial z} = -\frac{\partial \tau_{zx}}{\partial x} - \frac{\partial \tau_{zy}}{\partial y} = -\phi(z)\nabla^2\nabla^2 w. \tag{11.164}$$

Since

$$\int_{\frac{h}{2}}^{-\frac{h}{2}} \phi(z)\, dz == \frac{Eh^3}{12(1-\mu^2)}. \tag{11.165}$$

we get by integrating equation (11.164)

$$\sigma_z\left(\frac{h}{2}\right) - \sigma_z\left(-\frac{h}{2}\right) = \frac{Eh^3}{12(1-\mu^2)}\nabla^2\nabla^2 w. \tag{11.166}$$

Since the load is applied on the upper surface for which $z = -\frac{h}{2}$, then

$$\sigma_z\left(-\frac{h}{2}\right) = -q(x,y), \quad \sigma_z\left(\frac{h}{2}\right) = 0. \tag{11.167}$$

Hence, from equations (11.166) and (11.167), we get

$$\nabla^2\nabla^2 w = \frac{q(x,y)}{D}, \tag{11.168}$$

where

$$D = \frac{Eh^3}{12(1-\mu^2)}, \tag{11.169}$$

which is called the *flexural rigidity* of the plate.

Equation (11.168) is the differential equation of the plate in terms of its vertical deflection function $w(x, y)$ and is known as *equation of Sophie Germain* who first obtained it in 1815.

Bending and Torsional Moments

We isolate from the plate an element with dimension dx, dy, h. Consider a strip of width dz at depth z from the middle plane as shown in *Fig 11.7*.

The elemental strip is acted upon by forces $\sigma_x\, dy\, dz$ along x-axis and $\sigma_x\, dxdz$ along y-axis. The moments of these forces about y- and x-axes are

$$\left.\begin{aligned} M_x\, dy &= \int_{-\frac{h}{2}}^{\frac{h}{2}} \sigma_x\, z\, dy\, dz, \quad \text{i.e.,} \quad M_x = \int_{-\frac{h}{2}}^{\frac{h}{2}} \sigma_x\, z\, dz \\ M_y\, dx &= \int_{-\frac{h}{2}}^{\frac{h}{2}} \sigma_y\, z\, dx\, dz, \quad \text{i.e.,} \quad M_y = \int_{-\frac{h}{2}}^{\frac{h}{2}} \sigma_y\, z\, dz \end{aligned}\right\}. \tag{11.170}$$

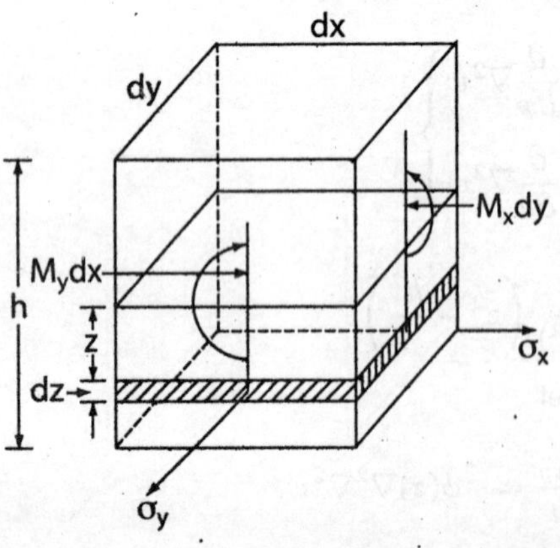

Fig 11.7

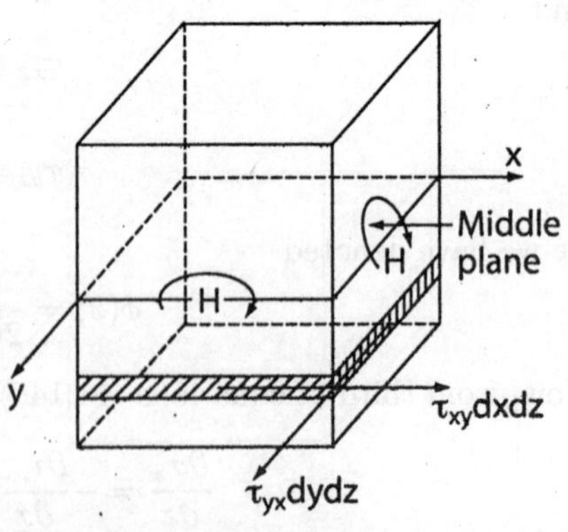

Fig 11.8

Substituting here the expressions for σ_x and σ_y from equation (11.158) and integrating, we get

$$\left.\begin{aligned} M_x &= -D\left(\frac{\partial^2 w}{\partial x^2} + \mu\frac{\partial^2 w}{\partial y^2}\right) \\ M_y &= -D\left(\frac{\partial^2 w}{\partial y^2} + \mu\frac{\partial^2 w}{\partial x^2}\right) \end{aligned}\right\}. \tag{11.171}$$

There will be shearing forces acting on the faces of elemental strip due to shear stresses τ_{xz}, τ_{yz} given by equations (11.162). The total shearing forces will be given by (*Fig 11.8*)

$$\left.\begin{aligned} N_x &= \int_{-\frac{h}{2}}^{\frac{h}{2}} \tau_{xz}\,dz \\ N_y &= \int_{-\frac{h}{2}}^{\frac{h}{2}} \tau_{yz}\,dz \end{aligned}\right\}. \tag{11.172}$$

Substituting here the expressions for shearing stresses from equations (11.162) and integrating, we find

$$\left.\begin{aligned} N_x &= -D\frac{\partial}{\partial x}\nabla^2 w \\ N_y &= -D\frac{\partial}{\partial y}\nabla^2 w \end{aligned}\right\}. \tag{11.173}$$

The shearing stresses $\tau_{xy} = \tau_{yx}$, as is seen in equation (11.158), vary directly as z and hence, the resultant of the system of elementary forces shown in *Fig 11.8* vanishes. They will reduce to a couple with moment

$$H = \int_{-\frac{h}{2}}^{\frac{h}{2}} \tau_{yx}\, z\,dz = -D(1-\mu)\frac{\partial^2 w}{\partial x\,\partial y}. \tag{11.174}$$

The shearing stresses τ_{xy} acting on the face normal to y-axis will also create a couple with moment $-H$ since $\tau_{xy} = \tau_{yx}$. These moments are called *twisting moments*. This is taken as positive when the rotation as seen from the side of positive x-axis is clockwise. Due care has been taken of this sign convention in getting equation (11.174).

We, therefore, conclude that the deflection function $w(x, y)$ of a thin elastic plate under the action of a load function $q(x, y)$ on its upper surface is determined by solving the differential equation (11.168) under suitable boundary conditions. After knowing the function $w(x, y)$, bending moments, shearing forces and twisting moment can be determined by using equations (11.171), (11.173) and (11.174) respectively.

Boundary Conditions

1. If the edge of a plate be hinged or simply supported, the deflection and the bending moment at the support should vanish. Suppose, the edge $x = a$ is simply supported. Then, for $x = a$,

$$w = 0 \quad \text{and} \quad M_x = -D\left(\frac{\partial^2 w}{\partial x^2} + \mu\frac{\partial^2 w}{\partial y^2}\right) = 0.$$

Since $w = 0$ along the edge $x = a$,

$$\frac{\partial w}{\partial y} \quad \text{and} \quad \frac{\partial^2 w}{\partial y^2}$$

also should vanish along that edge. Hence the boundary conditions reduce to

$$w = 0 \quad \text{and} \quad \frac{\partial^2 w}{\partial x^2} = 0 \text{ for } x = a. \tag{11.175}$$

Similarly, if the edge $y = b$ is simply-supported, the necessary boundary conditions will be

$$w = 0 \quad \text{and} \quad \frac{\partial^2 w}{\partial y^2} = 0 \text{ for } y = b. \tag{11.176}$$

2. If the edge $x = a$ is clamped, then the deflection and the angle of inclination of the tangent to the middle plane at the support should vanish. Thus

$$\text{for } x = a,\ w = 0 \quad \text{and} \quad \frac{\partial w}{\partial x} = 0. \tag{11.177}$$

Similarly, if the edge $y = b$ is clamped, then

$$w = 0 \quad \text{and} \quad \frac{\partial w}{\partial y} = 0 \text{ for } y = b. \tag{11.178}$$

3. If any edge, say $x = a$, is free, then the forces M_x, H, N_x will be zero. This being a difficult situation, Kirchhoff proposed the boundary conditions in the form

$$M_x = 0 \quad \text{and} \quad N_x + \frac{\partial H}{\partial y} = 0 \text{ for } x = a.$$

That is, for $x = a$

$$\left.\begin{aligned} \frac{\partial^2 w}{\partial x^2} + \mu \frac{\partial^2 w}{\partial y^2} &= 0 \\ \text{and} \quad \frac{\partial}{\partial x}\left[\frac{\partial^2 w}{\partial x^2} + (2-\mu)\frac{\partial^2 w}{\partial y^2}\right] &= 0 \end{aligned}\right\} \tag{11.179}$$

Vibration of Plates

In the case of free vibration of plates

$$q(x, y) = -\rho h \frac{\partial^2 w}{\partial t^2},$$

where ρ is the density of the plate material and t is the time. Then the basic equation (11.168) assumes the form

$$\nabla^2 \nabla^2 w = -\frac{\rho h}{D}\frac{\partial^2 w}{\partial t^2}. \tag{11.180}$$

In the case of forced vibration of plates under an external load $F(x, t)$ per unit area, the governing equation of motion is

$$\nabla^2 \nabla^2 w = -\frac{\rho h}{D}\frac{\partial^2 w}{\partial t^2} + F(x, t)$$

$$\text{or,} \quad \frac{\partial^2 w}{\partial t^2} + \frac{D}{\rho h}\nabla^2 \nabla^2 w = \frac{F(x, t)}{\rho h}. \tag{11.181}$$

11.20 Plane Problems of Elasticity

In many practical situations, equations of elasticity are simplified assuming that the whole phenomenon takes place in a plane. Such problems of elasticity are called *plane problems* which fall into two categories.

Consider a thin plate of thickness h loaded over lateral surface by forces parallel to its faces (*Fig 11.9*). Suppose that the width of the plate along OZ-axis is small. If the load is within the critical value, the plate will not deflect along OZ-axis. The faces of the plate will be free of stress. Thus

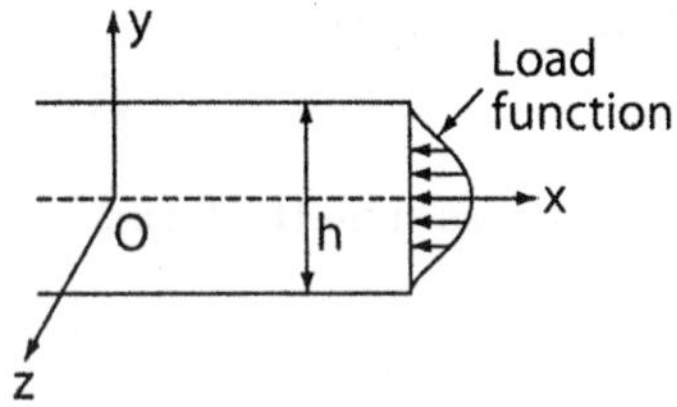

Fig 11.9

$$\sigma_z = \tau_{xz} = \tau_{yz} = 0. \tag{11.182}$$

Such problems are called *plane stress problems.* Since the thickness of the plate is small, the stresses in equation (11.182) are also neglected within the plate. But this may not be true in practice for the stress τ_{xz} and τ_{yz}, particularly when the load is non-uniform along the edge. In such cases the stresses are replaced by their average values across the thickness. Such problems are called *generalised plane stress problems.* The remaining stress components σ_x, τ_{xy} and σ_y vary slightly along OZ-axis and are taken to be independent of coordinate z. Thus, from equations (11.151), we find that for plane stress problem

$$\left.\begin{aligned}\epsilon_{xx} &= \frac{1}{E}(\sigma_x - \mu\sigma_y)\\ \epsilon_{xy} &= \frac{2(1+\mu)}{E}\tau_{xy}\\ \epsilon_{yy} &= \frac{1}{E}(\sigma_y - \mu\sigma_x)\\ \text{and}\quad \epsilon_{zz} &= -\frac{\mu}{E}(\sigma_x + \sigma_y)\end{aligned}\right\}. \tag{11.183}$$

The second group of plane problems arise when the strain ϵ_{zz} is zero. The displacements u, v do not depend on the coordinate z. Hence, from equations (11.150), we have

$$\epsilon_{yz} = \epsilon_{zx} = \epsilon_{zz} = 0. \tag{11.184}$$

These relations with the use of equations (11.151) yield

$$\sigma_z = \mu(\sigma_x + \sigma_y), \quad \tau_{xz} = \tau_{yz} = 0. \tag{11.185}$$

These kinds of problems are called *plane strain problems.* An example of such a problem is a long dam as shown in the *Fig 11.10*.

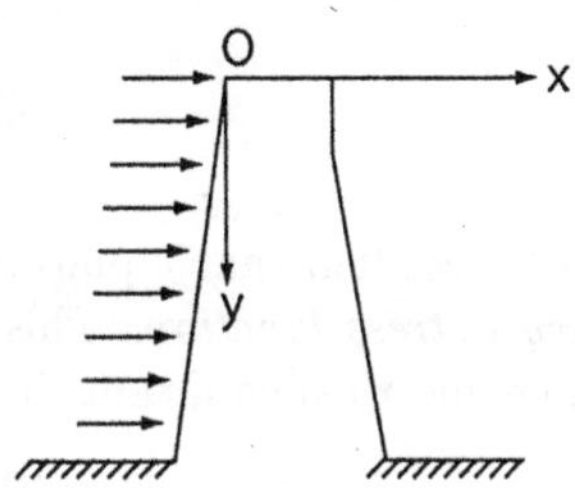

Fig 11.10

Equations (11.185) along with equations (11.151) yield

$$\begin{aligned}\epsilon_{xx} &= \frac{1}{E}\left[(1-\mu^2)\sigma_x - \mu(1+\mu)\sigma_y\right]\\ \epsilon_{yy} &= \frac{1}{E}\left[(1-\mu^2)\sigma_y - \mu(1+\mu)\sigma_x\right]\\ \epsilon_{xy} &= \frac{2(1+\mu)}{E}\tau_{xy}.\end{aligned}$$

With the notations

$$E_1 = \frac{E}{1-\mu^2} \quad \text{and} \quad \mu_1 = \frac{\mu}{1-\mu} \tag{11.186}$$

above relations assume the forms

$$\left.\begin{aligned} \epsilon_{xx} &= \frac{1}{E_1}(\sigma_x - \mu_1\sigma_y) \\ \epsilon_{yy} &= \frac{1}{E_1}(\sigma_y - \mu_1\sigma_x) \\ \epsilon_{xy} &= \frac{2(1+\mu_1)}{E_1}\tau_{xy}. \end{aligned}\right\}. \tag{11.187}$$

Thus the two problems can be considered to be one and the same, if we replace E by E_1 and μ by μ_1 in the case of plane strain problem.

We will consider the case when the body forces are constant throughout the volume of the body. In particular, taking y-axis vertically downward as in *Fig 11.10* and considering only gravity force, the components of body forces are

$$X = 0 \quad \text{and} \quad Y = g. \tag{11.188}$$

Then, equations (11.148) of equilibrium in plane problems become

$$\left.\begin{aligned} \frac{\partial \sigma_x}{\partial x} + \frac{\partial \tau_{xy}}{\partial y} &= 0 \\ \frac{\partial \tau_{xy}}{\partial x} + \frac{\partial \sigma_y}{\partial y} &= -\rho g \end{aligned}\right\}. \tag{11.189}$$

We introduce a function $\phi(x, y)$, such that

$$\left.\begin{aligned} \sigma_x = \frac{\partial^2 \phi}{\partial x^2}, \quad \sigma_y &= \frac{\partial^2 \phi}{\partial y^2} \\ \text{and} \quad \tau_{xy} &= -\frac{\partial^2 \phi}{\partial x \partial y} - \rho g x \end{aligned}\right\}. \tag{11.190}$$

It can be verified that equations (11.189) are identically satisfied. The function $\phi(x, y)$ is called *Airy's stress function.* This function is determined from the condition of compatibility. Cauchy's geometrical equations in this case are

$$\epsilon_{xx} = \frac{\partial u}{\partial x}, \quad \epsilon_{yy} = \frac{\partial v}{\partial y}, \quad \epsilon_{xy} = \frac{\partial u}{\partial y} + \frac{\partial v}{\partial x}. \tag{11.191}$$

It can easily be verified that

$$\frac{\partial^2}{\partial x^2}(\epsilon_{xx}) + \frac{\partial^2}{\partial y^2}(\epsilon_{yy}) - \frac{\partial^2}{\partial x \partial y}(\epsilon_{xy}) = 0. \tag{11.192}$$

Substituting the values of ϵ_{xx}, ϵ_{yy}, ϵ_{xy} from equation (11.183) in terms of stresses, equation (11.192) is reduced to

$$\nabla^2(\sigma_x + \sigma_y) = 0. \tag{11.193}$$

It may be noted that we have used the equation (11.189) of equilibrium in deriving equation (11.193). Substituting the expressions for σ_x and σ_y from equation (11.190) into equation (11.193), we find

$$\nabla^2\nabla^2\phi = 0. \tag{11.194}$$

Therefore, the stress function $\phi(x, y)$ satisfies the biharmonic equation. Hence, for solving any plane problem, we have to find the function $\phi(x, y)$ under certain boundary conditions and then to find the stresses by using equation (11.190).

SCHRÖDINGER'S EQUATION

11.21 Introduction

Physical phenomena like photo-electric effect, Compton effect, stable position of atom and many others could not be explained by laws of classical mechanics. This necessitated introduction of new ideas in the realm of atomic physics.

Max Planck introduced the idea of quantization of matter in 1900 to explain the observed energy distribution of black-body radiation. He assumed that radiation is a dynamical process of emission in the form of discrete packets called photons having energy

$$E = h\nu, \tag{11.195}$$

where ν = frequency and h = Planck's constant.

On the basis of Planck's quantum hypothesis, Niels Bohr proposed in 1913 a model of hydrogen atom. He assumed that, in an atom, only those orbits for which the relation

$$p = \frac{nh}{2\pi}, \quad n = 1, 2, \ldots \tag{11.196}$$

holds good are possible. Here p represents the angular momentum of the electron in the orbit.

By analogy with radiation, the French physicist Louis de Broglie predicted in 1924 the dual nature of matter. He assumed that every particle in motion behaves like a wave of definite wave length. Now, for a photon of zero rest mass, the angular momentum is

$$p = \frac{E}{c}, \quad \text{i.e.,} \quad E = pc. \tag{11.197}$$

From equations (11.195) and (11.197), we get

$$pc = h\nu = \frac{hc}{\lambda}, \quad \text{i.e.,} \quad \lambda = \frac{h}{p}, \tag{11.198}$$

where λ is the wavelength.

Thus, according to de Broglie, a particle of mass m moving with velocity v behaves like a wave of wavelength

$$\lambda = \frac{h}{p} = \frac{h}{mv}. \tag{11.199}$$

We use the notations

$$\hbar = \frac{h}{2\pi}, \quad k = \frac{2\pi}{\lambda}, \quad \omega = 2\pi\nu \tag{11.200}$$

and write equations (11.195) and (11.198) as

$$\left.\begin{aligned} E &= h\nu = 2\pi\hbar\nu = \hbar\omega \\ \text{and} \quad p &= \frac{h}{\lambda} = \frac{2\pi\hbar}{\lambda} = \hbar k \end{aligned}\right\}, \tag{11.201}$$

where ω = angular frequency and k = magnitude of wave vector.

The wave character of electrons was experimentally demonstrated by American physicists C.J. Davission and L.H. Germer in 1927.

The concepts of de Broglie were extended by Erwin Schrödinger in 1926 in a more general way to develop a wave equation for describing the motion of quantum systems. The space-time behaviour of such systems are determined by the laws of probability.

The basic characteristics of waves and particles are different. Because, a wave of finite wave length and frequency extends in infinite space and for infinite duration; whereas a particle is located at a definite point in space at any given instant of time. In order to bring about a compatability between these two, we have to localize the wave in a finite region of space by the superposition of waves of different wavelengths. In doing, so, we have introduced two kinds of velocity: *phase velocity* and *group velocity.* The phase velocity v_p of a wave is the velocity with which a crest or a through of the wave is propagated in a medium. It is determined by the equation

$$v_p = \frac{\omega}{k} = \lambda\nu. \tag{11.202}$$

In the case of light wave, the value of the phase velocity may even exceed the velocity of light in vacuum.

The velocity with which the resultant of the waves moves in the form a group of waves is called *group velocity.* It is determined by

$$v_g = \frac{d\omega}{dk}. \tag{11.203}$$

The value of the group velocity is always less than that of light in vacuum.

De Broglie proposed that the particle velocity v should be taken equal to the group velocity v_g of the associated wave.

11.22 Derivation of Equation

A differential equation for the wave associated with a particle in motion cannot be developed from first principles. We shall develop it from the classical wave equation with the use of de Broglie's wave properties of matter.

Let us consider de Broglie waves associated with a particle in motion such that the function $\psi(x, y, z, t)$, called *wave function,* represents the displacements of these waves at any time t and at any position (x, y, z) in space. We are already aquainted with classical waves, so that the function $\psi(x, y, z, t)$ satisfies the differential equation

$$\nabla^2\psi = \frac{1}{v^2}\frac{\partial^2\psi}{\partial t^2}, \tag{11.204}$$

where v is the velocity.

The amplitude of the wave at any point (x, y, z) is assumed to be periodic in t. Then, we can assume the solution in the form

$$\psi(x, y, z, t) = \phi(x, y, z)\, e^{-i\omega t}. \tag{11.205}$$

Substituting equation (11.205) into (11.204), we get

$$\nabla^2\phi = -\frac{\omega^2}{v^2}\phi. \tag{11.206}$$

Since, we have

$$\frac{\omega}{v} = \frac{2\pi\nu}{\lambda\nu} = \frac{2\pi}{\lambda} \tag{11.207}$$

equation (11.206) can be written as

$$\nabla^2\phi + \frac{4\pi^2}{\lambda^2}\phi = 0. \tag{11.208}$$

Now, the wavelength of the waves associated with the particle is

$$\lambda = \frac{h}{p} = \frac{h}{mv}. \tag{11.209}$$

Using this relation, we write equation (11.208) as

$$\nabla^2\phi + \frac{4\pi^2 m^2 v^2}{h^2}\phi = 0. \tag{11.210}$$

Now, let E = total energy of the particle and V = potential energy of the particle. Then, kinetic energy of the particle is given by

$$\tfrac{1}{2}mv^2 = E - V. \tag{11.211}$$

This gives

$$m^2v^2 = 2m(E - V) \tag{11.212}$$

In view of this, equation (11.210) becomes

$$\nabla^2\phi + \frac{8\pi^2 m}{h^2}\phi(E - V) = 0$$

$$\text{or,}\quad \nabla^2\phi + \frac{2m}{\hbar^2}\phi(E - V) = 0 \tag{11.213}$$

(since $\hbar = \dfrac{h}{2\pi}$).

Equation (11.213) is Schrödinger's time independent equation. We shall now find the differential equation which should be satisfied by the wave function $\psi(x, y, z, t)$. We multiply equation (11.213) by $e^{-i\omega t}$, and get

$$-(\nabla^2\phi)e^{-i\omega t} + \frac{2m}{\hbar^2}V\phi e^{-i\omega t} = \frac{2m}{\hbar^2}E\phi e^{-i\omega t}. \tag{11.214}$$

Now,

$$(\nabla^2\phi)e^{-i\omega t} = \nabla^2(\phi e^{-i\omega t}) = \nabla^2\psi \tag{11.215}$$

and

$$E\phi e^{-i\omega t} = E\psi = \frac{E}{-i\omega}\frac{\partial}{\partial t}(\psi) = \frac{ih}{2\pi}\frac{\partial\psi}{\partial t} = i\hbar\frac{\partial\psi}{\partial t} \tag{11.216}$$

(since $E = h\nu = \frac{\lambda}{2\pi}\omega$).

Using equations (11.215) and (11.216), we get from the equation (11.214)

$$-\nabla^2\psi + \frac{2m}{\hbar^2}V\psi = \frac{2m}{\hbar^2}\left(\frac{ih}{2\pi}\right)\frac{\partial\psi}{\partial t}$$

$$\text{or,}\quad \left(-\frac{\hbar^2}{2m}\nabla^2 + V\right)\psi = i\hbar\frac{\partial\psi}{\partial t}. \tag{11.217}$$

Equation (11.217) is Schrödinger's time-dependent equation.

The operator

$$\left(-\frac{\hbar^2}{2m}\nabla^2 + V\right) \tag{11.218}$$

is known as *Hamiltonian operator* and is denoted by $\hat{H}$. The operator

$$\hat{E} = i\hbar\frac{\partial}{\partial t} \tag{11.219}$$

is called the *energy operator.* Thus, the equation (11.217) can be written as

$$\hat{H}\psi = \hat{E}\psi. \tag{11.220}$$

If, in particular case, the potential energy V is independent of time, we can separate the three-dimensional Schrödinger equation (11.217) into time-dependent and time-independent parts. In this case, we assume

$$\psi(x,y,z,t) = \phi(x,y,z)\cdot f(t). \tag{11.221}$$

Substituting the function (11.221) into equation (11.217), we get

$$-\frac{\hbar^2}{2m}\left[\frac{\partial^2\phi}{\partial x^2} + \frac{\partial^2\phi}{\partial y^2} + \frac{\partial^2\phi}{\partial z^2}\right]f(t) + V\phi f = i\hbar\phi\frac{df}{dt}. \tag{11.222}$$

Dividing both sides by ϕf, we get

$$-\frac{\hbar^2}{2m}\left[\frac{\partial^2\phi}{\partial x^2} + \frac{\partial^2\phi}{\partial y^2} + \frac{\partial^2\phi}{\partial z^2}\right]\frac{1}{\phi} + V = i\hbar\frac{1}{f}\frac{df}{dt}. \tag{11.223}$$

Now, from equation (11.216), we have

$$E\psi = i\hbar\frac{\partial\psi}{\partial t}. \tag{11.224}$$

Substituting here the equation (11.221), we get

$$E\phi f = i\hbar\phi\frac{df}{dt}$$

which gives

$$E = i\hbar\frac{1}{f}\frac{df}{dt}. \tag{11.225}$$

It's solution is

$$f(t) = Ae^{-iEt/\hbar}. \tag{11.226}$$

This is the time dependent part of the solution of three-dimensional Schrödinger's equation. Now, in view of equation (11.225), equation (11.223) can be written as

$$-\frac{\hbar^2}{2m}\frac{1}{\phi}\nabla^2\phi + V = E \quad \text{i.e.,} \quad -\frac{\hbar^2}{2m}\nabla^2\phi + V\phi = E\phi$$

$$\text{or,} \quad \left(-\frac{\hbar^2}{2m}\nabla^2 + V\right)\phi = E\phi. \tag{11.227}$$

This is the three-dimensional time-independent Schrödinger's equation. It is also called three-dimensional steady state Schrödinger's equation.

11.23 Physical Significance of Wave Function

According to Max Born, the quantity

$$P = |\psi(x, y, z, t)|^2 \tag{11.228}$$

gives the probability of finding the particle surrounding the point (x, y, z) at time t, and is called the *probability density.* Since the probability of finding a particle in space at any time must be real, we take

$$\psi^*\psi \text{ for } |\psi|^2, \tag{11.229}$$

where ψ^* is complex conjugate of ψ. Thus, the probability that the particle will be found in a volume element $d\tau$ surrounding the point (x, y, z) at any time t is given by

$$Pd\tau = \psi^*\psi d\tau. \tag{11.230}$$

Since the particle is certainly to be found somewhere in space, the total probability must be unity. Hence

$$\int Pd\tau = \int \psi^*\psi d\tau = \int |\psi^2|\, d\tau = 1 \tag{11.231}$$

which means that the function $\psi(x, y, z, t)$ can be normalised. Any wave function satisfying the condition (11.231) is said to be normalisd. If the function ψ is not normalised in a particular case, we make it normalised by multiplying with a constant A, so that

$$\int |A\psi|^2\, d\tau = 1, \quad \text{i.e.,} \quad |A|^2 = \frac{1}{\int |\psi|^2\, d\tau} \tag{11.232}$$

which determines the unknown constant A.

11.24 Schrödinger's Equation for ψ^*

The wave function $\psi(x, y, z, t)$ being a complex one, we assume

$$\psi(x, y, z, t) = \phi_1(x, y, z, t) + i\phi_2(x, y, z, t) \tag{11.233}$$

where both ϕ_1, ϕ_2 are real functions.

We substitute the function in (11.233) into the Schrödinger's time-dependent equation (11.217). Then, we get

$$-\frac{\hbar^2}{2m}\nabla^2(\phi_1+i\phi_2)+V(\phi_1+i\phi_2)=i\hbar\frac{\partial}{\partial t}(\phi_1+i\phi_2)$$

$$\text{i.e.,}\quad \left(-\frac{\hbar^2}{2m}\nabla^2+V\right)\phi_1+i\left(-\frac{\hbar^2}{2m}\nabla^2+V\right)\phi_2=i\hbar\frac{\partial\phi_1}{\partial t}-\hbar\frac{\partial\phi_2}{\partial t}. \tag{11.234}$$

Equating real and imaginary parts, we get the following two equations

$$\left(-\frac{\hbar^2}{2m}\nabla^2+V\right)\phi_1=-\hbar\frac{\partial\phi_2}{\partial t} \tag{11.235}$$

$$\text{and}\quad \left(-\frac{\hbar^2}{2m}\nabla^2+V\right)\phi_2=\hbar\frac{\partial\phi_1}{\partial t}. \tag{11.236}$$

Now, the complex conjugate of ψ is

$$\psi^*=\phi_1(x,y,z,t)-i\phi_2(x,y,z,t). \tag{11.237}$$

We multiply equation (11.236) by $-i$ and add the result with equation (11.235). We obtain

$$\left(-\frac{\hbar^2}{2m}\nabla^2+V\right)(\phi_1-i\phi_2)=-i\hbar\frac{\partial}{\partial t}\left(\phi_1+\frac{\phi_2}{i}\right)=-i\hbar\frac{\partial}{\partial t}(\phi_1-i\phi_2)$$

$$\text{i.e.,}\quad \left(-\frac{\hbar^2}{2m}\nabla^2+V\right).\psi^*=-i\hbar\frac{\partial\psi^*}{\partial t} \tag{11.238}$$

which is the equation for $\psi^*(x,y,z,t)$.

11.25 Essential Conditions for Wave Functions

The wave function ψ must satisfy the following conditions so that it becomes physically meaningful.

1. It must be continuous and single-valued function.

Single-valuedness is required for avoiding any ambiguity in predictions. Continuity of the function is necessary, since the probability of finding the particle in a given volume element should vary continuously from point to point.

2. $\int|\psi|^2\,d\tau$ must be finite ($d\tau$ = volume element). This is necessary, since the probability of finding the particle in a given volume should be finite. Above integral is taken over all space.

3. The partial derivatives

$$\frac{\partial\psi}{\partial x},\ \frac{\partial\psi}{\partial y},\ \frac{\partial\psi}{\partial z}$$

should be continuous, single-valued and finite for all values of x,y,z,t.

This is required, since the above partial derivatives are related to the momenta of the particle. The momenta should be continous functions, except at a point where the potential energy is infinite.

4. The product $\psi\psi^*$, where ψ^* is the complex conjugate of ψ, is always a real quantity.

11.26 Probability Current Density

We consider a finite volume τ enclosed by a surface Ω. The total probability that a particle will be found at any time t within this volume of space is

$$\int_\tau P d\tau = \int_\tau \psi^* \psi d\tau.$$

Hence, the rate of change of this total probability is

$$\frac{\partial}{\partial t}\int_\tau P d\tau = \frac{\partial}{\partial t}\int_\tau \psi^* \psi d\tau = \int_\tau \left(\frac{\partial \psi^*}{\partial t}\psi + \psi^* \frac{\partial \psi}{\partial t}\right) d\tau. \tag{11.239}$$

Now, from equation (11.217) we have

$$\frac{\partial \psi}{\partial t} = \frac{1}{i\hbar}\left(-\frac{\hbar^2}{2m}\nabla^2 + V\right)\psi \tag{11.240}$$

and, from equation (11.238) we have

$$\frac{\partial \psi^*}{\partial t} = -\frac{1}{i\hbar}\left(-\frac{\hbar^2}{2m}\nabla^2 + V\right)\psi^*. \tag{11.241}$$

Hence,

$$\begin{aligned}\frac{\partial \psi^*}{\partial t}\psi + \psi^*\frac{\partial \psi}{\partial t} &= -\frac{1}{i\hbar}\left[\left(-\frac{\hbar^2}{2m}\nabla^2\psi^* + V\psi^*\right)\psi - \psi^*\left(-\frac{\hbar^2}{2m}\nabla^2\psi + V\psi\right)\right] \\ &= -\frac{i\hbar}{2m}\left(\psi\nabla^2\psi^* - \psi^*\nabla^2\psi\right).\end{aligned} \tag{11.242}$$

From equations (11.239) and (11.242) we get

$$\int_\tau \frac{\partial P}{\partial t} d\tau = -\frac{i\hbar}{2m}\int_\tau (\psi\nabla^2\psi^* - \psi^*\nabla^2\psi) d\tau. \tag{11.243}$$

Now, we know from vector analysis, that for two scalar functions ϕ and ψ, the following relations exist:

$$\left.\begin{aligned} \nabla\cdot(\phi\nabla\psi) &= \phi\nabla^2\psi + \nabla\phi\cdot\nabla\psi \\ \text{and}\quad \nabla\cdot(\psi\nabla\phi) &= \psi\nabla^2\phi + \nabla\phi\cdot\nabla\psi \end{aligned}\right\}. \tag{11.244}$$

From these equations, we get

$$\psi\nabla^2\phi - \phi\nabla^2\psi = \nabla\cdot(\psi\nabla\phi - \phi\nabla\psi). \tag{11.245}$$

Using the relation in (11.245), we get from equation (11.243)

$$\int_\tau \frac{\partial P}{\partial t} d\tau = -\frac{i\hbar}{2m}\int_\tau \nabla\cdot(\psi\nabla\psi^* - \psi^*\nabla\psi) d\tau. \tag{11.246}$$

We define the probability current density as

$$S = \frac{i\hbar}{2m}(\psi\nabla\psi^* - \psi^*\nabla\psi). \tag{11.247}$$

Then, equation (11.246) yields

$$\int_\tau \left(\frac{\partial P}{\partial t} + \nabla \cdot S \right) d\tau = 0. \tag{11.248}$$

Since the volume τ is arbitrary, the integrand must vanish. Hence,

$$\frac{\partial P}{\partial t} + \nabla \cdot S = 0. \tag{11.249}$$

Equation (11.249) reminds us of the equation of continuity in electrodynamics. Thus, this equation expresses the conservation of probability density in the same way as the equation of continuity expresses the conservation of charge density.

EXERCISE

1. Find the distribution of matter which produces a potential $\mu\,(x^2 + y^2 + z^2 - 2ax)$ at all points inside the sphere $x^2 + y^2 + z^2 - 2ax = 0$, and zero potential at all points outside it.
2. What distribution of matter will produce potential

$$\frac{M}{a}\left(1 - \frac{x}{3a}\right) \quad \text{or,} \quad \frac{M}{r}\left(1 - \frac{ax}{3r^2}\right).$$

   according as r is less or greater than a?
3. Show that the family of surfaces defined by $x^2 + y^2 =$ constant can be a family of equipotential surfaces in free space, and find the law of potential.
4. Derive the differential equation of vibration of a string subjected to gravity loading and viscous drag which is $c\dfrac{\partial u}{\partial t}$ per unit length of string.
5. Prove that, the torsional vibration of a circular shaft can be described by the equation

$$\frac{\partial^2 \theta}{\partial x^2} = \frac{\rho}{G}\frac{\partial^2 \theta}{\partial t^2},$$

   where $G =$ shear rigidity.
6. Starting from the beam equation, show that forced transverse vibration of a beam is described by

$$\frac{\partial^4 w}{\partial x^4} + \frac{\rho A}{\text{EI}}\frac{\partial^2 w}{\partial t^2} = \frac{F(x,t)}{\text{EI}},$$

   where EI $=$ flexural rigidity, $A =$ cross-sectional area of the beam, $F(x,t) =$ external load/unit length.
7. Derive the one-dimensional heat equation

$$\frac{\partial I}{\partial t} = k\frac{\partial^2 T}{\partial x^2}$$

   for a laterally insulated rod.
8. Show that, electromagnetic field given by

$$\mathbf{E} = [0, 0, \cos(y - ct)]$$
$$\text{and} \quad \mathbf{H} = [\cos(y - ct), 0, 0]$$

   satisfies Maxwell's equation.

9. Verify that Maxwell's equation for free space are satisfied by

$$\mathbf{B} = \frac{1}{c}\frac{\partial}{\partial t} \operatorname{grad} \phi \times \mathbf{k}$$

$$\mathbf{E} = \frac{\mathbf{k}}{c^2}\frac{\partial^2 \phi}{\partial t^2} + \frac{\partial}{\partial z} \operatorname{grad} \phi,$$

where $\mathbf{k}$ = unit vector along z-axis, and ϕ satisfies wave equation

$$\nabla^2 \phi - \frac{1}{c^2}\frac{\partial^2 \phi}{\partial t^2} = 0.$$

10. A line load p is uniformly distributed along the radius of a circular plate clamped along the boundary. Show that the problem reduces to solving the equation

$$\nabla^2 \nabla^2 w = \begin{cases} \dfrac{p}{\mathrm{RD}\epsilon} \\ \\ 0 \end{cases}$$

where ϵ is arbitrarily small and is to be defined with the boundary conditions

$$w = 0 \quad \text{and} \quad \frac{\partial w}{\partial r} = 0 \text{ for } r = R.$$

ANSWERS

1. $\rho = -\dfrac{3\mu}{2\pi}$ inside and $\rho = 0$ outside the sphere.

2. $\rho = 0$ for $r < a$ and $\rho = \dfrac{9Max}{4\pi}\left(\dfrac{1}{r} - \dfrac{1}{r^3}\right)$ for $r > 0$.

3. $V = A \log\left(x^2 + y^2\right) + B$; A, B are constants.

4. $\dfrac{\partial^2 u}{\partial x^2} = \dfrac{\gamma}{Tg}\dfrac{\partial^2 u}{\partial t^2} + \dfrac{c}{T}\dfrac{\partial u}{\partial t}$, γ = weight/unit length of the string.

Chapter 12

Special Functions

We have seen that the study of physical processes leads to various partial differential equations. These are reduced to ordinary differential equations during their solution process. While solving these ordinary differential equations, we come across many interesting functions each of which renders some speciality of its own. We call these functions as *Special functions.* We shall discuss some of them.

HYPERGEOMETRIC FUNCTIONS

12.1 General Solutions

We consider the differential equation

$$x(x-1)y'' + [(\alpha+\beta+1)x-\gamma]y' + \alpha\beta y = 0, \tag{12.1}$$

where α, β, γ are three parameters. We shall see later the necessary condition to be imposed on γ.

We find that, equation (12.1) can be written as

$$y'' + P(x)y' + Q(x)y = 0, \tag{12.2}$$

where

$$P(x) = \frac{(\sigma+\beta+1)x-\gamma}{x(x-1)} \tag{12.3}$$

and

$$Q(x) = \frac{\alpha\beta}{x(x-1)}. \tag{12.4}$$

This shows that, $x = 0$, $x = 1$ and $x = \infty$ are the singularities of the differential equations. We, therefore, seek the solution by series method around all the three singularities.

(A) *Solution around* $x = 0$: We assume the solution

$$y = \sum_{k=0}^{\infty} A_k x^{r+k}, \tag{12.5}$$

where r is a constant.

Substituting equation (12.5) along with its first and second derivatives in equation (12.1), we get

$$\sum_{k=0}^{\infty} A_k (r+k)(r+k-1)x^{r+k} - \sum_{k=0}^{\infty} A_k (r+k)(r+k-1)x^{r+k-1} + (\alpha+\beta+1)\sum_{k=0}^{\infty} A_k (r+k)x^{r+k} - \gamma\sum_{k=0}^{\infty} A_k (r+k)x^{r+k-1} + \alpha\beta\sum_{k=0}^{\infty} A_k x^{r+k} = 0. \tag{12.6}$$

Equating to zero the coefficient of x^{r-1}, we get the indicial equation

$$A_0 r(r-1) + \gamma A_0 r = 0,$$

which gives (since $A_0 \neq 0$)

$$r = 0 \quad \text{and} \quad r = 1-\gamma. \tag{12.7}$$

Putting $r = 0$ in equation (12.6), we get

$$\sum_{k=0}^{\infty} A_k k(k-1)x^k - \sum_{k=0}^{\infty} A_k k(k-1)x^{k-1} + (\alpha+\beta+1)\sum_{k=0}^{\infty} A_k k x^k - \gamma\sum_{k=0}^{\infty} A_k k x^{k-1} + \alpha\beta\sum_{k=0}^{\infty} A_k x^k = 0. \tag{12.8}$$

We equate to zero the coefficients of $x^0, x^1, x^2, \ldots$ etc., and get

$$\gamma A_1 - \alpha\beta A_0 = 0$$

$$2(\gamma+1)A_2 - (\alpha+1)(\beta+1)A_1 = 0$$

....

....

$$(k+1)(\gamma+k)A_{k+1} - (\alpha+k)(\beta+k)A_k = 0. \tag{12.9}$$

From equation (12.8), we get the recursion formula

$$A_{k+1} = \frac{(\alpha+k)(\beta+k)}{(k+1)(\gamma+k)} A_k. \tag{12.10}$$

From this recursion formula, we find the coefficients $A_1, A_2, \ldots$ (assuming $A_0 = 1$),

$$A_1 = \frac{\alpha\beta}{1\cdot\gamma}$$

$$A_2 = \frac{(\alpha+1)(\beta+1)}{2(\gamma+1)} A_1 = \frac{\alpha(\alpha+1)\beta(\beta+1)}{1\cdot 2\gamma(\gamma+1)}$$

....

....

$$A_{k+1} = \frac{(\alpha+k)(\beta+k)}{(k+1)(\gamma+k)} A_k = \frac{\alpha(\alpha+1)(\alpha+2)\cdots(\alpha+k)\beta(\beta+1)\cdots(\beta+k)}{1\cdot 2\cdots(k+1)\gamma(\gamma+1)(\gamma+2)\cdots(\gamma+k)}$$

We introduce the following notations of Pochhamber:

$$\left.\begin{aligned}(\alpha)_k &= \alpha(\alpha+1)(\alpha+2)\cdots(\alpha+k-1)\\ (\beta)_k &= \beta(\beta+1)(\beta+2)\cdots(\beta+k-1)\\ (\gamma)_k &= \gamma(\gamma+1)(\gamma+2)\cdots(\gamma+k-1)\end{aligned}\right\}. \tag{12.11}$$

It may be noted that, with the above notations, $\left(\frac{1}{2}\right)_5$ means

$$\tfrac{1}{2}\left(\tfrac{1}{2}+1\right)\left(\tfrac{1}{2}+2\right)\left(\tfrac{1}{2}+3\right)\left(\tfrac{1}{2}+4\right).$$

Hence

$$A_k = \frac{(\alpha)_k\,(\beta)_k}{k!\,(\gamma)_k}. \tag{12.12}$$

We write the particular solution for $r = 0$, as

$$y_1 = \sum_{k=0}^{\infty} \frac{(\alpha)_k\,(\beta)_k}{k!\,(\gamma)_k}\,x^k. \tag{12.13}$$

It is now obvious that γ should not be a negative integer. The series in equation (12.13) is known as *Hypergeometric series,* since for $\alpha = 1$ and $\beta = \gamma$, the series reduces to

$$y_1 = 1 + x + x^2 + \cdots = \frac{1}{1-x},$$

which is a geometric series. The hypergeometric series is convergent for $|x| < 1$, which is obvious from D'Alemberts ratio test.

The function

$$\sum_{k=0} \frac{(\alpha)_k\,(\beta_k)}{k!\,(\gamma)_k}\,x^k \tag{12.14}$$

is called *Hypergeometric function* and is denoted as $F(\alpha, \beta, \gamma; x)$. Thus, the solution for $r = 0$ is written as

$$y_1 = F(\alpha,\, \beta,\, \gamma;\, x). \tag{12.15}$$

Now, the solution for $r = 1 - \gamma$ is assumed as

$$y_2 = x^{1-\gamma} \sum_{k=0}^{\infty} A_k\, x^k. \tag{12.16}$$

We substitute the function (12.16) into the differential equation (12.1) and proceeding as before, we get the recursion formula,

$$A_{k+1} = \frac{(\alpha'+k)\,(\beta'+k)}{(k+1)\,(\gamma'+k)}\,A_k \tag{12.17}$$

where

$$\left.\begin{aligned}\alpha' &= \alpha-\gamma+1\\ \beta' &= \beta-\gamma+1\\ \text{and}\quad \gamma' &= 2-\gamma\end{aligned}\right\}. \tag{12.18}$$

Hence, the solution (12.16) is written as

$$\begin{aligned} y_2 &= x^{1-\gamma} F(\alpha', \beta', \gamma'; x) \\ &= x^{1-\gamma} F(\alpha-\gamma+1, \beta-\gamma+1, 2-\gamma; x). \end{aligned} \tag{12.19}$$

The solution (12.19) converges for $|x| < 1$ provided $2-\gamma \neq 0, -1, -2, \ldots$.

The two solutions y_1 and y_2 are independent so that, the solution of the equation (12.1) can be written as

$$\begin{aligned} y &= c_1 y_1 + c_2 y_2 \\ &= c_1 F(\alpha, \beta, \gamma; x) + c_2 x^{1-\gamma} F(\alpha-\gamma+1, \beta-\gamma+1, 2-\gamma; x). \end{aligned} \tag{12.20}$$

If $\gamma = 1$, the two solutions become identical, since

$$\begin{aligned} y_1 &= F(\alpha, \beta, \gamma; x) = F(\alpha, \beta, 1; x) \\ \text{and} \quad y_2 &= x^{1-\gamma} F(\alpha-\gamma+1, \beta-\gamma+1, 2-\gamma; x) = F(\alpha, \beta, 1; x). \end{aligned}$$

In this case, the second solution is to be found by the method discussed in chapter 4.

(B) *Solution around* $x = 1$: We make a change of variable, $z = 1 - x$.

In view of this, equation (12.1) becomes

$$z(z-1)\frac{d^2y}{dz^2} + [(\alpha+\beta+1)z - \gamma']\frac{dy}{dz} + \alpha\beta y = 0, \tag{12.21}$$

where

$$\gamma' = \alpha + \beta + 1 - \gamma. \tag{12.22}$$

Thus, equation (12.21) is similar to equation (12.1) if γ in the latter equation is replaced by γ' given in (12.22). A similar procedure as above results in the roots of indicial equation as

$$r = 0 \quad \text{and} \quad r = 1 - \gamma' = \gamma - \alpha - \beta. \tag{12.23}$$

Corresponding to these values, the two solutions are given by

$$\begin{aligned} y_1 &= F(\alpha, \beta, \gamma'; z) \\ &= F(\alpha, \beta, \alpha+\beta+1-\gamma; 1-x) \end{aligned} \tag{12.24}$$

$$\begin{aligned} \text{and} \quad y_2 &= z^{1-\gamma'} F(\alpha-\gamma'+1, \beta-\gamma'+1, 2-\gamma'; z) \\ &= (1-x)^{\gamma-\alpha-\beta} F(\gamma-\alpha, \gamma-\beta, \gamma-\alpha-\beta+1; 1-x) \end{aligned} \tag{12.25}$$

so that the complete solution is

$$y = c_1 y_1 + c_2 y_2. \tag{12.26}$$

(C) *Solution around* $x = \infty$: We substitute $x = \dfrac{1}{z}$. Then the differential equation (12.1) is transformed into

$$z^2(1-z)\frac{d^2y}{dz^2} + \left[2z(1-z) - (\alpha+\beta+1)z + \gamma z^2\right]\frac{dy}{dz} + \alpha\beta y = 0. \tag{12.27}$$

We seek the solution of equation (12.27) in the form

$$y = \sum_{k=0}^{\infty} a_k z^{r+k}. \tag{12.28}$$

Substituting the function (12.28) into equation (12.27), we get

$$\sum_{k=0}^{\infty} a_k \left[(r+k)(r+k-1)+(2-\alpha-\beta-1)(r+k)+\alpha\beta\right] z^{r+k}$$
$$=\sum_{k=0}^{\infty} a_k \left[(r+k)(r+k-1)+(2-\gamma)(r+k)\right] z^{r+k+1}. \tag{12.29}$$

Equating the coefficients of z^r to zero, we have

$$a_0\left[r(r-1)+(1-\alpha-\beta)r+\alpha\beta\right]=0,$$

which gives $r=\alpha$ and $r=\beta$.

For $r=\alpha$, the recursion formula for coefficients a_{k+1} is given by

$$a_{k+1}[(\alpha+k+1)(\alpha+k)+(1-\alpha-\beta)(\alpha+k+1)+\alpha\beta]$$
$$=a_k\left[(\alpha+k)(\alpha+k-1)+(2-\gamma)(\alpha+k)\right]$$
$$\text{i.e.,}\quad a_{k+1}=\frac{(\alpha+k)(\alpha+k+1-\gamma)}{(k+1)(k+\alpha-\beta+1)}a_k. \tag{12.30}$$

This gives

$$a_1=\frac{\alpha(\alpha+1-\gamma)}{1\cdot(\alpha-\beta+1)}a_0$$
$$a_2=\frac{(\alpha+1)(\alpha-\gamma+2)a_1}{2\cdot(\alpha-\beta+2)}=\frac{\alpha(\alpha+1)(\alpha-\gamma+1)(\alpha-\gamma+2)}{1\cdot 2\cdot(\alpha-\beta+1)(\alpha-\beta+2)}a_0$$

....

....

$$a_k=\frac{(\alpha+k-1)(\alpha-\gamma+k)}{k\cdot(\alpha-\beta+k)}a_{k-1}=\frac{(\alpha)_k(\xi)_k}{k!(\eta)_k}a_0 \tag{12.31}$$

where

$$\left.\begin{aligned}\xi&=\alpha-\gamma+1\\ \eta&=\alpha-\beta+1\end{aligned}\right\}. \tag{12.32}$$

Hence, the solution for $r=\alpha$ is (taking $a_0=1$)

$$y_1=z^{\alpha}\sum_{k=0}^{\infty}\frac{(\alpha)_k(\xi)_k}{k!(\eta)_k}z^k=x^{-\alpha}F\left(\alpha,\xi,\eta;\frac{1}{x}\right)$$
$$\text{i.e.,}\quad y_1=x^{-\alpha}F\left(\alpha,\alpha-\gamma+1,\alpha-\beta+1;\frac{1}{x}\right). \tag{12.33}$$

Similarly, for $r=\beta$, we get the solution

$$y_2=x^{-\beta}F\left(\beta,\beta-\gamma+1,\beta-\alpha+1;\frac{1}{x}\right). \tag{12.34}$$

Hence, the complete solution about $x=\infty$ is

$$y=c_1y_1+c_2y_2, \tag{12.35}$$

where y_1, y_2 are given by equations (12.33) and (12.34).

12.2 Elementary Functions in Terms of Hyper-geometric Functions

Many elementary functions can be expressed in terms of hypergeometric function. We shall see a few of them.

(1) If $\alpha = -n$, $\beta = \gamma$ and $x = -x$, we find from equation (12.13)

$$\begin{aligned} F(-n, \beta, \beta; -x) &= 1 + \frac{(-n)(-x)}{1} + \frac{(-n)(-n+1)(-x)^2}{1.2} + \cdots \\ &= 1 + \frac{nx}{1} + \frac{n(n-1)}{2!}x^2 + \cdots \\ &= (1+x)^n. \end{aligned} \tag{12.36}$$

If we change $x \longrightarrow (x-1)$, we get

$$F(-n, \beta, \beta; 1-x) = x^n. \tag{12.37}$$

(2) If $\alpha = \beta = 1, \gamma = 2, x = -x$, then

$$\begin{aligned} F(1,1,2;-x) &= 1 + \frac{1.1}{1.2}(-x) + \frac{(1.2)\,(1.2)}{(1.2)(2.3)}(-x)^2 + \cdots \\ &= 1 - \frac{x}{2} + \frac{x^2}{3} - \frac{x^3}{4} + \cdots \\ &= \frac{1}{x}\left(x - \frac{x^2}{2} + \frac{x^3}{3} - \frac{x^4}{4} + \cdots\right) \\ &= \frac{1}{x}\log(1+x). \end{aligned} \tag{12.38}$$

(3) If $\alpha = \gamma = 1$, $\beta = n$ and $x \longrightarrow \frac{x}{n}$,

$$\begin{aligned} F\left(1, n, 1; \frac{x}{n}\right) &= 1 + \frac{1 \cdot n}{1 \cdot 1} \cdot \frac{x}{n} + \frac{1 \cdot 2 \cdot n(n+1)}{(1 \cdot 2)(1 \cdot 2)} \cdot \left(\frac{x}{n}\right)^2 + \cdots \\ &= 1 + \frac{x}{1!} + \frac{x^2}{2!} \cdot \frac{n+1}{n} + \frac{x^3}{3!} \cdot \frac{(n+1)(n+2)}{n^2} + \cdots. \end{aligned} \tag{12.39}$$

In the limit as $n \longrightarrow \infty$, we get

$$\begin{aligned} \lim_{n \longrightarrow \infty} F\left(1, n, 1; \frac{x}{n}\right) &= 1 + \frac{x}{1!} + \frac{x^2}{2!} + \frac{x^3}{3!} + \cdots \\ &= e^x. \end{aligned} \tag{12.40}$$

(4) If $\gamma = \frac{1}{2}$, and $x \longrightarrow -\frac{x^2}{4\alpha\beta}$, we have

$$F\left(\alpha, \beta, \frac{1}{2}; -\frac{x^2}{4\alpha\beta}\right) = 1 - \frac{\alpha \cdot \beta}{1 \cdot \frac{1}{2}} \cdot \frac{x^2}{4\alpha\beta} + \frac{\alpha(\alpha+1)(\beta)(\beta+1)}{1 \cdot 2 \cdot \frac{1}{2} \cdot \frac{3}{2}} \cdot \frac{x^4}{2^4\alpha^2\beta^2}.$$

In the limit as $\alpha \longrightarrow \infty$ and $\beta \longrightarrow \infty$, we get

$$\begin{aligned} \lim_{\alpha, \beta \longrightarrow \infty} F\left(\alpha, \beta, \frac{1}{2}; -\frac{x^2}{4\alpha\beta}\right) &= 1 - \frac{x^2}{1 \cdot 2} + \frac{x^4}{2 \cdot 4 \cdot 1 \cdot 3} - \cdots \\ &= \sum_{k=0}^{\infty} \frac{(-1)^k}{(2k)!} x^{2k} = \cos x. \end{aligned} \tag{12.41}$$

12.3 Differentiation of Hypergeometric Functions

We have, from expression (12.14)

$$F(\alpha,\beta,\gamma;x)=\sum_{k=0}^{\infty}\frac{(\alpha)_k(\beta)_k}{k!(\gamma)_k}x^k.$$

Hence

$$\frac{d}{dx}F(\alpha,\beta,\gamma;x)=\sum_{k=1}^{\infty}\frac{(\alpha)_k\,(\beta)_k}{(k-1)!\,(\gamma)_k}x^{k-1}=\sum_{k=0}^{\infty}\frac{(\alpha)_{k+1}\,(\beta)_{k+1}}{k!\,(\gamma)_{k+1}}x^k. \tag{12.42}$$

Now, according to notations (12.11)

$$\begin{aligned}(\alpha)_{k+1}&=\alpha(\alpha+1)(\alpha+2)\cdots(\alpha+k)\\&=\alpha(\alpha+1)_k.\end{aligned}$$

Similarly

$$\begin{aligned}(\beta)_{k+1}&=\beta(\beta+1)_k\\ \text{and}\quad (\gamma)_{k+1}&=\gamma(\gamma+1)_k.\end{aligned}$$

Hence, equation (12.42) can be written as

$$\begin{aligned}\frac{d}{dx}F(\alpha,\beta,\gamma;x)&=\frac{\alpha\beta}{\gamma}\sum_{k=0}^{\infty}\frac{(\alpha+1)_k\,(\beta+1)_k}{k!\,(\gamma+1)_k}x^k\\&=\frac{\alpha\beta}{\gamma}F(\alpha+1,\beta+1,\gamma+1;\,x).\end{aligned} \tag{12.43}$$

In general we can get

$$\frac{d^m}{dx^m}F(\alpha,\beta,\gamma;x)=\frac{(\alpha)_m\,(\beta)_m}{(\gamma)_m}F(\alpha+m,\beta+m,\gamma+m;x). \tag{12.44}$$

12.4 Integral Representation of Hypergeometric Functions

We have

$$F(\alpha,\beta,\gamma;x)=\sum_{k=0}^{\infty}\frac{(\alpha)_k\,(\beta)_k}{k!(\gamma)_k}x^k. \tag{12.45}$$

Now

$$\begin{aligned}
\frac{(\beta)_k}{(\gamma)_k} &= \frac{\beta(\beta+1)(\beta+2)\cdots(\beta+k-1)}{\gamma(\gamma+1)(\gamma+2)\cdots(\gamma+k-1)} \\
&= \frac{1\cdot 2\cdot 3\cdots\beta(\beta+1)\cdots(\beta+k-1)}{1\cdot 2\cdot 3\cdots\gamma(\gamma+1)\cdots(\gamma+k-1)}\cdot\frac{\Gamma(\gamma)}{\Gamma(\beta)} \\
&= \frac{\Gamma(\beta+k)}{\Gamma(\gamma+k)}\cdot\frac{\Gamma(\gamma)}{\Gamma(\beta)} \quad \text{[see equation (7.42)]} \\
&= \frac{\Gamma(\beta+k)}{\Gamma(\gamma+k)}\cdot\frac{\Gamma(\gamma)}{\Gamma(\beta)}\cdot\frac{\Gamma(\gamma-\beta)}{\Gamma(\gamma-\beta)} \\
&= \frac{\Gamma(\gamma)}{\Gamma(\beta)\Gamma(\gamma-\beta)}\cdot\frac{\Gamma(\gamma-\beta)\Gamma(\beta+k)}{\Gamma(\gamma+k)} \\
&= \frac{\Gamma(\gamma)}{\Gamma(\beta)\Gamma(\gamma-\beta)}\cdot B(\gamma-\beta,\beta+k) \quad \text{[see equation (7.46)]} \\
&= \frac{\Gamma(\gamma)}{\Gamma(\beta)\Gamma(\gamma-\beta)}\int_0^1 (1-t)^{\gamma-\beta-1}\cdot t^{\beta+k-1}\,dt \\
&= \frac{1}{B(\beta,\gamma-\beta)}\int_0^1 (1-t)^{\gamma-\beta-1}\cdot t^{\beta+k-1}\,dt.
\end{aligned} \tag{12.46}$$

Hence

$$\begin{aligned}
F(\alpha,\beta,\gamma;x) &= \frac{1}{B(\beta,\gamma-\beta)}\cdot\sum_{k=0}^{\infty}\frac{(\alpha)_k}{k!}x^k\int_0^1 (1-t)^{\gamma-\beta-1}\cdot t^{\beta+k-1}\,dt \\
&= \frac{1}{B(\beta,\gamma-\beta)}\cdot\int_0^1 (1-t)^{\gamma-\beta-1}\cdot t^{\beta-1}\left[\sum_{k=0}^{\infty}\frac{(\alpha)_k}{k!}(xt)^k\right]dt.
\end{aligned} \tag{12.47}$$

Now

$$\begin{aligned}
\sum_{k=0}^{\infty}\frac{(\alpha)_k}{k!}(xt)^k &= 1+\frac{\alpha(xt)}{1!}+\frac{\alpha(\alpha+1)(xt)^2}{2!}+\cdots \\
&= (1-xt)^{-\alpha}.
\end{aligned} \tag{12.48}$$

In view of equation (12.48), equation (12.47) assumes the form

$$F(\alpha,\beta,\gamma;x) = \frac{1}{B(\beta,\gamma-\beta)}\int_0^1 (1-t)^{\gamma-\beta-1}\,t^{\beta-1}(1-xt)^{-\alpha}\,dt. \tag{12.49}$$

Equation (12.49) gives the integral representation of hypergeometric function.
If, in particular $x=1$,

$$\begin{aligned}
F(\alpha,\beta,\gamma;1) &= \frac{1}{B(\beta,\gamma-\beta)}\cdot\int_0^1 (1-t)^{\gamma-\beta-1}\,t^{\beta-1}\,(1-t)^{-\alpha}\,dt \\
&= \frac{1}{B(\beta,\gamma-\beta)}\int_0^1 t^{\beta-1}\cdot(1-t)^{\gamma-\alpha-\beta-1}\,dt \\
&= \frac{B(\beta,\gamma-\alpha-\beta)}{B(\beta,\gamma-\beta)} = \frac{\Gamma(\gamma)\Gamma(\gamma-\alpha-\beta)}{\Gamma(\gamma-\alpha)\Gamma(\gamma-\beta)}.
\end{aligned} \tag{12.50}$$

This is known as *Gauss's formula*.

12.5 Confluent Hypergeometric Functions

In the hypergeometric differential equation (12.1), if we replace x by $\frac{x}{\beta}$, we get

$$x\left(\frac{x}{\beta}-1\right)\frac{d^2y}{dx^2}+\left[\left(\frac{\alpha}{\beta}+1+\frac{1}{\beta}\right)x-\gamma\right]\frac{dy}{dx}+\alpha y=0. \tag{12.51}$$

If now we make $\beta \longrightarrow \infty$, equation (12.51) becomes

$$x\frac{d^2y}{dx^2}+(\gamma-x)\frac{dy}{dx}-\alpha y=0. \tag{12.52}$$

Equation (12.52) is called *confluent hypergeometric differential equation.* It has two singularities—one at $x=0$ (regular) and the other at $x=\infty$ (irregular). Hence, we can obtain a series solution about $x=0$ in the form

$$y=\sum_{k=0}^{\infty} A_k x^{r+k}. \tag{12.53}$$

Putting the expression (12.53) in equation (12.52), we get

$$\sum_{k=0}^{\infty} A_k\left[(r+k)(r+k-1)+\gamma(r+k)\right]x^{r+k-1}=\sum_{k=0}^{\infty} A_k\left[(r+k)+\alpha\right]x^{r+k}. \tag{12.54}$$

We get the indicial equation as

$$r(r-1)+\gamma r=0,$$

which gives

$$r=0 \quad \text{and} \quad r=1-\gamma. \tag{12.55}$$

For $r=0$, equating coefficients of x^{r+k}, we get

$$A_{k+1}[(k+1)k+\gamma(k+1)]=A_k(k+\alpha),$$

which gives

$$A_{k+1}=\frac{k+\alpha}{(k+1)(k+\gamma)}A_k \tag{12.56}$$

We find the coefficients as

$$A_1=\frac{\alpha}{1\cdot\gamma}A_0$$

$$A_2=\frac{\alpha+1}{2(1+\gamma)}A_1=\frac{\alpha(\alpha+1)}{1\cdot 2\cdot\gamma(1+\gamma)}A_0$$

$$\cdots\cdots\cdots\cdots\cdots\cdots\cdots\cdots$$

$$A_k=\frac{\alpha(\alpha+1)\cdots(\alpha+k-1)}{1\cdot 2\cdots k\cdot\gamma(\gamma+1)\cdots(\gamma+k-1)}A_0$$

$$\text{i.e.,}\quad A_k=\frac{(\alpha)_k}{k!(\gamma)_k}\cdot A_0. \tag{12.57}$$

Hence, the solution is (putting $A_0 = 1$)

$$y_1 = 1 + \frac{\alpha}{1!\gamma}x + \frac{\alpha(\alpha+1)}{2!\gamma(\gamma+1)}x^2 + \cdots$$
$$= \sum_{k=0}^{\infty} \frac{(\alpha)_k}{k!(\gamma)_k}x^k. \tag{12.58}$$

The function in (12.58) is known as *confluent hypergeometric function of first kind.* It is denoted as $F(\alpha, \gamma; x)$. We find that

$$F(\alpha,\gamma;x) = \lim_{\beta\longrightarrow\infty} F\left(\alpha,\beta,\gamma;\frac{x}{\beta}\right)$$
$$= \lim_{\beta\longrightarrow\infty}\left[1 + \frac{\alpha\beta}{1!\gamma}\cdot\frac{x}{\beta} + \frac{\alpha(\alpha+1)(\beta)(\beta+1)}{2!\gamma(\gamma+1)}\cdot\left(\frac{x}{\beta}\right)^2 + \cdots\right]$$
$$= 1 + \frac{\alpha}{1!\gamma}x + \frac{\alpha(\alpha+1)x^2}{2!\gamma(\gamma+1)} + \cdots. \tag{12.59}$$

Now, for $r = 1-\gamma$, we seek the solution as

$$y_2 = x^{1-\gamma}\sum_{k=0}^{\infty} A_k x^k. \tag{12.60}$$

Proceeding as before, we get the recursion formula

$$A_{k+1} = \frac{\alpha' + k}{(k+1)(\gamma'+k)}\cdot A_k \tag{12.61}$$

where

$$\left.\begin{aligned} \alpha' &= \alpha - \gamma + 1 \\ \text{and}\quad \gamma' &= 2-\gamma \end{aligned}\right\}. \tag{12.62}$$

Hence, we write the solution as

$$y_2 = x^{1-\gamma}\, F(\alpha',\gamma';x)$$
$$= x^{1-\gamma}\, F(\alpha-\gamma+1, 2-\gamma; x). \tag{12.63}$$

Thus the complete solution is given by

$$y = c_1y_1 + c_2y_2$$
$$= c_1F(\alpha,\gamma;x) + c_2x^{1-\gamma}F(\alpha-\gamma+1,2-\gamma;x). \tag{12.64}$$

The function $F(\alpha-\gamma+1, 2-\gamma; x)$ is known as *confluent hypergeometric function of second kind.*

It is obvious that, for the particular solution $y_1(x)$, $\gamma \neq 0, -1, -2, \ldots$ and for the second particular solution $y_2(x)$, $2-\gamma \neq 0$. Hence, the two solutions are linearly independent for $\gamma \neq 0, \pm1, \pm2, \ldots$ By D'Alembert's ratio test, it is found that the series converges for every value of x.

12.6 Derivative of Confluent Hypergeometric Functions

Proceeding as in Section 12.3, we can have

$$\frac{d}{dx}F(\alpha,\gamma;x)=\frac{\alpha}{\gamma}F(\alpha+1,\gamma+1;x). \tag{12.65}$$

In general, we have

$$\frac{d^m}{dx^m}F(\alpha,\gamma;x)=\frac{(\alpha)_m}{(\gamma)_m}F(\alpha+m,\gamma+m;x). \tag{12.66}$$

12.7 Integral Representation of Confluent Hypergeometric Functions

We have

$$\begin{aligned}
F(\alpha,\gamma;x) &= \sum_{k=0}^{\infty}\frac{(\alpha)_k}{k!\,(\gamma)_k}x^k=\frac{\Gamma(\gamma)}{\Gamma(\alpha)}\sum_{k=0}^{\infty}\frac{\Gamma(\alpha+k)}{k!\,\Gamma(\gamma+k)}x^k\\
&= \frac{\Gamma(\gamma)}{\Gamma(\alpha)\Gamma(\gamma-\alpha)}\sum_{k=0}^{\infty}\frac{\Gamma(\alpha+k)\Gamma(\gamma-\alpha)}{k!\,\Gamma(\gamma+k)}\cdot x^k\\
&= \frac{\Gamma(\gamma)}{\Gamma(\alpha)\Gamma(\gamma-\alpha)}\sum_{k=0}^{\infty}B(\alpha+k,\gamma-\alpha)\frac{x^k}{k!}\\
&= \frac{\Gamma(\gamma)}{\Gamma(\alpha)\Gamma(\gamma-\alpha)}\sum_{k=0}^{\infty}\frac{x^k}{k!}\int_0^1 t^{\alpha+k-1}(1-t)^{\gamma-\alpha-1}dt\\
&= \frac{\Gamma(\gamma)}{\Gamma(\alpha)\Gamma(\gamma-\alpha)}\int_0^1 t^{\alpha-1}(1-t)^{\gamma-\alpha-1}\left\{\sum_{k=0}^{\infty}\frac{(xt)^k}{k!}\right\}dt\\
&= \frac{\Gamma(\gamma)}{\Gamma(\alpha)\Gamma(\gamma-\alpha)}\int_0^1 t^{\alpha-1}(1-t)^{\gamma-\alpha-1}e^{xt}\,dt.
\end{aligned} \tag{12.67}$$

Equation (12.67) is the integral representation of confluent hypergeometric function. If we change the variable, $t=1-s$, we get

$$\begin{aligned}
F(\alpha,\gamma;x) &= \frac{\Gamma(\gamma)}{\Gamma(\alpha)\Gamma(\gamma-\alpha)}\int_0^1 e^{x(1-s)}s^{\gamma-\alpha-1}(1-s)^{\alpha-1}\,ds\\
&= e^x\frac{\Gamma(\gamma)}{\Gamma(\alpha)\Gamma(\gamma-\alpha)}\int_0^1 e^{-xs}s^{\gamma-\alpha-1}(1-s)^{\alpha-1}\,ds.
\end{aligned} \tag{12.68}$$

But, if we replace $x\longrightarrow -x$, and $\alpha\longrightarrow\gamma-\alpha$ in the equation (12.67), we get

$$F(\gamma-\alpha,\gamma;-x)=\frac{\Gamma(\gamma)}{\Gamma(\gamma-\alpha)\Gamma(\alpha)}\int_0^1 e^{-xt}t^{\gamma-\alpha-1}(1-t)^{\alpha-1}\,dt. \tag{12.69}$$

From equations (12.68) and (12.69), we get

$$F(\alpha,\gamma;x)=\frac{\Gamma(\gamma)}{\Gamma(\alpha)\Gamma(\gamma-\alpha)}\cdot e^x\cdot\frac{\Gamma(\gamma-\alpha)\Gamma(\alpha)}{\Gamma(\gamma)}F(\gamma-\alpha,\gamma;-x)$$

$$\text{i.e.,}\quad F(\alpha,\gamma;x)=e^x\,F(\gamma-\alpha,\gamma;-x). \tag{12.70}$$

In particular, if $x = 0$, we have from equation (12.67), $F(\alpha, \gamma; 0) = 1$. This gives

$$\frac{\Gamma(\gamma)}{\Gamma(\alpha)\Gamma(\gamma-\alpha)} \cdot \int_0^1 t^{\alpha-1}(1-t)^{\gamma-\alpha-1}\,dt = 1$$

$$\text{i.e.,} \quad \frac{\Gamma(\gamma)}{\Gamma(\alpha)\Gamma(\gamma-\alpha)} \cdot B(\alpha, \gamma-\alpha) = 1. \tag{12.71}$$

Again, if $\gamma = \alpha$,

$$F(\alpha, \alpha; x) = \sum_{k=0}^{\infty} \frac{x^k}{k!} = e^x. \tag{12.72}$$

BESSEL'S FUNCTION

12.8 Bessel's Differential Equation

The differential equation of the form

$$x^2 \frac{d^2y}{dx^2} + x\frac{dy}{dx} + \left(x^2 - \nu^2\right) y = 0 \tag{12.73}$$

is known as Bessel's differential equation. The equation has a non-essential singularity at $x = 0$ and hence we can find its solution in a power series. We assume, therefore

$$y = \sum_{k=0}^{\infty} a_k x^{r+k}. \tag{12.74}$$

We substitute the function in (12.74) along with its first and second derivatives into the equation (12.73) and get

$$\sum_{k=0}^{\infty} a_k \left[(r+k)(r+k-1) + (r+k)\right] x^{r+k} + x^2 \sum_{k=0}^{\infty} a_k x^{r+k} - \nu^2 \sum_{k=0}^{\infty} a_k x^{r+k} = 0. \tag{12.75}$$

Equating to zero the coefficients of $x^r, x^{r+1}, \ldots$, we get

$$\begin{aligned}
a_0 \left[r(r-1) + r - \nu^2\right] &= 0 \\
a_1 \left[(r+1)r + (r+1) - \nu^2\right] &= 0 \\
a_2 \left[(r+2)(r+1) + (r+2) - \nu^2\right] + a_0 &= 0 \\
a_3 \left[(r+3)(r+2) + (r+3) - \nu^2\right] + a_1 &= 0 \\
\cdots\cdots\cdots\cdots\cdots\cdots\cdots\cdots & \cdots \\
\cdots\cdots\cdots\cdots\cdots\cdots\cdots\cdots & \cdots
\end{aligned}$$

From these relations, we get

$$\left.\begin{aligned} a_0\left(r^2-\nu^2\right) &= 0 \\ a_1\left[(r+1)^2-\nu^2\right] &= 0 \\ a_2\left[(r+2)^2-\nu^2\right]+a_0 &= 0 \\ a_3\left[(r+3)^2-\nu^2\right]+a_1 &= 0 \\ \ldots\ldots\ldots\ldots\ldots\ldots\ldots &\ \ldots \\ \ldots\ldots\ldots\ldots\ldots\ldots\ldots &\ \ldots \end{aligned}\right\} \tag{12.76}$$

Assuming $a_0 \neq 0$, we get from equations (12.76) $r^2 - \nu^2 = 0$, i.e., $r = \nu$, and $r = -\nu$.
For $r = \nu$: Equations (12.76) give

$$a_1 = a_3 = \cdots a_{2k-1} = 0, \quad k = 1, 2, \ldots \tag{12.77}$$

and

$$a_2 = -\frac{a_0}{(\nu+2)^2-\nu^2}$$

$$a_4 = -\frac{a_2}{(\nu+4)^2-\nu^2} = \frac{a_0}{\left[(\nu+2)^2-\nu^2\right]\left[(\nu+4)^2-\nu^2\right]}$$

In general

$$\begin{aligned} a_{2k} &= \frac{(-1)^k a_0}{\left[(\nu+2)^2-\nu^2\right]\left[(\nu+4)^2-\nu^2\right]\cdots\left[(\nu+2k)^2-\nu^2\right]} \\ &= \frac{(-1)^k a_0}{(4\nu+4)(8\nu+16)\cdots(4k\nu+4k^2)} \\ &= \frac{(-1)^k a_0}{2^{2k}k!\,(\nu+1)(\nu+2)\cdots(\nu+k)} \\ &= \frac{(-1)^k\Gamma(\nu+1)a_0}{2^{2k}\Gamma(k+1)\Gamma(\nu+k+1)}, \quad k = 1, 2, \ldots. \end{aligned} \tag{12.78}$$

Hence, we write the particular solution for $r = \nu$ as

$$J_\nu(x) = a_0 2^\nu \Gamma(\nu+1) \sum_{k=0}^{\infty} \frac{(-1)^k}{\Gamma(k+1)\Gamma(\nu+k+1)} \left(\frac{x}{2}\right)^{\nu+2k} \tag{12.79}$$

For $r = -\nu$: Following similar procedure as above, we get for $r = -\nu$ the particular solution as

$$J_{-\nu}(x) = b_0 2^{-\nu} \Gamma(-\nu+1) \sum_{k=0}^{\infty} \frac{(-1)^k}{\Gamma(k+1)\Gamma(-\nu+k+1)} \left(\frac{x}{2}\right)^{-\nu+2k}. \tag{12.80}$$

We assume

$$a_0 = \frac{1}{2^\nu \Gamma(\nu+1)} \tag{12.81}$$

$$\text{and} \quad b_0 = \frac{1}{2^{-\nu}\Gamma(-\nu+1)}. \tag{12.82}$$

Then the two particulars solutions of the Bessel's equation are written as

$$J_\nu(x) = \sum_{k=0}^{\infty} \frac{(-1)^k}{\Gamma(k+1)\Gamma(\nu+k+1)} \left(\frac{x}{2}\right)^{\nu+2k} \tag{12.83}$$

and
$$J_{-\nu}(x) = \sum_{k=0}^{\infty} \frac{(-1)^k}{\Gamma(k+1)\Gamma(-\nu+k+1)} \left(\frac{x}{2}\right)^{-\nu+2k}. \tag{12.84}$$

The functions in (12.83) and (12.84) are called *Bessel's function* of first kind of order ν and $-\nu$ respectively. Thus, we write the complete solution as

$$y(x) = AJ_\nu(x) + BJ_{-\nu}(x), \tag{12.85}$$

where A, B are constants.

When ν is an integer $= n$, we find from equation (12.84) that the denominator contains $\Gamma(-n+k+1)$, which is $= \pm\infty$ for values of $k = 0, 1, \ldots, (n-1)$. Thus, for ν as an integral value, solution (12.84) actually begins from $k = n$ and is given by

$$J_{-\nu}(x) = \sum_{k=\nu}^{\infty} \frac{(-1)^k}{\Gamma(k+1)\Gamma(k-\nu+1)} \left(\frac{x}{2}\right)^{-\nu+2k}. \tag{12.86}$$

Putting $k = \nu + m$, we get

$$\begin{aligned} J_{-\nu}(x) &= \sum_{m=0}^{\infty} \frac{(-1)^{\nu+m}}{\Gamma(\nu+m+1)\Gamma(m+1)} \left(\frac{x}{2}\right)^{\nu+2m} \\ &= (-1)^\nu \sum_{m=0}^{\infty} \frac{(-1)^m}{\Gamma(m+1)\Gamma(\nu+m+1)} \left(\frac{x}{2}\right)^{\nu+2m} \end{aligned}$$

i.e.,
$$J_{-\nu} = (-1)^\nu J_\nu(x). \tag{12.87}$$

Equation (12.87) shows that, for integral values of ν, the two solutions $J_\nu(x)$ and $J_{-\nu}(x)$ are not independent of each other. Hence, in this case the general solution given in (12.85) is no more valid. We shall seek for a second solution for integral values of $\nu = n$. Now, the first solution for $\nu = n > 0$ is

$$\begin{aligned} J_n(x) &= \sum_{k=0}^{\infty} \frac{(-1)^k}{k!(n+k)!} \left(\frac{x}{2}\right)^{n+2k} \\ &= \frac{x^n}{2^n\, n!} \left[1 - \frac{x^2}{2(2n+2)} + \frac{x^4}{2.4(2n+2)(2n+4)} - \cdots\right]. \end{aligned} \tag{12.88}$$

In particular, for $n = 1$,

$$\begin{aligned} J_1(x) &= \sum_{k=0}^{\infty} \frac{(-1)^k}{k!\,(k+1)!} \left(\frac{x}{2}\right)^{2k+1} \\ &= \frac{1}{0!\,1!}\frac{x}{2} - \frac{1}{1!\,2!}\left(\frac{x}{2}\right)^3 + \frac{1}{1!\,2!\,3!}\left(\frac{x}{2}\right)^5 - \cdots. \end{aligned} \tag{12.89}$$

Again, for $n = 0$

$$J_0(x) = \sum_{k=0}^{\infty} \frac{(-1)^k}{(k!)^2} \left(\frac{x}{2}\right)^{2k} = 1 - \frac{x^2}{2^2} + \frac{x^4}{2^2 4^2} - \frac{x^6}{2^2 4^2 6^2} + \cdots \tag{12.90}$$

The functions $J_0(x)$ and $J_1(x)$ are graphically shown in *Fig 12.1(a)*.

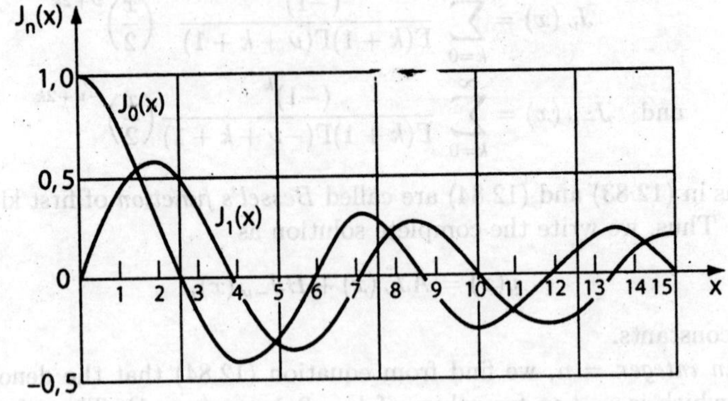

Fig 12.1(a)

For $\nu = m+\frac{1}{2}, m = 0, 1, 2, \ldots$, the Bessel's functions can be expressed in terms of elementary functions.

In particular, if $\nu = \frac{1}{2}$, we write from equation (12.83),

$$J_{\frac{1}{2}}(x) = \sum_{k=0}^{\infty} \frac{(-1)^k}{\Gamma(k+1)\Gamma\left(k+\frac{3}{2}\right)} \cdot \left(\frac{x}{2}\right)^{\frac{1}{2}+2k} \tag{12.91}$$

Using Legendre's duplication formula (7.98) (where $m \longrightarrow k+1$), we have

$$2^{2k+1}\Gamma(k+1)\Gamma\left(k+\frac{3}{2}\right) = (2k+1)!\sqrt{\pi}.$$

Hence, from equation (12.91)

$$J_{\frac{1}{2}}(x) = \sqrt{\frac{2}{\pi x}} \sum_{k=0}^{\infty} \frac{(-1)^k x^{2k+1}}{(2k+1)!}, \quad \text{i.e.,} \quad J_{\frac{1}{2}}(x) = \sqrt{\frac{2}{\pi x}} \sin x. \tag{12.92}$$

Similarly, for $\nu = -\frac{1}{2}$,

$$J_{-\frac{1}{2}}(x) = \sum_{k=0}^{\infty} \frac{(-1)^k}{\Gamma(k+1)\Gamma\left(k+\frac{1}{2}\right)} \left(\frac{x}{2}\right)^{-\frac{1}{2}+2k}$$

$$= \sqrt{\frac{2}{\pi x}} \sum_{k=0}^{\infty} \frac{(-1)^k x^{2k}}{(2k)!} \quad \text{[using equation (7.98)]}$$

$$\text{i.e.,} \quad J_{-\frac{1}{2}}(x) = \sqrt{\frac{2}{\pi x}} \cos x. \tag{12.93}$$

Functions $J_{1/2}(x)$ and $J_{-1/2}(x)$ are shown in *Fig 12.1(b)*.

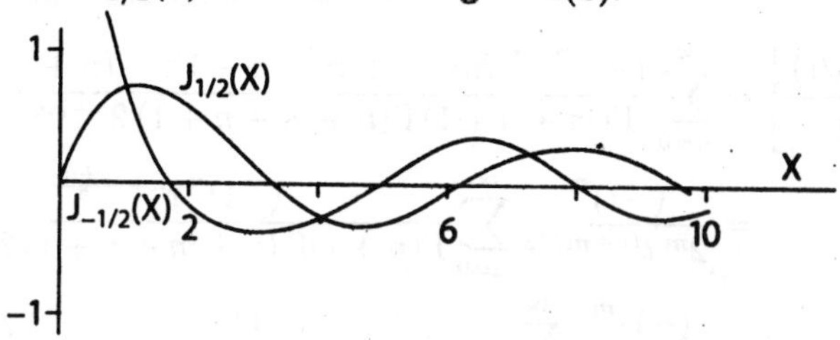

12.9 Bessel's Function in Terms of Hypergeometric Functions

The Bessel's functions of first kind and of order ν and $-\nu$ can be expressed in terms of hypergeometric functions.

We have

$$J_\nu(x) = \sum_{k=0}^{\infty} \frac{(-1)^k}{\Gamma(k+1)\Gamma(\nu+k+1)}\left(\frac{x}{2}\right)^{\nu+2k}$$

$$= \frac{1}{\Gamma(\nu+1)}\left(\frac{x}{2}\right)^{\nu} \sum_{k=0}^{\infty} \frac{(-1)^k\Gamma(\nu+1)}{k!\Gamma(\nu+k+1)}\left(\frac{x}{2}\right)^{2k}$$

$$= \frac{1}{\Gamma(\nu+1)}\left(\frac{x}{2}\right)^{\nu} \sum_{k=0}^{\infty} \frac{(-1)^k}{k!(\nu+1)(\nu+2)\cdots(\nu+k)}\left(\frac{x}{2}\right)^{2k}$$

$$= \lim_{\alpha,\beta\longrightarrow\infty} \frac{1}{\Gamma(\nu+1)}\left(\frac{x}{2}\right)^{\nu} \times$$

$$\sum_{k=0}^{\infty} \frac{\alpha(\alpha+1)\cdots(\alpha+k-1)\beta(\beta+1)\cdots(\beta+k-1)}{1\cdot 2\cdots k(\nu+1)(\nu+2)\cdots(\nu+k)}\left(\frac{-x^2}{4\alpha\beta}\right)^k$$

i.e., $$J_\nu(x) = \lim_{\alpha,\beta\longrightarrow\infty}\left[\frac{1}{\Gamma(\nu+1)}\left(\frac{x}{2}\right)^{\nu} F\left(\alpha,\beta,\nu+1;-\frac{x^2}{4\alpha\beta}\right)\right]. \tag{12.94}$$

Similarly

$$J_{-\nu}(x) = \lim_{\alpha,\beta\longrightarrow\infty}\left[\frac{1}{\Gamma(-\nu+1)}\left(\frac{x}{2}\right)^{-\nu} F\left(\alpha,\beta,-\nu+1;-\frac{x^2}{4\alpha\beta}\right)\right] \tag{12.95}$$

12.10 Recurrence Relations

We substitute $x^2 = t$ in equation (12.83). Then, we get

$$\frac{J_\nu(\sqrt{t})}{t^{\frac{\nu}{2}}} = \sum_{k=0}^{\infty} \frac{(-1)^k t^k}{\Gamma(k+1)\Gamma(\nu+k+1)2^{\nu+2k}}.$$

Differentiating m-times with respect to t

$$\frac{d^m}{dt^m}\left[\frac{J_\nu(\sqrt{t})}{t^{\frac{\nu}{2}}}\right] = \sum_{k=m}^{\infty} \frac{(-1)^k k(k-1)\cdots(k-m+1)t^{k-m}}{\Gamma(k+1)\Gamma(\nu+k+1)2^{\nu+2k}}.$$

Taking $k = m+n$,

$$\frac{d^m}{dt^m}\left[\frac{J_\nu(\sqrt{t})}{t^{\frac{\nu}{2}}}\right] = \sum_{n=0}^{\infty} \frac{(-1)^{m+n}(m+n)(m+n-1)\cdots(n+1)\,t^n}{\Gamma(m+n+1)\,\Gamma(\nu+m+n+1)\,2^{\nu+2m+2n}}$$

$$= \frac{(-1)^m}{2^m\,t^{(\nu+m)/2}} \sum_{n=0}^{\infty} \frac{(-1)^n\,t^{n+\frac{(\nu+m)}{2}}}{\Gamma(n+1)\Gamma(\nu+m+n+1)\,2^{\nu+m+2n}}$$

$$= \frac{(-1)^m}{2^m x^{\nu+m}} \sum_{n=0}^{\infty} \frac{(-1)^n}{\Gamma(n+1)\,\Gamma\{(\nu+m)+n+1\}}\left(\frac{x}{2}\right)^{(\nu+m)+2n}$$

Hence, we get

$$\frac{d^m}{d(x^2)^m}\left[\frac{J_\nu(x)}{x^\nu}\right] = \frac{(-1)^m}{2^m}\frac{J_{\nu+m}(x)}{x^{\nu+m}}$$

$$\text{or,}\quad \frac{d^m}{(x\,dx)^m}\left[\frac{J_\nu(x)}{x^\nu}\right] = (-1)^m\frac{J_{\nu+m}(x)}{x^{\nu+m}}. \tag{12.96}$$

Again, putting $x^2 = t$, we can write equation (12.83) in the form

$$t^{\frac{\nu}{2}} J_\nu\left(\sqrt{t}\right) = \sum_{k=0}^{\infty} \frac{(-1)^k\,t^{\nu+k}}{\Gamma(k+1)\Gamma(\nu+k+1)\,2^{\nu+2k}}.$$

Differentiating m-times with respect to t, we get

$$\frac{d^m}{dt^m}\left[t^{\frac{\nu}{2}} J_\nu\left(\sqrt{t}\right)\right] = \sum_{k=0}^{\infty} \frac{(-1)^k(\nu+k)(\nu+k-1)\cdots(\nu+k-m+1)\,t^{\nu+k-m}}{\Gamma(k+1)\Gamma(\nu+k+1)\,2^{\nu+2k}}$$

$$= \frac{t^{\frac{(\nu-m)}{2}}}{2^m} \sum_{k=0}^{\infty} \frac{(-1)^k\,t^{\frac{(\nu-m)}{2}+k}}{\Gamma(k+1)\Gamma(\nu+k-m+1)\,2^{\nu-m+2k}}.$$

$$= \frac{x^{\nu-m}}{2^m} \sum_{k=0}^{\infty} \frac{(-1)^k}{\Gamma(k+1)\Gamma\{(\nu-m)+k+1\}}\left(\frac{x}{2}\right)^{(\nu-m)+2k}$$

$$= \frac{x^{\nu-m}}{2^m} J_{\nu-m}(x).$$

Hence

$$\frac{d^m}{d(x^2)^m}\left[x^\nu J_\nu(x)\right] = \frac{x^{\nu-m}}{2^m} J_{\nu-m}(x)$$

$$\text{or,}\quad \frac{d^m}{(x\,dx)^m}\left[x^\nu J_\nu(x)\right] = x^{\nu-m} J_{\nu-m}(x). \tag{12.97}$$

Equations (12.96) and (12.97) are recurrence relations for $J_\nu(x)$.
Putting $m = 1$ in these equations, we get

$$\frac{1}{x}\frac{d}{dx}\left[\frac{J_\nu(x)}{x^\nu}\right] = -\frac{J_{\nu+1}(x)}{x^{\nu+1}} \tag{12.98}$$

$$\text{and}\quad \frac{1}{x}\frac{d}{dx}\left[x^\nu J_\nu(x)\right] = x^{\nu-1} J_{\nu-1}(x). \tag{12.99}$$

From equations (12.98) and (12.99), we get

$$\nu J_\nu(x) - xJ'_\nu(x) = xJ_{\nu+1}(x) \tag{12.100}$$

$$\text{and}\quad \nu J_\nu(x) + xJ'_\nu(x) = xJ_{\nu-1}(x). \tag{12.101}$$

Adding and subtracting these equations we get

$$\nu J_\nu(x) = \frac{x}{2}\left[J_{\nu+1}(x) + J_{\nu-1}(x)\right] \tag{12.102}$$

$$\text{and}\quad J'_\nu(x) = \frac{1}{2}\left[J_{\nu-1}(x) - J_{\nu+1}(x)\right]. \tag{12.103}$$

Putting $\nu = 0$ in equation (12.100), we have

$$J'_0(x) = -J_1(x). \tag{12.104}$$

We substitute $\nu = \frac{1}{2}$ in equation (12.96) and use the relation (12.92). Then we get

$$J_{m+\frac{1}{2}}(x) = (-1)^m x^{m+\frac{1}{2}} \frac{d^m}{(x\,dx)^m}\left[\frac{J_{\frac{1}{2}}(x)}{x^{\frac{1}{2}}}\right]$$

$$= (-1)^m x^{m+\frac{1}{2}} \sqrt{\frac{2}{\pi}} \frac{d^m}{(x\,dx)^m}\left(\frac{\sin x}{x}\right). \tag{12.105}$$

For $m = 1$, equation (12.105) yields

$$J_{\frac{3}{2}}(x) = \sqrt{\frac{2}{\pi x}}\left(\frac{\sin x}{x} - \cos x\right). \tag{12.106}$$

Again, we put $\nu = -\frac{1}{2}$ in equation (12.97). Then, we get

$$J_{-(m+\frac{1}{2})} = x^{m+\frac{1}{2}} \frac{d^m}{(x\,dx)^m}\left[x^{-\frac{1}{2}} J_{-\frac{1}{2}}(x)\right].$$

Using equation (12.93)

$$J_{-(m+\frac{1}{2})}(x) = \sqrt{\frac{2}{\pi}}\, x^{m+\frac{1}{2}} \frac{d^m}{(x\,dx)^m}\left(\frac{\cos x}{x}\right). \tag{12.107}$$

For $m = 1$, equation (12.107) yields

$$J_{-\frac{3}{2}}(x) = \sqrt{\frac{2}{\pi x}}\left(-\frac{\cos x}{x} - \sin x\right). \tag{12.108}$$

If we put $\nu = \frac{3}{2}$ in equation (12.102) and use equation (12.107), we get

$$J_{\frac{5}{2}}(x) = \frac{3}{x} J_{\frac{3}{2}}(x) - J_{\frac{1}{2}}(x) \tag{12.109}$$

$$= \sqrt{\frac{2}{\pi x}}\left[\frac{3\sin x}{x^2} - \frac{3\cos x}{x} - \sin x\right] \tag{12.110}$$

$$= \sqrt{\frac{2}{\pi x}}\left[\frac{3 - x^2}{x^2}\sin x - \frac{3}{x}\cos x\right]. \tag{12.111}$$

Similarly, if we put $\nu = -\frac{3}{2}$ in equation (12.102) and use the relations in (12.108), we get

$$J_{-\frac{5}{2}}(x) = \sqrt{\frac{2}{\pi x}}\left[\frac{3}{x}\sin x + \frac{3-x^2}{x^2}\cos x\right] \tag{12.112}$$

In this way, we can get the values of Bessel's function of half-odd orders of n.

12.11 Generating Function of Bessel's Functions

We have the Maclaurin series:

$$e^{\frac{xt}{2}} = 1 + \frac{xt}{1!\,2} + \frac{x^2t^2}{2!\,2^2} + \frac{x^3t^3}{3!\,2^3} + \cdots = \sum_{r=0}^{\infty}\frac{x^r t^r}{r!\,2^r} \tag{12.113}$$

$$\text{and } e^{-\frac{xt}{2}} = 1 - \frac{1}{1!\,2}\left(\frac{x}{t}\right) + \frac{1}{2!\,2^2}\left(\frac{x}{t}\right)^2 - \cdots = \sum_{S=0}^{\infty}\frac{(-1)^S}{S!\,2^S}\left(\frac{x}{t}\right)^S \tag{12.114}$$

The series in equation (12.113) is convergent for all values of x and t. The series in (12.114) however, has got a singular point at $t = 0$. We, therefore, assume that $t \neq 0$.

Multiplying these series, we get

$$e^{\frac{x}{2}\left(t-\frac{1}{t}\right)} = \left[\sum_{r=0}^{\infty}\frac{x^r t^r}{r!\,2^r}\right]\left[\sum_{S=0}^{\infty}\frac{(-1)^s}{S!\,2^S}\left(\frac{x}{t}\right)^S\right] = \sum_{r=0}^{\infty}\sum_{S=0}^{\infty}\frac{(-1)^S x^{r+S}\,t^{r-S}}{r!\,S!\,2^{r+S}}. \tag{12.115}$$

We assume that

$$\left.\begin{aligned} r - S &= n, \quad \text{if } r > S\\ \text{and}\quad r - S &= -n, \quad \text{if } r < S\end{aligned}\right\}. \tag{12.116}$$

Then, the coefficient of t^n in equation (12.115) is given by

$$\begin{aligned}&\sum_{S=0}^{\infty}\frac{x^{n+S}}{(n+S)!\,2^{n+S}}\times\frac{(-1)^S x^S}{S!\,2^S} \quad [r = n+S \text{ from equation (12.116)}]\\ &= \sum_{S=0}^{\infty}\frac{(-1)^S}{S!\,(n+S)!}\left(\frac{x}{2}\right)^{n+2S} = J_n(x), \quad n \geq 0.\end{aligned} \tag{12.117}$$

Similarly, the coefficient of t^{-n} in equation (12.115) is given by

$$\begin{aligned}&\sum_{r=0}^{\infty}\frac{x^r}{r!\,2^r}\times\frac{(-1)^{r+n}x^{r+n}}{(r+n)!\,2^{r+n}} \quad [S = r+n \text{ from equation (12.116)}]\\ &= \sum_{r=0}^{\infty}(-1)^n\frac{(-1)^r}{r!\,(r+n)!}\left(\frac{x}{2}\right)^{n+2r}\\ &= (-1)^n J_n(x) = J_{-n}(x) \quad [\text{by equation (12.87)}].\end{aligned} \tag{12.118}$$

Thus, for integral values of n, the coefficients of t^n and t^{-n} in the expansion of the function

$$f(x,t) = e^{\frac{x}{2}\left(t-\frac{1}{t}\right)}, \tag{12.119}$$

for $t \neq 0$ are given by $J_n(x)$ and $J_{-n}(x)$ respectively. Hence, we can write the function $f(x,t)$ in (12.115) as

$$f(x,t) = e^{\frac{x}{2}\left(t-\frac{1}{t}\right)} = \sum_{n=-\infty}^{\infty} J_n(x)\, t^n. \tag{12.120}$$

Thus, it is possible to calculate Bessel's function of any order n (integer) with the use of equation (12.120). The function $f(x,t)$ in (12.119) is, therefore, called *generating function* for $J_n(x)$.

12.12 Bessel's Function of Second Kind

We have seen in Section 12.8 that the second solution of Bessel's equation obtained by series method is not independent for integral values of ν. Here we intend to find the required solution. We consider the differential equation

$$x^2\frac{d^2y}{dx^2} + x\frac{dy}{dx} + \left[x^2 - (n-\epsilon)^2\right] y = 0, \tag{12.121}$$

where $0 < \epsilon < 1$, and $n = 0$ or an integer. Two linearly independent solutions of this equation are $J_{n-\epsilon}(x)$ and $J_{-(n-\epsilon)}(x)$. A linear combination of these two solutions in the form

$$Y_{(n-\epsilon)}(x) = \frac{1}{\epsilon}\left[(-1)^n J_{-(n-\epsilon)}(x) - J_{n-\epsilon}(x)\right] \tag{12.122}$$

for arbitrary value of ϵ will also be a solution of the equation (12.121). The function

$$Y_n(x) = \lim_{\epsilon \longrightarrow 0} Y_{n-\epsilon}(x), \tag{12.123}$$

if the limit exists, will be the solution of Bessel's equation for integral value of $\nu = n$. Now, using equation (12.83) and (12.84) we write,

$$\frac{(-1)^n J_{-(n-\epsilon)}(x)}{\epsilon} = (-1)^n \sum_{k=0}^{\infty} \frac{(-1)^k}{\epsilon\Gamma(k+1)\Gamma(-n+\epsilon+k+1)}\left(\frac{x}{2}\right)^{-n+\epsilon+2k}$$

Splitting the interval in two parts, we write the RHS as

$$(-1)^n \sum_{k=0}^{n-1}[\cdots] + (-1)^n \sum_{k=n}^{\infty}[\cdots].$$

For the second part, we put $k = n + m$. Then this part becomes

$$(-1)^n \sum_{m=0}^{\infty} \frac{(-1)^{n+m}}{\epsilon\Gamma(n+m+1)\Gamma(\epsilon+m+1)}\left(\frac{x}{2}\right)^{n+\epsilon+2m}$$

$$= \sum_{k=0}^{\infty} (-1)^k \left(\frac{x}{2}\right)^{n+2k} \frac{1}{\epsilon\Gamma(n+k+1)\Gamma(\epsilon+k+1)}\left(\frac{x}{2}\right)^{\epsilon}.$$

Hence

$$\begin{aligned}\frac{(-1)^n J_{-(n-\epsilon)}(x)}{\epsilon} &= (-1)^n \sum_{k=0}^{\infty} \frac{(-1)^k}{\epsilon\Gamma(k+1)\Gamma(-n+\epsilon+k+1)}\left(\frac{x}{2}\right)^{-n+\epsilon+2k} \\ &\quad + \sum_{k=0}^{\infty} (-1)^k \left(\frac{x}{2}\right)^{n+2k} \frac{1}{\epsilon\Gamma(n+k+1)\Gamma(\epsilon+k+1)}\left(\frac{x}{2}\right)^{\epsilon}\end{aligned} \tag{12.124}$$

Also we have

$$\frac{1}{\epsilon}J_{n-\epsilon}(x)=\sum_{k=0}^{\infty}(-1)^k\left(\frac{x}{2}\right)^{n+2k}\frac{1}{\epsilon\Gamma(k+1)\Gamma(n-\epsilon+k+1)}\left(\frac{x}{2}\right)^{-\epsilon} \tag{12.125}$$

We now determine the following limits using L'Hospital's rule.

1.
$$\begin{aligned}
&\lim_{\epsilon\to 0}\frac{1}{\epsilon\Gamma(-n+\epsilon+k+1)}\\
&=\lim_{\epsilon\to 0}\frac{1}{\epsilon\Gamma[1-(n-\epsilon-k)]}\\
&=\lim_{\epsilon\to 0}\frac{\Gamma(n-\epsilon-k)\sin\{(n-\epsilon-k)\pi\}}{\epsilon\pi}\quad\text{[using equation (7.69)]}\\
&=\lim_{\epsilon\to 0}-\Gamma(n-\epsilon-k)\cos\{(n-\epsilon-k)\pi\}\\
&=-\Gamma(n-k)\cos\{(n-k)\pi\}\\
&=(-1)^{n-k+1}\Gamma(n-k).
\end{aligned} \tag{12.126}$$

2.
$$\begin{aligned}
&\lim_{\epsilon\to 0}\frac{1}{\epsilon\Gamma(\epsilon+k+1)}\left(\frac{x}{2}\right)^{\epsilon}\\
&=\lim_{\epsilon\to 0}\left[\frac{-\Gamma'(\epsilon+k+1)}{\{\Gamma(\epsilon+k+1)\}^2}\left(\frac{x}{2}\right)^{\epsilon}+\frac{1}{\Gamma(\epsilon+k+1)}\left(\frac{x}{2}\right)^{\epsilon}\log\frac{x}{2}\right]\\
&=-\frac{\Gamma'(k+1)}{\{\Gamma(k+1)\}^2}+\frac{1}{\Gamma(k+1)}\log\frac{x}{2}.
\end{aligned} \tag{12.127}$$

3.
$$\begin{aligned}
&\lim_{\epsilon\to 0}\frac{1}{\epsilon\Gamma(n-\epsilon+k+1)}\left(\frac{x}{2}\right)^{-\epsilon}\\
&=\lim_{\epsilon\to 0}\left[\frac{\Gamma'(n-\epsilon+k+1)}{\{\Gamma(n-\epsilon+k+1)\}^2}\left(\frac{x}{2}\right)^{-\epsilon}-\frac{1}{\Gamma(n-\epsilon+k+1)}\left(\frac{x}{2}\right)^{-\epsilon}\log\frac{x}{2}\right]\\
&=\frac{\Gamma'(n+k+1)}{\{\Gamma(n+k+1)\}^2}-\frac{1}{\Gamma(n+k+1)}\log\frac{x}{2}.
\end{aligned} \tag{12.128}$$

From equations (12.122)–(12.128) we get

$$
\begin{aligned}
Y_n(x) &= \lim_{\epsilon \longrightarrow 0} \frac{1}{\epsilon}\left[(-1)^n J_{-(n-\epsilon)}(x) - J_{n-\epsilon}(x)\right] \\
&= (-1)^n \sum_{k=0}^{n-1} \frac{(-1)^k (-1)^{n-k+1}\Gamma(n-k)}{\Gamma(k+1)}\left(\frac{x}{2}\right)^{-n+2k} \\
&\quad + \sum_{k=0}^{\infty} (-1)^k \left(\frac{x}{2}\right)^{n+2k} \frac{1}{\Gamma(n+k+1)}\left[-\frac{\Gamma'(k+1)}{\{\Gamma(k+1)\}^2} + \frac{1}{\Gamma(k+1)}\log\frac{x}{2}\right] \\
&\quad - \sum_{k=0}^{\infty} (-1)^k \left(\frac{x}{2}\right)^{n+2k} \frac{1}{\Gamma(k+1)}\left[\frac{\Gamma'(n+k+1)}{\{\Gamma(n+k+1)\}^2} - \frac{1}{\Gamma(n+k+1)}\log\frac{x}{2}\right].
\end{aligned}
\tag{12.129}
$$

Using the relation [see equation (7.118)]

$$
\frac{\Gamma'(k+1)}{\Gamma(k+1)} = -C + \sum_{j=1}^{k} \frac{1}{j},
$$

where C = Euler's constant = 0.5772156, we write the equation (12.129) as

$$
\begin{aligned}
Y_n(x) = & -\sum_{k=0}^{n-1} \frac{\Gamma(n-k)}{\Gamma(k+1)}\left(\frac{x}{2}\right)^{-n+2k} + \sum_{k=0}^{\infty} (-1)^k \left(\frac{x}{2}\right)^{n+2k} \times \\
& \frac{1}{\Gamma(k+1)\Gamma(n+k+1)}\left[2\log\frac{x}{2} + 2C - \sum_{j=1}^{k}\frac{1}{j} - \sum_{j=1}^{n+k}\frac{1}{j}\right] \\
= & -\sum_{k=0}^{n-1} \frac{(n-k-1)!}{k!}\left(\frac{x}{2}\right)^{-n+2k} + \sum_{k=0}^{\infty} \frac{(-1)^k}{k!\,(n+k)!}\left(\frac{x}{2}\right)^{n+2k} \times \\
& \left[2\log\frac{x}{2} + 2C - \sum_{j=1}^{k}\frac{1}{j} - \sum_{j=1}^{n+k}\frac{1}{j}\right].
\end{aligned}
$$

Hence

$$
\begin{aligned}
Y_n(x) = & \; 2J_n(x)\left(\log\frac{x}{2} + C\right) - \sum_{k=0}^{n-1} \frac{(n-k-1)!}{k!}\left(\frac{x}{2}\right)^{-n+2k} \\
& - \left(\frac{x}{2}\right)^n \sum_{k=0}^{\infty} \frac{(-1)^k}{k!\,(n+k)!}\left(\frac{x}{2}\right)^{2k} \times \\
& \left[2\left(1 + \frac{1}{2} + \frac{1}{3} + \cdots + \frac{1}{k}\right) + \frac{1}{k+1} + \frac{1}{k+2} + \cdots + \frac{1}{k+n}\right].
\end{aligned}
\tag{12.130}
$$

The function $Y_n(x)$ given by equation (12.130) is called *Bessel's function* of second kind of order n. In particular, if $n = 0$, we have

$$Y_0(x) = 2\left[J_0(x)\left(\log\frac{x}{2} + C\right) + \frac{1}{(1!)^2}\left(\frac{x}{2}\right)^2 - \frac{1}{(2!)^2}\left(1 + \frac{1}{2}\right)\left(\frac{x}{2}\right)^4 + \frac{1}{(3!)^2}\left(1 + \frac{1}{2} + \frac{1}{3}\right)\left(\frac{x}{2}\right)^6 - \dots\right]. \quad (12.131)$$

All the functions $J_n(x)$ and $Y_n(x)$ have an infinity of real roots. The first few roots of some of the Bessel's function are provided in the Table 12.1 below.

Table 12.1: Zeros of $J_n(x)$

Root Number	$J_0(x)$	$J_1(x)$	$J_2(x)$	$J_3(x)$
(1)	2.4048	3.8317	5.1356	6.3802
(2)	5.5201	7.0156	8.4172	9.7610
(3)	8.6537	10.1735	11.6198	13.0152
(4)	11.7915	13.3237	14.7960	16.2235
(5)	14.9309	16.4706	17.9598	19.4094
(6)	18.0711	19.6159	21.1170	22.5827

A graph of the function $\frac{1}{\pi} Y_0(x)$ is provided in the (*Fig 12.2*) along with the graph for $J_0(x)$. It is to be noted that $Y_n(x) \longrightarrow \infty$ as $x \longrightarrow 0$.

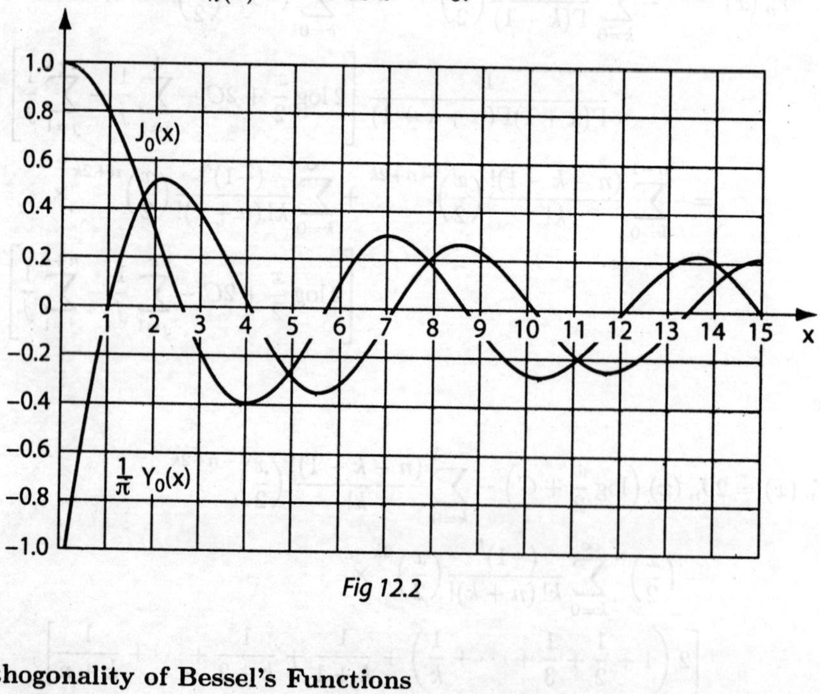

Fig 12.2

12.13 Orthogonality of Bessel's Functions

Bessel's functions of first kind has the following property of orthogonality:

If λ and μ are two roots of $J_\nu(x) = 0$, then

$$\int_0^1 x\, J_\nu(\lambda x)\, J_\nu(\mu x) = \begin{cases} 0, & \text{for } \lambda \neq \mu \\ \frac{1}{2}\{J_\nu(\lambda)\}^2, & \text{for } \lambda = \mu. \end{cases}$$

Proof. The differential equation satisfied by $J_\nu(x)$ is

$$x^2 \frac{d^2y}{dx^2} + x\frac{dy}{dx} + (x^2 - \nu^2)\, y = 0. \tag{12.132}$$

Changing the variable as $x \longrightarrow \lambda r$, and $y \longrightarrow u = J_\nu(\lambda r)$, the equation is reduced to

$$r^2 \frac{d^2u}{dr^2} + r\frac{du}{dr} + (\lambda^2 r^2 - \nu^2)\, u = 0. \tag{12.133}$$

Similarly, if we make another change of variable $x \longrightarrow \mu r$, and $y \longrightarrow v = J_\nu(\mu r)$, we get

$$r^2 \frac{d^2v}{dr^2} + r\frac{dv}{dr} + (\mu^2 r^2 - \nu^2)\, v = 0. \tag{12.134}$$

Multiplying equation (12.133) by $\frac{v}{r}$ and equation (12.134) by $\frac{u}{r}$, and subtracting, we get

$$r\left[v\frac{d^2u}{dr^2} - u\frac{d^2v}{dr^2}\right] + \left(v\frac{du}{dr} - u\frac{dv}{dr}\right) + (\lambda^2 - \mu^2)\, ruv = 0$$

or,
$$\frac{d}{dr}\left[r\left(v\frac{du}{dr} - u\frac{dv}{dr}\right)\right] + (\lambda^2 - \mu^2)\, ruv = 0$$

or,
$$\frac{d}{dr}\left[r\{\lambda J_\nu'(\lambda r)\, J_\nu(\mu r) - \mu J_\nu(\lambda r)\, J_\nu'(\mu r)\}\right] = (\mu^2 - \lambda^2)\, r J_\nu(\lambda r)\, J_\nu(\mu r).$$

Integrating within limit 0 to 1

$$\lambda J_\nu'(\lambda) J_\nu(\mu) - \mu J_\nu(\lambda) J_\nu'(\mu) = (\mu^2 - \lambda^2)\int_0^1 r J_\nu(\lambda r) J_\nu(\mu r)\, dr. \tag{12.135}$$

We know, λ and μ are roots of $J_\nu(x) = 0$ so that $J_\nu(\lambda) = J_\nu(\mu) = 0$.
Hence, from equation (12.135), we get

$$\int_0^1 r J_\nu(\lambda r) J_\nu(\mu r) dr = 0, \text{ for } \lambda \neq \mu. \tag{12.136}$$

If $\lambda = \mu$, we assume that λ is a root of $J_\nu(x) = 0$ so that $J_\nu(\lambda) = 0$; we make $\mu \longrightarrow \lambda$.
Then, from equation (12.135) we have

$$\begin{aligned}\lim_{\mu \longrightarrow \lambda} \int_0^1 r J_\nu(\lambda r) J_\nu(\mu r) dr &= \lim_{\mu \longrightarrow \lambda} \frac{\lambda J_\nu'(\lambda) J_\nu(\mu)}{\mu^2 - \lambda^2} \\ &= \lim_{\mu \longrightarrow \lambda} \frac{\lambda J_\nu'(\lambda) J_\nu'(\mu)}{2\mu} \quad \text{[by Hospital's rule]}.\end{aligned}$$

Hence

$$\int_0^1 r\{J_\nu(\lambda r)\}^2 dr = \frac{1}{2}\{J'_\nu(\lambda)\}^2 = \frac{1}{2}\{J_{\nu+1}(\lambda)\}^2. \tag{12.137}$$

Combining equations (12.136) and (12.137), we get

$$\int_0^1 rJ_\nu(\lambda r)J_\nu(\mu r)\,dr = \begin{cases} 0, & \text{for } \lambda \neq \mu \\ \frac{1}{2}\left\{J_{\nu+1}(\lambda)^2\right\}^2, & \text{for } \lambda = \mu. \end{cases} \tag{12.138}$$

Equation (12.138) expresses the orthogonality of Bessel's functions of first kind with *weighting function* $= r$. This property enables us to expand any function in the interval $[0, 1]$. The series thus obtained is called *Fourier-Bessel series.* Thus, we can expand a function defined in $[0, 1]$ as

$$f(x) = \sum_{k=0}^{\infty} A_k J_\nu(\lambda_k x), \tag{12.139}$$

where the coefficients are obtained from

$$A_k = \frac{2}{\{J_{\nu+1}(\lambda_k)\}^2}\int_0^1 xf(x)J_\nu(\lambda_k x)\,dx. \tag{12.140}$$

□

12.14 Neumann's Function

Neumann's function is defined as

$$N_\nu(x) = \frac{1}{\sin\nu\pi}\left[J_\nu(x)\cos\nu\pi - J_{-\nu}(x)\right] \tag{12.141}$$

for any non-integral value of ν. Thus, for $\nu = \frac{1}{2}$, and $\nu = -\frac{1}{2}$, we have

$$N_{\frac{1}{2}}(x) = -\sqrt{\frac{2}{\pi x}}\cos x \tag{12.142}$$

$$\text{and} \quad N_{-\frac{1}{2}}(x) = \sqrt{\frac{2}{\pi x}}\sin x. \tag{12.143}$$

Above values have been obtained from equation (12.141) with the use of equations (12.92) and (12.93). If, however, $\nu = n$, an integer, the value of $N_\nu(x)$ becomes indeterminate. In this case, we find $N_\nu(x)$ in the limit as $\nu \longrightarrow n$ with the use of L'Hospital's rule. Thus, we get

$$N_n(x) = \lim_{\nu \to n} N_\nu(x) = \lim_{\nu \to n} \frac{\dfrac{\partial J_n(x)}{\partial \nu}\cos\nu\pi - \pi J_\nu(x)\sin\nu\pi - \dfrac{\partial J_{-\nu}(x)}{\partial \nu}}{\pi\cos\nu\pi}$$

$$\text{i.e.,} \quad N_n(x) = \frac{1}{\pi}\left[\left(\frac{\partial J_\nu}{\partial \nu}\right)_{\nu=n} - (-1)^n\left(\frac{\partial J_{-\nu}}{\partial \nu}\right)_{\nu=n}\right]. \tag{12.144}$$

It can be proved that, for integral values of n,

$$N_n(x) = \frac{1}{\pi}Y_n(x). \tag{12.145}$$

The solutions $J_\nu(x)$ and $N_\nu(x)$ of the Bessel's equation are always linearly independent so that the general solution can be written as

$$y(x) = c_1 J_\nu(x) + c_2 N_\nu(x) \tag{12.146}$$

for all values of ν including integral values.

In place of $N_\nu(x)$, we can take $J_{-\nu}(x)$ for non-integral vlaues of ν and $Y_n(x)$ for integral and also zero values of ν.

Graphical representation of Neumann's function for $\nu = 0, 1$ and 2 is shown in *Fig 12.3.*

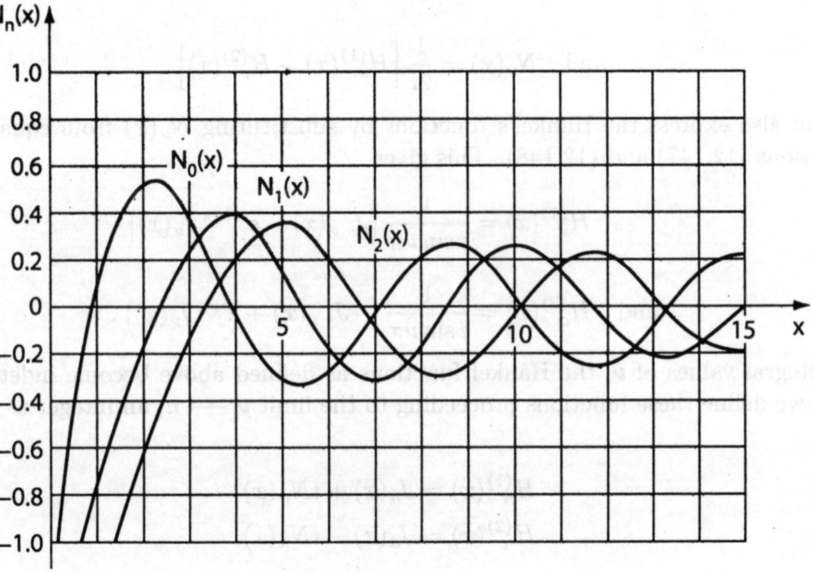

Fig 12.3

12.15 Hankel's Functions

Hankel's functions are defined as

$$H_\nu^{(1)}(x) = J_\nu(x) + iN_\nu(x) \tag{12.147}$$

$$\text{and} \quad H_\nu^{(2)}(x) = J_\nu(x) - iN_\nu(x) \tag{12.148}$$

for any non-integral value of ν. These functions are called *Bessel's function* of third kind. Thus, for $\nu = \frac{1}{2}$ and $\nu = -\frac{1}{2}$ with the use of equations (12.147) and (12.148), we have

$$H_{\frac{1}{2}}^{(1)}(x) = \sqrt{\frac{2}{\pi x}} \frac{e^{ix}}{i} \tag{12.149}$$

$$H_{-\frac{1}{2}}^{(1)}(x) = \sqrt{\frac{2}{\pi x}}\, e^{ix} \tag{12.150}$$

$$H^{(2)}_{\frac{1}{2}}(x) = -\sqrt{\frac{2}{\pi x}}\frac{e^{-ix}}{i} \tag{12.151}$$

and $$H^{(2)}_{-\frac{1}{2}}(x) = \sqrt{\frac{2}{\pi x}}\, e^{-ix}. \tag{12.152}$$

From equations (12.147) and (12.148), we can express $J_\nu(x)$ and $N_\nu(x)$ in terms of Hankel's functions in the following manner:

$$J_\nu(x) = \frac{1}{2}\left[H^{(1)}_\nu(x) + H^{(2)}_\nu(x)\right] \tag{12.153}$$

and $$N_\nu(x) = \frac{1}{2i}\left[H^{(1)}_\nu(x) - H^{(2)}_\nu(x)\right]. \tag{12.154}$$

We can also express the Hankel's functions by substituting $N_\nu(x)$ from equation (12.141) into equations (12.147) and (12.148). This gives

$$H^{(1)}_\nu(x) = \frac{1}{i\sin\nu\pi}\left[J_{-\nu}(x) - e^{-i\nu\pi}J_\nu(x)\right] \tag{12.155}$$

and $$H^{(2)}_\nu(x) = \frac{1}{i\sin\nu\pi}\left[-J_{-\nu}(x) + e^{i\nu\pi}J_\nu(x)\right]. \tag{12.156}$$

For integral values of ν, the Hankel functions as defined above become indeterminate. In this case, we define these functions proceeding to the limit $\nu \longrightarrow n$, an integer as before. Thus we get

$$H^{(1)}_n(x) = J_n(x) + iN_n(x) \tag{12.157}$$

$$H^{(2)}_n(x) = J_n(x) - iN_n(x), \tag{12.158}$$

where $N_n(x)$ is given by equation (12.145).

12.16 Modified Bessel's Functions

We consider the differential equation

$$x^2\frac{d^2y}{dx^2} + x\frac{dy}{dx} + \left(k^2x^2 - \nu^2\right)y = 0, \tag{12.159}$$

where k = constant, and ν = non-integral in general.

With the substitution $kx = z$, this equation is reduced to

$$z^2\frac{d^2y}{dz^2} + z\frac{dy}{dz} + \left(z^2 - \nu^2\right)y = 0, \tag{12.160}$$

which is Bessel's equation of order ν.

Hence, its general solution is

$$y = AJ_\nu(z) + BN_\nu(z), \tag{12.161}$$

A, B being constants. Going back to old variable x, equation (12.161) is written as

$$y = AJ_\nu(kx) + BN_\nu(kx). \tag{12.162}$$

If we take, $k = i$, the differential equation (12.159) takes the form

$$x^2\frac{d^2y}{dx^2} + x\frac{dy}{dx} - \left(x^2 + \nu^2\right)y = 0 \tag{12.163}$$

and its general solution is given as [putting $k = i$ in equation (12.162)]

$$y = AJ_\nu(ix) + BN_\nu(ix). \tag{12.164}$$

Now, replacing x by ix in equation (12.83), we get

$$i^{-\nu}J_\nu(ix) = \sum_{k=0}^{\infty}\frac{1}{\Gamma(k+1)\Gamma(\nu+k+1)}\left(\frac{x}{2}\right)^{\nu+2k} \tag{12.165}$$

The function

$$I_\nu(x) = \frac{J_\nu(ix)}{i^\nu} \tag{12.166}$$

is called *modified Bessel's function of first kind and of order* ν. If $\nu \neq$ an integer, the two linearly independent solutions of equation (12.163) are given by $I_\nu(x)$ and $I_{-\nu}(x)$. But, if n = an integer, they are not linearly independent, and hence we require a second solution. This second particular solution which will be linearly independent with $I_\nu(x)$ for all values of ν (including integral values) is called *modified Bessel's function of second kind and of order* ν. This is given by

$$K_\nu(x) = \frac{\pi i}{2}\, e^{\frac{\nu\pi i}{2}}\, H_\nu^{(1)}(ix). \tag{12.167}$$

The function in (12.167) is also known as *Macdonald function*.

Hence the general solution of equation (12.163) is

$$y = AI_\nu(x) + BK_\nu(x), \tag{12.168}$$

where ν may have integral as well as non-integral values.

For $\nu = \frac{1}{2}$ and $\nu = -\frac{1}{2}$, we have these relations:

$$I_{\frac{1}{2}}(x) = \sqrt{\frac{2}{\pi x}}\sinh x \tag{12.169}$$

$$I_{-\frac{1}{2}}(x) = \sqrt{\frac{2}{\pi x}}\cosh x \tag{12.170}$$

$$K_{\frac{1}{2}}(x) = k_{-\frac{1}{2}}(x) = \sqrt{\frac{\pi}{2x}}e^{-x}. \tag{12.171}$$

For $\nu = 0$, we have

$$I_0(x) = \sum_{k=0}^{\infty}\frac{1}{(k!)^2}\left(\frac{x}{2}\right)^{2k} \tag{12.172}$$

and

$$K_0(x) = -I_0(x)\left(\log\frac{x}{2} + C\right) + \sum_{k=0}^{\infty}\frac{1}{(k!)^2}\left(\frac{x}{2}\right)^{2k}\left(1 + \frac{1}{2} + \frac{1}{3} + \cdots\frac{1}{k}\right). \tag{12.173}$$

The general character of these functions is shown in *Fig 12.3(b)*.

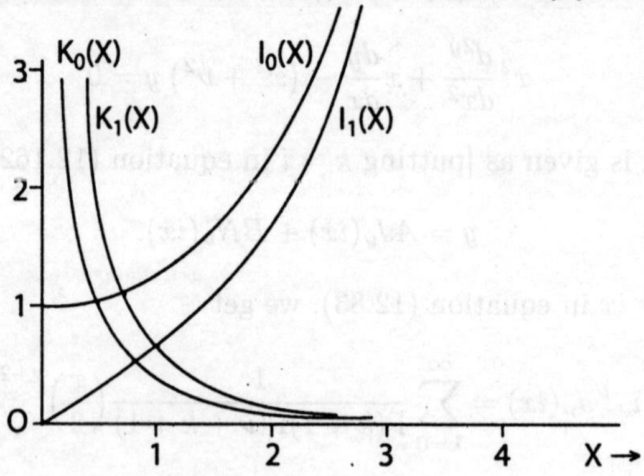

Fig 12.3(b)

For both the functions $I_\nu(x)$ and $K_\nu(x)$, following recurrence relations exist:

$$xI'_\nu(x) - \nu I_\nu(x) = xI_{\nu+1}(x) \tag{12.174}$$
$$xI'_\nu(x) + \nu I_\nu(x) = xI_{\nu-1}(x) \tag{12.175}$$
$$2I'_\nu(x) = I_{\nu+1}(x) + I_{\nu-1}(x) \tag{12.176}$$
$$2\nu I_\nu(x) = x\left[I_{\nu-1}(x) - I_{\nu+1}(x)\right]. \tag{12.177}$$

Similarly

$$xK'_\nu(x) - \nu K_\nu(x) = -xK_{\nu+1}(x) \tag{12.178}$$
$$xK'_\nu(x) + \nu K_\nu(x) = -xK_{\nu-1}(x) \tag{12.179}$$
$$-2K'_\nu(x) = K_{\nu+1}(x) + K_{\nu-1}(x) \tag{12.180}$$
$$2\nu K_\nu(x) = x\left[K_{\nu+1}(x) - K_{\nu-1}(x)\right]. \tag{12.181}$$

In particular, if $\nu = 0$, we have

$$I'_0(x) = I_1(x), \quad \text{and} \quad K'_0(x) = -K_1(x). \tag{12.182}$$

12.17 Kelvin's Functions

We consider the differential equation

$$\frac{d^2y}{dx^2} + \frac{1}{x}\frac{dy}{dx} - iy = 0, \tag{12.183}$$

which is obtained from equation (12.163) for $\nu = 0$ and $x \longrightarrow \sqrt{i}\,x$.

Equation (12.183) is a modified Bessel's equation of order zero. Hence, its solution is given by

$$y = AI_0\left(\sqrt{i}\,x\right) + BK_0\left(\sqrt{i}\,x\right). \tag{12.184}$$

The solution in (12.184) is a complex function. Kelvin introduced the following notations:

$$I_0\left(\sqrt{i}\,x\right) = \operatorname{Ber} x + i \operatorname{Bei} x$$
$$\text{and} \quad K_0\left(\sqrt{i}\,x\right) = \operatorname{Ker} x + i \operatorname{Kei} x. \tag{12.185}$$

Now, we replace x by $\sqrt{i}\,x$ in equation (12.165) and get

$$I_0\left(\sqrt{i}\,x\right) = \sum_{k=0}^{\infty} \frac{1}{(k!)^2}\left(\frac{\sqrt{i}\,x}{2}\right)^{2k} = 1 + \frac{ix^2}{2^2} + \frac{i^2x^4}{2^2 4^2} + \frac{i^3x^6}{2^2\,4^2\,6^2} + \cdots. \tag{12.186}$$

Separating real and imaginary parts, we have

$$\operatorname{Ber} x = 1 - \frac{x^4}{2^2\,4^2} + \frac{x^8}{2^2\,4^2\,6^2\,8^2} \tag{12.187}$$

and

$$\operatorname{Bei} x = \frac{x^2}{2^2} - \frac{x^6}{2^2\,4^2\,6^2} + \frac{x^{10}}{2^2 4^2\,6^2\,8^2\,10^2} - \cdots \tag{12.188}$$

From equations (12.187) and (12.188), it can be easily verified that

$$\left.\begin{aligned} \int_0^x x \operatorname{Ber} x\,dx &= x\frac{d}{dx}(\operatorname{Bei} x) \\ \int_0^x x \operatorname{Bei} x\,dx &= -x\frac{d}{dx}(\operatorname{Ber} x) \end{aligned}\right\}. \tag{12.189}$$

Again, replacing x by $\sqrt{i}\,x$ in equation (12.173), we get

$$\begin{aligned} K_0\left(\sqrt{i}\,x\right) &= -I_0\left(\sqrt{i}\,x\right)\left(\log\frac{\sqrt{i}\,x}{2} + C\right) \\ &\quad + \sum_{k=0}^{\infty}\frac{1}{(k!)^2}\left(\frac{\sqrt{i}x}{2}\right)^{2k}\left[1 + \frac{1}{2} + \cdots + \frac{1}{k}\right]. \end{aligned} \tag{12.190}$$

Now, $\sqrt{i}\,x = e^{\frac{i\pi}{4}}x$, hence

$$\log\sqrt{i}\,x = \frac{i\pi}{4} + \log x.$$

Thus, equation (12.190) with the use of equation (12.185) yields

$$\begin{aligned} K_0\left(\sqrt{i}\,x\right) &= (\operatorname{Ber} x + i\operatorname{Bei} x)\left(\log 2 - \frac{i\pi}{4} - \log x - C\right) \\ &\quad + \left[\frac{ix^2}{2^2} + \frac{i^2x^4}{2^2\cdot 4^2}\left(1+\frac{1}{2}\right) + \frac{i^3x^6}{2^2\cdot 4^2\cdot 6^2}\left(1+\frac{1}{2}+\frac{1}{3}\right) + \cdots\right] \end{aligned}$$

Separating real and imaginary parts

$$\begin{aligned} \operatorname{Ker} x &= \operatorname{Ber} x\,(\log 2 - \log x - C) + \frac{\pi}{4}\operatorname{Bei} x \\ &\quad - \frac{x^4}{2^2 4^2}\left(1+\frac{1}{2}\right) + \frac{x^8}{2^2\cdot 4^2\cdot 6^2\cdot 8^2}\left(1+\frac{1}{2}+\frac{1}{3}+\frac{1}{4}\right) - \cdots \end{aligned} \tag{12.191}$$

and

$$\begin{aligned} \operatorname{Kei} x &= \operatorname{Bei} x\,(\log 2 - \log x - C) - \frac{\pi}{4}\operatorname{Ber} x \\ &\quad + \frac{x^2}{2^2} - \frac{x^6}{2^2 4^2 6^2}\left(1+\frac{1}{2}+\frac{1}{3}\right) + \cdots \end{aligned} \tag{12.192}$$

It can be easily verified that

$$\int_0^x x\,\mathrm{Ker}\,x\,dx = x\frac{d}{dx}(\mathrm{Kei}\,x) \tag{12.193}$$

and
$$\int_0^x x\,\mathrm{Kei}\,x\,dx = -x\frac{d}{dx}(\mathrm{Ker}\,x) \tag{12.194}$$

12.18 Spherical Bessel's Functions

We consider the differential equation

$$r^2\frac{d^2R}{dr^2} + 2r\frac{dR}{dr} + \left[r^2 - n(n+1)\right]R = 0. \tag{12.195}$$

With the substitution

$$R(r) = \frac{z(r)}{\sqrt{r}} \tag{12.196}$$

equation (12.195) reduces to

$$r^2\frac{d^2z}{dr^2} + r\frac{dz}{dr} + \left[r^2 - \left(n+\frac{1}{2}\right)^2\right]z = 0. \tag{12.197}$$

Equation (12.197) is a Bessel's equation of order $\left(n+\frac{1}{2}\right)$ which has got solution

$$z(r) = AJ_{\left(n+\frac{1}{2}\right)}(r) + BJ_{-\left(n+\frac{1}{2}\right)}(r). \tag{12.198}$$

Spherical Bessel's functions are defined as

$$j_n(x) = \sqrt{\frac{\pi}{2x}}\,J_{n+\frac{1}{2}}(x) \tag{12.199}$$

and
$$y_n(x) = (-1)^{n+1}\sqrt{\frac{\pi}{2x}}\,J_{-\left(n+\frac{1}{2}\right)}(x). \tag{12.200}$$

Now, using equation (12.83), we have

$$J_{n+\frac{1}{2}}(x) = \sum_{k=0}^{\infty}\frac{(-1)^k}{\Gamma(k+1)\Gamma\left(n+k+\frac{3}{2}\right)}\left(\frac{x}{2}\right)^{n+2k+\frac{1}{2}}. \tag{12.201}$$

From Legendre's duplication formula, we have

$$\Gamma\left(n+k+\frac{3}{2}\right) = \frac{\sqrt{\pi}\,\Gamma(2n+2k+2)}{2^{2n+2k+1}\,\Gamma(n+k+1)}. \tag{12.202}$$

Hence, equation (12.201) becomes

$$J_{n+\frac{1}{2}}(x) = \sum_{k=0}^{\infty}(-1)^k\frac{(n+k)!\,2^{2n+2k+1}}{\sqrt{\pi}\,k!\,(2n+2k+1)!}\left(\frac{x}{2}\right)^{n+2k+\frac{1}{2}}. \tag{12.203}$$

Thus

$$j_n(x) = \sqrt{\frac{\pi}{2x}}J_{n+\frac{1}{2}}(x) = 2^n x^n\sum_{k=0}^{\infty}(-1)^k\frac{(n+k)!}{k!\,(2n+2k+1)!}x^{2k} \tag{12.204}$$

For $n = 0$

$$j_0(x) = \sum_{k=0}^{\infty} (-1)^k \frac{x^{2k}}{(2k+1)!} = \frac{1}{x} \sum_{k=0}^{\infty} (-1)^k \frac{x^{2k+1}}{(2k+1)!}$$

$$\text{i.e.,} \quad j_0(x) = \frac{\sin x}{x}. \tag{12.205}$$

Similarly, we can find

$$j_1(x) = \frac{\sin x}{x^2} - \frac{\cos x}{x}. \tag{12.206}$$

Similar expansion can be found for $y_n(x)$. In particular,

$$\text{for } n = 0, \quad y_0(x) = -\frac{\cos x}{x} \tag{12.207}$$

$$\text{and for } n = 1, \quad y_1(x) = -\frac{\cos x}{x^2} - \frac{\sin x}{x}. \tag{12.208}$$

Again, using the recurrence relations of Bessel's function of first kind, we get the recurrence relations of Spherical Bessel's function $j_n(x)$ as

$$j_{n+1}(x) + j_{n-1}(x) = \frac{2n+1}{x} j_n(x) \tag{12.209}$$

$$n j_{n-1}(x) - (n+1) j_{n+1}(x) = (2n+1) j_n'(x) \tag{12.210}$$

$$\frac{d}{dx}\left[x^{n+2} j_n(x)\right] = x^{n+1} j_{n-1}(x) \tag{12.211}$$

$$\text{and} \quad \frac{d}{dx}\left[x^{-n} j_n(x)\right] = -x^{-n} j_{n+1}(x). \tag{12.212}$$

EXAMPLE 12.1. *Show that*

(i) $\cos(x \sin\theta) = J_0(x) + 2\sum_{n=1}^{\infty} J_{2n}(x) \cos 2n\theta;$

(ii) $\sin(x \sin\theta) = 2\sum_{n=1}^{\infty} J_{2n-1}(x) \sin(2n-1)\theta.$

Solution: We have from the generating function [see equation (12.120)]

$$\exp\left[\frac{x}{2}\left(t - \frac{1}{t}\right)\right] = \sum_{n=-\infty}^{\infty} J_n(x) t^n. \tag{1}$$

Let $t = e^{i\theta}$, so that

$$t - \frac{1}{t} = 2i \sin\theta. \tag{2}$$

Substituting equation (2) in equation (1)

$$\exp(ix \sin\theta) = \sum_{n=-\infty}^{\infty} J_n(x) e^{in\theta}. \tag{3}$$

Equating real parts, we get

$$\cos(x \sin\theta) = \sum_{n=-\infty}^{\infty} J_n(x) \cos n\theta. \tag{4}$$

Now

$$\begin{aligned} J_n(x)t^n + J_{-n}(x)t^{-n} &= J_n(x)\left[e^{in\theta} + (-1)^n e^{-in\theta}\right] \\ &= 2J_n(x)\cos n\theta, \quad \text{for } n \text{ even} \\ &= 2iJ_n(x)\sin n\theta, \quad \text{for } n \text{ odd} \end{aligned} \tag{5}$$

Hence, from equations (4) and (5), we get

$$\cos(x\sin\theta) = J_0(x) + 2\sum_{n=1}^{\infty} J_{2n}\cos 2n\theta. \tag{6}$$

Similarly, equating imaginary parts in equation (3), we get

$$\sin(x\sin\theta) = 2\sum_{n=1}^{\infty} J_{2n-1}(x)\sin(2n-1)\theta. \tag{7}$$

If in particular, $\theta = \dfrac{\pi}{2}$, i.e., $t = i$, we get from equation (6)

$$\cos x = J_0(x) + 2\sum_{n=1}^{\infty}(-1)^n J_{2n}(x).$$

EXAMPLE 12.2. *Prove that*

$$\int_0^x x^\nu J_{\nu-1}(x)\,dx = x^\nu J_\nu(x).$$

Solution: We have, from equation (12.99)

$$\frac{d}{dx}\left[x^\nu J_\nu(x)\right] = x^\nu J_{\nu-1}(x).$$

Hence

$$\int_0^x x^\nu J_{\nu-1}(x)\,dx = x^\nu J_\nu(x).$$

EXAMPLE 12.3. *Show that*

$$\begin{vmatrix} J_\nu(x) & J'_\nu(x) \\ J_{-\nu}(x) & J'_{-\nu}(x) \end{vmatrix} = -\frac{2\sin\nu\pi}{\pi x}. \tag{1}$$

Solution: Let us denote, $y_1 = J_\nu(x)$ and $y_2 = J_{-\nu}(x)$.

Now, y_1, y_2 are two solutions of Bessel's equation

$$x^2\frac{d^2y}{dx^2} + x\frac{dy}{dx} + (x^2 - \nu^2)y = 0$$

$$\text{i.e.,}\quad \frac{d}{dx}\left(x\frac{dy}{dx}\right) + \left(x - \frac{\nu^2}{x}\right)y = 0. \tag{2}$$

Hence, we have

$$\frac{d}{dx}(xy_1') + \left(x - \frac{\nu^2}{x}\right) y_1 = 0 \tag{3}$$

$$\text{and} \quad \frac{d}{dx}(xy_2') + \left(x - \frac{\nu^2}{x}\right) y_2 = 0. \tag{4}$$

Multiplying (3) by $-y_2$, and (4) by y_1, and adding

$$\frac{d}{dx}[x(y_1y_2' - y_2y_1')] = 0,$$

which gives

$$x(y_1y_2' - y_2y_1') = \text{constant} = C, \tag{5}$$

i.e., $y_1y_2' - y_2y_1' = \dfrac{C}{x}$.

Equating coefficients of $\frac{1}{x}$ on both sides

$$\frac{1}{2^\nu\Gamma(\nu+1)} \cdot \frac{-1}{2^{-\nu}\Gamma(-\nu+1)}\{\nu + (-\nu)\} = C.$$

$$\therefore \quad C = \frac{-2\nu}{\Gamma(\nu+1)\Gamma(-\nu+1)} = \frac{-2}{\Gamma(\nu)\Gamma(1-\nu)} = -\frac{2\sin\nu\pi}{\pi}.$$

Hence

$$y_1y_2' - y_2y_1' = -\frac{2\sin\nu\pi}{\pi x} \tag{6}$$

Note 3. This shows that the Wronskian of the two solutions of the Bessel's equation is not zero. Hence the solutions are linearly independent (for non-integral values of ν).

Note 4. If we divide both sides of (6) by $\{J_\nu(x)\}^2$, we get

$$\frac{y_1y_2' - y_2y_1'}{y_1^2} = -\frac{2\sin\nu\pi}{\pi x y_1^2}$$

$$\text{i.e.,} \quad \frac{d}{dx}\left(\frac{y_2}{y_1}\right) = -\frac{2\sin\nu\pi}{\pi x y_1^2}$$

$$\text{i.e.,} \quad \frac{d}{dx}\left[\frac{J_{-\nu}(x)}{J_\nu(x)}\right] = -\frac{2\sin\nu\pi}{\pi x\{J_\nu(x)\}^2}.$$

EXAMPLE 12.4. *Prove that* $4J_\nu''(x) = J_{\nu-2}(x) - 2J_\nu(x) + J_{\nu+2}(x)$.

Solution: We have, from equation (12.103)

$$2J_\nu'(x) = J_{\nu-1}(x) - J_{\nu+1}(x). \tag{1}$$

Differentiating with respect to x,

$$2J_\nu''(x) = J_{\nu-1}'(x) - J_{\nu+1}'(x). \tag{2}$$

Replacing ν by $\nu - 1$ in equation (1)

$$2J_{\nu-1}'(x) = J_{\nu-2}(x) - J_\nu(x). \tag{3}$$

Replacing ν by $\nu+1$ in equation (1)

$$2J'_{\nu+1}(x) = J_\nu(x) - J_{\nu+2}(x). \tag{4}$$

From equations (3) and (4)

$$2\left[J'_{\nu-1}(x) - J'_{\nu+1}(x)\right] = J_{\nu-2}(x) - 2J_\nu(x) + J_{\nu+2}(x). \tag{5}$$

From equations (2) and (5), we get

$$4J''_\nu(x) = J_{\nu-2}(x) - 2J_\nu(x) + J_{\nu+2}(x).$$

EXAMPLE 12.5. *Solve:* $x\dfrac{d^2y}{dx^2} + a\dfrac{dy}{dx} + bxy = 0.$

Solution: We substitute $y = x^\alpha u(z)$, $z = x\sqrt{b}$, where $\alpha = \dfrac{(1-a)}{2}$.

Then

$$\frac{dy}{dx} = \alpha x^{\alpha-1}u(z) + \sqrt{b}x^\alpha u'(z)$$
$$\frac{d^2y}{dx^2} = \alpha(\alpha-1)x^{\alpha-2}u(z) + 2\alpha x^{\alpha-1}\sqrt{b}u'(z) + bx^\alpha u''(z).$$

The given equation is transformed to

$$z^2u'' + zu' + (z^2 - \alpha^2)\,u = 0,$$

which is a Bessel's equation of order α.

Hence, the solution is

$$u(z) = AJ_\alpha(z) + BN_\alpha(z), \tag{1}$$

i.e., $y = Ax^\alpha J_\alpha\left(x\sqrt{b}\right) + Bx^\alpha N_\alpha\left(x\sqrt{b}\right)$.

EXAMPLE 12.6. *Solve:* $x\dfrac{d^2y}{dx^2} + a\dfrac{dy}{dx} + bx^ky = 0.$

Solution: We substitute

$$\left.\begin{aligned} y &= x^\alpha u(z) \\ \text{and}\quad z &= Ax^\lambda \end{aligned}\right\}. \tag{1}$$

Then

$$\frac{dy}{dx} = \alpha x^{\alpha-1}u(z) + A\lambda x^{\alpha+\lambda-1}u(z) \tag{2}$$

$$\frac{d^2y}{dx^2} = (A\lambda)^2x^{\alpha+2\lambda-2}u''(z) + A\lambda(\alpha+\lambda-1)x^{\alpha+\lambda-2}u'(z)$$
$$+ A\lambda\alpha x^{\alpha+\lambda-2}u'(z) + \alpha(\alpha-1)x^{\alpha-2}u(z). \tag{3}$$

Hence the given equation

$$x\frac{d^2y}{dx^2} + a\frac{dy}{dx} + bx^ky = 0$$

is transformed into

$$(A\lambda)^2 u'' x^{\alpha+2\lambda-1} + A\lambda u'(\alpha+\alpha+\lambda-1+a)x^{\alpha+\lambda-1} + u\left[\alpha(\alpha-1)+a\alpha+bx^{k+1}\right]x^{\alpha-1} = 0$$

i.e., $$\lambda^2 z^2 u'' + \lambda z u'(2\alpha+\lambda-1+a) + \left[\alpha^2-\alpha+a\alpha+bx^{k+1}\right]u = 0$$

or, $$z^2 u''(z) + z u'(z)\left[\frac{2\alpha+\lambda-1+a}{\lambda}\right] + \left[\frac{\alpha^2-\alpha+a\alpha+bx^{k+1}}{\lambda^2}\right]u = 0. \tag{4}$$

We set

$$2\alpha+\lambda-1+a = \lambda, \text{ which gives } \alpha = \frac{1-a}{2}. \tag{5}$$

We set

$$z^2 = A^2 x^{2\lambda} = \frac{b}{\lambda^2} x^{k+1},$$

which gives

$$\lambda = \frac{k+1}{2} \quad \text{and} \quad A = \pm\frac{2\sqrt{b}}{k+1}. \tag{6}$$

Then

$$\frac{\alpha^2-\alpha+a\alpha}{\lambda^2} = \frac{1}{\lambda^2}\left[\left(\frac{1-a}{2}\right)\left(-1+\frac{1-a}{2}\right)+a\left(\frac{1-a}{2}\right)\right]$$

$$= -\frac{(1-a)^2}{4\lambda^2} = -\left(\frac{1-a}{1+k}\right)^2 = -p^2 \text{ (say)}. \tag{7}$$

Then equation (4) becomes

$$z^2 u''(z) + z u'(z) + (z^2-p^2)\,u(z) = 0.$$

Its solution is

$$u(z) = A' J_p(z) + B' J_{-p}(z). \tag{8}$$

Going back to old variables

$$y(x) = A' x^\alpha J_p(Ax^\lambda) + B' J_{-p}(Ax^\lambda),$$

where A, λ, p are given by equations (6) and (7).

EXAMPLE 12.7. *Determine an expression for* $\int x^2 J_2(x)\,dx$.

Solution: $I = \int x^2 J_2(x)\,dx = \int x^3\left[x^{-1} J_2(x)\right]dx.$

Integrating by parts

$$I = x^3 \int x^{-1} J_2(x)\,dx - \int 3x^2 \left\{\int x^{-1} J_2(x)\,dx\right\} dx$$

$$= x^3 \left[-x^{-1} J_1(x)\right] + \int 3x^2 \left[x^{-1} J_1(x)\right] dx$$

[using equation (12.98) for $\nu = 1$]

$$I = -x^2 J_1(x) + 3 \int x\, J_1(x)\,dx$$

$$= -x^2 J_1(x) + 3\left[x \int J_1(x)\,dx - \int 1 \cdot \left\{\int J_1(x) dx\right\} dx\right]$$

$$= -x^2 J_1(x) - 3x J_0(x) + 3\int J_0(x)\,dx \quad \text{[using equation (12.104)]}$$

12.19 An Application of Bessel's Function

We have discussed the problem of stability of struts in chapter 4 under different loading conditions. Here we shall consider the stability of high columns under the action of self-weight.

Let ρ = weight/unit length of the column; EI = constant flexural rigidity of the column and h = height of the column.

The column is free at the top end where we take the origin of coordinate, and is fixed at bottom.

At any section x from top, the compressive force is $P = \rho x$. The shear force is

$$Q = P\sin\theta \simeq P\tan\theta = \rho x \frac{dy}{dx}, \tag{12.213}$$

where θ is the angle made by the trangent at the section with the vertical.

Shear force due to bending is

$$Q = \text{EI}\frac{d^3y}{dx^3}. \tag{12.214}$$

From equations (12.213) and (12.214)

$$\text{EI}\frac{d^3y}{dx^3} = -\rho x\frac{dy}{dx} \quad \text{or,} \quad \text{EI}\frac{d^3y}{dx^3} + \rho x \frac{dy}{dx} = 0. \tag{12.215}$$

We take new variable

$$\xi = \frac{x}{h} \tag{12.216}$$

and denote

$$w = \frac{dy}{d\xi}, \quad \frac{\rho h^3}{\text{EI}} = \lambda^2. \tag{12.217}$$

Then equation (12.215) is reduced to

$$\frac{d^2w}{d\xi^2} + \lambda^2 \xi w = 0. \tag{12.218}$$

We make a change of variable

$$w = \xi^\alpha u(z), \quad z = A\xi^\beta, \tag{12.219}$$

where A, α, β are constants to be determined.

From equation (12.219) we find the derivatives of w:

$$\frac{dw}{d\xi} = \alpha\xi^{\alpha-1}u(z) + \xi^{\alpha}u'(z)\cdot A\beta\xi^{\beta-1} \tag{12.220}$$

$$\frac{d^2w}{d\xi^2} = \alpha(\alpha-1)\xi^{\alpha-2}u(z) + \alpha\xi^{\alpha-1}u'(z)\cdot A\beta\xi^{\beta-1} + A\beta(\alpha+\beta-1)\xi^{\alpha+\beta-2}u'(z) + A\beta\xi^{\alpha+\beta-1}\cdot u''(z)\cdot A\beta\xi^{\beta-1}. \tag{12.221}$$

Using equations (12.218) and (12.219) and dividing by $\xi^{\alpha-2}$, we get

$$(A\beta)^2u''(z)\xi^{2\beta} + A\beta\xi^{\beta}\left[\alpha + (\alpha+\beta-1)\right]u'(z) + \left[\alpha(\alpha-1) + \lambda^2\xi^3\right]u(z) = 0$$

$$\text{i.e.,}\quad z^2u''(z) + \frac{1}{\beta}(2\alpha+\beta-1)zu'(z) + \frac{1}{\beta^2}\left[\alpha(\alpha-1) + \lambda^2\xi^3\right]u(z) = 0. \tag{12.222}$$

We determine α, β such that

$$\frac{1}{\beta}(2\alpha+\beta-1) = 1, \quad \text{i.e.,} \quad \alpha = \frac{1}{2} \tag{12.223}$$

and

$$\frac{\lambda^2}{\beta^2}\xi^3 = z^2 = A^2\xi^{2\beta}, \quad \text{i.e.,} \quad \beta = \frac{3}{2} \quad \text{and} \quad A = \frac{\lambda}{\beta} = \frac{2\lambda}{3}.$$

Hence

$$\frac{1}{\beta^2}\alpha(\alpha-1) = -\frac{1}{9}. \tag{12.224}$$

In view of equations (12.223) and (12.224), equation (12.222) becomes

$$z^2u''(z) + zu'(z) + \left(z^2 - \frac{1}{9}\right)u(z) = 0. \tag{12.225}$$

Equation (12.225) is a Bessel's equation of first kind and order $\frac{1}{3}$. Hence, its general solution is

$$u(z) = CJ_{\frac{1}{3}}(z) + DJ_{-\frac{1}{3}}(z). \tag{12.226}$$

In terms of old variables, equation (12.226) becomes

$$w(\xi) = C\xi^{\frac{1}{2}}J_{\frac{1}{3}}\left(\frac{2}{3}\lambda\xi^{\frac{3}{2}}\right) + D\xi^{\frac{1}{2}}J_{-\frac{1}{3}}\left(\frac{2}{3}\lambda\xi^{\frac{3}{2}}\right). \tag{12.227}$$

The boundary conditions are

$$\left.\begin{aligned} &\text{for } x = 0, \text{ moment } M = 0, \text{ i.e., } \frac{d^2y}{dx^2} = 0 \\ \text{and} \quad &\text{for } x = h, \text{ rotation } \theta = 0, \text{ i.e., } \frac{dy}{dx} = 0 \end{aligned}\right\}. \tag{12.228}$$

In terms of the function $w(\xi)$, equations (12.228) are as under:

$$\left.\begin{aligned} &\text{for } \xi = 0, \quad \frac{dw}{d\xi} = 0 \\ \text{and} \quad &\text{for } \xi = 1, \quad w = 0 \end{aligned}\right\}. \tag{12.229}$$

Now, using formula (12.83) and (12.84), we have

$$F_1 = \xi^{\frac{1}{2}} J_{\frac{1}{3}}\left(\frac{2}{3}\lambda\xi^{\frac{3}{2}}\right) = \sum_{k=0}^{\infty} \frac{(-1)^k \xi^{\frac{1}{2}}}{\Gamma(k+1)\Gamma\left(k+\frac{4}{3}\right)} \left(\frac{1}{3}\lambda\xi^{\frac{3}{2}}\right)^{2k+\frac{1}{3}} \tag{12.230}$$

and $$F_2 = \xi^{\frac{1}{2}} J_{-\frac{1}{3}}\left(\frac{2}{3}\lambda\xi^{\frac{3}{2}}\right) = \sum_{k=0}^{\infty} \frac{(-1)^k \xi^{\frac{1}{2}}}{\Gamma(k+1)\Gamma\left(k+\frac{2}{3}\right)} \left(\frac{1}{3}\lambda\xi^{\frac{3}{2}}\right)^{2k-\frac{1}{3}}. \tag{12.231}$$

Boundary conditions (12.229) give

$$CF_1'\left(\frac{2}{3}\lambda\xi^{\frac{3}{2}}\right) + DF_2'\left(\frac{2}{3}\lambda\xi^{\frac{3}{2}}\right) = 0, \quad \text{for } \xi = 0 \tag{12.232}$$

and $$CF_1\left(\frac{2}{3}\lambda\xi^{\frac{3}{2}}\right) + DF_2\left(\frac{2}{3}\lambda\xi^{\frac{3}{2}}\right) = 0, \quad \text{for } \xi = 1. \tag{12.233}$$

Recalling that

$$\frac{d}{dx}\left[J_\nu(\alpha x)\right] = -\frac{v}{x} J_\nu(\alpha x) + \alpha J_{\nu-1}(\alpha x)$$

we find

$$F_1' = \lambda J_{-\frac{3}{2}}\left(\frac{2}{3}\lambda\xi^{\frac{3}{2}}\right)$$

which $\longrightarrow \infty$ as $\xi \longrightarrow 0$.

The first condition in equation (12.229) gives

$$C = 0. \tag{12.234}$$

From condition (12.233), we get

$$F_2\left(\frac{2}{3}\lambda\right) = 0, \text{ since } D \neq 0. \tag{12.235}$$

The minimum value of the argument for which condition (12.235) is satisfied is found to be

$$\frac{2}{3}\lambda = 1.866, \tag{12.236}$$

which gives, from (12.217)

$$\rho_{cr} = \lambda^2 \frac{\mathrm{EI}}{h^3} = 7.84 \frac{\mathrm{EI}}{h^3}. \tag{12.237}$$

LEGENDRE'S FUNCTIONS

12.20 Legendre's Differential Equation

The differential equation

$$(1-x^2)\frac{d^2y}{dx^2} - 2x\frac{dy}{dx} + \lambda y = 0, \tag{12.238}$$

where λ is a constant, is known as *Legendre's differential equation*.

The differential equation has no singularity at $x = 0$ and so we can develop a solution in power series about $x = 0$.

We assume the solution in the form

$$y(x) = \sum_{k=0}^{\infty} a_k x^{r+k}. \tag{12.239}$$

Substituting equation (12.239) into (12.238) and re-arranging terms, we get

$$\sum_{k=0}^{\infty} a_k (k+r)(k+r-1)x^{r+k-2}$$
$$-\sum_{k=0}^{\infty} a_k \left[(k+r)(k+r-1) + 2(k+r) - \lambda\right] x^{r+k} = 0. \tag{12.240}$$

Equating the coefficients of x^{r-2} to zero we get $a_0 r(r-1) = 0$, which gives $r = 0$, and $r = 1$ (since $a_0 \neq 0$).

Equating the coefficients of x^{r-1} to zero we get $a_1(r+1)r = 0$, which gives $a_1 = 0$, for $r = 1$, but a_1 is arbitrary for $r = 0$.

The recursion formula for the coefficients is given by

$$a_{k+2} = \frac{[(k+r)(k+r-1) + 2(k+r) - \lambda]}{(k+r+2)(k+r+1)} a_k. \tag{12.241}$$

Hence, *for* $r = 0$

$$a_{k+2} = \frac{k(k-1) + 2k - \lambda}{(k+2)(k+1)} a_k = \frac{k(k+1) - \lambda}{(k+2)(k+1)} a_k. \tag{12.242}$$

Since $a_0 \neq 0$, and a_1 is arbitrary for $r = 0$, we can find all the coefficients in terms of these two in the following manner.

For k even,

$$k = 0, \quad a_2 = -\frac{\lambda}{2!} a_0$$

$$k = 2, \quad a_4 = \frac{2.3 - \lambda}{4.3} a_2 = \frac{6 - \lambda}{4.3} a_2 = -\frac{\lambda(6-\lambda)}{4!} a_0$$

$$k = 4, \quad a_6 = \frac{4.5 - \lambda}{6.5} a_4 = -\frac{\lambda(6-\lambda)(20-\lambda)}{6!} a_0$$

.... etc.

Similarly, for *k odd,*

$$k = 1, \quad a_3 = \frac{1.2 - \lambda}{3.2} a_1 = \frac{2 - \lambda}{3!} a_1$$

$$k = 3, \quad a_5 = \frac{3.4 - \lambda}{5.4} a_3 = \frac{(2-\lambda)(12-\lambda)}{5!} a_1$$

$$k = 5, \quad a_7 = \frac{5.6 - \lambda}{7.6} a_5 = \frac{(2-\lambda)(12-\lambda)(30-\lambda)}{7!} a_1$$

.... etc.

Thus, the solution of the differential equation is

$$y(x) = a_0 y_1(x) + a_1 y_2(x), \tag{12.243}$$

where

$$y_1(x) = 1 - \frac{\lambda}{2!}x^2 - \frac{\lambda(6-\lambda)}{4!}x^4 - \frac{\lambda(6-\lambda)(20-\lambda)}{6!}x^6 - \cdots \tag{12.244}$$

$$\text{and} \quad y_2(x) = x + \frac{2-\lambda}{3!}x^3 + \frac{(2-\lambda)(12-\lambda)}{5!}x^5 + \cdots \tag{12.245}$$

Now, *for* $r = 1$, the recursion formula (12.241) becomes

$$a'_{k+2} = \frac{(k+1)k + 2(k+1) - \lambda}{(k+3)(k+2)}a'_k = \frac{(k+2)(k+1) - \lambda}{(k+3)(k+2)}a'_k. \tag{12.246}$$

For $r = 1$, $a'_0 \neq 0$, but $a'_1 = 0$. Hence all the odd coefficients are zero.
The even coefficients are:

$$k = 0, \quad a'_2 = \frac{2.1 - \lambda}{3.2}a'_0 = \frac{2-\lambda}{3!}a'_0$$

$$k = 2, \quad a'_4 = \frac{4.3 - \lambda}{5.4}a'_2 = \frac{(2-\lambda)(12-\lambda)}{5!}a'_0$$

$$k = 4, \quad a'_6 = \frac{6.5 - \lambda}{7.6}a'_4 = \frac{(2-\lambda)(12-\lambda)(30-\lambda)}{7!}a'_0$$

etc.

Hence, for $r = 1$, the solution is

$$y(x) = a'_0\left[x + \frac{2-\lambda}{3!}x^3 + \frac{(2-\lambda)(12-\lambda)}{5!}x^5 + \cdots\right] \tag{12.247}$$

$$= a'_0 y_2(x) \text{ [from equation (12.245)]}.$$

This shows that the particular solution $y_2(x)$ is not independent as obtained before. Hence we take the solution for $r = 0$ as

$$y(x) = a_0 y_1(x) = a_0\left[1 - \frac{\lambda}{2!}x^2 - \frac{\lambda(6-\lambda)}{4!}x^4 - \frac{\lambda(6-\lambda)(20-\lambda)}{6!}x^6 - \cdots\right]. \tag{12.248}$$

The series converges for $|x| < 1$ and diverges for $|x| = 1$. This can be tested by D'Alembert's ratio test.

Let us now develop a solution which is valid in the region $-1 \leq x \leq 1$. For this purpose, we assume

$$\lambda = n(n+1), \tag{12.249}$$

where n is a positive integer or zero. In such a case, the point at infinity is a regular singularity of Legendre's differential equation (12.238). We assume the solution in the form

$$y = \sum_{k=0}^{\infty} a_k x^{r-k}. \tag{12.250}$$

Substituting the solution function into the differential equation (12.238), and rearranging the terms, we get

$$\sum_{k=0}^{\infty} a_k(r-k)(r-k-1)x^{r-k-2}$$
$$+\sum_{k=0}^{\infty} a_k\left[\{n(n+1)-(r-k)(r-k-1)-2(r-k)\}\right]x^{r-k}=0. \tag{12.251}$$

Equating to zero the coefficient of the highest power of x, i.e., x^r, we get the indicial equation

$$a_0[n(n+1)-r(r-1)-2r]=0$$
$$\text{or,}\quad a_0[(n-r)(n+r+1)]=0,$$

which gives

$$r=n \text{ and } r=-(n+1). \tag{12.252}$$

From equation (12.251) the recursion formula is obtained as

$$a_k=-\frac{(r-k+2)(r-k+1)}{n(n+1)-(r-k)(r-k+1)}a_{k-2}. \tag{12.253}$$

For $r=n$, this recursion formula becomes

$$a_k=-\frac{(n-k+2)(n-k+1)}{n(n+1)-(n-k)(n-k+1)}a_{k-2}$$
$$\text{i.e.,}\quad a_k=-\frac{(n-k+2)(n-k+1)}{k(2n-k+1)}a_{k-2}. \tag{12.254}$$

Putting $k=1$ in equation (12.251), we get

$$a_1\{n(n+1)-(r-1)(r-2)-2(r-1)\}=0$$
$$\text{i.e.,}\quad a_1\{n(n+1)-r(r-1)\}=0,$$

which gives

$$a_1=0. \tag{12.255}$$

Putting $k=2,3,4,\ldots$ in equation (12.254),

$$a_2=-\frac{n(n-1)}{2(2n-1)}a_0$$
$$a_4=-\frac{(n-2)(n-3)}{4(2n-3)}a_2=\frac{n(n-1)(n-2)(n-3)}{2.4.(2n-1)(2n-3)}\cdot a_0$$

....

....

In view of equation (12.255)

$$a_1=a_3=a_5=\cdots=0\cdots$$

Hence, for $r = n$, the solution is (putting $a_0 = 1$)

$$y_1(x) = x^n \left[1 - \frac{n(n-1)}{2(2n-1)}x^{-2} + \frac{n(n-1)(n-2)(n-3)}{2.4.(2n-1)(2n-3)}x^{-4} - \cdots\right]. \tag{12.256}$$

Again, for $r = -(n+1)$, the recursion formula (12.253) becomes

$$a_k = \frac{(n+k)(n+k-1)}{k(2n+k+1)}a_{k-2}. \tag{12.257}$$

Putting $k = 2, 3, 4 \ldots$ in equation (12.257), we find the coefficients as

$$a_2 = \frac{(n+1)(n+2)}{2(2n+3)}a_0$$

$$a_4 = \frac{(n+3)(n+4)}{4(2n+5)}a_2 = \frac{(n+1)(n+2)(n+3)(n+4)}{2.4.(2n+3)(2n+5)} \cdot a_0$$

....

....

and as before, $a_1 = a_3 = \cdots = 0$.

Hence, the solution for $r = -(n+1)$ is given by

$$y_2(x) = x^{-n-1}\left\{1 + \frac{(n+1)(n+2)}{2(2n+3)}x^{-2} + \frac{(n+1)(n+2)(n+3)(n+4)}{2.4.(2n+3)(2n+5)}x^{-4} + \cdots\right\}. \tag{12.258}$$

The two solutions $y_1(x)$, $y_2(x)$ are linearly independent, so that the general solution of Legendre's equation is

$$y(x) = Ay_1(x) + By_2(x). \tag{12.259}$$

Both the solutions $y_1(x)$, $y_2(x)$ are series in descending powers of x and do not terminate. However

$$\lim_{k\to\infty}\left|\frac{y_{k+2}}{y_k}\right| = \frac{1}{x^2},$$

which converges for $|x| > 1$.

We take

$$A = \frac{(2n)!}{2^n(n!)^2} \tag{12.260}$$

and write the solution $Ay_1(x) = P_n(x)$, where

$$P_n(x) = \sum_{k=0}^{N} \frac{(-1)^k(2n-2k)!}{2^n k!(n-k)!(n-2k)!}x^{n-2k}$$

with

$$N = \begin{cases} \dfrac{n}{2}, & \text{for } n \text{ even} \\[2ex] \dfrac{(n-1)}{2}, & \text{for } n \text{ odd.} \end{cases} \tag{12.261}$$

The function $P_n(x)$ is called *Legendre's polynomial* of first kind of degree n. The values of $P_n(x)$ for $n = 0, 1, 2, 3, \ldots$ are calculated as under:

$$P_0(x) = 1$$
$$P_1(x) = x$$
$$P_2(x) = \tfrac{1}{2}\,(3x^2 - 1)$$
$$P_3(x) = \tfrac{1}{2}\,(5x^3 - 3x)$$
$$P_4(x) = \tfrac{1}{8}\,(35x^4 - 30x^2 + 3)$$
$$P_5(x) = \tfrac{1}{8}\,(63x^5 - 70x^3 + 15x)$$
$$\ldots\ldots\ldots\ldots\ldots\ldots \text{etc.}$$

These functions are shown graphically in the *Fig 12.4.*

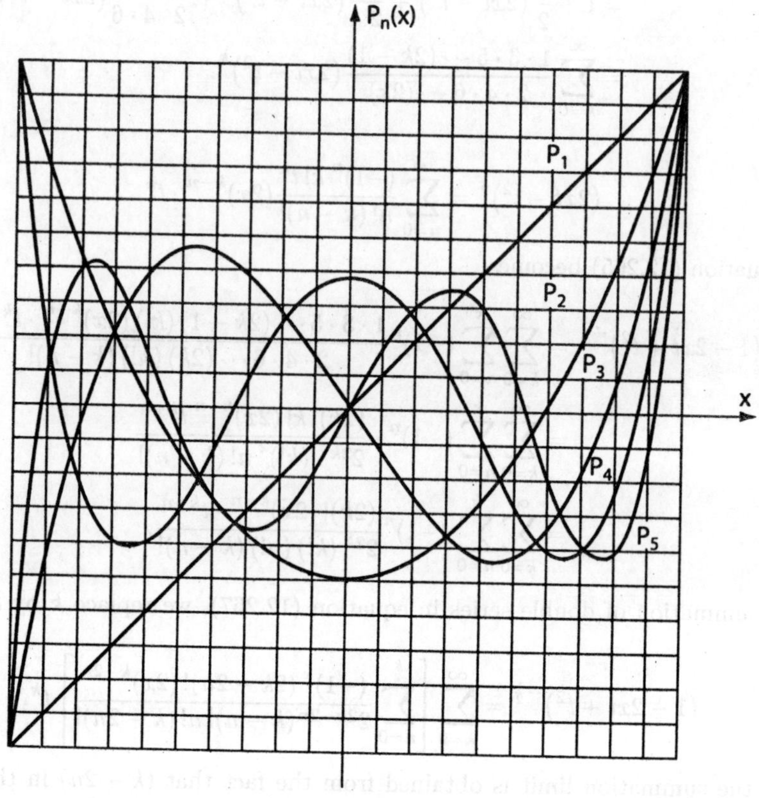

Fig 12.4

We take

$$B = \frac{2^n(n!)}{(2n+1)!} \tag{12.262}$$

and write

$$Q_n(x) = By_2(x). \tag{12.263}$$

The function $Q_n(x)$ is called Legendre's polynomial of second kind. These functions are seldom used in solving physical problems.

12.21 Generating Function for $P_k(x)$

Legendre's polynomial $P_k(x)$ is the coefficient of t^n in the expansion of

$$(1-2xt+t^2)^{-\frac{1}{2}}$$

that is

$$(1-2xt+t^2)^{-\frac{1}{2}} = \sum_{k=0}^{\infty} P_k(x)t^k. \tag{12.264}$$

We now prove this.

The function in the left-hand side of equation (12.264) can be expanded binomially. Thus

$$\begin{aligned}(1-2xt+t^2)^{-\frac{1}{2}} &= \{1-(2xt-t^2)\}^{-\frac{1}{2}} \\ &= 1+\frac{1}{2}(2xt-t^2)+\frac{1.3}{2.4}(2xt-t^2)^2+\frac{1\cdot 3\cdot 5}{2\cdot 4\cdot 6}(2xt-t^2)^3+\cdots \\ &= \sum_{k=0}^{\infty}\frac{1\cdot 3\cdot 5\cdots(2k-1)}{2\cdot 4\cdot 6\cdots(2k)}(2xt-t^2)^k. \end{aligned} \tag{12.265}$$

Now

$$(2xt-t^2)^k = \sum_{n=0}^{\infty}\frac{(-1)^n k!\, t^k}{n!\,(k-n)!}(2x)^{k-n}\cdot t^n. \tag{12.266}$$

Thus, equation (12.265) becomes

$$\begin{aligned}(1-2xt+t^2)^{-\frac{1}{2}} &= \sum_{k=0}^{\infty}\sum_{n=0}^{k}(-1)^n\frac{1\cdot 3\cdot 5\cdots(2k-1)\,(k!)\,(2x)^{k-n}\cdot t^{k+n}}{2\cdot 4\cdot 6\cdots(2k)\,(n!)\,(k-n)!} \\ &= \sum_{k=0}^{\infty}\sum_{n=0}^{k}(-1)^n\cdot\frac{(2k)!\,k!\,(2x)^{k-n}\, t^{k+n}}{2^{2k}\cdot(k!)^2\, n!\,(k-n)!} \\ &= \sum_{k=0}^{\infty}\sum_{n=0}^{k}(-1)^n\frac{(2k)!\,(2x)^{k-n}\cdot t^{k+n}}{2^{2k}\,(k!)\,(n!)\,(k-n)!}. \end{aligned} \tag{12.267}$$

For the summation of double series in equation (12.267), we replace k by $k-n$ and sum over n, i.e.,

$$(1-2xt+t^2)^{-\frac{1}{2}} = \sum_{k=0}^{\infty}\left[\sum_{n=0}^{\frac{k}{2}}\frac{(-1)^k\,(2k-2n)!\,(2x)^{k-2n}}{2^{2k-2n}(k-n)!\,n!\,(k-2n)!}\right]t^k, \tag{12.268}$$

where $\frac{k}{2}$ in the summation limit is obtained from the fact that $(k-2n)$ in the denominator should be positive, since

$$\frac{k}{2} = \begin{cases}\frac{1}{2}k, & \text{for } k \text{ even} \\ \frac{1}{2}(k-1), & \text{for } k \text{ odd.}\end{cases}$$

The series in equation (12.268) is identical to that in the r.h.s. of equation (12.256). Hence, we get

$$(1-2xt+t^2)^{-\frac{1}{2}} = \sum_{k=0}^{\infty} P_k(x)t^k. \tag{12.269}$$

Followings are the direct consequences of the equation (12.269)

(1) $$P_k(1) = 1. \tag{12.270}$$

If we put $x = 1$, we have

$$\left(1 - 2t + t^2\right)^{-\frac{1}{2}} = \sum_{k=0}^{\infty} P_k(1)t^k$$

$$\text{i.e.,} \quad (1-t)^{-1} = \sum_{k=0}^{\infty} P_k(1)t^k.$$

Hence, $P_k(1) = \text{coefficient of } t^k = 1$.

(2) $$P_k(-1) = (-1)^k P_k(1).$$

If we put $x = -1$, we have

$$\left(1 + 2t + t^2\right)^{-\frac{1}{2}} = \sum_{k=0}^{\infty} P_k(-1)t^k$$

$$\text{or,} \quad (1+t)^{-1} = \sum_{k=0}^{\infty} P_k(-1)t^k.$$

Hence, $P_k(-1) = \text{coefficient of } t^k = (-1)^k$.
Thus

$$P_k(-1) = (-1)^k = (-1)^k P_k(1). \tag{12.271}$$

(3) $$P_{2k}(0) = \frac{(-1)^k \cdot (2k)!}{2^{2k} \cdot k!}. \tag{12.272}$$

If we put $x = 0$, we get

$$\left(1 + t^2\right)^{-\frac{1}{2}} = \sum_{k=0}^{\infty} P_k(0)t^k.$$

Hence

$$\sum_{k=0}^{\infty} P_k(0)t^k = \left(1+t^2\right)^{-\frac{1}{2}}$$

$$= 1 - \frac{t^2}{2} + \frac{3}{8}t^4 + \cdots + (-1)^k \frac{1 \cdot 3 \cdot 5 \cdots (2k-1)}{2^k k!} t^{2k} + \cdots$$

Equating the coefficients of t^{2k},

$$P_{2k}(0) = (-1)^k \frac{1 \cdot 3 \cdot 5 \cdots (2k-1)}{2^k \cdot k!} = \frac{(-1)^k \cdot (2k)!}{2^{2k} \cdot (k!)^2}.$$

Similarly, equating the coefficient of $t^{2k} + 1$, we find $P_{2k+1}(0) = 0$.

(4) $P_k(-x) = (-1)^k P_k(x)$.
We have

$$\left(1 - 2xt + t^2\right)^{-\frac{1}{2}} = \sum_{k=0}^{\infty} P_k(x)t^k. \tag{12.273}$$

39

Changing x to $-x$, and t to $-t$, we get

$$(1-2xt+t^2)^{-\frac{1}{2}} = \sum_{k=0}^{\infty} P_k(-x)(-t)^k$$

$$\text{i.e.,} \quad \sum_{k=0}^{\infty} P_k(x)t^k = \sum_{k=0}^{\infty} (-1)^k P_k(-x)t^k$$

Equating coefficients of t^k,

$$P_k(x) = (-1)^k P_k(-x)$$

$$\text{or,} \quad P_k(-x) = (-1)^k P_k(x). \tag{12.274}$$

(5) If (r_1, θ_1) and (r_2, θ_2) be the polar coordinates of two points, then

$$\frac{1}{r} = \sum_{k=0}^{\infty} \frac{1}{r_1}\left(\frac{r_2}{r_1}\right)^k P_k(\cos\theta), \tag{12.275}$$

where r = distance between the points, and $\theta = \theta_1 - \theta_2$.
We have

$$r = (r_1^2 - 2r_1 r_2 \cos\theta + r_2^2)^{\frac{1}{2}}$$

$$= r_1\left[1 - 2\frac{r_2}{r_1}\cos\theta + \left(\frac{r_2}{r_1}\right)^2\right]^{\frac{1}{2}}.$$

Hence

$$\frac{1}{r} = \frac{1}{r_1}\left[1 - 2\frac{r_2}{r_1}\cos\theta + \left(\frac{r_2}{r_1}\right)^2\right]^{-\frac{1}{2}}$$

$$= \frac{1}{r_1}\sum_{k=0}^{\infty} P_k(\cos\theta)\left(\frac{r_2}{r_1}\right) \quad \left[\text{putting } x = \cos\theta, t = \frac{r_2}{r_1}\right].$$

Hence the result.

12.22 Recurrence Relations for $P_k(x)$

We denote

$$\phi(x,t) = (1-2xt+t^2)^{-\frac{1}{2}}.$$

Then

$$\frac{\partial \phi}{\partial x} = \frac{t}{1-2xt+t^2}\phi \tag{12.276}$$

and

$$\frac{\partial \phi}{\partial t} = \frac{x-t}{1-2xt+t^2}\phi. \tag{12.277}$$

We have

$$\sum_{k=0}^{\infty} P_k(x)t^k = (1-2xt+t^2)^{-\frac{1}{2}} = \phi(x,t).$$

(1) Differentiating partially with respect to t, we have

$$\sum_{k=0}^{\infty} kP_k(x)t^{k-1} = \frac{\partial\phi}{\partial t} = \frac{x-t}{1-2xt+t^2}\cdot\phi$$

[using equation (12.277)]

$$\text{or,}\quad \left(1-2xt+t^2\right)\sum_{k=0}^{\infty} kP_k(x)t^{k-1} = (x-t)\phi(x,t) = (x-t)\sum_{k=0}^{\infty} P_k(x)t^k.$$

Equating the coefficients of t^k on both sides

$$(k+1)P_{k+1}(x) - 2xkP_k(x) + (k-1)P_{k-1}(x) = xP_k(x) - P_{k-1}(x)$$

$$\text{or,}\quad (2k+1)xP_k(x) = (k+1)P_{k+1}(x) + kP_{k-1}(x). \tag{12.278}$$

(2) Now, from equation (12.276) and (12.277), we get

$$t\frac{\partial\phi}{\partial t} + (t-x)\frac{\partial\phi}{\partial x} = 0. \tag{12.279}$$

Since

$$\phi(x,t) = \sum_{k=0}^{\infty} P_k(x)t^k,$$

we have

$$\left.\begin{aligned}\frac{\partial\phi}{\partial x} &= \sum_{k=0}^{\infty} P_k'(x)t^k\\ \frac{\partial\phi}{\partial t} &= \sum_{k=0}^{\infty} kP_k(x)t^{k-1}\end{aligned}\right\}. \tag{12.280}$$

From equations (12.279) and (12.280), we get

$$t\sum_{k=0}^{\infty} kP_k(x)t^{k-1} + (t-x)\sum_{k=0}^{\infty} P_k'(x)t^k = 0.$$

Equating coefficients of t^k, we get

$$kP_k(x) - xP_k'(x) + P_{k-1}'(x) = 0. \tag{12.281}$$

(3) Again, we have

$$\frac{\partial\phi}{\partial x} = \frac{t}{(1-2xt+t^2)}\phi \quad\text{or,}\quad \sum_{k=0}^{\infty} P_k'(x)t^k = \frac{t}{(1-2xt+t^2)}\sum_{k=0}^{\infty} P_k(x)t^k.$$

$$\text{or,}\quad (1-2xt+t^2)\sum_{k=0}^{\infty} P_k'(x)t^k = t\sum_{k=0}^{\infty} P_k(x)t^k.$$

Equating coefficients of t^k on both sides

$$P'_k(x) - 2xP'_{k-1}(x) + P'_{k-2}(x) = P_{k-1}(x).$$

Replacing k by $k+1$,

$$P'_{k+1}(x) + P'_{k-1}(x) = 2xP'_k(x) + P_k(x). \tag{12.282}$$

(4) Differentiating equation (12.278) with respect to x, we get

$$(2k+1)xP'_k(x) + (2k+1)P_k(x) = (k+1)P'_{k+1}(x) + kP'_{k-1}(x). \tag{12.283}$$

Multiplying equation (12.282) by $(2k+1)$, and equation (12.283) by 2 and subtracting one from the other, we get

$$P'_{k+1}(x) - P'_{k-1}(x) = (2k+1)P_k(x). \tag{12.284}$$

(5) Differentiating equation (12.278) with respect to x and adding to the result k-times equation (12.282) we get

$$P'_{k+1}(x) = (1+k)P_k(x) + xP'_k(x). \tag{12.285}$$

Thus, we have found some recurrence relations which can again be used to develop some more relations.

12.23 Rodrigue's Formula

The Legendre's functions $P_k(x)$ are given by

$$P_k(x) = \frac{1}{2^k\, k!}\left(\frac{d}{dx}\right)^k \left(x^2-1\right)^k \tag{12.286}$$

which is known as *Rodrigue's formula.*

In order to prove this formula, we substitute

$$z = \left(x^2-1\right)^k. \tag{12.287}$$

Then

$$\frac{dz}{dx} = 2kx\left(x^2-1\right)^{k-1} \quad \text{or,} \quad \left(1-x^2\right)\frac{dz}{dx} + 2kxz = 0.$$

Differentiating this $(k+1)$ times with respect to x and using Leibnitz's theorem, we get (using $D = \frac{d}{dx}$)

$$\left(1-x^2\right)D^{k+2}(z) - (k+1)2xD^{k+1}(z) - \frac{k(k+1)}{2}\cdot 2D^k(z)$$
$$+2k\left[xD^{k+1}(z) + (k+1)D^k(z)\right] = 0$$

or,
$$\left(1-x^2\right)D^{k+2}(z) - 2xD^{k+1}(z) + k(k+1)D^k(z) = 0$$

or,
$$\left(1-x^2\right)\frac{d^2y}{dx^2} - 2x\frac{dy}{dx} + k(k+1)y = 0, \tag{12.288}$$

where

$$y = D^k(z) = \frac{d^k z}{dx^k}.$$

Equation (12.288) is a Legendre's equation whose convergent solution for k, a positive integer, is $P_k(x)$. Hence, we conclude that

$$P_k(x) = Ay = A\frac{d^k}{dx^k}\left(x^2-1\right)^k. \tag{12.289}$$

To determine the unknown constant A, we write

$$P_k(x) = A\frac{d^k}{dx^k}(x-1)^k(x+1)^k.$$

Using Leibnitz' theorem $P_k(x) = A[k!\,(x+1)^k$ + terms containing powers of $(x-1)]$
Putting $x = 1$,

$$P_k(1) = Ak!\,2^k,$$

$$\text{or,}\quad 1 = Ak!\,2^k;\ \text{thus } A = \frac{1}{2^k\,k!}. \tag{12.290}$$

Hence we write from equations (12.289) and (12.290)

$$P_k(x) = \frac{1}{2^k\,k!}\frac{d^k}{dx^k}\left(x^2-1\right)^k.$$

12.24 Orthogonality of Legendre's Polynomials

Legendre's polynomials are a set of orthogonal functions in the interval $[-1, 1]$, weighting function being unity. Thus

$$\int_{-1}^{1} P_m(x)P_n(x)dx = \frac{2}{2n+1}\delta_{mn}, \tag{12.291}$$

where δ_{mn} = Kronecker's delta.

Let us now prove it.

Since the Legendre's polynomials satisfy the differential equation

$$\left(1-x^2\right)y'' - 2xy' + k(k+1)y = 0$$

$$\text{i.e.,}\quad \frac{d}{dx}\left[\left(1-x^2\right)y'\right] + k(k+1)\,y = 0,$$

we can write for the functions $P_m(x)$, $P_n(x)$ the following two differential equations:

$$\frac{d}{dx}\left[\left(1-x^2\right)P_m'(x)\right] + m(m+1)P_m(x) = 0 \tag{12.292}$$

$$\text{and}\quad \frac{d}{dx}\left[\left(1-x^2\right)P_n'(x)\right] + n(n+1)P_n(x) = 0. \tag{12.293}$$

Multiplying equation (12.292) by $P_n(x)$ and (12.293) by $P_m(x)$ and then subtracting, we get

$$\frac{d}{dx}\left[\left(1-x^2\right)\left\{P_m(x)\,P_n'(x) - P_n(x)P_m'(x)\right\}\right] + \left[n(n+1) - m(m+1)\right]P_m(x)\cdot P_n(x) = 0.$$

Integrating from -1 to 1, we get

$$(1-x^2)\left\{P_m(x)\,P_n'(x) - P_n(x)\,P_m'(x)\right\}_{-1}^{1} + (n-m)(n+m+1)\int_{-1}^{1} P_m(x)P_n(x)\,dx = 0.$$

The first term is zero because of the factor $(1-x^2)$.
Hence, we get

$$(n-m)(n+m+1)\int_{-1}^{1} P_m(x)P_n(x)\,dx = 0,$$

which gives

$$\int_{-1}^{1} P_m(x)P_n(x)\,dx = 0, \quad \text{for } m \neq n. \tag{12.294}$$

For $m = n$, we evaluate this integral using generating function. We have

$$(1-2xt+t^2)^{-\frac{1}{2}} = \sum_{k=0}^{\infty} P_k(x)t^k.$$

Squaring both sides

$$(1-2xt+t^2)^{-1} = \left[\sum_{k=0}^{\infty} P_k(x)t^k\right]^2.$$

Integrating from -1 to 1, we have

$$\int_{-1}^{1} \frac{dx}{1-2xt+t^2} = \int_{-1}^{1}\left[\sum_{k=0}^{\infty} t^k P_k(x)\right]^2 dx. \tag{12.295}$$

By virtue of equation (12.294), the integrals containing cross terms in the series on RHS of equation (12.295) will vanish. Hence, we have

$$\int_{-1}^{1} \frac{dx}{1-2xt+t^2} = \sum_{k=0}^{\infty} t^{2k}\int_{-1}^{1} [P_k(x)]^2\,dx. \tag{12.296}$$

For the integral in the LHS, we put $z = 1-2xt+t^2$.
Then

$$\begin{aligned}\int_{-1}^{1} \frac{dx}{1-2xt+t^2} &= \frac{1}{2t}\int_{(1-t)^2}^{(1+t)^2} \frac{dz}{z} = \frac{1}{2t}\Big(\log z\Big)_{(1-t)^2}^{(1+t)^2}\\ &= \frac{1}{t}\log\frac{1+t}{1-t} = \frac{1}{t}[\log(1+t) - \log(1-t)]\\ &= 2\sum_{k=0}^{\infty} \frac{t^{2k}}{2k+1} \quad [\text{using series for } \log(1\pm x)]\end{aligned} \tag{12.297}$$

Hence, we get

$$\sum_{k=0}^{\infty} t^{2k}\int_{-1}^{1} [P_k(x)]^2 dx = \sum_{k=0}^{\infty} \frac{2}{2k+1}t^{2k},$$

which gives, on equating coefficients of t^{2k},

$$\int_{-1}^{1} [P_k(x)]^2 \, dx = \frac{2}{2k+1}. \tag{12.298}$$

Combining equations (12.294) and (12.298), we get

$$\int_{-1}^{1} P_m(x) P_n(x) \, dx = \begin{cases} 0, & \text{for } m \neq n \\ \frac{2}{2n+1}, & \text{for } m = n, \end{cases}$$

which proves the orthogonality condition.

This orthogonality property of Legendre's polynomials can be used successfully to expand any continuous or piece-wise continuous function defined in the interval $[-1, 1]$ in terms of Legendre's polynomials. The expansion thus obtained is called *Fourier-Legendre's* expansion. Thus, a function $f(x)$ defined in $[-1, 1]$ can be written as

$$f(x) = \Sigma A_k P_k(x). \tag{12.299}$$

The unknown constants A_k are obtained from

$$\int_{-1}^{1} f(x) P_k(x) \, dx = A_k \int_{-1}^{1} [P_k(x)]^2 \, dx = A_k \cdot \frac{2}{2k+1}.$$

Hence

$$A_k = \left(k + \frac{1}{2}\right) \int_{-1}^{1} f(x) P_k(x) \, dx, \tag{12.300}$$

which uniquely determines the coefficients A_k.

It is to be noted that above expansion is valid, if

(1) the function $f(x)$ is piece-wise smooth in the open interval $(-1, 1)$, and

(2) the integral $\int_{-1}^{1} [f(x)]^2 \, dx$ has a definite value.

We, however, do not offer any proof for the above necessary conditions.

12.25 Associated Legendre's Function

The Legendre's differential equation, as we know, is

$$\left(1 - x^2\right) \frac{d^2 y}{dx^2} - 2x \frac{dy}{dx} + k(k+1) y = 0. \tag{12.301}$$

Differentiating successively m-times this equation with the help of Leibnitz theorem, we have

$$\left[\left(1 - x^2\right) D^{m+2} - 2(m+1) x D^{m+1} + \{k(k+1) - m(m+1)\} D^m\right] y = 0, \tag{12.302}$$

where, $D = \frac{d}{dx}$.

We make the transformation

$$z = \left(1 - x^2\right)^{\frac{m}{2}} D^m(y). \tag{12.303}$$

Then

$$\frac{dz}{dx} = (1-x^2)^{\frac{m}{2}} D^{m+1}(y) - mx(1-x^2)^{\frac{m}{2}-1} D^m(y). \tag{12.304}$$

$$\frac{d^2z}{dx^2} = (1-x^2)^{\frac{m}{2}} D^{m+2}(y) - 2mx(1-x^2)^{\frac{m}{2}-1} D^m(y)$$

$$- m(1-x^2)^{\frac{m}{2}-1} D^m(y) + 2m\left(\frac{m}{2}-1\right) x^2(1-x^2)^{\frac{m}{2}-2} D^m. \tag{12.305}$$

In view of equations (12.303)–(12.305), equation (12.302) becomes

$$(1-x^2)\frac{d^2z}{dx^2} - 2x\frac{dz}{dx} + \left[k(k+1) - \frac{m^2}{1-x^2}\right] z = 0. \tag{12.306}$$

Equation (12.306) is known as *associated Legendre's equation.*

If $y_1(x) = P_k(x)$ and $y_2(x) = Q_k(x)$ are the solutions of Legendre's equation, then the solutions of associated Legendre's equation are given by

$$P_k^m(x) = (1-x^2)^{\frac{m}{2}} D^m P_k(x) \tag{12.307}$$

$$\text{and} \quad Q_k^m(x) = (1-x^2)^{\frac{m}{2}} D^m Q_k(x). \tag{12.308}$$

The functions $P_k^m(x)$ and $Q_k^m(x)$ are called *associated Legendre's polynomial* of rank m. Some associated Legendre's polynomials are

$$P_1^1(x) = (1-x^2)^{\frac{1}{2}}$$

$$P_2^1(x) = 3x(1-x^2)^{\frac{1}{2}}$$

$$P_3^1(x) = \frac{3}{2}(5x^2-1)(1-x^2)^{\frac{1}{2}}$$

$$P_3^2(x) = 15x(1-x^2)$$

$$P_3^3(x) = 15(1-x^2)^{\frac{3}{2}}$$

....

....

The associated Legendre's polynomials obey some recurrence relations. We find some of these relations.

(1) From equation (12.284) we have

$$P'_{k+1}(x) - P'_{k-1}(x) = (2k+1)P_k(x), \tag{12.309}$$

$$\text{i.e.,} \quad D\{P_{k+1}(x) - P_{k-1}(x)\} = (2k+1)P_k(x).$$

Differentiating m-times with respect to x

$$D^{m+1}\{P_{k+1}(x) - P_{k-1}(x)\} = (2k+1)D^m\{P_k(x)\}.$$

Multiplying by $(1-x^2)^{\frac{(m+1)}{2}}$ we get

$$P_{k+1}^{m+1}(x) - P_{k-1}^{m+1}(x) = (2k+1)(1-x^2)^{\frac{1}{2}} P_k^m(x). \tag{12.310}$$

(2) Differentiating equation (12.278) m-times with respect to x, we get

$$\begin{aligned}(2k+1)xD^m\{P_k(x)\} + (2k+1)mD^{m-1}\{P_k(x)\}\\ = (k+1)D^m\{P_{k+1}(x)\} + kD^m\{P_{k-1}(x)\}.\end{aligned} \tag{12.311}$$

Differentiating equation (12.309) $(m-1)$-times with respect to x

$$D^m\{P_{k+1}(x) - P_{k-1}(x)\} = (2k+1)D^{m-1}\{P_k(x)\}. \tag{12.312}$$

Multiplying equation (12.312) by m, and adding the result with equation (12.311), we have

$$\begin{aligned}(2k+1)xD^m\{P_k(x)\} = (k+m)D^{m-1}\{P_{k-1}(x)\}\\ + (k-m+1)D^m\{P_{k+1}(x)\}.\end{aligned} \tag{12.313}$$

Multiplying equation (12.313) by $(1-x^2)^{\frac{m}{2}}$ we get the recurrence relation

$$(2k+1)xP_k^m(x) = (k+m)P_{k-1}^m(x) + (k-m+1)P_{k+1}^m(x). \tag{12.314}$$

Thus, using the recurrence relations for Legendre's polynomials, we can find those for associated Legendre's polynomials.

The associated Legendre's polynomials of same rank are also orthogonal with weighting function unity. It can be proved that

$$\int_{-1}^{1} P_r^m(x)P_s^m(x)\,dx = \frac{2}{2s+1}\cdot\frac{(s+m)!}{(s-m)!}\delta_{rs}, \tag{12.315}$$

where

$$\delta_{rs} = \begin{cases} 0, & \text{for } r \neq s\\ 1, & \text{for } r = s.\end{cases}$$

EXAMPLE 12.8. *Express in terms of Legendre's polynomial the function*

$$f(x) = x^4 - 2x^3 + 3x^2 + 5x - 9.$$

Solution: We have

$$P_4(x) = \frac{1}{8}\left(35x^4 - 30x^2 + 3\right).$$

Hence

$$x^4 = \frac{8}{35}P_4(x) + \frac{6}{7}x^2 - \frac{3}{35}.$$

Similarly

$$x^3 = \frac{2}{5}P_3(x) + \frac{3}{5}x;\quad x^2 = \frac{2}{3}P_2(x) + \frac{1}{3};\quad x = P_1(x).$$

Hence

$$\begin{aligned}f(x) &= x^4 - 2x^3 + 3x^2 + 5x - 9\\ &= \frac{8}{35}P_4(x) + \frac{6}{7}x^2 - \frac{3}{35} - \frac{4}{5}P_3(x) - \frac{6}{5}x + 3x^2 + 5x - 9\\ &= \frac{8}{35}P_4(x) + \frac{27}{7}\left[\frac{2}{3}P_2(x) + \frac{1}{3}\right] + \frac{19}{5}P_1(x) - \frac{4}{5}P_3(x) - \left(9 + \frac{3}{35}\right)\\ &= \frac{8}{35}P_4(x) - \frac{4}{5}P_3(x) + \frac{18}{7}P_2(x) + \frac{19}{5}P_1(x) - \frac{39}{5}.\end{aligned}$$

EXAMPLE 12.9. *Prove the Christoffel's expansion*

$$P_k'(x) = (2k-1)P_k(x) + (2k-5)P_{k-3}(x) + (2k-9)P_{k-5}(x) + \cdots$$

Solution: We have the recurrence formula equation (12.284)

$$P_{k+1}'(x) - P_{k-1}'(x) = (2k+1)P_k(x)$$

$$\text{i.e.,}\quad P_k'(x) = P_{k-2}'(x) + (2k-1)P_{k-1}(x) \quad [\text{changing } k \longrightarrow k-1]$$

$$\text{or,}\quad P_k'(x) = P_{k-4}'(x) + (2k-5)P_{k-3}(x) + (2k-1)P_{k-1}(x).$$

Proceeding successively

$$P_k'(x) = (2k-1)P_{k-1}(x) + (2k-5)P_{k-3}(x) + (2k-9)P_{k-5}(x) + \cdots$$

EXAMPLE 12.10. *Show that* $\int_{-1}^{1} [P_k'(x)]^2\, dx = k(k+1)$.

Solution: Using Christoffel's expansion, the given integral is

$$= \int_{-1}^{1} dx \Big[(2k-1)P_{k-1}(x) + (2k-5)P_{k-3}(x) + \cdots\Big]^2. \tag{1}$$

The last term in equation (1) will be

$$\left.\begin{aligned} &3P_1(x), && \text{for } k \text{ even} \\ \text{and}\quad &P_0(x), && \text{for } k \text{ odd} \end{aligned}\right\} \tag{2}$$

The cross-terms in the integral will vanish due to orthogonality property. For other terms, we use the equation (12.298). Then we get

$$\int_{-1}^{1} [P_k'(x)]^2\, dx = (2k-1)^2 \cdot \frac{2}{2(k-1)+1} + (2k-5)^2 \frac{2}{2(k-3)+1} + \cdots \tag{3}$$

The last term in equation (3) will be

$$\left.\begin{aligned} &= 3^2 \cdot \frac{2}{2+1}, && \text{for } k \text{ even} \\ &= \frac{2}{2.0+1}, && \text{for } k \text{ odd} \end{aligned}\right\}. \tag{4}$$

Thus, for k even, the RHS of equation (3) is

$$= 2[(2k-1) + (2k-5) + \cdots + 3] = k(k+1)$$

Similarly, for k odd, the RHS of equation (3) is

$$= 2[(2k-1) + (2k-5) + \cdots + 1] = k(k+1).$$

EXAMPLE 12.11. *Find the Fourier-Legendre's expansion for the function*

$$f(x) = \begin{cases} 0, & \textit{for } -1 < x < 0 \\ \dfrac{\sqrt{x}}{2}, & \textit{for } 0 < x < 1. \end{cases}$$

Solution: We assume

$$f(x) = \sum_{k=0}^{\infty} A_k P_k(x).$$

Then coefficients are given by

$$A_k = \left(k+\frac{1}{2}\right)\int_{-1}^{1} f(x)P_k(x)dx \quad \text{[equation (12.300)]}$$

Hence

$$A_0 = \frac{1}{2}\int_0^1 \frac{\sqrt{x}}{2}\cdot P_0(x)\,dx = \frac{1}{2}\int_0^1 \frac{\sqrt{x}}{2}\,dx = \frac{1}{6}$$
$$A_1 = \frac{3}{2}\int_0^1 \frac{\sqrt{x}}{2}\cdot P_1(x)\,dx = \frac{3}{2}\int_0^1 \frac{\sqrt{x}}{2}\,(x)\,dx = \frac{3}{10}$$
$$A_2 = \frac{5}{2}\int_0^1 \frac{\sqrt{x}}{2}\cdot P_2(x)\,dx = \frac{5}{2}\int_0^1 \frac{\sqrt{x}}{2}\cdot\frac{1}{2}\left(3x^2-1\right) = \frac{5}{42}$$
$$A_3 = \frac{7}{2}\int_0^1 \frac{\sqrt{x}}{2}\cdot P_3(x)\,dx = \frac{7}{2}\int_0^1 \frac{\sqrt{x}}{2}\,dx\cdot\frac{1}{2}\left(5x^3-3x\right)\,dx = -\frac{7}{90}.$$

....

Thus

$$f(x) = \frac{1}{6}P_0(x) + \frac{3}{10}P_1(x) + \frac{5}{42}P_2(x) - \frac{7}{90}P_3(x) + \cdots.$$

EXAMPLE 12.12. *Express $P_k(\cos\theta)$ as a series of cosine of θ.*

Solution: We have

$$\left(1-2xt+t^2\right)^{-\frac{1}{2}} = \sum_{k=0}^{\infty} P_k(x)t^k. \tag{1}$$

Putting $x = \cos\theta$,

$$\left(1-2t\cos\theta+t^2\right)^{-\frac{1}{2}} = \sum_{k=0}^{\infty} P_k(\cos\theta)t^k. \tag{2}$$

Now

$$\cos\theta = \frac{1}{2}\left(e^{i\theta}+e^{-i\theta}\right). \tag{3}$$

Hence, LHS of equation (2) becomes

$$\left[1-2t\cdot\frac{1}{2}\left(e^{i\theta}+e^{-i\theta}\right)+t^2\right]^{-\frac{1}{2}} = \left(1-te^{i\theta}\right)^{-\frac{1}{2}}\left(1-te^{i\theta}\right)^{-\frac{1}{2}}$$
$$= \left[\sum_{k=0}^{\infty}\frac{(2k)!}{2^{2k}\,(k!)^2}t^k e^{ik\theta}\right]\left[\sum_{k=0}^{\infty}\frac{(2k)!}{2^{2k}\,(k!)^2}t^k e^{-ik\theta}\right] \tag{4}$$

The coefficients of t^k in equation (4) are

$$\frac{(2k)!}{2^{2k}(k!)^2}\left[\left(e^{ik\theta}+e^{-ik\theta}\right)+\frac{2k}{2(2k-1)}\left\{e^{(k-2)i\theta}+e^{-(k-2)i\theta}\right\}\right.$$
$$\left.+\frac{1}{2}\cdot\frac{3}{4}\cdot\frac{2k(2k-2)}{(2k-1)(2k-3)}\left\{e^{(k-4)i\theta}+e^{-(k-4)i\theta}\right\}+\cdots\right]$$
$$=\frac{(2k)!}{2^{2k}(k!)^2}\cdot 2\left[\cos k\theta+\frac{k}{2k-1}\cos(k-2)\theta+\frac{1.3}{2!}\frac{k(k-1)}{(2k-1)(2k-3)}\cos(k-4)\theta+\cdots\right]. \quad (5)$$

Thus, from equations (2) and (5), we get

$$P_k(\cos\theta)=\frac{(2k)!}{2^{2k-1}(k!)^2}$$
$$\left[\cos k\theta+\frac{k}{2k-1}\cos(k-2)\theta+\frac{1.3}{2!}\frac{k(k-1)}{(2k-1)(2k-3)}\cos(k-4)\theta+\cdots\right].$$

Note. The k-th term in binomial expansion of $(1-x)^{-\frac{1}{2}}$ is

$$\frac{1.3\cdots(2k-1)}{2.4\cdots(2k)}x^k=\frac{1.3\cdots(2k-1)}{2.4\cdots(2k)}\cdots\frac{2.4\cdots(2k)}{2.4\cdots(2k)}x^k$$
$$=\frac{(2k)!}{2^{2k}(k!)^2}x^k.$$

EXAMPLE 12.13. *Prove that* $\sum_{k=0}^{n}(2k+1)P_k(x)=P'_{n+1}(x)+P'_{n-1}(x)$. *Also show that*

$$P'_n(x)=\sum_{k=1}^{N}(2n-4k+3)P_{n-2k+1}(x)$$

where $N=\frac{n}{2}$, *for n even, and* $N=\frac{n+1}{2}$ *for n odd.*

Solution: We have the recurrence relation

$$(2n+1)P_n(x)=P'_{n+1}(x)-P'_{n-1}(x). \quad (1)$$

Putting $n=0,1,2,\ldots$ in equation (1), we get

$$P_0(x)=P'_1(x)-P'_{-1}(x)$$
$$3P_1(x)=P'_2(x)-P'_0(x)$$
$$\cdots\cdots\cdots\cdots\cdots\cdots\cdots\cdots$$
$$(2n+1)P_n(x)=P'_{n+1}(x)-P'_{n-1}(x).$$

Adding all these, we get $\left[\text{since } -P'_{-1}(x)=0, P'_0(x)=0\right]$

$$P_0(x)+3P_1(x)+5P_2(x)+\cdots(2n+1)P_n(x)=P'_{n+1}(x)+P'_{n-1}(x).$$

We change n to $(n-2k+1)$ in equation (1). Then

$$P'_{n-2k+2}(x)-P'_{n-2k}(x)=(2n-4k+3)P_{n-2k+1}(x).$$

Putting $k = 1, 2, \ldots, N$,

$$\begin{aligned} P'_n(x) - P'_{n-2}(x) &= (2n-1)P_{n-1}(x) \\ P'_{n-2}(x) - P'_{n-4}(x) &= (2n-5)P_{n-3}(x) \\ &\cdots\cdots\cdots \\ P'_2(x) - P'_0(x) &= 3P_1(x). \end{aligned}$$

Adding all the terms

$$P'_n(x) = \sum_{k=0}^{\infty}(2n - 4k + 3)P_{n-2k+1}(x),$$

where

$$N = \begin{cases} \dfrac{n}{2}, & \text{for } n \text{ even} \\ \dfrac{1}{2}(n+1), & \text{for } n \text{ odd.} \end{cases}$$

EXAMPLE 12.14. *Prove that* (i) $P_{2k}(-x) = P_{2k}(x)$; (ii) $P_{2k+1}(-x) = -P_{2k+1}(x)$.

Solution: We have from equation (12.274)

$$P_k(-x) = (-1)^k P_k(x). \tag{1}$$

Changing k to $2k$

$$P_{2k}(-x) = (-1)^{2k} P_{2k}(x) = P_{2k}(x). \tag{2}$$

Similarly, changing k to $(2k+1)$,

$$P_{2k+1}(-x) = (-1)^{2k+1} P_{2k+1}(x) = -P_{2k+1}(x).$$

EXAMPLE 12.15. *Prove that* $nP_n(x) = xP'_n(x) - P'_{n-1}(x)$.

Solution: Differentiating recurrence relation (12.278) with respect to x,

$$x(2n+1)P'_n(x) + (2n+1)P_n(x) = (n+1)P'_{n+1}(x) + nP'_{n-1}(x). \tag{1}$$

Multiplying recurrence relation (12.284) by $(n+1)$

$$\begin{aligned} (n+1)P'_{n+1}(x) - (n+1)P'_{n-1}(x) &= (n+1)(2n+1)P_n(x) \\ \text{or,}\quad (n+1)(2n+1)P_n(x) &= (n+1)P'_{n+1}(x) - (n+1)P'_{n-1}(x). \end{aligned} \tag{2}$$

Subtracting equation (2) from equation (1) and using equation (12.281)

$$\begin{aligned} x(2n+1)P'_n(x) - n(2n+1)P_n(x) &= (2n+1)P'_{n-1}(x) \\ \text{or,}\quad nP_n(x) &= xP'_n(x) - P'_{n-1}(x). \end{aligned} \tag{3}$$

HERMITE'S POLYNOMIALS

12.26 Hermite's Differential Equation

The differential equation

$$\frac{d^2y}{dx^2} - 2x\frac{dy}{dx} + 2\lambda y = 0 \quad (\lambda = \text{constant}) \tag{12.316}$$

is known as *Hermite's differential equation.*

The solution of the differential equation can be obtained by the generalised power series method of Frobenius, since there is no singularity in the finite plane. Thus we assume

$$y = \sum_{k=0}^{\infty} a_k x^{r+k} \quad (a_0 \neq 0). \tag{12.317}$$

Substituting the solution function (12.317) along with its derivatives into equation (12.316), we get, after re-arrangement of terms

$$\sum_{k=0}^{\infty} a_k(r+k)(r+k-1)x^{r+k-2} - 2\sum_{k=0}^{\infty} a_k(r+k-\lambda)x^{r+k} = 0. \tag{12.318}$$

Equating to zero the coefficient of x^{r-2}, we have, $a_0 r(r-1) = 0$ which gives $r = 0$, $r = 1$.
Equating to zero the coefficient of x^{r-1}, we have

$$a_1(r+1)r = 0,$$

which gives $a_1 \neq 0$ for $r = 0$.

The general recursion formula is

$$a_{k+2} = \frac{2(r+k-\lambda)}{(r+k+2)(r+k+1)}a_k. \tag{12.319}$$

For $r = 0$

$$a_{k+2} = \frac{2(k-\lambda)}{(k+2)(k+1)}a_k. \tag{12.320}$$

Thus, for even values of k, we have

$$k = 0, \quad a_2 = -\frac{2\lambda}{2.1}a_0 = -\frac{2\lambda}{2!}a_0$$

$$k = 2, \quad a_4 = -\frac{2(\lambda-2)}{4.3}a_2 = +\frac{2^2\lambda(\lambda-2)}{4!}a_0$$

$$k = 4, \quad a_6 = -\frac{2(\lambda-4)}{6.5}a_4 = -\frac{2^3\lambda(\lambda-2)(\lambda-4)}{6!}a_0$$

....

The general term is

$$a_{2n} = (-2)^n \frac{\lambda(\lambda-2)(\lambda-4)\cdots(\lambda-2n+2)}{(2n)!}a_0. \tag{12.321}$$

For odd values of k, we have

$$k=1, \quad a_3=-\frac{2(\lambda-1)}{3.2}a_1$$
$$k=3, \quad a_5=\frac{-2(\lambda-3)}{5.4}a_3=\frac{2^2(\lambda-1)(\lambda-3)}{5!}a_1$$
$$k=5, \quad a_7=\frac{-2(\lambda-5)}{7.6}a_5=-\frac{2^3(\lambda-1)(\lambda-3)(\lambda-5)}{7!}a_1.$$

The general term is

$$a_{2n+1}=\frac{(-2)^n}{(2n+1)!}(\lambda-1)(\lambda-3)(\lambda-5)\cdots(\lambda-2n+1)a_1\cdots \tag{12.322}$$

The two **particular** solutions for $r=0$ are given by (putting $a_0=1$ and $a_1=1$)

$$y_1(x)=1-\frac{2\lambda}{2!}x^2+\frac{2^2\lambda(\lambda-2)}{4!}x^4-\frac{2^3\lambda(\lambda-2)(\lambda-4)}{6!}x^6+\cdots \tag{12.323}$$

and

$$y_2(x)=x\left[1-\frac{2(\lambda-1)}{3!}x^2+\frac{2^2(\lambda-1)(\lambda-3)}{5!}x^4-\cdots\right]. \tag{12.324}$$

For $r=1, a_1=0$

The general recursion formula becomes

$$a_{k+2}=\frac{2(k+1-\lambda)}{(k+3)(k+2)}\cdot a_k. \tag{12.325}$$

Thus, the coefficients are

$$k=0, \quad a_2=\frac{2(1-\lambda)}{3.2}a_0=-\frac{2(\lambda-1)}{3!}a_0$$
$$k=2, \quad a_4=\frac{2(3-\lambda)}{5.4}a_2=\frac{2^2(\lambda-1)(\lambda-3)}{5!}a_0$$
$$k=4, \quad a_6=\frac{2(5-\lambda)}{7.6}a_4=-\frac{2^3(\lambda-1)(\lambda-3)(\lambda-5)}{7!}a_0$$
$$\cdots\cdots\cdots\cdots\cdots\cdots\cdots\cdots\cdots$$

Hence, the solution is (putting $a_0=1$)

$$y_3(x)=x\left[1-\frac{2(1-\lambda)}{3!}x^2+\frac{2^2(\lambda-1)(\lambda-3)}{5!}x^4+\cdots\right]. \tag{12.326}$$

This is clear that $y_3(x)$ is the same solution as $y_2(x)$ in equation (12.324). Hence, solution $y_1(x)$ and $y_2(x)$ for the case when $r=0$ are not linearly independent. We, therefore, take the solutions as

$$y=\begin{cases} y_1(x), & \text{for } r=0 \\ y_3(x), & \text{for } r=1 \end{cases}$$

so that the general solution for the differential equation (12.316) is taken as

$$y=Ay_1(x)+By_3(x), \tag{12.327}$$

where A, B are constants.

Now, when λ is a positive even integer, the series for $y_1(x)$ becomes a finite polynomial of degree λ when $\lambda = 2k-2$; when λ is a positive odd integer, the series for $y_3(x)$ becomes a finite polynomial of degree λ when $\lambda = 2k-1$.

In order to obtain a single series which will be valid for both even and odd values of λ, we assume a solution in terms of a series with descending powers of x as

$$y = \sum_{k=\lambda}^{0} a_k x^k. \tag{12.328}$$

Substituting this into the differential equation (12.316), we get the recursion formula as

$$a_{k+2} = -\frac{2(\lambda - k)}{(k+2)(k+1)} a_k, \quad \text{or,} \quad a_k = -\frac{(k+2)(k+1)}{2(\lambda - k)} a_{k+2}. \tag{12.329}$$

From this recursion formula, we get (putting $\lambda = k$)

$$a_{k-2} = -\frac{k(k-1)}{2.2} a_k$$

$$a_{k-4} = -\frac{(k-2)(k-3)}{2.4} a_{k-2} = \frac{k(k-1)(k-2)(k-3)}{2^2 \cdot 2 \cdot 4} a_k$$

$$\cdots\cdots\cdots\cdots\cdots\cdots\cdots\cdots\cdots\cdots\cdots\cdots$$

$$a_{k-2m} = \frac{(-1)^m k(k-1)(k-2)\cdots(k-2m+1)}{2^m \cdot 2 \cdot 4 \cdots (2m)} \cdot a_k \cdots \tag{12.330}$$

Now

$$\begin{aligned} k(k-1)(k-2)&\cdots(k-2m+1) \\ &= k(k-1)(k-2)\cdots(k-2m+1) \cdot \frac{(k-2m)(k-2m-1)\cdots 3\cdot 2\cdot 1}{(k-2m)(k-2m-1)\cdots 3\cdot 2\cdot 1} \\ &= \frac{k!}{(k-2m)!}. \end{aligned} \tag{12.331}$$

Again

$$2 \cdot 4 \cdots (2m) = 2^m m!. \tag{12.332}$$

Hence, the general term is written as

$$a_{k-2m} = (-1)^m \frac{k!}{2^{2m} m!(k-2m)!} a_k. \tag{12.333}$$

Thus, the solution of the differential equation is

$$y(x) = a_k \sum_{m=0}^{N} (-1)^m \frac{k!}{2^{2m}\, m!\, (k-2m)!} x^{k-2m}, \tag{12.334}$$

where

$$N = \begin{cases} \dfrac{k}{2}, & \text{for } k \text{ even} \\[2ex] \dfrac{k-1}{2}, & \text{for } k \text{ odd.} \end{cases}$$

We take the value of $a_k = 2^k$. Then the particular solution is written as

$$y(x) = \sum_{m=0}^{N} (-1)^m \frac{k!}{m!\,(k-2m)!}(2x)^{k-2m}. \tag{12.335}$$

The series in the RHS of equation (12.335) is known as Hermite's polynomial of degree k and is denoted by $H_k(x)$. Thus

$$H_k(x) = \sum_{m=0}^{N} (-1)^m \frac{k!}{m!\,(k-2m)!}\,(2x)^{k-2m}. \tag{12.336}$$

12.27 Generating Function

Let us consider the function

$$u(x) = e^{-x^2}. \tag{12.337}$$

Differentiating we get

$$\frac{du}{dx} = -2xe^{-x^2}, \quad \text{i.e.,} \quad \frac{du}{dx} + 2xu = 0.$$

Differentiating $(m+1)$ times with the help of Leibnitz formula ($D = \frac{d}{dx}$)

$$\left[D^{m+2} + 2xD^{m+1} + 2(m+1)D^m\right] u = 0. \tag{12.338}$$

Let

$$D^m(u) = (-1)^m e^{-x^2} y. \tag{12.339}$$

Then

$$D^{m+1}(u) = (-1)^m e^{-x^2}\,(y' - 2xy) \tag{12.340}$$

$$\text{and} \quad D^{m+2}(u) = (-1)^m e^{-x^2}\left(y'' - 4xy' + 4x^2y - 2y\right). \tag{12.341}$$

Using these relations into equation (12.338), we get

$$y'' - 2xy' + 2my = 0,$$

which is Hermite's equation. Hence, its solution is

$$\begin{aligned} y = H_m(x) &= (-1)^m e^{x^2} D^m(u) \\ &= (-1)^m e^{x^2} \frac{d^m}{dx^m}\left(e^{-x^2}\right). \end{aligned} \tag{12.342}$$

Equation (12.342) is the Rodrigue's formula for Hermite's polynomials.
In particular, we have

$$\begin{aligned} H_0(x) &= 1 \\ H_1(x) &= 2x \\ H_2(x) &= 4x^2 - 2 \\ H_3(x) &= 8x^3 - 12x \\ H_4(x) &= 16x^4 - 48x^2 + 12 \\ H_5(x) &= 32x^5 - 160x^3 + 120x \end{aligned}$$

....

The function defined by

$$h_m(x) = e^{-\frac{x^2}{2}} H_m(x) = (-1)^m e^{\frac{x^2}{2}} \frac{d^m}{dx^m}\left(e^{-x^2}\right) \tag{12.343}$$

is called *Hermite's function*, which satisfies *Weber's differential equation*

$$y'' + (1 + x^2 + 2m)\, y = 0. \tag{12.344}$$

12.28 Recurrence Relations

Hermite's polynomials can also be generated from the function

$$\phi(t, x) = e^{-t^2+2tx}, \tag{12.345}$$

such that

$$\phi(t, x) = e^{-t^2+2tx} = \sum_{k=0}^{\infty} \frac{1}{k!} H_k(x) t^k. \tag{12.346}$$

Now, from equation (12.345), we get

$$\frac{\partial \phi}{\partial x} = 2te^{-t^2+2tx}$$

$$\frac{\partial \phi}{\partial t} = 2(x-t)e^{-t^2+2tx},$$

which give

$$\frac{\partial \phi}{\partial x} - 2t\phi = 0 \tag{12.347}$$

$$\text{and} \quad \frac{\partial \phi}{\partial t} + 2(t-x)\phi = 0. \tag{12.348}$$

Again, from equation (12.346), we have

$$\frac{\partial \phi}{\partial x} = \sum_{k=0}^{\infty} \frac{1}{k!} H'_k(x) t^m$$

$$\text{and} \quad \frac{\partial \phi}{\partial t} = \sum_{k=0}^{\infty} \frac{1}{(k-1)!} H_k(x) t^{k-1},$$

which give, in view of equations (12.347) and (12.348)

$$\sum_{k=0}^{\infty} \frac{1}{k!} H'_k(x) t^k - 2t \sum_{k=0}^{\infty} \frac{1}{k!} H_k(x) t^k = 0 \tag{12.349}$$

$$\text{and} \quad \sum_{k=0}^{\infty} \frac{1}{(k-1)!} H_k(x) t^{k-1} + 2(t-x) \sum_{k=0}^{\infty} \frac{1}{k!} H_k(x) t^k = 0. \tag{12.350}$$

From equation (12.349) we have one recurrence formula

$$H'_k(x) - 2kH_{k-1}(x) = 0. \tag{12.351}$$

From equation (12.350), we get another recurrence formula

$$H_{k+1}(x) - 2xH_k(x) + 2kH_{k-1}(x) = 0. \tag{12.352}$$

The third recurrence relation is obtained by adding above two equations. Thus, we get

$$H_{k+1}(x) + H_k'(x) - 2xH_k(x) = 0. \tag{12.353}$$

Again, from equation (12.351) we have

$$H_k'(x) = 2kH_{k-1}(x).$$

Differentiating this successively $(k-1)$-times and using each time this recurrence relation, we get

$$\frac{d^m}{dx^m}H_k(x) = 2^m k(k-1)\cdots(k-m+1)H_{k-m}(x).$$

For $k = m$,

$$\frac{d^k}{dx^k}H_k(x) = 2^k(k!). \tag{12.354}$$

12.29 Orthogonality of Hermite's Polynomials

We consider the integral

$$I = \int_{-\infty}^{\infty} e^{-x^2} H_k(x)H_m(x)dx = \int_{-\infty}^{\infty} (-1)^m H_k(x)\frac{d^m}{dx^m}\left(e^{-x^2}\right) dx.$$

Integrating by parts

$$I = (-1)^m \left[H_k(x)\cdot\frac{d^{m-1}}{dx^{m-1}}\left(e^{-x^2}\right)\Big|_{-\infty}^{\infty} - \int_{-\infty}^{\infty} H_k'(x)\frac{d^{m-1}}{dx^{m-1}}\left(e^{-x^2}\right) dx\right].$$

The first part will always contain the factor e^{-x^2} due to which it will vanish. Hence

$$\begin{aligned} I &= -(-1)^m \int_{-\infty}^{\infty} H_k'(x)\frac{d^{m-1}}{dx^{m-1}}\left(e^{-x^2}\right) dx \\ &= (-1)^{m-1}\cdot 2k\int_{-\infty}^{\infty} H_{k-1}(x)\frac{d^{m-1}}{dx^{m-1}}\left(e^{-x^2}\right) dx \quad \text{[using equation (12.351)]}. \end{aligned}$$

Continuing the integration, we get

$$I = (-1)^{m-k}2^k(k!)\int_{-\infty}^{\infty}\frac{d^{m-k}}{dx^{m-k}}\left(e^{-x^2}\right) dx. \tag{12.355}$$

For $k < m$,

$$I = (-1)^{m-k}2^k(k!)\left[\frac{d^{m-k-1}}{dx^{m-k-1}}\left(e^{-x^2}\right)\right]_{-\infty}^{\infty} = 0.$$

Thus, the Hermite's polynomials are orthogonal with weighting function e^{-x^2} in the interval $(-\infty, +\infty)$. For $k = m$, we get from equation (12.355)

$$I = \int_{-\infty}^{\infty} e^{-x^2}\left[H_m(x)\right]^2 dx = 2^m(m!)\int_{-\infty}^{\infty} e^{-x^2} dx = 2^m(m!)\sqrt{\pi}. \tag{12.356}$$

This property of orthogonality helps us to expand a function $f(x)$ defined in $(-\infty, +\infty)$ in terms of Hermite's polynomials. Thus a function

$$f(x) = \sum_{k=0}^{\infty} A_k H_k(x), \tag{12.357}$$

where the coefficients A_k are given by

$$A_k = \frac{1}{2^k k! \sqrt{\pi}} \int_{-\infty}^{\infty} e^{-x^2} H_k(x) f(x)\, dx. \tag{12.358}$$

EXAMPLE 12.16. *Prove that $H_k(-x) = (-1)^k H_k(x)$.*

Solution: We have from equation (12.336)

$$H_k(x) = \sum_{k=0}^{N} (-1)^m \frac{k!}{m!\,(k-2m)!} (2x)^{k-2m}.$$

Then

$$\begin{aligned} H_k(-x) &= \sum_{k=0}^{N} (-1)^m \frac{k!}{m!\,(k-2m)!} (-2x)^{k-2m} \\ &= \sum_{k=0}^{N} (-1)^m (-1)^{k-2m} \frac{k!}{m!(k-2m)!} (2x)^{k-2m} \\ &= (-1)^k \sum_{k=0}^{N} (-1)^m \frac{k!}{m!(k-2m)!} (2x)^{k-2m} \\ &= (-1)^k H_k(x). \end{aligned}$$

EXAMPLE 12.17. *Show that*

(i) $e \cos 2x = \sum_{k=0}^{\infty} (-1)^k \frac{H_{2k}(x)}{(2k)!}$; (ii) $e \sin 2x = \sum_{k=0}^{\infty} (-1)^k \frac{H_{2k+1}(x)}{(2k+1)!}$.

Solution: We have from equation (12.346)

$$e^{-t^2+2tx} = \sum_{k=0}^{\infty} \frac{1}{k!} H_k(x) t^k.$$

Putting $t = i$, we get

$$e \cdot e^{2ix} = \sum_{k=0}^{\infty} \frac{1}{k!} H_k(x)(i)^k.$$

Breaking the RHS into even and odd terms

$$e(\cos 2x + i \sin 2x) = \sum_{k=0}^{\infty} (-1)^k \left[\frac{H_{2k}(x)}{(2k)!} + i \frac{H_{2k+1}(x)}{(2k+1)!} \right].$$

Equating real and imaginary parts

$$e\cos 2x = \sum_{k=0}^{\infty}(-1)^k \frac{H_{2k}(x)}{(2k)!}$$

and $$e\sin 2x = \sum_{k=0}^{\infty}(-1)^k \frac{H_{2k+1}(x)}{(2k+1)!}.$$

EXAMPLE 12.18. *Expand the function*

$$f(x) = \begin{cases} -1, & \text{for } x < 0 \\ 0, & \text{for } x = 0 \\ 1, & \text{for } x > 0 \end{cases}$$

in terms of Hermite's polynomial.

Solution: Let

$$f(x) = \sum_{k=0}^{\infty} A_k H_k(x).$$

Then

$$A_k = \frac{1}{2^k(k!)\sqrt{\pi}}\int_{-\infty}^{\infty} e^{-x^2} H_k(x) f(x)\,dx.$$

Since the function is odd, the coefficients $A_{2k} = 0$. Hence

$$A_{2k+1} = \frac{1}{2^{2k}(2k+1)!\sqrt{\pi}}\int_0^{\infty} e^{-x^2} H_{2k+1}(x)\,dx. \tag{1}$$

Using the recurrence relation (12.353), we get

$$A_{2k+1} = \frac{1}{2^{2k}(2k+1)!\sqrt{\pi}}\left[2\int_0^{\infty} x e^{-x^2} H_{2k}(x)\,dx - \int_0^{\infty} e^{-x^2} H'_{2k}(x)\,dx\right]. \tag{2}$$

The first integral on RHS upon integration by parts is

$$H_{2k}(x)\cdot\int_0^{\infty} 2xe^{-x^2}\,dx + \int_0^{\infty} H'_{2k}(x)e^{-x^2}\,dx$$

$$= -e^{-x^2}H_{2k}(x)\Big|_0^{\infty} + \int_0^{\infty} H'_{2k}(x)e^{-x^2}\,dx. \tag{3}$$

From equations (2) and (3), we get

$$A_{2k+1} = \frac{H_{2k}(0)}{2^{2k}(2k+1)!\sqrt{\pi}}. \tag{4}$$

Now, we have

$$e^{-t^2+2tx} = \sum_{k-0}^{\infty}\frac{1}{k!}H_k(x)t^k. \tag{5}$$

Putting $x = 0$,

$$e^{-t^2} = \sum_{k=0}^{\infty}\frac{1}{k!}H_k(0)t^k. \tag{6}$$

Comparing coefficients of t^{2k} on both sides, we get

$$H_{2n}(0) = (-1)^n \frac{2n!}{n!}. \tag{7}$$

Hence, from equation (4), we have

$$A_{2k+1} = (-1)^k \frac{1}{2^{2k}(2k+1)k!\sqrt{\pi}}. \tag{8}$$

Therefore,

$$\begin{aligned} f(x) &= \frac{1}{\sqrt{\pi}} \sum_{k=0}^{\infty} \frac{(-1)^k}{2^{2k}(2k+1)k!} H_{2k+1}(x) \\ &= \frac{1}{\sqrt{\pi}} \left[H_1(x) - \frac{1}{12} H_3(x) + \frac{1}{160} H_5(x) - \cdots \right]. \end{aligned}$$

LAGUERRE'S FUNCTION

12.30 Laguerre's Differential Equation

The differential equation

$$x\frac{d^2y}{dx^2} + (1-x)\frac{dy}{dx} + ny = 0, \tag{12.359}$$

where n = constant, is known as Laguerre's differential equation. It has a non-essential singularity at $x = 0$ and so we can find its solution by the method of series. We assume

$$y = \sum_{k=0}^{\infty} a_k x^{r+k}. \tag{12.360}$$

Substituting equation (12.360) into (12.359), we get

$$\sum_{k=0}^{\infty} a_k(r+k)(r+k-1)x^{r+k-1} + (1-x)\sum_{k=0}^{\infty} a_k(r+k)x^{r+k-1} + n\sum_{k=0}^{\infty} a_k x^{r+k} = 0$$

$$\text{or,} \quad \sum_{k=0}^{\infty} a_k(r+k)^2 x^{r+k-1} - \sum_{k=0}^{\infty} a_k(r+k-n)x^{r+k} = 0.$$

Equating to zero the coefficients of x^{k-1}, we have $a_0r^2 = 0$, which gives $r = 0$. The general recursion formula is

$$a_{k+1} = \frac{k-n}{(k+1)^2} a_k. \tag{12.361}$$

Thus

$$\begin{aligned}
&\text{for } k=0, \quad a_1 = -na_0 \quad = (-1)na_0\\
&\text{for } k=1, \quad a_2 = \frac{1-n}{2^2}a_1 = (-1)^2\frac{n(n-1)}{(2!)^2}a_0\\
&\text{for } k=2, \quad a_3 = \frac{2-n}{3^2}a_2 = (-1)^3\frac{n(n-1)(n-2)}{(3!)^2}a_0\\
&\dots\dots\dots\dots\dots\dots\dots\dots\dots\dots\\
&\dots\dots\dots\dots\dots\dots\dots\dots\dots\dots
\end{aligned}$$

Hence

$$\begin{aligned}
y(x) &= \sum_{k=0}^{\infty} a_k x^k\\
&= a_0\left[1 - nx + \frac{n(n-1)}{(2!)^2}x^2 - \frac{n(n-1)(n-2)}{(3!)^2}x^3 + \cdots\right]. \qquad (12.362)
\end{aligned}$$

Since the roots of indicial equation are equal, the second solution has to be obtained by the method discussed in chapter 3. We are, however, interested in the first solution (12.362) which reduces to a polynomial when n is a positive integer. We set $a_0 = 1$; then equation (12.362) for n, a positive integer, is written as

$$L_n(x) = \sum_{k=0}^{n} (-1)^k \frac{n!}{(k!)^2(n-k)!}x^k \qquad (12.363)$$

which is called *Laguerre's polynomial.*

The function in (12.363) is also written in descending powers of x in the form

$$L_n(x) = \frac{(-1)^n}{n!}\left[x^n - \frac{n^2}{1!}x^{n-1} + \frac{n^2(n-1)^2}{2!}x^{n-2} - \cdots\right] \qquad (12.364)$$

From this expansion, we write

$$\begin{aligned}
L_0(x) &= 1\\
L_1(x) &= 1 - x\\
L_2(x) &= 1 - 2x + \frac{x^2}{2}\\
L_3(x) &= 1 - 3x + \frac{3}{2}x^2 - \frac{x^3}{6}\\
L_4(x) &= 1 - 4x + 3x^2 - \frac{2x^3}{3} + \frac{x^4}{24}\\
&\dots\dots\dots\dots\dots\dots\dots\dots\\
&\dots\dots\dots\dots\dots\dots\dots\dots
\end{aligned}$$

12.31 Generating Function

The generating function for Laguerre's polynomial is

$$\phi(x,t) = \frac{\exp[-xt/(1-t)]}{1-t} \tag{12.365}$$

so that

$$\frac{1}{1-t}\exp\left(\frac{-xt}{1-t}\right) = \sum_{k=0}^{\infty} L_k(x)t^k. \tag{12.366}$$

The LHS of equation (12.366) can be written as

$$\frac{1}{1-t}\exp\left(\frac{-xt}{1-t}\right) = \frac{1}{1-t}\sum_{k=0}^{\infty}\frac{(-1)^k x^k t^k}{k!(1-t)^k} \quad \text{[using expansion for } e^z\text{]}$$

$$= \sum_{k=0}^{\infty}\frac{(-1)^k x^k t^k}{k!}(1-t)^{-(k+1)}$$

$$= \sum_{k=0}^{\infty}\frac{(-1)^k x^k t^k}{k!}\sum_{p=0}^{\infty}\frac{(k+1)(k+2)\cdots(k+p)t^p}{p!}$$

$$= \sum_{k=0}^{\infty}(-1)^k\frac{x^k}{k!}\sum_{p=0}^{\infty}\frac{1\cdot 2\cdot 3\cdots k(k+1)\cdots(k+p)t^{k+p}}{1\cdot 2\cdot 3\cdots k(p!)}$$

$$= \sum_{k=0}^{\infty}(-1)^k\frac{x^k}{k!}\sum_{p=0}^{\infty}\frac{(k+p)!}{k!\,p!}t^{k+p}.$$

We put $k+p=n$, then we get

$$\frac{1}{1-t}\exp\left(\frac{-xt}{1-t}\right) = \sum_{n=0}^{\infty}t^n\sum_{k=0}^{\infty}\frac{n!}{(k!)^2(n-k)!}x^k$$

$$= \sum_{k=0}^{\infty}t^n L_n(x) \quad \text{[by equation (12.363)]}.$$

Thus, we have proved the relation (12.365).
The Rodrigue's formula for the Laguerre's polynomial is

$$L_n(x) = \frac{e^x}{n!}\frac{d^n}{dx^n}\left(x^n e^{-x}\right). \tag{12.367}$$

To prove this, let us differentiate equation (12.366) n-times with respect to t. Writing

$$-\frac{xt}{1-t} = x - \frac{x}{1-t},$$

so that

$$\exp\left(-\frac{xt}{1-t}\right) = e^x\cdot\exp\left(-\frac{x}{1-t}\right),$$

we get

$$n!L_n(x) + \frac{(n+1)!}{1!}L_{n+1}(x)t + \frac{(n+2)!}{2!}L_{n+2}(x)t^2 + \cdots = e^x\frac{d^n}{dt^n}\left[\frac{1}{1-t}\exp\left(\frac{-x}{1-t}\right)\right].$$

In the limit $t \longrightarrow 0$, we get

$$\begin{aligned} n!L_n(x) &= e^x \frac{d^n}{dt^n}\left[\sum_{k=0}^{\infty} \frac{(-1)^k x^k}{k!(1-t)^{k+1}}\right]_{t=0} \\ &= e^x \left[\sum_{k=0}^{\infty} (-1)^k \frac{(k+1)(k+2)\cdots(k+n)}{k!(1-t)^{k+n+1}} x^k\right]_{t=0} \\ &= e^x \sum_{k=0}^{\infty} (-1)^k \frac{(k+1)(k+2)\cdots(k+n)}{k!} x^k. \end{aligned} \tag{12.368}$$

Now

$$\begin{aligned} \frac{d^n}{dx^n}\left(e^{-x}x^n\right) &= \frac{d^n}{dx^n}\left[x^n \sum_{k=0}^{\infty} (-1)^k \frac{x^k}{k!}\right] \\ &\doteq \sum_{k=0}^{\infty} (-1)^k \frac{(n+k)(n+k-1)\cdots(k+1)}{k!} x^k. \end{aligned} \tag{12.369}$$

From equations (12.368) and (12.369), we get

$$n!L_n(x) = e^x \frac{d^n}{dx^n}\left(x^n e^{-x}\right)$$

which gives the Rodrigue's formula equation (12.367).

12.32 Recurrence Relations

We have the generating function

$$\phi(x,t) = \frac{1}{1-t}\exp\left(\frac{-xt}{1-t}\right). \tag{12.370}$$

Hence

$$\frac{\partial \phi}{\partial x} = -\frac{t}{(1-t)^2}\exp\left(\frac{-xt}{1-t}\right) \tag{12.371}$$

$$\text{and} \quad \frac{\partial \phi}{\partial t} = \frac{1}{(1-t)^2}\left(1-\frac{x}{1-t}\right)\exp\left(\frac{-xt}{1-t}\right). \tag{12.372}$$

From equation (12.371) we get

$$(1-t)\frac{\partial \phi}{\partial x} + t\phi = 0. \tag{12.373}$$

From equation (12.372) we get

$$(1-t)^2\frac{\partial \phi}{\partial t} + (x+t-1)\phi = 0. \tag{12.374}$$

Again by equation (12.366)

$$\phi(x,t) = \sum_{k=0}^{\infty} L_k(x)t^k. \tag{12.375}$$

Hence

$$\frac{\partial \phi}{\partial x} = \sum_{k=0}^{\infty} L_k'(x)t^k \tag{12.376}$$

and
$$\frac{\partial \phi}{\partial t} = \sum_{k=0}^{\infty} kL_k(x)t^{k-1}. \tag{12.377}$$

From equation (12.373) using equations (12.376) and (12.377), we get

$$(1-t)\sum_{k=0}^{\infty} L_k'(x)t^k + t\sum_{k=0}^{\infty} L_k(x)t^k = 0.$$

Equating coefficients of t^k to zero, we get

$$L_k'(x) - L_{k-1}'(x) + L_{k-1}(x) = 0$$

$$\text{or,} \quad L_{k+1}'(x) = L_k'(x) - L_k(x). \tag{12.378}$$

This is one recurrence relation.
Again, from equation (12.374) using equations (12.376) and (12.377) we get

$$(1-t)^2\sum_{k=0}^{\infty} kL_k(x)t^{k-1} + (x+t-1)\sum_{k=0}^{\infty} L_k(x)t^k = 0.$$

Equating to zero the coefficients of t^k, we have

$$(k+1)L_{k+1}(x) - 2kL_k(x) + (k-1)L_{k-1}(x) + (x-1)L_k(x) + L_{k-1}(x) = 0$$

$$\text{or,} \quad (k+1)L_{k+1}(x) + kL_{k-1}(x) = (2k+1-x)L_k(x). \tag{12.379}$$

This is the second recurrence relation.
Now, differentiating equation (12.379) with respect to x

$$(k+1)L_{k+1}'(x) + kL_{k-1}'(x) = (2k+1-x)L_k'(x) - L_k(x).$$

Putting the value of $L_{k-1}'(x)$ and $L_{k+1}'(x)$ from equation (12.378), we get

$$xL_k'(x) = kL_k(x) - kL_{k-1}(x). \tag{12.380}$$

This is the third recurrence relation.

12.33 Laguerre's Functions

The functions defined by

$$l_k(x) = e^{-\frac{x}{2}} L_k(x) \tag{12.381}$$

are called *Laguerre's functions.* These functions are orthogonal in the interval $(0, \infty)$. We can prove that

$$I = \int_0^{\infty} l_k(x) l_m(x)\, dx = \begin{cases} 0, & \text{for } m \neq n \\ 1, & \text{for } m = n. \end{cases} \tag{12.382}$$

Let us now prove it. We have

$$I = \int_0^\infty l_k(x) l_m(x)\, dx = \int_0^\infty e^{-x} L_k(x) L_m(x)\, dx$$

$$(m!)I = \int_0^\infty L_k(x) \frac{d^m}{dx^m}\left(x^m e^{-x}\right) dx \quad \text{[using (12.367)]}$$

$$(m!)I = L_k(x) \frac{d^{m-1}}{dx^{m-1}}\left(x^m e^{-x}\right)\Big|_0^\infty - \int L_k'(x) \frac{d^{m-1}}{dx^{m-1}}\left(x^m e^{-x}\right) dx.$$

The first term vanishes due to presence of factors like $x^n e^{-x}$, $n > 0$. Thus, continuing the integration by parts k-times, for $k \le m$, we get

$$I = \frac{(-1)^k}{m!} \int_0^\infty \frac{d^k}{dx^k}[L_k(x)] \frac{d^{m-k}}{dx^{m-k}}\left(x^m e^{-x}\right) dx. \tag{12.383}$$

For $k < m$, integrating once more

$$I = (-1)^{k+1} \frac{1}{m!} \int_0^\infty \frac{d^{k+1}}{dx^{k+1}} L_k(x) \frac{d^{m-k-1}}{dx^{m-k-1}}\left(x^m e^{-x}\right) dx$$

$$= 0, \text{ since } \frac{d^{k+1}}{dx^{k+1}} L_k(x) = 0.$$

For $k = m$,

$$I = \frac{1}{m!} \int_0^\infty e^{-x} \{L_k(x)\}^2 dx$$

$$= (-1)^k \frac{1}{m!} \int_0^\infty \frac{d^k}{dx^k} L_k(x) \left[x^k e^{-x}\right] dx \quad \text{[from equation (12.383)]}$$

$$I = \frac{1}{m!} \int_0^\infty x^k e^{-x}\, dx \quad \left[\because \frac{d^k}{dx^k} L_k(x) = (-1)^k, \text{ from equation (12.364)}\right]$$

$$\text{or,} \quad I = \frac{k!}{m!} = 1 \quad \text{[since } k = m\text{]}.$$

Hence

$$\int_0^\infty l_k(x) l_m(x)\, dx = \begin{cases} 0, & \text{for } m \ne k \\ 1, & \text{for } m = k. \end{cases}$$

12.34 Associated Laguerre's Polynomials

We consider the Laguerre's differential equation

$$x \frac{d^2 y}{dx^2} + (1 - x) \frac{dy}{dx} + ny = 0. \tag{12.384}$$

Since the function $L_n(x)$ is its solution, we have

$$x L_n''(x) + (1 - x) L_n'(x) + n L_n(x) = 0. \tag{12.385}$$

Differentiating equation (12.385) k-times

$$x\frac{d^{k+2}}{dx^{k+2}}L_n(x)+(k+1-x)\frac{d^{k+1}}{dx^{k+1}}L_n(x)+(n-k)\frac{d^k}{dx^k}L_n(x)=0. \tag{12.386}$$

We substitute

$$L_n^k(x)=\frac{d^k}{dx^k}L_n(x). \tag{12.387}$$

The functions $L_n^k(x)$ are called *associated Laguerre's polynomials* of degree $(n-k)$. In view of equations (12.387), equation (12.386) reduces to

$$x\frac{d^2z}{dx^2}+(k+1-x)\frac{dz}{dx}+(n-k)z=0, \tag{12.388}$$

where we have set $z=L_n^k(x)$.

Equation (12.388) is known as *associated Laguerre's equation.* The constants n,k are positive integers.

We have the generating function for Laguerre's polynomial as

$$\frac{1}{1-t}\exp\left(\frac{-xt}{1-t}\right)=\sum_{n=0}^{\infty}L_n(x)t^n.$$

Differentiating this k-times with respect to x, we get

$$\frac{(-t)^k}{(1-t)^{k+1}}\exp\left(\frac{-tx}{1-t}\right)=\sum_{n=0}^{\infty}L_n^k(x)t^n. \tag{12.389}$$

Thus, the LHS of equation (12.389) is the generating function for associated Laguerre's polynomials.

From Rodrigue's formula (12.297) for Laguerre's polynomials, we can easily get the same for associated Laguerre's polynomials in the form

$$L_n^k(x)=\frac{1}{n!}e^x x^{-k}\frac{d^n}{dx^n}\left(e^{-x}x^{n+k}\right). \tag{12.390}$$

The functions defined by

$$Y_{n,k}(x)=e^{-\frac{x}{2}}x^{\frac{k}{2}}L_n^k(x) \tag{12.391}$$

are called *associated Laguerre's functions.* These functions are orthogonal in the interval $(0,\infty)$, and satisfy the following orthogonality conditions:

$$\int_0^\infty Y_{n,k}(x)Y_{m,k}(x)\,dx=\begin{cases}0, & \text{for } m\neq n\\ \dfrac{(k+n)!}{n!}, & \text{for } m=n.\end{cases} \tag{12.392}$$

Due to the orthogonality of Laguerre's functions in (12.381), we can expand a piece-wise continuous function $f(x)$ defined in the interval $(0,\infty)$ in terms of Laguerre's functions. Thus

$$f(x)=\sum_{n=0}^{\infty}a_n l_n(x), \tag{12.393}$$

where the constants a_n are given by

$$a_n = \int_0^\infty l_n(x) f(x)\, dx. \tag{12.394}$$

EXAMPLE 12.19. *Expand the function $f(x) = e^{-\alpha x}$ in terms of Laguerre's polynomials.*

Solution: Let

$$e^{-\alpha x} = \sum_{n=0}^{\infty} a_n L_n(x),$$

then

$$\begin{aligned} a_n &= \int_0^\infty e^{-x} L_n(x) e^{-\alpha x}\, dx \\ &= \frac{1}{n!}\int_0^\infty e^{-\alpha x} \frac{d^n}{dx^n}\left(x^n e^{-x}\right) dx \quad \text{[using equation (12.367)]} \\ &= \frac{1}{n!}\int_0^\infty x^n e^{-(\alpha+1)x}\, dx \quad \text{[integrating by parts].} \end{aligned}$$

Putting $(\alpha + 1)x = y$, we get

$$a_n = \frac{\alpha^n}{n!(\alpha+1)^{n+1}} \int_0^\infty y^n e^{-y}\, dy = \frac{\Gamma(n+1)\alpha^n}{n!(\alpha+1)^{n+1}}.$$

Thus

$$e^{-\alpha x} = \frac{1}{\alpha+1} \sum_{n=0}^{\infty} 1 \left(\frac{\alpha}{\alpha+1}\right)^n L_n(x).$$

For $\alpha = 1$,

$$\begin{aligned} e^{-x} &= \frac{1}{2}\sum_{n=0}^{\infty} \frac{1}{2^n} L_n(x) \\ &= \frac{1}{2}\left[L_0(x) + \frac{1}{2}L_1(x) + \frac{1}{4}L_2(x) + \frac{1}{8}L_3(x) + \cdots\right]. \end{aligned}$$

CHEBYSHEV'S POLYNOMIALS

12.35 Chebyshev's Polynomials of First Kind

Let us consider the differential equation

$$\left(1 - x^2\right) \frac{d^2 y}{dx^2} - x\frac{dy}{dx} + n^2 y = 0, \tag{12.395}$$

where n is a positive integer. In order to solve this differential equation, we substitute

$$x = \cos\phi. \tag{12.396}$$

Then

$$\frac{dy}{dx} = \frac{dy}{d\phi} \cdot \frac{d\phi}{dx} = -\frac{1}{\sin\phi} \cdot \frac{dy}{d\phi}$$

$$\text{and} \quad \frac{d^2y}{dx^2} = \frac{d}{d\phi}\left(\frac{dy}{dx}\right) \cdot \frac{d\phi}{dx} = \frac{1}{\sin^2\phi}\left(\frac{d^2y}{d\phi^2} - \frac{\cos\phi}{\sin\phi} \cdot \frac{dy}{d\phi}\right). \tag{12.397}$$

In view of these relations, equation (12.395) reduces to

$$\frac{d^2y}{d\phi^2} + n^2 y = 0. \tag{12.398}$$

Its solutions are

$$\left.\begin{aligned} y_1 &= \cos n\phi \\ y_2 &= \sin n\phi \end{aligned}\right\}. \tag{12.399}$$

Now, by D'Moivre's theorem, we have

$$\cos n\phi + i \sin n\phi = (\cos\phi + i \sin\phi)^n.$$

Equating real and imaginary parts

$$\cos n\phi = \cos^n\phi - {}^2C_n \cos^{n-2}\phi \sin^2\phi + {}^4C_n \cos^{n-4}\phi \sin^4\phi - \cdots$$

$$\text{and} \quad \sin n\phi = {}^1C_n \cos^{n-1}\phi \sin\phi - {}^3C_n \cos^{n-3}\phi \sin^3\phi + \cdots$$

Thus, the solutions (12.399) can be written as

$$y_1 = \cos\left(n\cos^{-1}x\right) = x^n - {}^2C_n\, x^{n-2}\left(1-x^2\right) + {}^4C_n\, x^{n-4}\left(1-x^2\right)^2 - \cdots \tag{12.400}$$

$$y_2 = \sin\left(n\cos^{-1}x\right) = {}^1C_n\, x^{n-1}\left(1-x^2\right)^{\frac{1}{2}} - {}^3C_n\, x^{n-3}\left(1-x^2\right)^{\frac{3}{2}} + \cdots \tag{12.401}$$

From equation (12.400), it is clear that $y_1(x)$ is a polynomial of degree n. This polynomial is known as *Chebyshev's polynomial of first kind.* Thus, we write

$$\begin{aligned} T_n(x) &= \cos\left(n\cos^{-1}x\right) \\ \text{or,} \quad T_n(\cos\phi) &= \cos n\phi. \end{aligned} \tag{12.402}$$

Rodrigue's formula for Chebyshev's polynomials of first kind is given by

$$T_n(x) = \frac{2^n n!}{(2n)!}\sqrt{x^2-1}\,\frac{d^n}{dx^n}\left[\left(x^2-1\right)^{n-\frac{1}{2}}\right]. \tag{12.403}$$

It is obvious that

$$T_n(-x) = (-1)^n T_n(x), \quad T_n(1) = 1 \quad \text{and} \quad T_n(-1) = (-1)^n. \tag{12.404}$$

For different values of n, we have from equation (12.403)

$$\left.\begin{aligned} T_0(x) &= 1 \\ T_1(x) &= x \\ T_2(x) &= 2x^2 - 1 \\ T_3(x) &= 4x^3 - 3x \\ T_4(x) &= 8x^4 - 8x^2 + 1 \\ T_5(x) &= 16x^5 - 20x^3 + 5x \\ T_6(x) &= 32x^6 - 48x^4 + 18x^2 - 1 \\ &\cdots\cdots\cdots\cdots\cdots\cdots \end{aligned}\right\}. \tag{12.405}$$

The nature of these functions are shown in *Fig 12.5.*

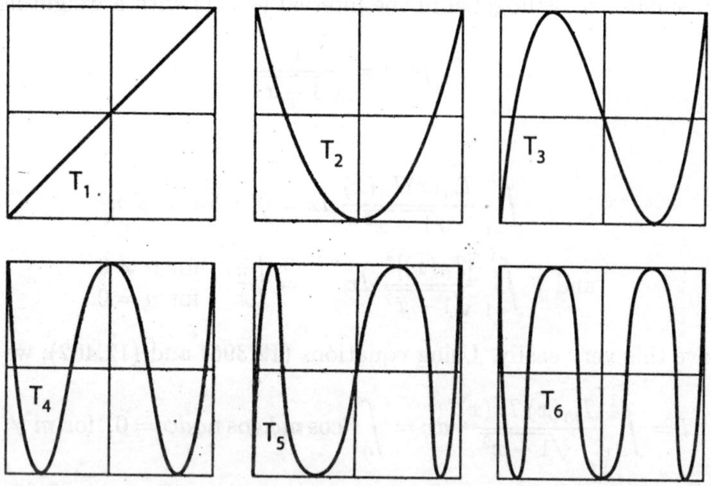

Fig 12.5

The *generating function* of Chebyshev's polynomials of first kind is given by

$$\psi(t, x) = \frac{1 - tx}{1 - 2tx + t^2} \tag{12.406}$$

so that

$$\sum_{k=0}^{\infty} T_k(x) t^k = \frac{1 - tx}{1 - 2tx + t^2}. \tag{12.407}$$

To prove this, we have

$$\begin{aligned}
\sum_{k=0}^{\infty} T_k(x)t^k &= \sum_{k=0}^{\infty} (\cos k\phi) t^k = \operatorname{Re} \sum_{k=0}^{\infty} \left(e^{i\phi} t\right)^k \text{ [Re means `real part of']} \\
&= \operatorname{Re}\left[\frac{1}{1 - e^{i\phi}t}\right] = \operatorname{Re}\left[\frac{1}{(1 - t\cos\phi) + it\sin\phi}\right] \\
&= \operatorname{Re}\left[\frac{(1 - t\cos\phi) - it\sin\phi}{1 - 2t\cos\phi + t^2}\right] \\
&= \frac{1 - t\cos\phi}{1 - 2t\cos\phi + t^2} \\
&= \frac{1 - tx}{1 - 2tx + t^2}.
\end{aligned}$$

Thus, we have proved the relation (12.407). Now we intend to find a recurrence relation. From equation (12.402), we have by changing n to $(n + 1)$,

$$\begin{aligned}
T_{n+1}(\cos\phi) = \cos(n+1)\phi &= \cos n\phi \cos\phi - \sin n\phi \cdot \sin\phi \\
&= 2\cos n\phi\cos\phi - [\cos n\phi\cos\phi + \sin n\phi\sin\phi] \\
&= 2\cos n\phi\cos\phi - \cos(n-1)\phi
\end{aligned}$$

$$\text{or,}\quad T_{n+1}(x) = 2xT_n(x) - T_{n-1}(x). \tag{12.408}$$

Equation (12.408) is the recurrence relation for Chebyshev's polynomials of first kind. These polynomials are orthogonal in the interval $(-1,1)$ with a weighting function

$$\rho(x) = \frac{1}{\sqrt{1-x^2}}. \tag{12.409}$$

Thus

$$\int_{-1}^{1} \frac{T_m(x)T_n(x)}{\sqrt{1-x^2}}\,dx = 0, \quad \text{for } m \neq n \tag{12.410}$$

$$\text{and} \quad \int_{-1}^{1} \frac{[T_n(x)]^2}{\sqrt{1-x^2}}\,dx = \begin{cases} \frac{\pi}{2}, & \text{for } n \neq 0 \\ \pi, & \text{for } n = 0. \end{cases} \tag{12.411}$$

We can prove this very easily. Using equations (12.396) and (12.402), we write

$$I = \int_{-1}^{1} \frac{T_m(x)\,T_n(x)}{\sqrt{1-x^2}}\,dx = \int_0^{\pi} \cos m\phi \cos n\phi d\phi = 0, \text{ for } m \neq n.$$

If $m = n$, we have

$$I = \int_0^{\pi} \cos^2 n\phi d\phi.$$

If now, $n = 0$,

$$I = \int_0^{\pi} d\phi = \pi.$$

If $n \neq 0$, then

$$I = \frac{1}{2}\int_0^{\pi} (1 + \cos 2n\phi)\,d\phi = \frac{\pi}{2}.$$

Thus we have established the orthogonality of Chebyshev's polynomials of first kind. Hence, a piece-wise continuous function $f(x)$ defined in $(-1,1)$ can be expanded in terms of these series like

$$f(x) = \sum_{k=0}^{\infty} A_k T_k(x). \tag{12.412}$$

Then the coefficients A_k are determined from

$$A_0 = \frac{1}{\pi}\int_{-1}^{1} \frac{f(x)\,dx}{\sqrt{1-x^2}} \tag{12.413}$$

$$A_k = \frac{2}{\pi}\int_{-1}^{1} \frac{T_k(x)\,f(x)\,dx}{\sqrt{1-x^2}} \quad (k = 1, 2, \ldots). \tag{12.414}$$

12.36 Chebyshev's Polynomials of Second Kind

We consider the differential equation

$$(1-x^2)\frac{d^2y}{dx^2} - 3x\frac{dy}{dx} + n(n+2)y = 0, \tag{12.415}$$

where n is a positive integer. The solution of this differential equation is a polynomial which is called *Chebyshev's polynomials of second kind* and is denoted by $U_n(x)$. We now find this polynomial.

We substitute

$$y(x) = \frac{V(x)}{\sqrt{1-x^2}}. \tag{12.416}$$

Then

$$\frac{dy}{dx} = \frac{1}{\sqrt{1-x^2}} \cdot \frac{dV}{dx} + \frac{Vx}{(1-x^2)^{\frac{3}{2}}} \tag{12.417}$$

and

$$\frac{d^2y}{dx^2} = \frac{1}{\sqrt{1-x^2}} \left[\frac{d^2V}{dx^2} + \frac{2x}{1-x^2} \frac{dV}{dx} + V\left(\frac{1}{1-x^2} + \frac{3x}{1-x^2} \right) \right]. \tag{12.418}$$

Substituting these relations into equation (12.415) we get after rearrangements of terms

$$(1-x^2) \frac{d^2V}{dx^2} - x\frac{dV}{dx} + (n+1)^2 V = 0. \tag{12.419}$$

Thus, $V(x)$ is the solution of Chebyshev's equation so that, we can write

$$V(x) = T_{n+1}(x). \tag{12.420}$$

Hence, from equation (12.416), we have

$$U_n(x) = \frac{T_{n+1}(x)}{\sqrt{1-x^2}}. \tag{12.421}$$

Now, from equation (12.401), we have

$$\begin{aligned} T_{n+1}(x) &= T_{n+1}(\cos\phi) = \sin\left[(n+1)\cos^{-1}x\right] \\ &= {}^1C_{n+1}x^n (1-x^2)^{\frac{1}{2}} - {}^3C_{n+1}x^{n-2}(1-x^2)^{\frac{3}{2}} + \cdots \end{aligned} \tag{12.422}$$

Hence

$$\begin{aligned} U_n(x) &= \frac{T_{n+1}(x)}{(1-x^2)^{\frac{1}{2}}} \\ &= {}^1C_{n+1}x^n - {}^3C_{n+1}x^{n-2}(1-x^2) + {}^5C_{n+1}x^{n-4}(1-x^2)^2 - \cdots \end{aligned} \tag{12.423}$$

which is a polynomial of degree n.

Rodrigue's formula for Chebyshev's polynomials of second kind is given by

$$U_n(x) = \frac{2^n(n+1)!}{(2n+1)!} \frac{1}{\sqrt{x^2-1}} \frac{d^n}{dx^n}\left[(x^2-1)^{n+\frac{1}{2}}\right]. \tag{12.424}$$

It is obvious that

$$U_n(-x) = (-1)^n U_n(x); \quad U_n(1) = n+1 \quad \text{and} \quad U_n(-1) = (-1)^n(n+1). \tag{12.425}$$

41

For different values of n, we have

$$\left.\begin{aligned}U_0(x)&=1\\U_1(x)&=2x\\U_2(x)&=4x^2-1\\U_3(x)&=8x^3-4x\\U_4(x)&=16x^4-12x^2+1\\U_5(x)&=32x^5-32x^3+6x\\U_6(x)&=64x^6-80x^4+24x^2-1\\&\cdots\cdots\cdots\cdots\\&\cdots\cdots\cdots\cdots\end{aligned}\right\}. \tag{12.426}$$

The *generating function* of Chebyshev's polynomials of second kind is given by

$$\psi(t,x)=\frac{1}{1-2tx+t^2} \tag{12.427}$$

so that

$$\sum_{n=0}^{\infty}U_n(x)t^n=\frac{1}{1-2tx+t^2}. \tag{12.428}$$

We now prove it. We have from equation (12.421)

$$\begin{aligned}\sum_{n=0}^{\infty}U_n(x)t^n&=\sum_{n=0}^{\infty}\frac{T_{n+1}(x)}{\sqrt{1-x^2}}t^n=\frac{1}{t\sin\phi}\sum_{n=0}^{\infty}\sin\{(n+1)\phi\}t^{n+1}\\&=\frac{1}{t\sin\phi}\,\mathrm{Im}\sum_{n=0}^{\infty}\left(e^{i\phi}\cdot t\right)^{n+1}\quad\text{[Im means `imaginary part of']}\\&=\frac{1}{t\sin\phi}\,\mathrm{Im}\left[\frac{e^{i\phi}\cdot t}{1-e^{i\phi}t}\right]=\frac{1}{t\sin\phi}\,\mathrm{Im}\left[\frac{t\cos\phi+it\sin\phi}{(1-t\cos\phi)-it\sin\phi}\right]\\&=\frac{1}{1-2t\cos\phi+t^2}=\frac{1}{1-2tx+t^2}.\end{aligned}$$

Thus, we have proved the relation (12.428).
Now, from equation (12.421), we get for $n\longrightarrow n+1$.

$$\begin{aligned}U_{n+1}(x)&=\frac{T_{n+2}(x)}{\sqrt{1-x^2}}=\frac{\sin\{(n+2)\phi\}}{\sin\phi}=\frac{1}{\sin\phi}\Big[\sin\{(n+1)+1\}\phi\Big]\\&=\frac{1}{\sin\phi}\Big[\sin(n+1)\phi\cos\phi+\cos(n+1)\phi\sin\phi\Big]\\&=\frac{1}{\sin\phi}\Big[2\sin(n+1)\phi\cdot\cos\phi-\{\sin(n+1)\phi\cos\phi-\cos(n+1)\phi\sin\phi\}\Big]\\&=2\cos\phi\cdot\frac{\sin(n+1)\phi}{\sin\phi}-\frac{\sin(n-1)\phi}{\sin\phi}\\&=2xU_n(x)-U_{n-1}(x).\end{aligned} \tag{12.429}$$

Equation (12.429) gives the recurrence relation for these polynomials.

Again, from equation (12.402), we get

$$T_{n+1}(x) = \cos(n+1)\phi.$$

Differentiating with respect to x

$$\frac{d}{dx}T_{n+1}(x) = -(n+1)\sin(n+1)\phi\left(-\frac{1}{\sin\phi}\right) = (n+1)U_n(x). \tag{12.430}$$

Equation (12.430) determines a relation between two kinds of polynomials of Chebyshev. The Chebyshev's polynomials of second kind are orthogonal in the interval $(-1, 1)$ with weighting function as

$$\rho(x) = \sqrt{1-x^2}. \tag{12.431}$$

Thus

$$\int_{-1}^{1}\sqrt{1-x^2}U_m(x) = \begin{cases} 0, & \text{for } m \neq n \\ \frac{\pi}{2}, & \text{for } m = n. \end{cases}$$

Therefore, any piece-wise continuous function in $(-1, 1)$ can be expanded in the series like

$$f(x) = \sum_{n=0}^{\infty} A_k U_k(x), \tag{12.432}$$

where

$$A_k = \frac{2}{\pi}\int_{-1}^{1}\sqrt{1-x^2}U_k(x)f(x)\,dx. \tag{12.433}$$

12.37 Chebyshev's Polynomial in Terms of Hypergeometric Function

We have the differential equation for Chebyshev's polynomials as

$$\left(1-x^2\right)\frac{d^2y}{dx^2} - x\frac{dy}{dx} + n^2y = 0. \tag{12.434}$$

We change the variable as

$$x = 1 - 2z. \tag{12.435}$$

Then the equation (12.434) becomes

$$\left(z^2 - z\right)\frac{d^2y}{dz^2} + \left(z - \frac{1}{2}\right)\frac{dy}{dz} - n^2y = 0. \tag{12.436}$$

Equation (12.436) corresponds to equation (12.1) for

$$\alpha = n, \quad \beta = -n \quad \text{and} \quad \gamma = \frac{1}{2}. \tag{12.437}$$

Hence, the solution for equation (12.434) can be written as

$$y(z) = F\left(n, -n, \frac{1}{2}; z\right)$$

$$\text{i.e.,} \quad y(x) = F\left(n, -n; \frac{1}{2}; \frac{1-x}{2}\right). \tag{12.438}$$

This equation gives the Chebyshev's polynomial of first kind in terms of Hyper-geometric function.

Similarly, we can find that Chebyshev's polynomials of second kind are expressed in terms of Hyper-geometric functions in the form

$$U_n(x) = (n+1)F\left(n+2, -n, \frac{3}{2}; \frac{1-x}{2}\right). \tag{12.439}$$

EXERCISE

1. Show that $\tan^{-1}(x) = xF\left(\frac{1}{2}, 1, \frac{3}{2}; -x^2\right)$.
2. (a) $\frac{1}{x}(e^x - 1) = F(1, 2; x)$;
   (b) $1 - 2x + \frac{x^2}{2} = F(-2, 1; x)$.
3. Evaluate $\int x^4 J_1(x)\, dx$.
4. Prove that $I_{n-1}(x) - I_{n+1}(x) = \frac{2n}{x} I_n(x)$.
5. Prove that
   (a) $\cos x = J_0(x) - 2J_2(x) + 2J_4(x) - \cdots$;
   (b) $\sin x = 2J_1(x) - 2J_3(x) + 2J_5(x) - \cdots$.
6. Solve the differential equation $y'' + (ax+b)^m y = 0$ in terms of Bessel's functions.
   [*Hints*: Put $(m+2)a \cdot z = 2(ax+b)^{\frac{m}{2}+1}$]
7. Show that $\displaystyle\int_{-1}^{1} \left[\frac{d}{dx} P_n(x)\right]^2 dx = n(n+1)$.
8. Prove that $\displaystyle\int_0^{\pi} P_n(\cos\theta) \cos n\theta\, d\theta = \frac{(2n)!}{2^n (n!)^2}\pi$.
9. Expand the function

$$f(x) = \begin{cases} 0, & \text{for } -1 \le x < 0 \\ 1, & \text{for } 0 < x \le 1 \end{cases}$$

   in terms of Legendre's polynomials.
10. Show that $x^4 = \frac{1}{35}\left[8P_4(x) + 30P_2(x) + 7P_0(x)\right]$.
11. Show that $\displaystyle\int_{-1}^{1} xP_n(x)P_{n-1}(x)\, dx = \frac{2n}{4n^2 - 1}$.
12. Prove the orthogonality of Associated Legendre's Polynomials

$$\int_{-1}^{1} P_k^m(x) P_r^m(x)\, dx = \frac{2}{2r+1} \frac{(r+m)!}{(r-m)!} \delta_{k,r}.$$

13. Prove that
   (a) $\displaystyle\frac{1}{e} \cosh 2x = \sum_{n=0}^{\infty} \frac{1}{(2n)!} H_{2n}(x)$;
   (b) $\displaystyle\frac{1}{e} \sinh 2x = \sum_{n=0}^{\infty} \frac{1}{(2n+1)!} H_{2n+1}(x)$.
14. Show that $\displaystyle\int_{-\infty}^{\infty} H_n\left(x + \frac{a}{2}\right) e^{-x^2}\, dx = \sqrt{\pi} a^n$.

15. Prove that $\int_0^\infty x^m e^{-x} L_n(x)\,dx = (-1)^n n!\,\delta_{m,n}$.

16. Derive the recurrence relation, $\frac{d}{dx}L_n^k(x) - \frac{d}{dx}L_{n-1}^k(x) + L_{n-1}^k(x) = 0$.

ANSWERS

3. $(4x^3 - 16x)J_1(x) - (x^4 - 8x^2)J_0(x)$.

6. $y(z) = (ax+b)^{\frac{1}{2}}\left[AJ_\lambda(z) + BY_\lambda(z)\right],\ \lambda = \frac{1}{m+2}$.

9. $f(x) = \frac{1}{2}P_0(x) + \frac{3}{4}P_1(x) - \frac{7}{16}P_3(x) + \frac{11}{32}P_5(x) + \cdots$

Chapter 13

Partial Differential Equations

13.1 Introduction

Equations containing partial derivatives of a function of two or more variables are called *partial differential equations* (PDE). The order of a partial differential equation is that of the derivative of highest order present in the equation. Thus, the equation

$$x^2\frac{\partial u}{\partial x}+y^2\frac{\partial u}{\partial y}=u, \tag{13.1}$$

is of order one, whereas, the equation

$$\frac{\partial^2 u}{\partial x^2}+xy\frac{\partial^2 u}{\partial x\partial y}+\frac{\partial^2 u}{\partial y^2}=0 \tag{13.2}$$

is of order two.

Partial differential equations arise in the study of geometrical and physical problems. We have already seen in chapter 11 some of the physical processes which are described by partial differential equations.

In this chapter, we intend to discuss the methods for solving partial differential equations confining ourselves to functions of two variables. We will use the following notations:

$$\left.\begin{array}{lll} u_x=\dfrac{\partial u}{\partial x}, & u_y=\dfrac{\partial u}{\partial y} & \\[2ex] u_{xx}=\dfrac{\partial^2 u}{\partial x^2}, & u_{xy}=\dfrac{\partial^2 u}{\partial x\partial y}, & u_{yy}=\dfrac{\partial^2 u}{\partial y^2} \end{array}\right\}, \tag{13.3}$$

where $u=u(x,y)$ is a function of two variables x,y.

GENERAL METHODS

13.2 Linear PDE of Order One

Equations of the type

$$P(x,y)\,u_x+Q(x,y)\,u_y=R(x,y)\,u, \tag{13.4}$$

are linear equations of order one.

For the general solution of such equations, the problem is reduced to an auxiliary system, called *Lagrange system* of ordinary differential equations

$$\frac{dx}{P} = \frac{dy}{Q} = \frac{du}{Ru}. \tag{13.5}$$

Case I. *P, Q, R are all constants.*

From the relation

$$\frac{dx}{P} = \frac{dy}{Q},$$

we obtain

$$Py - Qx = \text{constant} = A \text{ (say)}. \tag{13.6}$$

From the relation

$$\frac{dx}{P} = \frac{du}{Ru}, \tag{13.7}$$

we obtain

$$\ln u = \frac{R}{P}x + \ln B, \quad B = \text{constant}$$

$$\text{or,} \quad u = Be^{\frac{Rx}{P}}. \tag{13.8}$$

The constant B in equation (13.8) can be expressed as a function

$$B = \phi(Py - Qx) \tag{13.9}$$

in view of equation (13.6).

Hence the general solution is

$$u(x, y) = e^{\frac{Rx}{P}}\phi(Py - Qx). \tag{13.10}$$

EXAMPLE 13.1. *Solve:* $2u_x + 3u_y = 4u$.

Solution: Here, $P = 2$, $Q = 3$, $R = 4$.

By equation (13.10), the general solution is

$$u(x, y) = e^{2x}\phi(2y - 3x).$$

Case II. *P, Q, R are all functions.*

In this case, we are to solve the system of equations (13.5) taking pairs and eliminating the constants of integration. The method is explained with the help of examples.

EXAMPLE 13.2. *Solve:* $xy\, u_x - x^2 u_y = yu$.

Solution: Here $P = xy$, $Q = -x^2$, $R = y$.

Hence, equation (13.5) yields

$$\frac{dx}{xy} = \frac{dy}{-x^2} = \frac{du}{yu}. \tag{1}$$

From first pair, we have

$$x\,dx + y\,dy = 0 \quad \text{or,} \quad x^2 + y^2 = A. \tag{2}$$

From the first and the third relations in (1), we get

$$\frac{dx}{x} = \frac{du}{u} \quad \text{i.e.,} \quad \ln u = \ln x + \ln B$$

$$\text{or,} \quad u = Bx, \quad B = \text{constant}. \tag{3}$$

From equations (2) and (3), we write

$$u(x, y) = x\phi\left(x^2 + y^2\right), \tag{4}$$

where ϕ is an arbitrary function.

EXAMPLE 13.3. *Solve:* $x u_x + u_y = -uy$.

Solution: The auxiliary equation is

$$\frac{dx}{x} = \frac{dy}{1} = \frac{du}{-uy}. \tag{1}$$

The relation

$$\frac{dx}{x} = \frac{dy}{1},$$

yields

$$y = \ln\frac{x}{A} \quad \text{or,} \quad xe^{-y} = A = \text{constant}. \tag{2}$$

The relation

$$\frac{dy}{1} = \frac{du}{-uy}$$

yields

$$-\frac{y^2}{2} = \ln\left(\frac{u}{B}\right) \quad \text{or,} \quad u = Be^{-\frac{y^2}{2}}. \tag{3}$$

From equations (2) and (3)

$$u(x, y) = \phi(xe^{-y})\, e^{-\frac{y^2}{2}}.$$

Case III. *P, Q are constants, but* $R = 0$.

In this case, equation (13.4) assumes the form

$$Pu_x + Qu_y = 0. \tag{13.11}$$

We introduce the new variables

$$\xi = Px + Qy, \quad \beta = Qx - Py. \tag{13.12}$$

Then

$$u_x = \frac{\partial u}{\partial \xi}\frac{\partial \xi}{\partial x} + \frac{\partial u}{\partial \beta}\frac{\partial \beta}{\partial x} = P\frac{\partial u}{\partial \xi} + Q\frac{\partial u}{\partial \beta} \tag{13.13}$$

$$\text{and} \quad u_y = \frac{\partial u}{\partial \xi}\frac{\partial \xi}{\partial y} + \frac{\partial u}{\partial \beta}\frac{\partial \beta}{\partial y} = Q\frac{\partial u}{\partial \xi} - P\frac{\partial u}{\partial \beta}. \tag{13.14}$$

Hence, the equation (13.11) assumes the form

$$(P^2 + Q^2)\frac{\partial u}{\partial \xi} = 0 \quad \text{or,} \quad \frac{\partial u}{\partial \xi} = 0,$$

whose general solution is

$$u = \phi(\beta) = \phi(Qx - Py). \tag{13.15}$$

13.3 Linear PDE of Order Two (Constant Coefficients)

Linear partial differential equations of order two are of the form

$$Au_{xx} + Bu_{xy} + Cu_{yy} + Du_x + Eu_y + Fu = G, \tag{13.16}$$

where the coefficients $A, B, \ldots$ may be constants or functions of x, y.

These equations are classified as elliptic, parabolic or hyperbolic depending on the value of expression $B^2 - 4AC$. That is, the equation is

$$\left.\begin{array}{l} \text{(i) elliptic,} \quad\ \ \text{if } B^2 - 4AC < 0 \\ \text{(ii) parabolic, if } B^2 - 4AC = 0 \\ \text{(iii) hyperbolic, if } B^2 - 4AC > 0 \end{array}\right\}. \tag{13.17}$$

As an example, consider the equation

$$u_{xx} - 4u_{yy} + u_y = 0. \tag{13.18}$$

Here, $A = 1$, $B = 0$, $C = -4$.
Hence, $B^2 - 4AC = 16 > 0$.
Therefore, the equation (13.18) is hyperbolic.
Let the coefficients $A, B, \ldots, F$ be constants assuming real values only.
We introduce the operators

$$D_x = \frac{\partial}{\partial x}, \quad D_y = \frac{\partial}{\partial y}, \tag{13.19}$$

so that

$$\frac{\partial^2}{\partial x^2} = D_x^2, \quad \frac{\partial^2}{\partial x \partial y} = D_x D_y, \quad \frac{\partial^2}{\partial y^2} = D_y^2. \tag{13.20}$$

Then, equation (13.16) can be expressed as

$$\Phi(D_x, D_y)\, u = G(x, y). \tag{13.21}$$

Let us find the general solution of equation

$$\Phi(D_x, D_y)\, u = 0, \tag{13.22}$$

We assume that, $D = E = F = 0$, so that

$$\Phi(D_x, D_y)\, u \equiv (AD_x^2 + BD_x D_y + CD_y^2)\, u = 0. \tag{13.23}$$

We make the change of variable as $\xi = y + mx$ and assume the solution of equation (13.22) in the form

$$u(x, y) = \phi(y + mx) = \phi(\xi), \tag{13.24}$$

where ϕ is an arbitrary function. Then we have

$$D_x u = \frac{\partial u}{\partial x} = \frac{d\phi}{d\xi}\frac{\partial \xi}{\partial x} = m\frac{d\phi}{d\xi}$$

$$D_y u = \frac{\partial u}{\partial y} = \frac{d\phi}{d\xi}\frac{\partial \xi}{\partial y} = \frac{d\phi}{d\xi}.$$

Substituting these relations into equation (13.23) we get

$$(Am^2 + Bm + C)\frac{d^2\phi}{d\xi^2} = 0$$

which yields

$$Am^2 + Bm + C = 0. \tag{13.25}$$

Equation (13.25) yields two values m_1, m_2 for m.
If $m_1 \neq m_2$, the general solution is

$$u(x, y) = f_1(y + m_1 x) + f_2(y + m_2 x), \tag{13.26}$$

where f_1, f_2 are arbitrary functions.
If $m_1 = m_2 = m$ and each are real, the solution is

$$u(x, y) = f_1(y + mx) + x f_1(y + mx). \tag{13.27}$$

Let the equation be written as

$$\Phi(D_x, D_y) = (a_1 D_x + b_1 D_y - c_1)(a_2 D_x + b_2 D_y - c_2)\, u = 0. \tag{13.28}$$

Then the solution of equation will be the sum of solutions of

$$\left.\begin{aligned} a_1 D_x + b_1 D_y &= c_1 \\ a_2 D_x + b_2 D_y &= c_2 \end{aligned}\right\}. \tag{13.29}$$

Following equation (13.10), solutions of these equations are

$$\left.\begin{aligned} & e^{c_1 x/a_1}\phi_1(a_1 y - b_1 x) \\ \text{and} \quad & e^{c_2 x/a_2}\phi_2(a_2 y - b_2 x) \end{aligned}\right\} \tag{13.30}$$

Therefore, the solution of equation (13.28) will be

$$u(x, y) = e^{c_1 \frac{x}{a_1}}\phi_1(a_1 y - b_1 x) + e^{c_2 \frac{x}{a_2}}\phi_2(a_2 y - b_2 x). \tag{13.31}$$

EXAMPLE 13.4. *Solve:* $u_{xx} - 3u_{xy} + 2u_{yy} = 0$.

Solution: The given equation can be written as

$$(D_x - D_y)(D_x - 2D_y)\, u = 0.$$

Hence, $m_1 = 1$, $m_2 = 2$.
Therefore, the general solution is

$$u(x, y) = f_1(y + x) + f_2(y + 2x).$$

EXAMPLE 13.5. *Solve:* $u_{xx} - u_{yy} + 3u_x + u_y + 2u = 0$.

Solution: The equation can be written as

$$(D_x^2 - D_y^2 + 3D_x + D_y + 2)\,u = 0$$
$$\text{or,}\quad (D_x + D_y + 1)(D_x - D_y + 2)\,u = 0$$

which is split into two equations

$$(D_x + D_y + 1)\,u = 0 \qquad (1)$$
$$(D_x - D_y + 2)\,u = 0. \qquad (2)$$

For equation (1), $a = 1$, $b = 1$, $c = -1$. Hence its solution is

$$e^{-x} f_1(y - x). \qquad (3)$$

For equation (2), $a = 1$, $b = -1$, $c = -2$. Hence its solution is

$$e^{-2x} f_2(y + x). \qquad (4)$$

Thus, the general solution of the given equation is

$$u(x, y) = e^{-x} f_1(y - x) + e^{-2x} f_2(y + x). \qquad (5)$$

13.4 Particular Solutions

Consider the PDE of second order

$$\Phi(D_x, D_y)\,u = G(x, y), \qquad (13.32)$$

where the coefficients are constants. The general solution of equation (13.32) will consist of complimentary function corresponding to the solution of equation

$$\Phi(D_x, D_y) = 0 \qquad (13.33)$$

and the particular integral given by

$$\frac{1}{\Phi(D_x, D_y)} G(x, y). \qquad (13.34)$$

We now seek to find the particular integral corresponding to different types of function $G(x, y)$.

Let $\Phi(D_x, D_y)\,u = (D_x - m_1 D_y)(D_y - m_2 D_y)\,u$.

Then the particular integral is

$$\frac{1}{(D_x - m_1 D_y)(D_x - m_2 D_y)} G(x, y). \qquad (13.35)$$

Let

$$\frac{1}{D_x - m_2 D_y} G(x, y) = \theta(x, y). \qquad (13.36)$$

Then

$$(D_x - m_2 D_y)\theta = G(x, y)$$
$$\text{i.e.,}\quad \frac{\partial \theta}{\partial x} - m_2 \frac{\partial \theta}{\partial y} = G(x, y). \qquad (13.37)$$

The auxiliary system for this equation is

$$\frac{dx}{1} = \frac{dy}{-m_2} = \frac{d\theta}{G(x,y)}. \tag{13.38}$$

From the 1st and the 2nd relations

$$y + m_2 x = \text{constant} = c. \tag{13.39}$$

Equation (13.39) yields

$$y = c - m_2 x. \tag{13.40}$$

The first and third relations can be written as

$$\frac{dx}{1} = \frac{d\theta}{G(x, c - m_2 x)}$$

which yields

$$\theta = \int G(x, c - m_2 x)\, dx. \tag{13.41}$$

After integration, the constant is to be replaced by $(y + m_2 x)$ according to equation (13.39). Knowing $\theta(x,y)$, the second particular integral

$$\frac{1}{D_x - m_1 D_y} \cdot \theta(x,y), \tag{13.42}$$

can be obtained following the above procedure.

EXAMPLE 13.6. *Solve:* $u_{xx} - 3u_{xy} + 2u_{yy} = ye^x$.

Solution: The equation is written as

$$(D_x - D_y)(D_x - 2D_y)\, u = ye^x.$$

The complimentary function is given by example 13.4.
Let the first particular integral be

$$\frac{1}{D_x - 2D_y} ye^x = \theta(x,y). \tag{1}$$

Hence

$$\begin{aligned} \theta(x,y) &= \int G(x, c - m_2 x)\, dx = \int (c - 2x)\, e^x\, dx \quad (m_2 = 2) \\ &= ce^x - 2xe^x + e^x = (y + 2x)\, e^x - 2xe^x + e^x \\ &= ye^x + e^x. \end{aligned} \tag{2}$$

Similarly, let

$$\frac{1}{D_x - D_y} \theta(x,y) = \xi(x,y). \tag{3}$$

Then, following above procedure (here $m_1 = 1$),

$$\begin{aligned}\xi(x,y) &= \int \theta(x, c-x)\,dx = \int [(c-x)\,e^x + e^x]\,dx \\ &= ce^x - xe^x + e^x + e^x = (y+x)\,e^x - xe^x + 2e^x \\ &= ye^x + 2e^x. \end{aligned} \tag{4}$$

Thus, the particular integral of the given equation is the expression in the RHS of equation (4). Therefore, the general solution is

$$u(x,y) = f_1(y+x) + f_2(y+2x) + (y+2)\,e^x. \tag{5}$$

13.5 Particular Integrals for Some Functions

1. If $G(x,y) = e^{ax+by}$, then the particular integral of equation (13.32) is

$$\frac{1}{\Phi(a,b)} e^{ax+by}. \tag{13.43}$$

2. If $G(x,y) = \sin(ax+by)$, the particular integral of equation (13.32) is

$$-\frac{1}{\Phi(a,b)} \sin(ax+by). \tag{13.44}$$

3. If $G(x,y) = \cos(ax+by)$, the particular integral of equation (13.32) is

$$-\frac{1}{\Phi(a,b)} \cos(ax+by). \tag{13.45}$$

In all the above cases, it is necessary that

$$\Phi(a,b) \neq 0. \tag{13.46}$$

If $\Phi(a,b) = 0$, we write

$$\Phi(D_x, D_y) = \left(D_x - \frac{a}{b} D_y\right) F(D_x, D_y) \tag{13.47}$$

such that $F(a,b) \neq 0$. Then the particular integral corresponding to e^{ax+by} is given by

$$\frac{1}{F(a,b)} \frac{1}{D_x - \frac{a}{b} D_y} e^{ax+by} = \frac{1}{F(a,b)} x e^{ax+by}. \tag{13.48}$$

For the functions $\sin(ax+by)$ or $\cos(ax+by)$, the particular integral is obtained by the method of undetermined coefficients which is explained with examples.

4. If $G(x,y) = U(x,y)\,e^{ax+by}$, then the particular integral is given by

$$\frac{1}{\Phi(D_x, D_y)} U(x,y)\,e^{ax+by} = e^{ax+by} \frac{1}{\Phi(D_x + a, D_y + b)} U(x,y). \tag{13.49}$$

5. Let $G(x,y)$ be a polynomial in x, y. Then, for particular integral of

$$(D_x - mD_y)\, u = G(x,y),$$

we write,

$$D_x - mD_y = D_x \left(1 - m\frac{D_y}{D_x}\right)$$

so that

$$\frac{1}{D_x - mD_y} = \frac{1}{D_x}\left(1 - m\frac{D_y}{D_x}\right)^{-1} = \frac{1}{D_x}\left[1 + m\frac{D_y}{D_x} + (m\frac{D_y}{D_x})^2 + \cdots\right] \tag{13.50}$$

Similarly, we write

$$\frac{1}{aD_x + bD_y + c} = \frac{1}{c}\cdot\frac{1}{1 + (\frac{a}{c}\,D_x + \frac{b}{c}\,D_y)}$$
$$= \frac{1}{c}\left[1 - \left(\frac{a}{c}\,D_x + \frac{b}{c}\,D_y\right) + \left(\frac{a}{c}\,D_x + \frac{b}{c}\,D_y\right)^2 + \cdots\right]. \tag{13.51}$$

EXAMPLE 13.7. *Find the particular integral of*

$$u_{xx} - u_{xy} - 6u_{yy} = e^{2x+y} + \sin(2x - 3y).$$

Solution: The equation is written as

$$D_x^2 - D_xD_y - 6D_y^2 = e^{2x+y} + \sin(2x - 3y).$$

Hence

$$\Phi(D_x, D_y) = D_x^2 - D_xD_y - 6D_y^2. \tag{1}$$

For the function e^{2x+y}, $a = 2$, $b = 1$.

$$\therefore \quad \Phi(a,b) = 2^2 - 2\cdot 1 - 6\cdot 1^2 = -4 \neq 0.$$

Hence the particular integral is

$$-\tfrac{1}{4}\, e^{2x+y}. \tag{2}$$

For the function $\sin(2x - 3y)$, $a = 2$, $b = -3$.
Hence, $\Phi(a,b) = 2^2 - 2(-3) - 6(-3)^2 = -44 \neq 0$.
Therefore, the particular integral is $\frac{1}{44}\sin(2x - 3y)$.

EXAMPLE 13.8. *Find the particular integral of*

$$u_{xx} - 3u_{xy} + 2u_{yy} = e^{x+y} + \sin(2x + y).$$

Solution: The given equation is written as

$$\Phi(D_x, D_y)\, u = e^{x+y} + \sin(2x + y),$$

where

$$\Phi(D_x, D_y) = D_x^2 - 3D_xD_y + 2D_y^2. \tag{1}$$

For the function e^{x+y}, $a = 1$, $b = 1$.

Hence $\Phi(a,b) = 1^2 - 3 \cdot 1 \cdot 1 + 2 \cdot 1^2 = 0$.
We therefore write expression (1) as

$$\Phi(D_x, D_y) = (D_x - D_y)(D_x - 2D_y).$$

Hence the particular integral is

$$\frac{1}{D_x - D_y}\left[\frac{1}{D_x - 2D_y}e^{x+y}\right] = \frac{1}{D_x - D_y} \cdot \frac{1}{1-2.1}e^{x+y} = \frac{-1}{D_x - D_y}e^{x+y} = -xe^{x+y}.$$

For the function $\sin(2x+y)$, $a = 2$, $b = 1$.
Hence, $\Phi(a,b) = 2^2 - 3.2.1 + 2.1^2 = 0$.
We write the particular integral as

$$\bar{u}(x,y) = Cx\sin(2x+y) + Dx\cos(2x+y). \tag{2}$$

Evaluating the derivatives $\bar{u}_{xx}$, $\bar{u}_{xy}$ and $\bar{u}_{yy}$ of the expression (2) and substituting into equation (1), we find

$$-D\sin(2x+y) + C\cos(2x+y) = \sin(2x+y),$$

which yields

$$D = -1, \quad C = 0. \tag{3}$$

Hence, the particular integral is $= -x\cos(2x+y)$. Therefore, when $\Phi(a,b) = 0$, the particular integrals corresponding to $\sin(ax+by)$ or $\cos(ax+by)$ is to be sought in the form

$$Cx\sin(ax+by) + Dx\cos(ax+by).$$

EXAMPLE 13.9. *Solve:* $u_{xx} - u_{yy} + 3u_x + u_y + 2u = xe^{-y}$.

Solution: The equation can be written as

$$(D_x + D_y + 1)(D_x - D_y + 2)\,u = xe^{-y}. \tag{1}$$

Complimentary function is (see Example 13.5)

$$u(x,y) = e^{-x}f_1(y-x) + e^{-2x}f_2(y+x). \tag{2}$$

The particular integral is given by

$$\text{PI} = \frac{1}{(D_x + D_y + 1)(D_x - D_y + 2)}xe^{-y}. \tag{3}$$

Let

$$\frac{1}{D_x - D_y + 2}xe^{-y} = \theta(x,y). \tag{4}$$

Then

$$(D_x - D_y + 2)\theta = xe^{-y} \quad \text{i.e.,} \quad \frac{d\theta}{dx} - \frac{d\theta}{dy} = xe^{-y} - 2\theta. \tag{5}$$

Its auxiliary system is

$$\frac{dx}{1} = \frac{dy}{-1} = \frac{d\theta}{xe^{-y} - 2\theta}.$$

Then, from 1st and 2nd relation

$$y = c - x. \tag{6}$$

From 2nd and 3rd relation

$$\frac{d\theta}{dy} = 2\theta - xe^{-y}$$

$$\text{or,} \quad \frac{d\theta}{dy} - 2\theta = (y-c)\,e^{-y} \quad [\text{using equation (6)}]. \tag{7}$$

Multiplying both sides of equation (7) by the integrating factor e^{-2y} and integrating we find

$$\theta e^{-2y} = \int (y-c)\,e^{-3y}dy = (y-c)\frac{e^{-3y}}{-3} - \frac{1}{9}e^{-3y}.$$

Hence

$$\theta(x,y) = -\frac{1}{3}(y-c)\,e^{-y} - \frac{1}{9}e^{-y} = \frac{1}{3}xe^{-y} - \frac{1}{9}e^{-y}. \tag{8}$$

Again, let

$$\frac{1}{D_x + D_y + 1}\theta(x,y) = \xi(x,y). \tag{9}$$

Then

$$(D_x + D_y + 1)\xi(x,y) = \theta(x,y). \tag{10}$$

For the first part of $\theta(x,y)$, i.e., $\frac{1}{3}\,xe^{-y}$, the particular integral has the auxiliary system

$$\frac{dx}{1} = \frac{dy}{1} = \frac{d\xi}{\frac{1}{3}\,xe^{-y} - \xi}. \tag{11}$$

Then

$$y = x + c \tag{12}$$

and

$$\frac{d\xi}{dy} + \xi = \frac{1}{3}(y-c)\,e^{-y},$$

which yields

$$\xi e^{y} = \int \frac{1}{3}(y-c)\,dy$$

$$\text{i.e.,} \quad \xi e^{y} = \frac{1}{6}(y-c)^2 = \frac{1}{6}x^2 \quad \text{or,} \quad \xi(x,y) = \frac{1}{6}x^2e^{-y}. \tag{13}$$

Particular integral corresponding to second part of $\theta(x,y)$, i.e., $-\frac{1}{9}\,e^{-y}$ is obtained in the same way. Here we write, from 2nd and 3rd relations in equation (11)

$$\frac{d\xi}{dy} + \xi = -\frac{1}{9}e^{-y} \quad \text{or,} \quad \xi e^{y} = -\frac{1}{9}\int dy = -\frac{1}{9}y.$$

$$\therefore \quad \xi = -\frac{1}{9}ye^{-y}. \tag{14}$$

Hence, the general solution of the equation is

$$u(x,y) = e^{-x}f_1(y-x) + e^{-2x}f_2(y+x) + \frac{1}{6}x^2e^{-y} - \frac{1}{9}ye^{-y}.$$

EXAMPLE 13.10. *Solve:* $u_{xx} - u_{xy} - 2u_{yy} = e^{2x+y}\left(x^2 + y^2\right)$.

Solution: The equation can be written as

$$(D_x - 2D_y)(D_x + D_y)\, u = e^{2x+y}\left(x^2 + y^2\right). \tag{1}$$

The complimentary function is

$$u(x, y) = f_1(y + 2x) + f_2(y - x). \tag{2}$$

The particular integral is

$$\frac{1}{(D_x - 2D_y)(D_x + D_y)} e^{2x+y}\left(x^2 + y^2\right). \tag{3}$$

Recalling equation (13.49), here we have

$$U(x, y) = x^2 + y^2, \quad a = 2 \text{ and } b = 1. \tag{4}$$

Then the particular integral is

$$e^{2x+y} \frac{1}{(D_x - 2D_y)(D_x + D_y + 3)}\left(x^2 + y^2\right). \tag{5}$$

Let

$$\theta(x, y) = \frac{1}{D_x + D_y + 3}\left(x^2 + y^2\right). \tag{6}$$

Now

$$\begin{aligned}
\frac{1}{D_x + D_y + 3} &= \frac{1}{3} \cdot \frac{1}{1 + \frac{1}{3}(D_x + D_y)} = \frac{1}{3}\left[1 + \frac{1}{3}(D_x + D_y)\right]^{-1} \\
&= \frac{1}{3}\left[1 - \frac{1}{3}(D_x + D_y) + \frac{1}{9}(D_x + D_y)^2 + \cdots\right].
\end{aligned} \tag{7}$$

Hence

$$\begin{aligned}
\theta(x, y) &= \frac{1}{3}\left[\left(x^2 + y^2\right) - \frac{2}{3}(x + y) + \frac{16}{9}\right] \\
&= \frac{1}{27}\left(9x^2 + 9y^2 - 6x - 6y + 16\right).
\end{aligned} \tag{8}$$

Again

$$\begin{aligned}
\frac{1}{D_x - 2D_y}\theta(x, y) &= \frac{1}{D_x\left(1 - 2\frac{D_y}{D_x}\right)}\theta(x, y) \\
&= \frac{1}{D_x}\left[1 + 2\frac{D_y}{D_x} + 4\frac{D_y^2}{D_x^2} + \cdots\right]\theta(x, y) \\
&= \frac{1}{D_x}\left[\theta(x, y) + \frac{2}{D_x}\frac{\partial \theta}{\partial y} + \frac{4}{D_x^2}\frac{\partial^2 \theta}{\partial y^2} + \cdots\right].
\end{aligned} \tag{9}$$

Substituting for the function $\theta(x,y)$ from equation (8), we get

$$\frac{1}{D_x}\theta(x,y) = \frac{1}{27}\left(3x^3+9xy^2-3x^2-6xy+16x\right) \tag{10}$$

$$\frac{2}{D_x^2}\frac{\partial\theta}{\partial y} = \frac{2}{D_x^2}\left(\frac{18}{27}y-\frac{6}{27}\right) = 2\left(\frac{1}{3}x^2y-\frac{1}{9}x^2\right) \tag{11}$$

$$\frac{4}{D_x^3}\frac{\partial^2\theta}{\partial y^2} = \frac{4}{D_x^3}\left(\frac{18}{27}\right) = \frac{72}{27}\cdot\frac{x^3}{6} = \frac{4}{9}x^3. \tag{12}$$

Summing up the expressions (10), (11) and (12), we write the particular integral as

$$\frac{1}{27}e^{2x+y}\left(15x^3+9xy^2+18x^2y-9x^2-6xy+16x\right).$$

13.6 Cauchy Type Equations

Equations of the type

$$Ax^2u_{xx}+Bxy\,u_{xy}+Cy^2u_{yy}+Dx\,u_x+Ey\,u_y+Fu=G, \tag{13.52}$$

where $A, B, \ldots, F$ are constants and G may be a function of (x,y), is known as *Cauchy type equation.* The substitution

$$\xi=\ln x \quad\text{and}\quad \eta=\ln y \tag{13.53}$$

converts equation (13.52) into one with constant coefficients. We have

$$\frac{\partial}{\partial x}=\frac{\partial}{\partial\xi}\cdot\frac{d\xi}{dx}=\frac{1}{x}\frac{\partial}{\partial\xi}$$

$$\frac{\partial^2}{\partial x^2}=\frac{\partial}{\partial x}\left(\frac{1}{x}\frac{\partial}{\partial\xi}\right)=-\frac{1}{x^2}\frac{\partial}{\partial\xi}+\frac{1}{x}\frac{\partial}{\partial\xi}\left(\frac{\partial}{\partial\xi}\right)\frac{d\xi}{dx}=\frac{1}{x^2}\left(\frac{\partial^2}{\partial\xi^2}-\frac{\partial}{\partial\xi}\right).$$

Similarly

$$\frac{\partial}{\partial y}=\frac{1}{y}\frac{\partial}{\partial\eta},\quad \frac{\partial^2}{\partial y^2}=\frac{1}{y^2}\left(\frac{\partial^2}{\partial\eta^2}-\frac{\partial}{\partial\eta}\right).$$

In symbolic notations:

$$\left.\begin{array}{l} xD_x=D_\xi,\quad yD_y=D_\eta \quad xyD_xD_y=D_\xi D_\eta \\ x^2D_xD_x=D_\xi(D_\xi-1),\quad y^2D_yD_y=D_\eta(D_\eta-1) \end{array}\right\}. \tag{13.54}$$

With these transformations equation (13.52) assumes the form

$$Au_{\xi\xi}+Bu_{\xi\eta}+Cu_{\eta\eta}+(D-A)\,u_\xi+(E-C)\,u_\eta+F=G' \tag{13.55}$$

which is an equation with constant coefficients and hence can be solved by the methods we have discussed. The function $G'(\xi,\eta)$ is obtained from $G(x,y)$ by using the substitutions in equation (13.53).

EXAMPLE 13.11. *Solve:* $x^2u_{xx}-xy\,u_{xy}-2y^2u_{yy}+x\,u_x-2y\,u_y=\ln\left(\frac{y}{x}\right)-\frac{1}{2}.$

Solution: With new variables $\xi = \ln x$; $\eta = \ln y$, the given equation is transformed into

$$u_{\xi\xi} - u_{\xi\eta} - 2u_{\eta\eta} = \eta - \xi - \frac{1}{2}. \tag{1}$$

In symbolic notation, the equation (1) is written as

$$(D_\xi - 2D_\eta)(D_\xi + D_\eta)\, u = \eta - \xi - \frac{1}{2}. \tag{2}$$

The complimentary function is (see Example 13.10)

$$u(\xi, \eta) = f_1(2\xi + \eta) + f_2(\eta - \xi),$$

i.e.,

$$\begin{aligned} u(x, y) &= f_1\left(\ln x^2 y\right) + f_2\left(\ln \frac{y}{x}\right) \\ &= \phi_1(x^2 y) + \phi_2\left(\frac{y}{x}\right). \end{aligned} \tag{3}$$

The particular integrals of equation (1) can be written by inspection as

$$\text{PI} = \frac{1}{2}\xi^2\eta + \frac{1}{2}\xi\eta = \frac{1}{2}(\ln x)^2 \ln y + \frac{1}{2}\ln x \ln y. \tag{4}$$

Therefore, the general solution of the given equation is given by

$$u(x, y) = \phi_1(x^2 y) + \phi_2\left(\frac{y}{x}\right) + \frac{1}{2}(\ln x)^2 \ln y + \frac{1}{2}\ln x \ln y.$$

13.7 A Special Case of Linear PDE (Constant Coefficients)

We have assumed that the linear equations of second order leading to the expression $\Phi(D_x, D_y)$ has linear factors as in equation (13.28).

It may, however, so happen that the expression has no linear factors. In this case, we assume a solution

$$u(x, y) = Ae^{ax+by}, \tag{13.56}$$

where A, a, b are all constants.

Substituting the expression (13.56) into the differential equation

$$\Phi(D_x, D_y)\, u = 0 \tag{13.57}$$

we get

$$A\Phi(a, b)\, e^{ax+by} = 0. \tag{13.58}$$

Thus, the expression (13.56) satisfies the equation (13.57); only when

$$\Phi(a, b) = 0. \tag{13.59}$$

Since a, b are arbitrary, equation (13.59) will prove an infinite number of pairs (a, b). Therefore, the solution of equation (13.57) can be written as

$$u(x, y) = \sum_{m=1}^{\infty} A_m e^{a_m x + b_m y}. \tag{13.60}$$

If the RHS of equation (13.57) contains a function $G(x, y)$ instead of being zero, then we have to determine the particular integral following the previous methods.

EXAMPLE 13.12. *Solve:* $u_{xx} + 2u_{xy} - 2u_y + 3u = e^{x+y}\cos(x+2y)$.

Solution: In symbolic notation, the given equation is

$$\left(D_x^2 + 2D_xD_y - 2D_y + 3\right)\, u = e^{x+y}\cos(x+2y). \tag{1}$$

It is seen that

$$\Phi(D_x, D_y) \equiv D_x^2 + 2D_xD_y - 2D_y + 3 \tag{2}$$

has no linear factors. Hence we assume the complimentary function as

$$u(x,y) = \sum_{m=1}^{\infty} A_m e^{a_m x + b_m y}. \tag{3}$$

Substituting expression (3) into expression (2), we get

$$a_m^2 + 2a_m b_m - 2b_m + 3 = 0, \tag{4}$$

which is the essential requirement for the expression (3) to satisfy the PDE with RHS = 0. The particular integral, by equation (13.49), is given by

$$e^{x+y}\frac{1}{(D_x+1)^2 + 2(D_x+1)(D_y+1) - 2(D_y+1) + 3}\cos(x+2y)$$

$$= e^{x+y}\frac{1}{D_x^2 + 2D_xD_y + 4D_x + 3}\cos(x+2y). \tag{5}$$

Let

$$\left(D_x^2 + 2D_xD_y + 4D_x + 3\right)\left[P\cos(x+2y) + Q\sin(x+2y)\right] = \cos(x+2y). \tag{6}$$

After finding the derivatives in the LHS and comparing coefficients of $\cos(x+2y)$ and $\sin(x+2y)$ in both sides of equation (6), we get $-4P - Q = 0$, $-P + 4Q = 1$.
This yields, $P = -\frac{1}{17}$, $Q = \frac{4}{17}$.
Hence, the particular integral is

$$e^{x+y}\left[\frac{4}{17}\sin(x+2y) - \frac{1}{17}\cos(x+2y)\right]. \tag{7}$$

The general solution of the given equation is obtained by summing up the complimentary function in (3) and the particular integral in (7).
Note: The methods and procedures discussed so far regarding solution of second order PDEs with constant coefficients are equally applicable for higher order equations also, if their coefficients are constants. We have limited our discussion to equations of second order only for the sake of convenience.

13.8 Linear PDE of Order Two (Variable Coefficients)

The simplest types of linear PDE of order two with variable coefficients are

$$\left.\begin{aligned} u_{xx} &= P(x,y) \\ u_{xy} &= Q(x,y) \\ u_{yy} &= R(x,y) \end{aligned}\right\}. \tag{13.61}$$

All these equations can be directly integrated in two steps.
Consider, as an example, the equation

$$u_{xy} = x^2 + y^2. \tag{13.62}$$

Integrating once with respect to y

$$u_x = x^2 y + \frac{y^3}{3} + \phi_1(x). \tag{13.63}$$

Integrating once more with respect to x

$$u = \frac{1}{3}\left(x^3 y + y^3 x\right) + \psi_1(x) + \psi_2(y). \tag{13.64}$$

Sometimes, the order of the equation can be lowered with an introduction of a new function. In such cases, the problem reduces to solving linear PDE of order one.

EXAMPLE 13.13. *Solve:* $2x\,u_{xx} - y\,u_{xy} + 2u_x = 4xy^2$.

Solution: Substitute, $u_x = P(x, y)$. The given equation assumes the form

$$2x\frac{\partial P}{\partial x} - y\frac{\partial P}{\partial y} = 4xy^2 - 2P. \tag{1}$$

This is a linear first order PDE, whose auxialiary equation is

$$\frac{dx}{2x} = \frac{dy}{-y} = \frac{dP}{4xy^2 - 2P}. \tag{2}$$

From first two relations, we get

$$\ln x + \ln y^2 = \text{constant}, \quad \text{i.e.,} \quad xy^2 = c. \tag{3}$$

We assume the solution of equation (1) as

$$P = Axy^2 + By^2, \tag{4}$$

where A, B are constants.

Substituting the expression (4) into equation (1), we get

$$2x(Ay^2) - y(2Axy + 2By) + 2(Axy^2 + By^2) = 4xy^2$$

which yeilds, $A = 2$ and B undetermined.

Hence the solution (4) is written as

$$u_x = P = 2xy^2 + By^2.$$

In view of equation (3), the constant B can be replaced by an arbitrary function $\phi_1(xy^2)$. Therefore

$$u_x = 2xy^2 + y^2\phi_1(xy^2)$$

$$\text{or,} \quad u = x^2y^2 + \psi_1(xy^2) + \psi_2(y),$$

where

$$\frac{\partial}{\partial x}\psi_1(xy^2) = y^2\phi_1\left(xy^2\right).$$

EXAMPLE 13.14. *Solve:* $y\,u_{xy} - u_x = xy^2 \sin(xy)$.

Solution: Substitute, $u_x = P(x, y)$.

Then the given equation assumes the form

$$y\frac{\partial P}{\partial y} - P = xy^2 \sin(xy)$$

$$\text{or,}\quad \frac{1}{y^2}\left(y\frac{\partial P}{\partial y} - P\right) = x\sin(xy)$$

$$\text{or,}\quad \frac{\partial}{\partial y}\left(\frac{P}{y}\right) = x\sin(xy). \tag{1}$$

Integrating equation (1) with respect to y

$$\frac{P}{y} = -\cos(xy) + \phi_1(x)$$

$$\text{or,}\quad P = \frac{\partial u}{\partial x} = -y\cos(xy) + y\phi_1(x). \tag{2}$$

Integrating equation (2) with respect to x

$$u(x, y) = \sin(xy) + y\psi_1(x) + \psi_2(y), \tag{3}$$

where

$$\frac{\partial}{\partial x}\psi_1(x) = \phi_1(x). \tag{4}$$

EXAMPLE 13.15. *Solve:* $u_{xx} + 2y\,u_x - 8y^2 u = (2 + 3y)e^{x+y}$.

Solution: The equation in symbolic form is

$$(D_x - 2y)(D_x + 4y)u = (2 + 3y)e^{x+y}. \tag{1}$$

Let

$$(D_x + 4y)u = 0. \tag{2}$$

Considering y to be a constant, this yields

$$\frac{\partial u.}{\partial x} = -4yu, \quad \text{or,}\quad \frac{\partial u}{u} = -4y\partial x, \quad \text{i.e.,}\quad u = e^{-4xy}\phi_1(y). \tag{3}$$

Similarly, for the factor $(D_x - 2y)\,u = 0$,

$$u = e^{2xy}\phi_2(y). \tag{4}$$

Thus, the complimentary function is

$$u = e^{2xy}\phi_2(y) + e^{-4xy}\phi_1(y). \tag{5}$$

The particular integral is given by

$$\frac{1}{D_x^2 + 2yD_x - 8y^2}(2 + 3y)\,e^{x+y} = (2 + 3y)\frac{1}{1^2 + 2y \cdot 1 - y^2}e^{x+y}$$

$$= \frac{2 + 3y}{1 + 2y - 8y^2}e^{x+y}.$$

Hence the general solution is

$$u(x,y) = e^{-4xy}\phi_1(y) + e^{2xy}\phi_2(y) + \frac{2+3y}{1+2y-8y^2}e^{x+y}.$$

EXAMPLE 13.16. *Solve:* $x\,u_{xy} + y\,u_{yy} + u_y = 10x^3y.$

Solution: Substitute, $u_y = P(x,y)$.

Then the given equation assumes the form

$$x\frac{\partial P}{\partial x} + y\frac{\partial P}{\partial y} + P = 10x^3y. \tag{1}$$

The auxiliary system for this equation is

$$\frac{dx}{x} = \frac{dy}{y} = \frac{dP}{10x^3y - P}. \tag{2}$$

From the first two ratios

$$\frac{x}{y} = \text{constant} = C. \tag{3}$$

We now assume the solution of equation (1) as

$$P(x,y) = Ax^3y + \frac{B}{x}. \tag{4}$$

Substituting expression (4) into equation (1), we get $A = 2$. The constant B, in view of equation (3), can be taken as $\phi(\frac{y}{x})$.

Integrating

$$u(x,y) = x^3y^2 + \psi(\frac{y}{x}) + \psi_2(x),$$

where

$$\frac{\partial}{\partial y}\psi(\frac{y}{x}) = \frac{1}{x}\phi(\frac{y}{x}).$$

13.9 Canonical Forms

We recall the general type of linear second order PDF with variable coefficients

$$Au_{xx} + Bu_{xy} + Cu_{yy} + Du_x + Eu_y + Fu = G, \tag{13.65}$$

where $A, B, \ldots$ are all functions of (x,y).

We introduce the new variables,

$$\xi = \xi(x,y), \quad \eta = \eta(x,y). \tag{13.66}$$

Then, we have

$$u_x = u_\xi\xi_x + u_\eta\eta_x, \quad u_y = u_\xi\xi_y + u_\eta\eta_y \tag{13.67}$$

$$u_{xx} = u_{\xi\xi}\xi_x^2 + 2u_{\xi\eta}\xi_x\eta_x + u_{\eta\eta}\eta_x^2 + u_\xi\xi_{xx} + u_\eta\eta_{xx} \tag{13.68}$$

$$u_{xy} = u_{\xi\xi}\xi_x\xi_y + u_{\xi\eta}(\xi_x\eta_y + \xi_y\eta_x) + u_{\eta\eta}\eta_x\eta_y + u_\xi\xi_{xy} + u_\eta\eta_{xy} \tag{13.69}$$

$$u_{yy} = u_{\xi\xi}\xi_y^2 + 2u_{\xi\eta}\xi_y\eta_y + u_{\eta\eta}\eta_y^2 + u_\xi\xi_{yy} + u_\eta\eta_{yy}. \tag{13.70}$$

With these new variables, equation (13.65) will assume the form

$$\bar{A}u_{\xi\xi} + \bar{B}u_{\xi\eta} + \bar{C}u_{\eta\eta} + \bar{D}u_{\xi} + \bar{E}u_{\eta} + \bar{F}u = \bar{G}, \tag{13.71}$$

where

$$\bar{A} = A\xi_x^2 + B\xi_x\xi_y + C\xi_y^2 \tag{13.72}$$
$$\bar{B} = 2A\xi_x\eta_x + B(\xi_x\eta_y + \xi_y\eta_x) + 2C\xi_y\eta_y \tag{13.73}$$
$$\bar{C} = A\eta_x^2 + B\eta_x\eta_y + C\eta_y^2 \tag{13.74}$$
$$\bar{D} = A\xi_{xx} + B\xi_{xy} + C\xi_{yy} + D\xi_x + E\xi_y \tag{13.75}$$
$$\bar{E} = A\eta_{xx} + B\eta_{xy} + C\eta_{yy} + D\eta_x + E\eta_y \tag{13.76}$$
$$\text{and} \quad \bar{F} = F, \quad \bar{G} = G. \tag{13.77}$$

For canonical forms, we set $\bar{A} = 0$ and $\bar{C} = 0$.
From equation (13.72), $\bar{A} = 0$ yields

$$A\left(\frac{\xi_x}{\xi_y}\right)^2 + B\left(\frac{\xi_x}{\xi_y}\right) + C = 0,$$

i.e.,

$$\frac{\xi_x}{\xi_y} = \frac{1}{2A}\left(-B + \sqrt{B^2 - 4AC}\right). \tag{13.78}$$

Similarly, from equation (13.74), $\bar{C} = 0$ yields

$$\frac{\eta_x}{\eta_y} = \frac{1}{2A}\left(-B - \sqrt{B^2 - 4AC}\right). \tag{13.79}$$

Equations (13.78) and (13.79) are called *characteristic equations.* For $\xi(x, y) =$ constant

$$d\xi = \xi_x dx + \xi_x dy = 0$$

so that

$$\frac{dy}{dx} = -\frac{\xi_x}{\xi_y}. \tag{13.80}$$

Similarly, for $\eta(x, y) =$ constant

$$\frac{dy}{dx} = -\frac{\eta_x}{\eta_y}. \tag{13.81}$$

Equations (13.80) and (13.81) define characteristic curves from which we can get the new coordinates. It is to be noted that, in equations (13.78) and (13.79), opposite signs have been taken before the expression $\sqrt{B^2 - 4AC}$. Otherwise, the two coordinates will come out to be the same.

The canononical equation corresponding to equation (13.65) will have the forms:

1. $\left.\begin{aligned} u_{\xi\xi} - u_{\eta\eta} &= F(\xi, \eta, u, u_\xi, u_\eta) \\ u_{\xi\xi} &= F(\xi, \eta, u, u_\xi, u_\eta) \end{aligned}\right\}$ for hyperbolic types of equations. (13.82)

2. $u_{\xi\eta} = F(\xi, \eta, u, u_\xi, u_\eta)$ for parabolic types of equations. (13.83)

3. $u_{\xi\xi} + u_{\eta\eta} = F(\xi, \eta, u, u_\xi, u_\eta)$ for elliptic types of equations. (13.84)

Canonical forms of PDE are useful in order to study their solutions with the help of existing equations.

EXAMPLE 13.17. *Transform the* PDE, $u_{xx} + 2u_{xy} - 3u_{yy} = 0$, *into canonical form. What is the type of equation?*

Solution: Here $A = 1$, $B = 2$, $C = -3$, $D = E = F = 0$. Hence

$$B^2 - 4AC = 2^2 - 4 \cdot 1 \cdot (-3) = 16.$$

$$\therefore \quad \frac{dy}{dx} = -\frac{\xi_x}{\xi_y} = -\frac{1}{2A}\left(-B + \sqrt{B^2 - 4AC}\right) = -1.$$

$$\therefore \quad y + x = \text{constant} \tag{1}$$

is the one characteristic curves.

Similarly,

$$\frac{dy}{dx} = -\frac{\eta_x}{\eta_y} = -\frac{1}{2A}\left(-B - \sqrt{B^2 - 4AC}\right) = 3.$$

$$\therefore \quad y - 3x = \text{constant} \tag{2}$$

is the other characteristic curves.

For the canonical form, we take

$$\xi = y + x \quad \text{and} \quad \eta = y - 3x. \tag{3}$$

Then

$$\xi_x = 1, \quad \xi_y = 1, \quad \eta_x = -3, \quad \eta_y = 1. \tag{4}$$

All other higher derivatives are zero. Using equations (13.72)–(13.77), we find

$$\bar{A} = 1 \cdot 1^2 + 2 \cdot 1 \cdot 1 - 3 \cdot 1^2 = 0$$
$$\bar{B} = 2 \cdot 1 \cdot 1 \cdot (-3) + 2[1 \cdot 1 + 1 \cdot (-3)] - 2 \cdot 3 \cdot 1 \cdot 1 = -16$$
$$\bar{C} = 1 \cdot (-3)^2 + 2 \cdot (-3) \cdot 1 - 3 \cdot 1^2 = 0.$$

$\bar{D} = 0, \bar{E} = 0, \bar{F} = 0.$

Hence the transformed equation is

$$\bar{B} u_{\xi\eta} = 0 \quad \text{i.e.,} \quad u_{\xi\eta} = 0. \tag{5}$$

Comparing this equation with equation (13.83), it is seen that the transformed PDE is a parabolic type.

It may be noted here that the general solution of equation (5) can be directly written as

$$u(\xi, \eta) = \phi(\xi) + \psi(\eta). \tag{6}$$

In terms of old variables (x, y), this becomes

$$u(x, y) = \phi(y + x) + \psi(y - 3x). \tag{7}$$

SPECIAL METHODS

We now intend to discuss some special methods for solving PDEs. Here we will consider the necessary initial and boundary conditions attached with the PDE.

13.10 Linear PDE of Order One (Method of Characteristics)

Consider the equation

$$P(x,y)\,u_x + Q(x,y)\,u_y = R(x,y)\,u \tag{13.85}$$

with the condition

$$u(x,\alpha) = \phi(x), \quad -\infty < x < \infty. \tag{13.86}$$

We introduce two new coordinates ξ, η, such that ξ will change along the characteristic curves and η will change along the initial curve $y = \alpha$. This is shown in *Fig 13.1*.

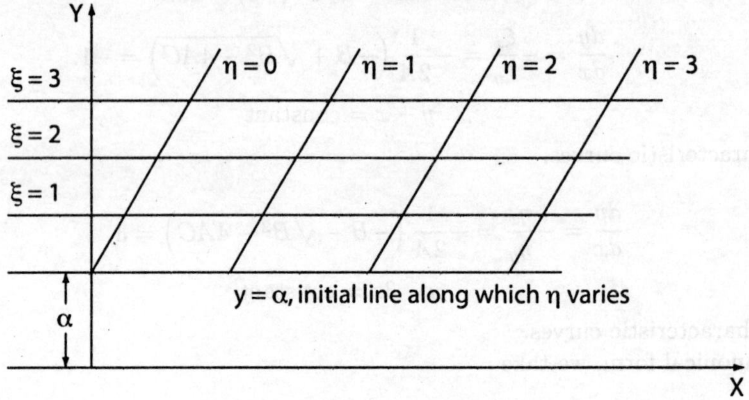

Fig 13.1

The characteristic curves $x(\xi)$, $y(\xi)$ are so chosen that

$$\frac{dx}{d\xi} = P(x,y), \quad \frac{dy}{d\xi} = Q(x,y). \tag{13.87}$$

Now, we have

$$\frac{du}{d\xi} = \frac{\partial u}{\partial x}\frac{dx}{d\xi} + \frac{\partial u}{\partial y}\frac{dy}{d\xi} = P(x,y)\,u_x + Q(x,y)\,u_y. \tag{13.88}$$

In view of equation (13.88), the PDE (13.85) assumes the form

$$\frac{du}{d\xi} = R(\xi,\eta)\,u, \tag{13.89}$$

which is a linear ordinary differential equation and can be solved by existing methods.

EXAMPLE 13.18. *Solve:* $x\,u_x + y\,u_y + 2u = 0$, $1 < y < \infty$ *under the condition*

$$u(x,1) = \sin x, \quad -\infty < x < \infty.$$

Solution: Here the new variable η changes along the initial line $y = 1$.

We have

$$\frac{dx}{d\xi} = x \quad \text{i.e.,} \quad x = Ae^{\xi}. \tag{1}$$

Similarly

$$\frac{dy}{d\xi} = y \quad \text{i.e.,} \quad y = Be^{\xi}, \tag{2}$$

where A, B are constants. To find these constants, let $x\big|_{\xi=0} = \eta$ and $y\big|_{\xi=0} = 1$.

Then, from equations (1) and (2), we get $A = \eta$, $B = 1$ so that

$$x = \eta e^{\xi}, \quad y = e^{\xi} \quad \text{i.e.,} \quad \frac{x}{\eta} = y \quad \text{or,} \quad \eta = \frac{x}{y}. \tag{3}$$

The given PDE now becomes

$$\left.\begin{aligned} &\frac{du}{d\xi} + 2u = 0 \\ &\text{with condition, } u\big|_{\xi=0} = \sin\eta \end{aligned}\right\} \tag{4}$$

Solution of equation (4) is

$$u(\xi, \eta) = (\sin\eta)\, e^{-2\xi}. \tag{5}$$

Going back to old variables using equation (3), get

$$u(x, y) = \frac{1}{y^2}\sin(\frac{x}{y}) \tag{6}$$

which is the required solution.

EXAMPLE 13.19. *Solve:* $xy\, u_x + x^2 u_y = yu$ *with the condition,* $u(x, 0) = \phi(x)$.

Solution: The given equation can be written as

$$y\, u_x + x\, u_y = \frac{y}{x} u. \tag{1}$$

We introduce two new variables ξ, η such that ξ changes along the characteristic curves $[x(\xi), y(\xi)]$ and η changes along the initial line $y = 0$.

Here

$$\frac{dx}{d\xi} = P(x, y) = y \quad \text{i.e.,} \quad x = y\xi + A \tag{2}$$

$$\text{and} \quad \frac{dy}{d\xi} = Q(x, y) = x \quad \text{i.e.,} \quad y = x\xi + B. \tag{3}$$

To determine the constants A and B, we let

$$x\big|_{\xi=0} = \eta \quad \text{and} \quad y\big|_{\xi=0} = 0. \tag{4}$$

These conditions yield from equations (2) and (3), $A = \eta$ and $B = 0$. Hence

$$x = y\xi + \eta \quad \text{and} \quad y = x\xi. \tag{5}$$

Equation (1) can now be written as

$$\frac{du}{d\xi} = \xi u, \quad \text{since } \frac{y}{x} = \xi,$$

which has the solution

$$u(\xi, \eta) = Ce^{\frac{\xi^2}{2}}. \tag{6}$$

Since, $u(0, \eta) = \phi(\eta)$, $C = \phi(\eta)$.

Therefore

$$u(\xi,\eta) = \phi(\eta)\, e^{\frac{\xi^2}{2}}. \tag{7}$$

Now, from equation (5)

$$\xi = \frac{y}{x}, \quad \eta = x - y\xi = x - \frac{y^2}{x}. \tag{8}$$

With the use of equation (8), the solution in (7) is written as

$$u(x,y) = \phi\left(x - \frac{y^2}{x}\right) \exp\left[\frac{1}{2}\left(\frac{y}{x}\right)^2\right] \tag{9}$$

which is the required solution.

13.11 D'Alemberts Method

This method provides a general solution of wave equation by transformation of variables.

Consider the one-dimensional wave equation

$$u_{tt} = c^2 u_{xx}. \tag{13.90}$$

We define new variables ξ, η such that

$$\xi = x + ct, \quad \eta = x - ct. \tag{13.91}$$

Then, we have

$$u_x = u_\xi + u_\eta, \quad u_t = c\,(u_\xi - u_\eta) \tag{13.92}$$

$$u_{xx} = u_{\xi\xi} + 2u_{\xi\eta} + u_{\eta\eta} \tag{13.93}$$

$$u_{tt} = c^2\,(u_{\xi\xi} - 2u_{\xi\eta} + u_{\eta\eta}). \tag{13.94}$$

Substituting expressions (13.93) and (13.94) into equation (13.90), we get

$$u_{\xi\eta} = 0, \tag{13.95}$$

whose general solution is

$$u(\xi,\eta) = f(\xi) + g(\eta), \tag{13.96}$$

where f, g are twice differentiable arbitrary functions. Returning to the original variables x, t, we write the solution function (13.96) as

$$u(x,t) = f(x+ct) + g(x-ct). \tag{13.97}$$

Let the initial conditions be given as

$$u(x,0) = \phi(x), \quad u_t(x,0) = 0. \tag{13.98}$$

Using these conditions in equation (13.97), we get

$$\phi(x) = f(x) + g(x) \tag{13.99}$$

$$0 = c\left(\frac{df}{dx} - \frac{dg}{dx}\right). \tag{13.100}$$

Integrating equation (13.100)

$$f(x) = g(x) + A, \tag{13.101}$$

where A is an arbitrary constant.

From equations (13.99) and (13.101)

$$f(x) = \frac{1}{2}\left[\phi(x) + A\right] \tag{13.102}$$

$$\text{and} \quad g(x) = \frac{1}{2}\left[\phi(x) - A\right]. \tag{13.103}$$

Replacing x by $x + ct$ in equation (13.102) and by $x - ct$ in equation (13.103), we get from equation (13.97)

$$u(x,t) = \frac{1}{2}\left[\phi(x+ct) + \psi(x-ct)\right] \tag{13.104}$$

which is the solution corresponding to the initial conditions in equation (13.98).

Consider now the following initial conditions:

$$u(x,0) = 0, \quad u_t(x,0) = \psi(x). \tag{13.105}$$

Then, from equation (13.97), we have

$$f(x) + g(x) = 0 \tag{13.106}$$

and

$$\psi(x) = c\left(\frac{df}{dx} - \frac{dg}{dx}\right). \tag{13.107}$$

Integrating equation (13.107)

$$f(x) - g(x) = \frac{1}{c}\int_0^x \psi(z)\,dz + B, \tag{13.108}$$

where B is the integration constant. Taking the limit $x \longrightarrow 0$ in equation (13.108), and using the relation (13.106), we find

$$B = 2f(0) = -2g(0). \tag{13.109}$$

Equation (13.108) along with equation (13.106) yields

$$f(x) = \frac{1}{2c}\int_0^x \psi(z)\,dz + f(0) \tag{13.110}$$

$$\text{and} \quad g(x) = -\frac{1}{2c}\int_0^x \psi(z)\,dz + g(0). \tag{13.111}$$

Changing the variable x to $x + ct$ in equation (13.110) and $x - ct$ in equation (13.111)

$$f(x+ct) = \frac{1}{2c}\int_0^{x+ct} \psi(z)\,dz + f(0) \tag{13.112}$$

$$\text{and} \quad g(x-ct) = -\frac{1}{2c}\int_0^{x-ct} \psi(z)\,dz + g(0). \tag{13.113}$$

Substituting these expressions into equation (13.97) and remembering the relation in (13.109), we get

$$u(x,t) = \frac{1}{2c}\left[\int_0^{x+ct} \psi(z)dz - \int_0^{x-ct} \psi(z)dz\right]$$

$$\text{or,}\quad u(x,t) = \frac{1}{2c}\int_{x-ct}^{x+ct} \psi(z)dz \tag{13.114}$$

which is the explicit solution corresponding to the initial conditions (13.105).

Let the initial conditions be given as

$$u(x,0) = \phi(x), \quad u_t(x,0) = \psi(x). \tag{13.115}$$

Since the equation (13.90) is linear, the solution corresponding to the initial conditions (13.115) can be obtained by summing up the solution functions in the above two cases. Thus, here we have

$$u(x,t) = \frac{1}{2}\left[\phi(x+ct) + \phi(x-ct) + \frac{1}{c}\int_{x-ct}^{x+ct} \psi(z)\,dz\right]. \tag{13.116}$$

EXAMPLE 13.20. *Solve:* $u_{tt} = u_{xx}$, $(-\infty < x < \infty, 0 < t < \infty)$ *given the initial conditions* $u(x,0) = \sin x$, $u_t(x,0) = \dfrac{1}{(1+x^2)}$.

Solution: Here, $\phi(x) = \sin x$, $\psi(x) = \dfrac{1}{(1+x^2)}$, $c = 1$.

By equation (13.116)

$$\begin{aligned} u(x,t) &= \frac{1}{2}\left[\sin(x+t) + \sin(x-t) + \int_{x-t}^{x+t} \frac{1}{1+z^2}dz\right] \\ &= \sin x \cos t + \frac{1}{2}\left[\tan^{-1} z\right]_{x-t}^{x+t} \\ &= \sin x \cos t + \frac{1}{2}\left[\tan^{-1}(x+t) - \tan^{-1}(x-t)\right] \\ &= \sin x \cos t + \frac{1}{2}\tan^{-1}\left[\frac{(x+t)-(x-t)}{1+(x+t)(x-t)}\right] \\ &= \sin x \cos t + \frac{1}{2}\tan^{-1}\left(\frac{2t}{1+x^2-t^2}\right). \end{aligned}$$

EXAMPLE 13.21. *For* $0 \le x < \infty$ *and* $0 < t < \infty$, *find a solution of the equation* $u_{tt} = c^2 u_{xx}$ *satisfying the conditions:*

$$u(x,0) = \frac{x^2}{1+x^2}, \quad u_t(x,0) = 0.$$

Solution: Here

$$\phi(x) = \frac{x^2}{1+x^2}, \quad \psi(x) = 0.$$

Since it is given that $0 \le x < \infty$, we extend the function $\phi(x)$ along the negative semi-axis and consider it to be odd there. Thus

$$\phi(x) = \frac{x^2}{1+x^2}, \quad \text{for } x \ge 0$$

$$\text{and}\quad \phi(x) = -\frac{x^2}{1+x^2}, \quad \text{for } x < 0.$$

Then the solution of the given equation is

$$u(x,t) = \frac{1}{2}\left[\frac{(x+ct)^2}{1+(x+ct)^2} + \frac{(x-ct)^2}{1+(x-ct)^2}\right], \quad \text{for } x \geq ct \tag{1}$$

$$= \frac{1}{2}\left[\frac{(x+ct)^2}{1+(x+ct)^2} - \frac{(ct-x)^2}{1+(ct-x)^2}\right], \quad \text{for } 0 \leq x \leq ct. \tag{2}$$

13.12 Fourier Method

In this method, the solution function of the equation is assumed as a product of functions each of which depends on a single variable only. After substituting this product into the original equation, we will get a number of ordinary differential equations with prescribed boundary conditions. After solving each of these ordinary differential equations, the solution of the original PDE is written as a series in the product of those functions. The method is also known as *Fourier method of separation of variables.*

We will study innumerable problems by this method in a later chapter, and, therefore, further discussion is not made here.

However, separation of variables cannot be done, if the given boundary condition is non-homogeneous. In such cases, we seek for a change of variable to make the boundary condition homogeneous.

Consider for example, the heat equation

$$u_t = \alpha^2 u_{xx} \quad (0 < x < L,\ 0 < t < \infty) \tag{13.117}$$

with boundary conditions

$$u(0,t) = \phi(t), u(L,t) = \psi(t). \tag{13.118}$$

We introduce a new function $w(x,t)$, such that

$$w(x,t) = u(x,t) - \left[\phi(t) + \frac{x}{L}\{\psi(t) - \phi(t)\}\right]. \tag{13.119}$$

Then, the original equation (13.117) in terms of $w(x,t)$ becomes

$$w_t = \alpha^2 w_{xx} - \frac{\partial P}{\partial t} \tag{13.120}$$

with boundary conditions

$$w(0,t) = 0 \quad \text{and} \quad w(L,t) = 0. \tag{13.121}$$

which are homogeneous.

Here, we have denoted

$$P(x,t) = \phi(t) + \frac{x}{L}\{\psi(t) - \phi(t)\}. \tag{13.122}$$

The original PDE has, however, changed to a non-homogeneous one due to which, it has to be solved by other methods.

If in equation (13.120), $\frac{\partial P}{\partial t} = 0$, which means that both the functions $\phi(t)$ and $\psi(t)$ in equation (13.118) are constant, it becomes a homogeneous one and therefore can be solved by Fourier's method of separation of variables.

13.13 Method of Integral Transforms

Linear PDEs can be transformed into ordinary differential equations by Fourier transforms and Laplace transforms. These integral transforms are also taken on given initial and boundary conditions. After finding the solution of the transformed ordinary differential equations with the transformed boundary conditions, inverse integral transforms are taken to obtain the solution of the original PDE.

The method will be discussed in detail in the chapters on Fourier transforms and Laplace transforms. However, here we make the following useful observations:

1. For mixed value problems of hyperbolic type, where the desired function has to satisfy certain initial as well as boundary conditions, the method of Laplace transform is very effective.
2. All functions do not have Fourier transforms. For example
$$f(x) = \text{constant}, \quad f(x) = e^x, \quad f(x) = \sin x \tag{13.123}$$
do not have Fourier transforms, since for these functions, the integral
$$\frac{1}{\sqrt{2\pi}} \int_{-\infty}^{\infty} f(x)\, e^{-i\xi x}\, dx \tag{13.124}$$
does not exist.
3. Compared to the Fourier transform, a wider class of functions have Laplace transforms.

13.14 Matrix Method for First Order PDE System

Let $u(x, y)$ and $v(x, y)$ be two functions connected by the system of linear PDE

$$\left.\begin{aligned} u_x + a_1 u_y + a_2 v_y &= 0 \\ v_x + b_1 u_y + b_2 v_y &= 0 \end{aligned}\right\}, \tag{13.125}$$

where a_1, a_2, b_1, b_2 are constants.

We write equation (13.125) in matrix notation as

$$\{U_x\} + [A]\,\{U_y\} = 0, \tag{13.126}$$

where

$$\{U_x\} = \begin{bmatrix} u_x \\ v_x \end{bmatrix} \tag{13.127}$$

$$\{U_y\} = \begin{bmatrix} u_y \\ v_y \end{bmatrix} \tag{13.128}$$

$$\text{and} \quad [A] = \begin{bmatrix} a_1 & a_2 \\ b_1 & b_2 \end{bmatrix}. \tag{13.129}$$

Let λ_1 and λ_2 be two eigenvalues of the matrix A. The corresponding eigenfunctions are denoted by $\{X_1\}$ and $\{X_2\}$.

We make a change to a new variable

$$\{W\} = \begin{bmatrix} w_1 \\ w_2 \end{bmatrix} \tag{13.130}$$

so that

$$\{U\} = [P]\{W\}, \tag{13.131}$$

where $[P]$ is a matrix whose columns are eigenvectors of matrix A, and

$$\{U\} = \begin{bmatrix} u \\ v \end{bmatrix}. \tag{13.132}$$

From equation (13.131), we have

$$\left.\begin{aligned} \{U_x\} &= [P]\{W_x\} \\ \text{and} \quad \{U_y\} &= [P]\{W_y\} \end{aligned}\right\}. \tag{13.133}$$

Substituting these relations into equation (13.126)

$$[P]\{W_x\} + [A][P]\{W_y\} = 0. \tag{13.134}$$

Multiplying both sides of this equation by $[P]^{-1}$,

$$\{W_x\} + [Q]\{W_y\} = 0, \tag{13.135}$$

where

$$[Q] = [P]^{-1}[A][P]. \tag{13.136}$$

Equation (13.135) will yield two uncoupled equations of the forms

$$\left.\begin{aligned} \frac{\partial w_1}{\partial x} + \alpha\frac{\partial w_1}{\partial y} &= 0 \\ \frac{\partial w_2}{\partial x} + \beta\frac{\partial w_2}{\partial y} &= 0 \end{aligned}\right\}, \tag{13.137}$$

which can be solved easily. Knowing the functions w_1, w_2 we can find the functions u, v from equation (13.132).

EXAMPLE 13.22. *Solve:* $u_x + u_y + v_y = 0$
$u_x + 4u_y + v_y = 0$
given the initial conditions $u(0, y) = \phi(y)$ *and* $v(0, y) = \psi(y)$.

Solution: Here, $[A] = \begin{bmatrix} 1 & 1 \\ 4 & 1 \end{bmatrix}$.

The eigenvalues of matrix A are given by $(1-\lambda)^2 - 4 = 0$, i.e., $\lambda_1 = -1$, $\lambda_2 = 3$.
The corresponding eigenvalues are:

$$\{X_1\} = \begin{Bmatrix} 1 \\ 2 \end{Bmatrix}, \quad \text{for } \lambda_1 = 3$$

$$\{X_2\} = \begin{Bmatrix} 1 \\ -2 \end{Bmatrix}, \quad \text{for } \lambda_2 = -1.$$

43

Hence

$$[P] = \begin{bmatrix} 1 & 1 \\ 2 & -2 \end{bmatrix}, \quad P^{-1} = \begin{bmatrix} \frac{1}{2} & \frac{1}{4} \\ \frac{1}{2} & -\frac{1}{4} \end{bmatrix}$$

$$[Q] = [P]^{-1}[A][P] = \begin{bmatrix} \frac{1}{2} & \frac{1}{4} \\ \frac{1}{2} & -\frac{1}{4} \end{bmatrix} \begin{bmatrix} 1 & 1 \\ 4 & 1 \end{bmatrix} \begin{bmatrix} 1 & 1 \\ 2 & -2 \end{bmatrix} = \begin{bmatrix} 3 & 0 \\ 0 & -1 \end{bmatrix}.$$

Hence, equations (13.137) in this case are:

$$\left.\begin{aligned} \frac{\partial w_1}{\partial x} + 3\frac{\partial w_1}{\partial y} = 0 \\ \text{and} \quad \frac{\partial w_2}{\partial x} - \frac{\partial w_2}{\partial y} = 0 \end{aligned}\right\}, \tag{1}$$

whose solutions are

$$\left.\begin{aligned} w_1(x, y) = F(y - 3x) \\ w_2(x, y) = G(y + x) \end{aligned}\right\}, \tag{2}$$

where F and G are arbitrary functions. Therefore

$$\begin{bmatrix} u \\ v \end{bmatrix} = \begin{bmatrix} 1 & 1 \\ 2 & -2 \end{bmatrix} \begin{bmatrix} w_1 \\ w_2 \end{bmatrix}$$

$$\left.\begin{aligned} \text{or,} \quad u(x, y) = w_1(x, y) + w_2(x, y) \\ \text{and} \quad v(x, y) = 2w_1(x, y) - 2w_2(x, y). \end{aligned}\right\}. \tag{3}$$

From equations (2) and (3), we get

$$\left.\begin{aligned} u(x, y) = F(y - 3x) + G(y + x) \\ v(x, y) = 2[F(y - 3x) - G(y + x)] \end{aligned}\right\}. \tag{4}$$

Using the boundary conditions, we have

$$\left.\begin{aligned} \phi(y) = F(y) + G(y) \\ \psi(y) = 2[F(y) - G(y)] \end{aligned}\right\} \tag{5}$$

Solving these, we get

$$\left.\begin{aligned} F(y) = \tfrac{1}{2}\,\phi(y) + \tfrac{1}{4}\,\psi(y) \\ G(y) = \tfrac{1}{2}\,\phi(y) - \tfrac{1}{4}\,\psi(y) \end{aligned}\right\}. \tag{6}$$

From equations (4) and (6), we get

$$u(x, y) = \frac{1}{2}\,\psi(y - 3x) + \frac{1}{4}\,\psi(y - 3x) + \frac{1}{2}\,\phi(x + y) - \frac{1}{4}\,\psi(x + y)$$

$$v(x, y) = \phi(y - 3x) + \frac{1}{2}\,\psi(y - 3x) - \phi(x + y) + \frac{1}{2}\,\psi(x + y).$$

13.15 Duhamel's Principle

Consider the heat equation

$$u_t = \alpha^2 u_{xx}, \quad (0 < x < L,\ 0 < t < \infty) \tag{13.138}$$

with the initial and boundary conditions

$$u(x,0) = \phi(x), \quad u(0,t) = 0, \quad u(L,t) = \psi(t), \tag{13.139}$$

where the function $\psi(t)$ is not a constant. The method of Fourier's separation of variables cannot be used here, since the boundary condition is non-homogeneous. Moreover, we cannot go for a change of variable to make the boundary condition homogeneous, since, in that case the original equation becomes non-homogeneous as discussed in Section 13.12.

Thus, the problem is a bit complicated one. However, if we take $\phi(t) = 1$, the problem becomes much easier and can be solved by Laplace transform or by Fourier method using a new variable

$$w(x,t) = \bar{u}(x,t) - \frac{x}{L}. \tag{13.140}$$

Solution of this problem, by Fourier method, is (see Chapter 18)

$$\bar{u}(x,t) = \frac{x}{L} + \sum_{m=1}^{\infty} A_m \sin\frac{m\pi x}{L} \exp\left[-\frac{m^2\pi^2\alpha^2}{L^2}t\right], \tag{13.141}$$

where

$$A_m = \frac{2}{L}\int_0^L \phi(x)\sin\frac{m\pi x}{L}\,dx \tag{13.142}$$

with $m = 1, 2, 3, \ldots, \infty$.

According to Duhamel's principle, the solution of the original equation defined in (13.138) and (13.139) will be given by

$$\begin{aligned} u(x,t) &= \int_0^t \psi(\tau)\frac{\partial \bar{u}(x,t-\tau)}{\partial \tau}d\tau \\ &= \int_0^t \bar{u}(x,t-\tau)\psi'(\tau)\,d\tau + \psi(0)\bar{u}(x,t). \end{aligned} \tag{13.143}$$

Thus, Duhamel's principle is very important in solving complex problems in terms of solution of the corresponding easier problem. If, in particular

$$\phi(x) = \frac{x(L-x)}{L^2} \quad \text{and} \quad \psi(t) = t$$

then

$$A_m = \frac{8}{m^3\pi^3}, \quad m = 1, 3, 5, \ldots$$

$$\bar{u}(x,t) = \frac{x}{L} + \frac{8}{\pi^3}\sum \frac{1}{m^3}\sin\frac{m\pi x}{L}\exp\left[-\frac{m^2\pi^2\alpha^2}{L^2}t\right].$$

Therefore

$$u(x,t) = \frac{x}{L} + \frac{8}{\pi^3}\sum \frac{1}{m^3}\sin\frac{m\pi x}{L}\left(\frac{m^2\pi^2\alpha^2}{L^2}\right)\left\{1-\exp\left[-\frac{m^2\pi^2\alpha^2}{L^2}t\right]\right\}$$
$$= \frac{x}{L} + \frac{8\alpha^2}{\pi L^2}\sum \frac{1}{m}\sin\frac{m\pi x}{L}\left\{1-\exp\left[-\frac{m^2\pi^2\alpha^2}{L^2}t\right]\right\}. \quad (13.144)$$

NON-LINEAR EQUATIONS

13.16 Elimination of Arbitrary Constants

Partial differential equations may be derived by the elimination of arbitrary constants from a given relation.

Consider, for example, the relation

$$u = (x-a)^2 + (y-b)^2. \quad (13.145)$$

Then, taking partial derivatives

$$u_x = \frac{\partial u}{\partial x} = 2(x-a), \quad u_y = \frac{\partial u}{\partial y} = 2(y-b),$$

which yields

$$(x-a) = \frac{u_x}{2}, \quad (y-b) = \frac{u_y}{2}. \quad (13.146)$$

Eliminating the arbitrary constants a, b from equations (13.145) and (13.146), we find

$$4u = u_x^2 + u_y^2, \quad (13.147)$$

which is a partial differential equation. We, therefore, can say that the expression (13.145) is the solution of the PDE (13.147).

Consider, again, the relation

$$u = ax + by + \phi(a,b) \quad (13.148)$$

where $\phi(a,b)$ is an arbitrary function of a and b. Then $u_x = a$, $u_y = b$, which, when substituted into equation (13.148) gives

$$u = x\,u_x + y\,u_y + \phi(u_x, u_y) \quad (13.149)$$

Therefore, the expression (13.148) is the solution of the PDE in (13.149), which was obtained by eliminating the arbitrary constants a and b.

13.17 Complete and Singular Solutions

We will consider only PDEs of order one. Let there be a non-linear PDE given by

$$\psi(x, y, u, u_x, u_y) = 0, \quad (13.150)$$

which is derived by eliminating the arbitrary constants a, b from the relation

$$\phi(x, y, u, a, b) = 0. \quad (13.151)$$

Then the expression (13.151) is called a *complete solution* of the PDE in (13.150). This complete solution represents a two parameter family of surfaces whose envelop, if it exists, are obtained by eliminating a, b from the equations

$$\phi = 0, \quad \frac{\partial \phi}{\partial a} = 0, \quad \frac{\partial \phi}{\partial b} = 0. \tag{13.152}$$

Let the equation of envelop thus obtained be

$$f(x, y, u) = 0. \tag{13.153}$$

If the relation in (13.153) satisfies the PDE in (13.150), it is called a *singular solution* of the PDE.

The singular solution may be obtained from the PDE (13.150) by eliminating $u_x = p$ and $u_y = q$ from the relations

$$\psi = 0, \quad \frac{\partial \psi}{\partial p} = 0, \quad \frac{\partial \psi}{\partial q} = 0. \tag{13.154}$$

We have denoted here u_x by p and u_y by q and will do so henceforth for the sake of convenience.

It may happen that one of the arbitrary constants, say b, is a function of a, i.e., $b = h(a)$. Then the complete solution,

$$\phi[x, y, u, a, h(a)] = 0, \tag{13.155}$$

represents a one-parameter family of surfaces. Its envelop will be found by eliminating a from

$$\phi = 0 \text{ and } \frac{\partial \phi}{\phi a} = 0, \tag{13.156}$$

and determining that part of the result which satisfies equation (13.150).

The totality of solutions obtained by varying the function $h(a)$ is called the *general solution.*

13.18 Special Types of Equations

We will consider the methods of solving the following types of equations:

Type I. Let the equation be given by

$$\psi(u_x, u_y) \equiv \psi(p, q) = 0. \tag{13.157}$$

Consider the relation

$$u = ax + h(a)y + c, \tag{13.158}$$

where a, c are arbitrary constants. Then

$$p = \frac{\partial u}{\partial x} = a, \quad q = \frac{\partial u}{\partial y} = h(a)$$

so that

$$q = h(p) \quad \text{i.e.,} \quad \psi(p, q) = 0. \tag{13.159}$$

Therefore, solution of equations (13.157) are given by (13.158). The singular solution is determined from equations

$$u - ax - h(a)y - c = 0, \quad x + h'(a)y = 0, \quad -1 = 0,$$

which shows that there is no singular solution. Thus, we conclude that equation (13.158) is the complete solution of equation (13.157), if

$$\psi[a, h(a)] = 0. \tag{13.160}$$

EXAMPLE 13.23. *Solve:* $pq = p + q$.

Solution: Here

$$\psi(p, q) \equiv pq - p - q = 0. \tag{1}$$

Its solution is given by

$$u = ax + h(a)y + c, \tag{2}$$

where

$$\psi[a, h(a)] = 0. \tag{3}$$

From equations (1) and (3), we write

$$a \cdot h(a) - a - h(a) = 0,$$

which yields

$$h(a) = \frac{a}{a-1}. \tag{4}$$

Substituting this value in equation (2), we get

$$u = ax + \frac{a}{a-1}y + c,$$

i.e., $(a-1)\,u = a(a-1)\,x + ay + d$ which is the complete solution.

Type II. Consider a non-linear PDE of the form

$$u = px + qy + \psi(p, q). \tag{13.161}$$

We have seen before that a complete solution of this equation is given by

$$u = ax + by + \psi(a, b), \tag{13.162}$$

where a, b are arbitrary constants and $\psi(a, b)$ is an arbitrary function.

Denoting equation (13.162) as

$$\phi(x, y, u, a, b) = 0, \tag{13.163}$$

the singular solution is obtained from

$$\phi = 0, \quad \frac{\partial \phi}{\partial a} = 0, \quad \frac{\partial \phi}{\partial b} = 0. \tag{13.164}$$

EXAMPLE 13.24. *Solve:* $u = px + qy + \frac{1}{2}p^2q^2$.

Solution: The complete solution is

$$u = ax + by + \frac{1}{2}a^2b^2. \tag{1}$$

Hence

$$\phi(x,y,u,a,b) \equiv u - ax - by - \frac{1}{2}a^2b^2 = 0.$$

$$\left.\begin{aligned} \frac{\partial\phi}{\partial a} &= -x - ab^2 = 0 \\ \frac{\partial\phi}{\partial b} &= -y - a^2b = 0 \end{aligned}\right\} \tag{2}$$

From equation (2), $x = -ab^2$, $y = -a^2b$.
Therefore

$$ax + by = -2a^2b^2. \tag{3}$$

From equations (1) and (3)

$$u = -2a^2b^2 + \frac{1}{2}a^2b^2 = -\frac{3}{2}a^2b^2. \tag{4}$$

Eliminating a, b from equation (2)

$$xy = a^3b^3 = \left(a^2b^2\right)^{\frac{3}{2}}. \tag{5}$$

From equations (4) and (5)

$$u = -\frac{3}{2}(xy)^{\frac{2}{3}}, \tag{6}$$

which is the singular solution.

Type III. Let a non-linear PDE is given in the form

$$\psi(u,p,q) = 0. \tag{13.165}$$

We assume the solution as

$$u = \phi(\xi), \quad \xi = x + ay, \tag{13.166}$$

where a is an arbitrary constant.
Then, we have

$$\left.\begin{aligned} p &= \frac{\partial u}{\partial x} = \frac{du}{d\xi}\frac{\partial\xi}{\partial x} = \frac{du}{d\xi} \\ q &= \frac{\partial u}{\partial y} = \frac{du}{d\xi}\frac{\partial\xi}{\partial y} = a\frac{du}{d\xi} \end{aligned}\right\}. \tag{13.167}$$

Substituting these relations into the original equation (13.165), we get

$$\psi\left(u, a, \frac{du}{d\xi}\right) = 0, \tag{13.168}$$

which is an ordinary differential equation of first order and can be solved by usual method.

EXAMPLE 13.25. $u^2\left(p^2 + q^2 - 1\right) = 1$.

Solution: Taking a new variable, $\xi = x + ay$, where a is an arbitrary constant, we have

$$p = \frac{du}{d\xi} \quad \text{and} \quad q = a\frac{du}{d\xi}. \tag{1}$$

From the given equation

$$p^2 + q^2 - 1 = \frac{1}{u^2} \quad \text{or,} \quad (1+a^2)\left(\frac{du}{d\xi}\right)^2 = \frac{1}{u^2} + 1 = \frac{1+u^2}{u^2}.$$

$$\therefore \quad \frac{du}{d\xi} = \frac{1}{\sqrt{1+a^2}}\frac{\sqrt{1+u^2}}{u} \quad \text{i.e.,} \quad \frac{u\,du}{\sqrt{1+u^2}} = \frac{d\xi}{\sqrt{1+a^2}}. \tag{2}$$

Integrating

$$\sqrt{1+u^2} = \frac{1}{\sqrt{1+a^2}}(\xi + b). \tag{3}$$

Squaring both sides, and substituting for ξ,

$$(1+a^2)\,(1+u^2) = (x + ay + b)^2, \tag{4}$$

which is the solution of the given equation.

Equation (4) is written as

$$\left.\begin{aligned} \phi(x,y,u,a,b) &\equiv (1+a^2)\,(1+u^2) - (x+ay+b)^2 = 0 \\ \therefore \quad \frac{\partial\phi}{\partial a} &\equiv (1+u^2)\,2a - 2y(x+ay+b) \qquad = 0 \end{aligned}\right\} \tag{5}$$

$$\frac{\partial\phi}{\partial b} \equiv -2(x+ay+b) = 0.$$

Solving these equations, we get $a + u^2 = 0$, which is the singular solution.

Type IV. Let a non-linear PDE be given in the form

$$\phi(x,p) - \psi(y,q) = 0. \tag{13.169}$$

The equation suggests that

$$\phi(x,p) = \psi(y,q), \tag{13.170}$$

which is possible only when each of $\phi(x,p)$ and $\psi(y,q)$ is a constant. Hence, we assume

$$\phi(x,p) = \lambda, \quad \psi(y,q) = \lambda, \tag{13.171}$$

where λ is an arbitrary constant.

Equation (13.171) can be solved for p and q to obtain

$$p = f(x,\lambda) \quad \text{and} \quad q = g(y,\lambda). \tag{13.172}$$

Since u is a function of x and y,

$$du = \frac{\partial u}{\partial x}dx + \frac{\partial u}{\partial y}dy = p\,dx + q\,dy. \tag{13.173}$$

Substituting from equation (13.172) the expressions for p and q into equation (13.173), we get

$$du = f(x, \lambda)\, dx + g(y, \lambda)\, dy$$

which upon integration, yields

$$u = \int f(x, \lambda)\, dx + \int g(y, \lambda)\, dy + \alpha. \tag{13.174}$$

Equation (13.174) is a complete solution of the given PDE. It contains two arbitrary constants λ and α.

EXAMPLE 13.26. *Solve:* $p^2 - q^2 = x - y$.

Solution: The given equation is written as

$$p^2 - x = q^2 - y = \lambda.$$

This gives, $p = \sqrt{x + \lambda}$, $q = \sqrt{y + \lambda}$.
Hence

$$du = p\, dx + q\, dy = \sqrt{x + \lambda}\, dx + \sqrt{y + \lambda}\, dy.$$

Integrating

$$u = \frac{2}{3}(x + \lambda)^{\frac{3}{2}} + \frac{2}{3}(y + \lambda)^{\frac{3}{2}} + \alpha \quad \text{or,} \quad 3(u - \alpha) = 2(x + \lambda)^{\frac{3}{2}} + 2(y + \lambda)^{\frac{3}{2}}$$

which is a complete solution of the given equation.
Here, λ and α are two arbitrary constants.

EXAMPLE 13.27. *Solve:* $3x^2y + \frac{1}{3}x^2q^2 = yp$.

Solution: The given equation can be written as

$$\frac{p}{3x^2} = 1 + \frac{1}{9}\frac{q^2}{y} \quad \text{(dividing both sides by } 3x^2y\text{)}.$$

Let

$$\frac{p}{3x^2} = \lambda \quad \text{and} \quad 1 + \frac{1}{9}\frac{q^2}{y} = \lambda.$$

Then

$$du = 3\lambda x^2\, dx + 3\sqrt{(\lambda - 1)\, y}\, dy.$$

Integrating

$$u = \lambda x^3 + 3\sqrt{\lambda - 1} \cdot \frac{2}{3} y^{\frac{3}{2}} + \alpha \quad \text{or,} \quad \left(u - \lambda x^3 - \alpha\right)^2 = 4(\lambda - 1)y^3,$$

which gives the complete solution.

Type V. Let the non-linear PDE be of the form

$$\psi(x, y, u, px, qy) = 0. \tag{13.175}$$

In such a case, we substitute

$$\xi = \ln x, \quad \eta = \ln y \tag{13.176}$$

so that

$$p = \frac{\partial u}{\partial x} = \frac{\partial u}{\partial \xi}\frac{d\xi}{dx} = \frac{1}{x}\frac{\partial u}{\partial \xi}$$

$$\text{and} \quad q = \frac{\partial u}{\partial y} = \frac{\partial u}{\partial \eta}\frac{d\eta}{dy} = \frac{1}{y}\frac{\partial u}{\partial \eta}.$$

Hence

$$px = \frac{\partial u}{\partial \xi} = \bar{p} \text{ (say)}, \quad qy = \frac{\partial u}{\partial \eta} = \bar{q} \text{ (say)}.$$

Equation (13.175) now assumes the form

$$\phi(\xi, \eta, u, \bar{p}, \bar{q}) = 0. \tag{13.177}$$

This equation may be arranged into one of the above four types and solved accordingly.

EXAMPLE 13.28. *Solve:* $2p^2x^2 - 3qyu = u^2$.

Solution: Substituting, $\xi = \ln x$, $\eta = \ln y$, the given equation is transformed into

$$2\left(\frac{\partial u}{\partial \xi}\right)^2 - 3u\frac{\partial u}{\partial \eta} - u^2 = 0. \tag{1}$$

This equation is now of type III. Assuming the solution in the form,

$$u = f(\xi + a\eta) = f(\theta), \quad \theta = \xi + a\eta,$$

we have

$$\frac{\partial u}{\partial \xi} = \frac{du}{d\theta}, \quad \frac{\partial u}{\partial \eta} = a\frac{du}{d\theta}.$$

Equation (1) assumes the form,

$$2\left(\frac{du}{d\theta}\right)^2 - 3ua\frac{du}{d\theta} - u^2 = 0.$$

$$\therefore \quad \frac{du}{d\theta} = \frac{1}{4}\left[3ua + \sqrt{9a^2u^2 + 8u^2}\right] \quad \text{or,} \quad 4\frac{du}{u} = \left(3a + \sqrt{8 + 9a^2}\right) d\theta.$$

Integrating

$$\ln u^4 = \left(3a + \sqrt{8 + 9a^2}\right)(\theta + \text{constant})$$

$$\text{or,} \quad \ln u^4 = \lambda(\xi + a\eta + b)$$

$$\text{i.e.,} \quad \ln u^4 = \lambda(\ln x + a \ln y + b), \tag{2}$$

where $\lambda = 3a + \sqrt{8 + 9a^2}$. Thus there are two arbitrary constants a, b in equation (2) which is the complete solution. It may be noted that there is no singular solution.

Type VI. Let the PDE be of the form

$$\psi\left(x, y, \frac{p}{u}, \frac{q}{u}\right) = 0. \tag{13.178}$$

In this case, we substitute

$$\xi = \ln u \quad \text{i.e.,} \quad u = e^{\xi}. \tag{13.179}$$

Then we have

$$p = \frac{\partial u}{\partial x} = \frac{\partial \xi}{\partial x}\frac{du}{d\xi} = u\frac{\partial \xi}{\partial x}.$$

Therefore

$$\frac{p}{u} = \frac{\partial \xi}{\partial x}, \quad \text{similarly } \frac{q}{u} = \frac{\partial \xi}{\partial y}. \tag{13.180}$$

Then the equation (13.178) assumes the form

$$\psi\left(x, y, \frac{\partial \xi}{\partial x}, \frac{\partial \xi}{\partial y}\right) = 0, \tag{13.181}$$

which may be put into one of the types II–IV and solved accordingly.

EXAMPLE 13.29. *Solve*: $p + q = u(\sin x + \cos y)$.

Solution: The given equation is written as

$$\frac{p}{u} + \frac{q}{u} = \sin x + \cos y. \tag{1}$$

Taking, $\xi = \ln u$, we have

$$\frac{p}{u} = \frac{\partial \xi}{\partial x}, \quad \frac{q}{u} = \frac{\partial \xi}{\partial y}. \tag{2}$$

Equation (1) assumes the form

$$\frac{\partial \xi}{\partial x} + \frac{\partial \xi}{\partial y} = \sin x + \cos y, \tag{3}$$

which is of type IV. Writing it as

$$\frac{\partial \xi}{\partial x} - \sin x = \cos y - \frac{\partial \xi}{\partial y} = \lambda \text{ (constant)},$$

we find

$$\frac{\partial \xi}{\partial x} = \lambda + \sin x, \quad \frac{\partial \xi}{\partial y} = \cos y - \lambda.$$

Hence

$$d\xi = \frac{\partial \xi}{\partial x}dx + \frac{\partial \xi}{\partial y}dy = (\lambda + \sin x)\,dx + (\cos y - \lambda)\,dy.$$

Integrating

$$\xi = (\lambda x - \cos x) + (\sin y - \lambda y) + \text{constant}.$$

Hence

$$u = \exp\left[\lambda(x - y) - \cos x + \sin y + \mu\right],$$

which is a complete solution.

Type VII. Let the PDE be of the form

$$pq = Ax^{\lambda}y^{\mu}u^{2\nu}, \tag{13.182}$$

which can be written as

$$\frac{pu^{-\nu}}{x^{\lambda}} \cdot \frac{q\bar{u}^{\nu}}{y^{\mu}} = A. \tag{13.183}$$

Here A, λ, μ, ν are real constants.
We define the new variables like

$$\xi = \frac{x^{\lambda+1}}{\lambda+1}, \quad \eta = \frac{y^{\mu+1}}{\mu+1}, \quad U = \frac{u^{1-\nu}}{1-\nu}. \tag{13.184}$$

Then, we have

$$\left.\begin{aligned} \frac{\partial U}{\partial \xi} &= \frac{\partial U}{\partial x}\frac{dx}{d\xi} = \bar{u}^{\nu}\frac{\partial u}{\partial x}\frac{1}{x^{\lambda}} = u^{-\nu}p\frac{1}{x^{\lambda}} \\ \frac{\partial U}{\partial \eta} &= \frac{\partial U}{\partial y}\frac{dy}{d\eta} = \bar{u}^{\nu}\frac{\partial u}{\partial y}\frac{1}{y^{\mu}} = u^{-\nu}q\frac{1}{y^{\mu}} \end{aligned}\right\}. \tag{13.185}$$

Substituting these expressions into equation (13.183) we get

$$\frac{\partial U}{\partial \xi} \cdot \frac{\partial U}{\partial \eta} = A, \tag{13.186}$$

which is of the type I. It's solution is $U = a\xi + b\eta + c$, with $ab = A$ so that, we can write

$$U = a\xi + \frac{A}{a}\eta + c. \tag{13.187}$$

Returning to the original variables

$$\frac{u^{1-\nu}}{1-\nu} = a\frac{x^{\lambda+1}}{\lambda+1} + \frac{Ay^{\mu+1}}{a(\mu+1)} + c. \tag{13.188}$$

Thus, equation (13.188) is a complete solution of the PDE in (13.182). It has no singular solution. From equation (13.188), it is observed that for the solution to exist, ν should not be equal to unity. Therefore, the method fails if $\nu = 1$.

EXAMPLE 13.30. *Solve:* $pq = 4xy\,u$.

Solution: Here, $A = 4, \lambda = 1, \mu = 1, \nu = \frac{1}{2}$.
A complete solution, by equation (13.188) is

$$\frac{u^{\frac{1}{2}}}{\frac{1}{2}} = a\frac{x^2}{2} + \frac{4y^2}{2a} + c$$

$$\text{or,} \quad u^{\frac{1}{2}} = \frac{ax^2}{4} + \frac{y^2}{a} + \frac{c}{2}$$

$$\text{or,} \quad 16u = \left(ax^2 + \frac{4y^2}{a} + 2c\right)^2$$

$$\text{or,} \quad 16a^2u = \left(a^2x^2 + 4y^2 + d\right)^2, \quad (2ac = d).$$

13.19 Charpit's Method

We have so far discussed about the method of solving particular types of non-linear PDEs. However, we may face with some non-linear PDEs which do not fall into any of the seven types. We discuss below a general method, known as *Charpit's method,* for solving non-linear PDEs of order one.

Let the non-linear PDE be given in the form

$$\psi(x, y, u, p, q) = 0. \tag{13.189}$$

Suppose that, it is possible to solve for p and q, such that

$$p = P(x, y, u), \quad q = Q(x, y, u), \tag{13.190}$$

using equation (13.189) and another equation

$$\phi(x, y, u, p, q) = 0, \tag{13.191}$$

which we are to find out.

Then, the total differential equation

$$du = P(x, y, u)\, dx + Q(x, y, u)\, dy \tag{13.192}$$

can be integrated to obtain a complete solution of equation (13.189).

To find the function ϕ in equation (13.191), we differentiate partially with respect to x and y both the functions ψ and ϕ. We get

$$\frac{\partial \psi}{\partial x} + \frac{\partial \psi}{\partial p}\frac{\partial p}{\partial x} + \frac{\partial \psi}{\partial q}\frac{\partial q}{\partial x} + p\frac{\partial \psi}{\partial u} = 0 \tag{13.193}$$

$$\frac{\partial \psi}{\partial y} + \frac{\partial \psi}{\partial p}\frac{\partial p}{\partial y} + \frac{\partial \psi}{\partial q}\frac{\partial q}{\partial y} + q\frac{\partial \psi}{\partial u} = 0 \tag{13.194}$$

$$\frac{\partial \phi}{\partial x} + \frac{\partial \phi}{\partial p}\frac{\partial p}{\partial x} + \frac{\partial \phi}{\partial q}\frac{\partial q}{\partial x} + p\frac{\partial \phi}{\partial u} = 0 \tag{13.195}$$

$$\frac{\partial \phi}{\partial y} + \frac{\partial \phi}{\partial p}\frac{\partial p}{\partial y} + \frac{\partial \phi}{\partial q}\frac{\partial q}{\partial y} + q\frac{\partial \phi}{\partial u} = 0. \tag{13.196}$$

The total differential equation

$$P(x, y, u)\, dx + Q(x, y, u)\, dy + R(x, y, u)\, du = 0 \tag{13.197}$$

is integrable if

$$P\left(\frac{\partial Q}{\partial u} - \frac{\partial R}{\partial y}\right) + Q\left(\frac{\partial R}{\partial x} - \frac{\partial P}{\partial u}\right) + R\left(\frac{\partial P}{\partial y} - \frac{\partial Q}{\partial x}\right) = 0. \tag{13.198}$$

Hence for the integrability of the total differential equation (13.192), there exists a relation between p and q in the form

$$\frac{\partial p}{\partial y} = \frac{\partial q}{\partial x}. \tag{13.199}$$

Thus we have a system of four equations for determining the function ϕ. Multiplying the first by $\frac{\partial \phi}{\partial p}$, second by $\frac{\partial \phi}{\partial q}$, third by $-\frac{\partial \psi}{\partial p}$, fourth by $-\frac{\partial \psi}{\partial q}$ and then adding with due care of equation (13.199), we get

$$A\frac{\partial \phi}{\partial p} + B\frac{\partial \phi}{\partial q} + C\frac{\partial \phi}{\partial x} + D\frac{\partial \phi}{\partial y} + E\frac{\partial \phi}{\partial u} = 0, \tag{13.200}$$

where A, B, ... are all functions given by

$$\left.\begin{aligned} A(x,y,u,p,q) &= \frac{\partial \psi}{\partial x} + p\frac{\partial \psi}{\partial u} \\ B(x,y,u,p,q) &= \frac{\partial \psi}{\partial y} + q\frac{\partial \psi}{\partial u} \\ C(x,y,u,p,q) &= -\frac{\partial \psi}{\partial p} \\ D(x,y,u,p,q) &= -\frac{\partial \psi}{\partial q} \\ E(x,y,u,p,q) &= -\left(p\frac{\partial \psi}{\partial p} + q\frac{\partial \psi}{\partial q}\right) \end{aligned}\right\} \tag{13.201}$$

Equation (13.200) is a linear partial differential equations in $\phi(x, y, u, p, q)$. The auxiliary system is

$$\frac{dp}{A} = \frac{dq}{B} = \frac{dx}{C} = \frac{dy}{D} = \frac{du}{E}. \tag{13.202}$$

Thus, any solution of this system involving p or q or both and containing an arbitrary constant may be taken for the function ϕ. However, for the solution in equation (13.190) to exist, we should have

$$\begin{bmatrix} \frac{\partial \psi}{\partial p} & \frac{\partial \psi}{\partial q} \\ \frac{\partial \phi}{\partial p} & \frac{\partial \phi}{\partial q} \end{bmatrix} \neq 0. \tag{13.203}$$

EXAMPLE 13.31. *Solve:* $p^2 + 2xp + q = e^{2y}$.

Solution: Here we have

$$\psi \equiv p^2 + 2xp + q - e^{2y} = 0. \tag{1}$$

Therefore

$$\frac{\partial \psi}{\partial x} = 2p, \quad \frac{\partial \psi}{\partial y} = -2e^{2y}, \quad \frac{\partial \psi}{\partial u} = 0, \quad \frac{\partial \psi}{\partial p} = 2(p+x), \quad \frac{\partial \psi}{\partial q} = 1,$$

$$A = 2p, \quad B = -2e^{2y}, \quad C = -2(p+x), \quad D = -1, \quad E = -\left[2p^2(1+x) + q\right].$$

The auxiliary equation is

$$\frac{dp}{2p} = \frac{dq}{-2e^{2y}} = \frac{dx}{-2(p+x)} = \frac{dy}{-1} = \frac{du}{-(2p^2 + 2p^2x + q)}. \tag{2}$$

From the relation

$$\frac{dq}{-2e^{2y}} = \frac{dy}{-1},$$

we have

$$q = e^{2y}. \tag{3}$$

From equation (1) with the use of equation (3)

$$p^2 + 2xp = 0. \tag{4}$$

Neglecting $p = 0$, we have $p = -2x$.
Hence

$$du = pdx + qdy = -2xdx + e^{2y}dy.$$

Integrating

$$u = -x^2 + \frac{1}{2}e^{2y} + c, \tag{5}$$

which is a complete solution.

EXERCISE

Find the general solution of the following partial differential equations:

1. (a) $3u_x + 4u_y = u$;
   (b) $2u_x + 3u_y = -5u$;
   (c) $y\,u_y - x\,u_x = u$;
   (d) $x\,u_x + y\,u_y = 3u$;
   (e) $4u_x - 3u_y = 0$.
2. (a) $u_{xx} - u_{xy} - 6u_{yy} = x + y$;
   (b) $u_{xx} - 5u_{xy} + 6u_{yy} = e^{2x+y} + \sin(x + y)$;
   (c) $u_{xx} - 2u_{xy} = e^{2x} + x^2 + y^2$.
3. (a) $u_{xx} - u_{xy} - 2u_{yy} + 6u_x - 9u_y + 5u = e^{2x+y}$;
   (b) $u_{xx} - u_{yy} + 3u_x - 3u_y = e^{2x+y}\sin(x + 2y)$;
   (c) $u_{xx} + u_{xy} + u_y - u = e^x + e^{-x}$;
   (d) $u_{xx} + u_{xy} + u_x + u_y + u = e^{-2x}\left(x^2 + 2y^2\right)$.
4. (a) $x^2u_{xx} - 4y^2u_{yy} - 4y\,u_y - u = x^2y^3 \ln y$;
   (b) $xy\,u_{xy} - y^2u_{yy} - 3x\,u_x + 2y\,u_y = xy$;
   (c) $x^2u_{xx} - 2xy\,u_{xy} - 3y^2u_{yy} + x\,u_x - 3y\,u_y = x^2y\sin(\ln x^2)$.
5. (a) $x\,u_{xx} + 2u_x = (4x + 4)\,e^{2x+3y}$;
   (b) $u_{yy} - x\,u_y = x\cos y + \sin y$;
   (c) $y\,u_{xy} - u_x = xy^2\sin(xy)$;
   (d) $2x\,u_{xx} - y\,u_{xy} + 2u_x = -6xy$.
6. Transform the following equations into canonical forms:
   (a) $x^2u_{yy} = y^2u_{xx}$;
   (b) $u_{xx} = 4u_{yy} - u_y$.
7. Solve: $u_{xy} = x^2y$, given the conditions, $u(x, 0) = x^2$, $u(1, y) = \cos y$.
8. Solve by the method of characteristics, the following equations:
   (a) $u_x + u_y = -2u$, $-\infty < x < \infty$, $0 < y < \infty$; given the condition, $u(x, 0) = \sin x$.
   (b) $x\,u_x + 2u_y + yu = 0$, $-\infty < x < \infty$, $0 < y < \infty$; given the condition, $u(x, 0) = \cos x$.

9. Solve by D'Alemberts method, the following equations:
   (a) $u_{tt} = \alpha^2 u_{xx}$, $-\infty < x < \infty$, $0 < t < \infty$; given the conditions,
$$u(x,0) = \frac{\sin x}{x}, \quad u_t(x,0) = 0.$$
   (b) $u_{tt} = u_{xx}$, $-\infty \le x < \infty$, $0 < t < \infty$ given the conditions
$$u(x,0) = \frac{x}{1+x^2}, \quad u_t(x,0) = \sin x.$$
   (c) $u_{tt} = \alpha^2 u_{xx}$, $0 \le x < \infty$, $0 < t < \infty$ given the conditions $u(x,0) = 0$, $u_t(x,0) = \sin x$.
10. Solve by separation of variables, the following equations:
   (a) $u_x = 4u_y$, with $u(0,y) = e^{-3y}$;
   (b) $u_{xx} = u_t$, with $u(1,t) = 0$ and $u_x(0,t) = e^{-t}$.
11. Solve the following system of equations:
   (a) $\left.\begin{aligned} u_t + 4v_x = 0 \\ v_t + u_x = 0 \end{aligned}\right\} \quad -\infty < x < \infty,\ 0 < t < \infty;$
   given the conditions, $u(x,0) = x^2$, $v(x,0) = \sin x$.
   (b) $\left.\begin{aligned} u_x + 2u_y + 3v_y = 0 \\ v_x + u_y + 4v_y = 0 \end{aligned}\right\} \quad -\infty < y < \infty,\ 0 < x < \infty;$
   given the conditions $u(x,0) = \phi(y)$, $v(x,0) = \psi(y)$.
12. Solve, by using Duhamel's principle, the heat equation
$$u_t = \alpha^2 u_{xx}, \quad (0 < x < L,\ 0 < t < \infty)$$
   given the boundary conditions
$$u(0,t) = 0, \quad u(1,t) = t^2 + t$$
   and the initial condition
$$u(x,0) = \sin\frac{\pi x}{L}.$$
13. Show that the equation
$$u_t = \alpha^2 u_{xx} - \xi u, \quad (\xi = \text{constant},\ 0 < x < L,\ 0 < t < \infty)$$
   with conditions
$$u(0,t) = 0, \quad u(L,t) = 0 \quad \text{and} \quad u(x,0) = \phi(x)$$
   can be transformed to the standard form of heat equation.
14. Solve the following non-linear equations
   (a) $pq = p^2 + q^2$;
   (b) $u = px + qy - 2p^2q$;
   (c) $u^2(16p^2 + 9q^2 - 4) = 4$;
   (d) $4(1 + u^3) = 9u^4pq$;
   (e) $p^2 - px + q = 0$;
   (f) $(1 - y^2)xq^2 + y^2p = 0$;
   (g) $x^2p^2 + yqu = u^2$;
   (h) $x^2p^2 + y^2q^2 = u$;
   (i) $p^2 + q^2 = u^2(x + y)$;
   (j) $p + q = u(e^x + e^y)$;
   (k) $p^2 + xy - q = \sin y$;
   (l) $y^2(p - q) = up^2$.

ANSWERS

1. (a) $e^{\frac{x}{3}}\phi(3y-4x)$;
(b) $e^{-\frac{5x}{2}}\phi(2y-3x)$;
(c) $\phi(xy, xu) = 0$;
(d) $u = y^3\phi\left(\frac{x}{y}\right)$;
(e) $u = \phi(3x+4y)$.

2. (a) $u = \phi_1(y-2x) + \phi_2(y+3x) + \frac{1}{2}x^2y + \frac{1}{3}x^3$;

(b) $u = \phi_1(y+2x) + \phi_2(y+3x) - xe^{2x+y} - \frac{1}{2}\sin(x+y)$;

(c) $u = \phi_1(y) + \phi_2(y+2x) + \frac{1}{4}e^{2x} + \frac{x^4}{12} + \frac{x^2y^2}{2} + \frac{4}{3}x^3y$.

3. (a) $u = e^{-5x}\phi_1(y-x) + e^{-x}\phi_2(y+2x) + \frac{1}{8}e^{2x+y}$;

(b) $u = \phi_1(y+x) + e^{-3x}\phi_2(y-x) - \frac{1}{18}e^{2x+y}[\sin(x+2y) + \cos(x+2y)]$;

(c) $u = e^{-x}\phi_1(y) + e^{x}\phi_2(y-x) + \frac{1}{2}xe^{x} - \frac{1}{2}xe^{-x}$;

(d) $u = \sum A_i e^{a_i x + b_i y} + \frac{1}{27}e^{-2x}(9x^2 + 18y^2 + 18x + 12y + 16)$.

4. (a) $u \sum A_i x^{a_i} y^{b_i} - \frac{1}{1225}x^2y^3(35\ln y - 24)$;

(b) $u = x^3\phi_1(\ln x) + \phi_2(\ln xy) - \frac{1}{2}xy\ln x$;

(c) $u = \phi_1(x^3y) + \phi_2\left(\frac{y}{x}\right) - \frac{1}{65}x^2y[4\cos(\ln x^2) + 7\sin(\ln x^2)]$.

5. (a) $u = e^{2x+3y} - \frac{1}{x}\phi_1(y) + \phi_2(y)$;
(b) $u = -\sin y + e^{xy}\phi_1(x) + \phi_2(x)$;
(c) $u = y\phi_1(x) + \phi_2(y) - \sin(xy)$;
(d) $u = -xy^2 + \phi_1(xy^2) + \phi_2(y)$.

6. (a) $u_{\xi\eta} = \frac{\xi u_\eta - \eta u_\xi}{2(\eta^2 - \xi^2)}, \xi = y^2 - x^2, \eta = y^2 + x^2$;

(b) $-16u_{\xi\eta} + u_\xi - 2u_\eta = 0, \xi = y + 2x, \eta = y - 2x$;

(c) $u = \frac{1}{6}x^3y^2 + \cos y - \frac{1}{6}y^2 - 1 + x^2$.

8. (a) $u = \sin(x-t)e^{-2t}$;
(b) $u = \cos(xe^{-t/2})e^{-t^2}$.

9. (a) $u = \dfrac{x\sin x\cos\alpha t - \alpha t\cos x\sin\alpha t}{x^2 - (\alpha t)^2}$;

(b) $u = \frac{1}{2}\left[\frac{x+t}{1+(x+t)^2} + \frac{x-at}{1+(x-at)^2}\right] + \sin x\sin t$;

(c) $u = \frac{1}{\alpha}\sin x\sin\alpha t$.

10. (a) $u = e^{-3x/4}e^{-3y}$;
(b) $u = (-\tan 1\cos x + \sin x)e^{-t}$.

11. (a) $u = [x^2 + 4t^2 + 2\cos x\sin 2t]$;
$v = [xt + \sin x\cos 2t]$;

44

(b) $u = -\frac{3}{4}[\psi(x-y) - \phi(x-y)] + \frac{1}{4}[3\psi(x-5y) + \phi(x-5y)]$

$v = \frac{1}{4}[\psi(x-y) - \phi(x-y)] + \frac{1}{4}[3\psi(x-5y) + \phi(x-5y)]$.

12. $u = \frac{1}{\lambda^2} \sin \frac{\pi x}{L} \left[2t + \left(1 - \frac{2}{\lambda^2}\right)\{1 - \exp(-\lambda^2 t)\}\right], \lambda^2 = \frac{n^2\alpha^2}{L^2}$.

14. (a) $u = ax + by + c, ab = a^2 + b^2$;

(b) $u = \frac{x\sqrt{y}}{2\sqrt{2}}$;

(c) $(1+u^2)(16a^2+9) = (x+ay+b)^2$, sing. soln. is $1+u^2 = 0$;

(d) $a(1+u^3) = (x+ay+b)^2$, sing. soln. is $1+u^3 = 0$;

(e) $u = \frac{1}{4}\left(x^2 + x\sqrt{x^2+4a}\right) + a\ln\left(x^2 + \sqrt{x^2+4a}\right) - ay + b$;

(f) $u = \frac{ax^2}{2} + \sqrt{a(y^2-1)}$;

(g) $\ln u^2 = \left(\sqrt{a^2+4} - a\right)(\ln x + a\ln y + b)$;

(h) $u = \frac{1}{4}(a\ln x + b\ln y + c)^2, a^2 + b^2 = 1$; sing. soln. is $u = 0$;

(i) $\ln u = \frac{2}{3}(a+x)^{3/2} + \frac{2}{3}(y-a)^{3/2} + b$; (k) $u = xy^2 + \frac{1}{20}y^5 + \cos y$;

(j) $u = \exp[e^x + \lambda x + e^y - \lambda y]$; (l) $u = \sqrt{\frac{y}{2}}x + \frac{2}{7}y^{7/2}$.

Chapter 14

Fourier Transforms

14.1 Definition

If a function $f(x)$ satisfies the Dirichlet's conditions and the integral,

$$\int_{-\infty}^{\infty} |f(x)|\, dx \leq M, \tag{14.1}$$

then the integral,

$$F(k) = \frac{1}{\sqrt{2\pi}} \int_{-\infty}^{\infty} f(x)\, e^{ikx}\, dx, \tag{14.2}$$

is defined as the *Fourier transform* of $f(x)$.

The integral

$$f(x) = \frac{1}{\sqrt{2\pi}} \int_{-\infty}^{\infty} F(k)\, e^{-ikx}\, dk \tag{14.3}$$

is defined as the *Fourier's inverse transform* of $F(k)$ which exists for all k. Equations (14.2) and (14.3) are called the *Fourier transform pair.* We can obtain these equations from the expression of Fourier integral which is written as

$$f(x) = \frac{1}{2\pi} \int_{-\infty}^{\infty} e^{-ikx}\, dk \left[\int_{-\infty}^{\infty} f(t)\, e^{ikt}\, dt\right]$$

$$\text{i.e.,}\quad f(x) = \frac{1}{\sqrt{2\pi}} \int_{-\infty}^{\infty} e^{-ikx}\, dk \left[\frac{1}{\sqrt{2\pi}} \int_{-\infty}^{\infty} f(t)\, e^{ikt}\, dt\right] = \frac{1}{\sqrt{2\pi}} \int_{-\infty}^{\infty} e^{-ikx} F(k)\, dk, \tag{14.4}$$

where

$$F(k) = \frac{1}{\sqrt{2\pi}} \int_{-\infty}^{\infty} f(t)\, e^{ikt}\, dt. \tag{14.5}$$

Thus, we get the Fourier integral transform of $f(x)$ as

$$\text{F.T. } f(x) = F(k) = \frac{1}{\sqrt{2\pi}} \int_{-\infty}^{\infty} f(x)\, e^{ikx}\, dx, \tag{14.6}$$

which is the equation (14.2). Fourier inverse transform is then given by equation (14.4), so that

$$\text{F.I.T. } f(x) = f(x) = \frac{1}{\sqrt{2\pi}} \int_{-\infty}^{\infty} F(k)\, e^{-ikx}\, dk. \tag{14.7}$$

14.2 Fourier Sine and Cosine Series

The Fourier transform of $f(x)$ given by equation (14.5) can be written as

$$F(k) = \frac{1}{\sqrt{2\pi}} \int_{-\infty}^{\infty} f(x) \{\cos kx + i \sin kx\} \, dx. \tag{14.8}$$

If $f(x)$ is an even function of x, i.e., $f(x) = f(-x)$, then

$$\int_{-\infty}^{\infty} f(x) \sin kx \, dx = 0. \tag{14.9}$$

Hence, we have from equation (14.8)

$$F(k) = \frac{1}{\sqrt{2\pi}} \int_{-\infty}^{\infty} f(x) \cos kx \, dx$$

$$\text{i.e.,} \quad F_c(k) = \sqrt{\frac{2}{\pi}} \int_{0}^{\infty} f(x) \cos kx \, dx, \tag{14.10}$$

which represents Fourier cosine transform. The inverse Fourier cosine transform of $F_c(k)$ is given by

$$f(x) = \sqrt{\frac{2}{\pi}} \int_{0}^{\infty} F_c(k) \cos kx \, dx. \tag{14.11}$$

If the function $f(x)$ is defined in the interval $(0, \pi)$ and vanishes elsewhere, then the equations (14.10) and (14.11) assume the forms

$$\left.\begin{aligned} F_c(k) &= \sqrt{\frac{2}{\pi}} \int_{0}^{\pi} f(x) \cos kx \, dx \\ f(x) &= \sqrt{\frac{2}{\pi}} \int_{0}^{\pi} F_c(k) \cos kx \, dx \end{aligned}\right\}. \tag{14.12}$$

These represent the finite Fourier cosine transform pair.

Similarly, if $f(x)$ is an odd function of x, i.e., $f(x) = -f(-x)$, then equation (14.8) assumes the form

$$F(k) = iF_s(k) = i\sqrt{\frac{2}{\pi}} \int_{0}^{\infty} f(x) \sin kx \, dx,$$

which gives

$$F_s(k) = \sqrt{\frac{2}{\pi}} \int_{0}^{\infty} f(x) \sin kx \, dx. \tag{14.13}$$

This is infinite Fourier sine transform. The inverse infinite Fourier sine transform of $F_s(k)$ is given by

$$f(x) = \sqrt{\frac{2}{\pi}} \int_{0}^{\infty} F_s(k) \sin kx \, dx. \tag{14.14}$$

If the function $f(x)$ is non-vanishing only in the interval $(0, \pi)$, then equations (14.13) and (14.14) assume the forms:

$$\left.\begin{aligned} F_s(k) &= \sqrt{\frac{2}{\pi}} \int_0^{\pi} f(x) \sin kx \, dx \\ f(x) &= \sqrt{\frac{2}{\pi}} \int_0^{\pi} F_s(k) \sin kx \, dx \end{aligned}\right\} \tag{14.15}$$

Equations in (14.15) constitute the finite sine Fourier transform pair.

EXAMPLE 14.1. *Find Fourier transform of the function*

$$f(x) = \begin{cases} 1, & \text{for } |x| < a \\ 0, & \text{for } |x| > a. \end{cases}$$

Solution: The Fourier transform is given by

$$F(k) = \frac{1}{\sqrt{2\pi}} \int_{-a}^{a} e^{ikx} \, dx = \frac{1}{\sqrt{2\pi}} \left(\frac{e^{ikx}}{ik} \right)_{-a}^{a} = \frac{1}{\sqrt{2\pi}} \frac{e^{ika} - e^{-ika}}{ik} = \sqrt{\frac{2}{\pi}} \cdot \frac{\sin ka}{k}.$$

EXAMPLE 14.2. *Find sine Fourier transform for the function*

$$f(x) = \begin{cases} x^2, & \text{for } 0 \le x < \pi \\ 0, & \text{elsewhere.} \end{cases}$$

Solution: The required transform is

$$\begin{aligned} F(k) &= \sqrt{\frac{2}{\pi}} \int_0^{\pi} x^2 \sin kx \, dx = \sqrt{\frac{2}{\pi}} \left[-\frac{x^2 \cos kx}{k} \Big|_0^{\pi} + \int_0^{\pi} \frac{2x}{k} \cos kx \right] \\ &= \sqrt{\frac{2}{\pi}} \left[-\frac{\pi^2}{k} \cos k\pi - \frac{2}{k^2} \int_0^{\pi} \sin kx \, dx \right] = \sqrt{\frac{2}{\pi}} \left[-\frac{\pi^2}{k} \cos k\pi + \frac{2}{k^3} \cos kx \Big|_0^{\pi} \right] \\ &= \sqrt{\frac{2}{\pi}} \left[-\frac{\pi^2}{k} (-1)^k - \frac{2}{k^3} \{1 - (-1)^k\} \right]. \end{aligned}$$

14.3 Properties of Fourier Transforms

Following are the basic properties of Fourier transforms:

1. *Linearity property.* The Fourier transform of a linear combination of a given number of functions is the linear combination of their Fourier transform.

Let

$$f(x) = \sum_{n=1}^{p} a_n f_n(x). \tag{14.16}$$

Then, FT of $f(x)$ is

$$F(k) = \sum_{n=1}^{p} a_n F_n(k), \tag{14.17}$$

where $F_1(k), F_2(k), \ldots$ are respectively the Fourier transforms of $f_1(x), f_2(x), \ldots$

Proof. From equation (14.16), Fourier transform of $f(x)$ is

$$F(k) = \sum_{n=1}^{p} a_n \left[\frac{1}{\sqrt{2\pi}} \int_{-\infty}^{\infty} f_n(x)\, e^{ikx}\, dx\right] = \sum_{n=1}^{p} a_n F_n(k).$$

Hence proved. □

2. *Change of scale property.* If $F(k)$ is the Fourier transform of $f(x)$, then the Fourier transform of $f(\lambda x)$, $\lambda =$ constant, is $\frac{1}{\lambda} F\left(\frac{k}{\lambda}\right)$.

Proof. Since $F(k)$ is the Fourier transform of $f(x)$, we have

$$F(k) = \frac{1}{\sqrt{2\pi}} \int_{-\infty}^{\infty} f(x)\, e^{ikx}\, dx. \tag{14.18}$$

If Fourier transform of $f(\lambda x)$ is $F_1(k)$, then

$$\begin{aligned} F_1(k) &= \frac{1}{\sqrt{2\pi}} \int_{-\infty}^{\infty} f(\lambda x)\, e^{ikx}\, dx = \frac{1}{\sqrt{2\pi}} \int_{-\infty}^{\infty} \frac{f(\xi)}{\lambda} e^{ik\xi/\lambda}\, d\xi, \quad \xi = \lambda x \\ &= \frac{1}{\sqrt{2\pi}\lambda} \int_{-\infty}^{\infty} f(\xi)\, e^{ik'\xi}\, d\xi, \quad k' = \frac{k}{\lambda}. \end{aligned}$$

Hence

$$F_1(k) = \frac{1}{\lambda} F(k') = \frac{1}{\lambda} F\left(\frac{k}{\lambda}\right) \tag{14.19}$$

Thus the property is proved. □

3. *Shifting of origin.* If $F(k)$ be the Fourier transform of $f(x)$, then the Fourier transform of $f(x-a)$ is given by $e^{ika} F(k)$.

Proof. We have

$$F(k) = \frac{1}{\sqrt{2\pi}} \int_{-\infty}^{\infty} f(x)\, e^{ikx}\, dx.$$

Hence, Fourier transform of $f(x-a)$ is

$$\begin{aligned} F_1(k) &= \frac{1}{\sqrt{2\pi}} \int_{-\infty}^{\infty} f(x-a)\, e^{ikx}\, dx = \frac{1}{\sqrt{2\pi}} \int_{-\infty}^{\infty} f(\xi)\, e^{ik(\xi+a)}\, d\xi, \quad \xi = x-a \\ &= e^{ika} \cdot \frac{1}{\sqrt{2\pi}} \int_{-\infty}^{\infty} f(\xi)\, e^{ik\xi}\, d\xi = e^{ika} F(k). \end{aligned} \tag{14.20}$$

Hence proved. □

4. *Modulation property.* If $F(k)$ be the Fourier transform of $f(x)$, then, the Fourier transform of $f(x)\cos bx$ is

$$F_1(k) = \frac{1}{2}\left[F(k+b) + F(k-b)\right]. \tag{14.21}$$

Proof. We are given

$$F(k) = \frac{1}{\sqrt{2\pi}} \int_{-\infty}^{\infty} f(x)\, e^{ikx}\, dx.$$

Then

$$\begin{aligned}
F_1(k) &= \frac{1}{\sqrt{2\pi}} \int_{-\infty}^{\infty} f(x) \cos bx \cdot e^{ikx}\, dx = \frac{1}{\sqrt{2\pi}} \int_{-\infty}^{\infty} f(x) \frac{e^{ibx} + e^{-ibx}}{2} \cdot e^{ikx}\, dx \\
&= \frac{1}{\sqrt{2\pi}} \int_{-\infty}^{\infty} \frac{1}{2} f(x) \left\{ e^{i(b+k)x} + e^{i(k-b)x} \right\} dx \\
&= \frac{1}{\sqrt{2\pi}} \left[\int_{-\infty}^{\infty} \frac{1}{2} f(x)\, e^{i(k+b)x}\, dx + \frac{1}{2} \int_{-\infty}^{\infty} f(x)\, e^{i(k-b)x}\, dx \right] \\
&= \frac{1}{2} \left[F(k+b) + F(k-b) \right].
\end{aligned}$$

Hence proved. □

5. *Convolution property.* The convolution of two functions $f(x)$ and $g(x)$ is denoted by f^*g and defined as

$$f^*g = \frac{1}{\sqrt{2\pi}} \int_{-\infty}^{\infty} f(t)\, g(x-t)\, dt. \tag{14.22}$$

If $F(k)$ and $G(k)$ be the Fourier transforms of $f(x)$ and $g(x)$ respectively, then the Fourier transform of the convolution of $f(x)$ and $g(x)$ is given by $F(k) \cdot G(k)$.

Proof. We are given

$$F(k) = \frac{1}{\sqrt{2\pi}} \int_{-\infty}^{\infty} f(x)\, e^{ikx}\, dx \tag{14.23}$$

$$\text{and} \quad G(k) = \frac{1}{\sqrt{2\pi}} \int_{-\infty}^{\infty} g(x)\, e^{ikx}\, dx. \tag{14.24}$$

Then Fourier transform of f^*g is [using the equation (14.22)]

$$\begin{aligned}
&= \frac{1}{\sqrt{2\pi}} \int_{-\infty}^{\infty} \left[\frac{1}{\sqrt{2\pi}} \int_{-\infty}^{\infty} f(t)\, g(x-t)\, dt \right] e^{ikx}\, dt \\
&= \frac{1}{2\pi} \int_{-\infty}^{\infty} f(t) \left[\int_{-\infty}^{\infty} g(x-t)\, e^{ikx}\, dx \right] dt \\
&= \frac{1}{\sqrt{2\pi}} \int_{-\infty}^{\infty} f(t) \left[e^{ikt}\, G(k) \right] dt \quad \text{[using property 3]} \\
&= G(k) \cdot \frac{1}{\sqrt{2\pi}} \int_{-\infty}^{\infty} f(t)\, e^{ikt}\, dt \\
&= G(k) \cdot F(k).
\end{aligned}$$

Hence proved. □

6. *Conjugate property.* If $F(k)$ be the Fourier transform of $f(x)$, then the Fourier transform of the complex conjugate of $f(x)$ is $F^*(-k)$.

Proof. We have

$$F(k) = \frac{1}{\sqrt{2\pi}} \int_{-\infty}^{\infty} f(x)\, e^{ikx}\, dx.$$

Taking complex conjugate on both sides

$$F^*(k) = \frac{1}{\sqrt{2\pi}} \int_{-\infty}^{\infty} f^*(x)\, e^{-ikx}\, dx.$$

Changing k by $-k$,

$$F^*(-k) = \frac{1}{\sqrt{2\pi}} \int_{-\infty}^{\infty} f^*(x)\, e^{ikx}\, dx.$$

Hence proved. □

14.4 Fourier Transforms for Functions of Many Variables

The Fourier transforms can be extended to functions of more than one variable. Thus, for a function $f(x, y)$ of two variables (x, y), the one-dimensional Fourier transform pairs are given by

$$\left.\begin{aligned} F(k, y) &= \frac{1}{\sqrt{2\pi}} \int_{-\infty}^{\infty} f(x, y)\, e^{ikx}\, dx \\ f(x, y) &= \frac{1}{\sqrt{2\pi}} \int_{-\infty}^{\infty} F(k, y)\, e^{-ikx}\, dk \end{aligned}\right\}. \tag{14.25}$$

Similarly, the sine and cosine transform pairs for $f(x, y)$ in one-dimension are

$$\left.\begin{aligned} F_s(k, y) &= \frac{1}{\sqrt{2\pi}} \int_{-\infty}^{\infty} f(x, y) \sin kx\, dx \\ f(x, y) &= \frac{1}{\sqrt{2\pi}} \int_{-\infty}^{\infty} F_s(k, y) \sin kx\, dk \end{aligned}\right\}. \tag{14.26}$$

and

$$\left.\begin{aligned} F_c(k, y) &= \frac{1}{\sqrt{2\pi}} \int_{-\infty}^{\infty} f(x, y) \cos kx\, dx \\ f(x, y) &= \frac{1}{\sqrt{2\pi}} \int_{-\infty}^{\infty} F_c(k, y) \cos kx\, dk \end{aligned}\right\}. \tag{14.27}$$

In case of two-dimensional Fourier transforms, we have

$$F(k, m) = \frac{1}{2\pi} \int_{-\infty}^{\infty} \int_{-\infty}^{\infty} f(x, y) e^{i(kx+my)}\, dx\, dy \tag{14.28}$$

and

$$f(x, y) = \frac{1}{2\pi} \int_{-\infty}^{\infty} \int_{-\infty}^{\infty} F(k, m) e^{-i(kx+my)}\, dk\, dm \tag{14.29}$$

In this way, we can go on extending the Fourier transform to any number of dimensions.

14.5 Parseval's Relations

If $F(k)$ and $G^*(k)$ are Fourier transforms of functions $f(x)$ and $g^*(x)$, the complex conjugate of $g(x)$, then

$$\int_{-\infty}^{\infty} f(x)\, g^*(x)\, dx = \int_{-\infty}^{\infty} F(k)\, G^*(k)\, dk. \tag{14.30}$$

As a particular case, if $f(x) = g(x)$, then

$$\int_{-\infty}^{\infty} |\, f(x)\,|^2\, dx = \int_{-\infty}^{\infty} |\, F(k)\,|^2\, dk. \tag{14.31}$$

Equations (14.30) and (14.31) are known as *Parsevals relations.*

Proof. We have

$$\begin{aligned}\int_{-\infty}^{\infty} f(x)\, g^*(x)\, dx &= \int_{-\infty}^{\infty} \left[\frac{1}{\sqrt{2\pi}} \int_{-\infty}^{\infty} F(k)\, e^{-ikx}\, dk\right] \left[\frac{1}{\sqrt{2\pi}} \int_{-\infty}^{\infty} G^*(m)\, e^{imx}\, dm\right] dx \\ &= \int_{-\infty}^{\infty} F(k)\, dk \int_{-\infty}^{\infty} G^*(m)\, dm \left[\frac{1}{2\pi} \int_{-\infty}^{\infty} e^{ix(m-k)}\, dx\right] \\ &= \int_{-\infty}^{\infty} F(k)\, dk \int_{-\infty}^{\infty} G^*(m)\, dm\, \delta(m-k) \qquad (14.32) \\ &= \int_{-\infty}^{\infty} F(k)\, G^*(k)\, dk \qquad \text{[using shifting property of delta function].}\end{aligned}$$

In equation (14.32), we have used

$$\frac{1}{2\pi} \int_{-\infty}^{\infty} e^{ix(m-k)}\, dx = \delta(m-k). \tag{14.33}$$

In fact,

$$\begin{aligned}\int_{-\infty}^{\infty} e^{ixk}\, dk &= \lim_{l \longrightarrow \infty} \frac{1}{ix} \left(e^{ixl} - e^{-ixl}\right) = 2\pi \lim_{l \longrightarrow \infty} \frac{1}{2\pi i x} \left(e^{ixl} - e^{-ixl}\right) \\ &= 2\pi \lim_{l \longrightarrow \infty} \frac{\sin xl}{\pi x} = 2\pi \delta(x), \qquad \text{[by definition of delta function]}\end{aligned}$$

so that

$$\frac{1}{2\pi} \int_{-\infty}^{\infty} e^{ixk}\, dk = \delta(x)$$

□

14.6 Evaluation of Some Integrals

We consider the function

$$f(x) = e^{-ax}, \quad (a > 0, x \geq 0). \tag{14.34}$$

Its cosine Fourier transform is

$$F_c(k) = \sqrt{\frac{2}{\pi}} \int_0^{\infty} e^{-ax} \cos kx\, dx = \sqrt{\frac{2}{\pi}} \frac{a}{a^2 + k^2}. \tag{14.35}$$

Similarly, the sine Fourier transform of $f(x)$ is

$$F_s(k) = \sqrt{\frac{2}{\pi}} \int_0^\infty e^{-ax} \sin kx \, dx = \sqrt{\frac{2}{\pi}} \cdot \frac{k}{a^2 + k^2}. \tag{14.36}$$

Since the function $f(x) = e^{-ax}$ is integrable in the interval $(0, \infty)$, we get by Fourier inverse transform

$$e^{-ax} = \sqrt{\frac{2}{\pi}} \int_0^\infty F_c(k) \cos kx \, dk = \frac{2}{\pi} \int_0^\infty \frac{a}{a^2 + k^2} \cos kx \, dx \quad \text{[using equation (14.35)]}$$

This gives

$$\int_0^\infty \frac{\cos kx}{a^2 + k^2} \, dk = \frac{\pi}{2a} e^{-ax}. \tag{14.37}$$

Similarly, by Fourier sine inverse transform, we have

$$e^{-ax} = \sqrt{\frac{2}{\pi}} \int_0^\infty F_s(k) \sin kx \, dk$$

$$\text{or,} \quad e^{-ax} = \frac{2}{\pi} \int_0^\infty \frac{k}{a^2 + k^2} \sin kx \, dk \quad \text{[using equation (14.36)]}$$

which gives

$$\int_0^\infty \frac{k \sin kx}{a^2 + k^2} \, dk = \frac{\pi}{2} e^{-ax}. \tag{14.38}$$

Thus we have obtained values of two integrals in equations (14.37) and (14.38) by Fourier transform method. We like to evaluate another integral given by

$$I = \int_0^\infty \frac{\sin^2 ax}{x^2} \, dx. \tag{14.39}$$

In example 14.1, we obtained Fourier transform of the function

$$f(x) = \begin{cases} 1, & \text{for } |x| < a \\ 0, & \text{for } |x| > a \end{cases} \tag{14.40}$$

as

$$F(k) = \sqrt{\frac{2}{\pi}} \frac{\sin ka}{k}. \tag{14.41}$$

According to Parseval's relation, we have

$$\int_{-\infty}^\infty |f(x)|^2 \, dx = \int_{-\infty}^\infty |F(k)|^2 \, dk$$

$$\text{or,} \quad \int_{-a}^a (1)^2 \, dx = \frac{2}{\pi} \int_{-\infty}^\infty \frac{\sin^2 ka}{k^2} \, dk \quad \text{[using equation (14.41)]}$$

$$\text{or,} \quad 2a = \frac{2}{\pi} \int_0^\infty \frac{\sin^2 ka}{k^2} \, dk.$$

Thus

$$\int_{-\infty}^\infty \frac{\sin^2 ka}{k^2} \, dk = \pi a, \quad \text{i.e.,} \quad \int_0^\infty \frac{\sin^2 ka}{k^2} \, dk = \frac{\pi a}{2}.$$

EXAMPLE 14.3. *Find Fourier transform of the function $f(x) = e^{-x^2/2}$.*

Solution: Fourier transform of the given function is

$$F(k) = \frac{1}{\sqrt{2\pi}} \int_{-\infty}^{\infty} e^{-x^2/2} e^{ikx}\, dx = \frac{1}{\sqrt{2\pi}} \int_{-\infty}^{\infty} e^{-\frac{1}{2}(x-ik)^2 - k^2/2}\, dx$$
$$= \frac{e^{-k^2/2}}{\sqrt{2\pi}} \int_{-\infty}^{\infty} e^{-z^2} \cdot \sqrt{2}\, dz, \quad \text{where } z = \frac{1}{\sqrt{2}}(x - ik)$$
$$= e^{k^2/2} \frac{1}{\sqrt{\pi}} \cdot \sqrt{\pi} = e^{-k^2/2}.$$

EXAMPLE 14.4. *Find Fourier cosine transform of the function $f(x) = e^{-x^2}$.*

Solution: Fourier cosine transform of the given function is

$$F_c(k) = \sqrt{\frac{2}{\pi}} \int_0^{\infty} e^{-x^2} \cos kx\, dx. \tag{1}$$

Now, we consider the integral

$$I = \int_0^{\infty} e^{-x^2} \cos kx\, dx. \tag{2}$$

Differentiating both sides with respect to k, we get

$$\frac{dI}{dk} = \int_0^{\infty} -xe^{-x^2} \sin kx\, dx = \frac{1}{2} \int_0^{\infty} \left(-2xe^{-x^2}\right) \sin kx\, dx$$
$$= \frac{1}{2} \int_0^{\infty} d\left(e^{-x^2}\right) \sin kx\, dx.$$

Integrating the RHS by parts, we get

$$\frac{dI}{dk} = \frac{1}{2} \left[\sin kx \left(e^{-x^2}\right)\Big|_0^{\infty} - \int_0^{\infty} ke^{-x^2} \cos kx\, dx\right]$$
$$= -\frac{1}{2} k \int_0^{\infty} e^{-x^2} \cos kx\, dx = -\frac{k}{2} I,$$

i.e., $\quad \dfrac{dI}{I} = -\dfrac{k}{2}\, dk,$

which gives

$$I = Ae^{-k^2/4}. \tag{3}$$

Now, from equation (2), we have, for $k = 0$

$$I = \int_0^{\infty} e^{-x^2}\, dx = \frac{\sqrt{\pi}}{2}. \tag{4}$$

From equations (3) and (4), we get

$$A = \frac{\sqrt{\pi}}{2}. \tag{5}$$

Hence, from equations (4) and (5)

$$I = \frac{\sqrt{\pi}}{2} e^{-k^2/4}. \tag{6}$$

Thus, from equations (1) and (6)

$$F_c(k) = \sqrt{\frac{2}{\pi}} I = \sqrt{\frac{2}{\pi}} \cdot \frac{\sqrt{\pi}}{2} e^{-k^2/4} = \frac{1}{\sqrt{2}} e^{-k^2/4}.$$

EXAMPLE 14.5. *Find the cosine Fourier transform of*

$$f(x) = \begin{cases} (1-x^2)^{n-\frac{1}{2}}, & \textit{for } 0 \le x < 1 \\ 0, & \textit{for } x > 1, \ (n = 0, 1, 2, \ldots) \end{cases}$$

and hence, find the value of the integral,

$$I = \int_0^\infty \frac{J_n(x)}{x^n} \cos kx \, dx.$$

Solution: The cosine Fourier transform of the given function is

$$F_c(k) = \sqrt{\frac{2}{\pi}} \int_0^1 \left(1-x^2\right)^{n-\frac{1}{2}} \cos kx \, dx, \quad n = 0, 1, 2, \ldots$$

Taking series expansion of $\cos kx$, we get on term-by-term integration

$$\begin{aligned} F_c(k) &= \sqrt{\frac{2}{\pi}} \int_0^1 \left(1-x^2\right)^{n-\frac{1}{2}} \sum_{m=0}^\infty (-1)^m \frac{(xk)^{2m}}{(2m)!} \, dx \\ &= \sqrt{\frac{2}{\pi}} \sum_{m=0}^\infty (-1)^m \frac{k^{2m}}{(2m)!} \int_0^1 x^{2m} \left(1-x^2\right)^{n-\frac{1}{2}} \, dx. \end{aligned} \tag{1}$$

Now

$$\begin{aligned} \int_0^1 x^{2m} \left(1-x^2\right)^{n-\frac{1}{2}} dx &= \frac{1}{2} \int_0^\infty z^{m-\frac{1}{2}} (1-z)^{n-\frac{1}{2}} \, dz, \quad \text{where } z = x^2 \\ &= \frac{1}{2} B\left(m + \frac{1}{2}, n + \frac{1}{2}\right) = \frac{1}{2} \frac{\Gamma(m+\frac{1}{2})\Gamma(n+\frac{1}{2})}{\Gamma(m+n+1)} \\ &= \frac{(2m-1)!!\,(2n-1)!!}{2^{m+n+1}(m+n)!} \pi. \end{aligned} \tag{2}$$

Since

$$\frac{(2m-1)!!}{(2m)!} = \frac{(2m-1)(2m-3)\cdots 3}{2m(2m-1)(2m-2)\cdots 2} = \frac{1}{2^m m!},$$

we get from equations (1) and (2)

$$\begin{aligned} F_c(k) &= \sqrt{\frac{\pi}{2}} (2n-1)!! \sum_{m=0}^\infty (-1)^m \frac{1}{m!\,(m+n)!} \frac{k^{2m}}{2^{n+2m}} \\ &= \sqrt{\frac{\pi}{2}} (2n-1)!! \sum_{m=0}^\infty \frac{(-1)^m}{m!\,(m+n)!} \frac{1}{k^n} \left(\frac{k}{2}\right)^{n+2m} \\ &= \sqrt{\frac{\pi}{2}} (2n-1)!! \frac{1}{k^n} J_n(k), \end{aligned} \tag{3}$$

where $J_n(k)$ is the Bessel function.

Taking the Fourier inverse transform, we have

$$f(x) = \sqrt{\frac{2}{\pi}} \int_0^\infty F_c(k) \cos kx \, dk = (2n-1)!! \int_0^\infty \frac{1}{k^n} J_n(k) \cos kx \, dk. \tag{4}$$

Thus, from the given function and equation (4), we get the value of the integral

$$\int_0^\infty \frac{J_n(z)}{z^n} \cos zx \, dz = \begin{cases} \dfrac{1}{(2n-1)!!}(1-x^2)^{n-\frac{1}{2}}, & \text{for } 0 \le x < 1 \\ 0, & \text{for } x > 1 \end{cases}$$

For $n = 0$, we have

$$\int_0^\infty J_0(z) \cos zx \, dx = (1-x^2)^{-\frac{1}{2}}, \text{ for } 0 \le x \le 1.$$

Note. We have used the notation

$$(2n-1)!! = (2n-1)(2n-3)\cdots 5.3.1 \tag{5}$$

14.7 Fourier Transforms of Derivatives

Let $F(k)$ be the Fourier transform of a function $f(x)$. We shall show that the Fourier transform of $\frac{df}{dx}$ is $-ikF(k)$.

Proof. We have

$$F(k) = \frac{1}{\sqrt{2\pi}} \int_{-\infty}^{\infty} f(x)\, e^{ikx} \, dx \tag{14.42}$$

Then, by definition, the Fourier transform of $f'(x) = \frac{df}{dx}$ is given by

$$F_1(k) = \frac{1}{\sqrt{2\pi}} \int_{-\infty}^{\infty} f'(x)\, e^{ikx} \, dx. \tag{14.43}$$

The integrand in equation (14.43) can be written as

$$\begin{aligned} &= \lim_{h \longrightarrow 0} \frac{f(x+h) - f(x)}{h} e^{ikx} \\ &= \lim_{h \longrightarrow 0} \left[\frac{f(x+h)}{h} e^{ik(x+h)} \cdot e^{-ikh} - \frac{f(x)}{h} e^{ikx} \right]. \end{aligned} \tag{14.44}$$

Substituting equation (14.44) into equation (14.43) and using equation (14.42), we get

$$\begin{aligned} F_1(k) &= \lim_{h \longrightarrow 0} \left[\frac{F(k)\, e^{-ikh}}{h} - \frac{F(k)}{h} \right] \\ &= F(k) \cdot \lim_{h \longrightarrow 0} \left(\frac{e^{-ikh} - 1}{h} \right) \\ &= F(k) \cdot \lim_{h \longrightarrow 0} \left(-ike^{-ikh} \right) \quad \text{[By L'Hospital's rule]} \\ &= -ikF(k). \end{aligned}$$

Hence proved. We can generalise this for the case of nth derivative of a function. Thus the Fourier transform of nth derivative of a function $f(x)$ denoted by $F_n(k)$ is given by

$$F_n(k) = (-ik)^n F(k), \tag{14.45}$$

provided the function $f(x)$ is such that $f(x)$ and all up to its $(n-1)$th derivatives $\longrightarrow 0$ as $x \longrightarrow \pm\infty$. □

14.8 Fourier Sine and Cosine Transforms of Derivatives

Let the Fourier sine and cosine transforms of a function $f(x)$ be given as

$$\left.\begin{aligned} F_s(k) &= \sqrt{\frac{2}{\pi}} \int_0^\infty f(x) \sin kx \, dx \\ F_c(k) &= \sqrt{\frac{2}{\pi}} \int_0^\infty f(x) \cos kx \, dx \end{aligned}\right\}. \tag{14.46}$$

We make the following assumptions:

(i) The function $f(x)$ is such that, $f(x) \longrightarrow 0$ as $x \longrightarrow \infty$;

(ii) All the derivatives of $f(x)$ also $\longrightarrow 0$ as $x \longrightarrow \infty$.

Then, the Fourier sine transform of first derivative of the function $f(x)$ is

$$\begin{aligned} F_{1s}(k) &= \sqrt{\frac{2}{\pi}} \int_0^\infty f'(x) \sin kx \, dx \\ &= \sqrt{\frac{2}{\pi}} \left[f(x) \sin kx \Big|_0^\infty - k \int_0^\infty f(x) \cos kx \, dx \right] \\ &= -kF_c(k). \end{aligned} \tag{14.47}$$

The first integral vanished due to the above assumptions.

Similarly, the Fourier cosine transform of first derivative of the function $f(x)$ is

$$\begin{aligned} F_{1c}(k) &= \sqrt{\frac{2}{\pi}} \int_0^\infty f'(x) \cos kx \, dx \\ &= \sqrt{\frac{2}{\pi}} \left[f(x) \cos kx \Big|_0^\infty + k \int_0^\infty f(x) \sin kx \, dx \right] \end{aligned}$$

$$\text{i.e.,} \quad F_{1c}(k) = kF_s(k) - \sqrt{\frac{2}{\pi}} f(0). \tag{14.48}$$

The Fourier sine transform of second derivative of the function $f(x)$ is

$$\begin{aligned}
F_{2s}(k) &= \sqrt{\frac{2}{\pi}}\int_0^\infty f''(x)\sin kx\,dx \\
&= \sqrt{\frac{2}{\pi}}\left[f'(x)\sin kx\Big|_0^\infty - \int_0^\infty kf'(x)\cos kx\,dx\right] \\
&= -k\sqrt{\frac{2}{\pi}}\int_0^\infty f'(x)\cos kx\,dx \\
&= -kF_{1c}(k) \\
&= -k^2F_s(k) + \sqrt{\frac{2}{\pi}}kf(0) \quad \text{[using equation (14.48)]}
\end{aligned} \tag{14.49}$$

Similarly, Fourier cosine transform of second derivative of $f(x)$ is

$$\begin{aligned}
F_{2c}(k) &= \sqrt{\frac{2}{\pi}}\int_0^\infty f''(x)\cos kx\,dx \\
&= \sqrt{\frac{2}{\pi}}\left[f'(x)\cos kx\,dx + \int_0^\infty kf'(x)\sin kx\,dx\right] \\
&= \sqrt{\frac{2}{\pi}}[-f'(0)] + kF_{1s}(k) \\
&= -\left[k^2F_c(k) + \sqrt{\frac{2}{\pi}}f'(0)\right] \quad \text{[using equation (14.47)]}
\end{aligned} \tag{14.50}$$

Case of Finite Interval. If the function $f(x)$ is defined in the finite interval $(0, \pi)$, then we have the sine and cosine Fourier transforms of $f(x)$ as

$$F_s(k) = \sqrt{\frac{2}{\pi}}\int_0^\pi f(x)\sin kx\,dx \tag{14.51}$$

$$\text{and} \quad F_c(k) = \sqrt{\frac{2}{\pi}}\int_0^\pi f(x)\cos kx\,dx. \tag{14.52}$$

Thus, the finite Fourier sine transform of first derivative of $f(x)$ is

$$\begin{aligned}
F_{1s}(k) &= \sqrt{\frac{2}{\pi}}\int_0^\pi f'(x)\sin kx\,dx \\
&= \sqrt{\frac{2}{\pi}}\left[f(x)\sin kx\Big|_0^\pi - \int_0^\pi kf(x)\cos kx\,dx\right] \\
&= -kF_c(k).
\end{aligned} \tag{14.53}$$

The finite Fourier cosine transform of first derivative of $f(x)$ is

$$\begin{aligned}F_{1c}(k) &= \sqrt{\frac{2}{\pi}}\int_0^\pi f'(x)\cos kx\,dx \\ &= \sqrt{\frac{2}{\pi}}\left[f(x)\cos kx\Big|_0^\pi + \int_0^\pi kf(x)\sin kx\,dx\right] \\ &= \sqrt{\frac{2}{\pi}}\left[(-1)^k f(\pi) - f(0)\right] + kF_s(k). \end{aligned} \tag{14.54}$$

Similarly, the finite Fourier sine transform of second derivative of $f(x)$ is

$$\begin{aligned}F_{2s}(k) &= \sqrt{\frac{2}{\pi}}\int_0^\pi f''(x)\sin kx\,dx \\ &= \sqrt{\frac{2}{\pi}}\left[f'(x)\sin kx\Big|_0^\pi - \int_0^\pi kf'(x)\cos kx\,dx\right] \\ \text{or,}\quad F_{2s}(k) &= -k\sqrt{\frac{2}{\pi}}\left[f(x)\cos kx\Big|_0^\pi + \int_0^\pi kf(x)\sin kx\,dx\right] \\ &= -k^2F_s(k) - k\sqrt{\frac{2}{\pi}}\left[(-1)^k f(\pi) - f(0)\right]. \end{aligned} \tag{14.55}$$

The finite Fourier cosine transform of second derivative of $f(x)$ is

$$\begin{aligned}F_{2c}(k) &= \sqrt{\frac{2}{\pi}}\int_0^\pi f''(x)\cos kx\,dx \\ &= \sqrt{\frac{2}{\pi}}\left[f'(x)\cos kx\Big|_0^\pi + \int_0^\pi kf'(x)\sin kx\,dx\right] \\ &= \sqrt{\frac{2}{\pi}}\left[(-1)^k f'(\pi) - f'(0)\right] + k\sqrt{\frac{2}{\pi}}\left[-\int_0^\pi kf(x)\cos kx\,dx\right] \\ &= \sqrt{\frac{2}{\pi}}\left[(-1)^k f'(\pi) - f'(0)\right] - k^2F_c(k). \end{aligned} \tag{14.56}$$

14.9 Use of Fourier Transforms in Solving Differential Equations

We consider the motion of a damped harmonic oscillator governed by the differential equation

$$\frac{d^2x}{dt^2} + 2\alpha\frac{dx}{dt} + \xi^2 x = f(t), \tag{14.57}$$

where $f(t)$ is any function such that

$$f(t) = \frac{1}{\sqrt{2\pi}}\int_{-\infty}^{\infty} F(\omega)\,e^{-i\omega t}\,d\omega \tag{14.58}$$

and

$$F(\omega) = \frac{1}{\sqrt{2\pi}}\int_{-\infty}^{\infty} f(t)\,e^{i\omega t}\,dt. \tag{14.59}$$

We take Fourier transform of both sides of equation (14.57) and get

$$\frac{1}{\sqrt{2\pi}}\int_{-\infty}^{\infty}\frac{d^2x}{dt^2}e^{i\omega t}\,dt+\frac{2\alpha}{\sqrt{2\pi}}\int_{-\infty}^{\infty}\frac{dx}{dt}e^{i\omega t}\,dt+\frac{\xi^2}{\sqrt{2\pi}}\int_{-\infty}^{\infty}x(t)\,e^{i\omega t}\,dt=\frac{1}{\sqrt{2\pi}}\int_{-\infty}^{\infty}f(t)\,e^{i\omega t}\,dt.$$

Using equations (14.45) and (14.59), we get

$$-\omega^2X(\omega)-2\alpha i\omega X(\omega)+\xi^2X(\omega)=F(\omega).$$

This gives

$$X(\omega)=\frac{F(\omega)}{\xi^2-\omega^2-2\alpha i\omega}. \tag{14.60}$$

where

$$X(\omega)=\frac{1}{\sqrt{2\pi}}\int_{-\infty}^{\infty}x(t)\,e^{i\omega t}\,dt. \tag{14.61}$$

Thus, by taking Fourier inverse transform, we get the solution of equation (14.57) in the form

$$x(t)=\frac{1}{\sqrt{2\pi}}\int_{-\infty}^{\infty}X(\omega)\,e^{-i\omega t}\,d\omega=\frac{1}{\sqrt{2\pi}}\int_{-\infty}^{\infty}\frac{F(\omega)\,e^{-i\omega t}}{\xi^2-\omega^2-2\alpha i\omega}\,d\omega\quad\text{[using equation (14.60)]} \tag{14.62}$$

14.10 Solution of Partial Differential Equations

Partial differential equations can also be solved by using Fourier transforms. We shall discuss a few boundary value problems in this context. Before that, we must define the Fourier transforms of partial derivatives of a function of two variables.

Let $f(x,y)$ be a function of two variables x, y such that

$$\left.\begin{aligned}&\text{(i) } f(x,y)\longrightarrow 0,\ \text{as } x\longrightarrow\pm\infty\\ &\text{and (ii) } \frac{\partial f}{\partial x}\longrightarrow 0,\ \text{as } x\longrightarrow\pm\infty\end{aligned}\right\}. \tag{14.63}$$

Then Fourier transform of $\dfrac{\partial f}{\partial x}$ is given by

$$F_1(k,y)=\frac{1}{\sqrt{2\pi}}\int_{-\infty}^{\infty}\frac{\partial f}{\partial x}e^{ikx}\,dx=-ikF(k,y). \tag{14.64}$$

The Fourier transform of $\dfrac{\partial^2 f}{\partial x^2}$ is given by

$$F_2(k,y)=\frac{1}{\sqrt{2\pi}}\int_{-\infty}^{\infty}\frac{\partial^2 f}{\partial x^2}e^{ikx}\,dx=-k^2F(k,y). \tag{14.65}$$

In general, the Fourier transform of $\dfrac{\partial^n f}{\partial x^n}$ is

$$F_n(k,y)=(-ik)^nF(k,y), \tag{14.66}$$

provided

$$\frac{\partial^{n-1} f}{\partial x^{n-1}} \longrightarrow 0, \text{ as } x \longrightarrow \pm\infty.$$

The corresponding relations for Fourier sine and cosine transforms are

$$F_{1c}(k,y) = \sqrt{\frac{2}{\pi}} \int_0^\infty \frac{\partial f}{\partial x} \cos kx \, dx = kF_s(k,y) - \sqrt{\frac{2}{\pi}} f(0,y). \tag{14.67}$$

$$F_{1s}(k,y) = \sqrt{\frac{2}{\pi}} \int_0^\infty \frac{\partial f}{\partial x} \sin kx \, dx = -kF_c(k,y). \tag{14.68}$$

$$F_{2c}(k,y) = \sqrt{\frac{2}{\pi}} \int_0^\infty \frac{\partial^2 f}{\partial x^2} \cos kx \, dx = -\left[k^2 F_c(k,y) + \sqrt{\frac{2}{\pi}} f'(0,y)\right] \tag{14.69}$$

and

$$F_{2s}(k,y) = \sqrt{\frac{2}{\pi}} \int_0^\infty \frac{\partial^2 f}{\partial x^2} \sin kx \, dx = -k^2 F_s(k,y) + \sqrt{\frac{2}{\pi}} k f(0,y). \tag{14.70}$$

Let us now consider heat flow through an infinite rod which is governed by the *Heat Conduction Equation* in one dimension

$$\frac{\partial u}{\partial t} = h^2 \frac{\partial^2 u}{\partial x^2}, \tag{14.71}$$

where $u(x,t)$ is the temperature function. We assume that both $u(x,t)$ and $\frac{\partial u}{\partial x}$ tend to zero as $x \longrightarrow \pm\infty$.

The initial condition is given by

$$u(x,0) = f(x), \tag{14.72}$$

where the function is such that its Fourier transform exists so that

$$F(k) = \frac{1}{\sqrt{2\pi}} \int_{-\infty}^\infty f(x) \, e^{ikx} \, dx. \tag{14.73}$$

We define the Fourier transform

$$U(k,t) = \frac{1}{\sqrt{2\pi}} \int_{-\infty}^\infty u(x,t) \, e^{ikx} \, dx. \tag{14.74}$$

Now, taking the Fourier transform on both sides of equation (14.71) we get

$$\frac{1}{\sqrt{2\pi}} \int_{-\infty}^\infty \frac{\partial u}{\partial t} e^{ikx} \, dx = \frac{h^2}{\sqrt{2\pi}} \int_{-\infty}^\infty \frac{\partial^2 u}{\partial x^2} e^{ikx} \, dx. \tag{14.75}$$

This gives

$$\frac{d}{dt} U(k,t) = -h^2 k^2 U(k,t). \tag{14.76}$$

Integrating, we get

$$U(k,t) = Ae^{-h^2 k^2 t}. \tag{14.77}$$

Now, from equation (14.74), we have, for $t = 0$,

$$U(k,0) = \frac{1}{\sqrt{2\pi}} \int_{-\infty}^{\infty} f(x)\, e^{ikx}\, dx \quad \text{[using equation (14.72)]}$$
$$= F(k) \quad \text{by equation (14.73).}$$

From equation (14.77), we get for $t = 0$,

$$U(k,0) = A. \tag{14.78}$$

Thus, $A = F(k)$ so that

$$U(k,t) = F(k)\, e^{-h^2k^2t}. \tag{14.79}$$

By taking inverse transform, we get

$$u(x,t) = \frac{1}{\sqrt{2\pi}} \int_{-\infty}^{\infty} \left[F(k)\, e^{-h^2k^2t}\right] e^{-ikx}\, dx. \tag{14.80}$$

Assuming

$$G(k) = e^{-h^2k^2t}, \tag{14.81}$$

we get from equation (14.80) by using convolution property

$$u(x,t) = \frac{1}{\sqrt{2\pi}} \int_{-\infty}^{\infty} f(t)\, g(x-t)\, dt, \tag{14.82}$$

where $g(x-t)$ is the Fourier inverse transform of $G(k)$ given by equation (14.81). Hence

$$g(x) = \frac{1}{\sqrt{2\pi}} \int_{-\infty}^{\infty} e^{-h^2k^2t} \cdot e^{-ikx}\, dk. \tag{14.83}$$

We substitute

$$\xi^2 = 2k^2h^2t, \text{ so that } \xi\, d\xi = 2kh^2t\, dk. \tag{14.84}$$

Then

$$g(x) = \frac{1}{\sqrt{2\pi}} \int_{-\infty}^{\infty} e^{-\xi^2/2} \exp\left[-\frac{i\xi}{h\sqrt{2t}}x\right] \frac{d\xi}{h\sqrt{2t}}$$
$$= \frac{1}{2h\sqrt{\pi t}} \int_{-\infty}^{\infty} e^{-\xi^2/2} e^{-i\alpha\xi}\, d\xi, \quad \text{where } \alpha = \frac{x}{h\sqrt{2t}}$$
$$= \frac{1}{h\sqrt{2t}} e^{-\alpha^2/2} = \frac{1}{h\sqrt{2t}} \exp\left(-\frac{x^2}{4h^2t}\right). \tag{14.85}$$

Hence, from equations (14.82) and (14.85), we have

$$u(x,t) = \frac{1}{\sqrt{2\pi}} \int_{-\infty}^{\infty} f(t) \frac{1}{h\sqrt{2t}} \exp\left[-\frac{(x-t)^2}{4h^2t}\right] dt$$
$$= \frac{1}{2h\sqrt{\pi t}} \int_{-\infty}^{\infty} f(t) \exp\left[-\frac{(x-t)^2}{4h^2t}\right] dt \tag{14.86}$$

which determines the temperature function in the finite bar for specific values of the function $f(x)$.

14.11 Solution of Wave Equation

We now consider the transverse vibration of a very long (infinite) elastic string fixed at both ends. The motion is governed by the one-dimensional wave equation,

$$\frac{\partial^2 \phi}{\partial t^2} = c^2 \frac{\partial^2 \phi}{\partial x^2} \tag{14.87}$$

where the function $\phi(x, t)$ determining the displacement of the string at any instant of time is such that

$$\phi(x,t) \longrightarrow 0 \quad \text{and} \quad \frac{\partial \phi}{\partial x} \longrightarrow 0 \text{ as } x \longrightarrow \pm\infty. \tag{14.88}$$

The initial conditions of motion are

$$\phi(x, 0) = f(x) \quad \text{and} \quad \left.\frac{\partial \phi}{\partial t}\right|_{t=0} = 0. \tag{14.89}$$

We define the following Fourier transform

$$F(k,t) = \frac{1}{\sqrt{2\pi}} \int_{-\infty}^{\infty} \phi(x,t)\, e^{ikx}\, dx. \tag{14.90}$$

Taking the Fourier transform of the wave equation (14.87), we get

$$\frac{1}{\sqrt{2\pi}} \frac{\partial^2}{\partial t^2} \int_{-\infty}^{\infty} \phi(x,t)\, e^{ikx}\, dx = c^2 \frac{1}{\sqrt{2\pi}} \int_{-\infty}^{\infty} \frac{\partial^2 \phi}{\partial x^2} e^{ikx}\, dx$$

$$\text{i.e.,} \quad \frac{d^2}{dt^2} F(k,t) = -c^2 k^2 F(k,t). \tag{14.91}$$

The general solution of this equation is

$$F(k,t) = A_1 e^{ikct} + A_2 e^{-ikct}, \tag{14.92}$$

where A_1, A_2 are unknown constants of integration to be determined from initial conditions in equation (14.89). From the condition $\phi(\dot{x}, 0) = f(x)$, we have from equation (14.90)

$$F(k,0) = \frac{1}{\sqrt{2\pi}} \int_{-\infty}^{\infty} \phi(x,0)\, e^{ikx}\, dx$$

$$\text{i.e.,} \quad A_1 + A_2 = \frac{1}{\sqrt{2\pi}} \int_{-\infty}^{\infty} f(x)\, e^{ikx}\, dx. \tag{14.93}$$

From the second initial condition, we get

$$\left.\frac{dF}{dt}\right|_{t=0} = \frac{1}{\sqrt{2\pi}} \int_{-\infty}^{\infty} \left(\frac{\partial \phi}{\partial t}\right)_{t=0} e^{ikx}\, dx = 0.$$

$$\text{i.e.,} \quad ikc(A_1 - A_2) = 0. \tag{14.94}$$

Solving equations (14.93) and (14.94), we get

$$A_1 = \frac{1}{2} F(k,0), \quad A_2 = \frac{1}{2} F(k,0). \tag{14.95}$$

Hence the solution equation (14.92) assumes the form

$$F(k,t) = \frac{1}{2}F(k,0)\left[e^{ikct} + e^{-ikct}\right]. \tag{14.96}$$

By taking the Fourier inverse transform of equation (14.90), we get the solution of the wave equation as

$$\phi(x,t) = \frac{1}{\sqrt{2\pi}}\int_{-\infty}^{\infty} F(k,t)\, e^{-ikt}\, dk = \frac{1}{\sqrt{2\pi}}\int_{-\infty}^{\infty} \frac{1}{2}F(k,0)\left[e^{ikct} + e^{-ikct}\right] e^{-ikx}\, dk$$

$$\text{or,}\quad \phi(x,t) = \frac{1}{\sqrt{2\pi}}\int_{-\infty}^{\infty} \frac{1}{2}F(k,0)\, e^{ik(ct-x)}\, dk + \frac{1}{\sqrt{2\pi}}\int_{-\infty}^{\infty} \frac{1}{2}F(k,0)\, e^{-ik(ct+x)}\, dk$$

$$= \frac{1}{2}[f(x-ct) + f(x+ct)] \quad \text{[using equation (14.90) for } t = 0]. \tag{14.97}$$

Here $f(x-ct)$ represents a wave travelling in the positive x-direction with speed c. At time $t + \Delta t$, position x becomes $x + c\Delta t$, then

$$\phi(x + c\Delta t, t + \Delta t) = f[x + c\Delta t - c(t + \Delta t)] = f(x - ct), \tag{14.98}$$

which shows that the wave repeats itself at $t + \Delta t$ and $x + \Delta x$. Similarly, $f(x+ct)$ represents a wave travelling in the negative $-x$ direction with speed c and repeats itself at $t + \Delta t$ and $x - \Delta x$.

14.12 Solution of Poisson's Equation

We consider three-dimensional Poisson's equation in vector form for the electrostatic potential function, that is

$$\nabla^2 \phi(\mathbf{r}) = -\frac{\rho(\mathbf{r})}{\epsilon}, \tag{14.99}$$

where $\phi(\mathbf{r})$ = potential function and $\rho(\mathbf{r})$ = charge density.

We define the following Fourier transforms in three-dimensions

$$\phi(\mathbf{k}) = \frac{1}{(2\pi)^{3/2}}\int_{-\infty}^{\infty} \phi(\mathbf{r})\, e^{i\mathbf{k}\cdot\mathbf{r}}\, d^3r, \tag{14.100}$$

where $d^3r = dx\, dy\, dz$.

The Fourier inverse transform of $\phi(\mathbf{k})$ is given by

$$\phi(\mathbf{r}) = \frac{1}{(2\pi)^{3/2}}\int_{-\infty}^{\infty} \phi(\mathbf{k})\, e^{-i\mathbf{k}\cdot\mathbf{r}}\, d^3k, \tag{14.101}$$

where d^3k represents triple integral in k-space. Now, taking Fourier transform of both sides of equation (14.99), we get

$$-k^2\phi(\mathbf{k}) = -\frac{p(\mathbf{k})}{\epsilon} \tag{14.102}$$

where we have assumed that

$$\phi(\mathbf{r}) \longrightarrow 0 \quad \text{and} \quad \frac{\partial \phi}{\partial \mathbf{r}} \longrightarrow 0, \quad \text{as } \mathbf{r} \longrightarrow \pm\infty. \tag{14.103}$$

Thus, equation (14.102) yields

$$\phi(\mathbf{k}) = \frac{p(\mathbf{k})}{k^2 \epsilon}, \tag{14.104}$$

where

$$p(\mathbf{k}) = \frac{1}{(2\pi)^{3/2}} \int_{-\infty}^{\infty} \rho(\mathbf{r}') e^{i\mathbf{k}\cdot\mathbf{r}'} d^3 r'. \tag{14.105}$$

Taking Fourier inverse transform of equation (14.104), we get the solution of Poisson's equation as

$$\begin{aligned} \phi(\mathbf{r}) &= \frac{1}{(2\pi)^{3/2}} \int_{-\infty}^{\infty} \frac{p(\mathbf{k})}{k^2 \epsilon} e^{-i\mathbf{k}\cdot\mathbf{r}} \, d^3 k \\ &= \frac{1}{(2\pi)^3} \int_{-\infty}^{\infty} \int_{-\infty}^{\infty} \frac{1}{k^2 \epsilon} \rho(\mathbf{r}') \, e^{-i\mathbf{k}\cdot(\mathbf{r}-\mathbf{r}')} d^3 k d^3 r'. \end{aligned} \tag{14.106}$$

Let us consider the integral

$$I = \int_{-\infty}^{\infty} \frac{1}{k^2} e^{-i\mathbf{k}\cdot(\mathbf{r}-\mathbf{r}')} d^3 k, \tag{14.107}$$

in k-space in spherical polar coordinates. Then

$$d^3 k = dx\, dy\, dz = k^2 \sin\theta \, d\theta d\phi \, dk. \tag{14.108}$$

Thus

$$\begin{aligned} I &= \int_0^{\infty} \int_0^{\pi} \int_0^{2\pi} \frac{1}{k^2} e^{-ik|\mathbf{r}-\mathbf{r}'|\cos\theta} . k^2 \sin\theta \, d\theta \, d\phi \, dk \\ &= 2\pi \int_0^{\infty} \int_0^{\pi} e^{-ik|\mathbf{r}-\mathbf{r}'|\cos\theta} \sin\theta \, d\theta \, dk. \end{aligned} \tag{14.109}$$

We substitute

$$\left.\begin{aligned} -ik\,|\,\mathbf{r}-\mathbf{r}'\,|\cos\theta &= z \\ \text{so that, } ik\,|\,\mathbf{r}-\mathbf{r}'\,|\sin\theta \, d\theta &= dz \end{aligned}\right\}. \tag{14.110}$$

Then integral in equation (14.109) takes the form

$$\begin{aligned} I &= 2\pi \left[\int \frac{e^z \, dz}{ik\xi} \right] dk, \quad \text{where } \xi = |\,\mathbf{r}-\mathbf{r}'\,| \\ &= 2\pi \int_0^{\infty} \left[\frac{e^{ik\xi} - e^{-ik\xi}}{ik\xi} \right] dk \\ &= 2\pi \int_0^{\infty} \frac{2\sin k\xi}{k\xi} dk = \frac{4\pi}{\xi} \int_0^{\infty} \frac{\sin k\xi}{k} dk \\ &= \frac{4\pi}{\xi} \cdot \frac{\pi}{2} = \frac{2\pi^2}{\xi} = \frac{2\pi^2}{|\,\mathbf{r}-\mathbf{r}'\,|}. \end{aligned} \tag{14.111}$$

Hence, from equations (14.106) and (14.111) we get

$$\phi(\mathbf{r}) = \frac{1}{(2\pi)^3 \epsilon} \int_{-\infty}^{\infty} \rho(\mathbf{r}') \frac{2\pi^2}{|\,\mathbf{r}-\mathbf{r}'\,|} d^3 r' = \frac{1}{4\pi\epsilon} \int_{-\infty}^{\infty} \frac{\rho(\mathbf{r}')}{|\,\mathbf{r}-\mathbf{r}'\,|} d^3 r' \tag{14.112}$$

which is in accordance with definition of potential.

14.13 Schrödinger Equation

We consider the motion of a free particle in one-dimension. The wave function $\psi(x,t)$ of the particle satisfies the Schrödinger differential equation

$$i\hbar\frac{\partial\psi}{\partial t}+\frac{\hbar^2}{2m}\frac{\partial^2\psi}{\partial x^2}=0, \tag{14.113}$$

where m = mass of the particle, $\hbar = h/2\pi$, h = Planck's constant.

The equation (14.113) is satisfied by a function

$$\psi(x,t)=Ae^{i(kx-\omega t)}, \tag{14.114}$$

where ω and k are constants, and ω = angular frequency, $k=\frac{2\pi}{\lambda}$ = magnitude of wave vector.

Now, if we substitute equation (14.114) into (14.113), we find that

$$\hbar\omega=\frac{\hbar^2k^2}{2m}, \tag{14.115}$$

which represents energy of a free particle given by

$$E=\frac{p^2}{2m},\quad p=\text{momentum} \tag{14.116}$$

in terms of Einstein's and de Broglie's following relations for matter waves

$$E=\hbar\omega \quad\text{and}\quad p=\hbar k. \tag{14.117}$$

Now, equation (14.114) represents a particular solution for a value of k and corresponding value of ω determined from equation (14.115). Hence we write this solution as

$$\psi_k(x,t)=A(k)\,e^{i(kx-\omega_k t)}. \tag{14.118}$$

The general solution of Schrödinger equation shall be a linear combination of all such solutions. Thus we write

$$\psi(x,t)=\int A(k)\,e^{i(kx-\omega_k t)}\,dk. \tag{14.119}$$

Knowing $A(k)$, we can find the value of $\psi(x,t)$ at any point and at any instant of time. Similarly, if we know $\psi(x,t)$, we can find $A(k)$ by the use of Fourier transform.

We write

$$\psi(x,t)=\frac{1}{\sqrt{2\pi}}\int A(k)\,e^{i(kx-\omega_k t)}\,dk=\frac{1}{\sqrt{2\pi}}\int\left[A(k)\,e^{-i\omega_k t}\right]e^{ikx}\,dx. \tag{14.120}$$

Hence, by inverse Fourier transform

$$A(k)\,e^{-i\omega_k t}=\frac{1}{\sqrt{2\pi}}\int\psi(x,t)\,e^{-ikx}\,dx,$$

so that

$$A(k)=\frac{1}{\sqrt{2\pi}}\int\psi(x,t)\,e^{-i(kx-\omega_k t)}\,dx. \tag{14.121}$$

$A(k)$ is called the *momentum wave function*, since the probability that the particle's momentum lies between $\hbar k$ and $\hbar(k+\Delta k)$ is

$$|A(k)|^2\,dk. \tag{14.122}$$

It can be proved that

$$\int |A(k)|^2\,dk = 1, \tag{14.123}$$

which shows that the momentum wave function $A(k)$, which is also the amplitude of particular plane wave corresponding to a definite k, is normalised to unity.

We can extend the above discussions in three dimensions. Thus, we can write

$$\psi(\mathbf{r},t) = \frac{1}{(2\pi)^{3/2}}\int A(\mathbf{k})\,e^{i(\mathbf{k}\cdot\mathbf{r}-\omega_k t)}\,d\mathbf{k}, \tag{14.124}$$

where $d\mathbf{k} = dk_x\,dk_y\,dk_z$.

The momentum wave function $A(\mathbf{k})$ is then obtained by inverse Fourier transform as

$$A(\mathbf{k}) = \frac{1}{(2\pi)^{3/2}}\int \psi(\mathbf{r},t)\,e^{-i(\mathbf{k}\cdot\mathbf{r}-\omega_k t)}\,d\mathbf{r}. \tag{14.125}$$

As an example, we consider a hydrogen atom in the ground state for which the wave function is given by

$$\psi(\mathbf{r}) = \left(\frac{1}{\pi a^3}\right)^{1/2} e^{-\mathbf{r}/a}, \tag{14.126}$$

where a = radius of first Bohr orbit.

Then the corresponding momentum wave function will be, by equation (14.125)

$$A(\mathbf{k}) = \frac{1}{(2\pi)^{3/2}}\left(\frac{1}{\pi a^3}\right)^{1/2}\int e^{-\mathbf{r}/a}e^{-i\mathbf{k}\cdot\mathbf{r}}\,d\mathbf{r}. \tag{14.127}$$

We choose k along z-axis, since r represents radial distance. Then

$$\mathbf{k}\cdot\mathbf{r} = kr\cos\theta. \tag{14.128}$$

Taking spherical polar coordinates, we have

$$d\mathbf{r} = r^2\sin\theta\,d\theta\,d\phi\,dr. \tag{14.129}$$

With these substitutions, equation (14.127) assumes the form,

$$\begin{aligned} A(\mathbf{k}) &= \frac{1}{\pi^2(2a)^{3/2}}\int_0^\infty\int_0^\pi\left[\exp\left(-\frac{r}{a}-ikr\cos\theta\right)r^2\right]dr\sin\theta\,d\theta\int_0^{2\pi}d\phi \\ &= \frac{2\pi}{\pi^2(2a)^{3/2}}\int_0^\infty\int_0^\pi \exp\left(-\frac{r}{a}\right)\exp(-ikr\cos\theta)r^2\sin\theta\,d\theta\,dr. \end{aligned} \tag{14.130}$$

We substitute $-ikr\cos\theta = z$, so that $ikr\sin\theta\,d\theta = dz$.

Hence, we get from equation (14.130)

$$\begin{aligned} A(\mathbf{k}) &= \frac{2}{\pi(2a)^{3/2}} \int_0^\infty \exp\left(-\frac{r}{a}\right) r\, dr \int \frac{e^z\, dz}{ik} \\ &= \frac{2}{\pi(2a)^{3/2}} \int_0^\infty \exp\left(-\frac{r}{a}\right) r\, dr \left[\frac{e^{-ikr\cos\theta}}{ik}\right]_0^\pi \\ &= \frac{4}{\pi k(2a)^{3/2}} \int_0^\infty \left(re^{-r/a}\right) \sin kr\, dr. \end{aligned} \tag{14.131}$$

Integrating by parts, and assuming two standard integrals,

$$\left.\begin{aligned} \int_0^\infty e^{-ax} \cos bx\, dx &= \frac{a}{a^2+b^2} \\ \text{and} \quad \int_0^\infty e^{-ax} \sin bx\, dx &= \frac{b}{a^2+b^2} \end{aligned}\right\}. \tag{14.132}$$

we finally get

$$A(\mathbf{k}) = \frac{4}{\pi k(2a)^{3/2}} \cdot \frac{2a^3 k}{\left(1+a^2k^2\right)^2} = \frac{(2a)^{3/2}}{\pi} \cdot \frac{1}{\left(1+a^2k^2\right)^2}. \tag{14.133}$$

EXERCISE

1. Find the Fourier cosine transform of $f(x) = \dfrac{1}{1+x^2}$.

2. Find the Fourier sine transform of $f(x) = \dfrac{e^{\lambda x}}{x}$.

3. Find the Fourier transform of

$$f(x) = \begin{cases} x, & \text{for } |x| < a \\ 0, & \text{for } |x| > a. \end{cases}$$

4. Calculate the Fourier transform of

$$f(t) = \begin{cases} \exp(-i\omega_0 t), & \text{for } -\dfrac{\alpha}{2} \le t \le \dfrac{\alpha}{2} \\ 0, & \text{for } |t| > \dfrac{\alpha}{2}. \end{cases}$$

5. Find the Fourier transform of

$$f(x) = \exp\left[-\alpha\,|x| + i\beta x\right].$$

6. Find the Fourier transform of

$$f(x) = \begin{cases} (1-x^2), & \text{for } |x| < 1 \\ 0, & \text{for } |x| > 1. \end{cases}$$

Hence evaluate the integral

$$\int_0^\infty \left(\frac{x \cos x - \sin x}{x^3} \right) \cdot \cos x \, dx.$$

7. Using the finite cosine transform, solve

$$\frac{\partial \phi}{\partial t} = \alpha \frac{\partial^2 \phi}{\partial x^2}, \quad \alpha = \text{constant}, \quad 0 < x < \pi$$

with the boundary conditions (a) $\phi(x, 0) = f(x)$ and (b) $\dfrac{\partial \phi}{\partial x} = 0$ at $x = 0$ and $x = \pi$.

8. Using the Fourier transform, solve the equation

$$\frac{\partial T}{\partial t} = h^2 \frac{\partial^2 T}{\partial x^2}$$

assuming that

$$T(x, y) \longrightarrow 0 \quad \text{and} \quad \frac{\partial T}{\partial x} \longrightarrow 0 \text{ as } x \longrightarrow \pm\infty$$

and the initial condition

$$T(x, 0) = \exp(-\alpha^2 x^2), \quad \alpha = \text{constant} > 0.$$

9. By use of the Fourier transform, solve the equation

$$\frac{\partial^2 u}{\partial t^2} = c^2 \frac{\partial^2 u}{\partial x^2},$$

when $u(x, t) \longrightarrow 0$ and $\dfrac{\partial u}{\partial x} \longrightarrow 0$ as $\longrightarrow \pm\infty$ with the initial conditions

$$u(x, 0) = f(x) \quad \text{and} \quad \left.\frac{\partial u}{\partial t}\right|_{t=0} = g(x).$$

10. By use of the Fourier transform, solve the Laplace's equation

$$\frac{\partial^2 \phi}{\partial x^2} + \frac{\partial^2 \phi}{\partial y^2} = 0,$$

when $\phi(x, y) \longrightarrow 0$ and $\dfrac{\partial \phi}{\partial x} \longrightarrow 0$ as $x \longrightarrow < \infty$ and $\phi(x, 0) = f(x)$ given.

11. By using Fourier sine transform, show that the solution of

$$\frac{\partial u}{\partial t} = \alpha^2 \frac{\partial^2 u}{\partial x^2}, \quad 0 < x < \infty, \quad 0 < t \leq \infty$$

with initial and boundary conditions $u(0, t) = A$, $u(x, 0) = 0$ is expressed as $Aerfc(x/2\alpha\sqrt{t})$.

12. Solve the integral equation

$$\int_0^\infty g(x) \sin kx = f(x)$$

where

$$f(x) = \begin{cases} \frac{\pi}{2} \sin x, & \text{for } 0 \le x \le \pi \\ 0, & \text{for } x > \pi \end{cases}$$

ANSWERS

1. $\sqrt{\frac{\pi}{2}} e^{-x}$.
2. $\sqrt{\frac{2}{\pi}} \tan^{-1} \frac{k}{\lambda}$.
3. $\sqrt{\frac{2}{\pi}} \frac{i}{k^2}(ak \cos ak - \sin ak)$.
4. $\frac{\alpha}{\sqrt{2\pi}} \frac{\sin \xi}{\xi}, \quad \xi = (k - \omega)\frac{\alpha}{2}$.
5. $\frac{1}{\sqrt{2\pi}} \frac{2\alpha}{\alpha^2 + (\beta - k)^2}$.
6. $\frac{4}{\sqrt{2\pi} k^3}(\sin k - k \cos k)$; $\frac{\pi}{4}(1 - x^2)$ for $|x| < 1$.
7. $\frac{\sin \pi x}{1 - x^2}$.

Chapter 15

Laplace Transforms

15.1 Definitions

A function $f(s)$ defined by the integral

$$f(s) = \int_0^\infty e^{-st} F(t) dt \tag{15.1}$$

is said to be the Laplace transforms of the function $F(t)$ and usually denoted as

$$f(s) = L[F(t)]. \tag{15.2}$$

The Laplace transform of $F(t)$ is said to exist if the integral in equation (15.1) converges for some value of s. Thus, for the existence of the Laplace transforms of a function $F(t)$, we impose on it two conditions, namely:

(1) The function $F(t)$ should be piecewise continuous. That is, the function $F(t)$ defined in the interval should be such that the interval can be sub-divided into finite number of intervals in each of which the function $F(t)$ is continuous and has finite right and left hand limits.

(2) The function $F(t)$ should be of exponential order. That is, there exists positive constants M and a, such that

$$|F(t)| \leq Me^{at} \text{ for all } t \geq 0. \tag{15.3}$$

Functions satisfying this condition are said to be of *exponential order as* $t \longrightarrow \infty$.

We now prove that the above conditions are sufficient.

15.2 Existence Theorem

If a function $F(t)$ is piecewise continuous in every finite interval $0 \leq t \leq t_0$ and is of exponential order a, the Laplace transforms exist for all $s > a$.

Proof. We have

$$L[F(t)] = \int_0^\infty e^{-st} F(t) dt = \int_0^{t_0} e^{-st} F(t) dt + \int_{t_0}^\infty e^{-st} F(t) dt. \tag{15.4}$$

The first integral in equation (15.4) exists since the function $F(t)$ is piecewise continuous in $0 \leq t \leq t_0$.

The second integral in equation (15.4) can be written as

$$\begin{aligned}\int_{t_0}^{\infty} e^{-st}F(t)dt &\leq \int_{t_0}^{\infty} \left| e^{-st}F(t) \right| dt \leq \int_{t_0}^{\infty} e^{-st} \left| F(t) \right| dt \\ &\leq \int_{t_0}^{\infty} e^{-st}Me^{at}dt, \quad \text{[since } F(t) \text{ is of exponential order]} \\ &= \frac{Me^{-(s-a)t_0}}{s-a}, \qquad s > a,\end{aligned}$$

which exists for $s > a$.

Thus, we conclude that the Laplace transforms exist for all $s > a$. □

15.3 Properties of Laplace Transforms

1. *Linearity property.* For a pair of constants a and b,

$$L[aF(t) + bG(t)] = aL[F(t)] + bL[G(t)]. \tag{15.5}$$

That is, the Laplace transforms operator is linear.

Proof. We have, by definition

$$\begin{aligned}L[aF(t) + bG(t)] &= \int_0^{\infty} [aF(t) + bG(t)]e^{-st}dt \\ &= a\int_0^{\infty} F(t)e^{-st}dt + b\int_0^{\infty} G(t)e^{-st}dt \\ &= aL[F(t)] + bL[G(t)].\end{aligned} \tag{15.6}$$

□

2. *First shifting property.* If $f(s)$ be the Laplace transform of $F(t)$, then the Laplace transform of $e^{\xi t}F(t)$ is $f(s-\xi)$.

Proof. We have, by definition

$$f(s) = L[F(t)] = \int_0^{\infty} e^{-st}F(t)dt. \tag{15.7}$$

Then

$$L\left[e^{\xi t}F(t)\right] = \int_0^{\infty} e^{\xi t}F(t)e^{-st}dt = \int_0^{\infty} e^{-(s-\xi)t}F(t)dt. \tag{15.8}$$

We put

$$s - \xi = p. \tag{15.9}$$

Then, from equation (15.8), we have

$$\begin{aligned}L\left[e^{\xi t}F(t)\right] &= \int_0^{\infty} e^{-p}F(t)dt = f(p), \quad \text{by definition} \\ &= f(s-\xi), \quad \text{by equation (15.9).}\end{aligned} \tag{15.10}$$

□

3. *Second shifting property.* If Laplace transform of $F(t)$ be $f(s)$, and if a function $G(t)$ is defined as

$$G(t) = \begin{cases} F(t-a), & t > a \\ 0, & t < a \end{cases} \tag{15.11}$$

then

$$L[G(t)] = e^{-as} L[F(t)]. \tag{15.12}$$

Proof. We have, by definition

$$\begin{aligned} L[G(t)] &= \int_0^\infty e^{-st} G(t)dt = \int_0^a e^{-st} G(t)dt + \int_a^\infty e^{-st} G(t)dt \\ &= \int_a^\infty e^{-st} F(t-a)dt, \quad \text{by equation (15.11).} \end{aligned} \tag{15.13}$$

We put $t = a + \xi$.
Then, from equation (15.13), we get

$$\begin{aligned} L[G(t)] &= \int_0^\infty e^{-s(a+\xi)} F(\xi)d\xi = e^{-sa} \int_0^\infty e^{-s\xi} F(\xi)d\xi \\ &= e^{-sa} L[F(t)], \quad \text{by definition.} \end{aligned}$$

□

4. *Scale change property.* If Laplace transform of $F(t)$ is $f(s)$, then Laplace transform of $F(at)$ is $\frac{1}{a} f\left(\frac{s}{a}\right)$.

Proof. We write, from definition,

$$\begin{aligned} L[F(at)] &= \int_0^\infty e^{-st} F(at)dt = \frac{1}{a} \int_0^\infty e^{-(s/a)\xi} F(\xi)d\xi, \quad \text{where } \xi = at \\ &= \frac{1}{a} f\left(\frac{s}{a}\right). \end{aligned} \tag{15.14}$$

We now consider some examples by direct method. □

EXAMPLE 15.1. *Find the Laplace transforms of the functions:* (a) $F(t) = 1$, (b) $F(t) = e^{at}$, (c) $F(t) = t^n$.

Solution:

(a) $L[F(t)] = \int_0^\infty e^{-st} F(t)dt = \int_0^\infty e^{-st} dt = -\frac{1}{s} e^{-st}\Big|_0^\infty = \frac{1}{s}.$

(b) $L[F(t)] = \int_0^\infty e^{-st} e^{at} dt = \int_0^\infty e^{-(s-a)t} dt = -\frac{1}{s-a} e^{-(s-a)t}\Big|_0^\infty = \frac{1}{s-a}, \; s > a.$

(c) $L[F(t)] = \int_0^\infty e^{-st} t^n dt.$

We substitute $\xi = st$.

Then

$$L[F(t)] = s^{\frac{1}{n+1}} \int_0^\infty \xi^n e^{-\xi} d\xi = s^{\frac{1}{n+1}} \Gamma(n+1). \tag{1}$$

If n is an integer, $\Gamma(n+1) = n!$
Hence equation (1) becomes

$$L[t^n] = \frac{n!}{s^{n+1}}.$$

EXAMPLE 15.2. *Find the Laplace transforms of the functions* (a) $\sin(at)$, (b) $\cosh(at)$.

Solution:

$$\text{(a) } L[\sin(at)] = \int_0^\infty e^{-st}\sin(at)dt = \left[\frac{e^{-st}}{a^2+s^2}\{-s\sin(at) - a\cos(at)\}\right]_0^\infty = \frac{a}{a^2+s^2},\ s > 0.$$

$$\text{(b) } L[\cosh(at)] = L\left[\frac{1}{2}\left(e^{at} + e^{-at}\right)\right] = \frac{1}{2}\left[\frac{1}{s-a} + \frac{1}{s+a}\right], \text{ [using result of example 15.1(b)]}$$

$$= \frac{s}{s^2-a^2}, \quad s > |a|.$$

EXAMPLE 15.3. *Find Laplace transform of the function*

$$F(t) = \begin{cases} \sin t, & 0 < t < \pi \\ 0, & t > \pi. \end{cases}$$

Solution:

$$L[F(t)] = \int_0^\infty e^{-st}F(t)dt = \int_0^\pi e^{-st}\sin t\, dt \quad [\because F(t) = 0, \text{ for } t > \pi]$$

$$= \left[\frac{e^{-st}}{s^2+1}(-s\sin t - \cos t)\right]_0^\pi = \frac{e^{-s\pi}+1}{s^2+1}.$$

We now consider some examples where the properties of Laplace transforms are used.

EXAMPLE 15.4. *Find Laplace transforms of the functions* (a) $(t^3+1)^2$; (b) $2e^{3t} - e^{-3t}$.

Solution:

$$\text{(a) } L\left[\left(t^3+1\right)^2\right] = L\left[t^6 + 2t^3 + 1\right] = L[t^6] + 2L[t^3] + L[1] = \frac{6!}{s^7} + 2\frac{3!}{s^4} + \frac{1}{s} = \frac{720}{s^7} + \frac{12}{s^4} + \frac{1}{s}.$$

$$\text{(b) } L\left[2e^{3t} - e^{-3t}\right] = 2L\left[e^{3t}\right] - L\left[e^{-3t}\right] = 2\frac{1}{s-3} - \frac{1}{s+3} = \frac{s+9}{s^2-9},\ s > 3.$$

EXAMPLE 15.5. *Find Laplace transforms of the functions:* (a) $(\sin^2 t)e^t$; (b) $e^{-3t}\cosh(2t)$.

Solution:

$$\text{(a) } L[\sin^2 t] = L\left[\frac{1}{2}(1 - \cos 2t)\right] = \frac{1}{2s} - \frac{1}{2}L[\cos 2t] = \frac{1}{2s} - \frac{1}{2}\frac{s}{s^2+4} = \frac{2}{s(s^2+4)}.$$

Hence

$$L\left[e^t \sin^2 t\right] = \frac{2}{(s-1)\left[(s-1)^2+4\right]}, \quad \text{by first shifting property.}$$

(b) $L[\cosh(2t)] = \dfrac{s}{s^2-4}$ [by example 15.2(b)]

Hence

$$L\left[e^{-3t}\cosh(2t)\right] = \frac{s+3}{(s+3)^2-4} \quad \text{[by first shifting property]}$$

$$= \frac{s+3}{s^2+6s+5}.$$

Theorem 15.1. *If $f(s)$ is the Laplace transform of the function $F(t)$, then the Laplace transform of $t^n F(t)$ is given by*

$$L\left[t^n F(t)\right] = (-1)^n \frac{d^n}{ds^n} f(s) \tag{15.15}$$

where n is a positive integer.

Proof. We have, by definition

$$L\left[t^n F(t)\right] = \int_0^\infty e^{-st} t^n F(t) dt. \tag{15.16}$$

Now, we can write

$$e^{-st} t^n = (-1)^n \frac{d^n}{ds^n}\left(e^{-st}\right). \tag{15.17}$$

Hence, in view of equation (15.17), equation (15.16) assumes the form

$$L\left[t^n F(t)\right] = \int_0^\infty (-1)^n \frac{d^n}{ds^n}\left(e^{-st}\right) F(t) dt = (-1)^n \frac{d^n}{ds^n} f(s).$$

Hence proved. □

For $n = 1$, we have

$$L[tF(t)] = -\frac{d}{ds} f(s).$$

EXAMPLE 15.6. *Find Laplace transform of $F(t) = t\cos \alpha t$.*

Solution: We have

$$L[\cos \alpha t] = \frac{s}{s^2+\alpha^2}.$$

Hence, by above theorem

$$L[t\cos \alpha t] = -\frac{d}{ds}\frac{s}{(s^2+\alpha^2)} = \frac{s^2-\alpha^2}{(s^2+\alpha^2)^2}.$$

Theorem 15.2. *If Laplace transform of $F(t)$ is $f(s)$, then the Laplace transform of $\dfrac{1}{t}F(t)$ is given by*

$$L\left[\frac{1}{t}F(t)\right] = \int_s^\infty f(x) dx. \tag{15.18}$$

Proof. We have, by definition

$$f(s) = \int_0^\infty e^{-st} F(t) dt. \tag{15.19}$$

Changing the variable s to ξ, we can write equation (15.19) as

$$f(\xi) = \int_0^\infty e^{-\xi t} F(t) dt. \tag{15.20}$$

Integrating both sides with respect to ξ from $\xi = s$ to $\xi = \infty$, we get

$$\int_s^\infty f(\xi) d\xi = \int_s^\infty d\xi \int_0^\infty e^{-\xi t} F(t) dt. \tag{15.21}$$

We can interchange the order of integration in the RHS of equation (15.21), since the variables ξ and t are independent. Hence,

$$\begin{aligned}\int_s^\infty f(\xi) d\xi &= \int_0^\infty F(t) dt \int_s^\xi e^{-\xi t} d\xi = -\int_0^\infty F(t) dt \left(\frac{e^{-\xi t}}{t} \right)_s^\infty \\ &= \int_0^\infty F(t) dt \frac{e^{-st}}{s} = \int_0^\infty e^{-st} \left\{ \frac{F(t)}{t} \right\} dt = L\left[\frac{F(t)}{t} \right].\end{aligned}$$

Hence, the theorem is proved. □

EXAMPLE 15.7. *Find the Laplace transform of the function*

$$F(t) = \frac{1}{t} \left(e^{-\alpha t} - e^{-\beta t} \right).$$

Solution: We have

$$L\left[e^{-\alpha t} \right] = \frac{1}{s+\alpha}.$$

Thus, by above theorem

$$L\left[\frac{1}{t} \left(e^{-\alpha t} - e^{-\beta t} \right) \right] = \int_s^\infty \left(\frac{1}{x+\alpha} - \frac{1}{x+\beta} \right) dx = \log \frac{x+\alpha}{x+\beta} \bigg|_s^\infty = \log \frac{s+\beta}{s+\alpha}.$$

Theorem 15.3. *If $F(t)$ is a periodic function with period T, then*

$$L[F(t)] = \frac{1}{1-e^{-sT}} \int_0^T e^{-st} F(t) dt. \tag{15.22}$$

Proof. We have, by definition

$$\begin{aligned}L[F(t)] &= \int_0^\infty e^{-st} F(t) dt = \int_0^T e^{-st} F(t) dt + \int_T^{2T} e^{-st} F(t) dt + \cdots \\ &= \sum_{n=0}^\infty \int_{nT}^{(n+1)T} e^{-st} F(t) dt, \quad s > 0.\end{aligned} \tag{15.23}$$

We substitute $\xi = t - nT$, so that $F(\xi + nT) = F(\xi)$, since $F(t)$ is periodic. Then from equation (15.23), we get

$$L[F(t)] = \sum_{n=0}^{\infty} e^{-nTs} \int_0^T e^{-s\xi} F(\xi) d\xi = (1 + e^{-sT} + e^{-2sT} + \cdots) \int_0^T e^{-s\xi} F(\xi) d\xi$$

$$= \frac{1}{1 - e^{-sT}} \int_0^T e^{-st} F(t) dt$$

Hence proved.

EXAMPLE 15.8. *Find Laplace transform of the function defined by*

$$F(t) = \begin{cases} \sin t, & 0 < t < \pi \\ 0, & \pi < t < 2\pi. \end{cases}$$

Solution: The given function is periodic with a period of 2π. Hence, by above theorem,

$$L[F(t)] = \frac{1}{1 - e^{-2\pi s}} \int_0^{2\pi} e^{-st} F(t) dt$$

$$\text{or,} \quad L[F(t)] = \frac{1}{1 - e^{-2\pi s}} \int_0^{\pi} e^{-st} \sin t \, dt. \tag{1}$$

Now,

$$\int_0^{\pi} e^{-st} \sin t \, dt = \left[-\frac{1}{s} e^{-st} \sin t \Big|_0^{\pi} + \frac{1}{s} \int_0^{\pi} e^{-st} \cos t \, dt \right] = \left[\frac{1}{s} \int_0^{\pi} e^{-st} \cos t \, dt \right]$$

$$= \frac{1}{s} \left[-\frac{1}{s} e^{-st} \cos t \Big|_0^{\pi} - \frac{1}{s} \int_0^{\pi} e^{-st} \sin t \, dt \right].$$

Hence

$$\left(1 + \frac{1}{s^2}\right) \int_0^{\pi} e^{-st} \sin t \, dt = \frac{1}{s^2} \left[e^{-\pi s} + 1 \right].$$

Thus

$$\int_0^{\pi} e^{-st} \sin t \, d\xi = \frac{1 + e^{-\pi s}}{s^2 + 1}. \tag{2}$$

From equations (1) and (2), we get

$$L[F(t)] = \frac{1 + e^{-\pi s}}{(1 - e^{-2\pi s})(s^2 + 1)} = \frac{1}{(1 - e^{-\pi s})(s^2 + 1)}.$$

15.4 Laplace Transforms of Derivatives

We consider a function $F(t)$ which has a Laplace transform $f(s)$ and the derivative $F'(t)$ which is piecewise continuous. The Laplace transform of derivative $F'(t)$ is defined as

$$L[F'(t)] = \int_0^{\infty} e^{-st} F'(t) dt. \tag{15.24}$$

Theorem 15.4. *If a function $F(t)$ is continuous and has a Laplace transform $f(s)$, then*

$$L[F'(t)] = sL[F(t)] - F(0), \tag{15.25}$$

where $F(0)$ = value of $F(t)$ for $t = 0$.

Proof. By definition, we have

$$L[F'(t)] = \int_0^{\infty} e^{-st} F'(t) dt. \tag{15.26}$$

Integrating by parts, we get

$$L[F'(t)] = e^{-st}F(t)\Big|_0^{\infty} + s\int_0^{\infty} e^{-st}F(t)dt. \tag{15.27}$$

We assume that the quantity

$$e^{-st}F(t) \longrightarrow 0 \text{ as } t \longrightarrow \infty. \tag{15.28}$$

In view of equation (15.28), we get from equation (15.27)

$$L[F'(t)] = sL[F(t)] - F(0). \tag{15.29}$$

We can extend this for second-order derivative.
Thus

$$\begin{aligned} L[F''(t)] &= sL[F'(t)] - F'(0) = s[sL[F(t)] - F(0)] - F'(0), \quad \text{[by equation (15.29)]} \\ &= s^2L[F(t)] - sF(0) - F'(0). \end{aligned} \tag{15.30}$$

In a similar manner, we can generalise for higher-order derivatives. However, we must demand that, for Laplace transform of $F^n(t)$ to exist, all the derivatives $F_1'(t), F''(t), \ldots, F^{n-1}(t)$ are continuous, $F^n(t)$ is piecewise continuous and $e^{-st}F^{n-1}(t) \longrightarrow 0$ as $t \longrightarrow \infty$ for all n.
Then, we can write

$$L\left[F^n(t)\right] = s^nL[F(t)] - s^{n-1}F(0) - s^{n-2}F'(0) - \cdots - F^{n-1}(0). \tag{15.31}$$

We can use this formula to find Laplace transforms of function in a very simple way, avoiding the process of complicated integrations. □

EXAMPLE 15.9. *Using the transform of a derivative, find the Laplace transform of $F(t) = t^2$.*

Solution: We have, $F(t) = t^2, F'(t) = 2t, F''(t) = 2, F'''(t) = 0.$
Now

$$\begin{aligned} L[F'''(t)] &= s^3L[F(t)] - s^2F(0) - sF'(0) - F''(0) \\ \text{or,} \quad L[0] &= s^3L[F(t)] - 2. \end{aligned}$$

Hence

$$L[F(t)] = \frac{2}{s^3}.$$

EXAMPLE 15.10. *Find the Laplace transform of $F(t) = t\cos t$.*

Solution: We have

$$\left.\begin{aligned} F(t) &= t\cos t \\ F'(t) &= \cos t - t\sin t \\ F''(t) &= -2\sin t - t\cos t \end{aligned}\right\}. \tag{1}$$

Using the transform of second derivative

$$L[F''(t)] = s^2 L[F(t)] - sF(0) - F'(0) = s^2 L[F(t)] - 1. \tag{2}$$

The LHS of equation (2) is

$$L[-2\sin t - t\cos t] = -2L[\sin t] - L[F(t)] = -\frac{2}{s^2+1} - L[F(t)]. \tag{3}$$

From equations (2) and (3), we get

$$-\frac{2}{s^2+1} - L[F(t)] = s^2 L[F(t)] - 1, \quad \text{or,} \quad (1+s^2)L[F(t)] = 1 - \frac{2}{s^2+1} = \frac{s^2-1}{s^2+1}.$$

Hence

$$L[F(t)] = \frac{s^2-1}{(s^2+1)^2}.$$

15.5 Laplace Transforms of Integrals

Theorem 15.5. *If a function $F(t)$ possesses a Laplace transform $f(s)$, then*

$$L\left\{\int_0^t F(\tau)d\tau\right\} = \frac{1}{s}f(s). \tag{15.32}$$

Proof. We assume that the function

$$G(t) = \int_0^t F(\tau)d\tau \tag{15.33}$$

is continuous and is of exponential order. Moreover, $G'(t)$ is piecewise continuous.
From equation (15.33), we get

$$G'(t) = F(t). \tag{15.34}$$

Hence, by equation (15.25), we have

$$L[G'(t)] = sL\left[G(t)\right] - G(0) = sL\left[G(t)\right], \text{ since } G(0) = 0.$$

Thus

$$L\left[G(t)\right] = \frac{1}{s}L\left[G'(t)\right] = \frac{1}{s}L\left[F(t)\right] = \frac{1}{s}f(s).$$

Hence proved. □

The Laplace transform of integrals have the following two consequences:
(1) If $L[F(t)] = f(s)$, then

$$L\left\{\int_0^t \frac{F(\tau)}{\tau}d\tau\right\} = \frac{1}{s}\int_s^\infty f(\xi)d\xi.$$

Proof. Let

$$G(t) = \frac{F(t)}{t}.$$

Then

$$L\left[G(t)\right] = L\left[\frac{F(t)}{t}\right] = \int_s^\infty f(\xi)d\xi \quad \text{[by theorem 15.2]}$$
$$= g(s), \quad \text{say.}$$

Hence

$$L\left\{\int_0^t G(\tau)d\tau\right\} = \frac{1}{s}g(s) \quad \text{[by theorem 15.5]}$$

Hence

$$L\left\{\int_0^t \frac{F(\tau)}{\tau}d\tau\right\} = \frac{1}{s}\int_s^\infty f(\xi)d\xi. \tag{15.35}$$

□

(2) If $L[F(t)] = f(s)$, then

$$\int_0^\infty \frac{f(t)}{t}dt = \int_0^\infty f(x)dx. \tag{15.36}$$

Proof. We have

$$L\left\{\frac{F(t)}{t}\right\} = \int_s^\infty f(\xi)d\xi \quad \text{[by theorem 15.2]}$$
$$= \int_s^0 f(\xi)d\xi + \int_0^\infty f(\xi)d\xi$$
$$= -\int_0^s f(\xi)d\xi + \int_0^\infty f(\xi)d\xi$$
$$= \int_0^\infty f(\xi)d\xi, \quad \text{if } s \longrightarrow 0+.$$

□

EXAMPLE 15.11. *The Laplace transform of a function $F(t)$ is*

$$f(s) = \frac{1}{s(s-a)}.$$

Find the function.

Solution: We know that

$$L\left(e^{at}\right) = \frac{1}{s-a}.$$

Then

$$L\left\{\int_0^t e^{a\tau}d\tau\right\} = \frac{1}{s(s-a)} \quad \text{[by theorem 15.5].}$$

Hence, the required function is

$$G(t) = \int_0^t e^{a\tau}d\tau = \frac{1}{a}e^{a\tau}\Big|_0^t = \frac{1}{a}\left(e^{at} - 1\right).$$

EXAMPLE 15.12. *Evaluate* $\int_0^\infty \frac{\sin t}{t} dt.$

Solution: We take $F(t) = \sin t$.

Hence

$$L[F(t)] = L[\sin t], \quad \text{or,} \quad f(s) = \frac{1}{s^2+1}. \tag{1}$$

Now

$$\int_0^\infty \frac{\sin t}{t} dt = \int_0^\infty \frac{F(t)}{t} dt = \int_0^\infty f(x)dx \quad \text{[by equation (15.36)]}$$
$$= \int_0^\infty \frac{dx}{x^2+1} = \tan^{-1} x \Big|_0^\infty = \frac{\pi}{2}.$$

15.6 Special Techniques for Finding Laplace Transforms

A. *Power series method.* Let a function $F(t)$ be expanded in a power series like

$$F(t) = \sum_{n=0}^{\infty} A_n t^n. \tag{15.37}$$

If $F(t)$ has no singularity and is of exponential order as $t \longrightarrow \infty$, then the Laplace transform of $F(t)$ is given by

$$f(s) = L[F(t)] = \sum_{n=0}^{\infty} A_n \frac{n!}{s^{n+1}}. \tag{15.38}$$

We consider some examples.

EXAMPLE 15.13. *Find the Laplae transform of* $F(t) = \sin\sqrt{t}$.

Solution: We have

$$\sin\sqrt{t} = \sqrt{t} - \frac{(\sqrt{t})^3}{3!} + \frac{(\sqrt{t})^5}{5!} - \cdots$$

Hence

$$L\left(\sin\sqrt{t}\right) = L\left(\sqrt{t}\right) - L\left[\frac{(\sqrt{t})^3}{3!}\right] + L\left[\frac{(\sqrt{t})^5}{5!}\right] - \cdots$$
$$= \frac{\Gamma(3/2)}{s^{3/2}} - \frac{\Gamma(5/2)}{3!\, s^{5/2}} + \frac{\Gamma(7/2)}{5!\, s^{7/2}} - \cdots$$
$$= \frac{\sqrt{\pi}}{s^{3/2}}\left[\frac{1}{2} - \frac{1}{3!}\cdot\frac{3}{2}\cdot\frac{1}{2}\cdot\frac{1}{s} + \frac{1}{5!}\cdot\frac{5}{2}\cdot\frac{3}{2}\cdot\frac{1}{2}\cdot\frac{1}{s^2} - \cdots\right]$$
$$= \frac{\sqrt{\pi}}{2s^{3/2}}\left[1 - \frac{1}{2^2 s} + \frac{1}{2!}\left(\frac{1}{2^2 s}\right)^2 - \frac{1}{3!}\left(\frac{1}{2^2 s}\right)^3 + \cdots\right]$$
$$= \frac{\sqrt{\pi}}{2s^{3/2}}\left(e^{-1/2^2 s}\right) = \frac{\sqrt{\pi}}{2s^{3/2}} e^{-s/4}.$$

EXAMPLE 15.14. *Find the Laplace transform of* $F(t) = t^{-1/2} erf\left(\sqrt{t}\right)$.

Solution: We have (ref. Chapter 20 for error integral)

$$F(t) = \frac{2t^{-1/2}}{\sqrt{\pi}} \int_0^{\sqrt{t}} e^{-x^2} dx = \frac{2t^{-1/2}}{\sqrt{\pi}} \int_0^{\sqrt{t}} \sum_{n=0}^{\infty} \frac{(-1)^n x^{2n}}{n!} dx$$

$$= \frac{2t^{-1/2}}{\sqrt{\pi}} \sum_{n=0}^{\infty} \frac{(-1)^n \left(\sqrt{t}\right)^{2n+1}}{(2n+1)n!} = \frac{2}{\sqrt{\pi}} \sum_{n=0}^{\infty} \frac{(-1^n)t^n}{n!\,(2n+1)}.$$

Hence

$$L[F(t)] = \frac{2}{\sqrt{\pi}} \sum_{n=0}^{\infty} \frac{(-1)^n}{(2n+1)} \frac{1}{s^{n+1}}. \tag{1}$$

Now

$$\tan^{-1}\left(\frac{1}{\sqrt{s}}\right) = \sum_{n=0}^{\infty} \frac{(-1)^n}{(2n+1)\left(\sqrt{s}\right)^{2n+1}}. \tag{2}$$

In view of equation (2), we write equation (1) as

$$L\left[t^{-1/2} erf\left(\sqrt{t}\right)\right] = \frac{2}{\sqrt{\pi s}} \tan^{-1} \frac{1}{\sqrt{s}}.$$

EXAMPLE 15.15. *Find the Laplace transform of*

$$F(t) = \frac{\sin \alpha t}{t}.$$

Solution:

$$L\left(\frac{\sin \alpha t}{t}\right) = L\left[\frac{1}{t}\left(\alpha t - \frac{\alpha^3 t^3}{3!} + \frac{\alpha^5 t^5}{5!} - \frac{\alpha^7 t^7}{7!}\right) + \cdots\right]$$

$$= L\left[\alpha - \frac{\alpha^3 t^2}{3!} + \frac{\alpha^5 t^4}{5!} - \frac{\alpha^7 t^6}{7!} + \cdots\right]$$

$$= \left[\frac{\alpha}{s} - \frac{\alpha^3}{3!}\frac{2!}{s^3} + \frac{\alpha^5}{5!}\frac{4!}{s^5} - \frac{\alpha^7}{7!}\frac{6!}{s^7} + \cdots\right]$$

$$= \left[\frac{\alpha}{s} - \frac{1}{3}\left(\frac{\alpha}{s}\right)^3 + \frac{1}{5}\left(\frac{\alpha}{s}\right)^5 - \frac{1}{7}\left(\frac{\alpha}{s}\right)^7 + \cdots\right].$$

$$= \sum_{n=0}^{\infty} (-1)^n \frac{1}{2n+1}\left(\frac{\alpha}{s}\right)^{2n+1} = \tan^{-1} \frac{\alpha}{s}.$$

EXAMPLE 15.16. *Find the Laplace transform of $F(t) = J_0(t)$, where $J_0(t)$ is Bessel's function of zero order.*

Solution: We have (see Chapter 12)

$$J_0(t) = \sum_{n=0}^{\infty} \frac{(-1)^n t^{2n}}{2^{2n}\, n!\, n!}.$$

Hence

$$L[J_0(t)] = \sum_{n=0}^{\infty} (-1)^n \frac{1}{2^{2n}\, n!\, n!} \frac{(2n)!}{s^{2n+1}}. \tag{1}$$

Remembering that

$$(2n)! = 2^n n! \cdot 1 \cdot 3 \cdot 5 \cdots (2n-1) \tag{2}$$

we find from equation (1)

$$L[J_0(t)] = \frac{1}{s}\left[1 + \sum_{n=1}^{\infty}(-1)^n \frac{1 \cdot 3 \cdot 5 \cdots (2n-1)}{2^n\, n!\, s^{2n}}\right]$$

$$= \frac{1}{s}\left(1 + \frac{1}{s^2}\right)^{-1/2} = \frac{1}{\sqrt{s^2+1}}.$$

B. *Method of differentiation.* Laplace transform of a function can be obtained by differentiating the Laplace transform of another known function.

EXAMPLE 15.17. *Find the Laplace transform of the function* $F(t) = t\sinh(at)$.

Solution: We know

$$L\left[\cosh(at)\right] = \frac{s}{s^2-a^2}, \quad \text{i.e.,} \quad \int_0^\infty e^{-st}\cosh(at)dt = \frac{s}{s^2-a^2}. \tag{1}$$

Differentiating both sides with respect to the parameter a,

$$\frac{d}{da}\int_0^\infty e^{-st}\cosh(at)dt = \frac{d}{da}\left(\frac{s}{s^2-a^2}\right), \quad \text{or,} \quad \int_0^\infty e^{-st} t\sinh(at)dt = \frac{2as}{(s^2-a^2)^2}.$$

Hence,

$$L[t\sinh(at)] = \frac{2as}{(s^2-a^2)^2}.$$

15.7 Some More Examples

EXAMPLE 15.18. *Find the Laplace transform of* $F(t) = \sin at - at\cos at$.

Solution: We have

$$F(t) = \sin at - at\cos at.$$

Hence

$$F'(t) = (a\cos at - a\cos at + a^2 t\sin at) = a^2 t\sin at.$$

Thus

$$L(F') = sL(F) - F(0) = sL(F).$$

Hence

$$L(F) = \frac{1}{s}L(F'). \tag{1}$$

Now

$$L(F') = L\left(a^2 t\sin at\right) = a^2 L(t\sin at). \tag{2}$$

Since

$$L(\sin at) = \frac{a}{s^2+a^2}.$$

$$\therefore \quad L(t\sin at) = -\frac{d}{ds}\left(\frac{a}{s^2+a^2}\right) = \frac{2as}{(s^2+a^2)^2}. \tag{3}$$

From equations (1), (2) and (3), we get

$$L(F) = \frac{a^2}{s} \cdot \frac{2as}{(s^2+a^2)^2} = \frac{2a^3}{(s^2+a^2)^2}.$$

EXAMPLE 15.19. *Find the Laplace transform of*

$$F(t) = \frac{\cos\sqrt{t}}{\sqrt{t}}.$$

Solution: Let $G(t) = \sin\sqrt{t}$.
Then

$$G'(t) = \frac{1}{2}\frac{\cos\sqrt{t}}{\sqrt{t}}, \quad G(0) = 0.$$

Hence

$$L[F(t)] = 2L[G'(t)]$$

$$\text{or,} \quad \frac{1}{2}L(F) = \left[sL\left(\sin\sqrt{t}\right)\right] = s\frac{\sqrt{\pi}}{2s^{3/2}}e^{-1/4s} \quad \text{[by example 15.13]}.$$

Thus,

$$L\left(\frac{\cos\sqrt{t}}{\sqrt{t}}\right) = \sqrt{\frac{\pi}{s}}e^{-1/4s}.$$

EXAMPLE 15.20. *Find Laplace transform of* $F(t) = \sinh(at)\sin(at)$.

Solution: We have

$$L(\sinh at) = \frac{a}{s^2-a^2}, \quad s > |a| \tag{1}$$

$$= f(s), \quad \text{say}.$$

Then

$$\begin{aligned} L\left(e^{iat}\sinh at\right) &= f(s-ia) \quad \text{[by first shifting property]} \\ &= \frac{a}{(s-ia)^2-a^2} \quad \text{[from equation (1)]} \\ &= \frac{a}{(s^2-2a^2)-2ias} \\ &= \frac{a\left(s^2-2a^2+2ias\right)}{(s^2-2a^2)-(2ias)^2}. \end{aligned}$$

Thus

$$L[(\cos at + i\sin at)\sinh(at)] = \frac{a(s^2-2a^2)+2ia^2s}{(s^2-2a^2)^2+4a^2s^2}.$$

Hence

$$L[\sin(at)\sinh(at)] = \frac{2a^2s}{(s^2-2a^2)^2+4a^2s^2} = \frac{2a^2s}{s^4+4a^4}.$$

Also,

$$L[\cos(at)\sinh(at)] = \frac{a\left(s^2-2a^2\right)}{s^4+4a^4}.$$

15.8 Inverse Laplace Transforms

If $f(s)$ be the Laplace transform of a function $F(t)$, then $F(t)$ is called *inverse Laplace transform* of $f(s)$. We write it as

$$F(t) = L^{-1}[f(s)]. \tag{15.39}$$

For examples

$$L^{-1}\left(\frac{1}{s}\right) = 1, \qquad L^{-1}\left(\frac{1}{s-a}\right) = e^{at}$$

$$L^{-1}\left(\frac{s}{s^2+a^2}\right) = \cos at, \qquad L^{-1}\left(\frac{1}{s^2-a^2}\right) = \frac{1}{a}\sinh(at)$$

...., etc.

15.9 Properties of Inverse Laplace Transforms

(1) *Linearity property.* If $L^{-1}\{f_1(s)\} = F_1(t)$ and $L^{-1}\{f_2(s)\} = F_2(t)$, then for two constants a and b,

$$L^{-1}\{af_1(s) + bf_2(s)\} = aF_1(t) + bF_2(t). \tag{15.40}$$

EXAMPLE 15.21. *Find inverse Laplace transforms of the function*

$$f(s) = \frac{3s}{s^2+2} - \frac{4s}{s^2-16}.$$

Solution:

$$L^{-1}\left(\frac{3s}{s^2+2}\right) = 3L^{-1}\left(\frac{s}{s^2+2}\right) = 3\cos\left(\sqrt{2}t\right) \tag{1}$$

$$L^{-1}\left(\frac{4s}{s^2-16}\right) = 4L^{-1}\left(\frac{s}{s^2-16}\right) = 4\cosh(4t). \tag{2}$$

Hence

$$L^{-1}\{f(s)\} = 3\cos\left(\sqrt{2}t\right) - 4\cosh(4t).$$

(2) *First shifting property.* If $L^{-1}\{f(s)\} = F(t)$, then

$$L^{-1}\{f(s-a)\} = e^{at}F(t). \tag{15.41}$$

EXAMPLE 15.22. *Find inverse Laplace transform of*

$$f(s) = \frac{s+3}{s^2+4s+20}.$$

Solution:

$$f(s) = \frac{s+3}{s^2+4s+20} = \frac{(s+2)+1}{(s+2)^2+(4)^2}.$$

Hence

$$L^{-1}\{f(s)\} = L^{-1}\left\{\frac{s+2}{(s+2)^2+(4)^2}\right\} + L^{-1}\left\{\frac{1}{(s+2)^2+(4)^2}\right\}$$
$$= e^{-2t}L^{-1}\left(\frac{s}{s^2+(4)^2}\right) + e^{-2t}L^{-1}\left(\frac{1}{s^2+4^2}\right)$$
$$= e^{-2t}\left(\cos 4t + \frac{1}{4}\sin 4t\right).$$

(3) *Second shifting property.* If $L^{-1}\{f(s)\} = F(t)$, then

$$L^{-1}\left\{e^{-as}f(s)\right\} = \begin{cases} F(t-a), & \text{for } t > a \\ 0, & \text{for } t < a \end{cases} \tag{15.42}$$

(4) *Change of scale property.* If $L^{-1}\{f(s)\} = F(t)$, then

$$L^{-1}\{f(as)\} = \frac{1}{a}F\left(\frac{t}{a}\right). \tag{15.43}$$

(5) *Inverse of Derivatives.* If $L^{-1}\{f(s)\} = F(t)$, then

$$L^{-1}\left\{\frac{d^n}{ds^n}f(s)\right\} = (-1)^n t^n F(t), \quad (n = 1, 2, \ldots). \tag{15.44}$$

(6) *Inverse transforms of integrals.* If $L^{-1}\{f(s)\} = F(t)$, then

$$L^{-1}\left[\int_s^\infty f(\xi)d\xi\right] = \frac{F(t)}{t}. \tag{15.45}$$

(7) *Multiplication by s.* If $L^{-1}\{f(s)\} = F(t)$, then

$$L^{-1}\{sf(s)\} = \frac{dF}{dt}. \tag{15.46}$$

(8) *Division by s.* If $L^{-1}\{f(s)\} = F(t)$, then

$$L^{-1}\left\{\frac{f(s)}{s}\right\} = \int_0^t F(\tau)d\tau. \tag{15.47}$$

EXAMPLE 15.23. *Find inverse Laplace transforms of*

(a) $\dfrac{s}{(s+a)^3}$; (b) $\dfrac{2s+3}{(s-1)^3}$; (c) $\dfrac{s}{(s^2-a^2)^2}$.

Solution:

(*a*) $$L^{-1}\left\{\frac{s}{(s+a)^3}\right\} = e^{-at}L^{-1}\left\{\frac{s-a}{s^3}\right\} = e^{-at}\left[L^{-1}\left(\frac{1}{s^2}\right) - aL^{-1}\left(\frac{1}{s^3}\right)\right]$$
$$= e^{-at}\left[t - \frac{at^2}{2}\right]$$

(*b*) $$L^{-1}\left\{\frac{2s+3}{(s-1)^3}\right\} = L^{-1}\left\{\frac{2(s-1)+5}{(s-1)^3}\right\} = 2L^{-1}\left\{\frac{1}{(s-1)^2}\right\} + 5L^{-1}\left\{\frac{1}{(s-1)^3}\right\}$$
$$= e^t\left[2L^{-1}\left(\frac{1}{s^2}\right) + 5L^{-1}\left(\frac{1}{s^3}\right)\right] = e^t\left(2t + \frac{5}{2}t^2\right).$$

(c) We know that

$$\frac{d}{ds}\left\{\frac{1}{s^2-a^2}\right\} = \frac{-2s}{(s^2-a^2)^2}.$$

Hence

$$\begin{aligned} L^{-1}\left\{\frac{s}{(s^2-a^2)^2}\right\} &= -\frac{1}{2}L^{-1}\left[\frac{d}{ds}\left(\frac{1}{s^2-a^2}\right)\right] \\ &= -\frac{1}{2}(-1)tL^{-1}\left(\frac{1}{s^2-a^2}\right) \quad \text{[by equation (15.44)]} \\ &= \frac{t}{2a}\sinh(at). \end{aligned}$$

EXAMPLE 15.24. *Find inverse Laplace transform of*

$$f(s) = \frac{4e^{-2s}}{s^2-16}.$$

Solution: We have

$$L^{-1}\left(\frac{4}{s^2-16}\right) = \sinh(4t).$$

Hence

$$L^{-1}\{f(s)\} = L^{-1}\left\{\frac{4e^{-2s}}{s^2-16}\right\} = \sinh\{4(t-2)\}, \quad \text{[by equation (15.42)]}$$

for $t > 2$.

Theorem 15.6. *If $f_1(s)$ and $f_2(s)$ are respectively the Laplace transforms of $F_1(t)$ and $F_2(t)$, then*

$$L[G(t)] = f_1(s)f_2(s)$$
$$\text{or,} \quad G(t) = L^{-1}\{f_1(s)f_2(s)\}, \tag{15.48}$$

where $G(t)$ is defined as

$$G(t) = \int_0^t F_1(\xi)F_2(t-\xi)d\xi = \int_0^t F_2(\xi)F_1(t-\xi)d\xi. \tag{15.49}$$

This is known as convolution or Faltung theorem. As an example, let

$$G(t) = \int_0^t J_0(\xi)J_0(t-\xi)d\xi,$$

where $J_0(\xi)$ is the Bessel's function of order zero.

Then, the Laplae transform of $G(t)$ is

$$\begin{aligned} L[G(t)] = L[J_0(t)]L[J_0(t)] &= \frac{1}{(s^2+1)^{1/2}}\cdot\frac{1}{(s^2+1)^{1/2}} \quad \text{[see example (15.16)].} \\ &= \frac{1}{s^2+1}. \end{aligned}$$

Hence

$$G(t) = L^{-1}\left(\frac{1}{s^2+1}\right) = \sin t$$

so that

$$\int_0^t J_0(\xi)J_0(t-\xi)d\xi = \sin t.$$

EXAMPLE 15.25. *Find inverse Laplace transform of*

$$f(s) = \frac{1}{s(s^2+1)^2}.$$

Solution: We have

$$f(s) = \frac{1}{s(s^2+1)^2} = \frac{s}{s^2(s^2+1)^2} = \frac{1}{s^2}\cdot\frac{s}{(s^2+1)^2} = f_1(s)\cdot f_2(s),$$

where

$$f_1(s) = \frac{1}{s^2}, \quad f_2(s) = \frac{s}{(s^2+1)^2}.$$

$$L^{-1}\{f_1(s)\} = L^{-1}\left(\frac{1}{s^2}\right) = t \tag{1}$$

$$L^{-1}\{f_2(s)\} = L^{-1}\left\{\frac{s}{(s^2+1)^2}\right\} = t\sin t. \tag{2}$$

Hence, by convolution theorem,

$$L^{-1}\{f_1(s)f_2(s)\} = \int_0^t (t-\xi)\xi\sin\xi d\xi = t\int_0^t \xi\sin\xi d\xi - \int_0^t \xi^2\sin\xi d\xi. \tag{3}$$

Now,

$$I_1 = \int_0^t \xi\sin\xi d\xi = -\xi\cos\xi\Big|_0^t + \int_0^t \cos\xi = -t\cos t + \sin t. \tag{4}$$

$$\begin{aligned} I_2 = \int_0^t \xi^2\sin\xi d\xi &= -\xi^2\cos\xi\Big|_0^t + \int_0^t 2\xi\cos\xi d\xi \\ &= -t^2\cos t + 2\left[\xi\sin\xi\Big|_0^t - \int_0^t \sin\xi d\xi\right] \\ &= -t^2\cos t + 2t\sin t + 2(\cos t - 1). \end{aligned} \tag{5}$$

From equations (3), (4) and (5), we get

$$\begin{aligned} L^{-1}\{f_1(s)f_2(s)\} &= t(-t\cos t + \sin t) + t^2\cos t - 2t\sin t - 2(\cos t - 1) \\ &= 2 - t\sin t - 2\cos t. \end{aligned}$$

15.10 Partial Fraction Method for Finding the Inverse

We consider the case when $f(s)$ is expressed as the ratio of two polynomials $P(s)$ and $Q(s)$ in the form

$$f(s) = \frac{P(s)}{Q(s)}, \tag{15.50}$$

where $Q(s)$ is of greater degree than $P(s)$. We shall see how to find the Laplace inverse transforms in the different cases.

Case I. *$Q(s)$ has distinct roots.*

Let $Q(s)$ have distinct roots $a_1, a_2, \ldots, a_n$. Then

$$Q(s) = (s - a_1)(s - a_2)\ldots(s - a_n).$$

In this case, we can express $f(s)$ as

$$f(s) = \frac{P(s)}{Q(s)} = \sum_{k=1}^{n} \frac{A_k}{s - a_k}, \tag{15.51}$$

where A_ks are constants. Any of these constants, say A_j, is found by multiplying both sides of equation (15.50) by $(s - a_j)$ and proceeding to limit $s \longrightarrow a_j$.

Thus,

$$A_j = \lim_{s \longrightarrow a_j} \frac{P(s)}{Q(s)}(s - a_j) \tag{15.52}$$

$$\text{or,} \quad A_j = P(a_j) \lim_{s \longrightarrow a_j} \frac{s - a_j}{Q(s)}$$

$$= P(a_j) \frac{1}{Q'(a_j)} \quad \text{[by L'Hospital's rule].}$$

Since

$$L^{-1}\left(\frac{A_j}{s - a_j}\right) = A_j e^{a_j t} = \frac{P(a_j)}{Q'(a_j)} e^{a_j t}, \tag{15.53}$$

we can write

$$L^{-1}\left\{\frac{P(s)}{Q(s)}\right\} = \sum_{k=1}^{n} \frac{P(a_k)}{Q'(a_k)} e^{a_k t}. \tag{15.54}$$

The summation in the RHS of equation (15.54) is called *Heaviside's expansion.*

The constants, A_j, can also be conveniently obtained from equation

$$A_j = \lim_{s \longrightarrow a_j} \left[\frac{P(s)}{Q(s)/(s - a_j)}\right]. \tag{15.55}$$

EXAMPLE 15.26. *Find inverse Laplace transform of*

$$f(s) = \frac{s - 1}{(s + 3)(s^2 + 3s + 2)}.$$

Solution: Here, $Q(s) = (s+3)(s^2+3s+2) = (s+3)(s+2)(s+1)$.

Hence we write

$$f(s) = \frac{A_1}{s+1} + \frac{A_2}{s+2} + \frac{A_3}{s+3}.$$

Using equation (15.55), we have

$$A_1 = \lim_{s \longrightarrow -1} \frac{s-1}{(s+2)(s+3)} = \frac{-2}{1.2} = -1$$

$$A_2 = \lim_{s \longrightarrow -2} \frac{s-1}{(s+1)(s+3)} = \frac{-3}{(-1)(1)} = 3$$

$$A_3 = \lim_{s \longrightarrow -3} \frac{s-1}{(s+1)(s+2)} = \frac{-4}{(-2)(-1)} = -2.$$

Hence

$$L^{-1}\{f(s)\} = -e^{-t} + 3e^{-2t} - 2e^{-3t}.$$

Case II. *$Q(s)$ has repeated roots.*

Let $Q(s)$ have one root a_1 repeated m times and other roots are $a_2, a_3, \ldots, a_p$ so that $n = m + p$.

Then

$$Q(s) = (s-a_1)^m (s-a_2)(s-a_3)\cdots(s-a_p).$$

Hence, we can express $f(s)$ as

$$f(s) = \frac{P(s)}{Q(s)} = \sum_{k=1}^{m} \frac{B_k}{(s-a_1)^k} + \sum_{j=2}^{p} \frac{A_j}{s-a_j}. \tag{15.56}$$

Proceeding as before, the constants A_js are found from

$$A_j = \frac{P(a_j)}{[Q(s)/(s-a_j)]_{s=a_j}} \quad (j = 2, 3, \ldots, p). \tag{15.57}$$

Multiplying both sides of equation (15.56) by $(s-a_1)^m$ and proceeding to limit $s \longrightarrow a_1$, we find

$$B_m = \frac{P(s)(s-a_1)^m}{Q(s)}, \quad \text{for } s = a_1$$

$$\text{i.e.,} \quad B_m = \frac{P(a_1)}{[Q(s)/(s-a_1)^m]_{s=a_1}}. \tag{15.58}$$

To find other constants B_k, $(k = 1, 2, \ldots, m-1)$, we write equation (15.56) as

$$\frac{P(s)}{Q(s)} = \frac{B_1}{s-a_1} + \frac{B_2}{(s-a_1)^2} + \cdots + \frac{B_m}{(s-a_1)^m} + \psi(s). \tag{15.59}$$

We multiply both sides of this equation by $(s-a_1)^m$ and get

$$\begin{aligned}\phi(s) &= \frac{P(s)(s-a_1)^m}{Q(s)} \\ &= B_1(s-a_1)^{m-1} + B_2(s-a_1)^{m-2} + \cdots + B_{m-1}(s-a_1) + B_m + \psi(s)(s-a_1)^m.\end{aligned}$$

Differentiating both sides with respect to s and putting $s = a_1$, we find

$$\frac{d}{ds}\{\phi(s)\} = B_{m-1}. \tag{15.60}$$

Thus, differentiating $(m - r)$ times and putting $s = a_1$, we get

$$B_r = \frac{1}{(m-r)!}\left[\frac{d^{m-r}}{ds^{m-r}}\{\phi(s)\}\right]_{s=a_1}. \tag{15.61}$$

It may be noted that in equation (15.61)

$$\phi(s) = \frac{P(s)}{[Q(s)/(s-a_1)^m]}. \tag{15.62}$$

EXAMPLE 15.27. *Find the inverse Laplace transform of*

$$f(s) = \frac{s^2+1}{(s+2)^2(s^2-5s+6)}.$$

Solution: Here, $P(s) = s^2 + 1$, $Q(s) = (s+2)^2(s-2)(s-3)$.
Hence

$$f(s) = \frac{P(s)}{Q(s)} = \frac{B_1}{s+2} + \frac{B_2}{(s+2)^2} + \frac{A_1}{s-2} + \frac{A_2}{s-3}.$$

The coefficients are

$$A_1 = \left[\frac{s^2+1}{(s+2)^2(s-3)}\right]_{s=2} = \frac{5}{4^2\cdot(-1)} = -\frac{5}{16}$$

$$A_2 = \left[\frac{s^2+1}{(s+2)^2(s-2)}\right]_{s=3} = \frac{10}{5^2\cdot 1} = \frac{2}{5}$$

$$B_2 = \left[\frac{s^2+1}{(s-2)(s-3)}\right]_{s=-2} = \frac{5}{(-4)(-5)} = \frac{1}{4}$$

$$B_1 = \frac{1}{1!}\frac{d}{ds}\left[\frac{s^2+1}{(s+2)(s-3)}\right]_{s=-2} = \left[\frac{-5s^2+10s+5}{\{(s-2)(s-3)\}^2}\right]_{s=-2} = \frac{1}{80}.$$

Hence, we have

$$f(s) = \frac{1}{80}\cdot\frac{1}{s+2} - \frac{1}{4}\cdot\frac{1}{(s+2)^2} - \frac{5}{16}\cdot\frac{1}{s-2} + \frac{2}{5}\cdot\frac{1}{s-3}.$$

Thus, we can write

$$L^{-1}\{f(s)\} = \frac{1}{80}e^{-2t} - \frac{1}{4}te^{-2t} - \frac{5}{16}e^{2t} + \frac{2}{5}e^{3t}.$$

Case III. *$Q(s)$ has an unrepeated quadratic factor.*

Let $Q(s)$ have a quadratic factor of the form $\{(s-a)^2 + b^2\}$. Then we can write

$$\frac{P(s)}{Q(s)} = \frac{B_1 s + B_2}{(s-a)^2 + b^2} + \frac{A_1}{s-a_1} + \frac{A_2}{s-a_2} + \cdots \tag{15.63}$$

We can find the constants $A_1, A_2, \ldots$ by using the equation (15.57) as before.
To find B_1 and B_2, we multiply both sides of equation (15.63) by $\{(s-a)^2+b^2\}$, and get

$$\phi(s) = B_1 s + B_2 + \{(s-a)^2+b^2\}\psi(s), \tag{15.64}$$

where we have denoted $\phi(s)$ and $\psi(s)$ as

$$\phi(s) = \frac{P(s)}{Q(s)/\{(s-a)^2+B^2\}}. \tag{15.65}$$

$$\psi(s) = \frac{A_1}{s-a_1} + \frac{A_2}{s-a_2} + \cdots \tag{15.66}$$

Now we proceed to the limit $s \longrightarrow (a+ib)$, and get from equation (15.64)

$$B_1(a+ib) + B_2 = \{\phi(s)\}_{s=a+ib}. \tag{15.67}$$

The equation (15.67) will give B_1 and B_2 on equating the real and imaginary parts.

EXAMPLE 15.28. *Find the inverse Laplace transform of*

$$f(s) = \frac{s}{(s-3)(s^2+4)}.$$

Solution: Let

$$\frac{s}{(s-3)(s^2+4)} = \frac{B_1 s + B_2}{s^2+4} + \frac{A_1}{s-3}. \tag{1}$$

Then

$$A_1 = \left[\frac{s}{s^2+4}\right]_{s=3} = \frac{3}{13} \tag{2}$$

and

$$B_1(2i) + B_2 = \left[\frac{s}{s-3}\right]_{s=2i} = \frac{2i}{2i-3}$$

$$\text{or,} \quad 2B_1 i + B_2 = \frac{2i(2i+3)}{(2i-3)(2i+3)} = \frac{-4+6i}{-13}.$$

Hence

$$B_1 = -\frac{3}{13}, \quad B_2 = \frac{4}{13}. \tag{3}$$

Thus we have

$$f(s) = \frac{1}{13}\left[\frac{-3s}{s^2+4} + \frac{4}{s^2+4} + \frac{3}{s-3}\right].$$

Hence inverse Laplace transform is

$$L^{-1}\{f(s)\} = \frac{1}{13}[-3\cos 2t + 2\sin 2t + 3e^{3t}].$$

Case IV. *$Q(s)$ has repeated quadratic factor.*

Let, as before, $Q(s)$ have quadratic factor $\{(s-a)^2+b^2\}$ repeated twice. Then, we can write

$$\frac{P(s)}{Q(s)} = \frac{C_1 s + C_2}{\{(s-a)^2+b^2\}^2} + \frac{B_1 s + B_2}{(s-a)^2+b^2} + \psi(s), \tag{15.68}$$

where as before

$$\psi(s) = \frac{A_1}{s-a_1} + \frac{A_2}{s-a_2} + \cdots \tag{15.69}$$

We multiply both sides of equation (15.68) by $\{(s-a)^2+b^2\}^2$ and get

$$\phi(s) = C_1 s + C_2 + (B_1 s + B_2)\{(s-a)^2+b^2\} + \psi(s)[(s-a)^2+b^2]^2. \tag{15.70}$$

where

$$\phi(s) = \frac{P(s)}{Q(s)/[(s-a)^2+b^2]^2} \tag{15.71}$$

Now we proceed to the limit $s \longrightarrow (a+ib)$ and get from (15.68)

$$C_1(a+ib) + C_2 = [\phi(s)]_{s=a+ib}. \tag{15.72}$$

Equating real and imaginary parts in (15.72), we can find C_1 and C_2.

To find B_1 and B_2, we differentiate equation (15.70) with respect to s and proceed to the limit $s \longrightarrow (a+ib)$.

Then we get

$$C_1 + 2[B_1(a+ib) + B_2](ib) = \frac{d}{ds}[\phi(s)]_{s=a+ib}. \tag{15.73}$$

Equating real and imaginary parts of equation (15.73), we can find B_1 and B_2, since C_1 is already known.

Some function $F(t)$ and their Laplace transforms $f(s)$ are given in Table 15.1.

EXAMPLE 15.29. *Find inverse Laplace transform of*

$$f(s) = \frac{s^2-b^2}{(s^2+b^2)^2}.$$

Solution: Let

$$f(s) = \frac{s^2-b^2}{(s^2+b^2)^2} = \frac{C_1 s + C_2}{(s^2+b^2)^2} + \frac{B_1 s + B_2}{(s^2+b^2)}. \tag{1}$$

By equation (15.72), putting $a=0$, we get

$$C_1(ib) + C_2 = (ib)^2 - b^2, \quad \therefore \quad C_1 = 0, \quad C_2 = -2b^2. \tag{2}$$

Again, by equation (15.73)

$$2\{B_1(ib) + B_2\}(ib) = \frac{d}{ds}(s^2-b^2)\bigg|_{s=ib}.$$

This gives

$$B_1 = 0 \quad \text{and} \quad B_2 = 1. \tag{3}$$

Thus

$$f(s) = \frac{-2b^2}{(s^2+b^2)^2} + \frac{1}{s^2+b^2} = \frac{1}{b}\left[-\frac{2b^3}{(s^2+b^2)^2} + \frac{b}{s^2+b^2}\right]. \tag{4}$$

Hence

$$L^{-1}\{f(s)\} = \frac{1}{b}[-\sin bt + bt\cos bt + \sin bt] = t\cos bt \quad \text{[see example 15.18]}$$

15.11 Applications of Laplace Transforms

A. *Evaluation of integrals.*

We know that, Laplace transform of a given function $F(t)$ is

$$f(s) = \int_0^\infty e^{-st} F(t)dt. \tag{15.74}$$

Proceeding to the limit $s \longrightarrow 0$ in equation (15.74) we get

$$f(0) = \int_0^\infty F(t)dt. \tag{15.75}$$

Equations (15.74) and (15.75) may be used to evaluate definite integrals.

EXAMPLE 15.30. *Evalulate the following integrals:*

(a) $\int_0^\infty te^{-st} \sin t\, dt$, (b) $\int_0^\infty J_0(t)dt$, (c) $\int_0^\infty e^{-x^2} dx$.

Solution: (a) We have

$$L(t \sin t) = -\frac{d}{ds}\left(\frac{1}{s^2+1}\right), \quad \text{or,} \quad \int_0^\infty e^{-st} t \sin t\, dt = \frac{2s}{(s^2+1)^2}.$$

Putting $s = 1$, we get

$$\int_0^\infty te^{-t} \sin t\, dt = \frac{1}{2}.$$

(b) From example 15.16, we have

$$L[J_0(t)] = \frac{1}{\sqrt{s^2+1}}, \quad \text{or,} \quad \int_0^\infty e^{-st} J_0(t)dt = \frac{1}{\sqrt{s^2+1}}.$$

Putting $s = 0$, $\int_0^\infty J_0(t)dt = 1$.

(c) Let

$$F(t) = \int_0^\infty e^{-tx^2} dx. \tag{1}$$

Then

$$L[F(t)] = \int_0^\infty e^{-st} \left[\int_0^\infty e^{-tx^2} dx\right] dt$$

$$\text{or,} \quad L[F(t)] = \int_0^\infty \left[\int_0^\infty e^{-st}\left(e^{-tx^2}\right) dt\right] dx = \int_0^\infty L\left[e^{-tx^2}\right] dx$$

$$= \int_0^\infty \frac{1}{s+x^2} dx = \frac{1}{\sqrt{s}} \tan^{-1} \frac{x}{\sqrt{s}}\bigg|_0^\infty = \frac{\pi}{2\sqrt{s}}. \tag{2}$$

Hence

$$F(t) = L^{-1}\left(\frac{\pi}{2\sqrt{s}}\right) = \frac{\pi}{2} L^{-1}\left(\frac{1}{\sqrt{s}}\right). \tag{3}$$

Since

$$L(t^n) = \frac{\Gamma(n+1)}{s^{n+1}}, \quad t^n = \Gamma(n+1)L^{-1}\left(\frac{1}{s^{n+1}}\right). \tag{4}$$

Putting $n = -\frac{1}{2}$, we get from equation (4)

$$\frac{1}{\sqrt{t}} = \Gamma\left(\frac{1}{2}\right) L^{-1}\left(\frac{1}{\sqrt{s}}\right). \tag{5}$$

From equations (4) and (5), we get

$$F(t) = \frac{\pi}{2} \cdot \frac{1}{\sqrt{t}\Gamma(\frac{1}{2})} = \frac{\sqrt{\pi}}{2\sqrt{t}}. \tag{6}$$

Thus, from equations (1) and (6)

$$\int_0^\infty e^{-tx^2} dx = \frac{\sqrt{\pi}}{2\sqrt{t}}. \tag{7}$$

Putting $t = 1$, we get

$$\int_0^\infty e^{-x^2} dx = \frac{\sqrt{\pi}}{2}. \tag{8}$$

EXAMPLE 15.31. *Using the method of Laplace transform, show that*

$$\int_0^\infty \cos x^2 dx = \frac{1}{2}\sqrt{\frac{\pi}{2}}.$$

Solution: Let

$$F(t) = \int_0^\infty \cos(tx^2) dx. \tag{1}$$

Then

$$L[F(t)] = \int_0^\infty e^{-st}\left[\int_0^\infty \cos(tx^2)dx\right] dt = \int_0^\infty \left[\int_0^\infty e^{-st}\cos(tx^2)dt\right] dx$$

$$= \int_0^\infty L(\cos tx^2)dx = \int_0^\infty \frac{s}{s^2+x^4}dx. \tag{2}$$

We substitute

$$x^2 = s\tan\phi, \text{ so that, } dx = \frac{s\sec^2\phi d\phi}{2\sqrt{s\tan\phi}}.$$

Then

$$\int_0^\infty \frac{s}{s^2+x^4}dx = \int_0^{\pi/2} \frac{s}{s^2(1+\tan^2\phi)} \cdot \frac{s\sec^2\phi}{2\sqrt{s\tan\phi}}d\phi$$

$$= \int_0^{\pi/2} \frac{1}{2\sqrt{s}}(\cos\phi)^{1/2}(\sin\phi)^{-1/2}d\phi$$

$$= \frac{1}{2\sqrt{s}} \frac{\Gamma(\frac{1}{4})\Gamma(\frac{3}{4})}{2\Gamma(1)}$$

$$= \frac{1}{4\sqrt{s}} \frac{\pi}{\sin\frac{\pi}{4}} \quad \left[\because \Gamma(m)\Gamma(1-m) = \frac{\pi}{\sin m\pi}\right]$$

$$= \frac{\pi}{2\sqrt{2s}}. \tag{3}$$

Thus, from equations (2) and (3)

$$L[F(t)] = \frac{\pi}{2\sqrt{2s}}.$$

Hence

$$F(t) = L^{-1}\left(\frac{\pi}{2\sqrt{2s}}\right) = \frac{\pi}{2\sqrt{2}} \cdot \frac{1}{\sqrt{\pi t}} = \frac{1}{2}\sqrt{\frac{\pi}{2t}}. \tag{4}$$

From equations (1) and (4), we get

$$\int_0^\infty \cos tx^2 dx = \frac{1}{2}\sqrt{\frac{\pi}{2t}}.$$

Putting $t = 1$,

$$\int_0^\infty \cos x^2 dx = \frac{1}{2}\sqrt{\frac{\pi}{2}}.$$

B. *Solution of differential equations.* Ordinary as well as partial differential equations can be solved by the methods of Laplace transforms. We shall consider some examples in order to highlight this technique.

EXAMPLE 15.32. *Solve the differential equation*

$$\frac{d^2y}{dx^2} - \lambda^2 y = 0, \quad \lambda \neq 0$$

with the conditions $y(0) = 0, y'(0) = 1$.

Solution: We have

$$\frac{d^2y}{dx^2} - \lambda^2 y = 0.$$

Taking Laplace transform on both sides

$$L\left(\frac{d^2y}{dx^2}\right) - \lambda^2 L(y) = 0. \tag{1}$$

Let

$$L(y) = f(s). \tag{2}$$

Then from equations (1) and (2), we have

$$\left[s^2 f(s) - sy(0) - y'(0)\right] - \lambda^2 f(s) = 0. \tag{3}$$

Using the given boundary conditions,

$$f(s)\left[s^2 - \lambda^2\right] = 1.$$

Hence

$$f(s) = \frac{1}{s^2 - \lambda^2} = \frac{1}{(s-\lambda)(s+\lambda)} \quad \text{or,} \quad f(s) = \frac{1}{2\lambda}\left[\frac{1}{s-\lambda} - \frac{1}{s+\lambda}\right]. \tag{4}$$

Taking inverse Laplace transform on both sides of equation (4), we have

$$y(x) = \frac{1}{2\lambda}\left[L^{-1}\left(\frac{1}{s-\lambda}\right) - L^{-1}\left(\frac{1}{s+\lambda}\right)\right] = \frac{1}{2\lambda}\left[e^{\lambda x} - e^{-\lambda x}\right].$$

EXAMPLE 15.33. *Solve the differential equation*

$$\frac{d^2y}{dt^2} + 4\frac{dy}{dt} + 3y = 0$$

with the conditions $y(0) = 1, \dot{y}(0) = 0$.

Solution: We have

$$\frac{d^2y}{dt^2} + 4\frac{dy}{dt} + 3y = 0. \tag{1}$$

Taking Laplace transform on both sides,

$$\{s^2 f(s) - sy(0) - \dot{y}(0)\} + 4\{sf(s) - y(0)\} + 3f(s) = 0, \tag{2}$$

where

$$L(y) = f(s). \tag{3}$$

Using initial conditions, we get from equation (2)

$$f(s) = \frac{s+4}{s^2+4s+3} = \frac{s+4}{(s+3)(s+1)}, \quad \text{or,} \quad f(s) = \frac{A}{s+3} + \frac{B}{s+1}. \tag{4}$$

By the partial fraction method, we find

$$A = \left[\frac{s+4}{s+1}\right]_{s=-3} = -\frac{1}{2}, \quad B = \left[\frac{s+4}{s+3}\right]_{s=-1} = \frac{3}{2}.$$

Hence, from equations (4) and (5)

$$f(s) = -\frac{1}{2}\frac{1}{s+3} + \frac{3}{2}\frac{1}{s+1}.$$

Taking inverse Laplace transform, we have

$$y(t) = -\frac{1}{2}e^{-3t} + \frac{3}{2}e^{-t}.$$

EXAMPLE 15.34. *Solve the Bessel's equation of zero order*

$$t\frac{d^2y}{dt^2} + \frac{dy}{dt} + ty = 0,$$

with the conditions that $y(0) = 1$. *Assume that* $y(t)$ *and its derivatives have transforms.*

Solution: We have

$$t\frac{d^2y}{dt^2} + \frac{dy}{dt} + ty = 0. \tag{1}$$

Taking Laplace transform on both sides

$$-\frac{d}{ds}[s^2 f(s) - sy(0) - y'(0)] + [sf(s) - y(0)] - \frac{d}{ds}[f(s)] = 0, \tag{2}$$

where we have taken

$$f(s) = L[y(t)]. \tag{3}$$

Using the initial condition $y(0) = 1$, we get from equation (2)

$$(s^2+1)f'(s) = -sf(s), \quad \text{or,} \quad \frac{f'(s)}{f(s)} = -\frac{s}{s^2+1}, \quad \text{i.e.,} \quad \frac{df}{f} = \frac{-sds}{s^2+1},$$

which gives on integration

$$f(s) = \frac{A}{\sqrt{s^2+1}}. \tag{4}$$

Here A is the integration constant. Now, from equation (4), we write

$$\begin{aligned} f(s) &= \frac{A}{s}\left(1+\frac{1}{s^2}\right)^{-1/2} = \frac{A}{s}\left[1-\frac{1}{2}\cdot\frac{1}{s^2}+\frac{1}{2^2}\cdot\frac{3}{2!}\cdot\frac{1}{s^4}+\cdots\right] \\ &= A\sum_{k=0}^{\infty}\frac{(-1)^k(2k)!}{(2^k k!)^2 s^{2k+1}}. \end{aligned} \tag{5}$$

Hence

$$y(t) = L^{-1}\{f(s)\} = A\sum_{k=0}^{\infty}\frac{(-1)^k t^{2k}}{(2^k k!)^2}. \tag{6}$$

Since $y(0) = 1$, we get from equation (6), $A = 1$.
Hence

$$y(t) = \sum_{k=0}^{\infty}\frac{(-1)^k t^{2k}}{(2^k k!)^2}. \tag{7}$$

Note: We have, in (5), used the relation (2) of example 15.16.

EXAMPLE 15.35. *A particle moves on a vertical z-axis under the force of gravity, the resistance of the medium being proportional to its velocity. Its equation of motion is*

$$m\frac{d^2z}{dt^2} = mg - k\frac{dz}{dt}.$$

Solve the equation under the condition

$$z(0) = 0 \quad and \quad \left.\frac{dz}{dt}\right|_{t=0} = V_0.$$

Solution: We have

$$m\frac{d^2z}{dt^2} = mg - k\frac{dz}{dt}, \quad \text{or,} \quad \frac{d^2z}{dt^2} + b\frac{dz}{dt} = g, \quad \text{where } b = \frac{k}{m}. \tag{1}$$

Taking Laplace transform on both sides

$$[s^2f(s) - sz(0) - z'(0)] + b[sf(s) - z(0)] = \frac{g}{s},$$

where $f(s) = L[z(t)]$.
Hence

$$\begin{aligned} f(s) &= \frac{g+V_0 s}{s^2(s+b)} \quad \text{[using initial conditions]} \\ &= \frac{A}{s}+\frac{B}{s^2}+\frac{C}{s+b} \quad \text{(say).} \end{aligned}$$

Then,

$$C = \left.\frac{g + V_0 s}{s^2}\right|_{s=-b} = -\frac{bV_0 - g}{b^2} \tag{2}$$

$$B = \left.\frac{g + V_0 s}{s + b}\right|_{s=0} = \frac{g}{b} \tag{3}$$

$$A = \frac{d}{ds}\left[\frac{g + V_0 s}{s + b}\right]_{s=0} = \left.\frac{bV_0 - g}{(s + b)^2}\right|_{s=0} = \frac{bV_0 - g}{b^2}. \tag{4}$$

Thus,

$$f(s) = \frac{bV_0 - g}{b^2}\left[\frac{1}{s} - \frac{1}{s + b}\right] + \frac{g}{b}\cdot\frac{1}{s^2}.$$

Taking the inverse transform, we get

$$L^{-1}[f(s)] = z(t) = \frac{bV_0 - g}{b^2}\left[1 - e^{-bt}\right] + \frac{g}{b}t,$$

where $b = \frac{k}{m}$.

C. *Solution of partial differential equations.* Let $\phi(x, t)$ be a function of two variables. The Laplace transform of the function $\phi(x, t)$ and its partial derivatives with respect to the variable t are given by the following relations.

1. $$L\left(\frac{\partial \phi}{\partial t}\right) = sf(x, s) - \phi(x, 0), \tag{15.76}$$

   where $f(x, s) = L\{\phi(x, t)\}$ with respect to variable t.

2. $$L\left(\frac{\partial^2 \phi}{\partial t^2}\right) = s^2 f(x, s) - s\phi(x, 0) - \phi'(x, 0), \tag{15.77}$$

   where $\phi'(x, 0) = \left.\frac{\partial \phi}{\partial t}\right|_{x=0}$.

3. $$L\left(\frac{\partial \phi}{\partial x}\right) = \frac{df}{dx}. \tag{15.78}$$

4. $$L\left(\frac{\partial^2 \phi}{\partial x^2}\right) = \frac{d^2 f}{dx^2}. \tag{15.79}$$

Now we consider some examples.

EXAMPLE 15.36. *Solve the wave equation*

$$\frac{\partial^2 \phi}{\partial t^2} = c^2 \frac{\partial^2 \phi}{\partial x^2},$$

under the conditions that

$$\phi(x, 0) = 0, \quad \phi'(x, 0) = 0, \qquad x > 0$$

$$\text{and} \quad \phi(0, t) = t, \quad \lim_{x \longrightarrow \infty} \phi(x, t) = 0, \quad t \geq 0.$$

Solution: We have

$$\frac{\partial^2 \phi}{\partial t^2} = c^2 \frac{\partial^2 \phi}{\partial x^2}. \tag{1}$$

Taking Laplace transform on both sides

$$s^2 f(x,s) - s\phi(x,0) - \phi'(x,0) = c^2 \frac{d^2 f}{dx^2},$$

which gives on using initial conditions

$$\frac{d^2 f}{dx^2} = \frac{s^2}{c^2} f. \tag{2}$$

Its general solution is

$$f(x,s) = Ae^{\xi x} + Be^{-\xi x}, \quad \text{where } \xi = \frac{s}{c}. \tag{3}$$

Since, $\lim_{x \longrightarrow \infty} \phi(x,t) = 0$, the function $f(x,s)$ should also vanish at infinity. Thus, $A = 0$. Then, from equation (3), we have

$$f(x,s) = Be^{-\xi x}. \tag{4}$$

To find the unknown constant B, we use the boundary condition

$$\phi(0,t) = t, \quad \text{i.e.,} \quad L\{\phi(0,t)\} = \frac{1}{s^2}, \quad \text{i.e.,} \quad f(0,s) = \frac{1}{s^2}. \tag{5}$$

Putting $x = 0$ in equation (4), and comparing with equation (5), we get

$$B = \frac{1}{s^2}. \tag{6}$$

Hence

$$f(x,s) = \frac{1}{s^2} e^{-\xi x}. \tag{7}$$

Taking inverse transform,

$$\phi(x,t) = L^{-1}\left\{\frac{1}{s^2} e^{-\frac{s}{c}x}\right\}.$$

By using second shifting property equation (15.42)

$$\phi(x,t) = \begin{cases} t - \dfrac{x}{c}, & \text{for } t > \dfrac{x}{c} \\ 0, & \text{for } t < \dfrac{x}{c}. \end{cases}$$

EXAMPLE 15.37. *Solve the equation*

$$\frac{\partial \phi}{\partial t} = h^2 \frac{\partial^2 \phi}{\partial x^2},$$

under the given conditions, $\phi(0,t) = 0$, $\phi(l,t) = 0$ *and* $\phi(x,0) = \sin(\pi x/l)$.

Solution: We have

$$\frac{\partial \phi}{\partial t} = h^2 \frac{\partial^2 \phi}{\partial x^2}. \tag{1}$$

Taking Laplace transform on both sides

$$sf(x,s) - \phi(x,0) = h^2 \frac{d^2 f}{dx^2}, \tag{2}$$

where

$$f(x,s) = L\{\phi(x,t)\}. \tag{3}$$

Using initial condition, $\phi(x,0) = \sin(\pi x/l)$, we get from equation (2)

$$\frac{d^2 f}{dx^2} - \frac{sf}{h^2} = -\frac{1}{h^2} \sin \frac{\pi x}{l}. \tag{4}$$

The solution of homogeneous equation in (4) is given by

$$f(x,s) = Ae^{(\sqrt{s}/h)x} + Be^{-(\sqrt{s}/h)x}. \tag{5}$$

Assuming the particular integral as

$$f = c \sin \frac{\pi x}{l} \tag{6}$$

we find

$$-c\left[\frac{\pi^2}{l^2} + \frac{s}{h^2}\right] = -\frac{1}{h^2}$$

which gives

$$c = \frac{1}{(\pi^2 h^2/l^2) + s}. \tag{7}$$

From equations (5) and (7), we get the complete solution of equation (4) as

$$f(x,s) = Ae^{(\sqrt{s}/h)x} + Be^{-(\sqrt{s}/h)x} + \frac{1}{(\pi^2 h^2/l^2) + s} \sin \frac{\pi x}{l}. \tag{8}$$

From boundary conditions, we have

$$\phi(0,t) = 0 \quad \text{and} \quad \phi(l,t) = 0$$

$$\text{i.e.,} \quad f(0,s) = 0 \quad \text{and} \quad f(l,s) = 0. \tag{9}$$

Using these conditions in (8), we find $A + B = 0$ and $Ae^{(\sqrt{s}/h)l} + Be^{-(\sqrt{s}/h)l} = 0$.
This gives $A = B = 0$.
Hence, from equation (8), we have

$$f(x,s) = \sin \frac{\pi x}{l} \frac{1}{(\pi h/l)^2 + s}.$$

Taking inverse Laplace transform, we get

$$\phi(x,t) = \sin \frac{\pi x}{l} e^{-(\pi h/l)^2 t}.$$

EXAMPLE 15.38. *A semi-infinite solid rod ($x > 0$) has its initial temperature zero everywhere. A constant heat flux is applied at the face $x = 0$. Find the temperature at any point of the rod, given that*

$$L^{-1}\left\{\frac{e^{-\lambda\sqrt{s}}}{s^{3/2}}\right\} = \sqrt{\frac{t}{\pi}}e^{-\lambda^2/4t^2} - \frac{\lambda}{2}erfc\left(\frac{\lambda}{2\sqrt{t}}\right),$$

where

$$erfc(x) = \frac{2}{\sqrt{\pi}}\int_x^\infty e^{-t^2}\,dt.$$

Solution: The heat conduction equation is

$$\frac{\partial \phi}{\partial t} = h^2\frac{\partial^2 \phi}{\partial x^2}. \tag{1}$$

Taking Laplace transforms of both sides, we get

$$sf(x,s) - \phi(x,0) = h^2\frac{d^2 f}{dx^2}, \tag{2}$$

where

$$f(x,s) = L\{\phi(x,t)\}. \tag{3}$$

Initial and boundary conditions are

$$\phi(x,0) = 0 \tag{4}$$

and

$$-k\frac{\partial \phi}{\partial x}\bigg|_{x=0} = Q, \tag{5}$$

where k is the thermal conductivity.

Using condition (4), we get from equation (2)

$$\frac{d^2 f}{dx^2} - \frac{s}{h^2}f = 0. \tag{6}$$

Its solution is

$$f(x,s) = Ae^{\sqrt{s}x/h} + Be^{-\sqrt{s}x/h}. \tag{7}$$

Since temperature, i.e., $\phi(x,t)$ is finite as $x \longrightarrow \infty$, $f(x,s)$ should also remain finite. Hence $A = 0$. Equation (7) takes the form,

$$f(x,s) = Be^{-\sqrt{s}x/h}. \tag{8}$$

Therefore,

$$\frac{df}{dx} = -\frac{\sqrt{s}}{h}Be^{-\sqrt{s}x/h}. \tag{9}$$

For $x = 0$,

$$\frac{df}{dx}\bigg|_{x=0} = -\frac{\sqrt{s}}{h}B. \tag{10}$$

From equation (5) on taking Laplace transform

$$-kL\left\{\frac{\partial \phi}{\partial x}\bigg|_{x=0}\right\} = QL(1) = \frac{Q}{s}, \quad \text{or,} \quad \frac{df}{dx}\bigg|_{x=0} = -\frac{Q}{ks}. \tag{11}$$

From equations (10) and (11), we have

$$-\frac{\sqrt{s}}{h}B = -\frac{Q}{ks}, \quad \text{i.e.,} \quad B = \frac{Qh}{ks^{3/2}}. \tag{12}$$

Hence, from equations (8) and (12)

$$f(x,s) = \frac{Qh}{k}\cdot\frac{e^{-\sqrt{s}x/h}}{s^{3/2}} = \frac{Qh}{k}\cdot\frac{e^{-\lambda\sqrt{s}}}{s^{3/2}}, \quad \text{where } \lambda = \frac{x}{h}. \tag{13}$$

Taking inverse Laplace transform

$$\phi(\lambda,t) = \frac{Qh}{k}L^{-1}\left[\frac{e^{-\lambda\sqrt{s}}}{s^{3/2}}\right] = \frac{Qh}{k}\left[\sqrt{\frac{t}{\pi}}e^{-\lambda^2/4t^2} - \frac{\lambda}{2}erfc\left(\frac{\lambda}{2\sqrt{t}}\right)\right].$$

Putting $\lambda = \frac{x}{h}$, we finally get

$$\phi(x,t) = \frac{Qh}{k}\left[\sqrt{\frac{t}{\pi}}\exp\left\{-\frac{x^2}{4h^2t^2}\right\} - \frac{x}{2h}erfc\left(\frac{x}{2h\sqrt{t}}\right)\right].$$

Table 15.1: Laplace Transforms

	$F(t)$	$f(s) = L[F(t)]$		$F(t)$	$f(s) = L[F(t)]$
1.	k	$\frac{k}{s}$	13.	$t\sin at$	$\frac{2as}{(s^2+a^2)^2}$
2.	t^n ($n > 0$, integer)	$\frac{n!}{s^{n+1}}$, $s > 0$	14.	$\sin at - at\cos at$	$\frac{2a^3}{(s^2+a^2)^2}$
3.	t^{k-1} ($k > 0$)	$\frac{\Gamma(k)}{s^k}$	15.	$t\cos at$	$\frac{s^2-a^2}{(s^2+a^2)^2}$
4.	e^{at}	$\frac{1}{s-a}$, $s > a$	16.	$\sin at + at\cos at$	$\frac{2as^2}{(s^2+a^2)^2}$
5.	$t^{n-1}\,e^{at}$	$\frac{(n-1)!}{(s-a)^n}$ $(n = 1, 2, \ldots)$	17.	$e^{\lambda t}\sinh at$	$\frac{a}{(s-\lambda)^2-a^2}$
6.	$t^{k-1}\,e^{at}$	$\frac{\Gamma(k)}{(s-a)^k}$, $k > 0$	18.	$e^{\lambda t}\cosh at$	$\frac{s-\lambda}{(s-\lambda)^2-a^2}$
7.	$\sin at$	$\frac{a}{s^2+a^2}$, $s > 0$	19.	$t\sinh at$	$\frac{2as}{(s^2-a^2)^2}$
8.	$\cos at$	$\frac{s}{s^2+a^2}$, $s > 0$	20.	$t\cosh at$	$\frac{s^2+a^2}{(s^2-a^2)^2}$
9.	$\sinh at$	$\frac{a}{s^2-a^2}$, $s > \lvert a \rvert$	21.	$t^2\sin at$	$\frac{2a(3s^2-a^2)}{(s^2+a^2)^3}$
10.	$\cosh at$	$\frac{s}{s^2-a^2}$, $s > \lvert a \rvert$	22.	$t^2\cos at$	$\frac{2s(s^2-3a^2)}{(s^2+a^2)^3}$
11.	$e^{\lambda t}\sin at$	$\frac{a}{(s-\lambda)^2+a^2}$	23.	$t^2\sinh at$	$\frac{2a(3s^2+a^2)}{(s^2-a^2)^3}$
12.	$e^{\lambda t}\cos at$	$\frac{s-\lambda}{(s-\lambda)^2+a^2}$	24.	$t^2\cosh at$	$\frac{2s(s^2+3a^2)}{(s^2-a^2)^3}$

EXERCISE

1. Find Laplace transforms of each of the following functions:
   (a) $3\cosh(at) - 5\sinh(at)$;
   (b) $2\sin^2(at)$;
   (c) $(e^{at} - 1)/a$;
   (d) $e^{2t} + 4e^{-3t}$.

2. Find Laplace transforms of:
   (a) $e^{-2t}\sin 2t$;
   (b) $e^t(t+2)^2$;
   (c) $e^{3t}\cosh(2t)$;
   (d) $e^t(\cos t - 3\sin 2t)$.

3. Using the theorem of Laplace transform of derivatives, find
   (a) $L(t^2e^{at})$;
   (b) $L(t^2\sin at)$;
   (c) $L(te^t\cos t)$;
   (d) $L\left[\sin 2t(t^2 - 4t)\right]$.

4. Find the Laplace transform of the following periodic functions:

$$\text{(a)}\ F(t) = \begin{cases} \dfrac{h}{a}t, & \text{for } 0 < t < a \\ \dfrac{h}{a}(2a - t), & \text{for } a < t < 2a \end{cases} \qquad \text{(b)}\ f(t) = \begin{cases} 1, & \text{for } 0 < t < \dfrac{T}{2} \\ 0, & \text{for } \dfrac{T}{2} < t < T \end{cases}$$

5. Show that

$$\text{(a)}\ L\{erf(\sqrt{t})\} = \frac{1}{s\sqrt{s+1}}; \qquad \text{(b)}\ L\left\{erfc\left(\frac{a}{2\sqrt{t}}\right)\right\} = \frac{1}{s}e^{-a\sqrt{s}};$$

$$\text{(c)}\ L\{J_1(t)\} = 1 - \frac{s}{\sqrt{s^2+1}}$$

$J_1(t) =$ Bessel's function of order one.

[*Hints*: $-J_1(t) = \dfrac{d}{dt}\{J_0(t)\}$]

6. Find inverse Laplace transforms of:

$$\text{(a)}\ \frac{s^2+4}{(s-1)(s^2+5s+4)}; \qquad \text{(d)}\ \frac{s^2-3s+2}{s^2(s-1)^2(s^2-5s+4)};$$

$$\text{(b)}\ \frac{s^2+2s+1}{(s-1)(s^2+2s-3)}; \qquad \text{(e)}\ \frac{s^2}{(s^2+1)(s^2+9)};$$

$$\text{(c)}\ \frac{8}{s^2(s-2)}; \qquad \text{(f)}\ \frac{2}{(s^2+4)^2(s^2+1)^2}.$$

7. Show that

$$L^{-1}\left[\frac{1}{1+\sqrt{1+s}}\right] = -1 + erf(\sqrt{t}) + \frac{e^{-t}}{\sqrt{\pi t}}.$$

8. Using the convolution theorem, find the inverse Laplace transforms of:

$$\text{(a)}\ \frac{1}{(s^2+a^2)^2}; \qquad \text{(b)}\ \frac{1}{s^2(s^2+a^2)}.$$

9. Using the convolution theorem, show that

$$\text{(a)}\ B(m,n) = \frac{\Gamma(m)\Gamma(n)}{\Gamma(m+n)};$$

(b) $L^{-1}\left[\frac{f(s)}{(s+a)^2}\right] = \int_0^t xe^{-ax}F(t-x)dx.$

10. Show that

(a) $\int_0^\infty t^2 e^{-t} \sin t \, dt = 0;$ (c) $\int_0^\infty \frac{\sin tx}{\sqrt{x}} dx = \sqrt{\frac{\pi}{2t}}, \; t > 0;$

(b) $\int_0^\infty e^{-t} erf(\sqrt{t}) dt = \frac{1}{\sqrt{2}};$ (d) $\int_0^\infty \frac{\cos x}{1+x^4} dx = \frac{\pi}{2e}.$

11. Solve the following differential equations by Laplace transform method:

(a) $\frac{d^2y}{dt^2} + 7\frac{dy}{dt} + 12y = 0, \; y(0) = 2, \; y'(0) = 0;$

(b) $\frac{d^2\phi}{dt^2} + k^2\phi = A \sin \omega t, \; \phi(0) = \phi'(0) = 0, \; \omega \neq k;$

(c) $\frac{d^2y}{dt^2} + 4\frac{dy}{dt} + 4y = 4\sin 2t, \; y(0) = 1, \; y'(0) = 0;$

(d) $x\frac{d^2y}{dx^2} + (2x+3)\frac{dy}{dx} + (x+3)y = ae^{-x}.$

12. Solve the partial differential equations by Laplace transform method

(a) $\frac{\partial^2\phi}{\partial x^2} - \frac{\partial^2\phi}{\partial t^2} = xt$, under the conditions

$\phi = \frac{\partial\phi}{\partial t} = 0$ for $t = 0$;

(b) $\frac{\partial^2\phi}{\partial t^2} = c^2\frac{\partial^2\phi}{\partial x^2}, \; (t > 0, \; x > 0)$

under the conditions, (i) $\phi = \phi' = 0$, for $t = 0$ and (ii) $\phi(0, t) = A \sin at$. The function $\phi(x, t)$ should be finite for $x \longrightarrow \infty$.

13. The differential equation that describes the deflection of an elastic beam under uniform load is

$$\frac{d^4y}{dx^4} = \frac{q}{EI},$$

where EI = flexural rigidity and q = load intensity.

Find $y(x)$ under the conditions (i) $y = y'' = 0$, for $x = 0$ and (ii) $y = y'' = 0$, for $x = l$, where l is the length of beam.

14. A semi-infinite rod $(x > 0)$ is initially at zero temperature. At time $t > 0$, a constant temperature source ϕ_0 is applied and maintained at face $x = 0$. Show that the temperature at any point of the rod at any time $t > 0$ is given by

$$\phi_0 \left\{1 - erf\left(\frac{x}{2h\sqrt{t}}\right)\right\}.$$

15. Solve the integral equation:

$$F(t) = at + \int_0^t F(\xi) \sin(t - \xi) d\xi.$$

ANSWERS

1. (a) $\dfrac{3s-5a}{s^2-a^2}, \quad s>a;$
(b) $\dfrac{4a^2}{s(s^2+4a^2)}, \quad s>0;$
(c) $\dfrac{1}{s(s-a)}, \quad s>0;$
(d) $\dfrac{5(s-1)}{(s-2)(s+3)}, \quad s>2.$

2. (a) $\dfrac{2}{(s+2)^2-4};$
(b) $\dfrac{2}{(s-1)^3}+\dfrac{2}{(s-1)^2}+\dfrac{4}{(s-1)};$
(c) $\dfrac{s-3}{s^2-6s+5};$
(d) $\dfrac{s-1}{s^2-2s}-\dfrac{6}{s^2-4s}.$

3. (a) $\dfrac{2}{(s-a)^3};$
(b) $\dfrac{2a(3s^2-a^2)}{(s^2+a^2)^3};$
(c) $\dfrac{s(s-2)}{(s^2-2s+2)^2};$
(d) $\dfrac{4(s^2-4s^3-16s-4)}{(s^2+4)^3}.$

4. (a) $\dfrac{h}{as^2}\tanh\left(\dfrac{as}{2}\right);$
(b) $\dfrac{1}{s}\dfrac{e^{-2s}}{(1-e^{-4s})}.$

6. (a) $\dfrac{1}{2}e^t-\dfrac{5}{6}e^{-t}+\dfrac{4}{3}e^{-4t};$
(b) $\dfrac{1}{4}e^{-3t}+\dfrac{3}{4}e^t+te^t;$
(c) $-4t-2+2e^{2t};$
(d) $\dfrac{7}{8}+\dfrac{t}{2}+\left(\dfrac{8}{3}+\dfrac{t}{3}\right)e^t+\dfrac{1}{72}e^{4t};$
(e) $\dfrac{1}{8}\cos t(2\sin 2t+\sin 4t)-\dfrac{1}{8}\sin t(\cos 4t+2\cos 2t-6);$
(f) $\dfrac{8}{27}(\cos 2t-\cos t)+\dfrac{1}{72}(\sin 2t-2t\cos 2t)+\dfrac{1}{9}(\sin t-t\cos t).$

8. (a) $\dfrac{1}{2a^2}[\sin at-t\cos at];$
(b) $\dfrac{1}{a^3}(at-\sin at).$

11. (a) $y(t)=2e^{-3t};$
(b) $\phi(t)=\dfrac{A}{\omega^2-k^2}\left(\dfrac{\omega}{k}\sin kt-\sin\omega t\right);$
(c) $y(t)=\left(\dfrac{3}{2}+3t\right)e^{-2t}-\dfrac{1}{2}\cos 2t;$
(d) $y(x)=\left(A+\dfrac{a}{3}t\right)e^{-t}, \quad A=\text{constant}.$

12. (a) $\phi(x,t)=-\dfrac{1}{6}xt^3;$
(b) $A\sin a\left(t-\dfrac{x}{c}\right)$, for $t>\dfrac{x}{c};$

13. $y(x)=\dfrac{q}{24EI}\left(x^4-2x^3l+xl^3\right).$

15. $a+t\left(1+\dfrac{1}{6}t^2\right).$

Chapter 16

Delta Functions and Green's Functions

DIRAC'S DELTA FUNCTION

16.1 Definitions

Dirac's delta functions are used to get a precise notation for dealing with quantities involving a certain kind of infinity. In one dimension, the quantity $\delta(x)$ depending on a parameter x is such that (i) $\delta(x)$ is zero everywhere except at $x = 0$ and (ii) At $x = 0$, $\delta(x)$ is so large that the integral of the function over an interval containing the point $x = 0$ is equal to unity. Thus,

$$\left.\begin{aligned} &\delta(x) = 0, \text{ for } x \neq 0 \\ \text{and } &\int_{-\infty}^{\infty} \delta(x)dx = 1 \end{aligned}\right\}. \tag{16.1}$$

As an example, consider a beam of length l supported at its ends. An impulse load $F_0(t)$ of high value is thrust upon it at point where we take origin. About this origin we consider a small region $(-\epsilon, \epsilon)$. The exact shape of the load function in this domain does not matter; but the integral of the function in this domain is unity. Then, in the limit $\epsilon \longrightarrow 0$, this function becomes delta function, so that we represent the impulsive load function as $F_0(t)\delta(x)$.

The delta function is not a mathematical function in usual sense, and hence, it does not have a definite value for each point of its domain.

The delta function is defined by giving a rule for integration of its product with another function $f(x)$ in the following manner:

$$\int_{-\infty}^{\infty} f(x)\delta(x)dx = f(0). \tag{16.2}$$

With a shift of origin of coordinates, equation (16.2) can be written as

$$\int_{-\infty}^{\infty} f(x)\delta(x-a)dx = f(a), \tag{16.3}$$

which is valid for any continuous function. Since, from definition, $\delta(x - a) = 0$ everywhere except at $x = a$, we can replace the function $f(x)$ by its value at a while integrating. Thus, from second of equation (16.1) the result in (16.3) directly follows.

Equation (16.3) shows that the process of multiplying a function of x by $\delta(x - a)$ and then integrating over all x is same as the process of substituting a for x.

The range of integration need not be from $-\infty$ to ∞. It may be over any region containing the critical point at which the δ function does not vanish. Thus, we may not even mention the limits of integration when it would be understood that the limits of integration are taken suitably.

16.2 Alternative Expressions for Delta Functions

There are many other ways of expressing the delta functions. We mention a few of them.

(1) Delta functions may be defined as the differential coefficient $\xi'(x)$ of *Heaviside unit step* function which is given by

$$\xi(x) = \begin{cases} 0, & \text{for } x < 0 \\ 1, & \text{for } x > 0. \end{cases} \tag{16.4}$$

In order to verify that $\delta(x)$ is equivalent to $\xi'(x)$, we substitute $\xi'(x)$ for $\delta(x)$ in the equation (16.2) and integrate it within limits $(-a, b)$, where a and b are two positive integers. Then we get, integrating by parts,

$$\begin{aligned}\int_{-a}^{b} f(x)\xi'(x)dx &= \xi(x)f(x)\Big|_{-a}^{b} - \int_{-a}^{b} f'(x)\xi(x)dx \\ &= f(b) - \int_{0}^{b} f'(x)dx \quad \text{[using equation (16.4)]} \\ &= f(0)\end{aligned}$$

which is in agreement with equation (16.2).

(2) Delta function can be expressed as

$$\delta(x) = \lim_{\lambda \longrightarrow \infty} \frac{\sin \lambda x}{\pi x}. \tag{16.5}$$

Now

$$\lim_{x \longrightarrow 0} \frac{\sin \lambda x}{\pi x} = \lim_{x \longrightarrow 0} \frac{\lambda \cos \lambda x}{\pi} = \frac{\lambda}{\pi}. \tag{16.6}$$

Thus, at $x = 0$

$$\frac{\sin \lambda x}{\pi x} = \frac{\lambda}{\pi}.$$

For other values of x, the value of the function oscillates with a period $2\pi/\lambda$. Moreover

$$\int_{-\infty}^{\infty} \frac{\sin \lambda x}{\pi x} dx = 1. \tag{16.7}$$

Thus, the function defined by equation (16.5) has all the properties of a delta function.

(3) Delta function can also be expressed in terms of Gaussian function as

$$\sigma(x) = \lim_{\sigma \longrightarrow 0} \frac{1}{\sigma\sqrt{2\pi}} \exp\left(-\frac{x^2}{2\sigma^2}\right), \quad \sigma > 0. \tag{16.8}$$

For $x = 0$, the value of Gaussian function is $\dfrac{1}{\sigma\sqrt{2\pi}}$, which tends to infinity as $\sigma \longrightarrow 0$.

Thus, at $x = 0$, the function is infinitely peaked for $\sigma \longrightarrow 0$. The function has a width $= 2\sigma$ which $\longrightarrow 0$ as $\sigma \longrightarrow 0$. Moreover

$$\int_{-\infty}^{\infty} \frac{1}{\sigma\sqrt{2\pi}} \exp\left(-\frac{x^2}{2\sigma^2}\right) = 1. \tag{16.9}$$

Hence, in the limit $\sigma \longrightarrow 0$, the Gaussian function has all the properties of a delta function.

(4) A very commonly used expression for delta function is

$$\delta(x) = \frac{1}{2\pi}\int_{-\infty}^{\infty} e^{i\theta x} d\theta. \tag{16.10}$$

To verify this, we consider the integral

$$\int_{-\infty}^{\infty} e^{i\theta x} d\theta = \lim_{\lambda \longrightarrow \infty} \int_{-\lambda}^{\lambda} e^{i\theta x} d\theta = \lim_{\lambda \longrightarrow \infty} \frac{e^{ix\lambda} - e^{-ix\lambda}}{ix} = 2\pi \lim_{\lambda \longrightarrow \infty} \frac{e^{ix\lambda} - e^{-ix\lambda}}{2i(\pi x)}$$

$$= 2\pi \lim_{\lambda \longrightarrow \infty} \frac{\sin \lambda x}{\pi x} = 2\pi\delta(x) \quad \text{[by equation (16.5)]}.$$

16.3 Equations Satisfied by Delta Function

There are a number of equations which are satisfied by delta functions. These are

(i) $$\delta(-x) = \delta(x) \tag{16.11}$$

(ii) $$x\delta(x) = 0 \tag{16.12}$$

(iii) $$\delta(ax) = \frac{1}{a}\delta(x), \quad a > 0 \tag{16.13}$$

(iv) $$\delta\left(x^2 - a^2\right) = \frac{1}{2a}\{\delta(x+a) + \delta(x-a)\}, \quad a > 0 \tag{16.14}$$

(v) $$f(x)\delta(x-a) = f(a)\delta(x-a) \tag{16.15}$$

(vi) $$\int \delta(a-x)dx\delta(x-b) = \delta(a-b) \tag{16.16}$$

The meaning of these equations is that both sides of any of these equations will give same results when used as a multiplying factor in an integrand. We prove these equations.

Equation (16.11) states that the delta function is an even function of x which is obvious. To verify equation (16.12), we take a continuous function $f(x)$, and find that

$$\int f(x)x\delta(x)dx = 0 \tag{16.17}$$

This shows that $x\delta(x)$ as a factor in an integrand is equivalent to zero. To prove equation (16.13), we consider

$$\int_{-\infty}^{\infty} f(x)\delta(ax)dx$$

which is equal to $f(0)\int_{-\infty}^{\infty} \delta(ax)dx.$

Putting $\xi = ax$

$$\int_{-\infty}^{\infty} f(x)\delta(ax)dx = f(0)\int_{-\infty}^{\infty} \frac{\delta(\xi)}{a}d\xi = \frac{f(0)}{a} \tag{16.18}$$

But, we have

$$\int_{-\infty}^{\infty} f(x)\delta(x)dx = f(0). \tag{16.19}$$

Comparing equations (16.18) and (16.19), we get (16.13).

Equation (16.14) shows that the delta function has two components $(x-a)$ and $(x+a)$. Hence

$$\begin{aligned}\delta\left(x^2-a^2\right) &= \delta[(x-a)(x+a)]\\ &= \frac{1}{|x+a|}\delta(x-a)+\frac{1}{|x-a|}\delta(x+a) \quad \text{[using equation (16.13)]}\end{aligned}$$

Hence, considering an arbitrary continuous function $f(x)$, we can write

$$\begin{aligned}\int f(x)\delta\left(x^2-a^2\right)dx &= \int \frac{f(x)}{|x+a|}\delta(x-a)dx + \int \frac{f(x)}{|x-a|}\delta(x+a)dx\\ &= \frac{1}{|2a|}[f(a)+f(-a)] \quad \text{[using equation (16.13)]}\end{aligned} \tag{16.20}$$

The right-hand side of equation (16.20) can be written as

$$\frac{1}{|2a|}\left[\int f(x)\delta(x-a)dx + \int f(x)\delta(x+a)dx\right]$$

Hence, we get

$$\delta\left(x^2-a^2\right) = \frac{1}{|2a|}[\delta(x-a)+\delta(x+a)].$$

Equation (16.15) is also obvious, since $\delta(x-a)$ is not zero for $x=a$ but is zero for all other values of x, the product $f(x)\delta(x-a)$ will remain non-zero for $x=a$ and will result in $f(a)\delta(x-a)$.

To verify equation (16.16), we consider any continuous function $f(a)$. Taking a as variable of integration, we have

$$\begin{aligned}\int f(a)da\int \delta(a-x)dx\delta(x-b) &= \int \delta(x-b)dx\int f(a)\delta(a-x)da\\ &= \int \delta(x-b)dxf(x) \quad \text{[using equation (16.3) by interchanging } a \text{ and } x]\\ &= \int f(a)da\delta(a-b).\end{aligned} \tag{16.21}$$

Thus, both the sides of equation (16.16) are equivalent when multiplied to an integrand with a as a variable of integration. Similarly, considering a function $f(b)$ with b as a variable of integration, we can prove the equivalence in equation (16.16).

16.4 Some Illustrations of Delta Functions

(1) Let us consider the differentiation of $\log x$, for which we have the usual formula

$$\frac{d}{dx}(\log x) = \frac{1}{x}. \tag{16.22}$$

The function $1/x$ is not defined in the neighbourhood of the point $x = 0$. Hence, in order to make this function well-defined in the neighbourhood of $x = 0$, we impose the extra condition that its integral from $-\epsilon$ to ϵ vanishes, where ϵ is an infinitesimally small positive number. But using this condition in equation (16.22) we find that the left-hand side equals $\log(-1)$ whereas the right-hand side vanishes. This leads to the conclusion that the equation (16.22) is not correct. Now, if we take principal values, $\log x$ has a purely imaginary term $i\pi$ for negative values of x. This purely imaginary term vanishes discontinuously as x passes through the value zero. On differentiation, this purely imaginary term gives $-i\pi\delta(x)$. Thus, the equation (16.22) is corrected as

$$\frac{d}{dx}(\log x) = \frac{1}{x} - i\pi\delta(x). \tag{16.23}$$

(2) Delta function is used very often to represent the condition of orthonormality of a system of functions. Thus, for the functions $\phi_n(x)$ belonging to a continuous spectrum, the orthonormality conditions is described by

$$\int_{-\infty}^{\infty} \phi_n^*(x)\phi_n(x') = \delta(x - x'). \tag{16.24}$$

Similar expression exists for functions corresponding to a discrete spectrum. Let us determine it.

Since the functions $\phi_n(x)$ constitute an orthonormal set, we have

$$\int_{-\infty}^{\infty} \phi_m^*(x)\phi_n(x)dx = \delta_{mn}. \tag{16.25}$$

Now, we can expand any well-behaved function $\psi(x)$ in terms of functions $\phi_n(x)$ as

$$\psi(x) = \sum_{n=1}^{\infty} A_n\phi_n(x). \tag{16.26}$$

In order to determine the coefficients A_n, we multiply equation (16.26) by $\phi_m^*(x)$ and integrate. Then we have

$$\begin{aligned}\int_{-\infty}^{\infty} \phi_m^*(x)\psi(x)dx &= \sum_{n=1}^{\infty} A_n \int_{-\infty}^{\infty} \phi_m^*(x)\phi_n(x)dx \\ &= \sum_{n=1}^{\infty} A_n\delta_{mn} \quad \text{[using equation (16.25)]} \\ &= A_m. \end{aligned} \tag{16.27}$$

Substituting the value of A_m from equation (16.27) into equation (16.26), we get

$$\begin{aligned}\psi(x) &= \sum_{n=1}^{\infty} \left[\int_{-\infty}^{\infty} \phi_m^*(x')\psi(x')dx'\right] \phi_n(x) \\ &= \int_{-\infty}^{\infty} \psi(x')dx' \left[\sum \phi_m^*(x')\phi_n(x)\right]. \end{aligned} \tag{16.28}$$

From equation (16.3), we can write

$$\psi(x) = \int_{-\infty}^{\infty} \psi(x')dx'\delta(x - x'). \tag{16.29}$$

Hence, from equations (16.28) and (16.29), we get

$$\delta(x - x') = \sum_{n=1}^{\infty} \phi_m^*(x')\phi_n(x). \tag{16.30}$$

Thus, equation (16.30) gives the orthonormality condition for a system of functions $\phi_n(x)$ corresponding to a discrete spectrum.

16.5 Delta Functions in Three-Dimensions

The concept of delta functions are easily extended in three dimensions. The three-dimensional Dirac delta function is defined as

$$\left.\begin{aligned} \delta(\mathbf{r}) = 0, \quad \text{for } \mathbf{r} \neq 0 \\ \text{and} \int \delta(\mathbf{r})d^3\mathbf{r} = 1 \end{aligned}\right\} \tag{16.31}$$

where

$$\left.\begin{aligned} \delta(\mathbf{r}) = \delta(x)\delta(y)\delta(z) \\ \text{and} \quad d^3\mathbf{r} = dxdydz \end{aligned}\right\}.$$

Equations (16.2) and (16.3) in three dimensions are generalised in the form

$$\int \delta(\mathbf{r})f(\mathbf{r})d^3\mathbf{r} = f(0) \tag{16.32}$$

$$\text{and} \quad \int f(\mathbf{r})\delta(\mathbf{r} - \mathbf{r}')d^3\mathbf{r} = f(\mathbf{r}'). \tag{16.33}$$

We now intend to find expressions for Dirac delta functions in cylindrical and spherical coordinates.

Cylindrical Coordinates

In this system, the coordinates are r, ϕ, z which are connected with Cartesian coordinates in the following manner:

$$\left.\begin{aligned} x = r\cos\phi, \quad y = r\sin\phi, \quad z = z \\ \text{and} \quad dxdydz = rdrd\phi dz \end{aligned}\right\}. \tag{16.34}$$

Any function $f(x, y, z)$ is transformed to a function $f(r, \phi, z)$. Then, using equation (16.34) we write

$$\int f(r, \phi, z)\delta(\mathbf{r} - \mathbf{r}')rdrd\phi dz = f(r', \phi', z') \tag{16.35}$$

which is possible only when

$$\delta(\mathbf{r}-\mathbf{r}') = \frac{\delta(r-r')\delta(\phi-\phi')\delta(z-z')}{r} \tag{16.36}$$

Equation (16.36) is the explicit form of δ function. In particular, if we consider polar coordinates in two-dimension, $z' = 0$. Then equation (16.35) gives

$$\int_{r=0}^{\infty}\int_{\phi=0}^{2\pi} f(r,\phi)\delta(\mathbf{r}-\mathbf{r}')rdrd\phi = \int_{r=0}^{\infty} f(r)\cdot(2\pi r)\delta(r) = f(r'),$$

which is possible only if

$$\delta(x)\delta(y) = \frac{\delta(r)}{2\pi r}. \tag{16.37}$$

Spherical Coordinates

In this system of coordinates

$$\left.\begin{aligned} x &= r\sin\theta\cos\phi \\ y &= r\sin\theta\sin\phi \\ z &= r\cos\theta \end{aligned}\right\} \tag{16.38}$$

where

$$0 \le \theta \le \pi, 0 \le \phi \le 2\pi \quad \text{and} \quad 0 \le r \le \infty. \tag{16.39}$$

Elemental volume is

$$dxdydz = r^2 \sin\theta d\theta d\phi dr. \tag{16.40}$$

Hence, equation (16.35) becomes

$$\int f(r,\theta,\phi)\delta(\mathbf{r}-\mathbf{r}')r^2\sin\theta d\theta d\phi dr = f(r',\theta',\phi') \tag{16.41}$$

which is possible only when

$$\delta(\mathbf{r}-\mathbf{r}') = \frac{\delta(r-r')\delta(\theta-\theta')\delta(\phi-\phi')}{r^2\sin\theta} \tag{16.42}$$

Equation (16.42) is the explicit form of δ function in spherical coordinates.
In particular, if the function $f(r,\theta,\phi)$ is independent of θ and ϕ, equation (16.41) gives

$$\int_{r=0}^{\infty} f(r)r^2\delta(\mathbf{r}-\mathbf{r}')dr \int_0^{\pi} \sin\theta d\theta \int_0^{2\pi} d\phi = f(r')$$

$$\text{or,} \quad 4\pi\int_0^{\infty} f(r)r^2\delta(\mathbf{r}-\mathbf{r}')dr = f(r')$$

Hence

$$\delta(\mathbf{r}-\mathbf{r}') = \frac{\delta(r-r')}{4\pi r^2} \tag{16.43}$$

16.6 Integral Transforms of Delta Functions

We know, that Fourier transform of a function $\psi(t)$ is defined as

$$\text{F.T.}\{\psi(t)\} = \frac{1}{\sqrt{2\pi}} \int_{-\infty}^{\infty} \psi(t) e^{-i\xi t} dt$$

Hence

$$\text{F.T.}\{\delta(t-a)\} = \frac{1}{\sqrt{2\pi}} \int_{-\infty}^{\infty} \delta(t-a) e^{-i\xi t} dt$$

$$= \frac{1}{\sqrt{2\pi}} e^{-i\xi a} \quad \text{[using equation (16.3)]}$$

In particular, if $a = 0$

$$\text{F.T.}\{\delta(t)\} = \frac{1}{\sqrt{2\pi}} \tag{16.44}$$

Similarly, Fourier inverse transform of $\delta(\xi - \xi_0)$ is

$$\int_{-\infty}^{\infty} \frac{1}{\sqrt{2\pi}} \delta(\xi - \xi_0) e^{i\xi t} d\xi = \frac{1}{\sqrt{2\pi}} e^{i\xi_0 t} \tag{16.45}$$

Again, the Laplace transform of a function $\psi(t)$ is given by

$$L\{\psi(t)\} = \int_0^{\infty} \psi(t) e^{-\xi t} dt \tag{16.46}$$

Hence

$$L\{\delta(t-a)\} = \int_0^{\infty} \delta(t-a) e^{-\xi t} dt = e^{-\xi a}. \tag{16.47}$$

In particular, if $a = 0$

$$L\{\delta(t)\} = 1 \tag{16.48}$$

EXAMPLE 16.1. *Show that, for the function*

$$f(x) = \frac{\sigma}{2} \sec h^2 \sigma(x-a), \quad \sigma > 0,$$

the value of $\int_{-\infty}^{\infty} f(x) dx = 1$ *for all values of* σ.
Also show that

$$\lim_{\sigma \longrightarrow \infty} f(x) = \delta(x-a).$$

Solution: The given integral

$$I = \int_{-\infty}^{\infty} f(x) dx = \int_{-\infty}^{\infty} \frac{\sigma}{2} \sec h^2 \sigma(x-a) dx.$$

$$= \int_0^{\infty} \sigma \sec h^2 \sigma(x-a) dx \quad \text{[since the function is even]}$$

$$= \int_0^{\infty} \sec h^2 \xi d\xi = \tanh \xi \Big|_0^{\infty} \quad [\xi = \sigma(x-a)]$$

i.e., $$I = \left[\frac{e^{\xi} - e^{-\xi}}{e^{\xi} + e^{-\xi}}\right]_0^{\infty} = \lim_{p \longrightarrow \infty} \left(\frac{e^p - e^{-p}}{e^p + e^{-p}}\right) = \lim_{p \longrightarrow \infty} \frac{1 - e^{-2p}}{1 + e^{-2p}} = 1.$$

Hence the integral is independent of σ.
Again

$$\lim_{\sigma \longrightarrow \infty} f(x) = \lim_{\sigma \longrightarrow \infty} \frac{\sigma}{2} \sec h^2 \sigma(x-a)$$

Putting $\xi = x - a$

$$\lim_{\sigma \longrightarrow \infty} f(x) = \lim_{\sigma \longrightarrow \infty} \frac{\sigma}{2} \frac{4}{(e^{\sigma\xi} + e^{-\sigma\xi})^2} = \lim_{\sigma \longrightarrow \infty} \frac{2\sigma}{e^{2\sigma\xi} + 2 + e^{-2\sigma\xi}}$$

$$= \lim_{\sigma \longrightarrow \infty} \frac{2}{2\xi\,(e^{2\sigma\xi} - e^{-2\sigma\xi})} \quad \text{[Hospital's rule]}$$

$$= 0.$$

Moreover, we have seen that

$$\int_{-\infty}^{\infty} f(x)dx = 1.$$

Hence, we can write

$$\lim_{\sigma \longrightarrow \infty} f(x) = \delta(x-a)$$

EXAMPLE 16.2. *Show that the function*

$$f(x) = A \exp\left\{-\frac{|x-a|}{\sigma}\right\}$$

satisfies the differential equation

$$f''(x) = \frac{1}{\sigma^2} f(x) - \frac{2}{\sigma}\sigma(x-a)$$

Solution: For $(x-a) > 0$,

$$f'(x) = -\frac{A}{\sigma} f(x) = -\frac{A}{\sigma} \exp\left\{-\frac{(x-a)}{\sigma}\right\}$$

and for $x - a < 0$,

$$f'(x) = \frac{A}{\sigma} f(x) = \frac{A}{\sigma} \exp\left\{\frac{(x-a)}{\sigma}\right\}.$$

Hence, at $x = a$, $f'(x)$ has a discontinuity of $-2A/\sigma$. Hence

$$f''(x) = \frac{A}{\sigma^2} f(x) - \frac{2A}{\sigma}\delta(x-a).$$

GREEN'S FUNCTION

16.7 Definition

Let us consider an inhomogeneous differential equation

$$L\phi(x) = f(x) \tag{16.49}$$

where $L =$ a differential operator and $f(x) =$ source function.

The differential equation (16.49) is supplemented by some boundary conditions which may be homogeneous or non-homogeneous. Assuming the source function to be unity, let $G(x, x')$ be the solution of the above differential equation considering appropriate boundary conditions. This function $G(x, x')$ is called *Green's function.* Thus we write

$$LG(x, x') = -\delta(x - x') \tag{16.50}$$

where $\delta(x - x')$ is the Dirac's delta function.

After determining the Green's function, the solution of equation (16.49) is obtained as

$$\phi(x) = -\int_a^b G(x, x')f(x')dx' \tag{16.51}$$

where $a \leq x' < b$. That the equation (16.51) is actually the solution of equation (16.49) can easily be verified. We operate by L both sides of equation (16.51) and get

$$L\phi(x) = -L\left[\int_a^b G(x, x')f(x')dx'\right] = -\int_a^b LG(x, x')f(x')dx'$$

$$= \int_a^b \delta(x - x')f(x')dx' \quad \text{[using equation (16.50)]}$$

i.e., $L\phi(x) = f(x)$.

Hence the function given by equation (16.51) satisfies the equation (16.49).

The most important properties of Green's function $G(x, x')$ are that

(i) It is symmetric; i.e.,

$$G(x, x') = G(x'x) \tag{16.52}$$

(ii) It is continuous everywhere including the point $x = x'$.

(iii) Its derivative with respect to x, i.e., $\frac{dG}{dx}$ is discontinuous at $x = x'$.

(iv) It satisfies the equation (16.50) for $a \leq x \leq b$.

(v) It is characteristic of the operator L and the prescribed boundary conditions.

Let us determine the Green's function for Laplace operator. We know that the potential of a unit point charge satisfies the equation

$$\nabla^2\left(\frac{1}{|\mathbf{r} - \mathbf{r}'|}\right) = -4\pi\delta(\mathbf{r} - \mathbf{r}') \tag{16.53}$$

which can be written as

$$\nabla^2\left(\frac{1}{4\pi|\mathbf{r} - \mathbf{r}'|}\right) = -\delta(\mathbf{r} - \mathbf{r}') \tag{16.54}$$

Comparing equation (16.54) with equation (16.50), we find that

$$G(\mathbf{r}, \mathbf{r}') = \frac{1}{4\pi\,|\,\mathbf{r} - \mathbf{r}'\,|} \tag{16.55}$$

Thus, equation (16.55) gives the explicit simple form of Green's function for Laplace operator.

16.8 Solution of Differential Equations

We have mentioned that the solution of a differential equation can be obtained in terms of the corresponding Green's function by using equation (16.51). We now generalise this for three-dimensional problems.

Let us consider the inhomogeneous differential equation

$$\nabla^2\phi(\mathbf{r}) = f(\mathbf{r}) \tag{16.56}$$

where the function $\phi(\mathbf{r})$ is defined in a region of volume V bounded by a closed surface S. Let the Green's function corresponding to $f(\mathbf{r}) = -\delta(\mathbf{r} - \mathbf{r}')$ is $G(\mathbf{r}, \mathbf{r}')$, so that

$$\nabla^2 G(\mathbf{r}, \mathbf{r}') = -\delta(\mathbf{r} - \mathbf{r}') \tag{16.57}$$

We multiply equation (16.56) by $G(\mathbf{r}, \mathbf{r}')$, subtract from the result $\phi(\mathbf{r}')$ times the equation (16.57) and, integrate throughout the volume V. Then we have

$$\iiint\limits_V [G(\mathbf{r}, \mathbf{r}')\nabla^2\phi(\mathbf{r}') - \phi(\mathbf{r}')\nabla^2 G(\mathbf{r}, \mathbf{r}')]\, d\mathbf{r}' = \iiint\limits_V [G(\mathbf{r}, \mathbf{r}')f(\mathbf{r}') + \phi(\mathbf{r}')\delta(\mathbf{r} - \mathbf{r}')]\, d\mathbf{r}'. \tag{16.58}$$

Using Green's theorem, this equation can be written as (n is unit normal to surface S)

$$\iint\limits_S \left[G(\mathbf{r}, \mathbf{r}')\frac{\partial\phi}{\partial n} - \phi\frac{\partial G(\mathbf{r}, \mathbf{r}')}{\partial n}\right] \cdot d\mathbf{S} = \iiint\limits_V G(\mathbf{r}, \mathbf{r}')f(\mathbf{r}')d\mathbf{r}' + \phi(\mathbf{r}). \tag{16.59}$$

If the surface S be made infinitely large, the integral in the left-hand side of equation (16.59) vanishes. Hence we get

$$\phi(\mathbf{r}) = -\iiint\limits_V G(\mathbf{r}, \mathbf{r}')f(\mathbf{r}')d\mathbf{r}' \tag{16.60}$$

In particular, let us consider the Poisson's equation of electrostatics

$$\nabla^2\phi(\mathbf{r}) = -\frac{\rho(\mathbf{r})}{\epsilon_0} \tag{16.61}$$

where $\rho(\mathbf{r})$ = the charge density in space, $\phi(\mathbf{r})$ = electrostatic potential and ϵ_0 = permittivity of the medium in space.

Then, using the Green's function given in equation (16.55), we write the solution of equation (16.61) as

$$\phi(\mathbf{r}) = \frac{1}{4\pi\epsilon_0}\iiint\limits_V \frac{\rho(\mathbf{r}')}{|\,\mathbf{r} - \mathbf{r}'\,|}d\mathbf{r}' \tag{16.62}$$

16.9 Expansion of Green's Function

Green's function can be expanded in terms of normalised eigenfunctions of the operator and the corresponding eigenvalues. Let $\phi_n(x)$ be normalised eigenfunctions of the differential operator L, and λ_n be the corresponding eigenvalues. Then , for $a \le x \le b$

$$L\phi_n(x) = \lambda_n \phi_n(x). \tag{16.63}$$

Since $\phi_n(x)$ constitute normalised orthogonal system of functions, we have

$$\int_a^b \phi_m(x)\phi_n(x)dx = \delta_{mn} \tag{16.64}$$

The Green's function for the operator L satisfies the differential equation

$$LG(x, x') = -\delta(x - x') \tag{16.65}$$

This Green's function $G(x, x')$ can be expanded in the form

$$G(x, x') = \sum_{n=1}^{\infty} A_n(x')\phi_n(x) \tag{16.66}$$

We now determine the expansion coefficients $A_n(x')$. We operate by L both sides of equation (16.66) and get

$$LG(x, x') = \sum_{n=1}^{\infty} A_n(x')L\phi_n(x)$$

Using equations (16.63) and (16.65), we have

$$-\delta(x - x') = \sum_{n=1}^{\infty} \lambda_n A_n(x')\phi_n(x) \tag{16.67}$$

Multiplying both sides of equation (16.67) by $\phi_m(x)$ and integrating within limits a to b, we get

$$-\int_a^b \phi_m \delta(x - x')dx = \sum_{n=1}^{\infty} \lambda_n A_n(x') \int_a^b \phi_m(x)\phi_n dx$$

$$\text{or,} \quad -\phi_m(x') = \sum_{n=1}^{\infty} \lambda_n A_n(x')\delta_{mn} \quad \text{[using equation (16.64)]}$$

$$= \lambda_m A_m(x').$$

Hence

$$A_m(x') = -\frac{\phi_m(x')}{\lambda_m}. \tag{16.68}$$

Thus, the expansion of Green's function in equation (16.66) becomes

$$G(x, x') = -\sum_{m=1}^{\infty} \frac{1}{\lambda_m}\phi_m(x')\phi_m(x) \tag{16.69}$$

Equation (16.69) is known as *bilinear formula for Green's function.*

16.10 Application of Green's Function

We consider Helmholtz equation in one dimension

$$\frac{d^2\phi}{dx^2} + \xi^2\phi(x) = f(x), \quad 0 \le x < a \tag{16.70}$$

and find its solution using Green's function. Let the boundary conditions be given by

$$\phi(0) = 0 \quad \text{and} \quad \phi(a) = 0. \tag{16.71}$$

Obviously, our first job is to find out the Green's function corresponding to unit source function and satisfying the boundary conditions. Thus, for Green's function $G(x, x')$, we are to solve

$$\frac{d^2G}{dx^2} + \xi^2 G(x, x') = -\delta(x - x') \tag{16.72}$$

$$\text{i.e.,} \quad \frac{d}{dx}\left(\frac{dG}{dx}\right) + \xi^2 G(x, x') = -\delta(x - x') \tag{16.73}$$

Integrating this equation between the limits $x' - \epsilon$ to $x' + \epsilon$ (ϵ is very small), we get

$$\int_{x'-\epsilon}^{x'+\epsilon} \frac{d}{dx}\left(\frac{dG}{dx}\right) dx + \xi^2 \int_{x'-\epsilon}^{x'+\epsilon} G(x, x')dx = -\int_{x'-\epsilon}^{x'+\epsilon} \delta(x - x')dx$$

Since $G(x, x')$ is a continuous function of x, the second integral vanishes in the limit as $\epsilon \longrightarrow 0$. Hence, we have

$$\left(\frac{dG}{dx}\right)_{x'-\epsilon}^{x'+\epsilon} = -1, \quad \text{i.e.,} \quad \left(\frac{dG}{dx}\right)_{x'+\epsilon} - \left(\frac{dG}{dx}\right)_{x'-\epsilon} = -1. \tag{16.74}$$

Thus, there is a discontinuity in the value of dG/dx at $x = x'$. We, therefore, split the interval $(0, a)$ in two parts like $(0, x')$ and (x', a) and assume two separate Green's functions $G_1(x, x')$ and $G_2(x, x')$ both of which satisfy the homogeneous equations

$$\frac{d^2G_1}{dx^2} + \xi^2 G_1 = 0, \quad \text{for } 0 \le x < x' \tag{16.75}$$

$$\text{and} \quad \frac{d^2G_2}{dx^2} + \xi^2 G_2 = 0, \quad \text{for } x' \le x < a \tag{16.76}$$

The boundary conditions to be satisfied are

$$\left.\begin{aligned} G_1(x, x') &= 0, \quad \text{for } x = 0 \\ \text{and} \quad G_2(x, x') &= 0, \quad \text{for } x = a \end{aligned}\right\}. \tag{16.77}$$

Moreover, at the point $x = x'$, the condition for continuity gives

$$G_1(x, x') = G_2(x, x'), \quad \text{for } x = x'. \tag{16.78}$$

Now if $p(x)$ and $q(x)$ are solutions of corresponding homogeneous equations, then

$$\frac{d^2p}{dx^2} + \xi^2 p(x) = 0, \quad \text{for } 0 \le x \le x' \tag{16.79}$$

$$\text{and} \quad \frac{d^2q}{dx^2} + \xi^2 q(x) = 0, \quad \text{for } x' \le x \le a \tag{16.80}$$

The boundary conditions to be satisfied by $p(x)$ and $q(x)$ are

$$p(0) = 0 \quad \text{and} \quad q(a) = 0 \tag{16.81}$$

Solutions of equations (16.79)–(16.81) are readily obtained as

$$p(x) = C_1 \sin \xi x, \quad 0 \le x \le x' \tag{16.82}$$

$$\text{and} \quad q(x) = C_2 \sin \xi(a - x), \quad x' \le x \le a \tag{16.83}$$

Now, comparing the set of equations (16.75)–(16.77) with the set (16.79)–(16.81), we have

$$G_1(x, x') = A \sin \xi x, \quad 0 \le x \le x' \tag{16.84}$$

$$\text{and} \quad G_2(x, x') = B \sin \xi(a - x), \quad x' \le x \le a \tag{16.85}$$

where A and B are constants.

Using the condition (16.78), we get

$$A \sin \xi x' = B \sin \xi(a - x') \tag{16.86}$$

Condition equation (16.74) in the limit $\epsilon \longrightarrow 0$ gives

$$-B\xi \cos \xi(a - x') - A\xi \cos \xi x' = -1 \tag{16.87}$$

Solving the system of equations (16.86) and (16.87), we get

$$\left.\begin{aligned} A &= \frac{\sin \xi(a - x')}{\xi \sin \xi a} \\ \text{and} \quad B &= \frac{\sin \xi x'}{\xi \sin \xi a} \end{aligned}\right\}. \tag{16.88}$$

Hence, the Green's function is

$$G_1(x, x') = \frac{\sin \xi x \sin \xi(a - x')}{\xi \sin \xi a}, \quad \text{for } 0 \le x \le x' \tag{16.89}$$

$$G_2(x, x') = \frac{\sin \xi x' \sin \xi(a - x)}{\xi \sin \xi a}, \quad \text{for } x' \le x \le a \tag{16.90}$$

Thus, the solution of equation (16.70) is given by

$$\begin{aligned} \phi(x) &= -\int_0^x G_1(x, x') f(x') dx' - \int_x^a G_2(x, x') f(x') dx' \\ &= -\frac{\sin \xi x}{\xi \sin \xi a} \int_0^x \sin \xi(a - x') f(x') dx' - \frac{\sin \xi(a - x)}{\xi \sin \xi a} \int_x^a \sin \xi x' f(x') dx' \end{aligned} \tag{16.91}$$

16.11 Non-homogeneous Boundary Conditions

We have discussed about the solution of one-dimensional Helmholtz equation (16.70) with homogeneous boundary conditions equation (16.71). But in general, the boundary conditions may be non-homogeneous and of the type

$$\left.\begin{aligned} \alpha_1 \phi(0) + \alpha_2 \phi'(0) &= \beta_1 \\ \alpha_3 \phi(a) + \alpha_4 \phi'_{,}(a) &= \beta_2 \end{aligned}\right\}. \tag{16.92}$$

In such cases, we split the boundary-value problems into two parts. The first part constitutes the solutions of non-homogeneous differential equation

$$\frac{d^2\phi_1}{dx^2}+\xi^2\phi_1(x)=f(x) \tag{16.93}$$

with homogeneous boundary conditions

$$\left.\begin{aligned}\alpha_1\phi_1(0)+\alpha_2\phi_1'(0)=0\\ \alpha_3\phi_1(a)+\alpha_4\phi_1'(a)=0\end{aligned}\right\}. \tag{16.94}$$

The second part constitutes the solution of homogeneous differential equation

$$\frac{d^2\phi_2}{dx^2}+\xi^2\phi_2(x)=0 \tag{16.95}$$

with non-homogeneous boundary conditions equation (16.92).

Then the complete solution would be given by

$$\phi(x)=\phi_1(x)+\phi_2(x). \tag{16.96}$$

The solution $\phi(x)$ of the first part can be obtained by the method of Green's function as discussed. For the second part, the solution of equation (16.95) is given by

$$\phi_2(x)=C\cos\xi x+D\sin\xi x \tag{16.97}$$

where C, D are constants which can be determined from boundary conditions in equation (16.92). Thus, the complete solution in equation (16.96) is determined.

The above discussion concerning one-dimensional Helmholtz's equation holds good for any other type of differential equation.

As an example, we consider the Helmholtz equation (16.70) with the boundary conditions

$$\phi(0)=\alpha \quad\text{and}\quad \phi'(a)=\beta. \tag{16.98}$$

We assume the solution as

$$\phi(x)=\phi_1(x)+\phi_2(x) \tag{16.99}$$

where $\phi_1(x)$ is the solution of equation (16.70) with homogeneous boundary conditions given by equation (16.81). This function $\phi_1(x)$ is represented by equation (16.91). The function $\phi_2(x)$ is the solution of homogeneous equation

$$\frac{d^2\phi_2}{dx^2}+\xi^2\phi_2=0 \tag{16.100}$$

subject to the non-homogeneous boundary conditions in equation (16.98).

Now the solution of equation (16.100) is

$$\phi_2(x)=A\cos\xi+B\sin\xi x \tag{16.101}$$

Using the boundary conditions in equation (16.98), we get

$$\alpha=A$$

$$\beta=-A\xi\sin\xi a+B\xi\cos\xi a$$

$$\text{or,}\quad B=\frac{\beta+\alpha\xi\sin\xi a}{\xi\cos\xi a}. \tag{16.102}$$

Hence, the solution in equation (16.100) becomes

$$\phi_2(x) = \alpha \cos \xi x + \frac{\beta + \alpha\xi \sin \xi a}{\xi \cos \xi a} \sin \xi x \tag{16.103}$$

Therefore, the complete solution of equation (16.70) with boundary conditions in equation (16.98) is obtained by adding the function $\phi_2(x)$ in equation (16.103) with the function $\phi_1(x)$ in equation (16.91).

EXERCISE

1. If $\delta_n(x) = \frac{n}{2} \sec h^2 nx$, show that

$$\int_{-\infty}^{\infty} \delta_n(x) dx = 1.$$

2. Show that

$$\lim_{\xi \longrightarrow \infty} \frac{\xi}{\sqrt{\pi}} e^{-\xi^2 x^2} = \delta(x)$$

3. Show that, for Legendre's Polynomials $P_n(x)$

$$\frac{1}{2} \sum_{n=0}^{\infty} (2n+1) P_n(x) P_n(x') = \delta(x - x'), \quad x \geq -1,$$

and $x' \leq 1$.

4. Find the Green's function for the differential equation

$$\frac{d^2\phi}{dx^2} - \alpha^2 \phi = 0, \quad 0 \leq x \leq a$$

with boundary conditions

$$\phi(0) = 0, \quad \text{and} \quad \left. \frac{1}{\phi} \frac{d\phi}{dx} \right|_{x=a} = -\alpha$$

5. Find the Green's function for the differential equation

$$\frac{d}{dx} \left\{ (1 - x^2) \frac{d\phi}{dx} \right\} = f(x), \quad |x| \leq 1$$

taking $\phi(x)$ to be finite at $x = \pm 1$.

6. Show that the Green's function for the equation

$$x^2 \frac{d^2\phi}{dx^2} + 2x \frac{d\phi}{dx} = f(x)$$

subject to the boundary conditions $\phi(\alpha) = 0$ and $\phi(\beta) = 0$ with $\alpha > 0, \beta > 0$ is given by

$$G(x, x') = -\frac{1}{\beta - \alpha} \left(1 - \frac{\alpha}{x}\right) \left(1 - \frac{\beta}{x'}\right), \quad \alpha \leq x < x'$$

$$G(x, x') = -\frac{1}{\beta - \alpha} \left(1 - \frac{\beta}{x}\right) \left(1 - \frac{\alpha}{x'}\right), \quad x' \leq x \leq \beta$$

ANSWERS

4. $$G(x,x') = \begin{cases} e^{-\alpha x}\sinh(\alpha x'), & 0 \le x' < x \\ e^{-\alpha x'}\sinh(\alpha x), & x \le x' \le a \end{cases}$$

5. $$G(x,x') = \begin{cases} -\dfrac{1}{2}\ln\left[(1-x)(1+x') + \ln 2 - \dfrac{1}{2}\right], & \text{for } -1 \le x < x' \\ -\dfrac{1}{2}\ln\left[(1+x)(1-x') + \ln 2 - \dfrac{1}{2}\right], & \text{for } x' \le x \le 1. \end{cases}$$

Chapter 17

Tensor Analysis

17.1 Introduction

The concept of a tensor was introduced by Gauss, Reimann and Christoffel while developing the subject of differential geometry. Ricci and his student Levi-Civita introduced Tensor analysis as a systematic branch of mathematics in a publication in 1901. Einstein popularised the subject by making an extensive use of tensors in his General Relativity theory. Today, the tensor analysis has emerged as an invaluable tool in many branches of theoretical physics.

The fundamental postulate of physics is that the laws of nature should be valid for all reference frames. The subject matter of tensor analysis has emerged as the study of the behaviour of a physical law from one system of coordinates to any other.

A *tensor* is a set of quantities called the coordinates of *components* whose *properties* are independent of the coordinate system. For two different coordinate systems, the components of a tensor obey specific transformation laws.

17.2 Notations and Symbols

Consider N quantities denoted by $x^1, x^2, \ldots, x^N$ which will be called *coordinates* of a point. The upper indices $1, 2, \ldots,$ which we shall call *superscripts* are simply identification labels and not indices of powers. The expression

$$\sum_{i=1}^{N} a_i x^i = a_1 x^1 + a_2 x^2 + \cdots + a_N x^N \tag{17.1}$$

representing the summation of N-terms will be written simply $a_i x^i$; the letter i attached with a is called *subscript.* If the subscript and superscript are numerical, then it will represent a single term, but not summation.

Repeat index in any expression is called a *dummy index,* and it can be replaced by any other letter. Thus, we can write, $a_i x^i = a_k x^k$.

If, in equation (17.1), $N = 4$, then

$$a_k x^k = a_1 x^1 + a_2 x^2 + a_3 x^3 + a_4 x^4 \tag{17.2}$$

Again, if ϕ is a function of $x^1, x^2, \ldots, x^N$, then

$$d\phi = \frac{\partial \phi}{\partial x^1} dx^1 + \frac{\partial \phi}{\partial x^2} dx^2 + \cdots + \frac{\partial \phi}{\partial x^N} dx^N \tag{17.3}$$

which is written as

$$d\phi = \frac{\partial \phi}{\partial x^i} dx^i \tag{17.4}$$

The Kronecker Delta is defined and denoted by

$$\delta^i_j = \begin{cases} 1, & \text{if } i = j \\ 0, & \text{if } i \neq j \end{cases} \tag{17.5}$$

Thus, $\delta^1_1 = 1, \delta^2_2 = 1, \delta^2_3 = 0$, etc.

Also

$$\delta^i_k x^k = x^i \tag{17.6}$$

since all other terms will vanish. That is why δ^i_k is called a *substitution factor*.

The superscripts are used to denote the components of a *contravariant tensor*. The subscripts are used to denote the components of a *covariant tensor*. The components of a *mixed tensor* is denoted by both superscripts and subscripts. Thus

(1) a^i are components of a contravariant tensor.

(2) B_{ij} are components of a covariant tensor.

(3) x^k_{ij} are components of a mixed tensor.

The *rank* of a tensor is the number of indices appearing in the symbol representing the components of a tensor (or a tensor itself). Thus, the ranks of above three tensors are respectively one, two and three. However, if an index appears as a subscript and again as a superscript, then it will be ignored. Thus C^{ik}_{jkl} is a mixed tensor of rank three (not five).

17.3 N-Dimensional Spacc

An ordered set of N variables $x^1, x^2, \ldots, x^N$ is called the coordinates of a *point*.

The *N-Dimensional space,* denoted by V_N, is the set of all these points.

A *subspace* V_M of the space V_N is defined for $M < N$ if the N variables $x^1, x^2, \ldots, x^N$ are given functions of M parameters. Considering a space V_N, the N-equations

$$\bar{x}^i = \phi^i\left(x^1, x^2, \ldots, x^N\right), \quad (i = 1, 2, \ldots, N) \tag{17.7}$$

where ϕ^i are single-valued continuous and differentiable functions of coordinates define a new coordinate system $\bar{x}^1, \bar{x}^2, \ldots, \bar{x}^N$.. It is essential that the N-functions ϕ^i are independent. The equation (17.7) are said to define a *transformation of coordinates.*

Differentiating equation (17.7) with respect to x^j, we get

$$d\bar{x}^i = \sum_{j=1}^{N} \frac{\partial \phi^i}{\partial x^j} dx^j = \sum_{j=1}^{N} \frac{\partial \bar{x}^i}{\partial x^j} dx^j = \frac{\partial \bar{x}^i}{\partial x^j} dx^j, \text{ (by summation convention)} \tag{17.8}$$

If the Jacobian determinant formed with the partial derivatives $\partial \bar{x}^i / \partial x^j$ does not vanish, then the inverse transformation exists, so that we can obtain

$$x^i = \psi^i\left(\bar{x}^1, \bar{x}^2, \ldots, \bar{x}^N\right), \quad (i = 1, 2, \ldots, N) \tag{17.9}$$

17.4 Contravariant and Covariant Vector

A set of N functions A^i of N coordinates x^i are said to form the components of a *contravariant vector,* if they transform according to the equation

$$\bar{A}^i = \frac{\partial \bar{x}^i}{\partial x^j} A^j \tag{17.10}$$

where $\bar{x}^i$ refers to the new coordinate system.

Multiplying equations (17.10) by $\partial x^k / \partial \bar{x}^i$ and summing over index i from 1 to N, we have

$$\frac{\partial x^k}{\partial \bar{x}^i} \bar{A}^i = \frac{\partial x^k}{\partial \bar{x}^i} \frac{\partial \bar{x}^i}{\partial x^j} A^j = \frac{\partial x^k}{\partial x^j} A^j = \delta^k_j A^j = A^k.$$

Thus, the solution of equations (17.10) is

$$A^j = \frac{\partial x^j}{\partial \bar{x}^i} \bar{A}^i \tag{17.11}$$

Comparing equations (17.8) and (17.10), we find that the differentials dx^j form the components of a contravariant vector.

A set of N functions A_i of the coordinates x^i are said to form the components of a *covariant vector,* if they transform according to the equation

$$\bar{A}_i = \frac{\partial x^j}{\partial \bar{x}^i} A_j \tag{17.12}$$

where $\bar{x}^i$ refers to the new coordinate system.

Multiplying equation (17.12) by $\partial \bar{x}^i / \partial x^k$ and summing over the index i from 1 to N, we have

$$\frac{\partial \bar{x}^i}{\partial x^k} \bar{A}_i = \frac{\partial \bar{x}^i}{\partial x^k} \frac{\partial x^j}{\partial \bar{x}^i} A_j = \frac{\partial x^j}{\partial x^k} A_j = \delta^j_k A_j = A_k \tag{17.13}$$

Thus, the solution of equation (17.12) is

$$A_j = \frac{\partial \bar{x}^i}{\partial x^j} \bar{A}_i \tag{17.14}$$

From equations (17.11) and (17.14), we find that

$$A^j B_j = \left(\frac{\partial x^j}{\partial \bar{x}^i} \bar{A}^i \right) \left(\frac{\partial \bar{x}^t}{\partial x^j} \bar{B}_t \right) = \delta^t_i \bar{A}^i \bar{B}_t = \bar{A}^t \bar{B}_t$$

Hence

$$A^j B_j = \bar{A}^t \bar{B}_t = \text{invariant} \tag{17.15}$$

Thus, if a function ϕ of coordinates x^i is transformed to a function ψ of new coordinates $\bar{x}^i$ such that $\phi = \psi$, then the function ϕ is called an *invariant* or a *scalar.*

EXAMPLE 17.1. *In rectangular Cartesian coordinates, the components of acceleration vectors are* $(\ddot{x}, \ddot{y})$. *Find these components in polar coordinates.*

Solution: In polar coordinates, $x = r\cos\theta, y = r\sin\theta$. This gives

$$\left.\begin{aligned} dx &= dr\cos\theta - (r\sin\theta)d\theta \\ dy &= dr\sin\theta + (r\cos\theta)d\theta \end{aligned}\right\} \tag{1}$$

On solving these we have

$$\left.\begin{aligned} dr &= (\cos\theta)dx + (\sin\theta)dy \\ d\theta &= \left(-\frac{1}{r}\sin\theta\right)dx + \left(\frac{1}{r}\cos\theta\right)dy. \end{aligned}\right\} \tag{2}$$

We put

$$\left.\begin{aligned} (x, y) &= (x^1, x^2), (r, \theta) = (\bar{x}^1, \bar{x}^2) \\ (\ddot{x}, \ddot{y}) &= (A^1, A^2), (a_r, a_\theta) = (\bar{A}^1, \bar{A}^2) \end{aligned}\right\} \tag{3}$$

where a_r, a_θ are components of acceleration vectors in polar coordinates.

Now, from (2) using (3), we have

$$d\bar{x}^1 = (\cos\theta)dx^1 + (\sin\theta)dx^2$$

$$d\bar{x}^2 = \left(-\frac{1}{r}\sin\theta\right)dx^1 + \left(\frac{1}{r}\cos\theta\right)dx^2$$

Hence

$$\left.\begin{aligned} \frac{\partial\bar{x}^1}{\partial x^1} &= \cos\theta, \quad \frac{\partial\bar{x}^1}{\partial x^2} = \sin\theta \\ \frac{\partial\bar{x}^2}{\partial x^1} &= -\frac{1}{r}\sin\theta, \quad \frac{\partial\bar{x}^2}{\partial x^2} = \frac{1}{r}\cos\theta \end{aligned}\right\} \tag{4}$$

Thus

$$a_r = \bar{A}^1 = \frac{\partial\bar{x}^1}{\partial x^1}A^1 + \frac{\partial\bar{x}^1}{\partial x^2}A^2 = (\cos\theta)\ddot{x} + (\sin\theta)\ddot{y} \tag{5}$$

and

$$a_\theta = \bar{A}^2 = \frac{\partial\bar{x}^2}{\partial x^1}A^1 + \frac{\partial\bar{x}^2}{\partial x^2}A^2 = \left(-\frac{1}{r}\sin\theta\right)\ddot{x} + \left(\frac{1}{r}\cos\theta\right)\ddot{y} \tag{6}$$

From (1)

$$\dot{x} = \frac{dx}{dt} = (\cos\theta)\dot{r} - r(\sin\theta)\dot{\theta}$$

$$\ddot{x} = (\cos\theta)\ddot{r} - 2(\sin\theta)\dot{\theta}\dot{r} - r(\cos\theta)\dot{\theta}^2 - r(\sin\theta)\ddot{\theta} \tag{7}$$

$$\dot{y} = (\sin\theta)\dot{r} + r(\cos\theta)\dot{\theta}$$

$$\ddot{y} = (\sin\theta)\ddot{r} + 2(\cos\theta)\dot{\theta}\dot{r} - r(\sin\theta)\dot{\theta}^2 + r(\cos\theta)\ddot{\theta} \tag{8}$$

Using (7) and (8), we get from (5) and (6)

$$a_r = \ddot{r} - r\dot{\theta}^2, \quad a_\theta = \ddot{\theta} + \frac{2}{r}\dot{r}\dot{\theta}.$$

EXAMPLE 17.2. *The components of a covariant vector in rectangular Cartesian systems are*

$$A_1 = \frac{y}{x}, \quad A_2 = \frac{x}{y}$$

Find these in polar coordinates.

Solution: Let the components in polar coordinates be $\bar{A}_1, \bar{A}_2$. Then, we have

$$\bar{A}_1 = \frac{\partial x^1}{\partial \bar{x}^1} A_1 + \frac{\partial x^2}{\partial \bar{x}^1} A_2 \tag{1}$$

$$\bar{A}_2 = \frac{\partial x^1}{\partial \bar{x}^2} A_1 + \frac{\partial x^2}{\partial \bar{x}^2} A_2 \tag{2}$$

From equation (1) of previous example, we get

$$\left.\begin{aligned} dx^1 &= d\bar{x}^1 \cos\theta - (r\sin\theta) d\bar{x}^2 \\ dx^2 &= d\bar{x}^1 \sin\theta + (r\cos\theta) d\bar{x}^2 \end{aligned}\right\}$$

Hence

$$\left.\begin{aligned} \frac{\partial x^1}{\partial \bar{x}^1} &= \cos\theta, \quad \frac{\partial x^1}{\partial \bar{x}^2} = -r\sin\theta \\ \frac{\partial x^2}{\partial \bar{x}^1} &= \sin\theta, \quad \frac{\partial x^2}{\partial \bar{x}^2} = +r\cos\theta \end{aligned}\right\}$$

Using these relations, we get from (1) and (2)

$$\bar{A}_1 = \cos\theta\left(\frac{y}{x}\right) + \sin\theta\left(\frac{x}{y}\right) = \sin\theta + \cos\theta$$

$$\bar{A}_2 = -r\sin\theta\left(\frac{y}{x}\right) + r\cos\theta\left(\frac{x}{y}\right) = -r\frac{\sin^2\theta}{\cos\theta} + r\frac{\cos^2\theta}{\sin\theta}$$

17.5 Higher Rank Tensors

We have seen the transformation laws of contravariant and covariant vectors which are tensors of rank one. Now we shall discuss the same for higher rank tensors.

Consider N^2 functions A^{ij} which are defined in $x^1, x^2, \ldots, x^N$ coordinate system. If due to a change in coordinate system to $\bar{x}^1, \bar{x}^2, \ldots, \bar{x}^N$, the quantities A^{ij} transform according to the equation

$$\bar{A}^{ij} = \frac{\partial \bar{x}^i}{\partial x^k}\frac{\partial \bar{x}^j}{\partial x^m} A^{km} \tag{17.16}$$

then we call A^{ij} the components of a *contravariant tensor* of *second order*.

Similarly, if the N^2 functions A_{ij} transform according to the law

$$\bar{A}_{ij} = \frac{\partial x^k}{\partial \bar{x}^i}\frac{\partial x^m}{\partial \bar{x}^j} A_{km} \tag{17.17}$$

we call A_{ij} the components of a *covariant tensor* of *second order*.

Again, if we have N^2 functions A^i_j which transform according to the equation

$$\bar{A}^i_j = \frac{\partial \bar{x}^i}{\partial x^k}\frac{\partial x^m}{\partial \bar{x}^j} A^k_m \tag{17.18}$$

we call A^i_j the components of a *mixed tensor* of second order.

A set of N^{s+p} quantities

$$A^{k_1k_2\cdots k_s}_{m_1m_2\cdots m_p} \tag{17.19}$$

is said to be the components of a *mixed* tensor of $(s+p)$th order, if they transform according to the equation

$$\bar{A}^{k_1k_2\cdots k_s}_{m_1m_2\cdots m_p} = \frac{\partial \bar{x}^{k_1}}{\partial x^{t_1}} \cdots \frac{\partial \bar{x}^{k_s}}{\partial x^{t_s}} \frac{\partial x^{q_1}}{\partial \bar{x}^{m_1}} \cdots \frac{\partial x^{q_p}}{\partial \bar{x}^{m_p}} A^{t_1t_2\cdots t_s}_{q_1q_2\cdots q_p} \tag{17.20}$$

17.6 Symmetric and Skew-Symmetric Tensors

The order of indices in a tensor is important. In general

$$A^{ij} \neq A^{ji} \tag{17.21}$$

If, however

$$\left.\begin{aligned} B^{ij}_k &= -B^{ji}_k \\ \text{or,}\quad B^k_{ij} &= -B^k_{ji} \end{aligned}\right\} \tag{17.22}$$

then we say that the tensors B^{ij}_k or B^k_{ij} are *skew symmetric.*

Thus, if two contravariant or two covariant indices are interchanged, each component of a skew symmetric tensor alters in sign but not in magnitude.

When all the indices of a tensor can be interchanged without altering it, the tensor is called *symmetric.*

If a tensor is symmetric with respect to two indices it will remain so in any other coordinate system with respect to those two indices. If we assume that A^{ij} is symmetric, then

$$A^{ij} = A^{ji} \tag{17.23}$$

Now, by equation (17.16), we have

$$\bar{A}^{ij} = \frac{\partial \bar{x}^i}{\partial x^k}\frac{\partial \bar{x}^j}{\partial x^m} A^{km} = \frac{\partial \bar{x}^j}{\partial x^m}\frac{\partial \bar{x}^i}{\partial x^k} A^{mk} = \bar{A}^{ji}.$$

Thus

$$\bar{A}^{ij} = \bar{A}^{ji} \tag{17.24}$$

Therefore, we have proved the property of symmetry for a contravariant tensor of second order in another coordinate system.

Similarly, it can be shown that the property of skew symmetry is independent of the coordinate system.

A symmetric tensor of second order has at most $\frac{1}{2}(N+1)N$ different components. If A^{ij} is symmetric, then

$$A^{12} = A^{21}, \quad A^{13} = A^{31}, \quad \ldots, \quad \text{etc.}$$

The total number of elements $= N^2$.

There are quantities A^{ii} (no summation) of N-numbers.

Thus, the total number of symmetric set $= N^2 - N$.

Hence, maximum number of different component is

$$\tfrac{1}{2}(N^2 - N) + N = \tfrac{1}{2}N(N+1)$$

Similarly, a skew-symmetric tensor A^{ij} of second order has at most $\frac{1}{2}N(N-1)$ different components. Because, in this case

$$A^{11} = A^{22} = A^{33} = \cdots = 0$$
$$\text{and} \quad A^{12} = -A^{21}, \quad A^{13} = -A^{31}, \cdots \text{etc.}$$

TENSOR ALGEBRA

17.7 Addition (Subtraction) of Tensors

Two tensors are said to be of *same type* if they have same number of contravariant and covariant indices. Thus A^i_{jk} and B^i_{jk} are same types of tensors. These tensors can be added and subtracted to produce a single tensor. Thus, the *sum* and *difference* of tensors A^i_{jk} and B^i_{jk} are respectively the tensors

$$C^i_{jk} = A^i_{jk} + B^i_{jk}; \quad D^i_{jk} = A^i_{jk} - B^i_{jk} \tag{17.25}$$

Let us consider the covariant tensor A_{ij} of second order. We can write it as

$$A_{ij} = \tfrac{1}{2}(A_{ij} + A_{ji}) + \tfrac{1}{2}(A_{ij} - A_{ji}) \tag{17.26}$$

Now

$$\tfrac{1}{2}(A_{ij} + A_{ji}) \quad \text{is symmetric}$$
$$\text{and} \quad \tfrac{1}{2}(A_{ij} - A_{ji}) \quad \text{is skew-symmetric.}$$

Thus, any covariant tensor of second order is the sum of a symmetric and a skew-symmetric tensor. We can show that this is also true for a contravariant tensor of second order.

17.8 Multiplication of Tensors

The *outer product* of two tensors denoted by

$$B^{ij}_{klm} \quad \text{and} \quad C^{qrs}_{uvwx}$$

is defined as

$$A^{ijqrs}_{klmuvwx} = B^{ij}_{klm} C^{qrs}_{uvwx} \tag{17.27}$$

The division of one tensor of rank greater than zero by another tensor of rank greater than zero is not defined.

17.9 Contraction of Tensors

The process by which we can reduce the number of covariant and contravariant indices of a mixed tensor by one is known as *contraction*. As an example, we consider a mixed tensor A^{ij}_{klm}. By the equation (17.20), we write

$$\bar{A}^{st}_{pqr} = \frac{\partial \bar{x}^s}{\partial x^i} \frac{\partial \bar{x}^t}{\partial x^j} \frac{\partial x^k}{\partial \bar{x}^p} \frac{\partial x^l}{\partial \bar{x}^q} \frac{\partial x^m}{\partial \bar{x}^r} A^{ij}_{klm} \tag{17.28}$$

Hence

$$\bar{A}^{sr}_{pqr} = \frac{\partial \bar{x}^s}{\partial x^i}\frac{\partial \bar{x}^r}{\partial x^j}\frac{\partial x^k}{\partial \bar{x}^p}\frac{\partial x^l}{\partial \bar{x}^q}\frac{\partial x^m}{\partial \bar{x}^r}A^{ij}_{klm} = \frac{\partial \bar{x}^s}{\partial x^i}\frac{\partial x^k}{\partial \bar{x}^p}\frac{\partial x^l}{\partial \bar{x}^q}\delta^m_j A^{ij}_{klm} = \frac{\partial \bar{x}^s}{\partial x^i}\frac{\partial x^k}{\partial \bar{x}^p}\frac{\partial x^l}{\partial \bar{x}^q}A^{im}_{klm} \tag{17.29}$$

Thus, we have obtained a mixed tensor whose covariant and contravariant indices have been reduced by one. Hence, the process of contraction enables us to obtain a tensor of order $(r-2)$ from a mixed tensor of order r.

17.10 Inner Product of Tensors

The process of combining outer multiplication and contraction to produce new tensors is called *inner multiplication* of two tensors. The resulting tensor is called the *inner product* of the two tensors.

For example, we consider two tensors A^{ij}_k and B^p_{qrs}. The inner product of these two tensors can be written as

$$C^{ij}_{qrs} = A^{ij}_k B^k_{qrs} \tag{17.30}$$

17.11 Quotient Law

We know that a set of functions forming the components of a tensor should satisfy the transformation equations of the type equation (17.20). This is the direct method, although practically not always feasible, to ascertain whether a set of functions form the components of a tensor.

The *quotient law* is a simple indirect method to ascertain whether a set of functions form the components of a tensor. The quotient law states that—

If there are N^r functions of coordinates x^i forming the components of a tensor of order r, then the inner product of these functions with an *arbitrary* tensor is itself a tensor. As an example, consider the set of N^3 functions A^{ijk}. These will form the components of a tensor of the type indicated by the indices, if

$$A^{ijk}B^r_{ij} = C^{rk} \tag{17.31}$$

provided that (i) B^r_{ij} is an *arbitrary* tensor and (ii) C^{rk} is a tensor.

In a new $\bar{x}^i$ coordinate system, equation (17.31) assumes the form

$$\bar{C}^{rk} = \bar{A}^{ijk}\bar{B}^r_{ij} \tag{17.32}$$

Using equation (17.20) we get

$$\frac{\partial \bar{x}^r}{\partial x^l}\frac{\partial \bar{x}^k}{\partial x^m}C^{lm} = \bar{A}^{ijk}\frac{\partial \bar{x}^r}{\partial x^u}\frac{\partial x^s}{\partial \bar{x}^i}\frac{\partial x^t}{\partial \bar{x}^j}B^u_{st} \tag{17.33}$$

Substituting equation (17.31) into equation (17.33) and making appropriate change of indices, that is

$$C^{lm} = A^{stm}B^l_{st} \tag{17.34}$$

we get

$$\frac{\partial \bar{x}^r}{\partial x^l}\left[\bar{A}^{ijk}\frac{\partial x^s}{\partial \bar{x}^i}\frac{\partial x^t}{\partial \bar{x}^j} - \frac{\partial \bar{x}^k}{\partial x^m}A^{stm}\right]B^l_{st} = 0 \tag{17.35}$$

We multiply the equation (17.35) by $\partial x^{\alpha}/\partial \bar{x}^{r}$, and sum over r, we get

$$\left[\bar{A}^{ijk}\frac{\partial x^s}{\partial \bar{x}^i}\frac{\partial x^t}{\partial \bar{x}^j} - \frac{\partial \bar{x}^k}{\partial x^m}A^{stm}\right]B^{\alpha}_{st} = 0 \tag{17.36}$$

Since, B^{α}_{st} is arbitrary, we must have

$$\bar{A}^{ijk}\frac{\partial x^s}{\partial \bar{x}^i}\frac{\partial x^t}{\partial \bar{x}^j} = \frac{\partial \bar{x}^k}{\partial x^m}A^{stm} \tag{17.37}$$

Inner multiplication of this equation by

$$\frac{\partial \bar{x}^r}{\partial x^s}\frac{\partial \bar{x}^g}{\partial x^t}$$

yields the result

$$\bar{A}^{rgk} = \frac{\partial \bar{x}^r}{\partial x^s}\frac{\partial \bar{x}^g}{\partial x^t}\frac{\partial \bar{x}^k}{\partial x^m}A^{stm} \tag{17.38}$$

Thus, A^{stm} is a tensor of third order having all contravariant indices. It is important to note that the tensor B^{l}_{st} should be arbitrary without having symmetric and skew symmetric properties. Otherwise, we cannot get the equation (17.37).

17.12 Conjugate Tensors

The covariant tensor A_{ij} and the contravariant tensor B^{ij} are said to be *conjugate* tensors, if they satisfy the equation

$$A_{ij}B^{ij} = \delta^k_j \tag{17.39}$$

A tensor of second order has a conjugate only if the determinant formed by its components is not zero.

THE LINE ELEMENT

17.13 The Metric Tensor

Let us consider two neighbouring points x^i and $x^i + dx^i$ in the space V_N. The distance between these two points is given by the quadratic differential form

$$ds^2 = g_{ij}dx^i dx^j \tag{17.40}$$

where it is assumed that (i) the g_{ij} are functions of x^i, (ii) g_{ij} is symmetric covariant tensor, i.e., $g_{ij} = g_{ji}$; (iii) $g = \det|g_{ij}| \neq 0$.

In this case, the space is said to be *Riemannian* space, and g_{ij}, the symmetric covariant tensor of second order is called the *fundamental tensor* or the *metric tensor* of the Riemannian space. The quadratic form in equation (17.40) is called the *metric.*

In Euclidean space V_3, referred to rectangular Cartesian coordinates, we write

$$ds^2 = (dx^1)^2 + (dx^2)^2 + (dx^3)^2 \tag{17.41}$$

so that

$$g_{11} = g_{22} = g_{33} = 1, \quad g_{ij} = 0 \tag{17.42}$$

In cylindrical coordinates (r, θ, z), we have

$$ds^2 = dr^2 + r^2 d\theta^2 + dz^2$$

which, upon writing $r = x^1, \theta = x^2, z = x^3$, becomes

$$ds^2 = (dx^1)^2 + (x^1)^2 (dx^2)^2 + (dx^3)^2 \tag{17.43}$$

Thus, here

$$g_{11} = 1, g_{22} = (x^1)^2, g_{33} = 1, g_{ij} = 0 \tag{17.44}$$

Similarly, in spherical coordinates (r, θ, ϕ)

$$ds^2 = dr^2 + r^2 d\theta^2 + (r \sin\theta)^2 d\phi^2 = (dx^1)^2 + (x^1)^2 (dx^2)^2 + (x^1 \sin x^2)^2 (dx^3)^2 \tag{17.45}$$

so that

$$g_{11} = 1, g_{22} = (x^1)^2, g_{33} = (x^1 \sin x^2)^2, g_{ij} = 0 \tag{17.46}$$

We say that ds^2 is *positive-definite,* if it is positive for all the values of the differentials dx^i, not all equal to zero. Thus, from the equations (17.41)–(17.46), we find that, the matrix of a Euclidean space is positive-definite in all the coordinate systems.

In the discussion of special relativity theory, we consider a V_4 space, where the line element ds is given by

$$ds^2 = -(dx^1)^2 - (dx^2)^2 - (dx^3)^2 + c^2 (dx^4)^2 \tag{17.47}$$

Thus the metric is positive for curves along which x^1, x^2, x^3 are constants and negative for curves along which x^4 is constant. In order that the distance ds between two points is real, we write the equation (17.40) as

$$ds^2 = e g_{ij} dx^i dx^j \tag{17.48}$$

where the factor e takes the values $+1$ or -1 so that ds^2 is always positive. This factor is called *indicator.*

17.14 Conjugate of Metric Tensors

Let us consider $\det | g_{ij} |$ whose value is g. Let G^{ji} be the cofactor of g_{ij}. Since g_{ij} is symmetric

$$g_{ij} = g_{ji}, \quad \text{hence } G^{ji} = G^{ij}$$

Then the normalised cofactor, denoted by g^{ij} is

$$g^{ij} = \frac{1}{g} G^{ij}$$

Thus we have

$$g^{ij} g_{jk} = \delta^i_k \tag{17.49}$$

so that by equation (17.39), g^{ij} and g_{ij} are conjugate tensors. But, we have to prove that g^{ij} is a contravariant tensor of second order. To do so, we introduce an arbitrary contravariant vector A^i, and put

$$B_i = g_{ij} A^j \tag{17.50}$$

We multiply equation (17.50) by g^{ik}, then

$$g^{ik}B_i = g^{ik}g_{ij}A^j = \delta^k_j A^j = A^k \tag{17.51}$$

Since A^i is arbitrary, so must be B_i. Hence, from quotient law, g^{ik} is a contravariant tensor of second order.

The tensors g_{ij} and g^{ij} are called *conjugate tensor.*

EXAMPLE 17.3. *Find g^{ij} in spherical coordinate system.*

Solution: In spherical coordinate, we have

$$g_{11} = 1, \quad g_{22} = \left(x^1\right)^2, \quad g_{33} = \left(x^1 \sin x^2\right)^2, \quad g_{ij} = 0$$

Hence

$$g = \det|g_{ij}| = g_{11}g_{22}g_{33} = \left(x^1\right)^2 \left(x^1 \sin x^2\right)^2$$

Cofactor of $g_{11} = g_{22}g_{33}$.
Hence

$$g^{11} = \frac{1}{g}g_{22}g_{33} = \frac{1}{g_{11}} = 1$$

Similarly

$$g^{22} = \frac{1}{g_{22}} = \frac{1}{\left(x^1\right)^2}; \quad g^{33} = \frac{1}{g_{33}} = \frac{1}{\left(x^1 \sin x^2\right)^2}$$

EXAMPLE 17.4. *Find g^{ij}, if $ds^2 = (dx^1)^2 + 2\cos\xi dx^1 dx^2 + (dx^2)^2$.*

Solution: Here, $g_{11} = 1, g_{22} = 1, g_{12} = g_{21} = \cos\xi$.
Hence

$$g = \begin{vmatrix} g_{11} & g_{12} \\ g_{21} & g_{22} \end{vmatrix} = \begin{vmatrix} 1 & \cos\xi \\ \cos\xi & 1 \end{vmatrix} = \sin^2\xi.$$

Thus,

$$g^{11} = \frac{g_{22}}{g} = \frac{1}{\sin^2\xi}, \quad g^{21} = g^{12} = \frac{-\cos\xi}{\sin^2\xi}, \quad g^{22} = \frac{1}{\sin^2\xi}$$

17.15 Length of a Curve

Let a curve be given by the equation

$$x^i = x^i(t) \tag{17.52}$$

where t is the parameter. Then the length of the curve between the points corresponding to $t = t_0$ and $t = t_1$ is defined by

$$s = \int_{t_0}^{t_1} \sqrt{e g_{ij} \frac{dx^i}{dt}\frac{dx^j}{dt}}\, dt \tag{17.53}$$

If, at all points of the curve

$$g_{ij}\frac{dx^i}{dt}\frac{dx^j}{dt} = 0, \tag{17.54}$$

the curve is called a *null line.*

In Riemannian space whose metric is positive-definite, no real null line exists. But in V_4, there exists some real curves, which are null lines. These are given by

$$\left.\begin{aligned} x^1 &= c\int r\cos\theta\cos\phi dt, \quad x^2 = c\int r\cos\theta\sin\phi dt \\ x^3 &= c\int r\sin\theta dt, \qquad\quad x^4 = \int r dt \end{aligned}\right\} \tag{17.55}$$

where r, θ, ϕ are functions of t.

17.16 Magnitude of a Vector

The *magnitude* A of a contravariant vector A^i is defined by

$$(A)^2 = e g_{ij} A^i A^j \tag{17.56}$$

The magnitude A is invariant.

The magnitude B of a covariant vector B_i is defined by

$$(B)^2 = e g^{ij} B_i B_j \tag{17.57}$$

If A or B equals to one, then the corresponding vector is called a *unit vector.*

In the Euclidean space under Cartesian coordinates, equation (17.56) gives the magnitude of a vector (a, b, c) as $\sqrt{a^2+b^2+c^2}$ which is expected.

17.17 Associate Tensors

Let us consider a contravariant vector A^j. We define its *associate,* the covariant vector A_i given by

$$A_i = g_{ij} A^j \tag{17.58}$$

Similarly, we say that the vector B^i is associate to the vector B_j, if

$$B^i = g^{ij} B_j \tag{17.59}$$

The process in equation (17.58) is called *lowering* the superscript, and that in equation (17.59) is called *raising* the subscripts. These processes may be performed on higher rank tensor also.

17.18 Angle between Two Vectors

We consider unit vectors A^j and B^i. Then the angle θ between them is defined by

$$\cos\theta = g_{ij} A^i B^j = A_j B^j = g^{jk} A_j B_k = A^k B_k \tag{17.60}$$

In Euclidean V_3 space under Cartesian coordinate system, this equation corresponds to

$$\cos\theta = ll' + mm' + nn' \tag{17.61}$$

where (l, m, n) and (l', m', n') are unit vectors.

If the metric of the Riemann space is positive-definite, equation (17.60) always defines a real angle. In fact, for all real values of λ and μ, the magnitude of the vector

$$\lambda A^i + \mu B^i \text{ is } \geq 0.$$

That is

$$g_{ij}(\lambda A^i + \mu B^i)(\lambda A^j + \mu B^j) \geq 0$$

which reduces to

$$\lambda^2 + 2\lambda\mu\cos\theta + \mu^2 \geq 0 \tag{17.62}$$

The discriminant of this quadratic should be negative or zero. Hence

$$\cos^2\theta - 1 \leq 0, \quad \text{or,} \quad \cos^2\theta \leq 1 \tag{17.63}$$

Hence θ is real.

If A^i and B^i are not unit vectors, then the angle between them is given by

$$\cos\theta = \left(g_{ij}A^iB^j\right)/(AB) \tag{17.64}$$

where A and B are magnitudes of the vectors A^i and B^j taken to be real and positive.

Two vectors are said to be *orthogonal* to one another if the angle between them is a right angle. In this case, equation (17.64) yields $g_{ij}A^iB^j = 0$.

TENSOR CALCULUS

17.19 Christoffel Symbols

The derivative of a tensor does not always lead to another tensor. Let us consider the tensor of rank one

$$A^j = \frac{\partial x^j}{\partial \bar{x}^i}\bar{A}^i \tag{17.65}$$

and differentiate it. Then we get

$$\frac{\partial A^j}{\partial x^k} = \frac{\partial x^j}{\partial \bar{x}^i}\frac{\partial \bar{x}^m}{\partial x^k}\frac{\partial \bar{A}^i}{\partial \bar{x}^m} + \frac{\partial^2 x^j}{\partial x^k \partial \bar{x}^k}\bar{A}^i \tag{17.66}$$

The second term in the right-handside shows that the derivative of A^j is not a tensor. The aim of the tensor calculus is to develop expressions involving derivatives of a tensor which are the components of a tensor.

Two *Christoffel symbols* of the first and second kinds defined respectively by

$$[ij, k] = \frac{1}{2}\left(\frac{\partial g_{ik}}{\partial x^j} + \frac{\partial g_{jk}}{\partial x^i} - \frac{\partial g_{ij}}{\partial x^k}\right) \tag{17.67}$$

and

$$\begin{Bmatrix} l \\ ij \end{Bmatrix} = g^{lk}[ij, k] \tag{17.68}$$

which are formed from the fundamental tensor g_{ij} are directly involved in this development. We, therefore, like to investigate into them.

In the two symbols, i, j, k are regarded as subscripts and l as a superscript when summation convention is used.

From the definitions of the symbols, it is clear that both of them are symmetric with respect to the indices i and j. That is

$$[ij, k] = [ji, k] \tag{17.69}$$

and

$$\begin{Bmatrix} l \\ ij \end{Bmatrix} = \begin{Bmatrix} l \\ ji \end{Bmatrix} \tag{17.70}$$

Since, $g_{ij} = g_{ji}$, it follows from equation (17.67) that

$$\frac{\partial g_{ik}}{\partial x^j} = [ij, k] + [kj, i] \tag{17.71}$$

Again, the inner multiplication of equation (17.68) by g_{lm} yields

$$g_{lm}\begin{Bmatrix} l \\ ij \end{Bmatrix} = g_{lm}g^{lk}[ij, k] = \delta^k_m[ij, k] = [ij, m] \tag{17.72}$$

Note. The Christoffel symbol of the second kind is also denoted by Γ^l_{ij}. In this case, it must be remembered that Γ^l_{ij} is just a symbol, but not a tensor. Similarly, the Christoffel symbol of the first kind is also denoted by $\Gamma_{ij,k}$ which is not a tensor.

EXAMPLE 17.5. *Find the Christoffel symbols when*

$$ds^2 = \left(dx^1\right)^2 + u\left(dx^2\right)^2 + \left(dx^3\right)^2$$

where u is a function of x^1 and x^2.

Solution: Here, $g_{11} = 1, g_{22} = u(x^1, x^2), g_{33} = 1, g_{ij} = 0 (i \neq j)$.

Hence, $g^{11} = 1, g^{22} = 1/u, g^{33} = 1, g^{ij} = 0 (i \neq j)$.

Thus, only derivatives of g_{22} and g^{22} are non-zero, all other derivatives vanish. The non-zero symbols of first kind are:

$$\Gamma_{12,2} = \frac{1}{2}\frac{\partial g_{22}}{\partial x^1} = \frac{1}{2}\frac{\partial u}{\partial x^1} = \frac{1}{2}u_1 \text{ (say)}$$

$$\Gamma_{22,1} = -\frac{1}{2}\frac{\partial g_{22}}{\partial x^1} = -\frac{1}{2}\frac{\partial u}{\partial x^1} = -\frac{1}{2}u_1 \text{ (say)}$$

$$\Gamma_{22,2} = \frac{1}{2}\left(\frac{\partial g_{22}}{\partial x^2} + \frac{\partial g_{22}}{\partial x^2} - \frac{\partial g_{22}}{\partial x^2}\right) = \frac{1}{2}\frac{\partial g_{22}}{\partial x^2} = \frac{1}{2}u_2 \text{ (say)}$$

Similarly, the non-zero symbols of second kind are:

$$\Gamma^2_{12} = g^{22}\Gamma_{12,2} = \frac{1}{2}\frac{u_1}{u}$$

$$\Gamma^1_{22} = g^{11}\Gamma_{22,1} = 1\left(-\frac{1}{2}u_1\right) = -\frac{1}{2}u_1$$

$$\Gamma^2_{22} = g^{22}\Gamma_{22,2} = \frac{1}{u}\left(\frac{1}{2}u_2\right) = \frac{u_2}{2u}$$

EXAMPLE 17.6. *Show that* $\begin{Bmatrix} i \\ ij \end{Bmatrix} = \frac{\partial}{\partial x^j} \left\{ \log \sqrt{|g|} \right\}$.

Solution: We know that $g = \det | g_{ij} |$.

The cofactor of g_{km} of the determinant is gg^{km}. The derivative of the determinant with respect to x^j is the sum of the products of derivatives of each element by its cofactor. Hence

$$\frac{\partial g}{\partial x^j} = gg^{km} \frac{\partial g_{km}}{\partial x^j} \tag{1}$$

Now

$$\begin{Bmatrix} i \\ ij \end{Bmatrix} = g^{ik}[ij, k] = \frac{1}{2} g^{ik} \left(\frac{\partial g_{ik}}{\partial x^j} + \frac{\partial g_{jk}}{\partial x^i} - \frac{\partial g_{ij}}{\partial x^k} \right) \tag{2}$$

Since g_{ij} is symmetric, the summation in (2) yields

$$\begin{Bmatrix} i \\ ij \end{Bmatrix} = \frac{1}{2} g^{ik} \frac{\partial g_{ik}}{\partial x^j} \tag{3}$$

Thus, from (1) and (3), we have

$$\begin{Bmatrix} i \\ ij \end{Bmatrix} = \frac{1}{2} \frac{1}{g} \frac{\partial g}{\partial x^j} = \frac{\partial}{\partial x^j} \{\log \sqrt{g}\}$$

Now, g may be negative. Hence, in general

$$\begin{Bmatrix} i \\ ij \end{Bmatrix} = \frac{\partial}{\partial x^j} \left\{ \log \sqrt{|g|} \right\}$$

EXAMPLE 17.7. *By the use of definition of Christoffel symbol of the second kind, show that*

$$\frac{\partial g^{mk}}{\partial x^l} = -g^{mi} \begin{Bmatrix} k \\ ij \end{Bmatrix} - g^{ki} \begin{Bmatrix} m \\ il \end{Bmatrix}$$

Solution: Since the tensors g_{ij} and g^{ij} are conjugate, we have

$$g_{ij} g^{ik} = \delta_j^k$$

Differentiating with respect to x^l, we get

$$\frac{\partial g^{ik}}{\partial x^l} g_{ij} + \frac{\partial g_{ij}}{\partial x^l} g^{ik} = 0$$

Inner multiplication by g^{jm} yields

$$\frac{\partial g^{mk}}{\partial x^l} + g^{jm} g^{ik} \frac{\partial g_{ij}}{\partial x^l} = 0$$

Hence

$$\begin{aligned} \frac{\partial g^{mk}}{\partial x^l} &= -g^{jm} g^{ik} \{[il, j] + [jl, i]\} \quad \text{[using equation (17.71)]} \\ &= -g^{ki} \begin{Bmatrix} m \\ il \end{Bmatrix} - g^{jm} \begin{Bmatrix} k \\ jl \end{Bmatrix} \quad \text{[using equation (17.68)]} \\ &= -g^{ki} \begin{Bmatrix} m \\ il \end{Bmatrix} - g^{mi} \begin{Bmatrix} k \\ il \end{Bmatrix} \end{aligned}$$

17.20 Transformation Laws of Christoffel Symbols

The fundamental tensor g_{ij} transforms according to the equation

$$\bar{g}_{km} = \frac{\partial x^i}{\partial \bar{x}^k}\frac{\partial x^j}{\partial \bar{x}^m} g_{ij} \tag{17.73}$$

Differentiating with respect to $\bar{x}^n$, we get

$$\frac{\partial \bar{g}_{km}}{\partial \bar{x}^n} = \frac{\partial}{\partial \bar{x}^n}\left(\frac{\partial x^i}{\partial \bar{x}^k}\right)\frac{\partial x^j}{\partial \bar{x}^m} g_{ij} + \frac{\partial x^i}{\partial \bar{x}^k}\frac{\partial}{\partial \bar{x}^n}\left(\frac{\partial x^j}{\partial \bar{x}^m}\right) g_{ij} + \frac{\partial x^i}{\partial \bar{x}^k}\frac{\partial x^j}{\partial \bar{x}^m}\frac{\partial g_{ij}}{\partial \bar{x}^n} \tag{17.74}$$

Since g_{ij} is symmetric, $g_{ij} = g_{ji}$, hence we can write equation (17.74) as

$$\frac{\partial \bar{g}_{km}}{\partial \bar{x}^n} = \left(\frac{\partial^2 x^i}{\partial \bar{x}^n \partial \bar{x}^k}\frac{\partial x^j}{\partial \bar{x}^m} + \frac{\partial x^i}{\partial \bar{x}^k}\frac{\partial^2 x^j}{\partial \bar{x}^n \partial \bar{x}^m}\right) g_{ij} + \frac{\partial x^i}{\partial \bar{x}^k}\frac{\partial x^j}{\partial \bar{x}^m}\frac{\partial g_{ij}}{\partial \bar{x}^n} \tag{17.75}$$

We obtain two similar equations by cyclic interchange of indices k, m and n. Thus, for the change $(k, m, n) \longrightarrow (n, k, m)$, we have

$$\frac{\partial \bar{g}_{nk}}{\partial \bar{x}^m} = \left(\frac{\partial^2 x^i}{\partial \bar{x}^m \partial \bar{x}^n}\frac{\partial x^j}{\partial \bar{x}^k} + \frac{\partial x^j}{\partial \bar{x}^n}\frac{\partial^2 x^i}{\partial \bar{x}^m \partial \bar{x}^k}\right) g_{ij} + \frac{\partial x^i}{\partial \bar{x}^n}\frac{\partial x^j}{\partial \bar{x}^k}\frac{\partial g_{ij}}{\partial \bar{x}^m} \tag{17.76}$$

and for the change $(n, k, m) \longrightarrow (m, n, k)$

$$\frac{\partial \bar{g}_{mn}}{\partial \bar{x}^k} = \left(\frac{\partial^2 x^i}{\partial \bar{x}^k \partial \bar{x}^m}\frac{\partial x^j}{\partial \bar{x}^n} + \frac{\partial x^j}{\partial \bar{x}^m}\frac{\partial^2 x^i}{\partial \bar{x}^k \partial \bar{x}^n}\right) g_{ij} + \frac{\partial x^i}{\partial \bar{x}^m}\frac{\partial x^j}{\partial \bar{x}^n}\frac{\partial g_{ij}}{\partial \bar{x}^k} \tag{17.77}$$

Now in terms of new coordinates $\bar{x}^i$, we may write the Christoffel symbol of first kind as

$$[\overline{km, n}] = \frac{1}{2}\left(\frac{\partial \bar{g}_{kn}}{\partial \bar{x}^m} + \frac{\partial \bar{g}_{mn}}{\partial \bar{x}^k} - \frac{\partial \bar{g}_{km}}{\partial \bar{x}^n}\right) \tag{17.78}$$

Substituting equations (17.75)–(17.77) into equation (17.78), we get, with appropriate changes of dummy indices

$$[\overline{lm, n}] = [ij, k]\frac{\partial x^i}{\partial \bar{x}^l}\frac{\partial x^j}{\partial \bar{x}^m}\frac{\partial x^k}{\partial \bar{x}^n} + \frac{\partial x^i}{\partial \bar{x}^n}\frac{\partial^2 x^j}{\partial \bar{x}^l \partial \bar{x}^m} g_{ij} \tag{17.79}$$

This shows that the symbol $[ij, k]$ transforms like a tensor only if the second term in the RHS of equation (17.79) vanishes. If this term vanishes we say that the transformation is *affine*.

Equation (17.79) gives the transformation law for the Christoffel symbol of the first kind. We shall now develop the transformation law for the Christoffel symbol of the second kind.

We have the transformation law of the contravariant fundamental tensor as

$$\bar{g}^{np} = g^{rs}\frac{\partial \bar{x}^n}{\partial x^r}\frac{\partial \bar{x}^p}{\partial x^s} \tag{17.80}$$

By the inner multiplication of both sides of equation (17.79) with the corresponding sides of equation (17.80), we have

$$\left\{\begin{matrix} \bar{p} \\ lm \end{matrix}\right\} = \left\{\begin{matrix} s \\ ij \end{matrix}\right\}\frac{\partial \bar{x}^p}{\partial x^s}\frac{\partial x^i}{\partial \bar{x}^l}\frac{\partial x^j}{\partial \bar{x}^m} + \frac{\partial \bar{x}^p}{\partial x^j}\frac{\partial^2 x^j}{\partial \bar{x}^l \partial \bar{x}^m} \tag{17.81}$$

This is the transformation law for the Christoffel symbol of the second kind.

Now, the inner multiplication of equation (17.81) by $\partial x^r/\partial \bar{x}^p$ yields

$$\left\{\begin{matrix} \bar{p} \\ lm \end{matrix}\right\} \frac{\partial x^r}{\partial \bar{x}^p} = \left\{\begin{matrix} s \\ ij \end{matrix}\right\} \frac{\partial x^r}{\partial \bar{x}^p} \frac{\partial \bar{x}^p}{\partial x^s} \frac{\partial x^i}{\partial \bar{x}^l} \frac{\partial x^j}{\partial \bar{x}^m} + \frac{\partial x^r}{\partial \bar{x}^p} \frac{\partial \bar{x}^p}{\partial x^j} \frac{\partial^2 x^j}{\partial \bar{x}^l \partial \bar{x}^m}$$

$$\text{or,} \quad \frac{\partial^2 x^r}{\partial \bar{x}^l \partial \bar{x}^m} = \left\{\begin{matrix} \bar{p} \\ lm \end{matrix}\right\} \frac{\partial x^r}{\partial \bar{x}^p} - \left\{\begin{matrix} r \\ ij \end{matrix}\right\} \frac{\partial x^i}{\partial \bar{x}^l} \frac{\partial x^j}{\partial \bar{x}^m} \tag{17.82}$$

17.21 Covariant Differentiation of Vectors

We shall now investigate the character of the partial derivatives of a contravariant vector. We have seen in equation (17.66) that the partial derivatives of a contravariant vector A^j do not form a tensor. However, if we can eliminate the second term in the RHS of the equation (17.66), then the quantity $\partial A^j/\partial x^k$ will have tensor character.

Now, we have

$$\frac{\partial A^k}{\partial x^j} = \frac{\partial \bar{A}^i}{\partial \bar{x}^n} \frac{\partial \bar{x}^n}{\partial x^j} \frac{\partial x^k}{\partial \bar{x}^i} + \bar{A}^i \frac{\partial^2 x^k}{\partial \bar{x}^i \partial \bar{x}^n} \frac{\partial \bar{x}^n}{\partial x^j} \tag{17.83}$$

Using the expression for second partial derivative of x^k from equation (17.82) we get

$$\frac{\partial A^k}{\partial x^j} = \frac{\partial \bar{A}^i}{\partial \bar{x}^n} \frac{\partial \bar{x}^n}{\partial x^j} \frac{\partial x^k}{\partial \bar{x}^i} + \bar{A}^i \frac{\partial \bar{x}^n}{\partial x^j} \left[\left\{\begin{matrix} \bar{p} \\ in \end{matrix}\right\} \frac{\partial x^k}{\partial \bar{x}^p} - \left\{\begin{matrix} k \\ rs \end{matrix}\right\} \frac{\partial x^r}{\partial \bar{x}^i} \frac{\partial x^s}{\partial \bar{x}^n} \right] \tag{17.84}$$

Using the transformation law

$$A^k = \bar{A}^i \frac{\partial x^k}{\partial \bar{x}^i}$$

with appropriate change of dummy indices, we can reduce the equation (17.84) to

$$\frac{\partial A^k}{\partial x^j} + \left\{\begin{matrix} k \\ rj \end{matrix}\right\} A^r = \left[\frac{\partial \bar{A}^i}{\partial \bar{x}^n} + \left\{\begin{matrix} \bar{i} \\ rn \end{matrix}\right\} \bar{A}^r \right] \frac{\partial \bar{x}^n}{\partial x^j} \frac{\partial x^k}{\partial \bar{x}^i} \tag{17.85}$$

We introduce the notation

$$A^k_{,m} = \frac{\partial A^k}{\partial x^m} + \left\{\begin{matrix} k \\ rm \end{matrix}\right\} A^r \tag{17.86}$$

Then the equation (17.85) can be written as

$$A^k_{,j} = \bar{A}^i_{,n} \frac{\partial \bar{x}^n}{\partial x^j} \frac{\partial x^k}{\partial \bar{x}^i} \tag{17.87}$$

which shows that $A^k_{,j}$ is a mixed tensor of second order. This tensor is called the *covariant derivative* of A^k with respect to x^j.

We shall now find a similar expression for the derivative of a covariant vector. By transformation law, we have

$$\bar{A}_k = A_j \frac{\partial x^j}{\partial \bar{x}^k} \tag{17.88}$$

Differentiating this with respect to $\bar{x}^k$, we have

$$\frac{\partial \bar{A}_k}{\partial \bar{x}^k} = \frac{\partial A_j}{\partial x^m} \frac{\partial x^m}{\partial \bar{x}^l} \frac{\partial x^j}{\partial \bar{x}^k} + A_j \frac{\partial^2 x^j}{\partial \bar{x}^k \partial \bar{x}^l} \tag{17.89}$$

We eliminate the second term of RHS by using the equation (17.82), change the dummy indices as required and use the equation (17.88) and finally get

$$\frac{\partial \bar{A}_k}{\partial \bar{x}^l} - \begin{Bmatrix} m \\ kl \end{Bmatrix} \bar{A}_m = \left[\frac{\partial A_j}{\partial x^n} - \begin{Bmatrix} r \\ jn \end{Bmatrix} \right] A_r \frac{\partial x^j}{\partial \bar{x}^k} \frac{\partial x^n}{\partial \bar{x}^l} \tag{17.90}$$

With the notation

$$A_{m,n} = \frac{\partial A_m}{\partial x^n} - \begin{Bmatrix} r \\ mn \end{Bmatrix} A_r \tag{17.91}$$

the equation (17.90) is reduced to

$$\bar{A}_{k,l} = A_{j,n} \frac{\partial x^j}{\partial \bar{x}^k} \frac{\partial x^n}{\partial \bar{x}^l} \tag{17.92}$$

The equation (17.92) shows that $A_{j,n}$ is a covariant tensor of second order. It is called the *covariant derivative* of A_j with respect to x^n.

The above relations for the covariant derivative may be extended to the case of a general mixed tensor of higher order.

EXAMPLE 17.8. *Show that the covariant derivatives of fundamental tensors vanish.*

Solution: We differentiate g_{ij} with respect to x^k. Then

$$g_{ij,k} = \frac{\partial g_{ij}}{\partial x^k} - \begin{Bmatrix} l \\ ik \end{Bmatrix} g_{lj} - \begin{Bmatrix} l \\ jk \end{Bmatrix} g_{il} = \frac{\partial g_{ij}}{\partial x^k} - \frac{\partial g_{ij}}{\partial x^k} = 0 \quad \text{[using equation (17.71)]} \tag{1}$$

Again

$$g^{ij}_{,k} = \frac{\partial g^{ij}}{\partial x^k} + \begin{Bmatrix} i \\ lk \end{Bmatrix} g^{lj} + \begin{Bmatrix} j \\ lk \end{Bmatrix} g^{il} = \frac{\partial g^{ij}}{\partial x^k} - \frac{\partial g^{ij}}{\partial x^k} = 0 \quad \text{[using the result of example 17.7]} \tag{2}$$

17.22 ∇-Operator Definitions

We consider a contravariant vector A^i and a covariant vector A_i. Then, by equation (17.86)

$$A^i_{,i} = \frac{\partial A^i}{\partial x^i} + \begin{Bmatrix} i \\ ri \end{Bmatrix} A^r = \frac{\partial A^i}{\partial x^i} + A^r \frac{\partial}{\partial x^r} \{\log \sqrt{g}\} \quad \text{[see Example 17.6]}$$

Thus

$$A^i_{,i} = \frac{1}{\sqrt{g}} \frac{\partial}{\partial x^r} \{\sqrt{g} A^r\}, \quad (g > 0) \tag{17.93}$$

This invariant is called *divergence* of the contravariant vector A^i and is denoted by div A^i. Hence

$$\operatorname{div} A^i = \frac{1}{\sqrt{g}} \frac{\partial}{\partial x^i} \{\sqrt{g} A^i\}, \quad (g > 0) \tag{17.94}$$

The divergence of a covariant vector A_i is defined by

$$\operatorname{div} A_i = g^{ik} A_{i,k} \tag{17.95}$$

Let us now consider an invariant ϕ. We define the covariant derivative of an invariant as the familiar partial derivative. Then

$$\phi_{,i} = \frac{\partial \phi}{\partial x^i} \tag{17.96}$$

This being a covariant vector, its covariant derivative with respect to x^j is given by [from equation (17.91)]

$$(\phi_{,i})_{,j} = \frac{\partial^2 \phi}{\partial x^i \partial x^j} - \begin{Bmatrix} r \\ ij \end{Bmatrix} \frac{\partial \phi}{\partial x^r} \tag{17.97}$$

The Laplacian of ϕ is defined as the divergence of the covariant vector $(\phi_{,i})_{,j}$ and is denoted by $\nabla^2\phi$. Hence

$$\nabla^2 \phi = g^{ik}(\phi_{,i})_{,k} = g^{ik}\left(\frac{\partial^2 \phi}{\partial x^i \partial x^k} - \begin{Bmatrix} r \\ ik \end{Bmatrix} \frac{\partial \phi}{\partial x^r}\right) \tag{17.98}$$

The *curl* of the covariant vector A_i is defined as

$$\operatorname{curl} A_i = A_{i,j} - A_{j,i} \tag{17.99}$$

From equation (17.91), we can write

$$\operatorname{curl} A_i = \left[\frac{\partial A_i}{\partial x^j} - \begin{Bmatrix} r \\ ij \end{Bmatrix} A_r\right] - \left[\frac{\partial A_j}{\partial x^i} - \begin{Bmatrix} r \\ ji \end{Bmatrix} A_r\right] = \frac{\partial A_i}{\partial x^j} - \frac{\partial A_j}{\partial x^i} \tag{17.100}$$

EXAMPLE 17.9. ***Find div A^i in spherical coordinates.***

Solution: From equation (17.46), we have

$$g_{11} = 1, \quad g_{22} = (x^1)^2, \quad g_{33} = (x^1 \sin x^2)^2, \quad g_{ij} = 0.$$

Hence

$$g = g_{11} g_{22} g_{33} = (x^1)^4 (\sin$$

Thus

$$\sqrt{g} = (x^1)^2 \sin x^2 \tag{1}$$

By equation (17.94), we have

$$\operatorname{div} A^i = \frac{1}{\sqrt{g}}\left[\frac{\partial}{\partial x^1}(\sqrt{g}A^1) + \frac{\partial}{\partial x^2}(\sqrt{g}A^2) + \frac{\partial}{\partial x^3}(\sqrt{g}A^3)\right] \tag{2}$$

Using equation (1), we find

$$\operatorname{div} A^i = \frac{1}{(x^1)^2 \sin x^2}\left[A^1 \frac{\partial}{\partial x^1}\left\{(x^1)^2 \sin x^2\right\} + A^2 \frac{\partial}{\partial x^2}\left\{(x^1)^2 \sin x^2\right\} + A^3 \frac{\partial}{\partial x^3}\left\{(x^1)^2 \sin x^2\right\}\right]$$

$$+ \frac{\partial A^1}{\partial x^1} + \frac{\partial A^2}{\partial x^2} + \frac{\partial A^3}{\partial x^3}$$

$$\text{or,} \quad \operatorname{div} A^i = \left(\frac{2}{x^1}A^1 + \frac{\partial A^1}{\partial x^1}\right) + \left[(\cot x^2)A^2 + \frac{\partial A^2}{\partial x^2}\right] + \frac{\partial A^3}{\partial x^3}$$

$$= \frac{2}{x^1}A^1 + (\cot x^2)A^2 + \frac{\partial A^1}{\partial x^1} + \frac{\partial A^2}{\partial x^2} + \frac{\partial A^3}{\partial x^3}$$

EXAMPLE 17.10. *Show that div A_j = div A^j, and hence find div A_j in spherical coordinates.*

Solution: We have

$$\begin{aligned}\text{div}\, A_j &= g^{jk} A_{j,k} \quad \text{(by definition)} \\ &= \left(g^{jk} A_j\right)_{,k} = \left(A^k\right)_{,k} = \text{div}\, A^j\end{aligned}$$

Thus, to find div A_j, we are to construct A^j and write

$$\text{div}\, A_j = \text{div}\, A^j \tag{1}$$

In spherical coordinates

$$g^{11} = 1, \quad g^{22} = \frac{1}{(x^1)^2}, \quad g^{33} = \frac{1}{(x^1 \sin x^2)^2}, \quad g^{ij} = 0 \tag{2}$$

Now, from (1) we have

$$\text{div}\, A_j = \text{div}\, A^j = \frac{1}{\sqrt{g}} \frac{\partial}{\partial x^r} \{\sqrt{g} A^r\} \quad \text{[from equation (17.93)]}$$

$$== \frac{1}{\sqrt{g}} \frac{\partial}{\partial x^r} \{\sqrt{g} g^{rk} A_k\} \tag{3}$$

Putting the values of g^{rk} from (2) and $\sqrt{g} = (x^1)^2 \sin x^2$ into equation (3), we find

$$\text{div}\, A_j = \left(\frac{2A_1}{x^1} + \frac{\partial A_1}{\partial x^1}\right) + \left[A_2 \frac{\cos x^2}{(\sin x^2)(x^1)^2} + \frac{1}{(x^1)^2} \frac{\partial A_2}{\partial x^2}\right] + \frac{1}{(x^1 \sin x^2)^2} \frac{\partial A_3}{\partial x^3}$$

$$\text{or,} \quad \text{div}\, A_j = \frac{\partial A_1}{\partial x^1} + \frac{1}{(x^1)^2} \frac{\partial A_2}{\partial x^2} + \frac{1}{(x^1 \sin x^2)^2} \frac{\partial A_3}{\partial x^3} + \frac{2}{x_1} A_1 + \frac{\cot x^2}{(x^1)^2} A_2$$

17.23 Intrinsic Derivatives

We consider a tensor

$$A^{ijk\cdots}_{pqr\cdots} \tag{17.101}$$

whose components are functions of t along a curve $x^i = x^i(t)$. The *intrinsic derivative* of this tensor is defined as

$$\frac{\partial}{\partial t} A^{ijk\cdots}_{pqr\cdots} = A^{ijk\cdots}_{pqr\cdots,a} \frac{dx^a}{dt} \tag{17.102}$$

Hence, the intrinsic derivative of a tensor is a tensor of same order and type.

If ϕ is an invariant, then

$$\frac{\partial \phi}{\partial t} = \frac{\partial \phi}{\partial x^k} \frac{dx^k}{dt} = \frac{d\phi}{dt} \tag{17.103}$$

Hence, the intrinsic derivative of an invariant equals its total derivative.

For a vector A^j

$$\frac{\partial A^j}{\partial t} = \left(A^j_{,k}\right) \frac{dx^k}{dt} = \left[\frac{\partial A^j}{\partial x^k} + \begin{Bmatrix} j \\ rk \end{Bmatrix} A^r\right] \frac{dx^k}{dt} = \frac{dA^j}{dt} + \begin{Bmatrix} j \\ rk \end{Bmatrix} A^r \frac{dx^k}{dt} \tag{17.104}$$

Similarly, for a vector A_j

$$\frac{\partial A_j}{\partial t} = \frac{dA_j}{dt} - \begin{Bmatrix} r \\ jk \end{Bmatrix} A_r \frac{dx^k}{dt} \tag{17.105}$$

CURVATURE TENSOR

17.24 Riemann-Christoffel Tensor

We consider an arbitrary covariant vector A_j. It's covariant derivative is

$$A_{j,n} = \frac{\partial A_j}{\partial x^n} - \begin{Bmatrix} k \\ jn \end{Bmatrix} A_k \tag{17.106}$$

We take covariant derivatives of equation (17.106). Then we have

$$A_{j,np} = \frac{\partial}{\partial x^p}(A_{j,n}) - \begin{Bmatrix} l \\ jp \end{Bmatrix} A_{l,n} - \begin{Bmatrix} l \\ np \end{Bmatrix} A_{j,l} \tag{17.107}$$

Using equation (17.106) in equation (17.107) with necessary change of indices, we get

$$A_{j,np} = \frac{\partial}{\partial x^p}\left[\frac{\partial A_j}{\partial x^n} - \begin{Bmatrix} k \\ jn \end{Bmatrix} A_k\right] - \begin{Bmatrix} l \\ jp \end{Bmatrix}\left[\frac{\partial A_l}{\partial x^n} - \begin{Bmatrix} k \\ ln \end{Bmatrix} A_k\right] - \begin{Bmatrix} l \\ np \end{Bmatrix}\left[\frac{\partial A_j}{\partial x^l} - \begin{Bmatrix} k \\ jl \end{Bmatrix} A_k\right] \tag{17.108}$$

We shall get a similar expression for $A_{j,pn}$ by interchanging n and p in equation (17.108). Thus

$$A_{j,pn} = \frac{\partial}{\partial x^n}\left[\frac{\partial A_j}{\partial x^p} - \begin{Bmatrix} k \\ jn \end{Bmatrix} A_k\right] - \begin{Bmatrix} l \\ jn \end{Bmatrix}\left[\frac{\partial A_l}{\partial x^p} - \begin{Bmatrix} k \\ lp \end{Bmatrix} A_k\right] - \begin{Bmatrix} l \\ np \end{Bmatrix}\left[\frac{\partial A_j}{\partial x^l} - \begin{Bmatrix} k \\ jl \end{Bmatrix} A_k\right] \tag{17.109}$$

Subtracting equation (17.109) from equation (17.108), we have

$$A_{j,np} - A_{j,pn} = \left[\frac{\partial}{\partial x^n}\begin{Bmatrix} k \\ jp \end{Bmatrix} - \frac{\partial}{\partial x^p}\begin{Bmatrix} k \\ jn \end{Bmatrix} + \begin{Bmatrix} l \\ jp \end{Bmatrix}\begin{Bmatrix} k \\ ln \end{Bmatrix} - \begin{Bmatrix} l \\ jn \end{Bmatrix}\begin{Bmatrix} k \\ lp \end{Bmatrix}\right] A_k \tag{17.110}$$

Since A_k is an arbitrary tensor, and $(A_{j,np} - A_{j,pn})$ is a tensor, it follows from quotient law that the expression within the bracket in equation (17.110) is a mixed tensor of fourth order. It is of contravariant order one and of covariant order three.

The equation (17.110) is generally written as

$$A_{j,np} - A_{j,pn} = R^k_{jnp} A_k \tag{17.111}$$

where R^k_{jnp} is called the *Riemann-Christoffel Tensor* and is given by

$$R^k_{jnp} = \frac{\partial}{\partial x^n}\begin{Bmatrix} k \\ jp \end{Bmatrix} - \frac{\partial}{\partial x^p}\begin{Bmatrix} k \\ jn \end{Bmatrix} + \begin{Bmatrix} l \\ jp \end{Bmatrix}\begin{Bmatrix} k \\ ln \end{Bmatrix} - \begin{Bmatrix} l \\ jn \end{Bmatrix}\begin{Bmatrix} k \\ lp \end{Bmatrix} \tag{17.112}$$

The Riemann-Christoffel tensor is formed exclusively from the metric tensor g_{ij} and its derivatives, and does not depend on the choice of A_k. If in Euclidean space, the Cartesian coordinate system is taken, then all the g_{ii} are constant so that all the components of R^k_{jnp} will vanish.

From equation (17.112), we find that

$$R^k_{jnp} = -R^k_{jpn} \tag{17.113}$$

That is, the Riemann-Christoffel tensor R^k_{jnp} is skew symmetric with respect to the indices n and p. It can also be shown that the sum of the Riemann-Christoffel tensors obtained by changing the indices j, n, p in cyclic order is zero. That is

$$R^k_{jnp} + R^k_{npj} + R^k_{pjn} = 0 \tag{17.114}$$

17.25 Curvature Tensor

The covariant *curvature tensor* is defined by

$$R_{rjnp} = g_{rk} R^k_{jnp} \tag{17.115}$$

Using the equation (17.112) in the RHS of equation (17.115), we have

$$R_{rjnp} = \frac{\partial}{\partial x^n}\left[g_{rk}\begin{Bmatrix} k \\ jp \end{Bmatrix}\right] - \frac{\partial g_{rk}}{\partial x^n}\begin{Bmatrix} k \\ jn \end{Bmatrix} - \frac{\partial}{\partial x^p}\left[g_{rk}\begin{Bmatrix} k \\ jn \end{Bmatrix}\right] + \frac{\partial g_{rk}}{\partial x^p}\begin{Bmatrix} k \\ jn \end{Bmatrix}$$
$$+ g_{rk}\begin{Bmatrix} l \\ jp \end{Bmatrix}\begin{Bmatrix} k \\ ln \end{Bmatrix} - g_{rk}\begin{Bmatrix} l \\ jn \end{Bmatrix}\begin{Bmatrix} k \\ lp \end{Bmatrix} \tag{17.116}$$

Using equations (17.71) and (17.72), we reduce this expression to the form

$$R_{rjnp} = \frac{\partial}{\partial x^n}[jp, r] - \frac{\partial}{\partial x^p}[jn, r] + \begin{Bmatrix} k \\ jn \end{Bmatrix}[rp, k] - \begin{Bmatrix} k \\ jp \end{Bmatrix}[rn, k] \tag{17.117}$$

By contracting the Riemann-Christoffel tensor R^k_{jnp} with respect to the indices k and p, we get from equation (17.112), *Ricci's tensor*

$$R_{jn} = \frac{\partial}{\partial x^n}\begin{Bmatrix} k \\ jk \end{Bmatrix} - \frac{\partial}{\partial x^k}\begin{Bmatrix} k \\ jn \end{Bmatrix} + \begin{Bmatrix} l \\ jk \end{Bmatrix}\begin{Bmatrix} k \\ ln \end{Bmatrix} - \begin{Bmatrix} l \\ jn \end{Bmatrix}\begin{Bmatrix} k \\ lk \end{Bmatrix} \tag{17.118}$$

It follows that $R_{jn} = R_{nj}$; that is, Ricci's tensor is symmetric.

The *curvature invariant* is defined by

$$R = g^{jn} R_{jn} \tag{17.119}$$

A space is called an *Einstein space*, if

$$R_{ij} = I g_{ij} \tag{17.120}$$

where I is an invariant.

Inner multiplication of equation (17.120) by g^{ij} gives, $R = NI$, and hence

$$R_{ij} = \frac{R}{N} g_{ij} \tag{17.121}$$

The *Einstein tensor* is defined by

$$G^i_j = g^{ik} R_{jk} - \tfrac{1}{2} R \delta^i_j \tag{17.122}$$

A covariant differentiation of this equation gives

$$G^i_{j,i} = g^{ik} R_{jk,i} - \tfrac{1}{2} R_{,i}\delta^i_j = g^{ik} R_{jk,i} - \tfrac{1}{2} R_{,j}, \quad \text{i.e.,} \quad G^i_{j,i} = 0 \tag{17.123}$$

which is an important equation in the theory of relativity.

17.26 Geodesics

We consider a curve in N-dimensional space defined by the equation

$$x^i = x^i(t) \tag{17.124}$$

where t is an independent parameter. The distance between two points $P(t_1)$ and $Q(t_2)$ lying on the curve is given by

$$s = \int_{t_1}^{t_2} \left[\phi\left(x^i, \dot{x}^i, t\right)\right]^{1/2} dt \tag{17.125}$$

where

$$\phi = g_{ij}\frac{dx^i}{dt}\frac{dx^j}{dt} \quad \text{and} \quad \dot{x}^i = \frac{dx^i}{dt} \tag{17.126}$$

The curve along which the integral equation (17.125) is *stationary* (a minimum) for infinitesimal variations is called *geodesics*. Thus, for a geodesics, we have

$$\delta s = \delta\int_{t_1}^{t_2} \sqrt{\phi}dt = 0 \tag{17.127}$$

Using Euler's equation for the extremum of functionals, we write

$$\frac{\partial\left(\sqrt{\phi}\right)}{\partial x^i} - \frac{d}{dt}\left(\frac{\partial\left(\sqrt{\phi}\right)}{\partial \dot{x}^i}\right) = 0$$

$$\text{or,} \quad -\frac{1}{2\sqrt{\phi}}\frac{\partial\phi}{\partial x^i} + \frac{d}{dt}\left[\frac{1}{2\sqrt{\phi}}\frac{\partial\phi}{\partial\dot{x}^i}\right] = 0$$

$$\text{or,} \quad -\frac{1}{\sqrt{\phi}}\frac{\partial\phi}{\partial x^i} + \frac{1}{\sqrt{\phi}}\frac{d}{dt}\left(\frac{\partial\phi}{\partial\dot{x}^i}\right) + \frac{\partial\phi}{\partial\dot{x}^i}\frac{d}{dt}\left(\frac{1}{\sqrt{\phi}}\right) = 0$$

$$\text{i.e.,} \quad \frac{d}{dt}\left(\frac{\partial\phi}{\partial\dot{x}^i}\right) - \frac{\partial\phi}{\partial x^i} - \frac{1}{2\phi}\frac{\partial\phi}{\partial\dot{x}^i}\frac{d\phi}{dt} = 0 \tag{17.128}$$

Let us now choose the arc-length s as the parameter t. Then, we have

$$\left.\begin{aligned} \dot{x}^i &= \frac{dx^i}{dt} = \frac{dx^i}{ds} \\ \phi &= g_{ij}\frac{dx^i}{ds}\frac{dx^j}{ds} \end{aligned}\right\} \tag{17.129}$$

Hence

$$\left.\begin{aligned} \frac{\partial\phi}{\partial\dot{x}^i} &= 2g_{ij}\dot{x}^j \\ \frac{\partial\phi}{\partial x^i} &= \frac{\partial g_{mk}}{\partial x^i}\frac{dx^m}{ds}\frac{dx^k}{ds} \end{aligned}\right\} \tag{17.130}$$

Moreover, since s is an arbitrary parameter

$$\frac{d\phi}{ds} = 0 \tag{17.131}$$

Hence, from equation (17.128) we get

$$\frac{d}{ds}\left(\frac{\partial \phi}{\partial \dot{x}^i}\right) - \frac{\partial \phi}{\partial x^i} = 0$$

$$\text{or,} \quad \frac{d}{ds}\left(2g_{ij}\dot{x}^j\right) - \frac{\partial g_{mk}}{\partial x^i}\dot{x}^m \dot{x}^k = 0$$

$$\text{or,} \quad g_{ij}\frac{d^2x^j}{ds^2} + \frac{\partial g_{ij}}{\partial x^l}\frac{dx^l}{ds}\frac{dx^j}{ds} - \frac{1}{2}\frac{\partial g_{mk}}{\partial x^i}\frac{dx^m}{ds}\frac{dx^k}{ds} = 0 \tag{17.132}$$

Using the Christoffel symbol of first kind, we write equation (17.132) in the form

$$g_{ij}\frac{d^2x^j}{ds^2} + [mk, i]\frac{dx^m}{ds}\frac{dx^k}{ds} = 0 \tag{17.133}$$

Inner multiplication of the equation (17.133) by $g^{i\lambda}$ gives

$$g^{i\lambda}g_{ij}\frac{d^2x^j}{ds^2} + g^{i\lambda}[mk, i]\frac{dx^m}{ds}\frac{dx^k}{ds} = 0, \quad \text{i.e.,} \quad \frac{d^2x^\lambda}{ds^2} + \begin{Bmatrix} \lambda \\ mk \end{Bmatrix}\frac{dx^m}{ds}\frac{dx^k}{ds} = 0 \tag{17.134}$$

This is the equation of the geodesics in the Riemann-space.

Now, it is possible to introduce a coordinate system such that, at a particular point P, the Christoffel symbols vanish. Such a coordinate system is called *geodesic coordinates* and the point P is called the *pole*. Hence, under geodesic coordinate system, the equation of the geodesics simplifies to

$$\frac{d^2x^\lambda}{ds^2} = 0 \tag{17.135}$$

In Euclidean space under Cartesian coordinate system, we have $g_{ii} = 1, g_{ij} = 0$ and also the Christoffel symbol vanishes. Thus, equation (17.135) determines the geodesic lines in this case also.

EXERCISE

1. The components of velocities in (x, y) coordinates are $(\dot{x}, \dot{y})$. Show that in polar coordinates they are $(\dot{r}, \dot{\theta})$.

2. In (x, y, z) coordinates, the components of a covariant vector are $A_1 = x, A_2 = y, A_3 = z$. Find its components in spherical coordinates (r, θ, ϕ).

3. Prove that, δ^i_j is a mixed tensor.

4. If A_{ij} is a skew-symmetric tensor, prove that $\left(\delta^i_j\delta^k_l + \delta^i_l\delta^k_j\right)A_{ik} = 0$.

5. If A^{ij} and B^k are tensors, prove that $A^{ij}B_j$ is also a tensor.

6. If $A_jB^{jk} = C^k$, where B^{jk} is an arbitrary contravariant tensor and C^k is a contravariant vector, then A_j is a covariant vector.

7. Show that the inner products (a) $\phi = A_jB^j$ and (b) $C_i = A_{ij}B^j$ are tensors.

8. Find the values of g^{ik} in cylindrical polar coordinates.

9. Calculate the Christoffel symbols in cylindrical polar coordinates.

10. Calculate the Christoffel symbols corresponding to the metric $ds^2 = (dx^1)^2 + G(dx^2)^2$, where G is a function of x^1, x^2.

11. Evaluate div A^j and $\nabla^2\phi$ in cylindrical polar coordinates.

12. Show that, the covariant derivative of δ^i_j is zero.

13. If A^{ijk} is a skew-symmetric tensor, show that $\frac{1}{\sqrt{g}}\frac{\partial}{\partial x^k}\left(\sqrt{g}A^{ijk}\right)$ is a tensor.

14. Show that $\frac{\partial}{\partial t}\left(\frac{dx^i}{dt}\right) = \frac{d^2x^i}{dt^2} + \begin{Bmatrix} i \\ jk \end{Bmatrix}\frac{dx^j}{dt}\frac{dx^k}{dt}$.

15. Prove that, $R^k_{knp} = 0$.

16. Prove that, $R_{pqrs} + R_{prsq} + R_{psqr} = 0$.

ANSWERS

2. $\bar{A}_1 = r\sin^2\theta\cos^2\phi + r\sin^2\theta\sin^2\phi + r\cos^2\theta$

$\bar{A}_2 = r^2\sin\theta\cos\theta\cos^2\phi + r^2\sin\theta\cos\theta\sin^2\phi - r^2\sin\theta\cos\theta$

$\bar{A}_3 = 0.$

8. $g^{11} = 1,\ g^{22} = \frac{1}{(x^1)^2},\ g^{33} = 1.$

9. $\Gamma_{12,2} = x^1,\ \Gamma_{22,1} = -x^1$, others are zero.

10. $\Gamma_{12,2} = \frac{1}{2}\frac{\partial G}{\partial x^1},\ \Gamma_{22,1} = -\frac{1}{2}\frac{\partial G}{\partial x^1},\ \Gamma_{22,2} = \frac{1}{2}\frac{\partial G}{\partial x^2}.$

11. $\operatorname{div} A^j = \frac{A^1}{x^1} + \frac{\partial A^1}{\partial x^1} + \frac{\partial A^2}{\partial x^2} + \frac{\partial A^3}{\partial x^3}$

$$\nabla^2\phi = \frac{\partial^2\phi}{(\partial x^1)^2} + \frac{1}{(x^1)^2}\frac{\partial^2\phi}{(\partial x^2)^2} + \frac{\partial^2\phi}{(\partial x^3)^2} + \frac{1}{x^1}\frac{\partial\phi}{\partial x^1}.$$

Chapter 18

Boundary Value Problems

In Chapter 11, we have discussed about the various types of partial differential equations appearing in different types of problems of mathematical physics. Here we like to find solutions of those partial differential equations under pre-assigned initial and boundary conditions by the Fourier's method of separation of variables.

LAPLACE'S EQUATION

18.1 Laplace's Equation in Cartesian Coordinates

In three-dimensional Cartesian coordinates, the Laplace's equation is

$$\frac{\partial^2 \phi}{\partial x^2} + \frac{\partial^2 \phi}{\partial y^2} + \frac{\partial^2 \phi}{\partial z^2} = 0 \tag{18.1}$$

where $\phi(x, y, z)$ represents, say potential function. We assume the solution in the form

$$\phi(x, y, z) = X(x)Y(y)Z(z) \tag{18.2}$$

Substituting the solution function equation (18.2) into equation (18.1), and dividing the result by XYZ, we get

$$\frac{X''}{X} + \frac{Y''}{Y} + \frac{Z''}{Z} = 0 \tag{18.3}$$

Since each term in equation (18.3) is a function of one variable, their sum can be zero only if each of them is a constant. Hence

$$\frac{d^2 X}{dx^2} = -\alpha^2 X \tag{18.4}$$

$$\frac{d^2 Y}{dy^2} = -\beta^2 Y \tag{18.5}$$

$$\text{and} \quad \frac{d^2 Z}{dz^2} = \gamma^2 Z \tag{18.6}$$

such that

$$\gamma^2 = \alpha^2 + \beta^2 \tag{18.7}$$

The solutions of ordinary differential equations (18.4)–(18.7) are

$$X(x) = A_1 \cos \alpha x + A_2 \sin \alpha x \tag{18.8}$$

$$Y(y) = A_3 \cos \beta y + A_4 \sin \beta y \tag{18.9}$$

$$Z(z) = A_5 e^{\gamma z} + A_6 e^{-\gamma z} \tag{18.10}$$

so that the general solution of Laplace's equation may be written as

$$\phi(x, y, z) = \sum (A_1 \cos \alpha x + A_2 \sin \alpha x)(A_3 \cos \beta y + A_4 \sin \beta y)(A_5 e^{\gamma z} + A_6 e^{-\gamma z}) \tag{18.11}$$

where $A_1, A_2, \ldots$ are arbitrary constants. These constants of integration along with the parameters α, β are determined from boundary conditions of specific problems.

EXAMPLE 18.1. *Determine the potential within a rectangular box having dimension (a, b, c) along (x, y, z) directions with the following boundary conditions:*

(i) $$\phi(x, y, z) = 0, \text{ for } x = 0, y = 0, z = 0 \tag{1}$$

(ii) $$\phi(x, y, z) = 0, \text{ for } x = a \text{ and } y = b \tag{2}$$

(iii) $$\phi(x, y, z) = f(x, y), \text{ for } z = c \tag{3}$$

Solution: Using the conditions at $x = 0$ and $x = a$, we find from equation (18.8) that

$$A_1 = 0 \quad \text{and} \quad \sin \alpha a = 0, \quad \text{i.e.,} \quad \alpha = \frac{m\pi}{a} \tag{4}$$

Similarly, using conditions at $y = 0$ and $y = b$, we find from equation (18.9) that

$$A_3 = 0, \quad \text{and} \quad \sin \beta b = 0, \quad \text{i.e.,} \quad \beta = \frac{n\pi}{b} \tag{5}$$

Again, $\phi(x, y, z) = 0$ at $z = 0$, which gives

$$A_5 + A_6 = 0, \quad \text{i.e.,} \quad A_5 = -A_6 \tag{6}$$

Hence, we can write the solution as

$$\phi(x, y, z) = \sum_{m=1}^{\infty} \sum_{n=1}^{\infty} A_{mn} \sin \frac{m\pi x}{a} \sin \frac{n\pi y}{b} \sinh \gamma_{mn} z \tag{7}$$

where

$$\gamma_{mn} = \pi \left(\frac{n^2}{b^2} + \frac{m^2}{a^2} \right)^{1/2} \tag{8}$$

m, n being positive integers.

Now, using the condition equation (3), we get

$$f(x, y) = \sum_{m=1}^{\infty} \sum_{n=1}^{\infty} A_{mn} \sin \frac{m\pi x}{a} \sin \frac{n\pi y}{b} \sinh \gamma_{mn} c \tag{9}$$

which is just a double Fourier series for $f(x, y)$.

Hence, the coefficients A_{mn} are given by

$$A_{mn} = \frac{4}{ab \sinh(\gamma_{mn} c)} \int_0^a \int_0^b f(x, y) \sin \frac{m\pi x}{a} \sin \frac{n\pi y}{b} dx dy \tag{10}$$

18.2 Laplace's Equation in Two-Dimensional Cartesian Coordinates

In two-dimensional Cartesian coordinates, Laplace's equation assumes the form

$$\frac{\partial^2\phi}{\partial x^2}+\frac{\partial^2\phi}{\partial y^2}=0. \tag{18.12}$$

Proceeding exactly as before, we find the solution in the form

$$\phi(x,y)=\sum(A_1\cos\alpha x+A_2\sin\alpha x)(A_3e^{\alpha y}+A_4e^{-\alpha y}) \tag{18.13}$$

Let us consider an example.

EXAMPLE 18.2. *Determine an expression for steady-state flow of heat in a finite rectangular plate of dimension* $a\times b$ *under the following boundary conditions.* (i) *Temperature* $=0$, *for* $x=0$ *and* $x=a$; (ii) *Temperature* $=0$, *for* $y=0$; *and* (iii) *Temperature* $=f(x)$, *for* $y=b$.

Solution: Laplace equation (18.12) describes the steady state temperature distribution in two dimensions.

Since $\phi(x,y)=0$, for $x=0$, and $x=a$, we get from equation (18.13)

$$A_1=0,\quad \text{and}\quad \sin\alpha a=0,\quad \text{i.e.,}\quad \alpha=\frac{m\pi}{a} \tag{1}$$

Again, since $\phi(x,y)=0$ for $y=0$, we get from equation (18.13)

$$A_3+A_4=0,\quad \text{i.e.,}\quad A_3=-A_4 \tag{2}$$

Thus the solution of Laplace equation (18.12) under the given boundary conditions is

$$\phi(x,y)=\sum_{m=1}^{\infty}A_m\sin\frac{m\pi x}{a}\sinh\beta_m y \tag{3}$$

$$=\sum_{m=1}^{\infty}A_m\sin\frac{m\pi x}{a}\sinh\frac{m\pi y}{a} \tag{4}$$

Using the condition at $y=b$, we get

$$f(x)=\sum_{m=1}^{\infty}A_m\sin\frac{m\pi x}{a}\sinh\frac{m\pi b}{a} \tag{5}$$

Hence

$$A_m=\frac{2}{a\sinh\dfrac{m\pi b}{a}}\int_0^a f(x)\sin\frac{m\pi x}{a}dx \tag{6}$$

18.3 Laplace's Equation in Spherical Polar Coordinates

Laplace's equation in spherical polar coordinates (r,θ,ϕ) takes the form

$$\frac{1}{r^2\sin\theta}\left[\frac{\partial}{\partial r}\left(r^2\sin\theta\frac{\partial\psi}{\partial r}\right)+\frac{\partial}{\partial\theta}\left(\sin\theta\frac{\partial\psi}{\partial\theta}\right)+\frac{\partial}{\partial\phi}\left(\frac{1}{\sin\theta}\frac{\partial\psi}{\partial\phi}\right)\right]=0 \tag{18.14}$$

The axis of symmetry is taken along $\theta = 0$.
We assume the solution of equation (18.14) in the form

$$\psi(r, \theta, \phi) = R(r)P(\theta)Q(\phi) \tag{18.15}$$

Substituting equation (18.15) into equation (18.14) and dividing by RPQ, we get

$$\frac{\sin\theta}{R}\frac{d}{dr}\left(r^2\frac{dR}{dr}\right) + \frac{1}{P}\frac{d}{d\theta}\left(\sin\theta\frac{dP}{d\theta}\right) + \frac{1}{\sin\theta}\frac{1}{Q}\frac{d^2Q}{d\phi^2} = 0$$

$$\text{or,}\quad \frac{1}{R}\frac{d}{dr}\left(r^2\frac{dR}{dr}\right) = -\frac{1}{P\sin\theta}\frac{d}{d\theta}\left(\sin\theta\frac{dP}{d\theta}\right) - \frac{1}{\sin^2\theta}\frac{1}{Q}\frac{d^2Q}{d\phi^2} = 0. \tag{18.16}$$

The LHS of equation (18.16) is a function of r only, whereas the RHS is a function of θ and ϕ. Hence, the two sides can be equal if and only if each side is a constant, say λ. Thus

$$\frac{1}{R}\frac{d}{dr}\left(r^2\frac{dR}{dr}\right) = \lambda, \quad \text{or,} \quad r^2\frac{d^2R}{dr^2} + 2r\frac{dR}{dr} - \lambda R = 0 \tag{18.17}$$

To solve this equation, we assume

$$R(r) = Ar^n, \quad A \text{ and } n \text{ are constants} \tag{18.18}$$

Substituting equation (18.18) into equation (18.17), we get

$$[n(n+1) - \lambda]R = 0$$

which gives (since $R \neq 0$)

$$\lambda = n(n+1) \tag{18.19}$$

Thus, equation (18.17) assumes the form

$$r^2\frac{d^2R}{dr^2} + 2r\frac{dR}{dr} - n(n+1)R = 0 \tag{18.20}$$

We assume a series solution of this equation (18.20) in the form

$$R(r) = \sum A_k r^k \tag{18.21}$$

On substitution in equation (18.20), we get

$$k(k+1) - n(n+1) = 0 \tag{18.22}$$

which is possible, if $k = n$, or, $k = -(n+1)$.
Hence the general solution of equation (18.20) is

$$R(r) = \sum\left(A_n r^n + \frac{B_n}{r^{n+1}}\right) \tag{18.23}$$

Now, let us consider the equation (18.16) which gives

$$-\frac{1}{P\sin\theta}\cdot\frac{d}{d\theta}\left(\sin\theta\frac{dP}{d\theta}\right) - \frac{1}{\sin^2\theta}\frac{1}{Q}\frac{d^2Q}{d\phi^2} = n(n+1)$$

$$\text{or,}\quad \frac{1}{Q}\frac{d^2Q}{d\phi^2} = -n(n+1)\sin^2\theta - \frac{\sin\theta}{P}\frac{d}{d\theta}\left(\sin\theta\frac{dP}{d\theta}\right) \tag{18.24}$$

Again, we find that the LHS of equation depends only on variable ϕ, and the RHS depends only on variable θ. This is possible only if both the sides are equal to a constant, say $-m^2$, m being an integer. Then we get from equation (18.24)

$$\frac{1}{Q}\frac{d^2Q}{d\phi^2} = -m^2, \quad \text{or,} \quad \frac{d^2Q}{d\phi^2} + m^2Q = 0 \tag{18.25}$$

Its solution is

$$Q(\phi) = Ce^{\pm im\phi} \tag{18.26}$$

where C is a constant of integration.

Lastly, we get from equation (18.24) the following differential equation for $P(\theta)$:

$$\sin\theta\frac{d}{d\theta}\left(\sin\theta\frac{dP}{d\theta}\right) + \{n(n+1)\sin^2\theta - m^2\}P = 0 \tag{18.27}$$

We substitute

$$\cos\theta = x, \quad dx = -\sin\theta d\theta \tag{18.28}$$

Equation (18.27) assumes the form

$$(1-x^2)\frac{d}{dx}\left[(1-x^2)\frac{dP}{dx}\right] + \left[(1-x^2)n(n+1) - m^2\right]P = 0$$

$$\text{or,}\quad \frac{d}{dx}\left[(1-x^2)\frac{dP}{dx}\right] + \left[n(n+1) - \frac{m^2}{1-x^2}\right]P = 0. \tag{18.29}$$

Equation (18.29) is associated Legendre's equation having the solution

$$P(x) = A'_nP_n^m(x) + B'_nQ_n^m(x)$$

$$\text{i.e.,}\quad P(\cos\theta) = A'_nP_n^m(\cos\theta) + B'_nQ_n^m(\cos\theta) \tag{18.30}$$

where $P_n^m(x), Q_n^m(x)$ are associated Legendre polynomials of order m and degree n, and A'_n, B'_n are constants to be determined from boundary conditions.

Thus, the general solution of equation (18.14) is obtained from equations (18.23), (18.26) and (18.30) by summing up for all n in the form

$$\psi(r,\theta,\phi) = \sum_{n=0}^{\infty}\left(A_nr^n + B_nr^{-(n+1)}\right)\left[A'_nP_n^m(x) + B'_nQ_n^m(x)\right]\cdot e^{\pm im\phi} \tag{18.31}$$

where, as before, $x = \cos\theta$.

Particular Case

If there is axial symmetry about z-axis, the function $\psi(r,\theta,\phi)$ will not depend on ϕ. In this case $m = 0$, and equation (18.29) assumes the form

$$\frac{d}{dx}\left[(1-x^2)\frac{dP}{dx}\right] + n(n+1)P = 0 \tag{18.32}$$

which is a Legendre's differential equation having solution in terms of Legendre's polynomials

$$P_n(x) = P_n(\cos\theta) \tag{18.33}$$

Then the solution of Laplace equation (18.14) is written as

$$\psi(r,\theta) = \sum_{n=0}^{\infty} \left(A_n r^n + B_n r^{-(n+1)}\right) P_n(\cos\theta) \tag{18.34}$$

The coefficients A_n, B_n are to be determined from boundary conditions, so that the series in equation (18.34) is unique. On the symmetry axis $z = r$ and $\theta = 0$, so that

$$\psi(z = r) = \sum_{n=0}^{\infty} \left(A_n r^n + B_n r^{-(n+1)}\right) \tag{18.35}$$

will be valid for positive z; for negative z, each term is to be multiplied by a factor $(-1)^n$. Thus, if the function $\psi(z)$ is known at an arbitrary point z on the symmetry axis, then the solution at any point in space can be obtained by multiplying each term of series in equation (18.35) by $P_n(\cos\theta)$.

EXAMPLE 18.3. *A ring of radius a has total charge q. Find potential at a point P at a distance z from centre along normal to the plane of the ring. Also find potential at any other point in space so that $r > a$.*

Solution: Let the charge density on the ring be σ, so that

$$\sigma = \frac{q}{2\pi a} \tag{1}$$

We consider a small segment $ad\theta$ of the ring. Then, potential at point P due to this elemental charge is

$$dV = \frac{\sigma a d\theta}{\sqrt{a^2 + z^2}} = \frac{q}{2\pi}\frac{d\theta}{\sqrt{a^2 + z^2}} \tag{2}$$

Hence, total potential is

$$\begin{aligned}
V &= \int_0^{2\pi} \frac{q\,d\theta}{2\pi\sqrt{a^2 + z^2}} = q\left(a^2 + z^2\right)^{-1/2} \\
&= \frac{q}{z}\left(1 + \frac{a^2}{z^2}\right)^{-1/2}, \qquad z > a \\
&= \frac{q}{z}\left[1 - \frac{1}{2}\left(\frac{a}{z}\right)^2 + \frac{\frac{1}{2}\cdot\frac{3}{2}}{2}\left(\frac{a}{z}\right)^4 - \cdots\right] \\
&= \frac{q}{z}\sum_{n=0}^{\infty}(-1)^n \frac{1\cdot 3\cdot 5\cdots(2n-1)}{2\cdot 4\cdot 6\cdots 2n}\left(\frac{a}{z}\right)^{2n} \\
&= q\sum_{n=0}^{\infty} \frac{(2n)!}{2^{2n}(n!)^2}\frac{a^{2n}}{z^{2n+1}}.
\end{aligned} \tag{3}$$

Hence, potential at any other point in space for $r > a$ is given by

$$V = q\sum_{n=0}^{\infty} \frac{(2n)!}{2^{2n}(n!)^2}\frac{a^{2n}}{r^{2n+1}} P_{2n}(\cos\theta)$$

EXAMPLE 18.4. *The potential function on the surface of a sphere of radius a having no charge at its centre is given by*

$$f(\theta) = \begin{cases} +V, & \text{for } 0 \le \theta \le \pi/2 \\ -V, & \text{for } \pi/2 \le \theta \le \pi \end{cases}.$$

Find the potential function inside the sphere.

Solution: Since there is no charge at the centre, the potential must be finite there. Hence, in equation (18.34), $B_n = 0$, so that

$$\psi(r,\theta) = \sum_{n=0}^{\infty} A_n r^n P_n(\cos\theta) \tag{1}$$

Using condition on the surface

$$f(\theta) = \sum_{n=0}^{\infty} A_n a^n P_n(\cos\theta) \tag{2}$$

Hence the coefficients A_n are given by

$$A_n = \frac{2n+1}{2a^n} \int_0^{\pi} f(\theta) P_n(\cos\theta) \sin\theta d\theta \tag{3}$$

$$\text{i.e.,} \quad A_n = \frac{2n+1}{2a^n} \left[\int_0^1 P_n(x) V dx - \int_{-1}^{0} P_n(x) V dx \right] \tag{4}$$

Now, $P_n(x)$ is odd about $x = 0$, when n is odd; also for n even $P_n(x)$ is even. Thus only odd coefficients in equation (4) will be non-zero. Hence

$$A_n = \frac{(2n+1)V}{a^n} \int_0^1 P_n(x) dx \tag{5}$$

Using expressions for $P_1(x), P_3(x), \ldots$ provided in chapter 12, we get from equation (5)

$$A_1 = \frac{3V}{2a}, \quad A_3 = \frac{7V}{8a^3}, \quad A_5 = \frac{11V}{16a^5}, \ldots \tag{6}$$

Hence, from (1) and (6), we get

$$\psi(r,\theta) = V \left\{ \frac{3}{2}\left(\frac{r}{a}\right) P_1(\cos\theta) - \frac{7}{8}\left(\frac{r}{a}\right)^3 P_3(\cos\theta) + \frac{11}{16}\left(\frac{r}{a}\right)^5 P_5(\cos\theta) - \cdots \right\} \tag{7}$$

EXAMPLE 18.5. *A conducting sphere of radius a is placed in a uniform electric field E_0. Show that potential at any point in space exterior to the sphere is given by*

$$\phi(r,\theta) = -E_0 r \cos\theta \left(1 - \frac{a^3}{r^3}\right)$$

Assume that the sphere is grounded.

Solution: We take centre of the sphere as origin and axis of z along the direction of E_0; the potential function will be independent of azimuthal angle ϕ. Thus, using equation (18.34), we write

$$\phi(r,\theta) = \sum_{n=0}^{\infty}\left(A_n r^n + \frac{B_n}{r^{n+1}}\right)P_n(\cos\theta) \tag{1}$$

Boundary conditions of the problem are:

(i) $\phi(r,\theta) = 0$, for $r = a$ (since sphere is grounded);

(ii) $\phi(r,\theta) = -E_0 z = -E_0 r\cos\theta$, for $z \longrightarrow \infty$.

Using condition (ii), we have (since for $z \longrightarrow \infty, B_n = 0$)

$$-E_0 r\cos\theta = \sum_{n=0}^{\infty} A_n r^n P_n(\cos\theta), \quad \text{or,} \quad -E_0 r P_1(\cos\theta) = \sum_{n=0}^{\infty} A_n r^n P_n(\cos\theta) \tag{2}$$

Equating coefficients of $P_1(\cos\theta), P_2(\cos\theta), \ldots$ on both sides, we get

$$A_0 = 0, \quad A_1 = -E_0, \quad A_n = 0 \quad (n > 1) \tag{3}$$

Using condition (i), we get

$$\sum_{n=0}^{\infty}\left(A_n a^n + \frac{B_n}{a^{n+1}}\right)P_n(\cos\theta) = 0. \tag{4}$$

Multiplying equation (4) by $P_m(\cos\theta)$ and integrating between the limit -1 and $+1$, we get

$$A_n\frac{2a^n}{2n+1} + \frac{2B_n}{a^{n+1}}\frac{1}{2n+1} = 0 \quad \text{[using orthogonality condition]}$$

which gives

$$B_n = -A_n a^{2n+1} \tag{5}$$

In view of equation (3), we get from (5)

$$B_0 = 0, \quad B_1 = -A_1 a^3 = E_0 a^3, \quad B_n = 0 \quad (n > 1).$$

Hence, potential at any point outside the sphere is

$$\phi(r,\theta) = -E_0 r\cos\theta + E_0\frac{a^3}{r^2}\cos\theta = -E_0 r\cos\theta\left(1 - \frac{a^3}{r^3}\right).$$

18.4 Laplace's Equation in Cylindrical Coordinates

Laplace's equation in cylindrical coordinates (r, θ, z) is

$$\frac{1}{r}\frac{\partial}{\partial r}\left(r\frac{\partial\phi}{\partial r}\right) + \frac{1}{r^2}\frac{\partial^2\phi}{\partial\theta^2} + \frac{\partial^2\phi}{\partial z^2} = 0 \tag{18.36}$$

We assume a solution in the form

$$\phi(r,\theta,z) = R(r)P(\theta)Z(z) \tag{18.37}$$

Substituting this into equation (18.36) and dividing by RPZ, we get

$$\frac{1}{Z}\frac{d^2Z}{dz^2} = -\frac{1}{R}\frac{d^2R}{dr^2} - \frac{1}{rR}\frac{dR}{dr} - \frac{1}{r^2P}\frac{d^2P}{d\theta^2} \tag{18.38}$$

The LHS being a function of z only, the equation (18.38) can be satisfied only when both sides are equal to a constant, say λ^2. Thus,

$$\frac{1}{Z}\frac{d^2Z}{dz^2} = \lambda^2, \quad \text{or,} \quad \frac{d^2Z}{dz^2} - \lambda^2 Z = 0 \tag{18.39}$$

Its solution is given by

$$Z(z) = c_1 e^{\lambda z} + c_2 e^{-\lambda z} \tag{18.40}$$

where c_1, c_2 are integration constants.

Again, taking the RHS of equation (18.38) equal to λ^2, we have

$$-\frac{1}{R}\frac{d^2R}{dr^2} - \frac{1}{rR}\frac{dR}{dr} - \frac{1}{r^2P}\frac{d^2P}{d\theta^2} = \lambda^2$$

which gives

$$\frac{r^2}{R}\frac{d^2R}{dr^2} + \frac{r}{R}\frac{dR}{dr} + \lambda^2 r^2 = -\frac{1}{P}\frac{d^2P}{d\theta^2} \tag{18.41}$$

Arguing as before, we have each side of equation (18.41) equal to a constant, say m^2. Then

$$-\frac{1}{P}\frac{d^2P}{d\theta^2} = m^2, \quad \text{or,} \quad \frac{d^2P}{d\theta^2} + m^2 P = 0. \tag{18.42}$$

Its solution is

$$P(\theta) = c_3 \cos m\theta + c_4 \sin m\theta \tag{18.43}$$

where c_3, c_4 are constants of integration.

Lastly, from equation (18.41), we get

$$\frac{r^2}{R}\frac{d^2R}{dr^2} + \frac{r}{R}\frac{dR}{dr} + \lambda^2 r^2 = m^2, \quad \text{or,} \quad r^2\frac{d^2R}{dr^2} + r\frac{dR}{dr} + (\lambda^2 r^2 - m^2)R = 0 \tag{18.44}$$

We substitute

$$\lambda r = x \tag{18.45}$$

Then equation (18.44) is transformed to

$$x^2\frac{d^2R}{dx^2} + x\frac{dR}{dx} + (x^2 - m^2)R = 0 \tag{18.46}$$

This is Bessel's differential equation of order m. For non-integral values of m, its solution is

$$R(x) = A_1 J_m(x) + A_2 J_{-m}(x)$$

$$\text{i.e.,} \quad R(\lambda r) = A_1 J_m(\lambda r) + A_2 J_{-m}(\lambda r) \tag{18.47}$$

If m is an integer, the solution of equation (18.46) is given in terms of Bessel's functions of first and second kinds of order m. Thus

$$R(\lambda r) = A_1 J_m(\lambda r) + A_2 N_m(\lambda r) \tag{18.48}$$

Now, if the function $P(\theta)$ in equation (18.43) is required to be single-valued, then m takes up only integral values. Then, the general solution is

$$\phi(r,\theta,z) = \sum_{m=0}^{\infty} \left(C_1 e^{\lambda z} + C_2 e^{-\lambda z}\right)\left(A_m \cos m\theta + B_m \sin m\theta\right) J_m(\lambda r) \tag{18.49}$$

where all the constants are to be determined from boundary conditions in specific problems.

Particular Case

We assume that the function ϕ in equation (18.36) is independent of the coordinate z. Then the Laplace's equation assumes the form

$$\frac{1}{r}\frac{\partial}{\partial r}\left(r\frac{\partial \phi}{\partial r}\right) + \frac{1}{r^2}\frac{\partial^2 \phi}{\partial \theta^2} = 0 \tag{18.50}$$

Assuming the solution in the form

$$\phi(r,\theta) = R(r)P(\theta) \tag{18.51}$$

and substituting it into equation (18.50), we get

$$\frac{P}{r}\frac{d}{dr}\left(r\frac{dR}{dr}\right) + \frac{R}{r^2}\frac{d^2P}{d\theta^2} = 0$$

Multiplying by $\frac{r^2}{PR}$, we get

$$\frac{r}{R}\frac{d}{dr}\left(r\frac{dR}{dr}\right) = -\frac{1}{P}\frac{d^2P}{d\theta^2} = k^2 \text{ (say)} \tag{18.52}$$

Thus, we get the following two differential equations

$$r^2\frac{d^2R}{dr^2} + r\frac{dR}{dr} - k^2R = 0 \tag{18.53}$$

$$\text{and} \quad \frac{d^2P}{d\theta^2} + k^2P = 0 \tag{18.54}$$

Solution of equation (18.53) is given by

$$R(r) = Ar^k + Br^{-k}, \quad k \neq 0 \tag{18.55}$$

For $k = 0$, equation (18.53) becomes

$$r^2\frac{d^2R}{dr^2} + r\frac{dR}{dr} = 0, \quad \text{i.e.,} \quad \frac{d}{dr}\left(r\frac{dR}{dr}\right) = 0$$

Integrating

$$r\frac{dR}{dr} = A_0$$

Hence

$$R(r) = A_0 \log r + B_0 \tag{18.56}$$

Again, the solution of equation (18.54) is given by

$$P(\theta) = C\cos k\theta + D\sin k\theta \tag{18.57}$$

For $k = 0$, equation (18.54) reduces to

$$\frac{d^2P}{d\theta^2} = 0$$

which gives

$$P(\theta) = C_0\theta + D_0 \tag{18.58}$$

Thus, we get the solution

$$\phi_0(r,\theta) = (A_0\log r + B_0)(C_0\theta + D_0), \quad \text{for } k = 0 \tag{18.59}$$

and

$$\phi_k(r,\theta) = (A_k r^k + B_k r^{-k})(C_k\cos k\theta + D_k\sin k\theta), \quad \text{for } k \neq 0. \tag{18.60}$$

The solutions of Laplace's equation in cylindrical coordinates when the function ϕ does not depend on z are called *circular harmonics.*

The general solution of Laplace's equation in cylindrical coordinates is obtained from equation (18.59) and equation (18.60) by summing up solutions for all possible values of k in the following form

$$\phi(r,\theta) = a_0\log r + \sum_{k=1}^{\infty} r^k(a_k\cos k\theta + b_k\sin k\theta) + \sum_{k=1}^{\infty} r^{-k}(c_k\cos k\theta + d_k\sin k\theta) + e_0 \tag{18.61}$$

where e_0 is constant.

EXAMPLE 18.6. *The potential on the side and bottom surface given by $z = 0$ of a circular cylinder of radius a and height h is zero. The potential on the surface $z = H$ is given to be $f(r,\theta)$. Determine the potential at any point inside the cylinder.*

Solution: Since the potential ϕ is zero for $z = 0$, we get, from equation (18.40)

$$Z(z) = \sinh\lambda z \tag{1}$$

Again, the potential function ϕ should be single-valued with respect to θ. Hence

$$P(\theta) = C\cos m\theta + D\sin m\theta \tag{2}$$

where m is an integer.

The function $R(\lambda r)$ in equation (18.47) should be finite at $r = 0$. Hence $A_2 = 0$.

Thus the solution is given by

$$\phi(r,\theta,z) = \sum_{m=0}^{\infty} \sinh\lambda z(A_m\cos m\theta + B_m\sin m\theta)J_m(\lambda r) \tag{3}$$

Since potential vanishes at $r = a$, we get

$$J_m(\lambda_{mn}a) = 0 \tag{4}$$

$$\text{i.e.,} \quad \lambda_{mn} = \frac{1}{a}\mu_{mn}, \quad n = 1, 2, 3, \cdots \tag{5}$$

where μ_{mn} are the roots of

$$J_m(\mu_{mn}) = 0. \tag{6}$$

Thus, the solution (3) takes the form

$$\phi(r, \theta, z) = \sum_{m=0}^{\infty}\sum_{n=1}^{\infty} \sinh(\lambda_{mn}z)(A_{mn}\cos m\theta + B_{mn}\sin m\theta)J_m(\lambda_{mn}r). \tag{7}$$

The potential at $z = H$ is given as $f(r, \theta)$. Hence

$$f(r, \theta) = \sum_{m=0}^{\infty}\sum_{n=1}^{\infty} \sinh(\lambda_{mn}H)(A_{mn}\cos m\theta + B_{mn}\sin m\theta)J_m(\lambda_{mn}r) \tag{8}$$

This is a Fourier series in θ and Fourier-Bessel series in r. Hence the coefficients are given by

$$A_{mn} = \xi\int_0^{2\pi}\cos m\theta d\theta\int_0^a f(r, \theta)rJ_m(\lambda_{mn}r)dr \tag{9}$$

$$B_{mn} = \xi\int_0^{2\pi}\sin m\theta d\theta\int_0^a f(r, \theta)rJ_m(\lambda_{mn}r)dr \tag{10}$$

where

$$\xi = \frac{2\text{cosech}\,(\lambda_{mn}H)}{\pi a^2 J_{m+1}^2(\lambda_{mn}a)} \tag{11}$$

EXAMPLE 18.7. *The upper half of a long cylinder of radius R is at potential V_1 and the lower half is at potential V_2. Find potential at any point inside the cylinder.*

Solution: This is a symmetric problem, and hence the solution given in equation (18.61) is valid. Since the potential is finite at $r = 0$, we get

$$a_0 = c_k = d_k = 0 \tag{1}$$

Hence

$$\phi(r, \theta) = e_0 + \sum_{k=1}^{\infty} r^k(a_k\cos k\theta + b_k\sin k\theta) \tag{2}$$

Using condition at lateral surface, we have

$$\text{for } r = R, \phi(r, \theta) = f(\theta) = \begin{cases} V_1, & \text{for } \pi > \theta > 0 \\ V_2, & \text{for } 2\pi > \theta > \pi \end{cases} \tag{3}$$

Thus, from (2), we have

$$f(\theta) = e_0 + \sum_{k=1}^{\infty} R^k(a_k\cos k\theta + b_k\sin k\theta) \tag{4}$$

This is just Fourier expansion for $f(\theta)$. Hence the coefficients are given by

$$e_0 = \frac{1}{2\pi}\int_0^{2\pi} f(\theta)d\theta = \frac{1}{2\pi}\left[\int_0^{\pi} V_1 d\theta + \int_{\pi}^{2\pi} V_2 d\theta\right] \quad \text{[using equation (3)]}$$

$$= \frac{1}{2}(V_1 + V_2). \tag{5}$$

$$a_k = \frac{1}{\pi R^k}\int_0^{2\pi} f(\theta)\cos k\theta d\theta = \frac{1}{\pi R^k}\left[\int_0^{\pi} V_1 \cos k\theta d\theta + \int_{\pi}^{2\pi} V_2 \cos k\theta d\theta\right] = 0. \tag{6}$$

$$b_k = \frac{1}{\pi R^k}\left[\int_0^{\pi} V_1 \sin k\theta d\theta + \int_{\pi}^{2\pi} V_2 \sin k\theta d\theta\right] = \frac{1}{\pi R^k}\left[(V_1 - V_2)\frac{1-\cos k\pi}{k}\right]$$

$$= \frac{2(V_1 - V_2)}{\pi k R^k}, \quad \text{for } k = \text{odd} \tag{7}$$

$$= 0, \quad \text{for } k = \text{even}. \tag{8}$$

Hence, the potential at any point is

$$\phi(r,\theta) = \frac{1}{2}(V_1 + V_2) + \frac{2}{\pi}(V_1 - V_2)\sum_{k=1}^{\infty}\left(\frac{r}{R}\right)^k \frac{\sin k\theta}{k}$$

where $k = 1, 3, 5, \ldots$

EXAMPLE 18.8. *An infinitely long rectangular plate of width π has its long parallel edges at temperature $0°$ and the base at ϕ_0. Show that the steady state temperature is given by*

$$\phi = \frac{4\phi_0}{\pi}\left\{e^{-y}\sin x + \frac{1}{3}e^{-3y}\sin 3x + \cdots\right\}$$

Solution: The steady-state temperature in two-dimensional Cartesian coordinates is given by equation (18.12), so that we write

$$\phi(x,y) = \sum (A_1 \cos \alpha x + A_2 \sin \alpha x)(A_3 e^{-\alpha y} + A_4 e^{-\alpha y}) \tag{1}$$

The plate being infinite in y-direction

$$\phi(x,y) \longrightarrow 0, \quad \text{as } y \longrightarrow \infty; \text{ hence } A_3 = 0 \tag{2}$$

Again, $\phi(x,y) = 0$, at $x = 0$

Hence

$$A_1 = 0. \tag{3}$$

Thus, equation (1) becomes

$$\phi(x,y) = Ae^{-\alpha y}\sin \alpha x \tag{4}$$

Now, $\phi(x,y) = 0$, for $x = \pi$; hence

$$\sin \alpha\pi = 0, \quad \text{i.e.,} \quad \alpha = 1, 2, 3, \ldots \tag{5}$$

Thus, the general solution (1) assumes the form

$$\phi(x,y) = \sum_{n=1}^{\infty} Ae^{-ny}\sin(nx) \tag{6}$$

Lastly, we have, from the condition at base

$$\phi(x,y) = \phi_0, \quad \text{for } y = 0.$$

Hence, from (6), we have

$$\phi_0 = \sum_{n=1}^{\infty} A_n \sin(nx)$$

which is just a Fourier sine series. Hence

$$\begin{aligned} A_n &= \frac{2}{\pi}\int_0^{\pi} \phi_0 \sin(nx)dx = -\frac{2\phi_0}{\pi}\cdot\frac{1}{n}\left[(-1)^n - 1\right] \\ &= \frac{4\phi_0}{n\pi}, \quad \text{for } n = \text{odd}. \end{aligned} \tag{7}$$

Thus, from equation (6), we get by using (7)

$$\phi(x,y) = \frac{4\phi_0}{\pi}\sum_{n=1,3,\ldots}^{\infty} e^{-ny}\frac{\sin(nx)}{n}.$$

which is the required result.

WAVE EQUATION

18.5 Vibration of Strings

Transverse free vibration of strings stretched horizontally between two fixed points is described by the equation

$$\frac{\partial^2 u}{\partial t^2} = c^2\frac{\partial^2 u}{\partial x^2} \tag{18.62}$$

We assume the solution in the form

$$u(x,t) = X(x)T(t) \tag{18.63}$$

and substitute it into equation (18.62). Then we get

$$XT'' = c^2 X'' T, \quad \text{or,} \quad \frac{X''}{X} = \frac{T''}{c^2 T} = \text{constant} = -\lambda^2 \text{ (say)} \tag{18.64}$$

Hence

$$X'' = -\lambda^2 X \tag{18.65}$$

$$\text{and} \quad T'' = -c^2\lambda^2 T. \tag{18.66}$$

Let the boundary conditions be given by

$$u(0,t) = u(l,t) = 0 \tag{18.67}$$

where l = length of the string. Then, for the function $X(x)$, the boundary conditions are

$$X(0) = X(l) = 0 \tag{18.68}$$

Now, from equation (18.65), we have

$$X'' + \lambda^2 X = 0 \tag{18.69}$$

which gives

$$X(x) = A\cos\lambda x + B\sin\lambda x \tag{18.70}$$

Using the boundary conditions in equation (18.68), we find

$$A = 0, \quad \text{and} \quad \sin\lambda l = 0, \quad \text{i.e.,} \quad \lambda_n = \frac{n\pi}{l}$$

so that general solution of equation (18.69) is

$$X_n(x) = B_n \sin\frac{n\pi x}{l}, \quad n = 1, 2, 3, \ldots \tag{18.71}$$

For $\lambda = \lambda_n$, the equation (18.66) becomes

$$T'' + c^2\lambda_n^2 T = 0$$

which gives

$$T_n(t) = C_n\cos(c\lambda_n t) + D_n\sin(c\lambda_n t)$$

$$\text{or,} \quad T_n(t) = C_n\cos\frac{cn\pi t}{l} + D_n\sin\frac{cn\pi t}{l} \tag{18.72}$$

Hence, the complete solution of equation (18.62) is

$$u(x,t) = \sum_{n=1}^{\infty}\left(A'_n\cos\frac{cn\pi t}{l} + B'_n\sin\frac{cn\pi t}{l}\right)\sin\frac{n\pi x}{l} \tag{18.73}$$

The unknown coefficients A'_n, B'_n are to be determined from initial conditions. Let

$$\left.\begin{aligned} u(x,0) &= f(x) \\ \text{and} \quad \left.\frac{\partial u}{\partial t}\right|_{t=0} &= g(x) \end{aligned}\right\} \tag{18.74}$$

be the initial conditions imposed. Then, from equation (18.73), using these conditions, we get

$$u(x,0) = \sum_{n=1}^{\infty} A'_n \sin\frac{n\pi x}{l} = f(x) \tag{18.75}$$

and

$$\left.\frac{\partial u}{\partial t}\right|_{t=0} = \sum_{n=1}^{\infty} B'_n\frac{cn\pi}{l}\sin\frac{n\pi x}{l} = g(x) \tag{18.76}$$

which are just the Fourier sine series. Hence, the coefficients are given by

$$A'_n = \frac{2}{l}\int_0^l f(x)\sin\frac{n\pi x}{l}dx \tag{18.77}$$

and

$$B'_n\frac{cn\pi}{l} = \frac{2}{l}\int_0^l g(x)\sin\frac{n\pi x}{l}dx$$

i.e.,

$$B'_n = \frac{2}{cn\pi}\int_0^l g(x)\sin\frac{n\pi x}{l}dx \tag{18.78}$$

Thus, we have obtained the complete solution of the problem.

Particular Cases

(1) Suppose that the functions $f(x)$ and $g(x)$ are given as

$$f(x) = a\sin\frac{\pi x}{l}; \quad g(x) = 0 \tag{18.79}$$

Hence, from equation (18.77) and equation (18.78), we get

$$\begin{aligned} A'_n &= \frac{2}{l}\int_0^l a\sin\frac{\pi x}{l}\cdot\sin\frac{n\pi x}{l}dx \\ &= a, \quad \text{for } n = 1 \\ &= 0, \quad \text{for } n > 1 \\ \text{and} \quad B'_n &= 0. \end{aligned}$$

Hence,

$$u(x,t) = a\sin\frac{\pi x}{l}\cdot\cos\frac{c\pi t}{l} \tag{18.80}$$

(2) If position of the string is initially given by

$$f(x) = a\sin^3\frac{\pi x}{l}$$

and is released from rest, then $g(x) = 0$.
Hence, $B'_n = 0$.
From equation (18.77), we get

$$A'_n = \frac{2}{l}\int_0^l a\left(\sin\frac{\pi x}{l}\right)^3 \sin\frac{n\pi x}{l}dx = \frac{2a}{4l}\int_0^l \left[3\sin\frac{\pi x}{l} - \sin\frac{3\pi x}{l}\right]\sin\frac{n\pi x}{l}dx$$

$$\left.\begin{aligned} \text{or,} \quad A'_n &= \frac{3a}{4}, \quad \text{for } n = 1 \\ &= -\frac{a}{4}, \quad \text{for } n = 3 \\ &= 0, \quad \text{for } n = 2 \text{ and } n > 3 \end{aligned}\right\} \tag{18.81}$$

Hence, the complete solution is

$$u(x,t) = \frac{a}{4}\left[3\cos\frac{c\pi t}{l}\sin\frac{\pi x}{l} - \cos\frac{3c\pi t}{l}\sin\frac{3\pi x}{l}\right] \tag{18.82}$$

(3) The string is plucked at the point $x = \xi$ with amplitude h, and released. Then

$$\left.\begin{aligned} f(x) &= \frac{h}{\xi}x, \quad \text{for } 0 \le x \le \xi \\ &= \frac{h}{(l-\xi)}(l-x), \quad \text{for } \xi \le x \le l \end{aligned}\right\}. \tag{18.83}$$

and

$$g(x) = 0 \tag{18.84}$$

Hence

$$B'_n = 0 \tag{18.85}$$

and

$$\begin{aligned} A'_n &= \frac{2}{l}\left[\int_0^{\xi}\left(\frac{h}{\xi}x\right)\sin\frac{n\pi x}{l}dx + \int_{\xi}^{l}\frac{h(l-x)}{l-\xi}\sin\frac{n\pi x}{l}dx\right] \\ &= \frac{2h}{l\xi}\int_0^{\xi} x\sin\frac{n\pi x}{l}dx + \frac{2h}{l(l-\xi)}\int_{\xi}^{l}(l-x)\sin\frac{n\pi x}{l}dx \\ &= \frac{2hl^2}{n^2\pi^2\xi(l-\xi)}\cdot\sin\frac{n\pi\xi}{l}. \end{aligned} \tag{18.86}$$

Hence, the complete solution is

$$u(x,t) = \sum_{n=1}^{\infty}\frac{2hl^2}{n^2\pi^2\xi(l-\xi)}\sin\frac{n\pi\xi}{l}\cos\frac{cn\pi t}{l}\sin\frac{n\pi x}{l} \tag{18.87}$$

If, again, the string is plucked at mid-point, then

$$\xi = \frac{l}{2}$$

Hence

$$\sin\frac{n\pi\xi}{l} = \sin\frac{n\pi}{2} = \begin{cases} 0, & \text{for even } n \\ (-1)^{n-1}, & \text{for odd } n \end{cases}$$

and

$$\frac{2hl^2}{\xi(l-\xi)} = \frac{2hl^2}{(l/2)(l/2)} = 8h$$

Thus, the solution equation (18.87) becomes

$$u(x,t) = \frac{8h}{\pi^2}\sum_{n=1,3,\ldots}^{\infty}\frac{(-1)^{n-1}}{n^2}\cos\frac{cn\pi t}{l}\sin\frac{n\pi x}{l} \tag{18.88}$$

(4) Left end is fixed, but right end moves with displacement $a\sin\omega t$.
In this case, the general solution is

$$u(x,t) = [A\cos(c\lambda t) + B\sin(c\lambda t)]\,(C\sin\lambda x + D\cos\lambda x) \tag{18.89}$$

The condition

$$u(0,t) = 0 \text{ gives } D = 0$$

The condition for right end gives

$$a\sin\omega t = [A\cos(c\lambda t) + B\sin(c\lambda t)]C\sin\lambda l$$

This will be satisfied only if

$$A = 0, \quad \omega = c\lambda \quad \text{and} \quad a = BC\sin\lambda l \tag{18.90}$$

Hence, the resulting solution is

$$u(x,t) = \frac{a}{\sin\lambda l}\sin(\omega t)\sin\frac{\omega}{c}x = a\,\text{cosec}\left(\frac{\omega l}{c}\right)\sin\omega t\sin\frac{\omega}{c}x. \tag{18.91}$$

18.6 Forced Vibration of Strings

Let the string be acted upon by an external load/unit length given by

$$\frac{F(x,t)}{\rho} = A\sin\omega t \tag{18.92}$$

We expand the RHS function in equation (18.92) in Fourier series as

$$A\sin\omega t = \sum_{n=1}^{\infty} F_n(t)\sin\frac{n\pi x}{l} \tag{18.93}$$

Then

$$F_n(t) = \frac{2}{l}\int_0^l A\sin\omega t\sin\frac{n\pi x}{l}dx = \frac{2A}{n\pi}[1-(-1)^n]\sin\omega t, \quad n = 1, 2, \ldots \tag{18.94}$$

It may be noted that we have accepted boundary conditions in equation (18.67). Now, we assume that

$$u(x,t) = \sum_{n=1}^{\infty} T_n(t)\sin\frac{n\pi x}{l} \tag{18.95}$$

Substituting equations (18.94) and (18.95) into the equation of forced vibration given by

$$\frac{\partial^2 u}{\partial t^2} = c^2\frac{\partial^2 u}{\partial x^2} + \frac{F(x,t)}{\rho} \tag{18.96}$$

we get

$$\sum_{n=1}^{\infty}\left[T_n'' + \frac{c^2n^2\pi^2}{l^2}T_n - \frac{2A}{n\pi}\{1-(-1)^n\}\sin\omega t\right]\frac{\sin n\pi x}{l} = 0$$

This gives

$$T_n'' + \omega_n^2 T_n = \alpha_n\sin\omega t \tag{18.97}$$

where

$$\omega_n = \frac{cn\pi}{l}, \quad \alpha_n = \frac{2A}{n\pi}\{1-(-1)^n\} \quad (n = 1, 2, 3, \ldots) \tag{18.98}$$

Solving equation (18.97), we get

$$T_n(t) = A_n\cos\omega_n t + B_n\sin\omega_n t + \frac{\alpha_n}{\omega_n^2-\omega^2}\sin\omega t \tag{18.99}$$

Here, $\omega_n \neq \omega$. We now use conditions in equation (18.74). Then we have

$$u(x,0) = \sum_{n=1}^{\infty} T_n(0)\sin\frac{n\pi x}{l} = f(x) \tag{18.100}$$

and

$$\left.\frac{\partial u}{\partial t}\right|_{t=0} = \sum_{n=1}^{\infty} T_n'(0)\sin\frac{n\pi x}{l} = g(x) \tag{18.101}$$

Using equation (18.99), we get from equations (18.100) and (18.101)

$$f(x) = \sum_{n=1}^{\infty} A_n \sin \frac{n\pi x}{l} \tag{18.102}$$

and

$$g(x) = \sum_{n=1}^{\infty} \left[\omega_n B_n + \frac{\omega \alpha_n}{\omega_n^2 - \omega^2}\right] \sin \frac{n\pi x}{l} \tag{18.103}$$

The coefficients A_n, B_n are thus given by

$$A_n = \frac{2}{l} \int_0^l f(x) \sin \frac{n\pi x}{l} dx \tag{18.104}$$

$$B_n = \frac{2}{l\omega_n} \int_0^l g(x) \sin \frac{n\pi x}{l} dx - \frac{\omega \alpha_n}{\omega_n (\omega_n^2 - \omega^2)} \tag{18.105}$$

After finding A_n and B_n from equations (18.104) and (18.105), we get final expression for $T_n(t)$ in equation (18.99), hence the complete solution is easily obtained by using the equation (18.95).

18.7 Longitudinal Vibration of Rods

Free longitudinal vibrations of rods are described by the partial differential equation

$$\frac{\partial^2 u}{\partial t^2} = c^2 \frac{\partial^2 u}{\partial x^2} \tag{18.106}$$

where

$$c^2 = \frac{E}{\rho}, E = \text{Young's modulus.} \tag{18.107}$$

The boundary conditions may be:
(i) both ends of the rod are fixed, i.e.,

$$u(0, t) = u(l, t) = 0 \tag{18.108}$$

(ii) one end fixed, and the other is free; at free end there is no strain. Hence

$$u(0, t) = 0, \quad \left.\frac{\partial u}{\partial x}\right|_{x=l} = 0 \tag{18.109}$$

(iii) both ends are free, i.e.,

$$\frac{\partial u}{\partial x} = 0, \quad \text{for } x = 0 \text{ and } x = l \tag{18.110}$$

We seek the solution of equation (18.106), in the form

$$u(x, t) = X(x)T(t) \tag{18.111}$$

On substitution of equation (18.111) into equation (18.106), we get

$$X'' + \lambda^2 X = 0 \tag{18.112}$$

$$\text{and} \quad T'' + c^2\lambda^2 T = 0 \tag{18.113}$$

Solution of equation (18.112) is given by

$$X(x) = A\cos\lambda x + B\sin\lambda x \tag{18.114}$$

Considering the boundary conditions in equation (18.109), we have

$$X(0) = 0; \quad X'(l) = 0. \tag{18.115}$$

Using these boundary conditions, we get

$$A = 0, \quad \text{and} \quad B\lambda\cos\lambda l = 0 \tag{18.116}$$

which gives

$$\cos\lambda l = 0, \quad \text{i.e.,} \quad \lambda_n = \frac{(2n+1)\pi}{2l} \quad (n = \text{integer}). \tag{18.117}$$

Hence

$$X_n(x) = \sin\left[\frac{(2n+1)\pi x}{2l}\right] \tag{18.118}$$

Now, from equation (18.113), we get for $\lambda = \lambda_n$

$$T_n(t) = C_n\cos(c\lambda_n t) + D_n\sin(c\lambda_n t) \quad \text{for } n = 0, 1, 2, \ldots \tag{18.119}$$

Thus, the complete solution is written as

$$u(x,t) = \sum_{n=0}^{\infty}[C_n\cos(c\lambda_n t) + D_n\sin(c\lambda_n t)]\sin\lambda_n x \tag{18.120}$$

For the determination of the constants C_n, D_n, let us assume the following initial conditions

$$u(x,0) = f(x) \tag{18.121}$$

$$\text{and} \quad \left.\frac{\partial u}{\partial t}\right|_{t=0} = g(x) \tag{18.122}$$

Using these conditions, we get from equation (18.120)

$$f(x) = \sum_{n=0}^{\infty} C_n\sin\lambda_n x \tag{18.123}$$

$$\text{and} \quad g(x) = \sum_{n=0}^{\infty} D_n c\lambda_n\sin\lambda_n x \tag{18.124}$$

These are just Fourier series for functions $f(x)$ and $g(x)$. Hence these coefficients are

$$C_n = \frac{\int_0^l f(x)\sin\lambda_n x dx}{\int_0^l \sin^2\lambda_n x dx} \quad (n = 0, 1, 2, \ldots) \tag{18.125}$$

and

$$D_n = \frac{\int_0^l g(x) \sin \lambda_n x dx}{c\lambda_n \int_0^l \sin^2 \lambda_n x dx} \quad (n = 0, 1, 2, \ldots) \tag{18.126}$$

Using the value of the integral

$$\int_0^l \sin^2 \lambda_n x dx = \frac{1}{2} \int_0^l (1 - \cos 2\lambda_n x) dx = \frac{l}{2} \tag{18.127}$$

we get, from equations (18.125) and (18.126)

$$C_n = \frac{2}{l} \int_0^l f(x) \sin \lambda_n x dx \quad (n = 0, 1, 2, \ldots) \tag{18.128}$$

$$D_n = \frac{2}{c\lambda_n l} \int_0^l g(x) \sin \lambda_n x dx = \frac{4}{(2n+1)c\pi} \int_0^l g(x) \sin \lambda_n x dx \quad (n = 0, 1, 2, \ldots) \tag{18.129}$$

Thus, the complete solution is given by equation (18.120), where coefficients C_n, D_n are to be obtained by using equations (18.128) and (18.129). It is obvious that equation (18.120) represents harmonic vibration with frequency

$$\omega_n = c\lambda_n = \frac{(2n+1)c\pi}{2l} = \frac{(2n+1)\pi}{2l} \sqrt{\frac{E}{\rho}} \tag{18.130}$$

and corresponding period of vibration is

$$T = \frac{2\pi}{\omega_n} = \frac{4l}{(2n+1)} \sqrt{\frac{\rho}{E}} \tag{18.131}$$

The fundamental tone of vibration is given for $n = 0$, when the frequency and period of vibrations are

$$\omega_0 = \frac{\pi}{2l} \sqrt{\frac{E}{\rho}}; \quad T_0 = 4l \sqrt{\frac{\rho}{E}} \tag{18.132}$$

EXAMPLE 18.9. *A rod of cross-sectional area A is fixed at one end. The other end is pulled by a force P along its length and is suddenly released. Find its motion.*

Solution: The initial strain due to force P is

$$\epsilon = \frac{P}{AE} \tag{1}$$

Thus, the initial conditions in equations (18.121) and (18.122) take the forms

$$u(x, 0) = f(x) = \epsilon x, \quad \frac{\partial u}{\partial t} = g(x) = 0 \tag{2}$$

Hence from equation (18.129),

$$D_n = 0 \tag{3}$$

From equation (18.128)

$$C_n = \frac{2\epsilon}{l}\int_0^l x\sin\frac{(2n+1)\pi x}{2l}dx = \frac{8\epsilon l}{(2n+1)^2\pi^2}(-1)^n, \quad n = 0, 1, 2, \ldots \tag{4}$$

Hence, the expression for longitudinal displacement is

$$u(x,t) = \frac{8\epsilon l}{\pi^2}\sum_{n=0}^{\infty}\frac{(-1)^n}{(2n+1)^2}\cos\left[\frac{c(2n+1)\pi t}{2l}\right]\sin\left[\frac{(2n+1)\pi x}{2l}\right]. \tag{5}$$

Maximum amplitudes at the end $x = l$ are

$$\text{for } n = 0, \text{ amplitude} = \frac{8\epsilon l}{\pi^2},$$

$$\text{for } n = 1, \text{ amplitude} = \frac{8}{9}\frac{\epsilon l}{\pi^2},$$

which shows that the maximum amplitude decreases very fast with increase of n.

EXAMPLE 18.10. *The end ($x = 0$) of a beam of length l moves with a constant velocity v. Suddenly it is held at this end firmly. Discuss its motion.*

Solution: The initial conditions in equation (18.121) and (18.122) in this case are

$$u(x,0) = 0 \quad \text{and} \quad \left.\frac{\partial u}{\partial t}\right|_{t=0} = v \tag{1}$$

Putting $f(x) = 0$ and $g(x) = v$, we get from equations (18.128) and (18.129)

$$C_n = 0 \tag{2}$$

and

$$D_n = \frac{4v}{(2n+1)c\pi}\int_0^l \sin\frac{(2n+1)\pi x}{2l}dx = \frac{8vl}{c\pi^2(2n+1)^2}, \quad n = 0, 1, 2, \ldots \tag{3}$$

Hence the expression for longitudinal displacement of the beam is

$$u(x,t) = \frac{8vl}{c\pi^2}\sum_{n=0}^{\infty}\frac{1}{(2n+1)^2}\sin\left\{\frac{c(2n+1)\pi t}{2l}\right\}\sin\left\{\frac{(2n+1)\pi x}{2l}\right\} \tag{4}$$

The longitudinal displacement at $x = l$ and at $t = l/c$, is

$$u = \frac{8vl}{c\pi^2}\left(1 + \frac{1}{9} + \frac{1}{25} + \cdots\right) = \frac{8vl}{c\pi^2}\cdot\frac{\pi^2}{8} = \frac{vl}{c}.$$

EXAMPLE 18.11. *A rod of length l is fixed at the end $x = 0$ and free at the end $x = l$. It is subjected to a sinusoidal load $P = P_0\sin\omega t$ at its free end. Discuss its motion.*

Solution: In this case, we seek the solution of equation (18.106) in the form

$$u(x,t) = U(x)\sin\omega t \tag{1}$$

Then, for the function $U(x)$, we get the equation

$$\frac{d^2U}{dx^2} + \frac{\omega^2}{c^2}U = 0 \tag{2}$$

which gives

$$U(x) = A\sin\frac{\omega}{c}x + B\cos\frac{\omega}{c}x \tag{3}$$

Since, $u(x,t) = 0$, for $x = 0$, we get from (3), $B = 0$. Hence

$$U(x) = A\sin\frac{\omega}{c}x$$

Thus, solution (1) takes up the form

$$u(x,t) = A\sin\frac{\omega}{c}x\sin\omega t \tag{4}$$

At $x = l$, the strain due to longitudinal vibration is

$$\left.\frac{\partial u}{\partial x}\right|_{x=l} = A\frac{\omega}{c}\cos\left(\frac{\omega}{c}l\right)\sin\omega t \tag{5}$$

If cross-sectional area of rod $= F$, and Young's modulus $= E$, then

$$EF\cdot\left.\frac{\partial u}{\partial x}\right|_{x=l} = P_0\sin\omega t \tag{6}$$

From equations (5) and (6), we find the unknown constant A as

$$A = \frac{cP_0}{EF\omega}\sec\left(\frac{\omega l}{c}\right) \tag{7}$$

Thus, the displacement function is obtained from equations (4) and (7) as

$$u(x,t) = \frac{cP_0}{EF\omega}\sec\left(\frac{\omega l}{c}\right)\sin\frac{\omega x}{c}\sin\omega t \tag{8}$$

The amplitude of vibration at $x = l$, is

$$U(l) = \frac{cP_0}{EF\omega}\tan\frac{\omega l}{c} \tag{9}$$

For

$$\frac{\omega l}{c} = \frac{\pi}{2}, \quad \text{i.e.,} \quad \omega = \frac{\pi c}{2l} \tag{10}$$

the amplitude tends to ∞ when resonance occurs. This value of ω corresponds to minimum value of frequency for free vibration, which is [from equation (18.130)]

$$\omega = \frac{(2n+1)}{2l}c\pi$$

so that for $n = 0$, $\omega = \dfrac{\pi c}{2l}$.

From equation (9), we find that

$$\text{for } \frac{\omega l}{c} = n\pi, \quad U(l) = 0 \tag{11}$$

That is, the free end does not move.

EXAMPLE 18.12. *A rod of length l is fixed at the end $x = 0$ and free at the end $x = l$ where a body of mass m is attached. It is subjected to sinusoidol load $P = P_0 \sin \omega t$ at free end. Discuss the motion.*

Solution: Proceeding as before, we get the solution

$$u(x,t) = \left(A \sin \frac{\omega}{c} x + B \cos \frac{\omega}{c} x\right) \sin \omega t \tag{1}$$

Since at $x = 0, u(x,t) = 0$, we get

$$B = 0 \tag{2}$$

Hence

$$u(x,t) = A \sin \frac{\omega}{c} x \sin \omega t \tag{3}$$

At the free end $x = l$, normal force acting due to displacements caused by vibration is

$$EF \frac{\partial u}{\partial x}\bigg|_{x=l} = EFA \frac{\omega}{c} \cos\left(\frac{\omega l}{c}\right) \sin \omega t \tag{4}$$

At the free end $x = l$, inertial force due to motion of the mass m is

$$m \frac{\partial^2 u}{\partial t^2}\bigg|_{x=l} = -m\omega^2 A \sin\left(\frac{\omega l}{c}\right) \sin \omega t \tag{5}$$

At the free end the external applied load is $P_0 \sin \omega t$. Hence for equilibrium by D'Alembert's principle

$$EFA \frac{\omega}{c} \cos\left(\frac{\omega l}{c}\right) \sin \omega t = P_0 \sin \omega t + m\omega^2 A \sin\left(\frac{\omega l}{c}\right) \sin \omega t$$

$$\text{or,} \quad A\left[EF \frac{\omega}{c} \cos\left(\frac{\omega l}{c}\right) - m\omega^2 \sin\left(\frac{\omega l}{c}\right)\right] = P_0$$

Hence

$$A = \frac{P_0}{EF \frac{\omega}{c} \cos\left(\frac{\omega l}{c}\right) - m\omega^2 \sin\left(\frac{\omega l}{c}\right)} \tag{6}$$

From (3) and (6), the amplitude of vibration at free end $x = l$, is

$$U(l) = \frac{P_0 \sin\left(\frac{\omega l}{c}\right)}{EF \frac{\omega}{c} \cos\left(\frac{\omega l}{c}\right) - m\omega^2 \sin\left(\frac{\omega l}{c}\right)} = \frac{P_0}{EF \frac{\omega}{c} \cot\left(\frac{\omega l}{c}\right) - m\omega^2} \tag{7}$$

This amplitude tends to infinity when

$$m = \frac{EF}{\omega c} \cot\left(\frac{\omega l}{c}\right) \tag{8}$$

Equation (8) determines the critical value of the mass attached in terms of ω.

18.8 Forced Longitudinal Vibration of Rods

We consider a rod of length l hung from the fixed ceiling at $x = 0$ so that it remains under gravity. The external force/unit volume is

$$F(x,t) = \rho \cdot g \tag{18.133}$$

ρ being the density of the material of rod. The equation of vibration in this case is

$$\frac{\partial^2 u}{\partial t^2} = c^2 \frac{\partial^2 u}{\partial x^2} + g \tag{18.134}$$

We assume the following boundary and initial conditions

$$u(0,t) = 0, \quad \left.\frac{\partial u}{\partial x}\right|_{x=l} = 0 \tag{18.135}$$

and

$$u(x,0) = f(x), \quad \left.\frac{\partial u}{\partial x}\right|_{t=0} = \theta(x) \tag{18.136}$$

Let

$$g = \sum_{n=0}^{\infty} g_n \sin \lambda_n x \tag{18.137}$$

Then

$$g_n = \frac{2}{l}\int_0^l g \sin \lambda_n x \, dx = \frac{2g}{l\lambda_n} \tag{18.138}$$

where as in Section 18.7,

$$\lambda_n = \frac{(2n+1)\pi}{2l}, \quad n = 0, 1, 2, \ldots \tag{18.139}$$

We assume the solution of equation (18.134) in the form

$$u(x,t) = \sum_{n=0}^{\infty} T_n(t) \sin \lambda_n x \tag{18.140}$$

Substituting this into equation (18.134) and using equations (18.137) and (18.138), we get

$$\sum_{n=0}^{\infty} \left[T_n'' + c^2\lambda_n^2 T_n - \frac{2g}{l\lambda_n} \right] \sin \lambda_n x = 0$$

which gives

$$T_n'' + c^2\lambda_n^2 T_n - \frac{2g}{l\lambda_n} = 0, \quad (n = 0, 1, 2, \ldots) \tag{18.141}$$

Solution of this equation is given by

$$T_n = A_n \cos(c\lambda_n t) + B_n \sin(c\lambda_n t) + \frac{2g}{lc^2\lambda_n^3} \tag{18.142}$$

Using the boundary conditions in equation (18.136), we have

$$\left.\begin{aligned} u(x,0) &= \sum_{n=0}^{\infty} T_n(0)\sin\lambda_n x = f(x) \\ \text{and}\quad \left.\frac{\partial u}{\partial t}\right|_{t=0} &= \sum_{n=0}^{\infty} T_n'(0)\sin\lambda_n x = \theta(x) \end{aligned}\right\}. \tag{18.143}$$

Hence

$$T_n(0) = A_n + \frac{2g}{lc^2\lambda_n^3} = \frac{2}{l}\int_0^l f(x)\sin\lambda_n x\,dx \tag{18.144}$$

$$\text{and}\quad T_n'(0) = B_n c\lambda_n = \frac{2}{l}\int_0^l \theta(x)\sin\lambda_n x\,dx \tag{18.145}$$

Equations (18.144) and (18.145) provide the values for unknown coefficients A_n and B_n when the functions $f(x)$ and $\theta(x)$ are given.

If in particular the initial conditions are as considered in example 18.9, we get

$$A_n = \frac{(-1)^n 8\epsilon l}{(2n+1)^2\pi^2} - \frac{16gl^2}{c^2(2n+1)^3\pi^3}, \quad \text{for } n = 0,1,2,\ldots \tag{18.146}$$

and $B_n = 0$. Thus, the complete solution is obtained by using equations (18.140), (18.142) and (18.146).

18.9 Vibration of Rectangular Membranes

The vibration of membranes in rectangular Cartesian coordinates is governed by the wave equation

$$\frac{\partial^2 u}{\partial t^2} = c^2\left(\frac{\partial^2 u}{\partial x^2} + \frac{\partial^2 u}{\partial y^2}\right) \tag{18.147}$$

where $c^2 = T/\rho$.

T = tension in the membrane and ρ = surface density. Let the membrane be bounded by the region

$$0 \le x \le a, \quad 0 \le y \le b \tag{18.148}$$

The boundary condition is given by

$$u(x,y,t) = 0 \tag{18.149}$$

along the contour of the membrane.

We assume the solution of equation (18.147) in the form

$$u(x,y,t) = \phi(x,y)T(t) \tag{18.150}$$

Substituting this into equation (18.147) and separating the variables, we get

$$\phi T'' = c^2\left(\frac{\partial^2\phi}{\partial x^2} + \frac{\partial^2\phi}{\partial y^2}\right)\cdot T$$

which gives

$$\frac{1}{\phi}\left(\frac{\partial^2\phi}{\partial x^2}+\frac{\partial^2\phi}{\partial y^2}\right)=\frac{T''}{c^2T}=-\lambda^2 \quad \text{(say)} \tag{18.151}$$

Hence, we get the following two equations

$$\frac{\partial^2\phi}{\partial x^2}+\frac{\partial^2\phi}{\partial y^2}+\lambda^2\phi=0 \tag{18.152}$$

and

$$T''+c^2\lambda^2T=0 \tag{18.153}$$

If in equation (18.151), we accept $+\lambda^2$ in place of $-\lambda^2$, then the solution of equation (18.153) will not be periodic and so there will not be any vibration.

Now, we assume the solution of equation (18.152) in the form

$$\phi(x,y)=X(x)Y(y) \tag{18.154}$$

Substituting this into equation (18.152), we get

$$X''Y+XY''+\lambda^2XY=0$$

$$\text{i.e.,}\quad \frac{X''}{X}=-\frac{Y''+\lambda^2Y}{Y}=\text{constant}=-k^2 \quad \text{(say)} \tag{18.155}$$

which gives

$$X''+k^2X=0 \tag{18.156}$$

$$\text{and}\quad Y''+\xi^2Y=0 \tag{18.157}$$

where

$$\xi^2=\lambda^2-k^2 \tag{18.158}$$

If in equation (18.155), we take the constant equal to $+k^2$ instead of $-k^2$, the boundary conditions along $x=0$ and $x=a$ cannot be satisfied after solving the equation (18.156) with k^2 replaced by $-k^2$. Readers may verify this.

Now, solutions of equations (18.156) and (18.157) are given by

$$X(x)=A\cos kx+B\sin kx \tag{18.159}$$

$$Y(y)=C\cos\xi x+D\sin\xi x \tag{18.160}$$

where A, B, C, D are constants to be determined from the following boundary conditions

$$X(0)=0,\quad X(a)=0 \tag{18.161}$$

$$\text{and}\quad Y(0)=0,\quad Y(b)=0 \tag{18.162}$$

Using these conditions, we get

$$A=C=0 \tag{18.163}$$

and

$$\sin ka=0,\quad \sin\xi b=0 \tag{18.164}$$

which give

$$ka=m\pi \quad\text{and}\quad \xi b=n\pi \tag{18.165}$$

where m, n are integers.

Thus, we write, taking $B = D = 1$,

$$X_m(x) = \sin\frac{m\pi x}{a} \tag{18.166}$$

$$Y_n(y) = \sin\frac{n\pi y}{b} \tag{18.167}$$

where $m = 1, 2, 3, \ldots, n = 1, 2, 3, \ldots$

Hence, the solution of equation (18.152) is

$$\phi_{mn}(x, y) = \sin\frac{m\pi x}{a}\sin\frac{n\pi y}{b} \tag{18.168}$$

where, by virtue of equations (18.158) and (18.165)

$$\lambda_{mn}^2 = \xi^2 + k^2 = \pi^2\left(\frac{m^2}{a^2} + \frac{n^2}{b^2}\right) \tag{18.169}$$

For each value of λ_{mn} given by equation (18.169), we find the solution of equation (18.153) as

$$T_{mn}(t) = A_{mn}\cos(c\lambda_{mn}t) + B_{mn}\sin(c\lambda_{mn}t) \tag{18.170}$$

Hence, the complete solution of the wave equation (18.149) is written as

$$u(x, y, t) = \sum_{m=1}^{\infty}\sum_{n=1}^{\infty}\left[A_{mn}\cos(c\lambda_{mn}t) + B_{mn}\sin(c\lambda_{mn}t)\right]\sin\frac{m\pi x}{a}\cdot\sin\frac{n\pi y}{b} \tag{18.171}$$

The unknown constants A_{mn}, B_{mn} are to be determined from initial conditions. Let these initial conditions be given by

$$u(x, y, 0) = f(x, y) \tag{18.172}$$

$$\text{and}\quad \left.\frac{\partial u}{\partial t}\right|_{t=0} = g(x, y) \tag{18.173}$$

Using these conditions, we get from equation (18.171)

$$f(x, y) = u(x, y, 0) = \sum_{m=1}^{\infty}\sum_{n=1}^{\infty} A_{mn}\sin\frac{m\pi x}{a}\sin\frac{n\pi y}{b} \tag{18.174}$$

$$\text{and}\quad g(x, y) = \left.\frac{\partial u}{\partial t}\right|_{t=0} = \sum_{m=1}^{\infty}\sum_{n=1}^{\infty} c\lambda_{mn}B_{mn}\sin\frac{m\pi x}{a}\sin\frac{n\pi y}{b} \tag{18.175}$$

These are just Fourier's expansions in double series. Hence the coefficients are given by

$$A_{mn} = \frac{4}{ab}\int_0^a\int_0^b f(x, y)\sin\frac{m\pi x}{a}\sin\frac{n\pi y}{b}dydy \tag{18.176}$$

$$B_{mn} = \frac{4}{abc\lambda_{mn}}\int_0^a\int_0^b g(x, y)\sin\frac{m\pi x}{a}\sin\frac{n\pi y}{b}dxdy \tag{18.177}$$

Thus, after finding A_{mn}, B_{mn}, the solution given by equation (18.171) becomes complete.

Frequency of free vibration of the membrane is given by [using equation (18.169)]

$$\omega_{mn} = c\lambda_{mn} = \pi c\left(\frac{m^2}{a^2}+\frac{n^2}{b^2}\right)^{1/2} \tag{18.178}$$

In particular, if $a = b$,

$$\omega_{mn} = \frac{\pi c}{a}\sqrt{m^2+n^2} \tag{18.179}$$

For fundamental tone, $m = n = 1$. Hence

$$\omega_{11} = \frac{\pi c}{a}\sqrt{2} \tag{18.180}$$

The corresponding time period is

$$\frac{2\pi}{\omega_{11}} = \frac{2\pi a}{\pi c\sqrt{2}} = \sqrt{2}\frac{a}{c} \tag{18.181}$$

18.10 Free Vibrations of Circular Membranes

We consider a circular membrane of radius a. In polar coordinates

$$x = r\cos\theta, \quad y = r\sin\theta \tag{18.182}$$

so that the wave equation (18.147) assumes the form

$$\frac{\partial^2 u}{\partial t^2} = c^2\left(\frac{\partial^2 u}{\partial r^2}+\frac{1}{r}\frac{\partial u}{\partial r}+\frac{1}{r^2}\frac{\partial^2 u}{\partial \theta^2}\right) \tag{18.183}$$

Let us consider the particular case, when the displacements of the membrane are independent of angle θ. Then equation (18.183) takes the form

$$\frac{\partial^2 u}{\partial t^2} = c^2\left(\frac{\partial^2 u}{\partial r^2}+\frac{1}{r}\frac{\partial u}{\partial r}\right) \tag{18.184}$$

The vibration governed by this equation is called *radial vibration.* The boundary condition is

$$u(a,t) = 0 \tag{18.185}$$

and the initial conditions are

$$\left.\begin{aligned} u(r,0) &= f(r) \\ \text{and} \quad \left.\frac{\partial u}{\partial t}\right|_{t=0} &= g(r) \end{aligned}\right\} \tag{18.186}$$

We assume the solution of equation (18.184) in the form

$$u(r,t) = R(r)T(t) \tag{18.187}$$

On substitution this into equation (18.184), we get

$$RT'' = c^2\left(R''T+\frac{1}{r}R'T\right)$$

$$\text{or,}\quad \frac{1}{R}\left(R''+\frac{1}{r}R'\right)=\frac{T''}{c^2T}=\text{constant}=-\lambda^2\ (\text{say}) \tag{18.188}$$

Thus, we get the following two differential equations for determining $R(r)$ and $T(t)$:

$$R''+\frac{1}{r}R'+\lambda^2R=0 \tag{18.189}$$

$$\text{and}\quad T''+c^2\lambda^2T=0 \tag{18.190}$$

Equation (18.189) is the Bessel's equation whose solutions are given by Bessel's function of zero order. Thus,

$$R(r)=AJ_0(\lambda r)+BY_0(\lambda r) \tag{18.191}$$

where A, B are constants.

Now, the function $Y_0(\lambda r)$ becomes infinite at $r \longrightarrow 0$. Hence, we accept $B=0$, so that the solution equation (18.191) becomes

$$R(r)=AJ_0(\lambda r) \tag{18.192}$$

Using the boundary condition in equation (18.185), we get from equation (18.192)

$$J_0(\lambda a)=0 \tag{18.193}$$

which gives

$$\lambda_n=\frac{\mu_n}{a}\quad (n=1,2,\ldots) \tag{18.194}$$

where μ_n are the roots of $J_0(\mu)=0$.

Hence, particular solution equation (18.192) becomes (putting $A=1$)

$$R_n(r)=J_0\left(\frac{\mu_n r}{a}\right),\quad (n=1,2,\ldots) \tag{18.195}$$

Again, for $\lambda=\lambda_n$, the solution of equation (18.190) is given by

$$T_n(t)=C_n\cos(c\lambda_n t)+D_n\sin(c\lambda_n t) \tag{18.196}$$

Hence, the complete solution of equation (18.184) is given by

$$u(r,t)=\sum_{n=1}^{\infty}\left[C_n\cos(c\lambda_n t)+D_n\sin(c\lambda_n t)\right]J_0(\lambda_n r) \tag{18.197}$$

where λ_n is given by equation (18.194).

We now use the initial conditions in equation (18.186) in order to determine the constant C_n and D_n. Thus, we get

$$f(r)=u(r,0)=\sum_{n=1}^{\infty}C_nJ_0(\lambda_n r) \tag{18.198}$$

$$\text{and}\quad g(r)=\left.\frac{\partial u}{\partial t}\right|_{t=0}=\sum_{n=1}^{\infty}D_nc\lambda_nJ_0(\lambda_n r) \tag{18.199}$$

These are Bessel-Fourier series for the functions $f(r)$ and $g(r)$. Hence the coefficients are

$$C_n = \frac{2}{a^2 J_1^2(\mu_n)} \int_0^a r f(r) J_0(\lambda_n r) dr \tag{18.200}$$

$$D_n = \frac{2}{c\lambda_n a^2 J_1^2(\mu_n)} \int_0^a r g(r) J_0(\lambda_n r) dr \tag{18.201}$$

Thus, the solution of radial vibration problem of circular membranes is completely solved. The frequency of radial vibration is

$$\omega_n = c\lambda_n = \frac{c}{a}\mu_n \tag{18.202}$$

18.11 Vibration of Circular Membranes (General Case)

We now consider the general equation (18.183) for the free vibration of circular membranes. We assume the solution in the form

$$u(r, \theta, t) = \phi(r, \theta) T(t) \tag{18.203}$$

and substitute this into equation (18.183).

Then we have

$$\frac{1}{\phi}\left(\frac{\partial^2 \phi}{\partial r^2} + \frac{1}{r}\frac{\partial \phi}{\partial r} + \frac{1}{r^2}\frac{\partial^2 \phi}{\partial \theta^2}\right) = \frac{T''}{c^2 T} = \text{constant} = -\lambda^2$$

which gives the following two equations

$$T'' + c^2\lambda^2 T = 0 \tag{18.204}$$

and

$$\frac{\partial^2 \phi}{\partial r^2} + \frac{1}{r}\frac{\partial \phi}{\partial r} + \frac{1}{r^2}\frac{\partial^2 \phi}{\partial \theta^2} = -\lambda^2 \phi \tag{18.205}$$

For the solution of equation (18.205), we assume

$$\phi(r, \theta) = R(r) P(\theta) \tag{18.206}$$

On substitution of this into equation (18.205), we get

$$-\frac{r^2}{R}\left(R'' + \frac{1}{r}R' + \lambda^2 R\right) = \frac{P''}{P} = \text{constant} = -\xi^2 \tag{18.207}$$

The constant ξ^2 in equation (18.207) has been taken negative, because the function $P(\theta)$ should be periodic with period 2π. For this reason, ξ should be an integer. Hence replacing ξ by n, we get following two equations from equation (18.207)

$$P'' + n^2 P = 0 \tag{18.208}$$

$$r^2 R'' + rR' + \left(\lambda^2 r^2 - n^2\right) R = 0 \tag{18.209}$$

Solution of equation (18.208) is

$$P(\theta) = A\cos n\theta + B\sin n\theta \tag{18.210}$$

and of equation (18.209) is

$$R(r) = CJ_n(\lambda r) + DY_n(\lambda r) \tag{18.211}$$

Since, for $r \longrightarrow 0$, $Y_n(\lambda r) \longrightarrow \infty$, we reject this term and take $D = 0$. Along the boundary $r = a$, there is no displacement of the membrane. Hence,

$$R(a) = CJ_n(\lambda a) = 0 \tag{18.212}$$

Thus,

$$\lambda_{mn} = \frac{1}{a}\mu_{nm} \tag{18.213}$$

where μ_{nm} are the roots of $J_n(\mu) = 0$. Taking $C = 1$, we find from equation (18.211)

$$R_{nm}(r) = J_n(\lambda_{nm} r) = J_n\left(\mu_{nm}\frac{r}{a}\right) \tag{18.214}$$

where

$$n = 0, 1, 2, \ldots; \quad m = 1, 2, 3, \ldots \tag{18.215}$$

Again, for each value of $\lambda = \lambda_{nm}$ given by equation (18.213), we find the solution of equation (18.204) in the form

$$T_{nm}(t) = C_{nm}\cos(c\lambda_{nm}t) + D_{nm}\sin(c\lambda_{nm}t) \tag{18.216}$$

Thus, the final solution of the wave equation is

$$u(r, \theta, t) = \sum_{n=0}^{\infty}\sum_{m=1}^{\infty}\left[C_{nm}\cos(c\lambda_{nm}t) + D_{nm}\sin(c\lambda_{nm}t)\right](A_n\cos n\theta + B_n\sin n\theta)\, J_n\left(\mu_{nm}\frac{r}{a}\right) \tag{18.217}$$

The unknown constants are to be determined from initial conditions.

18.12 Three-Dimensional Wave Equations (Cartesian System)

We consider a wave in three-dimensional Cartesian coordinates which is governed by the differential equation

$$\frac{\partial^2 u}{\partial x^2} + \frac{\partial^2 u}{\partial y^2} + \frac{\partial^2 u}{\partial z^2} = \frac{1}{c^2}\frac{\partial^2 u}{\partial t^2} \tag{18.218}$$

Let the region in which the wave is considered be defined by the planes

$$\left.\begin{array}{ll} x = 0, & x = x_0 \\ y = 0, & y = y_0 \\ z = 0, & z = z_0 \end{array}\right\}. \tag{18.219}$$

We assume the solution of equation (18.218) in the form

$$u(x, y, z, t) = X(x)Y(y)Z(z)T(t) \tag{18.220}$$

Substituting this into equation (18.218) and dividing both sides by $XYZT$, we get

$$\frac{1}{X}\frac{d^2X}{dx^2} + \frac{1}{Y}\frac{d^2Y}{dy^2} + \frac{1}{Z}\frac{d^2Z}{dz^2} = \frac{1}{c^2T}\frac{d^2T}{dt^2} \tag{18.221}$$

All the variables are separated; hence each of the four terms in equation (18.221) should be constants.

Thus

$$\frac{1}{X}\frac{d^2X}{dx^2} = -k_1^2 \quad \text{(say)}, \quad \text{i.e.,} \quad \frac{d^2X}{dx^2} + k_1^2 X = 0 \tag{18.222}$$

which gives

$$X(x) = A_1 \cos k_1 x + A_2 \sin k_1 x \tag{18.223}$$

Similarly,

$$Y(y) = B_1 \cos k_2 x + B_2 \sin k_2 x \tag{18.224}$$

$$Z(z) = C_1 \cos k_3 x + C_2 \sin k_3 x \tag{18.225}$$

$$\text{and} \quad T(t) = D_1 \cos(kct) + D_2 \sin(kct) \tag{18.226}$$

where

$$k^2 = k_1^2 + k_2^2 + k_3^2 \tag{18.227}$$

We assume the boundary condition

$$u(x, y, z, t) = 0 \begin{cases} \text{for } x = 0, & \text{and } x = x_0 \\ \text{for } y = 0, & \text{and } y = y_0 \\ \text{for } z = 0, & \text{and } z = z_0 \end{cases} \tag{18.228}$$

Using these conditions in equations (18.223)–(18.225), we get

$$A_1 = B_1 = C_1 = 0 \tag{18.229}$$

and

$$k_1 = \frac{m\pi}{x_0}, \quad k_2 = \frac{n\pi}{y_0}, \quad k_3 = \frac{p\pi}{z_0} \tag{18.230}$$

where m, n and p are integers like $1, 2, 3, \ldots$

Hence the general solution can be written as

$$u(x, y, z, t) = \sum_{m=1}^{\infty}\sum_{n=1}^{\infty}\sum_{p=1}^{\infty}[A_{mnp}\cos(k_{mnp}ct) + B_{mnp}\sin(k_{mnp}ct)] \sin\frac{m\pi x}{x_0} \sin\frac{n\pi y}{y_0} \sin\frac{p\pi z}{z_0} \tag{18.231}$$

where

$$k_{mnp} = \pi\left(\frac{m^2}{x_0^2} + \frac{n^2}{y_0^2} + \frac{p^2}{z_0^2}\right)^{1/2} \tag{18.232}$$

Let the initial conditions be given as

$$\left.\begin{aligned} u(x, y, z, 0) &= f(x, y, z) \\ \text{and} \quad \left.\frac{\partial u}{\partial t}\right|_{t=0} &= g(x, y, z) \end{aligned}\right\}. \tag{18.233}$$

Then, from equation (18.231) and (18.233), we get

$$f(x, y, z) = \sum\sum\sum A_{mnp} \sin\frac{m\pi x}{x_0} \sin\frac{n\pi y}{y_0} \sin\frac{p\pi z}{z_0} \tag{18.234}$$

$$g(x, y, z) = \sum\sum\sum B_{mnp}(k_{mnp}c) \sin\frac{m\pi x}{x_0} \sin\frac{n\pi y}{y_0} \sin\frac{p\pi z}{z_0} \tag{18.235}$$

These are just Fourier expansions of functions $f(x,y,z)$ and $g(x,y,z)$ in triple series. Hence, the coefficients are given by

$$A_{mnp} = \frac{8}{x_0 y_0 z_0} \int_0^{x_0} \int_0^{y_0} \int_0^{z_0} f(x,y,z) \sin\frac{m\pi x}{x_0} \sin\frac{n\pi x}{y_0} \sin\frac{p\pi x}{z_0} dxdydz \tag{18.236}$$

$$B_{mnp} = \frac{8}{x_0 y_0 z_0 (k_{mnp}c)} \int_0^{x_0} \int_0^{y_0} \int_0^{z_0} g(x,y,z) \sin\frac{m\pi x}{x_0} \sin\frac{n\pi x}{y_0} \sin\frac{p\pi x}{z_0} dxdydz \tag{18.237}$$

Thus, the solution of three-dimensional wave equation in Cartesian coordinates is complete.

18.13 Three-Dimensional Waves in Spherical Coordinates

The wave equation in spherical polar coordinate is

$$\frac{\partial^2 u}{\partial r^2} + \frac{2}{r}\frac{\partial u}{\partial r} + \frac{1}{r^2}\frac{\partial^2 u}{\partial \theta^2} + \frac{\cot\theta}{r^2}\frac{\partial u}{\partial \theta} + \frac{1}{r^2 \sin^2\theta}\frac{\partial^2 u}{\partial \phi^2} = \frac{1}{c^2}\frac{\partial^2 u}{\partial t^2} \tag{18.238}$$

We assume the solution as

$$u(r,\theta,\phi,t) = R(r)F(\theta)G(\phi)T(t) \tag{18.239}$$

and substitute this into equation (18.238). Then we get

$$\frac{R''}{R} + \frac{2}{r}\frac{R'}{R} + \frac{1}{r^2}\frac{F''}{F} + \frac{\cot\theta}{r^2}\frac{F'}{F} + \frac{1}{r^2\sin^2\theta}\frac{G''}{G} = \frac{1}{c^2}\frac{T''}{T} \tag{18.240}$$

Hence we take

$$\frac{1}{c^2}\frac{T''}{T} = -\lambda^2 = \text{constant}, \quad \text{or,} \quad T'' + c^2\lambda^2 T = 0 \tag{18.241}$$

Its solution is

$$T(t) = A\cos(c\lambda t) + B\sin(c\lambda t) \tag{18.242}$$

Now, LHS of equation (18.240) being also equal to $-\lambda^2$, we have after multiplying by $r^2\sin^2\theta$

$$r^2\sin^2\theta\left(\frac{R''}{R} + \frac{2}{r}\frac{R'}{R}\right) + \sin^2\theta\left(\frac{F''}{F} + \cot\theta\frac{F'}{F} + \lambda^2 r^2\right) = -\frac{G''}{G} = k^2 \tag{18.243}$$

Thus, we get

$$\frac{G''}{G} = -k^2, \quad \text{i.e.,} \quad G'' + k^2 G = 0 \tag{18.244}$$

Its solution is

$$G(\phi) = C\cos k\phi + D\sin k\phi \tag{18.245}$$

Equating LHS of equation (18.243) to k^2, we get after dividing by $\sin^2\theta$

$$\left(\frac{r^2}{R}R'' + 2r\frac{R'}{R} + \lambda^2 r^2\right) = -\left(\frac{F''}{F} + \cot\theta\frac{F'}{F} - \frac{k^2}{\sin^2\theta}\right) = +n^2 \tag{18.246}$$

Thus, we get two equations:

$$\frac{F''}{F} + \cot\theta\frac{F'}{F} + \left(n^2 - \frac{k^2}{\sin^2\theta}\right) = 0 \tag{18.247}$$

$$\text{and} \quad r^2 R'' + 2rR' + \left(\lambda^2 r^2 - n^2\right)R = 0. \tag{18.248}$$

Putting

$$n^2 = p(p+1) \tag{18.249}$$

we write equation (18.247) as

$$\frac{1}{\sin\theta}\frac{d}{d\theta}\left(\sin\theta\frac{dF}{d\theta}\right) + \left[p(p+1) - \frac{k^2}{\sin^2\theta}\right]F = 0 \tag{18.250}$$

We substitute

$$\cos\theta = \mu \tag{18.251}$$

Then equation (18.250) is transformed into

$$\frac{d}{d\mu}\left[(1-\mu^2)\frac{dF}{d\mu}\right] + \left[p(p+1) - \frac{k^2}{1-\mu^2}\right]F = 0 \tag{18.252}$$

This is Associated Legendre's equation whose solution is

$$F(\theta) = a_p P_p^k(\mu) + b_p Q_p^k(\mu) = a_p P_p^k(\cos\theta) + b_p Q_p^k(\cos\theta) \tag{18.253}$$

For finite solution at $\cos\theta \longrightarrow 0$, the coefficients $b_p = 0$. Hence the solution for $F(\theta)$ can be written as

$$F(\theta) = a_p P_p^k(\cos\theta) \tag{18.254}$$

Lastly, the radial equation (18.248) can be written as

$$\frac{d^2R}{dr^2} + \frac{2}{r}\frac{dR}{dr} + \left\{\lambda^2 - \frac{p(p+1)}{r^2}\right\}R = 0. \tag{18.255}$$

We substitute

$$R(r) = \frac{S(r)}{\sqrt{r}} \tag{18.256}$$

Then the equation (18.255) is reduced to

$$\frac{d^2S}{dr^2} + \frac{1}{r}\frac{dS}{dr} + \left\{\lambda^2 - \frac{(p+\frac{1}{2})^2}{r^2}\right\}S = 0 \tag{18.257}$$

which is the Bessel's differential equation. Its solution is

$$S(r) = J_{p+1/2}(r)$$

Hence

$$R(r) = \frac{S(r)}{\sqrt{r}} = \frac{1}{\sqrt{r}}J_{p+1/2}(r) \tag{18.258}$$

Thus the four functions $R(r), F(\theta), G(\phi)$ and $T(t)$ are determined by equations (18.258), (18.254), (18.245) and (18.242) respectively. The solution of the wave equation will be obtained by multiplying these four functions.

Note. The solutions in equation (18.242) and (18.245) may also be written in the forms

$$\left.\begin{aligned} T(t) &= \exp(\pm ic\lambda t) \\ G(\phi) &= \exp(\pm ik\phi) \end{aligned}\right\}. \tag{18.259}$$

18.14 Three-Dimensional Waves in Cylindrical Coordinates

Three-dimensional wave equation in cylindrical coordinates is

$$\frac{\partial^2 u}{\partial r^2}+\frac{1}{r}\frac{\partial u}{\partial r}+\frac{1}{r^2}\frac{\partial^2 u}{\partial \theta^2}+\frac{\partial^2 u}{\partial z^2}=\frac{1}{c^2}\frac{\partial^2 u}{\partial t^2} \tag{18.260}$$

We assume the solution in the form

$$u(r,\theta,z,t)=R(r)F(\theta)Z(z)T(t) \tag{18.261}$$

Substituting this into equation (18.260), we get

$$\frac{1}{R}\left(R''+\frac{R'}{r}\right)+\frac{1}{r^2}\frac{F''}{F}+\frac{Z''}{Z}=\frac{1}{c^2}\frac{T''}{T}=-\lambda^2 \tag{18.262}$$

Thus, we can take

$$\frac{1}{c^2}\frac{T''}{T}=-\lambda^2, \quad \text{i.e.,} \quad T''+\lambda^2c^2T=0 \tag{18.263}$$

whose solution is

$$T(t)=A_1\cos(\lambda ct)+B_1\sin(\lambda ct) \tag{18.264}$$

We also can take

$$\frac{Z''}{Z}=-\alpha^2, \tag{18.265}$$

$$\text{i.e.,} \quad Z''+\alpha^2Z=0 \tag{18.266}$$

whose solution is

$$Z(z)=A_2\cos\alpha z+B_2\sin(\alpha z) \tag{18.267}$$

From equation (18.262) on using equation (18.265) we have

$$\frac{1}{R}\left(R''+\frac{R'}{r}\right)+\frac{1}{r^2}\frac{F''}{F}=-\lambda^2+\alpha^2, \quad \text{i.e.,} \quad r^2\frac{R''}{R}+r\frac{R'}{R}+(\lambda^2-\alpha^2)\,r^2=-\frac{F''}{F}=\beta^2 \tag{18.268}$$

Here again the variables are separated.
Hence we have

$$\frac{F''}{F}=-\beta^2, \quad \text{i.e.,} \quad F''+\beta^2F=0 \tag{18.269}$$

whose solution is

$$F(\theta)=A_3\cos\beta\theta+B_3\sin\beta\theta \tag{18.270}$$

Lastly, we find the radial equation as

$$r^2\frac{R''}{R}+r\frac{R'}{R}+(\lambda^2-\alpha^2)r^2=\beta^2, \quad \text{i.e.,} \quad \frac{d^2R}{dr^2}+\frac{1}{r}\frac{dR}{dr}+\left(\xi^2-\frac{\beta^2}{r^2}\right)R=0 \tag{18.271}$$

where

$$\xi^2=\lambda^2-\alpha^2 \tag{18.272}$$

Equation (18.271) is a Bessel's equation of order β with argument (ξr).

Hence, its solution

$$R = A_4 J_\beta(\xi r) + B_4 Y_\beta(\xi r) \tag{18.273}$$

Since $Y_\beta(\xi r) \longrightarrow \infty$ for $r \longrightarrow 0$, we accept $B_4 = 0$. Thus

$$R(r) = A_4 J_\beta(\xi r) \tag{18.274}$$

Thus the final solution of equation (18.260) is obtained by multiplying the expressions in equations (18.264), (18.267), (18.270) and (18.274).

In exponential form, the solution may be written as

$$u(r, \theta, z, t) = Ce^{\pm i\beta\theta} \cdot e^{\pm i\alpha z} J_\beta(\xi r) e^{\pm i\lambda ct} \tag{18.275}$$

18.15 Flexural Vibration of Beams

We recall the differential equation for bending of beams under a normal load intensity of $q(x, y)$ as

$$\frac{d^2}{dx^2}\left(EI\frac{d^2w}{dx^2}\right) = q(x) \tag{18.276}$$

where $w(x)$ is the vertical deflection function and EI is the flexural rigidity of the beam.

For free vibration of the beam, the load function $q(x, y)$ is to be replaced by inertia load. If m be the mass/unit length of the beam

$$q(x) = -m\frac{\partial^2 w}{\partial t^2} \tag{18.277}$$

so that equation (18.276) assumes the form

$$\frac{\partial^2}{\partial x^2}\left(EI\frac{\partial^2 w}{\partial x^2}\right) + m\frac{\partial^2 w}{\partial t^2} = 0 \tag{18.278}$$

If in particular, the cross-section of the beam remains constant, equation (18.278) can be written as

$$\frac{\partial^2 w}{\partial t^2} + \frac{EI}{m}\frac{\partial^4 w}{\partial x^4} = 0 \tag{18.279}$$

For the solution of equation (18.279), we assume

$$w(x, t) = X(x)T(t) \tag{18.280}$$

Substituting this into equation (18.279), we get

$$\frac{T''}{T} = -\frac{EI}{m}\frac{X^{IV}}{X} = -\xi^2 = \text{constant}$$

so that, we have two equations

$$T'' + \xi^2 T = 0 \tag{18.281}$$

$$\text{and} \quad X^{IV} - \lambda^4 X = 0 \tag{18.282}$$

where

$$\lambda^4 = \frac{m\xi^2}{EI} \tag{18.283}$$

Solution of equation (18.282) can be written as

$$X(x) = AR(x) + BS(x) + CU(x) + DV(x) \tag{18.284}$$

where A, B, C, D are constants and R, S, U, V are Krilov's functions defined by

$$\left.\begin{aligned} R(x) &= \tfrac{1}{2}(\cosh \lambda x + \cos \lambda x) \\ S(x) &= \tfrac{1}{2}(\sinh \lambda x + \sin \lambda x) \\ U(x) &= \tfrac{1}{2}(\cosh \lambda x - \cos \lambda x) \\ V(x) &= \tfrac{1}{2}(\sinh \lambda x - \sin \lambda x) \end{aligned}\right\}. \tag{18.285}$$

The functions are related in the following manner

$$\left.\begin{aligned} R(x) &= \frac{1}{\lambda}\frac{dS}{dx} \\ S(x) &= \frac{1}{\lambda}\frac{dU}{dx} \\ U(x) &= \frac{1}{\lambda}\frac{dV}{dx} \\ V(x) &= \frac{1}{\lambda}\frac{dR}{dx} \end{aligned}\right\}. \tag{18.286}$$

It is also to be noted that, at $x = 0$, $R = 1$, $S = U = V = 0$.

The derivatives of the function $X(x)$ in equation (18.284) can be written with the use of relations in equation (18.286) as

$$\left.\begin{aligned} X' &= \lambda[AV(x) + BR(x) + CS(x) + DU(x)] \\ X'' &= \lambda^2[AU(x) + BV(x) + CR(x) + DS(x)] \\ X''' &= \lambda^2[AS(x) + BU(x) + CV(x) + DR(x)] \end{aligned}\right\}. \tag{18.287}$$

Equation (18.281) has the solution

$$T(t) = G \cos \xi t + H \sin \xi t \tag{18.288}$$

which shows that ξ represents the frequency of free vibration.

Boundary Conditions

For a a simply supported end, $w = 0$ and $M =$ bending moment $= EI\dfrac{\partial^2 w}{\partial x^2} = 0$

That is

$$X = 0 \quad \text{and} \quad X'' = 0 \tag{18.289}$$

For a clamped end, $w = 0$ and rotation

$$\frac{\partial w}{\partial x} = 0, \quad \text{i.e.,} \quad X = 0 \text{ and } X' = 0 \tag{18.290}$$

For a free end, both bending moment and shear force Q are zero. Thus

$$EI\frac{\partial^2 w}{\partial x^2} = 0 \quad \text{and} \quad EI\frac{\partial^3 w}{\partial x^3} = 0, \quad \text{i.e.,} \quad X'' = 0 \text{ and } X''' = 0 \tag{18.291}$$

For a free end carrying a mass M_0, the bending moment is zero, and the inertial force due to mass M_0 is equal to the shear force. Thus

$$X'' = 0 \quad \text{and} \quad -M_0 \frac{\partial^2 w}{\partial t^2} = EI \frac{\partial^3 w}{\partial x^2}$$

The second condition gives

$$\begin{aligned} -M_0 X T'' &= EI X''' T \\ \text{or,} \quad M_0 \xi^2 X T &= EI X''' T \quad \text{[using equation (18.281)]} \\ \text{or,} \quad M_0 \xi^2 X &= EI X''' \end{aligned} \tag{18.292}$$

For an elastically supported end, the bending moment is zero, and the shear force is equal to the reaction at the support. If k be the coefficient of rigidity of the elastic support, then

$$Q = kw \quad \text{or,} \quad EIX''' = kX$$

Hence the boundary conditions are

$$X'' = 0 \quad \text{and} \quad EIX''' = kX \tag{18.293}$$

EXAMPLE 18.13. *A uniform beam of length L is clamped at one end and elastically supported at the other end. The stiffness coefficient of the elastic support is k. Find an expression for determining the frequency of natural vibration of the beam.*

Solution: Taking the origin of coordinates at the clamped end, the boundary conditions are written as

$$\text{For } x = 0, \quad X = 0 \text{ and } X' = 0 \tag{1}$$

$$\text{For } x = L, \quad X'' = 0 \text{ and } EIX''' = kX \tag{2}$$

Using conditions (1) in equations (18.284) and (18.287)

$$A = 0 \quad \text{and} \quad B = 0 \tag{3}$$

Using conditions (2), we find

$$\begin{aligned} C\,R(\lambda L) + D\,S(\lambda L) &= 0 \\ EI[C\,V(\lambda L) + D\,R(\lambda L)] &= k[C\,U(\lambda L) + D\,V(\lambda L)] \end{aligned}$$

Eliminating C and D from these equations we get

$$\begin{vmatrix} R(\lambda L) & S(\lambda L) \\ EI\,V(\lambda L) - k\,U(\lambda L) & EI\,R(\lambda L) - k\,V(\lambda L) \end{vmatrix} = 0$$

Using equations (18.285) and simplifying, we get

$$\sinh \lambda L \cos \lambda L - \sinh \lambda L \cosh \lambda L = \frac{EI}{k}$$

which is the required equation for determining the natural frequency of vibration.

18.16 Beams with Initial and Boundary Conditions

Let, apart from the two boundary conditions at each end of the beam, the following initial conditions be given:

$$\text{At } t = 0, \quad w = f(x) \quad \text{and} \quad \frac{\partial w}{\partial t} = g(x) \tag{18.294}$$

If, in particular, both ends are simply supported, then

$$\text{for } x = 0 \quad \text{and} \quad x = L, \quad X = X'' = 0$$

which yields $A = C = 0$ and

$$\left.\begin{aligned} BS(\lambda L) + DV(\lambda L) = 0 \\ BV(\lambda L) + DS(\lambda L) = 0 \end{aligned}\right\}. \tag{18.295}$$

Eliminating B and D, we get

$$[S(\lambda L) + V(\lambda L)]\,[S(\lambda L) - V(\lambda L)] = 0$$

which, on simplification, gives

$$\sinh \lambda L \sin \lambda L = 0 \tag{18.296}$$

Since, $\sinh \lambda L \neq 0$, we have

$$\sin \lambda L = 0, \quad \text{i.e.,} \quad \lambda L = n\pi \quad (n = 1, 2, \ldots) \tag{18.297}$$

Thus, using equation (18.283), the frequency of vibration is given by

$$\xi_n = \frac{n^2\pi^2}{L^2}\sqrt{\frac{EI}{m}} \tag{18.298}$$

The eigenfunctions are given by

$$X = X_n = B\left[S(\lambda_n x) - \frac{V(\lambda_n L)}{S(\lambda_n L)} V(\lambda_n x)\right] = B_n \sin \lambda_n x \tag{18.299}$$

Here we have used the condition equation (18.297).
Hence the deflection function of the beam is given by

$$w(x,t) = \sum_{n=1}^{\infty} (G_n \cos \xi_n t + H_n \sin \xi_n t) \sin \lambda_n x \tag{18.300}$$

Using the initial conditions in equation (18.294)

$$\left.\begin{aligned} f(x) &= \sum_{n=1}^{\infty} G_n \sin \frac{n\pi x}{L} \\ g(x) &= \frac{\pi^2}{L^2}\sqrt{\frac{EI}{m}} \sum_{n=1}^{\infty} n^2 H_n \sin \frac{n\pi x}{L} \end{aligned}\right\}. \tag{18.301}$$

From these equations, we can evaluate the Fourier coefficients G_n and H_n when the functions $f(x)$ and $g(x)$ are explicitly given.

18.17 Beams with Axial Load

When the beam is acted upon by a compressive axial load P which remains constant during the process of vibration, equation (18.279) assumes the form

$$\frac{\partial^2 w}{\partial t^2} + \frac{P}{m}\frac{\partial^2 w}{\partial x^2} + \frac{EI}{m}\frac{\partial^4 w}{\partial x^4} = 0 \tag{18.302}$$

Assuming the solution in the form of equation (18.280), we get

$$T'' = -\xi^2 T \tag{18.303}$$

and

$$\frac{P}{m}X'' + \frac{EI}{m}X^{IV} = \xi^2 X \tag{18.304}$$

The first equation has the solution given in equation (18.288). The second equation is written as

$$X^{IV} + \beta^2 X'' - \lambda^4 X = 0 \tag{18.305}$$

where

$$\beta^2 = \frac{P}{EI} \quad \text{and} \quad \lambda^4 = \frac{m\xi^2}{EI} \tag{18.306}$$

Substituting $X = e^{\alpha x}$, we find the characteristic equation as

$$\alpha^4 + \alpha^2\beta^2 - \lambda^4 = 0$$

so that

$$\left(\alpha^2 + \beta^2/2\right)^2 = \lambda^4 + \frac{\beta^4}{4}$$

Taking

$$\left.\begin{aligned} \alpha_1 &= \sqrt{-\beta^2/2 + \sqrt{\lambda^4 + \beta^4/4}} \\ \alpha_2 &= \sqrt{\beta^2/2 + \sqrt{\lambda^4 + \beta^4/4}} \end{aligned}\right\} \tag{18.307}$$

the solution of equation (18.305) is written as

$$X(x) = A\sinh\alpha_1 x + B\cosh\alpha_1 x + C\sin\alpha_2 x + D\cos\alpha_2 x \tag{18.308}$$

In the particular case when the ends of the beam are simply supported, we have

$$X = X'' = 0, \quad \text{for } x = 0 \text{ and } x = L$$

These conditions yield, $B = D = 0$, and

$$\left.\begin{aligned} A\sinh\alpha_1 L + C\sin\alpha_2 L &= 0 \\ A\alpha_1^2\sinh\alpha_1 L - C\alpha_2^2\sin\alpha_2 L &= 0 \end{aligned}\right\} \tag{18.309}$$

Eliminating A and C

$$(\alpha_1^2 + \alpha_2^2)\sinh\alpha_1 L\sin\alpha_2 L = 0$$

$$\sin\alpha_2 L = 0, \quad \text{i.e.,} \quad \alpha_2 L = n\pi \quad (n = 1, 2, 3, \ldots)$$

Substituting α_2 from equation (18.307)

$$L\sqrt{\beta^2/2+\sqrt{\lambda^4+\beta^4/4}}=n\pi$$

$$\text{i.e.,}\quad \lambda^4=\left(\frac{n^2\pi^2}{L^2}-\frac{\beta^2}{2}\right)^2-\frac{\beta^4}{4}=\left(\frac{n\pi}{L}\right)^4-\beta^2\left(\frac{n\pi}{L}\right)^2$$

Putting the values of λ^4 and β^2, we find

$$\xi_n=\frac{n^2\pi^2}{L^2}\left[\frac{EI}{m}\left(1-\frac{PL^2}{n^2\pi^2EI}\right)\right]^{1/2} \tag{18.310}$$

which determines the frequency of vibrations. For tensile load P is negative and so the value of ξ_n increases. The frequency of vibration becomes zero when

$$P=\frac{n^2\pi^2}{L^2}EI \tag{18.311}$$

which is the critical value of the axial load.

Note. When the ends of the beam are simply supported, we can assume the solution of equation (18.282) and (18.305) directly as

$$X(x)=\sum_{n=1}^{\infty}A_n\sin\frac{n\pi x}{L} \tag{18.312}$$

since this sequence of functions always satisfies the boundary conditions

$$X=X''=0 \text{ for } x=0 \text{ and } x=L$$

Substituting the solution function equation (18.312) into those equations, we can determine the frequency directly and easily. But for other types of boundary conditions, the method described is to be followed.

DIFFUSION EQUATION

The diffusion or heat flow equation through solids is given as (ref. Chapter 1 for its derivation)

$$\nabla^2\phi=\frac{1}{h^2}\frac{\partial\phi}{\partial t} \tag{18.313}$$

where ϕ is non-steady state temperature and h^2 is a constant called *Diffusivity*.

In a steady state temperature distribution field

$$\frac{\partial\phi}{\partial t}=0 \tag{18.314}$$

and the problem reduces to solving Laplace's equation which we have already discussed. Here we intend to discuss about the solution of equation (18.313) under various situations.

18.18 Heat Flow through Bars

We consider a bar of length l having uniform cross-sectional area. We assume origin of coordinates at left end of the bar with x-axis taken along its length.

Equation (18.313) in this one-dimensional problem reduces to

$$\frac{\partial \phi}{\partial t} = h^2 \frac{\partial^2 \phi}{\partial x^2} \tag{18.315}$$

Initial and boundary conditions considered are

$$\text{(i)} \quad \phi(0,t) = \phi(l,t) = 0 \tag{18.316}$$

i.e., ends of the bar are kept at zero temperature and

$$\text{(ii)} \quad \phi(x,0) = f(x) \tag{18.317}$$

i.e., initial temperature distribution function is $f(x)$ which is given. Moreover

$$\phi \neq \infty \text{ for } t \longrightarrow \infty \tag{18.318}$$

For the solution of equation (18.315), we assume

$$\phi(x,t) = X(x)T(t) \tag{18.319}$$

On substitution this into equation (18.315), we get

$$\frac{X''}{X} = \frac{T'}{h^2 T} = -\lambda^2 \text{ (constant)} \tag{18.320}$$

which give two differential equations

$$X'' + \lambda^2 X = 0 \tag{18.321}$$

$$\text{and} \quad T' + \lambda^2 h^2 T = 0 \tag{18.322}$$

The solutions of these equations are

$$\left.\begin{aligned} X(x) &= A\cos\lambda x + B\sin\lambda x \\ T(t) &= A' e^{-\lambda^2 h^2 t} \end{aligned}\right\} \tag{18.323}$$

Using the boundary conditions in equation (18.316), we have

$$X(0) = A = 0$$

$$\text{and} \quad X(l) = B\sin\lambda l = 0$$

which gives

$$\lambda_n = \frac{n\pi}{l} \quad (n = 1, 2, 3, \ldots) \tag{18.324}$$

Hence, from equation (18.323), we get (putting $B = 1$)

$$X_n(x) = \sin\frac{n\pi x}{l} \quad (n = 1, 2, \ldots) \tag{18.325}$$

For $\lambda = \lambda_n$, equation (18.323) gives

$$T_n(t) = A_n e^{-\lambda_n^2 h^2 t} \quad (n = 1, 2, \ldots) \tag{18.326}$$

Hence, the complete solution of equation (18.315) is

$$\phi(x,t) = \sum_{n=1}^{\infty} A_n \sin \frac{n\pi x}{l} \exp\left\{-\frac{n^2\pi^2h^2}{l^2}t\right\}. \tag{18.327}$$

From equation (18.326), it is clear that λ^2 in equation (18.320) was taken negative for satisfaction of condition equation (18.318).

Using the initial condition in equation (18.317), we have

$$f(x) = \sum_{n=1}^{\infty} A_n \sin \frac{n\pi x}{l} \tag{18.328}$$

which is just the Fourier expansion of function $f(x)$.

Hence the coefficients A_n are given by

$$A_n = \frac{2}{l}\int_0^l f(x) \sin \frac{n\pi x}{l} dx \quad (n = 1, 2, \cdots) \tag{18.329}$$

Thus the complete solution is obtained. We now consider other types of boundary and initial conditions.

Case I. *Ends of the Bar are insulated.*

In this case the boundary conditions are

$$\frac{\partial \phi}{\partial x} = 0, \quad \text{for } x = 0 \text{ and for } x = l \tag{18.330}$$

Let other conditions in equation (18.317) and (18.318) remain unchanged. For equation (18.323), we have

$$\frac{\partial}{\partial x}X = -A\lambda \sin \lambda x + B\lambda \cos \lambda x \tag{18.331}$$

Using the conditions in equation (18.330), we get

$$B = 0 \quad \text{and} \quad \sin \lambda l = 0$$

which give

$$\lambda = \frac{n\pi}{l} \quad (n = 0, 1, 2, \ldots) \tag{18.332}$$

Hence

$$X_n(x) = A_n \cos \frac{n\pi x}{l} \quad (n = 0, 1, 2, \ldots) \tag{18.333}$$

Hence, solution of equation (18.315) is

$$\phi(x,t) = A_0 + \sum_{n=1}^{\infty} A_n \cos \frac{n\pi x}{l} \exp\left\{-\frac{n^2\pi^2h^2}{l^2}t\right\} \tag{18.334}$$

The coefficients are given by

$$A_0 = \frac{1}{l}\int_0^l f(x)dx \tag{18.335}$$

$$\text{and} \quad A_n = \frac{2}{l}\int_0^l f(x)\cos\frac{n\pi x}{l}dx \tag{18.336}$$

Case II. *Ends of the Bar are at constant temperatures.*

In this case, the boundary conditions are

$$\left.\begin{aligned}\phi(0,t) &= \text{constant} = P\\ \phi(l,t) &= \text{constant} = Q\end{aligned}\right\}. \tag{18.337}$$

We assume the solution in the form

$$\phi(x,t) = \sum_{n=1}^{\infty} T_n(t)\sin\frac{n\pi x}{l} \tag{18.338}$$

where

$$T_n(t) = \frac{2}{l}\int_0^l \phi(x,t)\sin\frac{n\pi x}{l}dx \tag{18.339}$$

Integrating equation (18.339) by parts, we get

$$\frac{l}{2}T_n(t) = \left[-\phi(x,t)\frac{l}{n\pi}\cos\frac{n\pi x}{l}\right]_{x=0}^{x=l} + \left[\frac{l^2}{n^2\pi^2}\frac{\partial\phi}{\partial x}\sin\frac{n\pi x}{l}\right]_{x=0}^{x=l} - \frac{l^2}{n^2\pi^2}\int_0^l \frac{\partial^2\phi}{\partial x^2}\sin\frac{n\pi x}{l}dx$$

In the first term in the RHS, we use the conditions in equation (18.337), the second term vanishes, and in the third term we use from equation (18.315)

$$\frac{\partial^2\phi}{\partial x^2} = \frac{1}{h^2}\frac{\partial\phi}{\partial t}$$

and finally get

$$\frac{1}{2}T_n = \frac{1}{n\pi}\left[P-(-1)^n Q\right] - \frac{l}{n^2\pi^2h^2}\int_0^l \frac{\partial\phi}{\partial t}\sin\frac{n\pi x}{l}dx \tag{18.340}$$

Now, from equation (18.339), we get

$$\frac{d}{dt}(T_n) = \frac{2}{l}\int_0^l \frac{\partial\phi}{\partial t}\sin\frac{n\pi x}{l}dx \tag{18.341}$$

From equations (18.340) and (18.341), we get

$$\frac{1}{2}T_n = f(n) - \frac{l^2}{2h^2n^2\pi^2}T'_n \tag{18.342}$$

where

$$f(n) = \frac{1}{n\pi}\left[P-(-1)^n Q\right] \tag{18.343}$$

Equation (18.342) yields

$$T_n' + \xi_n^2 T_n = 2\xi_n^2 f(n) \tag{18.344}$$

where

$$\xi_n^2 = \left(\frac{n\pi h}{l}\right)^2 \tag{18.345}$$

Solution of equation (18.344) is given by

$$T_n(t) = A_n e^{-\xi_n^2 t} + 2f(n) \tag{18.346}$$

Let the initial condition be given by

$$\phi(x,0) = \psi(x) \tag{18.347}$$

Then, from equation (18.338) we have

$$\psi(x) = \sum_{n=1}^{\infty} T_n(0) \sin\frac{n\pi x}{l} \tag{18.348}$$

Hence

$$T_n(0) = \frac{2}{l}\int_0^l \psi(x)\sin\frac{n\pi x}{l}dx$$

$$\text{or,}\quad A_n + 2f(n) = \frac{2}{l}\int_0^l \psi(x)\sin\frac{n\pi x}{l}dx \quad \text{[using equation (18.346)]}$$

Hence

$$A_n = \frac{2}{l}\int_0^l \psi(x)\sin\frac{n\pi x}{l}dx - \frac{2}{n\pi}\left[P - (-1)^n Q\right] \tag{18.349}$$

Thus, coefficients A_n in equation (18.346) are determined. The complete solution is then given by the series in equation (18.338).

EXAMPLE 18.14. *A bar of length l is heated so that both its ends are at zero temperature; the initial temperature being given by*

$$\phi(x,0) = \alpha x(l-x)/l^2, \quad \alpha = \textit{constant}.$$

Find temperature at any point of the bar.

Solution: The boundary conditions being as given in equation (18.316), we will get the solution as in equation (18.327), where unknown constants A_n are given by

$$A_n = \frac{2}{l}\int_0^l f(x)\sin\frac{n\pi x}{l}dx \quad (n = 1, 2, \ldots)$$

where

$$f(x) = \frac{\alpha x(l-x)}{l^2}$$

Hence

$$A_n = \frac{2\alpha}{l^2}\int_0^l x\sin\frac{n\pi x}{l}dx - \int_0^l \frac{2\alpha}{l^3}x^2\sin\frac{n\pi x}{l}dx$$

i.e.,

$$A_n = \frac{4\alpha}{n^3\pi^3}[1-(-1)^n]$$

Substituting this in equation (18.327), we get

$$\phi(x,t) = \frac{4\alpha}{\pi^3}\sum_{n=1}^{\infty}\frac{1}{n^3}\{1-(-1)^n\}\sin\frac{n\pi x}{l}\exp\left\{-\frac{n^2\pi^2h^2}{l^2}t\right\}$$

The function $\phi(x,t)$ vanishes for $n = 2, 4, \ldots$; hence for odd values of $n = 1, 3, \ldots$, the solution is

$$\phi(x,t) = \frac{8\alpha}{\pi^3}\sum_{n=1,2,3,\ldots}^{\infty}\frac{1}{n^3}\sin\frac{n\pi x}{l}\exp\left\{-\frac{n^2\pi^2h^2}{l^2}t\right\}$$

18.19 Heat Flow in Two Dimensions (Cartesian System)

In two-dimensional Cartesian coordinates, the heat flow equation is

$$\frac{\partial^2\phi}{\partial x^2} + \frac{\partial^2\phi}{\partial y^2} = \frac{1}{h^2}\frac{\partial\phi}{\partial t} \tag{18.350}$$

We intend to study the heat flow of a rectangular plate bounded by the region $0 \le x \le a$ and $0 \le y \le b$.

Let the boundary conditions are

$$\left.\begin{aligned}\phi(x,y,t) &= 0, \text{for } x = 0, \text{ and } x = a\\ \phi(x,y,t) &= 0, \text{for } y = 0, \text{ and } y = b\end{aligned}\right\} \tag{18.351}$$

and the initial conditions are

$$\left.\begin{aligned}\phi(x,y,0) &= f(x,y)\\ \phi(x,y,t) &\longrightarrow 0 \text{ as } t \longrightarrow \infty\end{aligned}\right\} \tag{18.352}$$

We assume the solution in the form

$$\phi(x,y,t) = X(x)Y(y)T(t) \tag{18.353}$$

Substituting this into equation (18.350) and dividing both sides by XYT, we get

$$h^2\left[\frac{X''}{X} + \frac{Y''}{Y}\right] = \frac{T'}{T} = -k^2 \quad \text{(say)} \tag{18.354}$$

which gives the following two equations:

$$T' + k^2T = 0 \tag{18.355}$$

$$\text{and} \quad \frac{X''}{X} + \frac{Y''}{Y} = -\frac{k^2}{h^2} \tag{18.356}$$

The solution of equation (18.355) is

$$T(t) = Ae^{-k^2t} \tag{18.357}$$

Equation (18.356) may be written as

$$\frac{X''}{X}+\frac{k^2}{h^2}=-\frac{Y''}{Y}=\xi^2 \quad \text{(say)} \tag{18.358}$$

Then we get the equation for $Y(y)$ in the form

$$Y''+\xi^2 Y=0 \tag{18.359}$$

which has the solution

$$Y(y)=C\cos\xi y+D\sin\xi y \tag{18.360}$$

The other equation from equation (18.358) is

$$X''+\left(\frac{k^2}{h^2}-\xi^2\right)X=0 \tag{18.361}$$

whose solution is

$$X(x)=P\cos\eta x+Q\sin\eta x \tag{18.362}$$

where

$$\eta^2=\frac{k^2}{h^2}-\xi^2 \tag{18.363}$$

Using the boundary conditions in equation (18.351), we get from equations (18.360) and (18.362) $C=0$, $P=0$ and

$$\sin\xi b=0, \quad \sin\eta a=0$$

giving

$$\xi=\frac{m\pi}{b} \quad \text{and} \quad \eta=\frac{n\pi}{a} \tag{18.364}$$

where $m=1,2,\ldots;n=1,2,\ldots$

Thus, from equation (18.363) we get

$$k_{mn}^2=h^2\left(\xi^2+\eta^2\right)=h^2\pi^2\left(\frac{m^2}{b^2}+\frac{n^2}{a^2}\right) \tag{18.365}$$

Hence, the solution of equation (18.350) becomes

$$\phi(x,y,t)=\sum_{m=1}^{\infty}\sum_{n=1}^{\infty}A_{mn}\sin\frac{n\pi x}{a}\sin\frac{m\pi y}{b}\exp\left(-k_{mn}^2t\right) \tag{18.366}$$

Using the initial condition in equation (18.352), we have

$$f(x,y)=\sum_{m=1}^{\infty}\sum_{n=1}^{\infty}A_{mn}\sin\frac{n\pi x}{a}\sin\frac{m\pi y}{b} \tag{18.367}$$

which gives

$$A_{mn}=\frac{4}{ab}\int_0^a\int_0^b f(x,y)\sin\frac{n\pi x}{a}\sin\frac{m\pi y}{b}dxdy \tag{18.368}$$

Thus the heat flow through rectangular plates is completely determined.

18.20 Heat Flow in a Circular Plate

In polar coordinates, the heat flow equation (18.350) is transformed into

$$\frac{\partial \phi}{\partial t} = h^2 \left(\frac{\partial^2 \phi}{\partial r^2} + \frac{1}{r}\frac{\partial \phi}{\partial r} + \frac{1}{r^2}\frac{\partial^2 \phi}{\partial \theta^2} \right) \tag{18.369}$$

We assume that initial and boundary conditions are not dependent on θ. Then the solution of equation (18.369) will also be independent of θ, so that the equation of heat flow in this case becomes

$$\frac{\partial \phi}{\partial t} = h^2 \left(\frac{\partial^2 \phi}{\partial r^2} + \frac{1}{r}\frac{\partial \phi}{\partial r} \right) \tag{18.370}$$

We assume the solution

$$\phi(r,t) = R(r)T(t) \tag{18.371}$$

which on substitution into equation (18.370) gives

$$RT' = h^2 \left(R''T + \frac{1}{r}R'T \right)$$

$$\text{i.e.,} \quad \frac{1}{R}\left(R'' + \frac{R'}{r} \right) = \frac{T'}{h^2 T} = \text{constant} = -\lambda^2 \tag{18.372}$$

Hence

$$T' + h^2\lambda^2 T = 0 \tag{18.373}$$

and

$$R'' + \frac{1}{r}R' + \lambda^2 R = 0 \tag{18.374}$$

Equation (18.374) is Bessel's equation having solution

$$R(r) = AJ_0(\lambda r) + BY_0(\lambda r) \tag{18.375}$$

The function $Y_0(\lambda r) \longrightarrow \infty$ for $r \longrightarrow 0$; hence we accept $B = 0$. Thus, .

$$R(r) = AJ_0(\lambda r) \tag{18.376}$$

Let the boundary condition be given by

$$\left. \frac{\partial \phi}{\partial r} \right|_{r=a} = 0 \tag{18.377}$$

This means that the lateral surface is insulated. Using this condition, we get from equation (18.376)

$$J_0'(\lambda a) = 0 \tag{18.378}$$

Hence

$$\lambda_n = \frac{\mu_n}{a} \tag{18.379}$$

where μ_n are the roots of equation $J_0'(\mu) = 0$.

Thus,

$$R_n(r) = J_0(\lambda_n r) = J_0\left(\frac{\mu_n}{a} r \right) \tag{18.380}$$

where $n = 1, 2, 3, \ldots$

Now, for $\lambda = \lambda_n$, we get the solution of equation (18.373) as

$$T_n(t) = A_n \exp\left(-h^2\lambda_n^2 t\right) \quad (n = 1, 2, \ldots) \tag{18.381}$$

Thus, we get the solution of equation (18.370) as

$$\phi(r,t) = \sum_{n=1}^{\infty} A_n J_0(\lambda_n r) \exp\left\{-h^2\lambda_n^2 t\right\} \tag{18.382}$$

Let the initial condition be given as

$$\phi(r,0) = f(r) \tag{18.383}$$

Using this condition, we get from equation (18.382)

$$f(r) = \sum_{n=1}^{\infty} A_n J_0(\lambda_n r) \tag{18.384}$$

which gives

$$A_n = \frac{2}{a^2 J_1^2(\mu_n)} \int_0^a r f(r) J_0(\lambda_n r) dr \tag{18.385}$$

where $n = 1, 2, 3, \ldots$

Thus, the solution of heat flow problem in circular plates is complete.

Let us consider other types of boundary conditions.

Case I. *Let $\phi(r,t) = 0$, for $r = a$. Then, from equation* (18.376), *we have*

$$J_0(\lambda a) = 0 \tag{18.386}$$

so that

$$\lambda_n = \frac{\mu_n}{a} \tag{18.387}$$

where μ_n are the roots of

$$J_0(\mu) = 0 \tag{18.388}$$

Using the initial condition as before, we find the unknown constants A_n from equation (18.385).

Case II. *Let the boundary condition be given by*

$$\left.\frac{\partial\phi}{\partial r}\right|_{r=a} + \alpha\phi(a,t) = 0 \quad (\alpha = \textit{constant}). \tag{18.389}$$

Using this condition, we get from equation (18.376)

$$\lambda J_0'(\lambda a) + \alpha J_0(\lambda a) = 0$$

$$\text{i.e.,} \quad \lambda a J_0'(\lambda a) + \alpha a J_0(\lambda a) = 0 \tag{18.390}$$

Hence, $\mu = \lambda a$ is the root of equation

$$\mu J_0'(\mu) + \alpha a J_0(\mu) = 0 \tag{18.391}$$

Let

$$\lambda_n = \frac{\mu_n}{a} \tag{18.392}$$

Then

$$R_n(r) = J_0(\lambda_n r) = J_0\left(\frac{\mu_n}{a} r\right) \quad (n = 1, 2, \ldots) \tag{18.393}$$

For $\lambda = \lambda_n$, the solution for $T_n(t)$ is given by equation (18.381). Using the initial condition as before, the unknown constants A_n are obtained as

$$A_n = \frac{2}{a^2 \xi(\mu)} \int_0^a r f(r) J_0(\lambda_n r) dr \tag{18.394}$$

where

$$\xi(\mu) = J_0'^2(\mu_n) + J_0^2(\mu_n) \tag{18.395}$$

Case III. *Let the boundary condition be given by*

$$\phi(a, t) = \phi_0 \quad (constant) \tag{18.396}$$

In this case, we make a change of variables

$$\phi(r, t) = \psi(r, t) + \phi_0 \tag{18.397}$$

so that the heat flow equation (18.397) assumes the form

$$\frac{\partial \psi}{\partial t} = h^2 \left(\frac{\partial^2 \psi}{\partial r^2} + \frac{1}{r} \frac{\partial \psi}{\partial r} \right) \tag{18.398}$$

with the boundary condition

$$\psi(r, t) = 0, \quad \text{for } r = a \tag{18.399}$$

Let us assume that the initial condition in terms of $\phi(r, t)$ was given as

$$\phi(r, 0) = \phi_0 + f(r) \tag{18.400}$$

This condition in terms of the function $\psi(r, t)$ assumes the form

$$\psi(r, 0) = \phi(r, 0) - \phi_0 = f(r) \tag{18.401}$$

Thus, proceeding exactly as before, we can solve the problem.

Hence, if the boundary condition is non-homogeneous, we can, by a proper change of variables, reduce the problem to that having homogeneous boundary condition.

BIHARMONIC EQUATIONS

18.21 Deflections of Rectangular Plates

Consider a plate $a \times b$ simply supported at the edges under the action of an arbitrary load function $q(x, y)$. Then, we are to solve the equation

$$\nabla^2 \nabla^2 w = \frac{q(x, y)}{D} \tag{18.402}$$

under the boundary conditions

$$\left.\begin{aligned} w = 0, \ \frac{\partial^2 w}{\partial x^2} = 0, \text{for } x = 0 \text{ and } x = a \\ w = 0, \ \frac{\partial^2 w}{\partial y^2} = 0, \text{for } y = 0 \text{ and } x = b \end{aligned}\right\} \tag{18.403}$$

Consistent with these boundary conditions, we assume the solution of equation (18.402) as

$$w(x, y) = \sum_{m=1}^{\infty} \sum_{n=1}^{\infty} A_{mn} \sin \frac{m\pi x}{a} \sin \frac{n\pi y}{b} \tag{18.404}$$

We expand the load function $q(x, y)$ in double Fourier series as

$$q(x, y) = \sum_{m=1}^{\infty} \sum_{n=1}^{\infty} B_{mn} \sin \frac{m\pi x}{a} \sin \frac{n\pi y}{b} \tag{18.405}$$

where the coefficients B_{mn} are given by

$$B_{mn} = \frac{4}{ab} \int_0^a \int_0^b q(x, y) \sin \frac{m\pi x}{a} \sin \frac{n\pi y}{b} dxdy \tag{18.406}$$

Differentiating the expression equation (18.404) and substituting its derivatives in the left side and expression equation (18.405) in the right side of equation (18.402), we find

$$D\pi^4 A_{mn} \left(\frac{m^2}{a^2} + \frac{n^2}{b^2} \right)^2 = B_{mn} \tag{18.407}$$

Case I. *Let $q(x, y) = q =$ constant. Then, from equation* (18.406)

$$B_{mn} = \frac{4q}{ab} \int_0^a \sin \frac{m\pi x}{a} dx \int_0^b \sin \frac{n\pi y}{b} dy \quad or, \quad B_{mn} = \frac{16q}{\pi^2 mn} \tag{18.408}$$

where

$$\left.\begin{aligned} m = 1, 3, 5, \ldots, \infty \\ n = 1, 3, 5, \ldots, \infty \end{aligned}\right\} \tag{18.409}$$

From equations (18.407) and (18.408)

$$A_{mn} = \frac{16q}{D\pi^6 \left(m^2/a^2 + n^2/b^2\right)^2 mn} \tag{18.410}$$

Thus, the deflection function $w(x, y)$ given by the expression equation (18.404) is now defined. We write it in the form

$$w(x, y) = \frac{16q}{D\pi^6} \sum_{m=1}^{\infty} \sum_{n=1}^{\infty} \frac{\sin \frac{m\pi x}{a} \sin \frac{n\pi y}{b}}{mn \left(m^2/a^2 + n^2/b^2\right)^2} \tag{18.411}$$

The maximum deflection occurs at the centre of the plate where $x = a/2$, $y = b/2$. This yields

$$w_{\max} = \frac{16q}{D\pi^6} \sum_{m=1}^{\infty} \sum_{n=1}^{\infty} \frac{(-1)^{\frac{m+n}{2}-1}}{mn \left(m^2/a^2 + n^2/b^2\right)^2} \tag{18.412}$$

Denoting the sum of the series as β, we can write

$$w_{\max} = \frac{16q}{D\pi^6} \beta \tag{18.413}$$

Case II. *Let a concentrated point load P act on the plate at the point (ξ, η). Considering an infinitesimal area $dxdy$ around the point, the distributed load over this area will be*

$$q = \frac{P}{dxdy} \tag{18.414}$$

Substituting this into equation (18.406) and noting that q is zero everywhere except in an elemental area surrounding the point (ξ, η), we find

$$B_{mn} = \frac{4P}{ab} \sin \frac{m\pi\xi}{a} \sin \frac{n\pi\eta}{b} \tag{18.415}$$

The coefficients A_{mn} of the series in equation (18.404) is obtained from equations (18.407) and (18.415) in the form

$$A_{mn} = \frac{4P \sin \frac{m\pi\xi}{a} \sin \frac{n\pi\eta}{b}}{D\pi^4 ab \left(m^2/a^2 + n^2/b^2\right)^2} \tag{18.416}$$

Hence, the equation of the deflected middle surface of the plate is

$$w(x, y) = \frac{4P}{D\pi^4 ab} \sum_{m=1}^{\infty} \sum_{n=1}^{\infty} \frac{\sin \frac{m\pi\xi}{a} \sin \frac{n\pi\xi}{b}}{\left(m^2/a^2 + n^2/b^2\right)} \sin \frac{m\pi x}{a} \sin \frac{n\pi y}{b} \tag{18.417}$$

It is to be noted that moments and shearing forces acting in the plate can be determined once the deflection function is known.

18.22 Free Vibration of Rectangular Plates

Consider a rectangular plate simply supported along its edges. Free vibration of the plate is governed by the equation

$$\nabla^2 \nabla^2 w + \frac{\rho h}{D} \frac{\partial^2 w}{\partial t^2} = 0 \tag{18.418}$$

Assuming the solution in the form

$$w(x, y, t) = W(x, y)T(t) \tag{18.419}$$

and substituting the same into equation (18.418), we get

$$\left(\nabla^2\nabla^2 W\right) T + \frac{\rho h}{D} W T'' = 0 \tag{18.420}$$

This yields two equations like

$$T'' + \xi^2 T = 0 \tag{18.421}$$

and

$$\nabla^2\nabla^2 W = \frac{\rho h \xi^2}{D} W \tag{18.422}$$

where ξ^2 is a constant which determines the spectrum of frequencies.

In order to determine the function $W(x, y)$ satisfying the simply supported conditions at $x = 0, x = a$ and $y = 0, y = b$, we assume as before

$$W(x, y) = \sum_{m=1}^{\infty}\sum_{n=1}^{\infty} A_{mn} \sin\frac{m\pi x}{a} \sin\frac{n\pi y}{b} \tag{18.423}$$

Substituting expression equation (18.423) into equation (18.422), we get

$$\left(\frac{m\pi}{a}\right)^4 + 2\left(\frac{m\pi}{a}\right)^2\left(\frac{n\pi}{b}\right)^2 + \left(\frac{n\pi}{b}\right)^4 = \frac{\rho h \xi^2}{D} \quad \text{or,} \quad \xi^2 = \frac{D\pi^4}{\rho h}\left(\frac{m^2}{a^2} + \frac{n^2}{b^2}\right)^2 \tag{18.424}$$

Therefore, the spectrum of free vibrations are given by

$$\xi_{mn} = \pi^2\left(\frac{m^2}{a^2} + \frac{n^2}{b^2}\right)\sqrt{\frac{D}{\rho h}} \tag{18.425}$$

The lowest frequency of vibration is given for $m = 1$ and $n = 1$, so that

$$\xi_{11} = \pi^2\left(\frac{1}{a^2} + \frac{1}{b^2}\right)\sqrt{\frac{D}{\rho h}} \tag{18.426}$$

If the side a of the plate is very large compared to the other side b, so that $b/a \longrightarrow 0$, then the lowest frequency of vibration is

$$\xi_{11} = \frac{\pi^2}{b^2}\sqrt{\frac{D}{\rho h}} \tag{18.427}$$

18.23 General Solution of Biharmonic Equations

We seek to find the general solution of the biharmonic equation for stress function $\phi(x, y)$

$$\nabla^2\nabla^2\phi = 0 \tag{18.428}$$

in rectangular Cartesian coordinates. We assume the solution in the form

$$\phi(x, y) = X(x)Y(y) \tag{18.429}$$

and substitute the same into equation (18.428). We get

$$X^{IV}Y + 2X''Y'' + XY^{IV} = 0 \tag{18.430}$$

In order to eliminate the function X from here, we take

$$\left.\begin{aligned} X^{IV} &= \lambda^4 X \\ X'' &= -\xi^2 X \end{aligned}\right\} \tag{18.431}$$

where λ^4 and ξ^2 are constants.

Equation (18.431) shows that the function X satisfies two differential equations simultaneously. For this purpose, the constants λ^4 and ξ^2 cannot be arbitrary, but they should be related. To find this relation, we differentiate twice the second of equation (18.431) and get

$$X^{IV} = -\xi^2 X'' = -\xi^2(-\xi^2 X) = \xi^4 X \tag{18.432}$$

Comparing the first of equation (18.431) with equation (18.432), we find

$$\xi^4 = \lambda^4, \quad \text{i.e.,} \quad \xi = \lambda \tag{18.433}$$

Thus, the system of equations (18.431) can be written as

$$\left.\begin{aligned} X^{IV} &= \xi^4 X \\ X'' &= -\xi^2 X \end{aligned}\right\} \tag{18.434}$$

Substituting these relations into equation (18.430)

$$Y^{IV} - 2\xi^2 Y'' + \xi^4 Y = 0 \tag{18.435}$$

Equation (18.435) determines the function $Y(y)$ whereas the function $X(x)$ is to be determined from second of equation (18.434), i.e.,

$$X'' + \xi^2 X = 0 \tag{18.436}$$

The general solution of this equation is

$$X(x) = A\cos\xi x + B\sin\xi x \tag{18.437}$$

where A, B are constants.

We seek the solution of equation (18.435) in the form

$$Y(y) = e^{\beta y} \tag{18.438}$$

The characteristic equation is obtained as

$$\beta^4 - 2\xi^2\beta^2 + \xi^4 = 0$$

which yields $\beta = \pm\xi$ having two double roots. Hence the general solution of equation (18.435) is written as

$$Y(y) = C\cosh\xi y + D\sinh\xi y + Fy\cosh\xi y + Gy\sinh\xi y \tag{18.439}$$

where C, D, F, G are constants.

Thus the particular solution of equation (18.428) will be

$$\begin{aligned}\phi(x, y) &= X(x)Y(y) \\ &= (A\cos\xi x + B\sin\xi x)(C\cosh\xi y + D\sinh\xi y + Fy\cosh\xi y + Gy\sinh\xi y) \qquad (18.440)\end{aligned}$$

Since the biharmonic equation is linear, the general solution will be obtained by summing up solutions of type (18.440).

The constants are to be determined from the relevant boundary conditions.

Considering the case of bending of a rectangular strip whose length l is considerably greater than the height h and which is acted upon by normal load $q(x)$ on the upper surface (see Fig 18.1), the boundary conditions are

$$\sigma_x = \frac{\partial^2\phi}{\partial y^2} = 0, \quad \text{for } x = 0 \text{ and } x = l \qquad (18.441)$$

$$\sigma_y = \frac{\partial^2\phi}{\partial x^2} = q(x); \quad \tau_{xy} = -\frac{\partial^2\phi}{\partial x\partial y} = 0, \quad \text{for } y = h \qquad (18.442)$$

$$\sigma_y = \frac{\partial^2\phi}{\partial x^2} = 0; \quad \tau_{xy} = -\frac{\partial^2\phi}{\partial x\partial y} = 0, \quad \text{for } y = 0. \qquad (18.443)$$

The first condition yields

$$A = 0, \quad B = 1 \quad \text{and} \quad \xi = \frac{m\pi}{l} \quad (m = 1, 2, \ldots) \qquad (18.444)$$

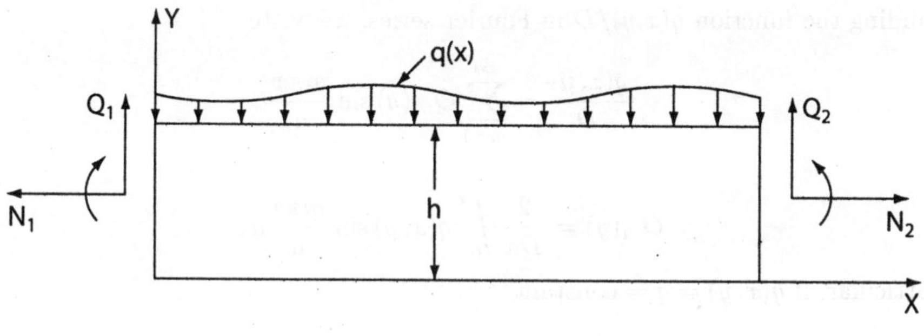

Fig 18.1

The third condition gives

$$C = 0, \quad D\xi + F = 0 \qquad (18.445)$$

The second condition yields

$$-\xi^2 \sin\xi x(D\sinh\xi h + Fh\cosh\xi h + Gh\sinh\xi h) = q(x) \qquad (18.446)$$

and

$$\frac{1}{\xi}(F\cosh\xi h + G\sinh\xi h) + D\cosh\xi h + Fh\sinh\xi h + Gh\cosh\xi h = 0 \qquad (18.447)$$

Multiplying both sides of equation (18.446) by $\sin\xi x$ and integrating from zero to l, we get

$$D\sinh\xi h + Fh\cosh\xi h + Gh\sinh\xi h = -\frac{2}{\xi^2 l}\int_0^l q(x)\sin\xi x\,dx \qquad (18.448)$$

From equations (18.445), (18.447) and (18.448), we can determine the three constants D, F, G so that complete solution for the stress function $\phi(x, y)$ is now available.

The method described is due to Filon. It can be used in the same way, when the upper and lower faces of the strip are acted upon by external shear loads also. Thus, it is possible to introduce arbitrary loads on the longer sides of the strip.

18.24 Rectangular Plates—Levy's Solution

Consider a rectangular plate whose opposite edge $x = 0$ and $x = a$ are simply supported and other two edges $y = 0$ and $y = b$ are arbitrarily supported or may even be free. Then we have the boundary conditions

$$w = 0, \quad \frac{\partial^2 w}{\partial x^2} = 0, \quad \text{for } x = 0 \text{ and } x = a \tag{18.449}$$

Consistent with those boundary conditions, we assume the solution of equation (18.402) as

$$w(x, y) = \sum_{m=1}^{\infty} Y_m(y) \sin \frac{m\pi x}{a} \tag{18.450}$$

Feeding this expression into the plate equation, we get

$$\sum_{m=1}^{\infty} \left[Y_m^{IV} - 2\left(\frac{m\pi}{a}\right)^2 Y_m'' + \left(\frac{m\pi}{a}\right)^4 Y_m \right] \sin \frac{m\pi x}{a} = \frac{q(x, y)}{D} \tag{18.451}$$

Expanding the function $q(x.y)/D$ in Fourier series, we write

$$\frac{q(x, y)}{D} = \sum_{m=1}^{\infty} Q_m(y) \sin \frac{m\pi x}{a} \tag{18.452}$$

where

$$Q_m(y) = \frac{2}{Da} \int_0^a q(x, y) \sin \frac{m\pi x}{a} dx \tag{18.453}$$

In particular, if $q(x, y) = q = \text{constant}$

$$Q_m = \frac{2q}{Da} \int_0^a \sin \frac{m\pi x}{a} dx = \frac{2q}{m\pi D}, \quad m = 1, 3, 5, \ldots \tag{18.454}$$

With the use of the expression equation (18.452), we get from equation (18.451)

$$Y_m^{IV} - 2\left(\frac{m\pi}{a}\right)^2 Y_m'' + \left(\frac{m\pi}{a}\right)^4 Y_m = Q_m \tag{18.455}$$

General solution of this equation can be written in the form

$$Y_m(y) = A_m \cosh \frac{m\pi y}{a} + B_m \sinh \frac{m\pi y}{a} + C_m y \cosh \frac{m\pi y}{a} + D_m y \sinh \frac{m\pi y}{a} + \bar{Q}_m \tag{18.456}$$

where $\bar{Q}_m$ is the particular solution corresponding to the function $Q_m(y)$.

For uniformly distributed load

$$Q_m(y) = \frac{2q}{m\pi D}, \quad \text{hence } \bar{Q}_m = \frac{2qa^4}{D(m\pi)^5} \tag{18.457}$$

The coefficients A_m, B_m, C_m, D_m are to be determined from the boundary conditions along the edges $y = 0$ and $y = b$. If, in particularly, these edges are rigidly fixed, then

$$w = 0, \quad \frac{\partial w}{\partial y} = 0, \quad \text{for } y = 0 \text{ and } y = b\cdot$$

i.e.,

$$Y_m = 0, \quad Y_m' = 0, \quad \text{for } y = 0 \text{ and } y = b \tag{18.458}$$

These conditions yield the following equations:

$$A_m = -\bar{Q}_m, \quad C_m = -\frac{m\pi b}{a} B_m$$

$$B_m \left(\sinh \frac{m\pi b}{a} - \frac{m\pi b}{a} \cosh \frac{m\pi b}{a} \right) + D_m b \sinh \frac{m\pi b}{a} = \bar{Q}_m \left(\cosh \frac{m\pi b}{a} - 1 \right)$$

$$-B_m \left(\frac{m\pi}{a} \right)^2 b \sinh \frac{m\pi b}{a} + D_m \left(\sinh \frac{m\pi b}{a} + b \cosh \frac{m\pi b}{a} \right) = \bar{Q}_m \frac{m\pi}{a} \sinh \frac{m\pi b}{a}$$

which can be solved for the coefficients.

18.25 Circular Plates

Here we introduce the polar coordinates (r, θ) so that

$$x = r \cos \theta, \quad y = r \sin \theta$$

The Laplacian operator becomes

$$\nabla^2 = \frac{\partial^2}{\partial x^2} + \frac{\partial^2}{\partial y^2} = \frac{\partial^2}{\partial r^2} + \frac{1}{r} \frac{\partial}{\partial r} + \frac{1}{r^2} \frac{\partial^2}{\partial \theta^2} \tag{18.459}$$

Hence the plate equation is written as

$$\left(\frac{\partial^2}{\partial r^2} + \frac{1}{r} \frac{\partial}{\partial r} + \frac{1}{r^2} \frac{\partial^2}{\partial \theta^2} \right) \left(\frac{\partial^2 w}{\partial r^2} + \frac{1}{r} \frac{\partial w}{\partial r} + \frac{1}{r^2} \frac{\partial^2 w}{\partial \theta^2} \right) = \frac{q(r, \theta)}{D} \tag{18.460}$$

If, in particular, the load is independent of angle θ, then the deflection function will not depend on θ. Equation (18.460) in this case assumes the form

$$\frac{d^4 w}{dr^4} + \frac{2}{r} \frac{d^3 w}{dr^3} - \frac{1}{r^2} \frac{d^2 w}{dr^2} + \frac{1}{r^3} \frac{dw}{dr} = \frac{q(r)}{D} \tag{18.461}$$

Expressions for moments and shearing forces are

$$M_r = -D \left[\frac{\partial^2 w}{\partial r^2} + \mu \left(\frac{1}{r} \frac{\partial w}{\partial r} + \frac{1}{r^2} \frac{\partial^2 w}{\partial \theta^2} \right) \right] \tag{18.462}$$

$$M_\theta = -D \left[\frac{1}{r} \frac{\partial w}{\partial r} + \frac{1}{r^2} \frac{\partial^2 w}{\partial \theta^2} + \mu \frac{\partial^2 w}{\partial r^2} \right] \tag{18.463}$$

$$N_r = -D \frac{\partial}{\partial r} \nabla^2 w, \quad N_\theta = -D \frac{1}{r} \nabla^2 w \tag{18.464}$$

$$H = -D (1 - \mu) \frac{\partial}{\partial r} \left(\frac{1}{r} \frac{\partial w}{\partial \theta} \right) \tag{18.465}$$

Boundary Conditions

For freely supported plate

$$w = 0, \quad M_r = 0 \text{ for } r = a \tag{18.466}$$

and for rigidly clamped boundary

$$w = 0, \quad \frac{\partial w}{\partial r} = 0, \quad \text{for } r = a \tag{18.467}$$

where a = radius of the circular plate.

18.26 Symmetrically Loaded Circular Plates

In this case all the physical quantities are independent of θ.

The basic plate equation (18.461) is recalled here

$$\frac{d^4w}{dr^4} + \frac{2}{r}\frac{d^3w}{dr^3} - \frac{1}{r^2}\frac{d^2w}{dr^2} + \frac{1}{r^3}\frac{dw}{dr} = \frac{q(r)}{D} \tag{18.468}$$

Moments and shear forces are given by

$$\left.\begin{aligned} M_r &= -D\left(\frac{d^2w}{dr^2} + \frac{\mu}{r}\frac{dw}{dr}\right) \\ M_\theta &= -D\left(\frac{1}{r}\frac{dw}{dr} + \mu\frac{d^2w}{dr^2}\right) \\ H &= 0 \end{aligned}\right\} \tag{18.469}$$

For the solution of equation (18.468), we assume

$$w(r) = r^\alpha \tag{18.470}$$

and get the characteristic equation

$$\alpha^2(\alpha - 2)^2 = 0$$

so that $\alpha = 0, 0, 2, 2$.

Hence the general solution of equation (18.468) can be written as

$$w(r) = A_1 + B_1 r^2 + C_1 \ln r + D_1 r^2 \ln r + \bar{w} \tag{18.471}$$

where $\bar{w}$ is the particular integral. If the load distribution is uniform so that $q(r) = q =$ constant, then

$$\bar{w} = \frac{qr^4}{64D} \tag{18.472}$$

Then, in this case

$$w(r) = A_1 + B_1 r^2 + C_1 \ln r + D_1 r^2 \ln r + \frac{qr^4}{64D} \tag{18.473}$$

If the plate is solid without having a hole in its central part, then deflections, and moments at the centre should have finite values whereas rotation should be zero at the centre $r = 0$. Hence, in equation (18.473), we take $C_1 = D_1 = 0$, so that

$$w(r) = A_1 + B_1 r^2 + \frac{qr^4}{64D} \tag{18.474}$$

Case I. *Simply-supported plate.*

Let the circular plate be simply-supported along its boundary. For this, the conditions are

$$w = 0, \quad M_r = -D\left(\frac{d^2w}{dr^2} + \frac{\mu}{r}\frac{dw}{dr}\right) = 0 \quad \text{for } r = a$$

These conditions yield

$$\left.\begin{aligned} A_1 &= \frac{qa^4(5+\mu)}{64(1+\mu)D} \\ B_1 &= -\frac{qa^2(6+2\mu)}{64(1+\mu)D} \end{aligned}\right\} \tag{18.475}$$

With the use of these values for A_1 and B_1 the deflection function equation (18.474) is written as

$$w(r) = \frac{q}{64D}\left[(a^2-r^2)^2 + \frac{4a^2(a^2-r^2)}{1+\mu}\right] \tag{18.476}$$

By equation (18.469), the bending moments are obtained as

$$\left.\begin{aligned} M_r &= \frac{q}{16}\left[(a^2-r^2)(3+\mu)\right] \\ M_\theta &= \frac{q}{16}\left[a^2(3+\mu) - r^2(1+3\mu)\right] \end{aligned}\right\} \tag{18.477}$$

Case II. *Clamped plate.*

In this case, the boundary conditions are

$$w = 0, \quad \frac{dw}{dr} = 0, \quad \text{for } r = a$$

These conditions yield

$$A_1 = \frac{q}{64D}a^4, \quad B_1 = -\frac{2q}{64D}a^2 \tag{18.478}$$

Hence, the deflection function is

$$w(r) = \frac{q}{64D}\left(a^2 - r^2\right)^2 \tag{18.479}$$

The bending moments are obtained as

$$\left.\begin{aligned} M_r &= \frac{q}{16}\left[a^2(1+\mu) - r^2(3+\mu)\right] \\ M_\theta &= \frac{q}{16}\left[a^2(1+\mu) - r^2(1+3\mu)\right] \end{aligned}\right\} \tag{18.480}$$

For symmetrically loaded plates, sometimes it is convenient to get the solution in terms of the angle of rotation. Denoting

$$\psi = \frac{dw}{dr} \tag{18.481}$$

the vertical shear N (dropping the suffix r) from equation (18.464) can be written as

$$N = -D\frac{d}{dr}\left(\frac{d^2w}{dr^2} + \frac{1}{r}\frac{dw}{dr}\right) \quad \text{or,} \quad -\frac{N}{D} = \frac{d}{dr}\left[\frac{1}{r}\frac{d}{dr}(\psi r)\right] \tag{18.482}$$

Integrating this twice

$$\psi = A_1 r + \frac{B_1}{r} - \frac{1}{Dr}\int\left[r\int N\,dr\right]dr \tag{18.483}$$

where the constants A_1, B_1 of integration are to be determined from boundary conditions.

Considering a concentrated point load P at the centre of the plate, the vertical shearing force for a small portion at the centre is

$$N = \frac{P}{2\pi r} \tag{18.484}$$

Substituting the value of N into equation (18.483) and integrating twice, we get

$$\psi = A_1 r + \frac{B_1}{r} - \frac{P}{4\pi D}r\left(\ln r - \frac{1}{2}\right)$$

which can be written as

$$\psi = Ar + \frac{B}{r} - \frac{P}{4\pi D}r\ln\left(\frac{r}{a}\right) \tag{18.485}$$

Now, at the centre, $\psi = 0$ for $r = 0$. Since

$$\lim_{r\longrightarrow 0} r\ln\frac{r}{a} = 0 \tag{18.486}$$

the constant $B = 0$. Therefore

$$\psi = Ar - \frac{P}{4\pi D}r\ln\left(\frac{r}{a}\right) \tag{18.487}$$

Integrating

$$w = C + Ar^2 - \frac{Pr^2}{8\pi D}\left(\ln\frac{a}{r} + \frac{1}{2}\right) \tag{18.488}$$

If *the plate is clamped* along the boundary

$$w = 0,\ \psi = \frac{dw}{dr} = 0,\quad \text{for } r = a$$

This yields

$$A = 0,\quad C = \frac{Pa^2}{16\pi D} \tag{18.489}$$

Hence, the deflection function is

$$w(r) = \frac{P}{8\pi D}\left[\frac{1}{2}\left(a^2 - r^2\right) - r^2\ln\frac{a}{r}\right] \tag{18.490}$$

The bending moments are given by

$$\left.\begin{aligned} M_r &= \frac{P}{4\pi}\left[(1+\mu)\ln\frac{a}{r} - 1\right] \\ M_\theta &= \frac{P}{4\pi}\left[(1+\mu)\ln\frac{a}{r} - \mu\right] \end{aligned}\right\} \tag{18.491}$$

If the *plate is simply-supported* along the boundary, then

$$w = 0,\quad M_r = 0,\quad \text{for } r = a$$

which yields the values of the constants as

$$\left.\begin{aligned} A &= \frac{P}{16\pi D}\cdot\frac{4+2\mu}{1+\mu} \\ C &= \frac{Pa^2}{16\pi D}\cdot\frac{3+\mu}{1+\mu} \end{aligned}\right\} \tag{18.492}$$

Therefore, the deflection function is written as

$$w(r) = \frac{P}{8\pi D}\left[\frac{1}{2}\frac{3+\mu}{1+\mu}\left(a^2 - r^2\right) - r^2 \ln\frac{a}{r}\right] \tag{18.493}$$

The moments can be calculated easily from this expression.

EXAMPLE 18.15. *A semi-circular plate of radius a is clamped along the curved boundary and is simply-supported along the straight edge. Find the deflection of the plate under a uniform load q.*

Solution: We are to solve the equation

$$\nabla^2\nabla^2 w = \frac{q}{D} \tag{1}$$

under the boundary conditions

$$w = 0, \quad \frac{\partial w}{\partial r} = 0, \quad \text{for } r = a \tag{2}$$

and

$$w = 0, \quad \frac{\partial^2 w}{\partial \theta^2} = 0, \quad \text{for } \theta = 0 \text{ and } \pi \tag{3}$$

We consider a function $u(r,\theta)$, such that

$$w(r,\theta) = \frac{qa^4}{24D}\left(\frac{r^4}{a^4}\sin^4\theta - u\right) \tag{4}$$

It can be verified that the function $u(r,\theta)$ satisfies the equation

$$\nabla^2\nabla^2 u = 0 \tag{5}$$

with the boundary conditions

$$\left.\begin{aligned} u = 0, \quad & \frac{\partial^2 u}{\partial \theta^2} = 0, \text{for } \theta = 0 \\ u = 0, \quad & \frac{\partial^2 u}{\partial \theta^2} = 0, \text{for } \theta = \pi \\ u = \sin^4\theta, \quad & \frac{\partial u}{\partial r} = \frac{4}{a}\sin^4\theta, \text{for } r = a \end{aligned}\right\} \tag{6}$$

To determine the function $u(r,\theta)$, we take

$$u_m(r,\theta) = (A_m \cos\lambda\theta + B_m \sin\lambda\theta)\, R_m(r) \tag{7}$$

From the conditions at $\theta = 0$ and $\theta = \pi$, we get

$$A_m = 0, \quad \lambda = m, \quad \text{with } m = 1, 2, 3, \ldots$$

Since the deflection of the plate is bounded, we can find

$$R_m(r) = C_m r^m + D_m r^{m+2} \tag{8}$$

Hence

$$u(r,\theta) = \sum_{m=1}^{\infty} \left[P_m \left(\frac{r}{a}\right)^m + Q_m \left(\frac{r}{a}\right)^{m+2} \right] \sin m\theta \tag{9}$$

Using the boundary conditions at $r = a$,

$$\left.\begin{aligned} \sin^4\theta &= \sum_{m=1}^{\infty} (P_m + Q_m) \sin m\theta \\ \frac{4}{a}\sin^4\theta &= \sum_{m=1}^{\infty} \left[\frac{m}{a} P_m + \frac{m+2}{a} Q_m \right] \sin m\theta \end{aligned}\right\} \tag{10}$$

Multiplying equation (10) by $\sin m\theta$ $(0 \le \theta \le \pi)$ both sides and integrating, we get

$$P_m + Q_m = \frac{2}{\pi} \int_0^{\pi} \sin^4\theta \sin m\theta d\theta \tag{11}$$

and

$$mP_m + (m+2)Q_m = \frac{8}{\pi} \int_0^{\pi} \sin^4\theta \sin m\theta d\theta \tag{12}$$

Now

$$\begin{aligned} I_m &= \int_0^{\pi} \sin^4\theta \sin m\theta \\ &= \frac{1-(-1)^m}{8} \left[\frac{1}{2}\left(\frac{1}{m+4} + \frac{1}{m-4} \right) - 2\left(\frac{1}{m+2} + \frac{1}{m-2} \right) + \frac{3}{m} \right] \end{aligned} \tag{13}$$

Then, equations (11) and (12) are solved to get

$$P_m = \frac{m-2}{\pi} I_m, \quad Q_m = \frac{4-m}{\pi} I_m \tag{14}$$

Therefore the deflection function is known from equation (4) by substituting the expression for $u(r, \theta)$ from equation (9) with the use of equations (13) and (14).

EXAMPLE 18.16. *A solid circular plate of radius a is clamped along its boundary and carries a load P distributed over a central circle of radius b. Determine the moments at different regions.*

Solution: Here we have two regions in the plate. In region 1 $(b \le r \le a)$, there is no load, so that from equation (18.483)

$$\psi_1 = A_1 r + \frac{B_1}{r} \tag{1}$$

In region 2 $(0 \le r \le b)$

$$N = \frac{P}{2\pi r}$$

so that

$$\psi_2 = A_2 r + \frac{B_2}{r} - \frac{P}{4\pi D} r \ln \frac{r}{b} \tag{2}$$

There are unknown constants A_1, B_1, A_2 and B_2 which are to be determined from the conditions:

At $r = b, \psi_1 = \psi_2$ and $M_{r1} = M_{r2}$. (3)

At $r = 0, \psi_1 = 0$. (4)

At $r = a, \psi_2 = 0$. (5)

Equation (4) yields $B_1 = 0$. Hence

$$\psi_1 = A_1 r \tag{6}$$

Condition (3) gives

$$\left(\frac{d\psi_1}{dr} + \frac{\mu\psi_1}{r}\right)_{r=b} = \left(\frac{d\psi_2}{dr} + \frac{\mu\psi_2}{r}\right)_{r=b} \tag{7}$$

Since $\psi_1 = \psi_2$ at $r = b$, equation (7) gives

$$\left(\frac{d\psi_1}{dr}\right)_{r=b} = \left(\frac{d\psi_2}{dr}\right)_{r=b} \tag{8}$$

Conditions at (5), (3) and (8) yield

$$\left.\begin{aligned} &A_2 a + \frac{B_2}{a} - \frac{Pa}{4\pi D} \ln \frac{a}{b} = 0 \\ &A_1 b = A_2 b + \frac{B_2}{b} \\ &A_1 = A_2 - \frac{B_2}{b^2} - \frac{P}{4\pi D} \end{aligned}\right\} \tag{9}$$

From these equations we get

$$\left.\begin{aligned} A_1 &= \frac{P}{4\pi D}\left(\ln \frac{a}{b} + \frac{1}{2}\frac{b^2}{a^2} - \frac{1}{2}\right) \\ A_2 &= \frac{P}{4\pi D}\left(\ln \frac{a}{b} + \frac{1}{2}\frac{b^2}{a^2}\right) \\ B_2 &= -\frac{Pb^2}{8\pi D} \end{aligned}\right\} \tag{10}$$

Therefore

$$\left.\begin{aligned} \psi_1 &= A_1 r = \frac{Pr}{4\pi D}\left(\ln \frac{a}{b} + \frac{1}{2}\frac{b^2}{a^2} - \frac{1}{2}\right) \\ \psi_2 &= \frac{Pr}{4\pi D}\left(\ln \frac{a}{b} + \frac{1}{2}\frac{b^2}{a^2}\right) - \frac{Pb^2}{8\pi D r} - \frac{Pr}{4\pi D} \ln \frac{r}{b} \end{aligned}\right\} \tag{11}$$

In region 1, the moments are given by

$$\left.\begin{aligned} M_r &= \frac{P}{4\pi}\left[(1+\mu)\left(\ln \frac{a}{r} + \frac{1}{2}\frac{b^2}{a^2}\right) + \frac{b^2}{2r^2}(1-\mu) - 1\right] \\ M_\theta &= \frac{P}{4\pi}\left[(1+\mu)\left(\ln \frac{a}{r} + \frac{1}{2}\frac{b^2}{a^2}\right) - \frac{b^2}{2r^2}(1-\mu) - 1\right] \end{aligned}\right\} \tag{12}$$

In region 2, the moments are given by

$$M_r = M_\theta = \frac{P(1+\mu)}{4\pi}\left(\ln\frac{a}{b} + \frac{1}{2}\frac{b^2}{a^2} - \frac{1}{2}\right) \tag{13}$$

which is a constant.

SCHRÖDINGER'S EQUATION

18.27 Linear Harmonic Oscillator

Schrödinger's one-dimensional time-independent equation is

$$-\frac{\hbar^2}{2m}\frac{d^2\psi}{dx^2} + (V - E)\psi = 0 \tag{18.494}$$

We intend to use this equation to study the linear harmonic oscillator which finds practical application in many physical situations. In this case, the potential energy is

$$V(x) = \tfrac{1}{2}kx^2 \tag{18.495}$$

so that equation (18.494) assumes the form

$$\frac{d^2\psi}{dx^2} + \frac{2m}{\hbar^2}\left(E - \frac{1}{2}kx^2\right) = 0 \tag{18.496}$$

We make the following transformations

$$\left.\begin{aligned} \xi &= \alpha x \\ \alpha^4 &= \frac{mk}{\hbar^2} \\ \lambda &= \frac{2E}{\hbar}\sqrt{\frac{m}{k}} \end{aligned}\right\} \tag{18.497}$$

Then, the equation (18.496) is reduced to

$$\frac{d^2\psi}{d\xi^2} + (\lambda - \xi^2)\psi = 0 \tag{18.498}$$

We consider the case when

$$\lambda \equiv (2n+1), \quad n = 0, 1, 2, \ldots \tag{18.499}$$

Then equation (18.498) assumes the form

$$\frac{d^2\psi}{d\xi^2} + (1 - \xi^2 + 2n)\psi = 0 \tag{18.500}$$

which is *Weber's differential equation.* With the transformation

$$\psi(\xi) = e^{-\xi^2/2}\phi(\xi) \tag{18.501}$$

the equation changes to Hermite's differential equation

$$\frac{d^2\phi}{d\xi^2} - 2\xi\frac{d\phi}{d\xi} + 2n\phi = 0 \tag{18.502}$$

whose solutions are Hermite's polynomial (see Chapter 12) $H_n(\xi)$. Thus the general solution of the linear harmonic oscillator can be written as

$$\psi_n(x) = N_n e^{-\alpha^2 x^2/2} H_n(\alpha x) \tag{18.503}$$

The quantity n in this equation is called the *principal quantum number.*

From third of equation (18.497) and (18.499), we have

$$\frac{2E_n}{\hbar}\sqrt{\frac{m}{k}} = 2n + 1 \quad \text{or,} \quad E_n = \left(n + \frac{1}{2}\right)\hbar\omega_c \tag{18.504}$$

where $\omega_c = \sqrt{k/m}$ = classical angular frequency of the linear oscillator.

Equation (18.504) represents the eigen energy of the linear harmonic oscillator. For $n = 0$, this gives the ground state energy

$$E_0 = \tfrac{1}{2}\hbar\omega_c \tag{18.505}$$

which is known as *zero point energy.*

18.28 Motion of a Particle in a One-Dimensional Box

We consider one-dimensional motion of a particle with zero potential energy over a fixed distance between $x = 0$ to $x = a$. The potential energy is assumed to be infinite at the points $x = a$ and beyond. This situation is known as a *particle in one-dimensional box* assuming rigid walls at $x = 0$ and $x = a$. Thus, in this case

$$\left.\begin{aligned} V(x) &= 0, \quad \text{for } 0 < x < a \\ V(x) &= \infty, \quad \text{for } x \le 0 \text{ and } x \ge a \end{aligned}\right\} \tag{18.506}$$

so that Schrödinger's one-dimensional time-independent equation (18.494) becomes

$$\frac{d^2\psi}{dx^2} + \frac{2m}{\hbar^2}E\psi = 0 \quad \text{or,} \quad \frac{d^2\psi}{dx^2} + \xi^2\psi = 0 \tag{18.507}$$

where

$$\xi^2 = \frac{2mE}{\hbar^2} \tag{18.508}$$

The general solution of equation (18.507) is

$$\psi = A\cos\xi x + B\sin\xi x \tag{18.509}$$

Now during the motion, the particle has a total energy E which remains constant even after colliding with rigid walls at $x = (0, a)$ so that it cannot leave the region. Hence, the wave function ψ must satisfy the boundary conditions

$$\left.\begin{aligned} \psi(x) &= 0, \quad \text{for } x = 0 \\ \psi(x) &= 0, \quad \text{for } x = a \end{aligned}\right\} \tag{18.510}$$

Applying these boundary conditions in equation (18.507), we have

$$A = 0 \quad \text{and} \quad B \sin \xi a = 0 \tag{18.511}$$

Since $B \neq 0$, we get

$$\xi a = n\pi, \quad \text{i.e.,} \quad \xi = \frac{n\pi}{a}, \quad n = 1, 2, \ldots \tag{18.512}$$

We have omitted the value $n = 0$, for in that case the wave function does not exist. Hence,

$$\psi_n(x) = B \sin \frac{n\pi x}{a} \tag{18.513}$$

The constant B is determined from the normalising condition, i.e., the total probability that the particle is somewhere in the box should be unity. This requires

$$\int_0^a |\psi(x)|^2 \, dx = 1 \tag{18.514}$$

Using equation (18.513), we have

$$\int_0^a B^2 \sin^2 \left(\frac{n\pi x}{a}\right) dx = 1$$

which yields

$$B^2 \frac{a}{2} = 1, \quad \text{i.e.,} \quad B = \sqrt{\frac{2}{a}} \tag{18.515}$$

Thus, the normalised wave function is

$$\psi_n(x) = \sqrt{\frac{2}{a}} \sin \frac{n\pi x}{a} \tag{18.516}$$

Now, from equation (18.508), we have

$$\frac{2mE}{\hbar^2} = \xi^2 = \frac{n^2\pi^2}{a^2}$$

which gives

$$E_n = \frac{n^2\pi^2\hbar^2}{2ma^2} \tag{18.517}$$

This represents the eigenvalues of energy. For different values of n, equation (18.517) determines energy-levels of the particle. The ground-state energy-level of the particle is given by $n = 1$ and its value is

$$E_1 = \frac{\pi^2 \hbar^2}{2ma^2} \tag{18.518}$$

In view of this we can write equation (18.517) in the form

$$E_n = n^2 E_1 \tag{18.519}$$

The integer n corresponding to the energy level E_n is called *quantum number* and the wave function $\psi_n(x)$ corresponding to this quantum number is called *eigenfunction* of the particle.

18.29 A Particle in a Three-Dimensional Box

We consider a particle of mass m moving inside a rectangular box defined by the surfaces

$$\left.\begin{aligned} x=0, &\quad x=a \\ y=0, &\quad y=b \\ z=0, &\quad z=c \end{aligned}\right\} \tag{18.520}$$

We assume that there is no force acting on the particle inside the box.

The potential function $V(x, y, z)$ is zero inside the box and infinite outside it. The Schrödinger time independent wave equation in three-dimensional rectangular coordinate axes is

$$\frac{\partial^2\psi}{\partial x^2}+\frac{\partial^2\psi}{\partial y^2}+\frac{\partial^2\psi}{\partial z^2}+\frac{2m}{\hbar^2}E\psi=0 \tag{18.521}$$

We assume the solution of equation (18.521) in the form

$$\psi(x,y,z)=X(x)Y(y)Z(z) \tag{18.522}$$

Substituting this into the equation (18.521) and dividing by XYZ, we get

$$\frac{X''}{X}+\frac{Y''}{Y}+\frac{Z''}{Z}=-\frac{2mE}{\hbar^2} \tag{18.523}$$

The variables in this equation (18.523) being separated, each part in LHS should be a constant. Thus, we get three ordinary differential equations

$$\frac{d^2X}{dx^2}+\lambda^2X=0 \tag{18.524}$$

$$\frac{d^2Y}{dy^2}+\xi^2Y=0 \tag{18.525}$$

$$\frac{d^2Z}{dz^2}+\eta^2Z=0 \tag{18.526}$$

where

$$\lambda^2+\xi^2+\eta^2=\frac{2mE}{\hbar^2} \tag{18.527}$$

Solutions of these equations are

$$X(x)=A\cos\lambda x+B\sin\lambda x \tag{18.528}$$

$$Y(y)=C\cos\xi y+D\sin\xi y \tag{18.529}$$

$$Z(z)=F\cos\eta z+G\sin\eta z \tag{18.530}$$

where A, B, C, D, F, G are integration constants to be determined from the boundary conditions that $\psi(x, y, z) = 0$ on the surface of infinite potential, i.e.,

$$\psi(x,y,z)=0, \quad \text{for} \quad \left.\begin{aligned} x&=0\\ x&=a\end{aligned}\right\}, \quad \left.\begin{aligned} y&=0\\ y&=b\end{aligned}\right\}, \quad \left.\begin{aligned} z&=0\\ z&=c\end{aligned}\right\} \tag{18.531}$$

Using these boundary conditions, we get

$$A=C=F=0 \tag{18.532}$$

and, $\sin\lambda a = 0$, $\sin\xi b = 0$, $\sin\eta c = 0$, which give

$$\lambda = \frac{n_x\pi}{a}, \quad \xi = \frac{n_y\pi}{b}, \quad \eta = \frac{n_z\pi}{c} \tag{18.533}$$

where n_x, n_y, n_z are integers $1, 2, 3, \ldots$

Thus, the total wave functions inside the box are given by

$$\psi(x, y, z) = N \sin\frac{n_x\pi x}{a} \sin\frac{n_y\pi y}{b} \sin\frac{n_z\pi z}{c} \tag{18.534}$$

where $N = BDG$, a constant which can be determined from normalising condition

$$\int_0^a \int_0^b \int_0^c \psi\psi^* dxdydz = 1 \tag{18.535}$$

Substituting equation (18.534) into equation (18.535) and integrating within proper limits, we get

$$N^2\frac{abc}{8} = 1, \quad \text{i.e.,} \quad N = \sqrt{\frac{8}{abc}} \tag{18.536}$$

Thus, the total normalised wave functions are

$$\psi(x, y, z) = \sqrt{\frac{8}{abc}} \sin\frac{n_x\pi x}{b} \sin\frac{n_y\pi y}{b} \sin\frac{n_z\pi z}{c} \tag{18.537}$$

Eigenvalues of energy are obtained from equations (18.527) and (18.533) which give

$$\frac{n_x^2\pi^2}{a^2} + \frac{n_y^2\pi^2}{b^2} + \frac{n_z^2\pi^2}{c^2} = \frac{2mE}{\hbar^2} \quad \text{or,} \quad E_{n_x, n_y, n_z} = \frac{\pi^2\hbar^2}{2m}\left[\left(\frac{n_x}{a}\right)^2 + \left(\frac{n_y}{b}\right)^2 + \left(\frac{n_z}{c}\right)^2\right] \tag{18.538}$$

The integers n_x, n_y, n_z are called quantum numbers which assume values $1, 2, 3, \ldots$ The energy in the ground state is obtained for $n_x = n_y = n_z = 1$. Thus, from equation (18.538) we find the lowest possible energy as

$$E_{111} = \frac{\pi^2\hbar^2}{2m}\left[\frac{1}{a^2} + \frac{1}{b^2} + \frac{1}{c^2}\right] \tag{18.539}$$

If the particle is confined in a cubical box when $a = b = c = l$, the energy levels are given by

$$E_{n_x, n_y, n_z} = \frac{\pi^2\hbar^2}{2ml^2}\left(n_x^2 + n_y^2 + n_z^2\right) \tag{18.540}$$

The energy in the ground state in this case is

$$E_{111} = \frac{3\pi^2\hbar^2}{2ml^2} \tag{18.541}$$

We find from equation (18.540), that when the particle moves in a cubical box, eigenvalues of energy depend on sum of the squares of quantum numbers. For three different sets of values of (n_x, n_y, n_z) say like $(3, 1, 1)$, $(1, 3, 1)$ and $(1, 1, 3)$, each state has same energy $11\pi^2\hbar^2/2ml^2$, although there exist three different eigenfunctions defining three different stationary states. These states are said to be *degenerate*.

18.30 The Hydrogen Atom

The hydrogen atom is a two-body system consisting of a single electron moving around a proton which is regarded as stationary. In absence of any external force field, the potential energy of the system is

$$V = -\frac{Ze^2}{r} \tag{18.542}$$

where Ze = charge on the nucleus, e = charge on electron and r = distance between electron and nucleus. We intend to solve the wave equation for the motion of the electron.

The three-dimensional time-independent Schrödinger equation for the motion of the electron is

$$\nabla^2\psi + \frac{2m_0}{\hbar^2}(E - V)\psi = 0 \tag{18.543}$$

where m_0 is the rest mass of the electron and E is its total energy.

In spherical polar coordinates (r, θ, ϕ), this equation is written as

$$\frac{1}{r^2}\frac{\partial}{\partial r}\left(r^2\frac{\partial\psi}{\partial r}\right) + \frac{1}{r^2\sin\theta}\frac{\partial}{\partial\theta}\left(\sin\theta\frac{\partial\psi}{\sigma\theta}\right) + \frac{1}{r^2\sin^2\theta}\frac{\partial^2\psi}{\partial\phi^2} + \frac{2m_0}{\hbar^2}(E - V)\psi = 0 \tag{18.544}$$

We assume the solution of this equation as

$$\psi(r, \theta, \phi) = R(r)F(\theta)Q(\phi) \tag{18.545}$$

Substituting this into equation (18.544) and multiplying the results by

$$\frac{r^2\sin^2\theta}{RFQ} \tag{18.546}$$

we get

$$\frac{\sin^2\theta}{R}\frac{d}{dr}\left(r^2\frac{dR}{dr}\right) + \frac{\sin\theta}{F}\frac{d}{d\theta}\left(\sin\theta\frac{dF}{d\theta}\right) + \frac{1}{Q}\frac{d^2Q}{d\phi^2} + \frac{2m_0}{\hbar^2}(E - V)r^2\sin^2\theta = 0 \tag{18.547}$$

The third term in this equation is a function of ϕ only and other terms are independent of ϕ. Hence this term should be a constant, say, $-m^2$. Thus

$$\frac{1}{Q}\frac{d^2Q}{d\phi^2} = -m^2 \tag{18.548}$$

$$\text{or,}\quad \frac{d^2Q}{d\phi^2} + m^2Q = 0. \tag{18.549}$$

This is one of the three differential equations and is called *azimuthal wave equation.* Now using equation (18.548) in equation (18.549) and dividing the result by $\sin^2\theta$, we get

$$\frac{1}{R}\frac{d}{dr}\left(r^2\frac{dR}{dr}\right) + \frac{2m_0r^2}{\hbar^2}(E - V) = \frac{m^2}{\sin^2\theta} - \frac{1}{F\sin\theta}\frac{d}{d\theta}\left(\sin\theta\frac{dF}{d\theta}\right) \tag{18.550}$$

The LHS of this equation is a function of r only whereas the RHS is a function of θ only. Thus, each side should be equal to a constant, say $l(l+1)$. Then we get the following two differential equations.

$$\frac{1}{R}\frac{d}{dr}\left(r^2\frac{dR}{dr}\right)+\frac{2m_0r^2}{\hbar^2}(E-V)=l(l+1)$$

$$\text{or,}\quad \frac{1}{r^2}\frac{d}{dr}\left(r^2\frac{dR}{dr}\right)+\left[\frac{2m_0}{\hbar^2}(E-V)-\frac{l(l+1)}{r^2}\right]R=0 \tag{18.551}$$

This equation is called the *radial wave equation.* The other equation is obtained from equation (18.550) in the form

$$\frac{m^2}{\sin^2\theta}-\frac{1}{F\sin\theta}\frac{d}{d\theta}\left(\sin\theta\frac{dF}{d\theta}\right)=l(l+1)$$

$$\text{or,}\quad \frac{1}{\sin\theta}\frac{d}{d\theta}\left(\sin\theta\frac{dF}{d\theta}\right)+\left[l(l+1)-\frac{m^2}{\sin^2\theta}\right]F=0 \tag{18.552}$$

This equation is called the *polar wave equation.* Thus, the complete soluton of Schrödinger equation (18.543) depends on the solutions of three equations (18.549), (18.551) and (18.552). We now find the solutions of these equations.

(a) *Solution of Azimuthal equation*: The azimuthal equation is

$$\frac{d^2Q}{d\phi^2}+m^2Q=0 \tag{18.553}$$

whose solution is

$$Q(\phi)=Ae^{\pm im\phi} \tag{18.554}$$

where the arbitrary constant A is determined from the normalising condition

$$\int_0^{2\pi}Q^*Qd\phi=1 \tag{18.555}$$

This gives

$$A=\frac{1}{\sqrt{2\pi}} \tag{18.556}$$

Hence

$$Q(\phi)=\frac{1}{\sqrt{2\pi}}e^{\pm im\phi} \tag{18.557}$$

Now, the function $Q(\phi)$ should be single-valued, i.e., it should have same value at $\phi=0$ and $\phi=2\pi$. Hence, from equation (18.557), we get $1=e^{\pm 2\pi im}$, i.e.,

$$\cos 2\pi m\pm i\sin 2\pi m=1 \tag{18.558}$$

which is possible only when $m=0$, or a positive or a negative integer. Thus, we write the solution equation (18.557) as

$$Q_m(\phi)=\frac{1}{\sqrt{2\pi}}e^{im\phi} \tag{18.559}$$

where $m=0,\pm1,\pm2,\ldots$

The quantity m is called the *magnetic quantum number.*

(b) *Solution of Polar equation*: In the polar equation (18.552), we substitute

$$x=\cos\theta \tag{18.560}$$

Then the equation is transformed into

$$\frac{d}{dx}\left[(1-x^2)\frac{dF}{dx}\right]+\left[l(l+1)-\frac{m^2}{1-x^2}\right]F=0 \tag{18.561}$$

This is an Associated Legendre's equation whose solution is

$$F(\theta)=BP_l^m(x)=BP_l^m(\cos\theta) \tag{18.562}$$

where $P_l^m(\cos\theta)$ are Associated Legendre's polynomials of mth order and lth degree (see Chapter 12).

The constant B is obtained from the normalising condition given by

$$\int_0^{\pi} F^*F\sin\theta d\theta=1, \quad \text{or,} \quad \int_{-1}^{1} B^2P_l^m(x)P_l^m(x)dx=1. \tag{18.563}$$

This gives

$$B=\left[\frac{2l+1}{2}\frac{(l-m)!}{(l+m)!}\right]^{1/2} \tag{18.564}$$

Hence the solution of polar equation is

$$F_{lm}(\theta)=\left[\frac{2l+1}{2}\cdot\frac{(l-m)!}{(l+m)!}\right]^{1/2} P_l^m(\cos\theta) \tag{18.565}$$

From this solution we find a term $(l-m)!$ in the numerator. This shows that the solution exists only for those values of the parameter l determined by $(l-m)\geq 0$.

The parameter l is called the *orbital quantum number.*

(c) *Solution of radial equation*: In the radial equation (18.551) we substitute the expression for potential $V(r)$ from equation (18.542) and get

$$\frac{1}{r^2}\frac{d}{dr}\left(r^2\frac{dR}{dr}\right)+\frac{2m_0}{\hbar^2}\left[E+\frac{Ze^2}{r}-\frac{l(l+1)\hbar^2}{2m_0r^2}\right]R=0 \tag{18.566}$$

We introduce a dimensionless variable

$$\rho=\alpha r, \quad \text{i.e.,} \quad r=\rho/\alpha \tag{18.567}$$

Then this equation assumes the form

$$\frac{1}{\rho^2}\frac{d}{d\rho}\left(\rho^2\frac{dR}{d\rho}\right)+\frac{2m_0}{\alpha^2\hbar^2}\left[E+\frac{Ze^2\alpha}{\rho}-\frac{l(l+1)\alpha^2\hbar^2}{2m_0\rho^2}\right]R=0 \tag{18.568}$$

We like to solve this equation for the bound state of the electron when E is negative. With this, we choose α such that

$$-\frac{2m_0E}{\alpha^2\hbar^2}=\frac{1}{4}, \quad \text{i.e.,} \quad \alpha^2=-\frac{8m_0E}{\hbar^2} \tag{18.569}$$

We also substitute

$$\lambda=\frac{2m_0Ze^2}{\alpha\hbar^2}=\frac{2m_0Ze^2}{\hbar^2}\sqrt{-\frac{\hbar^2}{8m_0E}}, \quad \text{i.e.,} \quad \lambda=\frac{Ze^2}{\hbar}\sqrt{-\frac{m_0}{2E}} \tag{18.570}$$

55

The particular choice of α in equation (18.569) is arbitrary and has been taken for convenience. Thus, the equation (18.568) assumes the form

$$\frac{1}{\rho^2}\frac{d}{d\rho}\left(\rho^2\frac{dR}{d\rho}\right)+\left[\frac{\lambda}{\rho}-\frac{1}{4}-\frac{l(l+1)}{\rho^2}\right]R=0$$

$$\text{or,}\quad \frac{d^2R}{d\rho^2}+\frac{2}{\rho}\frac{dR}{d\rho}+\left[\frac{\lambda}{\rho}-\frac{1}{4}-\frac{l(l+1)}{\rho^2}\right]R=0. \tag{18.571}$$

For $\rho \longrightarrow \infty$, this equation takes the form

$$\frac{d^2R}{d\rho^2}-\frac{1}{4}R=0 \tag{18.572}$$

which has solution

$$R=c_1e^{\rho/2}+c_2e^{-\rho/2} \tag{18.573}$$

We take $c_1=0$, otherwise the solution R will be infinite for $\rho \longrightarrow \infty$.
Thus, the solution of equation (18.572) is (taking $c_2=1$)

$$R(\rho)=e^{-\rho/2} \tag{18.574}$$

Hence, for exact solution of equation (18.571) we assume

$$R(\rho)=e^{-\rho/2}P(\rho) \tag{18.575}$$

where $P(\rho)$ is a finite polynomial in ρ. Now, from equation (18.575), we get

$$\frac{dR}{d\rho}=\left(\frac{dP}{d\rho}-\frac{1}{2}P\right)e^{-\rho/2} \tag{18.576}$$

$$\frac{d^2R}{d\rho^2}=\left(\frac{d^2P}{d\rho^2}-\frac{dP}{d\rho}+\frac{1}{4}P\right)e^{-\rho/2} \tag{18.577}$$

Substituting equations (18.575)–(18.577) into equation (18.571) and arranging similar terms, we get

$$\frac{d^2P}{d\rho^2}+\left(\frac{2}{\rho}-1\right)\frac{dP}{d\rho}+\left[\frac{\lambda-1}{\rho}-\frac{l(l+1)}{\rho^2}\right]P=0 \tag{18.578}$$

This equation has a singularity at $\rho=0$ which is removable and hence can be solved by power series method. Thus we assume

$$P(\rho)=\sum_{k=0}^{\infty}a_k\rho^{k+s} \tag{18.579}$$

Substituting the function $P(\rho)$ and its derivatives from equation (18.579) into equation (18.578), we get

$$\sum_{k=0}^{\infty}(s+k)(s+k-1)a_k\rho^{s+k-2}+\left(\frac{2}{\rho}-1\right)\sum_{k=0}^{\infty}a_k(k+s)\rho^{k+s-1}$$
$$+\left[\frac{\lambda-1}{\rho}-\frac{l(l+1)}{\rho^2}\right]\sum_{k=0}^{\infty}a_k\rho^{s+k}=0$$

Collecting terms with same powers of ρ, we have

$$\sum_{k=0}^{\infty}[(s+k)(s+k+1)-l(l+1)]\,a_k\rho^{s+k-2}-\sum_{k=0}^{\infty}(s+k+1-\lambda)a_k\rho^{s+k-1}=0 \qquad (18.580)$$

Equating to zero the coefficient of lowest power of ρ, i.e., ρ^{s-2} for $k=0$, we get

$$a_0\,[s(s+1)-l(l+1)]=0 \qquad (18.581)$$

Assuming that $a_0 \neq 0$, equation (18.581) gives two values of s, i.e., $s=l$ or $s=-(l+1)$. We choose $s=l$, since for $s=-(l+1)$ the function $R(\rho)$ fails to be finite.

Now, we equate to zero the coefficients of ρ^{s+k-1} and get

$$[(s+k+1)(s+k+2)-l(l+1)]\,a_{k+1}=(s+k+1-\lambda)a_k$$

$$\text{or,}\quad a_{k+1}=\frac{s+k+1-\lambda}{(s+k+1)(s+k+2)-l(l+1)}\cdot a_k \qquad (18.582)$$

Equation (18.582) is the recursion formula for the coefficients in the series in equation (18.579). Putting the value of $s=l$ in equation (18.582) and simplyfying its denominator, we have

$$a_{k+1}=\frac{l+k+1-\lambda}{(k+1)(k+2l+2)}a_k \qquad (18.583)$$

Equation (18.583) shows that, for some value of k, the coefficients $a_{k+1}=0$, and by recursion formula, all other coefficients thereafter vanish. Thus, the series terminates to a finite polynomials, if

$$l+k+1-\lambda=0,\quad \text{i.e.,}\quad \lambda=l+k+1 \qquad (18.584)$$

Since k and l are both integers, we find from equation (18.584) that λ is also an integer. Hence we replace λ by n, so that

$$n=l+k+1 \qquad (18.585)$$

Again, since $k=0,1,2,\ldots$ and $l=0,1,2,\ldots$ we get $n=1,2,3,\ldots$ The maximum value which l can have is

$$\max l=n-(k+1)=n-1$$

so that

$$l=0,1,2,\ldots(n-1) \qquad (18.586)$$

Here k is called the *radial quantum number* and n the *principal quantum number.*

We now return to the solution of equation (18.578) which we assumed as in equation (18.579).

Putting $s=l$, we write in the form

$$P(\rho)=\rho^l L(\rho) \qquad (18.587)$$

where $L(\rho)$ is a polynomial. Substituting this into equation (18.578) and putting $\lambda=n$, we get

$$\rho^2\frac{d^2L}{d\rho^2}+[2\rho(l+1)-\rho^2]\frac{dL}{d\rho}+[l(l-1)+2l-l\rho+(n-1)\rho-l(l+1)]L=0$$

$$\text{or,}\quad \rho\frac{d^2L}{d\rho^2}+[(2l+1)+1-\rho]\frac{dL}{d\rho}+(n-l-1)L=0 \qquad (18.588)$$

This is identical with Associated Laguerre equation

$$\rho\frac{d^2L_q^p}{d\rho^2}+(p+1-\rho)\frac{dL_q^p}{d\rho}+(q-p)L_q^p=0 \tag{18.589}$$

where

$$p=2l+1 \quad \text{and} \quad q=n-l-1+p=n+l. \tag{18.590}$$

Hence the solution of equation (18.588) is

$$L(\rho)=L_{n+l}^{2l+1}(\rho) \tag{18.591}$$

the RHS functions being Laguerr's associated polynomials. Hence, from equation (18.585), (18.587) and (18.591), we get the solution for the radial equation as

$$R(\rho)=Ce^{-\rho/2}\rho^l L_{n+l}^{2l+1}(\rho) \tag{18.592}$$

The constant C is determined from the normalising condition

$$\int_0^\infty R^*(r)R(r)r^2\,dr=1 \tag{18.593}$$

Putting the function equation (18.592) into equation (18.593) and remembering that $\rho=\alpha r$, we can find

$$C=\left[\left(\frac{2Z}{na_0}\right)^3\frac{(n-l-1)!}{2n\{(l+n)!\}^3}\right]^{1/2} \tag{18.594}$$

Hence the radial wave function is

$$R(r)=\left[\left(\frac{2Z}{na_0}\right)^3\frac{(n-l-1)!}{2n\{(l+n)!\}^3}\right]^{1/2}\exp\left(\frac{-Zr}{na_0}\right)\left(\frac{2Zr}{na_0}\right)^l L_{n+l}^{2l+1}\left(\frac{2Zr}{na_0}\right) \tag{18.595}$$

Here a_0 = Bohr radius, $\alpha=\dfrac{2Z}{na_0}$, and the Laguerre's associated polynomials are defined as

$$L_n^k(r)=\frac{d^k}{dr^k}L_n(r) \tag{18.596}$$

where $L_n(r)$ are Laguerre's polynomial defined as

$$L_n(r)=\epsilon^r\frac{d^n}{dr^n}\left(r^n e^{-r}\right) \tag{18.597}$$

The complete wave function for the hydrogen atom is thus written as

$$\psi_{nlm}(r,\theta,\phi)=R_{nl}(r)F_{lm}(\theta)Q_m(\phi) \tag{18.598}$$

where

$$\left.\begin{aligned} n&=1,2,3,\ldots\\ l&=0,1,2,\ldots,(n-1)\\ m&=0,1,2,\ldots,l \end{aligned}\right\} \tag{18.599}$$

The functions $Q_m(\phi)$, $F_{lm}(\theta)$ and $R_{nl}(r)$ are given by equations (18.559), (18.565) and (18.595) respectively.

Now, putting $\lambda = n$ in equation (18.570), we get

$$n = \frac{Ze^2}{\hbar}\sqrt{-\frac{m_0}{2E}} \tag{18.600}$$

which gives

$$E_n = -\left(\frac{m_0 Z^2 e^4}{2\hbar^2}\right)\cdot\frac{1}{n^2} \tag{18.601}$$

This equation gives the values of energy called energy levels of the hydrogen atom. For ground state, $n = 1$, and the energy of hydrogen atom is

$$E_1 = -\frac{Z^2 m_0 e^4}{2\hbar^2} \tag{18.602}$$

Degeneracy of Energy Levels

Equation (18.600) shows that the energy level of the hydrogen atom depends on the principal quantum number n. We have also seen that, for a particular value of n, the value of l can vary from zero to $(n-1)$, and again, for each value of l, m can vary from $-l$ to l. Therefore, for the possible values of l as given in equation (18.599) the total number of values of m is

$$\sum_{l=0}^{n-1}(2l+1) = \frac{2(n)(n-1)}{2} + n = n^2 \tag{18.603}$$

Thus, for $n = 2$, we get four quantum states, for $n = 3$ we get nine and so on.

Degeneracy is a phenomenon where system belonging to the same energy level has a number of independent quantum states. The energy level containing more than one independent quantum states is called a *degenerate energy level.* The total number of independent quantum states is called the *degree of generacy*. In the present case, the degree of generacy is n^2.

EXERCISE

1. A homogeneous spherical shell is made of material of diffusivity h. The inner and outer radii of the shell are a and b respectively. The inner shell is kept at constant temperature T_a and the outer shell is kept at constant temperature T_b. Show that temperature of the shell is given by

$$T(r) = (T_a - T_b)\frac{ab}{b-a}\cdot\frac{1}{r} + \frac{T_b b - T_a a}{b-a}$$

2. A point charge q is in front of a conducting sphere at zero potential. Find the electric field at any point $(0, 0, a)$ outside the sphere of radius R.

3. A grounded sphere of radius a is in presence of an external charge q at a distance d from the centre. Show that, the ratio of the charge on the part of the sphere visible from q to that on the part of the sphere invisible from q is

$$\left(\frac{d+a}{d-a}\right)^{1/2}$$

4. A long uncharged conducting circular cylinder of radius a is placed in a uniform electric field E_0, with its axis perpendicular to the lines of force. Show that, the field induced by the presence of the cylinder is given by the potential

$$\phi = -\left(1 - \frac{a^2}{r^2}\right) E_0 r \cos\theta$$

5. Discuss the vibration of a homogeneous string fixed at ends $x = 0$ and $x = a$ given by the initial conditions as

$$u(x,0) = \frac{4hx(a-x)}{a^2}, \quad \frac{\partial u(x,0)}{\partial t} = 0$$

6. A homogeneous rectangular membrane $(0 \leq x \leq a; 0 \leq y \leq b)$ fixed along the contour vibrates. Find the free vibration law of the membrane, if the initial conditions are

$$u(x,y,0) = xy(a-x)(b-y) \quad \text{and} \quad \left.\frac{\partial u}{\partial t}\right|_{t=0} = 0.$$

7. A homogeneous square membrane of side a with edges fixed performs free vibration. Find its displacement function when the initial conditions are

$$u(x,y,0) = \sin\frac{\pi x}{a}\sin\frac{\pi y}{a} \quad \text{and} \quad \left.\frac{\partial u}{\partial t}\right|_{t=0} = \frac{1}{a}$$

8. Find the temperature function in a homogeneous rod whose lateral surface is thermally insulated. Ends of the rod of length l are maintained at zero temperature and, at $t = 0$, it is given that the temperature distribution is $x(l-x)/l^2$.

9. Find the temperature distribution function inside an infinite circular cylinder of radius R, if its surface is at a given temperature, i.e.,

$$\left. u(r,\theta)\right|_{r=R} = \sin\theta$$

10. Find the temperature distribution function $u(x,t)$ in a long homogeneous rod whose lateral surface is heat insulated. The initial condition being

$$u(x,0) = e^{-x^2}$$

11. Find the stationary temperature distribution in a rectangular plate $(0 \leq x \leq a, 0 \leq y \leq b)$ if the temperature along the boundary are given as

$$\psi(x,0) = \psi(x,b) = 0$$
$$\psi(0,y) = y(b-y)$$
$$\psi(a,y) = \sin\frac{\pi y}{b}.$$

12. A beam carrying a load P at its middle point is simply supported at both ends. Discuss the vibration of the beam, if the load is suddenly withdrawn.

13. A circular plate of radius a has a central hole of radius b. An uniformly distributed load acts along the ring-like zone. Find the deflection function when both the circular edges are clamped.

14. An elliptic plate under uniformly distributed load q is clamped along its boundary defined by

$$\frac{x^2}{a^2} + \frac{y^2}{b^2} - 1 = 0$$

Show that the deflection fnnction is given by

$$w(x,y) = \lambda \left(\frac{x^2}{a^2} + \frac{y^2}{b^2} - 1 \right)^2$$

where λ is a constant to be determined.

ANSWERS

2. Potential $V = -\frac{q}{r}\frac{R}{a} \sum \left(\frac{R^2}{ar} \right)^n P_n(\cos\theta)$.

5. $u(x,t) = \frac{32h}{\pi^3} \sum_{n=0}^{\infty} \frac{1}{(2n+1)^3} \cos\lambda ct \sin\lambda x \quad \text{where } \lambda = (2n+1)\pi/a.$

6. $u(x,y,t) = \frac{64a^2b^2}{\pi^6} \sum_{k=0}^{\infty} \sum_{n=0}^{\infty} \frac{\cos\lambda ct \sin\alpha_k x \sin\beta_n y}{(2k+1)^3(2n+1)^3} \quad \text{where } \alpha_k = \frac{\pi(2k+1)}{a}, \beta_k \frac{\pi(2n+1)}{b}$ and $\lambda = \pi\sqrt{\alpha^2 + \beta^2}$..

8. $u(x,t) = \frac{8}{\pi^3} \sum_{n=0}^{\infty} \frac{1}{(2n+1)^3} e^{-\lambda^2 c^2 t} \sin\lambda x \quad \text{where } \lambda = \frac{\pi}{l}(2n+1).$

9. $u(r,\theta) = \frac{r}{R} \sin\theta.$

10. $u(x,t) = \frac{1}{2c\sqrt{\pi t}} \int_0^{\infty} \exp\left[-\left(2\xi^2 + \frac{(x+\xi)^2 + (x-\xi)^2}{4c^2 t} \right) \right] d\xi.$

11. $u(x,t) = \sum_{k=1}^{\infty} \frac{\sin\lambda y}{\sinh\lambda a} \left[A_k \sinh\lambda(a-x) + B_k \sinh\lambda x \right] \quad \text{where } \lambda = \frac{\pi k}{b}, B_1 = 1, B_k = 0$ for $k \geq 2$ and $A_k = \frac{8b^2}{\pi^3(k^3)}$ for k odd, $A_k = 0$ for k even.

12. $u(x,t) = \frac{2Fl^3}{\pi^4 EI} \sum_{n=1}^{\infty} \frac{\sin(n\pi/2)}{n^4} \sin\lambda x \cos\lambda^2 c^2 t \quad \text{where } \lambda = \frac{n\pi}{l}.$

Chapter 19

Integral Equations

19.1 Introduction

Any equation containing the unknown function under an integral sign is called an *integral equation.* The general type of linear integral equations is of the form

$$y(x) = \phi(x) + \lambda \int_a^b k(x,\xi)y(\xi)d\xi, \tag{19.1}$$

where $k(x,\xi)$ and $\phi(x)$ are known functions. The function $k(x,\xi)$ is defined in the triangle

$$a \le x < b, \quad a \le \xi \le x \tag{19.2}$$

and the function $\phi(x)$ is defined in the interval $[a,b]$.

The function $y(x)$ is unknown and defined in the interval $[a,b]$. The function $k(x,\xi)$ is called the *Kernel* of the equation. The constant λ is a parameter which for different values, creates a family of equations.

We assume that the function $k(x,\xi)$ and $\phi(x)$ are continuous in their respective domains of definition. Then the solutions of the integral equations (19.1) exists and are continuous in $[a,b]$.

Integral equations are classified into three groups depending upon the nature of the kernel $k(x,\xi)$. These are given below.

(1) *Volterra's equations*: Integral equations of the type

$$y(x) = \phi(x) + \lambda \int_a^x k(x,\xi)y(\xi)d\xi \tag{19.3}$$

are known as *Volterra's equation of second kind.* The function $\phi(x)$ is known as *free term* of the equation. *Volterra's equation of first kind* are of the form

$$\int_a^x k(x,\xi)y(\xi)d\xi = \phi(x) \tag{19.4}$$

(2) *Abel's Integral equations*: Integral equations of the type

$$\phi(x) = \int_0^x \frac{y(\xi)d\xi}{(x-\xi)^\alpha}, \quad 0 < \alpha < 1 \tag{19.5}$$

are known as *Abel's integral equations.*

(3) *Fredholm's Integral equations*: Linear Fredholm's integral equations of second kind are of the form

$$y(x) + \int_a^b k(x,\xi)y(\xi)d\xi = \phi(x) \tag{19.6}$$

The linear integral equations of the type

$$\int_a^b k(x,\xi)y(\xi)d\xi = \phi(x) \tag{19.7}$$

are known as Fredholm's equations of first kind. The function $k(x,\xi)$ is assumed to be given in the square $a \le x, \xi \le b$ and the function $\phi(x)$ in $[a,b]$.

Integral equations appear in many physical processes. For example, Volterra's equations of second kind appear in describing the dynamics of various processes. The integral equations are more convenient and powerful tools than differential equations in solving the physical process, since the boundary conditions are built into them.

19.2 Volterra's Equations of Second Kind

We shall now consider different aspects of linear Volterra's integral equations of second kind. On the very outset, we mention that, any linear differential equation with given initial conditions can be reduced to a linear integral equation of Volterra's second kind.

As an example, we consider the differential equation

$$\phi'' - 2\phi' + \phi = 0 \tag{19.8}$$

with the condition $\phi(2) = 1, \phi'(2) = -2$.

We assume that

$$\phi''(x) = y(x) \tag{19.9}$$

Then, integrating equation (19.9) with the use of boundary conditions, we get

$$\phi'(x) = \int_2^x y(u)du + \phi'(2)$$

$$\text{i.e.,}\quad \phi'(x) = -2 + \int_2^x y(u)du \tag{19.10}$$

Integrating once more, we have

$$\begin{aligned}\phi(x) &= \phi(2) - 2\int_2^x ds + \int_2^x ds\int_2^s y(u)du \\ &= 1 - 2(x-2) + \int_2^x (x-u)y(u)du\end{aligned} \tag{19.11}$$

We substitute equations (19.9) to (19.11) into the given differential equation and get

$$y(x) - 2\left[-2 + \int_2^x y(u)du\right] + \left[5 - 2x + \int_2^x (x-u)y(u)du\right] = 0$$

$$\text{or,}\quad y(x) = 2x - 9 + \int_2^x (2 - x + u)y(u)du \tag{19.12}$$

Thus, $y(x) = \phi''(x)$ satisfies the integral equation (19.12). Hence, if $\phi(x)$ is the solution of the differential equation (19.8) under the given initial conditions, then $\phi''(x) = y(x)$ is a solution of the integral equation (19.12) under the same initial conditions, and conversely. We have noticed that the initial conditions are built into the integral equation.

EXAMPLE 19.1. *Reduce the following boundary value problem into an integral equation*

$$\phi''(x) - \phi'(x)\sin x + e^x\phi(x) = x$$

with $\phi(0) = 1$ *and* $\phi'(0) = -1$.

Solution: We substitute

$$\phi''(x) = y(x) \tag{1}$$

Integrating and using boundary condition, we get

$$\phi'(x) = \phi'(0) + \int_0^x y(t)dt = -1 + \int_0^x y(t)dt \tag{2}$$

Integrating once more, we have

$$\phi(x) = \phi(0) - x + \int_0^x du \int_0^u y(t)dt = 1 - x + \int_0^x (x-t)y(t)dt \tag{3}$$

Putting the values of ϕ, ϕ', ϕ'' from equations (1), (2), (3) into the original differential equation, we get

$$y(x) - \sin x\left[-1 + \int_0^x y(t)dt\right] + e^x\left[1 - x + \int_0^x (x-t)y(t)dt\right] = x$$

$$\text{or,}\quad y(x) = x - \sin x - e^x(1-x) + \int_0^x (\sin x - e^x)(x-t)y(t)dt$$

19.3 Solution of Volterra's Equation of Second Kind

A solution of an integral equation is a function which is continuous in the domain of definition of the free term, and which reduces the integral equation to an identity.

We shall discuss the various methods of solving Volterra's linear integral equations of second kind.

A. By Reducing to a Cauchy Problem

We assume that the Kernel $k(x,\xi)$ and the free term $\phi(x)$ have continuous derivatives. Then we can differentiate the original integral equation which would be reduced to an ordinary differential equation with some boundary conditions. Thus we reduce the integral equation to some Cauchy problem which can be solved by known methods. This technique is always effective when the Kernel $k(x,\xi)$ is a polynomial in $(x-\xi)$.

We consider some examples.

EXAMPLE 19.2. *Solve the integral equation*

$$y(x) = 4e^x - \int_0^x (x-\xi)y(\xi)d\xi$$

Solution: We differentiate the given integral equation successively twice, and get

$$y'(x) = 4e^x - \int_0^x y(\xi)d\xi \tag{1}$$

$$y''(x) = 4e^x - y(x) \tag{2}$$

Equation (2) gives

$$y'' + y = 4e^x \tag{3}$$

which is an ordinary differential equation.

Its solution is

$$y = A\cos x + B\sin x + 2e^x \tag{4}$$

The initial conditions are obtained from original integral equation and equation (1) as

$$\text{for } x = 0, \quad y = 4 \quad \text{and} \quad y' = 4 \tag{5}$$

Using the conditions in (5), we find the integration constants A and B as $A = 2, B = 2$. Hence, the solution of the given integral equation is

$$y = 2\cos x + 2\sin x + 2e^x \tag{6}$$

EXAMPLE 19.3. *Solve the integral equation*

$$y(x) = \sin x + \frac{1}{2}\int_0^x (x-\xi)^2 y(\xi)d\xi. \tag{7}$$

Solution: We differentiate the integral equation successively twice and get

$$y'(x) = \cos x + \int_0^x (x-\xi)y(\xi)d\xi \tag{1}$$

$$y''(x) = -\sin x + \int_0^x y(\xi)d\xi \tag{2}$$

$$y'''(x) = -\cos x + y \tag{3}$$

From equation (3), we have

$$y'''(x) - y(x) = -\cos x \tag{4}$$

We assume the solution of homogeneous equation $y''' - y = 0$ as $y = e^{\xi x}$.

Then, for ξ we get the cubic equation $\xi^3 - 1 = 0$ which gives

$$\xi = 1, \quad \xi = -\frac{1}{2} + i\frac{\sqrt{3}}{2}, \quad \xi = -\frac{1}{2} - i\frac{\sqrt{3}}{2}$$

Thus the complimentary function is

$$y_1(x) = Ae^x + Be^{-x/2}(\cos\alpha x + i\sin\alpha x) + Ce^{-x/2}(\cos\alpha x - i\sin\alpha x), \quad \alpha = \frac{\sqrt{3}}{2}$$

The particular integral of equation (4) is assumed as

$$y_2(x) = D_1\cos x + D_2\sin x$$

Substituting this into equation (4), and comparing the coefficients of $\sin x$ and $\cos x$ on both sides, we find

$$D_1 = D_2 = \frac{1}{2}$$

Thus the complete solution of the differential equation (4) is written as

$$\begin{aligned} y(x) &= y_1(x) + y_2(x) \\ &= Ae^x + Be^{-x/2}\left(\cos\frac{\sqrt{3}}{2}x + i\sin\frac{\sqrt{3}}{2}x\right) \\ &\quad + Ce^{-x/2}\left(\cos\frac{\sqrt{3}}{2}x - i\sin\frac{\sqrt{3}}{2}x\right) + \frac{1}{2}(\cos x + \sin x) \end{aligned}$$

Using the initial conditions

$$y(0) = 0 \quad \text{and} \quad y'(0) = 1,$$

we find the unknown constants A, B, C as

$$A = \frac{1}{6}, \quad B = C = \frac{1}{3}$$

Thus, the solution of the given integral equation is

$$y(x) = \frac{1}{6}\left[e^x + 3(\cos x + \sin x) + 4e^{-x}\cos\frac{\sqrt{3}}{2}x\right]$$

Note: The Leibnitz's formula for differentiating an integral is useful here and hence is given below:

If $F(x) = \int_{a(x)}^{b(x)} f(x,t)dt$, then

$$\frac{dF}{dx} = \frac{db}{dx}f(b,x) - \frac{da}{dx}f(a,x) + \int_a^b \frac{\partial f}{\partial x}dt$$

As an example, suppose

$$F(x) = \int_0^x \cos(x-t)y(t)dt$$

Hence, $b(x) = x, a(x) = 0, f(x,t) = \cos(x-t)y(t)$ so that

$$\frac{db}{dx} = 1, \quad \frac{da}{dx} = 0, \quad f(b,x) = f(x,x) = y(x)$$

Thus,

$$\frac{dF}{dx} = y(x) - \int_0^x \sin(x-t)y(t)dt$$

EXAMPLE 19.4. *Reduce the integral equation to an ordinary differential equation and solve it*

$$y(x) = \frac{1}{1+x^2} + \int_0^x \sin(x-\xi)y(\xi)d\xi$$

Solution: We have

$$y(x) = \frac{1}{1+x^2} + \int_0^x \sin(x-\xi)y(\xi)d\xi \tag{1}$$

Differentiating it successively twice, we get

$$y'(x) = \frac{-2x}{(1+x^2)^2} + \int_0^x \cos(x-\xi)y(\xi)d\xi \tag{2}$$

$$y''(x) = \frac{2(x^2-1)}{(1+x^2)^3} + y(x) - \int_0^x \sin(x-\xi)y(\xi)d\xi \tag{3}$$

Eliminating from equations (1) and (3) the integral part, we get

$$y''(x) = \frac{2(x^2-1)}{(1+x^2)^2} + \frac{1}{1+x^2}$$

Integrating, we get

$$y'(x) = \int \frac{2(x^2-1)}{(1+x^2)^3}dx + \int \frac{1}{1+x^2}dx = -\frac{2x}{(1+x^2)^3} + \tan^{-1}x + A$$

Integrating once more, we have

$$\begin{aligned} y(x) &= \int \frac{-2x}{(1+x^2)^3}dx + \int \tan^{-1}x + Ax + B \\ &= \frac{1}{1+x^2} + \left[x\tan^{-1}x - \int \frac{x}{1+x^2}\right] + Ax + B \\ &= \frac{1}{1+x^2} + x\tan^{-1}x - \frac{1}{2}\log(1+x^2) + Ax + B. \end{aligned}$$

The initial conditions are: $y(0) = 1$, $y'(0) = 0$. Hence, $A = B = 0$.
Thus, the required solution is

$$y(x) = \frac{1}{1+x^2} + x\tan^{-1}x - \frac{1}{2}\log(1+x^2)$$

B. Method of Successive Approximation

In this method, we assume an approximate solution $y_0(x)$ of the integral equation

$$y(x) = \phi(x) + \int_a^x k(x,\xi)y(\xi)d\xi \tag{19.13}$$

The subsequent approximations are obtained from

$$y_1(x) = \phi(x) + \int_a^x k(x,\xi)y_0(\xi)d\xi$$

$$y_2(x) = \phi(x) + \int_a^x k(x,\xi)y_1(\xi)d\xi$$

$$\cdots\cdots\cdots\cdots\cdots\cdots\cdots\cdots$$

$$\cdots\cdots\cdots\cdots\cdots\cdots\cdots\cdots$$

$$y_n(x) = \phi(x) + \int_a^x k(x.\xi)y_{n-1}(\xi)d\xi \tag{19.14}$$

The function $y_n(x)$ converges to the solution of the integral equation as $n \longrightarrow \infty$. The rate of this convergence depends much on the choice of the zeroth approximation $y_0(x)$ which we generally take to be the free term of the integral equation.

EXAMPLE 19.5. *Solve, by the method of successive approximations, the integral equation*

$$y(x) = \frac{x^2}{2} + x - \int_0^x y(\xi)d\xi$$

Solution: We take the zeroth approximation as

$$y_0(x) = \frac{x^2}{2} + x$$

Then, the successive approximations are

$$y_1 = x + \frac{x^2}{2} - \int_0^x (\xi + \xi^2/2)d\xi = x + \frac{x^2}{2} - \left(\frac{x^2}{2} + \frac{x^3}{6}\right) = x - \frac{x^3}{6}$$

$$y_2 = x + \frac{x^2}{2} - \int_0^x (\xi - \xi^3/6)\, d\xi = x + \frac{x^2}{2} - \left(\frac{x^2}{2} - \frac{x^4}{24}\right) = x + \frac{x^4}{24}$$

....

$$y_n = x + (-1)^n \frac{x^{2n+1}}{(2n+1)!}$$

Hence as $n \longrightarrow \infty$, $y_n = y = x$ which is the solution.

EXAMPLE 19.6. *Solve, by the method of successive approximations, the integral equation*

$$y(x) = 1 + \int_0^x xy(\xi)d\xi$$

Solution: We assume $y_0(x) = 1$.

Then

$$y_1(x) = 1 + \int_0^x x d\xi = 1 + x^2$$

$$y_2(x) = 1 + \int_0^x x\left[1 + \xi^2\right] d\xi = 1 + x\left(x + \frac{x^3}{3}\right) = 1 + x^2 + \frac{x^4}{3}$$

$$y_3(x) = 1 + \int_0^x x\left(1 + \xi^2 + \frac{\xi^4}{3}\right) d\xi = 1 + x\left(x + \frac{x^3}{3} + \frac{x^5}{15}\right)$$

$$\text{i.e.,}\quad y_3(x) = 1 + x\left(x + \frac{x^3}{3.1} + \frac{x^5}{5.3.1}\right)$$

Thus

$$y_n(x) = 1 + x\left(\frac{x}{1} + \frac{x^3}{1.3} + \frac{x^5}{1.3.5} + \cdots\right) = 1 + x\sum_{k=0}^{\infty} \frac{x^{2k+1}}{(2k+1)!!}$$

Hence, we get the solution in terms of an infinite series.

EXAMPLE 19.7. *Solve, by the method of successive approximation, the integral equation*

$$y(x) = 1 + \int_0^x (x-\xi)y(\xi)d\xi$$

Solution: We take $y_0 = 1$.

Then

$$y_1(x) = 1 + \int_0^x (x-\xi)d\xi = 1 + \frac{x^2}{2}$$

$$y_2(x) = 1 + \int_0^x (x-\xi)\left(1+\frac{\xi^2}{2}\right)d\xi = 1 + \frac{x^2}{2} + \frac{x^4}{24}$$

Thus

$$y_n(x) = 1 + \frac{x^2}{2!} + \frac{x^4}{4!} + \cdots$$

Hence, as $n \longrightarrow \infty$, $y_n(x) \longrightarrow \cosh x$, so that the solution of the integral equation is $y = \cosh x$.

C. Method of Laplace Transforms

We consider Volterra's linear integral equation of second kind

$$y(x) = \phi(x) + \int_0^x k(x-\xi)y(\xi)d\xi \tag{19.15}$$

where the Kernel $k(x,\xi) = k(x-\xi)$, i.e., it depends only on the difference of the arguments. Such types of integral equations are called *convolution equation*. These equations are conveniently solved by Laplace transform method.

Let

$$\left.\begin{aligned} L\{y(x)\} &= f(s) \\ L\{\phi(x)\} &= g(s) \\ L\{k(x)\} &= k(s) \end{aligned}\right\} \tag{19.16}$$

Then

$$L\left[\int_0^x k(x-\xi)y(\xi)d\xi\right] = k(s)f(s) \tag{19.17}$$

by the convolution or Faltung theorem. Hence taking Laplace transform of both sides of equation (19.15), we have

$$f(s) = g(s) + k(s)f(s) \quad \text{or,} \quad f(s) = \frac{g(s)}{1-k(s)} \tag{19.18}$$

Taking inverse Laplace transform, we get the solution of the integral equation as

$$y(x) = L^{-1}\left\{\frac{g(s)}{1-k(s)}\right\} \tag{19.19}$$

EXAMPLE 19.8. *Solve by Laplace transform method, the integral equation*

$$y(x) = \sin x + \int_0^x \cos(x-\xi)y(\xi)d\xi$$

Solution: We have

$$y(x) = \sin x + \int_0^x \cos(x-\xi)y(\xi)d\xi \tag{1}$$

Let

$$L\{y(x)\} = f(s) \tag{2}$$

Taking Laplace transform of both sides of equation (1), we get

$$f(s) = \frac{1}{s^2+1} + \frac{s}{s^2+1}\cdot f(s) \quad \text{or,} \quad f(s)\left[1 - \frac{s}{s^2+1}\right] = \frac{1}{s^2+1}$$

$$\text{or,} \quad f(s) = \frac{1}{s^2-s+1} = \frac{1}{(s-1/2)^2+(\sqrt{3}/2)^2} \tag{3}$$

Taking inverse Laplace transform of equation (3), we get

$$y(x) = \frac{2}{\sqrt{3}}e^{x/2}\sin\left(\frac{\sqrt{3}}{2}x\right)$$

EXAMPLE 19.9. *Solve the following integral equation by Laplace transform method*

$$y(x) = e^{2x} + \int_0^x (x-\xi)e^{x-\xi}y(\xi)d\xi$$

Solution: We have

$$y(x) = e^{2x} + \int_0^x (x-\xi)e^{x-\xi}y(\xi)d\xi \tag{1}$$

Here, $\phi(x) = e^{2x}, k(x) = xe^x$.
Hence

$$L\{\phi(x)\} = \frac{1}{s-2}, \quad L\{k(x)\} = \frac{1}{(s-1)^2} \tag{2}$$

Thus, taking Laplace transform of both sides of equation (1), we get

$$f(s) = \frac{1}{s-2} + \frac{1}{(s-1)^2}f(s) \tag{3}$$

where $f(s) = L\{y(x)\}$.
From equation (3), we get

$$f(s)\left[1 - \frac{1}{(s-1)^2}\right] = \frac{1}{s-2} \quad \text{or,} \quad f(s) = \frac{(s-1)^2}{s(s-2)^2} \tag{4}$$

Now, let

$$\frac{(s-1)^2}{s(s-2)^2} = \frac{A}{s} + \frac{B}{s-2} + \frac{C}{(s-2)^2}$$

Then

$$A = \left[\frac{(s-1)^2}{(s-2)(s-2)}\right]_{s=0} = \frac{1}{4}$$

$$B = \frac{d}{ds}\left[\frac{(s-1)^2}{s}\right]_{s=2} = \frac{3}{4}$$

$$C = \left[\frac{(s-1)^2}{s}\right]_{s=2} = \frac{1}{2}$$

Thus

$$f(s) = \frac{1}{4} \cdot \frac{1}{s} + \frac{3}{4} \cdot \frac{1}{s-2} + \frac{1}{2} \cdot \frac{1}{(s-2)^2} \tag{5}$$

Taking inverse Laplace transform of (5), we get

$$y(x) = \frac{1}{4} + \frac{3}{4}e^{2x} + \frac{1}{2}xe^{2x} = \frac{1}{4}\left[1 + e^{2x}(2x+3)\right]$$

EXAMPLE 19.10. *Solve the integral equation*

$$y(x) = 1 + \int_0^x \cos(x-\xi)\sin(x-\xi)y(\xi)d\xi$$

by the Laplace transform method.

Solution: We write the given integral equation in the form

$$y(x) = 1 + \frac{1}{2}\int_0^x \sin 2(x-\xi)y(\xi)d\xi \tag{1}$$

Taking Laplace transform of both sides, we get

$$f(s) = \frac{1}{s} + \frac{1}{2}\frac{2}{s^2+4} \cdot f(s) \tag{2}$$

where

$$f(s) = L\{y(x)\} \tag{3}$$

From (2) we find

$$f(s)\left[1 - \frac{1}{s^2+4}\right] = \frac{1}{s}, \quad \text{or,} \quad f(s) = \frac{s^2+4}{s(s^2+3)} = \frac{4}{3} \cdot \frac{1}{s} - \frac{1}{3} \cdot \frac{s}{s^2+3} \tag{4}$$

Taking inverse Laplace transform, we get

$$y = L^{-1}\{f(s)\} = \frac{4}{3} - \frac{1}{3}\cos\sqrt{3}x$$

19.4 Solution of Volterra's Equation of First Kind

We consider the Volterra's linear integral equation of first kind in the form

$$\int_a^x k(x,\xi)y(\xi)d\xi = \phi(x) \tag{19.20}$$

From equation (19.20) we find that a necessary condition for the existence of the solution is

$$\phi(a) = 0 \tag{19.21}$$

We can solve these types of equations by two methods. One direct method is the technique of Laplace transform. Another way is to convert the Volterra's equation of first kind to that of second kind.

For this purpose, we assume that

(i) the derivatives $\phi'(x)$ and $\frac{\partial k}{\partial x}$ are continuous, and

(ii) $k(x,x) \neq 0$ everywhere in the interval $[a,b]$.

Then, differentiating the equation (19.20), we have

$$k(x,x)y(x) + \int_0^x \frac{\partial k}{\partial x} y(\xi)d\xi = \phi'(x)$$

$$\text{or,} \quad y(x) + \frac{1}{k(x,x)} \int_0^x \frac{\partial k}{\partial x} y(\xi)d\xi = \frac{\phi'(x)}{k(x,x)} \tag{19.22}$$

which is a Volterra's equation of second kind. It has, therefore, a unique continuous solution.

EXAMPLE 19.11. *Solve the following Volterra's equations of first kind by reducing to that of second kind*

$$\int_0^x \sin(x-\xi)y(\xi)d\xi = 1 - \cos x$$

Solution: We have

$$\int_0^x \sin(x-\xi)y(\xi)d\xi = 1 - \cos x \tag{1}$$

Differentiating successively twice, we get

$$\int_0^x \cos(x-\xi)y(\xi)d\xi = \sin x$$

$$\text{and} \quad y(x) - \int_0^x \sin(x-\xi)y(\xi)d\xi = \cos x \tag{2}$$

Thus, we have reduced the equation of the first kind to that of the second kind.

In order to solve equation (2), we take Laplace transform of both sides. Then, we get

$$f(s) - \frac{1}{s^2+1} \cdot f(s) = \frac{s}{s^2+1}$$

which gives

$$f(s) = \frac{1}{s}$$

Taking inverse Laplace transform, we get $y(x) = 1$ as the solution.

Note: In this case, we could solve by directly taking Laplace transform of equation (1). Then we shall get

$$\frac{1}{s^2+1} \cdot f(s) = \frac{1}{s} - \frac{s}{s^2+1}$$

which gives $f(s) = 1/s$ and hence, $y(x) = 1$.

EXAMPLE 19.12. *Solve the integral equation*

$$\int_0^x (1 - x^2 + \xi^2)y(\xi)d\xi = \frac{x^2}{2}$$

by reducing it to Volterra's equation of second type.

Solution: We have

$$\int_0^x (1 - x^2 + \xi^2) y(\xi) d\xi = \frac{x^2}{2} \tag{1}$$

Differentiating equation (1), we get

$$y(x) - \int_0^x 2xy(\xi) d\xi = x \tag{2}$$

Let

$$\psi(x) = \int_0^x y(\xi) d\xi \tag{3}$$

Using equation (3), we get from equation (2)

$$\psi'(x) - 2x\psi(x) = x \text{ with } \psi(0) = 0$$

Hence $\psi(x) = \frac{1}{2}e^{x^2} - \frac{1}{2}$ so that $y(x) = \psi'(x) = xe^{x^2}$.

EXAMPLE 19.13. *Solve the integral equation*

$$\int_0^x (x - \xi) e^{x-\xi} y(\xi) d\xi = \frac{1}{2} e^{2x} - xe^x - \frac{1}{2}$$

by Laplace transform method.

Solution: We have

$$\int_0^x (x - \xi) e^{x-\xi} y(\xi) d\xi = \frac{1}{2} e^{2x} - xe^x - \frac{1}{2} \tag{1}$$

Taking Laplace transform of both sides of equation (1)

$$\frac{1}{(s-1)^2} f(s) = \frac{1}{2} \cdot \frac{1}{s-2} - \frac{1}{(s-1)^2} - \frac{1}{2s}$$

$$\text{or,} \quad f(s) = \frac{1}{2} \frac{(s-1)^2}{s-2} - 1 - \frac{(s-1)^2}{2s} = \frac{1}{2}\left[s + \frac{1}{s-2}\right] - 1 - \left[\frac{s}{2} - 1 + \frac{1}{2s}\right]$$

$$= \frac{1}{2}\left(\frac{1}{s-2} - \frac{1}{s}\right)$$

Taking inverse Laplace transform, we get

$$y(x) = L^{-1}\{f(s)\} = \tfrac{1}{2}\left(e^{2x} - 1\right)$$

19.5 Method of Resolvent Kernels

Consider the Volterra integral equation of second kind given by equation (19.3). In order to solve by successive approximation, we set $y_0(x) = \phi(x)$ as the zeroth approximation. Then

$$y_1(x) = \phi(x) + \lambda \int_a^x k(x, \xi) \phi(\xi) d\xi = \phi(x) + \lambda \int_a^x k_1(x, \xi) \phi(\xi) d\xi \tag{19.23}$$

where

$$k_1(x, \xi) = k(x, \xi). \tag{19.24}$$

Similarly,

$$y_2(x) = \phi(x) + \lambda \int_a^x k_1(x,\xi)\left[\phi(\xi) + \lambda \int_a^t k_1(t,\xi)\phi(\xi)dt\right] d\xi$$

$$= \phi(x) + \lambda \int_a^x k_1(x,\xi)\phi(\xi)d\xi + \lambda^2 \int_a^x k_2(x,\xi)\phi(\xi)d\xi \tag{19.25}$$

where

$$k_2(x,\xi) = \int_\xi^x k(x,t)k_1(t,\xi)dt \tag{19.26}$$

Proceeding in the same manner, we can obtain the expression for nth approximation in the form

$$y_n(x) = \phi(x) + \sum_{i=1}^{\infty} \lambda^i \int_a^x k_i(x,\xi)\phi(\xi)d\xi \tag{19.27}$$

where

$$k_i(x,\xi) = \int_\xi^x k(x,t)k_{i-1}(t,\xi)dt \tag{19.28}$$

with $i = 2, 3, \ldots$

The Kernels $k_i(x,\xi)$ are called *iterated Kernels.*

Consider the series

$$H(x,\xi;\lambda) = \sum_{i=1}^{\infty} \lambda^{i-1} k_i(x,\xi) \tag{19.29}$$

The Kernel $k(x,\xi)$ is assumed to be continuous in its domain of definition. Then, for any fixed value of λ, the series in equation (19.29) converges uniformly in $a \le x \le b$ and $a \le \xi \le x$ as $i \longrightarrow \infty$.

Hence as $i \longrightarrow \infty$, equation (19.27) can be written as

$$y(x) = \phi(x) + \lambda \int_a^x H(x,\xi;\lambda)\phi(\xi)d\xi \tag{19.30}$$

which is the solution of the integral equation.

The function $H(x,\xi;\lambda)$ is called the *resolvent Kernel.*

EXAMPLE 19.14. *Solve the integral equation*

$$y(x) = \phi(x) + \lambda \int_0^x e^{x-\xi} y(\xi)d\xi$$

Solution: Here $k(x,\xi) = k_1(x,\xi) = e^{x-\xi} = e^x \cdot e^{-\xi}$.

Then

$$k_2(x,\xi) = \int_\xi^x e^{x-t}e^{t-\xi}dt = e^{x-\xi}\int_\xi^x dt = e^{x-\xi}(x-\xi)$$

$$k_3(x,\xi) = \int_\xi^x e^{x-t}e^{t-\xi}(t-\xi)dt = e^{x-\xi}\int_\xi^x (t-\xi)dt = \frac{1}{2}e^{x-\xi}(x-\xi)^2$$

$$k_4(x,\xi) = \frac{1}{2}\int_\xi^x e^{x-t}e^{t-\xi}(t-\xi)^2dt = \frac{1}{2}e^{x-\xi}\int_\xi^x (t-\xi)^2dt = \frac{1}{6}e^{x-\xi}(x-\xi)^3$$

In general

$$k_n(x,\xi) = e^{x-\xi}\frac{(x-\xi)^{n-1}}{(n-1)!}, \quad n = 1, 2, \ldots$$

Therefore

$$H(x,\xi;\lambda) = \sum_{n=1}^{\infty} \lambda^{n-1} k_n(x,\xi) = e^{x-\xi}\frac{[\lambda(x-\xi)]^{n-1}}{(n-1)!} = e^{(1+\lambda)(x-\xi)}$$

Hence, the solution of the given equation is

$$y(x) = \phi(x) + \lambda \int_0^x e^{(1+\lambda)(x-\xi)}\phi(\xi)d\xi$$

In particular, if $\lambda = 2$ and $\phi(x) = \sin x$, then

$$\begin{aligned} y(x) &= \sin x + 2\int_0^x e^{3x}e^{-3\xi}\sin\xi d\xi = \sin x + 2e^{3x}\int_0^x e^{-3\xi}\sin\xi d\xi \\ &= \sin x + 2e^{3x}\left[e^{-3\xi}\frac{-3\sin\xi - \cos\xi}{9+1}\right]_0^x \\ &= \sin x + 2e^{3x}\left[\frac{e^{-3x}}{10}(-3\sin x - \cos x) + \frac{1}{10}\right] \\ &= \frac{e^{3x}}{5} - \frac{\cos x}{5} + \frac{2}{5}\sin x. \end{aligned}$$

19.6 A Special Type of Kernels

If the Kernel $k(x,\xi)$ of the Volterra equation of second kind has the form

$$k(x,\xi) = \frac{A(x)}{A(\xi)} \tag{19.31}$$

then the integral equation can be reduced to an ordinary first order linear differential equation.

In this case, we substitute

$$\left.\begin{aligned} y_1(x) &= \frac{y(x)}{A(x)} \\ \phi_1(x) &= \frac{\phi(x)}{A(x)} \end{aligned}\right\} \tag{19.32}$$

Then the integral equation assumes the form

$$y_1(x) = \phi_1(x) + \lambda\int_a^x y_1(\xi)d\xi \tag{19.33}$$

Defining

$$y_2(x) = \int_a^x y_1(\xi)d\xi \tag{19.34}$$

equation (19.33) can be written as

$$y_2'(x) - \lambda y_2(x) = \phi_1(x) \tag{19.35}$$

Since $y_2(a) = 0$, we can write the solution of equation (19.35) as

$$y_2(x) = e^{\lambda x} \int_a^x e^{-\lambda \xi} \phi_1(\xi) d\xi \tag{19.36}$$

Hence

$$y_1(x) = y_2'(x) = \phi_1(x) + \lambda \int_a^x e^{\lambda(x-\xi)} \phi_1(\xi) d\xi \tag{19.37}$$

$$\text{and} \quad y(x) = \phi(x) + \lambda \int_a^x e^{\lambda(x-\xi)} k(x,\xi) \phi(\xi) d\xi \tag{19.38}$$

The result of the example 19.14 can easily be verified with equation (19.38).

EXAMPLE 19.15. *Solve the integral equation*

$$y(x) = e^{-x} \cos x - \int_0^x \cos x \, e^{-(x-\xi)} y(\xi) d\xi$$

Solution: Here $\phi(x) = e^{-x} \cos x$, $\lambda = -1$, $k(x,\xi) = e^{-x} \cos x / e^{-\xi}$.
Hence by equation (19.38)

$$\begin{aligned}
y(x) &= e^{-x} \cos x - \int_0^x e^{-(x-\xi)} \left(e^{-x} \cos x e^{\xi}\right) e^{-\xi} \cos \xi d\xi \\
&= e^{-x} \cos x - e^{-2x} \cos x \int_0^x e^{\xi} \cos \xi d\xi \\
&= e^{-x} \cos x - e^{-2x} \cos x \left[\frac{\cos \xi + \sin \xi}{2}\right]_0^x \\
&= e^{-x} \cos x - e^{-2x} \cos x \tfrac{1}{2} (\cos x + \sin x - 1) \\
&= e^{-x} \cos x \left[1 - \frac{e^{-x}}{2} (\cos x + \sin x - 1)\right]
\end{aligned}$$

19.7 Abel Equation

The generalised Abel equations

$$\phi(x) = \int_0^x \frac{y(\xi) d\xi}{(x-\xi)^{\alpha}}, \quad 0 < \alpha < 1 \tag{19.39}$$

are Volterra integral equations of first kind. The Kernel $k(x,\xi)$ is a function of $(x-\xi)$. That is

$$k(x,\xi) = k(x-\xi) \tag{19.40}$$

Let there be two functions $P(x,\xi)$ and $Q(x,\xi)$ of the type

$$P(x,\xi) = p(x-\xi) \quad \text{and} \quad Q(x,\xi) = q(x-\xi) \tag{19.41}$$

Then Volterra's composition of first kind is defined by

$$\overset{*}{P}\overset{*}{Q} = \int_{\xi}^x P(x,t) Q(t,\xi) dt \tag{19.42}$$

If we put $t = \xi + z$, then we have

$$\overset{*}{P}\overset{*}{Q} = \int_0^{x-\xi} p(x-\xi-z)q(z)dz \quad \text{or,} \quad \overset{*}{P}\overset{*}{Q} = \psi(x-\xi) \tag{19.43}$$

where

$$\psi(u) = \int_0^u p(u-z)q(z)dz = p * q \tag{19.44}$$

The above idea is used in solving equations (19.39).
We take the function $(x-\xi)^{\alpha-1}$ for Volterra's compositon with the Kernel $(x-\xi)^{-\alpha}$.
Then

$$\int_\xi^x \frac{dt}{(x-t)^\alpha (t-\xi)^{1-\alpha}} = \int_0^1 \frac{dz}{(1-z)^\alpha z^{1-\alpha}} = \Gamma(\alpha)\Gamma(1-\alpha) = \frac{\pi}{\sin(\alpha\pi)} \tag{19.45}$$

Here we have used the substitution

$$t = \xi + (x-\xi)z \tag{19.46}$$

By taking composition of both sides of equation (19.39), we get

$$\frac{\pi}{\sin(\alpha\pi)} \int_0^x y(\xi)d\xi = \int_0^x \frac{\phi(\xi)}{(x-\xi)^{x-\alpha}} d\xi \tag{19.47}$$

From equation (19.47), we obtain by differentiating

$$y(x) = \frac{\sin(\alpha\pi)}{\pi}\frac{d}{dx}\left[\int_0^x \frac{\phi(\xi)}{(x-\xi)^{1-\alpha}} d\xi\right] = \frac{\sin(\alpha\pi)}{\pi}\left[\frac{\phi(0)}{x^{1-\alpha}} + \int_0^x \frac{\phi'(\xi)}{(x-\xi)^{1-\alpha}} d\xi\right] \tag{19.48}$$

For the particular integral equation

$$\phi(x) = \int_0^x \frac{y(\xi)d\xi}{\sqrt{x-\xi}} \tag{19.49}$$

we have $\alpha = 1/2$. Hence the solution is obtained directly from equation (19.48) in the form

$$y(x) = \frac{1}{\pi}\left[\frac{\phi(0)}{\sqrt{x}} + \int_0^x \frac{\phi'(\xi)}{\sqrt{x-\xi}} d\xi\right] \tag{19.50}$$

It is to be noted that Abel's equation (19.49) is encountered in particle dynamics when it is required to determine a curve, taken to be smooth, along which the time of descent of a heavy particle is a given function of its initial position.

19.8 Fredholm's Integral Equations

We consider the family of Fredholm's integral equations of second kind given by

$$y(x) - \lambda \int_a^b k(x,\xi)y(\xi)d\xi = \phi(x) \tag{19.51}$$

where λ is a numerical parameter, real or complex.

The equation (19.51) has a unique solution under the condition

$$\int_a^b \int_a^b | k(x,\xi) |^2 dx d\xi < \frac{1}{\lambda^2} \tag{19.52}$$

We can find the solution of equation (19.51) by the method of successive approximations in the same way as before. If the condition in equation (19.52) is satisfied, the nth approximation to the solution will converge to the exact solution when $n \longrightarrow \infty$.

EXAMPLE 19.16. *Solve the equation*

$$y(x) - \int_0^1 x\xi y(\xi) d\xi = 2x.$$

Solution: We have

$$y(x) - \int_0^1 x\xi y(\xi) d\xi = 2x$$

Thus, $\lambda = 1, k(x,\xi) = x\xi$.
Hence

$$\int_0^1 \int_0^1 | k(x,\xi) |^2 dx d\xi = \int_0^1 \int_0^1 x^2 \xi^2 dx d\xi = \frac{1}{9}$$

Thus the condition equation (19.52) is satisfied. We assume the zeroth approximation as

$$y_0(x) = 2x$$

Then

$$y_1(x) = 2x + \int_0^1 (x\xi) 2\xi d\xi = 2x\left(1 + \frac{1}{3}\right)$$

$$y_2(x) = 2x + \int_0^1 (x\xi) 2\xi \left(1 + \frac{1}{3}\right) d\xi = 2x\left(1 + \frac{1}{3} + \frac{1}{9}\right)$$

Thus,

$$y_n(x) = 2x + 2x\left(\frac{1}{3} + \frac{1}{3^2} + \cdots\right)$$

As $n \longrightarrow \infty$

$$y_n(x) = y(x) = 2x + 2x\frac{1/3}{1 - \frac{1}{3}} = 3x.$$

EXAMPLE 19.17. *Solve the equation*

$$y(x) = 2\sin x + \frac{1}{2\pi}\int_0^\pi \sin x \, y(\xi) \xi d\xi$$

Solution: Here $\lambda = \frac{1}{2\pi}, k(x,\xi) = (\sin x)\xi$.
Hence

$$\int_0^\pi \int_0^\pi | k(x,\xi) |^2 dx d\xi = \int_0^\pi \int_0^\pi \xi^2 (\sin x)^2 dx d\xi = \frac{\xi^3}{3}\bigg|_0^\pi \frac{1}{2}\left(\pi - \frac{\sin 2x}{2}\right)\bigg|_0^\pi = \frac{\pi^4}{6}$$

Thus the condition in equation (19.52) is satisfied.
We assume

$$y_0(x) = 2\sin x$$

Then

$$y_1(x) = 2\sin x + \frac{1}{2\pi}\int_0^{\pi} (\sin x)\xi(2\sin\xi)d\xi = 2\sin x + \sin x$$

$$y_2(x) = 2\sin x + \frac{1}{2\pi}\int_0^{\pi} (\sin x)\xi(2\sin\xi + \sin\xi)d\xi = 2\sin x + \sin x + \frac{1}{2}\sin x$$

Thus

$$y_n(x) = 2\sin x + \sin x\left(1 + \frac{1}{2} + \frac{1}{2^2} + \cdots\right)$$

Hence, as $n \longrightarrow \infty, y_n(x) \longrightarrow y(x)$.
The solution of the equation is

$$y(x) = 2\sin x + \sin x \cdot \frac{1}{1-\dfrac{1}{2}} = 4\sin x.$$

19.9 Fredholm's Singular Equations

If the Kernel $k(x,\xi)$ in the equation (19.51) is of the form

$$k(x,\xi) = \sum_{i=1}^{n} P_i(x)Q_i(\xi) \tag{19.53}$$

we say that the Fredholm's equation is singular. In such a case, the integral equation can be reduced to a system of linear algebraic equations. Let us see how.

We write the integral equation (19.51) as ($\lambda = 1$)

$$y(x) - \int_a^b \left[\sum_{i=1}^{n} P_i(x)Q_i(\xi)\right] y(\xi)d\xi = \phi(x) \tag{19.54}$$

With the substitution

$$R_i = \int_a^b Q_i(\xi)y(\xi) \quad (i = 1, 2, \ldots, n) \tag{19.55}$$

we can write equation (19.54) as

$$y(x) - \sum_{i=1}^{n} R_i P_i(x) = \phi(x) \tag{19.56}$$

We multiply both sides of equation (19.56) by $Q_i(x)$, $i = 1, 2, \ldots, n$ and integrate between the limits from a to b. Then we have

$$\int_a^b y(x)Q_i(x)dx - \sum_{i=1}^{n} R_i\left[\int_a^b P_i(x)Q_i(x)dx\right] = \int_a^b \phi(x)Q_i(x)dx \tag{19.57}$$

$$\text{or,} \quad R_i - \sum_{j=1}^{n} A_{ij} R_i = F_i, \quad i = 1, 2, \ldots, n \tag{19.58}$$

where

$$A_{ij} = \int_a^b Q_i(x) P_j(x) \tag{19.59}$$

$$\text{and} \quad F_i = \int_a^b \phi(x) Q_i(x) dx \tag{19.60}$$

Solving the system of equations (19.58), we find R_i and then get the solution of **the integral** equation from equation (19.56) as

$$y(x) = \phi(x) + \sum_{i=1}^{n} R_i P_i(x) \tag{19.61}$$

EXAMPLE 19.18. *Solve the integral equation*

$$y(x) - \int_0^1 (2x - \xi) y(\xi) d\xi = \cos 2\pi x$$

Solution: Here $k(x, \xi) = 2x - \xi$. Thus

$$P_1(x) = 2x, \quad P_2(x) = -1$$
$$Q_1(x) = 1, \quad Q_2(x) = x$$

Hence,

$$A_{11} = \int_0^1 P_1(x) Q_1(x) = \int_0^1 2x\, dx = 1$$
$$A_{21} = \int_0^1 P_1(x) Q_2(x) dx = \int_0^1 (2x) x\, dx = \frac{2}{3}$$
$$A_{12} = \int_0^1 P_2(x) Q_1(x) dx = \int_0^1 (-1) \cdot 1 dx = -1$$
$$A_{22} = \int_0^1 P_2(x) Q_2(x) dx = \int_0^1 (-1) \cdot x\, dx = -\frac{1}{2}$$
$$F_1 = \int_0^1 \phi(x) Q_1(x) dx = \int_0^1 \cos 2\pi x \cdot 1 dx = 0$$
$$F_2 = \int_0^1 \phi(x) Q_2(x) dx = \int_0^1 \cos 2\pi x \cdot x\, dx = 0$$

Thus, the system of equations (19.58) in this case become

$$R_i - \sum_{j=1}^{2} A_{ij} R_i = 0, \quad i = 1, 2, \ldots$$

Hence $R_i = 0$, i.e., $R_1 = R_2 = 0$.
Thus, the solution of the integral equation is

$$y(x) = \phi(x) + \sum R_i P_i(x) = \cos 2\pi x.$$

EXAMPLE 19.19. *Solve the integral equation*

$$y(x) - \frac{1}{\pi}\int_0^{2\pi} \cos x \sin \xi \, y(\xi) d\xi = \sin x.$$

Solution: Here $k(x,\xi) = \dfrac{\cos x}{\pi} \sin \xi$,

$$\phi(x) = \sin x.$$

Hence

$$P_1(x) = \frac{\cos x}{\pi}, \quad Q_1(\xi) = \sin \xi$$

$$\therefore \quad A_{11} = \int_0^{2\pi} P_1(x) Q_1(x) dx = \frac{1}{\pi}\int_0^{2\pi} \cos x \sin x \, dx = 0$$

$$F_1 = \int_0^{2n} \phi(x) Q_1(x) dx = \int_0^{2\pi} \sin x \sin x \, dx = \pi$$

Hence by equation (19.58), we have

$$R_1 - A_{11} R_1 = F_1, \quad \text{i.e.,} \quad R_1 = \pi$$

Thus, the solution of the integral equation is

$$y(x) = \phi(x) + R_1 P_1(x) = \sin x + \pi \frac{\cos x}{\pi} = \sin x + \cos x$$

EXAMPLE 19.20. *Solve the integral equation*

$$y(x) - \int_0^{\pi/4} \tan \xi \, y(\xi) d\xi = \cos^2 x$$

Solution: Here $P_1(x) = 1, Q_1(\xi) = \tan \xi, \phi(x) = \cos^2 x$.
Hence

$$A_{11} = \int_0^{\pi/4} P_1(x) Q_1(x) dx = \int_0^{\pi/4} \tan x \, dx = -\log(\cos x)\Big|_0^{\pi/4} = \log \sqrt{2}$$

$$F_1 = \int_0^{\pi/4} \phi(x) Q_1(x) dx = \int_0^{\pi/4} \cos^2 x \cdot \tan x \, dx = \frac{1}{2}\int_0^{\pi/4} \sin 2x \, dx = -\frac{1}{4}\cos 2x\Big|_0^{\pi/4} = \frac{1}{4}$$

Hence, by equation (19.58), we have

$$R_1 - A_{11} R_1 = F_1, \quad \text{i.e.,} \quad R_1 - \log \sqrt{2} R_1 = \frac{1}{4},$$

$$\text{or,} \quad R_1 = \frac{1}{4(1 - \log \sqrt{2})} = \frac{1}{2(2 - \log 2)}$$

Thus, the solution of the integral equation is

$$y(x) = \phi(x) + R_1 P_1(x) = \cos^2 x + \frac{1}{2(2 - \log 2)}.$$

Note: In such simple problems, we can proceed directly. That is writing

$$R = \int_0^{\pi/4} \tan\xi \, y(\xi) d\xi \tag{1}$$

we get the integral equation as

$$y(x) - R = \cos^2 x \tag{2}$$

Multiplying both sides of (1) by $\tan x$ and integrating, we find R, and so on.

19.10 Eigenvalues and Eigenfunctions

We consider Fredholm's family of homogeneous integral equations

$$y(x) - \lambda \int_a^b k(x,\xi) y(\xi) d\xi = 0. \tag{19.62}$$

The particular value or values of λ for which the equations (19.62) have got non-trival solutions are called *eigenvalues of the Kernel.* The number $\mu = 1/\lambda$ is called the *eigenvalue of the integral equation*. The non-trival solutions corresponding to the eigenvalues λ are called eigenfunctions.

If the Kernel $k(x,\xi)$ is singular, that is

$$k(x,\xi) = \sum_{i=1}^{n} P_i(x) Q_i(t)$$

then we can proceed in the same manner as in the previous article to find the eigenvalues and corresponding eigenfunctions. With the substitution as in equation (19.55), we write the equations (19.62) in the form

$$y(x) = \lambda \sum_{i=1}^{n} R_i P_i(x) \tag{19.63}$$

Multiplying both sides of equation (19.63) by $Q_i(x)$, $i = 1, 2, \ldots, n$ and integrating to the desired limits, we get

$$R_i - \lambda \sum_{j=1}^{n} A_{ij} R_i = 0 \tag{19.64}$$

which is a system of linear algebraic equations for finding λ. After getting λ, we get the eigenfunctions from equation (19.63).

We shall consider some examples.

EXAMPLE 19.21. *Find the eigenvalues and eigenfunctions of the integral equation*

$$y(x) - \lambda \int_0^1 (1 - x^2) y(\xi) d\xi = 0$$

Solution: Here the Kernel is $k(x,\xi) = (1 - x^2)$.

Hence $P_1(x) = 1 - x^2, Q_1(\xi) = 1$.

Thus

$$A_{11} = \int_0^1 P_1(x) Q_1(x) dx = \int_0^1 (1 - x^2) dx = \frac{2}{3}$$

By equation (19.64), we have

$$R_1 - \lambda A_{11} R_1 = 0, \quad \text{i.e.,} \quad 1 - \frac{2}{3}\lambda = 0 \quad [R_1 \neq 0]$$

Hence $\lambda = 3/2$.
The solution corresponding to this value of λ to the integral equation is, by equation (19.63)

$$y(x) = \lambda R_1 P_1(x) = c(1 - x^2), \quad c = \text{constant}.$$

EXAMPLE 19.22. *Find the eigenvalues and eigenfunctions of the integral equation*

$$y(x) - \lambda \int_0^{\pi} x \sin \xi y(\xi) d\xi = 0$$

Solution: Here, $P_1(x) = x, Q_1(x) = \sin x$.
Thus

$$A_{11} = \int_0^{\pi} P_1(x) Q_1(x) dx = \int_0^{\pi} x \sin x \, dx = \pi$$

By equation (19.64), we have

$$R_1(1 - \lambda A_{11}) = 0, \quad \text{i.e.,} \quad \lambda = \frac{1}{A_{11}} = \frac{1}{\pi}$$

The solution of the integral equation, i.e., the eigenfunction corresponding to this eigenvalue is

$$y(x) = \lambda R_1 P_1(x) = C\,|\,x\,|, \quad C = \text{constant}.$$

EXAMPLE 19.23. *Find eigenvalues and eigenfunctions of the integral equation*

$$y(x) - \lambda \int_0^1 \left(xe^{\xi} + 2\xi\right) y(\xi) d\xi = 0$$

Solution: Here, $P_1(x) = x$, $Q_1(\xi) = e^{\xi}$, $P_2(x) = 2$, $Q_2(\xi) = \xi$.
Hence

$$A_{11} = \int_0^1 P_1(x) Q_1(x) dx = \int_0^1 xe^x dx = 1$$

$$A_{21} = \int_0^1 Q_2(x) P_1(x) dx = \int_0^1 x^2 dx = \frac{1}{3}$$

$$A_{12} = \int_0^1 Q_1(x) P_2(x) dx = \int_0^1 2e^x dx = 2(e - 1)$$

$$A_{22} = \int_0^1 Q_2(x) P_2(x) dx = \int_0^1 2x \, dx = 1$$

By equation (19.64), we have (putting $\mu = 1/\lambda$)

$$\begin{bmatrix} \mu - A_{11} & -A_{12} \\ -A_{21} & \mu - A_{22} \end{bmatrix} \begin{bmatrix} R_1 \\ R_2 \end{bmatrix} = 0 \tag{1}$$

The characteristic equation is given by

$$\begin{vmatrix} \mu-1 & -2(e-1) \\ -\frac{1}{3} & \mu-1 \end{vmatrix} = 0$$

$$\text{or,} \quad (\mu-1)^2 = \frac{2}{3}(e-1), \quad \text{i.e.,} \quad \mu_{1,2} = 1 \pm \sqrt{\frac{2}{3}(e-1)} \tag{2}$$

By using equation (2), we get from equation (1)

$$\begin{bmatrix} \pm\sqrt{\alpha} & -2(e-1) \\ -\frac{1}{3} & \pm\sqrt{\alpha} \end{bmatrix} \begin{bmatrix} R_1 \\ R_2 \end{bmatrix} = 0 \quad \text{or,} \quad \begin{pmatrix} R_1 \\ R_2 \end{pmatrix} = C \begin{pmatrix} \pm 3\sqrt{\alpha} \\ 1 \end{pmatrix}$$

where

$$\alpha = \frac{2}{3}(e-1)$$

Hence, the corresponding eigenfunctions are

$$y_1(x) = \lambda_1 \left[R_1 P_1(x) + R_2 P_2(x)\right] = \frac{1}{1+\sqrt{\alpha}} \cdot C\left[3\sqrt{\alpha}\, x + 2\right] = C_1 \left[\sqrt{\frac{2}{3}(e-1)}\, x + \frac{2}{3}\right]$$

$$y_2(x) = \frac{1}{1-\sqrt{\alpha}}(-C)\left[3\sqrt{\alpha}x - 2\right] = C_2 \left[\sqrt{\frac{2}{3}(e-1)}\, x - \frac{2}{3}\right]$$

EXAMPLE 19.24. *Find the eigenvalues and eigenfunctions of the integral equation*

$$y(x) - \lambda \int_0^{\pi} \sin(x+\xi) y(\xi) d\xi = 0$$

Solution: Here $k(x,\xi) = \sin(x+\xi) = \sin x \cos\xi + \cos x \sin\xi$.

Hence

$$P_1(x) = \sin x, \quad Q_1(\xi) = \cos\xi$$
$$P_2(x) = \cos x, \quad Q_2(\xi) = \sin\xi$$

Then

$$A_{11} = \int_0^{\pi} Q_1(x) P_1(x) dx = \int_0^{\pi} \cos x \sin x\, dx = 0$$

$$A_{12} = \int_0^{\pi} Q_1(x) P_2(x) dx = \int_0^{\pi} \cos^2 x\, dx = \pi$$

$$A_{21} = \int_0^{\pi} Q_2(x) P_1(x) dx = \int_0^{\pi} \sin^2 x\, dx = \pi$$

$$A_{22} = \int_0^{\pi} Q_2(x) P_2(x) dx = \int_0^{\pi} \sin x \cos x\, dx = 0$$

By equation (19.64), we get putting $\mu = 1/\lambda$

$$\begin{bmatrix} \mu - A_{11} & -A_{12} \\ -A_{21} & \mu - A_{22} \end{bmatrix} \begin{bmatrix} R_1 \\ R_2 \end{bmatrix} = 0 \tag{1}$$

Hence, the characteristic equation is

$$\begin{vmatrix} \mu & -\pi \\ -\pi & \mu \end{vmatrix} = 0, \quad \text{i.e.,} \quad \mu^2 - \pi^2 = 0$$

which gives $\mu_{1,2} = \pm\pi$.

Using these values of $\mu_{1,2}$, we get from equation (1)

$$\begin{bmatrix} \pm\pi & -\pi \\ -\pi & \pm\pi \end{bmatrix} \begin{bmatrix} R_1 \\ R_2 \end{bmatrix} = 0, \quad \text{i.e.,} \quad \begin{bmatrix} R_1 \\ R_2 \end{bmatrix} = C \begin{bmatrix} 1 \\ \pm 1 \end{bmatrix}$$

Thus, the corresponding eigenfunctions are

$$y_{1,2}(x) = \lambda_{1,2}\left[R_1 P_1(x) + R_2 P_2(x)\right] = C\lambda_{1,2}(\sin x \pm \cos x) = C_1(\sin x \pm \cos x)$$

19.11 Fredholm's Theorems

Following Fredholm's theorems are given without any proof.

1. For real Kernels, Fredholm's non-homogeneous equation (19.51) has a unique solution for any function $\phi(x)$ when λ is not an eigenvalue of the Kernel $k(x, \xi)$.

If λ is an eigenvalue, then the non-homogeneous equation (19.51) has a solution if and only if

$$\int_a^b \phi(x) y(x) dx = 0$$

for every solution $y(x)$ of equation (19.51) with $k(x, \xi) \longrightarrow k(\xi, x)$.

2. Fredholm's homogneous equation

$$y(x) = \lambda \int_a^b k(x, \xi) y(\xi) d\xi$$

has at least one non-trivial solution for some value of λ which is an eigenvalue of the Kernel $k(x, \xi)$.

In this case, the solution $y(x)$ is the eigenfunction of $k(x, \xi)$ corresponding to the eigenvalue λ.

EXERCISE

1. Reduce the following boundary value problems into integral equations

   (a) $\phi'' + \phi = x^2, \phi(0) = 1, \phi'(0) = 0;$

   (b) $\phi''' + x\phi = e^x, \phi(0) = 1, \phi'(0) = \phi''(0) = 0;$

2. Solve the integral equations by reducing to differential equations

   (a) $y(x) = e^{-x}\cos x - \displaystyle\int_0^x \cos x e^{-(x-\xi)} y(\xi) d\xi;$

   (b) $y(x) = x - 1 + \displaystyle\int_0^x (x - \xi) y(\xi) d\xi;$

(c) $y(x) = \cosh x - \int_0^x \sinh(x - \xi) y(\xi) d\xi.$

3. Solve the integral equations by the method of successive approximations

(a) $y(x) = 1 - x^2 + \int_0^x x y(\xi) d\xi;$

(b) $y(x) = 1 - \int_0^x (x - \xi) y(\xi) d\xi;$

(c) $y(x) = 1 + \int_0^x \xi^2 y(\xi) d\xi.$

4. Solve the integral equations by Laplace transform method

(a) $y(x) = \frac{x^2}{2} + \int_0^x (x - \xi) y(\xi) d\xi;$

(b) $y(x) = 1 + \int_0^x \cosh(x - \xi) y(\xi) d\xi;$

(c) $y(x) = e^{2x} + \int_0^x (x - \xi) e^{x-\xi} y(\xi) d\xi.$

5. Solve the following Volterra's equations of first kind by reducing them to second kind

(a) $\int_0^x e^{x-\xi} y(\xi) d\xi = \frac{x^2}{2};$

(b) $\int_0^x (x - \xi)^2 y(\xi) d\xi = x^3;$

(c) $\int_0^x (2 + x^2 - \xi^2) y(\xi) d\xi = x^2.$

6. Solve the following Volterra's convolution equations of first kind by Laplace transform

(a) $\int_0^x \cos(x - \xi) y(\xi) d\xi = x \sin x;$

(b) $\int_0^x e^{x-\xi} \cos(x - \xi) y(\xi) d\xi = x e^x.$

7. Solve the following Fredholm's equations of second kind by method of successive approximation

(a) $y(x) + \frac{1}{\pi} \int_0^\pi \cos^2 \xi y(\xi) d\xi = 1;$

(b) $y(x) - \frac{1}{2} \int_0^1 y(\xi) d\xi = \sin \pi x;$

(c) $y(x) + \frac{1}{2\pi} \int_0^{2\pi} [\cos(x + \xi) + \cos(x - \xi)] y(\xi) d\xi = \cos x.$

8. Solve the following Fredholm's equations of second kind with singular Kernel

(a) $y(x) - \int_{-\pi}^\pi \left[\frac{1}{\pi} \sin x \sin \xi + \xi \right] y(\xi) d\xi = \sin 2x;$

(b) $y(x) - \int_{-1}^{1} (x\xi + x^2) y(\xi) d\xi = 0.$

9. Find eigenvalues and eigenfunctions of the following integral equations

(a) $y(x) - \lambda \int_0^1 (1 + 2x)\xi y(\xi) d\xi = 0;$

(b) $y(x) - \lambda \int_0^{\pi} \cos x \cos \xi y(\xi) d\xi = 0;$

10. Solve the integral equation $e^{-\alpha} = \int_0^{\infty} f(x) \cos \alpha x \, dx.$

ANSWERS

1. (a) $y(x) = x^2 - 1 - \int_0^x (x - \xi) y(\xi) d\xi;$ (b) $y(x) = e^x - x - \frac{1}{2} \int_0^x x(x - \xi)^2 y(\xi) d\xi.$

2. (a) $y(x) = \cos x \exp[-(x + \sin x)];$
(b) $y(x) = -e^{-x};$
(c) $y(x) = 1.$

3. (a) $y(x) = 1;$
(b) $y(x) = \sin x;$
(c) $y(x) = \frac{x^3}{e^3}.$

4. (a) $y(x) = \cosh x - 1;$
(b) $y(x) = 1 + \frac{2}{\sqrt{5}} e^{x/2} \sinh \frac{\sqrt{5}}{2} x;$
(c) $y(x) = \frac{1}{4} e^{2x}(2x + 3) + \frac{1}{4}.$

5. (a) $y(x) = x - \frac{x^2}{2};$
(b) $y(x) = 3;$
(c) $y(x) = x \exp\left(-x^2/2\right).$

6. (a) $y(x) = 2 \sin x;$
(b) $y(x) = e^x \left(1 + \frac{x^2}{2}\right);$

7. (a) $y(x) = \frac{2}{3};$
(b) $y(x) = \sin \pi x + \frac{2}{\pi};$
(c) $y(x) = \frac{2}{3} \cos x.$

8. (a) $y(x) = \sin 2x + \lambda \left(2 + \frac{\sin x}{\pi}\right) - \pi, \lambda = \text{constant};$
(b) $y(x) = 0.$

9. (a) $\lambda = \dfrac{6}{7}, y(x) = A(1+2x), A = \text{constant};$ (b) $\lambda = \dfrac{2}{\pi}, y(x) = A\cos x, A = \text{constant}.$

10. $f(x) = \dfrac{2}{\pi}\dfrac{1}{(1+x^2)}.$

Chapter 20

Special Integrals

In mathematical physics, we come across many special integrals each of which plays a very important role in solving specific problems. We here intend to discuss some of these special integrals.

20.1 Error Integrals

These integrals are defined as

$$\phi(x) = \frac{2}{\sqrt{\pi}} \int_0^x e^{-t^2}\, dt \tag{20.1}$$

and are used quite frequently in probability theory and statistics. The function $\phi(x)$ is called *error function* and is denoted as $erf(x)$.

These integrals can be expressed in terms of incomplete gamma functions. For this, we substitute

$$t^2 = y \tag{20.2}$$

Then

$$\phi(x) = \frac{2}{\sqrt{\pi}} \int_0^{x^2} \frac{1}{2} e^{-y} y^{-1/2}\, dy = \frac{1}{\sqrt{\pi}} \gamma\left(\frac{1}{2}, x^2\right) \quad \text{[by equation (7.133)]} \tag{20.3}$$

If we replace x by $-x$ in equation (20.1), we have

$$\phi(-x) = \frac{2}{\sqrt{\pi}} \int_0^{-x} e^{-t^2}\, dt = -\frac{2}{\sqrt{\pi}} \int_0^x e^{-t^2}\, dt = -\phi(x) \tag{20.4}$$

Now, using equation (7.138), the incomplete gamma function can be expanded in power series (putting $m = 1/2$ and replacing x by x^2). Thus we get

$$\phi(x) = \frac{2}{\sqrt{\pi}} \left(x - \frac{x^3}{3} + \frac{x^5}{5(2!)} - \cdots \right) \tag{20.5}$$

The error integral can also be expanded in terms of confluent hypergeometric functions. We have

$$\begin{aligned} \phi(x) &= \frac{2}{\sqrt{\pi}} \int_0^x e^{-t^2}\, dt = \frac{2}{\sqrt{\pi}} \int_0^x \left[\sum_{k=0}^{\infty} \frac{(-1)^k}{k!} t^{2k} \right] dt \\ &= \frac{2}{\sqrt{\pi}} \sum_{k=0}^{\infty} \frac{(-1)^k}{k!} \frac{x^{2k+1}}{2k+1} = \frac{2x}{\sqrt{\pi}} \sum_{k=0}^{\infty} \frac{1}{k!} \frac{(-x^2)^k}{2k+1} \end{aligned} \tag{20.6}$$

Now

$$\frac{1}{2k+1} = \frac{(2k-1)(2k-3)\cdots 3\cdot 1}{(2k+1)(2k-1)(2k-3)\cdots 3\cdot 1} = \frac{\frac{1}{2}\cdot\frac{3}{2}\cdot\frac{5}{2}\cdots\frac{2k-1}{2}}{\frac{3}{2}\cdot\frac{5}{2}\cdot\frac{7}{2}\cdots\frac{2k+1}{2}}$$

$$= \frac{\frac{1}{2}\left(\frac{1}{2}+1\right)\left(\frac{1}{2}+2\right)\cdots\left(\frac{1}{2}+k-1\right)}{\frac{3}{2}\left(\frac{3}{2}+1\right)\cdots\left(\frac{3}{2}+k-1\right)} \tag{20.7}$$

Thus, in view of equation (20.7), we write the equation (20.6) as

$$\phi(x) = \frac{2x}{\sqrt{\pi}}\sum_{k=0}^{\infty}\frac{\left(\frac{1}{2}\right)_k}{(k!)\left(\frac{3}{2}\right)_k}(-x^2)^k = \frac{2x}{\sqrt{\pi}}F\left(\frac{1}{2},\frac{3}{2};-x^2\right) \tag{20.8}$$

There is another type of error integral which is defined as

$$\psi(x) = \frac{2}{\sqrt{\pi}}\int_x^{\infty} e^{-t^2}\,dt \tag{20.9}$$

This function is called complimentary error function and is denoted as $erfc(x)$.
With the substitution as in equation (20.2), the integral in equation (20.9) becomes

$$\psi(x) = \frac{1}{\sqrt{\pi}}\int_x^{\infty} e^{-y}y^{-1/2}\,dy = \frac{1}{\sqrt{\pi}}\Gamma\left(\frac{1}{2},x^2\right) \tag{20.10}$$

Now

$$\phi(x) = \frac{2}{\sqrt{\pi}}\int_0^x e^{-t^2}\,dt = \frac{2}{\sqrt{\pi}}\int_0^{\infty} e^{-t^2}\,dt - \frac{2}{\sqrt{\pi}}\int_x^{\infty} e^{-t^2}\,dt = \frac{2}{\sqrt{\pi}}I - \psi(x) \tag{20.11}$$

where

$$I = \int_0^{\infty} e^{-t^2}\,dt \tag{20.12}$$

Let us now evaluate the integral in equation (20.12).
The integral in equation (20.12) can be taken as

$$I = \int_0^{\infty} e^{-x^2}\,dx = \int_0^{\infty} e^{-y^2}\,dy$$

Thus

$$I^2 = \int_0^{\infty} e^{-x^2-y^2}\,dx\,dy \tag{20.13}$$

We consider the integral equation (20.13) within the region of a circle $0 \le r \le R$. Thus, using polar coordinates

$$I^2 = \int_0^{2\pi}\left[\int_0^R e^{-r^2}r\,dr\right]d\theta = \int_0^{2\pi}\left(\frac{1}{2}-\frac{1}{2}e^{-R^2}\right)d\theta = \pi\left(1-e^{-R^2}\right) \tag{20.14}$$

Now the circle of radius R lies completely within the square $-R \leq x \leq R, -R \leq y \leq R$, which in turn lies within the circle of radius $2R$. The integrand $e^{-x^2-y^2}$ being positive everywhere in xy-plane, we can write

$$\pi\left(1-e^{-R^2}\right) \leq \int_{-R}^{R}\int_{-R}^{R} e^{-x^2-y^2}\,dx\,dy \leq \pi\left(1-e^{-4R^2}\right)$$

$$\text{i.e.,}\quad \pi\left(1-e^{-R^2}\right) \leq \left(\int_{-R}^{R} e^{-x^2}\,dx\right)^2 \leq \pi\left(1-e^{-4R^2}\right)$$

Proceeding to the limit $R \longrightarrow \infty$, we get

$$\int_{-\infty}^{\infty} e^{-x^2}\,dx = \sqrt{\pi} \quad \text{or,} \quad \int_{0}^{\infty} e^{-x^2}\,dx = \frac{\sqrt{\pi}}{2} \tag{20.15}$$

[since the integrand is an even function]

Substituting the value of the integral equation (20.15) in the equation (20.11), we get

$$\phi(x) = 1 - \psi(x) \tag{20.16}$$

20.2 Sine Integrals

The sine integrals are defined by

$$Si(x) = \int_0^x \frac{\sin t}{t}\,dt \tag{20.17}$$

For $x > 0$,

$$Si(-x) = -Si(x) \tag{20.18}$$

The Taylor series expansion for $\sin x$ is

$$\sin x = x - \frac{x^3}{3!} + \frac{x^5}{5!} - \cdots \tag{20.19}$$

Using equation (20.19), the integral in equation (20.17) is expressed as

$$Si(x) = x - \frac{x^3}{3.3!} + \frac{x^5}{5.5!} - \cdots \tag{20.20}$$

The sine integral can also be expressed in terms of confluent hypergeometric functions. We have

$$Si(x) = \int_0^x \frac{\sin t}{t}\,dt = \int_0^x \frac{e^{it}-e^{-it}}{2it}\,dt = \int_0^x \frac{e^{it}-1}{2it}\,dt - \int_0^x \frac{e^{-it}-1}{2it}\,dt$$

Putting $t = ip$ in the first integral and $t = -ip$ in the second, we get

$$Si(x) = \frac{1}{2i}\left[\int_0^{ix} \frac{e^p-1}{p}\,dp - \int_0^{-ix} \frac{e^p-1}{p}\,dp\right] \tag{20.21}$$

Now, from equation (12.59), we have

$$F(1,2;x) = \sum_{k=0}^{\infty} \frac{x^k}{(k+1)!} = \frac{e^x-1}{x} \tag{20.22}$$

Hence

$$\int_0^x \frac{e^t - 1}{t} dt = \int_0^x F(1, 2; t) dt = \lim_{\alpha \longrightarrow 0} \int_0^x F(1 + \alpha, 2; t) dt$$

Using relation equation (12.65), we get

$$\int_0^x \frac{e^t - 1}{t} dt = \lim_{\alpha \longrightarrow 0} \int_0^x \frac{1}{\alpha} \frac{dF}{dt}(\alpha, 1; t) dt = \lim_{\alpha \longrightarrow 0} \frac{F(\alpha, 1; x)}{\alpha - 0} = \left[\frac{\partial F(\alpha, 1; x)}{\partial \alpha}\right]_{\alpha=0} \tag{20.23}$$

In view of equation (20.23), we write equation (20.21) in the form

$$Si(x) = \frac{1}{2i}\left[\frac{\partial F(\alpha, 1; ix)}{\partial \alpha} - \frac{\partial F(\alpha, 1; -ix)}{\partial \alpha}\right]_{\alpha=0} \tag{20.24}$$

20.3 Cosine Integral

The cosine integral is defined as

$$Ci(x) = \gamma + \log x + \int_0^x \frac{\cos t - 1}{t} dt \tag{20.25}$$

where $\gamma = 0.5772$ is Euler's constant.

The Taylor's series expansion for $\cos x$ is

$$\cos x = 1 - \frac{x^2}{2!} + \frac{x^4}{4!} - \ldots \tag{20.26}$$

Thus, the cosine integral in equation (20.25) is expressed in power series as

$$Ci(x) = \gamma + \log x + \left(-\frac{x^2}{2.2!} + \frac{x^4}{4.4!} - \cdots\right) \tag{20.27}$$

For $x > 0$,

$$Ci(-x) = Ci(x) - i\pi \tag{20.28}$$

Now

$$\begin{aligned}\int_0^x \frac{\cos t - 1}{t} dt &= \int_0^x \frac{e^{it} - 1}{2t} + \int_0^x \frac{e^{-it} - 1}{2t} dt \\ &= \frac{1}{2}\left[\int_0^{ix} \frac{e^p - 1}{p} dp + \int_0^{-ix} \frac{e^{-p} - 1}{p} dp\right] \\ &= \frac{1}{2}\left[\frac{\partial F(\alpha, 1; ix)}{\partial \alpha} + \frac{\partial F(\alpha, 1; -ix)}{\partial \alpha}\right]_{\alpha=0}\end{aligned} \tag{20.29}$$

Thus

$$Ci(x) = \gamma + \log x + \frac{1}{2}\left[\frac{\partial F(\alpha, 1; ix)}{\partial \alpha} + \frac{\partial F(\alpha, 1; -ix)}{\partial \alpha}\right]_{\alpha=0} \tag{20.30}$$

20.4 Fresnel Integrals

The sine and cosine Fresnel integrals are defined as

$$C(x) = \int_0^x \cos\left(\frac{\pi}{2}t^2\right) dt \tag{20.31}$$

$$\text{and} \quad S(x) = \int_0^x \sin\left(\frac{\pi}{2}t^2\right) dt \tag{20.32}$$

For small values of x, these integrals are evaluated as

$$C(x) = x - \left(\frac{\pi}{2}\right)^2 \frac{x^5}{5.2!} + \left(\frac{\pi}{2}\right)^4 \frac{x^9}{9.4!} + \cdots \tag{20.33}$$

$$\text{and} \quad S(x) = \left(\frac{\pi}{2}\right) \frac{x^3}{3.1!} - \left(\frac{\pi}{2}\right)^3 \frac{x^7}{7.3!} + \left(\frac{\pi}{2}\right)^5 \frac{x^{11}}{11.5!} - \cdots \tag{20.34}$$

The series in equation (20.33) can be written as

$$C(x) = \sum_{k=0}^{\infty} (-1)^k \left(\frac{\pi}{2}\right)^{2k} \frac{x^{4k+1}}{(4k+1)(2k!)} \tag{20.35}$$

Similarly, the series in equation (20.34) can be written as

$$S(x) = \sum_{k=0}^{\infty} (-1)^{k+1} \left(\frac{\pi}{2}\right)^{2k-1} \frac{x^{4k-1}}{(4k-1)(2k-1)!} \tag{20.36}$$

For large values of x, both the above series diverge. For $x \geq 1.5$, we use the following expressions

$$C(x) \sim \frac{1}{2} + \frac{1}{\pi x} \sin\left(\frac{\pi}{2}x^2\right) \tag{20.37}$$

$$\text{and} \quad S(x) \sim \frac{1}{2} - \frac{1}{\pi x} \cos\left(\frac{\pi}{2}x^2\right) \tag{20.38}$$

These integrals can also be expressed in terms of Bessel's function of odd-half orders. For this purpose, we substitute

$$\frac{\pi}{2}t^2 = y \tag{20.39}$$

Then

$$C(z) = \frac{1}{2} \int_0^z \sqrt{\frac{2}{\pi y}} \cos y \, dy = \frac{1}{2} \int_0^z J_{-1/2}(y) dy \quad \text{[ref. equation (12.93)]} \tag{20.40}$$

Now, from equation (12.103), we have

$$2J'_\nu(x) = J_{\nu-1}(x) - J_{\nu+1}(x) \tag{20.41}$$

Putting $\nu = \frac{1}{2}, \frac{5}{2}, \frac{9}{2}, \cdots$ we get by successive use of equation (20.41)

$$\begin{aligned} J_{-1/2}(y) &= 2J'_{1/2}(y) + J_{3/2} = 2J'_{1/2}(y) + 2J'_{5/2}(y) + J_{7/2}(y) \\ &= 2J'_{1/2}(y) + 2J'_{5/2}(y) + 2J'_{9/2}(y) + J_{11/2}(y) \\ &\qquad \cdots\cdots\cdots\cdots\cdots\cdots\cdots\cdots \\ &\qquad \cdots\cdots\cdots\cdots\cdots\cdots\cdots\cdots \end{aligned} \tag{20.42}$$

Thus, from equation (20.40) and (20.42) we have

$$C(z) = \frac{1}{2}\int_0^z J_{-1/2}(y)dy = J_{1/2}(z) + J_{5/2}(z) + J_{9/2}(z) + \cdots \tag{20.43}$$

Similarly, we can find that

$$S(z) = J_{3/2}(z) + J_{7/2}(z) + J_{11/2}(z) + \cdots \tag{20.44}$$

The series in equations (20.43) and (20.44) are convergent.
We now evaluate the following Fresnel integrals

$$F_1 = \int_{-\infty}^{\infty} \sin(t^2)dt \tag{20.45}$$

$$\text{and} \quad F_2 = \int_{-\infty}^{\infty} \cos(t^2)dt \tag{20.46}$$

With the substitution

$$y = t^2 \tag{20.47}$$

the integrals in equations (20.45) and (20.46) are reduced to

$$F_1 = \int_0^{\infty} \frac{\sin y}{\sqrt{y}}\,dy \tag{20.48}$$

$$\text{and} \quad F_2 = \int_0^{\infty} \frac{\cos y}{\sqrt{y}}\,dy \tag{20.49}$$

Let us now consider the integral

$$I = \frac{2}{\sqrt{\pi}}\int_0^{\infty} e^{-x^2 t}\,dx \tag{20.50}$$

With the substitution $x = y/\sqrt{t}$, this integral becomes

$$I = \frac{2}{\sqrt{\pi}} \cdot \frac{1}{\sqrt{t}}\int_0^{\infty} e^{-y^2}dy = \frac{1}{\sqrt{t}} \quad \text{[using equation (20.15)]} \tag{20.51}$$

Hence

$$\frac{1}{\sqrt{t}} = \frac{2}{\sqrt{\pi}}\int_0^{\infty} e^{-x^2 t}\,dx \tag{20.52}$$

Using equation (20.52), integrals in equations (20.48) and (20.49) are written as

$$F_1 = \frac{2}{\sqrt{\pi}}\int_0^{\infty} dy \int_0^{\infty} e^{-x^2 y}\sin y\,dx \tag{20.53}$$

$$\text{and} \quad F_2 = \frac{2}{\sqrt{\pi}}\int_0^{\infty} dy \int_0^{\infty} e^{-x^2 y}\cos y\,dx \tag{20.54}$$

Changing the order of integration

$$F_1 = \frac{2}{\sqrt{\pi}}\int_0^{\infty} dx \int_0^{\infty} e^{-x^2 y}\sin y\,dy \tag{20.55}$$

$$\text{and} \quad F_2 = \frac{2}{\sqrt{\pi}}\int_0^{\infty} dx \int_0^{\infty} e^{-x^2 y}\cos y\,dy \tag{20.56}$$

The interior integrals can be evaluated by parts so that

$$F_1 = \frac{2}{\sqrt{\pi}} \int_0^\infty \frac{dx}{1+x^4} \tag{20.57}$$

and $$F_2 = \frac{2}{\sqrt{\pi}} \int_0^\infty \frac{x^2}{1+x^4}\, dx \tag{20.58}$$

With the substitution $x = 1/z$, we get

$$F_2 = \frac{2}{\sqrt{\pi}} \int_0^\infty \frac{dz}{1+z^4} \tag{20.59}$$

which is same as F_1. Thus

$$F_1 = F_2 = \frac{2}{\sqrt{\pi}} \int_0^\infty \frac{dz}{1+z^4} \tag{20.60}$$

To evaluate the integral in equation (20.60), we note that

$$\begin{aligned} \frac{1}{1+z^4} &= \frac{1}{2\sqrt{2}} \left[\frac{z+\sqrt{2}}{z^2+\sqrt{2}z+1} - \frac{z-\sqrt{2}}{z^2-\sqrt{2}z+1} \right] \\ &= \frac{1}{4\sqrt{2}} \left[\frac{(2z+\sqrt{2})+\sqrt{2}}{z^2+\sqrt{2}z+1} - \frac{(2z-\sqrt{2})-\sqrt{2}}{z^2-\sqrt{2}z+1} \right] \end{aligned}$$

Thus

$$\int \frac{dz}{1+z^4} = \frac{1}{4\sqrt{2}} \left[\log \frac{z^2+\sqrt{2}z+1}{z^2-2\sqrt{2}+1} \right] + \frac{1}{2\sqrt{2}} \left[\tan^{-1}\left(\sqrt{2}z+1\right) + \tan^{-1}\left(\sqrt{2}z-1\right) \right] \tag{20.61}$$

Using equation (20.61), we get from equation (20.60)

$$F_1 = F_2 = \sqrt{\frac{\pi}{2}} \tag{20.62}$$

20.5 Liouville Integral

Liouville integral is defined as

$$L(\lambda) = \int_0^\infty \int_0^\infty \exp\left[-\left(x + y + \frac{\lambda^3}{xy} \right) \right] x^{\frac{1}{3}-1} \cdot y^{\frac{2}{3}-1}\, dx\, dy \tag{20.63}$$

where the parameter λ is positive.

Differentiating with respect to the parameter λ, we get

$$\frac{dL}{d\lambda} = -3\lambda^2 \int_0^\infty \int_0^\infty \exp\left[-\left(x + y + \frac{\lambda^3}{xy} \right) \right] x^{\frac{1}{3}-1} \cdot y^{\frac{2}{3}-1} \frac{dx\, dy}{xy} \tag{20.64}$$

We substitute

$$z = \frac{\lambda^3}{xy} \quad \text{(considering } y = \text{constant)}$$

so that

$$\frac{dx}{x} = -\frac{dz}{z} \tag{20.65}$$

Then equation (20.64) becomes

$$\frac{dL}{d\lambda} = -3\int_0^\infty \int_0^\infty \exp\left[-\left(y+z+\frac{\lambda^3}{yz}\right)\right] y^{\frac{1}{3}-1} \cdot z^{\frac{2}{3}-1}\, dydz = -3L.$$

Hence, integrating we have

$$L(\lambda) = Ce^{-3\lambda}, \quad C = \text{constant} \tag{20.66}$$

Thus, for $\lambda = 0$,

$$L(0) = C. \tag{20.67}$$

From equation (20.63), we have for $\lambda = 0$

$$\begin{aligned} L(0) &= \int_0^\infty \int_0^\infty \exp\left[-(x+y)\right] x^{\frac{1}{3}-1} y^{\frac{2}{3}-1}\, dx\, dy \\ &= \int_0^\infty e^{-x} x^{\frac{1}{3}-1}\, dx \int_0^\infty e^{-y} y^{\frac{2}{3}-1}\, dy \\ &= \Gamma\left(\frac{1}{3}\right)\Gamma\left(\frac{2}{3}\right) \quad \text{[by definition]} \\ &= \frac{\pi}{\sin\frac{\pi}{3}} \quad \text{[ref. equation (7.69)]} \\ &= \frac{2\pi}{\sqrt{3}}. \end{aligned} \tag{20.68}$$

Thus, from equations (20.67) and (20.68), we get

$$C = \frac{2\pi}{\sqrt{3}} \tag{20.69}$$

Hence, from equations (20.66) and (20.69), we get finally

$$L(\lambda) = \frac{2\pi}{\sqrt{3}} e^{-3\lambda} \tag{20.70}$$

20.6 Weber Integral

The Weber integral is defined as

$$W = \int_0^\infty e^{-a^2u^2} J_t(bu) u^{t+1}\, du \tag{20.71}$$

where $a > 0, b > 0$ and $t > -1$.

Using the expression for Bessel's function, we write the integral in equation (20.71) as

$$\begin{aligned} W &= \int_0^\infty e^{-a^2u^2} u^{t+1} \sum_{k=0}^\infty \frac{(-1)^k}{\Gamma(k+1)\Gamma(t+k+1)} \left(\frac{bu}{2}\right)^{t+2k} du \\ &= \sum_{k=0}^\infty \frac{(-1)^k}{\Gamma(k+1)\Gamma(t+k+1)} \left(\frac{b}{2}\right)^{t+2k} \int_0^\infty e^{-a^2u^2} u^{2t+2k+1}\, du. \end{aligned} \tag{20.72}$$

We substitute

$$\xi = a^2u^2, \text{ so that } udu = \frac{d\xi}{2a^2} \tag{20.73}$$

Then we get from equation (20.72)

$$\begin{aligned} W &= \sum_{k=0}^{\infty} \frac{(-1)^k}{\Gamma(k+1)\Gamma(t+k+1)} \cdot \left(\frac{b}{2}\right)^{t+2k} \int_0^{\infty} \frac{1}{2a^{2t+2k+2}} \cdot e^{-\xi}\xi^{t+k}\, d\xi \\ &= \frac{b^t}{(2a^2)^{t+1}} \sum_{k=0}^{\infty} \frac{1}{\Gamma(k+1)\Gamma(t+k+1)} \left(-\frac{b^2}{4a^2}\right)^k \Gamma(t+k+1) \\ &= \frac{b^t}{(2a^2)^{t+1}} \sum_{k=0}^{\infty} \frac{1}{k!}\left(-\frac{b^2}{4a^2}\right)^k = \frac{b^t}{(2a^2)^{t+1}} \exp\left(-\frac{b^2}{4a^2}\right). \end{aligned} \tag{20.74}$$

20.7 Bessel's Integral

The Bessel's integral is defined as

$$\int_0^{\pi} \cos(m\phi - z\sin\phi)d\phi = \pi J_m(z) \tag{20.75}$$

We have proved in Chapter 12 (example 12.1) that

$$\cos(x\sin\phi) = J_0(x) + 2\sum_{n=1}^{\infty} J_{2n}(x)\cos 2n\phi \tag{20.76}$$

$$\text{and} \quad \sin(x\sin\phi) = 2\sum_{n=1}^{\infty} J_{2n-1}(x)\sin(2n-1)\phi \tag{20.77}$$

Multiplying equation (20.76) by $\cos m\phi$ and integrating the result from 0 to π, we get

$$\int_0^{\pi} \cos m\phi \cos(x\sin\phi)d\phi = J_0(x)\int_0^{\pi} \cos m\phi\, d\phi + 2\sum_{n=1}^{\infty} J_{2n}(x)\int_0^{\pi} \cos m\phi \cos 2n\phi\, d\phi \tag{20.78}$$

Now

$$\int_0^{\pi} \cos m\phi\, d\phi = 0 \tag{20.79}$$

and

$$\int_0^{\pi} \cos m\phi \cos 2n\phi\, d\phi = \begin{cases} 0, & \text{for } m \neq 2n \\ \pi/2, & \text{for } m = 2n \end{cases} \tag{20.80}$$

In view of equations (20.79) and (20.80), equation (20.78) becomes

$$\int_0^{\pi} \cos m\phi \cos(x\sin\phi)d\phi = \frac{\pi}{2}\{1 + (-1)^m\}J_m(x) \tag{20.81}$$

Again, multiplying equation (20.77) by $\sin m\phi$ and integrating from 0 to π, we get

$$\int_0^{\pi} \sin m\phi \sin(x\sin\phi)d\phi = 2\sum_{n=1}^{\infty} J_{2n-1}\int_0^{\pi} \sin m\phi \sin(2n-1)\phi\, d\phi \tag{20.82}$$

Now

$$\int_0^\pi \sin m\phi \sin(2n-1)\phi \, d\phi = \begin{cases} 0, & \text{for } m \neq 2n-1 \\ \pi/2, & \text{for } m = 2n-1 \end{cases} \tag{20.83}$$

Using equation (20.83), we write equation (20.82) in the form

$$\int_0^\pi \sin m\phi \sin(x \sin\phi) d\phi = \frac{\pi}{2}\{1-(-1)^m\} J_m(x) \tag{20.84}$$

Adding equations (20.81) and (20.84), we get the Bessel's integral

$$\int_0^\pi \cos(m\phi - x\sin\phi) d\phi = \pi J_m(x) \tag{20.85}$$

20.8 Laplace's Integrals

Laplace's integrals are defined in terms of Legendre polynomials in the forms

$$P_n(x) = \frac{1}{\pi}\int_0^\pi \left\{x + \sqrt{x^2-1}\cos\phi\right\}^n d\phi \tag{20.86}$$

and

$$P_n(x) = \frac{1}{\pi}\int_0^\pi \frac{1}{\{x+\sqrt{x^2-1}\cos\phi\}^{n+1}} d\phi \tag{20.87}$$

We consider the integral

$$I = \int_0^\pi \frac{d\phi}{a - b\cos\phi}, \quad a > b > 0 \tag{20.88}$$

which can be evaluated by methods discussed in Chapter 9. Its value will be

$$I = \int_0^\pi \frac{d\phi}{a - b\cos\phi} = \frac{\pi}{\sqrt{a^2-b^2}} \tag{20.89}$$

We substitute

$$\left.\begin{aligned} a &= 1 - xt \\ b &= t\sqrt{x^2-1} \end{aligned}\right\} \tag{20.90}$$

Then, from equation (20.89), we get

$$\frac{\pi}{(1-2xt+t^2)^{1/2}} = \int_0^\pi \frac{d\phi}{1 - t\left(x+\sqrt{x^2-1}\cos\phi\right)} = \int_0^\pi \sum_{n=0}^\infty t^n \left(x + \sqrt{x^2-1}\cos\phi\right)^n d\phi$$

[using binomial expansion for $(1-z)^{-1}$, where $z = t\left(x+\sqrt{x^2-1}\cos\phi\right)$]

Hence

$$\frac{1}{(1-2xt+t^2)^{1/2}} = \frac{1}{\pi}\sum_{n=0}^\infty t^n \int_0^\pi \left(x+\sqrt{x^2-1}\cos\phi\right)^n d\phi$$

$$\text{or,} \quad \sum_{n=0}^\infty P_n(x) t^n = \sum_{n=0}^\infty t^n \cdot \frac{1}{\pi}\int_0^\pi \left(x+\sqrt{x^2-1}\cos\phi\right)^n d\phi.$$

Equating the coefficients of t^n on both sides, we get

$$P_n(x) = \frac{1}{\pi}\int_0^\pi \left(x + \sqrt{x^2-1}\cos\phi\right)^n d\phi \tag{20.91}$$

Replacing n by $-(n+1)$ in equation (20.91), we have

$$P_{-(n+1)}(x) = \frac{1}{\pi}\int_0^\pi \frac{d\phi}{\left(x + \sqrt{x^2-1}\cos\phi\right)^{n+1}} \tag{20.92}$$

Now, Legendre's differential equation (12.238) for $\lambda = n(n+1)$ remains invariant if we replace n by $-(n+1)$. Hence

$$P_{-(n+1)}(x) = P_n(x) \tag{20.93}$$

From equations (20.92) and (20.93), we have

$$P_n(x) = \frac{1}{\pi}\int_0^\pi \frac{d\phi}{\left(x + \sqrt{x^2-1}\cos\phi\right)^{n+1}} \tag{20.94}$$

20.9 Exponential Integrals

The exponential integrals are defined by

$$E(x) = \int_x^\infty \frac{e^{-t}}{t}\,dt \tag{20.95}$$

Let us consider the integral

$$I = \int_x^\lambda \frac{e^{-t}}{t}\,dt = \int_x^\lambda \frac{dt}{t} - \int_x^\lambda \frac{1-e^{-t}}{t}\,dt$$

i.e., $$I = \left[\log t - \sum_{k=1}^\infty (-1)^k \frac{t^k}{k(k!)}\right]_x^\lambda \quad \text{[using expansion for } e^{-t}] \tag{20.96}$$

Let, for $\lambda \longrightarrow \infty$, the expression on the RHS of equation (20.96) takes a value C. Then, we can write

$$E(x) = C - \log x - \sum_{k=1}^\infty (-1)^k \frac{x^k}{k(k!)} \tag{20.97}$$

In order to determine C, we consider the integral

$$\Gamma(x+1) = \int_0^\infty e^{-t} t^x\,dt$$

Hence

$$\Gamma'(x+1) = \int_0^\infty e^{-t}\log t \cdot t^x\,dt$$

or, $$\Gamma(x+1)\psi_1(x+1) = \int_0^\infty e^{-t}\log t \cdot t^x\,dt \quad \text{[using equation (7.113)]}$$

Hence, for $x = 0$,

$$\Gamma(1)\psi_1(1) = \int_0^\infty e^{-t}\log t\,dt \quad \text{or,} \quad \psi_1(1) = \int_0^\infty e^{-t}\log t\,dt \tag{20.98}$$

But, from equation (7.115), we have

$$\psi_1(1) = -\gamma \quad \text{(Euler's constant)} \tag{20.99}$$

Thus, from equations (20.98) and (20.99), we get

$$\int_0^\infty e^{-t}\log t\,dt = -\gamma \tag{20.100}$$

Now

$$\int_x^\infty e^{-t}\log t\,dt = -e^{-t}\log t\Big|_x^\infty + \int_x^\infty \frac{e^{-t}}{t}\,dt = e^{-x}\log x + E(x) \tag{20.101}$$

Again

$$\begin{aligned}\int_x^\infty e^{-t}\log t\,dt &= \int_0^\infty e^{-t}\log t\,dt - \int_0^x e^{-t}\log t\,dt \\ &= -\gamma - \int_0^x e^{-t}\log t\,dt \quad \text{[using equation (20.100)]}\end{aligned} \tag{20.102}$$

From equations (20.101) and (20.102), we get

$$e^{-x}\log x + E(x) = -\gamma - \int_0^x e^{-t}\log t\,dt$$

$$\text{or,} \quad C - (1 - e^{-x})\log x - \sum_{k=0}^{\infty}(-1)^k \frac{x^k}{k(k!)} = -\gamma - \int_0^x e^{-t}\log t\,dt \quad \text{[using equation (20.97)]}$$

Putting $x = 0$, we get

$$C = -\gamma \tag{20.103}$$

Hence, from equation (20.97), we get

$$E(x) = -\gamma - \log x - \sum_{k=0}^{\infty}(-1)^k \frac{x^k}{k(k!)} \tag{20.104}$$

If we put, in equation (20.95)

$$x = -\log y \quad \text{and} \quad t = -\log z \tag{20.105}$$

we get

$$E(-\log y) = -\int_0^y \frac{dz}{\log z} \tag{20.106}$$

which gives the value of logarithmic integral

$$li(x) = \int_0^x \frac{dz}{\log z} = -E(-\log x) \tag{20.107}$$

20.10 Dawson's Integral

Dawson's integral is defined by

$$F(x) = e^{-x^2} \int_0^x \exp(t^2)dt \tag{20.108}$$

An approximation to this integral due to Rybicki is

$$F(x) = \lim_{t \longrightarrow 0} \frac{1}{\sqrt{\pi}} \sum \frac{1}{n} \exp\{-(x-nt)^2\} \tag{20.109}$$

the summation being carried over odd-values of n.

The series in equation (20.109) converges very rapidly and accuracy increases exponentially for small values of t. Even for moderate values of t, the series in equation (20.109) gives accurate approximations.

Some values of the integral are given in the table 20.1.

Table 20.1: Values of Dawson's Integral

x	$F(x)$	x	$F(x)$
0.1	0.0993	1.1	0.5262
0.2	0.1948	1.2	0.5073
0.3	0.2826	1.3	0.4834
0.4	0.3599	1.4	0.4565
0.5	0.4244	1.5	0.4282
0.6	0.4748	1.6	0.3999
0.7	0.5105	1.7	0.3726
0.8	0.5321	1.8	0.3468
0.9	0.5407	1.9	0.3230
1.0	0.5381	2.0	0.3013

20.11 Heine's Integrals

Heine's two integrals are defined by

$$Q_n(x) = \int_0^\alpha \left\{x - \sqrt{x^2-1}\cosh\phi\right\}^n d\phi \tag{20.110}$$

and

$$Q_n(x) = \int_0^\infty \left\{x + \sqrt{x^2-1}\cosh\phi\right\}^{-n-1} d\phi \tag{20.111}$$

where $Q_n(x)$ are Legendre's polynomial of second kind, and

$$\alpha = \frac{1}{2}\log\frac{x+1}{x-1} \tag{20.112}$$

The above integrals are analogues of Laplace's integrals discussed before.

In order to deduce the above integrals, we assume the integral formula for $Q_n(x)$ in the form

$$Q_n(x) = \frac{1}{2^{n+1}} \int_{-1}^{1} \frac{(1-t^2)^n}{(x-t)^{n-1}}\, dt \tag{20.113}$$

We make the transformation (for $x > 1$)

$$t = x - \sqrt{x^2-1}e^{\phi} \tag{20.114}$$

so that as t varies from -1 to 1, ϕ varies from α to $-\alpha$, where α is given by equation (20.112). Substituting equation (20.114) into equation (20.113), we get

$$Q_n(x) = \frac{1}{2} \int_{-\alpha}^{\alpha} \left\{x - \sqrt{x^2-1}\cosh\phi\right\}^n\, d\phi$$

$$\text{i.e.,}\quad Q_n(x) = \int_{0}^{\alpha} \left\{x - \sqrt{x^2-1}\cosh\phi\right\}^n\, d\phi \tag{20.115}$$

since the integrand is an even function of ϕ. This integral formula is very useful in calculating the functions $Q_n(x)$. Thus,

$$Q_0(x) = \alpha = \frac{1}{2}\log\frac{x+1}{x-1} \tag{20.116}$$

$$Q_1(x) = x\alpha - \sqrt{x^2-1}\sinh\alpha = \frac{1}{2}x\log\frac{x+1}{x-1} - 1 \tag{20.117}$$

....

To deduce the second integral, we substitute

$$\left\{x + \sqrt{x^2-1}\cosh\xi\right\}\left\{x - \sqrt{x^2-1}\cosh\phi\right\} = 1 \tag{20.118}$$

so that as ϕ varies from 0 to α, ξ varies from zero to $+\infty$.

Denoting $\sqrt{x^2-1}$ by λ, we get from equation (20.118)

$$\lambda\cosh\phi = x - \frac{1}{x+\lambda\cosh\xi} \tag{20.119}$$

Differentiating

$$d\phi = \frac{\sinh\xi}{\sinh\phi}\frac{d\xi}{(x+\lambda\cosh\xi)^2} \tag{20.120}$$

Using equation (20.119), we get

$$(\lambda\sinh\phi)^2 = \lambda^2 - (\lambda\cosh\phi)^2 = (x^2-1) - \left[x^2 - \frac{2x}{x+\lambda\cosh\xi} + \frac{1}{(x+\lambda\cosh\xi)^2}\right]$$

$$= \frac{-(x+\lambda\cosh\xi)^2 + 2x(x+\lambda\cosh\xi) - 1}{(x+\lambda\cosh\xi)^2} = \frac{\lambda^2(\sinh\xi)^2}{(x+\lambda\cosh\xi)^2} \tag{20.121}$$

From equations (20.120) and (20.121)

$$d\phi = \frac{d\xi}{x+\sqrt{x^2-1}\cosh\xi} \tag{20.122}$$

Substituting equation (20.122) into equation (20.115), we get finally the second integral of Heine

$$Q_n(x) = \int_{0}^{\infty} \left\{x + \sqrt{x^2-1}\cosh\xi\right\}^{-n-1}\, d\xi \tag{20.123}$$

20.12 Elliptic Integrals

Incomplete elliptic integral of first kind is defined as

$$u = F(k,\phi) = \int_0^{\phi} \frac{d\theta}{\sqrt{1-k^2\sin^2\theta}} = \int_0^{x} \frac{d\xi}{\sqrt{(1-\xi^2)(1-k^2\xi^2)}} \tag{20.124}$$

where ϕ is called the amplitude of u written as $\phi = amu$ and $x = \sin\phi$. Here and below we take $0 < k < 1$.

Incomplete elliptic integral of second kind is defined as

$$E(k,\phi) = \int_0^{\phi} \sqrt{1-k^2\sin^2\theta}\, d\theta = \int_0^{x} \frac{\sqrt{1-k^2\xi^2}}{\sqrt{1-\xi^2}} d\xi \tag{20.125}$$

Incomplete integral of the third kind is defined as

$$\begin{aligned} \Pi(k,n,\phi) &= \int_0^{\phi} \frac{d\theta}{(1+n\sin^2\theta)\sqrt{1-k^2\sin^2\theta}} \\ &= \int_0^{x} \frac{d\xi}{(1+n\xi^2)\sqrt{(1-\xi^2)(1-k^2\xi^2)}} \end{aligned} \tag{20.126}$$

Writing $k^2 = \sin^2\alpha$, the graphical presentations of the incomplete integrals of first and second kinds are given in figures 20.1 and 20.2 respectively.

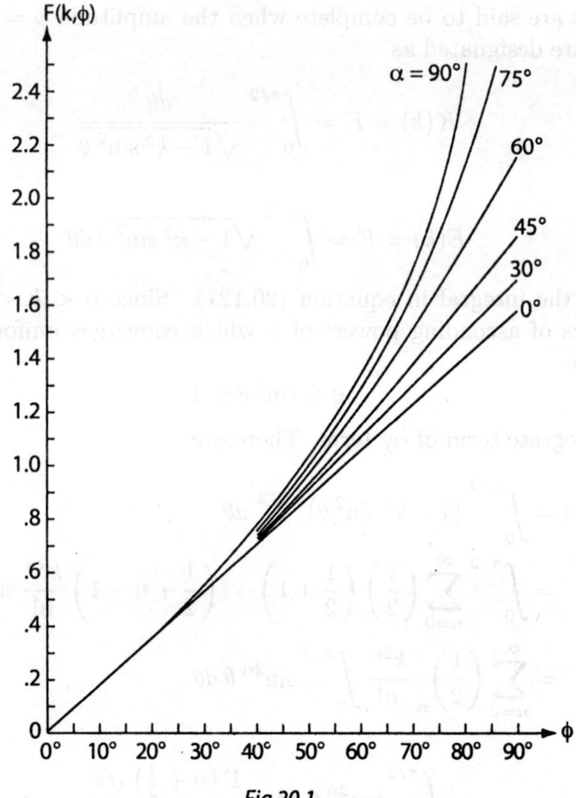

Fig 20.1

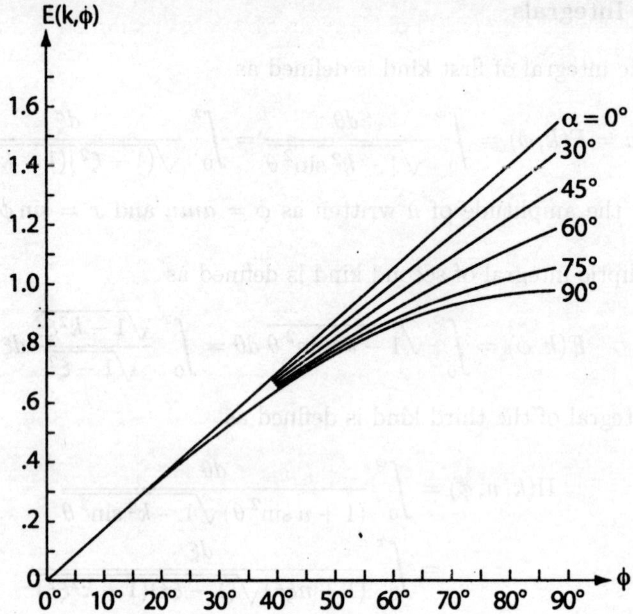

Fig 20.2

Some of the values of these integrals are also provided in Tables 20.2 and 20.3 respectively.

Elliptic integrals are said to be complete when the amplitude $\phi = \pi/2$ and so $x = 1$. The complete integrals are designated as

$$K(k) = K = \int_0^{\pi/2} \frac{d\theta}{\sqrt{1 - k^2 \sin^2 \theta}} \tag{20.127}$$

and

$$E(k) = E = \int_0^{\pi/2} \sqrt{1 - k^2 \sin^2 \theta}\, d\theta \tag{20.128}$$

Let us consider the integral in equation (20.127). Since $0 < k < 1$, we can expand the integrand as a series of ascending powers of k which converges uniformly with respect to θ, because, as we know

$$0 \le \sin^2 \theta \le 1 \tag{20.129}$$

Then we can integrate term of by term. Therefore,

$$\begin{aligned} K(k) &= \int_0^{\pi/2} \left(1 - k^2 \sin^2 \theta\right)^{-1/2} d\theta \\ &= \int_0^{\pi/2} \sum_{n=0}^{\infty} \left(\frac{1}{2}\right)\left(\frac{1}{2}+1\right)\cdots\left(\frac{1}{2}+n-1\right)\frac{k^{2n}}{n!} \sin^{2n}\theta\, d\theta \\ &= \sum_{n=0}^{\infty} \left(\frac{1}{2}\right)_n \frac{k^{2n}}{n!} \int_0^{\pi/2} \sin^{2n}\theta\, d\theta. \end{aligned} \tag{20.130}$$

Now

$$\int_0^{\pi/2} \sin^{2n}\theta\, d\theta = \frac{\Gamma\left(n + \frac{1}{2}\right)\sqrt{\pi}}{\Gamma(n+1)2} \tag{20.131}$$

Here we have used equation (7.91) of Chapter 7 by substituting $2n = 2n-1$. The right-hand side of equation (20.131) becomes

$$= \frac{\left(\frac{1}{2}\right)_n}{n!}\frac{\pi}{2} = \frac{\left(\frac{1}{2}\right)_n}{(1)_n}\frac{\pi}{2} \tag{20.132}$$

Hence

$$K(k) = \frac{\pi}{2}\sum_{n=0}^{\infty}\frac{\left(\frac{1}{2}\right)_n\left(\frac{1}{2}\right)_n}{(1)_n n!}(k^2)^n = \frac{\pi}{2}F\left(\frac{1}{2},\frac{1}{2},1;k^2\right) \tag{20.133}$$

Table 20.2: Incomplete Elliptic Integral of the First kind

ϕ \ α	0°	10°	20°	30°	40°	50°	60°	70°	80°	90°
0°	0.0000	0.0000	0.0000	0.0000	0.0000	0.0000	0.0000	0.0000	0.0000	0.0000
10°	0.1745	0.1746	0.1746	0.1748	0.1749	0.1751	0.1752	0.1753	0.1754	0.1754
20°	0.3491	0.3493	0.3499	0.3508	0.3520	0.3533	0.3545	0.3555	0.3561	0.3564
30°	0.5236	0.5243	0.5263	0.5294	0.5334	0.5379	0.5422	0.5459	0.5484	0.5493
40°	0.6981	0.6997	0.7043	0.7116	0.7213	0.7323	0.7436	0.7535	0.7604	0.7629
50°	0.8727	0.8756	0.8842	0.8982	0.9173	0.9401	0.9647	0.9876	1.0044	1.0107
60°	1.0472	1.0519	1.0660	1.0896	1.1226	1.1643	1.2126	1.2619	1.3014	1.3170
70°	1.2217	1.2286	1.2495	1.2853	1.3372	1.4068	1.4944	1.5959	1.6918	1.7354
80°	1.3963	1.4056	1.4344	1.4846	1.5597	1.6660	1.8125	2.0119	2.2653	2.4362
90°	1.5708	1.5828	1.6200	1.6858	1.7868	1.9356	2.165	2.5046	3.1534	∞

Table 20.3: Incomplete Elliptic Integral of the Second kind

ϕ \ α	0°	10°	20°	30°	40°	50°	60°	70°	80°	90°
0°	0.0000	0.0000	0.0000	0.0000	0.0000	0.0000	0.0000	0.0000	0.0000	0.0000
10°	0.1745	0.1745	0.1744	0.1743	0.1742	0.1740	0.1739	0.1738	0.1737	0.1736
20°	0.3491	0.3489	0.3483	0.3473	0.3462	0.3450	0.3438	0.3429	0.3422	0.3420
30°	0.5236	0.5229	0.5209	0.5179	0.5141	0.5100	0.5061	0.5029	0.5007	0.5000
40°	0.6981	0.6966	0.6921	0.6851	0.6763	0.6667	0.6575	0.6497	0.6446	0.6428
50°	0.8727	0.8698	0.8614	0.8483	0.8317	0.8134	0.7954	0.7801	0.7697	0.7660
60°	1.0472	1.0426	1.0290	1.0076	0.9801	0.9493	0.9184	0.8914	0.8728	0.8660
70°	1.2217	1.2149	1.1949	1.1632	1.1221	1.0750	1.0266	0.9830	0.9514	0.9397
80°	1.3963	1.3870	1.3597	1.3161	1.2590	1.1926	1.1225	1.0565	1.0054	0.9848
90°	1.5708	1.5589	1.5238	1.4675	1.3931	1.3055	1.2111	1.1184	1.0401	1.0000

Therefore, complete elliptic integrals of first kind can be expressed in terms of hypergeometric functions. Writing in the form of infinite series

$$K(k) = \frac{\pi}{2}\left[1 + \left(\frac{1}{2}\right)^2 k + \left(\frac{1.3}{2.4}\right)^2 k^2 + \left(\frac{1.3.5}{2.4.6}\right)^2 k^3 + \cdots\right] \tag{20.134}$$

Table 20.4: Complete Elliptic Integrals of First and Second kinds

α	K	E	α	K	E	α	K	E
0°	1.5708	1.5708	30°	1.6858	1.4675	60°	2.1565	1.2111
1°	1.5709	1.5707	31°	1.6941	1.4608	61°	2.1842	1.2015
2°	1.5713	1.5703	32°	1.7028	1.4539	62°	2.2132	1.1920
3°	1.5719	1.5697	33°	1.7119	1.4469	63°	2.2435	1.1826
4°	1.5727	1.5689	34°	1.7214	1.4397	64°	2.2754	1.1732
5°	1.5738	1.5678	35°	1.7312	1.4323	65°	2.3088	1.1638
6°	1.5751	1.5665	36°	1.7415	1.4248	66°	2.3439	1.1545
7°	1.5767	1.5649	37°	1.7522	1.4171	67°	2.3809	1.1453
8°	1.5785	1.5632	38°	1.7633	1.4092	68°	2.4198	1.1362
9°	1.5805	1.5611	39°	1.7748	1.4013	69°	2.4610	1.1272
10°	1.5828	1.5589	40°	1.7868	1.3931	70°	2.5046	1.1184
11°	1.5854	1.5564	41°	1.7992	1.3849	71°	2.5507	1.1096
12°	1.5882	1.5537	42°	1.8122	1.3765	72°	2.5998	1.1011
13°	1.5913	1.5537	43°	1.8256	1.3680	73°	2.6521	1.0927
14°	1.5946	1.5476	44°	1.8396	1.3594	74°	2.7081	1.0844
15°	1.5981	1.5442	45°	1.8541	1.3506	75°	2.7681	1.0764
16°	1.6020	1.5405	46°	1.8691	1.3418	76°	2.8327	1.0686
17°	1.6061	1.5367	47°	1.8848	1.3329	77°	2.9026	1.0611
18°	1.6105	1.5326	48°	1.9011	1.3238	78°	2.9786	1.0538
19°	1.6151	1.5283	49°	1.9180	1.3147	79°	3.0617	1.0468
20°	1.6200	1.5238	50°	1.9356	1.3055	80°	3.1534	1.0401
21°	1.6252	1.5191	51°	1.9539	1.2963	81°	3.2553	1.0338
22°	1.6307	1.5141	52°	1.9729	1.2870	82°	3.3699	1.0278
23°	1.6365	1.5090	53°	1.9927	1.2776	83°	3.5004	1.0223
24°	1.6426	1.5037	54°	2.0133	1.2681	84°	3.6519	1.0172
25°	1.6490	1.4981	55°	2.0347	1.2587	85°	3.8317	1.0127
26°	1.6557	1.4924	56°	2.0571	1.2492	86°	4.0528	1.0086
27°	1.6627	1.4864	57°	2.0804	1.2397	87°	4.3387	1.0053
28°	1.6701	1.4803	58°	2.1047	1.2301	88°	4.7427	1.0026
29°	1.6777	1.4740	59°	2.1300	1.2206	89°	5.4349	1.0008
30°	1.6858	1.4675	60°	2.1565	1.2111	90°	∞	1.0000

Complete elliptic integrals of second kind are denoted by

$$E(k) = \int_0^{\pi/2} \sqrt{1 - k^2 \sin^2\theta}\, d\theta = \int_0^1 \frac{\sqrt{1-k^2\xi^2}}{\sqrt{1-\xi^2}}\, d\xi \tag{20.135}$$

Proceeding as before, it can be shown that

$$E(k) = \frac{\pi}{2} F\left(-\frac{1}{2}, \frac{1}{2}, 1; k^2\right) \tag{20.136}$$

In the form of infinite series

$$E(k) = \frac{\pi}{2}\left[1 - \left(\frac{1}{2}\right)^2 k^2 - \left(\frac{1.3}{2.4}\right)^2 \frac{k^4}{3} - \left(\frac{1.3.5}{2.4.6}\right)^2 \frac{k^6}{5} - \dots\right] \tag{20.137}$$

Writing

$$k_1^2 = 1 - k^2 \tag{20.138}$$

the following integrals are defined

$$K' = K(k_1) = \int_0^{\pi/2} \left(1 - k_1^2 \sin^2\theta\right)^{-1/2} d\theta \tag{20.139}$$

$$\text{and} \quad E' = E(k_1) = \int_0^{\pi/2} \left(1 - k_1^2 \sin^2\theta\right)^{1/2} d\theta \tag{20.140}$$

Then there exists a relation between these integrals in the form

$$EK' + E'K - KK' = \frac{\pi}{2} \tag{20.141}$$

Equation (20.141) is known as *Legendre's relation.*

Some values of complete elliptic integrals of first and second kinds are provided in the Table 20.4.

Reduction to Standard Form

Consider the integral

$$I = \int_0^x \frac{a\,d\xi}{\left[(\xi^2 + a^2)(\xi^2 + b^2)\right]^{1/2}} \tag{20.142}$$

With the substitution, $\xi = b\tan\phi$, the integral is written as

$$I = \int_0^{b\tan\phi} \frac{ab\sec^2\phi\,d\phi}{\sqrt{a^2 + b^2\tan^2\phi}\,b\sec\phi} = \int_0^{b\tan\phi} \frac{d\phi}{\sqrt{1 - \dfrac{(a^2 - b^2)}{a^2}\sin^2\phi}} \tag{20.143}$$

which is of the form of elliptic integrals of first kind with $k^2 = (a^2 - b^2)/a^2 \quad (a > b)$.

Similarly, the integral

$$I = \int_0^x \frac{a\,d\xi}{\sqrt{(a^2 - \xi^2)(b^2 - \xi^2)}} \quad (a > b) \tag{20.144}$$

can be reduced to

$$I = \int_0^{b\sin\phi} \frac{d\phi}{\sqrt{1 - k^2\sin^2\phi}} \tag{20.145}$$

where $k^2 = b^2/a^2 < 1$, if we substitute $\xi = b\sin\phi$.

Thus, with appropriate substitutions, many integrals can be reduced to standard forms of elliptic integrals.

Landen's Transformation

The transformation

$$\tan\phi = \frac{\sin 2\phi_1}{k + \cos 2\phi_1} \quad \text{or,} \quad k\sin\phi = \sin(2\phi_1 - \phi) \tag{20.146}$$

is called *Landen's transformation.*

With this transformation, the incomplete elliptic integral of first kind defined by equation (20.124) can be reduced to

$$F(k,\phi) = \frac{2}{1+k}\int_0^{\phi_1} \frac{d\phi_1}{1-k_1^2\sin^2\phi_1} \tag{20.147}$$

where

$$k_1 = \frac{2\sqrt{k}}{1+k} \tag{20.148}$$

Equation (20.148) shows that $k < k_1 < 1$.

By successive application of Landen's transformation we can obtain a sequence of values k_n $(n = 1, 2, 3, \ldots)$ such that

$$k < k_1 < k_2 < \ldots < 1 \tag{20.149}$$

and hence

$$\lim_{n\longrightarrow\infty} k_n = 1.$$

Then it follows that

$$F(k,\phi) = \sqrt{\frac{k_1k_2k_3\ldots}{k}}\int_0^{\phi} \frac{d\theta}{\sqrt{1-\sin^2\theta}} = \sqrt{\frac{k_1k_2k_3\ldots}{k}}\ln\tan\left(\frac{\pi}{4}+\frac{\phi}{2}\right) \tag{20.150}$$

where

$$k_2 = \frac{2\sqrt{k_1}}{1+k_1},\quad k_3 = \frac{2\sqrt{k_2}}{1+k_2}, \cdots \text{so on} \tag{20.151}$$

and

$$\phi = \lim_{n\longrightarrow\infty} \phi_n.$$

With the use of equation (20.150), $F(k,\phi)$ can be computed.

EXAMPLE 20.1. *Show that*

$$\int_0^{\pi/2} \frac{dx}{\sqrt{\sin x}} = \sqrt{2}F\left(\sqrt{\frac{1}{2}}, \frac{\pi}{2}\right)$$

Hence prove that

$$F\left(\sqrt{\frac{1}{2}}, \frac{\pi}{2}\right) = \frac{1}{2}\sqrt{\frac{\pi}{2}}\,\frac{\Gamma(1/4)}{\Gamma(3/4)}$$

Solution: Let $x = \frac{\pi}{2} - y$. Then the given integral becomes

$$I = \int_0^{\pi/2} \frac{dy}{\sqrt{\cos y}} \tag{1}$$

The integral in equation (1) can be written as

$$I = \int_0^{\pi/2} \frac{dy}{\sqrt{1-2\sin^2\frac{y}{2}}} \tag{2}$$

Let

$$\sqrt{2}\sin\frac{y}{2} = \sin\phi \tag{3}$$

Then

$$\frac{1}{\sqrt{2}}\cos\frac{y}{2}\,dy = \cos\phi\,d\phi, \quad \text{i.e.,} \quad dy = \frac{\sqrt{2}\cos\phi\,d\phi}{\sqrt{1-\frac{1}{2}\sin^2\phi}} \tag{4}$$

Thus the integral in equation (2) becomes

$$I = \sqrt{2}\int_0^{\pi/2}\frac{d\phi}{\sqrt{1-\frac{1}{2}\sin^2\phi}} = \sqrt{2}F\left(\sqrt{\frac{1}{2}}, \frac{\pi}{2}\right) \tag{5}$$

Again, from the properties of gamma functions, we have

$$\int_0^{\pi/2}\frac{dx}{\sqrt{\sin x}} = \int_0^{\pi/2}\sin^{-1/2}x\,dx \tag{6}$$

whose value is

$$\frac{\Gamma(1/4)\Gamma(1/2)}{2\Gamma(3/4)} = \frac{\sqrt{\pi}}{2}\frac{\Gamma(1/4)}{\Gamma(3/4)}$$

Hence the result follows.

EXAMPLE 20.2. *Find the length of arc of the ellipse $x = a\sin\phi$, $y = b\cos\phi$, $a > b > 0$.*

Solution: We have $x = a\sin\phi$, $y = b\cos\phi$ so that

$$dx = a\cos\phi\,d\phi, \quad dy = -b\sin\phi\,d\phi$$

Hence

$$\begin{aligned} ds &= \sqrt{dx^2+dy^2} = \sqrt{a^2\cos^2\phi + b^2\sin^2\phi}\,d\phi \\ &= \sqrt{a^2-(a^2-b^2)\sin^2\phi}\,d\phi = a\sqrt{1-e^2\sin^2\phi}\,d\phi \end{aligned}$$

where

$$e = \sqrt{\frac{a^2-b^2}{a^2}} = \text{eccentricity of the ellipse}$$

Therefore, the total arc length of the ellipse is

$$4\int_0^{\pi/2} ds = 4a\int_0^{\pi/2}\sqrt{1-e^2\sin^2\phi}\,d\phi = 4aE\left(e, \frac{\pi}{2}\right)$$

EXAMPLE 20.3. *Evaluate the integral*

$$\int_0^1 \frac{dx}{\sqrt{(1+x^2)(1+2x^2)}}$$

Solution: Let $x = \tan\theta$. Then the given integral becomes

$$I = \int_0^{\pi/4} \frac{\sec^2\theta\, d\theta}{\sqrt{(1+\tan^2\theta)(1+2\tan^2\theta)}} = \int_0^{\pi/4} \frac{d\theta}{\sqrt{\cos^2\theta + 2\sin^2\theta}}$$

$$= \int_0^{\pi/4} \frac{d\theta}{\sqrt{2-\cos^2\theta}}$$

We put $\theta = \frac{\pi}{2} - \phi$. Then the integral in equation becomes

$$I = \frac{1}{\sqrt{2}} \int_{\pi/4}^{\pi/2} \frac{d\phi}{\sqrt{1-\frac{1}{2}\sin^2\phi}}$$

$$= \frac{1}{\sqrt{2}} \left[\int_0^{\pi/2} \frac{d\phi}{\sqrt{1-\frac{1}{2}\sin^2\phi}} - \int_0^{\pi/4} \frac{d\phi}{\sqrt{1-\frac{1}{2}\sin^2\phi}} \right]$$

$$= \frac{1}{\sqrt{2}} \left[F\left(\sqrt{\frac{1}{2}}, \frac{\pi}{2}\right) - F\left(\sqrt{\frac{1}{2}}, \frac{\pi}{4}\right) \right]$$

Some Integrals Reducible to Elliptic Type

1. $\displaystyle\int_x^\infty \frac{d\xi}{\sqrt{(\xi^2+a^2)(\xi^2+b^2)}}$ with $\xi = a\cot\phi$,

2. $\displaystyle\int_x^a \frac{d\xi}{\sqrt{(a^2-\xi^2)(\xi^2-b^2)}}$ with $\xi^2 = a^2 - (a^2-b^2)\sin^2\phi$,

3. $\displaystyle\int_a^x \frac{d\xi}{\sqrt{(\xi^2-a^2)(\xi^2-b^2)}}$ with $\sin^2\phi = \dfrac{\xi^2-a^2}{\xi^2-b^2}$,

4. $\displaystyle\int_x^\infty \frac{d\xi}{\sqrt{(\xi^2-a^2)(\xi^2-b^2)}}$ with $\xi = a\operatorname{cosec}\phi$,

5. $\displaystyle\int_b^x \frac{d\xi}{\sqrt{(\xi^2+a^2)(\xi^2-b^2)}}$ with $\xi = b\sec\phi$,

6. $\displaystyle\int_x^b \frac{d\xi}{\sqrt{(\xi^2+a^2)(\xi^2-b^2)}}$ with $\xi = b\cos\phi$,

7. $\displaystyle\int_a^b \frac{d\xi}{\sqrt{(\xi-\alpha)(\xi-\beta)(\xi-\gamma)}}$, $(\gamma > \beta > \alpha)$ with $\xi = t^2 + \gamma, t = \tan\phi$,

8. $\displaystyle\int_a^b \frac{d\xi}{\sqrt{(\xi^2+2\xi+\alpha)(\xi^2+2\xi+\beta)}}$, $(\alpha, \beta > 1)$ with $\xi + 1 = z, z = \sqrt{\beta-1}\tan\phi$,

9. $\displaystyle\int_0^x \frac{d\xi}{\sqrt{(1+\xi^4)}}$ with $\cos\phi = \dfrac{1-\xi^2}{1+\xi^2}$,

10. $\displaystyle\int_x^\infty \frac{d\xi}{\sqrt{(1+\xi^4)}}$ with $\cos\phi = \dfrac{\xi^2-1}{\xi^2+1}$.

20.13 Poisson's Integral

Poisson's integral is defined by

$$P_\nu(x) = \int_0^\pi \cos(x\cos\theta)(\sin\theta)^{2\nu}\,d\theta \tag{20.152}$$

where $\nu > -\frac{1}{2}$.

Let us now evaluate the integral.

We have

$$\frac{1}{\Gamma(\nu+k+1)} = \frac{1}{\Gamma\{(k+\frac{1}{2})+(\nu+\frac{1}{2})\}} = \frac{B\left(k+\frac{1}{2},\nu+\frac{1}{2}\right)}{\Gamma\left(k+\frac{1}{2}\right)\Gamma\left(\nu+\frac{1}{2}\right)}$$
$$= \frac{1}{\Gamma\left(k+\frac{1}{2}\right)\Gamma\left(\nu+\frac{1}{2}\right)}\int_0^1 x^{k-\frac{1}{2}}(1-x)^{\nu-\frac{1}{2}}\,dx$$

Substituting $x = t^2$, we get

$$\frac{1}{\Gamma(\nu+k+1)} = \frac{2}{\Gamma\left(k+\frac{1}{2}\right)\Gamma(\nu+\frac{1}{2})}\int_0^1 t^{2k}\left(1-t^2\right)^{\nu-\frac{1}{2}}\,dt$$
$$= \frac{1}{\Gamma\left(k+\frac{1}{2}\right)\Gamma\left(\nu+\frac{1}{2}\right)}\int_{-1}^1 t^{2k}\left(1-t^2\right)^{\nu-\frac{1}{2}}\,dt \tag{20.153}$$

Now, the Bessel's function of first kind is given by [ref. equation (12.79)]

$$J_\nu(x) = \sum_{k=0}^\infty \frac{(-1)^k}{\Gamma(k+1)\Gamma(\nu+k+1)}\left(\frac{x}{2}\right)^{\nu+2k}$$

Using equation (20.153), we get

$$J_\nu(x) = \sum_{k=0}^\infty \frac{(-1)^k(x/2)^{\nu+2k}}{\Gamma(k+1)\Gamma(k+\frac{1}{2})\Gamma(\nu+\frac{1}{2})}\int_{-1}^1 t^{2k}(1-t^2)^{\nu-\frac{1}{2}}\,dt$$

$$\text{or,}\quad J_\nu(x) = \frac{(x/2)^\nu}{\Gamma(\nu+\frac{1}{2})}\int_{-1}^1 (1-t^2)^{\nu-\frac{1}{2}}\sum_{k=0}^\infty \frac{(-1)^k(xt)^{2k}}{2^{2k}\Gamma(k+\frac{1}{2})\Gamma(k+1)}\,dt$$

Now, from Legendre's duplication formula (7.98), we have, for $m = k+\frac{1}{2}$

$$\sqrt{\pi}\Gamma(2k+1) = 2^{2k}\Gamma\left(k+\frac{1}{2}\right)\Gamma(k+1) \tag{20.154}$$

Hence, we get

$$J_\nu(x) = \frac{(x/2)^\nu}{\sqrt{\pi}\Gamma\left(\nu+\frac{1}{2}\right)}\int_{-1}^1 (1-t^2)^{\nu-\frac{1}{2}}\sum_{k=0}^\infty \frac{(-1)^k(xt)^{2k}}{\Gamma(2k+1)}\,dt$$
$$= \frac{(x/2)^\nu}{\sqrt{\pi}\Gamma(\nu+\frac{1}{2})}\int_{-1}^1 (1-t^2)^{\nu-\frac{1}{2}}\cos(xt)dt \tag{20.155}$$

Putting, $t = \cos\theta$, we get from equation (20.155)

$$J_\nu(x) = \frac{(x/2)^\nu}{\sqrt{\pi}\Gamma(\nu+\frac{1}{2})}\int_0^\pi \cos(x\cos\theta)(\sin\theta)^{2\nu}\,d\theta$$

Hence

$$P_\nu(x) = \int_0^\pi \cos(x\cos\theta)(\sin\theta)^{2\nu}\,d\theta = \sqrt{\pi}\Gamma\left(\nu+\frac{1}{2}\right)\frac{J_\nu(x)}{(x/2)^\nu} \tag{20.156}$$

20.14 Gauss Integral

The Gauss integral is defined as

$$G = \iint \frac{\cos(r,n)}{r^2}\,dS \tag{20.157}$$

where r = length of the radius vector joining a fixed point $A(\xi,\eta,\zeta)$ with a variable point $P(x,y,z)$ lying on a surface S, so that

$$r^2 = (x-\xi)^2 + (y-\eta)^2 + (z-\zeta)^2 \tag{20.158}$$

and $\cos(r,n)$ = cosine of the angle between the radius vector and normal to the surface S at the point P.

Now, let the direction cosines of the normal to the surface S be $\cos\lambda$, $\cos\mu$ and $\cos\nu$. Then, we have

$$\begin{aligned}\cos(r,n) &= \cos(x,r)\cos\lambda + \cos(y,r)\cos\mu + \cos(z,r)\cos\nu \\ &= \frac{x-\xi}{r}\cos\lambda + \frac{y-\eta}{r}\cos\mu + \frac{z-\xi}{r}\cos\nu\end{aligned} \tag{20.159}$$

In view of this relation, the Gauss integral assumes the form

$$\begin{aligned}G &= \iint\left(\frac{x-\xi}{r^3}\cos\lambda + \frac{y-\eta}{r^3}\cos\mu + \frac{z-\xi}{r^3}\cos\nu\right)dS \\ &= \iint\left(\frac{x-\xi}{r^3}dy\,dz + \frac{y-\eta}{r^3}dz\,dx + \frac{z-\xi}{r^3}dx\,dy\right)\end{aligned} \tag{20.160}$$

We now remember the following theorem of Ostrogradskii. It states: If the function P, Q, R along with their derivatives are continuous in a region bounded by a closed surface S, then the integral

$$\iint [P\,dy\,dx + Q\,dz\,dx + R\,dx\,dy] \tag{20.161}$$

will always vanish if

$$\frac{\partial P}{\partial x} + \frac{\partial Q}{\partial y} + \frac{\partial R}{\partial z} = 0 \tag{20.162}$$

If however, there is any singular point in that region, then the value of the integral equation (20.161) will have the same non-zero value on all the surfaces enclosing that singular point.

In equation (20.160), we set

$$P = \frac{x-\xi}{r^3} \tag{20.163}$$

$$Q = \frac{y-\eta}{r^3} \tag{20.164}$$

$$\text{and} \quad R = \frac{z-\xi}{r^3} \tag{20.165}$$

Then we get

$$\frac{\partial P}{\partial x} = \frac{1}{r^3} - \frac{3(x-\xi)^2}{r^5} \tag{20.166}$$

$$\frac{\partial Q}{\partial y} = \frac{1}{r^3} - \frac{3(y-\eta)^2}{r^5} \tag{20.167}$$

$$\text{and} \quad \frac{\partial R}{\partial z} = \frac{1}{r^3} - \frac{3(z-\zeta)^2}{r^5} \tag{20.168}$$

From these equations, we find that the condition equation (20.162) is satisfied. The functions P, Q, R and their derivatives are continuous everywhere in space except at the point $A(\xi, \eta, \zeta)$ which is a singular point. Hence, if the surface S does not enclose the point A, the value of the Gauss integral is zero. If the surface S encloses the point A, then the value of the Gauss integral will be same over all types of surfaces. Thus, we construct a sphere of radius R around the point A as centre. Then, the radius vector of the points on the surface of the sphere is constant and equal to R; and the direction of the normal to the surface at all points coincides with that of the radius vector so that

$$\cos(r, n) = 1 \tag{20.169}$$

Hence, the Gauss integral becomes

$$G = \iint \frac{dS}{R^2} = \frac{S}{R^2} = \frac{4\pi R^2}{R^2} = 4\pi \tag{20.170}$$

Physically, the Gauss integral is the measure of the solid angle under which the surface S is visible from the point A. Thus, when the point A is inside the surface, the total solid angle of visibility of the surface is 4π, and hence $G = 4\pi$. But when the point A is outside the surface S, the solid angles of visibility at different portions will have different signs and the net value will be zero, and hence $G = 0$.

In two-dimension, the Gauss integral is defined as

$$g = \int_L \frac{\cos(r, n)}{r}\, dl \tag{20.171}$$

where, as before,

r = length of radius vector joining a fixed point $A(\xi, \eta)$ with a variable point $P(x, y)$ on the curve L, and

$\cos(r, n)$ = cosine of the angle between the radius vector and the normal to the curve at P.

It can be proved that, when the point A is totally enclosed by the curve L, then $g = 2\pi$. But if the point A lies outside the closed curve L, the value of the integral $g = 0$. Physically, this integral gives a measure of the angle under which the curve L is visible from the point A.

20.15 Airy's Integral

The Airy's integrals are defined by

$$A_1(x) = \int_0^\infty \cos\,(t^3 + tx)\,dt \tag{20.172}$$

$$\text{and}\quad A_2(x) = \int_0^\infty \cos\,(t^3 - tx)\,dt \tag{20.173}$$

Both the integrals converge and can be expressed in terms of cylindrical functions. Without offering any proof, we give below the value of the integrals.

$$A_1(x) = \int_0^\infty \cos\,(t^3 + tx)\,dt = \frac{\sqrt{x}}{3} K_{1/3}\left(\frac{2x\sqrt{x}}{3\sqrt{3}}\right) \tag{20.174}$$

$$A_2(x) = \int_0^\infty \cos\,(t^3 - tx)\,dt = \frac{\pi\sqrt{x}}{3\sqrt{3}}\left[J_{-1/3}\left(\frac{2x\sqrt{x}}{3\sqrt{3}}\right) + J_{1/3}\left(\frac{2x\sqrt{x}}{3\sqrt{3}}\right)\right] \tag{20.175}$$

where $J_{-1/3}$, $J_{1/3}$ are Bessel's functions and $K_{1/3}$ is Kelvin's functions.

20.16 Poisson's Integral on Boundary Value Problem

We consider Laplace's equation in polar coordinates

$$\frac{1}{r}\frac{d}{dr}\left(r\frac{du}{dr}\right) + \frac{1}{r^2}\frac{d^2u}{d\theta^2} = 0 \tag{20.176}$$

where the function $u(r, \theta)$ is defined in a circular region of radius R and assumes a given value $f(\theta)$ on the boundarv of the circle. That is

$$u(R, \theta) = f(\theta) \tag{20.177}$$

The solution of this boundary value problem is given by the integral

$$u(r, \theta) = \frac{R^2 - r^2}{2\pi}\int_0^{2\pi} \frac{f(\xi)}{R^2 - 2Rr\cos(\theta - \xi) + r^2}\,d\xi \tag{20.178}$$

The integral in equation (20.178) is known as *Poisson's integral.* To prove this, we assume the solution of equation (20.176) in the form

$$u(r, \theta) = R(r)F(\theta) \tag{20.179}$$

Substituting this solution function into equation (20.176), we get

$$r \cdot \frac{rR'' + R'}{R} = -\frac{F''}{F} = \text{constant} = k^2 \text{ (say)} \tag{20.180}$$

Hence, we have

$$F'' + k^2F = 0 \tag{20.181}$$

Since the function $u(r, \theta)$ should be periodic with a period of 2π, the constant k^2 in equation (20.180) should be such that k's are integers, i.e., $k = 0, 1, 2, 3, \ldots$

Thus, the solution of equation (20.181) is given by

$$F(\theta) = A\cos k\theta + B\sin k\theta \tag{20.182}$$

where A, B are arbitrary constants.

Now, for the radial function $R(r)$, we get

$$r^2R'' + rR' - k^2R = 0 \tag{20.183}$$

For the solution of this equation, we assume

$$R(r) = Cr^\alpha \tag{20.184}$$

and substitute into equation (20.183). Then we get

$$C[\alpha(\alpha-1) + \alpha - k^2]r^\alpha = 0$$

which gives

$$\alpha^2 - k^2 = 0, \quad \text{i.e.,} \quad \alpha = \pm k \tag{20.185}$$

Thus the solution equation (20.184) can be written as

$$R(r) = Cr^k + Dr^{-k} \tag{20.186}$$

where C, D are arbitrary constants. Now, the solution function $u(r,\theta)$ should be finite at $r \longrightarrow 0$. Hence, we take $D = 0$. Thus

$$R(r) = Cr^k \tag{20.187}$$

From equations (20.182) and (20.187), we get the complete solution of equation (20.176) in the form

$$u(r,\theta) = \frac{a_0}{2} + \sum_{k=1}^{\infty}(a_k\cos k\theta + b_k\sin k\theta)\,r^k \tag{20.188}$$

Let us assume that the function $f(\theta)$ defined by equation (20.177) can be expanded in Fourier's series in the form

$$f(\theta) = \frac{A_0}{2} + \sum_{k=1}^{\infty}(A_k\cos k\theta + B_k\sin k\theta) \tag{20.189}$$

Then this series should converge absolutely and uniformly. Moreover, the coefficients are given by

$$\left.\begin{aligned} A_n &= \frac{1}{\pi}\int_0^{2\pi} f(\xi)\cos k\xi d\xi \\ B_n &= \frac{1}{\pi}\int_0^{2\pi} f(\xi)\sin k\xi d\xi \end{aligned}\right\} \tag{20.190}$$

where $k = 0, 1, 2, \ldots$

Now, let us consider the series

$$u(r,\theta) = \frac{A_0}{2} + \sum_{k=1}^{\infty}\left(\frac{r}{R}\right)^k(A_k\cos k\theta + B_k\sin k\theta) \tag{20.191}$$

Since $r \leq R$, i.e., $r/R \leq 1$, this series also should converge absolutely and uniformly. Moreover, the series in equation (20.191) for $r = R$ transforms to the series in equation (20.189). This shows that the series in equation (20.191) for $r = R$ satisfies the boundary condition equation (20.177) and also satisfies the Laplace's equation. Thus, the required solution function is given by equation (20.191). We now intend to show that this series is equivalent to the Poisson's integral.

Substituting the values of A_n, B_n from equation (20.190) into equation (20.191) and interchanging the summation and integration (which is permissible due to uniform convergence of the series), we get

$$u(r,\theta) = \frac{1}{\pi}\int_0^{2\pi} f(\xi)\left[\frac{1}{2} + \sum_{k=1}^{\infty}\left(\frac{r}{R}\right)^k \cos k(\theta - \xi)\right] d\xi \tag{20.192}$$

Now, let us consider the series

$$S = \frac{1}{2} + \sum_{n=1}^{\infty} z^n \tag{20.193}$$

where

$$z = re^{i\phi} = r(\cos\phi + i\sin\phi), \quad (r < 1) \tag{20.194}$$

Then

$$S = \frac{1}{2} + \sum_{n=1}^{\infty} z^n = \frac{1}{2} + \sum_{n=1}^{\infty} r^n(\cos n\phi + i\sin n\phi) \tag{20.195}$$

Again,

$$\begin{aligned} S &= \frac{1}{2} + \sum_{n=1}^{\infty} z^n = \frac{1}{2} + \frac{z}{1-z} \quad (\text{since } |z| < 1) \\ &= \frac{1+z}{2(1-z)} = \frac{(1 + r\cos\phi) + ir\sin\phi}{2(1 - r\cos\phi - ir\sin\phi)} = \frac{(1-r^2) + 2ri\sin\phi}{2(1 - 2r\cos\phi + r^2)} \end{aligned} \tag{20.196}$$

Equating real parts in equations (20.195) and (20.196), we have

$$\frac{1}{2} + \sum_{n=1}^{\infty} r^n \cos n\phi = \frac{1-r^2}{2(1 - 2r\cos\phi + r^2)} \tag{20.197}$$

Replacing $r \longrightarrow r/R$ and $\phi \longrightarrow (\theta - \xi)$, we get

$$\begin{aligned} \frac{1}{2} + \sum_{n=1}^{\infty}\left(\frac{r}{R}\right)^n \cos n(\theta - \xi) &= \frac{1 - (r/R)^2}{2\left[1 - 2\frac{r}{R}\cos(\theta - \xi) + \left(\frac{r}{R}\right)^2\right]} \\ &= \frac{R^2 - r^2}{2\left[R^2 - 2Rr\cos(\theta - \xi) + r^2\right]} \end{aligned} \tag{20.198}$$

Using this formula in equation (20.192), we find the solution of the boundary value problem as

$$u(r,\theta) = \frac{R^2 - r^2}{2\pi}\int_0^{2\pi} \frac{f(\xi)}{R^2 - 2Rr\cos(\theta - \xi) + r^2}\, d\xi \tag{20.199}$$

Thus, we have proved the Poisson's integral.

Note 1. Comparing imaginary parts in equations (20.195) and (20.196), we get another interesting formula

$$\sum_{n=1}^{\infty} r^n \sin n\phi = \frac{r \sin \phi}{1 - 2r \cos \phi + r^2} \quad (0 \le r < 1) \tag{20.200}$$

Note 2. If the given function $f(\theta) \equiv 1$, then from equation (20.190), we get

$$\frac{A_0}{2} = 1, \quad A_n = 0, \quad B_n = 0 \tag{20.201}$$

Hence, from equation (20.191), we have

$$u(r, \theta) \equiv 1 \tag{20.202}$$

Using these values in Poisson's integral, we find

$$1 = \frac{R^2 - r^2}{2\pi} \int_0^{2\pi} \frac{d\xi}{R^2 - 2Rr \cos(\theta - \xi) + r^2} \quad \left(\text{for } 0 \le \frac{r}{R} < 1\right) \tag{20.203}$$

20.17 Fourier's Integral for Mean Partial Sum

Let a function $f(x)$ be expanded in Fourier's series in the form

$$f(x) = \frac{a_0}{2} + \sum_{k=1}^{\infty} (a_k \cos kx + b_k \sin kx) \tag{20.204}$$

We denote the partial sum up to n-terms of this series as

$$S_n(x) = \frac{a_0}{2} + \sum_{k=1}^{n-1} (a_k \cos kx + b_k \sin kx) \tag{20.205}$$

Let the arithmetical mean of the partial sums be $\sigma_n(x)$. Then

$$\sigma_n(x) = \frac{1}{n} [S_0(x) + S_1(x) + \cdots + S_{n-1}(x)] \tag{20.206}$$

Using equation (20.205), we write

$$\sigma_n(x) = \frac{a_0}{2} + \sum_{k=1}^{n-1} \frac{n-k}{n} (a_k \cos kx + b_k \sin kx) \tag{20.207}$$

Now, for the partial sum $S_n(x)$, we have (ref. Chapter 10)

$$S_n(x) = \frac{1}{\pi} \int_{-\pi}^{\pi} f(x + \xi) \frac{\sin \left(n + \frac{1}{2}\right) \xi}{2 \sin \frac{\xi}{2}} \, d\xi \tag{20.208}$$

Hence, we get from equations (20.206) and (20.208)

$$\sigma_n(x) = \frac{1}{n\pi} \int_{-\pi}^{\pi} \frac{f(x + \xi)}{2 \sin \frac{\xi}{2}} \sum_{k=0}^{n-1} \sin \left(k + \frac{1}{2}\right) \xi d\xi \tag{20.209}$$

Now

$$2\sin\frac{\xi}{2}\sin\left(k+\frac{1}{2}\right)\xi = \cos k\xi - \cos(k+1)\xi \tag{20.210}$$

Hence

$$2\sin\frac{\xi}{2}\sum_{k=0}^{n-1}\sin\left(k+\frac{1}{2}\right)\xi = \sum_{k=0}^{n-1}\{\cos k\xi - \cos(k+1)\xi\} = 1-\cos n\xi = 2\sin^2\frac{n\xi}{2}$$

Thus, for $\xi \neq 0$, we get

$$\sum_{k=0}^{n-1}\sin\left(k+\frac{1}{2}\right)\xi = \frac{\sin^2\frac{n\xi}{2}}{\sin\frac{\xi}{2}} \tag{20.211}$$

Using this equation, we get from equation (20.209)

$$\sigma_n(x) = \frac{1}{n\pi}\int_{-\pi}^{\pi} f(x+\xi)\frac{\sin^2\frac{n\xi}{2}}{2\sin^2\frac{\xi}{2}}\,d\xi \tag{20.212}$$

This is Fourier's integral for mean partial sum.

Let $f(x) \equiv 1$, then $S_n(x) = 1$ for $n = 0, 1, 2, \cdots$ and so $\sigma_n(x) = 1$. Putting these values in equation (20.212), we have

$$1 = \frac{1}{n\pi}\int_{-\pi}^{\pi}\frac{\sin^2\frac{n\xi}{2}}{2\sin^2\frac{\xi}{2}}\,d\xi \tag{20.213}$$

Since the integrand is an even function we write

$$\frac{1}{2} = \frac{1}{n\pi}\int_{0}^{\pi}\frac{\sin^2\frac{n\xi}{2}}{2\sin^2\frac{\xi}{2}}\,d\xi \tag{20.214}$$

which can also be written as

$$\int_0^{\pi/2}\frac{\sin^2 nx}{\sin^2 x}\,dx = \frac{n\pi}{2} \tag{20.215}$$

Thus we obtain an interesting integral.

20.18 Fourier's Integrals

In chapter 10 we have seen that a function $f(x)$ satisfying Dirichlet's conditions can be expanded in Fourier's series. Here we shall see that the same function $f(x)$ defined in the interval $(-l, l)$ can be represented by the double integral

$$f(x) = \frac{1}{\pi}\int_0^{\infty} d\xi \int_{-\infty}^{\infty} f(t)\cos\xi(x-t)dt \tag{20.216}$$

This integral is known as *Fourier's integral.* We now establish this equation. Let the Fourier expansion of the function $f(x)$ be

$$f(x) = \frac{a_0}{2} + \sum_{k=1}^{\infty}\left(a_k\cos\frac{k\pi x}{l} + b_k\sin\frac{k\pi x}{l}\right) \tag{20.217}$$

where

$$a_k = \frac{1}{l}\int_{-l}^{l} f(t)\cos\frac{k\pi t}{l}\,dt \quad (k = 0, 1, 2, \cdots) \tag{20.218}$$

and

$$b_k = \frac{1}{l}\int_{-l}^{l} f(t)\sin\frac{k\pi t}{l}\,dt \quad (k = 1, 2, \cdots) \tag{20.219}$$

Substituting the values of the coefficients a_k, b_k into equation (20.217), we get

$$f(x) = \frac{1}{2l}\int_{-l}^{l} f(t)dt + \sum_{k=1}^{\infty}\frac{1}{l}\int_{-l}^{l} f(t)\cos\frac{k\pi}{l}(t-x)dt \tag{20.220}$$

Let us now determine the function in the interval $(-\infty, \infty)$. The equation (20.220) will be valid for any $l > |x|$. If we take $l \longrightarrow \infty$, the first integral in equation (20.220) obviously vanishes. For the other part, we take the variables (for $m = 1, 2, \cdots$)

$$\xi_m = \frac{m\pi}{l} \quad \text{and} \quad \Delta\xi_m = \xi_{m+1} - \xi_m = \frac{\pi}{l}$$

so that $\Delta\xi_m \longrightarrow 0$ as $l \longrightarrow \infty$. Then the equation (20.220) can be written as

$$f(x) = \lim_{l\longrightarrow\infty}\int_{-\infty}^{\infty}\frac{1}{\pi}\sum_{k=1}^{\infty}\Delta\xi_k\int_{-l}^{l} f(t)\cos\xi_k(t-x)dt$$

By definition of integral as the limit of sum, we have

$$f(x) = \frac{1}{\pi}\int_0^{\infty} d\xi\int_{-\infty}^{\infty} f(t)\cos\xi(t-x)dt \tag{20.221}$$

This is the Fourier integral.

If $f(x)$ is an odd function of x, then

$$f(-x) = -f(x) \tag{20.222}$$

In this case

$$\begin{aligned}
\int_{-\infty}^{\infty} f(t)\cos\xi(t-x)dt &= \int_{-\infty}^{0} f(t)\cos\xi(t-x)dt + \int_0^{\infty} f(t)\cos\xi(t-x)dt \\
&= -\int_{\infty}^{0} f(-t)\cos\xi(t+x)dt + \int_0^{\infty} f(t)\cos\xi(t-x)dt \\
&= \int_0^{\infty} f(t)\left[\cos\xi(t-x) - \cos\xi(t+x)\right]dt \\
&= 2\int_0^{\infty} f(t)\sin\xi x\sin\xi t.
\end{aligned} \tag{20.223}$$

Hence, the Fourier's integral equation (20.221) assumes the form

$$f(x) = \frac{2}{\pi}\int_0^{\infty} d\xi\int_0^{\infty} f(t)\sin\xi x\sin\xi t\,dt \tag{20.224}$$

Similarly, the Fourier's integral for an even function $f(x)$ when $f(-x) = f(x)$ is given by

$$f(x) = \frac{2}{\pi}\int_0^{\infty} d\xi \int_0^{\infty} f(t)\cos\xi x \cos\xi t \, dt \tag{20.225}$$

Fourier's integrals are useful in getting an analytical expressions of discontinuous functions defined throughout the infinite range $(-\infty, \infty)$.

For convenience, we split the integral in equation (20.221) into two parts, and write

$$f(x) = I_1 + I_2 \tag{20.226}$$

where

$$I_1 = \frac{1}{\pi}\int_0^{\infty}\left(\int_{-\infty}^{\infty} f(t)\cos\xi t\, dt\right)\cos\xi x d\xi \tag{20.227}$$

and

$$I_2 = \frac{1}{\pi}\int_0^{\infty}\left(\int_{-\infty}^{\infty} f(t)\sin\xi t\, dt\right)\sin\xi x d\xi \tag{20.228}$$

As an example, we consider a function $f(x)$ defined by

$$f(x) = \begin{cases} 0, & \text{for } x < 0 \\ e^{-\alpha x}, & \text{for } x > 0 \end{cases} \tag{20.229}$$

and determine its Fourier's integral.

In this case

$$\int_{-\infty}^{\infty} f(t)\cos\xi t\, dt = \int_{-\infty}^{0} f(t)\cos\xi t\, dt + \int_0^{\infty} f(t)\cos\xi t\, dt$$

$$= \int_0^{\infty} e^{-\alpha t}\cos\xi t\, dt = \frac{\alpha}{\alpha^2+\xi^2}$$

Hence

$$I_1 = \frac{1}{\pi}\int_0^{\infty}\frac{\alpha\cos\xi x}{\alpha^2+\xi^2}d\xi \tag{20.230}$$

Similarly, the integral

$$\int_{-\infty}^{\infty} f(t)\sin\xi t\, dt = \int_0^{\infty} f(t)\sin\xi t\, dt = \int_0^{\infty} e^{-\alpha t}\sin\xi t\, dt = \frac{\alpha\xi}{\alpha^2+\xi^2}$$

Hence

$$I_2 = \frac{1}{\pi}\int_0^{\infty}\frac{\alpha\xi\sin\xi x}{\alpha^2+\xi^2}d\xi \tag{20.231}$$

Thus

$$f(x) = I_1 + I_2 = \frac{\alpha}{\pi}\int_0^{\infty}\frac{\cos\xi x + \xi\sin\xi x}{\alpha^2+\xi^2}\, d\xi \tag{20.232}$$

As another example, let $f(x) = e^{-\alpha x}$ for $x \geq 0$ and $f(x) = f(-x)$ for $\alpha > 0$. The function being even, we have from equation (20.225)

$$f(x) = \frac{2}{\pi}\int_0^{\infty} d\xi\cos\xi x\int_0^{\infty} e^{-\alpha t}\cos\xi t\, dt = \frac{2}{\pi}\int_0^{\infty}\cos\xi x d\xi\left(\frac{\alpha}{\alpha^2+\xi^2}\right)$$

$$\text{i.e.,}\quad e^{-\alpha x} = \frac{2\alpha}{\pi}\int_0^{\infty}\frac{\cos\xi x}{\alpha^2+\xi^2}\, d\xi \tag{20.233}$$

20.19 Complex Form of Fourier's Integral

We can express the Fourier's integral in complex form in the following manner.

We write equation (20.221) as

$$f(x) = \frac{1}{2\pi}\int_{-\infty}^{\infty} d\xi\left[\int_{-\infty}^{\infty} f(t)\cos\xi(t-x)dt\right] \tag{20.234}$$

Since $\sin\xi(t-x)$ is an odd function of ξ, we can write

$$0 = \frac{1}{2\pi}\int_{-\infty}^{\infty} d\xi\left[\int_{-\infty}^{\infty} f(t)\sin\xi(t-x)dt\right] \tag{20.235}$$

From equations (20.234) and (20.235), we get

$$\begin{aligned} f(x) &= \frac{1}{2\pi}\int_{-\infty}^{\infty} d\xi\left[\int_{-\infty}^{\infty} f(t)\left\{\cos\xi(t-x)+i\sin\xi(t-x)\right\}dt\right] \\ &= \frac{1}{2\pi}\int_{-\infty}^{\infty} f(t)e^{i\xi(t-x)}\,dt \end{aligned}$$

$$\text{i.e.,}\quad f(x) = \frac{1}{2\pi}\int_{-\infty}^{\infty} e^{-i\xi x}\,d\xi\left[\int_{-\infty}^{\infty} f(t)e^{i\xi t}\,dt\right] \tag{20.236}$$

Equation (20.236) represents Fourier's integral in complex form. An alternative form may be obtained by changing $\xi \longrightarrow -\xi$. Then we get

$$f(x) = \frac{1}{2\pi}\int_{-\infty}^{\infty} e^{i\xi x}\,d\xi\left[\int_{-\infty}^{\infty} f(t)e^{-i\xi t}\,dt\right] \tag{20.237}$$

EXERCISE

1. Show that the asymptotic expansion of the error function can be expressed as

$$erf(x) \sim 1 - \frac{e^{-x^2}}{x\sqrt{x}}\left(1 - \frac{1}{2x^2} + \frac{1.3}{4x^4} - \frac{1.5}{8x^6} + \cdots\right)$$

2. Establish the relations:

(a) $\displaystyle\int_0^{\infty} e^{-ax}J_n(bx)\frac{dx}{x} = \frac{\left[\sqrt{a^2+b^2}-a\right]^n}{nb^n}$;

(b) $\displaystyle\int_0^{\infty} K_0(ax)\cos bx\,dx = \frac{\pi}{2}\left(a^2+b^2\right)^{-1/2}$;

(c) $\displaystyle\int_0^{\infty} J_n(x)\sin ax\,dx = \frac{\sin(n\sin^{-1}a)}{\sqrt{1-a^2}}$, $|a|<1$, $n>-2$.

3. Show that

(a) $\displaystyle\int_0^{\pi/2} \sqrt{\cos x}\,dx = 2\sqrt{2}E\left(\sqrt{\frac{1}{2}}\right) - \sqrt{2}K\left(\sqrt{\frac{1}{2}}\right)$;

(b) $\displaystyle\int_0^2 \frac{dx}{\sqrt{(4-x^2)(9-x^2)}} = \frac{1}{3}F\left(\frac{2}{3}, \frac{\pi}{2}\right);$

(c) $\displaystyle\int_1^\infty \frac{d\xi}{\sqrt{(\xi^2-1)(\xi^2+3)}} = \frac{1}{2}K\left(\frac{\sqrt{3}}{2}\right).$

4. A function $f(x)$ is defined in the following manner:

$$f(x) = \begin{cases} e^{-\lambda x}, & \text{for } x > 0 \\ f(-x), & \text{for } \lambda > 0 \end{cases}$$

Show that its Fourier's integral is given by

$$f(x) = \frac{2\lambda}{\pi}\int_0^\infty \frac{\cos \xi x}{\lambda^2 + \xi^2}\, d\xi$$

Chapter 21

Numerical Methods

21.1 Introduction

We frequently face with physics and engineering problems which cannot be solved by using existing analytical methods due to their complex nature. In such cases, we have to resort to some numerical techniques for solving the problems. With the advent of computers and availability of many useful software packages, the numerical methods have gained a high degree of popularity.

Numerical methods are founded on the branch of mathematics called *Numerical Analysis* which is a well-developed vast subject. We intend to discuss in this chapter some topics which we frequently require in solving physical problems. For an in-depth study of numerical techniques, advanced textbooks on the subject may be consulted.

21.2 Finite-Difference Operators

We consider a single-valued function $y = f(x)$ which is defined by a set of numerical values y_k corresponding to equally spaced values x_k of x. We define the following finite-difference operators which are useful in understanding the numerical methods.

(a) *Forward-Difference Operators*: Let the function $y = f(x)$ be defined by its values y_0, y_1, y_2, ..., y_k corresponding to equally spaced values x_0, x_1, ..., x_k of x. The common spacing between any two values of x be h, so that

$$\Delta x = x_{k+1} - x_k = h \tag{21.1}$$

The forward-difference of first order is defined by

$$\Delta f_k = f_{k+1} - f_k = f(x_{k+1}) - f(x_k) \tag{21.2}$$

where Δ is the forward-difference operator.

The forward-difference of second order is

$$\begin{aligned}\Delta^2 f_k &= \Delta(f_{k+1} - f_k) = \Delta f_{k+1} - \Delta f_k = (f_{k+2} - f_{k+1}) - (f_{k+1} - f_k)\\ &= f_{k+2} - 2f_{k+1} + f_k = f(x_{k+2}) - 2f(x_{k+1}) + f(x_k)\end{aligned} \tag{21.3}$$

Proceeding in this way we can get

$$\Delta^3 f_k = f_{k+3} - 3f_{k+2} + 3f_{k+1} - f_k \tag{21.4}$$

and so on. These differences are exhibited in Table 21.1.

Table 21.1: Forward-Differences

x	f	Δ	Δ^2	Δ^3	Δ^4
x_0	f_0				
		Δf_0			
x_1	f_1		$\Delta^2 f_0$		
		Δf_1		$\Delta^3 f_0$	
x_2	f_2		$\Delta^2 f_1$		$\Delta^4 f_0$
		Δf_2		$\Delta^3 f_1$	
x_3	f_3		$\Delta^2 f_2$		
		Δf_3			
x_4	f_4				

(b) *Backward-Difference Operators*: The first order backward-difference of the function $f(x)$ is defined as

$$\nabla f_k = f_k - f_{k-1} = f(x_k) - f(x_{k-1}) \tag{21.5}$$

where ∇ is the backward-difference operator.

The second order backward-difference of $f(x)$ is

$$\nabla^2 f_k = \nabla(\nabla f_k) = \nabla f_k - \nabla f_{k-1} = f_k - 2f_{k-1} + f_{k-2} \tag{21.6}$$

Similarly, the third order backward-difference of $f(x)$ is

$$\nabla^3 f_k = f_k - 3f_{k-1} + 3f_{k-2} - f_{k-3} \tag{21.7}$$

These differences are shown in the Table 21.2.

Table 21.2: Backward-Differences

x	f	∇	∇^2	∇^3	∇^4
x_0	f_0				
		∇f_1			
x_1	f_1		$\nabla^2 f_2$		
		∇f_2		$\nabla^3 f_3$	
x_2	f_2		$\nabla^2 f_3$		$\nabla^4 f_4$
		∇f_3		$\nabla^3 f_4$	
x_3	f_3		$\nabla^2 f_4$		
		∇f_4			
x_4	f_4				

(c) *Central-Difference Operator*: We denote

$$\left.\begin{aligned} f_{k+\frac{1}{2}} &= f(x_k + \Delta x/2) \\ f_{k-\frac{1}{2}} &= f(x_k - \Delta x/2) \end{aligned}\right\} \tag{21.8}$$

The first order central-difference of $f(x)$ is defined as

$$\delta f_k = f_{k+\frac{1}{2}} - f_{k-\frac{1}{2}} \tag{21.9}$$

where δ is the central difference operator.

The second order central difference is

$$\begin{aligned}\delta^2 f_k &= \delta(\delta f_k) = \delta\left(f_{k+\frac{1}{2}} - f_{k-\frac{1}{2}}\right)\\ &= \left(f_{k+\frac{1}{2}+\frac{1}{2}} - f_{k+\frac{1}{2}-\frac{1}{2}}\right) - \left(f_{k-\frac{1}{2}+\frac{1}{2}} - f_{k-\frac{1}{2}-\frac{1}{2}}\right)\\ &= f_{k+1} - 2f_k + f_{k-1}.\end{aligned} \tag{21.10}$$

Similarly, the third order central difference is

$$\delta^3 f_k = f_{k+\frac{3}{2}} - 3f_{k+\frac{1}{2}} + 3f_{k-\frac{1}{2}} - f_{k-\frac{3}{2}} \tag{21.11}$$

These differences are exhibited in Table 21.3.

Table 21.3: Central-Differences

x	f	δ	δ^2	δ^3	δ^4
x_0	f_0				
		$\delta f_{1/2}$			
x_1	f_1		$\delta^2 f_1$		
		$\delta f_{3/2}$		$\delta^3 f_{3/2}$	
x_2	f_2		$\delta^2 f_2$		$\delta^4 f_2$
		$\delta f_{5/2}$		$\delta^3 f_{5/2}$	
x_3	f_3		$\delta^2 f_3$		
		$\delta f_{7/2}$			
x_4	f_4				

21.3 Shift Operator

The shift operator E is defined by

$$E f_k = f_{k+1} \tag{21.12}$$

Thus, the shift operator E, operating on f_k gives the next higher tabulated value of $f(x)$. For each integer $n > 0$, the above definition of E implies that

$$E^n f_k = f_{k+n} \tag{21.13}$$

For non-integral value of n (say α), we define

$$E^\alpha f_k = f_{k+\alpha} = f(x_k + \alpha\Delta x) \tag{21.14}$$

which reduces to equation (21.12) for $\alpha = n$. Moreover, we find that

$$E^{-1} f_k = f_{k-1}; \quad E^{1/2} f_k = f_{k+\frac{1}{2}}; \quad E^{-1/2} f_k = f_{k-\frac{1}{2}} \tag{21.15}$$

The *averaging or mean operator* μ is defined as

$$\mu = \tfrac{1}{2}\left(E^{1/2} + E^{-1/2}\right) \tag{21.16}$$

Then

$$\mu f_k = \tfrac{1}{2}\left(E^{1/2} f_k + E^{-1/2} f_k\right) = \tfrac{1}{2}\left(f_{k+\frac{1}{2}} + f_{k-\frac{1}{2}}\right) \tag{21.17}$$

From equations (21.2) and (21.12)

$$\Delta f_k = f_{k+1} - f_k = Ef_k - f_k = (E-1)f_k$$

Hence

$$\Delta = E - 1 \quad \text{or,} \quad E = \Delta + 1 \tag{21.18}$$

Equation (21.18) gives the relation between shifting operator and forward difference operator. Similarly, from equations (21.5) and (21.15) we write

$$\nabla f_k = f_k - f_{k-1} = f_k - E^{-1}f_k = (1 - E^{-1})f_k$$

Hence

$$\nabla = 1 - E^{-1}, \quad \text{or,} \quad E^{-1} = 1 - \nabla \tag{21.19}$$

which gives the relation between shifting operator and backward-difference operator.

Some more symbolic relation of this type are given in examples and exercises.

EXAMPLE 21.1. *Show that* (a) $\delta = E^{1/2} - E^{-1/2}$, (b) $\mu^2 = 1 + \frac{1}{4}\delta^2$.

Solution: We have, by definition

$$\begin{aligned} \delta f_k &= f_{k+\frac{1}{2}} - f_{k-\frac{1}{2}} = E^{1/2}f_k - E^{-1/2}f_k \quad \text{[using equation (21.13)]} \\ &= \left(E^{1/2} - E^{-1/2}\right) f_k \end{aligned}$$

Hence

$$\delta = E^{1/2} - E^{-1/2} \tag{21.20}$$

(b) We have, by definition

$$\mu = \tfrac{1}{2}\left(E^{1/2} + E^{-1/2}\right)$$

Squaring both sides

$$\begin{aligned} \mu^2 &= \tfrac{1}{4}\left(E^{1/2} + E^{-1/2}\right)^2 = \tfrac{1}{4}(E + 2 + E^{-1}) = \tfrac{1}{4}\left[(E - 2 + E^{-1}) + 4\right] \\ &= \tfrac{1}{4}\left[\left(E^{1/2} - E^{-1/2}\right)^2 + 4\right] = \tfrac{1}{4}\left(\delta^2 + 4\right) \quad \text{[using equation (21.20)]} \end{aligned}$$

Hence

$$\mu^2 = 1 + \tfrac{1}{4}\delta^2 \tag{21.21}$$

EXAMPLE 21.2. *Show that* $\nabla = -\frac{1}{2}\delta^2 \left(1 + \frac{\delta^2}{4}\right)^{1/2}$.

Solution: From equation (21.19) we have

$$E^{-1} = 1 - \nabla$$

Substituting this relation into equation (21.20)

$$\delta = (1-\nabla)^{-1/2} - (1-\nabla)^{1/2} = \frac{1}{\sqrt{1-\nabla}} - \sqrt{1-\nabla} = \frac{\nabla}{\sqrt{1-\nabla}} \quad \text{or,} \quad \nabla^2 + \delta^2\nabla - \delta^2 = 0$$

The positive root of this equation gives

$$\nabla = -\frac{1}{2}\delta^2 + \delta\sqrt{1 + \frac{\delta^2}{4}}$$

Hence the result.

21.4 Differential Operators

Considering the function $y = f(x)$, we denote the differentiating operator as

$$D = \frac{d}{dx}, \quad D^2 = \frac{d^2}{dx^2}, \quad D^3 = \frac{d^3}{dx^3}, \quad \text{etc.} \tag{21.22}$$

Now

$$f_{k+1} = f(x_k + h)$$

which by using Taylor's series, can be written as

$$f_{k+1} = f_k + hf'_k + \frac{h^2}{2!}f''_k + \frac{h^3}{3!}f'''_k + \cdots \tag{21.23}$$

where the primes denote differentiation with respect to the independent variable x.

Using equations (21.12) and (21.22), we can write equation (21.23) in the form

$$Ef_k = \left[1 + hD + \frac{h^2}{2!}D^2 + \frac{h^3D^3}{3!} + \cdots\right] f_k = e^{hD} f_k$$

Hence

$$E = e^{hD} \tag{21.24}$$

Equation (21.24) gives the relation between the operator E to operator D.
Since $E = \Delta + 1$, we get from equation (21.24)

$$\Delta = E - 1 = e^{hD} - 1 \tag{21.25}$$

This gives

$$D = \frac{1}{h}\ln E = \frac{1}{h}\ln(\Delta + 1) \tag{21.26}$$

In series form, we can write

$$D = \frac{1}{h}\left(\Delta - \frac{\Delta^2}{2} + \frac{\Delta^3}{3} - \cdots\right) \tag{21.27}$$

Squaring both sides

$$D^2 = \frac{1}{h^2}\left(\Delta^2 - \Delta^3 + \frac{11}{12}\Delta^4 - \frac{5}{6}\Delta^5 - \cdots\right) \tag{21.28}$$

Multiplying equations (21.27) and (21.28) we get

$$D^3 = \frac{1}{h^3}\left(\Delta^3 - \frac{3}{2}\Delta^4 + \frac{7}{4}\Delta^5 - \cdots\right) \tag{21.29}$$

Similarly, other higher order differential operators can be obtained in terms of forward difference operator.

We can also express Δ in terms of the operator D by using equation (21.25). Expressing e^{hD} in series form, we write

$$\Delta = hD + \frac{h^2D^2}{2!} + \frac{h^3D^3}{3!} + \cdots \tag{21.30}$$

Squaring both sides, we get the second forward-difference in terms of the operator D. Thus

$$\Delta^2 = \left(hD + \frac{h^2D^2}{2!} + \frac{h^3D^3}{3!} + \cdots\right)^2 = h^2D^2 + h^3D^3 + \frac{7}{12}h^4D^4 + \cdots \tag{21.31}$$

Differential operator can also be related to the backward-difference operator by using equations (21.19) and (21.24).

Thus we write $E^{-1} = 1 - \nabla$ and $E^{-1} = e^{-hD}$, which yield

$$D = -\frac{1}{h}\ln E^{-1} = -\frac{1}{h}\ln(1-\nabla) = \frac{1}{h}\left(\nabla + \frac{1}{2}\nabla^2 + \frac{1}{3}\nabla^3 + \cdots\right) \tag{21.32}$$

Squaring we get

$$D^2 = \frac{1}{h^2}\left(\nabla^2 + \nabla^3 + \frac{11}{12}\nabla^4 + \frac{5}{6}\nabla^5 + \cdots\right) \tag{21.33}$$

Multiplying equations (21.32) and (21.33)

$$D^3 = \frac{1}{h^3}\left(\nabla^3 + \frac{3}{2}\nabla^4 + \frac{7}{4}\nabla^5 + \frac{15}{8}\nabla^6 + \cdots\right) \tag{21.34}$$

Thus we can get higher order differential operators in terms of the backward-difference operator.

The central-difference operator δ can also be expressed in terms of the differential operator D. Thus, from equations (21.20) and (21.24)

$$\begin{aligned}\delta &= E^{1/2} - E^{-1/2} = e^{hD/2} - e^{-hD/2}\\ &= \left(1 + \frac{hD}{2} + \frac{h^2D^2}{8} + \cdots\right) - \left(1 - \frac{hD}{2} + \frac{h^2D^2}{8} - \cdots\right)\\ &= hD + \frac{h^3D^3}{24} + \frac{h^5D^5}{1920} + \cdots\end{aligned} \tag{21.35}$$

Squaring both sides

$$\delta^2 = h^2D^2 + \frac{h^4D^4}{12} + \frac{h^6D^6}{360} + \cdots \tag{21.36}$$

The averaged central-difference

$$\mu\delta = \frac{1}{2}\left(E - E^{-1}\right) = \frac{1}{2}\left(e^{hD} - e^{-hD}\right)$$

$$\text{or,}\quad \mu\delta = \sinh(hD) \tag{21.37}$$

$$\text{or,}\quad \mu\delta = hD + \frac{h^3D^3}{3!} + \frac{h^5D^5}{5!} + \cdots \tag{21.38}$$

Equation (21.37) yields

$$D = \frac{1}{h}\sinh^{-1}(\mu\delta) = \frac{1}{h}\left(\mu\delta - \frac{\mu^3\delta^3}{6} + \frac{3}{40}\mu^5\delta^5 - \cdots\right) \tag{21.39}$$

EXAMPLE 21.3. *Show that* $\cosh(hD) = \frac{1}{2}(\delta^2 + 2)$. *Hence find* δ^2 *in terms of* D.

Solution: We have $\delta = E^{1/2} - E^{-1/2}$.

Hence $\delta^2 = E + E^{-1} - 2 = e^{hD} + e^{-hD} - 2$ or, $\delta^2 + 2 = 2\cosh(hD)$.

Then

$$\delta^2 = 2[\cosh(hD) - 1] = 2\left[1 + \frac{h^2D^2}{2!} + \frac{h^4D^4}{4!} + \frac{h^6D^6}{6!} + \cdots - 1\right]$$
$$= h^2D^2 + \frac{1}{12}h^4D^4 + \frac{1}{360}h^6D^6 + \cdots$$

21.5 Derivatives in Finite Difference Forms

Considering the function $y = f(x)$ defined by its values $f_0, f_1, \ldots$ at equispace points $x = x_0, x_1, \ldots$ we write the first derivative of the function at $x = x_k$ following equation (21.27) as

$$Df_k = \frac{1}{h}\left(\Delta - \frac{\Delta^2}{2} + \frac{\Delta^3}{3} - \cdots\right) f_k \tag{21.40}$$

Similarly, the second derivative is by equation (21.28)

$$D^2 f_k = \frac{1}{h^2}\left(\Delta^2 - \Delta^3 + \frac{11}{12}\Delta^4 - \cdots\right) f_k \tag{21.41}$$

Now, we cannot use all the terms in the infinite series in the r.h.s. of equations (21.40) and (21.41) and they are truncated by considering a few terms. *The sum of the omitted terms is called the truncation error and magnitude of the first truncated term is called the order of the magnitude of the error caused due to truncation.*

In equation (21.40), if we consider only the first term, then

$$Df_k = \frac{1}{h}\Delta f_k = \frac{1}{h}(f_{k+1} - f_k) \tag{21.42}$$

The first truncated term is $\frac{1}{2}\Delta^2 f_k$ which, by equation (21.31) is of order h^2. Hence the order of the magnitude of error $= h^2/h = h$; we express it as $e = 0(h)$.

Similarly, considering the first two terms in equation (21.40)

$$Df_k = \frac{1}{h}\left(\Delta f_k - \frac{1}{2}\Delta^2 f_k\right) = \frac{1}{h}\left[(f_{k+1} - f_k) - \frac{1}{2}(f_{k+2} - 2f_{k+1} + f_k)\right]$$
$$= \frac{1}{2h}(-f_{k+2} + 4f_{k+1} - 3f_k) \tag{21.43}$$

Here the first truncated term is $\Delta^3/3h$ which is of order h^2; hence the truncation error is $e = 0(h^2)$. Proceeding in the same way, the higher derivatives are expressed below:

(i) *For* $e = 0(h)$

$$D^2 f_k = \frac{1}{h^2}\Delta^2 f_k = \frac{1}{h^2}(f_{k+2} - 2f_{k+1} + f_k) \tag{21.44}$$
$$D^3 f_k = \frac{1}{h^3}\Delta^3 f_k = \frac{1}{h^3}(f_{k+3} - 3f_{k+2} + 3f_{k+1} - f_k) \tag{21.45}$$

and so on.

(ii) *For* $e = 0(h^2)$

$$D^2 f_k = \frac{1}{h^2}(\Delta^2 f_k - \Delta^3 f_k)$$
$$= \frac{1}{h^2}(f_{k+2} - 2f_{k+1} + f_k - f_{k+3} + 3f_{k+2} - 3f_{k+1} + f_k)$$
$$= \frac{1}{h^2}(-f_{k+3} + 4f_{k+2} - 5f_{k+1} + 2f_k) \quad (21.46)$$
$$D^3 f = \frac{1}{h^3}\left(\Delta^3 f_k - \frac{3}{2}\Delta^4 f_k\right)$$
$$= \frac{1}{h^3}\left[(f_{k+3} - 3f_{k+2} + 3f_{k+1} - f_k) - \frac{3}{2}(f_{k+4} - 4f_{k+3} + 6f_{k+2} - 4f_{k+1} + f_k)\right]$$
$$= \frac{1}{2h^3}(-3f_{k+4} + 14f_{k+3} - 24f_{k+2} + 18f_{k+1} - 5f_k) \quad (21.47)$$

and so on.

The derivatives can also be expressed in backward finite difference form. These are given below:

(i) *For* $e = 0(h)$

$$Df_k = \frac{1}{h}(f_k - f_{k-1}) \quad (21.48)$$
$$D^2 f_k = \frac{1}{h^2}(f_k - 2f_{k-1} + f_{k-2}) \quad (21.49)$$
$$D^3 f_k = \frac{1}{h^3}(f_k - 3f_{k-1} + 3f_{k-2} - f_{k-3}) \quad (21.50)$$

and so on.

(ii) *For* $e = 0(h^2)$

$$Df_k = \frac{1}{2h}(3f_k - 4f_{k-1} + f_{k-2}) \quad (21.51)$$
$$D^2 f_k = \frac{1}{h^2}(2f_k - 5f_{k-1} + 4f_{k-2} - f_{k-3}) \quad (21.52)$$
$$D^3 f_k = \frac{1}{2h^3}(5f_k - 18f_{k-1} + 24f_{k-2} - 14f_{k-3} + 3f_{k-4}) \quad (21.53)$$

and so on.

The derivatives in central differences with $e = 0(h^2)$ are expressed as

$$Df_k = \frac{1}{2h}(f_{k+1} - f_{k-1})$$
$$D^2 f_k = \frac{1}{h^2}(f_{k+1} - 2f_k + f_{k-1})$$

and so on.

The difference equations are used in finding approximate solutions of differential equations. This is discussed in a later section.

21.6 Numerical Interpolation

Consider the function $y = f(x)$ whose values at equally-spaced points $x = x_0, x_1, x_2, \ldots, x_k$ are given as $f_0, f_1, f_2, \ldots, f_k$. The values of the function at the point $x = x_i + \alpha h$ where $0 < \alpha < 1$, is not given but can be interpolated.

Thus, given a set of values $\{f_k(x)\}$ of a function $f(x)$ at a given set of points $\{x_k\}$, the method of finding the values of the function at a point lying within an interval is known as *Interpolation.*

(a) *Newton's Forward-Difference Interpolation*: For $0 < \alpha < 1$, and $0 \le i \le k$, we write the shifting operator formula by equation (21.14) as $f_{i+\alpha} = E^{\alpha} f_i$.

Substituting $E = 1 + \Delta$ here, we get

$$f_{i+\alpha} = (1+\Delta)^{\alpha} f_i \tag{21.54}$$

Now, using binomial theorem

$$(1+\Delta)^{\alpha} = 1 + \alpha\Delta + \frac{\alpha(\alpha-1)}{2!}\Delta^2 + \frac{\alpha(\alpha-1)(\alpha-2)}{3!}\Delta^3 + \cdots$$

Hence equation (21.54) can be written as

$$f_{i+\alpha} = \left[1 + \alpha\Delta + \frac{\alpha(\alpha-1)}{2!}\Delta^2 + \frac{\alpha(\alpha-1)(\alpha-2)}{3!}\Delta^3 + \cdots\right] f_i \tag{21.55}$$

Neglecting terms of order higher than Δ^3, we write

$$\begin{aligned} f_{i+\alpha} &= \left[1 + \alpha\Delta + \frac{\alpha(\alpha-1)}{2!}\Delta^2 + \frac{\alpha(\alpha-1)(\alpha-2)}{3!}\Delta^3\right] f_i \\ &= f_i + \alpha(f_{i+1} - f_i) + \frac{\alpha(\alpha-1)}{2}(f_{i+2} - 2f_{i+1} + f_i) \\ &\quad + \frac{\alpha(\alpha-1)(\alpha-2)}{6}(f_{i+3} - 3f_{i+2} + 3f_{i+1} - f_i) \end{aligned} \tag{21.56}$$

The first omitted term in this case is

$$\frac{\alpha(\alpha-1)(\alpha-2)(\alpha-3)}{4!}\Delta^4 f_i \tag{21.57}$$

Considering the maximum value of $\Delta^4 f_i$, which is of the order of h^4, the truncation error will be $e = 0(h^4)$ which is very small.

(b) *Newton's Backward-Difference Interpolation*: For $0 < \alpha < 1$ and $0 \le i \le k$, the shifting operator formula is written as $f_{i-\alpha} = E^{-\alpha} f_i$ which, upon using equation (21.19), assumes the form

$$f_{i-\alpha} = (1-\nabla)^{\alpha} f_i \tag{21.58}$$

Expanding binomially

$$f_{i-\alpha} = \left[1 - \alpha\nabla + \frac{\alpha(a-1)}{2!}\nabla^2 - \frac{\alpha(\alpha-1)(\alpha-2)}{3!}\nabla^3 + \cdots\right] f_i \tag{21.59}$$

Neglecting terms of order higher than ∇^3

$$
\begin{aligned}
f_{i-\alpha} &= f_i - \alpha\nabla f_i + \frac{\alpha(\alpha-1)}{2}\nabla^2 f_i - \frac{\alpha(\alpha-1)(\alpha-2)}{6}\nabla^3 f_i \\
&= f_i - \alpha(f_i - f_{i-1}) + \frac{\alpha(\alpha-1)}{2}(f_i - 2f_{i-1} + f_{i-2}) \\
&\quad - \frac{\alpha(\alpha-1)(\alpha-2)}{6}(f_i - 3f_{i-1} + 3f_{i-2} - f_{i-3})
\end{aligned}
\tag{21.60}
$$

Here also the truncation error will be $e = 0(h^4)$.

Newton's interpolation formulae are for equispaced arguments. The forward interpolation formula is applicable for interpolating near the beginning of the given set $\{x_k\}$, whereas the backward interpolation formula is applied to interpolate at points near the end of $\{x_k\}$.

EXAMPLE 21.4. *A table of values of x and $f(x)$ is provided below:*

x	0.30	0.32	0.34	0.36	0.38	0.40
$f(x)$	1.659	1.668	1.681	1.692	1.704	1.719

Calculate the values of $f(0.33)$ and $f(0.37)$.

Solution: (a) The point $x = 0.33$ lies close to the beginning of the table. Hence we use Newton's forward difference formula.

Here $i = 1$, $h = 0.02$, $\alpha = (0.33 - 0.32)/0.02 = 0.5$

$$\frac{\alpha(\alpha-1)}{2} = \frac{0.5(0.5-1)}{2} = -0.125.$$

$$\frac{\alpha(\alpha-1)(\alpha-2)}{6} = -0.125 \times \frac{(0.5-2)}{3} = 0.0625.$$

$$
\begin{aligned}
\Delta f_1 &= f_2 - f_1 = 1.681 - 1.668 = 0.013 \\
\Delta^2 f_1 &= f_3 - 2f_2 + f_1 = 1.692 - 2\times 1.681 + 1.668 = -0.002 \\
\Delta^3 f_1 &= f_4 - 3f_3 + 3f_2 - f_1 = 1.704 - 3\times 1.692 + 3\times 1.681 - 1.668 = -0.017.
\end{aligned}
$$

Therefore, by formula (21.56) with $e = 0(h^4)$

$$f(0.33) = 1.668 + 0.5\times 0.013 + 0.125\times 0.002 + 0.0625\times(-0.017) = 1.6737.$$

(b) Here $x = 0.37$ which is near the end of the table. Hence we apply Newton's backward interpolation formula given by equation (21.60) with $e = 0(h^4)$.

Thus, $i = 4, h = 0.02, \alpha = 0.5$.

$$
\begin{aligned}
\nabla f_4 &= f_4 - f_3 = 1.704 - 1.692 = 0.012 \\
\nabla^2 f_4 &= f_4 - 2f_3 + f_2 = 1.704 - 2\times 1.692 + 1.681 = 0.001 \\
\nabla^3 f_4 &= f_4 - 3f_3 + 3f_2 - f_1 = 1.704 - 3\times 1.692 + 3\times 1.681 - 1.668 = 0.003.
\end{aligned}
$$

Hence $f(0.37) = 1.704 - 0.5\times 0.012 - 0.125\times 0.001 - 0.0625\times 0.003 = 1.6977.$

(c) *Gauss's Forward-Difference Interpolation*: For interpolation near middle of the set $\{x_k\}$ of the given arguments we use Gauss's forward difference formula. The formula is assumed in the form

$$f(x_{i+\alpha}) = f_i + G_1\Delta f_i + G_2\Delta^2 f_{i-1} + G_3\Delta^3 f_{i-1} + G_4\Delta^4 f_{i-2} + \cdots \tag{21.61}$$

where the constants $G_1, G_2, \ldots$ are to be determined.

Now

$$\begin{aligned}\Delta^2 f_{i-1} &= \Delta^2 E^{-1} f_i = \Delta^2(1+\Delta)^{-1} f_i\\ &= \Delta^2(1-\Delta+\Delta^2-\Delta^3+\cdots)f_i\\ &= (\Delta^2-\Delta^3+\Delta^4-\Delta^5+\cdots)f_i\end{aligned} \tag{21.62}$$

Similarly

$$\Delta^3 f_{i-1} = \left(\Delta^3-\Delta^4+\Delta^5-\Delta^6+\cdots\right) f_i \tag{21.63}$$

$$\Delta^4 f_{i-2} = \left(\Delta^4-2\Delta^5+3\Delta^6-\cdots\right) f_i \tag{21.64}$$

Again, $f(x_{i+\alpha}) = f_{i+\alpha}$, which is expressed by equation (21.55). Substituting the values of $\Delta^2 f_{i-1}, \Delta^3 f_{i-1}$, etc. into equation (21.61) and comparing the result with equation (21.55), we get

$$\left.\begin{aligned}&G_1 = \alpha && G_2 = \alpha(\alpha-1)/2!\\ &G_3 = (\alpha+1)\alpha(\alpha-1)/3! && G_4 = (\alpha+1)\alpha(\alpha-1)(\alpha-2)/4!\end{aligned}\right\} \tag{21.65}$$

and so on.

Hence equation (21.61) is written as

$$\begin{aligned}f_{i+\alpha} = f_i + \alpha\Delta f_i + \frac{\alpha(\alpha-1)}{2!}\Delta^2 f_{i-1} + \frac{(\alpha+1)\alpha(\alpha-1)}{3!}\Delta^3 f_{i-1}\\ + \frac{(\alpha+1)\alpha(\alpha-1)(\alpha-2)}{4!}\Delta^4 f_{i-2} + \cdots\end{aligned} \tag{21.66}$$

(d) *Gauss's Backward Difference Interpolation*: Gauss's backward-difference formula is assumed in the form

$$f_{i+\alpha} = f_i + G_1\Delta f_{i-1} + G_2\Delta^2 f_{i-1} + G_3\Delta^3 f_{i-2} + G_4\Delta^4 f_{i-2} + \cdots \tag{21.67}$$

Proceeding as before, we can find

$$\left.\begin{aligned}&G_1 = \alpha && G_2 = \frac{\alpha(\alpha+1)}{2!}\\ &G_3 = \frac{(\alpha+1)\alpha(\alpha-1)}{3!} && G_4 = \frac{(\alpha+2)(\alpha+1)\alpha(\alpha-1)}{4!}\end{aligned}\right\} \tag{21.68}$$

and so on.

Hence Gauss's backward-difference formula is

$$\begin{aligned}f_{i+\alpha} = f_i + \alpha\Delta f_{i-1} + \frac{\alpha(\alpha+1)}{2!}\Delta^2 f_{i-1} + \frac{(\alpha+1)(\alpha)(\alpha-1)}{3!}\Delta^3 f_{i-2}\\ + \frac{(\alpha+2)(\alpha+1)\alpha(\alpha-1)}{4!}\Delta^4 f_{i-2} + \cdots\end{aligned} \tag{21.69}$$

(e) *Stirling's Interpolation Formula*: Stirling's interpolation formula is the average of Gauss's forward and backward difference formulae. Thus Stirling's interpolation formula is given by

$$f_{i+\alpha} = f_i + \frac{\alpha}{2}(\Delta f_i + \Delta f_{i-1}) + \frac{\alpha^2}{2}\Delta^2 f_{i-1} + \frac{\alpha(\alpha^2-1)}{12}(\Delta^3 f_{i-1} + \Delta^3 f_{i-2}) + \frac{\alpha^2(\alpha^2-1)}{24}\Delta^4 f_{i-2} + \cdots \tag{21.70}$$

EXAMPLE 21.5. *Calculate the value of $f(0.35)$ with $e = 0(h^3)$ using the table of values of x and $f(x)$ of the previous example.*

Solution: The argument lies between x_2 and x_3 close to the middle of the table and so we use Gauss's formula.

Here $i = 2, h = 0.02, \alpha = 0.5$.

Using Gauss's forward-difference formula with $e = 0(h^3)$,

$$\begin{aligned} f_{2+\alpha} &= f_2 + \alpha\Delta f_2 + \frac{\alpha(\alpha-1)}{2}\Delta^2 f_1 \\ &= f_2 + \alpha(f_3 - f_2) + \frac{\alpha(\alpha-1)}{2}(f_3 - 2f_2 + f_1) \\ &= 1.681 + 0.5(1.692 - 1.681) - 0.125(1.692 - 2 \times 1.681 + 1.668) \\ &= 1.681 + 0.0055 - 0.125(-0.002) = 1.68675. \end{aligned}$$

Using Gauss's backward-difference formula

$$\begin{aligned} f_{2+\alpha} &= f_2 + \alpha\Delta f_1 + \frac{\alpha(\alpha+1)}{2}\Delta^2 f_1 \\ &= f_2 + 0.5(f_2 - f_1) + 0.375(f_3 - 2f_2 + f_1) \\ &= 1.681 + 0.5(1.681 - 1.668) + 0.375(-0.002) \\ &= 1.68675 \end{aligned}$$

Using Stirlings' formula with $e = 0(h^3)$

$$\begin{aligned} f_{2+\alpha} &= f_2 + \frac{\alpha}{2}(\Delta f_2 + \Delta f_1) + \frac{\alpha^2}{2}\Delta^2 f_1 \\ &= f_2 + 0.25(f_3 - f_1) + 0.125(f_3 - 2f_2 + f_1) \\ &= 1.681 + 0.25(1.692 - 1.668) + 0.125(-0.002) \\ &= 1.68675 \end{aligned}$$

Thus we get the same value of interpolation at the point $x = 0.35$.

(f) *Lagrange's Interpolation Formula*: This formula is applicable for both equal and unequal intervals of arguments. Suppose, we are given the values of the function $f(x)$ as $f_0, f_1, f_2, \ldots, f_k$ corresponding to the arguments $x_0, x_1, x_2, \ldots, x_k$ which are not necessarily equally spaced. We assume the function $f(x)$ to be a polynomial of order k. This assumption is justified by a theorem due to Weierstrass which states:

If a function $f(x)$ is continuous in the interval $[x_0, x_k]$, then, for a given $\epsilon > 0$, there exists a polynomial $P(x)$ such that

$$|f(x) - P(x)| < \epsilon \text{ for all } x \text{ in } [x_0, x_k]$$

Thus we may write

$$\begin{aligned} f(x) = {} & A_0(x - x_1)(x - x_2)\cdots(x - x_k) \\ & + A_1(x - x_0)(x - x_2)\cdots(x - x_k) \\ & \cdots\cdots\cdots\cdots\cdots\cdots\cdots\cdots \\ & \cdots\cdots\cdots\cdots\cdots\cdots\cdots\cdots \\ & + A_k(x - x_0)(x - x_2)\cdots(x - x_{k-1}) \end{aligned} \tag{21.71}$$

where $A_0, A_1, \ldots, A_k$ are constants to be determined. Equation (21.71) is true for all values of x in $[x_0, x_k]$. Substituting $x = x_0$, we get

$$f(x_0) = f_0 = A_0(x_0 - x_1)(x_0 - x_2)\cdots(x_0 - x_k)$$

which yields

$$A_0 = \frac{f_0}{(x_0 - x_1)(x_0 - x_2)\cdots(x_0 - x_k)} \tag{21.72}$$

Similarly, substituting in equation (21.71) $x = x_1, x_2, \ldots, x_k$ respectively, we find

$$\begin{aligned} A_1 &= \frac{f_1}{(x_1 - x_0)(x_1 - x_2)\cdots(x_1 - x_k)} \\ A_2 &= \frac{f_2}{(x_2 - x_0)(x_2 - x_1)\cdots(x_2 - x_k)} \\ & \cdots\cdots\cdots\cdots\cdots\cdots\cdots\cdots \\ & \cdots\cdots\cdots\cdots\cdots\cdots\cdots\cdots \\ A_k &= \frac{f_k}{(x_k - x_0)(x_k - x_1)\cdots(x_k - x_{k-1})} \end{aligned} \tag{21.73}$$

Substituting these values of constants into equation (21.71) we get

$$f(x) = f_0P_0(x) + f_1P_1(x) + \cdots + f_kP_k(x) \tag{21.74}$$

where

$$P_j(x) = \prod_{i=0}^{k} \frac{x - x_i}{x_j - x_i} \tag{21.75}$$

for $j = 0, 1, 2, \ldots, k$ and $i \neq j$.

Equation (21.74) is the *Lagrange's Interpolation formula.*

EXAMPLE 21.6. *From the following table of values of x and $f(x)$ calculate the value of $f(2.53)$ by Lagrange's interpolation formula*

x	2.00	2.20	2.45	2.65	2.90
$f(x)$	0.3430	0.3040	0.2527	0.2216	0.1890

Solution: We are given $x = 2.53$, $x_0 = 2.00$, $x_1 = 2.20$, $x_2 = 2.45$, $x_3 = 2.65$ and $x_4 = 2.90$.

60

Then

$$x - x_0 = 0.53, \quad x - x_1 = 0.33, \quad x - x_2 = 0.08, \quad x - x_3 = -0.12, \quad x - x_4 = -0.37$$
$$x_0 - x_1 = -0.20, \quad x_0 - x_2 = -0.45, \quad x_0 - x_3 = -0.65, \quad x_0 - x_4 = -0.90$$
$$x_1 - x_2 = -0.25, \quad x_1 - x_3 = -0.45, \quad x_1 - x_4 = -0.70,$$
$$x_2 - x_3 = -0.20, \quad x_2 - x_4 = -0.45,$$
$$x_3 - x_4 = -0.25$$

By equation (21.75)

$$P_0(x) = \frac{(x-x_1)(x-x_2)(x-x_3)(x-x_4)}{(x_0-x_1)(x_0-x_2)(x_0-x_3)(x_0-x_4)} = \frac{(0.33)(0.08)(-0.12)(-0.37)}{(-0.20)(-0.45)(-0.65)(-0.90)}$$
$$= 0.02226$$

$$P_1(x) = \frac{(x-x_0)(x-x_2)(x-x_3)(x-x_4)}{(x_1-x_0)(x_1-x_2)(x_1-x_3)(x_1-x_4)} = \frac{(0.53)(0.08)(-0.12)(-0.37)}{(0.20)(-0.25)(-0.45)(-0.70)}$$
$$= -0.011953$$

$$P_2(x) = \frac{(x-x_0)(x-x_1)(x-x_3)(x-x_4)}{(x_2-x_0)(x_2-x_1)(x_2-x_3)(x_2-x_4)} = \frac{(0.53)(0.33)(-0.12)(-0.37)}{(0.45)(0.25)(-0.20)(-0.45)}$$
$$= 0.76696$$

$$P_3(x) = \frac{(x-x_0)(x-x_1)(x-x_2)(x-x_4)}{(x_3-x_0)(x_3-x_1)(x_3-x_2)(x_3-x_4)} = \frac{(0.53)(0.33)(0.08)(-0.37)}{(0.65)(0.45)(0.20)(-0.25)}$$
$$= 0.35398$$

$$P_4(x) = \frac{(x-x_0)(x-x_1)(x-x_2)(x-x_3)}{(x_4-x_0)(x_4-x_1)(x_4-x_2)(x_4-x_3)} = \frac{(0.53)(0.33)(0.08)(-0.12)}{(0.90)(0.70)(0.45)(0.25)}$$
$$= -0.0.02369$$

Hence by equation (21.74)

$$f(2.53) = 0.3430 \times 0.02226 + 0.3040 \times (-0.11953) + 0.2527 \times 0.76696$$
$$+ 0.2216 \times 0.35398 + 0.1890 \times (-0.02369)$$
$$= 0.00763 - 0.03634 + 0.19381 + 0.07844 - 0.00448$$
$$= 0.23906$$

21.7 Roots of Equations

In the study of many physical processes, we are encountered with the problem of finding the roots of equations which may be too complex to be solved by known methods. In such cases, we resort to finding an approximate value of the root with a certain degree of accuracy by numerical methods. There are many methods available in the literature out of which we will discuss only two.

Consider an equation $f(x) = 0$, where the function $f(x)$ may be algebraic or transcendental. We assume that the function $f(x)$ is continuous in an interval within which its root is being sought.

(a) *Regula Falsi Method*: Suppose, there exist two points $x = \alpha$ and $x = \beta$ at which the values of the function $f(x)$ are of opposite sign. Let $f(\alpha) < 0$ and $f(\beta) > 0$. Then, there exists a point $x = x_0$ at which $f(x) = 0$; that is x_0 is a root of the given equation. As a first approximation, x_0 may be obtained as the point of intersection of the chord joining points $[\alpha, f(\alpha)]$ and $[\beta, f(\beta)]$ with the x-axis. Thus

$$\frac{0 - f(\alpha)}{f(\beta) - f(\alpha)} = \frac{x_0 - \alpha}{\beta - \alpha}$$

which yields

$$x_0 = \frac{\alpha f(\beta) - \beta f(\alpha)}{f(\beta) - f(\alpha)} \tag{21.76}$$

If now $f(x_0) < 0$, the root of the given equation lies in the interval (x_0, β). If, however, $f(x_0) > 0$, the root of the given equation lies in the interval (α, x_0). Thus the process is to be repeated until we find the root with desired accuracy.

The method generally converges fast to a root. It depends on the choice of the interval (α, β)—narrower it is, faster will be the convergence.

EXAMPLE 21.7. *Find a real root of the equation* $x^2 - 2x + \ln x = 0$.

Solution: Here $f(x) = x^2 - 2x + \ln x$.

We find that $f(1.5) = -0.3445 < 0$ and $f(2) = 0.6931 > 0$. Hence $f(x) = 0$ has a root in the interval $(1.5, 2)$.

Putting $\alpha = 1.5$ and $\beta = 2$ in equation (21.76)

$$x_0 = \frac{1.5 \times f(2) - 2 \times f(1.5)}{f(2) - f(1.5)} = \frac{1.5 \times 0.6931 + 2 \times 0.3445}{0.6931 + 0.3445} = 1.666$$

$f(x_0) = -0.046 < 0.$

Hence, now $\alpha = 1.666$ and $\beta = 2$

$$x_1 = \frac{1.666 \times 0.6931 + 2 \times 0.046}{0.6931 + 0.046} = 1.7869$$

$f(x_1) = 0.1997 > 0.$

Hence, now $\alpha = 1.666$ and $\beta = 1.7869$

$$x_2 = \frac{1.666 \times 0.1997 + 1.7869 \times 0.046}{0.1997 + 0.046} = 1.6886$$

$f(x_2) = -0.00193 < 0.$

Hence, now $\alpha = 1.6886$ and $\beta = 1.7869$

$$x_3 = \frac{1.6886 \times 0.1997 + 1.7869 \times 0.00193}{0.1997 + 0.00193} = 1.68954$$

$f(x_3) = -0.000078.$

Hence the root $= 1.68954$.

(b) *Newton-Raphson Method*: Let α_0 be the exact root of equation $f(x) = 0$, where the function $f(x)$ is continuously differentiable. We guess as an approximate root $x = x_0$. Then $\alpha_0 = x_0 + h$ (say).

By Taylor's theorem

$$0 = f(\alpha_0) = f(x_0 + h) = f(x_0) + hf'(x_0) + \frac{h^2}{2!} f''(x_0) + \cdots$$

Neglecting terms of second and higher degree in h

$$0 = f(x_0) + hf'(x_0)$$

which yields

$$h = -\frac{f(x_0)}{f'(x_0)} \tag{21.77}$$

Then the next approximate root is

$$x_1 = x_0 + h = x_0 - \frac{f(x_0)}{f'(x_0)} \tag{21.78}$$

Equation (21.78) gives the root with error $e = 0(h^2)$. Continuing with the above iteration process, we can find the root as

$$x_2 = x_1 - \frac{f(x_1)}{f'(x_1)} \tag{21.79}$$

....

....

$$x_n = x_{n-1} - \frac{f(x_{n-1})}{f'(x_{n-1})} \tag{21.80}$$

till it converges to the exact root α. From equation (21.77) it is observed that if $f'(x_0) = 0$ or very near to zero, the value of h will be undefined or extremely large. In this case the method may not work. Either we make a new guess for x_0 or we may use the Taylor's series with $e = 0(h^2)$ neglecting the first derivative. Then we get

$$0 = f(x_0 + h) = f(x_0) + \frac{h^2}{2!} f''(x_0)$$

so that

$$h = \left[-\frac{2f(x_0)}{f''(x_0)}\right]^{1/2} \tag{21.81}$$

Thus, the next approximate root is given by

$$x_1 = x_0 + h = x_0 + \left[-\frac{2f(x_0)}{f''(x_0)}\right]^{1/2} \tag{21.82}$$

Further iteration process is to be continued using equation (21.82).

Let us define the function

$$P(x) = x - \frac{f(x)}{f'(x)}$$

which is called Newton-Raphson iteration function.

Then

$$P'(x) = 1 - \frac{\{f'(x)\}^2 - f(x)f''(x)}{\{f'(x)\}^2} = \frac{f(x)f''(x)}{\{f'(x)\}^2} \tag{21.83}$$

It can be proved that this method converges if $|P'(x)| < 1$, near the root

$$\text{i.e.,} \quad \frac{|f(x)f''(x)|}{\{f'(x)\}^2} < 1 \quad \text{near the root}$$

$$\text{i.e.,} \quad |f(x)f''(x)| < \{f'(x)\}^2 \tag{21.84}$$

Equation (21.84) gives the condition for convergence of the method. If α is a simple root of $f(x) = 0$, the method will converge rapidly.

It is easy to make the initial guess x_0 if the behaviour of the function or its graph is known. However, it may not be possible to have the graph readily. In such cases, we may have the following difficulties.

(1) The interval (α, β) where the desired root remains may be very small (Fig. 21.1). In such a case, if the initial guess made happens to be outside the interval, it would be difficult to find the root.

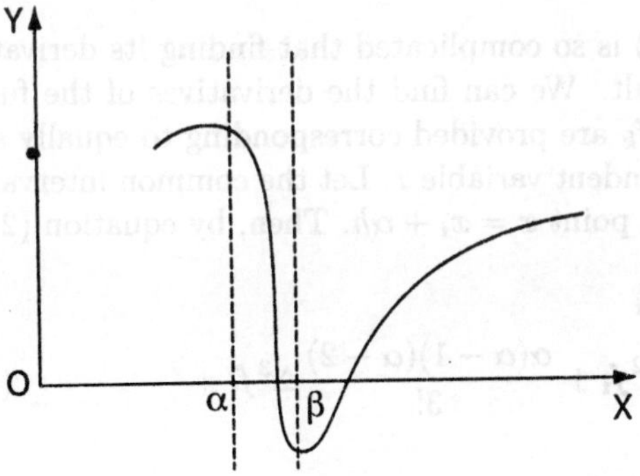

Fig 21.1

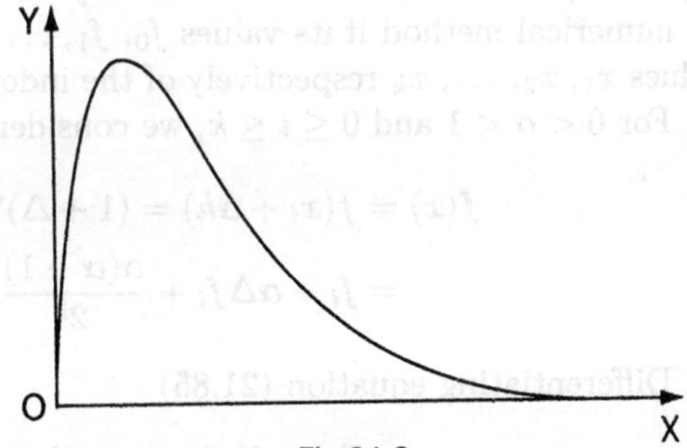

Fig 21.2

(2) It may happen that X-axis is an asymptote of the given function $f(x)$. In this case, the method will not converge. On the contrary, it would yield a large value of x where the function $f(x) \longrightarrow 0$. This value is actually not the root of equation. As an example, consider the function $f(x) = y\exp(-x)$ which is shown in Fig. 21.2. It can be found that the method diverges.

(3) If the initial guess x_0 is far away from the desired root, the method may yield repeated values of the root. This is known as *cycling* phenomenon. For example, if $f(x) = x^3 - x - 3$, cycling phenomenon will occur if initial guess $x_0 = 0$. However, with $x_0 = 2$, the method will converge.

EXAMPLE 21.8. *Find a real root of the equation $x + \sin x - 1 = 0$ by Newton-Raphson method.*

Solution: We have $f(x) = x + \sin x - 1$.

$\therefore\ f'(x) = 1 + \cos x$, which is zero for $x = \pi$.

We find that $f(0) = -1 < 0$ and $f(1) = \sin(1) = \sin(57.2598°) = 0.341471 > 0$.

We select $x_0 = 0.5$. Then $f(0.5) = -0.020579$, $f'(0.5) = 1.87758$.

Hence

$$h = -\frac{f(0.5)}{f'(0.5)} = 0.01096$$

For next iteration, $x_1 = x_0 + h = 0.51096$

$$f(x_1) = -0.0000249$$
$$f'(x_1) = 1.872275$$

Hence

$$h = \frac{-f(x_1)}{f'(x_1)} = 0.0000133$$

Thus $x_2 = x_1 + h = 0.510973$

$$f(x_2) = -0.000000644$$

Therefore $x_2 = 0.510973$ is the root.

21.8 Numerical Differentiation

Sometimes it may happen that a function $f(x)$ is so complicated that finding its derivative at a point by usual methods becomes very difficult. We can find the derivatives of the function by numerical method if its values $f_0, f_1, \ldots, f_k$ are provided corresponding to equally spaced values $x_1, x_2, \ldots, x_k$ respectively of the independent variable x. Let the common interval be h.

For $0 < \alpha < 1$ and $0 \le i \le k$, we consider a point $x = x_i + \alpha h$. Then, by equation (21.55)

$$f(x) = f(x_i + \alpha h) = (1+\Delta)^\alpha f_i$$
$$= f_i + \alpha\Delta f_i + \frac{\alpha(\alpha-1)}{2!}\Delta^2 f_i + \frac{\alpha(\alpha-1)(\alpha-2)}{3!}\Delta^3 f_i + \cdots \quad (21.85)$$

Differentiating equation (21.85)

$$\frac{df}{dx} = \frac{df}{d\alpha}\frac{d\alpha}{dx} = \frac{1}{h}\frac{df}{d\alpha}$$

$$\text{or,}\quad \frac{df}{dx} = \frac{1}{h}\left[\Delta f_i + \frac{2\alpha-1}{2}\Delta^2 f_i + \frac{3\alpha^2-6\alpha+2}{6}\Delta^3 f_i + \cdots\right] \quad (21.86)$$

Higher order derivatives may be found by repeating the differentiating. Thus

$$\frac{d^2 f}{dx^2} = \frac{1}{h^2}\left[\Delta^2 f_i + (\alpha-1)\Delta^3 f_i + \frac{6\alpha^2-18\alpha+11}{12}\Delta^4 f_i + \cdots\right] \quad (21.87)$$

$$\frac{d^3 f}{dx^3} = \frac{1}{h^3}\left[\Delta^3 f_i + \left(\alpha - \frac{3}{2}\right)\Delta^4 f_i + \cdots\right] \quad (21.88)$$

and so on. Above equations are based on Newton's forward-difference formula and hence are used when the point $x = x_i + \alpha h$ is near the beginning of the tabulated values. If this point is near the end of the table, we use Newton's backward difference formula. Thus, at the point $x = x_i - \alpha h$, by equation (21.59), we have

$$f(x) = f_i - \alpha\nabla f_i + \frac{\alpha(\alpha-1)}{2!}\nabla^2 f_i - \frac{\alpha(\alpha-1)(\alpha-2)}{3!}\nabla^3 f_i + \cdots \quad (21.89)$$

By repeated differentiation

$$\frac{df}{dx} = \frac{1}{h}\left(\nabla f_i - \frac{2\alpha - 1}{2}\nabla^2 f_i + \frac{3\alpha^2 - 6\alpha + 2}{6}\nabla^3 f_i - \cdots\right) \tag{21.90}$$

$$\frac{d^2 f}{dx^2} = \frac{1}{h^2}\left(\nabla^2 f_i - (\alpha - 1)\nabla^3 f_i + \frac{6\alpha^2 - 18\alpha + 11}{12}\nabla^4 f_i - \cdots\right) \tag{21.91}$$

$$\frac{d^3 f}{dx^3} = \frac{1}{h^3}\left(\nabla^3 f_i - \left(\alpha - \frac{3}{2}\right)\nabla^4 f_i \cdots\right) \tag{21.92}$$

and so on.

If the point x lies near the middle of the table, we use Stirling's formula. Thus, by equation (21.70), we write

$$f(x) = f_i + \frac{\alpha}{2}(\Delta f_i + \Delta f_{i-1}) + \frac{\alpha^2}{2}\Delta^2 f_{i-1} + \frac{\alpha(\alpha^2 - 1)}{12}(\Delta^3 f_{i-1} + \Delta^3 f_{i-2}) + \cdots \tag{21.93}$$

which yields

$$\frac{df}{dx} = \frac{1}{h}\left[\frac{\Delta f_i + \Delta f_{i-1}}{2} + \alpha\Delta^2 f_{i-1} + \frac{3\alpha^2 - 1}{12}(\Delta^3 f_{i-1} + \Delta^3 f_{i-2}) + \cdots\right] \tag{21.94}$$

$$\frac{d^2 f}{dx^2} = \frac{1}{h^2}\left[\Delta^2 f_{i-1} + \frac{\alpha}{2}(\Delta^3 f_{i-1} + \Delta^3 f_{i-2}) + \frac{6\alpha^2 - 1}{12}\Delta^4 f_{i-2} + \cdots\right] \tag{21.95}$$

$$\frac{d^3 f}{dx^3} = \frac{1}{h^3}\left[\frac{1}{2}(\Delta^3 f_{i-1} + \Delta^3 f_{i-2}) + \alpha\Delta^4 f_{i-2} + \cdots\right] \tag{21.96}$$

and so on.

EXAMPLE 21.9. *Calculate the values of* $\frac{df}{dx}$ *and* $\frac{d^2 f}{dx^2}$ *at the points* (i) $x = 0.33$, (ii) $x = 0.37$ *using the table in example* 21.4.

Solution: (i) Here we use Newton's forward-difference formula given by equations (21.86) and (21.87). Using directly the calculated values of example 21.4, we find $i = 1$, $h = 0.02$, $\alpha = 0.5$.

$$\therefore \quad \frac{2\alpha - 1}{2} = 0, \quad \alpha - 1 = -0.5,$$

$$\tfrac{1}{6}(3\alpha^2 - 6\alpha + 2) = -0.0416, \quad \tfrac{1}{12}(6\alpha^2 - 18\alpha + 11) = 0.2917$$

$$\Delta f_1 = 0.013, \quad \Delta^2 f_1 = -0.002, \quad \Delta^3 f_1 = -0.017$$

Hence

$$\left.\frac{df}{dx}\right|_{x=0.33} = \frac{1}{0.02}[0.013 + 0.0416 \times 0.017] = 0.68536, \quad e = 0(h^3)$$

$$\left.\frac{d^2 f}{dx^2}\right|_{x=0.33} = \frac{1}{(0.02)^2}[-0.002 - 0.5 \times (-0.017)] = 16.25, \quad e = 0(h^2)$$

(ii) Here we use Newton's backward-difference formula given by equations (21.90) and (21.91).

Here $i = 4, h = 0.02, \alpha = 0.5$

$$\nabla f_4 = 0.012, \quad \nabla^2 f_4 = 0.001, \quad \nabla^3 f_4 = 0.003$$

Hence

$$\left.\frac{df}{dx}\right|_{x=0.37} = \frac{1}{0.02}[0.012 - 0.0416 \times 0.003] = 0.59376, \quad e = 0(h^3)$$

$$\left.\frac{d^2 f}{dx^2}\right|_{x=0.37} = \frac{1}{(0.02)^2}[0.001 + 0.5 \times 0.003] = 6.25, \quad e = 0(h^2)$$

Unequal Intervals: In the case when the tabulated arguments $x_0, x_1, x_2, \ldots, x_k$ are not equally spaced, we have to use Lagrange's polynomial defined by equation (21.74). Hence we write

$$\frac{df}{dx} = f_0 P'(x) + f_1 P_1'(x) + \cdots + f_k P_k'(x) \tag{21.97}$$

$$\frac{d^2 f}{dx^2} = f_0 P''(x) + f_1 P_1''(x) + \cdots + f_k P_k''(x) \tag{21.98}$$

....

.... etc.

Here

$$P_j'(x) = \frac{d}{dx}\left[\prod_{i=0}^{k} \frac{x - x_i}{x_j - x_i}\right], \quad j = 0, 1, \ldots, k \text{ and } i \neq j \tag{21.99}$$

EXAMPLE 21.10. *Find the function $f(x)$ and its first derivative at the point $x = -3$ from the following data*

x	2	5	9	−11
$f(x)$	6	159	895	−1063

Solution: Here $x_0 = 2$, $x_1 = 5$, $x_2 = 9$ and $x_3 = -11$, so that the intervals of arguments are unequal. We use Lagrange's interpolation formula to find the function $f(x)$.

$$f_0 P_0(x) = f_0 \frac{(x - x_1)(x - x_2)(x - x_3)}{(x_0 - x_1)(x_0 - x_2)(x_0 - x_3)} = \frac{(x-5)(x-9)(x+11)}{(2-5)(2-9)(2+11)} \times 6$$

$$= \frac{2}{91}(x-5)(x-9)(x+11)$$

$$f_1 P_1(x) = f_1 \frac{(x - x_0)(x - x_2)(x - x_3)}{(x_1 - x_0)(x_1 - x_2)(x_1 - x_3)} = 159 \times \frac{(x-2)(x-9)(x+11)}{(5-2)(5-9)(5+11)}$$

$$= -\frac{53}{64}(x-2)(x-9)(x+11)$$

$$f_2 P_2(x) = f_2 \frac{(x - x_0)(x - x_1)(x - x_3)}{(x_2 - x_0)(x_2 - x_1)(x_2 - x_3)} = 895 \times \frac{(x-2)(x-5)(x+11)}{(9-2)(9-5)(9+11)}$$

$$= \frac{895}{560}(x-2)(x-5)(x+11)$$

$$f_3 P_3(x) = f_3 \frac{(x - x_0)(x - x_1)(x - x_2)}{(x_3 - x_0)(x_3 - x_1)(x_3 - x_2)} = -1063 \times \frac{(x-2)(x-5)(x-9)}{(-11-2)(-11-5)(-11-9)}$$

$$= \frac{1063}{4160}(x-2)(x-5)(x+9)$$

Therefore

$$
\begin{aligned}
f(x) &= f_0P_0(x) + f_1P_1(x) + f_2P_2(x) + f_3P_3(x) \\
&= \frac{2}{91}(x-5)(x-9)(x+11) - \frac{53}{64}(x-2)(x-9)(x+11) \\
&\quad + \frac{895}{560}(x-2)(x-5)(x+11) + \frac{1063}{4160}(x-2)(x-5)(x-9) \\
&= \frac{2}{91}\left(x^3 - 3x^2 - 109x + 495\right) - \frac{53}{64}\left(x^3 - 103x + 198\right) \\
&\quad + \frac{895}{560}\left(x^3 + 4x^2 - 67x + 110\right) + \frac{1063}{4160}\left(x^3 - 16x^2 + 73x - 90\right) \\
&= Ax^3 + Bx^2 + Cx + D
\end{aligned}
$$

where $A = 1.047596, B = 2.238474, C = -5.525521, D = -0.283620$

$$\frac{df}{dx} = 3Ax^2 + 2Bx + C$$

At $x = -3$,

$$\frac{df}{dx} = 27A - 6B + C = 9.328727$$

21.9 Numerical Integration

Consider the function $y = f(x)$ which is defined by its values $f_0, f_1, \ldots, f_k$ corresponding to equally spaced values $x_0, x_1, \ldots, x_k$ respectively of the independent variable x. Thus the interval $[x_0, x_k]$ is divided into k equal sub-intervals of size $h = (x_k - x_0)/k$. Then, any point x is written as $x = x_i + \alpha h$ for $0 < \alpha < 1$ and $0 \le i \le k$.

Then $dx = h d\alpha$ and

$$f(x) = f_i + \alpha\Delta f_i + \frac{\alpha(\alpha-1)}{2!}\Delta^2 f_i + \cdots \tag{21.100}$$

Considering the sub-interval $[x_i, x_{i+1}]$ where α varies from 0 to 1, we have

$$\int_{x_i}^{x_{i+1}} f(x)dx = \int_0^1 \left[f_i + \alpha\Delta f_i + \frac{\alpha(\alpha-1)}{2!}\Delta^2 f_i + \cdots\right] h d\alpha \tag{21.101}$$

(a) *Trapezoidal Rule*: Neglecting Δ^2 and higher-order terms in equation (21.101)

$$
\begin{aligned}
\int_{x_i}^{x_{i+1}} f(x)dx &= \int_0^1 [f_i + \alpha\Delta f_i]\, h d\alpha = h\left(f_i\alpha + \frac{\alpha^2}{2}\Delta f_i\right)\Bigg|_0^1 = h\left(f_i + \frac{1}{2}\Delta f_i\right) \\
&= h\left[f_i + \frac{1}{2}(f_{i+1} - f_i)\right] = \frac{h}{2}(f_{i+1} + f_i)
\end{aligned}
\tag{21.102}
$$

Putting $i = 0, 1, 2, \ldots, k$ and adding all these results we get

$$\int_{x_0}^{x_k} f(x)dx = \frac{h}{2}(f_0 + 2f_1 + 2f_2 + \cdots + f_k) \tag{21.103}$$

Equation (21.103) for integration is known as Trapezoidal rule and is actually the sum of the areas of k-number trapeziums obtained by joining the end points of ordinates $f_0, f_1, \ldots$ by straight lines.

Order of Error: In each sub-interval $[x_{i+1}, x_i]$, equation (21.102) has been obtained by truncating terms of Δ^2 (and of higher) which is of the order $0(h^2)$. Hence, error in equation (21.102) is $e = 0(h^3)$. The total error in the integration containing k sub-intervals is

$$k \times 0(h^3) = \frac{x_k - x_0}{h} \times 0(h^3) = 0(h^2)$$

where the interval $(x_k - x_0)$ has been assumed to be of order unity. Thus the integration value will be closer to its exact value when the step size h is small.

Bounds of Error: Considering the sub-interval $[x_i, x_{i+1}]$ and expanding $f(x)$ in Taylor's series around the point x_i,

$$\begin{aligned}\int_{x_i}^{x_{i+1}} f(x)dx &= \int_{x_i}^{x_{i+1}} \left[f_i + (x - x_i)f_i' + (x - x_i)^2 \frac{f_i''}{2} + \cdots \right] dx \\ &= \left[f_i x + \frac{(x - x_i)^2}{2} f_i' + \frac{(x - x_i)^3}{6} f_i'' + \cdots \right]_{x_i}^{x_{i+1}} \\ &= hf_i + \frac{h^2}{2} f_i' + \frac{h^3}{6} f_i'' + \cdots \qquad (21.104)\end{aligned}$$

Approximated value of the integral, by equation (21.102) is

$$\frac{h}{2}(f_i + f_{i+1}) = \frac{h}{2}\left(f_i + f_i + hf_i' + \frac{h^2}{2} f_i'' + \cdots \right). \qquad (21.105)$$

Truncation error in the sub-interval $[x_i, x_{i+1}]$

$$e = \frac{h}{2}(f_i + f_{i+1}) - \int_{x_i}^{x_{i+1}} f(x)dx = \frac{h^3}{12} f_i''$$

If

$$\begin{aligned} f''_{\text{max}} &= \max\{f_0'', f_1'', \ldots, f_k''\} \\ \text{and} \quad f''_{\text{min}} &= \min\{f_0'', f_1'', \ldots, f_k''\}\end{aligned}$$

of the function $f(x)$ in the interval of integration, then the truncation error in a single sub-interval is bounded by

$$\frac{h^3}{12} f''_{\text{min}} < e < \frac{h^3}{12} f''_{\text{max}} \qquad (21.106)$$

Since there are $k\,[= (x_k - x_0)/h]$ number of sub-intervals within the limit of integration, the total accumulated error E will have the bounds

$$\frac{kh^3}{12} f''_{\text{min}} < E < \frac{kh^3}{12} f''_{\text{max}} \quad \text{or,} \quad \frac{(x_k - x_0)}{12} h^2 f''_{\text{min}} < E < \frac{(x_k - x_0)}{12} h^2 f''_{\text{max}} \qquad (21.107)$$

(b) *Simpson's One-third Rule*: Let the number of sub-intervals k be an even integer.

Considering the sub-interval $[x_i, x_{i+2}]$ and neglecting Δ^3 and higher terms in equation (21.101), we write

$$\begin{aligned}\int_{x_i}^{x_{i+2}} f(x)dx &= \int_0^2 \left[f_i + \alpha\Delta f_i + \frac{\alpha(\alpha-1)}{2}\Delta^2 f_i\right] hd\alpha \\ &= h\left[f_i\alpha + \frac{\alpha^2}{2}\Delta f_i + \frac{1}{2}\left(\frac{\alpha^3}{3} - \frac{\alpha^2}{2}\right)\Delta^2 f_i\right]_0^2 \\ &= 2h\left[f_i + \Delta f_i + \frac{1}{6}\Delta^2 f_i\right] \\ &= 2h\left[f_i + (f_{i+1} - f_i) + \frac{1}{6}(f_{i+2} - 2f_{i+1} + f_i)\right] \\ &= \frac{h}{3}[f_i + 4f_{i+1} + f_{i+2}] \qquad (21.108)\end{aligned}$$

Putting $i = 0, 2, 4, \ldots, k/2$ in equation (21.108) and adding all these results, we get

$$\begin{aligned}&\int_{x_0}^{x_2} f(x)dx + \int_{x_2}^{x_4} f(x)dx + \cdots + \int_{x_{k-2}}^{x_k} f(x)dx \\ &\qquad = \frac{h}{3}[(f_0 + 4f_1 + f_2) + (f_2 + 4f_3 + f_4) + \cdots + (f_{k_2} + 4f_{k-1} + f_k)]\end{aligned}$$

or, $$\int_{x_0}^{x_k} f(x)dx = \frac{h}{3}[f_0 + 4f_1 + 2f_2 + 4f_3 + 2f_4 + \cdots + 4f_{k-1} + f_k] \qquad (21.109)$$

Equation (21.109) is Simpson's one-third rule of integration.

Order of Error: In evaluating the integral in the sub-interval $[x_i, x_{i+2}]$, we have neglected the terms

$$\begin{aligned}&\int_0^2 \left[\frac{\alpha(\alpha-1)(\alpha-2)}{3!}\Delta^3 f_i + \frac{\alpha(\alpha-1)(\alpha-2)(\alpha-3)}{4!}\Delta^4 f_i + \cdots\right] hd\alpha \\ &= \frac{h}{6}\left[\left(\frac{\alpha^4}{4} - \alpha^3 + \alpha^2\right)\Delta^3 f_i + \frac{1}{4}\left(\frac{\alpha^5}{5} - \frac{3}{2}\alpha^4 + \frac{11}{3}\alpha^3 - 3\alpha^2\right)\Delta^4 f_i\right]_0^2 \\ &= -\frac{h}{90}\Delta^4 f_i \qquad (21.110)\end{aligned}$$

Thus the error in equation (21.108) is $e = 0(h^5)$. Since there are $k/2$ pairs of sub-intervals, the total error in the integration formula equation (21.109) is of the order of $e = k/2 \times 0(h^5) = 0(h^4)$ which is much smaller compared to Trapezoidal rule. Simpson's formula has become much popular due to this reason.

Bounds of Error: Considering the interval $[x_i, x_{i+2}]$ and expanding the function $f(x)$ around the point x_i, we have

$$\begin{aligned}\int_{x_i}^{x_{i+2}} f(x)dx &= \int_{x_i}^{x_{i+2}} \left[f_i + (x - x_i)f_i' + (x - x_i)^2\frac{f_i''}{2} + \cdots\right] dx \\ &= \left[xf_i + \frac{(x-x_i)^2}{2}f_i' + \frac{(x-x_i)^3}{6}f_i'' + \frac{(x-x_i)^4}{24}f_i''' + \frac{(x-x_i)^5}{120}f_i^{IV} + \cdots\right]_{x_i}^{x_{i+2}} \\ &= 2hf_i + 2h^2 f_i' + \frac{4}{3}h^3 f_i'' + \frac{2}{3}h^4 f_i''' + \frac{4}{15}h^5 f_i^{IV} + \cdots \qquad (21.111)\end{aligned}$$

Approximated value of the integral by equation (21.108) is

$$
\begin{aligned}
&= \frac{h}{3}(f_i + 4f_{i+1} + f_{i+2}) \\
&= \frac{h}{3}\left[f_i + 4\left(f_i + hf_i' + \frac{h^2}{2}f_i'' + \frac{h^3}{6}f_i''' + \frac{h^4}{24}f_i^{IV} + \cdots\right)\right. \\
&\quad \left. + \left(f_i + 2hf_i' + 2h^2 f_i'' + \frac{4}{3}h^3 f_i''' + \frac{2}{3}h^4 f_i^{IV} + \cdots\right)\right] \\
&= 2hf_i + 2h^2 f_i' + \frac{4}{3}h^3 f_i'' + \frac{2}{3}h^4 f_i''' + \frac{5}{18}h^4 f_i^{IV} + \cdots
\end{aligned} \tag{21.112}
$$

Hence the truncation error in the sub-interval $[x_i, x_{i+2}]$ is

$$
\begin{aligned}
e &= \frac{h}{3}(f_i + 4f_{i+1} + f_{i+2}) - \int_{x_i}^{x_{i+2}} f(x)dx \\
&= \frac{h^5}{90} f_i^{IV} \quad \text{[using equations (21.111) and (21.112)]}
\end{aligned}
$$

Let

$$
f_{\max}^{IV} = \max\{f_0^{IV},\ f_1^{IV},\ \ldots,\ f_k^{IV}\}
$$

$$
\text{and} \quad f_{\min}^{IV} = \min\{f_0^{IV},\ f_1^{IV},\ \ldots,\ f_k^{IV}\}
$$

of the function $f(x)$ in the interval $[x_0, x_k]$.

Then the truncation error in a pair of sub-intervals is bounded by

$$
\frac{h^5}{90} f_{\min}^{IV} < e < \frac{h^5}{90} f_{\max}^{IV} \tag{21.113}
$$

Since there are $k/2 = (x_k - x_0)/2h$ number of pairs of sub-intervals, the total accumulated error E will have the bounds

$$
\frac{kh^5}{180} f_{\min}^{IV} < E < \frac{kh^5}{180} f_{\max}^{IV}, \quad \text{i.e.,} \quad \frac{(x_k - x_0)h^4}{180} f_{\min}^{IV} < E < \frac{(x_k - x_0)h^4}{180} f_{\max}^{IV} \tag{21.114}
$$

(c) *Simpson's Three-eighths Rule*: Let the number of sub-intervals k be divisible by three.

Considering the interval $[x_i, x_{i+3}]$ and neglecting Δ^3 and higher terms in equation (21.101), we write

$$
\begin{aligned}
\int_{x_i}^{x_{i+3}} f(x)dx &= \int_0^3 \left[f_i + \alpha\Delta f_i + \frac{\alpha(\alpha-1)}{2}\Delta^2 f_i + \frac{\alpha(\alpha-1)(\alpha-2)}{6}\Delta^3 f_i\right] h d\alpha \\
&= \int_0^3 \left[f_i\alpha + \frac{\alpha^2}{2}\Delta f_i + \frac{1}{2}\left(\frac{\alpha^3}{3} - \frac{\alpha^2}{2}\right)\Delta^2 f_i + \frac{1}{6}\left(\frac{\alpha^4}{4} - \alpha^3 + \alpha^2\right)\Delta^3 f_i\right]_0^3 \\
&= h\left[3f_i + \frac{9}{2}\Delta f_i + \frac{9}{4}\Delta^2 f_i + \frac{3}{8}\Delta^4 f_i\right] \\
&= h\left[3f_i + \frac{9}{2}(f_{i+1} - f_i) + \frac{9}{4}(f_{i+2} - 2f_{i+1} + f_i) + \frac{3}{8}(f_{i+3} - 3f_{i+2} + 3f_{i+1} - f_i)\right] \\
&= \frac{3h}{8}(f_i + 3f_{i+1} + 3f_{i+2} + f_{i+3})
\end{aligned} \tag{21.115}
$$

Putting $i = 0, 3, 6, \ldots, k/3$ in equation (21.115) and summing up all these results, we get

$$\int_{x_0}^{x_k} f(x)dx = \frac{3h}{8}\Big[(f_0 + 3f_1 + 3f_2 + f_3) + (f_3 + 3f_4 + 3f_5 + f_6) + \cdots + (f_{k-3} + 3f_{k-2} + 3f_{k-1} + f_k)\Big]$$
$$= \frac{3h}{8}(f_0 + f_k + 3f_1 + 3f_2 + 2f_3 + 3f_4 + 3f_5 + 2f_6 + \cdots + 3f_{k-2} + 3f_{k-1}) \qquad (21.116)$$

Equation (21.116) is Simpson's three-eighths rule for integration.

Order of Error: The first omitted term in evaluating the integral in equation (21.115) is

$$\int_0^3 \frac{\alpha(\alpha-1)(\alpha-2)(\alpha-3)}{4!}\Delta^4 f_i h d\alpha = \frac{\Delta^4 f_i h}{24}\left[\frac{\alpha^5}{5} - \frac{3}{2}\alpha^4 + \frac{11}{3}\alpha^3 - 3\alpha^2\right]_0^3$$
$$= -\frac{3h}{80}\Delta^4 f_i \qquad (21.117)$$

Thus the error in equation (21.115) is of the order of $0(h^5)$. Since there are $k/3$ intervals, the total error E is of the order of

$$E = \frac{k}{3} \times 0(h^5) = 0(h^4)$$

which is essentially the same as in Simpson's one-third rule.

Bounds of Error: Expanding the function $f(x)$ around the point x_i by Taylor's series , we have

$$\int_{x_i}^{x_{i+3}} f(x)dx = \int_{x_i}^{x_{i+3}}\left[f_i + (x - x_i)f_i' + \frac{(x-x_i)^2}{2}f_i'' + \frac{(x-x_i)^3}{6}f_i''' + \cdots\right]dx$$
$$= \left[xf_i + \frac{(x-x_i)^2}{2}f_i' + \frac{(x-x_i)^3}{6}f_i'' + \frac{(x-x_i)^4}{24}f_i''' + \frac{(x-x_i)^5}{120}f_i^{IV} + \cdots\right]_{x_i}^{x_i+3}$$
$$= 3hf_i + \frac{9h^2}{2}f_i' + \frac{9}{2}h^3 f_i'' + \frac{27}{8}h^4 f_i''' + \frac{81}{40}h^5 f_i^{IV} + \cdots \qquad (21.118)$$

Approximated value of the integral by equation (21.115) is

$$= \frac{3}{8}h(f_i + 3f_{i+1} + 3f_{i+2} + f_{i+3})$$
$$= \frac{3}{8}h\Bigg[f_i + 3\left(f_i + hf_i' + \frac{h^2}{2}f_i'' + \frac{h^3}{6}f_i''' + \frac{h^4}{24}f_i^{IV} + \cdots\right)$$
$$+ 3\left(f_i + 2hf_i' + 2h^2 f_i'' + \frac{4}{3}h^3 f_i''' + \frac{2}{3}h^4 f_i^{IV} + \cdots\right)$$
$$+ \left(f_i + 3hf_i' + \frac{9h^2}{2}f_i'' + \frac{9h^2}{2}f_i''' + \frac{27}{8}h^4 f_i^{IV} + \cdots\right)\Bigg]$$
$$= 3hf_i + \frac{9}{2}h^2 f_i' + \frac{9}{2}h^3 f_i'' + \frac{27}{8}h^4 f_i''' + \frac{33}{16}h^5 f_i^{IV} + \cdots \qquad (21.119)$$

Hence the truncation error in the interval $[x_i, x_{i+3}]$

$$e = \frac{3}{8}h\,(f_i + 3f_{i+1} + 3f_{i+2} + f_{i+3}) - \int_{x_i}^{x_{i+3}} f(x)dx$$
$$= \frac{3}{80}h^5 f_i^{IV} \quad \text{[using equations (21.118) and (21.119)]}$$

The truncation error e in the interval $[x_i, x_{i+3}]$ is thus bounded by

$$\frac{3}{80}h^5 f_{\min}^{IV} < e < \frac{3}{80}h^5 f_{\max}^{IV} \tag{21.120}$$

Therefore, the total accumulated error in the interval of integration will have the bounds

$$\frac{kh^5}{80} f_{\min}^{IV} < E < \frac{kh^5}{80} f_{\max}^{IV}$$
$$\text{i.e.,} \quad \frac{(x_k - x_0)h^4}{80} f_{\min}^{IV} < E < \frac{(x_k - x_0)h^4}{80} f_{\max}^{IV} \tag{21.121}$$

EXAMPLE 21.11. *Evaluate the integral*

$$\int_0^1 \frac{dx}{1+x^2}$$

by Simpson's one-third rule taking $k = 8$.

Solution: Here $h = (1-0)/8 = 0.125$ and $f(x) = \frac{1}{1+x^2}$.

The arguments and corresponding values of the function $f(x)$ are tabulated below:

x:	0.00	0.125	0.250	0.375	0.500	0.625	0.750	0.875	1.000
$f(x)$:	1.00000	0.98461	0.94117	0.87671	0.80000	0.71910	0.64000	0.56637	0.50000

$$S_1 = f_0 + f_8 = 1.50000$$
$$S_2 = f_1 + f_3 + f_5 + f_7 = 3.14679$$
$$S_3 = f_2 + f_4 + f_6 = 2.38117$$

By equation (21.109)

$$\int_0^1 \frac{dx}{1+x^2} = \frac{h}{3}[(f_0 + f_8) + 4\,(f_1 + f_3 + f_5 + f_7) + 2\,(f_2 + f_4 + f_6)]$$
$$= \frac{0.125}{3}[S_1 + 4S_2 + 2S_3]$$
$$= \frac{0.125}{3}[1.50000 + 12.58716 + 4.76234]$$
$$= 0.785395$$

Exact value of the integral

$$= \tan^{-1} x\Big|_0^1 = \tan^{-1} 1 = \frac{\pi}{4} = 0.785397$$

(d) *Boole's Rule*: Let the number of sub-intervals k be divisible by 4. Considering the interval $[x_i, x_{i+4}]$ and neglecting Δ^4 and higher terms in equation (21.101), we can get finally

$$\int_{x_i}^{x_{i+4}} f(x)dx = \frac{2h}{45}[7f_i + 32f_{i+1} + 12f_{i+2} + 32f_{i+3} + 7f_{i+4}] \tag{21.122}$$

Putting $i = 0, 4, 8, \ldots, k/4$ in equation (21.122) and summing up all these results, we get

$$\int_{x_0}^{x_k} f(x)dx = \frac{2h}{45}[(7f_0 + 32f_1 + 12f_2 + 32f_3 + 7f_4) + (7f_4 + 32f_5 + 12f_6 + 32f_7 + 7f_8) + \cdots] \tag{21.123}$$

Equation (21.123) is the Boole's formula for integration. The truncation error in equation (21.122) can be shown to be

$$e = \frac{8h^7}{945}f_i^{VI} \tag{21.124}$$

Thus, the total accumulated error in the interval of integration is

$$E = \frac{k}{4}\cdot\frac{8h^7}{945}f_i^{VI} = \frac{2(x_k - x_0)h^6}{945}f_i^{VI} \tag{21.125}$$

If $f_{\min}^{VI}$ and $f_{\max}^{VI}$ are respectively minimum and maximum values of $f^{VI}(x)$ at a point within the interval of integration, then the total accumulated error in this method has the bounds

$$\frac{2(x_k - x_0)h^6}{945}f_{\min}^{VI} < E < \frac{2(x_k - x_0)h^6}{945}f_{\max}^{VI} \tag{21.126}$$

EXAMPLE 21.12. *Evaluate the integral*

$$I = \int_0^1 e^{-x}\sin 4x dx$$

by the different methods, taking $k = 4$.

Solution: Here the uniform step size, $h = 0.25$. Then

$$\begin{aligned} x_0 &= 0.00, \quad f_0 = 0.00000 \\ x_1 &= 0.25, \quad f_1 = 0.65534 \\ x_2 &= 0.50, \quad f_2 = 0.55152 \\ x_3 &= 0.75, \quad f_3 = 0.06667 \\ x_4 &= 1.00, \quad f_4 = -0.27841 \end{aligned}$$

Using trapezoidal rule [equation (21.103)]

$$\begin{aligned} I &= \frac{0.25}{2}[0.00000 + 2\times(0.65534 + 0.55152 + 0.0667) - 0.27841] \\ &= 0.28358 \end{aligned}$$

By Simpson's one-third rule [equation (21.109)]

$$\begin{aligned} I &= \frac{0.25}{3}[0.00000 + 4\times 0.65534 + 2\times 0.55152 + 4\times 0.06667 - 0.27841] \\ &= 0.30939 \end{aligned}$$

Using Boole's formula [equation (21.123)]

$$I = \frac{2 \times 0.25}{45}[7 \times 0.00 + 32 \times 0.65534 + 12 \times 0.55152 + 32 \times 0.06667 - 7 \times 0.27841$$

$$= 0.30860$$

Exact value of the integral is

$$= e^{-x} \frac{-\sin 4x - 4\cos 4x}{1^2 + 4^2}\bigg|_0^1 = \frac{4e - \sin 4 - 4\cos 4}{17e} = 0.30825$$

Thus, Boole's formula gives the best result when the number of sub-intervals is kept the same.

21.10 Curve Fitting

In science and engineering, we very often require curve fitting of the experimental data collected in the form of the set of points (x_1, y_1), (x_2, y_2), ..., (x_k, y_k), where the arguments $\{x_k\}$ are distinct. The main aim is to find a function $y = f(x)$ which satisfies the above variables.

For this purpose, we choose a function containing some unknown coefficients which are to be determined from the best approximation that goes through (or very near) the given points. Considering a pair of points (x_i, y_i), the *error* or *deviation* caused due to the assumption of the function $y = f(x)$ is defined as

$$e_i = f(x_i) - y_i, \quad \text{for } 1 \leq i \leq k \tag{21.127}$$

The *root-mean-square* (rms) error is defined as

$$E = \left[\frac{1}{k}\sum_{i=1}^{k} e_i^2\right]^{1/2} \tag{21.128}$$

The best-fitting curve will be the one that minimizes the rms error E which will be a function of constants.

There are many possibilities of the type of function $y = f(x)$. We will consider a few of them.

(a) *Linear Function*: Consider the class of linear functions of the form

$$y = f(x) = Ax + B \tag{21.129}$$

Then the root-mean-square error is

$$E = \left[\frac{1}{k}\sum_{i=1}^{k} (Ax_i + B - y_i)^2\right]^{1/2} \tag{21.130}$$

which will be a minimum if and only if

$$k^2 E^2 = \sum_{i=1}^{k} (Ax_i + B - y_i)^2 \tag{21.131}$$

is minimum.

The expression in the right-hand side of equation (21.131) depends on the constants A and B, since x_i, y_i are known values. Denoting

$$F(A,B) = \sum_{i=1}^{k} (Ax_i + B - y_i)^2 \tag{21.132}$$

we write the conditions for minimum value of $F(A,B)$ as

$$\left.\begin{aligned} \frac{\partial F}{\partial A} &\equiv 2\sum_{i=1}^{k} (Ax_i + B - y_i)\, x_i = 0 \\ \frac{\partial F}{\partial B} &= 2\sum_{i=1}^{k} (Ax_i + B - y_i) = 0 \end{aligned}\right\} \tag{21.133}$$

Equation (21.133) can be put in the form

$$\left.\begin{aligned} A\Sigma x_i^2 + B\Sigma x_i - \Sigma x_i y_i &= 0 \\ A\Sigma x_i + kB - \Sigma y_i &= 0 \end{aligned}\right\} \tag{21.134}$$

which will determine the unknown constants A and B. Thus the line $y = Ax + B$, called the least-squares lines, is determined.

EXAMPLE 21.13. *Determine the least-square line $y = Ax + B$ for the data given below:*

x_i	-6	-2	0	2	6
y_i	-5.3	-3.5	-1.7	0.2	4.0

Solution: The calculations are given in the following table:

	x_i	y_i	x_i^2	$x_i y_i$
	-6	-5.3	36	31.8
	-2	-3.5	4	7.0
	0	-1.7	0	0
	2	0.2	4	0.4
	6	4.0	36	24
Sum:	0	-6.3	80	63.2

Equation (21.133) in this case takes the form

$$\left.\begin{aligned} 80A + 0 &= 63.2 \\ 0 + 6B &= -6.3 \end{aligned}\right\}$$

which yield $A = 0.79$, $B = -1.05$.

Thus the least-square line is $y = 0.79x - 1.05$.

(b) *Power Curve*: The least-square power curves are of the form

$$y = Ax^{\alpha} \tag{21.135}$$

where α is a known constant. Thus, only A is the unknown constant to be determined in this case.

Equation (21.132) in this case assumes the form

$$F(A) = \sum_{i=1}^{k} (Ax_i^\alpha - y_i)^2 \tag{21.136}$$

For minimum of $F(A)$

$$\frac{dF}{dA} \equiv 2\sum_{i=1}^{k} (Ax_i^\alpha - y_i)\, x_i^\alpha = 0$$

which yields

$$A = (\Sigma x_i^\alpha y_i) / (\Sigma x_i^{2\alpha}) \tag{21.137}$$

EXAMPLE 21.14. *Find the least-square power curve $y = Ax^2$ for the data:*

x_i	2.0	2.3	2.6	2.9	3.2
y_i	5.9	8.3	10.7	13.7	17.0

Also determine the RMS error.

Solution: Here $\alpha = 2$. Computations are shown in the table below:

x_i	y_i	x_i^2	x_i^4	$x_i^2 y_i$	$f(x_i)$	e_i	e_i^2
2.0	5.9	4.0	16.0	23.60	6.470	0.57	0.325
2.3	8.3	5.29	27.98	43.91	8.560	0.258	0.067
2.6	10.7	6.76	45.69	72.33	10.36	0.236	0.056
2.9	13.7	8.41	70.73	115.22	13.606	−0.09	0.009
3.2	17.0	10.24	104.86	174.08	16.566	−0.434	0.188
			265.26	429.14			0.645

Hence

$$A = \frac{429.14}{265.26} = 1.6178.$$

The least-square power curve is $y = 1.6178x^2$.

The RMS error $= \sqrt{\frac{0.645}{5}} = 0.359$.

(c) *Exponential Curve*: Suppose that we require to fit the given point (x_i, y_i), $i = 1, 2, \ldots, k$ to an exponential curve of the form

$$y = Ae^{Bx} \tag{21.138}$$

Taking logarithm of both sides

$$\ln y = \ln A + Bx$$

which can be expressed as

$$Y = BX + C \tag{21.139}$$

where

$$Y = \ln y, \quad X = x, \quad C = \ln A \tag{21.140}$$

Thus the problem reduces to finding a linear function in equation (21.139). Following equation (21.134), we write the conditions for minimum RMS error

$$\left.\begin{aligned} B\left(\Sigma X_i^2\right) + C\left(\Sigma X_i\right) &= \Sigma X_i Y_i \\ B(\Sigma X_i) + kC &= \Sigma Y_i \end{aligned}\right\} \tag{21.141}$$

After finding B and C, the value of A in equation (21.138) is to be found from the relation

$$C = \ln A, \quad \text{i.e.,} \quad A = e^c \tag{21.142}$$

In actual computations, we are to make transformation of the given data (x_i, y_i) to the new data set (X_i, Y_i) according to equation (21.140). This process is called *data linearization.*

EXAMPLE 21.15. *Find the exponential fit curve for the data set given below:*

x_i	0.5	1.5	2.5	3.5	4.5
y_i	1.75	2.82	4.30	5.75	7.5

Solution: Computations are given in the following table:

	$x_i = X_i$	y_i	$Y_i = \ln y_i$	X_i^2	X_iY_i
	0.5	1.75	0.5596	0.25	0.2798
	1.5	2.82	1.0367	2.25	1.5551
	2.5	4.30	1.4586	6.25	3.6465
	3.5	5.75	1.7492	12.25	6.1222
	4.5	7.5	2.0149	20.25	9.0671
Sum:	12.5		6.8190	41.25	20.6706

By equation (21.141)

$$41.25B + 12.5C = 20.6706$$
$$12.5B + 5C = 6.819$$

Solving we get $B = 0.3622$ and $C = -0.4583$.
Hence $A = e^c = 0.6324$.
Therefore, the exponential fit curve is

$$y = Ae^{Bx} = 0.6324\exp(0.3622x)$$

(d) *Parabolic Curve Fitting*: Suppose we are required to fit a given set of points $\{x_i, y_i\}$, $i = 1, 2, \ldots, k$ to a parabola given by

$$y = f(x) = Ax^2 + Bx + C \tag{21.143}$$

The RMS error in this case is

$$E = \left[\frac{1}{k}\sum_{i=1}^{k}\left(Ax_i^2 + Bx_i + C - y_i\right)\right]^{1/2} \tag{21.144}$$

Minimization of RMS error requires to find the minimum of

$$F(A,B,C) = \sum_{i=1}^{k} \left(Ax_i^2 + Bx_i + C - y_i\right)^2 \tag{21.145}$$

Conditions for the minimum of the function $F(A,B,C)$ are

$$\left.\begin{aligned} \frac{\partial F}{\partial A} &\equiv 2\sum_{i=1}^{k} \left(Ax_i^2 + Bx_i + C - y_i\right) x_i^2 = 0 \\ \frac{\partial F}{\partial B} &\equiv 2\sum_{i=1}^{k} \left(Ax_i^2 + Bx_i + C - y_i\right) x_i = 0 \\ \frac{\partial F}{\partial C} &\equiv 2\sum_{i=1}^{k} \left(Ax_i^2 + Bx_i + C\right) = 0 \end{aligned}\right\} \tag{21.146}$$

Equation (21.146) can be expressed as

$$\left.\begin{aligned} A\alpha + B\beta + C\gamma &= \Sigma x_i^2 y_i \\ A\beta + B\gamma + C\delta &= \Sigma x_i y_i \\ A\gamma + B\delta + kC &= \Sigma y_i \end{aligned}\right\} \tag{21.147}$$

where

$$\left.\begin{aligned} \alpha &= \sum_{i=1}^{k} x_i^4, \quad \beta = \sum_{i=1}^{k} x_i^3 \\ \gamma &= \sum_{i=1}^{k} x_i^2, \quad \delta = \sum_{i=1}^{k} x_i \end{aligned}\right\} \tag{21.148}$$

Equation (21.147) is a linear system in A, B, C and can be easily solved.

EXAMPLE 21.16. *Find the least-squares parabolic fit curve for the data set given below:*

x_i :	-3	-1	1	3
y_i :	-1	25	25	1

Solution: Here $k = 4$. Computations are given in the following table:

	x_i	y_i	x_i^2	x_i^3	x_i^4	$x_i y_i$	$x_i^2 y_i$
	-3	-1	9	-27	81	3	-9
	-1	25	1	-1	1	-25	25
	1	25	1	1	1	25	25
	3	1	9	27	81	3	9
Sum:	0	50	20	0	164	6	50

Hence $\alpha = 164, \beta = 0, \gamma = 20, \delta = 0$.

Set of equation (21.147) becomes

$$\left.\begin{aligned} 164A+0+20C&=50\\ 0+20B+0&=\ \ 6\\ 20A+0+4C&=50 \end{aligned}\right\}$$

Solving this system, we find $A=-3.125, B=0.300, C=28.125$.
Hence the desired parabola is

$$y=Ax^2+Bx+C=-3.125x^2+0.3x+28.125.$$

Note: The method can be extended to polynomials of higher degree. But the resulting curve may exhibit large deviations if the data set do not actually have a polynomial nature. It is risky to use polynomials of degree four or higher.

21.11 System of Linear Equations

Consider a system of n linear equations in n unknowns given by

$$\left.\begin{aligned} a_{11}x_1+a_{12}x_2+\cdots+a_{1n}x_n&=b_1\\ a_{21}x_1+a_{22}x_2+\cdots+a_{2n}x_n&=b_2\\ \cdots\cdots\cdots&\cdots\cdots\cdots\\ \cdots\cdots\cdots&\cdots\cdots\cdots\\ a_{n1}x_1+a_{n2}x_2+\cdots+a_{nn}x_n&=b_n \end{aligned}\right\} \tag{21.149}$$

In matrix notation, we write

$$[A]\{X\}=\{b\} \tag{21.150}$$

where

$$[A]=\begin{bmatrix} a_{11} & a_{12} & \cdots & a_{1n}\\ a_{21} & a_{22} & \cdots & a_{2n}\\ \cdots & \cdots & \cdots & \cdots\\ a_{n1} & a_{n2} & \cdots & a_{nn} \end{bmatrix} \tag{21.151}$$

$$\{X\}=\begin{bmatrix} x_1\\ x_2\\ \vdots\\ x_n \end{bmatrix};\quad \{b\}=\begin{bmatrix} b_1\\ b_2\\ \vdots\\ b_n \end{bmatrix}. \tag{21.152}$$

The unique solution of the system of equations will exist, if

$$\text{Det}\ |A|\neq 0 \tag{21.153}$$

That is, the matrix $[A]$ is invertible. We intend to consider the following three methods for solving the above system of linear equations.

(a) *Gaussian Elimination Method*: This is a direct method which yields the exact solution when n is not very large. We assume that, the diagonal elements

$$a_{ii} \neq 0, \quad i = 1, 2, \ldots, n \tag{21.154}$$

We multiply 1st equation by $-a_{i1}/a_{11}$ and add the result to the ith equation, where $i = 2, 3, \ldots, n$. We get the new system

$$\left.\begin{aligned} a_{11}x_1 + a_{12}x_2 + \cdots\cdots\cdots + a_{1n}x_n &= b_1 \\ a_{22}^{(1)}x_2 + \cdots + a_{2n}^{(1)}x_n &= b_2^{(1)} \\ \cdots\cdots\cdots\cdots\cdots\cdots \\ a_{n2}^{(1)}x_2 + \cdots + a_{nn}^{(1)}x_n &= b_n^{(1)} \end{aligned}\right\} \tag{21.155}$$

Again we assume that $a_{22}^{(1)} \neq 0$. We multiply the second equation by $-a_{i2}^{(1)}/a_{22}^{(1)}$ and add the result to the ith equation. Here $i = 3, 4, \ldots, n$. We get another system of equation in the form

$$\left.\begin{aligned} a_{11}x_1 + a_{12}x_2 + \cdots\cdots\cdots\cdots\cdots + a_{1n}x_n &= b_1 \\ a_{22}^{(1)}x_2 + \cdots\cdots\cdots + a_{2n}^{(1)}x_n &= b_2^{(1)} \\ a_{33}^{(2)}x_3 + \cdots + a_{3n}^{(2)}x_n &= b_3^{(2)} \\ \cdots\cdots\cdots\cdots\cdots\cdots \\ a_{n3}^{(2)}x_3 + \cdots + a_{nn}^{(2)}x_n &= b_n^{(2)} \end{aligned}\right\} \tag{21.156}$$

Proceeding in this way, we will be eliminating the first term of each of the equations going downwards. After $(n-1)$th operation, we will get the following reduced system of equations

$$\left.\begin{aligned} a_{11}x_1 + a_{12}x_2 + \cdots\cdots\cdots\cdots\cdots + a_{1n}x_n &= b_1 \\ a_{22}^{(1)}x_2 + \cdots\cdots\cdots + a_{2n}^{(1)}x_n &= b_2^{(1)} \\ a_{33}^{(2)}x_3 + \cdots + a_{3n}^{(2)}x_n &= b_3^{(2)} \\ \cdots\cdots\cdots\cdots\cdots\cdots \\ \cdots\cdots\cdots\cdots\cdots\cdots \\ a_{nn}^{(n-1)}x_n &= b_n^{(n-1)} \end{aligned}\right\} \tag{21.157}$$

The diagonal elements of this reduced system are called *pivots* which are all non-zero according to our assumption. The determinant of the coefficient matrix of equation (21.157) has the value

$$\text{Det}\,|A| = a_{11}a_{22}^{(1)}a_{33}^{(2)} \cdots a_{nn}^{(n-1)} \neq 0 \tag{21.158}$$

Hence the system of equations has a unique solution. We solve it by the so-called *back substitution method.* From last equation

$$x_n = b_n^{(n-1)}/a_{nn}^{(n-1)} \tag{21.159}$$

Putting this value of x_n into the second last equation, we get

$$x_{n-1} = \frac{b_{n-1}^{(n-2)} - a_{n-1,n}^{(n-2)}x_n}{a_{n-1,n-1}^{(n-2)}} \tag{21.160}$$

Proceeding in this way, we can find all the values of x_i $(i = 1, 2, \ldots, n)$. But here is a note of caution. While reducing the system to the upper triangular form, if the pivot of an equation (called pivotal equation) is found to be too small, then division by it will create a very large number which will greatly affect the elimination process and may yield unbelievable solutions. To avoid such situations, we find the equation which has numerically highest pivot and take that equation as the pivotal equation for the purpose of elimination. This is known as *pivoting process.*

EXAMPLE 21.17. *Solve the system of equations*

$$\begin{aligned} -5x_1 + 2x_2 - x_3 &= -1 \\ x_1 + 0 + 3x_3 &= 5 \\ 3x_1 + x_2 + 6x_3 &= 17 \end{aligned}$$

by Gaussian elimination method.

Solution: We note that the pivot of the first equation is numerically highest. Hence we write the system as

$$-5x_1 + 2x_2 - x_3 = -1 \tag{1}$$
$$x_1 + 0 + 3x_3 = 5 \tag{2}$$
$$3x_1 + x_2 + 6x_3 = 17 \tag{3}$$

We multiply equation (1) by $-1/(-5) = 1/5$ and add the result to other two equations and get

$$\frac{2}{5}x_2 + \frac{14}{5}x_3 = \frac{24}{5} \tag{4}$$
$$\frac{11}{5}x_2 + \frac{27}{5}x_3 = \frac{82}{5} \tag{5}$$

We re-arrange equations (4) and (5) taking the higher pivot as

$$\frac{11}{5}x_2 + \frac{27}{5}x_3 = \frac{82}{5} \tag{6}$$
$$\frac{2}{5}x_2 + \frac{14}{5}x_3 = \frac{24}{5} \tag{7}$$

Multiplying equation (6) by $-2/11$ and adding the result to equation (7), we get

$$\frac{20}{11}x_3 = \frac{20}{11} \tag{8}$$

From equation (8), $x_3 = 1$. Putting this value into equation (6), we find $x_2 = 5$. Putting $x_2 = 5$ and $x_3 = 1$ into equation (1), we get $x_1 = -2$. Hence the required solution is $x_1 = -2$, $x_2 = 5$ and $x_3 = 1$.

(b) *Jacobi Iteration Method*: This is an iterative method for solving a system of linear equations. Assuming that equation (21.154) is valid, we write the given system in the form

$$x_1 = \frac{1}{a_{11}}\left[b_1 - \sum_{i=2}^{n} a_{1i}x_i\right]$$

$$x_2 = \frac{1}{a_{22}}\left[b_2 - \sum_{\substack{i=1\\ i\neq 2}}^{n} a_{2i}x_i\right] \tag{21.161}$$

....

....

$$x_j = \frac{1}{a_{jj}}\left[b_j - \sum_{\substack{i=1\\ i\neq j}}^{n} a_{ji}x_i\right], \quad j = 1, 2, \ldots, n$$

Equations (21.161) are used for the iterative process starting with an initial approximate solutions $x_i^{(0)}, i = 1, 2, \ldots, n$. The successive iterations are given by

$$x_j^{(k+1)} = \frac{1}{a_{jj}}\left[b_j - \sum_{\substack{i=1\\ i\neq j}}^{n} a_{ji}x_i^{(k)}\right] \tag{21.162}$$

where $j = 1, 2, \ldots, n$ and $k = 0, 1, 2, \ldots$

This is an efficient method for solving large systems for which the coefficient matrix A is sparse. But sometimes the iteration equations may produce a divergent sequence of solutions.

Convergence: A matrix A of $n \times n$ dimension is said to be *strictly diagonally dominant*, if

$$|a_{k,k}| > \sum_{\substack{j=1\\ j\neq k}}^{n} |a_{k,j}|, \quad \text{for } k = 1, 2, \ldots, n \tag{21.163}$$

This means that, in each row of the matrix, the magnitude of the diagonal coefficient must exceed the sum of the magnitudes of other coefficients.

Equation (21.150) has a unique solution if the matrix A is strictly diagonally dominant. In this case the method converges for any initial approximation $x_i^{(0)}$.

(c) *Gauss-Seidel Method*: In this method, assuming $a_{ii} \neq 0$, we use the improved value of a variable as soon as it is available. Thus, starting from an approximate value $x_i^{(0)}$, we find $x_1^{(1)}$ by 1st of equations (21.161), use this value in 2nd equation to get $x_2^{(1)}$, then use these two values in the 3rd equation to get $x_3^{(1)}$ and so on. Thus, if $x_i^{(k)}$ is the kth approximation to x_i, then

$$x_i^{(k+1)} = \frac{1}{a_{ii}}\left[b_i - \sum_{j<i} a_{ij}x_j^{(k+1)} - \sum_{j>i} a_{ij}x_j^{(k)}\right] \tag{21.164}$$

where $k = 0, 1, 2, \ldots, i = 1, 2, \ldots, n$ and $j = 1, 2, \ldots, n$.

The method converges for any initial approximation $x_i^{(0)}$ if the coefficient matrix is strictly diagonally dominant. This method, in most of the cases, converges faster then Jacobi method.

EXAMPLE 21.18. *Solve the system of equations*

$$\begin{aligned} 4x_1 + 11x_2 - \ x_3 &= 33 \\ x_1 + \ x_2 + 4x_3 &= 12 \\ 8x_1 - \ 3x_2 + 2x_3 &= 20 \end{aligned}$$

by Jacobi and Gauss-Seidel methods.

Solution: We rewrite the equations as

$$8x_1 - 3x_2 + 2x_3 = 20 \tag{1}$$

$$4x_1 + 11x_2 - x_3 = 33 \tag{2}$$

$$x_1 + x_2 + 4x_3 = 12 \tag{3}$$

From equation (1)

$$x_1 = (20 + 3x_2 - 2x_3)/8 \tag{4}$$

From equation (2)

$$x_2 = (33 - 4x_1 + x_3)/11 \tag{5}$$

From equation (3)

$$x_3 = (12 - x_1 - x_2)/4 \tag{6}$$

We present the solution in tabular form by assuming initial values (1, 1, 1).

(a) *Jacobi Iteration*

k	x_1	x_2	x_3
1	2.625	2.727	2.500
2	2.897	2.273	1.662
3	2.937	2.097	1.742
4	2.850	2.090	1.741
5	2.848	2.122	1.765
6	2.855	2.125	1.757
7	2.857	2.121	1.755
8	2.856	2.121	1.756

Hence the solutions are $x_1 = 2.856, x_2 = 2.121, x_3 = 1.756$.

(b) *Gauss-Seidel Method*

k	x_1	x_2	x_3
1	2.625	2.136	1.809
2	2.849	2.128	1.756
3	2.859	2.120	1.755
4	2.856	2.121	1.756

Here we get the solutions faster.

21.12 Ordinary Differential Equations

Approximate solutions of differential equations (whether ordinary or partial) become necessary when we are unable to find closed-form solutions. In finding the approximate solution by meaningful numerical methods, we are to be careful about two things:

(i) the solution obtained should be sufficiently accurate so that it is acceptable for our meaningful purposes.

(ii) the numerical solution should not be *unstable* which means that there should not be large variations in the dependent variable from one step to the next resulting in oscillating solutions.

We intend to discuss first order equations only taking time (t) as the independent variable. Since higher order equations can be expressed as a set of first order equations, the methods discussed will be equally applicable to them.

(A) *Initial Value Problems*: Consider the initial value problem

$$\frac{dy}{dt} = f(t, y) \text{ with } y(a) = y_0 \tag{21.165}$$

where $[a, b]$ is the interval over which we want to find the solution. We subdivide the interval $[a, b]$ into M equal sub-intervals so that

$$\left.\begin{aligned} t_k &= a + hk \text{ for } k = 0, 1, 2, \ldots, M \\ \text{where } h &= \frac{b-a}{M} \end{aligned}\right\} \tag{21.166}$$

The value h is called the *step size*. Our aim is to generate a set of point $\{(t_k, y_k)\}$ that will satisfy the differential equation approximately. We discuss the following methods.

(a) *Euler's method*: Assuming $y, y', y'', \ldots$ to be continuous functions, we can use Taylor's theorem and write

$$y_{k+1} = y_k + hy_k' + \frac{h^2}{2!}y_k'' + \frac{h^3}{3!}y_k''' + \cdots \tag{21.167}$$

where primes denote differentiation with respect to time.

If the step size h is small enough, we can neglect the second and higher order terms in equation (21.167) and get

$$y_{k+1} = y_k + hy_k' \text{ with } e = 0(h^2) \tag{21.168}$$

which is *Euler's approximation*.

This method is simple and can also be used to solve higher-order equations.

(b) *Direct method*: In this method we replace the first derivative y' by difference equation. Thus, using forward differences by equation (21.43) we write

$$y_k' = \frac{d}{dt}(y_k) = Dy_k = \frac{1}{2h}(-y_{k+2} + 4y_{k+1} - 3y_k) \tag{21.169}$$

with $e = 0(h^2)$.

Substituting this directly into equation (21.165) and rearranging terms, we get

$$y_{k+2} = 4y_{k+1} - 3y_k - 2hf(t_k, y_k) \tag{21.170}$$

Putting $k = 0$, we find that

$$y_2 = 4y_1 - 3y_0 - 2hf(t_0, y_0) \tag{21.171}$$

which requires the value of unknown y_1. We can find this starting value of y_1 by using Euler's method, since both these methods have $e = 0(h^2)$.

For higher-order equations, we can express higher derivatives by using equations (21.47) and others in difference forms, substitute them into the given differential equation and form an equation of the type of equation (21.170). Then we can proceed to find the solution.

EXAMPLE 21.19. *Solve $y' + y = t^2$ over the interval $[0, 0.2]$ by Euler's method and Direct method. It is given that $y(0) = 1$ and $h = 0.05$.*

Solution: The given equation is written as

$$y' = t^2 - y \tag{1}$$

(i) *Euler's method*: By equation (21.168)

$$y_{k+1} = y_k + hy'_k = y_k + h\left(t_k^2 - y_k\right)$$

i.e.,

$$y_{k+1} = (1-h)y_k + ht_k^2 \tag{2}$$

Thus, for the first four steps we find

$$\begin{aligned}
t_0 &= 0; \quad y_0 = 1\\
t_1 &= 0.05; \quad y_1 = (1-0.05)\times 1 = 0.950\\
t_2 &= 0.10; \quad y_2 = (1-0.05)\times 0.950 + (0.05)\times(0.05)^2 = 0.9026\\
t_3 &= 0.15; \quad y_3 = (1-0.05)\times 0.9026 + (0.05)\times(0.10)^2 = 0.8580\\
t_4 &= 0.20; \quad y_4 = (1-0.05)\times 0.858 + (0.05)\times(0.15)^2 = 0.8162
\end{aligned}$$

(ii) *Direct method*: By equation (21.170)

$$y_{k+2} = 4y_{k+1} - 3y_k - 2hf(t_k, y_k) = 4y_{k+1} - 3y_k - 2h\left(t_k^2 - y_k\right)$$

$$\text{i.e.,} \quad y_{k+2} = 4y_{k+1} - (3-2h)y_k - 2ht_k^2 \tag{3}$$

Therefore, for the first four steps we get, using the value of y_1 from Euler's method

$$\begin{aligned}
t_0 &= 0; \quad y_0 = 1\\
t_1 &= 0.05; \quad y_1 = 0.950\\
t_2 &= 0.10; \quad y_2 = 4\times 0.950 - (3 - 2\times 0.05)\times 1 = 0.90\\
t_3 &= 0.15; \quad y_3 = 4\times 0.9 - (3 - 2\times 0.05)\times 0.95 - 2\times 0.05\times(0.05)^2 = 0.8447\\
t_4 &= 0.20; \quad y_4 = 4\times 0.8447 - (3 - 2\times 0.05)\times 0.9 - 2\times 0.05\times(0.1)^2 = 0.768
\end{aligned}$$

(c) *Taylor's method*: In this method, we retain higher-order terms in equation (21.167). Thus, considering first three terms, we write

$$y_{k+1} = y_k + hy'_k + \frac{h^2}{2}y''_k \tag{21.172}$$

with $e = 0(h^3)$.

The second derivative y''_k is to be obtained by differentiating the given differential equation

$$y' = f(t, y) \tag{21.173}$$

with respect to t. In doing so, due care is to be taken of the fact that $f(t, y)$ is a function of two variables t and y, where y itself is a function of t.

EXAMPLE 21.20. *Solve the problem in example 21.19 by Taylor's method.*

Solution: The given equation is

$$y' = t^2 - y \quad (1)$$

$$\therefore \quad y'' = 2t - y' \quad (2)$$

Equation (21.172) in this case becomes

$$y_{k+1} = y_k + hy'_k + \frac{h^2}{2}(2t_k - y'_k) \quad \text{[using equation (2)]}$$

$$\text{i.e.,} \quad y_{k+1} = y_k + \left(h - \frac{h^2}{2}\right)y'_k + h^2 t_k$$

$$= y_k + 0.04875 \times (t_k^2 - y_k) + 0.0025 t_k \quad \text{[using equation (1)]}$$

$$= 0.95125 y_k + 0.04875 t_k^2 + 0.0025 t_k$$

Hence, for the first four steps we find

$t_0 = 0; \quad y_0 = 1$

$t_1 = 0.05; \quad y_1 = 0.95125 \times 1 = 0.95125$

$t_2 = 0.10; \quad y_2 = (0.95125)^2 + 0.04875 \times (0.05)^2 + 0.0025 \times 0.05 = 0.9051$

$t_3 = 0.15; \quad y_3 = 0.95125 \times 0.9051 + 0.04875 \times (0.1)^2 + 0.0025 \times 0.1 = 0.8617$

$t_4 = 0.20; \quad y_4 = 0.95125 \times 0.8617 + 0.04875 \times (0.15)^2 + 0.0025 \times 0.15 = 0.8212.$

Note: The exact solution of the equation $y' = t^2 - y$ with $y(0) = 1$ can be found as

$$y = -e^{-t} + t^2 - 2t + 2.$$

The values of this function for the first four step sizes are

$$t_0 = 0; \quad y_0 = 1$$
$$t_1 = 0.05; \quad y_1 = 0.9512$$
$$t_2 = 0.10; \quad y_2 = 0.9051$$
$$t_3 = 0.15; \quad y_3 = 0.8617$$
$$t_4 = 0.20; \quad y_4 = 0.8212$$

This shows that Taylor's method is very accurate.

(d) *Runge-Kutta methods*: We have seen that Taylor's method yields very accurate results. But it has a big shortcoming that we require to evaluate higher order derivatives which may be complicated or inaccurate. *Runge-Kutta* methods require only the first derivative and still yield results as accurate as the Taylor's method.

We will consider the first order equation

$$y' = f(t, y) \text{ with } y(a) = y_0 \quad (21.174)$$

where $a \leq t \leq b$ and develop the Runge-Kutta method with $e = 0(h^3)$.

If ψ_k be an approximation to the slope of the function $y(t)$ in the interval $t_k < t \leq t_{k+1}$, then we can write

$$y_{k+1} = y_k + h\psi_k \quad (21.175)$$

For a good approximation of ψ_k, we assume

$$\psi_k = \alpha k_1 + \beta k_2 \tag{21.176}$$

where k_1, k_2 have the forms

$$\left.\begin{aligned} k_1 &= f(t_k, y_k) = f_k \\ k_2 &= f\left(t_k + ph, y_k + qhk_1\right) \end{aligned}\right\} \tag{21.177}$$

Here the quantities α, β, p, q are constants which will be determined later. Equation (21.177) suggests that

$$\Delta t = ph; \quad \Delta y = qhk_1 = qhf_k \tag{21.178}$$

Expanding k_2 in a Taylor's series and neglecting higher-order terms

$$\begin{aligned} k_2 &= f\left(t_k, y_k\right) + \frac{\partial}{\partial y} f\left(t_k, y_k\right) \Delta y + \frac{\partial}{\partial t} f\left(t_k, y_k\right) \Delta t \\ &= f_k + qhf_k \frac{\partial}{\partial y}\left(f_k\right) + ph\frac{\partial}{\partial t}\left(f_k\right) \end{aligned} \tag{21.179}$$

Here we have used equation (21.178). Using equations (21.176) to (21.179), we get from equation (21.175)

$$\begin{aligned} y_{k+1} &= y_k + h\psi_k = y_k + h\left(\alpha k_1 + \beta k_2\right) \\ &= y_k + (\alpha + \beta)hf_k + \beta h^2 \left[p\frac{\partial}{\partial t}\left(f_k\right) + qf_k\frac{\partial}{\partial y}\left(f_k\right)\right] \end{aligned} \tag{21.180}$$

with $e = 0(h^3)$.

Now, we can expand y_k in Taylor's series taking only three terms so that $e = 0(h^3)$

$$y_{k+1} = y_k + hy'_k + \frac{h^2}{2}y''_k = y_k + hf_k + \frac{h^2}{2}\frac{d}{dt}\left(f_k\right) \tag{21.181}$$

Since

$$\frac{df}{dt} = \frac{\partial f}{\partial y}\frac{\partial y}{\partial t} + \frac{\partial f}{\partial t}\frac{\partial t}{\partial t} = y'\frac{\partial f}{\partial y} + \frac{\partial f}{\partial t} \tag{21.182}$$

Equation (21.181) can be written as

$$y_{k+1} = y_k + hf_k + \frac{h^2}{2}\left[f_k\frac{\partial}{\partial y}\left(f_k\right) + \frac{\partial}{\partial t}\left(f_k\right)\right] \tag{21.183}$$

Comparing the coefficients of h, h^2 in equations (21.180) and (21.183), we get

$$\alpha + \beta = 1, \quad \beta p = \tfrac{1}{2} \quad \text{and} \quad \beta q = \tfrac{1}{2} \tag{21.184}$$

Thus, there are three equations with four unknowns. We generally choose $\beta = \frac{1}{2}$ or $\beta = 1$. *For* $\beta = \frac{1}{2}$

$$\alpha = \tfrac{1}{2}, \quad p = 1, \quad q = 1$$

Equations (21.175) to (21.177) yield

$$y_{k+1} = y_k + h(\alpha k_1 + \beta k_2) = y_k + \frac{h}{2}\left[f(t_k, y_k) + f\left(t_k + h, y_k + hf_k\right)\right] \tag{21.185}$$

For $\beta = 1$

$$\alpha = 0, \quad p = \tfrac{1}{2}, \quad q = \tfrac{1}{2}$$

Hence we find

$$y_{k+1} = y_k + h\beta k_2 = y_k + hf\left(t_k + \frac{h}{2}, y_k + \frac{h}{2}f_k\right) \tag{21.186}$$

Equations (21.185) and (21.186) can be used to evaluate y_{k+1} with $e = 0(h^3)$. The result will be of same accuracy as will be obtained by Taylor's method.

The Runge-Kutta methods with $e = 0(h^4)$ and $e = 0(h^5)$ can also be developed in the same way. However, the Runge-Kutta method with $e = 0(h^5)$ is most popular and is widely used. We give below the final results obtained by Runge and Kutta without going into the details:

$$y_{k+1} = y_k + \frac{h}{6}\left(\xi_k + 2\eta_k + 2\theta_k + \zeta_k\right), \quad e = 0(h^5) \tag{21.187}$$

where

$$\left.\begin{aligned} \xi_k &= f(t_k, y_k) \\ \eta_k &= f\left(t_k + \frac{h}{2}, y_k + \frac{h}{2}\xi_k\right) \\ \theta_k &= f\left(t_k + \frac{h}{2}, y_k + \frac{h}{2}\eta_k\right) \\ \zeta_k &= f\left(t_k + h, y_k + h\theta_k\right) \end{aligned}\right\} \tag{21.188}$$

EXAMPLE 21.21. *Solve the problem in example* 21.19 *by Runge-Kutta method with* $e = 0(h^5)$.

Solution: We have $f(t, y) = t^2 - y$; $h = 0.05$; $y_0 = 1$; $t_0 = 0$.

By equation (21.188)

$$\xi_0 = 0 - y_0 = -1.0$$

$$\eta_0 = \left(t_0 + \frac{h}{2}\right)^2 - \left(y_0 + \frac{h}{2}\xi_0\right) = 6.25 \times 10^{-4} - 0.975 = -0.974375$$

$$\theta_0 = \left(t_0 + \frac{h}{2}\right)^2 - \left(y_0 + \frac{h}{2}\eta_0\right) = 6.25 \times 10^{-4} - 0.97564 = -0.9750$$

$$\zeta_0 = (t_0 + h)^2 - (y_0 + h\theta_0) = 0.0025 - 0.95125 = -0.94875$$

Hence

$$\begin{aligned} y_1 &= y_0 + \frac{h}{6}(\xi_0 + 2\eta_0 + 2\theta_0 + \zeta_0) \\ &= 1.0 + \frac{0.05}{6}(-1.0 - 2 \times 0.974375 - 2 \times 0.975 - 0.94875) \\ &= 0.95127. \end{aligned}$$

Similarly, we calculate

$$\xi_1 = t_1^2 - y_1 = (0.05)^2 - 0.95127 = -0.94877$$

$$\eta_1 = \left(t_1 + \frac{h}{2}\right)^2 - \left(y_1 + \frac{h}{2}\xi_1\right)$$

$$= \left(0.05 + \frac{0.05}{2}\right)^2 - \left(0.95127 - \frac{0.05}{2} \times 0.94877\right) = -0.921925$$

$$\theta_1 = \left(t_1 + \frac{h}{2}\right)^2 - \left(y_1 + \frac{h}{2}\eta_1\right)$$

$$= \left(0.05 + \frac{0.05}{2}\right)^2 - \left(0.95127 - \frac{0.05}{2} \times 0.921925\right) = -0.922596$$

$$\zeta_1 = (t_1 + h)^2 - (y_1 + h\theta_1)$$

$$= (0.05 + 0.05)^2 - (0.95127 - 0.05 \times 0.922596) = -0.805140$$

Hence

$$y_2 = y_1 + \frac{h}{6}(\xi_1 + 2\eta_1 + 2\theta_1 + \zeta_1)$$

$$= 0.95127 + \frac{0.05}{6}(-0.94877 - 0.921925 \times 2 - 2 \times 0.922596 - 0.805140).$$

$$= 0.905912$$

In the same way we can find y_3 and y_4. The computational effort in this method is quite much, but we do not require higher-order derivatives.

Systems of Differential Equations

The Runge-Kutta methods can be successively used to solve systems of differential equations. Consider, for example, the initial value problem

$$\left.\begin{aligned} x' &= g(t, x, y) \\ y' &= f(t, x, y) \end{aligned}\right\} \tag{21.189}$$

with $x(t_0) = x_0$ and $y(t_0) = y_0$.

The Runge-Kutta formulas for $e = O(h^5)$ are

$$\left.\begin{aligned} x_{k+1} &= x_k + \frac{h}{6}(\alpha_k + 2\beta_k + 2\gamma_k + \delta_k) \\ y_{k+1} &= y_k + \frac{h}{6}(\xi_k + 2\eta_k + 2\theta_k + \zeta_k) \end{aligned}\right\} \tag{21.190}$$

where

$$\left.\begin{aligned} \alpha_k &= g(t_k,\ x_k,\ y_k) \\ \beta_k &= g\left(t_k + \frac{h}{2},\ x_k + \frac{h}{2}\alpha_k,\ y_k + \frac{h}{2}\xi_k\right) \\ \gamma_k &= g\left(t_k + \frac{h}{2},\ x_k + \frac{h}{2}\beta_k,\ y_k + \frac{h}{2}\eta_k\right) \\ \delta_k &= g\,(t_k + h,\ x_k + h\gamma_k,\ y_k + h\theta_k) \end{aligned}\right\} \tag{21.191}$$

and

$$\left.\begin{aligned}
\xi_k &= f(t_k,\ x_k,\ y_k)\\
\eta_k &= f\left(t_k+\frac{h}{2},\ x_k+\frac{h}{2}\alpha_k,\ y_k+\frac{h}{2}\xi_k\right)\\
\theta_k &= f\left(t_k+\frac{h}{2},\ x_k+\frac{h}{2}\beta_k,\ y_k+\frac{h}{2}\eta_k\right)\\
\zeta_k &= f\left(t_k+h,\ x_k+h\gamma_k,\ y_k+h\theta_k\right)
\end{aligned}\right\} \tag{21.192}$$

Higher-Order Differential Equations

Many problems in physics and engineering lead to differential equations involving higher derivatives. Consider the second-order initial value problem

$$x''(t)+Ax'(t)+Bx(t)=f(t) \tag{21.193}$$

with A, B constants and initial conditions

$$x(t_0)=x_0 \quad \text{and} \quad x'(t_0)=y_0 \tag{21.194}$$

Equation (21.193) can be replaced by a system of two first-order equations by the substitution

$$x'(t)=y(t) \tag{21.195}$$

so that

$$x''(t)=y'(t) \tag{21.196}$$

From equation (21.193) we write

$$x''(t)=f(t)-Ax'(t)-Bx(t)=f(t)-Ay(t)-Bx(t)$$

$$\text{i.e.,}\quad y'(t)=F(t,x,y)\ \text{(say)} \tag{21.197}$$

Thus, the differential equation (21.193) becomes a system

$$\left.\begin{aligned}
\frac{dx}{dt}&=y\\
\frac{dy}{dt}&=F(t,x,y)
\end{aligned}\right\} \tag{21.198}$$

with $x(t_0)=x_0$ and $y(t_0)=y_0$.

Now this system can be solved numerically by Runge-Kutta method with $e=0(h^5)$ using equations (21.190).

EXAMPLE 21.22. *Compute x_1, y_1 for the initial value problem*

$$\frac{d^2x}{dt^2}+4\frac{dx}{dt}+5x=0$$

with conditions $x(0)=3$ and $x'(0)=-5$. Take $h=0.1$.

Solution: The given equation is written as

$$x''(t) = -4x'(t) - 5x(t)$$

Hence, it is equivalent to the system

$$\left.\begin{aligned} \frac{dx}{dt} &= y \\ \frac{dy}{dt} &= -5x - 4y \end{aligned}\right\}$$

with $x_0 = 3$ and $y_0 = -5$. Therefore, here we have

$$g(t, x, y) = y; \quad f(t, x, y) = -5x - 4y, \quad h = 0.1$$

Using equations (21.191) and (21.192), we find

$$\begin{aligned}
\alpha_0 &= g(y_0) = y_0 = -5; \quad \xi_0 = -5 \times 3 - 4(-5) = 5 \\
\beta_0 &= y_0 + \frac{h}{2}\xi_0 = -5 + \frac{0.1}{2} \times 5 = -4.75 \\
\eta_0 &= -5\left(x_0 + \frac{h}{2}\alpha_0\right) - 4\left(y_0 + \frac{h}{2}\xi_0\right) \\
&= -5\left(3 - \frac{0.1}{2} \times 5\right) - 4\left(-5 + \frac{0.1}{2} \times 5\right) = 5.25 \\
\gamma_0 &= y_0 + \frac{h}{2}\eta_0 = -5 + \frac{0.1}{2} \times 5.25 = -4.7375 \\
\theta_0 &= -5\left(x_0 + \frac{h}{2}\beta_0\right) - 4\left(y_0 + \frac{h}{2}\eta_0\right) \\
&= -5\left(3 - \frac{0.1}{2} \times 4.75\right) - 4\left(-5 + \frac{0.1}{2} \times 5.25\right) = 5.1375 \\
\delta_0 &= y_0 + h\theta_0 = -5 + 0.1 \times 5.1375 = -4.486625 \\
\zeta_0 &= -5(x_0 + h\gamma_0) - 4(y_0 + h\theta_0) \\
&= -5(3 - 0.1 \times 4.7375) - 4(-5 + 0.1 \times 5.1375) = 5.31375
\end{aligned}$$

Therefore

$$\begin{aligned}
x_1 &= x_0 + \frac{h}{6}(\alpha_0 + 2\beta_0 + 2\gamma_0 + \delta_0) \\
&= 3 - \frac{0.1}{6}(5 + 2 \times 4.75 + 2 \times 4.7375 + 4.48625) = 2.52565 \\
y_1 &= y_0 + \frac{h}{6}(\xi_0 + 2\eta_0 + 3\theta_0 + \zeta_0) \\
&= -5 + \frac{0.1}{6}(5 + 2 \times 5.25 + 2 \times 5.1375 + 5.31375) = -4.48185
\end{aligned}$$

Note: In practice, we develop computer programs to calculate the sequence of data $\{x_k, y_k\}$ which may be very large.

(b) *Boundary Value Problems*: Consider the differential equations of the type

$$x'' = f(t, x, x') \tag{21.199}$$

62

for $a \le t \le b$ with the boundary conditions

$$x(a) = \alpha \quad \text{and} \quad x(b) = \beta \tag{21.200}$$

This is called a *boundary value problem.* This boundary value problem has a unique solution, if:

(i) $f(t, x, y)$ and its partial derivatives f_x, f_y are continuous on the region R defined by $a \le t \le b$; $-\infty < x < \infty$; $-\infty < y < \infty$ (here $y = x'$).

(ii) there exists a constant $M > 0$ for which

$$\frac{\partial f}{\partial x} > 0 \quad \text{and} \quad \left|\frac{\partial f}{\partial y}\right| \le M \tag{21.201}$$

We discuss the following two methods for solving the boundary value problems by numerical methods:

(a) *Linear Shooting Method*: Consider a linear boundary value problem

$$x'' = P(t)x'(t) + Q(t)x(t) + R(t) \tag{21.202}$$

with $x(a) = \alpha$ and $x(b) = \beta$.

Then

$$f(t, x, y) = P(t)y + Q(t)x + R(t)$$

Hence

$$\frac{\partial f}{\partial x} = Q(t) \quad \text{and} \quad \frac{\partial f}{\partial y} = P(t)$$

Thus, equation (21.202) will have a unique solution if f, $P(t)$, $Q(t)$ are continuous on R, and there exists a constant $M > 0$ such that

$$Q(t) > 0 \quad \text{and} \quad |P(t)| \le M \tag{21.203}$$

for all $a \le t \le b$.

The boundary value problem can be split into two initial value problems like

$$\text{(i)} \quad U'' = P(t)U'(t) + Q(t)U(t) + R(t) \tag{21.204}$$

$$\text{with conditions } U(a) = \alpha \text{ and } U'(a) = 0 \tag{21.205}$$

$$\text{(ii)} \quad V'' = P(t)V'(t) + Q(t)V(t) \tag{21.206}$$

$$\text{with conditions } V(a) = 0 \text{ and } V'(a) = 1 \tag{21.207}$$

Then, the solution of equation (21.202) is the linear combination of solutions of these two initial value problems; that is

$$x(t) = c_1U(t) + c_2V(t) \tag{21.208}$$

Here $U(t)$, $V(t)$ are unique solutions of equations (21.204) and (21.206) respectively, and c_1, c_2 are unknown constants to be determined from initial conditions. Thus

$$\text{for } t = a, \quad \alpha = c_1\alpha + 0, \quad \text{i.e.,} \quad c_1 = 1$$

and for $t = b$, $\beta = U(b) + c_2V(b)$

$$\text{i.e.,} \quad c_2 = \frac{\beta - U(b)}{V(b)} \tag{21.209}$$

Therefore, the solution given by equation (21.208) assumes the form

$$x(t) = U(t) + \frac{\beta - U(b)}{V(b)}V(t) \tag{21.210}$$

where Runge-Kutta method can be used to solve numerically the initial value problems given by equations (21.204) and (21.206). We also can use Taylor's method with $e = 0(h^3)$.

EXAMPLE 21.23. *Solve the boundary value problem*

$$x'' = 8x + t, \quad 0 \leq t \leq 1$$

with the conditions $x(0) = 1$ *and* $x(1) = 0$.

Solution: The given boundary value problem is split into following two initial value problems:

(1) $U'' = 8U + t$, with conditions $U(0) = 1.0, U'(0) = 0$

(2) $V'' = 8V$, with conditions $V(0) = 0.0, V'(0) = 1$

We divide the interval $[0,1]$ into five sub-intervals with step size $h = 0.2$.

For the first problem

$$U_0 = 1, \quad U_0' = 0, \quad U_0'' = 8U_0 = 8, \quad U_0''' = 8U_0' + 1 = 1$$

Then, we have

$$U_1 = U_0 + hU_0' + \frac{h^2}{2}U_0'' = 1 + \frac{(0.2)^2}{2} \times 8 = 1.16$$

$$U_1' = U_0' + hU_0'' + \frac{h^2}{2}U_0''' = 0.2 \times 8 + \frac{(0.2)^2}{2} \times 1 = 1.62$$

$$U_1'' = U_0'' + hU_0''' = 8 + 0.2 \times 1 = 8.2, \quad U_1''' = U_0''' = 1$$

$$U_2 = U_1 + hU_1' + \frac{h^2}{2}U_1'' = 1.16 + 0.2 \times 1.62 + \frac{(0.2)^2}{2} \times 8.2 = 1.648$$

$$U_2' = U_1' + hU_1'' + \frac{h^2}{2}U_1''' = 1.62 + 0.2 \times 8.2 + 0.02 \times 1 = 3.28$$

$$U_2'' = U_1'' + hU_1''' = 8.2 + 0.2 \times 1 = 8.4, \quad U_2''' = U_1''' = 1$$

$$U_3 = U_2 + hU_2' + \frac{h^2}{2}U_2'' = 1.648 + 0.2 \times 3.28 + 0.02 \times 8.4 = 2.472$$

Proceeding in the same way, we find

$$U_4 = 3.64, \quad U_5 = 5.16$$

For the second problem

$$V_0 = 0, \quad V_0' = 1, \quad V_0'' = 0, \quad V_0''' = 8$$

Then we have

$$V_1 = V_0 + hV_0' + \frac{h^2}{2}V_0'' = 0.2$$
$$V_1' = 1 + 0.02 \times 8 = 1.16$$
$$V_1'' = 0 + 0.2 \times 8 = 1.6, \quad V_1''' = V_0''' = 8$$
$$V_2 = V_1' + hV_1' + \frac{h^2}{2}V_1'' = 0.2 + 0.2 \times 1.16 + 0.02 \times 1.6 = 0.464$$
$$V_2' = V_1' + hV_1'' + \frac{h^2}{2}V_1''' = 1.16 + 0.2 \times 1.6 + 0.02 \times 8 = 1.64$$
$$V_2'' = V_1'' + hV_1''' = 1.6 + 0.2 \times 8 = 3.2$$
$$V_3 = V_2 + hV_2'' + \frac{h^2}{2}V_2''' = 0.464 + 0.2 \times 1.64 + 0.02 \times 3.2 = 0.856$$

Similarly

$$V_4 = 1.44, \quad V_5 = 2.28$$

The solution of the given boundary value problem is

$$x(t) = U(t) + cV(t)$$

where by equation (21.209)

$$c = \frac{0 - 5.16}{2.88} = -2.26315$$

The computed values are given in the following table

Table: Result of example 21.23

t	$U(t)$	$V(t)$	$x(t) = U(t) - 2.263V(t)$	Remarks
0.0	1.0	0.0	1.0	Accuracy of the results will improve if we take a large number of sub-intervals and solve the problem on computer.
0.2	1.16	0.2	0.707	
0.4	1.648	0.464	0.598	
0.6	2.472	0.856	0.535	
0.8	3.64	1.44	0.381	
1.0	5.16	2.28	0.00	

(b) *Finite-Difference Method*: In this method, we divide the interval $[a, b]$ into N-numbers of the sub-intervals each of step size $h = (b - a)/N$ so that

$$t_k = a + hk, \quad k = 0, 1, 2, \ldots, N$$

Using the central difference formulas with $e = 0(h^2)$, equation (21.202) is written as

$$\frac{1}{h^2}(x_{k+1} - 2x_k + x_{k-1}) = P_k \frac{1}{2h}(x_{k+1} - x_{k-1}) + Q_k x_k + P_k \tag{21.211}$$

where $P_k = P(t_k), Q_k = Q(t_k)$ and $R_k = R(t_k)$.

Arranging similar terms in equation (21.211), we get

$$\left(1 - \frac{h}{2}P_k\right) x_{k+1} - \left(2 + Q_k h^2\right) x_k + \left(1 + \frac{h}{2}P_k\right) x_{k-1} = h^2 R_k \tag{21.212}$$

where $k = 1, 2, 3, \ldots$

The system of equations obtained by putting $k = 1, 2, \ldots, N$ in equation (21.212) has the familiar tridiagonal form. Standard computer programs are available for solving large systems of this type. The numerical solution will be a set of discrete points $\{(t_k, x_k)\}$ which can be compared with the analytical solution when it is available.

If we use the forward-difference formulas with $e = 0(h)$, then equation (21.202) can be written as

$$\frac{1}{h^2}(x_{k+2} - 2x_{k+1} + x_k) = \frac{P_k}{h}(x_{k+1} - x_k) + Q_k x_k + R_k$$

$$\text{i.e.,}\quad x_{k+2} - (2 + hP_k)x_{k+1} + \left(1 + hP_k - h^2 Q_k x_k\right) = h^2 R_k \tag{21.213}$$

where $k = 0, 1, 2, \ldots$

EXAMPLE 21.24. *Solve the boundary value problem*

$$x'' = -3tx' + (1 - t^2)x + 2t^3$$

with conditions $x(0) = x(6) = 0$.

Solution: We divide the interval $(0, 6)$ into six equal sub-intervals of step size $h = 1$.

Here $P(t) = -3t, Q(t) = (1 - t^2)$ and $R(t) = 2t^3$.

By using equation (21.213) and remembering the given boundary conditions $x_0 = 0$ and $x_6 = 0$, we get the following system of equations

$$x_2 - 2x_1 = 0 \tag{1}$$

$$x_3 + x_2 - 2x_1 = 2 \tag{2}$$

$$x_4 + 4x_3 - 2x_2 = 16 \tag{3}$$

$$x_5 + 7x_4 = 54 \tag{4}$$

$$10x_5 + 4x_4 = 128 \tag{5}$$

From equations (1) and (2) $x_3 = 2$.

From equations (4) and (5) $x_5 = 10.3$, $x_4 = 6.24$.

From equation (3) using values of x_3 and x_4 $x_2 = -0.88$.

From equation (1), $x_2 = 2x_1$, i.e., $x_1 = \frac{1}{2}x_2 = -0.44$.

Thus we have obtained the solution set $(0, -0.44, -0.88, 2, 6.24, 10.3, 0)$

21.13 Partial Differential Equations

We are already aware that partial differential equations may be hyperbolic, parabolic and elliptic types. We intend to discuss the numerical solutions of these types of equations by the finite difference method.

A. *Hyperbolic Equation*: We consider the wave equation

$$\frac{\partial^2 u}{\partial t^2} = c^2 \frac{\partial^2 u}{\partial x^2}, \quad (0 < x < a; 0 < t < b) \tag{21.214}$$

with initial and boundary conditions

$$\left.\begin{aligned} u(0,t) = f(t), &\quad u(a,t) = g(t) \\ u(x,0) = \phi(x), &\quad \frac{\partial u}{\partial t}(x,0) = \psi(x) \end{aligned}\right\} \tag{21.215}$$

We take rectangular $x-t$ axes. We divide the interval $(0, a)$ along x-axis into N equal sub-intervals of step size $\Delta x = h$ and the interval $(0, b)$ along t-axis into M equal sub-intervals of step size $\Delta t = k$. Thus the region of definition of the function $u(x, t)$ is partitioned into a grid (Fig 21.3).

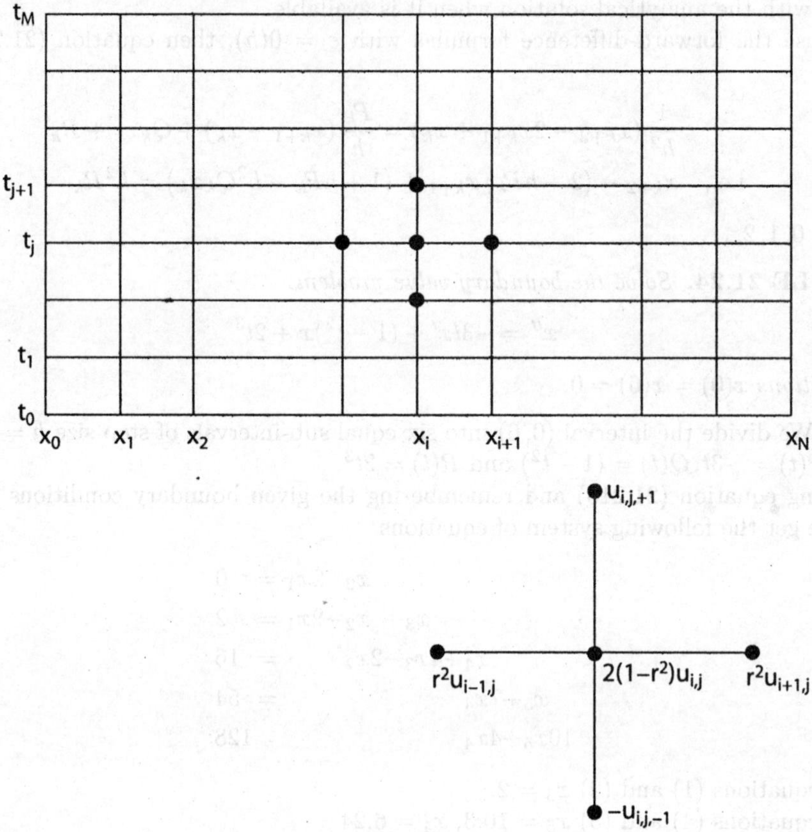

Fig 21.3

The central difference formulas for the second order partial derivatives in equation (21.214) are

$$\left.\begin{aligned} \frac{\partial^2 u}{\partial x^2} &\simeq \frac{1}{h^2}\left[u(x+h,t) - 2u(x,t) + u(x-h,t)\right] \\ \frac{\partial^2 u}{\partial t^2} &\simeq \frac{1}{k^2}\left[u(x,t+k) - 2u(x,t) + u(x,t-h)\right] \end{aligned}\right\} \tag{21.216}$$

Since the grid spacing is uniform along both axes,

$$\left.\begin{aligned} x_{i+1} &= x_i + h, \quad x_{i-1} = x_i - h \\ t_{j+1} &= t_j + k, \quad t_{j-1} = t_j - k \end{aligned}\right\} \tag{21.217}$$

where $i = 0, 1, \ldots, N$ and $j = 0, 1, \ldots, M$.

We use the approximation

$$u_{i,j} = u\left(x_i, t_j\right)$$

in equation (21.216) and write them as

$$\left.\begin{aligned}\frac{\partial^2 u}{\partial x^2} &\simeq \frac{1}{h^2}\left[u_{i+1,j} - 2u_{i,j} + u_{i-1,j}\right]\\ \frac{\partial^2 u}{\partial t^2} &\simeq \frac{1}{k^2}\left[u_{i,j+1} - 2u_{i,j} + u_{i,j-1}\right]\end{aligned}\right\} \tag{21.218}$$

Substituting these into equation (21.214) we get

$$\left.\begin{aligned}\frac{1}{k^2}\left[u_{i,j+1} - 2u_{i,j} + u_{i,j-1}\right] &= \frac{c^2}{h^2}\left[u_{i+1,j} - 2u_{i,j} + u_{i-1,j}\right]\\ \text{i.e.,}\quad u_{i,j+1} &= \frac{c^2k^2}{h^2}\left[u_{i+1,j} - 2u_{i,j} + u_{i-1,j}\right] + 2u_{i,j} - u_{i,j-1}\end{aligned}\right\} \tag{21.219}$$

The initial and boundary conditions in equation (21.215) can be written in difference form as

$$\left.\begin{aligned}u_{0,j} = f(t_j),\quad u_{N,j} = g(t_j)\\ u_{i,0} = \phi(x_i),\quad u_{i,1} = u_{i,0} + \psi(x_i)\end{aligned}\right\} \tag{21.220}$$

Defining $r = \dfrac{ck}{h}$, we write equation (21.219) in the form

$$u_{i,j+1} = r^2\left(u_{i+1,j} + u_{i-1,j}\right) + 2(1 - r^2)u_{i,j} - u_{i,j-1} \tag{21.221}$$

where $i = 1, 2, \ldots, N$, $j = 1, 2, \ldots, M$.

Equation (21.221) is used to find the values of $(j + 1)$th row.

While using this equation, it is of paramount importance to see that the method is stable. The stability in the method is assured, if

$$r = \frac{ck}{h} \leq 1$$

If, in particular, $r = 1$, equation (21.221) assumes the simple form

$$u_{i,j+1} = u_{i+1,j} + u_{i-1,j} - u_{i,j-1} \tag{21.222}$$

EXAMPLE 21.25. *The motion of a vibrating string is governed by the wave equation*

$$\frac{\partial^2 u}{\partial t^2} = 4\frac{\partial^2 u}{\partial x^2} \text{ for } 0 < x < 1 \text{ and } 0 < t < 0.5$$

with the initial and boundary conditions $u(0, t) = 0$ *and* $u(1, t) = 0$ *for* $0 \leq t \leq 0.5$

$$u(x, 0) = \phi(x) = \begin{cases}\dfrac{5x}{3}, & \text{for } 0 \leq x \leq \dfrac{3}{5}\\[2ex] \dfrac{5 - 5x}{2}, & \text{for } \dfrac{3}{5} \leq x \leq 1\end{cases}$$

and

$$\frac{\partial u}{\partial t}(x, 0) = \psi(x) = 0.$$

Taking the time step size $\Delta t = 0.05$, *calculate the displacements of the string at time* $t = 0.1$.

Solution: Here $k = 0.05$ and $c = 2$. We choose $h = 0.1$ so that

$$r = \frac{ck}{h} = \frac{2 \times 0.05}{0.1} = 1$$

Then equation (21.221) assumes the form

$$u_{i,j+1} = u_{i+1,j} + u_{i-1,j} - u_{i,j-1} \tag{1}$$

Initial condition

$$\frac{\partial u}{\partial t}(x, 0) = 0 \text{ gives } u_{i,1} = u_{i,0} \tag{2}$$

Using the given boundary condition we find

$$\begin{aligned} &u_{0,0} = 0.0, \quad u_{1,0} = 0.167, \quad u_{2,0} = 0.333, \quad u_{3,0} = 0.5, \quad u_{4,0} = 0.667, \quad u_{5,0} = 0.833 \\ &u_{6,0} = 1.0, \quad u_{7,0} = 0.75, \quad u_{8,0} = 0.5, \quad u_{9,0} = 0.25, \quad u_{10,0} = 0.0 \end{aligned} \tag{3}$$

Using equations (2) and (3), we find for $t = 0.05$

$$\begin{aligned} &u_{0,1} = u_{0,0} = 0.0, \quad u_{1,1} = u_{1,0} = 0.167 \quad u_{2,1} = u_{2,0} = 0.333, \\ &u_{3,1} = u_{3,0} = 0.50, \quad u_{4,1} = u_{4,0} = 0.667, \quad u_{5,1} = u_{5,0} = 0.833 \\ &u_{6,1} = u_{6,0} = 1.0, \quad u_{7,1} = u_{7,0} = 0.75, \quad u_{8,1} = u_{8,0} = 0.50, \\ &u_{9,1} = u_{9,0} = 0.25, \quad u_{10,1} = u_{10,0} = 0.0 \end{aligned} \tag{4}$$

Using equations (1), (3) and (4), we find for $t = 0.1$, the displacements of the string $u_{i,2}$ as under

$$\begin{aligned} u_{0,2} &= 0 \\ u_{1,2} &= u_{2,1} + u_{0,1} - u_{1,0} = 0.333 + 0.0 - 0.166 = 0.166 \\ u_{2,2} &= u_{3,1} + u_{1,1} - u_{2,0} = 0.50 + 0.166 - 0.333 = 0.333 \\ u_{3,2} &= u_{4,1} + u_{2,1} - u_{3,0} = 0.667 + 0.333 - 0.50 = 0.50 \\ u_{4,2} &= u_{5,1} + u_{3,1} - u_{4,0} = 0.833 + 0.5 - 0.667 = 0.667 \\ u_{5,2} &= u_{6,1} + u_{4,1} - u_{5,0} = 1.0 + 0.667 - 0.833 = 0.833 \\ u_{6,2} &= u_{7,1} + u_{5,1} - u_{6,0} = 0.75 + 0.833 - 1.0 = 0.583 \\ u_{7,2} &= u_{8,1} + u_{6,1} - u_{7,0} = 0.50 + 1.0 - 0.75 = 0.75 \\ u_{8,2} &= u_{9,1} + u_{7,1} - u_{8,0} = 0.25 + 0.75 - 0.50 = 0.50 \\ u_{9,2} &= u_{10,1} + u_{8,1} - u_{9,0} = 0.0 + 0.50 - 0.25 = 0.25 \\ u_{10,2} &= 0 \end{aligned}$$

B. *Parabolic Equation*: We consider the one-dimensional heat equation

$$\frac{\partial u}{\partial t} = c^2 \frac{\partial^2 u}{\partial x^2}, \quad \text{for } 0 < x < l \text{ and } 0 < t < T \tag{21.223}$$

with the initial condition

$$u(x, 0) = \phi(x) \quad \text{for } 0 \le x \le l \tag{21.224}$$

and the boundary conditions

$$\left.\begin{aligned} u(0, t) &= f(t) \text{ for } x = 0 \text{ and } 0 \le t \le T \\ u(l, t) &= g(t) \text{ for } x = l \text{ and } 0 \le t \le T \end{aligned}\right\} \tag{21.225}$$

Equation (21.223) models the diffusion problems for a rod of length l assuming no heat losses from the lateral surfaces. If the ends of the rod are kept at constant temperatures T_1 and T_2, then we will have the functions $f(t) = T_1$ and $g(t) = T_2$.

We choose the step size $\Delta x = h$ and the time step size $\Delta t = k$. Denoting the function $u(x,t)$ in difference notation as $u_{i,j}$, which represents temperature at $x = x_i$ and $t = t_j$, we write, by using forward differences

$$\frac{\partial u}{\partial t}(x_i, t_j) = \frac{1}{k}\left(u_{i,j+1} - u_{i,j}\right) \tag{21.226}$$

Similarly, keeping the time fixed, we write the second derivative with respect to x using central differences

$$\frac{\partial^2 u}{\partial x^2} = \frac{1}{h^2}\left(u_{i+1,j} - 2u_{i,j} + u_{i-1,j}\right) \tag{21.227}$$

Thus, the heat equation (21.223) in difference notation assumes the form (Fig 21.4)

$$\frac{1}{k}\left(u_{i,j+1} - u_{i,j}\right) = \frac{c^2}{h^2}\left(u_{i+1,j} - 2u_{i,j} + u_{i-1,j}\right)$$

$$\text{or,} \quad u_{i,j+1} = \frac{kc^2}{h^2}\left(u_{i+1,j} - 2u_{i,j} + u_{i-1,j}\right) + u_{i,j} \tag{21.228}$$

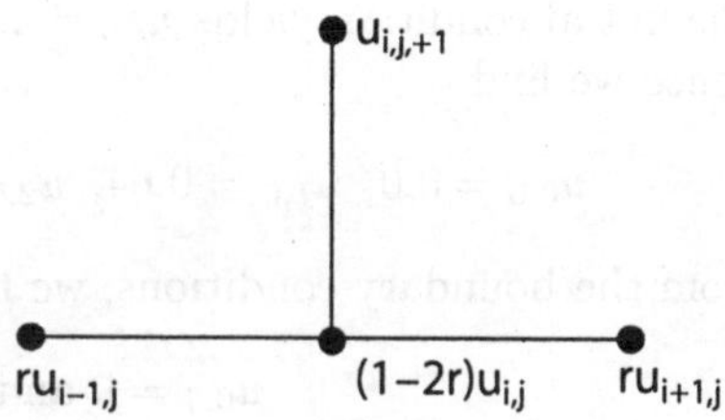

Fig 21.4

The initial and boundary conditions in difference notations are written as

$$u_{i,0} = \phi(x_i), \quad u_{0,j} = f(t_j), \quad u_{N,j} = g(t_j) \tag{21.229}$$

where N is the total number of steps on x-axis. Equations (21.228) and (21.229) are used to determine the values of the function $u(x,t)$ at other nodal points of the grid. However, it is important to see that the solution is stable. This means that any error made at one stage of the calculations should be eventually dampened out at a later stage. Denoting $r = kc^2/h^2$, equation (21.228) is stable, if and only if $0 \le r \le \frac{1}{2}$. That is, the step size k must satisfy the condition $k \le h^2/(2c^2)$.

If, in particular, $r = 1/2$, equation (21.228) assumes a simple form

$$u_{i,j+1} = \frac{1}{2}\left(u_{i+1,j} + u_{i-1,j}\right) \tag{21.230}$$

If an insulated boundary condition is imposed at $x = 0$, then

$$\frac{\partial u}{\partial t}(0,t) = 0, \quad \text{i.e.,} \quad \frac{1}{h}\left(u_{1,j} - u_{0,j}\right) = 0 \tag{21.231}$$

If the end $x = l$ is insulated, then, by using backward-difference, we have

$$\frac{1}{h}\left(u_{N,j} - u_{N-1,j}\right) = 0 \tag{21.232}$$

EXAMPLE 21.26. *A heat equation is given by*

$$\frac{\partial u}{\partial t} = \frac{\partial^2 u}{\partial x^2}, \quad 0 < x < 1 \textit{ and } 0 < t < 0.20$$

with initial condition at $t = 0$

$$u(x,0) = \phi(x) = 4x - 4x^2, \quad 0 \le x \le 1$$

and the boundary conditions

$$u(0,t) = 0 \text{ and } u(1,t) = 0, \quad 0 \le t \le 0.20.$$

Taking 5 *step sizes along x-axis, determine the temperature function at time* $t = 0.06$.

Solution: We are given $h = 0.2$ and we take $k = 0.02$.

Since $c = 1$, $r = 1/2$.

The points along x-axis are: $x_0 = 0.0$, $x_1 = 0.2$, $x_2 = 0.4$, $x_3 = 0.6$, $x_4 = 0.8$, $x_5 = 1.0$.

The points along t-axis are: $t_0 = 0.0$, $t_1 = 0.02$, $t_2 = 0.04$, ... $t_{10} = 0.20$.

The initial condition yields $u_{i,0} = 4x_i - 4x_i^2$.

Hence we find

$$u_{0,0} = 0.0,\ u_{1,0} = 0.64,\ u_{2,0} = 0.96,\ u_{3,0} = 0.96,\ u_{4,0} = 0.64,\ u_{5,0} = 0 \tag{1}$$

From the boundary conditions, we have

$$u_{0,j} = 0 \text{ and } u_{5,j} = 0 \quad (j = 0, 1, \ldots, 10) \tag{2}$$

By equation (21.230), we have

$$u_{i,j+1} = \tfrac{1}{2}(u_{i+1,j} + u_{i-1,j}) \tag{3}$$

Hence, at $t = 0.02$, when $j = 0$, we get

$$u_{i,1} = \tfrac{1}{2}(u_{i+1,0} + u_{i-1,0}) \tag{4}$$

From equation (4) we find by using values in (1) the following

$$\left.\begin{aligned}
u_{1,1} &= \tfrac{1}{2}(u_{2,0} + u_{0,0}) = \tfrac{1}{2}(0.96 + 0.0) = 0.48 \\
u_{2,1} &= \tfrac{1}{2}(u_{3,0} + u_{1,0}) = \tfrac{1}{2}(0.96 + 0.64) = 0.80 \\
u_{3,1} &= \tfrac{1}{2}(u_{4,0} + u_{2,0}) = \tfrac{1}{2}(0.64 + 0.96) = 0.80 \\
u_{4,1} &= \tfrac{1}{2}(u_{5,0} + u_{3,0}) = \tfrac{1}{2}(0.0 + 0.96) = 0.48 \\
u_{5,1} &= 0 \quad \text{[by equation (2)]}
\end{aligned}\right\} \tag{5}$$

Again, at $t = 0.04$, $j = 1$, so that equation (3) yields

$$u_{i,2} = \tfrac{1}{2}(u_{i+1,1} + u_{i-1,1}) \tag{6}$$

Hence, putting $i = 1, 2, \ldots$ and using equation (5), we get

$$\left.\begin{aligned}
u_{1,2} &= \tfrac{1}{2}(u_{2,1} + u_{0,1}) = \tfrac{1}{2}(0.80 + 0.0) = 0.40 \\
u_{2,2} &= \tfrac{1}{2}(u_{3,1} + u_{1,1}) = \tfrac{1}{2}(0.80 + 0.48) = 0.64 \\
u_{3,2} &= \tfrac{1}{2}(u_{4,1} + u_{2,1}) = \tfrac{1}{2}(0.48 + 0.80) = 0.64 \\
u_{4,2} &= \tfrac{1}{2}(u_{5,1} + u_{3,1}) = \tfrac{1}{2}(0.0 + 0.80) = 0.40 \\
u_{5,2} &= 0 \quad \text{[by equation (2)]}
\end{aligned}\right\} \tag{7}$$

Similarly, at $t = 0.06, j = 2$, so that equation (3) yields

$$u_{i,3} = \tfrac{1}{2}\left(u_{i+1,2} + \tfrac{1}{2}u_{i-1,2}\right) \tag{8}$$

Putting $i = 1, 2, \ldots$ and using equation (7), we get

$$\left.\begin{aligned}
u_{1,3} &= \tfrac{1}{2}\left(u_{2,2} + u_{0,2}\right) = \tfrac{1}{2}(0.64 + 0.0) = 0.32 \\
u_{2,3} &= \tfrac{1}{2}\left(u_{3,2} + u_{1,2}\right) = \tfrac{1}{2}(0.64 + 0.40) = 0.52 \\
u_{3,3} &= \tfrac{1}{2}\left(u_{4,2} + u_{2,2}\right) = \tfrac{1}{2}(0.4 + 0.64) = 0.52 \\
u_{4,3} &= \tfrac{1}{2}\left(u_{5,2} + u_{3,2}\right) = \tfrac{1}{2}(0.0 + 0.64) = 0.32 \\
u_{5,3} &= 0 \quad \text{[by equation (2)]}
\end{aligned}\right\} \tag{9}$$

Equation (9) give the value of temperature function $u(x,t)$ at $t = 0.06$. Computations can be advanced in the same way to find all the values of $u(x,t)$.

C. *Elliptic Equations*: As an example of elliptic partial differential equation, we consider Laplace's equation

$$\frac{\partial^2 u}{\partial x^2} + \frac{\partial^2 u}{\partial y^2} = 0 \tag{21.233}$$

within a rectangular region $(0 \le x \le a, 0 \le y \le b)$.

The boundary conditions may be

$$\text{(i)} \qquad u(x,y)\Big|_{\Gamma} = C \tag{21.234}$$

which is known as Dirichlet's problem

$$\text{(ii)} \qquad \frac{\partial u}{\partial n}(x,y)\Big|_{\Gamma} = C, \quad (n = \text{direction of outward normal}) \tag{21.235}$$

which is known as Neumann's problem

$$\text{(iii)} \qquad u(x,y)\Big|_{\Gamma_1} = C_1 \quad \text{and} \quad \frac{\partial u}{\partial x}\Big|_{\Gamma_2} = C_2 \tag{21.236}$$

where Γ_1, Γ_2 are two portions of the rectangular contour. This is known as mixed value problem. We divide the rectangular region into $N \times M$ squares of side h, i.e., $\Delta x = h$ and $\Delta y = h$ (Fig 21.5).

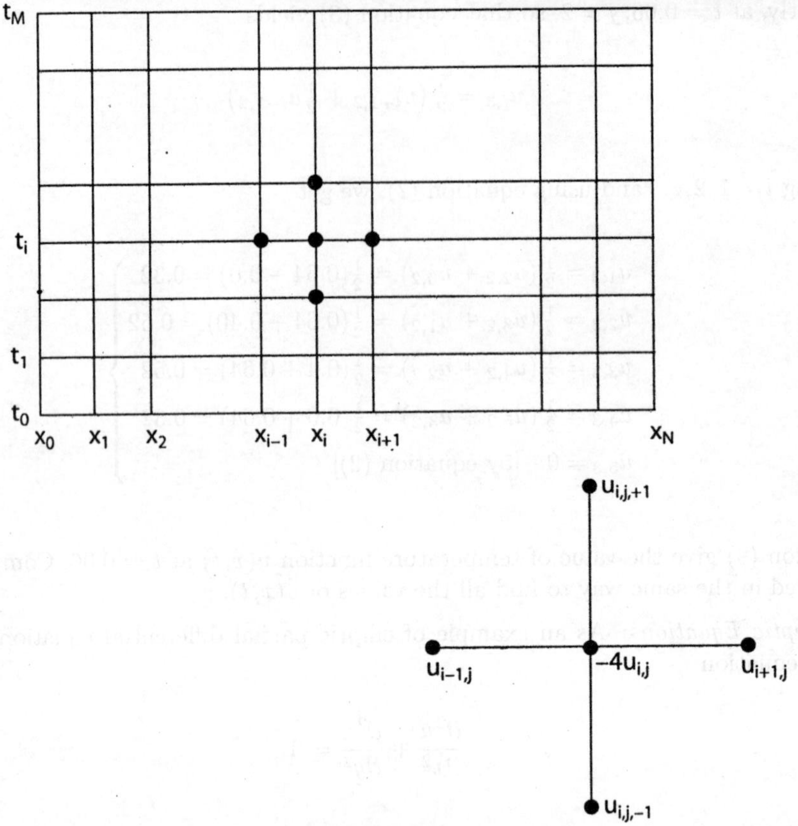

Fig 21.5

Then equation (21.233) can be written in difference notation using central differences as

$$\frac{1}{h^2}\left(u_{i+1,j} - 2u_{i,j} + u_{i-1,j}\right) + \frac{1}{h^2}\left(u_{i,j+1} - 2u_{i,j} + u_{i,j-1}\right) = 0$$

$$\text{i.e.,}\quad u_{i+1,j} + u_{i-1,j} + u_{i,j+1} + u_{i,j-1} - 4u_{i,j} = 0. \tag{21.237}$$

which is known as *five-point difference formula* for Laplace's equation. This formula gives the value of $u_{i,j}$ as the average of four neighbouring points. Considering Dirichlet's problem, we assume that the values of the function $u(x, y)$ are known at the grid points on the boundary.

Applying equation (21.237) at each of the interior points of the grid, we will obtain a system of linear algebraic equations which can be solved by known methods. This will yield approximate values of $u(x, y)$ at all of the interior points of the grid.

For example, let us consider a square region divided into 3×3 small squares (Fig. 21.6).

Applying equation (21.237), we get the following system of equations

$$\left.\begin{aligned} -4p_1 + p_2 + p_3 \qquad &= -u_{1,0} - u_{0,1} \\ p_1 - 4p_2 \qquad + p_4 &= -u_{2,0} - u_{3,1} \\ p_1 \qquad - 4p_3 + p_4 &= -u_{0,2} - u_{1,3} \\ p_2 + p_3 - 4p_4 &= -u_{2,3} - u_{3,2} \end{aligned}\right\} \tag{21.238}$$

Thus we have four equations for determining four unknowns p_1, p_2, p_3, p_4.

It is to be noted that, although the value of the step size h does not appear in the equation, accuracy of the solution much depends on it. For finer meshes, the system of algebraic equations will be very large when we resort to computers for the solution.

For the Neumann problem, suppose we are given the boundary condition

$$\frac{\partial u}{\partial x} = 0, \quad \text{for } x = a \tag{21.239}$$

In difference notation it can be written as

$$\frac{1}{2h}(u_{N+1,j} - u_{N-1,j}) \simeq 0$$

which yields

$$u_{N+1,j} \simeq u_{N-1,j} \tag{21.240}$$

Applying equation (21.237) for the point (x_N, y_j) on the boundary, we get

$$u_{N+1,j} + u_{N-1,j} + u_{N,j+1} + u_{N,j-1} - 4u_{N,j} = 0 \tag{21.241}$$

Substituting for $u_{N+1,j}$ from equation (21.240), we get

$$2u_{N-1,j} + u_{N,j+1} + u_{N,j-1} - 4u_{N,j} = 0 \tag{21.242}$$

Similarly, if the edge $y = b$ is fixed, then the boundary condition along the edge is

$$\frac{\partial u}{\partial y} = 0, \quad \text{for } y = b \tag{21.243}$$

For points on this boundary satisfying this condition, we will have the equation

$$2u_{i,M-1} + u_{i+1,M} + u_{i-1,M} - 4u_{i,M} = 0 \tag{21.244}$$

EXAMPLE 21.27. *Find an approximate solution to the boundary value problem*

$$\frac{\partial^2 u}{\partial x^2} + \frac{\partial^2 u}{\partial y^2} = 0$$

in the rectangle $0 \le x \le 3$, $0 \le y \le 3$ *with the conditions*

$$u(x,0) = 20, \quad u(x,3) = 120$$
$$\text{and} \quad u(0,y) = 60, \quad u(3,y) = 0$$

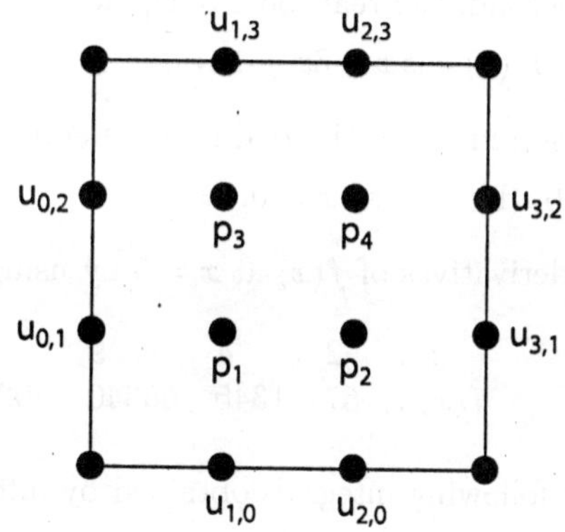

Fig 21.6

Solution: See Fig. 21.6 for the grid to be used. Using the given boundary conditions, we write equation (21.238) as

$$\begin{aligned}
-4p_1 + p_2 + p_3 \qquad &= -60 - 20 = -80 \\
p_1 - 4p_2 \qquad + p_4 &= -20 - 0 = -20 \\
p_1 \qquad - 4p_3 + p_4 &= -60 - 120 = -180 \\
p_2 + p_3 - 4p_4 &= 0 - 120 = -120
\end{aligned}$$

Solving this system of equation by Gauss's elimination method, we find $p_1 = 45$, $p_2 = 30$, $p_3 = 70$, $p_4 = 55$.

EXERCISE

1. Show that (a) $\Delta = \frac{\delta^2}{2} + \delta\left(1 + \frac{\delta^2}{4}\right)^{1/2}$, (b) $E^{-1/2} = \mu - \frac{\delta}{2}$.

2. Show that $D = \frac{\mu}{h}\left(\delta - \frac{\delta^3}{6} + \frac{\delta^5}{30} - \cdots\right)$.

3. From the following table of x and $f(x)$, find the value of $f(12)$

x:	10	15	20	25	30
$f(x)$:	0.10	0.15	0.20	0.255	0.03

4. Find $f(37.5)$ from the following table

x:	25	30	35	40	45	50
$f(x)$:	24.15	22.03	20.22	18.64	17.26	16.05

5. Find $f(15)$ from the following table by using Lagrange's interpolation formula

x:	10	12	14	16	18	20
$f(x)$:	2.42	1.94	1.50	1.11	0.79	0.54

6. Find by Regula falsi method, the real roots of equations
(a) $x - \ln\{2(x+1)\} = 0$, (b) $\cos x - 3x + 1 = 0$.

7. Find by Newton-Raphson method, the real roots of equations
(a) $2x - \log_{10} x - 7 = 0$, (b) $x + \tan x = 0$.

8. Find first and second derivatives of $f(x)$ at $x = 5$ by using the following table

x:	2	4	9	13
$f(x)$:	57	1345	66340	402052

9. Compare the values of following integrals obtained by different methods:

(a) $\int_0^2 (x + \sin x)e^x dx$; (b) $\int_0^{10} x^2 \exp(-x\sin x)dx$.

10. Find the least-squares parabolic fit curve $y = Ax^2 + Bx + C$ for the data set

x_i :	-3	0	2	4
y_i :	3	1	1	3

11. Find the curve fit $y = C\exp(Ax)$ by changing the variables $X = x$, $Y = \ln y$, $C = \exp(B)$ for the data set

x_i :	1	2	3	4	5
y_i :	0.6	1.9	4.3	7.6	12.6

12. Solve the following equations by Gauss elimination method and Gauss-Seidel method:

(a)
$$\begin{aligned} 3.4x - 5.8y + 9.4z &= 20 \\ 2.7x + 4.2y + 1.3z &= 15 \\ 2.0x - 5.1y + 5.4z &= 8 \end{aligned}$$

(b)
$$\begin{aligned} 2a_1 - 10a_2 + a_3 + a_4 &= -15 \\ 10a_1 - 2a_2 - a_3 - a_4 &= 3 \\ a_1 + a_2 + 2a_3 - 10a_4 &= 9 \\ a_1 + a_2 - 10a_3 + 2a_4 &= -27 \end{aligned}$$

13. Solve the initial value problem

$$2\frac{dy}{dt} + y = t, \quad 0 < t < 3$$

with $y(0) = 1$ by Euler's method taking step size $h = 0.25$. Compare your results by Taylor's method for the value of $y(0.75)$.

14. Solve by Runge-Kutta method with $e = 0(h^5)$

$$y' = 2ty^2, \quad 0 < t < 0.2$$

with $y(0) = 1$. Compare your results with the exact solution

$$y = \frac{1}{1 - t^2}.$$

15. Solve by Runge-Kutta method using equation (21.186)

$$2yy' = t^2, \quad 0 < t < 3$$

with $y(0) = 2$, taking $h = 0.4$. Find the value of $y(2)$.

16. Solve $\dfrac{d^2x}{dt^2} + x = 6\cos t$ with $x(0) = 2$ and $x'(0) = 3$ and compare your results with the exact solution

$$x = 2\cos t + 3\sin t + 3t\sin t.$$

17. Solve by linear shooting method

$$x'' = p(t)x' - q(t)x + 1, \quad 0 < t < 4$$

with $p(t) = \dfrac{2t}{1+t^2}$, $q(t) = \dfrac{2}{1+t^2}$ and $x(0) = 1.25$, $x(4) = -0.95$. Estimate the value of $x(1.6)$.

18. Solve the equation in Q.17 by finite difference method taking $h = 0.8$ and compare the value of $x(1.6)$.

19. Solve the wave equation

$$\frac{\partial^2 u}{\partial t^2} = 4\frac{\partial^2 u}{\partial x^2}, \quad 0 \le x \le 1 \text{ and } 0 \le t < 0.5$$

with $u(0,t) = u(1,t) = 0$, $u(x,0) = \sin \pi x$ and $\frac{\partial u}{\partial t}(x,0) = 0$.

Take $h = 0.2, k = 0.1$. Estimate $u(0.2, 0.4)$ and $u(0.4, 0.2)$.

20. Solve the diffusion equation

$$\frac{\partial u}{\partial t} = \frac{\partial^2 u}{\partial x^2}, \quad 0 < x < 1 \text{ and } 0 \le t \le 0.1 \tag{21.245}$$

with initial conditions

$$\begin{aligned} u(0,t) = 0, \quad u(1,t) = 0, \quad &\text{for } 0 \le t \le 0.1 \\ \text{and} \quad u(x,0) = \phi(x) = \sin \pi x, \quad &\text{for } 0 \le x \le 1. \end{aligned}$$

Taking $h = 0.2$ and $k = 0.02$, estimate the values of $u(0.6, 0.06)$ and $u(0.8, 0.04)$.

21. A harmonic function $u(x,y)$ defined in the rectangle $0 \le x \le 3$, $0 \le y \le 3$ has its values on the boundary as under

$$\begin{aligned} u(x,0) = 8 \quad &\text{and} \quad u(x,3) = 80 \\ u(0,y) = 60 \quad &\text{and} \quad \frac{\partial u}{\partial x}(3,y) = 0 \end{aligned}$$

Determine the system of six equations for solving the six unknowns (see Fig. 21.7)

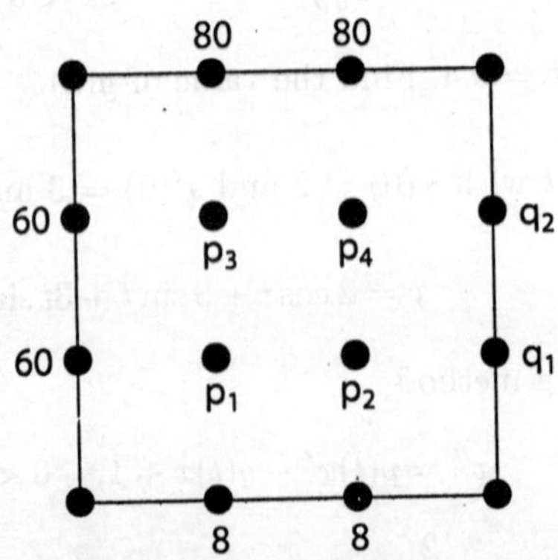

Fig 21.7

ANSWERS

3. 0.1206.

4. 19.41.

5. 1.29.

6. (a) 1.68, (b) 0.607.

7. (a) 3.789, (b) 2.029.

8. 476, 3530.

10. $y = 0.178x^2 - 0.192x + 0.851$.

11. $y = 0.363\exp(0.747x)$.

12. (a) 1, 2, 3; (b) 1, 2, 3, 0.

13. 0.7598, 0.8118.

15. 2.58.

17. 0.4969.

19. $-0.4755, 0.2939$.

21.
$$\begin{aligned} -4p_1 + p_2 + p_3 &= -68 \\ p_1 - 4p_2 + p_4 + q_1 &= -8 \\ 2p_2 + q_2 - 4q_1 &= -8 \\ p_1 + p_4 - 4p_3 &= -140 \\ p_2 + p_3 + q_2 - 4p_4 &= -80 \\ 2p_4 + q_1 - 4q_2 &= -80 \end{aligned}$$

Appendix A

Determinants

A *determinant* is a square array of quantities having some definite value. In general, a determinant of nth order is written as

$$\Delta = \begin{vmatrix} a_{11} & a_{12} & \cdots & a_{1n} \\ a_{21} & a_{22} & \cdots & a_{2n} \\ \cdots & \cdots & \cdots & \cdots \\ a_{n1} & a_{n2} & \cdots & a_{nn} \end{vmatrix} \tag{A.1}$$

The quantities of a determinant are called *elements*. The elements in the horizontal direction form *rows* and those in vertical direction form columns. The element in ith row and jth column will be denoted by a_{ij}.

The *minor of any element of a determinant is obtained by omitting the row and column containing the element.* Thus, the minor of b_2 in the determinant

$$\Delta = \begin{vmatrix} a_1 & b_1 & c_1 \\ a_2 & b_2 & c_2 \\ a_3 & b_3 & c_3 \end{vmatrix} \tag{A.2}$$

is $\begin{bmatrix} a_1 & c_1 \\ a_3 & c_3 \end{bmatrix}$.

The *cofactor* of any element a_{ij} is its minor multiplied by the factor $(-1)^{i+j}$. Thus, the cofactor of b_2 in (A.2) is

$$(-1)^{2+2} \begin{vmatrix} a_1 & c_1 \\ a_3 & c_3 \end{vmatrix} = \begin{vmatrix} a_1 & c_1 \\ a_3 & c_3 \end{vmatrix}$$

The cofactor of the element a_{ij} is denoted generally by A_{ij}.

The value of a determinant of order n by *Laplace's Development* is given by

$$\det A = \sum_{j=1}^{n} a_{ij} A_{ij} \text{ for any } i \tag{A.3}$$

Thus, the values of the determinant in (A.2) is

$$\det A = a_1 \begin{vmatrix} b_2 & c_2 \\ b_3 & c_3 \end{vmatrix} - b_1 \begin{vmatrix} a_2 & c_2 \\ a_3 & c_3 \end{vmatrix} + c_1 \begin{vmatrix} a_2 & b_2 \\ a_3 & b_3 \end{vmatrix}$$

Following the above rule, we shall obtain

$$\begin{vmatrix} b_2 & c_2 \\ b_3 & c_3 \end{vmatrix} = b_2c_2 - b_3c_2$$

and so on.

Some Properties of Determinant

1. The value of a determinant does not change if its rows and columns are interchanged. Thus

$$\Delta = \begin{vmatrix} a_1 & b_1 & c_1 \\ a_2 & b_2 & c_2 \\ a_3 & b_3 & c_3 \end{vmatrix} = \begin{vmatrix} a_1 & a_2 & a_3 \\ b_1 & b_2 & b_3 \\ c_1 & c_2 & c_3 \end{vmatrix} \tag{A.4}$$

2. If two rows or columns of a determinant are interchanged, the sign of the determinant is changed (the numerical value remains the same). Thus

$$\Delta = \begin{vmatrix} a_1 & b_1 & c_1 \\ a_2 & b_2 & c_2 \\ a_3 & b_3 & c_3 \end{vmatrix} = - \begin{vmatrix} a_2 & b_2 & c_2 \\ a_1 & b_1 & c_1 \\ a_3 & b_3 & c_3 \end{vmatrix} \tag{A.5}$$

3. If two rows or two columns of a determinant are identical, then the value of the determinant is zero. Thus

$$\Delta = \begin{vmatrix} a_1 & b_1 & c_1 \\ a_2 & b_2 & c_2 \\ a_1 & b_1 & c_1 \end{vmatrix} = 0 \tag{A.6}$$

4. If every element of a row or a column of a determinant is multiplied by a factor, then the determinant is multiplied by the same factor. Thus, if

$$\Delta = \begin{vmatrix} a_1 & b_1 & c_1 \\ a_2 & b_2 & c_2 \\ a_3 & b_3 & c_3 \end{vmatrix}$$

then

$$\begin{vmatrix} a_1 & \lambda b_1 & c_1 \\ a_2 & \lambda b_2 & c_2 \\ a_3 & \lambda b_3 & c_3 \end{vmatrix} = \lambda\Delta \tag{A.7}$$

5. A determinant remains unchanged, if λ times the elements of any row or column are added or subtracted from the corresponding elements of any row or column. Here λ is any given number. Thus

$$\Delta = \begin{vmatrix} a_1 & b_1 & c_1 \\ a_2 & b_2 & c_2 \\ a_3 & b_3 & c_3 \end{vmatrix} = \begin{vmatrix} a_1 + \lambda b_1 & b_1 & c_1 \\ a_2 + \lambda b_2 & b_2 & c_2 \\ a_3 + \lambda b_3 & b_3 & c_3 \end{vmatrix} \tag{A.8}$$

6. If, in a determinant, the elements a_{ij} and a_{ji} are equal, it is called a *symmetric* determinant A.

   Thus

$$\Delta = \begin{vmatrix} a & h & g \\ h & b & f \\ g & f & c \end{vmatrix} \tag{A.9}$$

   is *symmetrical.*

7. If, in a determinant, $a_{ij} = -a_{ji}$, for all i and j, then it is called *skew-symmetric.*

   Thus

$$\Delta = \begin{vmatrix} 0 & a & b \\ -a & 0 & c \\ -b & -c & 0 \end{vmatrix} \tag{A.10}$$

   is skew-symmetric.

8. The square of any determinant is a symmetric determinant.
9. The value of every skew-symmetric determinant of *odd order* is zero.
10. Every skew symmetric determinant of *even order* is a perfect square.
11. If A_{ij} and A_{ji} be the cofactors of the elements a_{ij} and a_{ji} respectively of a symmetric determinant, then $A_{ij} = A_{ji}$ for all i and j.
12. If A_{ij} and A_{ji} be the cofactors of the elements a_{ij} and a_{ji} respectively of a skew-symmetric determinant, then

$$A_{ij} = (-1)^{n-1} A_{ji} \tag{A.11}$$

   where n is the order of the determinant $|a_{ij}|$.

Appendix B

Jacobian

If ξ and η are functions of two variables x and y, i.e.,

$$\xi = \xi(x, y) \quad \text{and} \quad \eta = \eta(x, y)$$

then the determinant

$$\begin{vmatrix} \dfrac{\partial \xi}{\partial x} & \dfrac{\partial \xi}{\partial y} \\ \dfrac{\partial \eta}{\partial x} & \dfrac{\partial \eta}{\partial y} \end{vmatrix} \tag{B.1}$$

is called *differential determinant* or *Jacobian* and is denoted by

$$\frac{\partial(\xi, \eta)}{\partial(x, y)} \tag{B.2}$$

Thus, for three functions

$$\xi = \xi(x, y, z), \quad \eta = \eta(x, y, z), \quad \theta = \theta(x, y, z)$$

we write the Jacobian as

$$\frac{\partial(\xi, \eta, \theta)}{\partial(x, y, z)} = \begin{vmatrix} \dfrac{\partial \xi}{\partial x} & \dfrac{\partial \xi}{\partial y} & \dfrac{\partial \xi}{\partial z} \\ \dfrac{\partial \eta}{\partial x} & \dfrac{\partial \eta}{\partial y} & \dfrac{\partial \eta}{\partial z} \\ \dfrac{\partial \theta}{\partial x} & \dfrac{\partial \theta}{\partial y} & \dfrac{\partial \theta}{\partial z} \end{vmatrix} \tag{B.3}$$

Properties of Jacobians

1.

$$\frac{\partial(\xi, \eta)}{\partial(x, y)} \cdot \frac{\partial(x, y)}{\partial(\xi, \eta)} = 1 \tag{B.4}$$

The LHS is

$$
= \begin{vmatrix} \frac{\partial \xi}{\partial x} & \frac{\partial \xi}{\partial y} \\ \frac{\partial \eta}{\partial x} & \frac{\partial \eta}{\partial y} \end{vmatrix} \times \begin{vmatrix} \frac{\partial x}{\partial \xi} & \frac{\partial x}{\partial \eta} \\ \frac{\partial y}{\partial \xi} & \frac{\partial y}{\partial \eta} \end{vmatrix} = \begin{vmatrix} \frac{\partial \xi}{\partial x}\frac{\partial x}{\partial \xi} + \frac{\partial \xi}{\partial y}\frac{\partial y}{\partial \xi} & \frac{\partial \xi}{\partial x}\frac{\partial x}{\partial \eta} + \frac{\partial \xi}{\partial y}\frac{\partial y}{\partial \eta} \\ \frac{\partial x}{\partial \xi}\frac{\partial \eta}{\partial x} + \frac{\partial y}{\partial \xi}\frac{\partial \eta}{\partial y} & \frac{\partial x}{\partial \eta}\frac{\partial \eta}{\partial x} + \frac{\partial y}{\partial \eta}\frac{\partial \eta}{\partial y} \end{vmatrix}
$$

$$
= \begin{vmatrix} \frac{\partial \xi}{\partial \xi} & \frac{\partial \xi}{\partial \eta} \\ \frac{\partial \eta}{\partial \xi} & \frac{\partial \eta}{\partial \eta} \end{vmatrix} = \begin{vmatrix} 1 & 0 \\ 0 & 1 \end{vmatrix} = 1.
$$

2. If $u = u(x, y)$ and $v = v(x, y)$, where $x = x(\xi, \eta)$ and $y = y(\xi, \eta)$, then

$$
\frac{\partial(u, v)}{\partial(\xi, \eta)} = \frac{\partial(u, v)}{\partial(x, y)} \frac{\partial(x, y)}{\partial(\xi, \eta)} \tag{B.5}
$$

The RHS is

$$
= \begin{vmatrix} \frac{\partial u}{\partial x} & \frac{\partial u}{\partial y} \\ \frac{\partial v}{\partial x} & \frac{\partial v}{\partial y} \end{vmatrix} \times \begin{vmatrix} \frac{\partial x}{\partial \xi} & \frac{\partial x}{\partial \eta} \\ \frac{\partial y}{\partial \xi} & \frac{\partial y}{\partial \eta} \end{vmatrix} = \begin{vmatrix} \frac{\partial u}{\partial x}\frac{\partial x}{\partial \xi} + \frac{\partial u}{\partial y}\frac{\partial y}{\partial \xi} & \frac{\partial u}{\partial x}\frac{\partial x}{\partial \eta} + \frac{\partial u}{\partial y}\frac{\partial y}{\partial \eta} \\ \frac{\partial v}{\partial x}\frac{\partial x}{\partial \xi} + \frac{\partial v}{\partial y}\frac{\partial y}{\partial \xi} & \frac{\partial v}{\partial x}\frac{\partial x}{\partial \eta} + \frac{\partial v}{\partial y}\frac{\partial y}{\partial \eta} \end{vmatrix}
$$

$$
= \begin{vmatrix} \frac{\partial u}{\partial \xi} & \frac{\partial u}{\partial \eta} \\ \frac{\partial v}{\partial \xi} & \frac{\partial v}{\partial \eta} \end{vmatrix} = \frac{\partial(u, v)}{\partial(\xi, \eta)}
$$

Appendix C

Stirling's Approximations

We have

$$n! = \Gamma(n+1)\int_0^\infty t^n e^{-t}dt = \int_0^\infty e^{n\log t - t}\,dt \tag{C.1}$$

We substitute $t = n + \sqrt{n}\xi$, i.e., $dt = \sqrt{n}d\xi$.
Thus, from (C.1)

$$n! = \sqrt{n}\int_{-\sqrt{n}}^{\infty} \exp\left[n\log\left(n+\sqrt{n}\xi\right) - \left(n+\sqrt{n}\xi\right)\right]d\xi \tag{C.2}$$

Now

$$\log\left(n+\sqrt{n}\xi\right) = \log n + \log\left(1+\frac{\xi}{\sqrt{n}}\right) = \log n + \frac{\xi}{\sqrt{n}} - \frac{\xi^2}{2n} \quad \text{[for large } n\text{]} \tag{C.3}$$

Also, for large n, the lower limit of integral in (C.2) can be taken as $-\infty$. Thus, from (C.2) and (C.3)

$$n! = \sqrt{n}n^n e^{-n}\int_{-\infty}^{\infty} e^{-\xi^2/2}d\xi = \sqrt{2\pi}n^{n+\frac{1}{2}}e^{-n} \tag{C.4}$$

Hence

$$\log(n!) = \tfrac{1}{2}\log(2\pi) + \left(n+\tfrac{1}{2}\right)\log n - n \tag{C.5}$$

For large values of n, further approximation gives

$$\log(n!) = n\log n - n \tag{C.6}$$

Appendix D

Miscellaneous Functions

RIEMANN'S ZETA FUNCTION

D.1 Definition and Evaluation

Riemann's zeta functions are in general defined as

$$\xi(k) = \sum_{n=1}^{\infty} \frac{1}{n^k} = \frac{1}{1^k} + \frac{2}{2^k} + \frac{3}{3^k} + \cdots \quad \text{(D.1)}$$

Thus

$$\xi(2) = \sum_{n=1}^{\infty} \frac{1}{n^2} = \frac{1}{1^2} + \frac{1}{2^2} + \frac{1}{3^2} + \cdots$$

$$\xi(3) = \sum_{n=1}^{\infty} \frac{1}{n^3} = \frac{1}{1^3} + \frac{1}{2^3} + \frac{3}{3^3} + \cdots$$

$$\xi(4) = \sum_{n=1}^{\infty} \frac{1}{n^4} = \frac{1}{1^4} + \frac{2}{2^4} + \frac{3}{3^4} + \cdots$$

and so on.

The sum of the above series can be evaluated by the method of Fourier expansion. Let us evaluate the sum $\xi(2)$. For this purpose, we consider the function

$$f(x) = x^2, \quad \text{for } -\pi \leq x \leq \pi \quad \text{(D.2)}$$

The Fourier's expansion of the function $f(x)$ is given by

$$f(x) \equiv x^2 = \frac{\pi^2}{3} + \sum_{n=1}^{\infty} \frac{4}{n^2}(-1)^n \cos nx \quad \text{(D.3)}$$

This we have obtained in Example 10.5.

Putting $x = \pi$ in equation (D.3), we get

$$\pi^2 = \frac{\pi^2}{3} + \sum_{n=1}^{\infty} \frac{4}{n^2}(-1)^n \cos n\pi = \frac{\pi^2}{3} + \sum_{n=1}^{\infty} \frac{4}{n^2}(-1)^n(-1)^n = \frac{\pi^2}{3} + 4\xi(2)$$

Hence

$$\xi(2) = \frac{\pi^2}{6} \tag{D.4}$$

Similarly, in order to evaluate $\xi(4)$, we consider the function

$$f(x) = x^4 \quad \text{for } -\pi \le x \le \pi \tag{D.5}$$

Fourier's coefficients in this case are

$$a_0 = \frac{1}{2\pi}\int_{-\pi}^{\pi} x^4 dx = \frac{\pi^4}{5}$$

$$a_n = \frac{1}{\pi}\int_{-\pi}^{\pi} x^4 \cos nx dx = (-1)^n\left(\frac{8\pi^2}{n^2} - \frac{48}{n^4}\right).$$

$b_n = 0$, since the function $f(x)$ is even.

Hence

$$x^4 = \frac{\pi^4}{5} + \sum_{n=1}^{\infty}(-1)^n\left(\frac{8\pi^2}{n^2} - \frac{48}{n^4}\right)\cos nx$$

Putting $x = \pi$, we get

$$\pi^4 = \frac{\pi^4}{5} + \sum_{n=1}^{\infty}(-1)^n\left(\frac{8\pi^2}{n^2} - \frac{48}{n^4}\right)(-1)^n = \frac{\pi^4}{5} + \sum_{n=1}^{\infty}\left(\frac{8\pi^2}{n^2} - \frac{48}{n^4}\right)$$

$$\text{or,} \quad \frac{4}{5}\pi^4 = 8\pi^2\xi(2) - 48\xi(4) \tag{D.6}$$

Using the value of $\xi(2)$ from equation (D.4), we get

$$\frac{4}{5}\pi^4 = 8\pi^2\left(\frac{\pi^2}{6}\right) - 48\xi(4)$$

which gives

$$\xi(4) = \frac{\pi^4}{90} \tag{D.7}$$

Zeta functions are also expressed as

$$\xi(k) = \frac{1}{\Gamma(k)}\int_0^{\infty}\frac{x^{k-1}}{e^x - 1}dx \tag{D.8}$$

where $k > 1$.

Some special values of the function are

$$\left.\begin{aligned} &\xi(0) = -\frac{1}{2}; \quad \xi(1) = \infty \\ &\xi(-2n) = 0; \quad \xi(1-2n) = -\frac{B_{2n}}{2n} \end{aligned}\right\} \tag{D.9}$$

where $n = 1, 2, \ldots$ and B_{2n} are Bernoulli's numbers.

Riemann's zeta functions can be expressed in terms of poly-gamma functions. We know that the poly-gamma functions are given by [ref. equation (7.121)]

$$\psi_k(m) = (-1)^k\Gamma(k)\sum_{n=0}^{\infty}\frac{1}{(m+n)^k}, \quad k \ge 2$$

If we put $m = 1$, then

$$\psi_k(1) = (-1)^k \Gamma(k) \sum_{n=0}^{\infty} \frac{1}{(n+1)^k} = (-1)^k \Gamma(k) \sum_{n=1}^{\infty} \frac{1}{n^k} = (-1)^k \Gamma(k) \xi(k)$$

Hence

$$\xi(k) = (-1)^k \frac{1}{\Gamma(k)} \psi(1) \tag{D.10}$$

BERNOULLI POLYNOMIALS

D.2 Definition and Generation

Bernoulli polynomials are generated from the function

$$\frac{te^{xt}}{e^t - 1} = \sum_{n=0}^{\infty} B_n(x) \frac{t^n}{n!}, \quad |t| < 2\pi \tag{D.11}$$

If we express the polynomial $B_n(x)$ in the form

$$B_n(x) = \sum_{k=0}^{\infty} a_k x^k \tag{D.12}$$

then the coefficients a_k can be determined for different values of n. Thus

$$\begin{aligned}
&B_0(x) = 1\\
&B_1(x) = -\tfrac{1}{2} + x\\
&B_2(x) = \tfrac{1}{6} - x + x^2\\
&B_3(x) = 0 + \tfrac{1}{2}x - \tfrac{3}{2}x^2 + x^3\\
&B_4(x) = -\tfrac{1}{30} + 0 + x^2 - 2x^3 + x^4\\
&\dots\dots\dots\dots\dots\dots\dots\dots\\
&\dots\dots\dots\dots\dots\dots\dots\dots
\end{aligned} \tag{D.13}$$

For $x = 0$, $B_n(0) = B_n$ which are called Bernoulli numbers. Thus

$$B_0 = 1, \quad B_1 = -\tfrac{1}{2}, \quad B_2 = \tfrac{1}{6}, \quad B_3 = 0, \quad B_4 = -\tfrac{1}{30}, \quad B_5 = 0, \quad B_6 = \tfrac{1}{42}, \quad \dots \tag{D.14}$$

It is to be noted that, for $n = 1, 2, \dots$

$$B_{2n+1} = 0 \tag{D.15}$$

These numbers occur in series expansion of some circular functions. As an example

$$\tan x = x + \frac{x^3}{3} + \frac{2x^5}{15} + \dots + \frac{(-1)^{n-1} 2^{2n} (2^{2n} - 1)}{(2n)!} B_{2n} x^{2n-1} \tag{D.16}$$

for $|x| < \pi/2$.

Bernoulli functions are symmetric. Thus

$$\left.\begin{aligned} B_n(1-x) &= (-1)^n B_n(x) \\ (-1)^n B_n(-x) &= B_n(x) + nx^{n-1} \end{aligned}\right\} \quad (n = 0, 1, 2, \ldots) \tag{D.17}$$

The derivatives of the function are related by

$$B_n'(x) = nB_{n-1}(x) \quad (n = 1, 2, \ldots) \tag{D.18}$$

The integral of the functions $B_m(x)$ and $B_n(x)$ can be evaluated as

$$\int_0^1 B_m(x)B_n(x)dx = (-1)^{n-1}\frac{m!n!}{(m+n)!}B_{m+n} \tag{D.19}$$

There exists a recurrence relation for these functions in the form

$$B_n(x+1) - B_n(x) = nx^{n-1} \tag{D.20}$$

for $n = 0, 1, 2, \ldots$

In view of equation (D.15), the integral vanishes when $(m+n)$ is odd. The polynomials are, therefore, orthogonal, when $(m+n)$ is odd.

EULER POLYNOMIALS

D.3 Generation and Properties

Euler polynomials are generated from the function

$$\frac{2e^{xt}}{e^t+1} = \sum_{n=0}^{\infty} E_n(x)\frac{t^n}{n!}, \quad |\,t\,| < \pi \tag{D.21}$$

Expressing the polynomials $E_n(x)$ in the form

$$E_n(x) = \sum_{k=0}^{\infty} a_k x^k \tag{D.22}$$

we find for different values of n the following functions

$$\begin{aligned} E_0(x) &= 1 \\ E_1(x) &= -\tfrac{1}{2} + x \\ E_2(x) &= 0 - x + x^2 \\ E_3(x) &= \tfrac{1}{4} + 0 - \tfrac{3}{2}x^2 + x^3 \\ E_4(x) &= 0 + x + 0 - 2x^3 + x^4 \\ &\ldots\ldots\ldots\ldots\ldots\ldots\ldots\ldots \\ &\ldots\ldots\ldots\ldots\ldots\ldots\ldots\ldots \end{aligned}$$

From equation (D.21), putting $x = 0$,

$$\frac{2}{e^t + 1} = \sum_{n=0}^{\infty} E_n(0)\frac{t^n}{n!} \tag{D.23}$$

where $E_n = 2^n E_n(1/2)$, $(n = 0, 1, 2, \ldots)$, are called Euler numbers.

For $n = 1, 2, \ldots,$

$$E_n(0) = -E_n(1) \tag{D.24}$$

Some Euler numbers are

$$\left.\begin{array}{lll} E_0 = 1, & E_2 = -1, & E_3 = 0 \\ E_4 = 5, & E_5 = 0, & E_6 = 61 \\ \ldots\ldots\ldots & \ldots\ldots\ldots & \ldots\ldots\ldots \end{array}\right\} \tag{D.25}$$

For $n = 1, 2, \ldots$

$$E_{2n+1} = 0 \tag{D.26}$$

Euler numbers occur in series expansion of some circular functions. As an example

$$\sec x = 1 + \frac{x^2}{2} + \frac{5x^4}{24} + \frac{61x^6}{720} + \cdots + \frac{(-1)^n E_{2n}}{(2n)!}x^{2n} + \cdots \tag{D.27}$$

where $|\,x\,| < \pi/2$.

Euler functions have the following symmetry relations

$$\left.\begin{array}{r} E_n(1-x) = (-1)^n E_n(x) \\ (-1)^{n+1}E_n(-x) = E_n(x) - 2x^n \end{array}\right\} \tag{D.28}$$

for $n = 0, 1, 2, \ldots$

The derivatives of the functions are related by

$$E_n'(x) = nE_{n-1}(x) \tag{D.29}$$

for $n = 1, 2, 3, \ldots$

A recurrence relation for the functions exists in the form

$$E_n(x+1) + E_n(x) = 2x^n \tag{D.30}$$

for $n = 0, 1, 2, \ldots$

The integral of the functions can be evaluated as

$$\int_0^1 E_m(x)E_n(x)dx = (-1)^n 4\left(2^{m+n+2} - 1\right)\frac{m!n!}{(m+n+2)!}B_{m+n+2} \tag{D.31}$$

for $m, n = 0, 1, 2, \ldots$

When $m + n =$ odd, $B_{m+n+2} = 0$. Hence the integral vanishes. Therefore, the functions $E_m(x)$ and $E_n(x)$ are orthogonal when $(m + n)$ is odd.

JACOBI'S ELLIPTIC FUNCTIONS

D.4 Definition and Notations

An elliptic function is defined to be a doubly-periodic analytic function whose only possible singularities in the finite part of the plane are poles.

We recall the incomplete elliptic integral

$$u = F(k, \phi) = \int_0^\phi \frac{d\theta}{\sqrt{1 - k^2 \sin^2 \theta}} = \int_0^x \frac{d\xi}{\sqrt{(1 - \xi^2)(1 - k^2\xi^2)}} \tag{D.32}$$

where $\phi =$ amplitude of u and is written as $\phi = a\,mu$.

Jacobi's elliptic functions are defined by the equations

$$\left.\begin{aligned} sn\,u &= \sin(a\,mu) = \sin\phi = x \\ cn\,u &= \cos(a\,mu) = \cos\phi = \sqrt{1 - x^2} \\ dn\,u &= \sqrt{1 - k^2 sn^2 u} = \sqrt{1 - k^2 x^2} \end{aligned}\right\} \tag{D.33}$$

Derivatives of these functions are

$$\left.\begin{aligned} \frac{d}{du} &= (sn\,u) = cn\,u\,dn\,u \\ \frac{d}{du} &= (cn\,u) = -dn\,u\,sn\,u \\ \frac{d}{du} &= (dn\,u) = -k^2\,sn\,u\,cn\,u \end{aligned}\right\} \tag{D.34}$$

The reciprocals and quotients of Jacobi's elliptic functions are denoted by

$$\left.\begin{aligned} \frac{1}{sn\,u} &= n\,su; & \frac{1}{cn\,u} &= nc\,u; & \frac{1}{dn\,u} &= nd\,u \\ \frac{sn\,u}{cn\,u} &= sc\,u; & \frac{sn\,u}{dn\,u} &= sd\,u; & \frac{cn\,u}{dn\,u} &= cd\,u \\ \frac{cn\,u}{sn\,u} &= cs\,u; & \frac{dn\,u}{sn\,u} &= ds\,u; & \frac{dn\,u}{cn\,u} &= dc\,u \end{aligned}\right\} \tag{D.35}$$

These notations are due to Glaisher.

Appendix E

Distribution Functions

E.1 Definition

Distribution functions are important concepts in the theory of probability and are used in the study of diversified field of science and technology. These are functions of random variables, whose values depend on chance. The random variables may be discrete and continuous. Consider that ξ is a random variable and x is an arbitrary real number. The probability that ξ will take on a value $< x$, is called *distribution function of probabilities* of the random variable ξ. It is denoted as

$$F(x) = P(\xi < x). \tag{E.1}$$

If the random variable is discrete, that is, it can take on only a finite or countable set of values, the distribution function is also discrete. In this case, the collection of numbers $\{p_i\}$ satisfying

$$\left.\begin{aligned} P(\xi = x_i) = p_i > 0 \text{ for all } i \\ \text{and } \sum_{i=1}^{\infty} p_i = 1 \end{aligned}\right\} \tag{E.2}$$

is called the *probability mass function* for the random variable ξ.

The distribution function of ξ is given by

$$F(x) = P(\xi \leq x) = \sum_{x_i \leq x} p_i. \tag{E.3}$$

We may write the distribution function $F(x)$ in the form

$$F(x) = \sum_{i=1}^{\infty} p_i \varepsilon(x - x_i), \tag{E.4}$$

where the function $\varepsilon(x)$ is defined as

$$\varepsilon(x) = \begin{cases} 1, & \text{for } x \geq 0 \\ 0, & \text{for } x < 0. \end{cases} \tag{E.5}$$

Distribution functions have got the following properties:

(1) Every distribution function is a non-decreasing function.

(2) The distribution function is continuous on the left and satisfies the conditions

$$F(-\infty) = 0 \text{ and } F(+\infty) = 1. \tag{E.6}$$

Conversely, also every function satisfying the above conditions may be regarded as the distribution function of some random variable.

A random variable ξ is called *continuous*, if there exists a non-negative function $f(x)$ such that, for every real number x, we have

$$F(x) = \int_{-x}^{x} f(t)\,dt, \tag{E.7}$$

where $F(x)$ is the distribution function of random variable ξ. The function $f(x)$ is called the *probability density function* of the random variable ξ. Every non-negative real function $f(x)$ that is integrable over real axis $(-\infty, \infty)$ and satisfies the equality

$$\int_{-\infty}^{\infty} f(x)\,dx = 1, \tag{E.8}$$

is the probability density function of some continuous random variable. The condition in (E.8) is known as the *normalisation condition.*

If the density function is absolutely continuous and is differentiable for all values of x, then

$$f(x) = F'(x). \tag{E.9}$$

The probability density function $f(x)$ has the following properties:

(1) It is a non-decreasing function, i.e.,

$$f(x) \geq 0. \tag{E.10}$$

(2) For any two values of the argument x_1, x_2, it satisfies the equation

$$P\{x_1 \leq \xi \leq x_2\} = \int_{x_1}^{x_2} f(x)\,dx. \tag{E.11}$$

(3)

$$\int f(x)\,dx = 1. \tag{E.12}$$

EXAMPLE E.1. *The probability density function of a random variable ξ is defined as*

$$f(x) = \begin{cases} x, & \textit{for } 0 < x \leq 1 \\ 2 - x, & \textit{for } 1 \leq x \leq 2 \\ 0, & \textit{otherwise.} \end{cases}$$

Show that $f(x)$ is really a probability density function. Also find the distribution function.

Solution: The function $f(x)$ should satisfy the equation (E.8) for it to be a probability density function.

Now,

$$\int_{-\infty}^{\infty} f(x)\,dx = \int_{0}^{1} x\,dx + \int_{1}^{2} (2-x)\,dx = \frac{1}{2} + \left\{\left(4 - \frac{4}{2}\right) - \left(2 - \frac{1}{2}\right)\right\}$$
$$= \frac{1}{2} + \frac{1}{2} = 1.$$

Hence, $f(x)$ is a probability density function.

The distribution function $F(x)$ is given by equation (E.7). Hence,

$$\begin{aligned}
F(x) &= 0, && \text{for } x \le 0\\
F(x) &= \int_0^x t\,dt = \frac{x^2}{2}, && \text{for } 0 \le x \le 1\\
F(x) &= \int_0^x t\,dt + \int_1^x (2-t)\,dt\\
&= 2x - \frac{x^2}{2} - 1, && \text{for } 1 \le x \le 2\\
\text{and } F(x) &= 1, && \text{for } x \ge 2.
\end{aligned}$$

E.2 Moments of a Distribution Function

Let ξ be a discrete random variable with probability mass function

$$p_k = P\{\xi = x_k\},\ k = 1, 2, \ldots. \tag{E.13}$$

If

$$\sum_{k=1}^{\infty} |x_k| p_k \le \infty, \tag{E.14}$$

we say that the *expected value* or *mathematical expectation* of ξ exists, and write:

$$\mu = E(\xi) = \sum_{k=1}^{\infty} x_k p_k. \tag{E.15}$$

If the series in (E.14) does not converge, we say that $E(\xi)$ does not exist.

For example, the probability mass function of a random variable ξ in Bernoullian trials is given to be

$$P(\xi = x) = q^x p, \tag{E.16}$$

where p = probability of success and $q = 1 - p$ = probability of failure.

Then, expectation

$$\mu = E(\xi) = \sum_{x=0}^{\infty} x q^x p,$$

i.e.,

$$\begin{aligned}
E(\xi) &= 0 \cdot p + 1 \cdot qp + 2 \cdot q^2 p + \ldots\\
&= qp(1 + 2q + 3q^2 + \ldots).
\end{aligned}$$

$$\therefore\ E(\xi) = qp \cdot \frac{1}{(1-q)^2} = qp \cdot \frac{1}{p^2} = \frac{q}{p}. \tag{E.17}$$

If the random variable ξ is of continuous type, having probability density function $f(x)$, we say that the expectation of ξ exists and is equal to

$$E(\xi) = \int_{-\infty}^{\infty} x f(x)\,dx, \tag{E.18}$$

provided

$$\int_{-\infty}^{\infty} |x| f(x)\, dx < \infty. \tag{E.19}$$

For example, let ξ be a continuous random variable with the probability density function

$$f(x) = \begin{cases} \frac{2}{x^3}, & \text{for } x \geq 1 \\ 0, & \text{for } x < 1. \end{cases}$$

Then

$$\begin{aligned} \int_0^{\infty} |x| f(x)\, dx &= \int_1^{\infty} x\, f(x)\, dx \\ &= \int_1^{\infty} \frac{2}{x^2} dx < \infty. \end{aligned}$$

Hence, expectation of ξ exists and equals to

$$E(\xi) = \int_1^{\infty} \frac{2}{x^2} dx = 2.$$

We now consider another example, where the probability density function is given as

$$f(x) = \frac{1}{\pi} \cdot \frac{1}{1+x^2}, \quad -\infty < x \leq \infty.$$

Here, the integral

$$\int_{-\infty}^{\infty} |x| f(x)\, dx = \frac{1}{\pi} \int_{-\infty}^{\infty} \frac{|x|}{1+x^2} dx$$

diverges. Hence, $E(\xi)$ does not exist.

In fact,

$$E(\xi) = \int_{-\infty}^{\infty} x\, f(x)\, dx = \int_{-\infty}^{\infty} \frac{x\, dx}{\pi(1+x^2)} = 0.$$

Thus, it is advisable to check first whether the expectation of the random variable actually exists or not. Then only we should proceed to find its value.

We now mention, without offering any proof, the following theorem on expectation.

Theorem E.1. *If ξ be a non-negative random variable with distribution function $F(x)$, then*

$$E(\xi) = \int_0^{\infty} [1 - F(x)] dx \tag{E.20}$$

in the sense that, if one side exists, so will the other and the two sides are equal.

The expectation of the random variable ξ^r is called the *rth moment* of ξ around the origin. In the case of discrete random variable, we write:

$$\mu_r = \sum_{k=1}^{\infty} x_k^r p_k. \tag{E.21}$$

If the random variable is of continuous type, we write:

$$\mu_r = E(\xi^r) = \int_{-\infty}^{\infty} x^r f(x)\, dx, \tag{E.22}$$

where $f(x)$ = probability density function.

Suppose, the rth moment of ξ around origin exists.

Now,

$$|x|^{r-1} \leq |x|^r + 1. \tag{E.23}$$

Hence, writing $s = r - 1$, we have for existence of $E(\xi^s)$, the condition (E.19) gives

$$\int_{-\infty}^{\infty} |x|^s f(x)\, dx < \infty$$

$$\text{or, } \int_{k\leq 1} |x|^s f(x)\, dx + \int_{k>1} |x|^s f(x)\, dx < \infty, \quad [k = |x|^s]$$

$$\text{or, } \leq P\{|\xi|^s \leq 1\} + E|\xi|^r < \infty, \tag{E.24}$$

which is true, since rth moment exists.

Thus, we have the theorem:

Theorem E.2. *If the moment of order r exists for a random variable ξ, its moments of order $0 < s < r$ should also exist.*

Let now k be a positive integer and a be a constant. If $E\{(\xi - a)^k\}$ exists, we call it the moment of order k about the point a. Taking $E(\xi) = a = \mu$, we call $E\{(\xi - \mu)^k\}$ the central moment of order k and write:

$$\mu_k = E\{(\xi - \mu)^k\}. \tag{E.25}$$

For $k = 2$, we have

$$\mu_2 = E\{(\xi - \mu)^2\}, \tag{E.26}$$

which is called the *variance* of ξ and is denoted as

$$\sigma^2 = \text{var}\,(\xi) = E\{(\xi - \mu)^2\}. \tag{E.27}$$

The quantity σ is called the *standard deviation* of ξ. Equation (E.27) is also written as

$$\sigma^2 = \text{var}\,(\xi) = E(\xi^2) - \{E(\xi)\}^2. \tag{E.28}$$

The variance of a random variable has the following important properties:

$$\left.\begin{array}{ll} (1)\ \text{var}\,(\xi) = 0, & \text{if and only if } \xi \text{ degenerates} \\ (2)\ \text{var}\,(\xi) < E\{(\xi - a)^2\}, & \text{for any } a \neq E(\xi) \end{array}\right\}. \tag{E.29}$$

The random variable η defined by

$$\eta = \frac{\xi - E(\xi)}{\sqrt{\text{var}\,(\xi)}} = \frac{\xi - \mu}{\sigma}, \tag{E.30}$$

is called *standardised random variable*.

E.3 Probability Generating Function

Functions that generate probabilities are called *probability generating functions.* Let us consider a random variable ξ, such that

$$\left.\begin{aligned} p_k &= P\{\xi = k\},\ k = 0, 1, 2, \ldots \\ \text{with } &\sum_{k=0}^{\infty} p_k = 1 \end{aligned}\right\} \tag{E.31}$$

Then the function defined by

$$P(t) = \sum_{k=0}^{\infty} p_k t^k \tag{E.32}$$

is called the *probability generating function* of ξ. It is to be noted that the series in (E.32) converges for $|t| \leq 1$.

As an example, let ξ be a random variable such that $P\{\xi = k\} = pq^k$, where $q = 1 - p$ and $0 < p < 1$, and $k = 0, 1, 2, \ldots$.

Then

$$P(t) = \sum_{k=0}^{\infty} pq^k t^k = p\frac{1}{1 - tq}, \quad \text{for } |t| \leq 1.$$

Now,

$$P(1) = p\frac{1}{1-q} = 1.$$

Hence, the function $P(t)$ converges uniformly and absolutely for $|t| \leq 1$. It generates uniquely the probability for different values of t in $|t| \leq 1$.

E.4 Moment Generating Functions

Functions that generate moments of a random variable are called *moment generating functions.* The functions defined by

$$M(t) = E(e^{t\xi}) \tag{E.33}$$

is known as the *moment generating function* for the random variable ξ, provided the expectation on the RHS of (E.33) exists about the origin and the series

$$\sum e^{tx} P(\xi = x) \tag{E.34}$$

converges absolutely.

For example, let the probability density function of a random variable be given by

$$f(x) = \begin{cases} \frac{1}{2}\exp(-x/2), & \text{for } x > 0 \\ 0, & \text{otherwise.} \end{cases}$$

Then

$$M(t) = \frac{1}{2}\int_0^{\infty} e^{(t-1/2)x}\,dx = \frac{1}{1-2t}, \tag{E.35}$$

which converges absolutely for $t < \frac{1}{2}$.

Similarly, for a discrete random variable ξ, if the probability mass function is

$$P\{\xi = k\} = \begin{cases} \exp(-\lambda)\frac{\lambda^k}{k!}, & k = 0, 1, \dots \\ 0, & \text{otherwise}, \end{cases}$$

then

$$M(t) = E(e^{t\xi}) = e^{-\lambda}\sum_{k=0}^{\infty} e^{tk} \cdot \frac{\lambda^k}{k!} = e^{-\lambda}\sum_{k=0}^{\infty}\frac{(\lambda e^t)^k}{k!},$$

i.e.,

$$M(t) = e^{-\lambda} \cdot e^{\lambda e^t} = e^{-\lambda(1-e^t)}, \tag{E.36}$$

where t can have any value.

We note the following interesting property of the moment generating function.

If the moment generating function $M(t)$ of a random variable ξ exists for t defined in a certain region, then it can be expanded in Maclaurin's series like

$$M(t) = M(0) + \frac{M'(0)}{1!}t + \frac{M''(0)}{2!}t^2 + \dots \tag{E.37}$$

such that $E(\xi^k)$ is given by the coefficient of $t^k/k!$.

Thus, if the moment generating function $M(t)$ is given, we can find expectation of the random variable. For example, if the function $M(t)$ is given by equation (E.35), we find

$$M'(t) = \frac{2}{(1-2t)^2}; \text{ hence, } M'(0) = 2.$$

Again,

$$M''(t) = \frac{4 \cdot 2}{(1-2t)^3}; \text{ hence, } M''(0) = 8.$$

Thus,

$$E(\xi) = \text{coefficient of } \frac{t}{1!} = M'(0) = 2$$

$$E(\xi^2) = \text{coefficient of } \frac{t^2}{2!} = M''(0) = 8.$$

Conversely, if the moment generating function $M(t)$ for a random variable ξ exists for t in some defined region, then the derivatives of all order of $M(t)$ exist, and can be determined from

$$M^{(k)}(t)|_{t=0} = E(\xi^k), \text{ for } k > 0. \tag{E.38}$$

E.5 Some Discrete Distribution Functions

(1) Binomial Distribution

If the probability mass function of a random variable ξ is given by

$$p_k = P\{\xi = k\} = \binom{n}{k}p^k(1-p)^{n-k}, \tag{E.39}$$

where $0 \leq p \leq 1$ and $k = 0, 1, 2, \ldots, n$, we say that the random variable has a *binomial distribution* with parameter p.

Here, we find that

$$\sum_{k=0}^{n} p_k = \{p + (1-p)\}^n = 1, \tag{E.40}$$

which shows that the p_k's really define the probability mass function.

The distribution function is given by

$$F(x) = \sum_{k=0}^{n} \binom{n}{k} p^k (1-p)^{n-k} \varepsilon(x-k). \tag{E.41}$$

Expectation of ξ is

$$E(\xi) = \sum_{k=0}^{n} k \binom{n}{k} p^k (1-p)^{n-k} = np \sum_{k=0}^{n-1} \binom{n-1}{k-1} p^{k-1} (1-p)^{n-k} = np. \tag{E.42}$$

$$E(\xi^2) = E\{\xi(\xi-1) + \xi\} = \sum_{k=0}^{n} k(k-1) \binom{n}{k} p^k (1-p)^{n-k} + np = n(n-1)p^2 + np. \tag{E.43}$$

Hence, variance of ξ is

$$\text{var}\,(\xi) = E(\xi^2) - \{E(\xi)\}^2 = n(n-1)p^2 + np - n^2p^2 = np(1-p) = npq, \tag{E.44}$$

where $q = 1 - p$.

The probability generating function is

$$\begin{aligned} P(t) = \sum p_k t^k &= \sum_{k=0}^{n} \binom{n}{k} (pt)^k (1-p)^{n-k} \\ &= \{pt + (1-p)\}^n \text{ for } |t| \leq 1 = (q + pt)^n. \end{aligned} \tag{E.45}$$

The moment generating function is

$$\begin{aligned} M(t) = \sum_{k=0}^{n} e^{tk} p_k &= \sum_{k=0}^{n} e^{tk} \binom{n}{k} p^k (1-p)^{n-k} \\ &= [pe^{tk} + (1-p)]^n = (pe^{tk} + q)^n. \end{aligned} \tag{E.46}$$

(2) Poisson Distribution

The random variable ξ is said to have Poisson distribution, if its probability mass function is given by

$$p_k = P\{\xi = k\} = \frac{e^{-\lambda}\lambda^k}{k!}, \quad \lambda > 0 \text{ and } k = 0, 1, 2, \ldots. \tag{E.47}$$

Now,

$$\sum_{k=0}^{\infty} p_k = \sum_{k=0}^{\infty} e^{-\lambda} \frac{\lambda^k}{k!} = e^{-\lambda} \sum_{k=0}^{\infty} \frac{\lambda^k}{k!} = e^{-\lambda} \cdot e^{\lambda} = 1. \tag{E.48}$$

Thus, equation (E.47) really defines a probability mass function.

The distribution function is given by

$$F(x) = \sum_{k=0}^{\infty} e^{-\lambda} \frac{\lambda^k}{k!} \varepsilon(x-k). \tag{E.49}$$

The expectations of the random variable ξ are:

$$\begin{aligned} E(\xi) &= \sum_{k=0}^{\infty} k \cdot \frac{e^{-\lambda}\lambda^k}{k!} = \lambda e^{-\lambda} \sum_{m=0}^{\infty} \frac{\lambda^m}{m!}, \quad m = k-1 \\ &= \lambda e^{-\lambda} \cdot e^{\lambda} = \lambda \end{aligned} \tag{E.50}$$

$$\begin{aligned} E(\xi^2) &= E\{\xi(\xi-1)+\xi\} = \sum_{k=0}^{\infty} k(k-1)e^{-\lambda}\frac{\lambda^k}{k!} + E(\xi) \\ &= \lambda^2 e^{-\lambda} \sum_{n=0}^{\infty} \frac{\lambda^n}{n!} + \lambda, \quad n = k-2 \\ &= \lambda^2 e^{-\lambda} \cdot e^{\lambda} + \lambda = \lambda^2 + \lambda. \end{aligned} \tag{E.51}$$

The variance of the random variable is

$$\text{var}\,(\xi) = E(\xi^2) - \{E(\xi)\}^2 = \lambda^2 + \lambda - \lambda^2 = \lambda. \tag{E.52}$$

The probability generating function is given by

$$P(t) = \sum_{k=0}^{\infty} p_k t^k = \sum_{k=0}^{\infty} e^{-\lambda} \frac{(\lambda t)^k}{k!} = e^{-\lambda} \cdot e^{\lambda t} = e^{-\lambda(1-t)}, \ |t| \le 1. \tag{E.53}$$

The moment generating function is given by the equation (E.36).

E.6 Some Continuous Distribution Functions

(1) Rectangular Distribution

A random variable ξ is said to have a rectangular distribution on the interval (a, b), if its probability density function is given by

$$f(x) = \begin{cases} \frac{1}{b-a}, & \text{for } a \le x \le b \\ 0, & \text{elsewhere.} \end{cases} \tag{E.54}$$

Now,

$$\int_{-\infty}^{\infty} f(x)\,dx = \int_a^b \frac{dx}{b-a} = 1, \tag{E.55}$$

which shows that the function $f(x)$ in (E.54) really defines a probability density function. The distribution function of the random variable ξ is

$$F(x) = \int_{-\infty}^{x} f(x)\,dx,$$

i.e.,

$$F(x) = \begin{cases} 0, & \text{for } x < a \\ \frac{x-a}{b-a}, & \text{for } a \le x < b \\ 1, & \text{for } x = b. \end{cases} \tag{E.56}$$

Expectation of ξ is

$$E(\xi) = \int_0^\infty |x| f(x)\, dx = \int_a^b x \frac{dx}{b-a} = \frac{1}{2}(a+b). \tag{E.57}$$

Similarly,

$$E(\xi^2) = \int_a^b \frac{x^2}{b-a} dx = \frac{1}{3} \cdot \frac{b^3 - a^3}{b-a} = \frac{1}{3}(b^2 + ab + a^2). \tag{E.58}$$

Thus, the variance of ξ is

$$\text{var}(\xi) = E(\xi^2) - \{E(\xi)\}^2 = \frac{1}{3}(b^2 + ab + a^2) - \left(\frac{a+b}{2}\right)^2 = \frac{(b-a)^2}{12}. \tag{E.59}$$

The moment generating function of ξ is

$$M(t) = \int_a^b \frac{e^{tx}}{b-a} dx = \frac{1}{(b-a)t}(e^{tb} - e^{ta}),\ t \neq 0. \tag{E.60}$$

(2) Gamma Distribution Function

A random variable ξ is said to have gamma-type distribution with parameters α, β, if its probability density function is defined as

$$f(x) = \begin{cases} \frac{1}{\Gamma(\alpha)\beta^\alpha} x^{\alpha-1} e^{-x/\beta}, & \text{for } 0 < x < \infty \\ 0, & \text{for } x \le 0, \end{cases} \tag{E.61}$$

where $\alpha > 0, \beta > 0$.

Now,

$$\begin{aligned} \int_{-\infty}^\infty f(x)\, dx &= \int_0^\infty \frac{1}{\Gamma(\alpha)\beta^\alpha} x^{\alpha-1} e^{-x/\beta} dx \\ &= \frac{1}{\Gamma(\alpha)} \int_0^\infty z^{\alpha-1} e^{-z} dz\ [z = x/\beta] \\ &= \frac{1}{\Gamma(\alpha)} \cdot \Gamma(\alpha) = 1, \end{aligned} \tag{E.62}$$

which shows that the function in (E.61) really defines a probability density function.

The distribution function of the random variable ξ is given by

$$F(x) = \int_{-\infty}^x f(x)\, dx,$$

i.e.,

$$F(x) = \begin{cases} \dfrac{1}{\Gamma(\alpha)\beta^\alpha} \displaystyle\int_0^x t^{\alpha-1} e^{-t/\beta} dt, & x > 0 \\ 0, & \text{elsewhere.} \end{cases} \tag{E.63}$$

The moment generating function is

$$M(s) = \frac{1}{\Gamma(\alpha)\beta^\alpha} \int_0^\infty t^{\alpha-1} e^{t(s-1/\beta)} dt. \tag{E.64}$$

Substituting

$$z = t\left(\frac{1}{\beta} - s\right) = t\frac{(1-\beta s)}{\beta},$$

so that

$$dz = \frac{1-\beta s}{\beta} dt,$$

we get from (E.64)

$$\begin{aligned} M(s) &= \frac{1}{\Gamma(\alpha)} \cdot \frac{1}{(1-\beta s)^\alpha} \int_0^\infty z^{\alpha-1} e^{-z} dz, \; s < \frac{1}{\beta} \\ &= \frac{1}{(1-\beta s)^\alpha}, \text{ for } s < \frac{1}{\beta}. \end{aligned} \tag{E.65}$$

Hence, expectations are given by

$$E(\xi) = M'(s)\big|_{s=0} = \alpha\beta \tag{E.66}$$

$$E(\xi^2) = M''(s)\big|_{s=0} = \alpha(\alpha+1)\beta^2. \tag{E.67}$$

The variance of ξ is given by

$$\text{var}(\xi) = E(\xi^2) - \{E(\xi)\}^2 = \alpha(\alpha+1)\beta^2 - (\alpha\beta)^2 = \alpha\beta^2. \tag{E.68}$$

Particular cases

(i) For $\alpha = 1$, we get the *exponential distribution* with parameter β. Hence, in this case, the probability density function is

$$f(x) = \begin{cases} e^{-x/\beta} \cdot \beta^{-1}, & \text{for } x > 0 \\ 0, & \text{elsewhere.} \end{cases} \tag{E.69}$$

The distribution function is

$$F(x) = \frac{1}{\beta} \int_0^x e^{-t/\beta} dt, \text{ for } x > 0. \tag{E.70}$$

The expectation is

$$E(\xi) = \beta, \; E(\xi^2) = 2\beta^2. \tag{E.71}$$

The variance is

$$\text{var}(\xi) = \beta^2. \tag{E.72}$$

The moment generating function is

$$M(s) = (1 - \beta s)^{-1}, \text{ for } s < \frac{1}{\beta}. \tag{E.73}$$

(ii) For $\alpha = n/2$, (n is a positive integer), and $\beta = 2$, we get a *chi-square distribution* with n degrees of freedom whose probability density function is defined as

$$f(x) = \begin{cases} \dfrac{1}{\Gamma(n/2)\,(2)^{n/2}} e^{-x/2} x^{(n/2)-1}, & \text{for } 0 < x < \infty \\ 0, & \text{elsewhere.} \end{cases} \tag{E.74}$$

Thus, we have the following values:

$$E(\xi) = n, \; E(\xi^2) = n(n+2), \; \text{var}\,(\xi) = 2n. \tag{E.75}$$

The moment generating function is

$$M(s) = (1 - 2s)^{-n/2}, \text{ for } s < 1/2. \tag{E.76}$$

(3) Beta Distribution Function

A random variable ξ having the probability distribution function defined by

$$f(x) = \begin{cases} \dfrac{x^{\alpha-1}(1-x)^{\beta-1}}{B(\alpha,\beta)}, & \text{for } 0 < x < 1 \\ 0, & \text{elsewhere} \end{cases} \tag{E.77}$$

is said to have beta-type distribution with parameters $\alpha, \beta (\alpha > 0, \beta > 0)$.

It can easily be verified that

$$\int_{-\infty}^{\infty} f(x)\,dx = 1, \tag{E.78}$$

confirming that the function in (E.77) really defines a probability density function.

The distribution function is

$$F(x) = \int_{-\infty}^{x} f(x)\,dx = \begin{cases} 0, & \text{for } x \le 0 \\ \dfrac{1}{B(\alpha,\beta)} \displaystyle\int_0^x x^{\alpha-1}(1-x)^{\beta-1}dx, & \text{for } 0 < x < 1 \\ 1, & \text{for } x = 1. \end{cases} \tag{E.79}$$

For any positive integer n,

$$F(\xi^n) = \frac{1}{B(\alpha,\beta)} \int_0^x x^{n+\alpha-1}(1-x)^{\beta-1}dx = \frac{B(n+\alpha,\beta)}{B(\alpha,\beta)} = \frac{\Gamma(n+\alpha)\Gamma(\alpha+\beta)}{\Gamma(\alpha)\Gamma(n+\alpha+\beta)}. \tag{E.80}$$

Hence,

$$E(\xi) = \frac{B(1+\alpha,\beta)}{B(\alpha,\beta)} = \frac{\Gamma(1+\alpha)\Gamma(\alpha+\beta)}{\Gamma(\alpha)\Gamma(1+\alpha+\beta)}$$

$$= \frac{\alpha\Gamma(\alpha)\Gamma(\alpha+\beta)}{\Gamma(\alpha)\,(\alpha+\beta)\Gamma(\alpha+\beta)} \ [\because \ \Gamma(a+1) = a\Gamma(a)]$$

$$\text{i.e., } E(\xi) = \frac{\alpha}{\alpha+\beta}. \tag{E.81}$$

Similarly,

$$E(\xi^2) = \frac{\alpha(\alpha+1)}{(\alpha+\beta)\,(\alpha+\beta+1)}. \tag{E.82}$$

The variance of ξ is

$$\text{var}\,(\xi) = E(\xi^2) - \{E(\xi)\}^2 = \frac{\alpha(\alpha+1)}{(\alpha+\beta)\,(\alpha+\beta+1)} - \frac{\alpha^2}{(\alpha+\beta)^2}$$

$$= \frac{\alpha\beta}{(\alpha+\beta)^2(\alpha+\beta+1)}. \tag{E.83}$$

The moment generating function of ξ is

$$M(t) = \frac{1}{B(\alpha,\beta)} \cdot \int_0^1 e^{tx} x^{\alpha-1}(1-x)^{\beta-1}dx, \text{ for all } t$$

$$= \frac{1}{B(\alpha,\beta)} \int_0^1 \left(\sum_{k=0}^{\infty} \frac{(tx)^k}{k!}\right) x^{\alpha-1}(1-x)^{\beta-1}dx$$

$$= \frac{1}{B(\alpha,\beta)} \sum_{k=0}^{\infty} \frac{t^k}{k!} \int_0^1 x^{k+\alpha-1}(1-x)^{\beta-1}dx$$

$$= \sum_{k=0}^{\infty} \frac{t^k}{k!} E(\xi^k) = \sum_{k=0}^{\infty} \frac{t^k}{\Gamma(k+1)} \cdot \frac{\Gamma(\alpha+k)\Gamma(\alpha+\beta)}{\Gamma(\alpha+\beta+k)\Gamma(\alpha)}. \tag{E.84}$$

For $\alpha = \beta = 1$, we get rectangular distribution on $(0, 1)$. In that case,

$$E(\xi) = \tfrac{1}{2},\ E(\xi^2) = \tfrac{1}{3},\ \text{var}\,(\xi) = \tfrac{1}{12}.$$

(4) Cauchy Distribution Function

A random variable ξ is said to have a Cauchy distribution with parameters α, β if its probability density function is given by

$$f(x) = \frac{\alpha}{\pi} \cdot \frac{1}{\alpha^2 + (x-\beta)^2},\ \alpha > 0 \tag{E.85}$$

for $-\alpha < x < \infty$.

Now,

$$\int_{-\infty}^{\infty} f(x)\,dx = \frac{\alpha}{\pi} \int_{-\infty}^{\infty} \frac{dx}{\alpha^2 + (x-\beta)^2} = \frac{1}{\pi} \int_{-\infty}^{\infty} \frac{dz}{1+z^2},\ z = (x-\beta)/\alpha$$

$$= \frac{2}{\pi} \tan^{-1} z\Big|_0^{\infty} = 1, \tag{E.86}$$

which shows that the function in (E.85) really defines a probability density function. The distribution function of the random variable ξ is (for $\beta = 0$ and $\alpha = 1$)

$$F(x) = \frac{1}{2} + \frac{1}{\pi}\tan^{-1} x, \ -\infty < x \leq \infty. \tag{E.87}$$

Now,

$$E(\xi^k) = \frac{1}{\pi}\int_{-\infty}^{\infty} x^k \frac{dx}{1+x^2} = \frac{2}{\pi}\int_0^{\infty} \frac{x^k dx}{1+x^2}.$$

We substitute

$$z = \frac{1}{1+x^2}, \text{ i.e., } x^2 = \frac{1}{z} - 1; \ 2x\,dx = -\frac{1}{z^2}dz.$$

Then

$$E(\xi^k) = \frac{2}{\pi}\int_0^1 \left(\frac{1-z}{z}\right)^{(k-1)/2} \cdot \frac{1}{2z^2} z\,dz = \frac{1}{\pi}\int_0^1 z^{(1-k)/2-1}(1-z)^{(k-1)/2}dz, \tag{E.88}$$

which converges for $k < 1$ and diverges for $k \geq 1$. Thus, the moment generating function of the Cauchy random variable does not exist. This distribution is important in the study like classical dispersion theory, configuration of atomic spectra, etc.

(5) Normal Distribution Function

A random variable ξ is said to have a normal distribution with parameter $\sigma > 0$ and μ ($-\infty < \mu < \infty$), if its probability density function is given by

$$f(x) = \frac{1}{\sigma\sqrt{2\pi}} \exp\left\{\frac{-(x-\mu)^2}{2\sigma^2}\right\}. \tag{E.89}$$

Now,

$$\begin{aligned}\int_{-\infty}^{\infty} f(x)\,dx &= \frac{1}{\sigma\sqrt{2\pi}}\int_{-\infty}^{\infty} \exp\left\{-\frac{(x-\mu)^2}{2\sigma^2}\right\} dx \\ &= \frac{1}{\sqrt{2\pi}}\int_{-\infty}^{\infty} e^{-z^2/2}dz, \ \left(z = \frac{x-\mu}{\sigma}\right) = 1,\end{aligned} \tag{E.90}$$

which shows that the function in (E.89) really defines a probability density function. The distribution function is

$$F(x) = \frac{1}{\sigma\sqrt{2\pi}}\int_{-\infty}^{x} \exp\left\{-\frac{(x-\mu)^2}{2\sigma^2}\right\} dx. \tag{E.91}$$

The moment generating function is

$$
\begin{aligned}
M(t) &= \frac{1}{\sigma\sqrt{2\pi}}\int_{-\infty}^{\infty}\exp\left[tx-\frac{(x-\mu)^2}{2\sigma^2}\right]dx\\
&= \frac{1}{\sigma\sqrt{2\pi}}\int_{-\infty}^{\infty}\exp\left[tx-\frac{x^2}{2\sigma^2}+\frac{x\mu}{\sigma^2}-\frac{\mu^2}{2\sigma^2}\right]dx\\
&= \frac{1}{\sigma\sqrt{2\pi}}\int_{-\infty}^{\infty}\exp\left\{\mu t+\frac{\sigma^2t^2}{2}-\frac{(x-\mu-\sigma^2t)^2}{2\sigma^2}\right\}dx\\
&= \exp\left(\mu t+\frac{\sigma^2t^2}{2}\right)\frac{1}{\sigma\sqrt{2\pi}}\int_{-\infty}^{\infty}\exp\left\{-\frac{(x-\mu-\sigma^2t)^2}{2\sigma^2}\right\}dx\\
&= \exp\left(\mu t+\frac{\sigma^2t^2}{2}\right)\frac{1}{\sqrt{2\pi}}\int_{-\infty}^{\infty}e^{-z^2/2}dz,\ z=\frac{x-\mu-\sigma^2t}{\sigma}
\end{aligned}
$$

$$
\text{i.e., } M(t) = \exp\left(\mu t+\frac{\sigma^2t^2}{2}\right). \tag{E.92}
$$

Thus, moments of all orders exist.

Hence,

$$
E(\xi) = M'(t)\big|_{t=0} = (\mu+\sigma^2t)M(t)\big|_{t=0} = \mu \tag{E.93}
$$

$$
E(\xi^2) = M''(t)\big|_{t=0} = M(t)\{\sigma^2+(\mu+\sigma^2t)^2\}_{t=0} = \sigma^2+\mu^2 \tag{E.94}
$$

$$
\text{var}(\xi) = \sigma^2+\mu^2-\mu^2 = \sigma^2. \tag{E.95}
$$

The following problems are left for readers as exercises.

1. Show that, the function

$$
f(x) = \begin{cases} 0, & \text{for } x < a \\ ge^{-g(x-a)}, & \text{for } x \geq a\ (g > 0) \end{cases}
$$

can be the probability density function of a continuous type random variable.

2. A discrete random variable ξ has probability mass function

$$
P\{\xi = x\} = \binom{x+r-1}{x}p^r(1-p)^r,\ x = 0, 1, 2, \ldots,
$$

where r is a fixed positive integer ≥ 1 and $0 < p < 1$. (This is known as *Negative Binomial Distribution*).

Show that $\text{var}(\xi) = \dfrac{rq}{p}, q = 1-p$.

3. A random variable ξ has probability density function defined by

$$
f(x) = \begin{cases} \frac{1}{12}x^2(1-x), & \text{for } 0 < x < 1 \\ 0, & \text{elsewhere.} \end{cases}
$$

Show that $\text{var}(\xi) = 1/25$.

Also determine the expression for the moment generating function for ξ.

Appendix F

Functions of Two or More Variables

F.1 Definition

Let us consider three variables x, y, z. If for each given pair (x, y), we can determine one or more values of z, then the variable z is said to be a *a function of two variables* x and y, and we denote it by

$$z = f(x, y). \tag{F.1}$$

We call z *a dependent variable* and (x, y) the *independent variables.*

The function is called *single-valued,* if only one value of z corresponds to each pair (x, y) for which the function is defined. The function is said to be *multiple-valued,* if there are more than one value of z corresponding to each pair (x, y). A multiple-valued function, in general, can be considered as a collection of single-valued functions and hence, we will confine our discussions to single-valued functions only.

In Cartesian coordinate system, the function defined by equation (F.1) represents a surface. For each pair (x, y) corresponds to a point in the xy-plane and a perpendicular from this point equal to length z determined by the function will define a point in space. The collection of these points will describe a surface in three-dimensional space.

The pair of values for which the function is defined is called the *domain of definition* or simply, *domain* of the function. A rectangular domain may be defined by

$$\left.\begin{array}{l} a \leq x \leq b \\ c \leq y \leq d \end{array}\right\}. \tag{F.2}$$

This is shown in the figure (F.1).

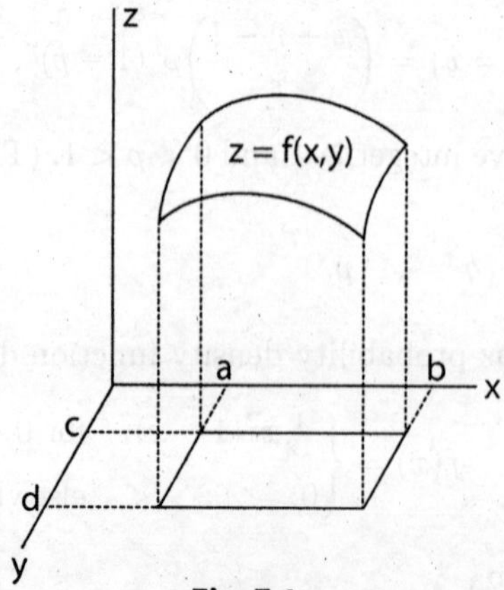

Fig. F.1

Limits

For the function $f(x, y)$, we say that l is the limit as $x \to a$ and $y \to b$, if for any positive number ε we can find some positive number δ [depending on ε and (a, b) in general] such that

$$|f(x, y) - l| < \varepsilon, \tag{F.3}$$

whenever $0 < (x - a) < \delta$ and $0 < (y - b) < \delta$ and we write:

$$\lim_{\substack{x \to a \\ y \to b}} f(x, y) = l. \tag{F.4}$$

In order that this limit exists, it must have the same value regardless of the path of approach of (x, y) to (a, b).

Continuity

The function $f(x, y)$ of two independent variables x and y is said to be *continuous* at a point (a, b), if for any positive number ε, however small, we can find a positive number δ [depending on ε and (a, b) in general] such that

$$|f(x, y) - f(a, b)| < \varepsilon, \tag{F.5}$$

whenever $|x - a| < \delta$ and $|y - b| < \delta$.

Thus, in order that $f(x, y)$ be continuous at (a, b), we must have the conditions: 1. $f(x, y)$ is defined at (a, b) and 2. $\lim_{\substack{x \to a \\ y \to b}} f(x, y) = f(a, b)$.

If a function is not continuous at a point (a, b), it is said to be *discontinuous* at that point which is then called a *point of discontinuity.* The function is said to be continuous in its domain of definition, if it is so at all the points in the domain.

Addition, subtraction and multiplication of continuous functions are continuous. Division of two continuous functions are continuous in the region, where the function in the denominator does not vanish.

A function which is continuous in a closed and bounded region is *uniformly continuous* in the region.

Let us consider the function

$$z = \begin{cases} xy\dfrac{x^2 - y^2}{x^2 + y^2}, & \text{when } x \neq 0, y \neq 0 \\ 0, & \text{when } x = 0, y = 0. \end{cases} \tag{F.6}$$

The function will be continuous at the origin, if

$$|z(h, k) - z(0, 0)| < \varepsilon \tag{F.7}$$

for $|h| < \delta$ and $|k| < \delta$.

Now,

$$|z(h, k) - z(0, 0)| = |hk|\frac{|h^2 - k^2|}{h^2 + k^2} = |hk|.$$

If we accept $\delta^2 = \varepsilon$, then the condition in (F.7) is satisfied. Hence, the function is continuous at $(0, 0)$.

F.2 Partial Derivatives

In a function of more than one variable, the ordinary derivative of the function w.r.t. one independent variable, considering all the remaining independent variables as constants, is called the *partial derivative*.

For the function $f(x, y)$, the partial derivative w.r.t. x is denoted by

$$\frac{\partial f}{\partial x} \text{ or, } f_x \text{ or, } f_x(x, y)$$

and that w.r.t. y by

$$\frac{\partial f}{\partial y} \text{ or, } f_y \text{ or, } f_y(x, y).$$

Analytically,

$$\left.\begin{aligned} \frac{\partial f}{\partial x} &= \lim_{\Delta x \to 0} \frac{f(x+\Delta x, y) - f(x, y)}{\Delta x} \\ \text{and } \frac{\partial f}{\partial y} &= \lim_{\Delta y \to 0} \frac{f(x, y+\Delta y) - f(x, y)}{\Delta y} \end{aligned}\right\}, \tag{F.8}$$

when the limits exist.

As an example, we consider the function

$$f(x, y) = 3x^3 + 2x^2y + 4y^3.$$

Then

$$f_x \equiv \frac{\partial f}{\partial x} = 9x^2 + 4xy$$

$$f_y \equiv \frac{\partial f}{\partial y} = 2x^2 + 12y^2.$$

At the point $x = 2, y = 1$,

$$f_x = 9(2)^2 + 4(2)(1) = 44$$

$$f_y = 2(2)^2 + 12(1)^2 = 20.$$

If a function has partial derivatives, which are continuous in a region, then the function itself must be continuous in that region. However, mere existence of the partial derivatives does not guarantee the continuity of the function. As an example, we consider

$$f(x, y) = \begin{cases} \dfrac{xy}{x^2 + y^2}, & \text{for } (x, y) \neq (0, 0) \\ 0, & \text{otherwise.} \end{cases}$$

Here

$$f_x \equiv \frac{\partial f}{\partial x} = \frac{(x^2 + y^2) \cdot y - xy(2x)}{(x^2 + y^2)^2} = \frac{y^3 - x^2y}{(x^2 + y^2)^2}.$$

Similarly,

$$f_y \equiv \frac{\partial f}{\partial y} = \frac{x^3 - xy^2}{(x^2 - y^2)^2}.$$

We find that the values of the partial derivatives cannot be determined at the origin. Hence, we obtain these values from first principles.

Thus,

$$f_x(0,0) = \lim_{h\to 0} \frac{f(h,0) - f(0,0)}{h} = 0$$

$$f_y(0,0) = \lim_{k\to 0} \frac{f(0,k) - f(0,0)}{k} = 0.$$

Thus, the partial derivatives exist at $(0,0)$.

Let us now determine the continuity of $f(x,y)$ at $(0,0)$. We assume that $x \to 0$ and $y \to 0$ along the line $y = mx$.

Then

$$\lim_{\substack{x\to 0\\ y\to 0}} f(x,y) = \lim \frac{mx^2}{x^2 + m^2x^2} = \frac{m}{1+m^2},$$

which has got different values depending upon m. Hence, the limit does not exist, so that the function is not continuous at $(0,0)$ even though the partial derivatives exist at the point.

Successive Partial Derivatives

Let $f(x,y)$ be a function of two variables having partial derivatives f_x and f_y at each point (x,y) in a region. These will, in general, be functions of (x,y) so that they may also have partial derivatives. Thus, we write:

$$\left.\begin{aligned} \frac{\partial}{\partial x}\left(\frac{\partial f}{\partial x}\right) &= \frac{\partial^2 f}{\partial x^2} = f_{xx} \\ \frac{\partial}{\partial y}\left(\frac{\partial f}{\partial y}\right) &= \frac{\partial^2 f}{\partial y^2} = f_{yy} \\ \frac{\partial}{\partial x}\left(\frac{\partial f}{\partial y}\right) &= \frac{\partial^2 f}{\partial x \partial y} = f_{xy} \\ \frac{\partial}{\partial y}\left(\frac{\partial f}{\partial x}\right) &= \frac{\partial^2 f}{\partial y \partial x} = f_{yx} \end{aligned}\right\}. \tag{F.9}$$

In general, $f_{yx} \neq f_{xy}$. *If, however, the functions* f_{xy} *and* f_{yx} *both exist for a particular set of values of* (x,y) *and are continuous, then* $f_{xy} = f_{yx}$.

Let us consider the function defined by equation (F.6). We find

$$z_x = \frac{\partial z}{\partial x} = y\left[\frac{x^2 - y^2}{x^2 + y^2} + x\frac{(x^2+y^2)\cdot 2x - (x^2 - y^2)\cdot 2x}{(x^2+y^2)^2}\right]$$

$$= y\left[\frac{x^2 - y^2}{x^2+y^2} + \frac{4x^2y^2}{(x^2+y^2)^2}\right].$$

Similarly,

$$z_y = \frac{\partial z}{\partial y} = x\left[\frac{x^2 - y^2}{x^2+y^2} - \frac{4x^2y^2}{(x^2+y^2)^2}\right].$$

Hence,

$$z_{xy}(0,0) = \lim_{h\to 0} \frac{z_y(h,0) - z_y(0,0)}{h} = \lim_{h\to 0} \frac{h}{h} = 1$$

$$\text{and } z_{yx}(0,0) = \lim_{k\to 0} \frac{z_x(0,k) - z_x(0,0)}{k} = \lim_{k\to 0} -\frac{k}{k} = -1.$$

Thus, we find that for the function considered above $z_{xy}(0,0) \neq z_{yx}(0,0)$.

F.3 Total Differential

We consider a function, $z = f(x,y)$, which has continued first order partial derivatives in a region.

Let Δx and Δy be increments given to x and y respectively. Then the increment in the function $z = f(x,y)$ is

$$\begin{aligned}\Delta z &= f(x+\Delta x, y+\Delta y) - f(x,y)\\ &= \{f(x+\Delta x, y+\Delta y) - f(x, y+\Delta y)\} + \{f(x, y+\Delta y) - f(x,y)\}.\end{aligned}$$

By mean value theorem,

$$\Delta z = \Delta x f_x(x+\theta_1\Delta x, y+\Delta y) + \Delta y f_y(x, y+\theta_2\Delta y),$$

where $0 < \theta_1 < 1$ and $0 < \theta_2 < 1$.

We assume that the partial derivatives f_x and f_y are continuous at (x,y). Then

$$f_x(x+\theta_1\Delta x, y+\Delta y) = f_x(x,y) + \varepsilon_1$$

$$f_y(x, y+\theta_2\Delta y) = f_y(x,y) + \varepsilon_2,$$

where $\varepsilon_1, \varepsilon_2$ approach to zero along with Δx and Δy.

Hence,

$$\Delta z = \frac{\partial f}{\partial x}\Delta x + \frac{\partial f}{\partial y}\Delta y + \varepsilon_1\Delta x + \varepsilon_2\Delta y. \tag{F.10}$$

When $\Delta x \to 0$ and $\Delta y \to 0$, we write, $\Delta z = dz$. Taking $\Delta x = dx$ and $\Delta y = dy$,

$$dz = \frac{\partial f}{\partial x}dx + \frac{\partial f}{\partial y}dy. \tag{F.11}$$

The expression in equation (F.11) is called the *total differential* or simply, *differential* of z. This can be generalised for a function of any number of variables.

F.4 Composite Functions

We consider the function, $z = f(x,y)$, where the independent variables x and y are functions of other two variables r and s, so that $x = g(r,s)$, $y = h(r,s)$.

Then z will be a function of r and s.

Such functions are called *composite functions.*

Assuming the differentiability of the functions f, g and h, we write by equation (F.10):

$$\frac{\partial z}{\partial r} = \lim_{\Delta r \to 0} \frac{\Delta z}{\Delta r} = \lim_{\Delta r \to 0} \left\{ \frac{\partial z}{\partial x} \cdot \frac{\Delta x}{\Delta r} + \frac{\partial z}{\partial y} \cdot \frac{\Delta y}{\Delta r} + \varepsilon_1 \frac{\Delta x}{\Delta r} + \varepsilon_2 \frac{\Delta y}{\Delta r} \right\}.$$

This yields

$$\frac{\partial z}{\partial r} = \frac{\partial z}{\partial x} \cdot \frac{\partial x}{\partial r} + \frac{\partial z}{\partial y} \cdot \frac{\partial y}{\partial r}. \tag{F.12}$$

Similarly, we can get

$$\frac{\partial z}{\partial s} = \frac{\partial z}{\partial x} \cdot \frac{\partial x}{\partial s} + \frac{\partial z}{\partial y} \cdot \frac{\partial y}{\partial s}. \tag{F.13}$$

These results can be generalised in the case of functions of many variables and are often called *chain rules.*

If x and y depend only on one variable r, so that z is a function of r only, then by equation (F.10),

$$\frac{dz}{dr} = \lim_{\Delta r \to 0} \frac{\Delta z}{\Delta r} = \lim_{\Delta r \to 0} \left\{ \frac{\partial z}{\partial x} \cdot \frac{\Delta x}{\Delta r} + \frac{\partial z}{\partial y} \cdot \frac{\Delta y}{\Delta r} + \varepsilon_1 \frac{\Delta x}{\Delta r} + \varepsilon_2 \frac{\Delta y}{\Delta r} \right\}$$

which yields

$$\frac{dz}{dr} = \frac{\partial z}{\partial x} \cdot \frac{dx}{dr} + \frac{\partial z}{\partial y} \cdot \frac{dy}{dr}. \tag{F.14}$$

F.5 Homogeneous Functions

A function $F(x_1, x_2, \ldots, x_n)$ of n variables is called a *homogenous function of degree p* if, for all values of the parameters λ and some constant p, we have the identity

$$F(\lambda x_1, \lambda x_2, \ldots, \lambda x_n) = \lambda^p F(x_1, x_2, \ldots, x_n). \tag{F.15}$$

For example, we consider the function

$$F(x, y) = ax^2 + 2hxy + by^2.$$

Then $F(\lambda x, \lambda y) = a(\lambda x)^2 + 2h(\lambda x)(\lambda y) + b(\lambda y)^2 = \lambda^2(ax^2 + 2hxy + by^2)$
$= \lambda^2 F(x, y)$.

Thus, the function is homogeneous of degree 2. There is an important theorem, known as *Euler's theorem,* on homogeneous functions. The theorem states:

If $F(x_1, x_2, \ldots, x_n)$ is a homogeneous function of degree p, then

$$x_1 \frac{\partial F}{\partial x_1} + x_2 \frac{\partial F}{\partial x_2} + \cdots + x_n \frac{\partial F}{\partial x_n} = pF. \tag{F.16}$$

In order to prove this, let us consider a homogeneous function $F(x, y)$ of two variables and of degree p. Then by definition, we have

$$F(\lambda x, \lambda y) = \lambda^p F(x, y) \tag{F.17}$$

for all values of the parameters λ and for some constant p.

Denoting $u = \lambda x$ and $v = \lambda y$, we write equation (F.17) as

$$F(u, v) = \lambda^p F(x, y). \tag{F.18}$$

Differentiating the left-hand side of equation w.r.t. to λ, we have

$$\frac{\partial F}{\partial \lambda} = \frac{\partial F}{\partial u} \cdot \frac{\partial u}{\partial \lambda} + \frac{\partial F}{\partial v} \cdot \frac{\partial v}{\partial \lambda} = \frac{\partial F}{\partial u} x + \frac{\partial F}{\partial v} y. \tag{F.19}$$

The derivative of the right-hand side of equation (F.18) is

$$\frac{\partial F}{\partial \lambda} = p\lambda^{p-1} F(x, y). \tag{F.20}$$

From equations (F.19) and (F.20), we get

$$x\frac{\partial F}{\partial u} + y\frac{\partial F}{\partial v} = p\lambda^{p-1} F.$$

Taking $\lambda = 1$, so that $u = x$ and $v = y$, we get

$$x\frac{\partial F}{\partial x} + y\frac{\partial F}{\partial y} = pF. \tag{F.21}$$

This proves the theorem.

F.6 Implicit Functions

An equation of the type

$$f(x, y) = 0 \tag{F.22}$$

defines, in general, one variable say y as a function of other variable x. Then y is called an *implicit function* of x.

Let f_x, f_y be continuous and y be a differentiable function of x. Then for small increments Δx and Δy of x and y, we have

$$f(x + \Delta x, y + \Delta y) - f(x, y) = 0$$

$$\text{or, } \{f(x + \Delta x, y + \Delta y) - f(x, y + \Delta y)\} + \{f(x, y + \Delta y) - f(x, y)\} = 0.$$

By the mean value theorem, this becomes

$$\Delta x f_x(x + \theta_1 \Delta x, y + \Delta y) + \Delta y f_y(x, y + \theta_2 \Delta y) = 0, \tag{F.23}$$

where $0 < \theta_1 < 1$ and $0 < \theta_2 < 1$.

Since f_x and f_y are continuous, we can write equation (F.23) in the form (after dividing by Δx)

$$(f_x + \varepsilon_1) + \frac{\Delta y}{\Delta x}(f_y + \varepsilon_2) = 0. \tag{F.24}$$

The function y being differentiable, $\Delta y \to 0$, when $\Delta x \to 0$. Thus, proceeding to these limits in equation (F.24), we get

$$f_x + \frac{dy}{dx} f_y = 0,$$

which yields

$$\frac{dy}{dx} = \frac{f_x}{f_y} \tag{F.25}$$

provided $f_y \neq 0$ at the point of consideration.

EXAMPLE F.1. *Investigate the continuity of the function*

$$f(x,y) = \begin{cases} \dfrac{x^2 - y^2}{x^2 + y^2}, & (x,y) \neq (0,0) \\ 0, & (x,y) = (0,0) \end{cases}$$

at the point $(0,0)$.

Solution: Assuming that $x \to 0$ and $y \to 0$ along the line $y = mx$, we have

$$\lim_{\substack{x\to 0\\ y\to 0}} f(x,y) = \lim_{\substack{x\to 0\\ y\to 0}} \frac{x^2 - m^2x^2}{x^2 + m^2x^2} = \frac{1-m^2}{1+m^2}.$$

Thus, the limit depends on the value of m, i.e., on the path of approach towards $(0,0)$. Hence, the function is not continuous at $(0,0)$.

Alternatively, since

$$\lim_{x\to 0}\left(\lim_{y\to 0}\frac{x^2-y^2}{x^2+y^2}\right) = \lim_{x\to 0}\left(\frac{x^2}{x^2}\right) = 1$$

$$\text{and } \lim_{y\to 0}\left(\lim_{x\to 0}\frac{x^2-y^2}{x^2+y^2}\right) = \lim_{y\to 0}\left(-\frac{y^2}{y^2}\right) = -1,$$

and not equal, the limit of the given function as $x \to 0$, $y \to 0$ does not exist. Hence, the function cannot be continuous at $(0,0)$.

EXAMPLE F.2. *If* $V = x^2 + y^2 + z^2$, *show that* $xV_x + yV_y + zV_z = 2V$.

Solution: We are given $V = x^2 + y^2 + z^2$.

$$\left.\begin{aligned} \therefore\ V_x &= \frac{\partial V}{\partial x} = 2x \\ V_y &= \frac{\partial V}{\partial y} = 2y \\ V_z &= \frac{\partial V}{\partial z} = 2z \end{aligned}\right\} \qquad (1)$$

Hence, $xV_x + yV_y + zV_z = (2x)x + (2y)y + (2z)z$
$= 2(x^2 + y^2 + z^2) = 2V.$

Alternatively, we find that for any constant λ,

$$\begin{aligned} V(\lambda x, \lambda y, \lambda z) &= (\lambda x)^2 + (\lambda y)^2 + (\lambda z)^2 = \lambda^2(x^2 + y^2 + z^2) \\ &= \lambda^2 V(x,y,z). \end{aligned}$$

Hence, V is a homogeneous function of order 2.

Then by Euler's theorem we directly get

$$x\frac{\partial V}{\partial x} + y\frac{\partial V}{\partial y} + z\frac{\partial V}{\partial z} = 2V.$$

Hence, the result.

EXAMPLE F.3. *If $u = f(xyz)$, show that* $x\dfrac{\partial u}{\partial x} = y\dfrac{\partial u}{\partial y} = z\dfrac{\partial u}{\partial z}$.

Solution: Let $xyz = t$. Then $u = f(t)$.

Hence

$$\left.\begin{aligned} \frac{\partial u}{\partial x} &= \frac{\partial u}{\partial t}\frac{\partial t}{\partial x} = f'(t)\cdot yz \\ \frac{\partial u}{\partial y} &= \frac{\partial u}{\partial t}\frac{\partial t}{\partial y} = f'(t)\cdot xz \\ \frac{\partial u}{\partial z} &= \frac{\partial u}{\partial t}\frac{\partial t}{\partial z} = f'(t)\cdot xy \end{aligned}\right\}. \tag{1}$$

Then

$$\left.\begin{aligned} x\frac{\partial u}{\partial x} &= f'(t)\cdot xyz \\ y\frac{\partial u}{\partial y} &= f'(t)\cdot xyz \\ z\frac{\partial u}{\partial z} &= f'(t)\cdot xyz \end{aligned}\right\}. \tag{2}$$

From equations (2), we find

$$x\frac{\partial u}{\partial x} = y\frac{\partial u}{\partial y} = z\frac{\partial u}{\partial z}.$$

EXAMPLE F.4. *If $u = \sin^{-1}\dfrac{x}{y} + \tan^{-1}\dfrac{y}{x}$, show that* $x\dfrac{\partial u}{\partial x} + y\dfrac{\partial u}{\partial y} = 0$.

Solution: We have $u = \sin^{-1}\dfrac{x}{y} + \tan^{-1}\dfrac{y}{x}$.

$$\therefore\ \frac{\partial u}{\partial x} = \frac{1}{\sqrt{1-\dfrac{x^2}{y^2}}}\cdot\frac{1}{y} + \frac{1}{\sqrt{1+\dfrac{y^2}{x^2}}}\cdot\left(-\frac{y}{x^2}\right) = \frac{1}{\sqrt{y^2-x^2}} - \frac{y}{x^2+y^2} \tag{1}$$

and

$$\frac{\partial u}{\partial y} = \frac{1}{\sqrt{1-\dfrac{x^2}{y^2}}}\cdot\left(\frac{x}{y^2}\right) + \frac{1}{\sqrt{1+\dfrac{y^2}{x^2}}}\cdot\frac{1}{x},$$

i.e.,

$$\frac{\partial u}{\partial y} = \frac{-x}{y\sqrt{y^2-x^2}} + \frac{x}{x^2+y^2}. \tag{2}$$

Using equations (1) and (2), we find

$$x\frac{\partial u}{\partial x} + y\frac{\partial u}{\partial y} = \left(\frac{x}{\sqrt{y^2-x^2}} - \frac{xy}{x^2+y^2}\right) + \left(\frac{-x}{\sqrt{y^2-x^2}} + \frac{xy}{x^2+y^2}\right) = 0.$$

EXAMPLE F.5. *If $V = z\tan^{-1}\dfrac{y}{x}$, show that*

$$\frac{\partial^2 V}{\partial x^2} + \frac{\partial^2 V}{\partial y^2} + \frac{\partial^2 V}{\partial z^2} = 0.$$

Solution: We have $V = z\tan^{-1}\dfrac{y}{x}$.

$$\therefore\ \frac{\partial V}{\partial x} = z\cdot\frac{1}{1+\dfrac{y^2}{x^2}}\cdot\left(-\frac{y}{x^2}\right) = \frac{-yz}{x^2+y^2}$$

$$\frac{\partial^2 V}{\partial x^2} = -yz\left[-\frac{1}{(x^2+y^2)^2}\right]\cdot 2x = \frac{2xyz}{(x^2+y^2)^2} \tag{1}$$

$$\frac{\partial V}{\partial y} = z\cdot\frac{1}{1+\dfrac{y^2}{x^2}}\cdot\frac{1}{x} = \frac{zx}{x^2+y^2}$$

$$\frac{\partial^2 V}{\partial y^2} = zx\cdot\left[-\frac{1}{(x^2+y^2)^2}\right]\cdot 2y = \frac{-2xyz}{(x^2+y^2)^2} \tag{2}$$

$$\frac{\partial V}{\partial z} = \tan^{-1}\frac{y}{x},\ \therefore\ \frac{\partial^2 V}{\partial z^2} = 0. \tag{3}$$

Using equations (1), (2) and (3), we get

$$\frac{\partial^2 V}{\partial x^2} + \frac{\partial^2 V}{\partial y^2} + \frac{\partial^2 V}{\partial z^2} = \frac{2xyz}{(x^2+y^2)^2} - \frac{2xyz}{(x^2+y^2)} + 0 = 0.$$

EXAMPLE F.6. *If $x = \rho\cos\theta$ and $y = \rho\sin\theta$, show that*

$$\left(\frac{\partial V}{\partial x}\right)^2 + \left(\frac{\partial V}{\partial y}\right)^2 = \left(\frac{\partial V}{\partial \rho}\right)^2 + \frac{1}{\rho^2}\left(\frac{\partial V}{\partial \theta}\right)^2.$$

Solution: Since $x = \rho\cos\theta$, hence,

$$\frac{\partial x}{\partial \rho} = \cos\theta,\ \frac{\partial x}{\partial \theta} = -\rho\sin\theta. \tag{1}$$

Similarly, $y = \rho\sin\theta$.

$$\therefore\ \frac{\partial y}{\partial \rho} = \sin\theta,\ \frac{\partial y}{\partial \theta} = \rho\cos\theta. \tag{2}$$

Now,

$$\frac{\partial V}{\partial \rho} = \frac{\partial V}{\partial x}\cdot\frac{\partial x}{\partial \rho} + \frac{\partial V}{\partial y}\cdot\frac{\partial y}{\partial \rho} = \frac{\partial V}{\partial x}\cos\theta + \frac{\partial V}{\partial y}\sin\theta \tag{3}$$

and

$$\frac{\partial V}{\partial \theta} = \frac{\partial V}{\partial x}\cdot\frac{\partial x}{\partial \theta} + \frac{\partial V}{\partial y}\cdot\frac{\partial y}{\partial \theta} = \frac{\partial V}{\partial x}(-\rho\sin\theta) + \frac{\partial V}{\partial y}\cos\theta.$$

$$\therefore\ \frac{1}{\rho}\frac{\partial V}{\partial \theta} = -\frac{\partial V}{\partial x}\sin\theta + \frac{\partial V}{\partial y}\cos\theta. \tag{4}$$

From equations (3) and (4), we have

$$\left(\frac{\partial V}{\partial \rho}\right)^2 + \frac{1}{\rho^2}\left(\frac{\partial V}{\partial \theta}\right)^2 = \left(\frac{\partial V}{\partial x}\cos\theta + \frac{\partial V}{\partial y}\sin\theta\right)^2 + \left(-\frac{\partial V}{\partial x}\sin\theta + \frac{\partial V}{\partial y}\cos\theta\right)^2$$

$$= \left(\frac{\partial V}{\partial x}\right)^2 + \left(\frac{\partial V}{\partial y}\right)^2.$$

EXAMPLE F.7. *If $V = f(u)$, u being a homogeneous function of degree n in x and y, show that*

$$x\frac{\partial V}{\partial x} + y\frac{\partial V}{\partial y} = nu\frac{dV}{du}.$$

Solution: Since $V = f(u)$, u being a function of x and y, we have

$$\frac{\partial V}{\partial x} = \frac{dV}{du}\cdot\frac{\partial u}{\partial x} \tag{1}$$

and

$$\frac{\partial V}{\partial y} = \frac{dV}{du}\cdot\frac{\partial u}{\partial y}. \tag{2}$$

From equations (1) and (2), we get

$$x\frac{\partial V}{\partial x} + y\frac{\partial V}{\partial y} = \frac{dV}{du}\left(x\frac{\partial u}{\partial x} + y\frac{\partial u}{\partial y}\right). \tag{3}$$

Since u is a homogeneous function of degree n, we have

$$x\frac{\partial u}{\partial x} + y\frac{\partial u}{\partial y} = nu. \tag{4}$$

From equations (3) and (4), we get

$$x\frac{\partial V}{\partial x} + y\frac{\partial V}{\partial y} = nu\frac{dV}{du}.$$

EXAMPLE F.8. *If $u = \sin^{-1}\left(\frac{x^2 + y^2}{x + y}\right)$, show that*

$$x\frac{\partial u}{\partial x} + y\frac{\partial u}{\partial y} = \tan u.$$

Solution: Let

$$V = \sin u = \frac{x^2 + y^2}{x + y}. \tag{1}$$

Obviously, V is a homogeneous function of degree one in x and y. Hence,

$$x\frac{\partial V}{\partial x} + y\frac{\partial V}{\partial y} = \sin u. \tag{2}$$

Now,

$$\frac{\partial V}{\partial x} = \frac{dV}{du}\cdot\frac{\partial u}{\partial x} = \cos u\frac{\partial u}{\partial x}$$
$$\frac{\partial V}{\partial y} = \frac{dV}{du}\cdot\frac{\partial u}{\partial y} = \cos u\frac{\partial u}{\partial y}.$$

Hence,

$$x\frac{\partial V}{\partial x} + y\frac{\partial V}{\partial y} = \cos u\left(x\frac{\partial u}{\partial x} + y\frac{\partial u}{\partial y}\right). \tag{3}$$

From equations (2) and (3), we have

$$\cos u\left(x\frac{\partial u}{\partial x}+y\frac{\partial u}{\partial y}\right)=\sin u.$$

$$\therefore\ x\frac{\partial u}{\partial x}+y\frac{\partial u}{\partial y}=\frac{\sin u}{\cos u}=\tan u.$$

EXAMPLE F.9. *If H be a homogeneous function in x, y, z of degree n and if*

$$u=(x^2+y^2+z^2)^{-\frac{1}{2}(n+1)},$$

show that

$$\frac{\partial}{\partial x}\left(H\frac{\partial u}{\partial x}\right)+\frac{\partial}{\partial y}\left(H\frac{\partial u}{\partial y}\right)+\frac{\partial}{\partial z}\left(H\frac{\partial u}{\partial z}\right)=0.$$

Solution: Since H is a homogeneous function of degree n in x, y, z, we have

$$x\frac{\partial H}{\partial x}+y\frac{\partial H}{\partial y}+z\frac{\partial H}{\partial z}=nH. \tag{1}$$

Again, we have

$$\frac{\partial}{\partial x}\left(H\frac{\partial u}{\partial x}\right)=H\frac{\partial^2 u}{\partial x^2}+\frac{\partial H}{\partial x}\cdot\frac{\partial u}{\partial x} \tag{2}$$

$$\frac{\partial}{\partial y}\left(H\frac{\partial u}{\partial y}\right)=H\frac{\partial^2 u}{\partial y^2}+\frac{\partial H}{\partial y}\cdot\frac{\partial u}{\partial y} \tag{3}$$

$$\frac{\partial}{\partial z}\left(H\frac{\partial u}{\partial z}\right)=H\frac{\partial^2 u}{\partial z^2}+\frac{\partial H}{\partial z}\cdot\frac{\partial u}{\partial z}. \tag{4}$$

Adding these, we get

$$\frac{\partial}{\partial x}\left(H\frac{\partial u}{\partial x}\right)+\frac{\partial}{\partial y}\left(H\frac{\partial u}{\partial y}\right)+\frac{\partial}{\partial z}\left(H\frac{\partial u}{\partial z}\right)$$
$$=H\left(\frac{\partial^2 u}{\partial x^2}+\frac{\partial^2 u}{\partial y^2}+\frac{\partial^2 u}{\partial z^2}\right)+\left(\frac{\partial H}{\partial x}\cdot\frac{\partial u}{\partial x}+\frac{\partial H}{\partial y}\cdot\frac{\partial u}{\partial y}+\frac{\partial H}{\partial z}\cdot\frac{\partial u}{\partial z}\right). \tag{5}$$

Now,

$$u=(x^2+y^2+z^2)^{-\frac{1}{2}(n+1)}.$$

Hence,

$$\frac{\partial u}{\partial x}=-\tfrac{1}{2}(n+1)\cdot(x^2+y^2+z^2)^{-\frac{1}{2}(n+1)-1}\cdot 2x$$
$$=-(n+1)r^{-(n+3)}\cdot x, \tag{6}$$

where

$$r^2=x^2+y^2+z^2. \tag{7}$$

Similarly,

$$\frac{\partial u}{\partial y}=-(n+1)r^{-(n+3)}\cdot y \tag{8}$$

$$\frac{\partial u}{\partial z}=-(n+1)r^{-(n+3)}\cdot z. \tag{9}$$

Putting the values from equations (6), (8), (9), we get

$$\frac{\partial H}{\partial x}\cdot\frac{\partial u}{\partial x}+\frac{\partial H}{\partial y}\cdot\frac{\partial u}{\partial y}+\frac{\partial H}{\partial z}\cdot\frac{\partial u}{\partial z}=-(n+1)r^{-(n+3)}\left(x\frac{\partial H}{\partial x}+y\frac{\partial H}{\partial y}+z\frac{\partial H}{\partial z}\right)$$
$$=-n(n+1)r^{-(n+3)}H \text{ [using (1)].} \tag{10}$$

Again,

$$\frac{\partial^2 u}{\partial x^2}=-(n+1)r^{-(n+3)}-x(n+1)\{-(n+3)\}\cdot r^{-(n+4)}\frac{x}{r}$$
$$=-(n+1)r^{-(n+3)}+(n+1)(n+3)x^2r^{-(n+5)}. \tag{11}$$

Similarly,

$$\frac{\partial^2 u}{\partial y^2}=-(n+1)r^{-(n+3)}+(n+1)(n+3)y^2r^{-(n+5)} \tag{12}$$
$$\frac{\partial^2 u}{\partial z^2}=-(n+1)r^{-(n+3)}+(n+1)(n+3)z^2r^{-(n+5)}. \tag{13}$$

Adding equations (11), (12) and (13), we have

$$\frac{\partial^2 u}{\partial x^2}+\frac{\partial^2 u}{\partial y^2}+\frac{\partial^2 u}{\partial z^2}=-3(n+1)r^{-(n+3)}+(n+1)(n+3)r^{-(n+5)}(x^2+y^2+z^2)$$
$$=-3(n+1)r^{-(n+3)}+(n+1)(n+3)r^{-(n+3)}$$
$$=n(n+1)r^{-(n+3)}. \tag{14}$$

Hence,

$$H\left(\frac{\partial^2 u}{\partial x^2}+\frac{\partial^2 u}{\partial y^2}+\frac{\partial^2 u}{\partial z^2}\right)=n(n+1)r^{-(n+3)}H. \tag{15}$$

Putting the values from equations (10) and (15) in the right side of equation (5), we get the result.

EXAMPLE F.10. *If $u=f(x^2+2yz, y^2+2zx)$, prove that*

$$(y^2-zx)\frac{\partial u}{\partial x}+(x^2-yz)\frac{\partial u}{\partial y}+(z^2-xy)\frac{\partial u}{\partial z}=0.$$

Solution: Let

$$\left.\begin{aligned}\phi&=x^2+2yz\\ \psi&=y^2+2zx\end{aligned}\right\}. \tag{1}$$

Then

$$\frac{\partial u}{\partial x}=\frac{\partial u}{\partial \phi}\cdot\frac{\partial \phi}{\partial x}+\frac{\partial u}{\partial \psi}\cdot\frac{\partial \psi}{\partial x}=\frac{\partial u}{\partial \phi}\cdot 2x+\frac{\partial u}{\partial \psi}\cdot 2z \tag{2}$$
$$\frac{\partial u}{\partial y}=\frac{\partial u}{\partial \phi}\cdot\frac{\partial \phi}{\partial y}+\frac{\partial u}{\partial \psi}\cdot\frac{\partial \psi}{\partial y}=\frac{\partial u}{\partial \phi}\cdot 2z+\frac{\partial u}{\partial \psi}\cdot 2y \tag{3}$$
$$\text{and }\frac{\partial u}{\partial z}=\frac{\partial u}{\partial \phi}\cdot\frac{\partial \phi}{\partial z}+\frac{\partial u}{\partial \psi}\cdot\frac{\partial \psi}{\partial z}=\frac{\partial u}{\partial \phi}\cdot 2y+\frac{\partial u}{\partial \psi}\cdot 2x. \tag{4}$$

Hence,

$$(y^2 - zx)\frac{\partial u}{\partial x} = 2(y^2 - zx)\left(x\frac{\partial u}{\partial \phi} + z\frac{\partial u}{\partial \psi}\right) \quad (5)$$

$$(x^2 - yz)\frac{\partial u}{\partial y} = 2(x^2 - yz)\left(z\frac{\partial u}{\partial \phi} + y\frac{\partial u}{\partial \psi}\right) \quad (6)$$

$$(z^2 - xy)\frac{\partial u}{\partial z} = 2(z^2 - xy)\left(y\frac{\partial u}{\partial \phi} + x\frac{\partial u}{\partial \psi}\right). \quad (7)$$

Adding both sides of equations (5), (6) and (7), we get the result.

EXAMPLE F.11. *Prove that $y = f(x + at) + g(x - at)$ satisfies the equation*

$$\frac{\partial^2 y}{\partial t^2} = a^2\frac{\partial^2 y}{\partial x^2},$$

where f and g are at least twice differentiable and a is a constant.

Solution: Let

$$u = x + at, v = x - at. \quad (1)$$

Then

$$y = f(u) + g(v). \quad (2)$$

Hence,

$$\frac{\partial y}{\partial t} = \frac{\partial y}{\partial u}\cdot\frac{\partial u}{\partial t} + \frac{\partial y}{\partial v}\cdot\frac{\partial v}{\partial t} = a\frac{\partial y}{\partial u} - a\frac{\partial y}{\partial v} = af'(u) - ag'(v). \quad (3)$$

Differentiating equation (3), we have

$$\frac{\partial^2 y}{\partial t^2} = \frac{\partial}{\partial t}\left(\frac{\partial y}{\partial t}\right) = \frac{\partial}{\partial t}\{af'(u) - ag'(v)\} = af''(u)\frac{\partial u}{\partial t} - ag''(v)\frac{\partial v}{\partial t}$$
$$= a^2 f''(u) + a^2 g''(v). \quad (4)$$

Again,

$$\frac{\partial y}{\partial x} = \frac{\partial y}{\partial u}\cdot\frac{\partial u}{\partial x} + \frac{\partial y}{\partial v}\cdot\frac{\partial v}{\partial x} = f'(u) + g'(v). \quad (5)$$

Hence,

$$\frac{\partial^2 y}{\partial x^2} = \frac{\partial}{\partial x}\{f'(u) + g'(v)\},$$

i.e.,

$$\frac{\partial^2 y}{\partial x^2} = f''(u)\frac{\partial u}{\partial x} + g''(v)\frac{\partial v}{\partial x} = f''(u) + g''(v). \quad (6)$$

From equations (4) and (6), we get

$$\frac{\partial^2 y}{\partial t^2} = a^2\frac{\partial^2 y}{\partial x^2}.$$

EXAMPLE F.12. *If u be a homogeneous function of x and y of degree n, prove that*

$$x^2\frac{\partial^2 u}{\partial x^2} + 2xy\frac{\partial^2 u}{\partial x \partial y} + y^2\frac{\partial^2 u}{\partial y^2} = n(n-1)u.$$

Solution: Since $u(x, y)$ is a homogeneous function of degree n, we have, for all values of t,

$$u(tx, ty) = t^n u(x, y). \tag{1}$$

Let $tx = \xi$ and $ty = \eta$. Differentiating both sides of equation (1) w.r.t. t, we have

$$\frac{\partial u}{\partial \xi} \cdot \frac{\partial \xi}{\partial t} + \frac{\partial u}{\partial \eta} \cdot \frac{\partial \eta}{\partial t} = nt^{n-1}u(x, y) \text{ or, } x\frac{\partial u}{\partial \xi} + y\frac{\partial u}{\partial \eta} = nt^{n-1}u(x, y). \tag{2}$$

Let

$$\phi = x\frac{\partial u}{\partial \xi},\ \psi = y\frac{\partial u}{\partial \eta}. \tag{3}$$

Then

$$\frac{\partial \phi}{\partial t} = \frac{\partial \phi}{\partial \xi} \cdot \frac{\partial \xi}{\partial t} + \frac{\partial \phi}{\partial \eta} \cdot \frac{\partial \eta}{\partial t} = x\frac{\partial}{\partial \xi}\left(x\frac{\partial u}{\partial \xi}\right) + y\frac{\partial}{\partial \eta}\left(x\frac{\partial u}{\partial \xi}\right) = x^2\frac{\partial^2 u}{\partial \xi^2} + xy\frac{\partial^2 u}{\partial \xi \partial \eta}. \tag{4}$$

Similarly,

$$\frac{\partial \psi}{\partial t} = \frac{\partial \psi}{\partial \xi} \cdot \frac{\partial \xi}{\partial t} + \frac{\partial \psi}{\partial \eta} \cdot \frac{\partial \eta}{\partial t} = x\frac{\partial}{\partial \xi}\left(y\frac{\partial u}{\partial \eta}\right) + y\frac{\partial}{\partial \eta}\left(y\frac{\partial u}{\partial \eta}\right) = xy\frac{\partial^2 u}{\partial \xi \partial \eta} + y^2\frac{\partial^2 u}{\partial \eta^2}. \tag{5}$$

Differentiating both sides of equation (2) w.r.t. t, we get

$$\frac{\partial}{\partial t}\left(x\frac{\partial t}{\partial \xi}\right) + \frac{\partial}{\partial t}\left(y\frac{\partial u}{\partial \eta}\right) = n(n-1)t^{n-2}u(x, y).$$

Using relations in (4) and (5), we have

$$\left(x^2\frac{\partial^2 u}{\partial \xi^2}\right) + xy\frac{\partial^2 u}{\partial \xi \partial \eta} + \left(xy\frac{\partial^2 u}{\partial \xi \partial \eta} + y^2\frac{\partial^2 u}{\partial \eta^2}\right) = n(n-1)t^{n-2}u. \tag{6}$$

Putting $t = 1$ so that $\xi = x$ and $\eta = y$ in equation (6), we get

$$x^2\frac{\partial^2 u}{\partial x^2} + 2xy\frac{\partial^2 u}{\partial x \partial y} + y^2\frac{\partial^2 u}{\partial y^2} = n(n-1)u.$$

EXAMPLE F.13. *If V be a function of r alone, where $r^2 = x^2 + y^2 + z^2$, show that*

$$\frac{\partial^2 V}{\partial x^2} + \frac{\partial^2 V}{\partial y^2} + \frac{d^2 V}{\partial z^2} = \frac{d^2 V}{dr^2} + \frac{2}{r} \cdot \frac{dV}{dr}.$$

Solution: Since $r^2 = x^2 + y^2 + z^2$, we have

$$\frac{\partial r}{\partial x} = \frac{x}{r},\ \frac{\partial r}{\partial y} = \frac{y}{r},\ \frac{\partial r}{\partial z} = \frac{z}{r}. \tag{1}$$

Now

$$\left.\begin{aligned} \frac{\partial V}{\partial x} &= \frac{\partial V}{\partial r} \cdot \frac{\partial r}{\partial x} = \frac{x}{r} \cdot \frac{dV}{dr} \\ \frac{\partial V}{\partial y} &= \frac{\partial V}{\partial r} \cdot \frac{\partial r}{\partial y} = \frac{y}{r} \cdot \frac{dV}{dr} \\ \frac{\partial V}{\partial z} &= \frac{\partial V}{\partial r} \cdot \frac{\partial r}{\partial z} = \frac{z}{r} \cdot \frac{dV}{dr} \end{aligned}\right\}. \tag{2}$$

We have used ordinary derivative of V w.r.t. r, since V is a function of r alone.

Hence,

$$\frac{\partial}{\partial x}\left(\frac{\partial V}{\partial x}\right) = \frac{\partial}{\partial r}\left(\frac{x}{r}\cdot\frac{dV}{dr}\right)\frac{\partial r}{\partial x} = \frac{x}{r}\cdot\frac{\partial}{\partial r}\left(\frac{x}{r}\cdot\frac{dV}{dr}\right)$$

$$\text{i.e., } \frac{\partial^2 V}{\partial x^2} = \frac{x}{r}\left[\frac{x}{r}\cdot\frac{d^2V}{dr^2} + \frac{dV}{dr}\left(\frac{1}{r}\cdot\frac{\partial x}{\partial r} - \frac{x}{r^2}\right)\right]$$

$$= \frac{x^2}{r^2}\cdot\frac{d^2V}{dr^2} + \frac{1}{r}\cdot\frac{dV}{dr} - \frac{x^2}{r^3}\cdot\frac{dV}{dr}. \quad (3)$$

Similarly,

$$\frac{\partial^2 V}{\partial y^2} = \frac{y^2}{r^2}\cdot\frac{d^2V}{dr^2} + \frac{1}{r}\cdot\frac{dV}{dr} - \frac{y^2}{r^3}\cdot\frac{dV}{dr} \quad (4)$$

$$\frac{\partial^2 V}{\partial z^2} = \frac{z^2}{r^2}\cdot\frac{d^2V}{dr^2} + \frac{1}{r}\cdot\frac{dV}{dr} - \frac{z^2}{r^3}\cdot\frac{dV}{dr}. \quad (5)$$

Adding equations (3), (4) and (5), we get

$$\frac{\partial^2 V}{\partial x^2} + \frac{\partial^2 V}{\partial y^2} + \frac{\partial^2 V}{\partial z^2} = \frac{1}{r^2}\cdot\frac{d^2V}{dr^2}(x^2+y^2+z^2) + \frac{3}{r}\cdot\frac{dV}{dr} - \frac{1}{r^3}(x^2+y^2+z^2)\frac{dV}{dr}$$

$$= \frac{d^2V}{dr^2} + \frac{2}{r}\cdot\frac{dV}{dr} \; [\text{since } r^2 = x^2+y^2+z^2].$$

EXAMPLE F.14. *If $f(x,y,z)=0$ defines z as an implicit function of x and y in a region of the xy-plane, prove that*

$$\frac{\partial z}{\partial x} = -\frac{f_x}{f_z} \; \textit{and} \; \frac{\partial z}{\partial y} = -\frac{f_y}{f_z}, \; f_z \neq 0.$$

Solution: Since z is a function of x and y, we have

$$dz = \frac{\partial z}{\partial x}dx + \frac{\partial z}{\partial y}dy. \quad (1)$$

Again, we have

$$f(x,y,z) = 0. \quad (2)$$

Hence,

$$df \equiv \frac{\partial f}{\partial x}dx + \frac{\partial f}{\partial y}dy + \frac{\partial f}{\partial z}dz = 0: \quad (3)$$

Using equation (1), we get

$$\frac{\partial f}{\partial x}dx + \frac{\partial f}{\partial y}dy + \frac{\partial f}{\partial z}\left(\frac{\partial z}{\partial x}dx + \frac{\partial z}{\partial y}dy\right) = 0$$

$$\text{or, } \left(\frac{\partial f}{\partial x} + \frac{\partial f}{\partial z}\cdot\frac{\partial z}{\partial x}\right)dx + \left(\frac{\partial f}{\partial y} + \frac{\partial f}{\partial z}\cdot\frac{\partial z}{\partial y}\right)dy = 0. \quad (4)$$

Since x and y are independent, coefficients of dx and dy should vanish. Hence,

$$\left.\begin{aligned}\frac{\partial f}{\partial x}+\frac{\partial f}{\partial z}\cdot\frac{\partial z}{\partial x}=0\\ \frac{\partial f}{\partial y}+\frac{\partial f}{\partial z}\cdot\frac{\partial z}{\partial y}=0\end{aligned}\right\}. \tag{5}$$

From equations (5), we find

$$\frac{\partial z}{\partial x}=-\frac{f_x}{f_z},\ \frac{\partial z}{\partial y}=-\frac{f_y}{f_z},\ \text{where } f_z\neq 0.$$

EXAMPLE F.15. *If* $z^3-xz-y=0$, *find* $\dfrac{\partial^2 z}{\partial x\partial y}$.

Solution: We have

$$z^3-zx-y=0. \tag{1}$$

Here, z is the dependent variable which depends on independent variables x and y. Differentiating equation (1) w.r.t. x, keeping y constant, we have

$$3z^2\frac{\partial z}{\partial x}-x\frac{\partial z}{\partial x}-z=0 \text{ or, } \frac{\partial z}{\partial x}=\frac{z}{3z^2-x}. \tag{2}$$

Again, differentiating equation (1) w.r.t. y, keeping x constant, we have

$$3z^2\frac{\partial z}{\partial y}-x\frac{\partial z}{\partial y}-1=0 \text{ or, } \frac{\partial z}{\partial y}=\frac{1}{3z^2-x}. \tag{3}$$

Differentiating this w.r.t. x, we get

$$\frac{\partial^2 z}{\partial x\partial y}=\frac{-1}{(3z^2-x)^2}\left(6z\frac{\partial z}{\partial x}-1\right). \tag{4}$$

Putting the value of $\dfrac{\partial z}{\partial x}$ from equation (2), we find

$$\frac{\partial^2 z}{\partial x\partial y}=\frac{1-6z\{z/(3z^2-x)\}}{(3z^2-x)^2}=-\frac{x+3z^2}{(3z^2-x)^3}.$$

F.7 Perfect Differential

We consider three functions $P(x,y,z)$, $Q(x,y,z)$ and $R(x,y,z)$ having continuous partial derivatives. The expression

$$\mu(P\,dx+Q\,dy+R\,dz) \tag{F.26}$$

is called an *exact or perfect differential,* if there exists a function $u(x,y,z)$, such that

$$\mu(P\,dx+Q\,dy+R\,dz)=du \tag{F.27}$$

for all values of dx, dy and dz. Here, $\mu(x,y,z)$ is a factor.

Since u is a function of (x, y, z), we have

$$du = \frac{\partial u}{\partial x}dx + \frac{\partial u}{\partial y}dy + \frac{\partial u}{\partial z}dz. \tag{F.28}$$

Comparing equations (F.27) and (F.28), we have

$$\frac{\partial u}{\partial x} = \mu P(x, y, z) \tag{F.29}$$

$$\frac{\partial u}{\partial y} = \mu Q(x, y, z) \tag{F.30}$$

$$\frac{\partial u}{\partial z} = \mu R(x, y, z). \tag{F.31}$$

Differentiating equation (F.29) w.r.t. y and (F.30) w.r.t. x and assuming that

$$\frac{\partial^2 u}{\partial x \partial y} = \frac{\partial^2 u}{\partial y \partial x},$$

we have

$$\mu\frac{\partial P}{\partial y} + P\frac{\partial \mu}{\partial y} = \mu\frac{\partial Q}{\partial x} + Q\frac{\partial \mu}{\partial x} \text{ or, } \mu\left(\frac{\partial P}{\partial y} - \frac{\partial Q}{\partial x}\right) = Q\frac{\partial \mu}{\partial x} - P\frac{\partial \mu}{\partial y}. \tag{F.32}$$

Similarly, we have

$$\mu\left(\frac{\partial Q}{\partial z} - \frac{\partial R}{\partial y}\right) = R\frac{\partial \mu}{\partial y} - Q\frac{\partial \mu}{\partial z} \tag{F.33}$$

$$\mu\left(\frac{\partial R}{\partial x} - \frac{\partial P}{\partial z}\right) = P\frac{\partial \mu}{\partial z} - R\frac{\partial \mu}{\partial x}. \tag{F.34}$$

Multiplying equation (F.32) by R, (F.33) by P, (F.34) by Q and adding, we get

$$P\left(\frac{\partial Q}{\partial z} - \frac{\partial R}{\partial y}\right) + Q\left(\frac{\partial R}{\partial x} - \frac{\partial P}{\partial z}\right) + R\left(\frac{\partial P}{\partial y} - \frac{\partial Q}{\partial x}\right) = 0. \tag{F.35}$$

If the factor function $\mu =$ constant, then

$$\frac{\partial \mu}{\partial x} = \frac{\partial \mu}{\partial y} = \frac{\partial \mu}{\partial z} = 0. \tag{F.36}$$

In this case, we have, from equations (F.32)-(F.34),

$$\left.\begin{aligned} \frac{\partial P}{\partial y} - \frac{\partial Q}{\partial x} &= 0 \\ \frac{\partial Q}{\partial z} - \frac{\partial R}{\partial y} &= 0 \\ \frac{\partial R}{\partial x} - \frac{\partial P}{\partial z} &= 0 \end{aligned}\right\}. \tag{F.37}$$

EXAMPLE F.16. *Determine whether $(2xy^2 + 3y\cos 3x)\,dx + (2x^2y + \sin 3x)\,dy$ is an exact differential of a function and if so, find the function.*

Solution: The given expression is

$$P\,dx + Q\,dy, \tag{1}$$

where

$$P = 2xy^2 + 3y\cos 3x \tag{2}$$

$$\text{and } Q = 2x^2y + \sin 3x. \tag{3}$$

Hence,

$$\frac{\partial P}{\partial y} = 4xy + 3\cos 3x \tag{4}$$

$$\text{and } \frac{\partial Q}{\partial x} = 4xy + 3\cos 3x. \tag{5}$$

From equations (4) and (5), we find that

$$\frac{\partial P}{\partial y} = \frac{\partial Q}{\partial x}. \tag{6}$$

Hence, the given expression is an exact differential.
Let

$$P\,dx + Q\,dy = du = \frac{\partial u}{\partial x}dx + \frac{\partial u}{\partial y}dy.$$

Then

$$\frac{\partial u}{\partial x} = P = 2xy^2 + 3y\cos 3x \tag{7}$$

$$\text{and } \frac{\partial u}{\partial y} = Q = 2x^2y + \sin 3x. \tag{8}$$

Integrating equation (7) w.r.t. x and (8) w.r.t. y, we have

$$u = x^2y^2 + y\sin 3x + f(y) + c_1 \tag{9}$$

$$u = x^2y^2 + y\sin 3x + g(x) + c_2. \tag{10}$$

Comparing equations (9) and (10), we find that

$$f(y) = 0,\ g(x) = 0. \tag{11}$$

Hence,

$$u(x, y) = x^2y^2 + y\sin 3x + c.$$

EXAMPLE F.17. *If* $u = f(y - z, z - x, x - y)$, *prove that*

$$\frac{\partial u}{\partial x} + \frac{\partial u}{\partial y} + \frac{\partial u}{\partial z} = 0.$$

Solution: Let

$$x_1 = y - z,\ x_2 = z - x,\ x_3 = x - y. \tag{1}$$

Then

$$u = f(x_1, x_2, x_3). \tag{2}$$

Hence,

$$\frac{\partial u}{\partial x} = \frac{\partial f}{\partial x_1}\cdot\frac{\partial x_1}{\partial x} + \frac{\partial f}{\partial x_2}\cdot\frac{\partial x_2}{\partial x} + \frac{\partial f}{\partial x_3}\cdot\frac{\partial x_3}{\partial x} = \frac{\partial f}{\partial x_1}\cdot 0 + \frac{\partial f}{\partial x_2}\cdot(-1) + \frac{\partial f}{\partial x_3}\cdot 1$$

$$= -\frac{\partial f}{\partial x_2} + \frac{\partial f}{\partial x_3}. \tag{3}$$

Similarly,

$$\frac{\partial u}{\partial y} = \frac{\partial f}{\partial x_1}\cdot\frac{\partial x_1}{\partial y} + \frac{\partial f}{\partial x_2}\cdot\frac{\partial x_2}{\partial y} + \frac{\partial f}{\partial x_3}\cdot\frac{\partial x_3}{\partial y} = \frac{\partial f}{\partial x_1} - \frac{\partial f}{\partial x_3} \tag{4}$$

and $$\frac{\partial u}{\partial z} = \frac{\partial f}{\partial x_1}\cdot\frac{\partial x_1}{\partial z} + \frac{\partial f}{\partial x_2}\cdot\frac{\partial x_2}{\partial z} + \frac{\partial f}{\partial x_3}\cdot\frac{\partial x_3}{\partial z} = -\frac{\partial f}{\partial x_1} + \frac{\partial f}{\partial x_2}. \tag{5}$$

Adding equations (3), (4) and (5), we get

$$\frac{\partial u}{\partial x} + \frac{\partial u}{\partial y} + \frac{\partial u}{\partial z} = \left(-\frac{\partial f}{\partial x_2} + \frac{\partial f}{\partial x_3}\right) + \left(\frac{\partial f}{\partial x_1} - \frac{\partial f}{\partial x_3}\right) + \left(-\frac{\partial f}{\partial x_1} + \frac{\partial f}{\partial x_2}\right) = 0.$$

EXAMPLE F.18. *If $u = f(ax^2 + 2hxy + by^2)$ and $v = \phi(ax^2 + 2hxy + by^2)$, show that*

$$\frac{\partial}{\partial y}\left(u\frac{\partial v}{\partial x}\right) = \frac{\partial}{\partial x}\left(u\frac{\partial v}{\partial y}\right).$$

Solution: Let

$$z = ax^2 + 2hxy + by^2. \tag{1}$$

Then

$$u = f(z),\ v = \phi(z). \tag{2}$$

Hence,

$$\frac{\partial v}{\partial x} = \frac{\partial v}{\partial z}\cdot\frac{\partial z}{\partial x} = \phi'(z)\,(2ax + 2hy) \tag{3}$$

$$\frac{\partial v}{\partial y} = \frac{\partial v}{\partial z}\cdot\frac{\partial z}{\partial y} = \phi'(z)\,(2hx + 2by). \tag{4}$$

Then

$$\frac{\partial}{\partial y}\left(u\frac{\partial v}{\partial x}\right) = \frac{\partial}{\partial y}[f(z)\phi'(z)\,(2ax + 2hy)] = \frac{\partial}{\partial y}[\psi(z)\,(2ax + 2hy)],\ \psi(z) = f(z)\phi'(z)$$

$$= \psi(z)2h + (2ax + 2hy)\psi'(z)\frac{\partial z}{\partial y} = \psi(z)2h + (2ax + 2hy)\,(2hx + 2by)\psi'(z). \tag{5}$$

Similarly,

$$\frac{\partial}{\partial x}\left(u\frac{\partial v}{\partial y}\right) = \frac{\partial}{\partial x}[f(z)\phi'(z)\,(2hx + 2by)] = \frac{\partial}{\partial x}[\psi(z)\,(2hx + 2by)]$$

$$= \psi(z)2h + (2hx + 2by)\psi'(z)\frac{\partial z}{\partial x} = \psi(z)2h + (2hx + 2by)\,(2ax + 2hy)\psi'(z). \tag{6}$$

From equations (5) and (6), we find that

$$\frac{\partial}{\partial y}\left(u\frac{\partial v}{\partial x}\right) = \frac{\partial}{\partial x}\left(u\frac{\partial v}{\partial y}\right).$$

EXAMPLE F.19. *If* $u = x\phi(x+y) + y\psi(x+y)$, *show that*

$$\frac{\partial^2 u}{\partial x^2} - 2\frac{\partial^2 u}{\partial x \partial y} + \frac{\partial^2 u}{\partial y^2} = 0.$$

Solution: Let $z = x + y$; so that

$$\frac{\partial z}{\partial x} = \frac{\partial z}{\partial y} = 1. \tag{1}$$

Then

$$u = x\phi(z) + y\psi(z). \tag{2}$$

Hence,

$$\begin{aligned}\frac{\partial u}{\partial x} &= x\phi'(z)\frac{\partial z}{\partial x} + \phi(z) + y\psi'(z)\frac{\partial z}{\partial x} \\ &= x\phi'(z) + y\psi'(z) + \phi(z).\end{aligned} \tag{3}$$

$$\begin{aligned}\frac{\partial^2 u}{\partial x^2} &= \frac{\partial}{\partial x}\{x\phi'(z) + y\psi'(z) + \phi(z)\} \\ &= x\phi''(z) + \phi'(z) + y\psi''(z) + \phi'(z) = x\phi'' + y\psi'' + 2\phi'.\end{aligned} \tag{4}$$

Similarly,

$$\frac{\partial u}{\partial y} = x\phi' + y\psi' + \psi \tag{5}$$

$$\frac{\partial^2 u}{\partial y^2} = x\phi'' + y\psi'' + 2\psi' \tag{6}$$

$$\frac{\partial^2 u}{\partial x \partial y} = \frac{\partial}{\partial x}(x\phi' + y\psi' + \psi) = x\phi'' + \phi' + y\psi'' + \psi'. \tag{7}$$

From equations (4), (6) and (7), we get

$$\begin{aligned}\frac{\partial^2 u}{\partial x^2} - 2\frac{\partial^2 u}{\partial x \partial y} + \frac{\partial^2 u}{\partial y^2} &= (x\phi'' + y\psi'' + 2\phi') - 2(x\phi'' + y\psi'' + \phi' + \psi') + (x\phi'' + y\psi'' + 2\psi') \\ &= 0.\end{aligned}$$

EXAMPLE F.20. *If* $f(u^2 - x^2, u^2 - y^2, u^2 - z^2) = 0$, *where* u *is a function of* x, y, z, *show that*

$$\frac{1}{x}\frac{\partial u}{\partial x} + \frac{1}{y}\frac{\partial u}{\partial y} + \frac{1}{z}\frac{\partial u}{\partial z} = \frac{1}{u}.$$

Solution: Since u is a function of x, y, z, we have

$$du = \frac{\partial u}{\partial x}dx + \frac{\partial u}{\partial y}dy + \frac{\partial u}{\partial z}dz. \tag{1}$$

Let

$$p = u^2 - x^2,\ q = u^2 - y^2,\ r = u^2 - z^2. \tag{2}$$

Then, we are given

$$f(p, q, r) = 0.. \tag{3}$$

Hence,

$$df \equiv \frac{\partial f}{\partial p}dp + \frac{\partial f}{\partial q}dq + \frac{\partial f}{\partial r}dr = 0$$

$$\text{or, } \frac{\partial f}{\partial p}(2udu - 2xdx) + \frac{\partial f}{\partial q}(2udu - 2ydy) + \frac{\partial f}{\partial r}(2udu - 2zdz) = 0$$

$$\text{or, } udu\left(\frac{\partial f}{\partial p} + \frac{\partial f}{\partial q} + \frac{\partial f}{\partial r}\right) - \left(x\frac{\partial f}{\partial p} + y\frac{\partial f}{\partial q} + z\frac{\partial f}{\partial r}\right) dx = 0. \tag{4}$$

Substituting the value of du from equation (1) into equation (4) and arranging the terms, we get

$$\begin{aligned} &\left\{u\frac{\partial u}{\partial x}\left(\frac{\partial f}{\partial p} + \frac{\partial f}{\partial q} + \frac{\partial f}{\partial r}\right) - x\frac{\partial f}{\partial p}\right\} dx \\ &+ \left\{u\frac{\partial u}{\partial y}\left(\frac{\partial f}{\partial p} + \frac{\partial f}{\partial q} + \frac{\partial f}{\partial r}\right) - y\frac{\partial f}{\partial q}\right\} dy \\ &+ \left\{u\frac{\partial u}{\partial z}\left(\frac{\partial f}{\partial p} + \frac{\partial f}{\partial q} + \frac{\partial f}{\partial r}\right) - z\frac{\partial f}{\partial r}\right\} dz = 0. \end{aligned} \tag{5}$$

Since x, y, z are independent, coefficients of dx, dy and dz in equation (5) should vanish. Hence,

$$\left.\begin{aligned} \frac{u}{x}\cdot\frac{\partial u}{\partial x}\left(\frac{\partial f}{\partial p} + \frac{\partial f}{\partial q} + \frac{\partial f}{\partial r}\right) = \frac{\partial f}{\partial p} \\ \frac{u}{y}\cdot\frac{\partial u}{\partial y}\left(\frac{\partial f}{\partial p} + \frac{\partial f}{\partial q} + \frac{\partial f}{\partial r}\right) = \frac{\partial f}{\partial q} \\ \frac{u}{z}\cdot\frac{\partial u}{\partial z}\left(\frac{\partial f}{\partial p} + \frac{\partial f}{\partial q} + \frac{\partial f}{\partial r}\right) = \frac{\partial f}{\partial r} \end{aligned}\right\}. \tag{6}$$

Adding these equations, we have

$$\left(\frac{u}{x}\cdot\frac{\partial u}{\partial x} + \frac{u}{y}\cdot\frac{\partial u}{\partial y} + \frac{u}{z}\cdot\frac{\partial u}{\partial z}\right)\left(\frac{\partial f}{\partial p} + \frac{\partial f}{\partial q} + \frac{\partial f}{\partial r}\right) = \left(\frac{\partial f}{\partial p} + \frac{\partial f}{\partial q} + \frac{\partial f}{\partial r}\right)$$

$$\text{or, } \frac{u}{x}\cdot\frac{\partial u}{\partial x} + \frac{u}{y}\cdot\frac{\partial u}{\partial y} + \frac{u}{z}\cdot\frac{\partial u}{\partial z} = 1.$$

$$\therefore \frac{1}{x}\cdot\frac{\partial u}{\partial x} + \frac{1}{y}\cdot\frac{\partial u}{\partial y} + \frac{1}{z}\cdot\frac{\partial u}{\partial z} = \frac{1}{u}.$$

F.8 Taylor's Theorem for the Functions of Two Variables

We consider a function $f(x, y)$ of two variables x and y. The theorem states:

If $f(x, y)$ possesses continuous partial derivatives up to nth order in a neighbourhood of a point (a, b), then for any point $(a + h, b + k)$,

$$f(a + h, b + k) = f(a, b) + Lf(a, b) + \frac{1}{2!}L^2 f(a, b) + \cdots + \frac{1}{n!}L^n f(a, b) + R_n, \tag{F.38}$$

where $L(\dots)$ is the differential operator

$$L(\dots) = h\frac{\partial \dots}{\partial x} + k\frac{\partial \dots}{\partial y} \tag{F.39}$$

and R_n is the remainder after n terms given by

$$R_n = \frac{1}{(n+1)!}\left(h\frac{\partial}{\partial x} + k\frac{\partial}{\partial y}\right)^{n+1} f(a+\theta h, b+\theta k) \tag{F.40}$$

with $0 < \theta < 1$.

If $R_n \to 0$ as $n \to \infty$ for all (x, y), equation (F.38) then represents an infinite series expansion of the function $f(x, y)$ in powers of $(x-a)$ and $(y-b)$ convergent in this region. This series is called a *Taylor's series* in two variables.

EXAMPLE F.21. *Expand the function* $f(x, y) = x^2y + 3y - 2$ *in powers of* $(x-1)$ *and* $(y+2)$.

Solution: Here $a = 1, b = -2, h = (x-1), k = (y+2)$ and $f(x, y) = x^2y + 3y - 2, f(1, -2) = -10$.

Then $f_x = 2xy, \quad f_x(1, -2) = -4$
$f_y = x^2 + 3, \quad f_y(1, -2) = 4$
$f_{xx} = 2y, \quad f_{xx}(1, -2) = -4$
$f_{xy} = 2x, \quad f_{xy}(1, -2) = 2$
$f_{yy} = 0, \quad f_{yy}(1, -2) = 0$
....

By Taylor's theorem, we have

$$\begin{aligned} f(x, y) &= f(1, -2) + [hf_x(1, -2) + kf_y(1, -2)] + \frac{1}{2!}[hf_x(1, -2) + kf_y(1, -2)]^2 + \ldots \\ &= -10 - 4(x-1) + 4(y+2) \\ &\quad + \frac{1}{2!}[(x-1)^2 f_{xx}(1, -2) + 2(x-1)(y+2) f_{xy}(1, -2) + (y+2)^2 f_{yy}] + \ldots \\ &= -10 - 4(x-1) + 4(y+2) - 2(x-1)^2 + 2(x-1)(y+2) + \ldots \end{aligned}$$

up to second degree terms.

Note.

$$\left(h\frac{\partial}{\partial x} + k\frac{\partial}{\partial y}\right)^2 = \left(h^2\frac{\partial^2}{\partial x^2} + 2hk\frac{\partial^2}{\partial x \partial y} + k^2\frac{\partial^2}{\partial y^2}\right)$$

$$\left(h\frac{\partial}{\partial x} + k\frac{\partial}{\partial y}\right)^3 = \left(h^3\frac{\partial^3}{\partial x^3} + 3h^2k\frac{\partial^3}{\partial x^2 \partial y} + 3hk^2\frac{\partial^3}{\partial x}\partial y^2 + k^3\frac{\partial^3}{\partial y^3}\right)$$

and so on.

If we differentiate the function $f(x, y)$ further, we have $f_{xxx} = 0$, $f_{yyy} = 0$, $f_{xyy} = 0$ and $f_{xxy} = 2$, i.e., $f_{xxy}(1, -2) = 2$.

Then the additional term in the expansion will be

$$\frac{1}{3!}[3h^2kf_{xxy}(1, -2)] = \frac{1}{3 \cdot 2}[3(x-1)^2(y+2) \cdot 2] = 3(x-1)^2(y+2).$$

The higher order derivatives being zero, the expansion is in a finite series.

EXAMPLE F.22. *Expand the function* $f(x, y) = e^x \log(1+y)$ *in powers of* x *and* y *up to third degree.*

Solution: We have $a = b = 0, h = x, k = y$.

$$f = e^x \log(1+y) \qquad \therefore f(0,0) = 0$$

$$f_x = e^x \log(1+y) \qquad \therefore f_x(0,0) = 0$$

$$f_{xx} = e^x \log(1+y) \qquad \therefore f_{xx}(0,0) = 0$$

$$f_{xy} = \frac{e^x}{1+y} \qquad \therefore f_{xy}(0,0) = 1.0$$

$$f_{xxy} = \frac{e^x}{1+y} \qquad \therefore f_{xxy}(0,0) = 1.0$$

$$f_{xxx} = e^x \log(1+y) \qquad \therefore f_{xxx}(0,0) = 0$$

$$f_y = \frac{e^x}{1+y} \qquad \therefore f_y(0,0) = 1.0$$

$$f_{yy} = -\frac{e^x}{(1+y)^2} \qquad \therefore f_{yy}(0,0) = -1.0$$

$$f_{yyy} = \frac{2e^x}{(1+y)^3} \qquad \therefore f_{yyy}(0,0) = 2.0$$

$$f_{xyy} = -\frac{e^x}{(1+y)^2} \qquad \therefore f_{xyy}(0,0) = -1.0.$$

By Taylor's theorem,

$$\begin{aligned} f(x,y) = f(0,0) &+ [x f_x(0,0) + y f_y(0,0)] + [x^2 f_{xx}(0,0) + 2xy f_{xy}(0,0) + y^2 f_{yy}(0,0)] \\ &+ [x^3 f_{xxx}(0,0) + 3x^2 y f_{xxy}(0,0) + 3xy^2 f_{xyy}(0,0) + y^3 f_{yyy}(0,0)]. \end{aligned}$$

Putting here the values from above, we get

$$f(x,y) = y + 2xy - y^2 + 3x^2 y - 3xy^2 + 2y^3.$$

F.9 Maxima and Minima of Functions of Two Variables

A function $f(x,y)$ is said to have a relative maxima (or minima) at a point (a,b) if the value of $f(a,b)$ is greater (or less) than any other value of $f(x,y)$ in the neighbourhood of the point (a,b). Thus, analytically we say that:

A function $f(x,y)$ is said to have a relative maximum (or minimum) value at a point (a,b) of the domain of $f(x,y)$ according as

$$\left.\begin{aligned} f(a+h,b+k) - f(a,b) \le 0 \\ \text{and } f(a+h,b+k) - f(a,b) \ge 0 \end{aligned}\right\} \tag{F.41}$$

for all h, k such that $0 < |h| < \delta, 0 < |k| < \delta$, where δ is a sufficiently small positive number.

A necessary condition that the function $f(x,y)$ has a relative maximum (or minimum) at a point (a,b) is that

$$f_x(a,b) = 0, \ f_y(a,b) = 0. \tag{F.42}$$

The point (a, b) satisfying equation (F.42) is called a *critical point*.
We define

$$\Delta = \{f_{xx} \cdot f_{yy} - (f_{xy})^2\}. \tag{F.43}$$

The point (a, b) in the domain of $f(x, y)$ is a relative maximum point if

$$\Delta(a, b) > 0 \text{ and } f_{xx}(a, b) < 0 \text{ (or, } f_{yy} < 0). \tag{F.44}$$

Similarly, the point (a, b) is a relative minimum point if

$$\Delta(a, b) > 0 \text{ and } f_{xx}(a, b) > 0 \text{ (or, } f_{yy} < 0). \tag{F.45}$$

If

$$\Delta(a, b) < 0, \tag{F.46}$$

the point (a, b) is neither a relative maximum point nor a relative minimum point for the function $f(x, y)$. Such points are called *Saddle points*.

If $\Delta(a, b) = 0$, no information is obtained and further analysis is necessary.

EXAMPLE F.23. *Investigate the maximum and minimum of the function* $f(x, y) = 3x^3 + 4x^2y - 3xy^2 - 4y$ *at the point* $(0, 0)$.

Solution: We have $f = 3x^3 + 4x^2y - 3xy^2 - 4y$.

$$\begin{aligned} f_x &= 9x^2 + 8xy - 3y^2, & f_x(0,0) &= 0 \\ f_y &= 4x^2 - 6xy - 4, & f_y(0,0) &= -4 \\ f_{xx} &= 18x + 8y, & f_{xx}(0,0) &= 0 \\ f_{xy} &= 8x - 6y, & f_{xy}(0,0) &= 0 \\ f_{yy} &= -6x, & f_{yy}(0,0) &= 0. \end{aligned}$$

Thus, we find that $f_x(0, 0) = 0$, but $f_y(0, 0) = -4 \neq 0$.
The necessary conditions given in equation (F.42) are not satisfied. Moreover,

$$\begin{aligned} \Delta(0, 0) &= f_{xx} \cdot f_{yy} - (f_{xy})^2 \Big|_{(0,0)} \\ &= 0. \end{aligned}$$

Hence, the given function has neither a maximum nor a minimum at the point $(0, 0)$.

EXAMPLE F.24. *Find the relative maxima and minima of the function*

$$f(x, y) = x^3 + y^3 - 3x - 12y + 20.$$

Solution: We have $f(x, y) = x^3 + y^3 - 3x - 12y + 20$.

$$\therefore\ f_x = 3x^2 - 3,\ f_y = 3y^2 - 12 \tag{1}$$

$$f_{xx} = 6x,\ f_{xy} = 0,\ f_{yy} = 6y,\ \Delta = 36xy. \tag{2}$$

The condition $f_x = 0$ gives

$$3x^2 - 3 = 0, \text{ i.e., } x = \pm 1 \tag{3}$$

and $f_y = 0$ gives

$$3y^2 - 12 = 0, \text{ i.e., } y = \pm 2. \tag{4}$$

Thus, the critical points are $A(1,2), B(-1,2), C(1,-2), D(-1,-2)$.
At point A, $\Delta = 36 \cdot 1 \cdot 2 = 72 > 0$ and f_{xx} or $f_{yy} > 0$.
Hence the point is a relative minimum point.
At point B, $\Delta = -72 < 0$ and hence, the point is a saddle point.
At point C, $\Delta = -72 < 0$ and hence, the point is a saddle point.
At point D, $\Delta = 72 > 0$ and $f_{xx} < 0, f_{yy} < 0$. Hence, the point is a relative maximum point.

F.10 Lagrange's Multiplier

Sometimes we need to determine the relative extremum of a function

$$u = f(x, y, z) \tag{F.47}$$

subject to the condition that x, y, z satisfy the equation

$$\phi(x, y, z) = 0. \tag{F.48}$$

This is known as *constraint equation.*

In this case, the *method of Lagrange's undetermined multiplier* is used.

Now, for the function $f(x, y, z)$ to have extremum points, we must have the necessary conditions

$$f_x = 0, f_y = 0, f_z = 0. \tag{F.49}$$

Hence,

$$du \equiv f_x dx + f_y dy + f_z dz = 0. \tag{F.50}$$

Differentiating the constraint equation (F.48), we have

$$\phi_x dx + \phi_y dy + \phi_z dz = 0. \tag{F.51}$$

Multiplying equation (F.51) by λ and adding the result to the equation (F.50), we get

$$(f_x + \lambda\phi_x)\, dx + (f_y + \lambda\phi_y)\, dy + (f_z + \lambda\phi_z)\, dz = 0. \tag{F.52}$$

The variables x, y, z being independent, we have

$$\left.\begin{aligned} f_x + \lambda\phi_x &= 0 \\ f_y + \lambda\phi_y &= 0 \\ f_z + \lambda\phi_z &= 0 \end{aligned}\right\}. \tag{F.53}$$

These equations along with the constraint equation (F.48) determine λ and the coordinates x, y, z of the extremum points. The factor λ is called *Lagrange's multiplier.*

From equations (F.53), we find that

$$\lambda = -\frac{f_x}{\phi_x} = -\frac{f_y}{\phi_y},$$

which yields

$$f_x\phi_y - f_y\phi_x = 0. \tag{F.54}$$

This is the necessary condition that the function $f(x, y, z)$ will have an extremum subject to the condition that $\phi(x, y, z) = 0$. The above condition holds if

$$f_z \neq 0 \text{ and } \phi_z \neq 0. \tag{F.55}$$

The method of Lagrange's multiplier can be generalised when there are more than one constraint function.

EXAMPLE F.25. *Find the maximum value of* $f(x, y, z) = x^2y^2z^2$ *subject to the condition* $x^2 + y^2 + z^2 = a^2$.

Solution: We consider the function

$$F(x, y, z) = x^2y^2z^2 + \lambda(x^2 + y^2 + z^2 - a^2). \tag{1}$$

For maximum of this function, we have

$$\left.\begin{aligned} F_x &\equiv 2xy^2z^2 + 2\lambda x = 0 \\ F_y &\equiv 2yz^2x^2 + 2\lambda y = 0 \\ F_z &\equiv 2zx^2y^2 + 2\lambda z = 0 \end{aligned}\right\}. \tag{2}$$

Since x, y, z are lying on a sphere of radius a with centre at origin, $x \neq 0, y \neq 0$ and $z \neq 0$. Hence, equations (2) yield

$$-\lambda = y^2z^2 = z^2x^2 = x^2y^2, \text{ i.e., } x^2 = y^2 = z^2. \tag{3}$$

By the constraint condition, we have

$$x^2 + y^2 + z^2 = a^2 \text{ or, } 3x^2 = a^2, \text{ i.e., } x = \pm\frac{a}{\sqrt{3}}.$$

Hence,

$$x = y = z = \pm\frac{a}{\sqrt{3}} \tag{4}$$

determines the point of maximum value of $f(x, y, z)$. Thus, the maximum value is

$$f_{\max} = \left(\pm\frac{a}{\sqrt{3}}\right)^2 \left(\pm\frac{a}{\sqrt{3}}\right)^2 \left(\pm\frac{a}{\sqrt{3}}\right)^2 = \frac{a^6}{27}.$$

EXAMPLE F.26. *Find the minimum value of* $f(x, y) = x^2 + y^2$ *subject to the constraint* $x^2 + 8xy + 7y^2 = 225$.

Solution: We are given

$$f(x, y) \equiv x^2 + y^2 \tag{1}$$

$$\phi(x, y) \equiv x^2 + 8xy + 7y^2 - 225 = 0. \tag{2}$$

Introducing a Lagrange's multiplier λ, we consider a function

$$F(x, y) = f(x, y) + \lambda\phi(x, y). \tag{3}$$

The conditions for minimum of $F(x, y)$ give

$$\left.\begin{aligned} f_x + \lambda\phi_x &\equiv 2x + \lambda(2x + 8y) = 0 \\ f_y + \lambda\phi_y &\equiv 2y + \lambda(8x + 14y) = 0 \end{aligned}\right\}. \tag{4}$$

These equation can be written as

$$\left.\begin{aligned} (\lambda + 1)x + 4\lambda y &= 0 \\ 4\lambda x + (1 + 7\lambda)y &= 0 \end{aligned}\right\}. \tag{5}$$

Since $x \neq 0, y \neq 0$, we have

$$\begin{vmatrix} \lambda+1 & 4\lambda \\ 4\lambda & 1+7\lambda \end{vmatrix} = 0 \text{ or, } (\lambda+1)(1+7\lambda) - 16\lambda^2 = 0$$
$$\text{or, } 7\lambda^2 + 8\lambda + 1 - 16\lambda^2 = 0$$
$$\text{or, } 9\lambda^2 - 8\lambda - 1 = 0.$$

$$\therefore\ \lambda = 1 \text{ or, } -\frac{1}{9}. \tag{6}$$

For $\lambda = 1$, from equation (5), we get $x = -2y$. Putting this value in equation (2),

$$(-2y)^2 + 8(-2y)y + 7y^2 = 225$$
$$\text{or, } -5y^2 = 225 \text{ or, } y^2 = -45,$$

which is not a real solution. Hence, this is discarded.

For $\lambda = -1/9$, from equation (5), we get

$$\frac{8}{9}x - \frac{4}{9}y = 0, \text{ i.e., } y = 2x.$$

Putting this value in equation (2),

$$x^2 + 8x(2x) + 7(2x)^2 = 225$$
$$\text{or, } 45x^2 = 225, \text{ i.e., } x^2 = 5.$$

Hence, $y^2 = 4x^2 = 20$, so that $x^2 + y^2 = 5 + 20 = 25$ is the required minimum value.

EXAMPLE F.27. *Find the maximum volume of the rectangular parallelepiped inscribed within the ellipsoid*

$$\frac{x^2}{a^2} + \frac{y^2}{b^2} + \frac{z^2}{c^2} = 1.$$

Solution: Let the lengths of sides of the parallelepiped inscribed within the ellipsoid symmetrical with the coordinate axes be $2x, 2y$ and $2z$. Then its volume is

$$V = 8xyz. \tag{1}$$

We are to determine maximum value of V subject to the constraint

$$\frac{x^2}{a^2} + \frac{y^2}{b^2} + \frac{z^2}{c^2} - 1 = 0. \tag{2}$$

We consider the function

$$f(x, y, z) = 8xyz + \lambda\left(\frac{x^2}{a^2} + \frac{y^2}{b^2} + \frac{z^2}{c^2} - 1\right). \tag{3}$$

For the extremum of this function, we have

$$f_x \equiv 8yz + 2\lambda\frac{x}{a^2} = 0 \tag{4}$$

$$f_y \equiv 8zx + 2\lambda\frac{y}{b^2} = 0 \tag{5}$$

$$f_z \equiv 8xy + 2\lambda\frac{z}{c^2} = 0. \tag{6}$$

From equations (4) and (5), we have

$$2\lambda = -\frac{8yza^2}{x} = -\frac{8zxb^2}{y},$$

which yields

$$\frac{x^2}{a^2} = \frac{y^2}{b^2}. \tag{7}$$

Similarly, from equations (5) and (6), we have

$$2\lambda = -\frac{8zxb^2}{y} = -\frac{8xyc^2}{z}, \text{ i.e., } \frac{y^2}{b^2} = \frac{z^2}{c^2}. \tag{8}$$

From equations (7) and (8), we get

$$\frac{x^2}{a^2} = \frac{y^2}{b^2} = \frac{z^2}{c^2}. \tag{9}$$

Using this relation in the constraint equation (2), we find

$$x = \pm\frac{a}{\sqrt{3}},\ y = \pm\frac{b}{\sqrt{3}},\ z = \pm\frac{c}{\sqrt{3}} \tag{10}$$

as the coordinates of the vertices of the parallelepiped.

The maximum volume is

$$V = 8xyz = \frac{8}{3\sqrt{3}}abc.$$

EXAMPLE F.28. *Prove that the maximum and minimum distances from the origin to the curve of intersection defined by*

$$\frac{x^2}{a^2} + \frac{y^2}{b^2} + \frac{z^2}{c^2} = 1 \text{ and } Ax + By + Cz = 0$$

can be obtained by solving for λ the equation

$$\frac{A^2a^2}{a^2-\lambda} + \frac{B^2b^2}{b^2-\lambda} + \frac{C^2c^2}{c^2-\lambda} = 0.$$

Solution: The distance from the origin to the curve of intersection is given by $x^2 + y^2 + z^2$. Thus, we are to determine extremum values of

$$f(x, y) = x^2 + y^2 + z^2 \tag{1}$$

subject to the constraint equations

$$\frac{x^2}{a^2} + \frac{y^2}{b^2} + \frac{z^2}{c^2} = 1 \tag{2}$$

$$\text{and } Ax + By + Cz = 0. \tag{3}$$

We consider the function

$$F(x, y) = x^2 + y^2 + z^2 - \lambda\left(\frac{x^2}{a^2} + \frac{y^2}{b^2} + \frac{z^2}{c^2} - 1\right) + \mu(Ax + By + Cz).$$

For the extremum of this function, we must have

$$F_x = 2x - \frac{2\lambda x}{a^2} + \mu A = 0 \tag{4}$$

$$F_y = 2y - \frac{2\lambda y}{b^2} + \mu B = 0 \tag{5}$$

$$F_z = 2z - \frac{2\lambda z}{c^2} + \mu C = 0. \tag{6}$$

Multiplying equations (4), (5), (6) by x, y, z in succession and adding, we get

$$2(x^2 + y^2 + z^2) - 2\lambda\left(\frac{x^2}{a^2} + \frac{y^2}{b^2} + \frac{z^2}{c^2}\right) + \mu(Ax + By + Cz) = 0.$$

This upon using the constraint equations (2) and (3) yields

$$x^2 + y^2 + z^2 = \lambda. \tag{7}$$

Thus, the maximum and minimum distances are given by the two values of λ. We are to determine λ.

From equations (4), (5) and (6), we find

$$x = -\frac{\mu A}{2\left(1 - \dfrac{\lambda}{a^2}\right)} = -\frac{\mu A a^2}{2(a^2 - \lambda)} \tag{8}$$

$$y = -\frac{\mu B}{2\left(1 - \dfrac{\lambda}{b^2}\right)} = -\frac{\mu B b^2}{2(b^2 - \lambda)} \tag{9}$$

$$z = -\frac{\mu C}{2\left(1 - \dfrac{\lambda}{c^2}\right)} = -\frac{\mu C c^2}{2(c^2 - \lambda)}. \tag{10}$$

Putting these values in the constraint equation (3), we have after dividing by $\mu/2$ throughout,

$$\frac{A^2a^2}{a^2 - \lambda} + \frac{B^2b^2}{b^2 - \lambda} + \frac{C^2c^2}{c^2 - \lambda} = 0, \tag{11}$$

which determines the value of λ.

Hence, the proposition is proved.

EXAMPLE F.29. *Find the area of the ellipse* $3x^2 + 2xy + y^2 = 1$.

Solution: The centre of the given ellipse is at the origin. In order to determine its area, we are to find semi-major and semi-minor axes of the ellipse. The distance of any point on the ellipse being $(x^2 + y^2)$, we are to find the extremum value of

$$f(x, y) = x^2 + y^2 \tag{1}$$

subject to the constraint

$$3x^2 + 2xy + y^2 = 1. \tag{2}$$

We consider the function

$$F(x, y) = x^2 + y^2 + \lambda(3x^2 + 2xy + y^2 - 1). \tag{3}$$

For its extremum, we must have

$$\left.\begin{aligned} 2x + 6\lambda x + 2\lambda y = 0 \\ 2y + 2\lambda x + 2\lambda y = 0 \end{aligned}\right\}. \tag{4}$$

or

$$\left.\begin{aligned} (2 + 6\lambda)x + 2\lambda y = 0 \\ 2\lambda x + (2 + 2\lambda)y = 0 \end{aligned}\right\}. \tag{5}$$

Since $(x, y) \neq (0, 0)$, we must have

$$\begin{aligned} \begin{vmatrix} 2+6\lambda & 2\lambda \\ 2\lambda & 2+2\lambda \end{vmatrix} &= 0 \text{ or, } (2+6\lambda)(2+2\lambda) - 4\lambda^2 = 0 \\ &\text{or, } 8\lambda^2 + 16\lambda + 4 = 0 \\ &\text{or, } 2\lambda^2 + 4\lambda + 1 = 0. \end{aligned} \tag{6}$$

Hence,

$$\lambda = \frac{1}{4}[-4 \pm \sqrt{16 - 8}] = \frac{1}{2}[-2 \pm \sqrt{2}]. \tag{7}$$

For $\lambda = \frac{1}{2}(-2 + \sqrt{2})$, we have from equation (5),

$$x = -\frac{2\lambda}{2 + 6\lambda} y = -\left[\frac{-2 + \sqrt{2}}{2 + 3(-2 + \sqrt{2})}\right] y = \frac{2 - \sqrt{2}}{-4 + 3\sqrt{2}} y$$

$$\text{i.e., } x = \frac{(2 - \sqrt{2})(3\sqrt{2} + 4)}{(3\sqrt{2})^2 - (4)^2} y = \frac{(2\sqrt{2} + 2)y}{2} = (\sqrt{2} + 1)y. \tag{8}$$

Putting this value into equation (2), we have

$$y^2[3(\sqrt{2} + 1)^2 + 2(\sqrt{2} + 1) + 1] = 1 \text{ or, } y^2(8\sqrt{2} + 12) = 1.$$

$$\therefore\ y^2 = \frac{1}{4(3 + 2\sqrt{2})}. \tag{9}$$

Then, from equation (8), we have

$$x^2 = (\sqrt{2} + 1)^2 y^2 = \frac{3 + 2\sqrt{2}}{4(3 + 2\sqrt{2})} = \frac{1}{4}. \tag{10}$$

Hence,

$$x^2 + y^2 = \frac{1}{4}\left(1 + \frac{1}{3 + 2\sqrt{2}}\right) = \frac{1}{4} \cdot \frac{4 + 2\sqrt{2}}{3 + 2\sqrt{2}} = \frac{1}{2} \cdot \frac{2 + \sqrt{2}}{(\sqrt{2} + 1)^2}. \tag{11}$$

Similarly, for $\lambda = -\frac{1}{2}(2 + \sqrt{2})$, we have

$$x = \frac{2 + \sqrt{2}}{2 - 3(2 + \sqrt{2})} y = \frac{2 + \sqrt{2}}{-(4 + 3\sqrt{2})} y = -\frac{(2 + \sqrt{2})(4 - 3\sqrt{2})}{(4)^2 - (3\sqrt{2})^2} y = \frac{-2 + 2\sqrt{2}}{-2} y.$$

Then

$$y^2[3(\sqrt{2}-1)^2 - 2(\sqrt{2}-1) + 1] = 1 \text{ or, } y^2(-8\sqrt{2}+12) = 1.$$

$$\therefore\ y^2 = \frac{1}{4(3-2\sqrt{2})}. \tag{12}$$

$$\therefore\ x^2 = (\sqrt{2}-1)^2 y^2 = \frac{3-2\sqrt{2}}{4(3-2\sqrt{2})} = \frac{1}{4}. \tag{13}$$

$$\therefore\ x^2 + y^2 = \frac{1}{4}\left[1 + \frac{1}{3-2\sqrt{2}}\right] = \frac{1}{4}\cdot\frac{4-2\sqrt{2}}{3-2\sqrt{2}} = \frac{1}{2}\cdot\frac{2-\sqrt{2}}{(\sqrt{2}-1)^2}. \tag{14}$$

If a, b are semi-major and semi-minor axes, the area of the ellipse

$$= \pi ab = \pi\sqrt{\frac{1}{2}\cdot\frac{2+\sqrt{2}}{(\sqrt{2}+1)^2}}\cdot\sqrt{\frac{1}{2}\cdot\frac{2-\sqrt{2}}{(\sqrt{2}-1)^2}} = \frac{\pi}{2}\cdot\frac{\sqrt{2}}{1} = \frac{\pi}{\sqrt{2}}.$$

Appendix G

Infinite Series

G.1 Definition

An expression of the form

$$a_1 + a_2 + a_3 + \cdots + a_n + \ldots,$$

where $a_1, a_2, \ldots$ have definite values, positive or negative, constant or variables, is called an *infinite series* and is denoted by Σa_n. Thus, we write:

$$\Sigma a_n = a_1 + a_2 + \cdots + a_n + \ldots. \quad \text{(G.1)}$$

The quantities $a_1, a_2, \ldots$, etc. are called *terms* of the infinite series. Let

$$S_n = a_1 + a_2 + \cdots + a_n, \quad \text{(G.2)}$$

where S_n is the sum of first n terms of the series.

If

$$\lim_{n \to \infty} S_n = S, \quad \text{(G.3)}$$

where S is a finite value, the series Σa_n is said to be *convergent* and S is called its sum.

If

$$\lim_{n \to \infty} S_n = \pm\infty, \quad \text{(G.4)}$$

the series Σa_n is said to be *divergent*.

If, however, the value of $\lim_{n \to \infty} S_n$ oscillates finitely or infinitely, the series Σa_n is said to be *oscillatory*.

Infinite series have the following basic properties:

(a) A series of positive terms is either convergent or divergent, but cannot oscillate.

(b) The nature of an infinite series remains unchanged by multiplication of each term by a constant.

(c) The nature of an infinite series is not affected by the inclusion or removal of a finite number of terms.

(d) If the terms of a convergent series of positive terms be rearranged, then the series remains convergent and its sum remains the same.

G.2 Two Useful Series

(a) Geometric series

The geometric series

$$\Sigma a_n = a + ax + ax^2 + \cdots + ax^n + \ldots, \tag{G.5}$$

where a is the first term and x is the *common ratio,* converges if $|x| < 1$, diverges if $|x| \geq 1$, oscillates finitely if $x = -1$ and oscillates infinitely if $x < -1$. When convergent, its sum is

$$S = a\frac{1}{1-x}. \tag{G.6}$$

Proof. Let $|x| < 1$, i.e., $-1 < x < 1$. Then

$$S_n = a(1 + x + x^2 + \cdots + x^{n-1}) = a\frac{1-x^n}{1-x}.$$

Since $|x| < 1, x^n \to 0$ as $n \to \infty$. Hence,

$$\lim_{n\to\infty} S_n = \frac{a}{1-x}.$$

Thus, the series is convergent and its sum is $\dfrac{a}{1-x}$.

If $x > 1$, then

$$S_n = a\frac{x^n - 1}{x-1} \text{ which} \to \infty \text{ as } n \to \infty..$$

Hence, the series is divergent when $x > 1$.

If $x = 1$, the series becomes $a(1 + 1 + 1 + 1 + \ldots \infty)$ so that $S_n = na$. This $S_n \to \infty$, when $n \to \infty$. Thus, the series is divergent, when $x = 1$.

If $x = -1$, the series becomes $a(1 - 1 + 1 - 1 + \ldots \infty)$ so that $S_n = a$ or 0, according as n is odd or even.

Hence, the series oscillates finitely, when $x = -1$.

If $x < -1$, then

$$S_n = a\frac{x^n - 1}{x-1}.$$

In this case, x^n oscillates between $-\infty$ and $+\infty$. Hence, the series oscillates infinitely when $x < -1$.

We, therefore, conclude that the *geometric series converges only when the common ratio x satisfies the condition $|x| < 1$.*

(b) p-series

The infinite series

$$\Sigma a_n = \frac{1}{1^p} + \frac{1}{2^p} + \cdots + \frac{1}{n^p} + \ldots \tag{G.7}$$

is convergent, if $p > 1$ and divergent, if $p \leq 1$.

Proof. The series is of positive terms and hence, its nature is not affected if the terms are grouped in any way.

Case I. $p > 1$.

In this case, $3^p > 2^p$, i.e., $\dfrac{1}{3^p} < \dfrac{1}{2^p}$.

Thus, we have

$$\frac{1}{2^p} + \frac{1}{3^p} < \frac{2}{2^p}.$$

Similarly,

$$\frac{1}{4^p} + \frac{1}{5^p} + \frac{1}{6^p} + \frac{1}{7^p} < \frac{4}{4^p}.$$

Hence, the series

$$\Sigma a_n < 1 + \frac{2}{2^p} + \frac{4}{4^p} + \ldots. \tag{G.8}$$

But the series in the RHS of equation (G.8) is a geometric series with common ratio

$$x = \frac{2}{2^p} < 1.$$

Hence, this comparing series is convergent. The given series must also be convergent for $p > 1$.

Case II. $p = 1$.

In this case,

$$\Sigma a_n = 1 + \frac{1}{2} + \frac{1}{3} + \cdots + \frac{1}{n} + \ldots.$$

Now

$$\frac{1}{3} + \frac{1}{4} > \frac{2}{4} = \frac{1}{2}$$

$$\frac{1}{5} + \frac{1}{6} + \frac{1}{7} + \frac{1}{8} > \frac{4}{8} = \frac{1}{2}$$

and so on.

Hence, the series

$$\Sigma a_n = 1 + \frac{1}{2} + \frac{1}{3} + \cdots + \frac{1}{n} + \ldots$$

$$> 1 + \frac{1}{2} + \frac{1}{2} + \frac{1}{2} + \ldots,$$

which is divergent. Hence, the given series is divergent for $p = 1$.

Case III. $p < 1$.

In this case,

$$\Sigma a_n = 1 + \frac{1}{2^p} + \frac{1}{3^p} + \cdots + \frac{1}{n^p} + \ldots$$

$$> 1 + \frac{1}{2} + \frac{1}{3} + \ldots \frac{1}{n} + \ldots,$$

which is divergent. Hence, the given series is divergent for $p < 1$.

G.3 Tests of Convergence

Various methods have been developed for testing the convergence of infinite series. We mention below some of the important methods without going for the proofs.

(1) A necessary condition for convergence of the series Σa_n is that

$$\lim_{n\to\infty} a_n = 0. \tag{G.9}$$

(2) If a series Σa_n of positive and monotonically decreasing terms be convergent, then it is necessary that

$$\lim_{n\to\infty} na_n = 0. \tag{G.10}$$

This is known as *Pringsheim's theorem.*

The condition in equation (G.10) is not sufficient.

(3) The alternating series

$$\Sigma a_n = a_1 - a_2 + a_3 - a_4 + \ldots$$

in which the terms are alternately positive and negative, is convergent, if

$$\left.\begin{array}{rl} & a_n > a_{n+1}, \text{ for all } n \\ \text{and} & \lim\limits_{n\to\infty} a_n = 0 \end{array}\right\}. \tag{G.11}$$

(4) *Comparison test*: If Σa_n and Σb_n be two infinite series of positive terms, and if

$$\lim_{n\to\infty} \frac{a_n}{b_n} = k, \tag{G.12}$$

where k is a finite positive non-zero number, then the series are either both convergent or both divergent.

(5) *D'Alembert's Ratio Test*: If Σa_n be a series of positive terms and if

$$\lim_{n\to\infty} \left(\frac{a_{n+1}}{a_n}\right) = l, \tag{G.13}$$

then the series Σa_n is

$$\left.\begin{array}{rl} & \text{convergent, if } l < 1 \\ \text{and} & \text{divergent, if } l > 1 \end{array}\right\}. \tag{G.14}$$

(6) *Raabe's test*: The series Σa_n of positive terms is convergent or divergent according as

$$\lim_{n\to\infty} \left\{ n \left(\frac{a_n}{a_{n+1}} - 1 \right) \right\} > 1 \text{ or } < 1. \tag{G.15}$$

The test fails if the limit = 1.

(7) *Logarithmic Test*: The series Σa_n of the positive terms is convergent or divergent according as

$$\lim_{n\to\infty} \left\{ n \log \left(\frac{a_n}{a_{n+1}} \right) \right\} > 1 \text{ or } < 1. \tag{G.16}$$

This test is generally applied when D'Alembert's ratio lest fails.

(8) *Cauchy's Root Test*: The series Σa_n of positive terms is convergent or divergent according as

$$\lim_{n\to\infty} (a_n)^{1/n} < 1 \text{ or } > 1. \tag{G.17}$$

The test fails when the limit $= 1$.

(9) *Absolute convergence*: The series Σa_n containing both positive and negative terms is called *absolutely convergent* if the series $|\Sigma a_n|$ be convergent.

The series Σa_n containing both positive and negative terms is convergent or non-convergent according as

$$\lim_{n\to\infty} \left|\frac{a_{n+1}}{a_n}\right| < 1 \text{ or } > 1. \tag{G.18}$$

The test fails if the limit $= 1$.

(10) *D'Morgan and Bertrand's Test*: The series Σa_n of positive terms is convergent or divergent according as

$$\lim_{n\to\infty} \left[(\log n)\left\{n\left(\frac{a_n}{a_{n+1}} - 1\right) - 1\right\}\right] > 1 \text{ or } < 1. \tag{G.19}$$

This test is generally applied when both D'Alembert's ratio test and Rabee's test fail.

(11) *Gauss's Test*: Let Σa_n be a series of positive terms. If we can express

$$\frac{a_n}{a_{n+1}} = 1 + \frac{\alpha}{n} + \frac{\beta_n}{n^p}(p > 1), \tag{G.20}$$

where $\{\beta_n\}$ is bounded sequence, then the series Σa_n converges, if $\alpha > 1$ and diverges, if $\alpha \leq 1$.

Following is an useful theorem for testing convergence or divergence of a series:

Theorem G.1. *Let* $\lim_{n\to\infty} n^p a_n = \lambda$.

Then the series Σa_n converges, if $p > 1$ and λ is finite and diverges if $p \leq 1$ and $\lambda \neq 0$ (may be infinite).

G.4 Power Series

An infinite series in which the successive terms are in ascending positive integral powers of a variable, say x, is called a power series in x. Thus,

$$\Sigma a_n x^n = a_0 + a_1 x + a_2 x^2 + \cdots + a_n x^n + \ldots, \tag{G.21}$$

where $a_0, a_1, \ldots$ are constants independent of x called *coefficients*, is a power series in x.

Every power series is convergent for $x = 0$. A power series is convergent either for all values of x or for a Cartesian range of values of x.

The *interval of convergence* of a power series in x is the range of values of x, such that the series converges for every value of x in this interval and does not converge for other values of x. Thus, if the power series in equation (G.21) converges for every $|x| < R$ but diverges for $|x| > R$, then $(-R, R)$ is the interval of convergence. R is called the *radius of convergence*.

If a power series is convergent in $(-\infty, \infty)$, the series is said to be *everywhere convergent*. *Every power series is absolutely convergent in its interval of convergence.*

The interval of power series is determined by ratio test and root test. Thus, for the series in equation (G.21),

$$\left.\begin{aligned} &\text{(i)} \quad \text{if } \lim_{n\to\infty}\left|\frac{u_{n+1}}{u_n}\right| = \lambda \\ &\text{(ii)} \quad \text{if } \lim_{n\to\infty}|u_n|^{1/n} = \lambda \end{aligned}\right\}, \tag{G.22}$$

where λ is a non-zero finite quantity, then the interval of convergence of the series is $(1/\lambda, 1/\lambda)$.

In the first case,

$$\frac{u_{n+1}}{u_n} = \frac{a_{n+1}x^{n+1}}{a_n x^n} = \frac{a_{n+1}}{a_n}x.$$

Hence,

$$\lim_{n\to\infty}\left|\frac{u_{n+1}}{u_n}\right| = \lim_{n\to\infty}\left|\frac{a_{n+1}}{a_n}\right||x| = \lambda|x|,$$

which is less or greater than 1 according as $|x| <$ or $> 1/\lambda$. Hence, the result follows. Similarly, in the second case we use Cauchy's root test. Thus,

$$\lim_{n\to\infty}|u_n|^{1/n} = \lim_{n\to\infty}|a_n x^n|^{1/n} = \lim_{n\to\infty}|a_n|^{1/n}|x| = \lambda|x|,$$

which is less or greater than 1 according as $|x| <$ or $> 1/\lambda$.

If, however, $|x| = 1/\lambda$, the test fails.

G.5 Properties of Power Series

We give below some important properties of power series without any proof.

(1) Two power series, if they converge, may be added and subtracted term by term.

(2) Two power series can be multiplied out term by term for values of x lying within the intervals of convergence of both the series.

(3) The quotient of two power series which are convergent is another power series, provided x remains within a sufficiently small interval in which the denominator does not vanish.

(4) Term by term limits are possible for every power series within its interval of convergence.

(5) If a function $f(x)$ is represented by a power series, i.e., $f(x) = \Sigma a_n x^n$, then $f(x)$ is continuous and differentiable for any number of times at every point x lying within its interval of convergence. The differential coefficients may be obtained by term by term differentiation.

(6) The integral of a function $f(x)$ represented by a power series $\Sigma a_n x^n$, over every interval lying within the interval of convergence, can be obtained by term by term integration.

(7) If two power series $\Sigma a_n x^n$ and $\Sigma b_n x^n$ both converge in the same interval $(-R, R)$ and both represent the same function $f(x)$, then

$$a_n = b_n \text{ for all } n. \tag{G.23}$$

EXAMPLE G.1. *Show that the series*

$$1 + \frac{1}{1\cdot 2} + \frac{1}{1\cdot 2\cdot 3} + \frac{1}{1\cdot 2\cdot 3\cdot 4} + \cdots$$

is convergent.

Solution: We have

$$a_1 = 1,\ a_2 = \frac{1}{1\cdot 2} = \frac{1}{2}$$
$$a_3 = \frac{1}{1\cdot 2\cdot 3} < \frac{1}{1\cdot 2\cdot 2} < \frac{1}{2^2}$$
$$a_4 = \frac{1}{1\cdot 2\cdot 3\cdot 4} < \frac{1}{1\cdot 2\cdot 2\cdot 2} = \frac{1}{2^3}$$
$$\dots\dots\dots\dots\dots\dots\dots\dots\dots\dots$$

and so on.

Hence, the given series is

$$< 1 + \frac{1}{2} + \frac{1}{2^2} + \frac{1}{2^3} + \dots. \tag{1}$$

The series in equation (1) is a geometric series with common ratio $\frac{1}{2} < 1$, so that it is convergent. Hence, by comparison test, the given series also must be convergent.

EXAMPLE G.2. *Test the convergence or divergence of the series whose nth term is* $\sqrt{n^2+1} - n$.

Solution: We are given

$$a_n = \sqrt{n^2+1} - n. \tag{1}$$

Let us consider a series whose nth term is

$$b_n = 1/n. \tag{2}$$

Then

$$\frac{a_n}{b_n} = n(\sqrt{n^2+1} - n) = \frac{n\{(n^2+1) - n^2\}}{\sqrt{n^2+1}+n} = \frac{1}{\sqrt{1+\frac{1}{n^2}}+1}. \tag{3}$$

Hence,

$$\lim_{n\to\infty}\left(\frac{a_n}{b_n}\right) = \lim_{n\to\infty}\frac{1}{\sqrt{1+\frac{1}{n^2}}+1} = \frac{1}{2}. \tag{4}$$

Thus, the series Σa_n and Σb_n are either both convergent or both divergent.

But the series $\Sigma b_n = \Sigma\frac{1}{n}$ is a p-series, where $p = 1$ and so is divergent. Hence, the given series by comparison test is divergent.

EXAMPLE G.3. *Test the convergence or divergence of the series*

(1) $\sum \dfrac{\log n}{2n^3 - 1}$ (2) $\dfrac{n+\sqrt{n}}{2n^3-1}$ (3) $\sin^3\left(\dfrac{1}{n}\right)$.

Solution:

(1) Since $\log n < n$ and $\dfrac{1}{2n^3-1} \le \dfrac{1}{n^3}$, we have

$$\frac{\log n}{2n^3-1} \le \frac{n}{n^3} = \frac{1}{n^2}. \tag{1}$$

Now, the series $\Sigma\dfrac{1}{n^2}$ converges, since $p = 2 > 1$ in p-series. Hence, the given series also converges.

(2) We have

$$\lim_{n\to\infty} n^2 a_n = \lim_{n\to\infty} n^2\left(\frac{n+\sqrt{n}}{2n^3-1}\right) = \lim_{n\to\infty}\frac{1+\frac{1}{\sqrt{n}}}{2-\frac{1}{n^3}} = \frac{1}{2}. \tag{2}$$

Hence, by Theorem G.1, the series converges.

(3) We have

$$\lim_{n\to\infty} n^3 a_n = \lim_{n\to\infty} n^3 \sin^3\left(\frac{1}{n}\right) = \lim_{n\to\infty}\left(\frac{\sin\frac{1}{n}}{\frac{1}{n}}\right)^3 = 1. \tag{3}$$

Hence, by Theorem G.1, where $p = 3 > 1$, the given series converges.

EXAMPLE G.4. *Test the convergence of the series*

$$\frac{x}{1}+\frac{1}{2}\cdot\frac{x}{3}+\frac{1\cdot 3}{2\cdot 4}\cdot\frac{x^5}{5}+\frac{1\cdot 3\cdot 5}{2\cdot 4\cdot 6}\cdot\frac{x^7}{7}+\ldots$$

Solution: Here

$$a_n = \frac{1\cdot 3\cdot 5\ldots(2n-3)}{2\cdot 4\cdot 6\ldots(2n-2)}\cdot\frac{x^{2n-1}}{2n-1}$$

$$\text{and } a_{n+1} = \frac{1\cdot 3\cdot 5\ldots(2n-1)}{2\cdot 4\cdot 6\ldots 2n}\cdot\frac{x^{2n+1}}{2n+1}.$$

Hence,

$$\begin{aligned}\frac{a_{n+1}}{a_n} &= \frac{1\cdot 3\cdot 5\ldots(2n-1)}{2\cdot 4\cdot 6\ldots 2n}\cdot\frac{x^{2n+1}}{2n+1}\cdot\frac{2\cdot 4\cdot 6\ldots(2n-2)}{1\cdot 3\cdot 5\ldots(2n-3)}\cdot\frac{2n-1}{x^{2n-1}}\\ &= \frac{(2n-1)^2}{2n(2n+1)}x^2. \end{aligned}\tag{1}$$

Then

$$\begin{aligned}\lim_{n\to\infty}\left(\frac{a_{n+1}}{a_n}\right) &= \lim_{n\to\infty}\frac{(2n-1)^2}{2n(2n+1)}x^2 = \lim_{n\to\infty}\frac{4n^2-4n+1}{4n^2+2n}\cdot x^2\\ &= \lim_{n\to\infty}\frac{1-\frac{1}{n}+\frac{1}{4n^2}}{1+\frac{2}{n}}x^2 = x^2. \end{aligned}\tag{2}$$

Thus, by D'Alembert's test, the series converges if $x^2 < 1$, i.e., $|x| < 1$ and diverges if $|x| > 1$.

The test fails if $x = 1$. In this case, we use Rabee's test. Thus,

$$\begin{aligned}\lim_{n\to\infty} n\left(\frac{a_n}{a_{n+1}}-1\right) &= \lim_{n\to\infty} n\left(\frac{4n^2+2n}{4n^2-4n+1}-1\right) = \lim_{n\to\infty}\frac{n(6n-1)}{4n^2-4n+1}\\ &= \frac{6}{4} = \frac{3}{2} > 1.\end{aligned}$$

Hence, the series is convergent.

EXAMPLE G.5. *Test for convergence the series*

$$\left(\frac{1}{2}\right)^2+\left(\frac{1\cdot 3}{2\cdot 4}\right)^2+\left(\frac{1\cdot 3\cdot 5}{2\cdot 4\cdot 6}\right)^2+\cdots+\left\{\frac{1\cdot 3\cdot 5\ldots(2n-1)}{2\cdot 4\cdot 6\ldots 2n}\right\}^2+\ldots.$$

Solution: In this case,

$$\lim_{n\to\infty}\left(\frac{a_{n+1}}{a_n}\right)=\lim_{n\to\infty}\left(\frac{2n+1}{2n+2}\right)^2=1. \tag{1}$$

Hence, D'Alembert's ratio test fails.
Again,

$$\lim_{n\to\infty} n\left(\frac{a_n}{a_{n+1}}-1\right)=\lim_{n\to\infty} n\left\{\left(\frac{2n+2}{2n+1}\right)^2-1\right\}=\lim_{n\to\infty}\frac{n(4n+3)}{4n^2+4n+1}=1. \tag{2}$$

Thus, Rabee's test also fails.
Now,

$$\frac{a_{n+1}}{a_n}=\left(\frac{2n+1}{2n+2}\right)^2=1-\frac{1}{n}+\frac{5-\frac{4}{n}}{4n^2+8n+4}. \tag{3}$$

The RHS of equation (3) can be written as

$$=1-\frac{1}{n}+\frac{\alpha_n}{n^2}, \tag{4}$$

where

$$\alpha_n=\frac{5-\frac{4}{n}}{4+\frac{8}{n}+\frac{4}{n^2}},$$

which tends to $\frac{5}{4}$ as $n\to 0$. Thus, the sequence $\{\alpha_n\}$ is bounded.

Hence, by Gauss's test, the series diverges.

EXAMPLE G.6. *Test the convergence of the series*

$$1+\frac{1+\alpha}{1+\beta}+\frac{(1+\alpha)(2+\alpha)}{(1+\beta)(2+\beta)}+\frac{(1+\alpha)(2+\alpha)(3+\alpha)}{(1+\beta)(2+\beta)(3+\beta)}+\ldots,$$

where α,β are constants.

Solution: Here,

$$a_n=\frac{(1+\alpha)(2+\alpha)\ldots(n+\alpha)}{(1+\beta)(2+\beta)\ldots(n+\beta)} \tag{1}$$

$$\text{and } a_{n+1}=\frac{(1+\alpha)(2+\alpha)\ldots(n+\alpha)(n+1+\alpha)}{(1+\beta)(2+\beta)\ldots(n+\beta)(n+1+\beta)}. \tag{2}$$

Hence,

$$\lim_{n\to\infty}\left(\frac{a_{n+1}}{a_n}\right)=\lim_{n\to\infty}\frac{n+1+\alpha}{n+1+\beta}=1. \tag{3}$$

Thus, D'Alembert's test fails.
Now,

$$\lim_{n\to\infty} n\left(\frac{a_n}{a_{n+1}}-1\right)=\lim_{n\to\infty} n\left(\frac{n+1+\beta}{n+1+\alpha}-1\right)=\lim_{n\to\infty}\frac{n(\beta-\alpha)}{n+1+\alpha}$$

$$=\lim_{n\to\infty}\frac{\beta-\alpha}{1+\frac{1}{n}+\frac{\alpha}{n}}=\beta-\alpha. \tag{4}$$

Thus, by Rabee's test, the series converges if $(\beta-\alpha)>1$ and diverges if $(\beta-\alpha)<1$.

EXAMPLE G.7. *Prove that the series*

$$x + \frac{x^2}{2} + \frac{x^3}{3} + \frac{x^4}{4} + \ldots (x \geq 0)$$

is convergent if $0 \leq x \leq 1$ *and divergent if* $x \geq 1$.

Solution: Here,

$$a_n = \frac{x^n}{n},\ a_{n+1} = \frac{x^{n+1}}{n+1}.$$

Hence,

$$\lim_{n\to\infty} \frac{a_{n+1}}{a_n} = \lim_{n\to\infty} \frac{n}{n+1} x = x. \tag{1}$$

Thus, the series converges if $x < 1$. When $x = 0$, the series obviously converges to zero. Hence, the series converges if $0 \leq x < 1$.

If $x > 1$, the series diverges.

For $x = 1$, we have by Rabee's test

$$\lim_{n\to\infty} n\left(\frac{a_n}{a_{n+1}} - 1\right) = \lim_{n\to\infty} n\left(\frac{n+1}{n} - 1\right) = 1.$$

Hence, the test fails. Let us use D'Morgan and Bertrand's test.

We have

$$\lim_{n\to\infty} \log n\left\{n\left(\frac{a_n}{a_{n+1}} - 1\right) - 1\right\} = \lim_{n\to\infty} \log n\left\{n\left(\frac{1}{n}\right) - 1\right\} = 0. \tag{2}$$

Thus, the series is divergent for $x = 1$. From above, we conclude that the series converges if $0 \leq x < 1$ and diverges if $x \geq 1$.

EXAMPLE G.8. *Test the convergence of the series*

$$1 + \frac{\alpha\cdot\beta}{1\cdot\gamma}x + \frac{\alpha(\alpha+1)\beta(\beta+1)}{1\cdot 2\cdot\gamma(\gamma+1)}x^2 + \frac{\alpha(\alpha+1)(\alpha+2)\beta(\beta+1)(\beta+2)}{1\cdot 2\cdot 3\cdot\gamma(\gamma+1)(\gamma+2)}x^3 + \ldots.$$

Solution: Here,

$$a_n = \frac{\alpha(\alpha+1)(\alpha+2)\ldots(\alpha+n-1)\beta(\beta+1)(\beta+2)\ldots(\beta+n-1)}{1\cdot 2\cdot 3\ldots n\cdot\gamma(\gamma+1)(\gamma+2)\ldots(\gamma+n-1)}x^n.$$

Hence,

$$\begin{aligned}\lim_{n\to\infty} \frac{a_{n+1}}{a_n} &= \lim_{n\to\infty} \frac{(\alpha+n)(\beta+n)}{(n+1)(\gamma+n)}x \\ &= \lim_{n\to\infty} \frac{\left(\frac{\alpha}{n}+1\right)\left(\frac{\beta}{n}+1\right)}{\left(1+\frac{1}{n}\right)\left(\frac{\gamma}{n}+1\right)}x = x.\end{aligned} \tag{1}$$

Hence, by D'Alembert's test, the series converges if $0 < x < 1$ and diverges if $x > 1$.

For $x = 1$, the test fails. We try Gauss's test.

We have, for $x = 1$,

$$\frac{a_n}{a_{n+1}} = \frac{(n+1)(\gamma+n)}{(\alpha+n)(\beta+n)} = \frac{n^2+(\gamma+1)n+\gamma}{n^2+(\alpha+\beta)n+\alpha\beta} = 1 + \frac{n(\gamma+1-\alpha-\beta)+\gamma-\alpha\beta}{n^2+(\alpha+\beta)n+\alpha\beta}$$

$$= 1 + \frac{\gamma+1-\alpha-\beta}{n} + \frac{\beta_n}{n^2}, \qquad (2)$$

where

$$\beta_n = \frac{A+B/n}{\left(1+\frac{\alpha}{n}\right)\left(1+\frac{\beta}{n}\right)}, \qquad (3)$$

where

$$\left.\begin{aligned} A &= \gamma-\alpha\beta-(\alpha+\beta)(\gamma+1-\alpha-\beta) \\ \text{and } B &= -\alpha\beta(\gamma+1-\alpha-\beta) \end{aligned}\right\}. \qquad (4)$$

We note that A and B are fixed quantities.
From equation (3), we find that

$$\lim_{n\to\infty} \beta_n = A. \qquad (5)$$

This implies that the sequence $\{\beta_n\}$ is bounded.
Hence, when $x = 1$, the given hypergeometric series converges if

$$\gamma+1-\alpha-\beta > 1, \text{ i.e., if } \gamma > \alpha+\beta \qquad (6)$$

and diverges if

$$\gamma \le \alpha+\beta. \qquad (7)$$

EXAMPLE G.9. *Prove that the series*

$$\frac{1+2}{1\cdot 2} + \left(\frac{2+2}{2\cdot 2}\right)^2 + \left(\frac{3+2}{2\cdot 3}\right)^3 + \cdots + \left(\frac{n+2}{2\cdot n}\right)^n + \ldots$$

is convergent.

Solution: Here,

$$a_n = \left(\frac{n+2}{2\cdot n}\right)^n = \left(\frac{1}{2}+\frac{1}{n}\right)^n. \qquad (1)$$

Hence,

$$(a_n)^{1/n} = \frac{1}{2}+\frac{1}{n}. \qquad (2)$$

Then

$$\lim_{n\to\infty} (a_n)^{1/n} = \lim_{n\to\infty}\left(\frac{1}{2}+\frac{1}{n}\right) = \frac{1}{2} < 1. \qquad (3)$$

Thus, by Cauchy's root test the series is convergent.

EXAMPLE G.10. *Test the convergence of the series*

$$\left(\frac{2^2}{1^2}-\frac{2}{1}\right)^{-1} + \left(\frac{3^3}{2^3}-\frac{3}{2}\right)^{-2} + \left(\frac{4^4}{3^4}-\frac{4}{3}\right)^{-3} + \ldots.$$

Solution: Here, the nth term is

$$a_n = \left\{\left(\frac{n+1}{n}\right)^{n+1} - \frac{n+1}{n}\right\}^{-n}. \tag{1}$$

Hence,

$$(a_n)^{1/n} = \left\{\left(\frac{n+1}{n}\right)^{n+1} - \frac{n+1}{n}\right\}^{-1} = \left(\frac{n+1}{n}\right)^{-1}\left\{\left(\frac{n+1}{n}\right)^{n} - 1\right\}^{-1}$$
$$= \left(1+\frac{1}{n}\right)^{-1}\left\{\left(1+\frac{1}{n}\right)^{n} - 1\right\}^{-1} \tag{2}$$

Now,

$$\left.\begin{aligned} \lim_{n\to\infty}\left(1+\frac{1}{n}\right)^{-1} &= 1 \\ \text{and } \lim_{n\to\infty}\left(1+\frac{1}{n}\right)^{n} &= e \end{aligned}\right\}. \tag{3}$$

Hence,

$$\lim_{n\to\infty}(a_n)^{1/n} = 1\cdot(e-1)^{-1} = \frac{1}{e-1}. \tag{4}$$

Since $e > 2$, $\frac{1}{e-1} < 1$.
Hence, by Cauchy's root test, the series is convergent.

EXAMPLE G.11. *Test the convergence of the series*

$$\sum_{n=2}^{\infty}\frac{1}{(\log n)^p}.$$

Solution: We have

$$a_n = \frac{1}{(\log n)^p} \tag{1}$$

$$\text{and } a_{n+1} = \frac{1}{\{\log(n+1)\}^p}. \tag{2}$$

Hence,

$$\frac{a_{n+1}}{a_n} = \left[\frac{\log n}{\log(n+1)}\right]^p = \left[\frac{\log n}{\log n + \log\left(1+\frac{1}{n}\right)}\right]^p$$
$$= \left[\frac{\log n}{\log n + \left(\frac{1}{n} - \frac{1}{2n^2} + \frac{1}{3n^3} - \dots\right)}\right]^p$$
$$= \left[\frac{1}{1 + \frac{1}{n\log n} - \frac{1}{2n^2\log n} + \frac{1}{3n^3+\log n} - \dots}\right]^p. \tag{3}$$

From equation (3) we find

$$\lim_{n\to\infty} \frac{a_{n+1}}{a_n} = 1. \tag{4}$$

Thus, D'Alembert's test fails. We apply Logarithmic test.
Now,

$$\log\left(\frac{a_n}{a_{n+1}}\right) = p\log\left\{1 + \frac{1}{n\log n} - \frac{1}{2n^2\log n} + \frac{1}{3n^3\log n} - \dots\right\}.$$

Hence,

$$\begin{aligned}\lim_{n\to\infty} n\log\left(\frac{a_n}{a_{n+1}}\right) &= \lim_{n\to\infty} np\log\left\{1 + \frac{1}{n\log n} - \frac{1}{2n^2\log n} + \dots\right\} \\ &= \lim_{n\to\infty} np\log(1+x),\end{aligned} \tag{5}$$

where

$$x = \frac{1}{n\log n} - \frac{1}{2n^2\log n} + \frac{1}{3n^3\log n} - \dots. \tag{6}$$

From equations (5) and (6), we get

$$\begin{aligned}\lim_{n\to\infty} n\log\left(\frac{a_n}{a_{n+1}}\right) &= \lim_{n\to\infty} np\left(x - \frac{x^2}{2} + \frac{x^3}{3} - \dots\right) \\ &= \lim_{n\to\infty} p\left(nx - \frac{nx^2}{2} + \frac{nx^3}{3} - \dots\right).\end{aligned} \tag{7}$$

Now,

$$\begin{aligned}\lim_{n\to\infty} nx &= \lim_{n\to\infty}\left(\frac{1}{\log n} - \frac{1}{2n\log n} + \frac{1}{3n^2\log n} - \dots\right) \\ &= 0.\end{aligned}$$

Similarly, other terms in equation (7) will also tend to zero when $n \to \infty$.
Since the limit is less than unity, the series diverges for all values of p.

EXAMPLE G.12. *Test the convergence of the series*

$$\frac{\sin\sqrt{1}}{1^{3/2}} - \frac{\sin\sqrt{2}}{2^{3/2}} + \frac{\sin\sqrt{3}}{3^{3/2}} - \dots.$$

Solution: We have

$$|a_n| = \left|\frac{\sin\sqrt{n}}{n^{3/2}}\right| < \frac{1}{n^{3/2}}. \tag{1}$$

Hence, each term of the given series is less than or equal to (in absolute value) the corresponding term of the series

$$\Sigma b_n = \frac{1}{1^{3/2}} + \frac{1}{2^{3/2}} + \frac{1}{3^{3/2}} + \dots. \tag{2}$$

But, the series in equation (2) is a p-series, where $p = 3/2 > 1$. Hence, the series Σb_n converges. Thus, the given series is absolutely convergent.

EXAMPLE G.13. *Find the interval of convergence of the series*

$$x - \frac{x^3}{3!} + \frac{x^5}{5!} - \ldots.$$

Solution: We have

$$a_n = (-1)^{n-1} \frac{x^{2n-1}}{(2n-1)!} \tag{1}$$

$$\text{and } a_{n+1} = (-1)^n \frac{x^{2n+1}}{(2n+1)!}. \tag{2}$$

Hence,

$$\lim_{n\to\infty} \left|\frac{a_{n+1}}{a_n}\right| = \lim_{n\to\infty} \left|(-1)^n \frac{x^{2n+1}}{(2n+1)!} \cdot \frac{(2n-1)!}{(-1)^{n-1}x^{2n-1}}\right| = \lim_{n\to\infty} \frac{x^2}{2n(2n+1)}$$
$$= 0 \text{ for all values of } x.$$

Hence, the series is absolutely convergent and its interval of convergence is $(-\infty, \infty)$.

EXAMPLE G.14. *Determine the domain of convergence of the series*

$$x - \frac{x^2}{2} + \frac{x^3}{3} - \ldots.$$

Solution: Here,

$$a_n = (-1)^{n-1} \frac{x^n}{n} \tag{1}$$

$$\text{and } a_{n+1} = (-1)^n \frac{x^{n+1}}{n+1}. \tag{2}$$

Hence,

$$\lim_{n\to\infty} \left|\frac{a_{n+1}}{a_n}\right| = \lim_{n\to\infty} \left|(-1)^n \frac{x^{n+1}}{n+1} \cdot \frac{n}{(-1)^{n-1}x^n}\right| = \lim_{n\to\infty} \left(\frac{1}{1+\frac{1}{n}}\right)|x| = |x|. \tag{3}$$

Thus, the series is absolutely convergent for $|x| < 1$, i.e., in the interval $-1 < x < 1$. For $|x| = 1$, the ratio test fails. Let $x = -1$. Then the given series becomes

$$\Sigma a_n = -\left(1 + \frac{1}{2} + \frac{1}{3} + \ldots\right), \tag{4}$$

which diverges. Hence, the given series diverges for $x = -1$.

When $x = 1$, the given series becomes

$$\Sigma a_n = 1 - \frac{1}{2} + \frac{1}{3} - \ldots, \tag{5}$$

which is convergent. Thus, the given series converges for $x = 1$. Hence, the series is convergent in the interval $-1 < x \leq 1$. The radius of convergence is 1.

G.6 Some Special Tests for Convergence

(1) *Alternative to Bertrand's Test*: The series Σa_n of positive terms converges or diverges according as

$$\lim_{n\to\infty}\left\{n\log\left(\frac{a_n}{a_{n+1}}\right)-1\right\}\log n > \text{ or } < 1. \tag{G.24}$$

(2) *Kummer's Test*: Let Σa_n be a series of positive terms and $\{b_n\}$ be a sequence of positive real numbers. We construct a sequence $\{w_n\}$, where

$$w_n = \frac{a_n}{a_{n+1}}b_n - b_{n+1}. \tag{G.25}$$

Then Σa_n converges, if $\underline{\lim}\, w_n > 0$ and Σa_n diverges, if $\overline{\lim}\, w_n < 0$, provided $\Sigma 1/b_n$ is a divergent series.

(3) *Cauchy's Condensation Test*: If $\phi(n) > 0$ for n positive integers and the sequence $\{\phi(n)\}$ is monotone decreasing, then the two series $\Sigma\phi(n)$ and $\Sigma h^n\phi(h^n)$, (h is a positive integer > 1) are either both convergent or both divergent.

(4) *Weierstrass M-test*: If a sequence of positive constants $M_1, M_2, M_3, \ldots$ can be found such that in some interval

$$\left.\begin{array}{r}\text{(i) } |a_n(x)| \le M_n,\ n = 1, 2, 3, \ldots\\ \text{and (ii) } \Sigma M_n \text{ converges}\end{array}\right\}, \tag{G.26}$$

then the series $\Sigma a_n(x)$ is uniformly and absolutely convergent in the interval.

(5) *Dirichlet's Test*: If the sequence $\{a_n\}$ is a monotone decreasing sequence of positive constants having limit zero and there exists a constant P such that, for $a \le x \le b$,

$$|u_1(x) + u_2(x) + \cdots + u_n(x)| < P \tag{G.27}$$

for all $n \ge N$, then the series $\Sigma a_n u_n(x)$ is uniformly convergent in $a \le x \le b$.

(6) *Integral Test*: If $f(x)$ is a positive, continuous and monotone decreasing function for $x \ge N$ and is such that

$$f(n) = a_n,\ n = N, N+1, N+2, \ldots, \tag{G.28}$$

then the series Σa_n converges or diverges according as

$$\int_N^\infty f(x)\,dx = \lim_{M\to\infty}\int_N^M f(x)\,dx \tag{G.29}$$

converges or diverges.

We generally take $N = 1$.

EXAMPLE G.15. *Test the convergence of the series*

$$1^p + \left(\frac{1}{2}\right)^p + \left(\frac{1\cdot 3}{2\cdot 4}\right)^p + \left(\frac{1\cdot 3\cdot 5}{2\cdot 4\cdot 6}\right)^p + \ldots.$$

Solution: Here

$$a_n = \left\{\frac{1\cdot 3\cdot 5\ldots(2n-1)}{2\cdot 4\cdot 6\ldots 2n}\right\}^p \tag{1}$$

$$\text{and } a_{n+1} = \left\{\frac{1\cdot 3\cdot 5\ldots(2n+1)}{2\cdot 4\cdot 6\ldots(2n+2)}\right\}^p. \tag{2}$$

Hence,

$$\lim_{n\to\infty} \frac{a_{n+1}}{a_n} = \lim_{n\to\infty} \left(\frac{2n+1}{2n+2}\right)^p = \lim_{n\to\infty} \left(\frac{2+\frac{1}{n}}{2+\frac{2}{n}}\right)^p = 1. \tag{3}$$

Thus, D'Alembert's test fails. Let us apply Logarithmic test.
Now,

$$\begin{aligned} n\log\frac{a_n}{a_{n+1}} &= n\log\left(\frac{1+\frac{1}{n}}{1+\frac{1}{2n}}\right)^p = np\left\{\log\left(1+\frac{1}{n}\right) - \log\left(1+\frac{1}{2n}\right)\right\} \\ &= np\left\{\left(\frac{1}{n} - \frac{1}{2n^2} + \frac{1}{3n^3} - \dots\right)\right\} - \left(\frac{1}{2n} - \frac{1}{2}\cdot\frac{1}{4n^2} + \frac{1}{3}\cdot\frac{1}{8n^3} - \dots\right) \\ &= np\left(\frac{1}{2n} - \frac{3}{8n^2} + \frac{7}{24n^3} - \dots\right) = p\left(\frac{1}{2} - \frac{3}{8n} + \frac{7}{24n^2} - \dots\right). \end{aligned} \tag{4}$$

Hence,

$$\lim_{n\to\infty} n\log\frac{a_n}{a_{n+1}} = \frac{p}{2}. \tag{5}$$

Thus, the series converges if $p/2 > 1$ and diverges if $p/2 < 1$.

When $p/2 = 1$, i.e., $p = 2$, the test fails. Now, let us apply alternative form of Bertrand's test. Then, for $p = 2$, we have from equation (4)

$$\begin{aligned} \left(n\log\frac{a_n}{a_{n+1}} - 1\right)\log n &= \left\{2\left(\frac{1}{2} - \frac{3}{8n} + \frac{7}{24n^2} - \dots\right) - 1\right\}\log n \\ &= \left(-\frac{3}{4n} + \frac{7}{24n^2} - \dots\right)\log n = \left(-\frac{3}{4} + \frac{7}{24n} - \dots\right)\frac{\log n}{n}. \end{aligned} \tag{6}$$

Since,

$$\lim_{n\to\infty} \frac{\log n}{n} = 0, \tag{7}$$

the whole expression in the RHS of equation (6) approaches zero as $n \to \infty$. Thus, the limit is less than 1 so that the series diverges for $p = 2$.

EXAMPLE G.16. *Test the series* $\sum_{n=1}^{\infty} \frac{1}{n^p}$ *by Cauchy's condensation method.*

Solution: Let

$$\phi(n) = \frac{1}{n^p}. \tag{1}$$

The sequence $\left\{\frac{1}{n^p}\right\}$ is monotone decreasing. We take a positive integer $h = 2 > 1$.
Then

$$\phi(h^n) = \frac{1}{(h^n)^p}. \tag{2}$$

Hence,

$$h^n\phi(h^n) = h^n\frac{1}{(h^n)^p} = \frac{1}{(h^{p-1})^n}. \tag{3}$$

For Cauchy's condensation test, we consider the following two series

$$\Sigma\phi(n) = \Sigma\frac{1}{n^p} \tag{4}$$

$$\text{and } \Sigma h^n\phi(h^n) = \Sigma\frac{1}{(h^{p-1})^n}. \tag{5}$$

These two series will either both converge or will both diverge.

Case I. $p > 1$.

In this case, $\frac{1}{h^{p-1}}$ is positive and less than 1. Hence, the geometric series

$$\sum_{n=1}^{\infty}\frac{1}{(h^{p-1})^n}$$

with common ratio $\frac{1}{h^{p-1}} < 1$ is convergent.

Therefore, the series $\Sigma\frac{1}{n^p}$ converges for $p > 1$.

Case II. $p < 1$.

In this case, $\frac{1}{h^{p-1}} = h^{1-p}$ is > 1. Thus, the common ratio of the geometric series (5) is > 1, so that the series diverges. Hence, the series (4) will also diverge when $p < 1$.

Case III. $p = 1$.

In this case, the series (5) becomes

$$= 1 + 1 + 1 + \ldots, \tag{6}$$

which obviously diverges. Hence, the series (4) when $p = 1$, i.e., the series $\Sigma\frac{1}{n}$, also diverges.

EXAMPLE G.17. *Test for uniform convergence of the series*
(a) $\sum\frac{\cos nx}{n^4}$, (b) $\sum\frac{1}{n^2+x^2}$, (c) $\sum\frac{\sin^2 nx}{2^n-1}$.

Solution:
(a) We have

$$|a_n(x)| = \left|\frac{\cos nx}{n^4}\right| \le \frac{1}{n^4}. \tag{1}$$

The series $\Sigma M_n = \Sigma\frac{1}{n^4}$ converges, since $p = 4 > 1$. Thus, the given series converges uniformly and absolutely for all x by Weierstrass M-test.

(b) We have

$$|a_n(x)| = \left|\frac{1}{n^2+x^2}\right| \le \frac{1}{n^2}. \tag{2}$$

The series $\Sigma M_n = \Sigma\frac{1}{n^2}$ converges, since it is a p-series with $p = 2 > 1$.
Hence, by M-test the given series converges uniformly and absolutely.
(c) We have

$$|a_n(x)| = \left|\frac{\sin^2 nx}{2^n-1}\right| \le \frac{1}{2^n-1}. \tag{1}$$

Bibliography

1. Arfken G, *Mathematical Methods for Physicists.* Academic Press, New York, 1970.
2. Ayres F, (Jr) *Differential Equations.* Schaum, New York, 1952.
3. Churchill R V, *Fourier Series and Boundary Value Problems.* McGraw-Hill, New York, 1963.
4. Copson E T, *Theory of Functions of a Complex Variable.* ELBS and Oxford University Press, 1962.
5. Dettman J W, *Mathematical Methods in Physics and Engineering.* McGraw-Hill, New York, 1962.
6. Duff G F, *Differential Equations of Applied Mathematics.* John Wiley & Sons Inc., New York, 1966.
7. Farlow S J, *Partial Differential Equations for Scientists and Engineers.* Dover Publications, New York, 1994.
8. Fixtengoltz G M, *Course on Differential and Integral Calculus* (in Russian). Vol. 3, "Nauka", Moscow, 1966.
9. Goreux S J, *Introduction to Tensor Calculus.* New Central Book Agency, Calcutta, 1992.
10. Kuznetsov D C, *Special Functions* (in Russian). "Buishaia Shkola", 1965.
11. Lindgren B W, *Vector Calculus.* Macmillan, New York, 1964.
12. Potter M C and Goldberg J, *Mathematical Methods.* Prentice-Hall of India, New Delhi, 2003.
13. Ramsey A S, *Theory of Newtonian Attractions.* Cambridge University Press, 1956.
14. Shokurov V, *Higher Mathematics* (Translated from Russian). Vol. 3, MIR Publishers, Moscow, 1988.
15. Silverman R A and Englewood Cliffs N J, *Complex Analysis with Applications.* Prentice-Hall Inc., 1974.
16. Sneddon I N, *Fourier Series.* Dover Publications, New York, 1961.
17. Spain B, *Tensor Calculus.* Radha Publishing House, Calcutta, 1988.
18. Spiegel M R, *Advanced Calculus.* Schaum, New York, 1963.
19. Spiegel M R, *Laplace Transforms.* Schaum, New York, 1965.
20. Spiegel M R, *Theory and Problems of Vector Analysis.* Schaum, New York, 1959.
21. Tikhonov A N and Samarskii A A, *Equations of Mathematical Physics.* Dover Publications, New York, 1990.
22. Tolstov G P, *Fourier Series* (in Russian). "Nauka", Moscow, 1960.
23. Tricomi F G, *Integral Equations.* Dover Publications, New York, 1985.
24. Watson G N, *Complex Integration and Cauchy's Theorem.* Cambridge, 1914.

